# TechOne: Basic Automotive Service and Maintenance

# TechOne: Basic Automotive Service and Maintenance

Don Knowles

Knowles Automotive Training
Moose Jaw, Saskatchewan
CANADA

Jack Erjavec, Series Editor

Professor Emeritus
Columbus State Community College
Columbus, Ohio

Australia • Brazil • Japan • Korea • Mexico • Singapore • Spain • United Kingdom • United States

**TechOne: Basic Automotive Service and Maintenance**
Don Knowles, Jack Erjavec,

Vice President,Technology and Trades SBU: Alar Elken

Editorial Director: Sandy Clark

Senior Acquisitions Editor: David Boelio

Developmental Editor: Matthew Thouin

Marketing Director: Dave Garza

Channel Manager: William Lawrenson

Marketing Coordinator: Mark Pierro

Production Director: Mary Ellen Black

Production Editor: Barbara L. Diaz

Art/Design Specialist: Cheri Plasse

Technology Project Manager: Kevin Smith

Technology Project Specialist: Linda Verde

Editorial Assistant: Kevin Rivenburg



Library of Congress Control Number: 2004051643

ISBN-13: 978-1-4018-5208-5

ISBN-10: 1-4018-5208-4

**Delmar**
Executive Woods
5 Maxwell Drive
Clifton Park, NY 12065
USA

Cengage Learning is a leading provider of customized learning solutions with office locations around the globe, including Singapore, the United Kingdom, Australia, Mexico, Brazil, and Japan. Locate your local office at **www.cengage.com/global**

Cengage Learning products are represented in Canada by Nelson Education, Ltd.

To learn more about Delmar, visit **www.cengage.com/delmar**

Purchase any of our products at your local bookstore or at our preferred online store **www.CengageBrain.com**



Printed in the United States of America
4 5 6 7 8 17 16 15 14 13

# Contents

## Section 7: Engine Control Systems 299



## Section 8: Emissions and Emission Systems 351

## Section 14: Suspension Systems 579



## Section 15: Steering Systems 621

# Preface

## THE SERIES

Welcome to Delmar's *TechOne*, a state-of-the-art series designed to respond to today's automotive instructor and student needs. *TechOne* offers current, concise information on ASE and other specific subject areas, combining classroom theory, diagnosis, and repair into one easy-to-use volume.

You'll notice several differences from a traditional textbook. First, a large number of short chapters divide complex material into chunks. Instructors can give tight, detailed reading assignments that students will find easier to digest. These shorter chapters can be taught in almost any order, allowing instructors to pick and choose the material that best reflects the depth, direction, and pace of their individual classes.

*TechOne* also features an art-intensive approach to suit today's visual learners—images drive the chapters. From drawings to photos, you will find more art to better understand the systems, parts, and procedures under discussion. Look also for helpful graphics that draw attention to key points in features like You Should Know and Interesting Fact.

Just as importantly, each *TechOne* starts off with a section on Safety and Communication, which stresses safe work practices, tool competence, and familiarity with workplace "soft skills," such as customer communication and the roles necessary to succeed as an automotive technician. From there, learners are ready to tackle the technical material in successive sections and test their knowledge with the ASE practice questions at the end of each chapter.

## THE SUPPLEMENTS

*TechOne* comes with an **Instructor's Manual** that includes answers to all chapter-end review questions and a complete correlation of the text to NATEF standards. A **CD-ROM**, included with each Instructor's Manual, contains **PowerPoint Slides** for classroom presentations, a **Computerized Testbank** with hundreds of questions to aid in creating tests and quizzes, and an electronic version of the Instructor's Manual. **Electronic Worksheets** are also available for lab based learning activities.

Flexibility is the key to *TechOne*. For those who would like to purchase jobsheets, Delmar's NATEF Standards Job Sheets are a good match. Topics cover the eight ASE subject areas and include:

- Engine Repair
- Automatic Transmissions and Transaxles
- Manual Drive Trains and Axles
- Suspension and Steering
- Brakes
- Electrical and Electronic Systems
- Heating and Air Conditioning
- Engine Performance

Plus,

- Advanced Engine Performance
- Fuels and Emissions

Visit **http://www.autoed.com** for a complete catalog.

## OTHER TITLES IN THIS SERIES

*TechOne* is Delmar's latest automotive series. We are excited to announce these future titles:

- Engine Repair
- Automatic Transmissions
- Suspension and Steering
- Heating and Air Conditioning
- Advanced Automotive Electronic Systems
- Advanced Engine Performance
- Automotive Fuels & Emissions

Check with your sales representative for availability.

## A NOTE TO THE STUDENT

There are now more computers on a car than aboard the first spacecraft, and even gifted backyard mechanics long ago turned their cars over to automotive professionals for diagnosis and repair. That's a statement about the nation's need for the knowledge and skills you'll develop as you continue your studies. Whether you eventually choose a career as a certified or licensed technician, service writer, manager, or automotive engineer—or even if you decide to open your own shop—hard work will give you the opportunity to become one of the 840,000 automotive professionals providing and maintaining safe and efficient automobiles on our roads. As a member of a technically-proficient, cutting-edge workforce, you'll fill a need, and, even better, you'll have a career to feel proud of.

Best of luck in your studies,

The Editors of Delmar, Cengage Learning

# About the Author

Don Knowles was employed as an automotive instructor at the Saskatchewan Institute of Applied Science and Technology (SIAST) Palliser Campus from 1961 to 1987. During his tenure at SIAST, he produced over 60 automotive training videos in cooperation with the SIAST audio visual department and wrote a computer-assisted instruction (CAI) program coordinated with each video. Don's specialty was compiling and teaching mobile training programs for experienced technicians.

Don has been writing automotive textbooks and other training materials since 1981. During this time he has written over 35 automotive textbooks, student manuals, and training CDs in automotive and medium/heavy truck technology. Don has also written many short articles for service bulletins.

Since 1987 Don has been owner/manager of Knowles Automotive Training. Knowles Automotive Training is an educational member of the Automotive Service Association (ASA). Don is ASE L1 certified and ASE master-certified in automotive and medium/heavy truck. He has been a member of the Society of Automotive Engineers (SAE) for over 25 years and is also a member of the Council of Advanced Automotive Trainers (CAAT) and the North American Council of Automotive Teachers (NACAT).

# Acknowledgments

The publisher would to thank the following reviewers, whose technical expertise was invaluable in creating this text:

John Cuprisin
Penn College of Technology
Williamsport, PA

Chris English
Blue Ridge Community College
Flat Rock, NC

Mario Schwarz
Santa Fe Community College
Gainesville, FL

Raymond Scow
Truckee Meadows Community College
Reno, NV

James Sipes
American River College
Sacramento, CA

Ted Sumners
Midland College
Midland, TX

# Features of the Text

*TechOne* includes a variety of learning aids designed to encourage student comprehension of complex automotive concepts, diagnostics, and repair. Look for these helpful features:

**Section Openers** provide students with a **Section Table of Contents** and **Objectives** to focus the learner on the section's goals.

**Interesting Facts** spark student attention with industry trivia or history. Interesting facts appear on the section openers and are then scattered throughout the chapters to maintain reader interest.

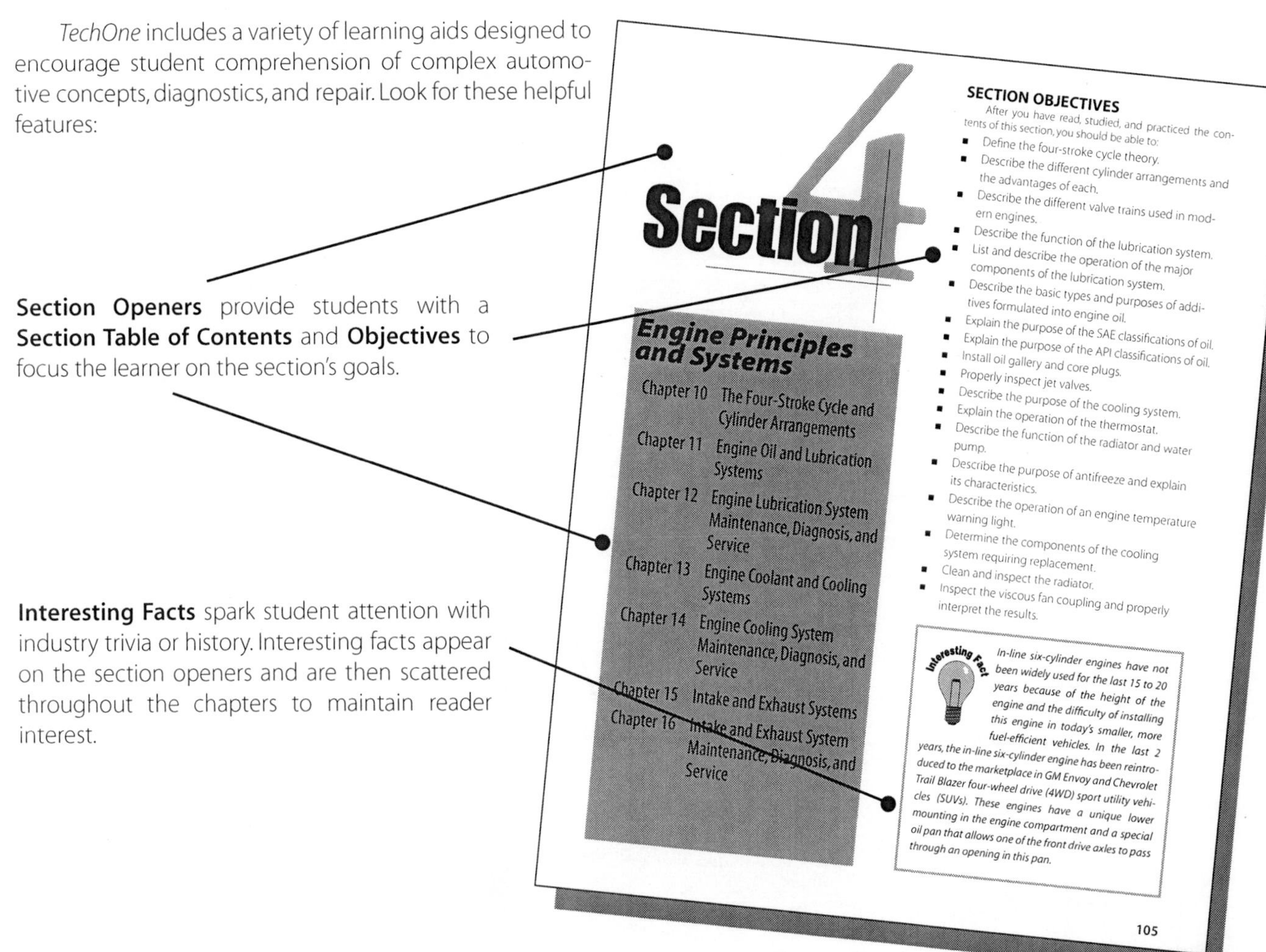

Section 4

**Engine Principles and Systems**

Chapter 10 The Four-Stroke Cycle and Cylinder Arrangements
Chapter 11 Engine Oil and Lubrication Systems
Chapter 12 Engine Lubrication System Maintenance, Diagnosis, and Service
Chapter 13 Engine Coolant and Cooling Systems
Chapter 14 Engine Cooling System Maintenance, Diagnosis, and Service
Chapter 15 Intake and Exhaust Systems
Chapter 16 Intake and Exhaust System Maintenance, Diagnosis, and Service

**SECTION OBJECTIVES**

After you have read, studied, and practiced the contents of this section, you should be able to:

- Define the four-stroke cycle theory.
- Describe the different cylinder arrangements and the advantages of each.
- Describe the different valve trains used in modern engines.
- Describe the function of the lubrication system.
- List and describe the operation of the major components of the lubrication system.
- Describe the basic types and purposes of additives formulated into engine oil.
- Explain the purpose of the SAE classifications of oil.
- Explain the purpose of the API classifications of oil.
- Install oil gallery and core plugs.
- Properly inspect jet valves.
- Describe the purpose of the cooling system.
- Explain the operation of the thermostat.
- Describe the function of the radiator and water pump.
- Describe the purpose of antifreeze and explain its characteristics.
- Describe the operation of an engine temperature warning light.
- Determine the components of the cooling system requiring replacement.
- Clean and inspect the radiator.
- Inspect the viscous fan coupling and properly interpret the results.

Interesting Fact

*In-line six-cylinder engines have not been widely used for the last 15 to 20 years because of the height of the engine and the difficulty of installing this engine in today's smaller, more fuel-efficient vehicles. In the last 2 years, the in-line six-cylinder engine has been reintroduced to the marketplace in GM Envoy and Chevrolet Trail Blazer four-wheel drive (4WD) sport utility vehicles (SUVs). These engines have a unique lower mounting in the engine compartment and a special oil pan that allows one of the front drive axles to pass through an opening in this pan.*

105

# Using Service Information

An **Introduction** orients readers at the beginning of each new chapter.

## Introduction

After students or technicians are familiar with the various sources of service information, they must understand how to access and use this information. When individuals are not familiar with the proper use of service information, they can waste valuable time searching for the required information. Students and technicians must also understand where to locate the best and quickest source for the information they require. Using brief or inadequate service information may lead to improper diagnostic and service procedures that do not locate and correct the cause of the customer's complaint.

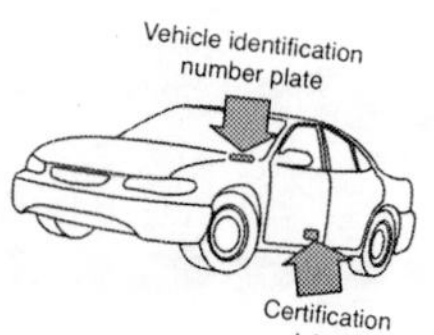


**Figure 1.** Vehicle identification number (VIN) location.

**You Should Know** informs the reader whenever special safety cautions, warnings, or other important points deserve emphasis.

**Technical Terms** are bolded in the text upon first reference and are defined.

### VEHICLE IDENTIFICATION NUMBER (VIN) INTERPRETATION

When completing a work order, ordering parts, or servicing a vehicle, certain facts must be known about the vehicle. The vehicle manufacturer, make, and model year must be known. The vehicle manufacturer and make are often on the nameplate attached to the exterior of the body. Examples of vehicle manufacturer and make are Chrysler Concorde and Honda Acura. The customer can also supply the necessary information regarding the vehicle manufacturer, make, and model year. This information is also on the vehicle registration.

The **vehicle identification number (VIN)** is a series of letters and numbers that identify specific vehicle parameters such as the vehicle make, model, model year, and size of engine. On modern vehicles, the VIN number is mounted on the top of the dash, and this number is visible through the lower left side of the windshield **(Figure 1)**.

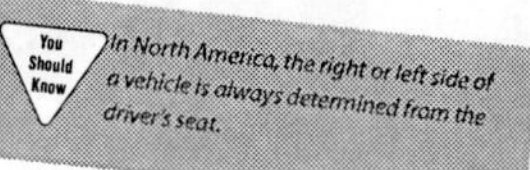

You Should Know
*In North America, the right or left side of a vehicle is always determined from the driver's seat.*


The VIN mounting location makes it more difficult for car thieves to change the VIN. An explanation of the VIN is provided in the vehicle manufacturer's service manual. The VIN explanation is usually included in the service manual

95

Chapter 2 The Automotive Business • 17

4. To promote public awareness of the iATN and its members.
5. To continually adapt to the changing needs of supporting iATN members.

iATN members agree to:

1. Uphold the highest standards of professionalism, competence, and integrity.
2. Pursue excellence through ongoing education and the exchange of knowledge and experience of fellow iATN members.
3. Maintain quality through the use of proper tools, equipment, parts, and procedures.
4. Promote public awareness of the importance of quality professional service and the advancing technology of the modern automobile.
5. Support and promote the mission and goals of the iATN.

The iATN is supported by a number of top companies, associations, and publications in the automotive industry. If an iATN member encounters an automotive problem, iATN suggests that the technician attempt to diagnose the problem using information from service manuals, data systems such as Mitchell-on-Demand, and technical hotlines. The technician can also access the iATN Fix Database. If the technician still cannot solve the problem, a problem file may be sent to iATN outlining the necessary vehicle information and the details of the problem. This problem file is sent via the Internet to many iATN members. Any iATN member who has experienced and solved the same problem can post a fix file to iATN that outlines the answer to the automotive problem and the required fix procedure. An iATN member may configure his or her account to only receive problem and fix files in specific areas of automotive repair such as engine electronics and fuel systems. The iATN offers many other member benefits such as live conferencing and technical forums. These forums include technical discussion forums, technical theory forums, technical tip forums, and tool and equipment forums.

A **Summary** concludes each chapter in short, bulleted sentences. **Review Questions** are structured in a variety of formats, including ASE style, challenging students to prove they've mastered the material.

## Summary

- Independent repair shops are privately owned, and they are not affiliated with vehicle manufacturers, automotive parts suppliers, or chain organizations.
- Specialty shops usually perform one type of automotive repair work.
- Quick-lube shops specialize in fast lubrication service.
- Automotive dealerships have a contract with one or more vehicle manufacturers to sell and service the manufacturer's vehicles.
- The ATRA distributes technical and business information to automotive transmission technicians and rebuilding shops.
- MACS Worldwide provides technician information, training, and communication for the automotive A/C industry.
- The ASA provides membership service such as the monthly *AutoInc.* magazine and a yearly Congress of Automotive Repair and Service.
- The iATN provides forums for the exchange of information in the automotive industry.

## Review Questions

1. In an automotive dealership, the vehicle manufacturer may set standards regarding all of these items *except:*
   A. the layout of the facility.
   B. new vehicle sales policies.
   C. accounting and financing practices.
   D. staff benefits policies.
2. When discussing dealership management, Technician A says customer service and satisfaction is extremely important. Technician B says communication between dealership departments is not a high priority. Who is correct?
   A. Technician A
   B. Technician B
   C. Both Technician A and Technician B
   D. Neither Technician A nor Technician B
3. All of these statements about the ATRA are true *except:*
   A. ATRA was founded in 1954.
   B. ATRA has members in 24 countries.
   C. ATRA distributes information on most automotive electronic systems.
   D. ATRA publishes customer bulletins.
4. When discussing the iATN, Technician A says the iATN has members in 136 countries. Technician B says the iATN is helpful when diagnosing automotive problems. Who is correct?
   A. Technician A
   B. Technician B
   C. Both Technician A and Technician B
   D. Neither Technician A nor Technician B

A **Bilingual Glossary** is found in the **Appendix** of every *TechOne* book, which offers Spanish translations of technical terms alongside their English counterparts. A comprehensive **Index** helps instructors and students pinpoint information in the text.

# Bilingual Glossary

**Aiming pads** Small projections on the front of some headlights to which headlight aligning equipment may be attached.
***Patines de alineación*** *Pequeñas proyecciones en la parte delantera de algunos faros a los cuales se puede conectar el equipo de alineación de faros.*

**Air bag diagnostic monitor (ASDM)** An automotive computer responsible for air bag system operation.
***Monitor de bolsa de aire (ASD)*** *Una computadora automotiva que es responsable por la operación del sistema de la bolsa de aire.*

**Airflow restriction indicator** An indicator located in the air cleaner or intake duct to display air cleaner restriction by the color of the indicator window.
***Indicadora de restricción del aire*** *Un indicador ubicado en el limpiador de aire o en el ducto de entrada que indica la restricción del aire por medio del color de la ventanilla del indicador.*

**Alternating current** flows in one direction and then in the opposite direction.
***Corriente alterna*** *fluye en una dirección y luego en la dirección opuesta.*

**American Petroleum Institute (API) rating** A universal engine oil rating that classifies oils according to the type of service for which the oil is intended.
***Evaluación del Instituto Americano de Petroleo (API)*** *Una evaluación universal del aceite automotivo que clasifica a los aceites según el tipo del servicio que se le requiere.*

**Amperes** A measurement for the amount of current flowing in an electric circuit.
***Amperes*** *Una medida de la cantidad del corriente que fluye en un circuito eléctrico.*

**Analog meter** A meter with a pointer and a scale to indicate a specific reading.
***Medidora análoga*** *Un medidor con una indicadora y una gama para indicar una lectura específica.*

**Analog voltage signal** A signal that is continuously variable within a specific range.
***Señal análoga del voltaje*** *Una señal que es variable continuamente en una gama específica.*

**Angular bearing load** A load applied at an angle somewhere between the horizontal and vertical positions.
***Carga de soporte angular*** *Una carga aplicada en un ángulo que se encuentra entre las posiciones horizontales y verticales.*

**Atom** The smallest particle of an element.
***Átomo*** *La partícula más pequeña de un elemento.*

**Automatic Transmission Rebuilders Association (ATRA)** provides technical information to transmission shops and technicians.
***Asociación de Reconstrutores de Transmisiones Automáticas (ATRA)*** *Provea la información técnica a los talleres de transmisiones y a los técnicos.*

**Automotive dealership** sells and services vehicles produced by one or more vehicle manufacturers.
***Sucursal automotivo*** *vende y repara los vehículos producidos por uno o varios fabricantes de automóviles.*

**Automotive Service Association (ASA)** promotes professionalism and excellence in the automotive repair industry through education, representation, and member services.
***Asociación de Servicio Automativo (ASA)*** *Promueve la profesionalidad y la excelencia en la industria de reparación automotiva por medio de la educación, la representación y los servicios de sus miembros.*

**Belleville spring** A diaphragm spring made from thin sheet metal that is formed into a cone shape.
***Resorte Belleville*** *Un resorte de tipo diafragma hecho de una hoja delgada de metal que es en la forma de un cono.*

679

# Index

691

# Safety and Communication

*In 2002, there were 78,034 general automotive repair businesses in the United States. These businesses employed 310,325 individuals and had total sales of $39.8 billion. In the same year there were over 21,000 automotive dealerships with service facilities, and the total sales of parts and labor in these facilities was $25.8 billion. Total motor vehicle registrations in the United States increased from 108,418,197 in 1970 to 216,308,623 in 1999.*

## SECTION OBJECTIVES

After you have read, studied, and practiced the contents of this section, you should be able to:

- Recognize shop hazards and take the necessary steps to avoid personal injury or property damage.
- Explain the purposes of the Occupational Safety and Health Act.
- Describe the harmful effects of carbon monoxide on the human body.
- Explain why asbestos dust in the shop air is a hazard.
- Describe the main purposes of the right-to-know laws.
- Describe the purpose of material safety data sheets (MSDS).
- Define an independent repair shop.
- Describe the main items listed on a work order.
- Explain employer to employee obligations.
- Describe employee to employer obligations.
- Describe the National Institute for Automotive Service Excellence (ASE) certification of automotive technicians.
- Explain National Automotive Technicians Education Foundation (NATEF) certification of automotive training programs.
- Complete job applications and resumes.
- Describe the use of electrical and electronic test equipment.
- Explain why smoking and drug or alcohol abuse is dangerous in the shop.
- Describe basic electrical, gasoline, and fire safety precautions.
- Describe the proper procedure for using a fire extinguisher.
- Explain the proper procedure for lifting heavy objects.
- Explain the necessary precautions when operating a vehicle lift.
- Describe hydraulic jack and safety stand safety precautions.
- Explain the necessary safety precautions when using power tools.

# General Shop Safety

## Introduction

Safety is extremely important in the automotive shop! The knowledge and practice of safety precautions prevent serious personal injury and expensive property damage. Automotive students and technicians must be familiar with shop hazards and shop safety rules. The first step in providing a safe shop is learning about shop hazards and safety rules. The second, and most important, step in this process is applying your knowledge of shop hazards and safety rules while working in the shop. In other words, you must actually develop safe working habits in the shop from your understanding of shop hazards and safety rules. When shop employees have a careless attitude toward safety, accidents are more likely to occur. All shop personnel must develop a serious attitude toward safety. The result of this serious attitude is that shop personnel will learn and adopt all shop safety rules.

Shop personnel must be familiar with their rights regarding hazardous waste disposal. These rights are explained in the right-to-know laws. Shop personnel must also be familiar with the types of hazardous materials in the automotive shop and the proper disposal methods for these materials according to state and federal regulations.

## OCCUPATIONAL SAFETY AND HEALTH ACT AND ENVIRONMENTAL PROTECTION AGENCY

The Occupational Safety and Health Act (OSHA) was passed by the United States government in 1970. The purposes of this legislation are:

1. To assist and encourage the citizens of the United States in their efforts to assure safe and healthful working conditions by providing research, information, education, and training in the field of occupational safety and health.
2. To assure safe and healthful working conditions for working men and women by authorizing enforcement of the standards developed under the Act.

Since approximately 25 percent of workers are exposed to health and safety hazards on the job, the OSHA is necessary to monitor, control, and educate workers regarding health and safety in the workplace.

The Environmental Protection Agency (EPA) is the Federal agency responsible for air and water quality in the United States. The EPA was established to compile and enforce regulations that apply to waste generated by service or manufacturing businesses. These regulations include waste disposal and landfills for household and industrial waste. The EPA is also responsible for vehicle emission standards and the enforcement of these standards as they apply to vehicle manufacturers and owners. In many states, vehicle owners must have a compulsory vehicle emission test at specific intervals.

## SHOP HAZARDS

Service technicians and students encounter many hazards in an automotive shop. When these hazards are known, basic shop safety rules and procedures must be followed to avoid personal injury. Some of the hazards in an automotive shop include the following:

1. Flammable liquids such as gasoline and paint must be handled and stored properly in approved, closed containers.
2. Flammable materials such as oily rags must be stored properly in closed containers to avoid a fire hazard.

3. Batteries contain a corrosive sulfuric acid solution and produce explosive hydrogen gas while charging.

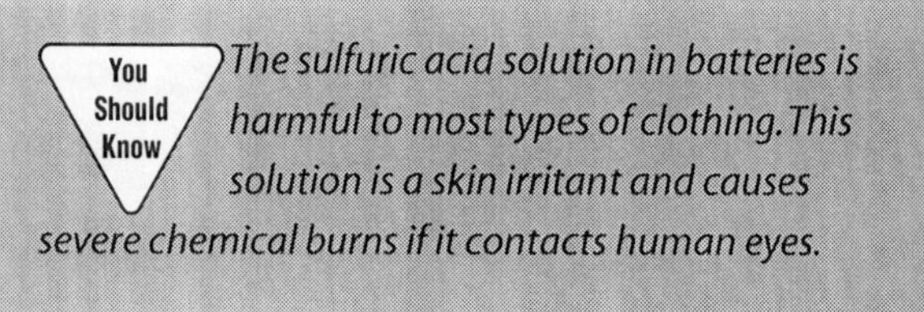


*The sulfuric acid solution in batteries is harmful to most types of clothing. This solution is a skin irritant and causes severe chemical burns if it contacts human eyes.*

4. Loose sewer and drain covers may cause foot or toe injuries.
5. Caustic liquids, such as those in hot cleaning tanks, are harmful to skin and eyes.
6. High-pressure air in the shop's compressed-air system can be very dangerous or fatal if it penetrates the skin and enters the bloodstream. High-pressure air released near the eyes causes eye injury.
7. Frayed cords on electrical equipment and lights may result in severe electrical shock.
8. Hazardous waste material, such as batteries, and caustic cleaning solution must be handled with the adequate personal protection **(Figure 1)**.

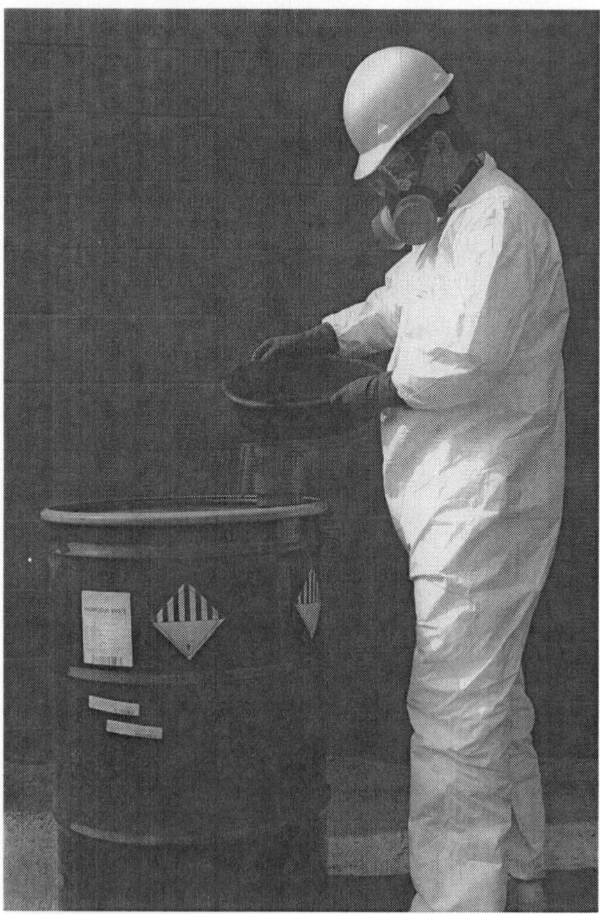

**Figure 1.** Recommended safety clothing and equipment is required when handling hazardous materials.

9. Carbon monoxide from vehicle exhaust is poisonous.
10. Loose clothing or long hair may become entangled in rotating parts on equipment or vehicles, resulting in serious injury.
11. Dust and vapors generated during some repair jobs are harmful. Asbestos dust, which may be released during brake lining service and clutch service, is a contributor to lung cancer.
12. High noise levels from shop equipment such as an air chisel may be harmful to the ears.
13. Oil, grease, water, or parts cleaning solutions on shop floors may cause someone to slip and fall, resulting in serious injury.
14. The incandescent bulbs used in some trouble lights may shatter when the light is dropped. This action may ignite flammable materials in the area and cause a fire. Many insurance companies now require the use of trouble lights with fluorescent bulbs in the shop.

## SHOP SAFETY RULES

The application of some basic shop rules helps prevent serious, expensive accidents. Failure to comply with shop rules may cause personal injury or expensive damage to vehicles and shop facilities. It is the responsibility of the employer and all shop employees to make sure that shop rules are understood and followed until these rules become automatic habits. The following basic shop rules should be followed.

1. Always wear safety glasses and other protective equipment that is required by a service procedure **(Figure 2)**. For example, a brake parts washer must be

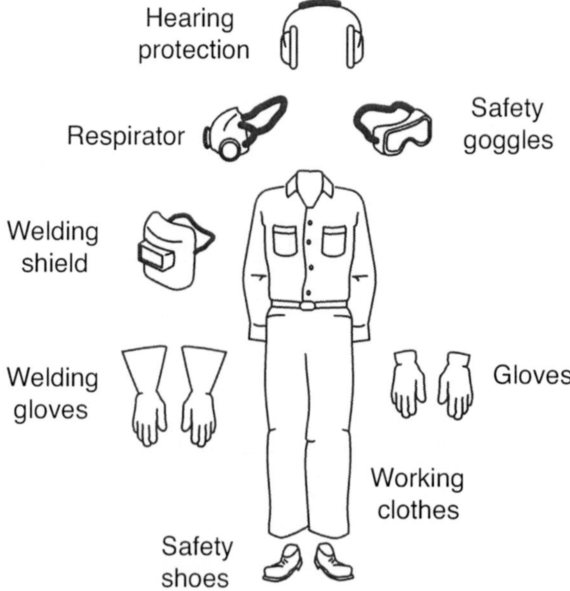


**Figure 2.** Shop safety equipment, including safety goggles, respirator, welding shield, proper work clothes, ear protection, welding gloves, work gloves, and safety shoes.

used to avoid breathing asbestos dust into the lungs. Asbestos dust is a known cause of lung cancer. This dust is encountered in manual transmission clutch facings and brake linings.
2. Tie long hair securely behind your head, and do not wear loose or torn clothing.
3. Do not wear rings, watches, or loose hanging jewelry. If jewelry such as a ring, metal watch band, or chain makes contact between an electrical terminal and ground, the jewelry becomes extremely hot, resulting in severe burns.
4. Do not work in the shop while under the influence of alcohol or drugs.
5. Set the parking brake when working on a vehicle. If the vehicle has an automatic transmission, place the gear selector in park unless a service procedure requires another selector position. When the vehicle is equipped with a manual transmission, position the gear selector in neutral with the engine running, or in reverse with the engine stopped.
6. Always connect a shop exhaust hose to the vehicle tailpipe, and be sure the shop exhaust fan is running. If it is absolutely necessary to operate a vehicle without a shop exhaust pipe connected to the tailpipe, open the large shop door to provide adequate ventilation.

**You Should Know** *Carbon monoxide in the vehicle exhaust may cause severe headaches and other medical problems. High concentrations of carbon monoxide may result in death!*

7. Keep hands, clothing, and wrenches away from rotating parts such as cooling fans. Remember that electric-drive fans may start turning at any time, even with the ignition off.
8. Always leave the ignition switch off unless a service procedure requires another switch position.
9. Do not smoke in the shop. If the shop has designated smoking areas, smoke only in these areas.
10. Store oily rags and other discarded combustibles in covered metal containers designed for this purpose.
11. Always use the wrench or socket that fits properly on the bolt. Do not substitute metric for United States Customary (USC) wrenches, or vice versa.
12. Keep tools in good condition. For example, do not use a punch or chisel with a mushroomed end. When struck with a hammer, a piece of the mushroomed metal could break off, resulting in severe eye or other injury.
13. Do not leave power tools running and unattended.
14. Serious burns may be prevented by avoiding contact with hot metal components, such as exhaust manifolds, other exhaust system components, radiators, and some air conditioning lines and hoses.
15. When a lubricant such as engine oil is drained, always wear heavy plastic gloves because the oil could be hot enough to cause burns.
16. Prior to getting under a vehicle, be sure the vehicle is placed securely on safety stands.
17. Operate all shop equipment, including lifts, according to the equipment manufacturer's recommended procedure. Do not operate equipment unless you are familiar with the correct operating procedure.
18. Do not run or engage in horseplay in the shop.
19. Obey all state and federal fire, safety, and environmental regulations.
20. Do not stand in front of or behind vehicles.
21. Always place fender, seat, and floor mat covers on a customer's vehicle before working on the car.
22. When one end of a vehicle is raised, place wheel chocks on both sides of the wheels remaining on the floor.
23. All shop employees must be familiar with the location of shop safety equipment.
24. Collect oil, fuel, brake fluid, and other liquids in the proper safety containers.
25. Use only approved cleaning fluids and equipment. Do not use gasoline to clean parts.
26. Be sure safety shields are in place on all rotating equipment.
27. All shop equipment must have regular scheduled maintenance and adjustment.
28. Some shops have safety lines around equipment. Always work within these lines when operating equipment.
29. Be sure the shop heating equipment is properly ventilated.
30. Post emergency numbers near the phone. These numbers should include a doctor, ambulance, fire department, hospital, and police.
31. Do not leave hydraulic jack handles where someone may trip over them.
32. Keep isles free of debris.
33. Inform the shop foreman of any safety dangers and suggestions for safety improvement.
34. Do not direct high-pressure air from an air gun against human skin or near the eyes.
35. All shop employees must wear proper footwear. Heavy-duty work boots or shoes with steel toes are the best footwear in an automotive shop.

**You Should Know** *High-pressure air may penetrate the skin and enter the blood stream. Air in the blood stream may be fatal! High-pressure air discharged near the eyes may cause serious eye damage.*

## SMOKING, ALCOHOL, AND DRUGS IN THE SHOP

Do not smoke when working in the shop. If the shop has designated smoking areas, smoke only in these areas. Do not smoke in customers' cars. A nonsmoker will not appreciate cigarette odor in the car. A spark from a cigarette or lighter may ignite flammable materials in the workplace. The use of drugs or alcohol must be avoided while working in the shop. Even a small amount of drugs or alcohol affects reaction time. In an emergency situation, slow reaction time may cause personal injury. If a heavy object falls off the workbench and your reaction time is slowed by drugs or alcohol, you may not be able to move your foot out of the way in time to avoid a foot injury. When a fire starts in the workplace and you are a few seconds slower operating a fire extinguisher because of alcohol or drug use, it could make the difference between extinguishing a fire and having expensive fire damage.

## AIR QUALITY

Vehicle exhaust contains small amounts of carbon monoxide, a colorless, odorless, poisonous gas. Weak concentrations of carbon monoxide in the shop air may cause nausea and headaches. Strong concentrations of carbon monoxide may be fatal. All shop personnel are responsible for air quality in the shop. Shop management is responsible for an adequate exhaust system to remove exhaust fumes from the maximum number of vehicles that may be running in the shop at the same time. Technicians should never run a vehicle in the shop unless a shop exhaust hose is installed on the tailpipe of the vehicle. The exhaust fan must be switched on to remove exhaust fumes.

If shop heaters or furnaces have restricted chimneys, they release carbon monoxide emissions into the shop air. Therefore, chimneys should be checked periodically for restriction and proper ventilation.

Monitors are available to measure the level of carbon monoxide in the shop. Some of these monitors read the amount of carbon monoxide present in the shop air; others provide an audible alarm if the concentration of carbon monoxide exceeds the danger level. Diesel exhaust contains some carbon monoxide, but particulate emissions are also present in the exhaust from these engines. Particulates are basically small carbon particles that can be harmful to the lungs.

The sulfuric acid solution in vehicle batteries is a corrosive, poisonous liquid. If a battery is charged with a fast charger at a high rate for a period of time, the battery becomes hot, and the sulfuric acid solution begins to boil. Under this condition, the battery may emit a strong sulfuric acid smell, and these fumes may be harmful to the lungs. If this happens, the battery charger should be turned off or the charging rate should be reduced considerably.

Some automotive clutch facings and brake linings contain asbestos. Never use compressed air to blow dirt from these components since this action disperses asbestos dust into the shop air where it may be inhaled by technicians and other people in the shop. A brake parts washer or a vacuum cleaner with a high efficiency particulate air (HEPA) filter must be used to clean the dust from these components.

Even though technicians take every precaution to maintain air quality in the shop, some undesirable gases may still reach the air. For example, exhaust manifolds may get oil on them during an engine overhaul. When the engine is started and these manifolds become hot, the oil burns off the manifolds and pollutes the shop air with oil smoke. Adequate shop ventilation must be provided to take care of this type of air contamination.

## SHOP SAFETY EQUIPMENT

Technicians must understand shop safety equipment and the use of this equipment. Technicians must also know the location of safety equipment. When the technician understands safety equipment and knows the location of this equipment, accidents such as fires may be extinguished quickly. If the technician is not familiar with the use of safety equipment or does not know its location in the shop, he or she may require a longer time to put the fire extinguisher into operation. This delay may allow a fire to get out of control and become an expensive disaster.

### Fire Prevention and Fire Extinguishers

An automotive shop is a dangerous place for a fire because a number of flammable liquids such as gasoline, engine oil, and transmission fluid are usually stored in the shop. The shop also contains a number of vehicles with gasoline or diesel fuel in the fuel tanks. Technicians must always practice fire prevention! For example, never turn on the ignition switch or crank the engine in a vehicle with the fuel line disconnected. If this action is taken, fuel will be discharged from the disconnected fuel line, and a spark from the ignition system may ignite this fuel. Oily rags must be stored in closed containers. Oily rags may generate enough heat to self-ignite and start a fire. This process is called spontaneous combustion. When oily rags are stored in closed containers, they cannot receive enough oxygen to support a fire.

Fire extinguishers are one of the most important pieces of safety equipment. All shop personnel must know the location of the fire extinguishers in the shop. If you have to waste time looking for an extinguisher after a fire starts, the fire could get out of control before you put the extinguisher into operation. Fire extinguishers should be located where they are easily accessible at all times.

A decal on each fire extinguisher identifies the type of chemical in the extinguisher and provides operating infor-

**Figure 3.** Types of fire extinguishers.

mation **(Figure 3)**. Shop personnel should be familiar with the following types of fires and fire extinguishers.

1. Class A fires are those involving ordinary combustible materials such as paper, wood, clothing and textiles. Water, foam, and multipurpose dry chemical fire extinguishers are used on these types of fires.
2. Class B fires involve the burning of flammable liquids such as gasoline, oil, paint, solvents, and greases. These fires may also be extinguished with multipurpose dry chemical fire extinguishers. In addition, fire extinguishers containing halogen, or halon, may be used to extinguish class B fires. The chemicals in this type of extinguisher attach to the hydrogen, hydroxide, and oxygen molecules to stop the combustion process almost instantly. However, the resultant gases from the use of halogen-type extinguishers are very toxic and harmful to the operator of the extinguisher. A water-type fire extinguisher will cause the fire to spread even more.
3. Class C fires involve the burning of electrical equipment such as wires, motors, and switches. These fires are extinguished with multipurpose dry chemical fire extinguishers. Water or foam fire extinguishers will conduct electricity and cause electrical shock.
4. Class D fires involve the combustion of metal chips, turnings, and shavings. Special dry chemical fire extinguishers are the only type of extinguisher recommended for these fires.

Additional information regarding types of extinguishers for various types of fires is provided in **Figure 4**. Some

| | Class of fire | Typical fuel involved | Type of extinguisher |
|---|---|---|---|
| Class A Fires (green) | For ordinary combustibles<br>Put out a class A fire by blowing its temperature or by coating the burning combustibles. | Wood<br>Paper<br>Cloth<br>Rubber<br>Plastics<br>Rubbish<br>Upholstery | Water*<br>Foam*<br>Multipurpose dry chemical |
| Class B Fires (red) | For flammable liquids<br>Put out a class B fire by smothering it. Use an extinguisher that gives a blanketing, flame-interrupting effect; cover whole flaming liquid surface. | Gasoline<br>Oil<br>Grease<br>Paint<br>Lighter fluid | Foam*<br>Carbon dioxide<br>Halogenated agent<br>Standard dry chemical<br>Purple K dry chemical<br>Multipurpose dry chemical |
| Class C Fires (blue) | For electrical equipment<br>Put out a class C fire by shutting off power as quickly as possible and by always using a nonconducting extinguisher agent to prevent electric shock. | Motors<br>Appliances<br>Wiring<br>Fuse boxes<br>Switchboards | Carbon dioxide<br>Halogenated agent<br>Standard dry chemical<br>Purple K dry chemical<br>Multipurpose dry chemical |
| Class D Fires (yellow) | For combustible metals<br>Put out a class D fire of metal chips, turnings, or shavings by smothering or coating with a specially designed extinguisher agent. | Aluminum<br>Magnesium<br>Potassium<br>Sodium<br>Titanium<br>Zirconium | Dry powder extinguisher and agents only |

*Cartridge-operated water, foam, and soda-acid types of extinguishers are no longer manufactured. These extinguishers should be removed from service when they become due for their next hydrostatic pressure test.

**Figure 4.** A guide to fire extinguisher selection.

multipurpose dry chemical fire extinguishers may be used on class A, class B, or class C fires.

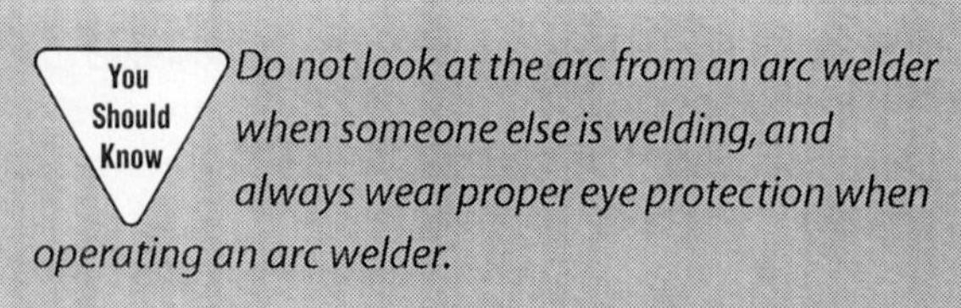


*Do not look at the arc from an arc welder when someone else is welding, and always wear proper eye protection when operating an arc welder.*

## Causes of Eye Injuries

Eye injuries may occur in various ways in the automotive shop. Some of the more common eye accidents are:

1. Thermal burns from excessive heat
2. Irradiation burns from excessive light, such as from an arc welder
3. Chemical burns from strong liquids such as battery electrolyte
4. Foreign material in the eye
5. Penetration of the eye by a sharp object
6. A blow from a blunt object

Wearing safety glasses and observing shop safety rules will prevent most eye accidents.

## Eyewash Fountains

If a chemical gets in your eyes, it must be washed out immediately to prevent a chemical burn. An eyewash fountain is the most effective way to wash the eyes. An eyewash fountain is similar to a drinking water fountain, but the eyewash fountain has water jets placed throughout the fountain top. Every shop should be equipped with some eyewash facility **(Figure 5)**. Be sure you know the location, and know how to use the eyewash fountain in the shop.

**Figure 5.** An eyewash fountain.

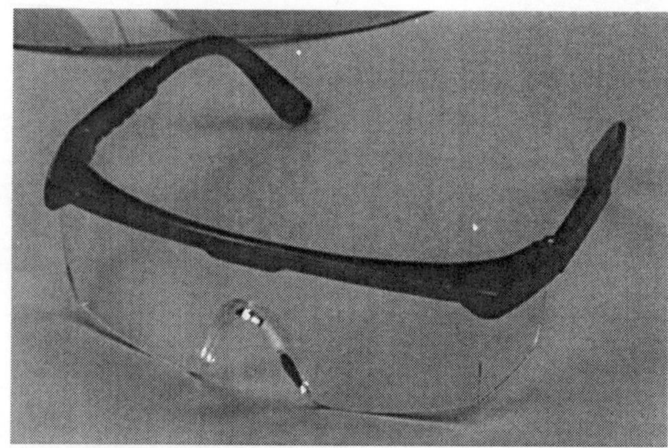

**Figure 6.** Safety glasses with side protection must be worn in the automotive shop.

## Safety Glasses and Face Shields

Wearing safety glasses or a face shield is one of the most important safety rules in an automotive shop. Face shields protect the face; safety glasses protect the eyes. When grinding, safety glasses must be worn, a face shield can be worn. Many shop insurance policies require the use of eye protection in the shop. Some automotive technicians have been blinded in one or both eyes because they did not bother to wear safety glasses. All safety glasses must be equipped with safety glass or plastic lenses, and they should provide some type of side protection **(Figure 6)**. When selecting a pair of safety glasses, they should feel comfortable on your face. If they are uncomfortable, you may tend to take them off, leaving your eyes unprotected. A face shield should be worn when handling hazardous chemicals or when using an air or electric grinder or buffer **(Figure 7)**.

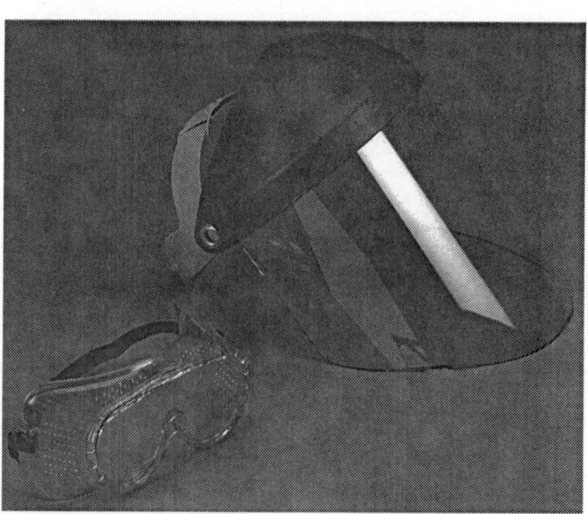

**Figure 7.** A face shield.

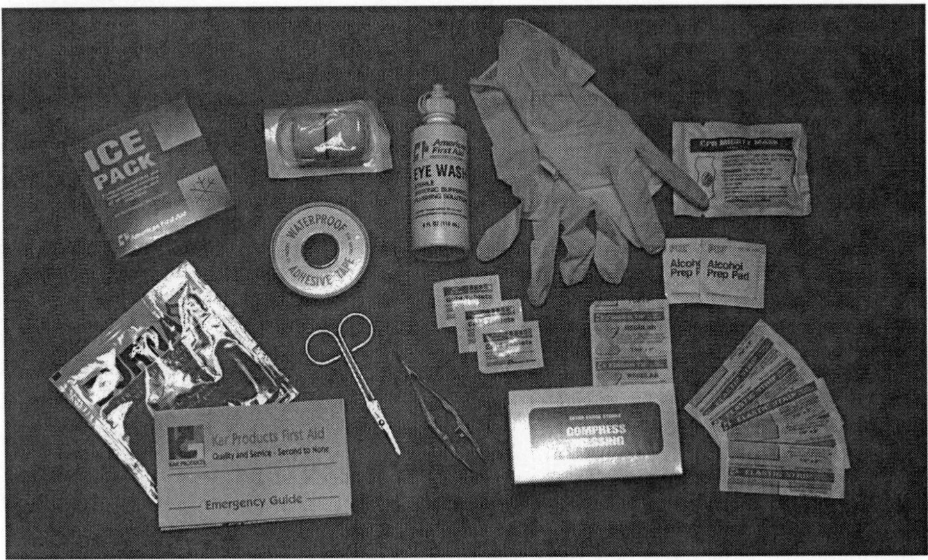


Figure 8. A first aid kit.

## First Aid Kits

First aid kits should be clearly identified and conveniently located **(Figure 8)**. These kits contain such items as bandages and ointment required for minor cuts. All shop personnel must be familiar with the location of first aid kits. At least one of the shop personnel should have basic first aid training. This person should be in charge of administering first aid and keeping first aid kits filled.

## SHOP LAYOUT

There are many different types of shops in the automotive service industry, including new car dealers, independent repair shops, specialty shops, service stations, and fleet shops.

The shop layout in any shop is important to maintain shop efficiency and contribute to safety. Every shop employee must be familiar with the location of all safety equipment in the shop. Shop layout includes bays for various types of repairs, space for equipment storage, and office locations. Most shops have specific bays for certain types of work, such as electrical repair, wheel alignment and tires, and machining **(Figure 9)**. Safety equipment such as fire extinguishers, first aid kits, and eyewash fountains must be in easily accessible locations, and the location of each piece of safety equipment must be clearly marked. Areas such as the parts department and the parts cleaning area must be located so they are easily accessible from all areas of the shop. The service manager's office should also be centrally located. All shop personnel should familiarize themselves with the shop layout, especially the location of

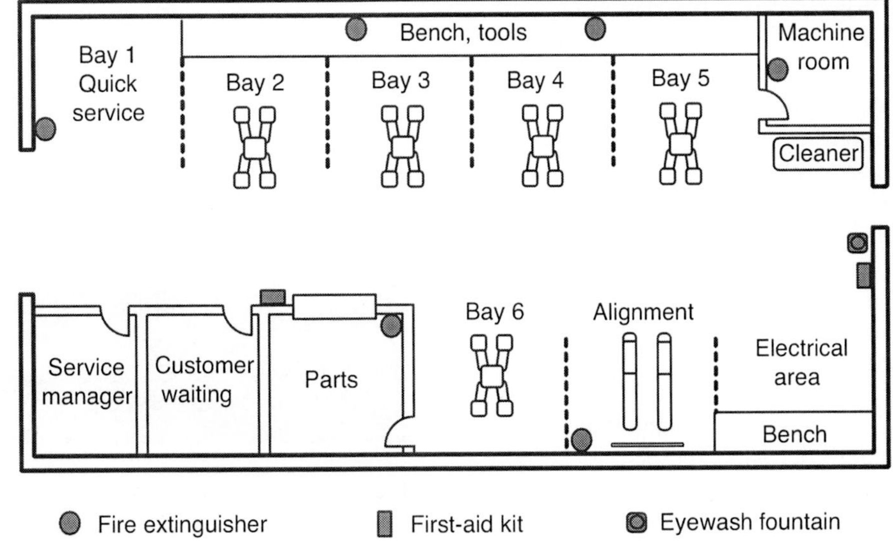


Figure 9. A shop layout.

safety equipment. If you know the exact fire extinguisher locations, you may put an extinguisher into operation a few seconds faster. Those few seconds could make the difference between a fire that is quickly extinguished and one that becomes out of control, causing extensive damage and personal injury.

The tools and equipment required for a certain type of work are stored in that specific bay. For example, the equipment for electrical and electronic service work is stored in the bay allotted to that type of repair. When certain bays are allotted to specific types of repair work, unnecessary equipment movement is eliminated. Each technician has his or her own tools on a portable roll cabinet that is moved to the vehicle being repaired. Special tools are provided by the shop, and these tools may be located on tool boards attached to the wall. Other shops may have a tool room where special tools are located. Adequate workbench space must be provided in those bays where bench work is required.

## HAZARDOUS WASTE DISPOSAL

Hazardous waste materials in automotive shops are chemicals or components that the shop no longer needs. These materials pose a danger to the environment and to people if they are disposed of in ordinary trash cans or sewers. However, it should be noted that no material is considered hazardous waste until the shop has finished using it and is ready to dispose of it. The Environmental Protection Agency (EPA) publishes a list of hazardous materials, which is included in the Code of Federal Regulations. Waste is considered hazardous if it is included on the EPA list of hazardous materials, or if it has one or more of these characteristics:

1. *Reactive.* Any material that reacts violently with water or other chemicals is considered hazardous. If a material releases cyanide gas, hydrogen sulphide gas, or similar gases when exposed to low-pH acid solutions, it is hazardous. A material that is **reactive** reacts with some other chemicals and gives off gas(es) during the reaction.
2. *Corrosive.* If a material burns the skin or dissolves metals and other materials, it is considered hazardous. A material that is **corrosive** causes another material to be gradually worn away by chemical action.
3. *Toxic.* Materials are hazardous if they leach one or more of eight heavy metals in concentrations greater than 100 times primary drinking water standard. A **toxic** substance is poisonous to animal or human life.
4. *Ignitable.* A liquid is hazardous if it has a flash point below 140°F (60°C). A solid is hazardous if it ignites spontaneously. A substance that is **ignitable** can be ignited spontaneously or by another source of heat or flame.

Federal and state laws control the disposal of hazardous waste materials. Every shop employee must be familiar with these laws. Hazardous waste disposal laws include the Resource Conservation and Recovery Act (RCRA). This law basically states that hazardous material users are responsible for hazardous materials from the time they become a waste until the proper waste disposal is completed.

Many automotive shops hire an independent hazardous waste hauler to dispose of hazardous waste material **(Figure 10)**. The shop owner or manager should have a written contract with the hazardous waste hauler. Rather than have hazardous waste material hauled to an approved hazardous waste disposal site, a shop may choose to recycle the material in the shop. An example of this would be a shop that has a machine to recycle used antifreeze, or a shop that is heated by using drained engine oil. Therefore, the user must store hazardous waste material properly and safely and be responsible for the transportation of this material until it arrives at an approved hazardous waste disposal site and is processed according to the law.

The RCRA controls these types of automotive waste:

1. Paint and body repair products waste
2. Solvents for parts and equipment cleaning
3. Batteries and battery acid
4. Mild acids used for metal cleaning and preparation
5. Waste oil, engine coolants, or antifreeze
6. Air-conditioning refrigerants
7. Engine oil filters

Never, under any circumstances, use these methods to dispose of hazardous waste material:

1. Pour hazardous wastes on weeds to kill them.
2. Pour hazardous wastes on gravel streets to prevent dust.
3. Throw hazardous wastes in a dumpster.
4. Dispose of hazardous wastes anywhere but an approved disposal site.
5. Pour hazardous wastes down sewers, toilets, sinks, or floor drains.

**Figure 10.** A hazardous waste hauler.

The right-to-know laws state that employees have a right to know when the materials they use at work are hazardous. The right-to-know laws started with the Hazard Communication Standard published by OSHA in 1983. This document was originally intended for chemical companies and manufacturers that required employees to handle hazardous materials in their work situation.

At the present time, most states have established their own right-to-know laws. Meanwhile, the federal courts have decided to apply these laws to all companies, including automotive service shops. Under the right-to-know laws, the employer has three responsibilities regarding the handling of hazardous materials by its employees.

First, all employees must be trained about the types of hazardous materials they will encounter in the workplace. Employees must be informed about their rights under legislation regarding the handling of hazardous materials. All hazardous materials must be properly labeled, and information about each hazardous material must be posted on material safety data sheets (MSDS), which are available from the manufacturer **(Figure 11)**.

The employer has a responsibility to place MSDS where they are easily accessible by all employees. The MSDS provide extensive information about the hazardous material, such as:

1. Chemical name
2. Physical characteristics
3. Protective equipment required for handling
4. Explosion and fire hazards
5. Other incompatible materials
6. Health hazards such as signs and symptoms of exposure, medical conditions aggravated by exposure, and emergency and first aid procedures
7. Safe handling precautions
8. Spill and leak procedures

Second, the employer has a responsibility to make sure that all hazardous materials are properly labeled. The label information must include health, fire, and reactivity hazards posed by the material, as well as the protective equipment necessary to handle the material. The manufacturer must supply all warning and precautionary information about hazardous materials, and this information must be read and understood by the employee before handling the material.

MATERIAL SAFETY DATA SHEET

PRODUCT NAME: KLEAN-A-KARB (aerosol) #- HPMS 102068
PRODUCT: 5078, 5931, 6047T
(page 1 of 2)

| 1. Ingredients | CAS # | ACGIH TLV | OSHA PEL | OTHER LIMITS | % |
|---|---|---|---|---|---|
| Acetone | 67-64-1 | 750ppm | 750ppm | | 2-5 |
| Xylene | 1330-20-7 | 100ppm | 100ppm | | 68-75 |
| 2-Butoxy Ethanol | 111-76-2 | 25ppm | 25ppm | (skin) | 3-5 |
| Methanol | 67-56-1 | 200ppm | 200ppm | | 3-5 |
| Detergent | - | NA | NA | | 0-1 |
| Propane | 74-98-6 | NA | 1000ppm | | 10-20 |
| Isobutane | 75-28-5 | NA | NA | 1000ppm | 10-20 |

2. PHYSICAL DATA : (without propellent)
Specific Gravity : 0.865
Vapor Pressure : ND
% Volatile : >99
Boiling Point : 176°F Initial
Evaporation Rate : Moderately Fast
Freezing Point : ND
Vapor Density : ND
Solubility: Partially soulable in water
pH :NA
Appearance and Odor: A clear colorless liquid, aromatic odor

3. FIRE AND EXPLOSION DATA
Flashpoint : –40°F
Method : TCC
Flammable Limits propellent
LEL: 1.8 UEL: 9.5
Extinguishing Media: $CO_2$, dry chemical, foam
Unusual Hazards : Aerosol cans may explode when heated above 120°F.

4. REACTIVITY AND STABILITY
Stability : Stable
Hazardous decomposition products: $CO_2$, carbon monoxide (thermal)
Materials to avoid: Strong oxidizing agents and sources of ignition

5. PROTECTION INFORMATION
Ventilation : Use mechanical means to insure vapor concentration is below TLV.
Respiratory: Use self-contained breathing apparatus above TLV.

Gloves : Solvent resistant
Eye and Face: Safety Glasses
Other Protective Equipment: Not normally required for aerosol product usage

**Figure 11.** Material safety data sheets (MSDS) inform employees about hazardous materials.

Third, employers are responsible for maintaining permanent files regarding hazardous materials. These files must include information on hazardous materials in the shop, proof of employee training programs, and information about accidents such as spills or leaks of hazardous materials. The employer's files must also include proof that employees' requests for hazardous material information such as MSDS have been met. A general right-to-know compliance procedure manual must be maintained by the employer.

## Summary

- The United States Occupational Safety and Health Act of 1970 assures safe and healthful working conditions and authorizes enforcement of safety standards.
- Many hazardous materials and conditions can exist in an automotive shop, including flammable liquids and materials, corrosive acid solutions, loose sewer covers, caustic liquids, high-pressure air, frayed electrical cords, hazardous waste materials, carbon monoxide, improper clothing, harmful vapors, high noise levels, and spills on shop floors.
- MSDS provide information regarding hazardous materials, labeling, and handling.
- The danger regarding hazardous conditions and materials may be avoided by eliminating shop hazards and applying the necessary shop rules and safety precautions.
- The automotive shop owner/management must supply the necessary shop safety equipment, and all shop personnel must be familiar with the location and operation of this equipment. Shop safety equipment includes gasoline safety cans, steel storage cabinets, combustible material containers, fire extinguishers, eyewash fountains, safety glasses and face shields, first aid kits, and hazardous waste disposal containers.

## Review Questions

1. While discussing shop hazards, Technician A says high-pressure air from an air gun may penetrate the skin. Technician B says air in the blood stream may be fatal. Who is correct?
   A. Technician A
   B. Technician B
   C. Both Technician A and Technician B
   D. Neither Technician A nor Technician B
2. While discussing hazardous waste disposal, Technician A says the right-to-know laws require employers to train employees regarding hazardous waste materials. Technician B says the right-to-know laws do not require employers to keep permanent records regarding hazardous waste disposal. Who is correct?
   A. Technician A
   B. Technician B
   C. Both Technician A and Technician B
   D. Neither Technician A nor Technician B
3. While discussing material safety data sheets (MSDS), Technician A says these sheets explain employers' and employees' responsibilities regarding hazardous waste handling and disposal. Technician B says these sheets contain specific information about hazardous materials. Who is correct?
   A. Technician A
   B. Technician B
   C. Both Technician A and Technician B
   D. Neither Technician A nor Technician B
4. While discussing hazardous material disposal, Technician A says certain types of hazardous waste material may be poured down a floor drain. Technician B says a shop is responsible for hazardous waste materials from the time they become waste until the proper waste disposal is completed. Who is correct?
   A. Technician A
   B. Technician B
   C. Both Technician A and Technician B
   D. Neither Technician A nor Technician B
5. According to health and safety inspection records, the percentage of workers who are exposed to health and safety hazards on the job is:
   A. 25 percent.
   B. 40 percent.
   C. 55 percent.
   D. 62 percent.
6. All of these statements about the Environmental Protection Agency (EPA) are true *except*:
   A. The EPA is responsible for air and water quality in the United States.
   B. The EPA was established to compile hazardous waste regulations.
   C. The EPA was established to compile vehicle manufacturing standards.
   D. The EPA was established to enforce hazardous waste regulations.
7. Technician A says loose clothing is dangerous because it may become entangled in rotating components on vehicles or shop equipment. Technician B says asbestos dust may be generated in the shop air during brake service. Who is correct?
   A. Technician A
   B. Technician B
   C. Both Technician A and Technician B
   D. Neither Technician A nor Technician B
8. All of these statements about the danger of wearing finger rings or metal watch bands in an automotive shop are true *except*:
   A. A ring or metal watch band may cause a chemical burn on the finger or arm.
   B. A ring or metal watch band can make electrical contact between an electrical terminal and ground.
   C. A ring or metal watch band may become very hot if high electric current flows through one of these components.
   D. A ring or metal watch band may cause severe burns to a finger or arm.
9. Technician A says when a vehicle is parked in the shop, the ignition switch should be left on the OFF position unless a service procedure requires another switch position. Technician B says oil rags should be stored in containers without lids. Who is correct?
   A. Technician A
   B. Technician B
   C. Both Technician A and Technician B
   D. Neither Technician A nor Technician B
10. All of these statements about carbon monoxide in the shop air are true *except*:
   A. Carbon monoxide causes headaches.
   B. Carbon monoxide may enter the shop air from vehicle exhaust.
   C. A restricted furnace chimney may cause carbon monoxide in the shop air.
   D. Carbon monoxide contributes to lung cancer.
11. The poisonous gas in vehicle exhaust is ________________ ________________.
12. Never direct high-pressure shop air from an air gun against human ____________ or ______________.
13. Wearing rings or other jewelry may cause severe ____________.
14. Breathing asbestos dust may cause ______________ ______________.
15. List four classes of fires and explain the type of fire extinguisher that should be used for each type of fire.
16. Explain two requirements related to location of safety equipment in the shop.
17. List five illegal methods of hazardous waste disposal.
18. List eight types of information found in MSDS related to hazardous materials.

# The Automotive Business

## *Introduction*

This chapter contains information regarding various types of automotive repair shops and the type of work done in each shop. As an automotive technology student, this information should assist you in deciding on the type of shop where you would like to work. Information regarding several automotive trade organizations is also included in this chapter. After you become a certified Automotive Technician, you will require frequent update training to learn about the fast-paced changes in Automotive Technology. One of the best ways to receive the necessary updated information is to join one or more trade organizations that supply current information.

## TYPES OF AUTOMOTIVE REPAIR SHOPS

It is important for students and technicians to understand the various types of automotive shops so they can decide which type of shop they would prefer to work in. This knowledge should lead to greater job satisfaction.

### Independent Repair Shops

Independent repair shops are the most common type of automotive repair facility **(Figure 1)**. Independent repair shops are privately owned and operated without being affiliated with a vehicle manufacturer, automotive parts manufacturer, or chain organization.

Independent repair shops may perform all types of repairs on a large variety of vehicles. In this type of operation, a wide variety of equipment is required, and the technicians must be knowledgeable regarding the diagnostic and repair procedures on many different vehicles. However, as automotive technology becomes more advanced, many of these shops are specializing in specific types of repairs such as undervehicle work that includes suspension, steering, wheel alignment and balancing, and brakes. Other independent shops may choose to specialize in repairs on specific makes of cars and light-duty trucks. Independent repair shops usually stock some of the more popular automotive parts, and they obtain most of the required parts from an automotive parts store or a dealership parts department.

### Specialty Shops

Specialty shops specialize in one type of automotive repair work. Specialty shops include muffler shops, transmission shops, fuel injection and tune-up shops, brake shops, and tire shops.

**Figure 1.** An independent automotive repair shop.

**Figure 2.** A specialty automotive shop may be part of a national or a regional automotive chain.

Specialty shops may be part of an automotive national or regional chain that has similar shops in other cities **(Figure 2)**. For example, many tire companies have a chain of tire stores across the United States. If specialty shops are part of an automotive chain, they are often an independently owned franchise. The owner must follow the policies of the chain organization. Specialty tire shops sell and install tires and usually specialize in undervehicle service such as wheel balancing and alignment, suspension repairs, and brake service. Muffler shops specialize in exhaust system repairs, but they may do other undervehicle repair work such as suspension and brakes.

## Quick-Lube Shops

Quick-lube shops specialize in lubrication work that consists of oil and filter changes and chassis lubrication. Most quick-lube shops specialize in performing this work quickly while the owner or driver waits in the shop waiting room. Some quick-lube shops use pits rather than vehicle lifts because using a pit allows them to perform faster lubrication work. The vehicle is driven over a service pit, and one technician in the pit drains the oil, changes the filter, and lubricates the chassis. Another technician working under the hood checks the fluid levels and installs the new oil. Quick-lube shops usually stock different brands of oil so they can supply the type of oil requested by the customer. Quick-lube shops usually perform a series of checks while performing lubrication work. These checks may include fluid levels, light operation, belt condition, wiper blade condition, and tire condition. Some quick-lube shops perform minor service work such as replacing belts, lights, and wiper blades.

*In 2003, the average independent automotive repair facility in the United States was a 5,854-square foot facility with 7 service bays. Each of these repair facilities employed an average of 3 technicians with over 5 years experience, and the average experienced technician's salary was $47,948.00.*

**Figure 3.** A dealership sells and services vehicles manufactured by one or more vehicle manufacturers.

## Automotive Dealerships

Automotive dealerships are usually independently owned, but they have a contract to market and service the vehicles from one or more vehicle manufacturer. Small dealerships may sell and service only one type of vehicle, whereas larger dealerships usually sell and service several vehicle makes **(Figure 3)**. The dealership must conform to certain standards set by the vehicle manufacturer. These standards may include regulations regarding the size and layout of the facility, new vehicle sales policies, accounting and financing practices, and service procedures.

The dealership usually has a general manager who is responsible for the entire operation. In some dealerships, the general manager may be the owner. The department managers are responsible to the general manager. Department managers usually include a business manager, sales manager, parts manager, and service manager. Business office staff may include accountants and a receptionist, and these personnel are responsible to the office manager. The sales manager is responsible for the sales personnel. In larger dealerships there may be separate managers for new vehicle sales and used vehicle sales. The parts manager is responsible for parts personnel, and the service manager is responsible for shop personnel including service writer(s), technicians, technician's assistant, and clean-up personnel. Dealership technicians are required to have a complete knowledge of the diagnostic and service procedures on the vehicles sold by the dealership, and the technician is required to perform warranty service on these vehicles.

# SUCCESSFUL DEALERSHIP MANAGEMENT

The top priority in all departments of the dealership must be customer service and satisfaction. When the sales department sells a customer a new vehicle, the customer may be very impressed with the professionalism and expertise of the sales personnel. However, if the customer takes this new vehicle to the service department and encounters service personnel that are indifferent to the customer's concerns regarding the vehicle, where do you think the customer will purchase his or her next vehicle? All dealership personnel must be proficient in public relations. All dealership staff must be positive and polite at all times when dealing with customers, and they must exhibit a concern for the customers and their vehicles.

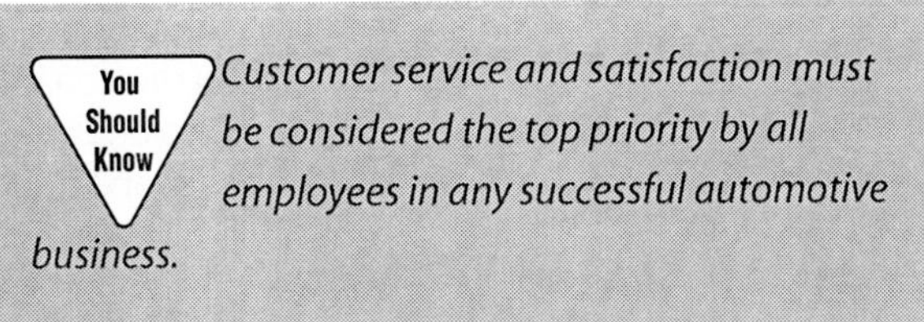

*You Should Know* — *Customer service and satisfaction must be considered the top priority by all employees in any successful automotive business.*

Dealerships must also place high priority on communication and cooperation.

For example, there must be excellent communication between all departments for the dealership to operate efficiently and profitably. Regular staff meetings are essential to voice concerns and provide communication between departments. Adequate communication between service personnel and customers is also essential. Service personnel should always explain in basic terms the problem with a customer's vehicle, and also the cost of repairs. Some dealerships have a policy of introducing their new vehicle purchasers to some of the service department personnel, including a technician. The customer appreciates personalized attention, and this provides the customer with a positive image of the dealership. It is easy to lose this personalized attention for the customer in large dealership operations. Small touches like calling a customer by name, improve public relations. Technicians must always place fender, seat, and floor mat covers on customer vehicles while performing service work on the vehicle.

## AUTOMATIC TRANSMISSION REBUILDERS ASSOCIATION (ATRA)

The Automatic Transmission Rebuilders Association (ATRA) was founded in 1954.

ATRA provides technical information to transmission technicians and rebuilding shops.

ATRA is the oldest and largest organization of independently owned transmission rebuilding firms with members in the United States, Canada, and twenty-two other countries. ATRA is dedicated to the welfare and improvement of the automatic transmission repair industry for the benefit of the motoring public. ATRA distributes technical and business information through bulletins, books, seminars, *GEARS* magazine, a yearly exposition, and Internet programs. Specific information may be downloaded from the ATRA website. This site may be accessed at http://www.atra-gears.com/.

The ATRA bookstore is exclusively for ATRA members and certified ATRA technicians. This bookstore supplies books related to business management including topics such as advertising, marketing, and taxes. Books, bulletins, technical manuals, and technical training videos are also available. The bookstore also markets ATRA member patches and chevrons. ATRA also provides customer bulletins on topics such as "What is a REMAN transmission?"

ATRA also provides technician certification tests in these areas: Transmission Rebuilder, Transmission Remove and Replace Technician, Transmission Diagnostician. These tests are provided at many locations throughout North America.

The *GEARS* magazine published by ATRA provides information on technical shop and office management, safety and EPA concerns, new products and new industry developments, and current interests.

ATRA provides technical seminars in many different locations. These seminars explain the popular transmissions used in vehicles built by the major manufacturers. These technical seminars provide transmission fixes, updates, and modifications to correct many transmission problems.

ATRA's yearly Powertrain Industry Exposition provides technical update information, new product information, demonstrations regarding new equipment or products, professional development, and networking opportunities.

*You Should Know* — *One of the best ways to keep up to date on automotive technology is to become a member of an automotive trade organization that provides information related to the area of automotive service in which you are working.*

## MOBILE AIR CONDITIONING SOCIETY WORLDWIDE (MACS WORLDWIDE)

The Mobile Air Conditioning Society Worldwide (MACS Worldwide) was founded in 1981. MACS provides technical and business information through the Automotive Cooling Journal, MACS Action, and MACS Service Reports.

MACS was founded to meet the automotive air conditioning (A/C) industry's need for comprehensive technical information, training, and communication. MACS Worldwide has over 1,600 members including A/C service shops, installers, distributors, component suppliers, and manufacturers. These members are located throughout the United States, Canada, and many other countries. The MACS website may be accessed at http://www.macsw.org.

MACS provides members with up-to-date technical and business information through a number of publications. These publications include *Automotive Cooling Journal, MACS Action,* and *MACS Service Reports.* The *Automotive Cooling Journal* is published monthly and provides reports on people, industry trends, new industry developments, technical information, and industry events such as trade shows and expositions. *MACS Action* is published bimonthly and provides members with information on marketing, business management, and association news. MACS

Service Reports are published monthly, and this technical publication provides A/C service bulletins, service tips, and specific technical information regarding A/C and cooling system service.

MACS provides training programs in these areas:

1. Automotive Air Conditioning, Diagnosis, and Service
2. Automotive Heating and Air Conditioning Test Preparation
3. Electronics and Electrical Control with an Introduction to Automatic Temperature Control
4. Guidelines for Automotive Air Conditioning Retrofit
5. Problem Servicing Mobile Air Conditioning Systems

MACS training programs are sold as a package that contains specific materials such as a student manual, instructor guide package, overhead transparencies, color slides, and a video. Instructors who wish to purchase MACS training programs numbered 1 through 3 above must have a MACS Technician Training Institute (TTI) training certificate. If an instructor holds a current state-approved teaching certificate, MACS will issue a MACS TTI certificate to the instructor upon submission of the instructor's credentials to MACS. When an instructor does not have a current state approved teaching certificate, the instructor must attend a seminar approved by MACS to obtain a MACS TTI certificate.

The MACS Technician Training and Certification Program for Refrigerant Recovery and Recycling is approved by the EPA, and is offered as a self-study program conducted by mail or in a classroom environment with an instructor. Instructors must meet specific criteria to become a MACS Certification Trainer/Proctor.

MACS sponsors an annual convention and trade show for members. This convention provides technical seminars, special speakers, and a display of new products and equipment at the trade show.

## AUTOMOTIVE SERVICE ASSOCIATION (ASA)

The Automotive Service Association (ASA) had its beginning in 1951 when a group of automotive shop owners recognized that the problems facing independent automotive repair businesses could be solved more efficiently through their association and cooperation. This recognition led to the formation of the Independent Garageman's Association of Texas that evolved through various organizations into the ASA. ASA promotes professionalism and excellence in the automotive repair industry through education, representation, and member services.

ASA presently has 12,000 member shops and individuals representing 65,000 professionals in many different countries. The purpose of ASA is to advance the professionalism and excellence in the automotive repair industry through education, representation, and member services. ASA has a mechanical division and a collision division. The ASA website may be accessed at http://www.asashop.org.

ASA publishes the monthly *AutoInc.* magazine. This magazine contains mechanical, collision, and management information. *AutoInc.* also includes a report from ASA's legislative office in Washington, D.C. ASA publishes monthly bulletins entitled *Mechanical Division Dispatch and Collision Repair Report.*

ASA sponsors the annual Congress of Automotive Repair and Service (CARS) for the mechanical division. This conference provides technical seminars, forums on current industry concerns such as inspection/maintenance (I/M) programs, and Automotive Management Institute AMI seminars. ASA also hosts an annual convention that provides division meetings, AMI seminars, and keynote speakers.

ASA operates an automotive hotline called Identifix. This hotline service provides diagnostic information via telephone to solve specific automotive problems. A hotline service such as Identifix employs experts in various areas of automotive expertise, and they have all the necessary service manuals and diagnostic computer software.

Identifix offers other services such as faxing wiring diagrams upon request. Identifix pubishes a bi-monthly Identifix Update bulletin containing many service tips.

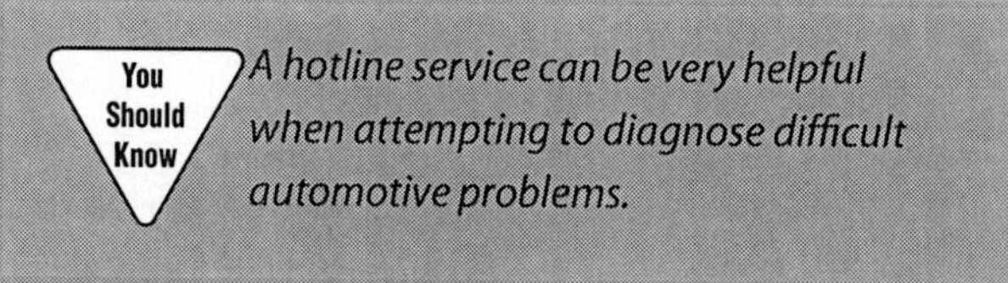


*A hotline service can be very helpful when attempting to diagnose difficult automotive problems.*

## INTERNATIONAL AUTOMOTIVE TECHNICIANS NETWORK (iATN)

The International Automotive Technicians Network (iATN) is a group of over 43,000 automotive technicians in 138 countries. These members share technician knowledge and information with other members via the Internet. iATN is the largest network of automotive technicians in the world.

The iATN purpose is to promote the continued growth, success, and image of the professional automotive technician by providing a forum for the exchange of knowledge and the promotion of education, professionalism, and integrity. The iATN website may be accessed at http://www.iatn.net.

The iATN goals are these:

1. To provide programs and services that promote the professional growth and effectiveness if iATN members through the exchange of knowledge.
2. To provide an interactive, technology-rich learning environment.
3. To increase the public's understanding, appreciation, and respect for the Automotive Service Industry.

4. To promote public awareness of the iATN and its members.
5. To continually adapt to the changing needs of supporting iATN members.

iATN members agree to:

1. Uphold the highest standards of professionalism, competence, and integrity.
2. Pursue excellence through ongoing education and the exchange of knowledge and experience of fellow iATN members.
3. Maintain quality through the use of proper tools, equipment, parts, and procedures.
4. Promote public awareness of the importance of quality professional service and the advancing technology of the modern automobile.
5. Support and promote the mission and goals of the iATN.

The iATN is supported by a number of top companies, associations, and publications in the automotive industry. If an iATN member encounters an automotive problem, iATN suggests that the technician attempt to diagnose the problem using information from service manuals, data systems such as Mitchell-on-Demand, and technical hotlines. The technician can also access the iATN Fix Database. If the technician still cannot solve the problem, a problem file may be sent to iATN outlining the necessary vehicle information and the details of the problem. This problem file is sent via the Internet to many iATN members. Any iATN member who has experienced and solved the same problem can post a fix file to iATN that outlines the answer to the automotive problem and the required fix procedure. An iATN member may configure his or her account to only receive problem and fix files in specific areas of automotive repair such as engine electronics and fuel systems. The iATN offers many other member benefits such as live conferencing and technical forums. These forums include technical discussion forums, technical theory forums, technical tip forums, and tool and equipment forums.

## Summary

- Independent repair shops are privately owned, and they are not affiliated with vehicle manufacturers, automotive parts suppliers, or chain organizations.
- Specialty shops usually perform one type of automotive repair work.
- Quick-lube shops specialize in fast lubrication service.
- Automotive dealerships have a contract with one or more vehicle manufacturers to sell and service the manufacturer's vehicles.
- The ATRA distributes technical and business information to automotive transmission technicians and rebuilding shops.
- MACS Worldwide provides technician information, training, and communication for the automotive A/C industry.
- The ASA provides membership service such as the monthly *AutoInc.* magazine and a yearly Congress of Automotive Repair and Service.
- The iATN provides forums for the exchange of information in the automotive industry.

## Review Questions

1. In an automotive dealership, the vehicle manufacturer may set standards regarding all of these items *except*:
   A. the layout of the facility.
   B. new vehicle sales policies.
   C. accounting and financing practices.
   D. staff benefits policies.
2. When discussing dealership management, Technician A says customer service and satisfaction is extremely important. Technician B says communication between dealership departments is not a high priority. Who is correct?
   A. Technician A
   B. Technician B
   C. Both Technician A and Technician B
   D. Neither Technician A nor Technician B
3. All of these statements about the ATRA are true *except*:
   A. ATRA was founded in 1954.
   B. ATRA has members in 24 countries.
   C. ATRA distributes information on most automotive electronic systems.
   D. ATRA publishes customer bulletins.
4. When discussing the iATN, Technician A says the iATN has members in 136 countries. Technician B says the iATN is helpful when diagnosing automotive problems. Who is correct?
   A. Technician A
   B. Technician B
   C. Both Technician A and Technician B
   D. Neither Technician A nor Technician B

5. Technician A says a wide variety of test equipment is required in an independent repair facility. Technician B says technicians working in an independent repair facility may be required to be knowledgeable regarding the diagnostic and service procedures on many different vehicles. Who is correct?
   A. Technician A
   B. Technician B
   C. Both Technician A and Technician B
   D. Neither Technician A nor Technician B
6. Technician A says an independent repair shop may specialize in specific types of repairs. Technician B says an independent repair shop may specialize in repairs on specific makes of vehicles. Who is correct?
   A. Technician A
   B. Technician B
   C. Both Technician A and Technician B
   D. Neither Technician A nor Technician B
7. All of these statements about a typical automotive dealership are true *except*:
   A. A general manager is responsible for the entire operation.
   B. The parts manager is responsible for the shop technicians and personnel.
   C. A large dealership may have a new vehicle sales manager and a used vehicle sales manager.
   D. The business manager is responsible for the office staff.
8. Technician A says all dealership personnel must be proficient in public relations. Technician B says the top priority in all dealership departments must be customer service and satisfaction. Who is correct?
   A. Technician A
   B. Technician B
   C. Both Technician A and Technician B
   D. Neither Technician A nor Technician B
9. The ASA promotes professionalism and excellence in the automotive repair industry through all of these methods *except*:
   A. test equipment sales.
   B. education.
   C. representation.
   D. member services.
10. Technician A says the ASA operates an automotive hotline service called Fix-It-Right. Technician B says an automotive hotline service provides diagnostic information via telephone to solve specific automotive problems. Who is correct?
    A. Technician A
    B. Technician B
    C. Both Technician A and Technician B
    D. Neither Technician A nor Technician B
11. Quick-lube shops may also perform ___________ service work.
12. The ATRA provides technician certification tests in these areas:
    1. ___________________________
    2. ___________________________
    3. ___________________________
13. The ASA publishes a monthly magazine entitled ___________________.
14. ATRA publishes a magazine entitled _____________.
15. Describe the advantages of ATRA membership for automotive technicians.
16. Describe the MACS publications and training programs for automotive technicians.
17. Explain the purposes of Identifix.
18. Describe the benefits of iATN membership for automotive technicians.

# Basic Shop Operation

## Introduction

This chapter discusses the following topics: job responsibilities for each shop position, completing and processing work orders, employer and employee obligations, job responsibilities for technicians, technician certification, certification of automotive training programs, job applications and resumes, and customer relations.

It is very important for technicians to understand the job responsibilities of each shop employee. When technicians have this understanding, they know the person to whom they are responsible in the shop operation. Technicians are often promoted to other positions in the shop. When technicians are familiar with the job responsibilities in the shop, they are better prepared to accept one of the management positions.

Technicians must also be familiar with work orders and the shop procedure for completing and processing these orders. A thorough understanding of employer and employee obligations helps the technician to meet employer expectations with excellent job performance. A technician must be familiar with job responsibilities to perform efficiently on the job.

Technicians must have a knowledge of ASE certification so they can choose their area(s) of certification. Technicians must also know the value of ASE certification. Technicians should be familiar with the NATEF certification for automotive training programs.

When applying for a job, technicians must know how to properly complete a job application and resume. Technicians must understand how to establish and maintain excellent customer relations.

### SERVICE WRITERS

The service writer is extremely important in maintaining excellent customer relations because the person in this position is usually the first person to greet the customer. In a typical automotive repair shop, the service writer meets the customers and completes the work orders.

The service writer should always be neatly and cleanly dressed, and he or she must be polite and professional. Photo identification with the service writer's name clearly displayed helps the customer to know the service writer. The service writer should also call the customer by name.

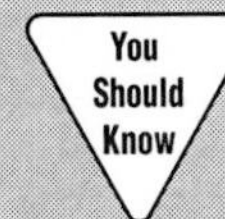


*All personnel in a successful automotive shop must place top priority on customer service and satisfaction.*

The service writer is responsible for completing the work order. The service writer must listen carefully to the customer's concerns. If the customer's problem is obvious from the operation of the vehicle with the engine running in the shop, the service writer does not need to question the vehicle owner any further regarding the problem. However, when the customer's complaint is not obvious from the operation of the vehicle while the engine is running in the shop, the service writer should politely question the customer regarding the problem. The service writer should ask the customer regarding the exact symptoms and/or sounds related to the problem with the customer's vehicle. The service writer should also find out from the customer if the problem occurs only at a certain vehicle speed or a specific temperature. The service writer must write an accurate description of the customer's problem and the necessary repairs on the work order to allow the service department technician to accurately diagnose and repair the problem and correct the customer's complaint.

If the customer requests an estimate for the vehicle repairs, the service writer should obtain the estimate as quickly as possible. The service writer should also be sure any other customer concerns are met. For example, the customer may require the service of the customer shuttle vehicle to get to his or her place of work. If the customer is going to wait for a quick repair, the service writer should be sure the customer knows where the customer waiting room is located in the building.

## CASHIERS

The cashier's job is very important because he or she is usually the last person to deal with the customer. The cashier presents the customer with the completed work order and collects the payment for the work completed.

It is extremely important for the cashier to appear well dressed and professional. The cashier must always remain polite and courteous. Sometimes customers are shocked by the cost of the repairs to their vehicle. The cashier must be willing to explain the cost of the repairs. In some cases this may require the assistance of someone from the service department, such as the service writer or service manager. In many shops, the cashier is responsible for entering the customer's name, address, and vehicle information in a computer database. This information is used to remind customers regarding scheduled maintenance or other special service offers from the service department.

## SERVICE MANAGER

The service manager is responsible for the complete shop operation. The service manager is responsible for the general operation, all shop personnel, and handling customer complaints.

Large automotive shops may have a shop foreman who works under the supervision of the service manager. Customer complaints are usually taken care of by the service manager. Therefore, the service manager must have excellent public relations skills. The service manager must also have good organizational skills and extensive experience in the automotive service industry. The service manager is responsible for implementing the vehicle manufacturer's policies on warranty and service procedures. The service manager completes the necessary arrangements and scheduling for technician training. In most shops, the service manager is responsible for hiring shop personnel. The service manager is responsible for communication and cooperation with other departments and personnel in the business.

## SHOP FOREMAN

The shop foreman is responsible for the quality of the repair work completed by the technicians. The shop foreman supervises the technicians and helps them with service problems that are difficult to diagnose.

Some shops may have a lead technician that performs this job. The shop foreman must have good public relations skills, extensive experience in automotive repair, and excellent diagnostic ability. The service manager and the shop foreman must be familiar with the latest technology on the vehicles sold by the dealership.

## REPAIR ORDERS

Repair orders may vary depending on the shop, but repair orders usually have this basic information:

1. Customer's name, address, and phone number(s)
2. Customer's signature
3. Vehicle make, model, year, and color
4. Vehicle identification number (VIN), **Figure 1**
5. Vehicle mileage
6. Engine displacement
7. Date and time
8. Service writer's code number
9. Work order number
10. Labor rate
11. Estimate of repair costs
12. Accurate and concise description of the vehicle problem

In many shops, the repair orders are completed on a computer terminal, and the computer may automatically write the vehicle repair history on the repair order if the vehicle has a previous repair history in the shop's computer system.

**Figure 1.** The VIN is visible through the driver's side of the windshield.

## TECHNICIANS AND REPAIR ORDERS

The repair order informs the technician regarding the problem(s) with the vehicle. In many shops the technician has to enter a starting time on the repair order. This may be done on a computer terminal or by inserting the work order into the time clock. The technician may also have to enter his code number on the work order to indicate who worked on the vehicle. The technician's code number on the work order is also used to pay the technician for the repair job. The technician must diagnose and repair the problem(s) indicated on the repair order. When the technician obtains parts from the parts department to complete the repair, the technician must present the order number. The parts personnel enter the parts and the cost on the repair order. In some shops the technician is required to enter the completed repairs on the work order. For example, the description of the problem on the work order may be "A/C system inoperative." If the technician replaced the A/C compressor fuse to correct the problem, he or she may enter "Replaced A/C compressor fuse" on the order. Some shops require the service writer or shop foreman to sign the work order when the repair job is successfully completed. The work order is routed back to the cashier who calculates all the charges on the work order including the appropriate taxes. Some shops add a miscellaneous charge on the work order. A typical miscellaneous charge is 10 percent of the total charges on the work order. This miscellaneous charge is to cover the cost of small items such as bolts, cotter pins, grease, lubricants, and sealers that are not entered separately on the work order.

In many shops the technicians work on a flat-rate basis. In these shops the technician is paid a flat rate for each repair. In a dealership, the flat-rate time is set by the vehicle manufacturer. Independent shops use generic flat-rate manuals published by firms such as Mitchell Publications. If the flat-rate time is 2.0 hours for completing a specific vehicle repair, the customer is charged for 2.0 hours, and the technician is paid for 2.0 hours even though he or she completed the repair in 1.5 hours. Conversely, if the technician takes 2.5 hours to complete the job, the technician is only paid for 2.0 hours and the customer is charged for 2.0 hours. The flat-rate time is usually entered on the work order.

## EMPLOYEE TO EMPLOYER OBLIGATIONS

The ever-increasing electronics content on today's vehicles requires that technicians are familiar with the latest electronics technology. There are many different ways to obtain training on the latest automotive technology, but it is absolutely essential. Automotive training may be obtained by these methods:

1. Obtaining training information, service manuals, or bulletins from OEMs, independent parts and component manufacturers, or independent suppliers of service manuals and training books. After the information is obtained, it is essential that you read and study it.
2. Join an organization dedicated to supplying information to automotive service technicians such as the ATRA.
3. Join an internet organization such as the iATN, where you can communicate with other technicians and obtain the answers to service problems.
4. Download information available on the Internet from automotive equipment manufacturers. Many of these manufacturers provide operator's manuals and other information on their equipment for downloading purposes.
5. Attend satellite training seminars available from some independent automotive training organizations and OEMs.
6. Attend training seminars in your location sponsored by equipment and parts manufacturers or OEMs.
7. Attend training seminars sponsored by the automotive department at your local college.

The successful automotive technician must be committed to life-long training, but the technician's employer must also be committed to assisting technicians employed in his or her shop to obtain the necessary training. This assistance may be financial, providing time off work to attend training seminars, or arranging the necessary training programs.

When you begin employment, you enter into a business agreement with your employer. A business agreement involves an exchange of goods or services that have value. Although the automotive technician may not have a written agreement with his or her employer, the technician exchanges time, skills, and effort for money paid by the employer. Both the employee and the employer have obligations. The automotive technician's obligations include the following:

1. Productivity: As an automotive technician, you have a responsibility to your employer to make the best possible use of time on the job. Each job should be done in a reasonable length of time. Employees are paid for their skills, effort, and time.
2. Quality: Each repair job should be a quality job! Work should never be done in a careless manner. Nothing improves customer relations like quality workmanship.
3. Teamwork: The shop staff are a team, and everyone including technicians and management personnel are team members. You should cooperate with and care about other team members. Each member of the team should strive for harmonious relations with fellow workers. Cooperative teamwork helps improve shop efficiency, productivity, and customer relations.

Customers may be "turned off" by bickering among shop personnel.

4. Honesty: Employers and customers expect and deserve honesty from automotive technicians. Honesty creates a feeling of trust among technicians, employers, and customers.
5. Loyalty: As an employee, you are obligated to act in the best interests of your employer, both on and off the job.
6. Attitude: Employees should maintain a positive attitude at all times. As in other professions, automotive technicians have days when it may be difficult to maintain a positive attitude. For example, there will be days when the technical problems on a certain vehicle are difficult to solve. However, a negative attitude certainly will not help the situation! A positive attitude has a positive effect on the job situation as well as on the customer and employer.
7. Responsibility: You are responsible for your conduct on the job and your work-related obligations. These obligations include always maintaining good workmanship and customer relations. Attention to details such as always placing fender, seat, and floor mat covers on customer vehicles prior to driving or working on the vehicle greatly improve customer relations.
8. Following directions: All of us like to do things "our way." Such action, however, may not be in the best interests of the shop, and as an employee you have an obligation to follow the supervisor's directions.
9. Punctuality and regular attendance: Employees have an obligation to be on time for work and to be regular in attendance on the job. It is very difficult for a business to operate successfully if it cannot count on its employees to be on the job at the appointed time.
10. Regulations: Automotive technicians should be familiar with all state and federal regulations pertaining to their job situation, such as the Occupational Safety and Health Act (OSHA) and hazardous waste disposal laws.

## EMPLOYER TO EMPLOYEE OBLIGATIONS

Employer to employee obligations include:

1. Wages: The employer has a responsibility to inform the employee regarding the exact amount of financial remuneration they will receive and when they will be paid.
2. Fringe benefits: A detailed description of all fringe benefits should be provided by the employer. These benefits may include holiday pay, sickness and accident insurance, and pension plans.
3. Working conditions: A clean, safe workplace must be provided by the employer. The shop must have adequate safety equipment and first aid supplies. Employers must be certain that all shop personnel maintain the shop area and equipment to provide adequate safety and a healthy workplace atmosphere.
4. Employee instruction: Employers must provide employees with clear job descriptions, and be sure that each worker is aware of his or her obligations.
5. Employee supervision: Employers should inform their workers regarding the responsibilities of their immediate supervisors and other management personnel.
6. Employee training: Employers must make sure that each employee is familiar with the safe operation of all the equipment that they are required to use in their job situation. Since automotive technology is changing rapidly, employers should provide regular update training for their technicians. Under the right-to-know laws, employers are required to inform all employees about hazardous materials in the shop. Employees should be familiar with MSDS, which detail the labeling and handling of hazardous waste and the health problems if exposed to hazardous waste.

## JOB RESPONSIBILITIES

An automotive technician has specific responsibilities regarding each job performed on a customer's vehicle. These job responsibilities include:

1. Do every job to the best of your ability. There is no place in the automotive service industry for careless workmanship! Automotive technicians and students must realize they have a very responsible job. During many repair jobs you, as a student or technician working on a customer's vehicle, actually have the customer's life and the safety of his or her vehicle in your hands. For example, if you are doing a brake job and leave the wheel nuts loose on one wheel, that wheel may fall off the vehicle at high speed. This could result in serious personal injury for the customer and others, plus extensive vehicle damage. If this type of disaster occurs, the individual who worked on the vehicle and the shop may be involved in a very expensive legal action. As a student or technician working on customer vehicles, you are responsible for the safety of every vehicle that you work on! Even when careless work does not create a safety hazard, it leads to dissatisfied customers who often take their business to another shop. Nobody benefits when that happens.
2. Treat customers fairly and honestly on every repair job. Do not install parts that are unnecessary to complete the repair job.
3. Use published specifications; do not guess at adjustments.
4. Follow the service procedures in the service manual provided by the vehicle manufacturer or an independent manual publisher.
5. When the repair job is completed, always be sure the customer's complaint has been corrected.
6. Do not be too concerned with work speed when you begin working as an automotive technician. Speed comes with experience.

## NATIONAL INSTITUTE FOR AUTOMOTIVE SERVICE EXCELLENCE (ASE)

The National Institute for Automotive Service Excellence (ASE) has provided voluntary testing and certification of automotive technicians on a national basis for many years. The ASE provides certification for technicians in various areas such as Automotive, Medium/Heavy Duty Truck, and Collision Repair.

The image of the automotive service industry has been enhanced by the ASE certification program. More than 415,000 technicians now have current certifications and work in a wide variety of automotive service shops. ASE provides certification in eight areas of automotive repair: engine repair, automatic transmissions/transaxles, manual drive train and axles, suspension and steering, brakes, electrical systems, heating and air conditioning, and engine performance.

A technician may take the ASE test and become certified in any or all of the eight areas. When a technician passes an ASE test in one of the eight areas, an Automotive Technician's shoulder patch is issued by ASE. If a technician passes all eight tests, he or she receives a Master Technician's shoulder patch **(Figure 2)**. Retesting at 5-year intervals is required to remain certified.

The certification test in each of the eight areas contains forty to eighty multiple-choice questions. The test questions are written by a panel of automotive service experts from various areas of automotive service including automotive instructors, service managers, automotive manufacturers' representatives, test equipment representatives, and certified technicians. The test questions are pretested and checked for quality by a national sample of technicians. On an ASE certification test, approximately 45 percent to 50 percent of the questions are Technician A and Technician B format, and the multiple-choice format is used in 40 percent to 45 percent of the questions. Less than 10 percent of ASE certification questions are an *except* format where the technician selects one incorrect answer out of four possible answers. ASE tests are designed to test the technician's understanding of automotive systems, testing and diagnostic knowledge, and ability to follow proper repair procedures.

**Figure 2.** ASE certification shoulder patches worn by automotive technicians and master technicians.

ASE regulations demand that each technician must have 2 years of working experience in the automotive service industry prior to taking a certification test or tests. However, relevant formal training may be substituted for 1 year of working experience. Contact ASE for details regarding this substitution.

ASE also provides certification tests in automotive specialty areas such as Advanced Engine Performance Specialist, Alternate Fuels Light Vehicle Compressed Natural Gas, Parts Specialist, Machinist, Cylinder Head Specialist, Machinist, Cylinder Block Specialist, and Machinist, Assembly Specialist.

Shops that employ ASE-certified technicians display an official ASE blue seal of excellence. This blue seal increases the customer's awareness of the shop's commitment to quality service and the competency of certified technicians.

## NATIONAL AUTOMOTIVE TECHNICIANS EDUCATION FOUNDATION (NATEF)

In 1978, the Industry Planning Council (IPC) and American Vocational Association (AVA) were concerned about the quality of automotive education across the United States. The IPC was composed of representatives from the automotive industry and vocational education. These two organizations directed a multi-year study which developed an automotive task list and an evaluation guide for automotive training programs. The ASE was selected to administer the evaluation of automotive training programs. NATEF was established as a separate foundation and an affiliate of ASE. NATEF evaluates and certifies automotive, autobody and medium/heavy truck training programs.

The evaluation of automotive training programs involves a comprehensive self-evaluation and an on-site evaluation. During the self-evaluation program, instructors, administrators, and advisory committee members rate the automotive program using these standards:

1. Purpose
2. Administration
3. Learning resources
4. Finances
5. Student services
6. Instruction
7. Equipment
8. Facilities
9. Instructional staff
10. Cooperative work agreements

The self-evaluation helps to identify areas in the automotive training program that require improvement according to national standards. After the self-evaluation is completed, the automotive program submits the self-evaluation materials with an application to NATEF. An on-site evaluation is scheduled by NATEF if the training program appears to meet the required standards. An Evaluation Team Leader (ETL) and technicians in the training program conduct the on-site evaluation. The ETL must be an educator with ASE master-technician certification and NATEF program evaluation training. The ETL, in cooperation with the evaluation team, submits a report to NATEF. If the automotive training program meets the required standards, the program receives NATEF certification. After NATEF certification is obtained, the automotive training program may display the ASE-NATEF logo **(Figure 3)**. Automotive training programs must be recertified every 5 years. More than 1,000 automotive training programs have been certified by NATEF. Autobody and Medium/Heavy Duty Truck training programs are also certified by NATEF.

In 1995, Ohio State University (OSU) conducted an extensive study regarding the effect of NATEF certification on student learning. OSU researchers concluded that NATEF certification has a significant positive effect on learning that takes place in automotive technician training programs.

**Figure 3.** The ASE-NATEF logo.

## JOB APPLICATION

The first impression that you make on a prospective employer is very important! Your letter of application and resume are very important when you are looking for employment. Some larger businesses have a standard job application form that must be filled out when applying for employment. This application form will request all the pertinent facts regarding your past employment experience, training, job preferences, special skills, and personal information. Always fill out the job application form completely and neatly.

Smaller businesses may request a resume and a letter of application from those seeking employment. The resume is a brief, one-page document that includes the following information.

1. Address and telephone number.
2. Work experience—List your previous work experience beginning with your most recent job. Most employers want a work history for the last 5 years. List the dates of employment, and include the names and addresses of previous employers.
3. References—List the names, addresses, and phone numbers of two or three previous employers or instructors who are familiar with your work and/or training. Always contact the people that you would like to use for a reference before listing their names.
4. Education—Be sure to include the names and addresses of schools or colleges that you attended, and state the diplomas and/or degrees that you obtained. List any areas of specialization. Include industry classes that you have attended, and list any special awards that you have received.
5. Special Skills—Include a description of any special skills that you have. For example, you may have specialized in electronic system diagnosis in a previous job, and as a result you have special skills in this area.
6. Hobbies and Special Interests—List any hobbies that you participate in. For example, you may be interested in building racing engines and automotive racing. Although this may not be directly related to your work as a technician, this hobby indicates your extensive interest and participation in the automotive industry.

The letter of application is a brief, one-page document that contains these items:

1. The name and address of the person who is responsible for conducting the hiring interview. This is usually the service manager or personnel manager.
2. The technician's name, address, and phone number.
3. The job title of the position for which you are applying, with an explanation of how you found out about this job opportunity.
4. An explanation of why you would like this position.
5. Reasons why you believe that you would be successful in this position.

6. An afffirmation of your dedication to excellent performance in this job position.

## CUSTOMER RELATIONS

When dealing with customers, always remember the two Ps: positive and polite. There may be some days when we do not feel positive and polite! Perhaps there are two vehicles in the shop with time-consuming, difficult diagnostic problems, and the vehicle owners are requesting their cars. Even though we do not feel positive and polite, we must maintain these attitudes when talking to customers. These attitudes will certainly improve our customer relations and bring customers back to the shop the next time their vehicle needs repairs.

**You Should Know** *Excellent customer relations are absolutely essential for the successful operation of an automotive shop. Many customers have taken their automotive repair business to another shop because they were treated with indifference, incompetence, or disrespect. Conversely, many repeat customers have been obtained by the courteous and positive attitude of service personnel.*

Employees at all levels in an automotive shop should be courteous and maintain a positive attitude. Perhaps the customer's vehicle is older than average and not in very good mechanical condition. In some cases, the customer may not realize the poor condition of their vehicle. The customer's vehicle is important to him or her, and we should respect that fact. Rather than being negative about the condition of the customer's vehicle, the technician and all service personnel should see this customer and vehicle as an opportunity for a considerable amount of service work.

Technicians and all service personnel should be interested in the customer's vehicle concerns, and these concerns must be corrected by the service work performed on the vehicle.

Automotive service personnel must be dedicated to quality workmanship. Repairs should not be done with the attitude that a repair is good enough to keep the vehicle running. Each repair should be the best possible repair to correct the customer's complaint. All service personnel must be committed to quality care of the customer's vehicle. This includes all personnel who will have any part in servicing the customer's vehicle. For example, in a large shop operation, quality care of customer vehicles applies to car jockeys who may drive these vehicles in to out of the shop. Do not smoke in customer vehicles because the customer may be a non-smoker who does not appreciate the odor of cigarette smoke in their vehicle. Always place seat covers and floor mat covers on the vehicle, and place covers on the fenders when working under the hood.

All personnel in an automotive service shop must be committed to teamwork. Teamwork means working together for the improvement of the shop operation. This may mean helping a co-worker lift a heavy component. Perhaps a co-worker is having trouble diagnosing a difficult problem, and you have encountered and corrected this problem on a previous service job. Teamwork means sharing diagnostic knowledge!

All automotive service personnel must be neat in appearance. Coveralls or smocks should be changed as often as necessary to maintain a neat, clean appearance.

## Summary

- Technicians must understand the various job descriptions in a typical automotive repair shop.
- Technicians must be familiar with work orders and how work orders are processed in a typical shop.
- Technicians must have a knowledge of employer to employee obligations.
- Technicians must understand employee to employer obligations.
- Technicians must be familiar with their job responsibilities to perform efficiently in an automotive shop.
- Technicians must understand the ASE certification of automotive technicians.
- Technicians must be familiar with NATEF certification of automotive training programs.
- Technicians must be able to complete job applications and resumes.
- Technicians must be able to establish and maintain good customer relations.

## Review Questions

1. Technician A says the service writer is responsible for meeting the customers. Technician B says the service writer is responsible to the service manager. Who is correct?
   A. Technician A
   B. Technician B
   C. Both Technician A and Technician B
   D. Neither Technician A nor Technician B
2. Technician A says that 4 years of work experience in an automotive shop are required prior to ASE certification. Technician B says that ASE provides certification in specialty areas such as Cylinder Head Specialist. Who is correct?
   A. Technician A
   B. Technician B
   C. Both Technician A and Technician B
   D. Neither Technician A nor Technician B
3. Technician A says a NATEF-certified automotive training program must be recertified every 3 years. Technician B says prior to NATEF certification of an automotive training program, a self-evaluation must be completed by the program staff. Who is correct?
   A. Technician A
   B. Technician B
   C. Both Technician A and Technician B
   D. Neither Technician A nor Technician B
4. In a large automotive repair shop, Technician A says the service manager is responsible to the shop foreman. Technician B says the service manager is responsible to the cashier. Who is correct?
   A. Technician A
   B. Technician B
   C. Both Technician A and Technician B
   D. Neither Technician A nor Technician B
5. Service writers should follow all of these job performance guidelines *except*:
   A. The service writer should be neatly and cleanly dressed.
   B. The service writer should deal with the customer quickly without discussing the customer's concerns.
   C. The service writer should be professional and polite.
   D. The service writer should call the customer by name.
6. Technician A says the service writer should ask the customer regarding the exact symptoms and/or sounds related to the customer's complaint. Technician B says the service writer should find out if the customer's vehicle complaint occurs at a specific vehicle speed or engine temperature. Who is correct?
   A. Technician A
   B. Technician B
   C. Both Technician A and Technician B
   D. Neither Technician A nor Technician B
7. All of these statements regarding the service manager's responsibilities are true *except*:
   A. The service manager is responsible for implementing the vehicle manufacturer's warrantee policies and recommended service procedures.
   B. The service manager is responsible for hiring shop personnel.
   C. The service manager is responsible to the shop foreman.
   D. The service manager is responsible for communication with other departments in the business.
8. Technician A says the repair order should contain an accurate and precise description of the vehicle problem. Technician B says the repair order should contain the customer's signature. Who is correct?
   A. Technician A
   B. Technician B
   C. Both Technician A and Technician B
   D. Neither Technician A nor Technician B
9. In an automotive dealership, flat-rate time is set by:
   A. the service manager.
   B. the shop foreman.
   C. the dealership manager.
   D. the vehicle manufacturer.
10. Technician A says a resume should contain personal medical records. Technician B says a resume should contain the job title for which you are applying. Who is correct?
    A. Technician A
    B. Technician B
    C. Both Technician A and Technician B
    D. Neither Technician A nor Technician B
11. In an automotive repair shop, the shop foreman supervises the ____________________.
12. In many automotive shops, technicians work on a ______________ ______________ basis.
13. Each ASE automotive certification test contains ______________ to ______________ multiple choice questions.
14. ASE automotive master technician status is obtained by passing ________________ ASE certification tests.
15. Explain ten employee to employer obligations.
16. Describe six employer to employee obligations.
17. Describe a technician's job responsibilities when working in an automotive shop.
18. Describe the information that should be included on a resume.

# Chapter 4 Tools and Equipment

## Introduction

Automotive technicians must use a variety of tools. Technicians must also be familiar with shop equipment, power tools, and special tools. This chapter provides information regarding the design and purpose of common hand tools. Information is also provided for basic shop equipment, power tools, and electronic equipment. Many precautions are provided that you need to know regarding the safe operation of shop equipment, power tools, and electronic equipment.

### COMMON HAND TOOLS

Technicians must be familiar with different types of hand tools and the proper use of these tools. Using the proper tool for the job often saves time and allows a technician to perform their work more efficiently. Improper tool use may cause vehicle component damage and injury to the technician.

#### Wrenches

A set of box-end and open-end wrenches are included in a basic technician's tool set. A box-end wrench **(Figure 1)** completely surrounds a nut or the head of a bolt and is less apt to slip and cause damage or injury. When there is very little space around a nut or bolt, it may be difficult to use a box-end wrench. An open-end wrench has an open, squared end that may be used where a box-end wrench will not fit. Open-end and box-end wrenches may have different sizes at either end.

Box-end wrenches are available in either six points or twelve points. Twelve-point box-end wrenches allow you to work in smaller areas compared to six-point wrenches. An open-end wrench is often the best tool for rotating a nut or holding a bolt head.

A technician's tool set should also include a set of combination wrenches that have an open-end wrench on one end and a box-end wrench on the opposite end.

Both ends of these wrenches are the same size, and the open end and box end can be used alternately on the same bolt or nut. On most older domestic vehicles, the nuts and bolts require wrenches with United States Customary (USC) sizes. The **United States Customary (USC)** system is a system of weights and measures used in the United States.

These nut and bolts are usually manufactured in increments of 1/16 of an inch. Most imported and newer domestic vehicles require metric-sized wrenches with increments of 1 millimeter (mm). The metric system of measurements may

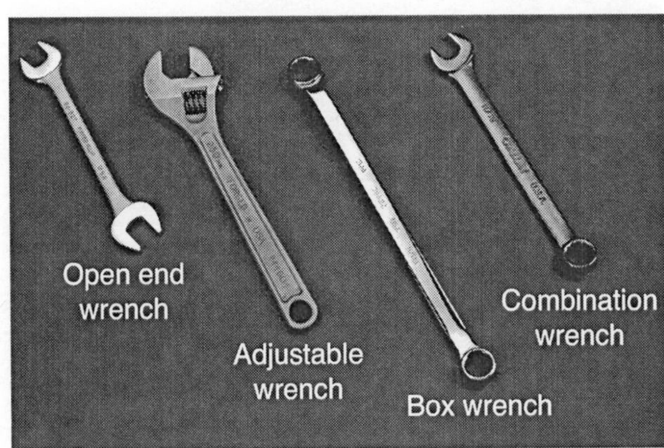


**Figure 1.** Common hand wrenches.

be referred to as the International System (SI). The **International system (SI)** is a system of weights and measures in which every measurement is multiplied or divided by 10 to obtain larger or smaller units. Technicians must have complete sets of both USC and metric wrenches.

> **You Should Know**
>
> *Metric and USC wrenches are not interchangeable. For example, a $^{9}/_{16}$-inch wrench is 0.20 inches larger than a 14 millimeter nut. If the $^{9}/_{16}$-inch wrench is used to turn or hold a 14-millimeter nut, the wrench may slip. This action may result in skinned knuckles for the technician and rounded-off shoulders on the nut. Only two sizes of metric and USC wrenches are the same size. Eight millimeter is the same size as $^{5}/_{16}$ inch, and 19 millimeter is the same size as $^{3}/_{4}$ inch.*

Allen wrenches are used to tighten or loosen small bolts or setscrews with a machined hex-shaped recess in the head of the bolt or screw. The proper size Allen wrench fits snugly into the recess of the bolt head.

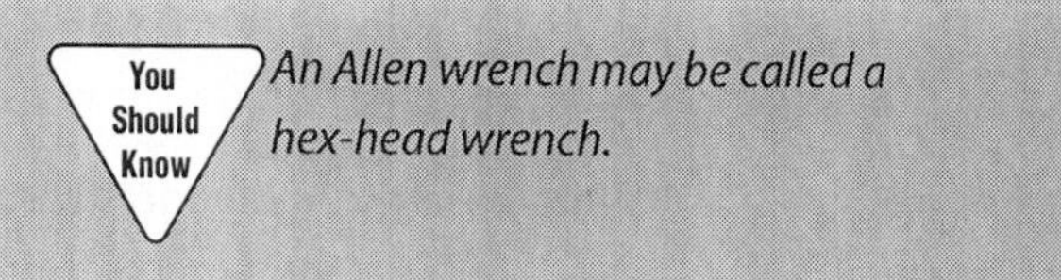

You
Should
Know
An Allen wrench may be called a
hex-head wrench.


To loosen or tighten line or tubing fittings, flare-nut wrenches should be used rather than open-end wrenches.

These fittings are usually made from soft metal that distorts easily. Using an open-end wrench on line or tubing fittings tends to round off the shoulders on the fitting. Flare-nut wrenches are like a box-end wrench with an opening in the box end. This opening allows the flare-nut wrench to be slid over the line and placed on the line fitting. A flare-nut wrench surrounds most of the line fitting, and thus prevents the wrench from slipping on the fitting.

## Ratchets and Sockets

A typical technician's tool set contains USC and metric socket sets in $^{1}/_{4}$-inch, $^{3}/_{8}$-inch, and $^{1}/_{2}$-inch drive. The drive size refers to the size of the square opening in the top of the socket. Each socket set will usually contain various extensions: a ratchet, and a long breaker bar. The ratchet allows you to tighten or loosen a bolt without removing and resetting the wrench after you have turned it **(Figure 2)**. The long breaker bar is used to provide increased leverage when loosening tight bolts.

Many sockets are designed with twelve points, but six-point and eight-point sockets are found in some socket sets. A six-point socket has stronger walls and improved grip compared to a twelve-point socket, but six-point sockets have half the positions of a twelve-point socket. Six-point sockets are mostly used on fasteners that are rusted or rounded. Eight-point sockets are available for use on square nuts or square-headed bolts. Universal joints are included in most socket sets. When loosening or tightening a fastener, a universal joint allows an angle between the socket and the ratchet and extension.

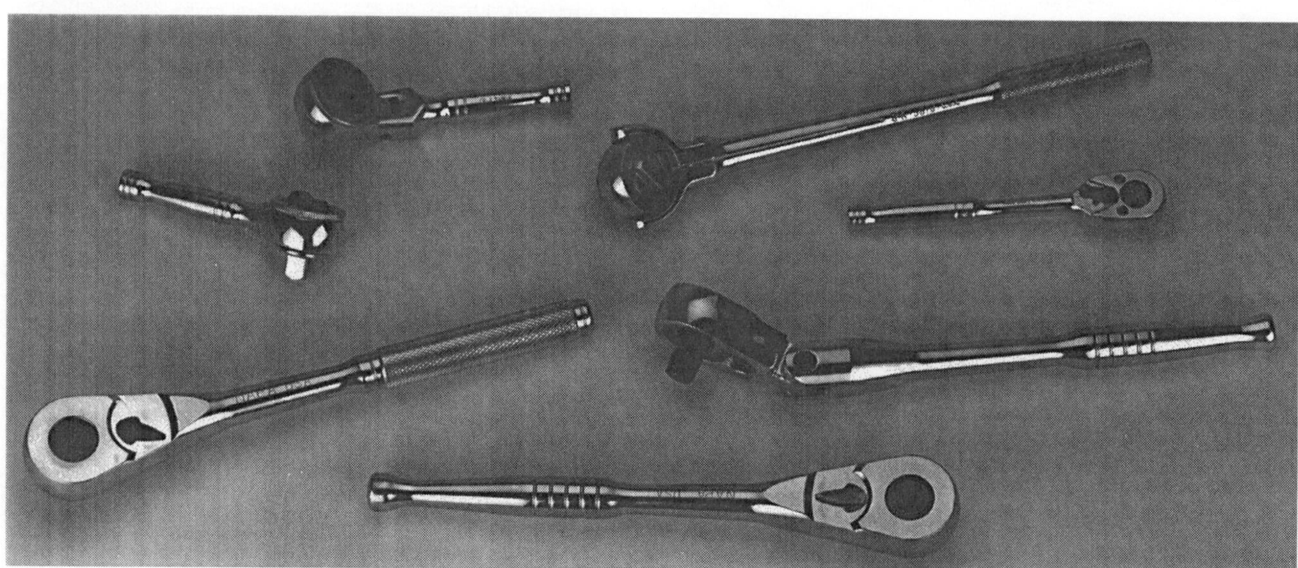

**Figure 2.** Various types of ratchets.

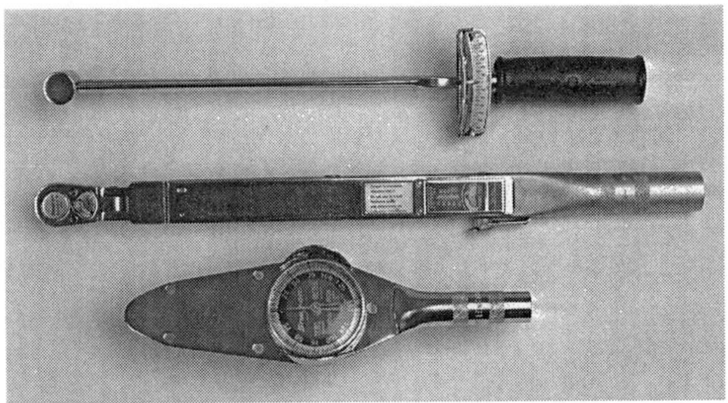

Figure 3. Three types of torque wrenches.

## Torque Wrenches

A torque wrench measures the tightness of a bolt or nut. Vehicle manufacturers provide torque specifications in their service manuals for most fasteners on various components.

The torque specifications are provided in foot-pounds (USC) or Newton-meters (metric). A foot-pound is the work or pressure accomplished by a force of 1 pound applied through a distance of 1 foot. A Newton-meter is the work or pressure accomplished by a force of 1 kilogram applied through a distance of 1 meter.

Torque wrench drive sizes are the same as socket drive sizes. Torque wrenches may be dial type, break-over type, torsion-bar type, and digital-reading type. When using a dial-type torque wrench, the bolt torque is indicated on a dial as the bolt is tightened. A break-over type torque wrench must be set to the specified torque. When the bolt torque reaches this setting, the wrench provides an audible click. On a torsion-bar torque wrench, the wrench bends as the bolt is tightened, and a pointer moves across a scale on the wrench to indicate the bolt torque. On a digital-reading torque wrench, the bolt torque is indicated on a digital reading on the wrench as the bolt is tightened **(Figure 3)**.

## Screwdrivers

The flat-tip screwdriver is the most common type **(Figure 4)**. Blade-type screwdrivers are available in many different sizes depending on the size of fastener to be loosened or tightened. On this type of screwdriver, the blade tip fits into a slot in the fastener head. A Phillips screwdriver has a cross point on the blade tip. This cross point has four surfaces that fit four matching recesses in the fastener head. The Phillips screwdriver provides more gripping power compared to a blade-type screwdriver **(Figure 5)**. A Torx-type screwdriver has a six-prong tip that fits snuggly into a matching recess in the fastener head **(Figure 6)**. Torx-type screwdrivers are available in different sizes. All screwdrivers are available in different lengths from 2-inch stubby screwdrivers to 12 inches or longer. An offset screwdriver is a blade-type screwdriver with the blade tips positioned at an 90° angle in relation to the shank of the screwdriver, and the blades at opposite ends of the screwdriver are at right angles to each other. This offset screwdriver may be used to loosen or tighten fasteners that do not have direct access above the fastener head.

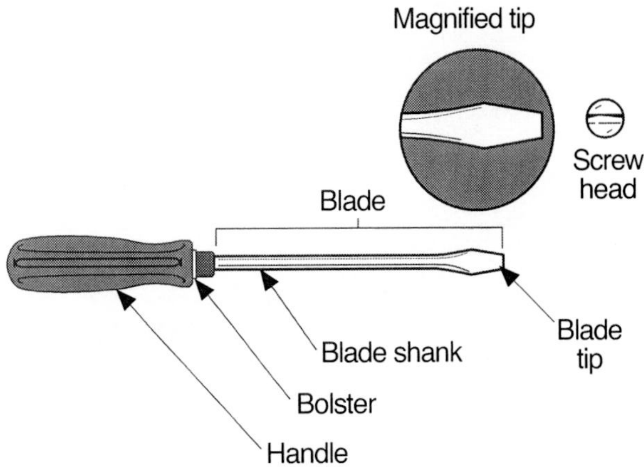


Figure 4. A flat-tip screwdriver.

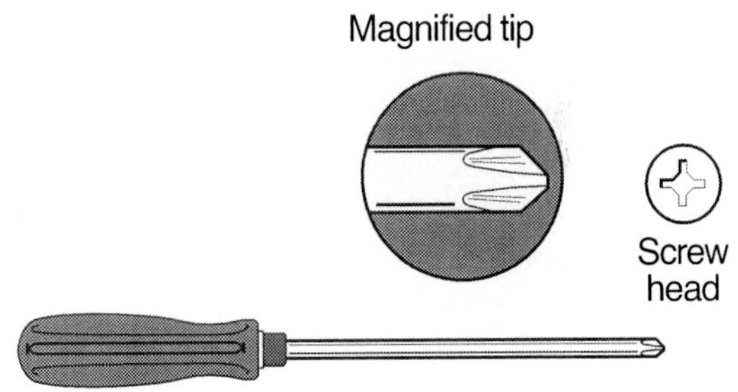


Figure 5. A Phillips-head screwdriver.

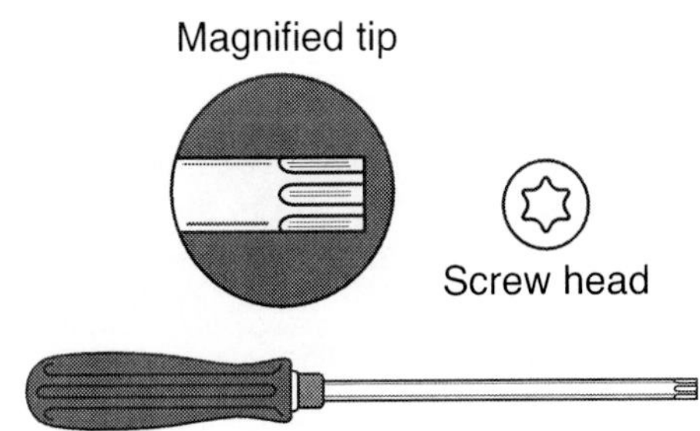


Figure 6. A Torx screwdriver.

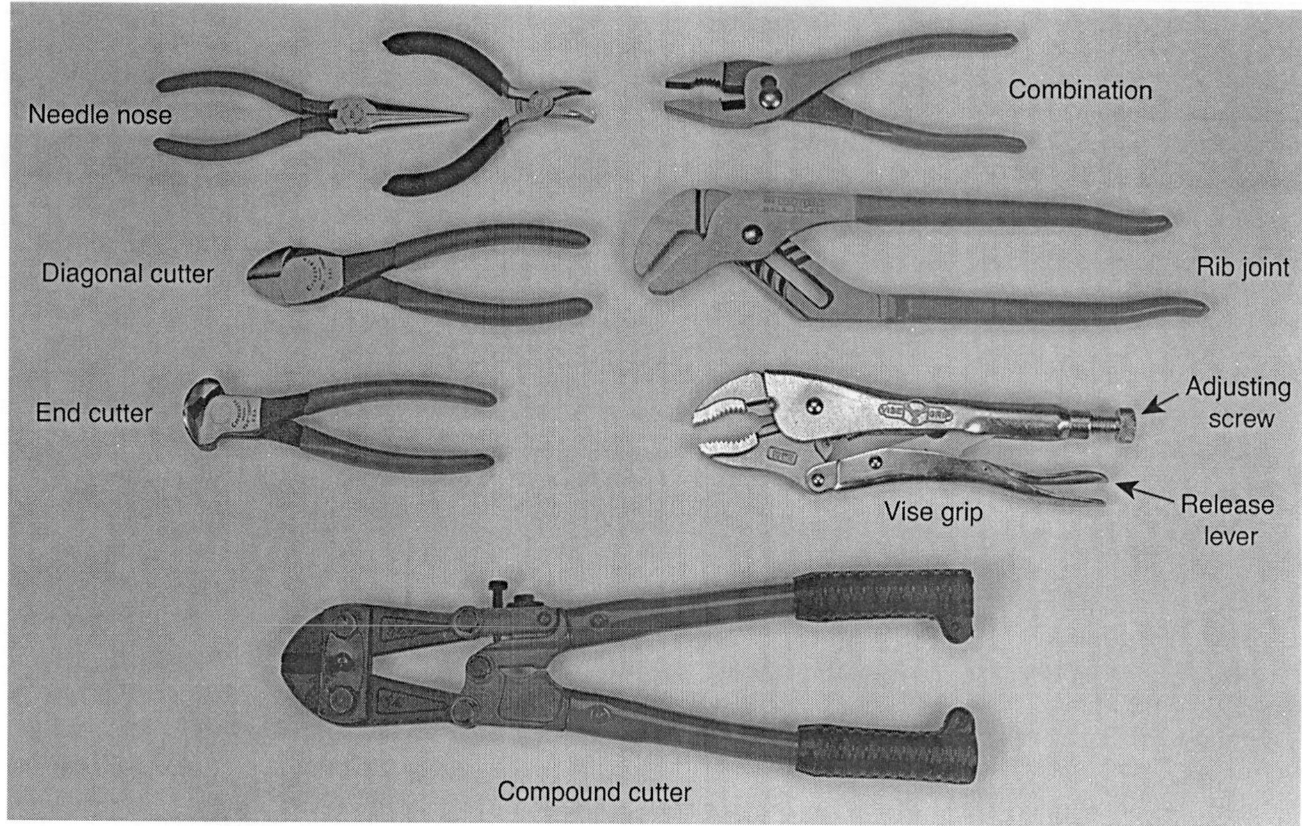


**Figure 7.** Various types of pliers.

## Pliers

Diagonal pliers have sharp cutting edges on the plier jaws. These pliers are used for cutting wire or removing cotter pins.

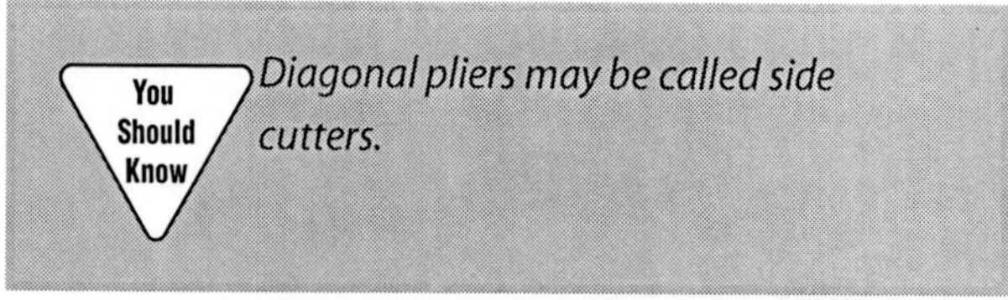


*Diagonal pliers may be called side cutters.*

Combination pliers have a slip joint between the two jaws. The slip joint can be set for either of two jaw openings. Other types of pliers include channel lock, needle nose, snapring, and vise grip **(Figure 7)**. Channel-lock pliers contain several channels that provide different jaw openings.

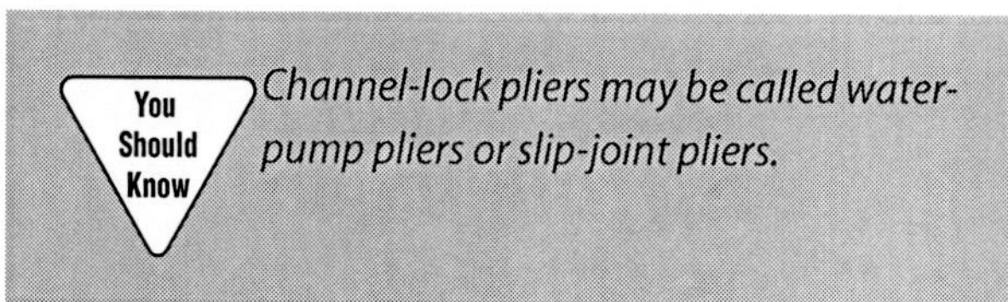


*Channel-lock pliers may be called water-pump pliers or slip-joint pliers.*

These pliers are used to grip large objects. Needle nose pliers are used to grip objects in a small opening. Snapring pliers have special jaws that grip the ends of snap rings to remove and replace these components. Vise-grip pliers are designed so the jaws may be locked onto a component.

## Hammers and Mallets

Ball-peen hammers are available in different weights. The average tool set contains at least two ball-peen hammers, one 8 ounce and possibly a 16 ounce. A soft-faced mallet is also required for tapping parts during removal or replacement. The soft-faced mallet may have plastic and lead or plastic and brass faces on the head of the mallet. When this type of mallet is used to tap components apart, it will not mark the component. If a ball-peen hammer is used to tap components apart, it will definitely mark and damage the components.

## Punches and Chisels

A tool set contains a variety of punches and chisels. Drift punches are used to remove roll pins **(Figure 8)**. Brass drift punches are available if component marking and damage is a concern. Tapered punches are used to align

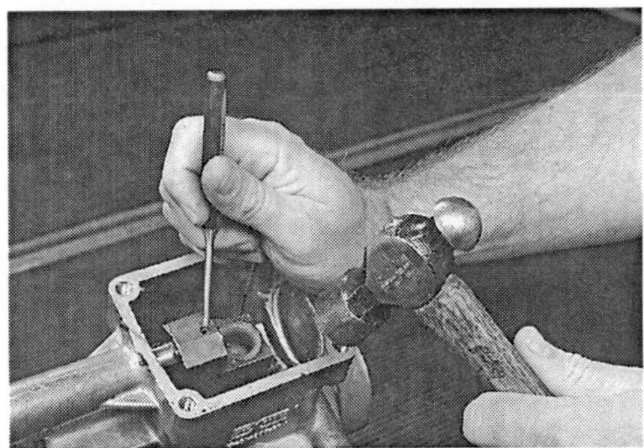

Figure 8. Using a punch to remove a retaining pin.

bolt holes. Center punches have a sharp tip that is used to indent a component in the proper location prior to drilling a hole in the component. This indent prevents drill wandering and component damage. Various types of chisels include: flat, cape, round-nose cape, and diamond point.

## FILES, TAPS, AND DIES

Types of files in a tool set may include: flat, half-round, round, and triangular **(Figure 9)**. Files may also be single cut or double cut. Single-cut files have the cutting grooves positioned diagonally across the face of the file. Double-cut files have the cutting grooves positioned diagonally in both directions across the face. Double-cut files are considered first cut or roughening files because they remove a great deal of metal. Single-cut files remove small amounts of metal, and they are considered finishing files.

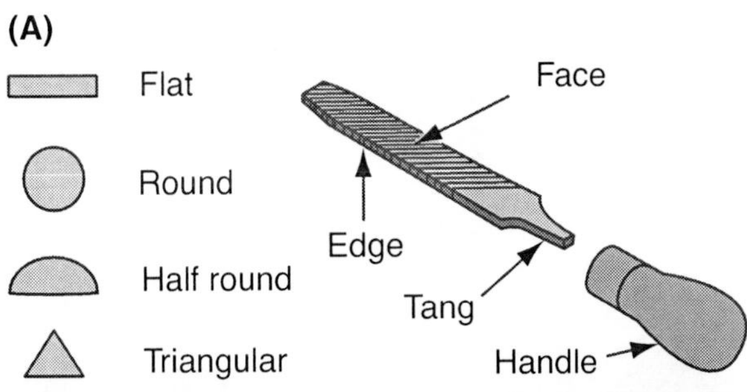


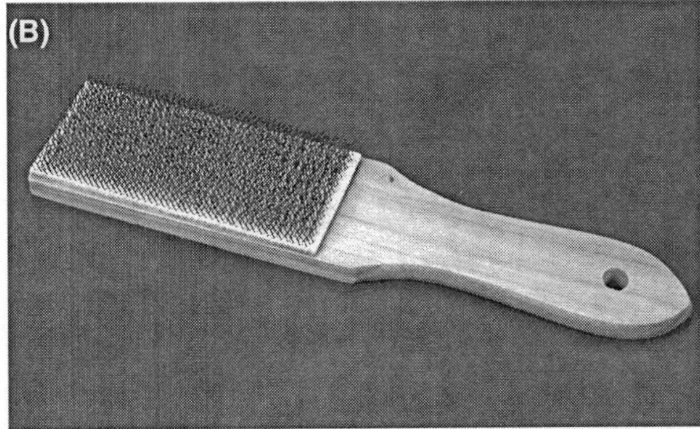


Figure 9. (A) Various types of files. (B) A file card.

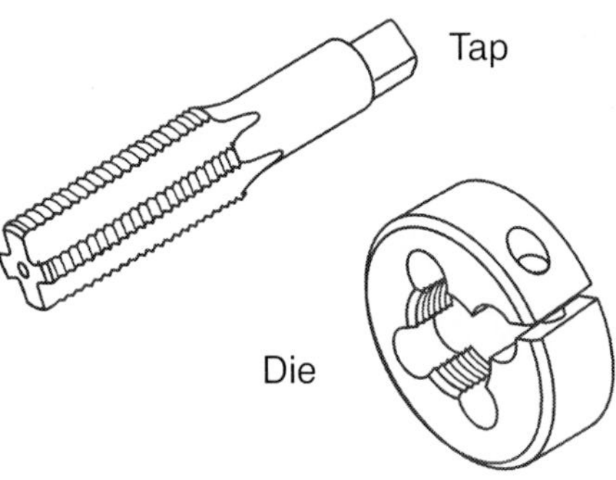


Figure 10. Tap and die.

The most common threading tools are taps and dies **(Figure 10)**. Taps and dies are available in USC and metric threads. USC taps and dies are also available in national fine (NF) or national coarse (NC). A tap is used to cut or restore internal threads in an opening. Openings in metal castings may be classified as blind holes or through holes. A blind hole bottoms in the casting, whereas a through hole extends through the casting **(Figure 11)**. A tapered tap is used in a through hole, and a bottoming tap is used to cut or restore threads in a blind hole **(Figure 12)**. A pitch gauge

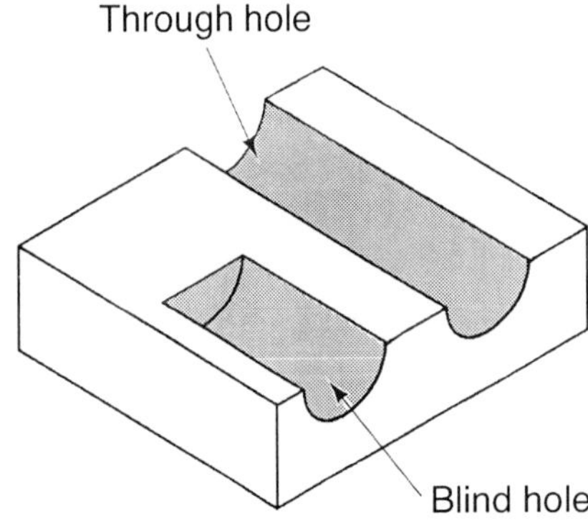


Figure 11. A comparison of through and blind holes.

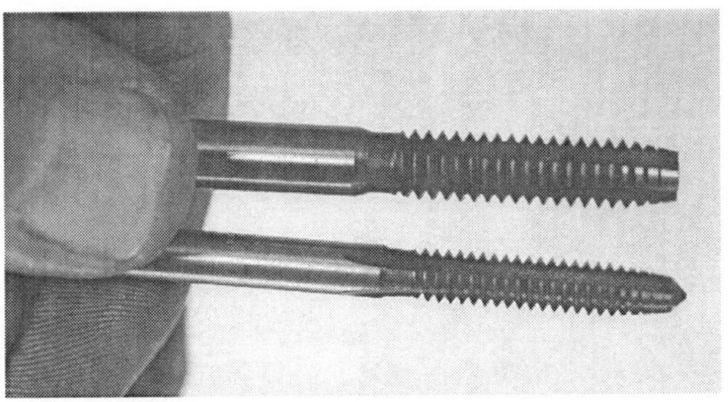

Figure 12. A bottoming tap and a tapered tap.

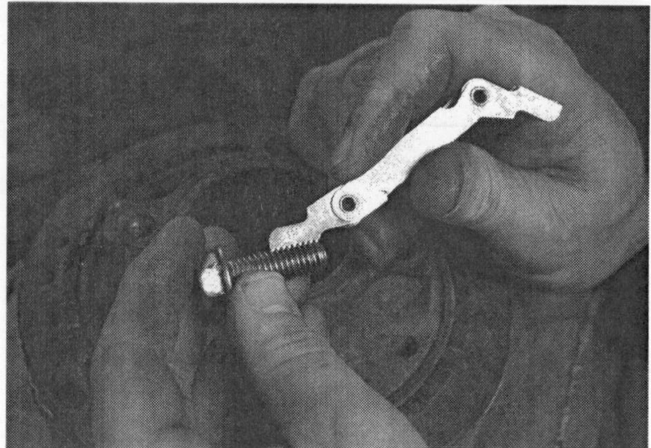

Figure 13. Using a pitch guage to determine the proper tap selection.

may be placed over the bolt threads to determine the proper tap **(Figure 13)**. The tap is rotated with a special handle that is designed to grasp the square top on the tap.

A die is a circular or hex-shaped tool that is designed to cut or restore external threads on a bolt, rod, or fastener. A die with the proper thread size must be selected. The die is mounted in a special turning handle.

A screw extractor may be used to extract broken bolts or studs. Screw extractors are available in various sizes, and a specific drill size is specified for each extractor. The end of the broken bolt or stud is center punched and drilled with the specified drill. When the screw extractor is rotated, it is designed to grip the drilled opening in the bolt. The same turning handle that is used to rotate a tap may be used to rotate the screw extractor and remove the broken bolt **(Figure 14)**.

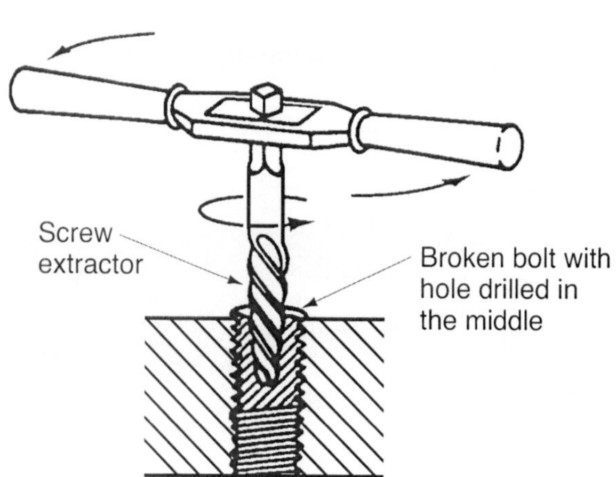


Figure 14. Using a screw extractor to remove a broken bolt.

## GEAR AND BEARING PULLERS

Many gears and bearings have a slight interference fit (press fit) when they are installed in a housing or on a shaft. For example, the inside diameter of a bore is 0.001 inch smaller than the outside diameter of the shaft that fits in the bore. When the shaft is installed in the bore, it must be pressed in to overcome the 0.001-inch interference fit. A **press fit** is present when a part is forced into an opening that is slightly smaller than the part itself to provide a tight fit.

This press fit prevents the parts from moving on each other. The removal of gears and bearings must be done carefully to prevent damage to these components and the bore or shaft where they are mounted. Bearing pullers are designed to fit over the outer diameter of the bearing or through the center opening in the bearing to pull the bearing from its mounting location. The bearing puller must have the right jaws or adapters to fit the bearing or gear properly so the puller will not slip off the component being pulled. Various pullers are illustrated in **Figure 15**.

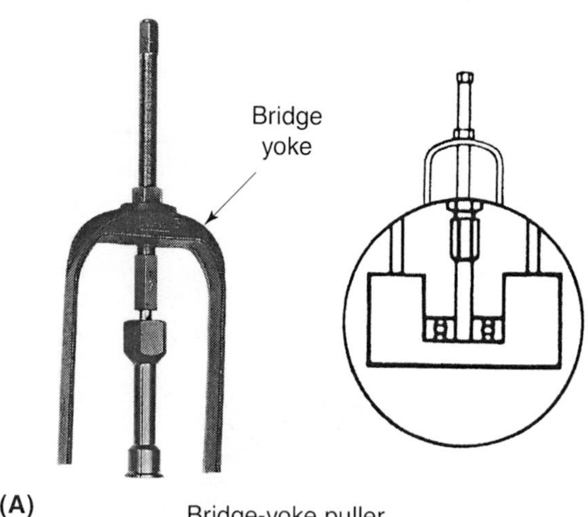


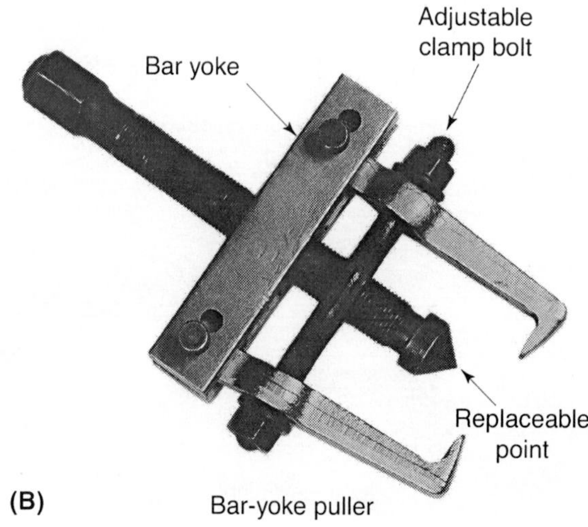


Figure 15. Types of bearing pullers. (A) A bridge-yoke puller. (B) A bar-yoke puller.

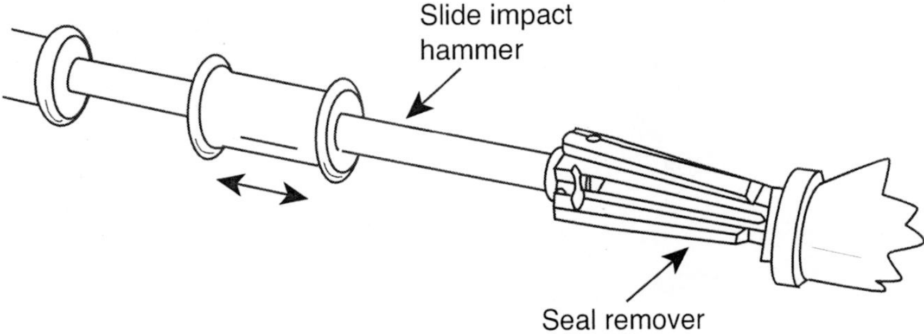


**Figure 16.** Using a slide hammer-type seal puller to remove a seal.

> **You Should Know** *Do not use a hammer to strike the hex-shaped end of a puller when the puller is tightened on a gear or bearing. This action may cause the puller to suddenly slip off the component, resulting in personal injury.*

## BUSHING AND SEAL PULLERS AND DRIVERS

Technicians have to use bushing and seal pullers and drivers when servicing components such as wheel hubs, transmissions, and differentials. Pullers are usually a threaded or slide-hammer type **(Figure 16)**. Bushings or seals may be damaged if the wrong tool is used for removal or installation. Seal drivers must fit the seal properly, and the seal must be started squarely into the housing **(Figure 17)**. Seal drivers are available in various diameters to fit squarely against the outside edge of different size seals. They also provide an internal recess that allows the puller to fit over a protruding shaft. The seal driver handle is tapped with a soft hammer to install the seal.

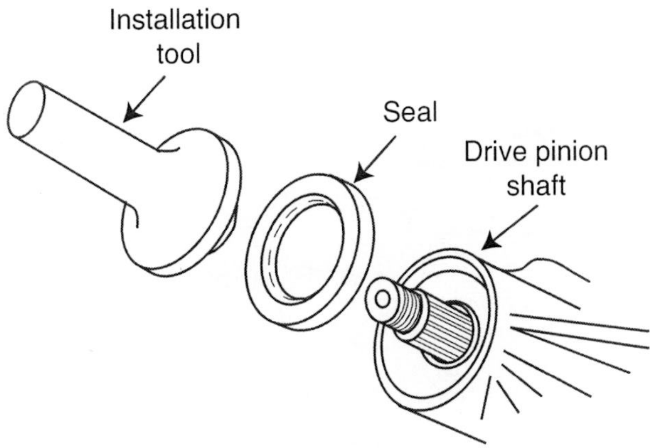


**Figure 17.** Using a seal driver to install a seal.

Be sure the housing is clean and free from burrs before installing the seal. The outer diameter of some seals is coated with a sealer to prevent leaks between the seal case and the housing. If the seal case is not coated, a special sealer may be placed only on the outer diameter of the seal case. After the seal is installed, lubricate the seal lips with the vehicle manufacturer's specified lubricant.

## POWER TOOLS AND SHOP EQUIPMENT

Power tools make a technician's job easier and increase production. Power tools operate faster and supply more torque compared to hand tools. Power tools also require increased safety measures. Power tools may be operated by air pressure (pneumatic), electricity, or hydraulic fluid. Pneumatic tools are commonly used by technicians because these tools have less weight, more torque, and require less maintenance than electric power tools. However, electric power tools are usually less expensive than pneumatic tools. Power can be supplied from a conventional wall socket to a power tool, but and air hose from the shop air supply must be connected to a pneumatic power tool.

> **You Should Know** *Improper use of power tools may cause serious personal injury. Never operate a power tool unless you are familiar with the manufacturer's recommended operating procedures.*

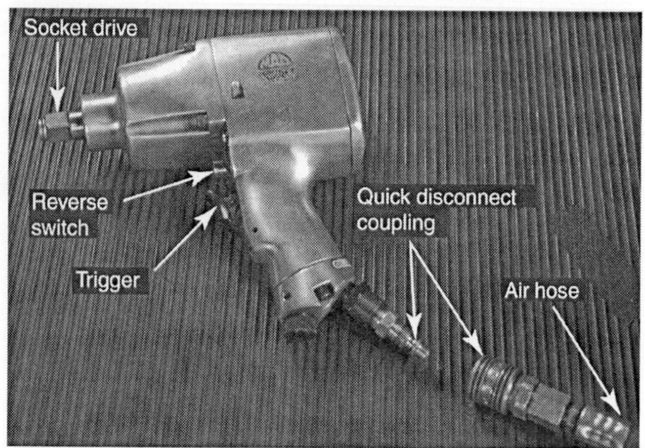


**Figure 18.** An air impact wrench.

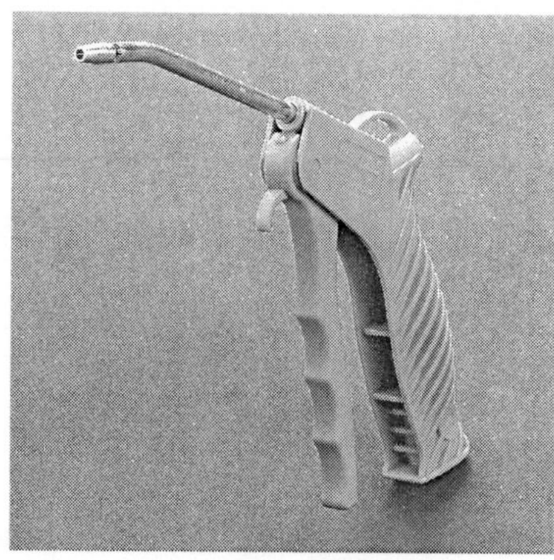

**Figure 19.** An OSHA-approved air blowgun.

## Impact Wrench

An impact wrench uses air pressure or electricity to loosen or tighten a nut or bolt with a hammering action **(Figure 18)**. Light-duty impact wrenches are available in drive sizes: of 1/4 inch, 3/8 inch, and 1/2 inch. Heavy-duty impact wrenches are available in 3/4-inch and 1-inch drives.

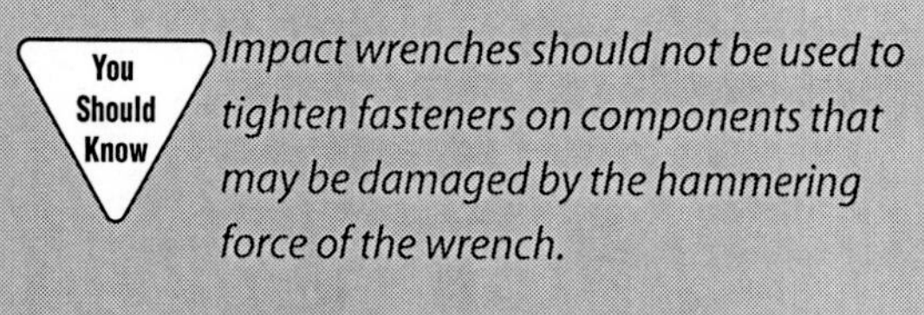


*Impact wrenches should not be used to tighten fasteners on components that may be damaged by the hammering force of the wrench.*

You Should Know

*Impact wrenches require the use of thick-walled sockets to withstand the hammering force of the wrench. If conventional sockets are used on an impact wrench, they may shatter, and this action can result in personal injury.*

Air ratchets are often used when loosening or tightening fasteners because they allow the technician to work faster than an ordinary ratchet. An air ratchet turns fasteners without a hammering force, and this tool can be used with conventional sockets. Air ratchets usually have a 3/8-inch drive. Air ratchets are not torque sensitive. Therefore, fasteners should be tightened snugly with an air ratchet and then tightened to the specified torque with a torque wrench.

## Blowgun

After parts are cleaned, they are blown off and dried with a blowgun. A blowgun snaps into the end of a shop air hose and directs airflow when a button or lever is pressed **(Figure 19)**. Always use OSHA-approved blowguns. Before using a blowgun, be sure the air-bleed holes in the side of the gun are not plugged.

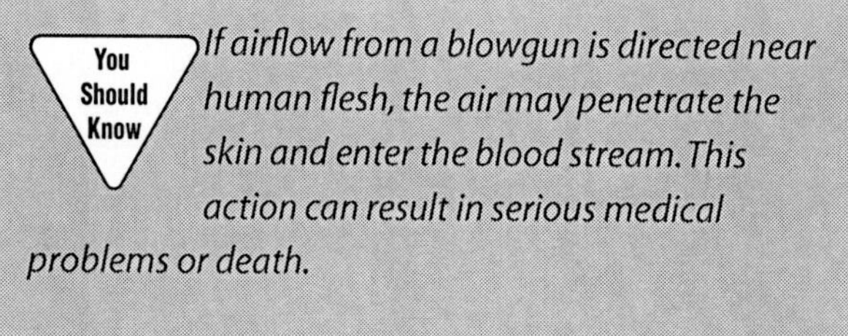


*If airflow from a blowgun is directed near human flesh, the air may penetrate the skin and enter the blood stream. This action can result in serious medical problems or death.*

## Bench Grinder

A bench grinder is usually bolted to the work bench **(Figure 20)**. The bench grinder must have shields and

**Figure 20.** A bench grinder.

guards in place, and always wear face protection when using the grinder. Bench grinders are classified by wheel size. Wheel sizes of 6 to 10 inches are commonly used in automotive shops. Bench grinders may be equipped with three different types of wheels.

1. A grinding wheel is used for sharpening tools or deburring metal components.
2. A wire wheel brush is used for buffing or general cleaning, such as rust or paint removal.
3. A buffing wheel is used for polishing and buffing.

## Trouble Light

A trouble light is used to supply adequate light in the immediate work area. A trouble light may be powered from a conventional wall socket or it may be battery powered. In some shops, the trouble lights are suspended from reels attached to the ceiling. In many areas, insurance regulations demand the use of fluorescent trouble lights. Incandescent trouble lights may shatter and burn if they are dropped, and this action can result in a fire. The bulb in an incandescent trouble light must be protected by a cage. Always keep the trouble light cord away from rotating components. An incandescent trouble light can burn carpet or upholstery. Be cautious when using this type of light inside a vehicle.

## Hydraulic Press

A hydraulic press is used to supply the necessary force to disassemble or reassemble press-fit components **(Figure 21)**. A hydraulic press uses a hydraulic cylinder and ram to remove and install precision-fit components from their mounting location.

**Figure 21.** A hydraulic press.

Although presses may be operated by hand, air pressure, or electricity, the hydraulic press is the most common. Most hydraulic presses are floor mounted, but smaller presses may be mounted on the bench or on a pedestal. A hydraulic cylinder and ram is mounted above the press bed. When the hydraulic pump is operated by hand, the ram extends against the work bed to exert pressure on the press-fit component. The component being pressed must be properly supported on the press bed to prevent component slipping. A shield must be in place around the component being pressed.

You Should Know *When operating a hydraulic press, never operate the pump handle until the reading on the pressure gauge exceeds the maximum pressure rating of the press. If this pressure is exceeded, some part of the press may suddenly break, causing severe personal injury.*

## Floor Jack

A floor jack is used to raise a vehicle off the ground. The hydraulic floor jack is the most common jack, but air-operated jacks are available **(Figure 22)**. A floor jack uses hydraulic pressure supplied to a hydraulic cylinder, ram, and lift pad to lift one end or one corner of a vehicle.

Floor jacks are rated according to the amount of weight they can lift safely. The jack handle is usually moved up and down to operate the hydraulic floor jack. The lift pad on the floor jack must be positioned under the manufacturer's specified lift point on the vehicle. A release lever on the jack handle is moved slowly to lower the floor jack. Do not leave a jack handle in the downward position where someone may trip over it.

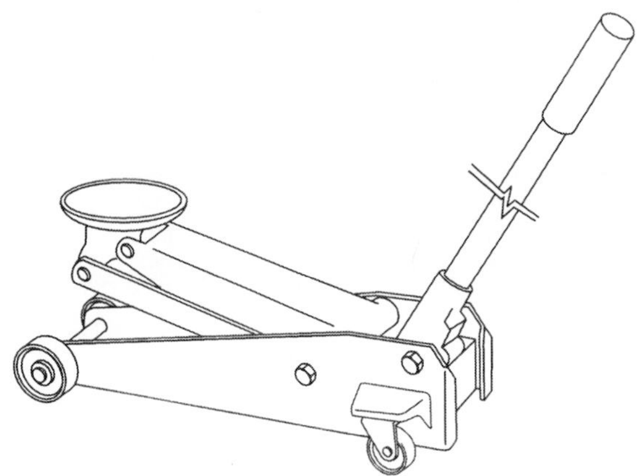

**Figure 22.** A hydraulic floor jack.

You Should Know *The maximum lifting capacity of a floor jack is usually written on the jack decal. Never lift a vehicle that is heavier than the weight rating of the floor jack. This action may cause the jack to break or collapse, resulting in personal injury and/or vehicle damage.*

*If the lift pad on a floor jack is not positioned under the vehicle manufacturer's specified lift point, undervehicle components may be damaged.*

## Safety Stands

After a vehicle is raised with a floor jack, it must be supported on safety stands before working under the vehicle **(Figure 23)**. The safety stand must be positioned under a strong structural part of the vehicle chassis. A service manual provides the location of vehicle lift and support points. Safety stands are rated according to the amount of weight they will support. Never support a vehicle on a safety stand if the weight supported exceeds the rating of the safety stand. When a vehicle is lowered onto safety stands, be sure all four legs on each safety stand remain in contact with the floor.

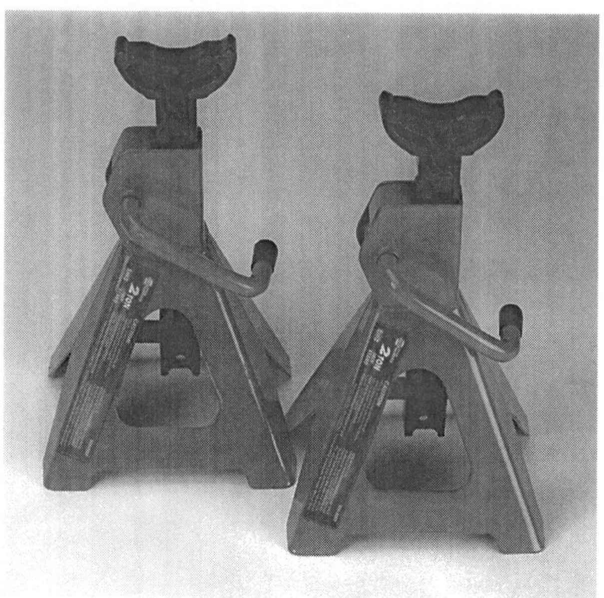

**Figure 23.** Safety stands are used to support a vehicle after it has been jacked up.

## Vehicle Lift (Hoist)

A vehicle lift raises a vehicle so the technician can work under it. The lift arms must be positioned under the vehicle manufacturer's recommended lift points before raising a vehicle **(Figure 24)**.

You Should Know *The maximum capacity of a vehicle lift is placed on an identification plate on the lift. Never lift a vehicle that is heavier than the maximum capacity of the lift. This action may bend the lift or cause the vehicle to slip off the lift, resulting in personal injury and/or vehicle damage.*

You Should Know *Always be sure the lift arms are securely positioned under the vehicle manufacturer's specified lift points before raising the vehicle. These lift points are illustrated in the service manual. If the lift arms are not positioned under the proper lift points, chassis components may be damaged.*

**Figure 24.** Vehicle lifts are used to raise a vehicle.

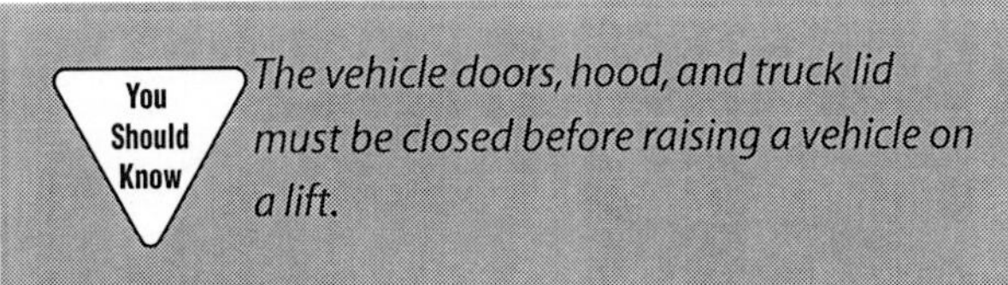


*The vehicle doors, hood, and truck lid must be closed before raising a vehicle on a lift.*

Some lifts have an electric motor which drives a hydraulic pump to create fluid pressure and force the lift upward. Other lifts use shop air pressure to force the lift upward. If shop air pressure is used, it is applied to hydraulic fluid in the lift cylinder. A control lever is placed near the lift, which supplies shop air pressure to the lift cylinder. On an electric lift, a switch near the lift turns the lift motor on. After a vehicle is raised, always be sure the safety lock is engaged. When the safety lock is released, a lever is used to slowly lower the lift. Before lowering a lift, be sure there are no tools, equipment, or people under the lift.

## Engine Lift

An engine lift, or hoist, is used to remove an engine through the hood opening **(Figure 25)**. A sling is used to attach the hoist to the engine.

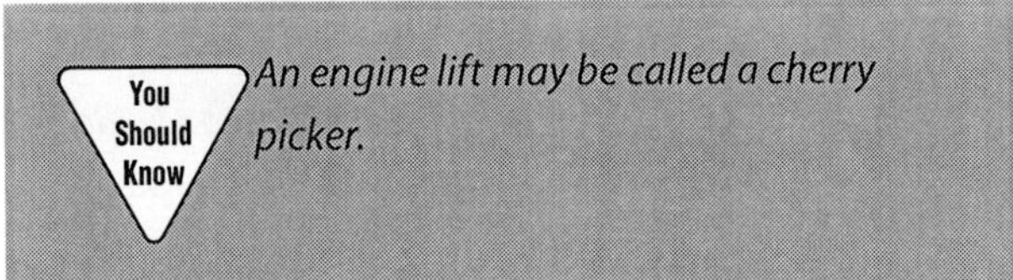


*An engine lift may be called a cherry picker.*

The sling attaching bolts must be strong enough to support the engine weight, and these holts must be threaded in far enough to prevent the bolts from stripping out. Most service manuals provide the proper locations for sling attachment.

The engine lift usually has an adjustable boom and legs. Adjust the legs out far enough to prevent the lift from tipping when the engine weight is supplied to the lift. Extending the boom lowers the lift capacity. After adjusting the boom and legs, be sure the lock pins in these components are properly installed and retained. When an engine is being lifted or lowered, do not place any part of your body under the lift. Once the engine is removed from the vehicle, immediately lower it onto the floor or install it on an engine stand.

**Figure 25.** An engine hoist.

## Tire Changers

Tire changers are used to mount and demount tires **(Figure 26)**. These changers may be used on most common tire sizes. A wide variety of tire changers are avail-

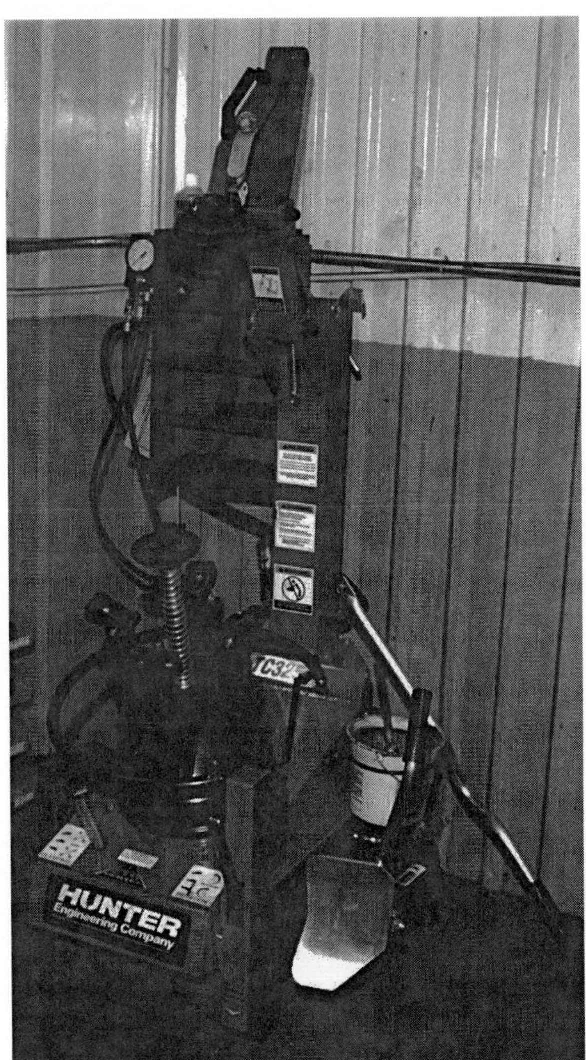


**Figure 26.** A typical tire changer.

able, and each one operates differently. Most tire changers are pneumatically powered. Always follow the procedure in the equipment operator's manual and the directions provided by your instructor.

## ELECTRICAL AND ELECTRONIC TEST EQUIPMENT

Technicians must be familiar with electrical and electronic test equipment. Using the proper electrical and electronic test equipment makes diagnosis faster and more accurate. When electrical or electronic equipment is used improperly, the circuit being diagnosed and the test equipment may be severely damaged.

### Circuit Testers

Circuit testers are used to diagnose defects in electrical circuits. A large variety of circuit testers are available for automotive testing, but the most common circuit tester is the 12-volt test light. A sharp probe is molded into the handle on a 12-volt test light, and the upper part of the handle is transparent. A 12-volt bulb is mounted in the transparent handle **(Figure 27)**, and the probe is connected to the bulb terminal. A ground clip is connected to the wire extending from the handle, and this wire is attached to the ground side of the bulb. In most test situations, the ground clip is connected to ground on the vehicle, and the probe is connected to an electric circuit to determine if voltage is available. The bulb is illuminated if voltage is available at the probe.

**You Should Know** *Do not use a conventional 12-volt test light to diagnose computer system wires or components. The current draw of these test lights may damage computer system components.*

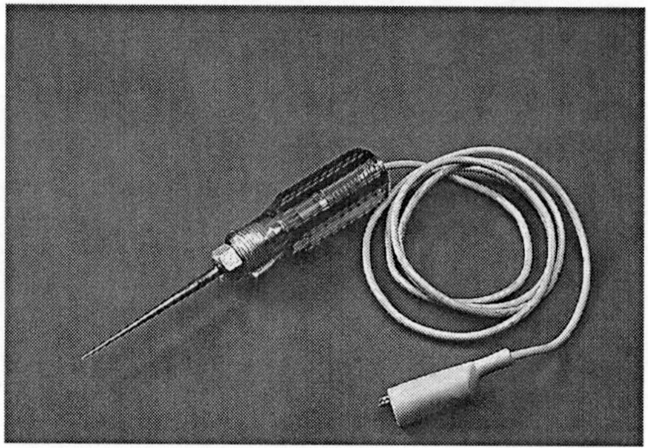

**Figure 27.** A 12-volt test light.

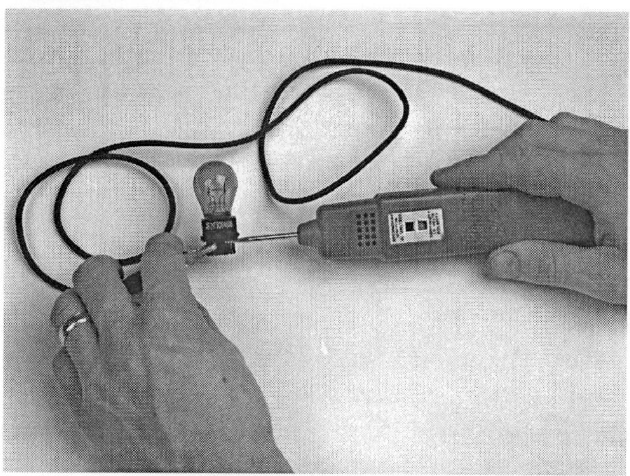

**Figure 28.** A self-powered test light.

High impedance test lights are available for diagnosing computer systems. A **high-impedance test light** contains a very small, high-resistance bulb. This type of test light should only be used to test computer systems when the vehicle manufacturer's service procedure recommends the use of such equipment.

**You Should Know** *If any type of circuit tester is used to diagnose an air bag system, accidental air bag deployment may occur. Use only the vehicle manufacturer's recommended equipment on these systems.*

A self-powered test light is similar in appearance to a 12-volt test light, but this test light has an internal battery that supplies power to the circuit being tested **(Figure 28)**.

In many automotive circuits, voltage is supplied to one end of the circuit, and the other end of the circuit is connected to ground. The ground side of the circuit may be called the negative side of the circuit, and the positive side of the circuit is connected to the battery positive terminal. When diagnosing an electrical circuit with a self-powered test light, the positive side of the circuit is disconnected, and the test light probe is connected to the circuit so the test light supplies voltage to the circuit in place of the battery. If the circuit is not open, the test light is illuminated because the other end of the circuit is connected to ground. When the circuit is open, the test light is not illuminated.

### Multimeter

**Multimeters** are small hand-held meters that provide the following readings in various scales: DC volts, AC volts, ohms, amperes, and milliamperes.

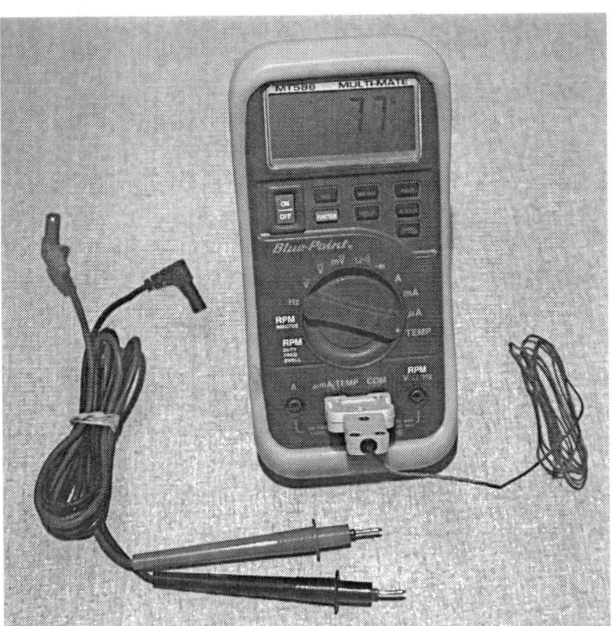

**Figure 29.** A digital multimeter.

Multimeters do not have heavy leads, so the maximum current flow reading on this type of meter is often 10 amperes. A control knob on the front of the meter is rotated to select the desired scale. Most multimeters are digital type, and some of these meters are auto-ranging which means the meter will automatically change to a higher scale if the reading goes above the highest value on the scale being used **(Figure 29)**. If the meter is not auto-ranging, the technician must manually select the next highest scale. If a multimeter is connected improperly, current flow through the meter may be excessive. Under this condition, an internal fuse will blow and protect the meter. **Digital meters** provide a digital reading, whereas **analog meters** have a moveable pointer that moves across various scales. Digital multimeters have higher resistance compared to analog meters.

Always use the type of meter recommended by the vehicle manufacturer. The meter leads are plugged into the appropriate terminals on the front of the meter, and the reading provided by the terminal position is indicated beside the terminal. The black test lead is usually plugged into the common (com) terminal. This terminal may be referred to as a ground terminal. Some multimeters have additional test capabilities such as diode condition, frequency or hertz (Hz), temperature, engine rpm, ignition dwell, and distributor condition.

## Tachometer

Analog tach-dwell meters are one of the most basic types of ignition test equipment. The colored tach-dwell meter lead is connected to the negative primary ignition coil terminal, and the black meter lead is connected to ground. A switch on the tach-dwell meter is used to select engine rpm or dwell. Ignition dwell is not adjustable on electronic-type distributor ignition (DI) or on electronic ignition (EI). Therefore, the dwell reading is not very useful for diagnostic purposes. The tachometer indicates engine rpm. A **distributor ignition (DI)** system uses a distributor to distribute spark from the coil secondary terminal to the spark plugs. An **electronic ignition (EI)** system does not have a distributor. This type of ignition system has a coil for each spark plug or pair of spark plugs.

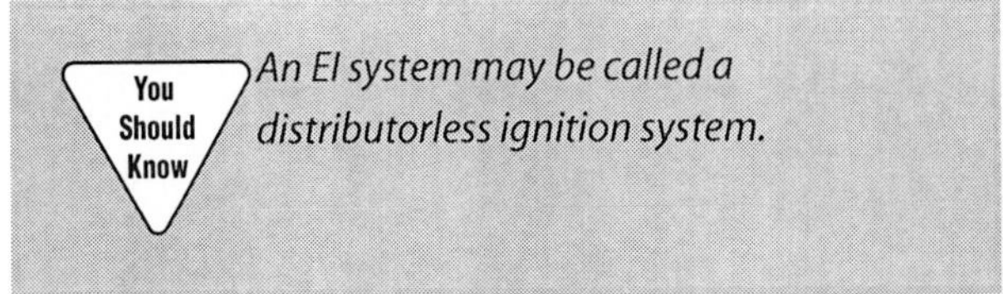


*An EI system may be called a distributorless ignition system.*

Most tachometers are digital. This type of tachometer usually has an **inductive pickup lead** that is connected over the number 1 spark plug wire. This type of tachometer is suitable for EI systems.

## Timing Light

A timing light is used to check ignition timing with the engine running **(Figure 30)**. Voltage is supplied to the timing light from two leads connected to the battery terminals with the correct polarity. Most timing lights have a lead with an inductive clamp that is positioned over the number 1 spark plug wire. When the timing light trigger switch is pulled on with the engine idling, the timing light emits a beam of light each time the number 1 spark plug fires. The engine timing marks are usually positioned on the crankshaft pulley or on the flywheel. A stationary pointer or notch is located above the rotating timing marks. The tim-

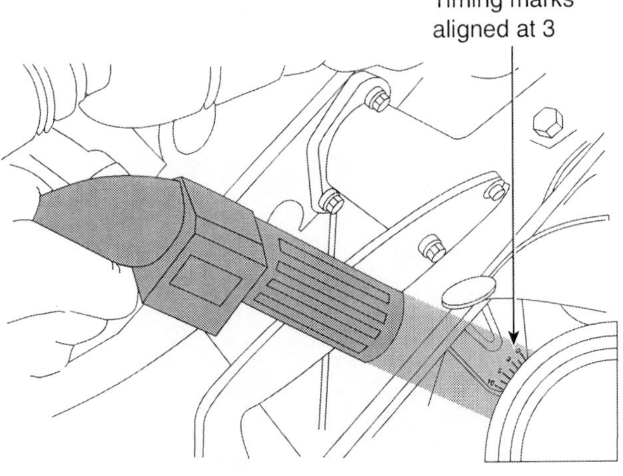


**Figure 30.** A timing light.

ing marks are lines on the crankshaft pulley or flywheel that indicate various degrees of crankshaft rotation. One line represents top dead center (TDC), and other lines represent specific degrees of crankshaft rotation before top dead center (BTDC). Some engines have degree lines on the timing marks that indicate after top dead center (ATDC). On other timing marks, the degree lines are on the pointer, and the crankshaft has only one notch for a timing mark.

Always complete all the vehicle manufacturer's recommended timing procedures before checking the ignition timing. For example, on some fuel-injected engines, an in-line timing connector must be disconnected to prevent the computer from supplying spark advance while checking the timing. With the engine idling at the specified rpm, the timing light beam is aimed at the timing marks. If necessary, rotate the distributor until the timing marks are located at the vehicle manufacturer's specified position. Be sure the distributor hold-down bolt is tightened to the specified torque after the timing is set. Timing adjustments are not possible on EI systems.

Many timing lights have an advance knob that allows the technician to check the spark advance. With the engine running at the specified higher speed, rotate the timing advance knob on the timing light until the timing marks come back to TDC. The degree scale around the advance knob indicates the degrees of spark advance. Some timing lights have a digital readout in place of the degree scale around the advance knob.

## Volt-Amp Tester

A volt-amp tester is used to test voltage and amperes in automotive circuits **(Figure 31)**. A volt-amp tester has the capability to perform a battery load test, a starter draw test, an alternator maximum output test, and an alternator voltage test.

A typical volt-amp tester contains an ammeter, voltmeter, and a carbon pile load. Many volt-amp testers provide digital readings. A carbon pile load is a stack of carbon disks, and a control knob on the volt-amp tester tightens these disks together as the knob is rotated clockwise. When this knob is rotated counterclockwise to the off position, the carbon disks are not contacting each other. Two heavy leads are connected from the carbon pile load to the battery terminals with the correct polarity. The red, positive lead is connected to the positive battery terminal, and the black, negative lead is connected to the negative battery terminal. The ammeter usually has an inductive clamp that fits over the wire in which the amperes is being measured. This type of clamp reads the current flow from the magnetic strength surrounding the wire.

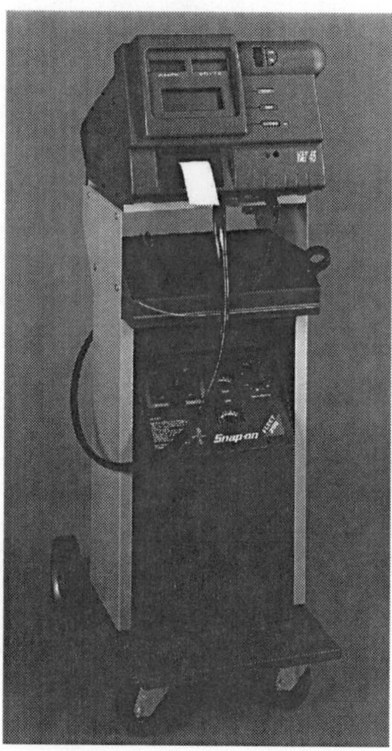


**Figure 31.** A volt-amp tester with a digital ammeter and a voltmeter.

## Scan Tool

A scan tool is used for diagnosing automotive computer systems **(Figure 32)**. On many vehicles, the scan tool is connected to a data link connector (DLC). Vehicles manufactured since 1996 are equipped with on-board diagnostic II (OBD II) systems. On these systems, the DLC is mounted under the dash. On some older vehicles, the DLC is mounted under the hood. On many vehicles, the various computers are interconnected by data links which are also connected to the DLC. Therefore, when the scan tool is connected to the DLC, it can be used to diagnose several different computer systems on the vehicle such as the engine computer, transmission computer, body computer, antilock brake system (ABS) computer, suspension computer, and air-conditioning (A/C) computer. Most scan tools have removable modules, and the proper module for the vehicle and system being tested must be inserted in the scan tool before connecting the tool.

**Figure 32.** A scan tool.

## Electronic Wheel Balancer

Electronic wheel balancers are used in most automotive shops **(Figure 33)**. Do not attempt to use this equipment until you have studied the operator's manual and your instructor has demonstrated the safe use of the balancer. The electronic wheel balancer is used to indicate the proper position and amount of wheel weight required to provide correct static and dynamic wheel balance.

You Should Know

*When using an electronic wheel balancer, always lower the safety shield over the tire and wheel before spinning the wheel. Failure to lower the safety shield may cause personal injury.*

## Exhaust Analyzers

Exhaust analyzers are very valuable diagnostic tools. Five-gas analyzers read the levels of carbon monoxide (CO),

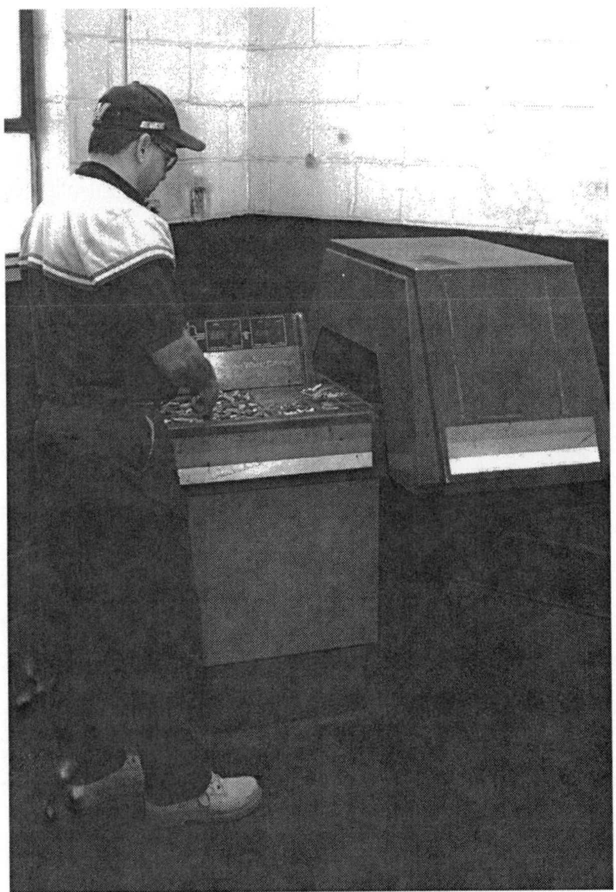

**Figure 33.** An electronic wheel balancer.

**Figure 34.** An exhaust gas analyzer.

hydrocarbons (HC), oxides of nitrogen ($NO_x$), carbon dioxide ($CO_2$), and oxygen ($O_2$), **Figure 34**.

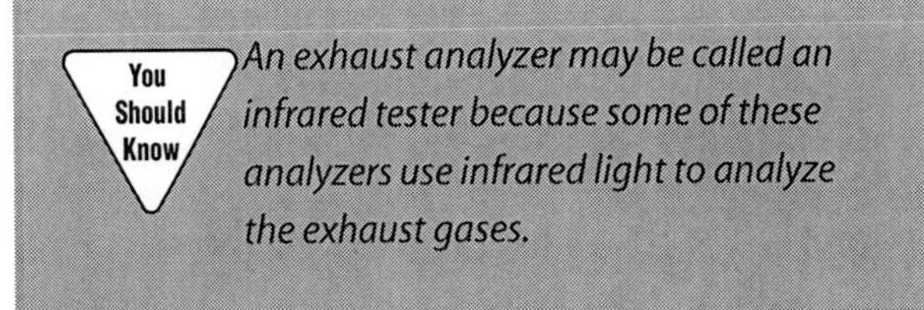

You Should Know

*An exhaust analyzer may be called an infrared tester because some of these analyzers use infrared light to analyze the exhaust gases.*

Some shops are equipped with four-gas analyzers which do not indicate $NO_x$ emissions. A pickup and hose assembly is connected from the vehicle tailpipe to the exhaust analyzer. When an exhaust analyzer is turned on, it performs an automatic warmup and calibration procedure. A filter on the analyzer removes water and other particles from the exhaust before they enter the analyzer. If the filter or hose is restricted, a warning light is illuminated on the analyzer. The engine is usually warmed up to normal operating temperature before performing an exhaust emission analysis.

The levels of CO and HC in the exhaust are a direct indication of engine performance. For example, a high HC reading may indicate a misfiring cylinder, and a high CO reading may be caused by a rich air-fuel ratio. A high $NO_x$ reading may be caused by a malfunctioning exhaust gas recirculation (EGR) system. The levels of $CO_2$ and $O_2$ are affected very little by the catalytic converter on the vehicle. A high $O_2$ level indicates a lean air-fuel ratio which may be caused by an intake manifold leak. Five-gas analyzers can be used to diagnose the following conditions: rich or lean air-fuel ratios, faulty injectors, catalytic converter malfunction, air pump malfunction, intake manifold leaks, improper evaporative (EVAP) system operation, improper EGR system operation, or defective engine conditions such as low cylinder compression or defective head gaskets.

The five-gas exhaust analyzer is also very useful when diagnosing vehicles that failed a compulsory emission test.

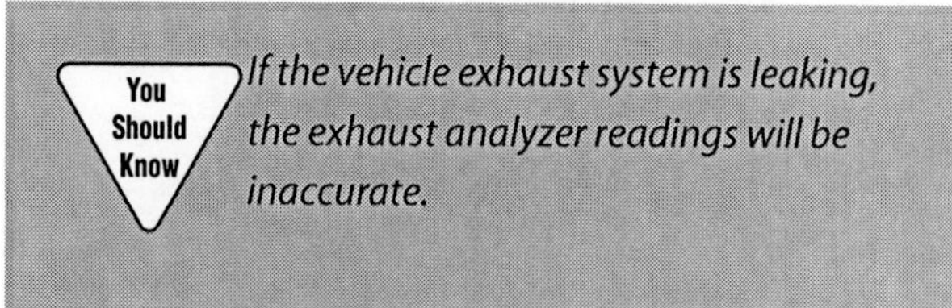


*If the vehicle exhaust system is leaking, the exhaust analyzer readings will be inaccurate.*

## Engine Analyzer

An engine analyzer contains all the necessary test equipment to perform a complete analysis of the engine and all the related engine systems **(Figure 35)**. A computer in the engine analyzer guides the technician through all the tests. Most engine analyzers contain the following test equipment and test capabilities: a cylinder output test, a pressure gauge, a vacuum gauge, a vacuum pump, a tachometer, a timing light/probe, a voltmeter, an ohmmeter, an ammeter, a carbon pile load, an oscilloscope, a scan tool, an exhaust/emissions analyzer, and a laboratory (lab) scope.

An engine analyzer will test the battery, starting system, charging system, primary and secondary ignition systems, fuel system, electronic control systems, emission levels, and engine condition. The analyzer is connected to

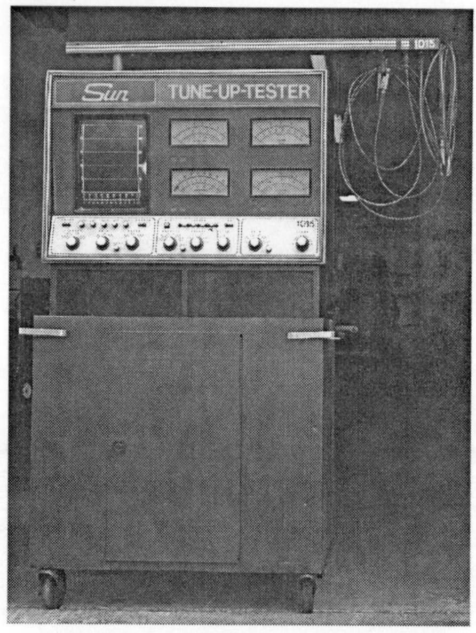


**Figure 35.** An engine analyzer.

these systems by a variety of leads, inductive clamps, probes, and connectors. The data received from these connections is processed by several computers in the analyzer. Many engine analyzers have vehicle specifications on disk or CD. The technician must enter the necessary information regarding the vehicle being analyzed. A keyboard is used to submit commands or information into the analyzer. Most engine analyzers perform a complete series of tests and record the results automatically. A printer connected to the analyzer will print out all test results. The analyzer compares all the test results to specifications and identifies any test results that are not within specification. Many analyzers also provide diagnostic assistance for problems indicated by the out-of-specification readings. The technician may also select any test function or functions separately. Engine analyzer functions vary depending on the equipment manufacturer. The technician must be familiar with the engine analyzer in their shop.

# *Summary*

- Wrenches include box end, open end, and combination type.
- Allen wrenches are hex-shaped tools that fit snugly into a matching recess in the fastener head.
- Flare-nut wrenches are like box-end wrenches with an opening cut through the box end.
- A ratchet allows the technician to tighten or loosen a fastener without removing and resetting the wrench after the fastener is rotated.
- Sockets may have twelve, eight, or six points.
- A universal joint allows an angle between the socket and ratchet or extension.
- A torque wrench measures the tightness of a bolt.
- Screwdrivers may be blade-type Phillips, Torx, or offset type.
- Types of pliers include combination, diagonal, and channel lock.

- A tap is used to thread openings in a casting (internal threads), and a die is used to cut or repair threads on a fastener (external threads).
- Gear and bearing pullers are used to remove gears or bearings that are mounted with a press fit.
- An impact wrench or air ratchet is used to remove and install fasteners.
- A blowgun is used to blow dry parts after the cleaning process.
- A bench grinder is used to grind or buff various components.
- A hydraulic press is used to remove and install press-fit components.
- A floor jack is used to raise one corner or one end of a vehicle.
- A vehicle lift is used to raise a vehicle so undercar work may be performed.
- An engine lift is used to lift an engine during engine removal and replacement.
- Electrical test equipment includes circuit testers, multimeters, tachometers, timing lights, volt-amp testers, and scan tools.
- Electronic equipment includes wheel balancers, exhaust analyzers, and engine analyzers.

# Review Questions

1. While discussing systems of weights and measures: Technician A says the international system (SI) is called the metric system. Technician B says newer domestic vehicles require metric-sized wrenches. Who is correct?
   A. Technician A
   B. Technician B
   C. Both Technician A and Technician B
   D. Neither Technician A nor Technician B
2. Breathing carbon monoxide may cause:
   A. arthritis.
   B. cancer.
   C. impaired vision.
   D. headaches.
3. While discussing hand tools: Technician A says flare-nut wrenches are used to loosen or tighten cylinder head bolts. Technician B says that flare-nut wrenches are used to loosen or tighten fuel line fittings. Who is correct?
   A. Technician A
   B. Technician B
   C. Both Technician A and Technician B
   D. Neither Technician A nor Technician B
4. All these statements about shop tools and equipment are true *except*:
   A. A tap is used to cut or restore the threads on a bolt.
   B. Single-cut files are considered finishing files.
   C. A scan tool may be used to diagnose automotive computer systems.
   D. A tire changer is usually pneumatically powered.
5. Technician A says Allen wrenches are hex-shaped. Technician B says USC sockets are available with a $^5/_8$ drive. Who is correct?
   A. Technician A
   B. Technician B
   C. Both Technician A and Technician B
   D. Neither Technician A nor Technician B
6. All of these statements about sockets are true *except*:
   A. Most sockets in a typical technician's tool set are designed with twelve points.
   B. Compared to a twelve-point socket, a six-point socket has improved grip on a fastener.
   C. Compared to a twelve-point socket, a six-point socket has stronger walls.
   D. Twelve-point and six-point sockets have the same number of installation positions on a fastener.
7. Technician A says torque wrench drive sizes are the same as socket drive sizes. Technician B says torque wrenches may be dial type or torsion bar type. Who is correct?
   A. Technician A
   B. Technician B
   C. Both Technician A and Technician B
   D. Neither Technician A nor Technician B
8. Technician A says double-cut files are considered to be finishing files. Technician B says single-cut files have cutting grooves positioned straight across the face of the file. Who is correct?
   A. Technician A
   B. Technician B
   C. Both Technician A and Technician B
   D. Neither Technician A nor Technician B
9. All of these statements about impact wrenches are true *except*:
   A. Conventional sockets should be used with a light-duty impact wrench.
   B. Impact wrenches may be operated by air pressure or electricity.
   C. Impact wrenches loosen or tighten bolts or nuts with a hammering action.
   D. A light-duty impact wrench may have a $^1/_2$-inch drive.

10. Technician A says a conventional 12-volt test light may be used to diagnose a computer circuit. Technician B says a conventional 12-volt test light may be used to diagnose an air bag circuit. Who is correct?
    A. Technician A
    B. Technician B
    C. Both Technician A and Technician B
    D. Neither Technician A nor Technician B
11. When a fastener is tightened to the torque setting on a break-over type torque wrench, the wrench provides an ____________ ____________.
12. A Torx screwdriver has an ____________ ____________ tip.
13. A bottoming tap is used to cut or restore threads in a ____________ hole.
14. Describe the design and purpose of an open-end wrench.
15. Explain how a self-powered test light may be used to test for an open circuit in a wire.
16. Explain the purpose of a scan tool, and describe the proper on-vehicle connection for this tool.
17. List the five gases indicated on a five-gas analyzer.
18. Explain the automotive systems and conditions that may be tested with an engine analyzer.

# Chapter 5 Tool and Equipment Safety

## *Introduction*

Each person in an automotive shop must follow proper, safe procedures when handling flammable liquids and operating shop equipment. When all personnel in the automotive shop follow these procedures, personal injury, vehicle damage, and property damage may be prevented. All shop personnel must also be familiar with different types of fires that may occur in an automotive shop, and they must also understand the proper type of fire extinguisher to use on various fires. It is essential that shop personnel know how to operate fire extinguishers and extinguish fires. Shop personnel must understand safe shop housekeeping procedures and safe vehicle operation in the shop.

## ELECTRICAL SAFETY

In the automotive shop you will be using electric drills, shop lights, wheel balancers, and wheel aligners. Electrical safety precautions must be observed on this equipment.

- Frayed cords on electrical equipment must be replaced or repaired immediately.
- All electrical cords from lights and electrical equipment must have a ground connection. The ground connector is the round terminal in a three-prong electrical plug. Do not use a two-prong adaptor to plug in a three-prong electrical cord. Three-prong electrical outlets should be mandatory in all shops.
- Do not leave electrical equipment running and unattended.

## GASOLINE SAFETY

Gasoline is a very explosive liquid! One exploding gallon of gasoline has a force equal to fourteen sticks of dynamite. It is the expanding vapors from gasoline that are extremely dangerous. These vapors are present even in cold temperatures. Vapors formed in gasoline tanks on cars are controlled, but vapors from a gasoline storage can may escape from the can, resulting in a hazardous situation. Therefore, gasoline storage containers must be placed in a well-ventilated space.

Approved gasoline storage cans have a flash-arresting screen at the outlet **(Figure 1)**. This screen prevents external ignition sources from igniting the gasoline within the can while the gasoline is being poured. Follow these safety precautions regarding gasoline containers.

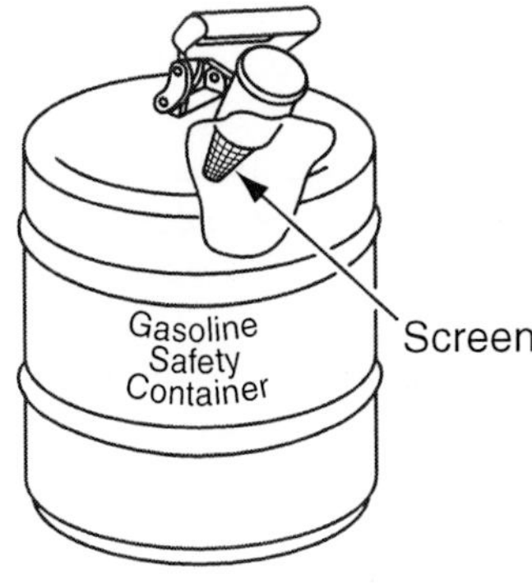


**Figure 1.** An approved gasoline container.

1. Always use approved gasoline containers that are painted red for proper identification.
2. Do not fill gasoline containers completely full. Always leave the level of gasoline at least one inch from the top of the container. This allows for expansion of the gasoline at higher temperatures. If gasoline containers are completely full, the gasoline will expand when the temperature increases. This expansion forces gasoline from the can and creates a dangerous spill.
3. If gasoline containers must be stored, place them in a well-ventilated area such as a storage shed. Do not store gasoline containers in your home or in the trunk of a vehicle.
4. When a gasoline container must be transported, be sure it is secured against upsets.
5. Do not store a partially filled gasoline container for long periods of time because it may give off vapors and produce a potential danger.
6. Never leave gasoline containers open except while filling or pouring gasoline from the container.
7. Do not prime an engine with gasoline while cranking the engine.
8. Never use gasoline as a cleaning agent.

## FIRE SAFETY

When fire safety rules are observed, personal injury and expensive fire damage to vehicles and property may be avoided.

- Familiarize yourself with the location and operation of all shop fire extinguishers.
- If a fire extinguisher is used, report it to management so the extinguisher can be recharged.
- Do not use any type of open flame heater to heat the work area.
- Do not turn the ignition switch on or crank the engine with a gasoline line disconnected.
- Store all combustible materials such as gasoline, paint, and oily rags in approved safety containers.
- Clean up gasoline, oil, or grease spills immediately.
- Always wear clean shop clothes. Do not wear oil-soaked clothes.
- Do not allow sparks and flames near batteries.
- Be sure that welding tanks are securely fastened in an upright position.
- Do not block doors, stairways, or exits.
- Do not smoke when working on vehicles.
- Do not smoke or create sparks near flammable materials or liquids.
- Store combustible shop supplies (such as paint) in a closed steel cabinet.
- Store gasoline in approved safety containers.
- If a gasoline tank is removed from a vehicle, do not drag the tank on the shop floor.
- Know the approved fire escape route from your classroom or shop to the outside of the building.
- If a fire occurs, do not open doors or windows. This action creates extra draft, which makes the fire worse.
- Do not put water on a gasoline fire because the water will make the fire worse.
- Call the fire department as soon as a fire begins, and then attempt to extinguish the fire.
- If possible, stand 6 to 10 feet from the fire and aim the fire extinguisher nozzle at the base of the fire with a sweeping action.
- If a fire produces a great deal of smoke in the room, remain close to the floor to obtain oxygen and avoid breathing smoke.
- If the fire is too hot or the smoke makes breathing difficult, get out of the building.
- Do not re-enter a burning building.
- Keep solvent containers covered except when pouring from one container to another. When flammable liquids are transferred from bulk storage, the bulk container should be grounded to a permanent shop fixture such as a metal pipe. During this transfer process, the bulk container should be grounded to the portable container **(Figure 2)**. These ground wires prevent the buildup of a static electric charge, which could cause a spark and a disastrous explosion. Always discard or clean empty solvent containers because fumes in these containers are a fire hazard.
- Familiarize yourself with different types of fires and fire extinguishers, and know the type of extinguisher to use on each fire.

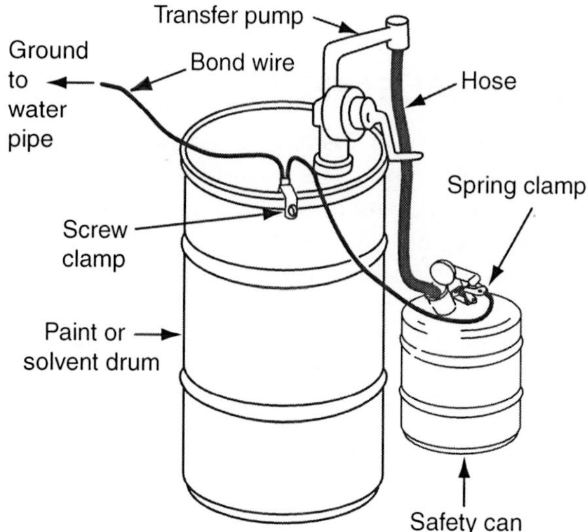


**Figure 2.** Safe procedures for flammable liquid transfer.

## USING A FIRE EXTINGUISHER

Everyone working in the shop must know how to operate the fire extinguishers. There are several different types of fire extinguishers, but their operation usually involves the following steps.

1. Get as close as possible to the fire without jeopardizing your safety.
2. Grasp the extinguisher firmly and aim the extinguisher at the fire.
3. Pull the pin from the extinguisher handle.
4. Squeeze the handle to dispense the contents of the extinguisher.
5. Direct the fire extinguisher nozzle at the base of the fire, and dispense the contents of the extinguisher with a sweeping action back and forth across the fire. Most extinguishers discharge their contents in 8 to 25 seconds.
6. Always be sure the fire is extinguished.
7. Always keep an escape route open behind you so a quick exit is possible if the fire becomes out of control.

## VEHICLE OPERATION

When driving a customer's vehicle, certain precautions must be observed to prevent accidents and maintain good customer relations.

1. Prior to driving a vehicle, make sure the brakes are operational and fasten the safety belt.
2. Check to be sure there is no person or object under the car before you start the engine.
3. If the vehicle is parked on a lift, be sure the lift is fully down and the lift arms, or components, are not in contact with the vehicle chassis.
4. Check to see if there are any objects directly in front of or behind the vehicle before driving away.
5. Always drive slowly in the shop, and watch carefully for personnel and other moving vehicles.
6. Make sure the shop door is up high enough so there is plenty of clearance between the top of the vehicle and the door.
7. Watch the shop door to be certain that it is not coming down as you attempt to drive under the door.
8. If a road test is necessary, wear your seat belt, obey all traffic laws, and never drive in a reckless manner.
9. Do not squeal tires when accelerating or turning corners.

If the customer observes that service personnel take good care of his or her car by driving carefully and installing fender, seat, and floor mat covers, the service department image is greatly enhanced in the customer's eyes. These procedures impress upon the customer that shop personnel respect the car. Conversely, if grease spots are found on upholstery or fenders after service work is completed, the customer will probably think the shop is careless, not only in car care but also in service work quality.

## HOUSEKEEPING SAFETY

Careful housekeeping habits prevent accidents and increase worker efficiency. Good housekeeping also helps impress upon the customer that quality work is a priority in this shop. Follow these housekeeping rules:

- Keep aisles and walkways clear of tools, equipment, and other items.
- Be sure all sewer covers are securely in place.
- Keep floor surfaces free of oil, grease, water, and loose material.
- Sweep up under a vehicle before lowering the vehicle on the lift.
- Proper trash containers must be conveniently located, and these containers should be emptied regularly.
- Access to fire extinguishers must be unobstructed at all times, and fire extinguishers should be checked for proper charge at regular intervals.
- Tools must be kept clean and in good condition.
- When not in use, tools must be stored in their proper location.
- Oily rags must be stored in approved, covered containers **(Figure 3)**. A slow generation of heat occurs from oxidation of oil on these rags. Heat may continue to be generated until the ignition temperature is reached. The oil and the rags then begin to burn, causing a fire. This action is called spontaneous combustion. However, if the oily rags are in an airtight, approved container, the fire cannot receive enough oxygen to cause burning.

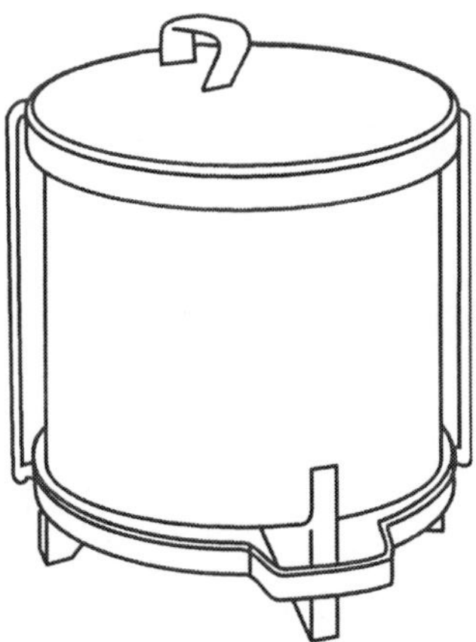

**Figure 3.** Dirty shop towels or rags must be kept in approved, closed containers.

Figure 4. Store combustible materials in an approved safety cabinet.

- Store paint, gasoline, and other flammable liquids in a closed steel cabinet **(Figure 4)**.
- Rotating components on equipment and machinery must have guards, and all shop equipment should have regular service and adjustment schedules.
- Keep the workbenches clean. Do not leave heavy objects, such as used parts, on the bench after you are finished with them.
- Keep parts and materials in their proper location.
- When not in use, creepers must not be left on the shop floor. Creepers should be stored in a specific location.
- The shop should be well lighted, and all lights should be in working order.
- Frayed electrical cords on lights or equipment must be replaced.
- Walls and windows should be cleaned regularly.
- Stairs must be clean, well lighted, and free of loose material.

## AIR BAG SAFETY

Technicians must be familiar with air bag safety rules. If air bag safety rules are not followed, expensive air bags may be accidentally deployed, and the technician may be injured.

1. When service is performed on any air bag system component, always disconnect the negative battery cable, isolate the cable end, and wait for the amount of time specified by the vehicle manufacturer before proceeding with the necessary diagnosis or service. The average waiting period is 2 minutes, but some vehicle manufacturers specify up to 10 minutes. Failure to observe this precaution may cause accidental air bag deployment and personal injury.
2. Replacement air bag system parts must have the same part number as the original part. Replacement parts of lesser or questionable quality must not be used. Improper or inferior components may result in inappropriate air bag deployment and injury to the vehicle occupants.
3. Do not strike or jar a sensor or an air bag system diagnostic monitor (ASDM). This may cause air bag deployment or make the sensor inoperative. Accidental air bag deployment may cause personal injury, and an inoperative sensor may result in air bag deployment failure, causing personal injury to vehicle occupants. An **air bag system diagnostic monitor (ASDM)** is the computer that operates the air bag system.
4. All sensors and mounting brackets must be properly torqued to ensure correct sensor operation before an air bag system is powered up. If sensor fasteners do not have the proper torque, improper air bag deployment may result in injury to vehicle occupants.
5. When working on the electrical system on an air bag-equipped vehicle, use only the vehicle manufacturer's recommended tools and service procedures. The use of improper tools or service procedures may cause accidental air bag deployment and personal injury. For example, do not use 12-volt or self-powered test lights when servicing the electrical system on an air-bag-equipped vehicle.

## LIFTING AND CARRYING

Many automotive service jobs require heavy lifting. Know your maximum weight lifting ability, and do not attempt to lift more than this weight. If a heavy part exceeds your weight lifting ability, have a coworker help with the lifting job. Follow these steps when lifting or carrying an object:

1. If the object is going to be carried, be sure your path is free from loose parts or tools.
2. Position your feet close to the object; position your back reasonably straight for proper balance.
3. Your back and elbows should be kept as straight as possible. Continue to bend your knees until your hands reach the best lifting location on the object.
4. Be certain the container is in good condition. If a container falls apart during the lifting operation, parts may drop out of the container and result in foot injury or part damage.
5. Maintain a firm grip on the object; do not attempt to change your grip while lifting is in progress.
6. Straighten your legs to lift the object, and keep the object close to your body. Use leg muscles rather than back muscles **(Figure 5)**.
7. If you have to change direction of travel, turn your whole body. Do not twist.

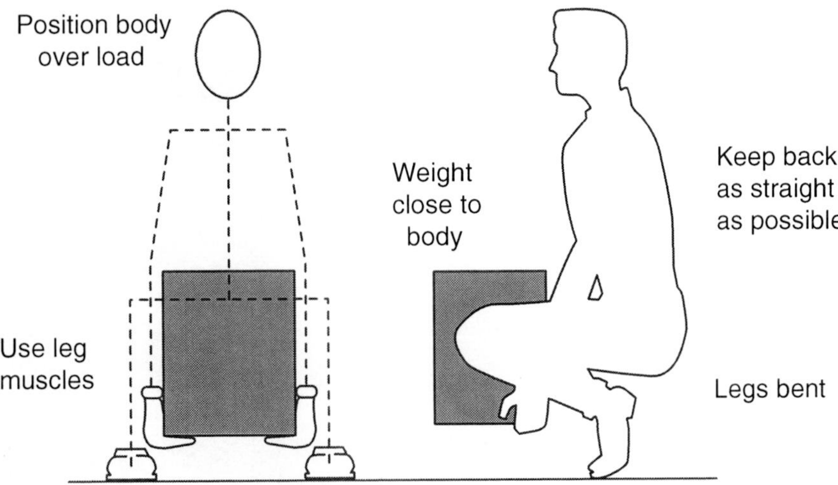


**Figure 5.** Use your leg muscles—never your back—to lift heavy objects.

8. Do not bend forward to place an object on a workbench or table. Position the object on the front surface of the workbench and slide it back. Do not pinch your fingers under the object while setting it on the front of the bench.
9. If the object must be placed on the floor or a low surface, bend your legs to lower the object. Do not bend your back forward because this movement strains back muscles.
10. When a heavy object must be placed on the floor, place suitable blocks under the object to prevent jamming your fingers.

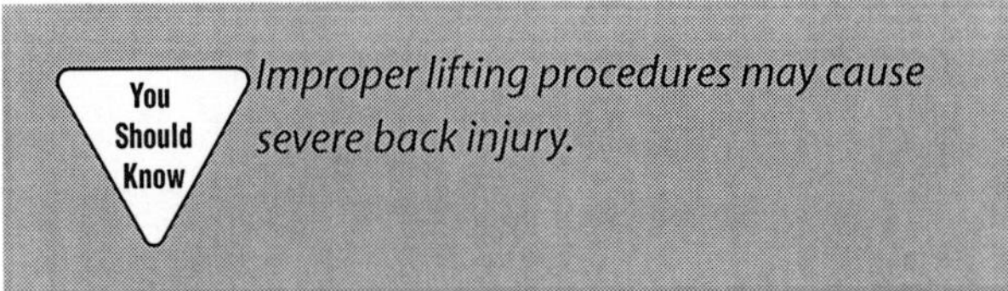


*Improper lifting procedures may cause severe back injury.*

## HAND TOOL SAFETY

Many shop accidents are caused by improper use and care of hand tools. Follow these safety steps when working with hand tools:

1. Maintain tools in good condition and keep them clean. Worn tools may slip and result in hand injury. If a hammer with a loose head is used, the head may fly off and cause personal injury or vehicle damage. If your hand slips off a greasy tool, it may cause some part of your body to hit the vehicle, causing injury.
2. Using the wrong tool for the job may damage the tool, fastener, or your hand if the tool slips. If you use a screwdriver as a chisel or pry bar, the blade may shatter causing serious personal injury.
3. Use sharp pointed tools with caution. Always check your pockets before sitting on the vehicle seat. A screwdriver, punch, or chisel in the back pocket may put an expensive tear in the upholstery. Do not lean over fenders with sharp tools in your pockets.
4. Tools that are intended to be sharp should be kept sharp. A sharp chisel, for example, will do the job faster with less effort.

## VEHICLE LIFT (HOIST) SAFETY

Special precautions and procedures must be followed when a vehicle is raised on a lift. Follow these steps when operating a lift:

1. Always be sure the lift is completely lowered before driving a vehicle on or off the lift.
2. Do not hit or run over lift arms and adaptors when driving a vehicle on or off the lift. Have a coworker guide you when driving a vehicle onto the lift. Do not stand in front of a lift with the car coming toward you.
3. Be sure the lift pads contact the car manufacturer's recommended lifting points shown in the service manual. If the proper lifting points are not used, components under the vehicle, such as brake lines, suspension components, or body parts, may be damaged. Failure to use the recommended lifting points may cause the vehicle to slip off the lift, resulting in severe vehicle damage and personal injury.
4. Before a vehicle is raised or lowered, close the doors, hood, and trunk lid.

5. When a vehicle has been lifted a short distance off the floor, stop the lift and check the contact between the hoist lift pads and the vehicle to be sure the lift pads are still on the recommended lifting points.
6. When a vehicle has been raised, be sure the safety mechanism is in place to prevent the lift from dropping accidentally.
7. Prior to lowering a vehicle, always make sure there are no objects, tools, or people under the vehicle.
8. Do not rock a vehicle on a lift during a service job.
9. When a vehicle is raised, removal of some heavy components may cause vehicle imbalance. For example, since front-wheel-drive cars have the engine and transaxle at the front of the vehicle, these cars have most of their weight on the front end. Removing a heavy rear-end component on these cars may cause the back end of the car to rise off the lift. If this happens, the vehicle could fall off the lift!
10. Do not raise a vehicle on a lift with people in the vehicle.
11. When raising pickup trucks or vans on a lift, remember these vehicles are higher than a passenger car. Be sure there is adequate clearance between the top of the vehicle and the shop ceiling, or components under the ceiling.
12. Do not raise a four-wheel-drive vehicle with a frame contact lift unless proper adaptors are used. Lifting a vehicle on a frame contact lift without the proper adaptors may damage axle joints.
13. Do not operate a front-wheel-drive vehicle that is raised on a frame contact lift. This may damage the front drive axles.

You Should Know

*When a vehicle is raised on a lift, the vehicle must be raised high enough to allow engagement of the lift locking mechanism. If the locking mechanism is not engaged, the lift may drop suddenly resulting in personal injury and/or vehicle damage.*

## HYDRAULIC JACK AND SAFETY STAND SAFETY

Accidents involving the use of floor jacks and safety stands may be avoided if these safety precautions are followed:

1. Never work under a vehicle unless safety stands are placed securely under the vehicle chassis and the vehicle is resting on these stands **(Figure 6)**.
2. Prior to lifting a vehicle with a floor jack, be sure that the jack lift pad is positioned securely under a recommended lifting point on the vehicle. Lifting the front end of a vehicle with the jack placed under a radiator support may cause severe damage to the radiator and support.

**Figure 6.** Safety stands.

3. Position the safety stands under a strong chassis member, such as the frame or axle housing. The safety stands must contact the vehicle manufacturer's recommended lifting points.
4. Since the floor jack is on wheels, the vehicle tends to move as it is lowered from a floor jack onto safety stands. Always be sure the safety stands remain under the chassis member during this operation, and be sure the safety stands do not tip. All the safety stand legs must remain in contact with the shop floor.
5. When the vehicle is lowered from the floor jack onto safety stands, remove the floor jack from under the vehicle. Never leave a jack handle sticking out from under a vehicle. Someone may trip over the handle and injure himself or herself.

## POWER TOOL SAFETY

Power tools use electricity, shop air, or hydraulic pressure as a power source. Careless operation of power tools may cause personal injury or vehicle damage. Follow these steps for safe power tool operation:

1. Do not operate power tools with frayed electrical cords.
2. Be sure the power tool cord has a proper ground connection.
3. Do not stand on a wet floor while operating an electric power tool.

4. Always unplug an electric power tool before servicing the tool.
5. Do not leave a power tool running and unattended.
6. When using a power tool on small parts, do not hold the part in your hand. The part must be secured in a bench vise or with locking pliers.
7. Do not use a power tool on a job where the maximum capacity of the tool is exceeded.
8. Be sure that all power tools are in good condition; always operate these tools according to the tool manufacturer's recommended procedure.
9. Make sure all protective shields and guards are in position.
10. Maintain proper body balance while using a power tool.
11. Always wear safety glasses or a face shield.
12. Wear ear protection.
13. Follow the equipment manufacturer's recommended maintenance schedule for all shop equipment.
14. Never operate a power tool unless you are familiar with the tool manufacturer's recommended operating procedure. Serious accidents occur from improper operating procedures.
15. Always make sure that the wheels are securely attached and are in good condition on the bench grinder.
16. Keep fingers and clothing away from grinding and buffing wheels. When grinding or buffing a small part, hold the part with a pair of locking pliers.
17. Always make sure the sanding or buffing disk is securely attached to the sander pad.
18. Special heavy-duty sockets must be used on impact wrenches. If ordinary sockets are used on an impact wrench, they may break and cause serious personal injury.

## COMPRESSED-AIR EQUIPMENT SAFETY

The shop air supply contains high-pressure air in the shop compressor and air lines. Serious injury or property damage may result from careless operation of compressed-air equipment. Follow these steps to improve safety.

1. Never operate an air chisel unless the tool is securely connected to the chisel with the proper retaining device.
2. Never direct a blast of air from an air gun against any part of your body. If air penetrates the skin and enters the bloodstream, it may cause very serious health problems and even death.
3. Safety glasses or a face shield should be worn for all shop tasks, including those tasks involving the use of compressed-air equipment.
4. Wear ear protection when using compressed-air equipment.
5. Always maintain air hoses and fittings in good condition. If an end suddenly blows off an air hose, the hose will whip around, possibly causing personal injury.
6. Use only air gun nozzles approved by OSHA.
7. Do not use an air gun to blow debris off clothing or hair.
8. Do not clean the workbench or floor with compressed air. This action may blow very small parts against your skin or into your eye. Small parts blown by compressed air may also cause vehicle damage. For example, if the car in the next stall has the air cleaner removed, a small part may find its way into the carburetor or throttle body. When the engine is started, this part will likely be pulled into the cylinder by engine vacuum, and the part will penetrate through the top of a piston.
9. Never spin bearings with compressed air because the bearing will rotate at extremely high speed. This may damage the bearing or cause it to disintegrate, causing personal injury.
10. All pneumatic tools must be operated according to the tool manufacturer's recommended operating procedure.
11. Follow the equipment manufacturer's recommended maintenance schedule for all compressed-air equipment.

## CLEANING EQUIPMENT SAFETY AND ENVIRONMENTAL CONSIDERATIONS

All technicians are required to clean parts during their normal work routines. Face shields and protective gloves must be worn while operating cleaning equipment. In most states, environmental regulations require that the runoff from steam cleaning must be contained in the steam cleaning system. This runoff cannot be dumped into the sewer system. Since it is expensive to contain this runoff in the steam cleaning system, the popularity of steam cleaning has decreased. The solution in hot and cold cleaning tanks may be caustic, and contact between this solution and skin or eyes must be avoided. Parts cleaning often creates a slippery floor, and care must be taken when walking in the parts cleaning area. The floor in this area should be cleaned frequently. When the cleaning solution in hot or cold cleaning tanks is replaced, environmental regulations require that the old solution be handled as hazardous waste. Use caution when placing aluminum or aluminum alloy parts in a cleaning solution. Some cleaning solutions will damage these components. Always follow the cleaning equipment manufacturer's recommendations.

### Parts Washers with Electromechanical Agitation

Some parts washers provide electromechanical agitation of the parts to provide improved cleaning action

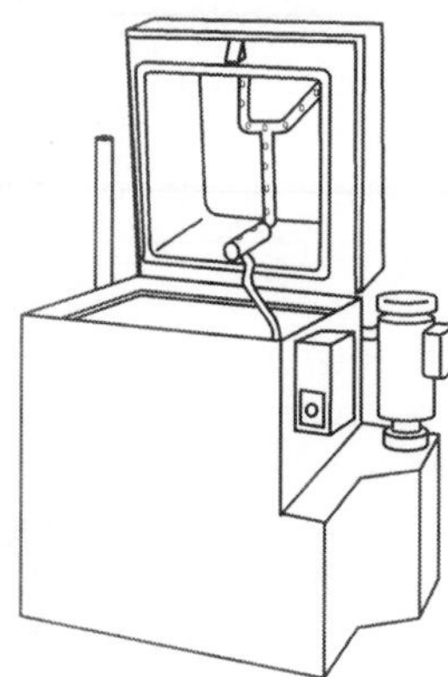

Figure 7. A parts washer with electromechanical agitation.

(**Figure 7**). These parts washers may be heated with gas or electricity. Various water-based hot tank cleaning solutions are available depending on the type of metals being cleaned. For example, Kleer-Flo Greasoff® number 1 powdered detergent is available for cleaning iron and steel. Non-heated electromechanical parts washers are also available, and these washers use cold cleaning solutions such as Kleer-Flo Degreasol® formulas.

Many cleaning solutions, such as Kleer-Flo Degreasol® 99R, contain no ingredients listed as hazardous by the EPA's RCRA Act. This cleaning solution is a blend of sulfur-free hydrocarbons, wetting agents, and detergents. Degreasol® 99R does not contain aromatic or chlorinated solvents, and it conforms to California's Rule 66 for clean air. Always use the cleaning solution recommended by the equipment manufacturer.

## Cold Parts Washer with Agitation Immersion Tank

Some parts washers have an agitator immersion chamber under the shelves that provides thorough parts cleaning. Folding work shelves provide a large upper cleaning area with a constant flow of solution from the dispensing hose (**Figure 8**). This cold parts washer operates on Degreasol® 99R cleaning solution.

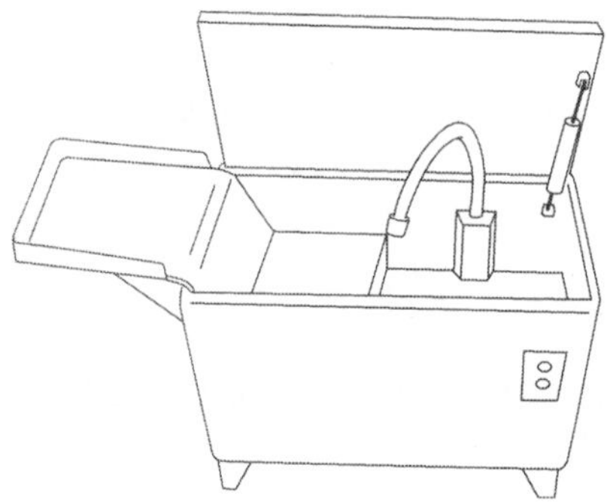

Figure 8. A cold parts washer with an agitator immersion tank.

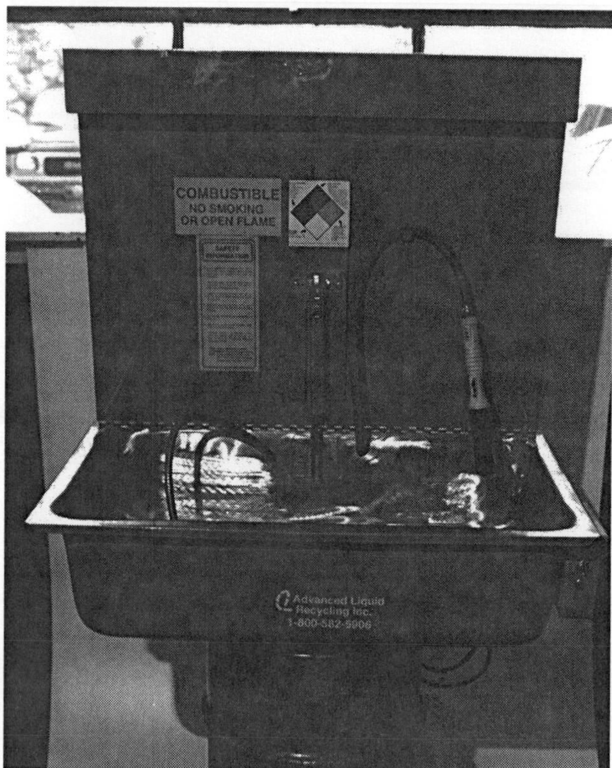


Figure 9. An aqueous parts cleaning tank.

## Aqueous Parts Cleaning Tank

The aqueous parts cleaning tank uses a water-based, environmentally friendly cleaning solution, such as Greasoff® 2, rather than traditional solvents. The immersion tank is heated and agitated for effective parts cleaning (**Figure 9**). A sparger bar pumps a constant flow of cleaning solution across the surface to push floating oils away, and an integral skimmer removes these oils. This action prevents floating surface oils from redepositing on cleaned parts.

## HANDLING SHOP WASTES

The shop is responsible for hazardous waste until such waste is delivered to a hazardous waste site. Many shops contract a hazardous waste hauler to transport hazardous waste from the shop to government-approved recyclers or hazardous waste disposal sites. Always hire a properly licensed waste hauler, and be sure you know how the waste hauler is disposing of shop wastes. Be sure to have a written contract with the hazardous waste hauler. The hazardous waste hauler fills out the necessary forms related to waste disposal and communicates with various state and federal agencies that are in charge of hazardous waste disposal regulations. Always keep all shipping bills from your hazardous waste hauler to prove you have recycled or disposed of hazardous waste material.

## Batteries

Batteries should be recycled by shipping them to a reclaimer or back to the battery distributor. When defective batteries are stored on site, they should be kept in water-tight, acid-resistant containers. Acid residue from batteries is hazardous because it is corrosive, and it may contain lead and other toxins. Inspect defective batteries for cracks and leaks. Spilled battery acid should be neutralized by covering it with baking soda or lime, and then clean up and dispose of all hazardous material.

## Oil

Used oil is usually hauled to an oil recycling facility. Place oil drip pans under vehicles with oil leaks so oil does not drip onto the storage area. In some states it is legal to burn used oil in a commercial space heater. State and local authorities must be contacted regarding regulations and permits.

## Oil Filters

Used oil filters should be allowed to drain into an appropriate drip pan for 24 hours. After the draining process, oil filters should be squashed and recycled.

## Solvents

Parts cleaning equipment that uses hazardous cleaning chemicals should be replaced with cleaning equipment that uses water-based degreasers. If hazardous chemicals are used in cleaning equipment, these chemicals must be recycled or disposed of as hazardous waste. Evaporation from spent cleaning chemicals is a contributor to ozone depletion and smog formation. Spent cleaning chemicals should be placed in closed, labelled containers and stored on drip pans or in diked areas. The storage area for waste materials should be covered to prevent rain from washing contaminants from stored materials into the ground water. This storage area may have to be fenced and locked if vandalism is a possibility.

## Liquids

Engine coolant should be collected and recycled in an approved coolant recycling machine. Other liquids such as brake fluid and transmission fluid should be labelled and stored in the same area as solvents. Used brake fluid or transmission fluid should be recycled or disposed of as hazardous waste material.

## Shop Towels

When dirty shop towels are stored on site, they should be placed in closed containers that are clearly marked, "contaminated shop towels only." Shop towels should be cleaned by a laundry service that has the capability to treat the waste water generated by cleaning these towels.

## Refrigerants

When servicing automotive air conditioning systems, it is illegal to vent refrigerants to the atmosphere. Certified equipment must be used to recover and recycle the refrigerant and recharge air conditioning systems. This service work must be performed by an EPA-certified technician.

# INTERPRETING MATERIAL SAFETY DATA SHEETS (MSDS)

The product manufacturer's name and address is provided at the top of the Material Safety Data Sheet (MSDS) **(Figure 10)**. The product manufacturer's phone number is also provided so it can be contacted in case of an emergency. The product name is provided with the chemical family name, plus any synonyms.

The ingredients in the hazardous material are listed in the **Ingredients section**. The **threshold limit value (TLV)** and the **permissible exposure limit (PEL)** recommended by OSHA are listed in this section. The TLV and PEL values are the permissible concentrations of the hazardous material in the air to which a person may be exposed daily without known harmful effects. These values are usually

MATERIAL SAFETY DATA SHEET

PRODUCT NAME: KLEAN-A-KARB (aerosol) #- HPMS 102068
PRODUCT: 5078, 5931, 6047T
(page 1 of 2)

| 1. Ingredients | CAS # | ACGIH TLV | OSHA PEL | OTHER LIMITS | % |
|---|---|---|---|---|---|
| Acetone | 67-64-1 | 750ppm | 750ppm | | 2-5 |
| Xylene | 1330-20-7 | 100ppm | 100ppm | | 68-75 |
| 2-Butoxy Ethanol | 111-76-2 | 25ppm | 25ppm | (skin) | 3-5 |
| Methanol | 67-56-1 | 200ppm | 200ppm | | 3-5 |
| Detergent | - | NA | NA | | 0-1 |
| Propane | 74-98-6 | NA | 1000ppm | | 10-20 |
| Isobutane | 75-28-5 | NA | NA | 1000ppm | 10-20 |

2. PHYSICAL DATA : (without propellent)
Specific Gravity : 0.865
Vapor Pressure : ND
% Volatile : >99
Boiling Point : 176°F Initial
Evaporation Rate : Moderately Fast
Freezing Point : ND
Vapor Density : ND
Solubility: Partially soulable in water
pH :NA
Appearance and Odor: A clear colorless liquid, aromatic odor

3. FIRE AND EXPLOSION DATA
Flashpoint : –40°F
Method : TCC
Flammable Limits propellent LEL: 1.8 UEL: 9.5
Extinguishing Media: $CO_2$, dry chemical, foam
Unusual Hazards : Aerosol cans may explode when heated above 120°F.

4. REACTIVITY AND STABILITY
Stability : Stable
Hazardous decomposition products: $CO_2$, carbon monoxide (thermal)
Materials to avoid: Strong oxidizing agents and sources of ignition

5. PROTECTION INFORMATION
Ventilation : Use mechanical means to insure vapor concentration is below TLV.
Respiratory: Use self-contained breathing apparatus above TLV.
Gloves : Solvent resistant
Eye and Face: Safety Glasses
Other Protective Equipment: Not normally required for aerosol product usage

**Figure 10.** A material safety data sheet (MSDS).

expressed in parts per million (PPM). The percentage column indicates the percentage of the ingredient in relation to the total weight or volume of the hazardous product.

The **Physical Data section** of the MSDS provides information about the hazardous material such as the appearance and odor of the material. This information could help emergency personnel to recognize a hazardous material. This section also provides other information such as specific gravity, boiling point, and solubility.

The **Fire and Explosion Data section** may help you to prevent or fight a fire involving the hazardous material. The flash point listed in this section indicates the lowest temperature at which a liquid gives off enough vapor to ignite. The proper type of fire extinguisher that should be used to put out a fire involving this material is listed in the extinguishing media information. This section also provides information regarding any materials that should be kept away from a fire involving this hazardous material.

The section on **Reactivity and Stability** provides information regarding the mixing of other material with the hazardous material. The stability classification indicates how the hazardous material resists chemical or physical change. This section also provides information about materials that may cause violent reactions when brought in contact with the hazardous material.

The section on **Protection Information** supplies information regarding the necessary protective equipment required when you are handling the hazardous material. This section also provides information regarding the proper ventilation procedure when dealing with the hazardous material.

Many MSDS also contain a section on leak and spill procedures. These procedures include the personal precautions to be observed if the hazardous material is spilled. This section also provides information about the proper procedure for disposal of the materials used in the clean up process.

## *Summary*

- Electrical safety precautions include replacing frayed electrical cords, assuring that all electrical equipment has a ground connection, and not leaving electrical equipment running and unattended.
- Gasoline must be stored only in approved gasoline containers.
- Fire safety precautions include never turning on an ignition switch or cranking an engine with a fuel line disconnected, and storing all combustible materials in approved containers.
- Shop employees must be familiar with proper fire extinguisher operation.
- Safe vehicle operating procedures include always checking for adequate brake operation before driving the vehicle and making sure the shop door is up high enough before attempting to drive into or out of the shop.
- Safe housekeeping procedures include keeping floors free from oil, grease, water, and loose material, and providing unobstructed access to fire extinguishers and other shop safety equipment.
- Air bag safety precautions include always disconnecting the vehicle battery and waiting until the proper amount of time has elapsed before servicing an air bag system component. Use only the vehicle manufacturer's recommended tools for servicing these systems.
- Shop employees must follow proper lifting and carrying procedures to avoid back injury.
- Proper hand tool safety includes keeping tools in good condition and using the proper tool for the repair job being performed.
- Lift safety precautions include making sure the lift arms are connecting the vehicle manufacturer's specified vehicle lift points, being sure the lift safety mechanism is in place after the vehicle is raised on a lift.
- Hydraulic jack and safety stand safety precautions include always placing the jack lift pad on the vehicle manufacturer's specified lift point, and never exceeding the weight lifting capacity of the jack.
- Power tool safety procedures include always being sure the power tool has a ground connection. Do not operate power tools with frayed electrical cords.
- Compressed-air equipment safety precautions include never directing compressed air against or near human flesh or eyes. Always maintain air hoses and fittings in good condition.
- Cleaning equipment safety precautions include wearing face shields and protective gloves when operating cleaning equipment, and keeping the floor clean and dry in the cleaning equipment area.
- Shop personnel must know the proper disposal procedures for all shop wastes.
- Shop personnel must be able to interpret MSDS.

## Review Questions

1. While lifting heavy objects in the automotive shop:
   A. bend your back to pick up the heavy object.
   B. place your feet as far as possible from the object.
   C. bend forward to place the object on the workbench.
   D. straighten your legs to lift an object off the floor.
2. While discussing power tool safety, Technician A says an electric power tool cord does not require a ground. Technician B says frayed electric cords should be replaced. Who is correct?
   A. Technician A
   B. Technician B
   C. Both Technician A and Technician B
   D. Neither Technician A nor Technician B
3. While operating hydraulic equipment safely in the automotive shop, remember that:
   A. Safety stands have a maximum weight capacity.
   B. The driver's door should be open when raising a vehicle on a lift.
   C. A lift does not require a safety mechanism to prevent lift failure.
   D. Four-wheel-drive vehicles should be lifted on a frame contact lift.
4. All these shop rules are correct *except*:
   A. USC tools may be substituted for metric tools.
   B. Foot injuries may be caused by loose sewer covers.
   C. Hands should be kept away from electric-drive cooling fans.
   D. Power tools should not be left running and unattended.
5. Technician A says the light-colored, flat connector is the ground connection in a three-prong electrical plug. Technician B says a two-prong adaptor may be used to plug in a three-prong end on an electrical cord. Who is correct?
   A. Technician A
   B. Technician B
   C. Both Technician A and Technician B
   D. Neither Technician A nor Technician B
6. All of these statements about gasoline containers are true *except*:
   A. Gasoline containers should be stored in a non-ventilated area.
   B. When transporting a filled gasoline container, it must be secured.
   C. Gasoline containers should be filled only to within 1 inch of the container top.
   D. Gasoline must be placed in approved gasoline containers.
7. When discussing shop safety and fires, Technician A says combustible shop supplies such as paint should be stored in a metal cabinet. Technician B says water should be used to extinguish a gasoline fire. Who is correct?
   A. Technician A
   B. Technician B
   C. Both Technician A and Technician B
   D. Neither Technician A nor Technician B
8. All of these statements about fire extinguishers and their use are true *except*:
   A. Direct the fire extinguisher nozzle at the base of a fire.
   B. To activate a fire extinguisher, pull the pin and rotate the handle.
   C. Many fire extinguishers discharge their contents in 8 to 25 seconds.
   D. When using a fire extinguisher to put out a fire, keep an escape route open behind you.
5. When discussing air bag system service and diagnosis, Technician A says the negative battery terminal should be disconnected and the vehicle manufacturer's specified time period elapsed before working on an air bag system. Technician B says all air bag sensor mounting bolts must be tightened to the specified torque to ensure proper sensor operation. Who is correct?
   A. Technician A
   B. Technician B
   C. Both Technician A and Technician B
   D. Neither Technician A nor Technician B
10. All of these statements about using a vehicle lift are true *except*:
    A. When raising a vehicle on a lift, the driver's door should be left open.
    B. The lift arms must contact the vehicle manufacturer's specified lift points.
    C. After the vehicle is raised, the lift safety mechanism must be in place.
    D. When lifted on a frame-contact lift, a four-wheel-drive vehicle requires special adaptors.
11. The round terminal in a three-prong electrical plug is the ______________ connection.
12. When servicing an air bag system, the technician must not ______________ or ______________ air bag sensors.
13. If battery acid is spilled, it may be neutralized by covering it with ______________ ______________.
14. Gasoline containers should be stored in a ______________ ______________ area.

15. Explain the proper safety precautions when filling gasoline containers.
16. Describe the proper fire extinguisher operating procedure.
17. Explain the necessary safety procedures when servicing air bag systems.
18. Describe the proper procedure for lifting heavy objects.

# Section 2

## Basic Shop Procedures, Measurements, and Fasteners



## SECTION OBJECTIVES

After you have read, studied, and practiced the contents of this section, you should be able to:

- Demonstrate the proper use of an impact wrench.
- Use various types of torque wrenches.
- Use gear, bearing, and seal pullers.
- Remove broken studs and screws.
- Remove damaged nuts.
- Use an acetylene torch for heating.
- Inspect and change respirator filters.
- Jump start a vehicle with a discharged battery.
- Understand USC and metric measurement systems.
- Perform measurements with a feeler gauge.
- Accurately measure components with an inside micrometer.
- Perform accurate measurements with a dial indicator.
- List four different types of fastener threads.
- Explain bolt diameter, pitch, length, thread depth, and grade marks.
- Describe the advantage of torque-to-yield bolts.
- Describe the proper tightening procedure for torque-to-yield bolts.

*The average age of vehicles in the United States increased from 7.8 years in 1990 to 9.2 years in 2002. This means there are more older vehicles on the highways, and these vehicles will require more service.*

# Chapter 6 Basic Shop Procedures

## Introduction

It is extremely important for automotive students and technicians to learn the proper use of shop equipment. Improper equipment use may cause personal injury, equipment damage, or vehicle damage. The first time students are required to use specific shop equipment, it is very important for them to learn the proper procedure for operating this equipment. When improper equipment operating procedures are used, it is more difficult to relearn the proper procedures.

### USING SHOP TOOLS AND EQUIPMENT

Technicians must be familiar with the use of shop tools and equipment. When technicians lack this knowledge, expensive tools and equipment may be damaged, and personal injury may occur. A knowledge of shop tools and equipment also allows technicians to work faster and more efficiently.

### CONNECTING AND USING AN IMPACT WRENCH

An impact wrench is a hand-held, reversible power tool for removing and tightening bolts and nuts. Impact wrenches may be powered by electricity or shop air pressure. Heavy-duty impact wrenches may deliver up to 450 foot-pounds (ft.-lb) (607.5 N•m) of torque. When an impact wrench is operating, the drive shaft on this tool may rotate at speeds from 2,000 to 14,000 rpm. When using an air-operated impact wrench, the first step is to connect a shop air hose to the wrench. A **female-type quick disconnect coupling** on the end of the air hose is pushed onto the matching **male-type quick disconnect coupling** fitting on the impact wrench **(Figure 1)**. To release the **quick disconnect coupling**, set the impact tool on the workbench, grasp the air hose, and pull the outer, spring-loaded housing on the female part of the coupling toward the air hose.

When using an electric impact wrench, the electrical cord on the wrench must be plugged into a 110 volt electrical outlet. Always be sure the impact wrench electrical cord has a proper ground connection. Some electric impact wrenches are powered by an internal, rechargeable

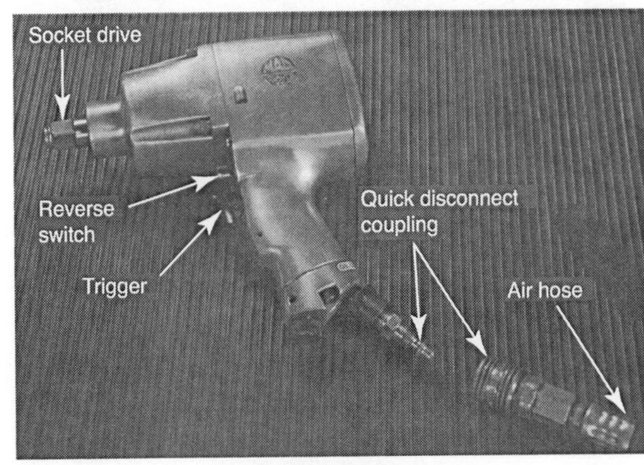


**Figure 1.** An air-operated impact wrench.

**Figure 2.** An electric impact wrench with an internal, rechargeable battery.

battery **(Figure 2)**. Battery-powered impact wrenches are convenient when working in areas where shop air pressure or electrical outlets are not available.

If an electrical cord on an impact wrench does not have a proper ground connection, electrical shock and personal injury may occur when using the wrench.

> **You Should Know** *Do not use an electric impact wrench with a frayed electrical cord or when standing on a wet floor. These actions may cause electrical shock and personal injury.*

The most common size of an impact wrench drive shaft is 1/2 inch (12.7 millimeter). Heavy-duty impact wrench sockets must be used on an impact wrench **(Figure 3)**.

> **You Should Know** *If an ordinary socket is used on an impact wrench, the socket may shatter, resulting in personal injury. Using the wrong size of socket on an impact wrench will damage the corners on the fastener being loosened or tightened.*

**Figure 3.** An impact wrench socket.

Be sure the impact wrench switch is set for the desired rotation. When the proper size of socket is installed on the impact wrench drive, the socket is then placed over the fastener to be loosened. The trigger on the impact wrench is pulled to activate the wrench. When the trigger is pulled and the impact wrench drive shaft encounters the turning resistance of the fastener, a small spring-loaded hammer inside the impact wrench strikes an anvil attached to the impact wrench drive shaft. Each impact of the hammer against the anvil moves the socket and fastener in the desired direction to loosen and remove the fastener. When removing wheel and tire assemblies, always chalk-mark one of the wheel studs in relation to the wheel rim so the wheel may be re-installed in the original direction to maintain wheel and tire balance.

When installing and tightening a nut or bolt, always be sure the threads are in good condition. Damaged threads must be repaired by with a tap or die. If damaged threads are not repaired, they may be stripped completely when the fastener is tightened, and this condition will not allow the fastener to be tightened to the specified torque. Many nuts are bidirectional, but other nuts must be installed in the proper direction. For example, many wheel nuts have a chamfer on the inner end that fits into tapered openings in the wheel rim to center the wheel **(Figure 4)**. These self-centering nuts must be installed with the chamfer facing toward the wheel. Other non-self-centering wheel nuts have a flat integral swivel washer on the inner side of the nut. This swivel washer must be installed so it is facing toward the wheel **(Figure 5)** In these applications, the wheel is centered by the hub design, and an O-ring

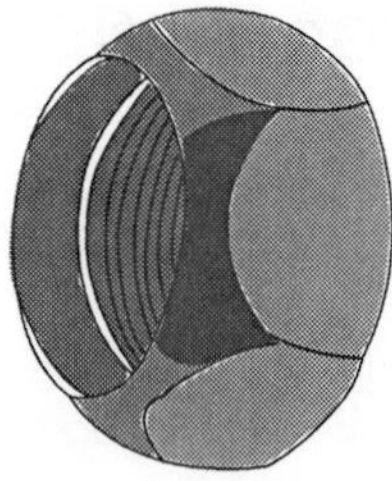

**Figure 4.** A wheel nut with a chamfered inner end.

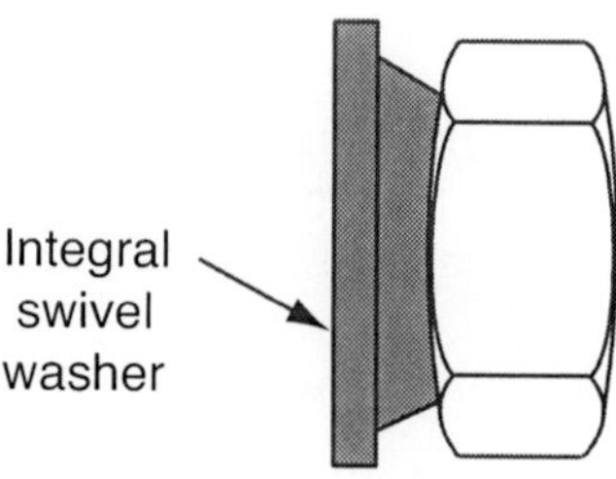


**Figure 5.** A non-self-centering wheel nut design.

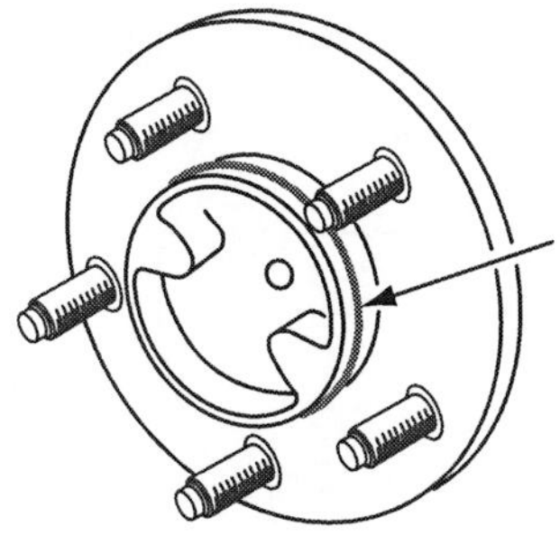

Figure 6. A hub O-ring that helps to provide wheel centering.

mounted on the hub also helps to center the wheel **(Figure 6)**. When installing a fastener, always determine if the fastener has a right-hand thread or a left-hand thread so you will know which direction to rotate the fastener. A fastener with a **right-hand thread** must be rotated clockwise when tightening the fastener. A fastener with a **left-hand thread** must be rotated counterclockwise to tighten the fastener.

Always start a fastener onto the threads by hand, and be sure the fastener rotates easily. If the fastener is not properly started onto the threads, the threads on the fastener will be cross threaded and ruined when the fastener is tightened. A **cross threaded** condition may also be defined as those threads that have been destroyed by starting a fastener onto its threads when the fastener is tipped slightly to one side and the threads on the fastener are not properly aligned with its matching threads. A cross threaded condition may also be referred to as stripped threads.

An impact wrench does not have torque-limiting capabilities. Therefore, fasteners can be overtightened with an impact wrench, and this action may damage the components mounted by the fasteners. For example, wheel studs are mounted in the hub flange and extend through the brake rotor or drum and the wheel openings **(Figure 7)**. If

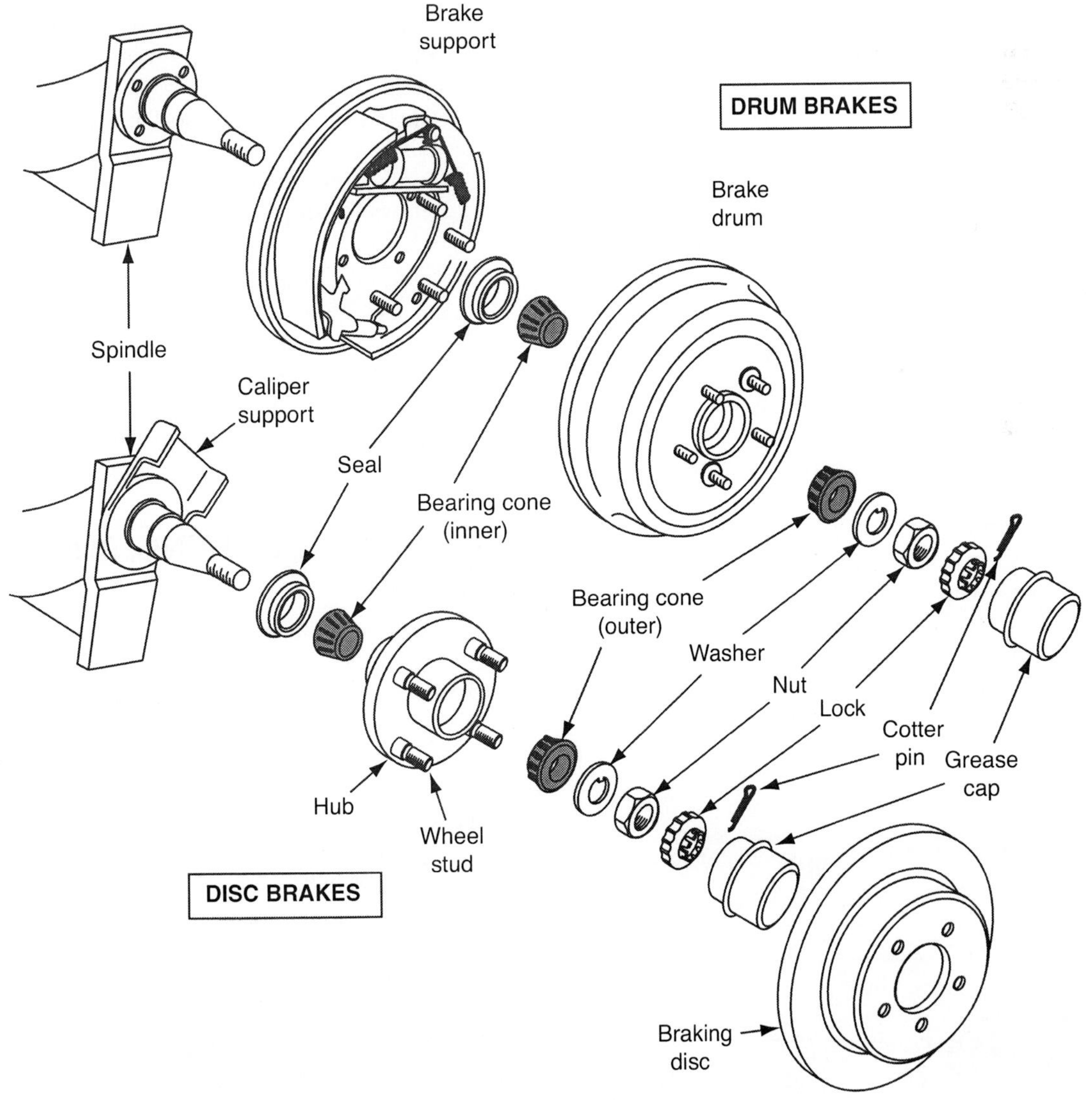


Figure 7. Wheel studs for disc and drum brakes.

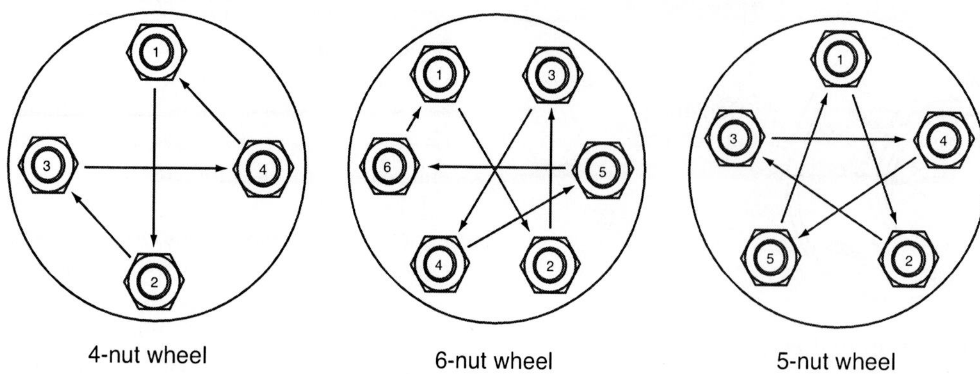


**Figure 8.** A wheel nut tightening sequence.

the wheel nuts are overtightened with an impact wrench, the brake rotor may be distorted, resulting in brake problems. To avoid this problem, tighten the fasteners snugly with the impact wrench, and then tighten the fasteners to the specified torque with a torque wrench. Some air-operated impact wrenches have an adjustable pressure that should be set to the low setting to avoid overtightening wheel nuts.

Some fasteners, such as wheel nuts, must be tightened in the proper sequence to avoid distorting the wheel rim **(Figure 8)**.

When air is compressed, moisture condenses in the air and tends to collect in the shop air lines. Develop a habit of oiling your air impact wrench daily. This action prevents internal rust, and extends the life of the impact wrench.

## USING A TORQUE WRENCH

A torque wrench allows the technician to tighten a fastener to the torque specified by the vehicle manufacturer. The specified torque provides the necessary clamping force between the components retained by the fasteners. Excessive torque may distort and warp components resulting in fluid leaks. Insufficient torque provides reduced clamping force. This condition may result in movement of the components retained by the fasteners, and this movement may cause damaged gaskets and leaks. Torque wrenches may be beam type, click type, dial type, or digital type **(Figure 9)**. Follow these steps when using a torque wrench:

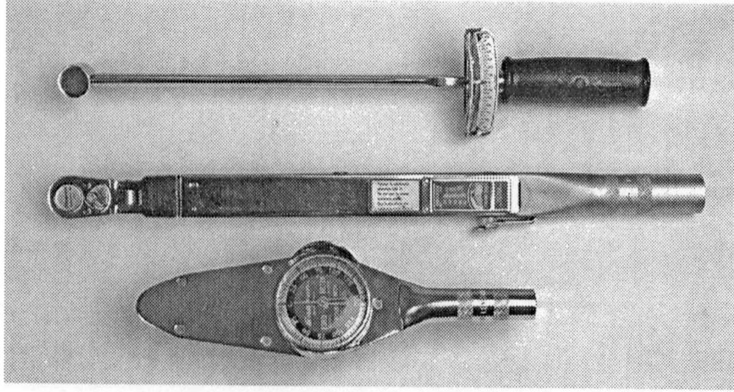

**Figure 9.** Three types of torque wrenches.

1. Obtain the torque specifications from the vehicle manufacturer's service manual.
2. Familiarize yourself with any special torque procedures in the service manual, such as proper fastener tightening sequence.
3. Divide the torque specifications by three.
4. When using a click-type torque wrench, set the specified torque on the wrench handle.
5. Install the proper size of socket on the torque wrench drive.
6. Maintain the torque wrench at a 90 degree angle to the fastener being tightened.
7. Tighten the fastener to $^1/_3$ of the specified torque.
8. Tighten the fasteners in proper sequence to $^1/_3$ of the specified torque if applicable.
9. Tighten the fastener(s) to $^2/_3$ of the specified torque.
10. Tighten the fastener(s) to within 10 ft.-lb (13 N•m) of the specified torque.
11. Tighten the fastener(s) to the specified torque.
12. Recheck the torque on the fastener(s).

## USING GEAR, BEARING, AND SEAL PULLERS

Many gears and bearings are mounted on their shafts with a precision press-fit. This press-fit prevents any motion between the gear or bearing and the shaft on which it is mounted.

The absence of motion between these components prevents wear on the mating surfaces between the gear or bearing and the shaft. Always select the proper puller for the job. The puller jaws must make proper contact on the gear, bearing, or seal to be pulled **(Figure 10)**. Some pullers have adjustable jaws to provide a proper jaw fit on various sizes of bearings, gears, or pulleys. Be sure the screw in the center of the puller does not damage the contacting surface on the shaft. For example, when pulling a crankshaft pulley, be sure the screw does not damage the threads in the center of the crankshaft. When pulling a gear or bearing, rotate the puller screw with steady, even pressure. Do not hammer or pry on the puller or the component being pulled.

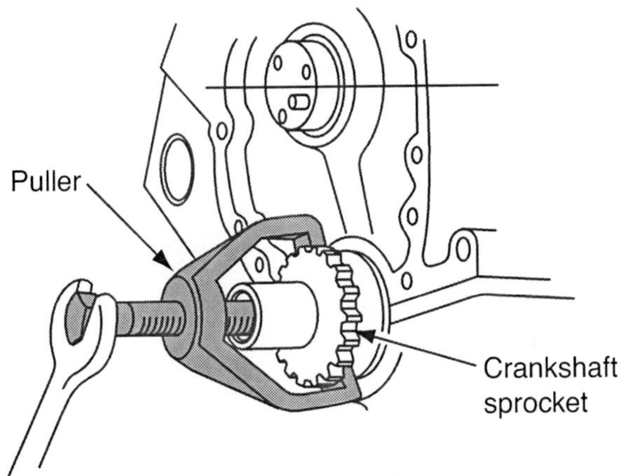


**Figure 10.** Using a puller to remove the crankshaft gear.

> **You Should Know** *When a puller is attached to a component, do not hammer on the puller or the component. This action may cause the puller to fly off the component resulting in personal injury.*

Some pullers have bolts that extend through the puller and thread into the gear being pulled **(Figure 11)**. Always be sure these bolts are threaded far enough into the gear so they do not pull out during the pulling operation. Some pullers have a slide hammer mounted on the puller shaft. The jaws of the puller are attached to the component to be pulled. In **Figure 12,** the jaws on a slide-hammer puller are attached to pull a rear axle seal on a rear-wheel-drive vehicle. The technician pulls the slide hammer quickly against the outer end of the puller, and the resulting force pulls the component to which the jaws are attached.

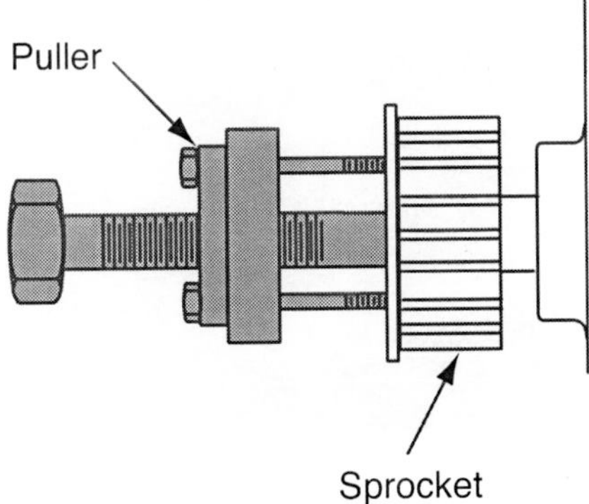


**Figure 11.** Using a puller with bolts extending from the puller into the gear being pulled.

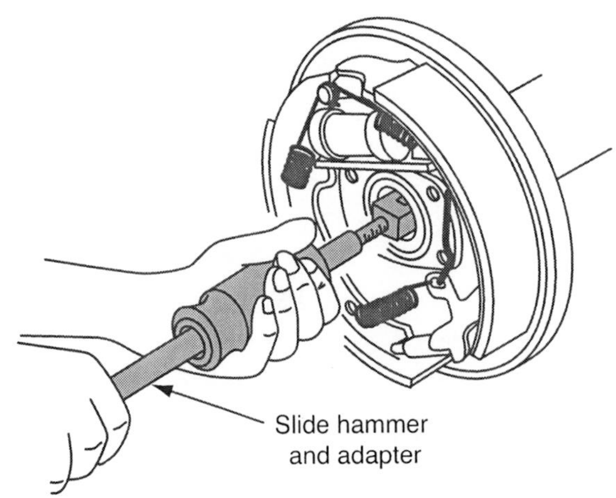


**Figure 12.** Using a slide hammer-type puller.

## REMOVING BROKEN STUDS AND SCREWS

Removing a broken stud or screw is usually a challenging and time consuming job for a technician. A great deal of time can be saved when the necessary precautions are taken to avoid breaking studs and screws. To avoid breaking fasteners, the first step is to spray all the fasteners with penetrating oil before attempting to remove the fasteners. This oil will work its way into the threads and dissolve rust and corrosion. The oil also lubricates the threads, which makes the fastener easier to remove. The second step to avoid breaking studs and screws is to always use the proper wrench. A box-end wrench or a socket grip the fastener more snugly compared to an open-end wrench. Thirdly, be sure you are attempting to turn the fastener in the proper direction. Remember that a few automotive fasteners have left-hand threads. In some cases, the fastener may still break even though you followed the proper steps to avoid breakage.

When a fastener will not loosen with normal force, do not apply excessive force and break the fastener. If a fastener does not loosen with normal force, try tightening the fastener slightly, and then attempt to loosen the fastener. Sometimes this tightening action helps to loosen the fastener. When a fastener refuses to loosen, try heating the fastener with an acetylene torch. If a nut will not loosen on a stud or bolt, heat the nut to expand it. When a cap screw will not loosen in a threaded block opening, heat the block casting in the area of the threaded opening. This heat may expand the block and allow the fastener to rotate.

> **You Should Know** *Do not spray penetrating oil on a fastener that has been heated and is still hot. Some of these oils are flammable, and this action could result in a sudden flame that ignites other combustible materials or causes personal injury.*

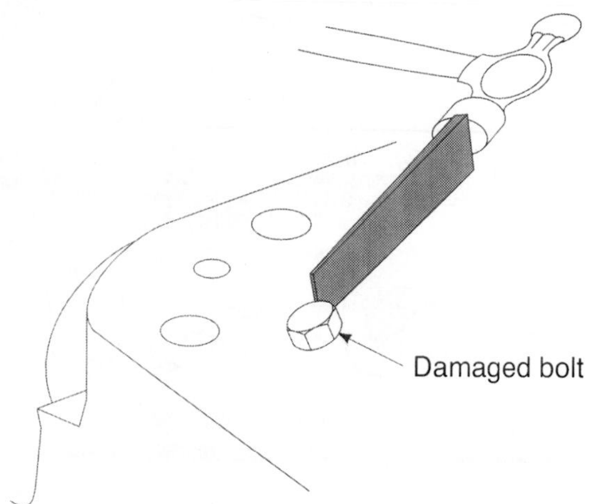


**Figure 13.** Using a hammer and chisel to loosen a fastener.

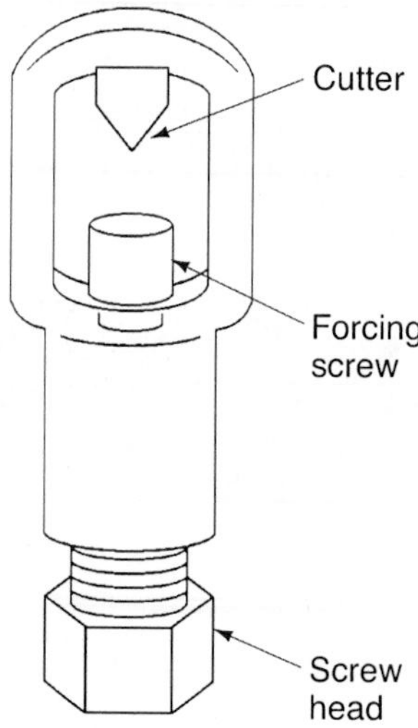


**Figure 15.** A nut splitter.

Use these methods to remove damaged or broken fasteners:

1. If the head on a fastener is damaged so a wrench slips on the head, a hammer and chisel may be used to tap on the side of the head and loosen the fastener **(Figure 13)**.
2. When a fastener does break off, if the fastener is sticking up above the casting it may be possible to install a stud remover over the fastener **(Figure 14)**. The stud remover is rotated with a $^1/_2$-inch breaker bar.
3. Sometimes a nut may be arc welded on top of a broken fastener. Allow the nut to cool, and then rotate the nut to remove the fastener.
4. When a nut is damaged, a nut splitter may be used to split the nut and remove it from the fastener **(Figure 15)**.
5. Broken fasteners may be removed with a left-hand drill bit and a reversible drill **(Figure 16)**.

**Figure 14.** A stud remover.

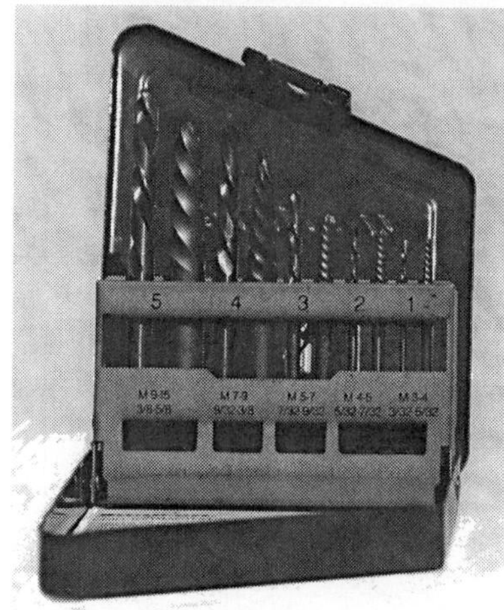


**Figure 16.** An extractor set with left-hand twist drill bits.

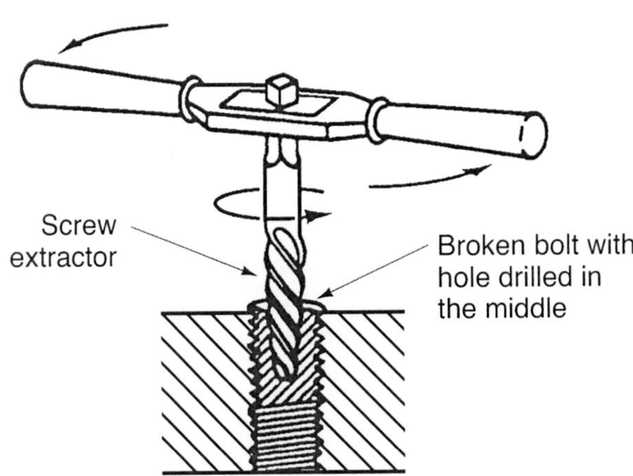


Figure 17. Using a screw extractor to remove a broken fastener.

6. Broken fasteners may be removed with a screw extractor. The first step in this process is to select the proper size drill bit for the size of extractor being used. Center punch the exact center of the broken fastener, and drill the fastener. The drilled hole must be deep enough so the extractor does not bottom in the hole. Insert the extractor into the hole and rotate it counterclockwise with a turning handle to remove the fastener **(Figure 17)**.

**You Should Know** *If a broken fastener will not loosen when attempting to rotate it with a screw extractor and turning handle, do not apply excessive force and break the extractor off in the fastener. Extractors are made from very hard, brittle steel, and the broken extractor will be very difficult to remove.*

7. If a broken fastener cannot be removed with a screw extractor, center punch it on the exact center of the broken fastener, and then drill all the way down through the fastener. Use progressively larger drill bits until all that remains of the fastener is a thin skin. Use a small metal pick to remove this thin skin from the threaded opening. Clean up the threaded opening with a bottoming tap.

## REMOVING DAMAGED NUTS

Nuts and fasteners located under the vehicle become rusted and corroded after a period of time. Always spray penetrating oil on nuts before attempting to loosen them. Directions on the can of penetrating oil will indicate the length of time you should wait before attempting to loosen the nut after the oil application. If the nut still will not loosen after it is sprayed with penetrating oil, heat the nut with an acetylene torch. Let the nut cool and then attempt to loosen it. If the corners of the nut are rounded off so a socket or box-end wrench slips on the nut, use a large pair of vise grips. Adjust the vise grips so they grip the nut tightly. Position your hand so that if the vise grips slip, your hand will not be injured. If the nut cannot be loosened using these methods, a nut splitter may be used to split the nut. The splitter is placed over the nut, and the screw on the splitter is rotated with a socket and ratchet until the nut is split. Do not rotate the screw on the splitter until the splitter damages the fastener threads. When installing nuts or cap screws in locations subjected to excessive heat and/or road splash, coat the threads with anti-seize compound before installation to prevent rust and corrosion on the fastener threads. The next technician that has to remove these fasteners will find they are easy to loosen.

## USING AN ACETYLENE TORCH FOR HEATING

Technicians are often required to use an acetylene torch for heating, cutting, or welding. The first basic step in using this equipment is to learn how to light and adjust the torch for heating. Acetylene welders have two pressurized tanks. One tank contains oxygen and the other tank contains acetylene. In this explanation, we are assuming that the gauges are properly mounted on the tanks and all the hoses and the torch body are installed correctly on the end of the hoses. Most acetylene welders have a green hose and a red hose. The green hose is connected to the oxygen gauge and the red hose is attached to the acetylene gauge. The torch body has an oxygen control valve that adjusts the flow of oxygen from the green hose into the torch and an acetylene control valve that adjusts the flow of acetylene from the red hose into the torch **(Figure 18)**. These valves are usually marked oxygen (OX) and acetylene (AC).

**You Should Know** *Acetylene gas is explosive! To avoid personal injury and possible explosions and fire, always use an acetylene welder under the supervision of your instructor until you are familiar with the use of this equipment.*

Follow these steps to light the acetylene torch.

1. Be sure the valves on the oxygen and acetylene cylinders are fully closed by turning them clockwise until they are bottomed.
2. Observe the pressure gauges on the acetylene and oxygen cylinders. The high-pressure gauge in each set

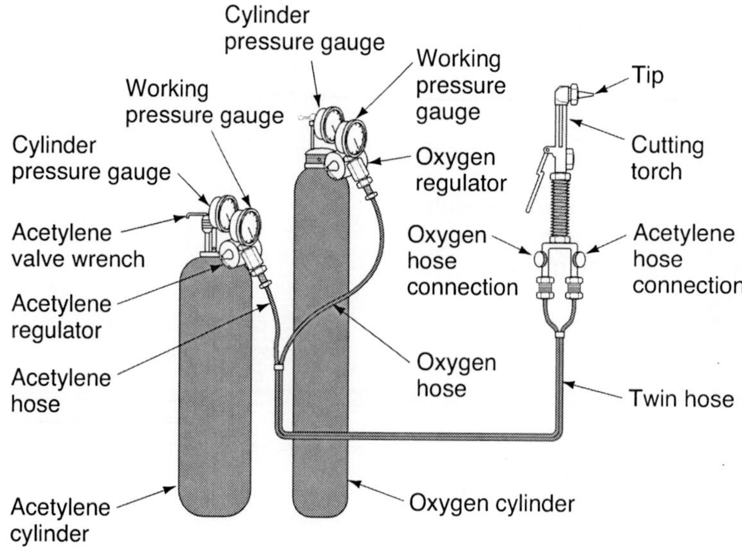


**Figure 18.** Acetylene welding components.

indicates the gas pressure in the cylinder and the low-pressure gauge indicates the pressure supplied to the torch. Be sure all the gauges indicate zero pressure.

3. If the gauges do not read zero, be sure the cylinder valves are closed and open the oxygen valve and the acetylene valve on the torch handle. Rotate the control knobs counterclockwise on both gauge sets until all the gauges read zero. Then close the oxygen and acetylene valves on the torch handle.
4. Install the proper end on the torch handle for the heating job you have to perform. Typically there are several heating tips with an acetylene torch. The smallest tip has a very small orifice designed for heating small components. The largest tip has a large orifice that is intended to produce a wider flame for heating larger components. The tips are threaded onto the end of the torch handle. Install the required tip snugly on the torch handle.
5. Close both valves on the torch handle and open the valve on the oxygen cylinder by rotating it counterclockwise. Turn the control knob on the oxygen gauge set clockwise until the low-pressure gauge indicates 10 psi supplied to the torch.
6. Open the valve on the acetylene cylinder by rotating it counterclockwise, and turn the control knob on the acetylene gauge set clockwise until the low-pressure gauge indicates 5 psi applied to the torch.
7. Put on a pair of welding gloves and a pair of welding goggles.

**You Should Know** *The flame from an acetylene torch produces dangerous ultraviolet light rays and is extremely hot. Always wear approved welding goggles and gloves when using this equipment. Never allow the flame near any part of your body, clothes, or any ignitable material. Always use a striker to light the torch. Do not use matches or a cigarette lighter to light an acetylene torch. This action places your hand near the flame, which could result in severe burns to your hand.*

8. Aim the torch handle and tip away from the acetylene welder components, any part of your body, or any combustible materials.
9. Open the torch acetylene valve slightly, and operate the striker to create a spark and light the torch. A **striker** has a spring-type striker bar with a flint attached under the tip of the bar. When the striker is squeezed, the flint moves across a rough surface to create a spark.
10. Turn the torch acetylene valve on slowly until the black, sooty flame turns to a yellow flame.
11. Turn the torch oxygen valve slowly until the flame changes to a blue color. Continue to slowly open the torch oxygen valve until you can see a small, distinct

inner cone right at the torch orifice in the center of the large flame. This type of flame is desirable for heating and it may be referred to as a **neutral flame.**

12. To shut the torch off, close the torch acetylene valve first to extinguish the flame, then turn the torch oxygen valve off.
13. Turn valves on the oxygen and acetylene cylinders clockwise until they are fully closed.
14. Open the acetylene and oxygen valves on the torch and rotate the control knobs counterclockwise on both the acetylene and oxygen gauge sets until these knobs feel loose. Both gauges should read 0 psi.
15. Close the acetylene and oxygen valves on the torch.

## CHECKING AND CHANGING RESPIRATOR FILTERS

A respirator is worn in the shop to protect the technician from hazardous airborne particles such as asbestos dust, paint spray, and solvent vapors. The respirator is held on the technician's face by straps attached to the respirator that fit behind the technician's head. As the technician breathes, air is taken into the respirator through two circular filters on each side of the respirator **(Figure 19)**. The filters keep hazardous particles from entering the technician's lungs. Air is expelled through an outlet in the lower center of the respirator.

> **You Should Know** *Be sure the respirator contains the proper filter for the hazardous material to which you are subjected. The respirator may not provide adequate protection if it has the wrong filter.*

**Figure 19.** A respirator.

The filter retainers on each side of the respirator are usually rotated counterclockwise to remove the retainers and filters. If there is any buildup of particles, dirt, or paint on the filters, they should be replaced with the proper filters for the hazardous material to which you are exposed. Be sure the new filters fit properly with no air gaps between the filters and their seats on the respirator. The straps must hold the respirator snugly on your head. Be sure the respirator provides an adequate seal on your face.

## JUMP-STARTING A VEHICLE WITH A DISCHARGED BATTERY

A technician is sometimes required to **jump start** a vehicle with a discharged battery. This seems like a basic operation, but if it is not done properly, very expensive computer damage may occur on the boost vehicle or the vehicle with the discharged battery.

**Jump-starting** a vehicle with a discharged battery may be called battery boosting.

> **You Should Know** *Connecting booster cables with wrong polarity may damage electronic equipment on the boost vehicle or the vehicle being boosted. Loose booster cable connections may create a spark causing the battery in either vehicle to explode, resulting in serious personal injury and paint damage on the vehicle.*

Always wear eye protection when connecting a booster battery to a discharged battery. Be sure the discharged battery is not frozen. Boosting a frozen battery may cause a battery explosion. Do not lean over the battery when connecting the booster cables. Do not allow any bumper or body contact between the boost vehicle and the vehicle being boosted. Follow these steps to connect the booster cables from the battery in the boost vehicle to the discharged battery:

1. Set the parking brake on both vehicles. Place the automatic transmission in park or a manual transmission in neutral.
2. Turn off all the electrical accessories on both vehicles.
3. Connect one end of the positive booster cable to the positive terminal of the discharged battery.
4. Connect the other end of the positive booster cable to the positive terminal of the battery in the boost vehicle.
5. Connect one end of the negative booster cable to the negative battery terminal in the boost vehicle.

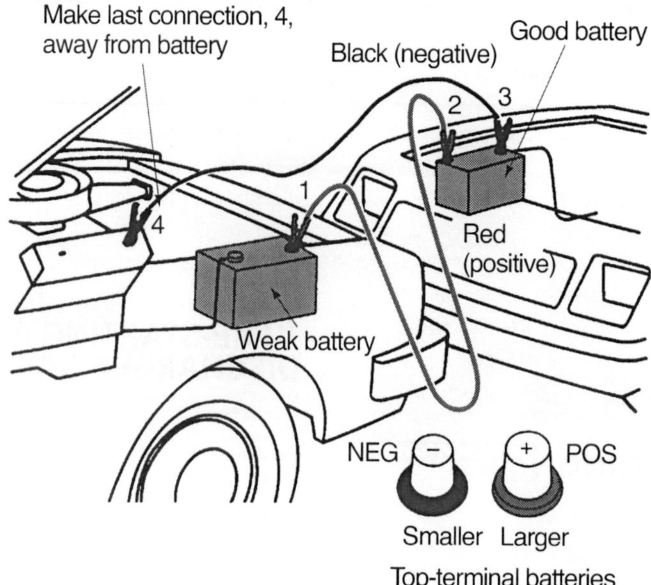


**Figure 20.** Proper jump starting connections.

6. Connect the other end of the negative booster cable to a good ground on the engine of the vehicle being boosted **(Figure 20)**.
7. Start the vehicle with the discharged battery.
8. Disconnect the negative booster cable from the engine in the vehicle with the discharged battery, and then disconnect the other end of this booster cable from the negative battery terminal in the boost vehicle.
9. Disconnect the end of the positive booster cable from the positive battery terminal in the boost vehicle, and then disconnect the other end of this cable from the positive battery terminal in the vehicle being boosted.

## Summary

- An impact wrench is a reversible power tool for removing and tightening bolts and nuts.
- Impact wrenches may be powered by electricity or shop air pressure.
- A torque wrench is used to tighten fasteners to the specified torque.
- Torque wrenches may be beam type, click type, or dial type.
- A puller is used to remove press-fit gears or bearings from their mounting location.
- Broken fasteners may be removed with a left-hand drill bit and a reversible drill or a screw extractor.
- Damaged nuts may be removed by splitting them with a nut splitter.
- When using an acetylene torch for heating, the oxygen pressure regulator should be set to 10 psi and the acetylene pressure regulator should be adjusted to 5 psi.
- Always wear welding gloves and goggles when using an acetylene torch.
- A respirator prevents hazardous materials in the shop air from entering the technician's lungs.
- When jump-starting a vehicle with a discharged battery, the booster cables must be connected with the proper polarity to the boost vehicle and the vehicle being boosted.
- When jump-starting a vehicle with a discharged battery, the negative booster cable must be connected to a good ground on the engine in the vehicle being boosted.

## Review Questions

1. Technician A says an impact wrench has torque-limiting capabilities. Technician B says the drive shaft on an impact wrench may turn at speeds between 2,000 and 14,000 rpm. Who is correct?
   A. Technician A
   B. Technician B
   C. Both Technician A and Technician B
   D. Neither Technician A nor Technician B
2. Technician A says fasteners may be overtightened and damaged with an impact wrench. Technician B says a fastener should be started onto its threads with an impact wrench. Who is correct?
   A. Technician A
   B. Technician B
   C. Both Technician A and Technician B
   D. Neither Technician A nor Technician B
3. Technician A says broken fasteners may be removed with a left-hand drill bit and a reversible drill. Technician B says broken fasteners may be removed with a screw extractor. Who is correct?
   A. Technician A
   B. Technician B
   C. Both Technician A and Technician B
   D. Neither Technician A nor Technician B
4. When jump-starting a vehicle with a discharged battery, Technician A says the negative booster cable should be connected to an engine ground in the vehicle being boosted. Technician B says when jump-starting a vehicle with a discharged battery, the bumpers should be touching on the boost vehicle and the vehicle being boosted. Who is correct?
   A. Technician A
   B. Technician B
   C. Both Technician A and Technician B
   D. Neither Technician A nor Technician B
5. All of these statements about electric and air-operated impact wrenches are true *except*:
   A. A female-type quick disconnect coupling is threaded into an air-operated impact wrench.
   B. Some electric-operated impact wrenches are powered by an internal, rechargeable battery.
   C. A technician may receive an electric shock from a 110-volt electric impact wrench with a defective ground connection.
   D. Electric and air-operated impact wrenches are reversible.
6. Technician A says a nut may become cross threaded if it is not started by hand on a fastener. Technician B says a brake rotor may be bent by overtightening the wheel nuts with an impact wrench. Who is correct?
   A. Technician A
   B. Technician B
   C. Both Technician A and Technician B
   D. Neither Technician A nor Technician B
7. When removing tight or seized fasteners, all of these precautionary steps should be taken to avoid breaking the fastener *except*:
   A. Spray the fasteners with penetrating oil.
   B. Be sure you are turning the fastener in the proper direction.
   C. Be sure you are using the proper wrench on the fastener.
   D. Apply excessive force to the wrench to loosen the fastener.
8. Technician A says broken fasteners may be removed by welding a nut on top of the fastener. Technician B says a damaged nut may be removed by splitting the nut with a nut splitter. Who is correct?
   A. Technician A
   B. Technician B
   C. Both Technician A and Technician B
   D. Neither Technician A nor Technician B
9. All of these statements about the use of acetylene welding equipment are true *except*:
   A. The red hose is connected to the acetylene regulator.
   B. Acetylene gas is poisonous.
   C. Typically the pressure on the low-pressure oxygen gauge should be set to 10 psi.
   D. Typically the pressure on the low-pressure acetylene gauge should be set to 5 psi.
10. The most common size of impact wrench drive shaft used in an automotive shop is:
   A $^1/_4$ inch.
   B. $^1/_2$ inch.
   C. $^3/_4$ inch.
   D. $^7/_8$ inch.
11. A torque wrench may be ____________ type, ____________ type, ____________ type, or ____________ type.
12. A puller is used to remove components that have a ____________ ____________.
13. When using an acetylene torch, the green hose is connected to the ____________ regulator.
14. After using an acetylene torch, the torch valves should be opened and the control knobs on the oxygen and acetylene gauges rotated ____________ until the gauges read ____________ psi.
15. Explain the precautions that must be observed when installing and tightening a fastener with an impact wrench.
16. Explain the results of insufficient and excessive fastener torque.
17. Describe the first step that should be taken when removing a fastener with seized, rusted threads.
18. When using an electric impact wrench, explain two precautions that must be observed to avoid electrical shock.

# Chapter 7 Measuring Systems, Measurements, and Fasteners

## Introduction

Measuring systems, precision measurements, and fasteners are explained in this chapter. Automotive technicians must have an understanding of measuring systems. The fasteners used on modern cars have metric-sized heads. However, many of the vehicle manufacturer's specifications are provided in USCS measurements. Therefore, technicians must be familiar with both the USCS and metric systems of measurement. Technicians also need to be familiar with the procedure for using measuring tools to perform precision measurements. Technicians must be able to identify bolt and nut strength markings so they always use an equivalent replacement fastener. An understanding of different types of fasteners is also a necessity for automotive technicians. For example, technicians must know that a number of automotive bolts require replacement each time they are removed. Many cylinder head bolts are an example of this type of fastener.

## MEASURING SYSTEMS

Two systems of weights and measures are commonly used in the United States. One system of weights and measures is the USCS. Common measurements are the inch, foot, yard, and mile. In this system, the quart and gallon are common measurements for volume, and ounce, pound, and ton are measurements for weight. A second system of weights and measures is referred to as the metric system. The basic USCS linear measurement is the yard, whereas the corresponding linear measurement in the metric system is the meter **(Figure 1)**. Each unit of measurement in the metric system is related to the other metric units by a factor of ten. Thus, every metric unit can be multiplied or divided by ten to obtain larger units (multiples) or smaller units (submultiples). For example, the meter can be divided by ten to obtain centimeters ($^1/_{100}$ meter) or millimeters ($^1/_{1000}$ meter).

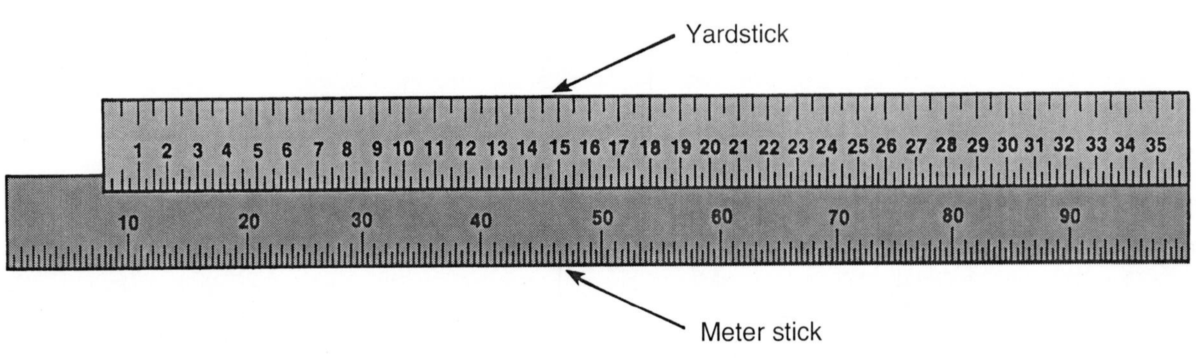


**Figure 1.** A meter is slightly longer than a yard.

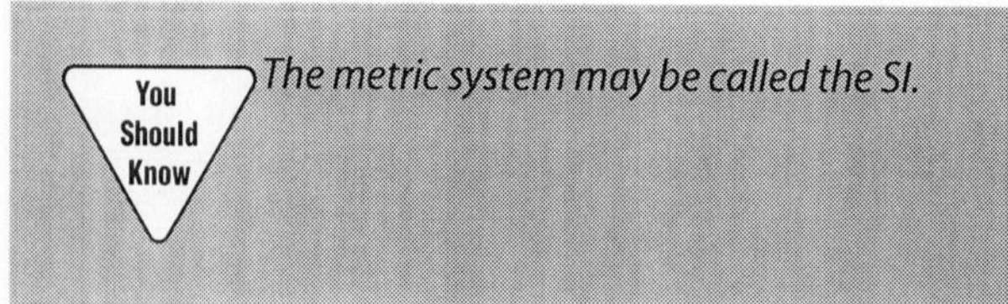


The U.S. government passed the Metric Conversion Act in 1975 in an attempt to move American industry and the general public to accept and adopt the metric system. The automotive industry has adopted the metric system, and in recent years, most bolts, nuts, and fittings on vehicles have been changed to metric. During the early 1980s, some vehicles had a mix of USC and metric bolts. Import vehicles have used the metric system for many years. Although the automotive industry has changed to the metric system, the general public in the United States has been slow to convert from the USCS to the metric system. One of the factors involved in this change is cost. What would it cost to change every highway distance and speed sign in the United States to read kilometers? The answer to that question is probably hundreds of millions or billions of dollars.

Service technicians must be able to work with both the USCS and metric systems. Some common equivalents between the metric and USCS are listed in **Figure 2**.

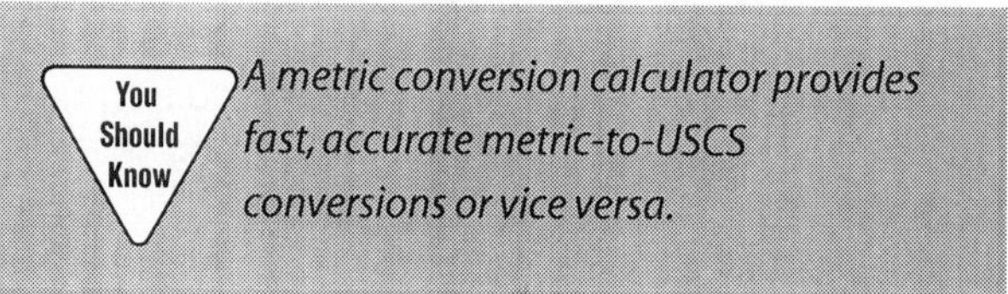


In the USCS, phrases such as ⅛ inch are used for measurements. The metric system uses a set of prefixes for this purpose. For example, the prefix kilo indicates 1,000; this prefix indicates there are 1,000 meters in a kilometer. Common prefixes in the metric system are listed in **Figure 3**.

1 meter (m) = 39.37 inches
1 centimeter (cm) = 0.3937 inch
1 millimeter (mm) = 0.03937 inch
1 inch = 2.54 cm
1 inch = 25.4 mm
1 ft. lb. = 1.35 newton meters (Nm)
1 in. lb. = 0.112 Nm
1 in. hg. = 3.38 kilopascals (kPa)
1 psi = 6.89 kPa
1 mile = 1.6 kilometer (m)
1 hp = 0.746 kilowatt (kW)
degrees F – 32 divided by 1.8 = degrees Celsius (C);
example, 212ºF – 32 = 180 divided by 1.8 = 100ºC

**Figure 2.** Metric and USCS equivalents.

| NAME | SYMBOL | MEANING |
|---|---|---|
| mega | M | one million |
| kilo | k | one thousand |
| hecto | h | one hundred |
| deca | da | ten |
| deci | d | one tenth of |
| centi | c | one hundredth of |
| milli | m | one thousandth of |
| micro | μ | one millionth of |

**Figure 3.** Prefixes in the metric system.

## Measurement of Mass

In the metric system, mass is measured in grams, kilograms, or tonnes. One thousand grams equal 1 kilogram. In the USCS, mass is measured in ounces, pounds, or tons. When converting pounds to kilograms, 1 pound equals 0.453 kilogram.

## Measurement of Length

In the metric system, length is measured in millimeters, centimeters, meters, or kilometers. For example, ten millimeters (mm) equal 1 centimeter (cm). In the USCS, length is measured in inches, feet, yards, or miles. When distance conversions are made between the two systems, some of the conversion factors are shown in **Figure 4**.

1 inch = 25.4 millimeters
1 foot = 30.48 centimeters
1 yard = 0.91 meter
1 mile = 1.60 kilometers

**Figure 4.** Measurements of length or distance.

## Measurement of Volume

In the metric system, volume is measured in milliliters, cubic centimeters, and liters. One cubic centimeter equals 1 milliliter. If a cube has a length, depth, and height of 10 centimeters (cm), the volume of the cube is 10 centimeter $\times$ 10 centimeter $\times$ 10 centimeter = 1,000 cm$^3$ = 1 liter. When volume conversions are made between the two systems, 1 cubic inch equals 16.38 cubic centimeters. If an engine has a displacement of 350 cubic inches, 350 $\times$ 16.38 = 5733 cubic centimeters, and 5733/1,000 = 5.7 liters.

# PRECISION MEASUREMENTS

Precision measurements are required in many areas of automotive service, such as engine repair, brake service, and steering diagnosis. If precision measurements are inaccurate, component operation is adversely affected and premature component failure may occur. Therefore, technicians must be familiar with these measurements.

## Measuring Space with a Feeler Gauge and Machinist's Rule

Feeler gauges are one of the most common automotive measuring tools. Many feeler gauge sets are thin strips of metal of varying precision thickness **(Figure 5)**. Wire-type feeler gauge sets contain round wires rather than metal strips **(Figure 6)**. A USCS set of round wire feeler gauges is marked in thousands of an inch, and this marking indicates the thickness of the gauge.

Wire feeler gauges are used to measure the spark plug gap between the center and ground electrode **(Figure 7)**. The spark from the ignition system jumps across this gap to ignite the air-fuel mixture in the cylinder. The vehicle manufacturer provides spark plug gap specifications. When installing new spark plugs or servicing used spark plugs, the gap must be measured and adjusted if necessary. When learning to use feeler gauges, it is helpful to practice measuring and adjusting the gaps on some used spark plugs. Select a typical feeler gauge size for measuring spark plug gaps, such as 0.045 inch. Attempt to insert the feeler gauge between the spark plug electrodes. Bend the ground electrode as required until the selected feeler gauge slides between the spark plug electrodes with a light push fit. Be careful not to damage the center electrode or spark plug insulator when bending the ground electrode.

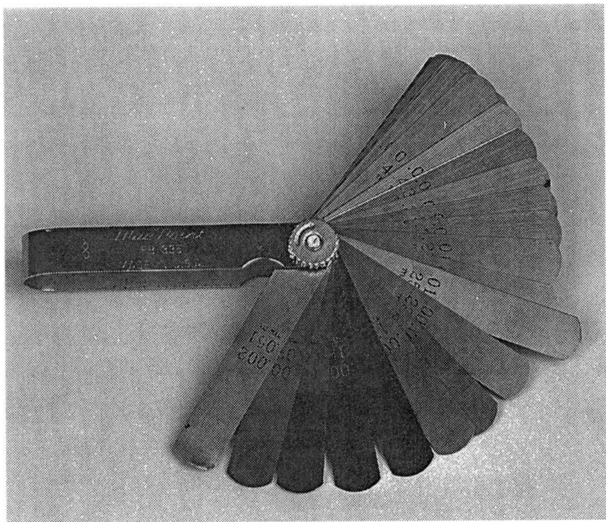

**Figure 5.** A strip-type feeler gauge set.

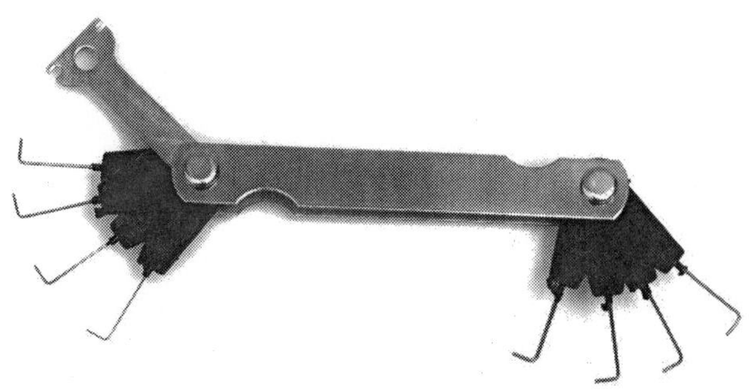

**Figure 6.** A round wire feeler gauge set.

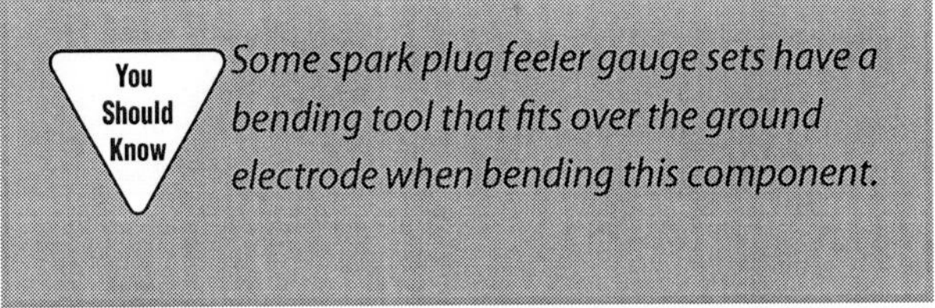


*Some spark plug feeler gauge sets have a bending tool that fits over the ground electrode when bending this component.*

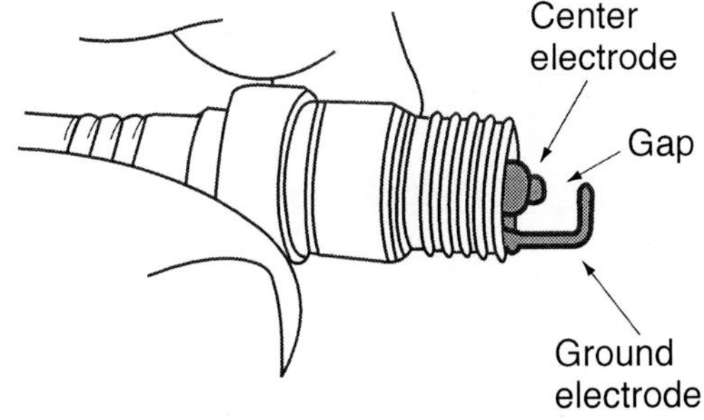


**Figure 7.** Measuring a spark plug gap with a round feeler gauge.

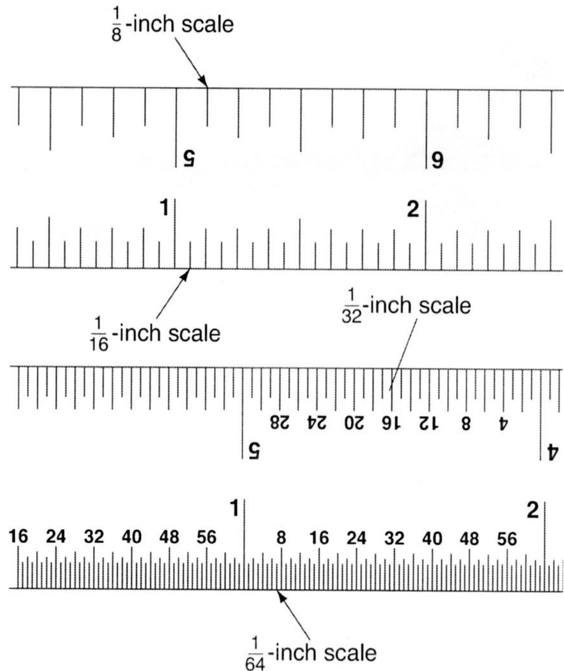


Figure 8. A machinist's rule.

A machinist's rule is commonly used for automotive measurements that do not require a close tolerance. A machinist's rule has four scales, two on each side of the rule. USCS scales include increments $^{1}/_{8}$, $^{1}/_{16}$, $^{1}/_{32}$, and $^{1}/_{64}$ inch **(Figure 8)**.

## Using an Outside Micrometer

When learning to use an outside micrometer, you can practice measurements on a used engine valve. When using an outside micrometer, insert two fingers through the micrometer frame and rotate the spindle with the thumb and forefinger on the same hand **(Figure 9)**. An

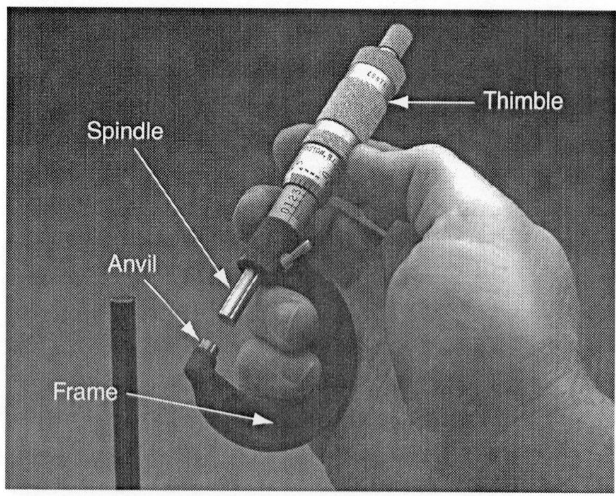


Figure 9. Proper hand position on a micrometer.

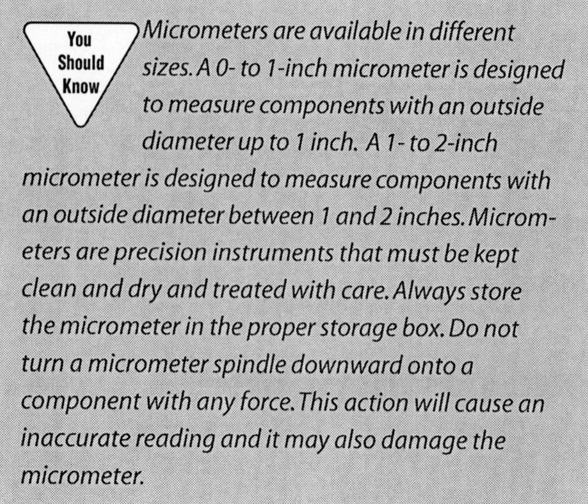


*Micrometers are available in different sizes. A 0- to 1-inch micrometer is designed to measure components with an outside diameter up to 1 inch. A 1- to 2-inch micrometer is designed to measure components with an outside diameter between 1 and 2 inches. Micrometers are precision instruments that must be kept clean and dry and treated with care. Always store the micrometer in the proper storage box. Do not turn a micrometer spindle downward onto a component with any force. This action will cause an inaccurate reading and it may also damage the micrometer.*

**outside micrometer** is designed to measure the outside diameter of various components.

Always use a clean shop towel to wipe any dust or dirt from the measuring surfaces on the micrometer spindle and anvil. Any dirt on these surfaces causes inaccurate readings.

Place the spindle and anvil over the valve stem and slowly rotate the thimble until the spindle and anvil contact the valve stem **(Figure 10)**. The spindle and anvil must be positioned at a right angle to the valve stem, and these components must be centered on the diameter of this stem. Move the micrometer with a slight rocking action as you turn the thimble the last few thousandths of an inch. Rotate the thimble with a light rotating force until the spin-

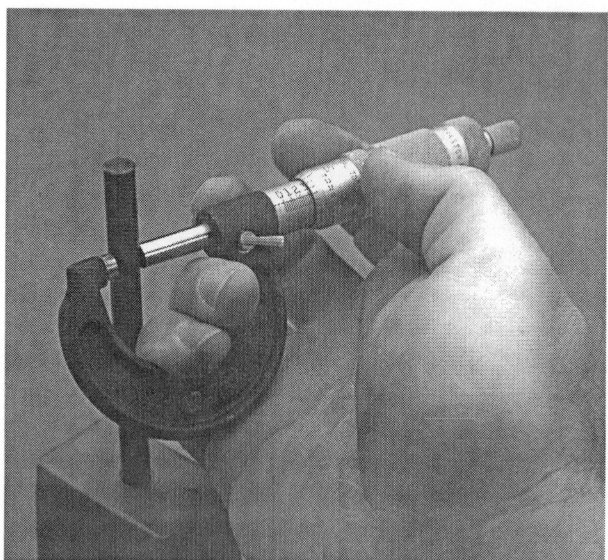

Figure 10. Positioning a micrometer to measure valve stem diameter.

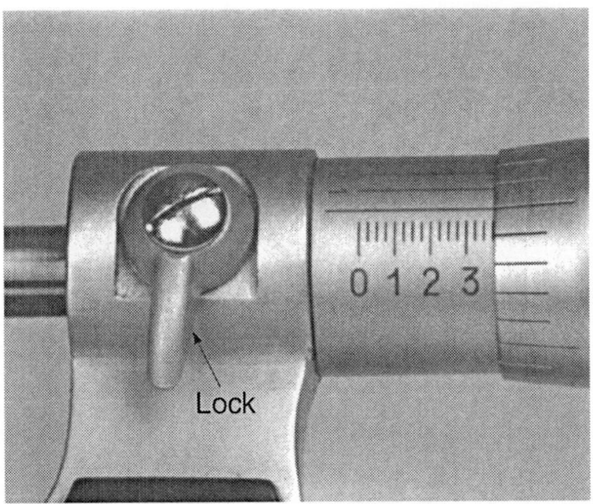


Figure 11. A micrometer with a lock lever.

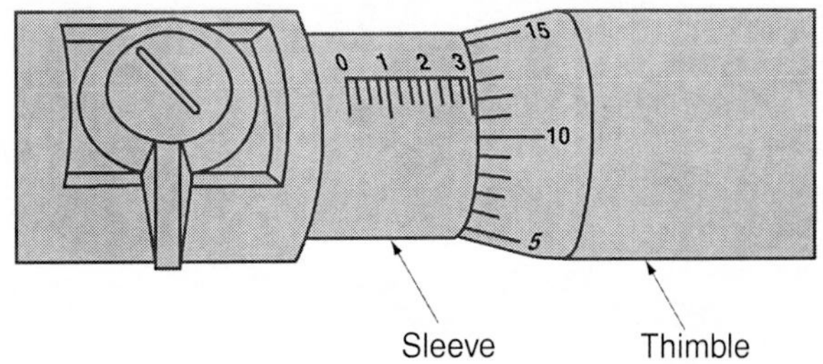


Figure 13. A micrometer reading.

dle and anvil contact the valve stem lightly and squarely. You will know by feel when the spindle and anvil are centered on the valve stem diameter. Some micrometers have a lock that prevents the thimble from moving and changing the reading after the micrometer is positioned properly. Flip the lock lever over and lock the micrometer reading **(Figure 11)**. With the reading locked on the micrometer, the spindle and anvil should slide over the valve stem with a slight drag.

Until you become used to the feel of a micrometer when it is properly positioned for component measurement, the thimble ratchet may be used to rotate the thimble **(Figure 12)**. When the anvil and spindle are properly positioned on a component with the correct force, the ratchet clicks and the spindle stops turning. This action prevents overtightening of the spindle on the component being measured.

## Interpreting a Micrometer Reading

After the reading from a component is locked on the micrometer, the next step is to interpret the reading. Each graduation on the sleeve represents 0.025 inch, and each number on the sleeve represents 0.100 inch. In **Figure 13**, 3 is exposed on the sleeve indicating 0.300 inch. Next, add the reading on the thimble to the reading on the sleeve. Each sleeve graduation indicates 0.001 inch. The thimble reading in Figure 8 is 0.013 inch. Therefore, the actual micrometer reading is 0.300 + 0.013 = 0.313 inch.

## Using a Dial Indicator

Dial indicators are usually sold in sets containing the dial indicator, a magnetic base, and other attaching arms **(Figure 14)**. Many automotive measurements, such as brake rotor runout and ball joint wear, require the use of a dial indicator. The magnetic base is used to hold the dial indicator onto a metal component. The dial indicator may also be attached with special vise grips or C-clamps.

When mounting a dial indicator, the support arm length should be kept to a minimum. Long support arms allow a

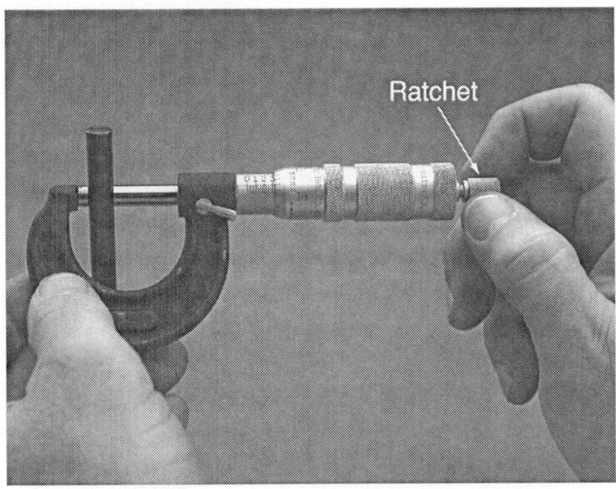


Figure 12. A ratchet on a micrometer thimble.

Figure 14. A dial indicator with a magnetic base.

Figure 15. A dial indicator installed to measure brake rotor runout.

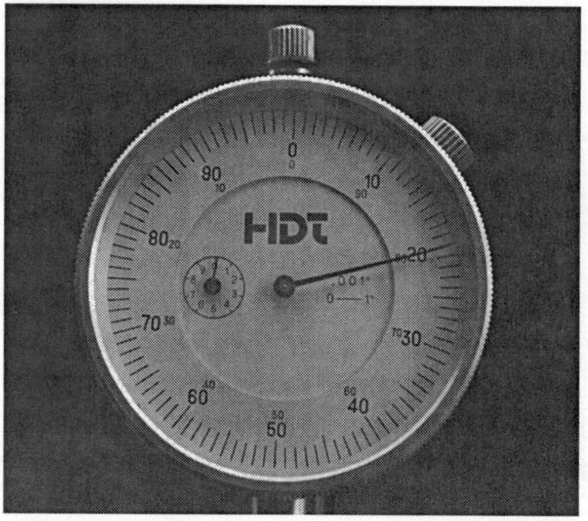


Figure 17. A dial indicator reading.

slight dial indicator movement which causes inaccurate readings. The dial indicator should be mounted so it is at a 90° angle to the component being measured. Be sure the dial indicator mounting is secure and does not allow any indicator movement. In **Figure 15**, the dial indictor is set up to measure brake rotor runout. Always be sure the brake rotor and the dial indicator stem and plunger are clean. Dirt on these surfaces causes inaccurate readings. In this figure, a dial indicator with a magnetic base is attached to the lower end of the front strut. Be sure the dial indicator is at a 90° angle to the brake rotor. Position the dial indicator plunger against the rotor surface so approximately one-half of the stem is sticking out of the indicator. After the plunger is properly positioned, rotate the lock to prevent dial indicator movement.

Gently grasp the movable dial indicator face and rotate this face until the zero position on the face is aligned with the dial indicator pointer **(Figure 16)**. Each dial indicator division indicates 0.001 inch. The pointer may swing to the right or left of zero when taking a dial indicator reading. Slowly rotate the brake rotor and observe the dial indicator. The pointer may move twenty divisions on right side of zero during part of a brake rotor rotation **(Figure 17)**. When the brake rotor is rotated for the remainder of a revolution, the dial indicator pointer may move to ten divisions to the left of zero. In this example, the total brake rotor runout is 0.020 + 0.010 = 0.030 inch. A small, separate pointer in the dial indicator face indicates the number of times the large gauge pointer makes a complete revolution.

## FASTENERS

There are many different types of fasteners used throughout the vehicle. The most common are threaded fasteners, including bolts, studs, screws, and nuts. When servicing automotive components, these fasteners must be inspected for thread damage, fillet damage, and stretch before they can be reused **(Figure 18)**. In addition, many

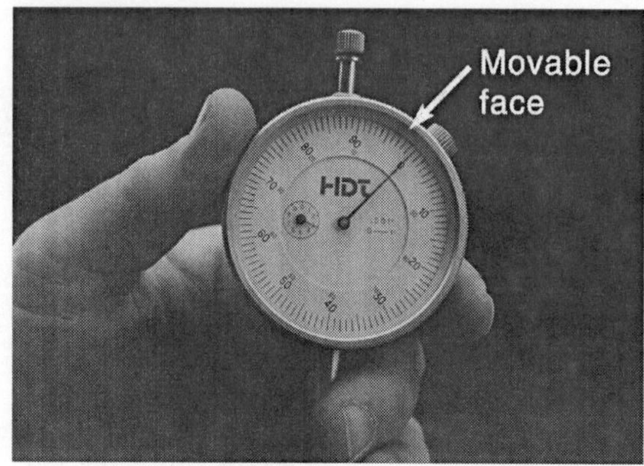


Figure 16. Zeroing the dial indicator.

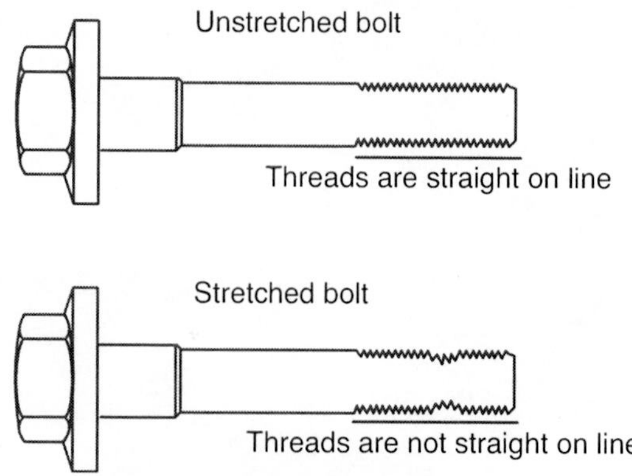


Figure 18. All bolts should be checked for stretch and other damage before reusing them.

threaded fasteners are not designed for reuse and the service manual should be referenced for the manufacturer's recommendations. If a threaded fastener requires replacement, select a fastener the same diameter, thread pitch, strength, and length as the original. All bolts in the same connection must be of the same grade. Use nut grades that match their respective bolts, and use the correct washers and pins as originally equipped. Torque the fasteners to the specified value, and use torque-to-yield bolts where specified.

Threaded fasteners used in automotive applications are classified by the Unified National Series using four basic categories: Unified National Coarse (UNC or NC), Unified National Fine (UNF or NF), Unified National Extrafine (UNEF or NEF), and Unified National Pipe Thread (UNPT or NPT).

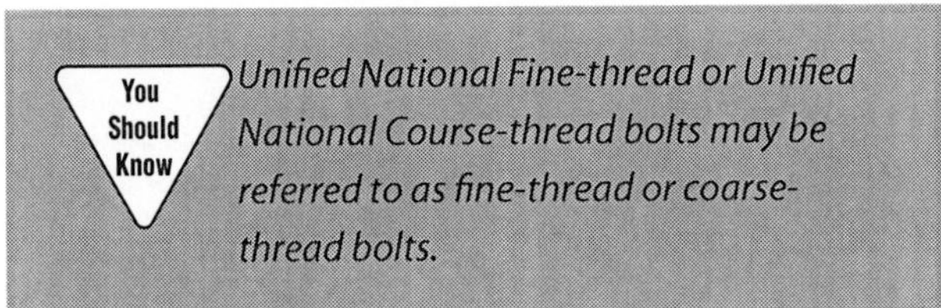


*Unified National Fine-thread or Unified National Course-thread bolts may be referred to as fine-thread or coarse-thread bolts.*

In recent years, the automotive industry has switched to the use of metric fasteners. Metric threads are classified as course or fine, as denoted by an SI or ISO lettering.

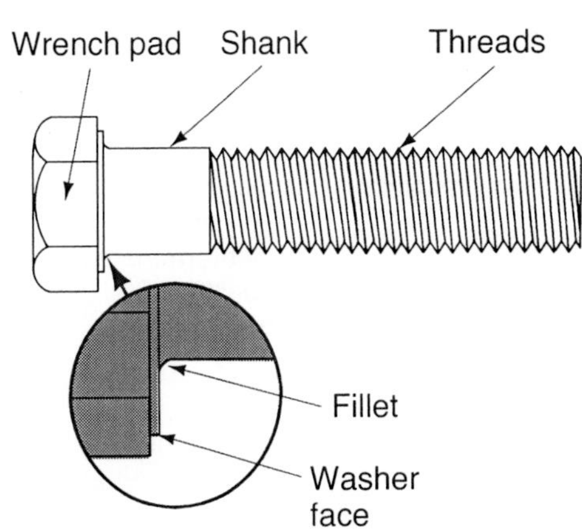


**Figure 19.** A typical bolt.

The most common type of threaded fastener used on the vehicle is the bolt **(Figure 19)**. To understand proper selection of a fastener, terminology must be defined **(Figure 20)**. The head of the bolt is used to torque the fastener. Several head designs are used, including hex, Torx, slot, and spline. **Bolt diameter** is the measure across the threaded area or shank.

The **pitch** in the USCS is the number of threads per inch. In the metric system, thread pitch is a measure of the distance (in millimeters) between two adjacent threads.

**Bolt length** is the distance from the bottom of the head to the end of the bolt. The bolt **grade** denotes its

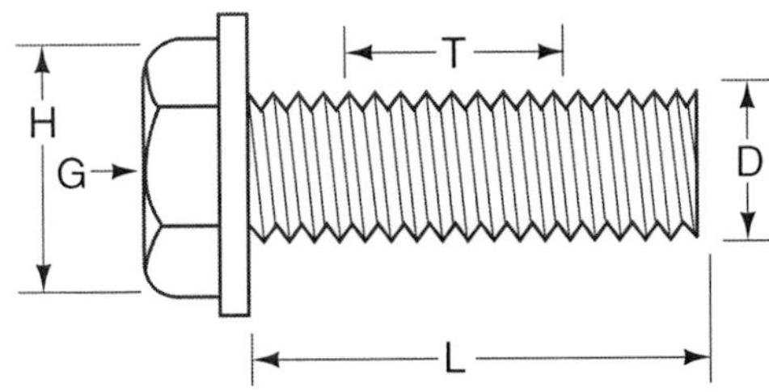


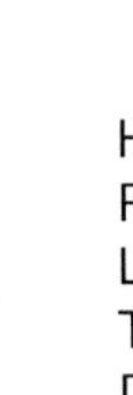

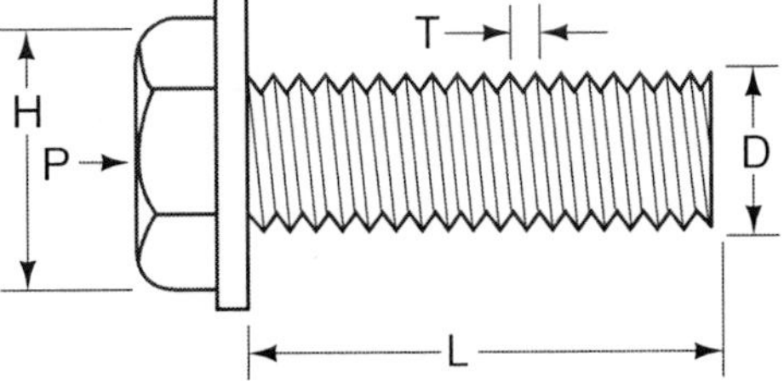


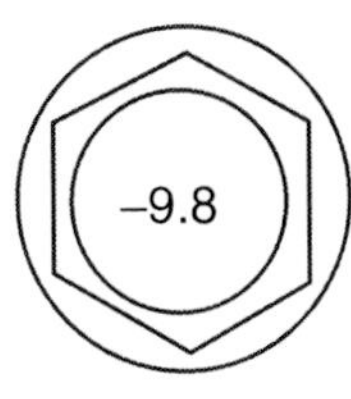


H = Head
G = Grade marking (bolt strength)
L = Length (inches)
T = Thread pitch (thread/inch)
D = Nominal diameter (inches)

A

H = Head
P = Property class (bolt strength)
L = Length (millimeters)
T = Thread pitch (thread/millimeter)
D = Nominal diameter (millimeter)

B

**Figure 20.** Bolt terminology.

| SAE grade markings | 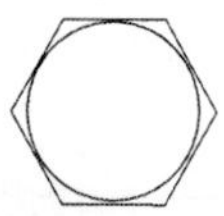 | 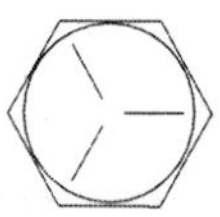 | 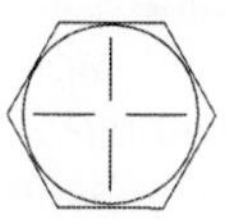 | 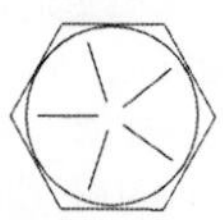 | 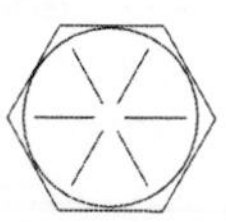 |
|---|---|---|---|---|---|
| Definition | No lines: unmarked indeterminate quality SAE grades 0-1-2 | 3 lines: common commercial quality Automotive and AN bolts SAE grade 5 | 4 lines: medium commercial quality Automotive and AN bolts SAE grade 6 | 5 lines: rarely used SAE grade 7 | 6 lines: best commercial quality NAS and aircraft screws SAE grade 8 |
| Material | Low carbon steel | Med. carbon steel tempered | Med. carbon steel quenched and tempered | Med. carbon alloy steel | Med. carbon alloy steel quenched and tempered |
| Tensile strength | 65,000 psi | 120,000 psi | 140,000 psi | 140,000 psi | 150,000 psi |

**Figure 21.** Bolt grade identification marks.

strength and is used to designate the amount of stress the bolt can withstand.

The grade of the bolt depends upon the material it is constructed from, bolt diameter, and thread depth. Grade marks are placed on the top of the head in the USCS to identify the bolt's strength **(Figure 21)**. In the metric system, the strength of the bolt is identified by a property class number on the head **(Figure 22)**. The larger the number, the greater the tensile strength. **Thread depth** is the height of the thread from its base to the top of its peak.

Like bolts, nuts are graded according to their tensile strength **(Figure 23)**. As discussed earlier, the nut grade must be matched to the bolt grade. The strength of the

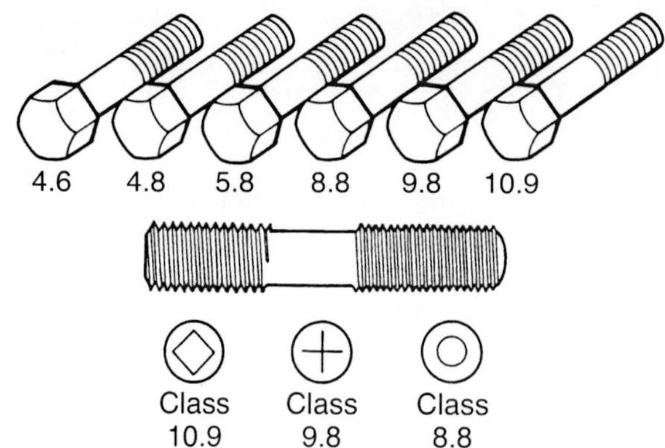


**Figure 22.** Metric bolt strength identification.

| Inch system | | Metric system | |
|---|---|---|---|
| Grade | Identification | Class | Identification |
| Hex nut grade 5 | 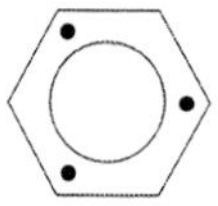 3 dots | Hex nut property grade 9 | 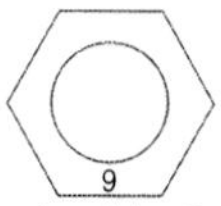 Arabic 9 |
| Hex nut grade 8 | 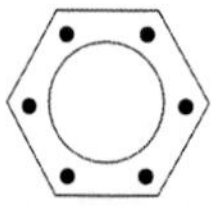 6 dots | Hex nut property grade 10 | 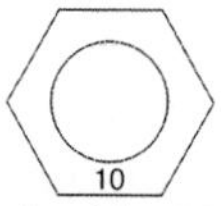 Arabic 10 |
| Increasing dots represent increasing strength. | | Can also have blue finish or paint dab on hex flat. Increasing numbers represent increasing strength. | |

**Figure 23.** Nut grade markings.

**STANDARD BOLT AND NUT TORQUE SPECIFICATIONS**

| Size Nut or Bolt | Torque (foot-pounds) | Size Nut or Bolt | Torque (foot-pounds) | Size Nut or Bolt | Torque (foot-pounds) |
|---|---|---|---|---|---|
| 1/4–20 | 7–9 | 7/16–20 | 57–61 | 3/4–10 | 240–250 |
| 1/4–28 | 8–10 | 1/2–13 | 71–75 | 3/4–16 | 290–300 |
| 5/16–18 | 13–17 | 1/2–20 | 83–93 | 7/8–9 | 410–420 |
| 5/16–24 | 15–19 | 9/16–12 | 90–100 | 7/8–14 | 475–485 |
| 3/8–16 | 30–35 | 9/16–18 | 107–117 | 1–8 | 580–590 |
| 3/8–24 | 35–39 | 5/8–11 | 137–147 | 1–14 | 685–695 |
| 7/16–14 | 46–50 | 5/8–18 | 168–178 | | |

**Figure 24.** Standard nut and bolt torque specifications.

connection is only as strong as the lowest grade used. For example, if a grade 8 bolt is used with a grade 5 nut, the connection is only a grade 5.

Proper torque of the fastener is important to prevent thread damage and to provide the correct clamping forces. The service manual provides the manufacturer's recommended torque value and tightening sequence for most fasteners used in the engine or other components. The amount of torque a fastener can withstand is based on its tensile strength **(Figure 24)**. In order to obtain proper torque, the fastener's threads must be cleaned and may require light lubrication.

## Torque-to-Yield Bolts

Automotive components such as engines and transaxles are designed with very close tolerances. These tolerances require an equal amount of clamping forces at mating surfaces. Normal head bolt torque values have a calculated 25-percent safety factor, that is, they are torqued to only 75 percent of the bolt's maximum proof load **(Figure 25)**. Using the chart, it can be seen that a small difference between torque values at the bolt head can result in a large difference in clamping forces.

Since torque is actually force used to turn a fastener against friction, the actual clamping forces can vary even at the same torque value. Up to about 25 ft.-lb (35 N•m) of torque, the clamping force is pretty constant. However, above this point, variation of actual clamping forces at the same torque value can be as high as 200 percent. This is due to variations in thread conditions or dirt and oil in some threads. Up to 90 percent of the torque is used up by friction, leaving 10 percent for the actual clamping. The result could be that some bolts are having to provide more clamping force than others, distorting the cylinder bores.

To compensate and correct for these factors, many manufacturers use torque-to-yield bolts. A **torque-to-yield bolt** is a bolt that has been tightened to a specified yield or stretch point.

The yield point of identical bolts does not vary much. A bolt that has been torqued to its yield point can be rotated

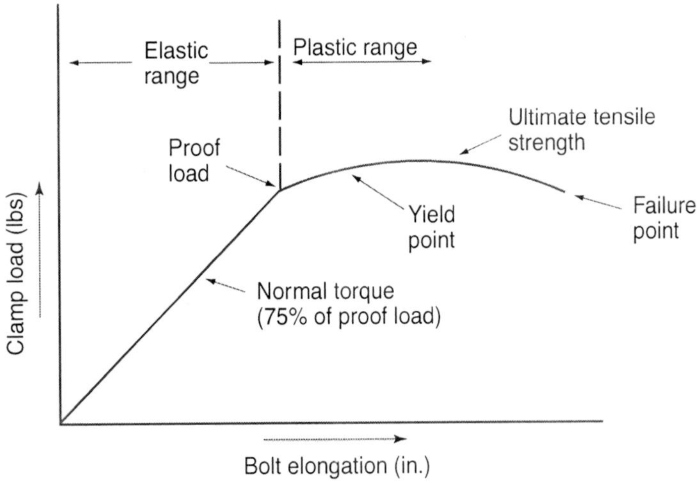


**Figure 25.** The relationship between proper clamp load and bolt failure.

an additional amount without any increase in clamping force. When a set of torque-to-yield fasteners is used, the torque is actually set to a point above the yield point of the bolt. This assures the set of fasteners will have an even clamping force.

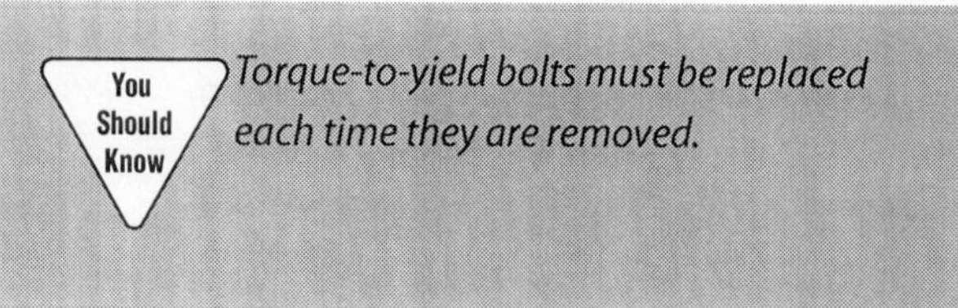

You Should Know
Torque-to-yield bolts must be replaced each time they are removed.


Manufacturers vary on specifications and procedures for securing torque-to-yield bolts. Always refer to the service manual for exact procedures. In most instances, a torque wrench is first used to tighten the bolts to their yield point. Next, the bolt is turned an additional number of degrees as specified in the service manual.

In the graph **(Figure 25)**, notice that a bolt can be elongated considerably at its yield point before it reaches its failure point. Also notice that the clamp load is consistent between the proof load and the failure point of the bolt. Bolts that are torqued to their yield points have been stretched beyond their elastic limit and require replacement whenever they are removed or loosened.

## Summary

- Every unit in the metric system can be divided or multiplied by ten to obtain smaller units or larger units.
- In the metric system, length is measured in millimeters, centimeters, meters, or kilometers.
- In the USC system, length is measured in inches, feet, yards, or miles.
- In the metric system, volume is measured in milliliters, cubic centimeters, and liters.
- In the USC system, volume is measured in cubic inches, quarts, and gallons.
- Spark plug gaps are measured with a round wire feeler gauge.
- A 1- to 2-inch outside micrometer is designed to measure components with an outside diameter between 1 and 2 inches.
- Each rotation of a USC micrometer thimble represents 0.025 inch, and each number on the sleeve indicates 0.100 inch.
- When reading a micrometer, the thimble reading is added to the sleeve reading to obtain the diameter of the component being measured.
- When mounting a dial indicator, the indicator stem should be at a 90 degree angle to the component being measured.
- The dial indicator must be securely mounted to eliminate movement of the indicator.
- Bolt strength in the USCS is indicated by grade marks on the bolt head.
- Torque-to-yield bolts provide a more uniform clamping force.
- Torque-to-yield bolts are tightened to the specified torque and then rotated a specific number of degrees.

## Review Questions

1. Technician A says in the metric system, 1,000 meters equals 1 kilometer. Technician B says 1 cubic inch is equal to 16.38 cubic centimeters. Who is correct?
   A. Technician A
   B. Technician B
   C. Both Technician A and Technician B
   D. Neither Technician A nor Technician B
2. Technician A says in the metric system, mass is measured in kilograms. Technician B says 1 kilogram equals 0.453 pounds. Who is correct?
   A. Technician A
   B. Technician B
   C. Both Technician A and Technician B
   D. Neither Technician A nor Technician B
3. Technician A says to bend the center electrode when adjusting spark plug gaps. Technician B says if a spark plug gap is adjusted properly, the specified feeler gauge should fit between the electrodes with a light drag. Who is correct?
   A. Technician A
   B. Technician B
   C. Both Technician A and Technician B
   D. Neither Technician A nor Technician B

4. Technician A says a 1- to 2-inch outside micrometer will measure the outside diameter of a shaft with a 0.500-inch diameter. Technician B says when reading an outside micrometer, the thimble reading must be subtracted from the sleeve reading.
   Who is correct?
   A. Technician A
   B. Technician B
   C. Both Technician A and Technician B
   D. Neither Technician A nor Technician B
5. All of these statements about the metric system of measurements are true *except*:
   A. The basic linear measurement is the meter.
   B. Each unit of measurement is related to the other units by a factor of 20.
   C. The prefix kilo indicates 1,000.
   D. The liter is a measurement for volume.
6. Technician A says a machinist's rule is used for automotive measurements that do not require close tolerance. Technician B says a feeler gauge may be used to measure the gap or clearance between two components. Who is correct?
   A. Technician A
   B. Technician B
   C. Both Technician A and Technician B
   D. Neither Technician A nor Technician B
7. All of these statements about using an outside micrometer are true *except*:
   A. The surfaces on the micrometer spindle and anvil must be clean.
   B. The spindle and anvil must be positioned at a right angle to the component being measured.
   C. The thimble should be turned with a light rotating force.
   D. A 1- to 2-inch micrometer will measure a component with a diameter of 2.25 inches.
8. When measuring the diameter of a shaft with a USCS micrometer, the number 4 is exposed on the sleeve but no other lines on the sleeve are visible after the 4, and the thimble reading is 0.020. The shaft diameter is:
   A. 0.380 inch.
   B. 0.400 inch.
   C. 0.420 inch.
   D. 0.440 inch.
9. While measuring brake rotor runout with a dial indicator, Technician A says the full length of the dial indicator stem should be sticking out of the indicator. Technician B says the dial indicator stem should be positioned at a 90 degree angle to the brake rotor. Who is correct?
   A. Technician A
   B. Technician B
   C. Both Technician A and Technician B
   D. Neither Technician A nor Technician B
10. All of these statements about fastener replacement are true *except*:
    A. Torque-to-yield bolts may be reused if they are removed.
    B. All bolts in the same connection must have the same grade.
    C. Nut grades must match their respective bolts.
    D. Fasteners must be tightened to the specified torque in the proper sequence.
11. When an outside micrometer is adjusted to measure the diameter of a round component, the spindle and anvil should slide over the component with a ______________ ______________.
12. The ratchet on a micrometer may be used to rotate the ______________.
13. If the sleeve reading on a micrometer is 0.150 inch and the thimble reading is 0.011 inch, the micrometer reading is ______________ inch.
14. A dial indicator may be mounted on a metal component with a ______________ base.
15. Describe the proper procedure for mounting a dial indicator to measure brake rotor runout.
16. After the dial indicator is properly mounted, describe the proper procedure for measuring the brake rotor runout.
17. Explain the advantage of torque-to-yield bolts, and give one example of where these bolts are used in a vehicle.
18. Describe the proper procedure for tightening torque-to-yield bolts.

## Service Information



## SECTION OBJECTIVES

After you have read, studied, and practiced the contents of this section, you should be able to:

- Explain the importance of service information as it relates to vehicle repair.
- Explain some of the information found in an owner's manual.
- Describe the general layout of a vehicle manufacturer's service manual.
- Explain the usual differences between a typical vehicle manufacturer's service manual and a generic service manual.
- Describe how electronic service information is stored and accessed.
- Explain the purpose of computer software.
- Explain the meaning of flash programming.
- Describe the purposes of service bulletins.
- Explain how a labor estimating guide is used when preparing a vehicle repair estimate.
- Describe the information on a typical emission control label.
- Interpret VINs.
- Locate the specified fluid capacities in a service manual.
- Find the vehicle specifications in a service manual.
- Describe two different types of maintenance schedules.
- List six different automotive service topics provided electronically on CDs or DVDs.

*General Motors published a service manual for 1946, 1947, and 1948 Chevrolet cars. This manual contained 177 pages. A General Motors Service manual for 2001 light-duty trucks contains five volumes with several hundred pages in each volume. An owner's manual for a 1954 Buick contained 40 pages, whereas an owner's manual for a 2002 GMC Sierra truck contains 505 pages. These figures confirm the tremendous expansion of automotive service information as vehicles have become increasingly more complex.*

# Service Information

## Introduction

Access to service information is absolutely essential when repairing modern vehicles. Without the necessary service information, technicians may waste a great deal of time diagnosing a problem that is explained in a service bulletin. In many cases it is impossible to diagnose electrical/electronic problems without a wiring diagram for the circuit being diagnosed. The technician must also have access to specifications for the vehicle being repaired. For example, without the proper torque specifications and procedures, expensive aluminum components in today's modern engines or transaxles may be warped and ruined.

### TYPES OF SERVICE INFORMATION

Vehicle manufacturers supply an owner's manual with each vehicle. This manual contains very basic service information such as using the jack supplied with the vehicle and changing a tire and wheel. The owner's manual also includes information related to checking fluid levels and tire pressure. Information regarding the type of lubricants required for the vehicle and necessary fluid capacities are also included in the owner's manual.

Service manuals are one of the most common types of service information. Service manuals are provided by vehicle manufacturers or aftermarket suppliers of automotive information and components. With the increasing vehicle complexity in recent years, service manuals have become more voluminous. Some vehicle manufacturer's service manuals for specific vehicles now contain five volumes with several hundred pages in each volume. Because of the large increase in the number of pages in service manuals, many vehicle manufacturers are now providing their service manuals on compact disks (CDs). Placing service manuals on CDs greatly reduces the storage space requirements for these manuals.

Vehicle manufacturers and some independent automotive service groups publish service bulletins. These bulletins contain service information related to the diagnosis and repair of a specific vehicle problem. Some service bulletins contain information regarding improved, modified parts that are required to eliminate a specific vehicle defect or complaint.

An increasing amount of service information is available electronically on CDs, computer disks, or computer software. Most vehicle manufacturers use computer software extensively to provide service information to their dealerships. The dealership computers can access and/or download service information from the vehicle manufacturer's computer system.

### OWNER'S MANUAL

The owner's manual provides valuable information for the vehicle owner. For example, this manual provides information regarding the proper use of vehicle restraints, such as seat belts, air bags, and child restraints. An index is included in the owner's manual so the desired information can be accessed quickly. Customer assistance and warranty information is also provided in the owner's manual. The amount of information in the owner's manual may vary depending on the vehicle make and model year. Vehicles have become more complex in recent years, and owner's manuals have also been expanded to include basic information regarding the proper operation of vehicle systems and components.

The owner's manual includes information regarding vehicle instrumentation and warning systems. The use of all the vehicle controls such as switches and levers are explained in the owner's manual. The owner's manual also provides many cautions so the driver/owner understands the results of improper system or vehicle operation. Maintenance schedules, fluid capacities, and the recommended lubricants and fluids are provided in the owner's manual. The owner's manual provides information on checking all the fluid levels and adding the proper fluids.

The owner's manual provides information regarding the towing capacity of the vehicle, and this manual also includes towing precautions. Basic service information is included in the owner's manual such as changing a tire, wiper blades, or light bulbs. The owner's manual also provides information on proper cleaning and maintenance of the vehicle interior and exterior. Valuable information regarding the location, identification, and replacement of electrical fuses is included in the owner's manual.

## MANUFACTURER'S SERVICE MANUAL

The service manual is one of the most important tools for today's technician. It provides information concerning component identification, service procedures, specifications, and diagnostic information. In addition, the service manual provides information concerning wiring harness connections and routing, wiring diagrams, component location, and fluid capacities.

The service manual provides an explanation of the VIN. The VIN information is essential when ordering parts. Most service manuals published by vehicle manufacturers now have a standard format **(Figure 1)**. The service manual usually provides illustrations to guide the technician through the service operation **(Figure 2)**. Always use the correct manual for the vehicle and system being serviced. Follow each step in the service procedure. Do not skip steps!

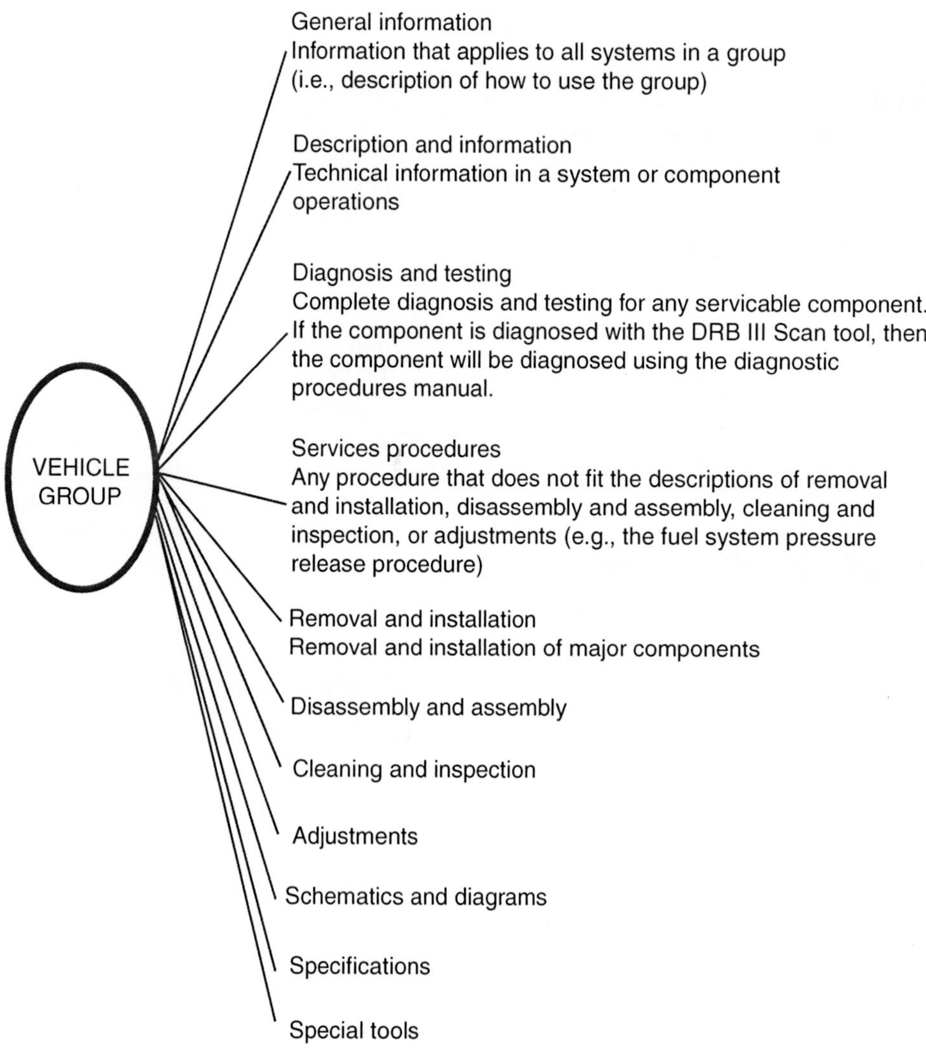
VEHICLE GROUP
General information
Information that applies to all systems in a group (i.e., description of how to use the group)
Description and information
Technical information in a system or component operations
Diagnosis and testing
Complete diagnosis and testing for any servicable component. If the component is diagnosed with the DRB III Scan tool, then the component will be diagnosed using the diagnostic procedures manual.
Services procedures
Any procedure that does not fit the descriptions of removal and installation, disassembly and assembly, cleaning and inspection, or adjustments (e.g., the fuel system pressure release procedure)
Removal and installation
Removal and installation of major components
Disassembly and assembly
Cleaning and inspection
Adjustments
Schematics and diagrams
Specifications
Special tools

**Figure 1.** A uniform service manual layout.

**Ball Joint Boot Replacement**

NOTE: The upper control arm ball joint, lower control arm ball joint, and knuckle upper ball joint are attached with the boot retainer to improve the sealing efficiency of the boot.

1. Remove the set ring and boot.

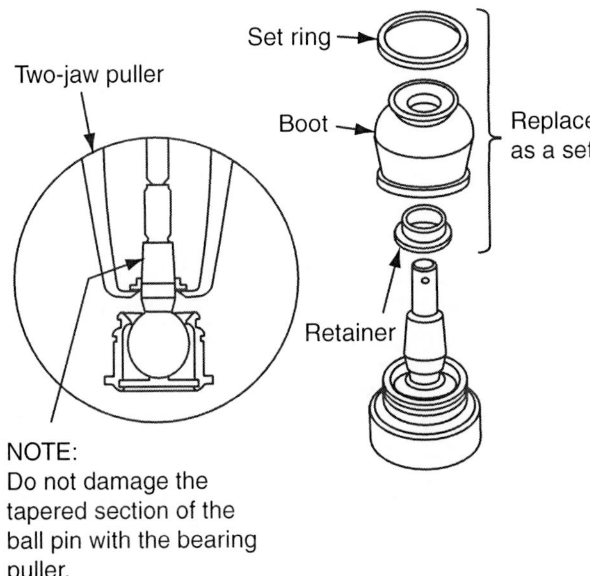


NOTE:
Do not damage the tapered section of the ball pin with the bearing puller.

2. Remove the retainer.

NOTE: The knuckle lower ball joint does not have a retainer.

3. Pack the interior of the boot and lip with grease.

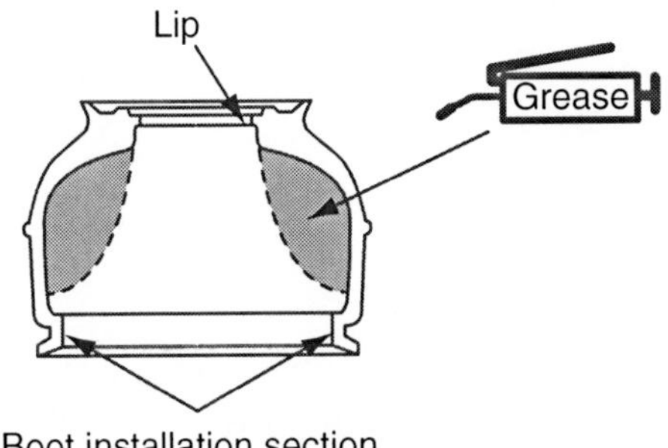


CAUTION: Do not contaminate the boot installation section with grease.

4. Wipe the grease off the sliding surface of the ball pin and pack with fresh grease.

5. Insert the new retainer lightly into the ball joint pin.

NOTE: When installing the ball joint, press the retainer into the ball joint pin.

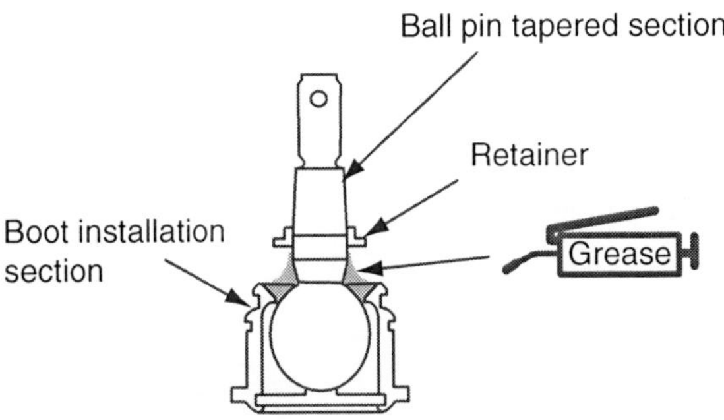


**CAUTION:**

**Keep grease off the boot installation section and the tapered section of the ball pin.**

**Do not allow dust, dirt, or other foreign materials to enter the boot.**

6. Install the boot in the groove of the boot installation section securely, then bleed air.

7. Adjust the special tool with the adjusting bolt until the end of the tool aligns with the groove on the boot.

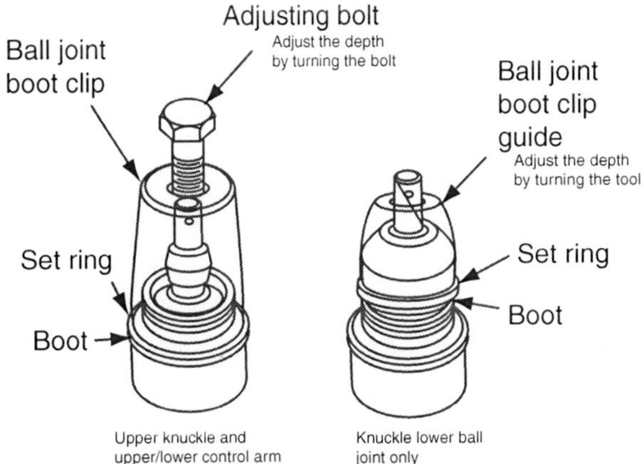


8. Slide the set ring over the tool and into position.

CAUTION: After installing the boot, check the ball pin tapered section and threads for grease contamination and wipe them if necessary.

**Figure 2.** Illustrations in a service manual guide the technician through service procedures.

| | | Tire size | Pressure | |
|---|---|---|---|---|
| | | | Front | Rear |
| Cold tire inflation pressure | For all roads, including full rated loads | P195/70R14 | 220 kPa (2.2 kgf/cm2, 32 psi) | 240 kPa (2.4 kgf/cm2, 34 psi) |
| | | P205/65R15 | 220 kPa (2.2 kgf/cm2, 32 psi) | 240 kPa (2.4 kgf/cm2, 34 psi) |
| | Optional inflation for reduced loads (1 to 4 passengers) | P195/70R14 | 180 kPa (1.8 kgf/cm2, 26 psi) | 180 kPa (1.8 kgf/cm2, 26 psi) |
| | | P205/65R15 | 180 kPa (1.8 kgf/cm2, 26 psi) | 180 kPa (1.8 kgf/cm2, 26 psi) |
| Vehicle height | Tire size | | Height | |
| | | | Front | Rear |
| | P195/70R14 | | 210 mm (8.27 in.) | 270 mm (10.63 in.) |
| | P205/65R15 | | 213 mm (8.39 in.) | 276 mm (10.87 in.) |
| Front wheel alignment | Toe-in (total) | | 0° +/- 0.2° (0 +/- 2mm, 0 +/- 0.08 in.) | |
| | Wheel angle | Tire size | Inside wheel | Outside wheel |
| | | P195/70R14 | 37°20' +/- 2° | 32°15' |
| | | P205/65R15 | 36°00' +/- 2° | 31°20' |
| | Camber | | -0°35' +/- 45' | |
| | Cross camber | | 45' or less | |
| | Caster | | 1°05' +/- 45' | |
| | Cross caster | | 45' or less | |
| | Steering axis inclination | | 13°00' +/- 45' | |
| Rear wheel alignment | Toe-in (total) | | 0.4° +/- 0.2° (4 +/- 2mm, 0.16 +/- 0.08in.) | |
| | Camber | | -0°15' +/- 45' | |
| | Cross camber | | 45' or less | |

**Figure 3.** A specification table.

Measurements such as torque, end play, and clearance specifications are located in or near the service manual text or procedural information. Specification tables are usually provided at the end of the procedural information or component area **(Figure 3)**.

Because the service manual is divided into a number of main component and system areas, a table of contents is provided at the front of the manual to provide quick access to the desired information. Each component area or system is covered in a section of the service manual **(Figure 4)**. At the beginning of each section in the service manual, a smaller table of contents guides the technician to the information regarding the specific system or component being serviced. The service manual may be divided into several volumes because of the extensive amount of information required to service today's vehicles.

| TABLE OF CONTENTS | SECTION NUMBER |
|---|---|
| GENERAL INFO. AND LUBE | |
| General Information | 0A |
| Maintenance and Lubrication | 0B |
| HEATING AND AIR COND. | |
| Heating and Vent. (nonA/C) | 1A |
| Air Conditioning System | 1B |
| V-5 A/C Compressor Overhaul | 1D3 |
| BUMPERS AND FRONT BODY PANELS | |
| Bumpers (See 10-4) | |
| Fr. End Body Panels (See 10-5) | |
| STEERING, SUSPENSION, TIRES, AND WHEELS | |
| Diagnosis | 3 |
| Wheel Alignment | 3A |
| Power Steering Gear & Pump | 3B1 |
| Front Suspension | 3C |
| Rear Suspension | 3D |
| Tires and Wheels | 3E |
| Steering Col. On-Vehicle Service | 3F |
| Steering Col. - Std. Unit Repair | 3F1 |
| Steering Col. - Tilt, Unit Repair | 3F2 |
| DRIVE AXLES | |
| Drive Axles | 4D |
| BRAKES | |
| General Info. - Diagnosis and On-Car Service | 5 |
| Compact Master Cylinder | 5A1 |
| Disc Brake Caliper | 5B2 |
| Drum Brake - Anchor Plate | 5C2 |
| Power Brake Booster Assembly | 5D2 |
| ENGINES | |
| General Information | 6 |
| 2.0 Liter I-4 Engine | 6A1 |
| 3.1 Liter V6 Engine | 6A3 |
| Cooling System | 6B |
| Fuel System | 6C |
| Engine Electrical - General | 6D |
| Battery | 6D1 |
| Cranking System | 6D2 |
| Charging System | 6D3 |
| Ignition System | 6D4 |
| Engine Wiring | 6D5 |
| Driveability & Emissions - Gen. | 6E |
| Driveability & Emissions - TBI | 6E2 |
| Driveability & Emissions - PFI | 6E3 |
| Exhaust System | 6F |

| TABLE OF CONTENTS | SECTION NUMBER |
|---|---|
| TRANSAXLE | |
| Auto.Transaxel On-Car Serv. | 7A |
| Auto. Trans. - Hydraulic Diagnosis | 3T40-HD |
| Auto. Trans. - Unit Repair | 3T40 |
| Man. Trans. On-Car Service | 7B |
| 5-Sp. 5TM40 Man. Trans. Unit Repair | 7B1 |
| 5-Sp. Isuzu Man. Trans. Unit Repair | 7B2 |
| Clutch | 7C |
| CHASSIS ELECTRICAL, INSTRUMENT PANEL, AND WIPER/WASHER | |
| Electrical Diagnosis | 8A |
| Lighting and Horns | 8B |
| Instrument Panel and Console | 8C |
| Windshield Wiper/Washer | 8E5 |
| ACCESSORIES | |
| Audio System | 9A |
| Cruise Control | 9B |
| Engine Block Heater | 9C |
| BODY SERVICE | |
| General Body Service | 10-1 |
| Stationary Glass | 10-2 |
| Underbody | 10-3 |
| Bumpers | 10-4 |
| Body Front End | 10-5 |
| Doors | 10-6 |
| Rear Quarters | 10-7 |
| Body Rear End | 10-8 |
| Roof & Convertible Top | 10-9 |
| Seats | 10-10 |
| Safety Belts | 10-11 |
| Body Wiring | 10-12 |
| Unibody Collision Repair | 11-1 |
| Welded Panel Replacement | 11-2 |
| INDEX | |
| Alphabetical Index | |

**Figure 4.** The table of contents directs you to the major systems and component areas in the service manual.

## TROUBLESHOOTING AND DIAGNOSTIC TABLES

Diagnostic information in each section of the service manual is usually provided in diagnostic procedure charts **(Figure 5)**.

**Diagnostic procedure charts** provide the necessary diagnostic steps in the proper order to diagnose specific problems in vehicle systems. The test results obtained in a specific diagnostic step guide the technician to the next appropriate step. When the technician follows the appropriate diagnostic procedure chart, unnecessary diagnostic steps are avoided.

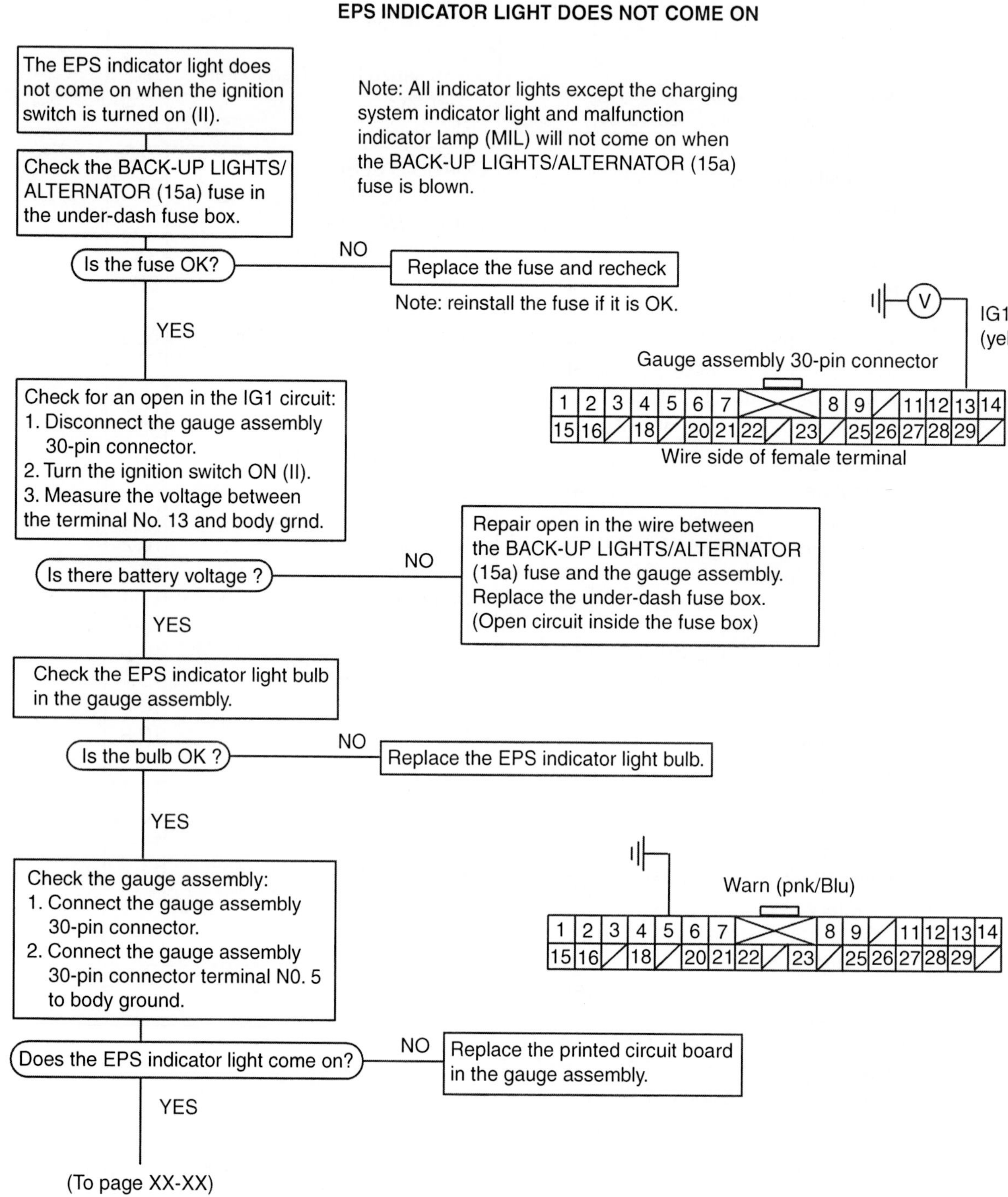


**Figure 5.** A typical diagnostic chart.

> **You Should Know** *When using diagnostic procedure charts in a service manual, do not skip steps in the procedure unless you are instructed to do so. Skipping steps in a diagnostic procedure may lead to inaccurate diagnosis and wasted time.*

## GENERIC SERVICE MANUALS

Generic service manuals are usually published by an independent automotive service organization or an aftermarket automotive component supplier. Generic service manuals are usually abbrieviated compared to vehicle manufacturer's service manuals. A generic service manual may include coverage of several years of a specific vehicle make. Some generic service manuals include coverage of several vehicle models supplied by the same manufacturer.

## ELECTRONIC SERVICE INFORMATION

Service and parts information can also be provided through computer services **(Figure 6)**. Computerized service information may be provided on computer disks or CDs. Service information on CDs can be stored and accessed more easily compared to information in service manuals. Computers may also be connected to a central database to obtain service information. Using the computer keyboard, light pen, mouse, or touch-sensitive screen, the technician selects choices from a series of menus on the computer monitor. When the desired information is accessed, it may be printed out for detailed study.

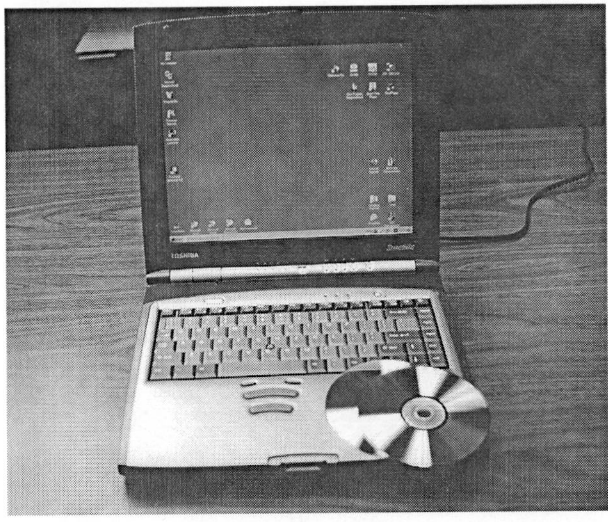

**Figure 6.** Computers are replacing printed service manuals in many shops.

## COMPUTER SOFTWARE

Computer software is the program in a computer that is responsible for proper operation of the system to which the computer is connected. A higher percentage of new-vehicle cost each year is required for the rapid increase in the complexity of computer software. The computer software contains very extensive information regarding the operation of the computer system under all possible operating conditions. When many computers are replaced, the necessary software must be downloaded from the appropriate scan tool or personal computer (PC) system **(Figure 7)**. This process is usually referred to as flash programming. **Flash programming** may also be used to reprogram on-board computers to correct specific computer system operational problems.

After a new computer is installed in a vehicle system, the scan tool is connected to a DLC which is usually located under the vehicle instrument panel **(Figure 8)**. Vehicles manufactured since 1996 have OBD II systems that have a 16-terminal DLC positioned under the driver's side of the instrument panel.

**Figure 7.** A scan tool.

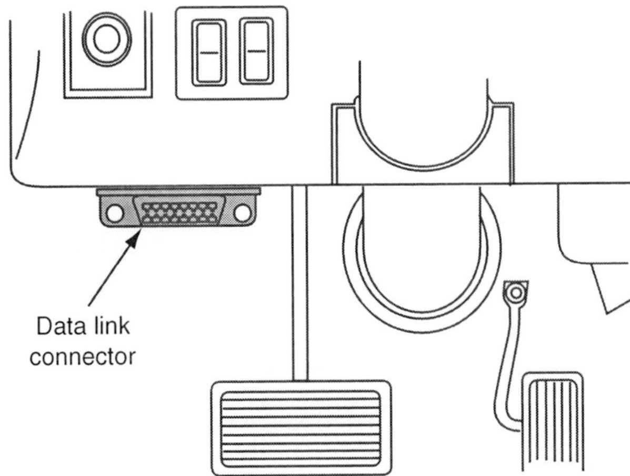


**Figure 8.** A data link connector (DLC).

You Should Know

*Never connect or disconnect the scan tool cable connection to the DLC with the ignition switch on. This action may cause computer damage or scan tool damage. The ignition switch must be turned off when connecting or disconnecting the scan tool.*

Data links that interconnect each computer on the vehicle are connected to the DLC to allow access to any vehicle computer. When the scan tool is connected to the DLC, the appropriate buttons on this tool are pressed to download the necessary software into the appropriate computer.

## SERVICE BULLETINS

Service procedures may be improved by the vehicle manufacturer at any time. Many service bulletins provide information regarding modified components and service procedures to correct specific problems on certain vehicles. When diagnosing automotive problems, technicians must have up-to-date service bulletin information to provide fast, accurate diagnosis. The vehicle manufacturer's service bulletins are made available to dealerships in printed form and/or on CD. Some vehicle manufacturers make their service bulletins available for sale to independent automotive shops. Some service bulletins provide up-to-date corrections for service manuals. If a significant number of corrections are required, a second edition of the manual may be published. When service information is provided on CD, the CDs are updated frequently to provide the latest information. Some independent publishers of automotive service information produce CDs that contain service bulletin information for many different vehicle makes and model years.

## LABOR ESTIMATING GUIDES

Many automotive repair shops provide the customer with an estimate of vehicle repair costs. Parts and labor costs are the two main expenses in any vehicle repair. The labor is the cost of the time required for the technician to repair the customer's vehicle. An estimating guide is used to provide an estimate of the labor costs required to repair a vehicle.

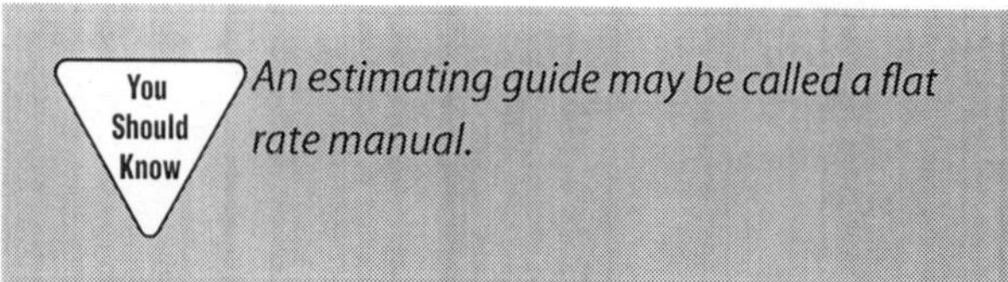

You Should Know

*An estimating guide may be called a flat rate manual.*

An estimating guide lists the time required to perform all the possible repair jobs on different vehicles. The service person greeting the customer finds the number of hours to complete the necessary repair on the vehicle, and this number of hours is then multiplied by the shop labor rate. For example, if the estimating guide indicates that 2.5 hours are required to repair the vehicle and the shop labor rate is $60.00 per hour, the estimated labor cost is 2.5 × 60, or $150.00.

The cost of automotive parts changes frequently. Therefore, when providing a parts cost estimate, it is more accurate to phone a parts store and obtain the exact price of the components required to repair the vehicle. The labor and parts costs plus all applicable taxes and miscellaneous charges are then added together to compile the parts and labor cost for repairing the customer's vehicle.

## VEHICLE SERVICE DECALS AND WARNING LABELS

Each vehicle has a number of warning labels. Many of these labels are adhered to the body structure in the engine compartment **(Figure 9)**. The warning labels pro-

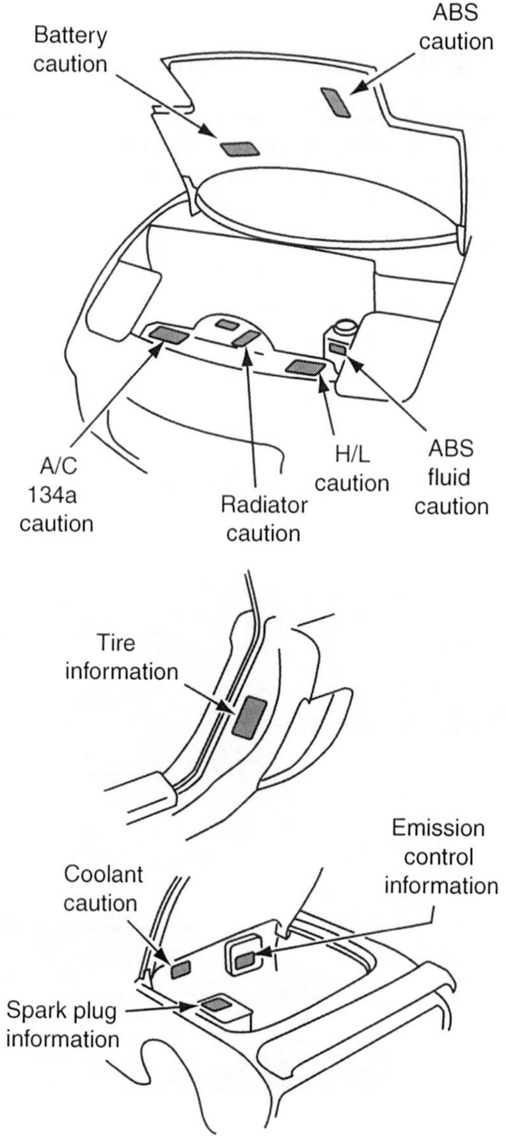


**Figure 9.** Vehicle warning labels.

vide warnings that must be observed when servicing various vehicle systems.

The emission control label is also mounted on the engine or in the engine compartment. This label provides valuable information regarding the engine size and the emission equipment on the engine. The emission control label also lists some engine specifications, such as spark plug type and gap, idle speed, and valve lash **(Figure 10)**.

The vacuum hose routing label is also mounted in the underhood area. This label illustrates the vacuum hose connections and routing for the vehicle **(Figure 11)**. The vacuum hose routing diagram is very helpful when improper vacuum hose connections are suspected.

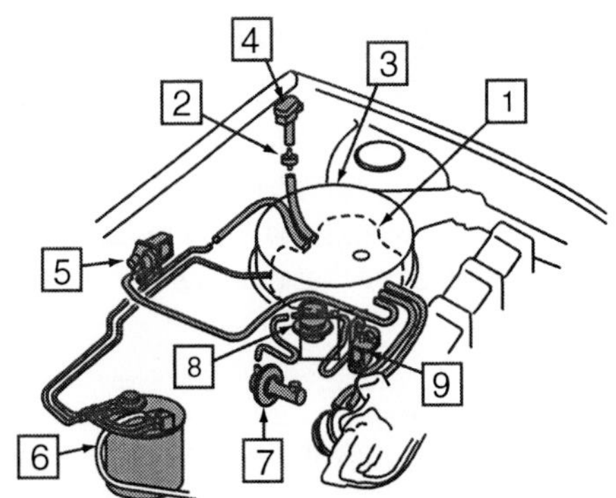


1 Throttle Body
2 Filter
3 Silencer
4 Pressure sensor
5 AC vacuum switch
6 Charcoal canister
7 EGR valve
8 EGR modulator
9 EGR VSV

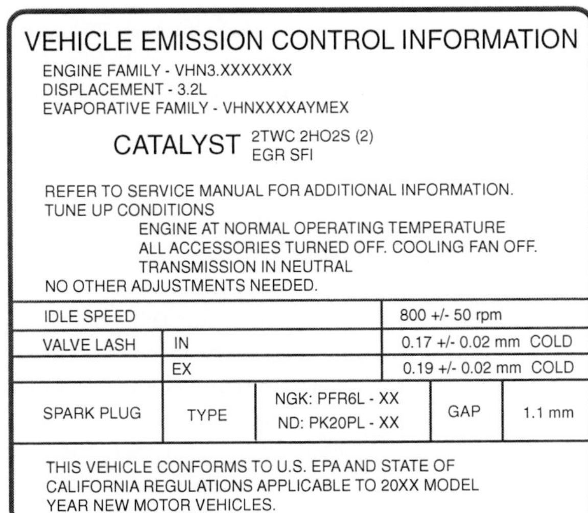

VEHICLE EMISSION CONTROL INFORMATION

ENGINE FAMILY - VHN3.XXXXXXX
DISPLACEMENT - 3.2L
EVAPORATIVE FAMILY - VHNXXXXAYMEX

CATALYST 2TWC 2HO2S (2) EGR SFI

REFER TO SERVICE MANUAL FOR ADDITIONAL INFORMATION.
TUNE UP CONDITIONS
ENGINE AT NORMAL OPERATING TEMPERATURE
ALL ACCESSORIES TURNED OFF. COOLING FAN OFF.
TRANSMISSION IN NEUTRAL
NO OTHER ADJUSTMENTS NEEDED.

| IDLE SPEED | | | 800 +/- 50 rpm | |
|---|---|---|---|---|
| VALVE LASH | IN | | 0.17 +/- 0.02 mm COLD | |
| | EX | | 0.19 +/- 0.02 mm COLD | |
| SPARK PLUG | TYPE | NGK: PFR6L - XX<br>ND: PK20PL - XX | GAP | 1.1 mm |

THIS VEHICLE CONFORMS TO U.S. EPA AND STATE OF CALIFORNIA REGULATIONS APPLICABLE TO 20XX MODEL YEAR NEW MOTOR VEHICLES.

**Figure 10.** An emission control label.

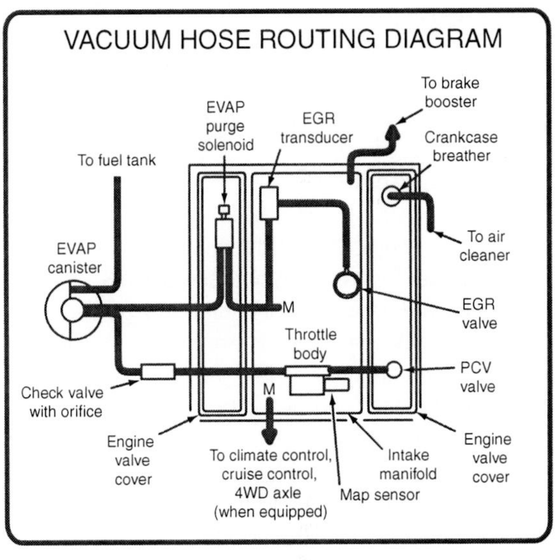


**Figure 11.** A vacuum hose routing diagram.

# Summary

- Sources of automotive service information include owner's manuals, service manuals, service bulletins, electronic media such as CDs and computer disks, computer software, and service bulletins.
- The owner's manual provides basic service and maintenance information.
- Service manuals contain information such as component identification, service procedures, specifications, and diagnostic information.
- Service manuals may be published by vehicle manufacturers or independent automotive service organizations.
- Generic service manuals may cover several model years of a certain vehicle.
- Compared to printed service manuals, CDs provide greatly increased storage capacity.
- Data for programming or reprogramming computers is stored in computer software and is downloaded from a scan tool or PC into an on-board automotive computer.
- Service bulletins contain valuable information regarding modified components and/or service procedures to correct specific vehicle problems.
- Labor estimating guides list the labor time required to perform specific repairs on various vehicles.
- Vehicle service decals provide essential information regarding emission systems and vacuum hose routing.
- Vehicle warning labels make owners and technicians aware of precautions that must be observed when operating or servicing a vehicle.

# Review Questions

1. Technician A says a vehicle owner's manual provides information about transaxle overhaul. Technician B says a vehicle owner's manual contains information about the type of oil that must be used in the engine. Who is correct?
   A. Technician A
   B. Technician B
   C. Both Technician A and Technician B
   D. Neither Technician A nor Technician B
2. Information about cleaning and maintenance of the vehicle interior and exterior is found in the:
   A. service manual.
   B. generic service manual.
   C. troubleshooting table.
   D. owner's manual.
3. Technician A says a CD has more information storage capacity than a printed service manual. Technician B says service bulletin information may be supplied on CDs. Who is correct?
   A. Technician A
   B. Technician B
   C. Both Technician A and Technician B
   D. Neither Technician A nor Technician B
4. Technician A says a labor estimating guide provides labor costs for specific vehicle repairs. Technician B says a labor estimating guide lists the cost of parts that are required for each vehicle repair. Who is correct?
   A. Technician A
   B. Technician B
   C. Both Technician A and Technician B
   D. Neither Technician A nor Technician B
5. All of these statements about owner's manuals are true *except*:
   A. The owner's manual includes information regarding tire changing.
   B. The owner's manual includes information regarding cooling system capacity.
   C. The owner's manual includes information regarding fuel injector service.
   D. The owner's manual includes information regarding the type of engine oil.
6. Technician A says a service bulletin may provide information about improved parts that are designed to correct a specific vehicle problem. Technician B says service bulletins may provide specific diagnostic and service information to correct a certain vehicle problem. Who is correct?
   A. Technician A
   B. Technician B
   C. Both Technician A and Technician B
   D. Neither Technician A nor Technician B
7. Technician A says the owner's manual contains vehicle warranty information. Technician B says the owner's manual contains information regarding vehicle warning systems. Who is correct?
   A. Technician A
   B. Technician B
   C. Both Technician A and Technician B
   D. Neither Technician A nor Technician B
8. Vehicle manufacturer's service manuals provide information on:
   A. proper cleaning and maintenance of the vehicle interior.
   B. diagnostic and service procedures.
   C. maintenance schedules.
   D. light bulb replacement.
9. Technician A says service manuals for a specific vehicle are limited to one volume. Technician B says some service manuals are available on CD. Who is correct?
   A. Technician A
   B. Technician B
   C. Both Technician A and Technician B
   D. Neither Technician A nor Technician B
10. All of these statements about generic service manuals are true *except*:
    A. Generic service manuals are usually abbreviated compared to other service manuals.
    B. Generic service manuals may include coverage of several years of a specific vehicle make.
    C. Generic service manuals may include coverage of several vehicle models supplied by the same manufacturer.
    D. Generic service manuals are usually published by vehicle manufacturers.
11. The emission control label provides information about the engine size and the ________________ equipment on the vehicle.
12. Many emission control labels provide some engine ________________.
13. The vacuum hose routing label is mounted in the ________________ area.
14. Specification tables are provided in the ________________ manual.
15. List ten information topics that are discussed in an owner's manual.
16. Describe how specific service information is accessed in a vehicle manufacturer's service manual.
17. Explain the meaning of flash programming as it relates to automotive computers.
18. Describe the purpose of a labor estimating guide.

# Chapter 9 Using Service Information

## Introduction

After students or technicians are familiar with the various sources of service information, they must understand how to access and use this information. When individuals are not familiar with the proper use of service information, they can waste valuable time searching for the required information. Students and technicians must also understand where to locate the best and quickest source for the information they require. Using brief or inadequate service information may lead to improper diagnostic and service procedures that do not locate and correct the cause of the customer's complaint.

### VEHICLE IDENTIFICATION NUMBER (VIN) INTERPRETATION

When completing a work order, ordering parts, or servicing a vehicle, certain facts must be known about the vehicle. The vehicle manufacturer, make, and model year must be known. The vehicle manufacturer and make are often on the nameplate attached to the exterior of the body. Examples of vehicle manufacturer and make are Chrysler Concorde and Honda Acura. The customer can also supply the necessary information regarding the vehicle manufacturer, make, and model year. This information is also on the vehicle registration.

The **vehicle identification number (VIN)** is a series of letters and numbers that identify specific vehicle parameters such as the vehicle make, model, model year, and size of engine. On modern vehicles, the VIN number is mounted on the top of the dash, and this number is visible through the lower left side of the windshield **(Figure 1)**.

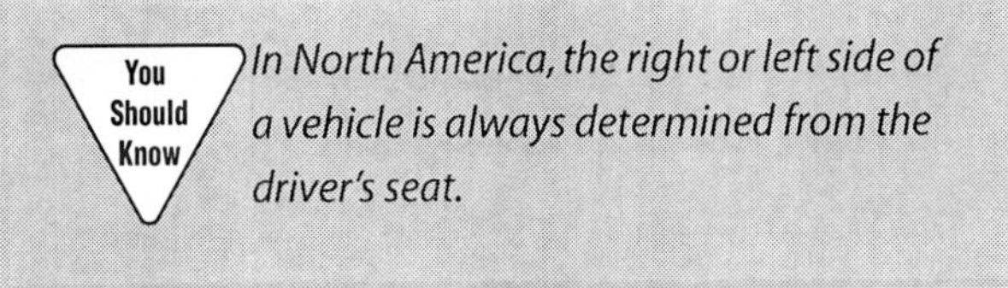


*In North America, the right or left side of a vehicle is always determined from the driver's seat.*

The VIN mounting location makes it more difficult for car thieves to change the VIN. An explanation of the VIN is provided in the vehicle manufacturer's service manual. The VIN explanation is usually included in the service manual

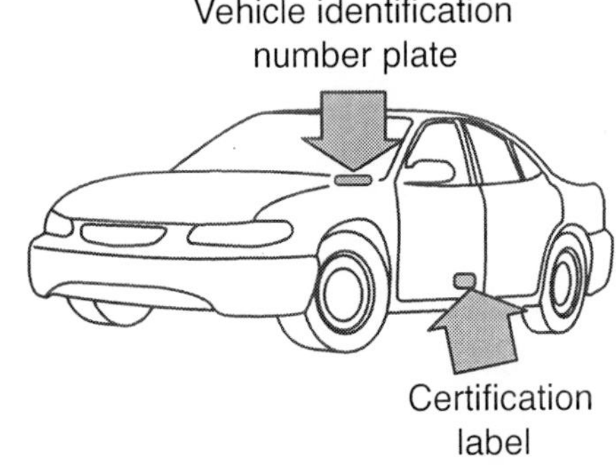


**Figure 1.** Vehicle identification number (VIN) location.

Introduction or General Information section. The Introduction or General Information section is the first section in a service manual. If the manual has more than one volume, the Introduction or General Information section is in volume one. A VIN for a 2002 PT Cruiser is explained as follows:

| *Position in VIN* | *Interpretation* | *Code = Description* |
|---|---|---|
| 1 | Country of Origin | 1 = Built in United States by DaimlerChrysler<br>3 = Built in Mexico by DaimlerChrysler De Mexico |
| 2 | Make | C = Chrysler |
| 3 | Vehicle Type | 4 = Multipurpose passenger vehicle without side air bags<br>8 = Multipurpose passenger vehicle with side air bags |
| 4 | Other | F = 1815–2267 kilograms (4000–5000 pounds) |
| 5 | Line | Y = Cruiser LHD |
| 5—Export | Line | E = Cruiser LHD<br>Z = Cruiser RHD |
| 6 | Series | 4 = High Line<br>5 = Premium<br>6 = Sport |
| 6—Export | Transmission | B = 4-Speed Automatic<br>N = 5-Speed Manual |
| 7 | Body Style | 8 = Hatchback |
| 8 | Engine | 9 = 2.0L 4-Cyl. Gasoline DOHC MPI<br>B = 2.4L 4-Cyl. Gasoline 16V DOHC |
| 8—Export | Engine | F = 1.6L 4-Cyl. 16V Gasoline SOHC<br>9 = 2.0L 4-Cyl. Gasoline DOHC MPI<br>U = 2.2L 4-Cyl. Turbo Diesel MPI |
| 9 | Check Digit | Verification of vehicle authenticity |
| 10 | Model Year | 2 = 2002 |
| 11 | Assembly Plant | T = Toluca Assembly<br>U = Graz Assembly |
| 12 through 17 | Vehicle Build Sequence | 6-digit number from assembly plant |

The tenth digit in the VIN indicates the model year. The check digit in the ninth position of the VIN is used by the vehicle manufacturer and government agencies to verify the authenticity of the vehicle. The ninth digit in the VIN helps to protect the consumer from theft and fraud. The formula for interpreting the ninth digit in the VIN is not provided to the general public.

Some imported vehicles have a slightly different VIN number interpretation **(Figure 2)**. On these vehicles, some of the information related to engine family and evaporative family is interpreted from numbers on the emission label. A series of letters and numbers for the engine family and the evaporative family are displayed on the emission label **(Figure 3)**. Interpretations of the engine family and evapo-

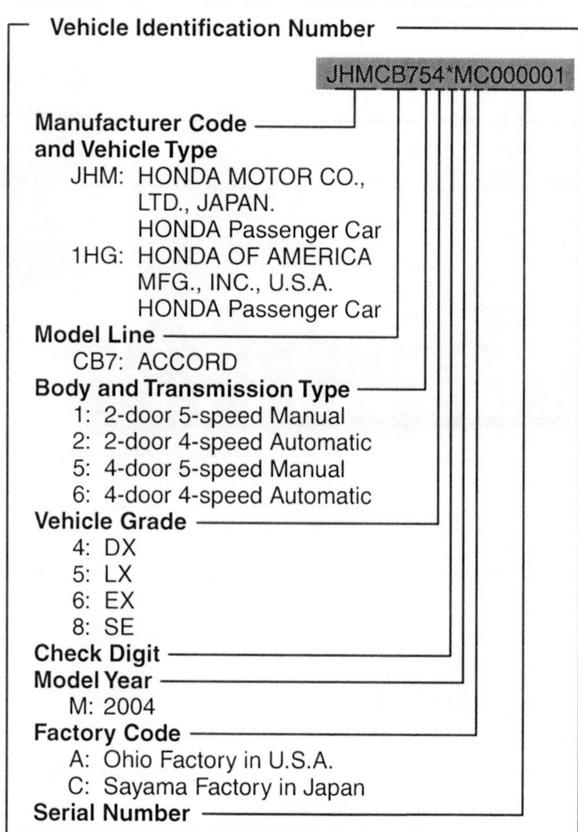


**Figure 2.** Vehicle identification number (VIN) interpretation.

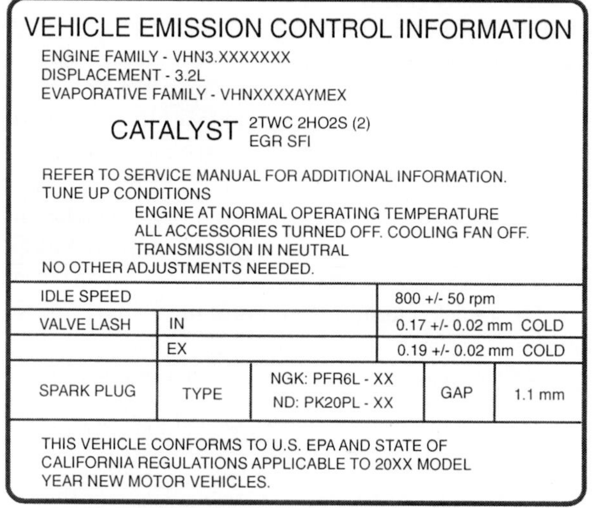


**Figure 3.** A vehicle emission control label.

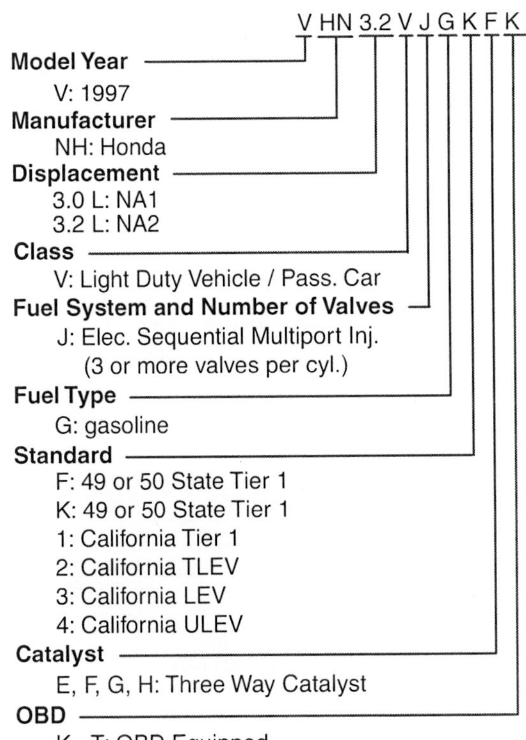


**Figure 4.** An engine family number interpretation.

rative family numbers are provided in the service manual **(Figure 4 and Figure 5)**.

In the engine family number the Standard digit indicates the emission standards the vehicle is designed to meet. If the vehicle has a forty-nine-state emission rating, the vehicle will meet emission standards in the forty-nine states other than California. A fifty-state emission designation indicates the vehicle will meet emission standards in all fifty states. Any of the California emission designations indicate the vehicle is designed to meet that specific California emission standard. Some of the abbreviation explanations regarding California emission standards are these: TLEV—Transitional low emission vehicle; LEV—Low emission vehicle; ULEV—Ultra-low emission vehicle.

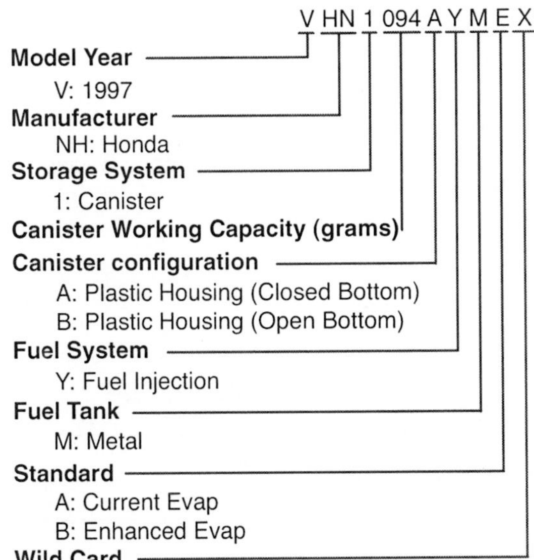


**Figure 5.** An evaporative family number intepretation.

## LOCATING CAPACITIES AND FLUID REQUIREMENTS IN A SERVICE MANUAL

A general index is located at the front of a service manual. This index directs the technician to the main sections in the manual **(Figure 6)**. The fluid capacities and requirements

| TABLE OF CONTENTS | SECTION NUMBER |
|---|---|
| **GENERAL INFOR. AND LUBE** | |
| General Information | 0A |
| Maintenance and Lubrication | 0B |
| **HEATING AND AIR CONDITIONING** | |
| Heating and Ventilation (Non-A/C) | 1A |
| Air Conditioning System | 1B |
| V-5 A/C Compressor Overhaul | 1D3 |
| **STEERING, SUSPENSION, TIRES, AND WHEELS** | |
| Diagnosis | 3 |
| Wheel Alignment | 3A |
| Power Steering Gear & Pump | 3B1 |
| Front Suspension | 3C |
| Rear Suspension | 3D |
| Tires and Wheels | 3E |
| Steering Column, On-Vehicle Service | 3F |
| Steering Column—Std., Unit Repair | 3F1 |
| Steering Column—Tilt, Unit Repair | 3F2 |
| **DRIVE AXLES** | |
| Drive Axles | 4D |
| **BRAKES** | |
| General Information—Diagnosis and On-car Service | 5 |
| Compact Master Cylinder | 5A1 |
| Disk Brake Caliper | 5B2 |
| Drum Brake—Anchor Plate | 5C2 |
| Power Brake Booster Assembly | 5D2 |
| **ENGINES** | |
| General Information | 6 |
| 2.0 Liter L-4 Engine | 6A1 |
| 3.1 Liter V6 Engine | 6A3 |
| Cooling System | 6B |
| Fuel System | 6C |
| Engine Electrical—General | 6D |
| Battery | 6D1 |
| Cranking System | 6D2 |
| Charging System | 6D3 |
| Ignition System | 6D4 |
| Engine Wiring | 6D5 |
| Driveability and Emissions—General | 6E |
| Driveability and Emissions—TBI | 6E2 |
| Driveability and Emissions—PFI | 6E3 |
| Exhaust System | 6F |
| **TRANSAXLE** | |
| Auto. Transaxle On-Car Service | 7A |
| Auto. Trans.—Hydraulic Diagnosis | 3T40-HD |
| Auto. Trans.—Unit Repair | 3T40 |
| Man. Trans.— On-Car Service | 7B |
| 5-Sp. 5TM40 Man. Trans. Unit Repair | 7B1 |
| 5-Sp. Isuzu Man. Trans. Unit Repair | 7B2 |
| Clutch | 7C |
| **CHASSIS ELECTRICAL, INSTRUMENT PANEL & WIPER/WASHER** | |
| Electrical Diagnosis | 8A |
| Lighting & Horns | 8B |
| Instrument Panel & Console | 8C |
| Windshield Wiper/Washer | 8E5 |

**Figure 6.** A service manual general index.

are often located in the Maintenance or Lubrication and Maintenance section of the vehicle manufacturer's service manual. In some service manuals, the fluid capacities are listed with other specifications at the end of each section. Fluid capacities are also provided in the owner's manual. Technicians must know the specified fluid capacities so they do not overfill or underfill various components. Typical capacities for a 2002 PT Cruiser are the following:

| *Description* | *Specification* |
|---|---|
| Fuel tank | 57L (15 gal.) |
| Engine oil—1.6L | 4.5L (4.8 qts.) with filter change |
| Engine oil—2.0L, 2.2L | 4.3L (4.5 qts.) with filter change |
| Engine oil—2.4L | 4.8L (5.0 qts.) with filter change |
| Cooling system | 7.0L (7.4 qts.) includes recovery bottle and heater |
| Automatic transaxle | 3.8L (4.0 qts.) service fill |
| Automatic transaxle | 8.1L (8.6 qts.) overhaul fill with converter empty |
| Manual transaxle, NV T350 | 2.4–2.7L (2.5–2.8 qts.) |

Fluid requirements are usually listed in the Lubrication and Maintenance section of the service manual. In some service manuals, the fluid requirements are listed with other specifications at the end of each section. These requirements may be provided in text form or in a chart. A typical lubrication requirement chart is provided in **Figure 7** and **Figure 8**.

The lubrication chart lists the engine oil grade and viscosity ratings at various temperatures. Figure 7 illustrates different labels and seals on oil containers. Engine oil with the proper American Petroleum Institute (API) rating and Society of Automotive Engineers (SAE) viscosity rating must be used. The **American Petroleum Institute (API) rating** is a universal engine oil rating that classifies oils according to the type of service for which the oil is intended. The **Society of Automotive Engineers (SAE) viscosity rating** is a universal rating for engine oil viscosity in relation to the atmospheric temperature in which the oil will be operating.

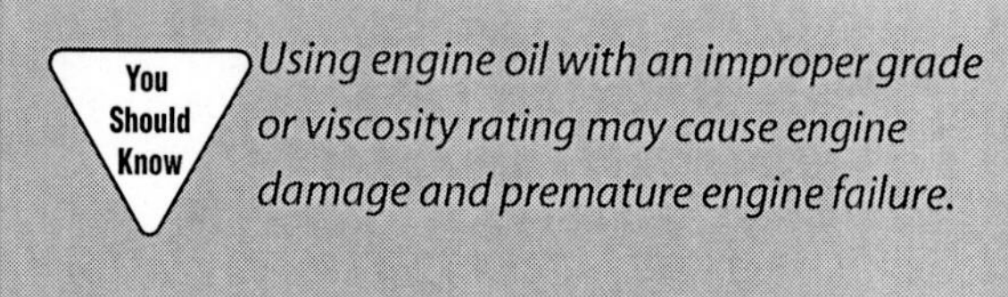


*Using engine oil with an improper grade or viscosity rating may cause engine damage and premature engine failure.*

Technicians must be familiar with engine coolant requirements. In recent years, most engines have aluminum cylinder heads and some engines also have aluminum cylinder blocks. Some radiators on these engines also have aluminum cores. In these cooling systems, most vehicle manufacturers recommend an ethylene glycol-based coolant with **hybrid organic additive technology (HOAT)**.

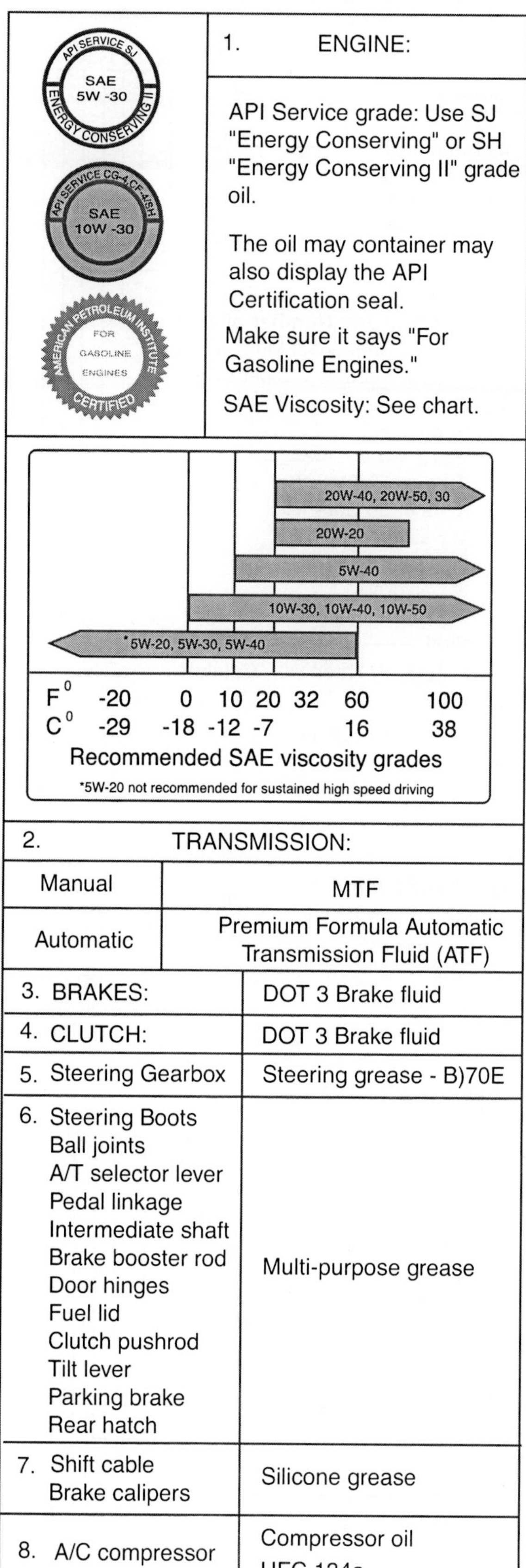


**Figure 7.** A lubrication chart.

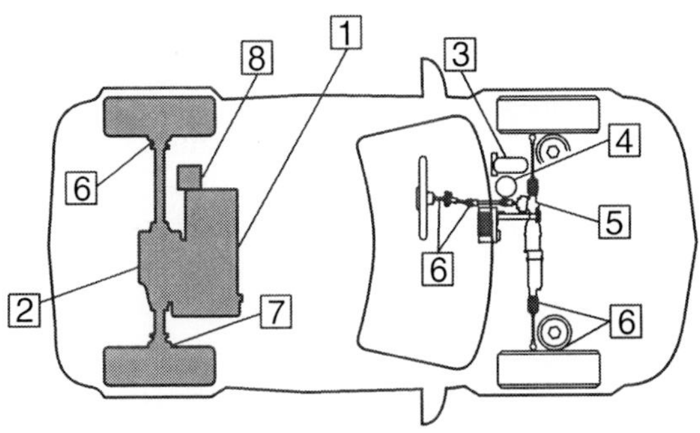


**Figure 8.** Vehicle lubrication points.

This type of coolant is green in color and contains special additives to minimize corrosion in the cooling system.

The coolant with a HOAT additive usually has a 5-year/100,000-mile warranty. In most climates where below-freezing temperatures are encountered, the coolant is a mixture of 50 percent ethylene glycol and 50 percent water. This mixture provides a coolant freeze point of −35°F (−37°C). The coolant specified in the vehicle manufacturer's service manual must be used.

**You Should Know** *Mixing orange- or magenta-colored antifreeze with green antifreeze containing a HOAT additive reduces the corrosion protection of the coolant and may cause premature cooling system corrosion plus water pump seal failure. If different types of antifreeze are mixed, the cooling system should be flushed and filled with the proper mixture of the vehicle manufacturer's specified coolant.*

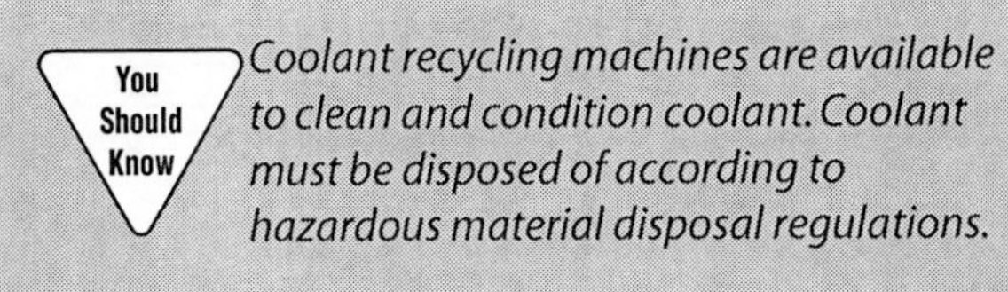

You Should Know
Coolant recycling machines are available to clean and condition coolant. Coolant must be disposed of according to hazardous material disposal regulations.


## LOCATING VEHICLE SPECIFICATIONS AND MAINTENANCE SCHEDULES IN A SERVICE MANUAL

Technicians must use proper specifications for the vehicle being serviced. For example, if the proper torque specifications are not used, expensive aluminum castings may be warped resulting in fluid leaks. In many service manuals, the specifications are located at the end of each section. The index for each section lists the page number where the specifications are located **(Figure 9)**. The specifications provide all the necessary measurements, adjustments, and torque values for the components being serviced **(Figure 10)**. Some service manuals provide

**STARTING SYSTEM**

| 5S-FE | |
|---|---|
| Description | ST - 2 |
| System circuit | ST - 3 |
| Operation | ST - 3 |
| Preparation | ST - 4 |
| Starter | ST - 5 |
| Starter relay | ST - 19 |
| Clutch start switch | ST - 19 |
| (M/T) | ST - 19 |
| Neutral start switch | ST - 19 |
| (A/T) | ST - 19 |
| Service specifications | ST - 20 |
| **3VZ-FE** | |
| Description | ST - 21 |
| System circuit | ST - 22 |
| Operation | ST - 22 |
| Preparation | ST - 23 |
| Starter | ST - 24 |
| Starter relay | ST - 38 |
| Clutch start switch | ST - 38 |
| (M/T) | ST - 38 |
| Neutral start switch | ST - 38 |
| (A/T) | ST - 38 |
| Service specifications | ST - 39 |

**Figure 9.** A service manual section index.

**SERVICE SPECIFICATIONS**
**SERVICE DATA**

| | | |
|---|---|---|
| Starter | Rated voltage and output power | 12V 1.4kW |
| | No-load characteristics (current) | 90A or less 11.5V |
| | No-load characteristics (RPM) | 3,000 rpm or more |
| | Brush length (STD) | 15.0 mm |
| | Brush length (MIN) | 10.0 mm |
| | Spring installed load | 18 - 24 N |
| | Commutator | |
| | Diameter (STD) | 30 mm |
| | Diameter (MIN) | 29 mm |
| | Undercut depth (STD) | 0.6 mm |
| | Undercut depth (MIN) | 0.2 mm |
| | Circle runout (MAX) | 0.05 mm |

**TORQUE SPECIFICATIONS**

| | | |
|---|---|---|
| Starter mounting bolt | 39 N·m | 29 ft.-lbf |

**Figure 10.** Manufacturer's specifications.

specifications at the beginning of each subheading within a section. Other service manuals insert the specifications at the appropriate place in each service procedure.

Technicians and service personnel must use maintenance schedules when servicing vehicles. Often customers are not aware that certain maintenance should be performed on their vehicle because of the mileage on the vehicle or the time since the last service interval. The service writer should check the vehicle mileage and past service record against the maintenance schedule to determine if routine maintenance is required on the vehicle. Performing the maintenance listed in the maintenance schedule will maintain safe and trouble-free operation of the vehicle.

The maintenance schedules are usually provided in the Lubrication and Maintenance section of the service manual. These maintenance schedules may be listed in charts (**Figure 11**). In some service manuals, the maintenance schedules are in text format. The maintenance schedules are also in the owner's manual. Some vehicle manufacturers provide maintenance schedules for normal vehicle operation and separate maintenance schedules for severe vehicle operating conditions. Some vehicle manufacturers define severe operating conditions as these:

1. Driving less than 5 miles (8 km) per trip, or in freezing temperatures, driving less than 10 miles (16 km) per trip
2. Driving in extremely hot temperatures above 90°F (32°C)
3. Extensive idling or long periods of stop-and-go driving
4. Trailer towing, driving with a car-top carrier, or driving continually in mountainous conditions

| Service at the indicated mileage or time whichever comes first. | Miles x 1,000 | 15 | 30 | 45 | 60 | 75 | 90 | 105 | 120 | Note |
|---|---|---|---|---|---|---|---|---|---|---|
| | km x 1,000 | 24 | 48 | 72 | 96 | 120 | 144 | 168 | 192 | |
| | months | 12 | 24 | 36 | 48 | 60 | 72 | 84 | 96 | |
| Replace engine oil | Replace every 7,5000 miles or 12 months | | | | | | | | | Capacity 5.0L. |
| Replace engine oil filter | | ● | ● | ● | ● | ● | ● | ● | ● | |
| Check engine oil and coolant | Check oil and coolant at each fuel stop | | | | | | | | | Check for leaks. |
| Replace air cleaner element | | | ● | | ● | | ● | | ● | |
| Inspect valve clearance | | | | | | | ● | | | Intake 0.15-0.19mm<br>Exhaust 0.17-0.21mm<br>Measure when cold |
| Replace spark plugs | | | | | | | | ● | | Gap: 1.0-1.1mm |
| Replace timing belt / inspect water pump | | | | | | | | ● | | |
| Inspect and adjust drive belts | | | ● | | ● | | ● | | ● | Check for cracks.<br>Check for belt deflection at center of belt.<br>Alternator 11.0-13.5mm<br>A/C Compressor 10-12mm |
| Replace fuel filter | | | | | ● | | | | ● | |
| Inspect idle speed | | | | | | | | ● | | 800 +/-rpm (MT neutral)<br>780 +/-rpm (AT neutral) |
| Replace engine coolant | | | | ● | | ● | | ● | | Capacity 12.0L |
| Replace transmission fluid | | | | | | | ● | | | Manual trans. MTF 2.65L<br>Automatic trans. ATF 2.9L |
| Inspect front and rear brakes | | ● | ● | ● | ● | ● | ● | ● | ● | Check brake pad thickness and movement.<br>Check caliper for leakage. |
| Replace brake fluid (including ABS) | | | | ● | | | ● | | | DOT3 brake fluid.<br>Check that fluid is between upper and lower marks in reservoir. |
| Check parking brake adjustment | | ● | ● | ● | ● | ● | ● | ● | ● | Engaged 10 to 14 notches |
| Rotate tires (Check inflation and condition at least once per month) | | Rotate every 7,500 miles | | | | | | | | Rotation method shown in owner's manual. |

**Figure 11.** A vehicle maintenance schedule.

| Service at the indicated mileage or time whichever comes first. | Miles x 1,000 | 15 | 30 | 45 | 60 | 75 | 90 | 105 | 120 | Note |
|---|---|---|---|---|---|---|---|---|---|---|
| | km x 1,000 | 24 | 48 | 72 | 96 | 120 | 144 | 168 | 192 | |
| | months | 12 | 24 | 36 | 48 | 60 | 72 | 84 | 96 | |
| **Visually inspect the following items** | | | | | | | | | | |
| Tie-rod ends, steering gear box, and boots | | | | | | | | | | Check for steering linkage looseness. Check condition of boots and fluid leakage. |
| Suspension components | | | | | | | | | | Check for bolt tightness. Check ball joint boots. |
| Driveshaft boots | | | | | | | | | | Check condition of boots |
| Brake hoses and lines (including ABS) | | | | | | | | | | Check for leakage. |
| All fluid levels and condition of fluid | | • | • | • | • | • | • | • | • | Check for levels, condition of fluid, and leakage. |
| Cooling system hoses and connections | | | | | | | | | | Check all hoses for leakage and damage. Check fan operation. |
| Exhaust system | | | | | | | | | | Check catalytic converter heat shield, exhaust pipe, and muffler for damage and leaks. |
| Fuel lines and connections | | | | | | | | | | Check for leaks. |
| Inspect air bag system | | 10 years after production | | | | | | | | |

**Figure 11.** A vehicle maintenance schedule, continued.

5. Driving on muddy, dusty, or icy roads on which salt has been applied

Maintenance schedules for severe operating conditions require more frequent vehicle servicing such as changing the oil, oil filter, and air filter at lower mileage intervals.

## USING GENERIC SERVICE MANUALS

Many generic service manuals have the same format as the vehicle manufacturer's service manuals. This format was explained previously in this chapter. Generic service manuals usually cover several models and model years. For example, one generic service manual covers 1985 through 1992 Volkswagen GTI, Golf, and Jetta. Compared to vehicle manufacturer's service manuals, generic manuals tend to be abbreviated.

Some generic service manual publishers produce a series of service manuals to cover specific vehicle classifications such as domestic cars, domestic light trucks and vans, and import cars trucks and vans **(Figure 12).** The series of manuals based on domestic cars has separate manuals covering each these topics: Engine Performance, Electrical, Chassis, Engine, Heating and Air Conditioning, and Electrical Component Locator. These generic service manuals cover several model years of domestic vehicles. The generic service manuals in the Engine Performance cate-

| MECHANICAL MANUALS | Page |
|---|---|
| Mech. Parts and Labor Estimating Manuals | 19 |
| Service and Repair Manuals: | |
| Domestic Cars | 20 |
| Domestic Light Trucks, and Vans | 22 |
| Import Cars, Light Trucks, and Vans | 23 |
| Medium and Heavy Duty Truck Manuals | 25 |
| Transmission Manuals | 26 |
| Specialty Manuals | 27 |
| Training Products | 27 |
| ASE Test Preparation Manuals | 28 |

**Figure 12.** Generic service manual categories.

| ENGINE PERFORMANCE | |
|---|---|
| DV84 | Engine performance: Dom. Cars 1982-84 (Includes procedures, illustrations, specifications, wiring and vacuum diagrams on fuel systems, computerized engine controls, ignition systems, and emission systems) |
| DV86 | Engine performance: Dom. cars 1985-86 (Same coverage as DV84 above) |
| DV88 | Engine performance: Dom. cars 1987-88 (Same coverage as DV84 above) |
| DV90 | Engine performance: Dom. cars 1989-90 (Same coverage as DV84 above) |
| DV91 | Engine performance: Dom. cars 1991 (Same coverage as DV84 above) |
| DV93 | Engine performance: Dom. cars 1992-93 (Same coverage as DV84 above) |
| DV95 | Engine performance: Dom. cars 1994-95 (Same coverage as DV84 above) |
| DTS96 | Engine performance: Dom. Cars, Light Trucks, & Vans 1996 |
| DTS97 | Engine performance: Dom. Cars, Light Trucks, & Vans 1997 |
| DTS98 | Engine performance: Dom. Cars, Light Trucks, & Vans 1998 |
| DTS99 | Engine performance: Dom. Cars, Light Trucks, & Vans 1999 |
| DTS00 | Engine performance: Dom. Cars, Light Trucks, & Vans 2000 |
| DTS01 | Engine performance: Dom. Cars, Light Trucks, & Vans 2001 |
| DTS02 | Engine performance: Dom. Cars, Light Trucks, & Vans 2002 |

**Figure 13.** Generic engine performance service manuals.

gory for domestic cars and light trucks and vans are listed in **Figure 13**.

Some publishers produce CDs or DVDs containing automotive service information. This information may also be available via the Internet. The automotive service information subjects available electronically from one publisher are listed in **Figure 14**. When one of these electronic service information systems is installed and opened in a computer, it begins with a simple home page from which the desired information is selected. The special features of OnDemand5 Repair are illustrated in **Figure 15** and the service information sections included in this category are listed in **Figure 16**.

| SOFTWARE and INFORMATION SYSTEMS | |
|---|---|
| Teamworks Package | 2 |
| Repair Package | 4 |
| Estimator Package | 8 |
| Internet .Com | 10 |
| Manager / Manager Plus | 12 |
| Transmission Package | 14 |
| Truck Package | 15 |
| Club Information | 16 |
| Vintage Service Information | 17 |

**Figure 14.** Generic service information available in electronic format.

- New, more powerful search engine
- Categories within Repair and Estimator are streamlined like never before! Now users can go from Repair to Estimator without having to reselect the category
- New fluid capacities
- Single click access to technical service bulletins
- Single click access to maintenance

**Figure 15.** Special features of Mitchell, OnDemand5 repair.

**ON DEMAND repair gives you more coverage.**

- Accessories
- Air Bags
- Air Conditioning & Heating
- Brakes
- Clutches
- Diagnostics
- Drive Axles
- Electrical
- Emission Controls
- Engine Cooling
- Engine Performance
- Maintenance
- Recalls
- Steering
- Suspension
- Technical Service Bulletins
- Transmission Servicing & Electronic Diagnosis
- Wheel Alignment
- Wiring Diagrams

**Figure 16.** Mitchel OnDemand5 repair information sections.

## Summary

- The VIN is a series of letters and numbers that identifies vehicle parameters such as the make, model, model year, and size of engine.
- The vehicle emission label provides information about the emission equipment on the vehicle and the emission standards that the vehicle is designed to meet.
- Fluid capacities and fluid requirements are usually provided in the service manual's Lubrication and Maintenance section.
- Using an engine oil or coolant other than the ones specified in the service manual may cause premature engine wear and cooling system corrosion and failure.
- In engines with aluminum cylinder heads and cylinder blocks, the cooling system requires a coolant with hybrid organic additive technology (HOAT).
- Most service manuals have a general index at the front to the manual that lists the sections in the manual.
- A sub-index at the beginning of each section identifies specific information within the section.
- In many service manuals, the specifications are located at the end of each section.
- Maintenance schedules are usually located in the Lubrication and Maintenance section of a service manual.
- Separate maintenance schedules may be provided for normal and severe operating conditions.
- Some generic service manuals have the same format as vehicle manufacturer's service manuals.
- Some generic manual publishers provide automotive service information on CDs, DVDs, or via the Internet.

## Review Questions

1. Technician A says the VIN number is located on the top of the dash on the left side. Technician B says the tenth digit in the VIN identifies the model year of the vehicle. Who is correct?
   A. Technician A
   B. Technician B
   C. Both Technician A and Technician B
   D. Neither Technician A nor Technician B
2. Technician A says the eighth digit in the VIN identifies the engine size. Technician B says the Check Digit is the ninth digit in the VIN. Who is correct?
   A. Technician A
   B. Technician B
   C. Both Technician A and Technician B
   D. Neither Technician A nor Technician B
3. Technician A says if a vehicle emission label indicates a forty-nine-state emission designation, the vehicle will meet California emission standards. Technician B says the TLEV abbreviation stands for top-level emission venue. Who is correct?
   A. Technician A
   B. Technician B
   C. Both Technician A and Technician B
   D. Neither Technician A nor Technician B
4. Technician A says the fluid capacities are usually provided in the Lubrication and Maintenance section of a service manual. Technician B says technicians must be familiar with fluid capacities and fluid requirements for the vehicle they are servicing. Who is correct?
   A. Technician A
   B. Technician B
   C. Both Technician A and Technician B
   D. Neither Technician A nor Technician B
5. When discussing VIN interpretation, Technician A says the first digit in the VIN indicates the vehicle country of origin. Technician B says the fourth digit in the VIN indicates vehicle type. Who is correct?
   A. Technician A
   B. Technician B
   C. Both Technician A and Technician B
   D. Neither Technician A nor Technician B
6. All of these statements about vehicle fluid capacities are true *except*:
   A. Fluid capacities are usually located in the service manual at the beginning of the Engine section.
   B. Fluid capacities are provided in the owner's manual.
   C. Technicians must know the proper fluid capacities to avoid overfilling various components and systems.
   D. In some service manuals, the fluid capacities are located in the Maintenance and Lubrication section.

7. All of these statements about engine oil ratings are true *except*:
   A. The API oil grade rating indicates the type of service for which the oil is intended.
   B. The SAE viscosity oil rating indicates the atmospheric temperature for which the oil is suitable.
   C. Engine oils with different API or SAE viscosity ratings may be mixed.
   D. An engine oil chart in the service manual or owner's manual indicates the proper oil grade and viscosity rating at different temperatures.
8. An engine coolant solution containing 50 percent ethylene glycol and 50 percent water provides protection against freezing to:
   A. 0°F.
   B. −12°F.
   C. −20°F.
   D. −35°F.
9. Technician A says a green-colored antifreeze contains hybrid organic additive technology. Technician B says an antifreeze containing hybrid organic additive technology has special additives to prevent cooling system leaks. Who is correct?
   A. Technician A
   B. Technician B
   C. Both Technician A and Technician B
   D. Neither Technician A nor Technician B
10. According to vehicle manufacturers, severe vehicle operating conditions that require different maintenance schedules include all of these operating conditions *except*:
    A. trailer towing or driving continuously in mountainous road conditions.
    B. extensive idling or long periods of stop-and-go driving.
    C. driving continuously at low elevations of sea level to 1,000 feet.
    D. driving in extremely hot temperatures above 90°F.
11. The last six digits in a VIN are assigned to the vehicle by the ____________ ____________.
12. The ninth digit in the VIN is not made available to the ____________ ____________.
13. A coolant with HOAT is specified for engines with ____________ cylinder heads and blocks.
14. A coolant with HOAT additive is ____________ in color.
15. Explain where the specifications are usually located in a service manual.
16. Define severe operating conditions as they apply to maintenance schedules.
17. Explain the difference between normal maintenance schedules and severe operation maintenance schedules.
18. Describe the basic differences between generic service manuals and vehicle manufacturer's service manuals.

# Section 4

## Engine Principles and Systems



## SECTION OBJECTIVES

After you have read, studied, and practiced the contents of this section, you should be able to:

- Define the four-stroke cycle theory.
- Describe the different valve trains used in modern engines.
- Describe the function of the lubrication system.
- Describe the basic types of oil additives.
- Explain the purpose of the SAE classifications of oil.
- Explain the purpose of the API classifications of oil.
- Describe the purpose of the cooling system.
- Explain the operation of the thermostat.
- Describe the purpose of antifreeze.
- Describe the operation of an engine temperature warning light.
- Clean and inspect the radiator.
- Inspect and diagnose the viscous fan coupling.
- Test the operation of electric cooling fans.
- Explain the operation of an airflow restrictor in the air cleaner.
- Describe the purpose of the intake manifold.
- Describe the operation and advantages of intake manifolds with dual runners.
- List the main exhaust system components.
- Service and replace air filters.
- Use a vacuum gauge to diagnose engine problems, and test the intake system for vacuum leaks.
- Perform an exhaust system restriction test.
- Test catalytic converters.

*In-line six-cylinder engines have not been widely used for the last 15 to 20 years because of the height of the engine and the difficulty of installing this engine in today's smaller, more fuel-efficient vehicles. In the last 2 years, the in-line six-cylinder engine has been reintroduced to the marketplace in GM Envoy and Chevrolet Trail Blazer four-wheel drive (4WD) sport utility vehicles (SUVs). These engines have a unique lower mounting in the engine compartment and a special oil pan that allows one of the front drive axles to pass through an opening in this pan.*

# Chapter 10 The Four-Stroke Cycle and Cylinder Arrangements

## Introduction

One of the many laws of physics utilized within the automotive engine is thermodynamics. **Thermodynamics** is the relationship between heat energy and mechanical energy.

The driving force of the engine is the expansion of gases. Gasoline (a liquid fuel) will change states to a gas if it is heated or burned. Gasoline must be mixed with oxygen before it can burn. In addition, the air-fuel mixture must be burned in a confined area in order to produce power. Gasoline that is burned in an open container produces very little power, but if the same amount of fuel is burned in an enclosed container, it will expand with force. When the air-fuel mixture burns, it also expands as the molecules of the gas collide with each other and bounce apart. Increasing the temperature of the gasoline molecules increases their speed of travel, causing more collisions and expansion.

Heat is generated by compressing the air-fuel mixture in the combustion chamber. Igniting the compressed mixture causes the heat, pressure, and expansion to multiply. This process releases the energy of the gasoline so it can produce work. The igniting of the mixture is a controlled burn, not an explosion. The controlled combustion releases the fuel energy at a controlled rate in the form of heat energy. The heat, and consequential expansion of molecules, increases the pressure inside the combustion chamber. Typically, the pressure works on top of a piston that is connected by a connecting rod to the crankshaft. As the piston is driven, it causes the crankshaft to rotate.

The engine produces torque, which is applied to the drive wheels. As the engine drives the wheels to move the vehicle, a certain amount of work is done. The rate of work being performed is measured in horsepower. **Torque** is a rotating force around a pivot point.

### ENGINE CYCLES

This section will define many of the terms used by automotive manufacturers to classify their engines. A stroke is the amount of piston travel from TDC to BDC measured in inches or millimeters.

For example, if the piston is at the top of its travel and is then moved to the bottom of its travel, one stroke has occurred. Another stroke occurs when the piston is moved from the bottom of its travel to the top again. A cycle is a sequence that is repeated. In the four-stroke engine, four strokes are required to complete one cycle.

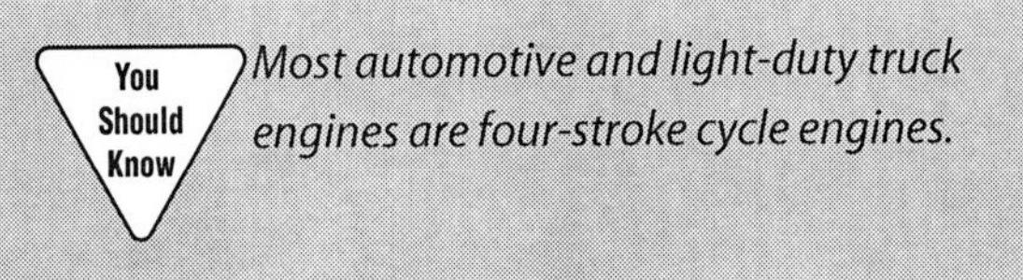


*Most automotive and light-duty truck engines are four-stroke cycle engines.*

The internal combustion engine must draw in an air-fuel mixture, compress the mixture, ignite the mixture, and then expel the exhaust. This is accomplished in four piston strokes **(Figure 1)**. The process of drawing in the air-fuel mixture is actually accomplished by atmospheric pressure pushing it into a low-pressure area created by the downward movement of the piston.

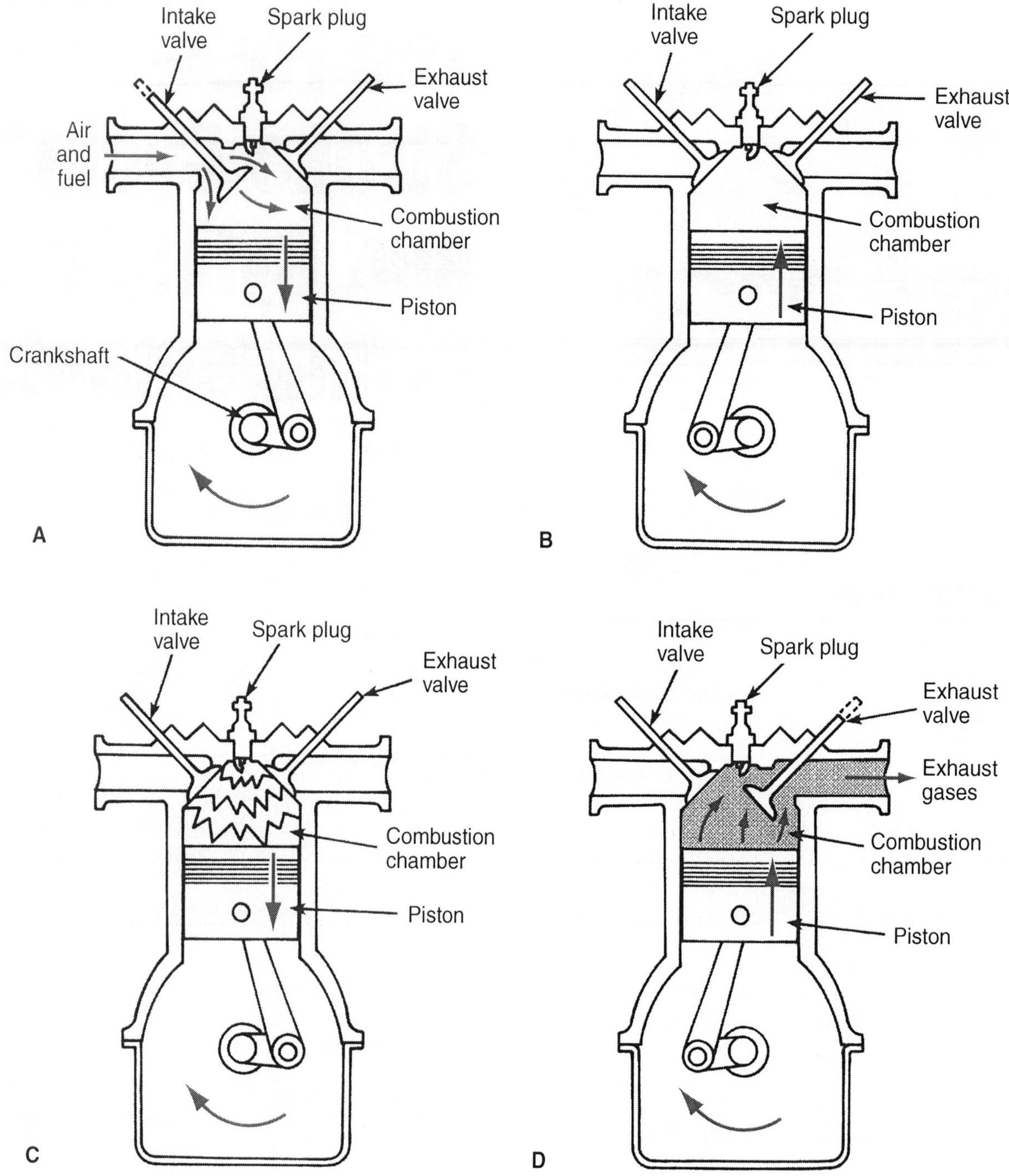


**Figure 1.** (A) Intake stroke. (B) Compression stroke. (C) Power stroke. (D) Exhaust stroke.

## THE INTAKE STROKE

The first stroke of the cycle is the intake stroke **(Figure 1A )**. As the piston moves down from TDC, the intake valve is opened so the vaporized air-fuel mixture can be pushed into the cylinder by atmospheric pressure. At **top dead center (TDC)**, the piston is at the very top of its stroke.

As the piston moves downward in its stroke, a vacuum is created (low pressure). Since high pressure moves toward a low pressure, the air-fuel mixture is pushed past the open intake valve and into the cylinder. After the piston reaches bottom dead center (BDC), the intake valve is closed and the stroke is completed. At **bottom dead center (BDC)**, the piston is at the very bottom of its stroke.

Closing the intake valve after BDC allows an additional amount of air-fuel mixture to enter the cylinder, increasing the volumetric efficiency of the engine. Even though the piston is at the end of its stroke and no more vacuum is created, the additional mixture enters the cylinder since it weighs more than air alone. **Volumetric efficiency** is the actual amount of air-fuel mixture entering the cylinders compared to the amount of air-fuel mixture that could enter the cylinders under ideal conditions.

## THE COMPRESSION STROKE

The compression stroke begins as the piston starts its travel back to TDC **(Figure 1B)**. The intake and exhaust valves are both closed, trapping the air-fuel mixture in the combustion chamber. The movement of the piston toward TDC compresses the mixture. As the molecules of the mixture are pressed tightly together, they begin to heat. When the piston reaches TDC, the mixture is fully compressed and a spark is induced in the cylinder by the ignition system. Compressing the mixture provides better burning and intense combustion.

**You Should Know** *Combustion chamber sealing is very important. If pressure can leak past the rings or valves during the compression stroke, cylinder pressure and engine power are reduced.*

## THE POWER STROKE

When the spark occurs at the spark plug electrodes in the compressed mixture, the rapid burning causes the molecules to expand, beginning the power stroke **(Figure 1C)**. The expanding molecules create a pressure above the piston and push it downward. The downward movement of the piston in this stroke is the only time the engine is productive concerning power output. During the power stroke, the intake and exhaust valves remain closed.

**You Should Know** *The effective power stroke is the shortest stroke in the four-stroke cycle. The extreme pressure created by combustion in the cylinder only lasts for approximately 25 degrees of crankshaft rotation. Therefore, timing of the combustion event is very important. The spark at the plug electrodes must occur at just the right instant in relation to piston position so the combustion of the air-fuel mixture creates maximum downward force on the piston.*

*An engine that uses a spark at the spark plug electrodes to ignite the air-fuel mixture may be referred to as an SI engine.*

## THE EXHAUST STROKE

The exhaust stroke of the cycle begins when the piston reaches BDC of the power stroke **(Figure 1D)**. Just prior to the piston reaching BDC, the exhaust valve is opened. The upward movement of the piston back toward TDC pushes out the exhaust gases from the cylinder past the exhaust valve and into the vehicle's exhaust system. As the piston approaches TDC, the intake valve opens and the exhaust valve is closed a few degrees after TDC. The degrees of crankshaft rotation when both the intake and exhaust valves are open is called **valve overlap**. During valve overlap, the incoming air-fuel mixture through the intake valve helps to purge any remaining exhaust gases from the cylinder. The cycle is then repeated again as the piston begins the intake stroke.

**You Should Know** *The duration of valve overlap is very important to achieve maximum engine power, performance, and smooth engine operation. During valve overlap, the remaining burned exhaust gases are purged out of the cylinder and fresh air-fuel mixture must be swept into the combustion chamber.*

Most engines in use today are referred to as **reciprocating**. Power is produced by the up-and-down movement of the piston in the cylinder. This linear motion is then converted to rotary motion by a crankshaft.

## DIESEL ENGINE PRINCIPLES

Diesel and gasoline engines have many similar components. However, the diesel engine does not use an ignition system consisting of spark plugs and coils. Instead of using a spark delivered by the ignition system, the diesel engine uses the heat produced by compressing air in the combustion chamber to ignite the fuel. Fuel injectors are used to supply fuel into the combustion chamber. The fuel is sprayed under pressure from the injector as the piston is completing its compression stroke. The temperature increase generated by compressing the air (approximately

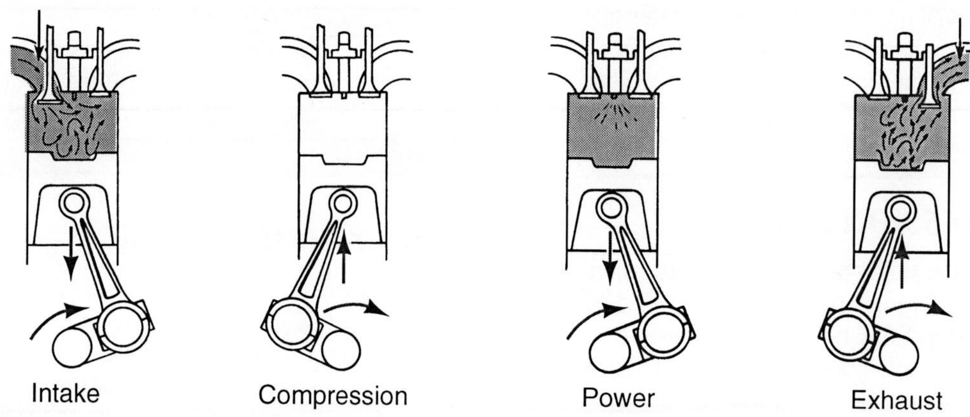


**Figure 2.** The four strokes of a diesel engine.

1,000°F) is sufficient to ignite the fuel as it is injected into the cylinder. This begins the power stroke **(Figure 2)**.

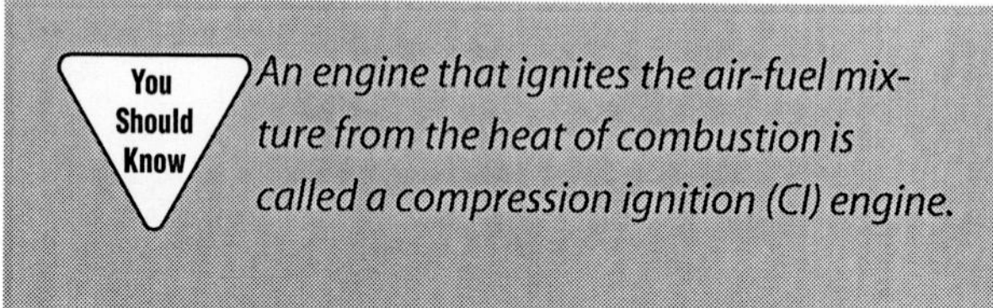


*An engine that ignites the air-fuel mixture from the heat of combustion is called a compression ignition (CI) engine.*

Since starting the diesel engine is dependent upon heating the intake air to a high enough level to ignite the fuel, a method of preheating the intake air is required to start a cold engine. Some manufacturers use glow plugs to accomplish this. Another method includes using a heater grid in the air intake system. **Glow plugs** are small, round, electrical devices positioned in a precombustion chamber. When the engine is cold, voltage is supplied to the glow plugs and the resulting current flow heats the glow plugs which heat the precombustion chambers.

In addition to ignition methods, there are other differences between the gasoline and diesel engines **(Figure 3)**. The combustion chambers of the diesel engine are designed to accommodate the different burning characteristics of diesel fuel.

| | Gasoline | Diesel |
|---|---|---|
| Intake | Air/fuel | Air |
| Compression | 8–10 to 1<br>130 psi<br>545°F | 13–25 to 1<br>400–600 psi<br>1,000°F |
| Air/fuel mixing point | Carburetor or before intake valve with fuel injection | Near TDC by injection |
| Combustion | Spark ignition | Compression ignition |
| Power | 464 psi | 1,200 psi |
| Exhaust | 1,300°–1,800°F<br>CO = 3% | 700°–900°F<br>CO = 0.5% |
| Efficiency | 22–28% | 32–38% |

**Figure 3.** Comparisons between gasoline and diesel engines.

There are three common combustion chamber designs:

- An open-type combustion chamber is located directly inside of the piston. The fuel is injected into the center of the chamber. Turbulence is produced by the shape of the chamber.
- A precombustion chamber is a smaller, second chamber connected to the main combustion chamber. Fuel is injected into the precombustion chamber where the combustion process is started. Combustion then spreads to the main chamber. Since the fuel is not injected directly on top of the piston, compression pressures do not need to be overcome by the injection pump and less fuel pressure is required.
- A turbulence combustion chamber is a chamber designed to create turbulence as the piston compresses the air. When the fuel is injected into the turbulent air, a more efficient burn is achieved.

Diesel engines can be either four-stroke or two-stroke designs. Two-stroke engines complete all cycles in two strokes of the piston, much like a gasoline two-stroke engine **(Figure 4)**.

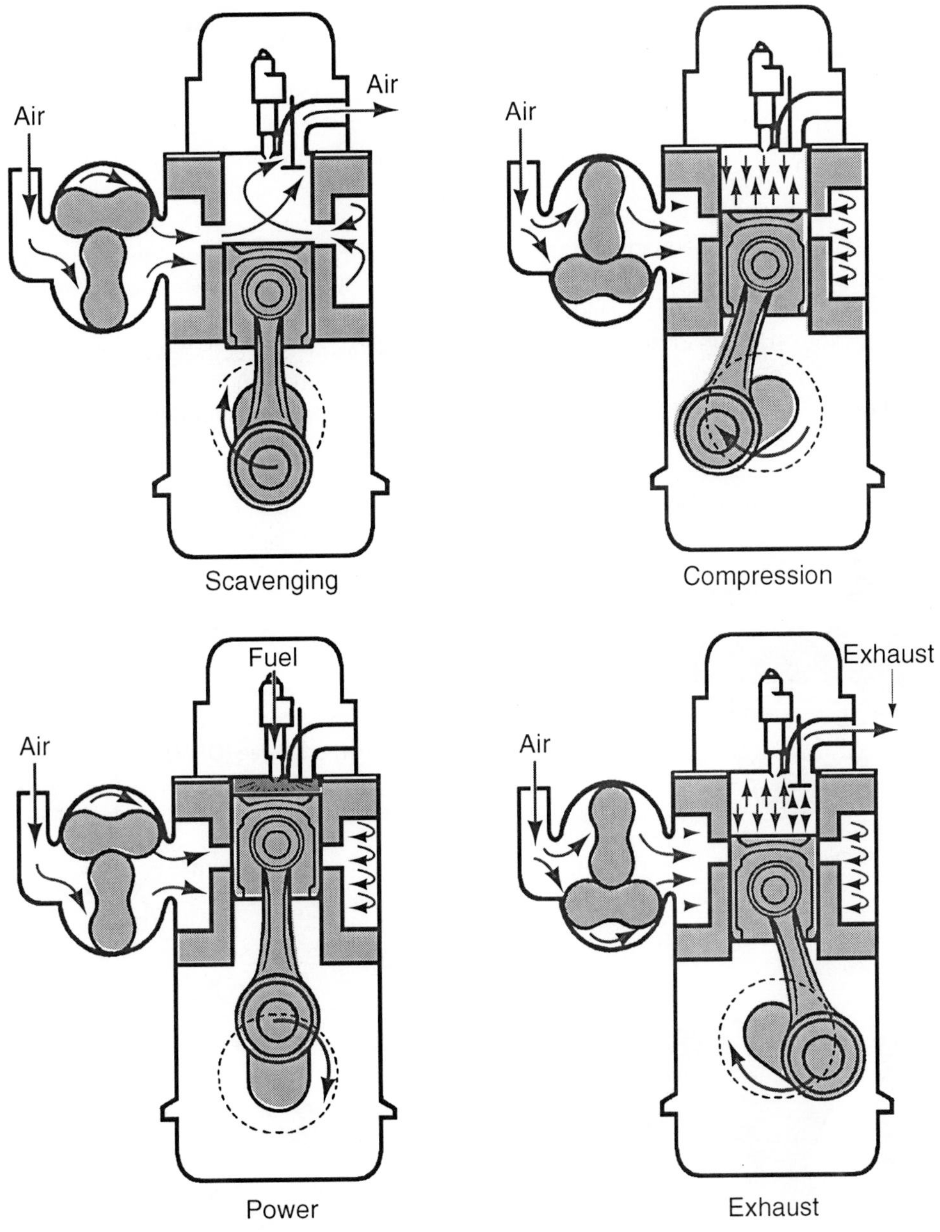


**Figure 4.** Some diesel engines are two-stroke cycle.

## CYLINDER ARRANGEMENTS

One cylinder would not be able to produce sufficient power to meet the demands of today's vehicles. Most automotive and truck engines use three, four, five, six, eight, ten, or twelve cylinders. The number of cylinders used by the manufacturer is determined by the amount of work required from the engine.

Vehicle manufactures attempt to achieve a balance between power, economy, weight, and operating characteristics. An engine having more cylinders generally runs smoother than those having three or four cylinders because there is less crankshaft rotation between power strokes. However, adding more cylinders increases the weight of the vehicle and the cost of production.

Engines are also classified by the arrangement of the cylinders. The cylinder arrangement used is determined by vehicle design and purpose. The most common engine designs are in-line and V-type.

The in-line engine places all of its cylinders in a single row **(Figure 5)**. Advantages of this engine design include ease of manufacturing and serviceability. The disadvantage of this engine design is the block height. Since the engine is tall, aerodynamic design of the vehicle is harder to achieve. Most manufacturers overcome this disadvantage by mounting the engine transversely in the engine compartment of front-wheel-drive vehicles. A **transversely** mounted engine is mounted sideways on the vehicle.

The V-type engine has two rows of cylinders set 60 degrees to 90 degrees from each other **(Figure 6)**. The V-type design allows for a shorter block height and improved vehicle aerodynamics. The length of the block is also shorter than in-line engines with the same number of cylinders.

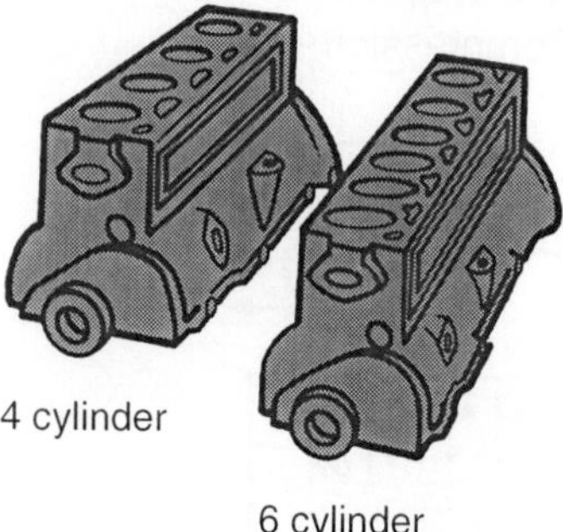


Figure 5. In-line engines.

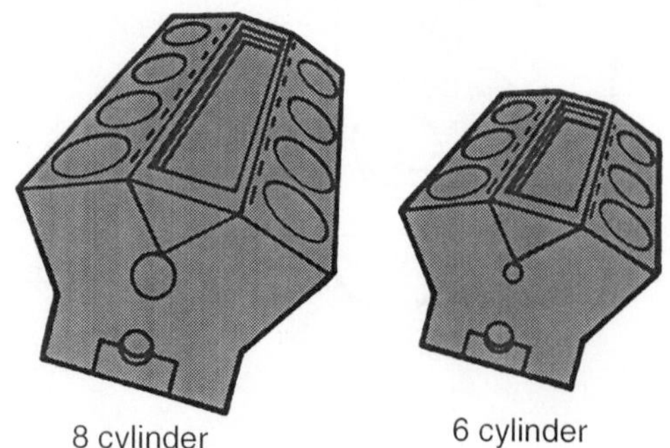


Figure 6. V-type engine design.

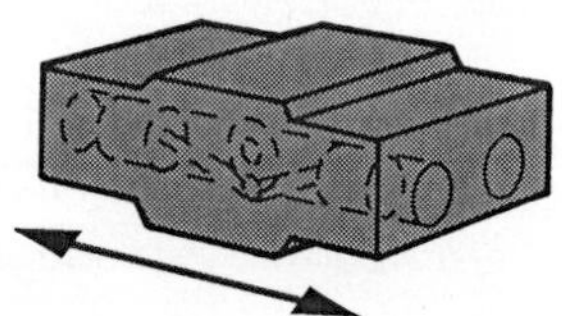

Figure 7. Opposed engine design.

Figure 8. A slant-type engine.

A common engine design used for rear-engine vehicles as well as the front-wheel-drive Subaru, is the opposed cylinder engine **(Figure 7)**. This engine design has two rows of cylinders directly across from each other. The main advantage of this engine design is the very small vertical height. An **opposed cylinder engine** has the piston banks mounted at 180 degrees in relation to each other.

Another engine design is the slant cylinder **(Figure 8)**. This design is similar to in-line engines except the cylinders are placed at a slant. This design reduces the height of the engine and allows for more aerodynamic vehicle designs.

## VALVE ARRANGEMENTS

Engines can also be classified by their valve train types. The three most commonly used valve trains are the:

**Overhead valve (OHV).** The intake and exhaust valves are located in the cylinder head while the camshaft and lifters are located in the engine block **(Figure 9)**. Valve train

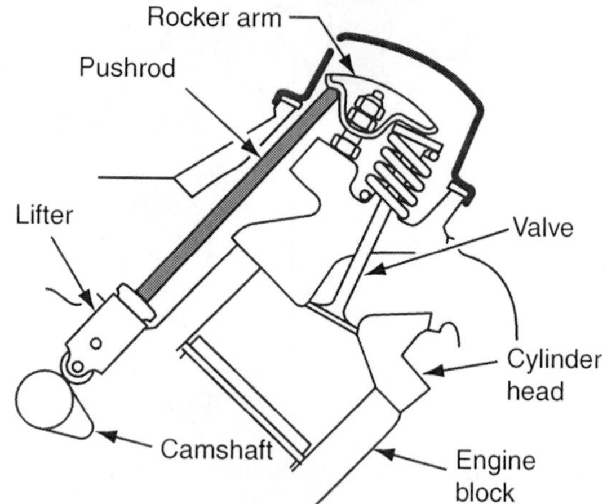


Figure 9. An overhead valve engine.

components of this design include lifters, pushrods, and rocker arms.

**Overhead cam (OHC).** The intake and exhaust valves are located in the cylinder head along with the camshaft **(Figure 10)**. The valves are operated directly by the camshaft and a follower, thereby eliminating many of the moving parts required in the OHV engine. If a single camshaft is used in each cylinder head, the engine is classified as a single overhead cam (SOHC) engine. This designation is used even if the engine has two cylinder heads with one camshaft each.

**Dual overhead cam (DOHC).** The DOHC uses separate camshafts for the intake and exhaust valves. A DOHC V-8 engine is equipped with a total of four camshafts **(Figure 11)**.

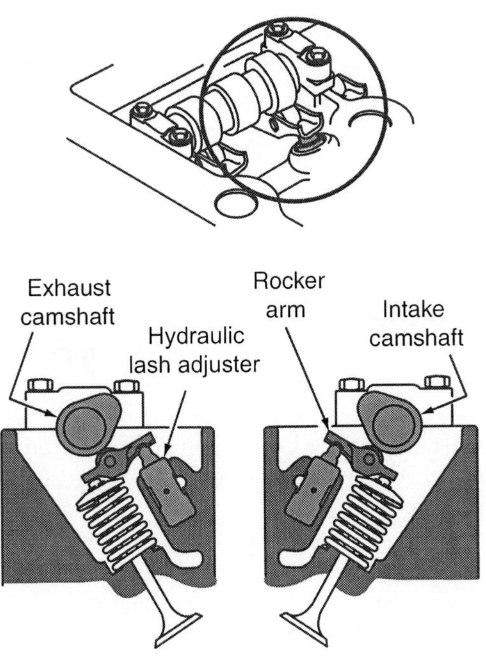


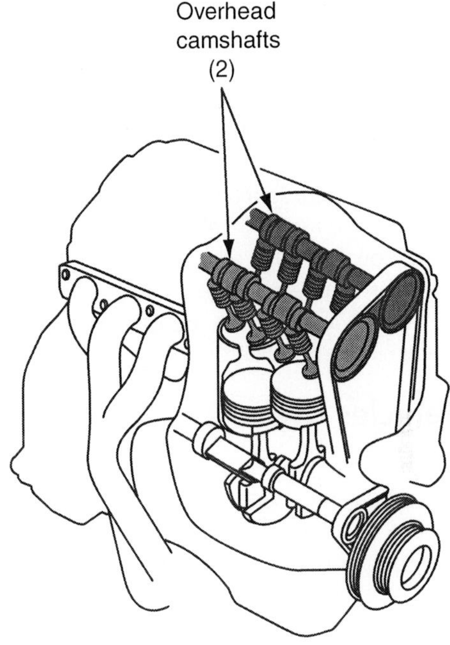


Figure 10. An overhead cam engine.

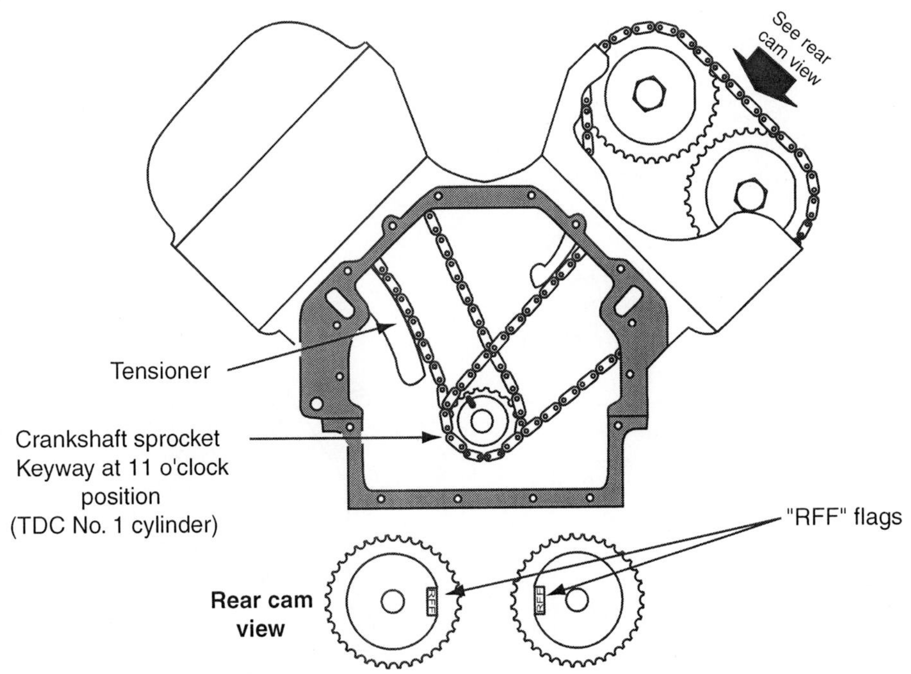


Figure 11. A dual overhead cam engine.

## VEHICLES WITH ALTERNATE POWER SOURCES

Most vehicle manufacturers in the world are working on the development of hybrid vehicles and fuel cell powered vehicles. Some hybrid vehicles are available in North America. **Hybrid vehicles** operate on two power sources such as a gasoline engine and an electric motor.

Fuel cell vehicles are driven electrically and the fuel cells produce electricity for the electric drive motor(s). Fuel cells use hydrogen and oxygen to produce electricity, so these vehicles require hydrogen fuel. Fuel cell vehicles are zero-emission vehicles. A significant number of fuel cell vehicles have been produced as concept cars by several vehicle manufacturers. Vehicle manufacturers indicate they will be marketing more fuel cell vehicles to selected fleet owners between now and 2010, and after this date fuel cell vehicles should be available to the general public. However, the first fuel cell vehicles will likely be marketed in areas that do not meet clean air standards.

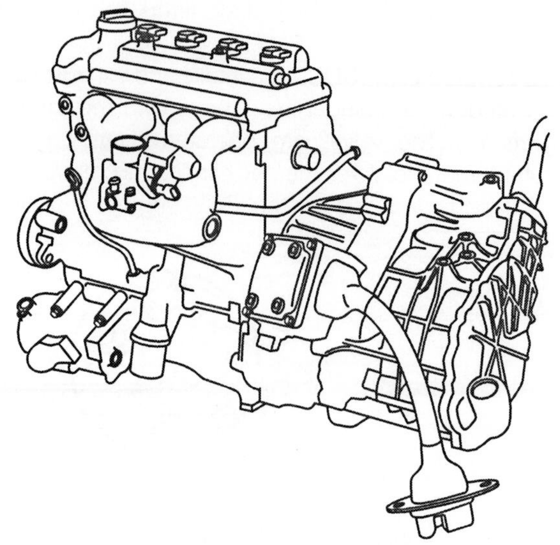

**Figure 13.** A hybrid power system with a gasoline engine and an electric propulsion motor.

### Hybrid Vehicles

In this section our objective is to provide a brief description of several alternate vehicle power sources. These power sources are being sold in very limited numbers or they are still in the research and development stage. In the 1990s, most vehicle manufacturers began developing electric vehicles, and these developments continue in the twenty-first century. The main reasons for this action is more stringent emission standards and the necessity of reducing our dependence on imported crude oil. As emission standards become more stringent, it becomes increasingly difficult and expensive to meet these standards with gasoline and diesel engines. The California Air Resources Board (CARB) established a low-emission vehicles/clean fuel program to further reduce mobile source emissions in California during the late 1990s. In this program, emission standards are established for five vehicle types: conventional vehicle (CV), transitional low-emission vehicle (TLEV), low-emission vehicle (LEV), ultra-low-emission vehicle (ULEV), and zero-emission vehicle (ZEV) **(Figure 12)**. The CARB has developed a sales-weighting and emissions credit system for introducing the TLEV, LEV, ULEV, and ZEV vehicles into the California market. With present technology, ZEV standards cannot be achieved with a gasoline-powered vehicle.

| | CV | TLEV | LEV | ULEV | ZEV |
|---|---|---|---|---|---|
| NMOG | 0.25 | 0.125 | 0.075 | 0.040 | 0.0 |
| CO | 3.4 | 3.4 | 3.4 | 1.7 | 0.0 |
| NOx | 0.4 | 0.4 | 0.2 | 0.2 | 0.0 |

**Figure 12.** California tailpipe emission standards in g/mile for passenger cars at 50,000 miles.

A hybrid vehicle has two different power sources. In most hybrid vehicles, the power sources are a small displacement gasoline or diesel engine and an electric motor. The Toyota Prius is a hybrid vehicle with a 1.5L DOHC, gasoline direct injection four-cylinder engine **(Figure 13)**. In a direct injection engine, the injectors deliver the gasoline directly into the combustion chamber. This type of injection system must operate at higher pressure so the fuel can be discharged out of the injectors into the high compression pressure in the combustion chambers. The engine is mounted transversely under the hood. The nickel/metal/hydride battery pack is installed behind the rear seat leaving adequate storage room in the trunk **(Figure 14)**.

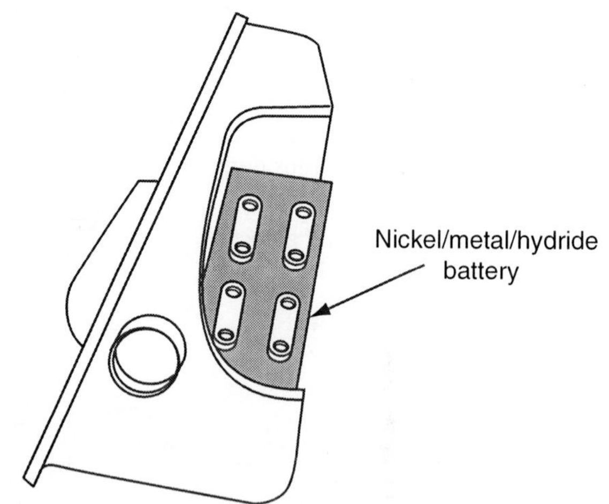


**Figure 14.** A nickel/metal hydride battery pack.

The generator/starter, power split device, propulsion motor/regenerator, and sprocket chain to the final drive are mounted in an aluminum case that is bolted to the back of the engine. This case is about the same length as a conventional four-speed automatic transaxle **(Figure 15)**. An automatic controller operates the system and determines whether power to the drive wheels is supplied from the electric motor, the gasoline engine, or both. Charging the batteries with an external charger is not necessary because the batteries are charged while the engine is running. The system operation may be summarized as follows:

1. During startup, low speed, and low speed deceleration, only the electric propulsion motor supplies power to the drive wheels.
2. During the normal speed range, power from the engine is divided by the power split device so that some of the engine power is supplied to the drive wheels and engine power is also supplied to the generator. The electric power supplied from the generator drives the electric propulsion motor, and power from this motor is also supplied to the drive wheels.
3. Full-throttle operation is the same as normal-speed driving operation except the battery also supplies power to the propulsion motor to maximize motor output to the drive wheels.
4. During deceleration, the wheels drive the propulsion motor, and this motor supplies power to recharge the batteries.
5. The battery state of charge is continually monitored by the controller, and power is supplied from the generator to recharge the battery whenever this action is required.
6. When the vehicle is brought to a stop, the engine is shut off.

A specially designed compact sedan is being developed for the hybrid drive system. The hybrid system is designed to meet ULEV emission standards and improve fuel economy. The Honda Insight is another hybrid car that is available.

## Fuel Cell-Powered Vehicles

A fuel cell-powered vehicle is an electric vehicle with some important differences. An electric motor supplies torque to the drive wheels, but the fuel cell produces and supplies electric power to the electric motor whereas in an electric vehicle this power is supplied by the batteries. Most of the vehicle manufacturers, in cooperation with some independent laboratories, are involved in fuel cell research and development programs. A number of prototype fuel cell vehicles have been produced. Several vehicle manufacturers are committed to having a fuel cell vehicle in very limited numbers by the year 2004.

Fuel cells electrochemically combine oxygen from the air with hydrogen from a hydrocarbon fuel to produce electricity. The oxygen and hydrogen are fed to the fuel cell as "fuel" for the electrochemical reaction. There are different types of fuel cells, but the most common type is the proton exchange membrane (PEM) fuel cell. Each individual fuel cell contains two electrodes separated by a membrane. Electrical power output from one individual fuel cell is very low, and so many fuel cells are contained in a fuel cell stack to supply enough electric energy to the electric drive motor **(Figure 16)**. Hydrogen is supplied to one of the elec-

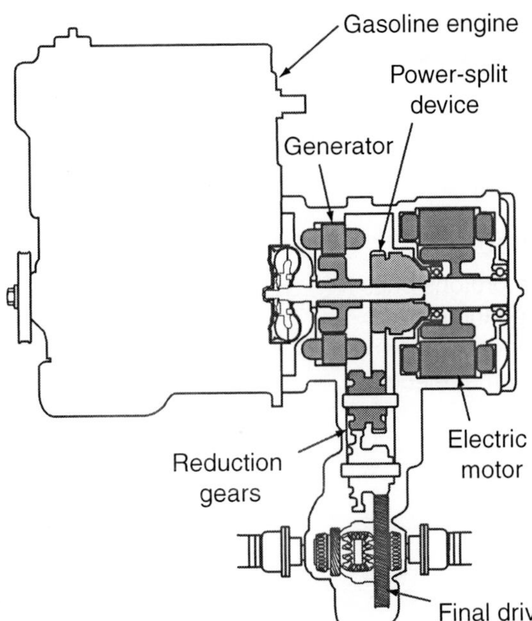


**Figure 15.** Hybrid power system internal components.

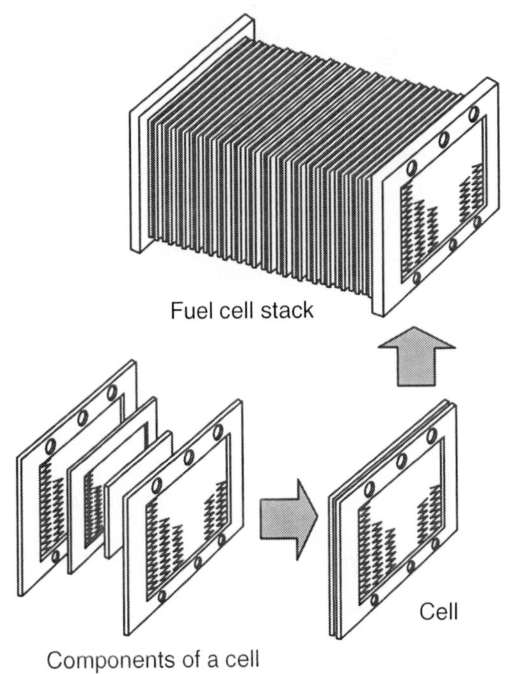


**Figure 16.** A fuel cell stack.

trodes in each fuel cell. This electrode is coated with a catalyst that separates the electrons and protons in the hydrogen. The movement of electrons supplies electrical power to the drive motor. The protons move through the proton exchange membrane to the other electrode. Oxygen from the air is supplied to this electrode, and oxygen from the air combines with the hydrogen protons to form water vapor, which is emitted from the vehicle **(Figure 17)**. The only emission from a fuel cell is water vapor ($H_2O$).

One of the major areas of research and development is the source of hydrogen fuel in fuel cell vehicles. One solution is to store hydrogen onboard the vehicle. This concept has several complications. Using hydrogen fuel would require a whole new refueling system in all areas of the country, which is very expensive. Gaseous hydrogen can be stored in large cylinders containing a hydride material something like steel wool. These cylinders would require a large storage space on the vehicle. Liquid hydrogen must be refrigerated to approximately –400°F to keep it in liquid form. Storing liquid hydrogen on a vehicle involves the use of a special double-walled insulated fuel tank. Storing liquid hydrogen on a vehicle also involves some safety concerns because as the fuel tank warms up, the pressure increases, and this may activate the pressure relief valve. This action discharges flammable hydrogen into the atmosphere creating a source of danger and pollution.

An onboard reformer may be used to extract hydrogen from liquid fuels, such as gasoline or methanol. The main disadvantage to this system is the space required by the reformer. Using methanol would also require a new refueling system across the country. If a fuel cell-powered vehicle is fueled directly with hydrogen, the exhaust emissions are almost zero. A fuel cell-powered vehicle with an onboard reformer to extract hydrogen from some other fuel does create a very small amount of tailpipe emissions because of the reformer action. Recent developments include a multireformer that operates on different types of fuels. Hydrogen can be obtained by electrolysis from water. However, this process requires a great deal of electrical energy. With present technology, this method of obtaining hydrogen is not an option. It appears that the first fuel cell vehicles offered to customers will have onboard reformers that extract hydrogen from some well-known fuel. The two major obstacles to be overcome in the development of fuel cell vehicles are cost and the onboard space required by the system components. Research and development is taking place at a rapid pace, and components are quickly being downsized and improved. As production of fuel cell vehicle components increases, the price will be reduced.

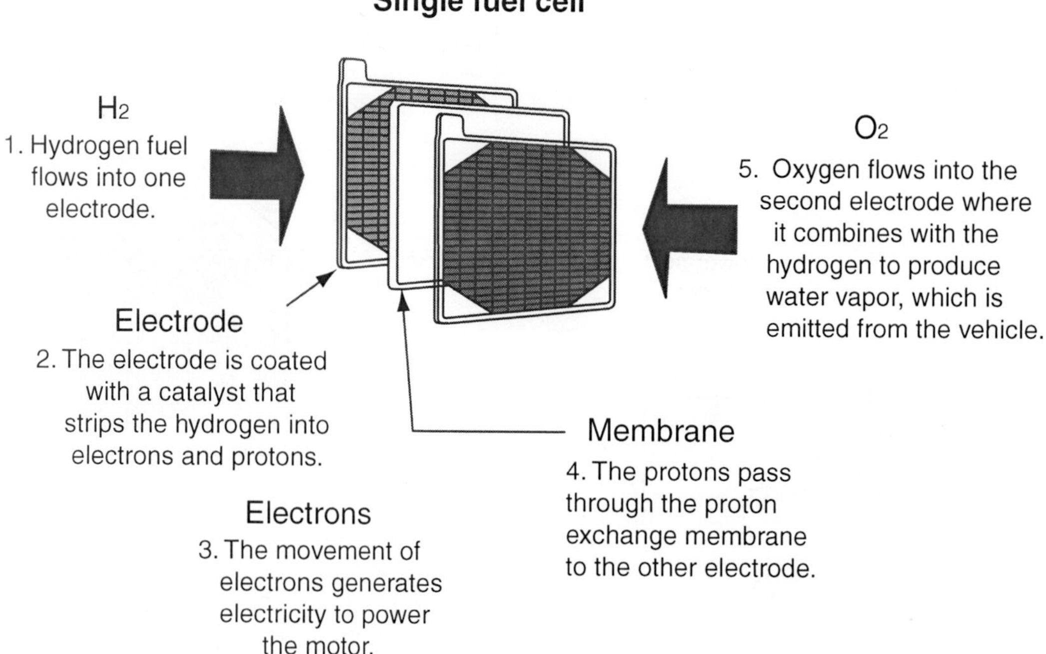
Single fuel cell
H2
1. Hydrogen fuel flows into one electrode.
Electrode
2. The electrode is coated with a catalyst that strips the hydrogen into electrons and protons.
Electrons
3. The movement of electrons generates electricity to power the motor.
Membrane
4. The protons pass through the proton exchange membrane to the other electrode.
O2
5. Oxygen flows into the second electrode where it combines with the hydrogen to produce water vapor, which is emitted from the vehicle.

**Figure 17.** Fuel cell operation.

## Summary


- Most automotive engines operate on the four-stroke cycle principle.
- A stroke is the piston movement from TDC to BDC.
- During the intake stroke, the piston is moving downward, the intake valve is open, and the exhaust valve is closed.
- During the compression stroke, the piston is moving upward and both valves are closed.
- During the power stroke, the piston is forced downward by combustion pressure and both valves remain closed. The exhaust valve opens near BDC on the power stroke.
- During the exhaust stroke, the piston is moving upward, the exhaust valve is open, and the intake valve is closed. The intake valve opens a few degrees before TDC on the exhaust stroke, and the exhaust valve closes a few degrees after TDC on the exhaust stroke.
- Valve overlap is the number of degrees that the crankshaft rotates when both valves are open at the same time and the piston is near TDC on the exhaust stroke.
- Three of the most common valve train designs are overhead valve (OHV), overhead cam (OHC), and dual-overhead cam (DOHC).
- Hybrid vehicles are powered by two different power sources, such as a gasoline engine and an electric propulsion motor.
- A fuel cell produces electricity when hydrogen and oxygen are supplied to the cell.


## Review Questions

1. Technician A says during the intake stroke, the intake and exhaust valves are open. Technician B says the spark plug fires when the piston is at BDC on the intake stroke. Who is correct?
   A. Technician A
   B. Technician B
   C. Both Technician A and Technician B
   D. Neither Technician A nor Technician B
2. Technician A says a stroke occurs when a piston moves from TDC to BDC. Technician B says the air is moved into the cylinder on the intake stroke by vacuum in the cylinder and atmospheric pressure. Who is correct?
   A. Technician A
   B. Technician B
   C. Both Technician A and Technician B
   D. Neither Technician A nor Technician B
3. Technician A says during the compression stroke, the intake and exhaust valves are closed. Technician B says during the compression stroke, the piston is moving upward in the cylinder. Who is correct?
   A. Technician A
   B. Technician B
   C. Both Technician A and Technician B
   D. Neither Technician A nor Technician B
4. Technician A says the intake valve opens when the piston is a few degrees before TDC on the exhaust stroke. Technician B says the exhaust valve closes when the piston is a few degrees after TDC on the exhaust stroke. Who is correct?
   A. Technician A
   B. Technician B
   C. Both Technician A and Technician B
   D. Neither Technician A nor Technician B
5. All of these statements about the power stroke in a four-cycle engine are true *except*:
   A. The power stroke is the shortest stroke in the four-stroke cycle.
   B. The extreme pressure created by combustion in the cylinder lasts for about 120 degrees of crankshaft rotation.
   C. During the power stroke, the expanding molecules in the combustion chamber force the piston downward.
   D. Leaking piston rings decrease combustion chamber pressure and engine power.
6. In a four-stroke cycle engine during valve overlap:
   A. the intake valve is closing.
   B. the exhaust valve is beginning to open.
   C. both valves are open at the same time.
   D. the piston is at TDC on the power stroke.

7. All of these statements about diesel engine principles are true *except*:
   A. The heat developed by compressing the air in the cylinder ignites the air-fuel mixture.
   B. Compared to a gasoline engine, the diesel engine has a lower compression ratio.
   C. Fuel injectors spray fuel into the combustion chamber or precombustion chamber.
   D. Glow plugs supply additional combustion chamber heat when starting a cold engine.
8. While discussing engine design, Technician A says a V-type engine has a 60 degree or a 90 degree angle between the two cylinder banks. Technician B says in an opposed cylinder engine has a 180 degree angle between the two cylinder banks. Who is correct?
   A. Technician A
   B. Technician B
   C. Both Technician A and Technician B
   D. Neither Technician A nor Technician B
9. While discussing engine design, Technician A says in an OHV engine, the camshaft is mounted on top of the cylinder head. Technician B says in an OHV engine, the valve lifters are mounted in the cylinder head. Who is correct?
   A. Technician A
   B. Technician B
   C. Both Technician A and Technician B
   D. Neither Technician A nor Technician B
10. All of these statements about hybrid vehicles are true *except*:
   A. A hybrid vehicle may be powered by a gasoline engine and an electric motor.
   B. A hybrid vehicle may be powered by a diesel engine and an electric motor.
   C. A computer determines which power source is connected to the drive wheels.
   D. Some current hybrid vehicles are powered by a gasoline engine and a diesel engine.
11. During valve overlap, both valves in a cylinder are ____________________.
12. A transverse-mounted engine is positioned ____________________ in the engine compartment.
13. In an OHC engine, the camshaft(s) are positioned ________________ the ____________________ ________________.
14. In a compression ignition engine, the air-fuel mixture is ignited by the ________________ of ________________.
15. Explain how an internal combustion engine produces power and torque.
16. Describe the piston and valve movement during the intake stroke.
17. Describe the piston and valve position during the power stroke.
18. Explain the purpose of valve overlap.

# Chapter 11 Engine Oil and Lubrication Systems

## Introduction

When the engine is operating, the moving parts generate heat due to friction. If this heat and friction were not controlled, the components would weld together. In addition, heat is created from the combustion process. It is the function of the engine's lubrication system to supply oil to the high friction and wear locations in the engine and to dissipate heat away from them **(Figure 1)**.

It is impossible to eliminate all of the friction within an engine, but a properly operating lubrication system helps

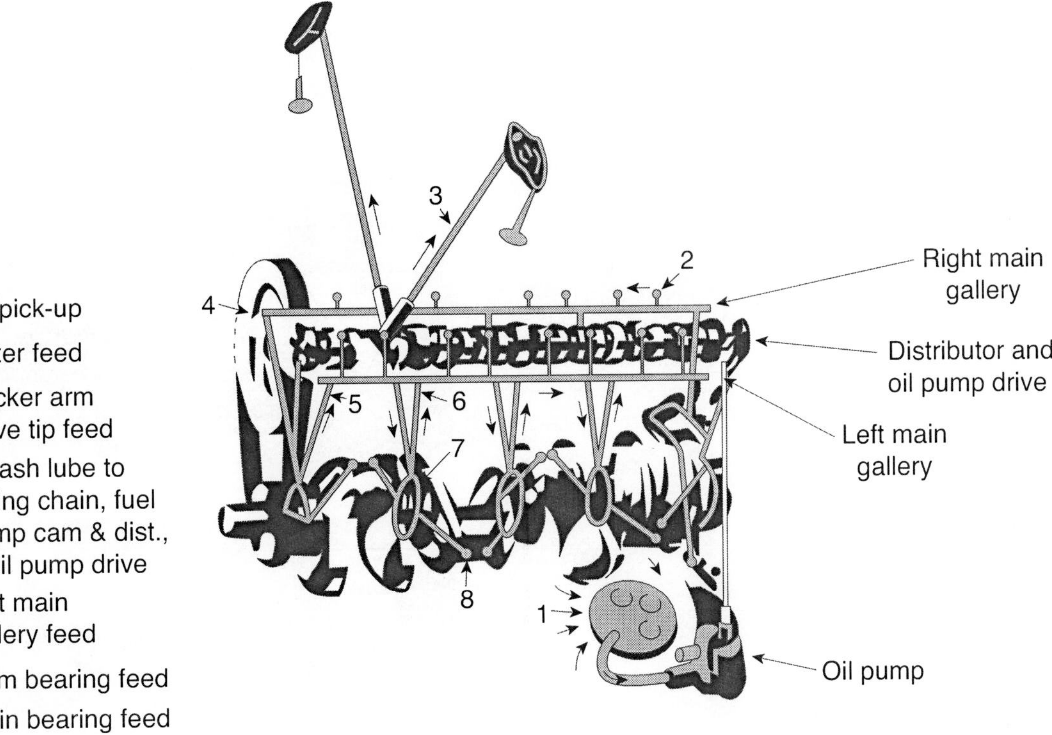


**Figure 1.** Oil flow through the engine lubrication system.

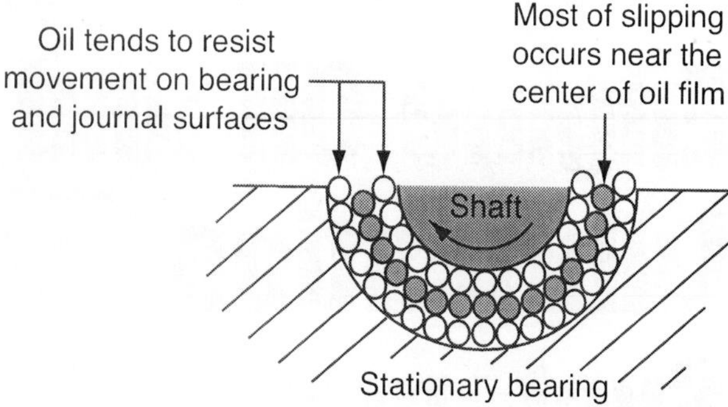


**Figure 2.** The oil film prevents metal-to-metal contact between engine compartments.

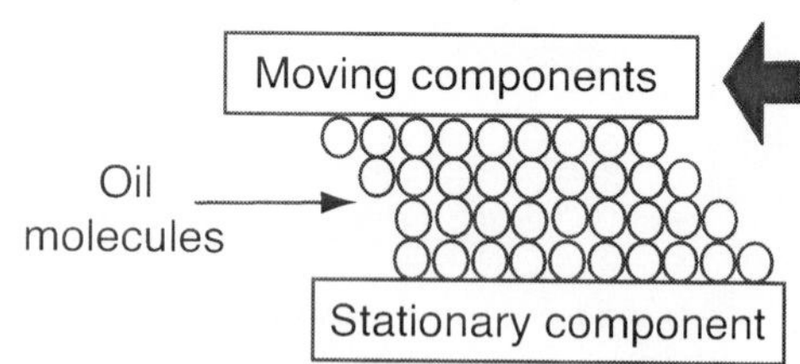


**Figure 3.** Oil molecules work as small bearings to reduce friction.

to reduce it. Lubrication systems provide an oil film to prevent moving parts from coming in direct contact with each other **(Figure 2)**. Oil molecules work as small bearings rolling over each other to reduce friction **(Figure 3)**. Another function of the lubrication system is to act as a shock absorber between the connecting rod and crankshaft. The purpose of engine oil is to lubricate, clean, seal, and cool.

Besides providing friction reduction, engine oil absorbs heat and transfers it to another area for cooling. As the oil flows through the engine, it conducts heat from the parts it comes in contact with. When the oil is returned to the oil pan, it is cooled by airflow over the pan. Other purposes of the oil include sealing the piston rings and washing away abrasive metal and dirt. To perform these functions, the engine lubrication system includes the following components:

- Engine oil
- Oil pan or sump
- Oil filter
- Oil pump
- Oil galleries

## ENGINE OIL RATING AND CLASSIFICATION

Engine oil must provide a variety of functions under all of the extreme engine operating conditions. To perform these tasks, additives are mixed with natural oil. A brief description of these additives follows:

- Antifoaming agents—These additives are included to prevent aeration of the oil. Aeration will result in low oil pump pressure and insufficient lubrication to the engine parts.
- Antioxidation agents—Heat and oil agitation result in oxidation. These additives work to prevent the buildup of varnish and to prevent the oil from breaking down into harmful substances that can damage engine bearings. **Oxidation** occurs when some of the oil combines with oxygen in the air to form an undesirable compound.
- Detergents and dispersants—These are added to prevent deposit buildup resulting from carbon, metal particles, and dirt. Detergents break up larger deposits and prevent smaller ones from grouping together. A dispersant is added to prevent carbon particles from grouping together.
- Viscosity index improver—As the oil increases in temperature it has a tendency to thin out. These additives prevent oil thinning.
- Pour point depressants—Prevent oil from becoming too thick to pour at low temperatures.
- Corrosion and rust inhibitors—Displace water from the metal surfaces and neutralize acid.
- Cohesion agents—Maintain a film of oil in high-pressure points to prevent wear. Cohesion agents are deposited on parts as the oil flows over them and remain on the parts as the oil is pressed out. **Cohesion** of the engine oil is the tendency of the oil molecules to remain on the friction surfaces of engine components.

Oil is rated by two organizations: the SAE and the API. The SAE has standardized oil viscosity ratings, while the API rates oil to classify its service or quality limits. These rating systems were developed to make proper selection of engine oil easier.

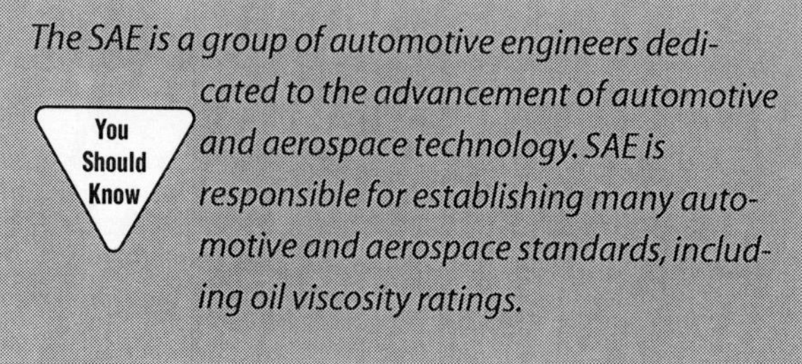


*The SAE is a group of automotive engineers dedicated to the advancement of automotive and aerospace technology. SAE is responsible for establishing many automotive and aerospace standards, including oil viscosity ratings.*

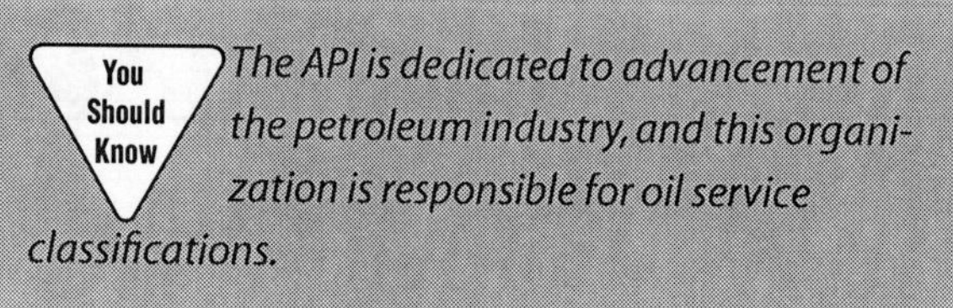


*The API is dedicated to advancement of the petroleum industry, and this organization is responsible for oil service classifications.*

The API classifies engine oils by a two-letter system **(Figure 4)**. The prefix letter is either listed as S-class or C-class to classify the oil usage. The second letter denotes various grades of oil within each classification and denotes the oil's ability to meet the engine manufacturer's warranty

**Figure 4.** Oil containers have identification labels for API and SAE ratings.

requirements **(Figure 5)**. One more oil classification has been developed by the API other than the ones listed in Figure 5. This classification is SL, and this oil meets the requirements for the latest automotive engines.

SAE ratings provide a numeric system of determining the oil's viscosity. The higher the number, the thicker or heavier the weight of the oil. For example, oil classified as SAE 50 is thicker than SAE 20. The thicker the oil, the slower it will pour. Thicker oils should be used when engine temperatures are high, but they can cause lubrication problems in cooler climates. Thinner oils will flow through the engine faster and easier at colder temperatures but may break down under higher temperatures or heavy engine loads. To provide a compromise between these conditions, multiviscosity oils have been developed. For example, an SAE 10W-30 oil has a viscosity of 10W at lower temperatures to provide easy flow. As the engine temperature increases, the viscosity of the oil actually increases to a viscosity of 30 to prevent damage resulting from excessively thin oil. The W in the designation refers to winter, which means the viscosity is determined at 0°F (–18°C). If there is no W, the viscosity is determined at 210°F (100°C).

Oils have recently been developed that may be

| API GASOLINE ENGINE DESIGNATION | |
|---|---|
| **SC** | Service typical of gasoline engines in 1964 through 1967. Oil designed for this service provides control of high- and low-temperature deposits, wear, dust, and corrosion in gasoline engines. |
| **SD** | Service typical of gasoline engines in 1968 through 1970. Oils designed for this service provide more protection against high- and low-temperature deposits, wear, rust, and corrosion in gasoline engines. SD oil can be used in engines requiring SC oil. |
| **SE** | Service typical of gasoline engines in automobiles and some trucks beginning in 1972. Oil designed for this service provides more protection against oil oxidation, high-temperature engine deposits, rust, and corrosion in gasoline engines. SE oil can be used in engines requring SC or SD oil. |
| **SF** | Service typical of gasoline engines in automobiles and some trucks beginning with 1980. SF oils provide increased oxidation stability and improved antiwear performance over oils that meet API designation SE. It also provides protection against engine deposits, rust, and corrosion. SF oils can be used in engines requiring SC, SD, or SE oils. |
| **SG** | Service typical of gasoline automobiles and light-duty trucks, plus CC classification diesel engines beginning in the late 1980s. SG oils provide the best protection against engine wear, oxidation, engine deposits, rust, and corrosion. It can be used in engines requiring SC, SD, or SF oils. |
| **SH** | This classification was replaced in 1997. When compared to SG motor oil, this oil produces 19% less engine sludge, 20% less cam wear, 5% less engine varnish, 21% improved oxidation, 7% less piston varnish, 3% less engine rust, and 13% less bearing wear. As you can see, this is a superior product to anything used in the past and should prolong the life of any engine. |
| **SJ** | This is the latest classification and it meets the same minimum requirements in areas of deposit control, resistance to rust, and oxidation as the older classifications. It also allows the engine to run with less friction, which results in better fuel economy. This oil classification was introduced in 1997 and replaces all earlier API categories. The companies are working on the next classification, which will be labeled SK, but it is not ready at this date. |

**Figure 5.** The API rating system indicates the quality of the oil and its ability to meet the manufacturer's requirements.

**SAE GRADES OF MOTOR OIL**

| Lowest Atmospheric Temperature Expected | Single-Grade Oils | Multigrade Oils |
|---|---|---|
| 32°F (0°C) | 20, 20W, 30 | 10W-30, 10W-40, 15W-40, 20W-40, 20W-50 |
| 0°F (–18°C) | 10W | 5W-30, 10W-30, 10W-40, 15W-40 |
| –15°F (–26°C) | 10W | 10W-30, 10W-40, 5W-30 |
| Below –15°C (–26°C) | 5W* | *5W-20, 5W-30 |

*SAE 5W and 5W-20 grade oils are not recommended for sustained high-speed driving.

**Figure 6.** The average atmospheric temperature in which the engine is operating determines the proper SAE oil grade.

labeled "energy conserving." These oils use friction modifiers to reduce friction and increase fuel economy. Selection of oil is based upon the type of engine and the conditions it will be running in. The API ratings must meet the requirements of the engine and provide protection under the normal expected running conditions. A main concern in selecting SAE ratings is ambient temperatures **(Figure 6)**. Always select oil that meets or exceeds the manufacturer's recommendations.

In recent years, synthetic oils have become increasingly popular. The advantage of synthetic oils is their stability over a wide temperature range. Before using these oils in an engine, refer to the vehicle's warranty information to confirm use of this oil does not void the warranty.

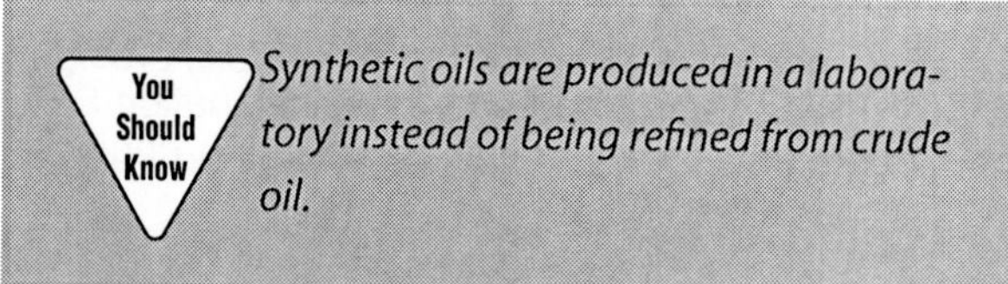


*Synthetic oils are produced in a laboratory instead of being refined from crude oil.*

## OIL PUMPS

There are two basic types of oil pumps: rotor and gear. Both types are positive displacement pumps. The rotary pump generally has a four-lobe inner rotor with a five-lobe outer rotor **(Figure 7)**. The outer rotor is driven by the inner rotor. As the lobes come out of mesh, a vacuum is created and atmospheric pressures above the oil level in the pan forces it into the pickup tube. The oil is trapped between the lobes as it is directed to the outlet. As the lobes come back into mesh, the oil is pressurized and expelled from the pump.

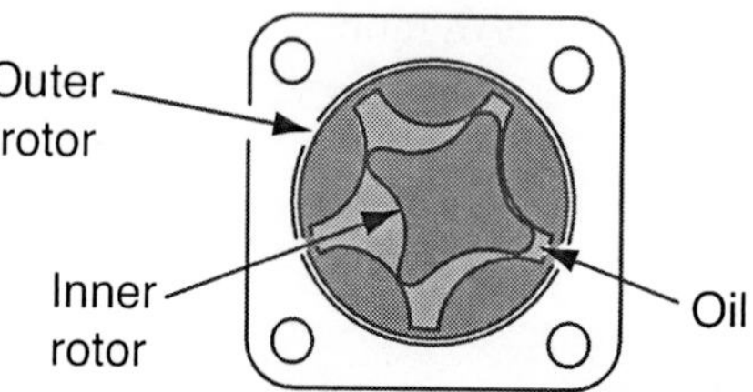


**Figure 7.** A typical rotor-type oil pump.

Gear pumps can use two gears riding in mesh with each other or use two gears and a crescent design **(Figure 8)**. Both types operate in the same manner as the rotor-type pump. The advantage of the rotor-type pump is its capability to deliver a greater volume of oil since the rotor cavities are larger.

In the past, most oil pumps were driven off the camshaft by a drive shaft fitting into the bottom of the dis-

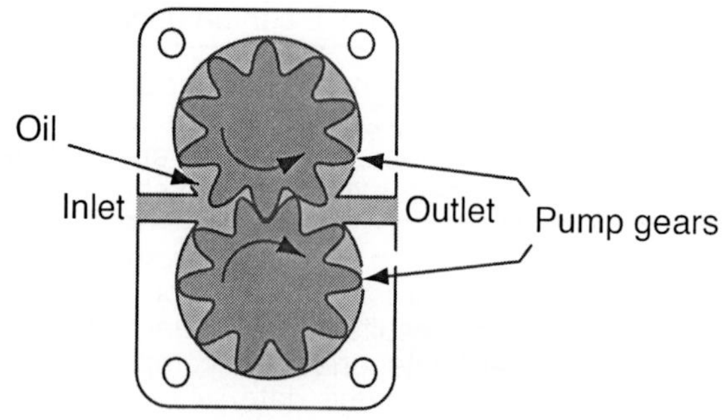


**Figure 8.** A typical gear-type oil pump.

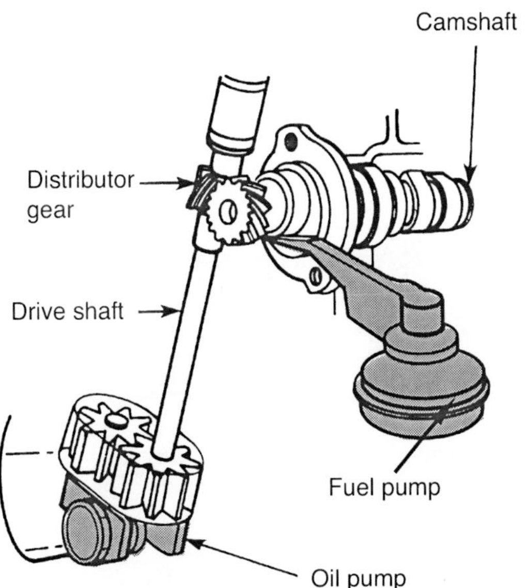


**Figure 9.** Many oil pumps are driven by a gear on the camshaft.

tributor shaft **(Figure 9)**. This is still a popular method for engines using distributor ignition systems. Many of today's engines do not use a distributor and drive the oil pump by the front of the crankshaft **(Figure 10)**.

Since oil pumps are positive displacement types, output pressures must be regulated to prevent excessive pressure buildup. A pressure relief valve opens to return oil to the sump if the specified pressure is exceeded **(Figure 11)**. A calibrated spring holds the valve closed, allowing pressure to increase. Once the oil pressure is great enough to overcome the spring pressure, the valve opens and returns the oil to the sump.

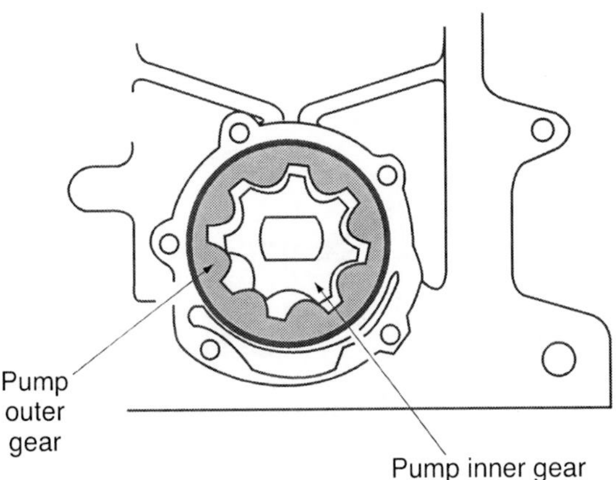


**Figure 10.** In many engines, the oil pump is located at the front of the engine block and driven directly by the crankshaft.

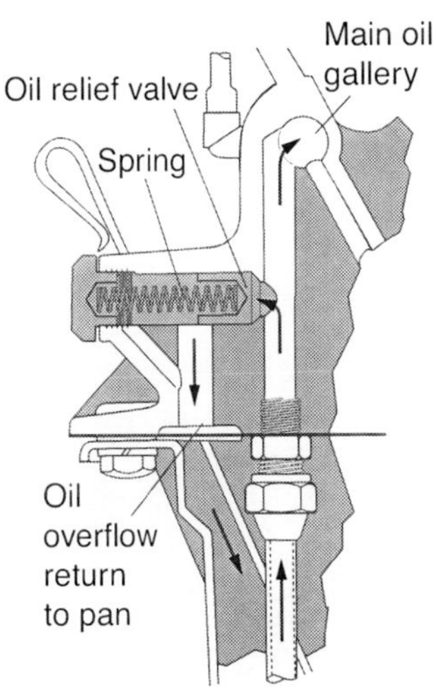


**Figure 11.** The oil pressure relief valve opens to return the oil to the sump if the oil pressure is excessive.

## OIL FILTER

As the oil flows through the engine, it works to clean dirt and deposits from the internal parts. These contaminants are deposited into the oil pan or sump with the oil. Since the oil pump pickup tube syphons the oil from the sump, it can also pick up these contaminants. The pickup tube has a screen mesh to prevent larger contaminants from being picked up and sent back into the engine. The finer contaminants must be filtered from the oil to prevent them from being sent with the oil through the engine. Oil filter elements have a pleated paper or fibrous material designed to filter out particles between 20 and 30 microns **(Figure 12)**. A micron has a very small diameter. The thick-

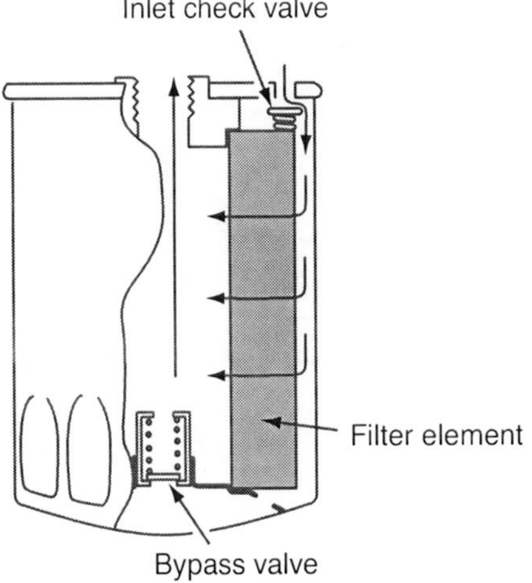


**Figure 12.** Oil flow through the oil filter.

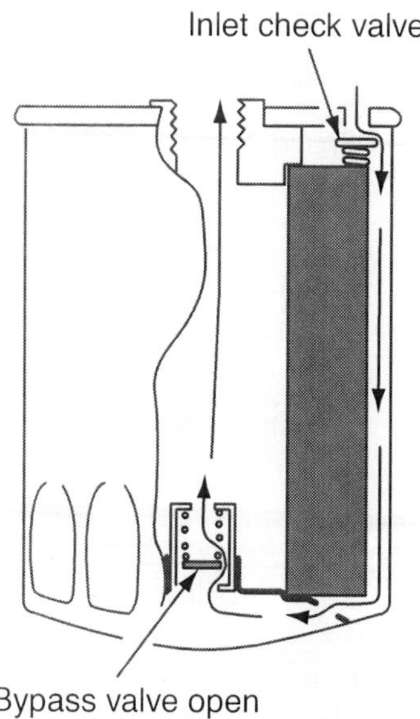


**Figure 13.** If the oil filter becomes plugged, the bypass valve opens to protect the engine.

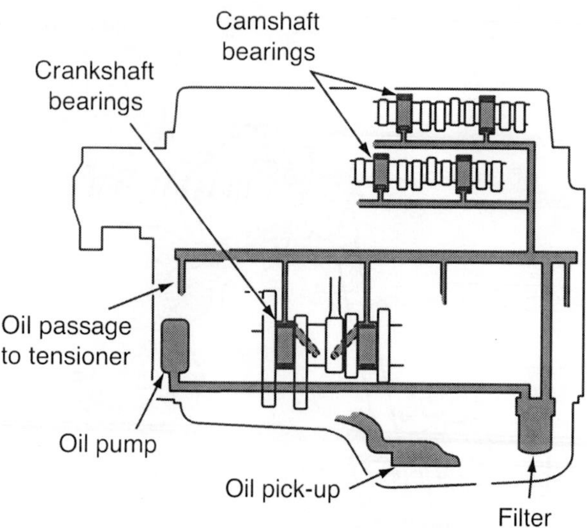


**Figure 14.** The lubrication system delivers oil to the high friction and wear areas of the engine.

ness of a human hair is usually about 60 to 90 microns, and 100 microns is equal to 0.004 inch. Most oil filters use a full filtration system that filters all the oil delivered by the pump. Under pressure from the pump, oil enters the filter on the outer areas of the element and works its way toward the center. In the event the filter becomes partially restricted and the pressure drop across the filter reaches a predetermined limit, a bypass valve opens and allows the oil to bypass the filter and enter the oil galleries. If this occurs, the oil will no longer be filtered **(Figure 13)**.

An inlet check valve is used to prevent oil drainback from the oil pump when the engine is shut off. The check valve is a rubber flap covering the inside of the inlet holes. When the engine is started and the pump begins to operate, the valve opens and allows oil to flow to the filter.

## LUBRICATION SYSTEM PURPOSE AND OPERATION

After the oil leaves the filter, it is directed through galleries to various parts of the engine. Oil galleries are drilled passages in the cylinder block and head. A main oil gallery is usually drilled the length of the block. From the main gallery, pressurized oil branches off to upper and lower portions of the engine **(Figure 14)**. Oil is directed to the crankshaft main bearings through the main bearing saddles. Passages drilled in the crankshaft then direct the oil to the connecting rod bearings **(Figure 15)**. Some manufacturers drill a small spit or squirt hole in the connecting rod to spray pressurized oil delivered to the bearings out and onto the cylinder walls **(Figure 16)**. When the spit hole aligns with the oil passage in the rod journal, it squirts the oil onto the cylinder wall.

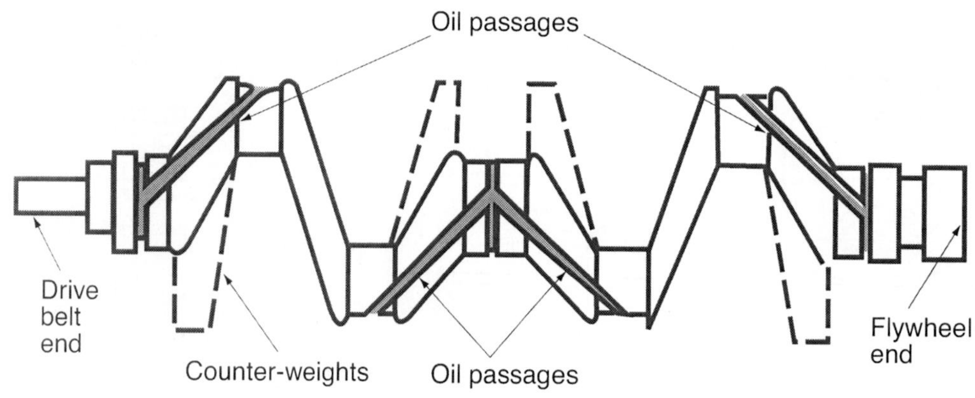


**Figure 15.** Oil passages drilled in the crankshaft supply lubrication to the bearings.

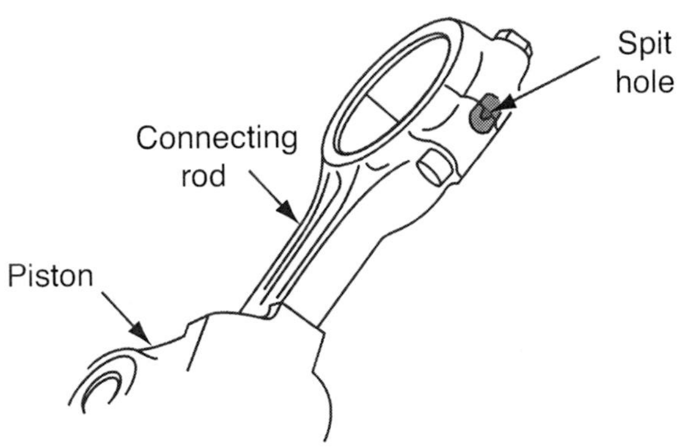


**Figure 16.** A spit hole in the connecting rod sprays oil onto the cylinder walls and helps to cool the piston.

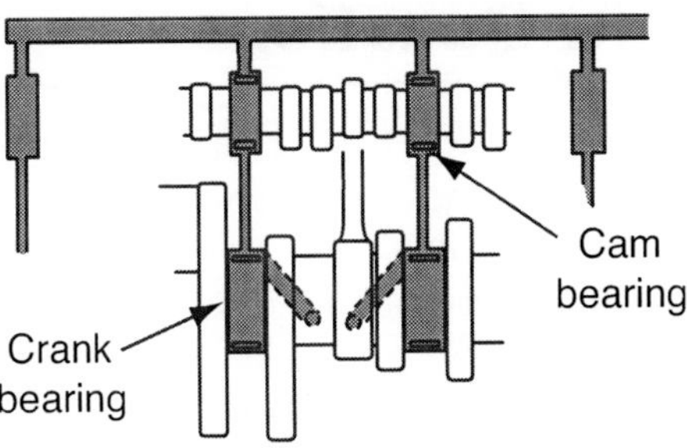


**Figure 17.** In an OHV engine, oil is supplied from the main bearings to the camshaft bearings.

Each camshaft bearing (on OHV engines) also receives oil from passages drilled in the block **(Figure 17)**. The valve train can receive oil through passages drilled in the block and then through a hollow rocker arm shaft. Engines using pushrods usually pump oil from the lifters through the pushrods to the valve train **(Figure 18)**. Oil sent to the cylinder head is returned to the sump through drain passages cast into the head and block.

Not all areas of the engine are lubricated by pressurized oil. The crankshaft rotates in the oil sump, creating some splash-and-throw effects. Oil that is picked up in this manner is thrown throughout the crankcase. This oil splashes onto the cylinder walls below the pistons. This provides lubrication and cooling to the piston pin and piston rings. In addition, oil that is forced out of the connecting rod bearings is thrown out to lubricate parts that are not fed pressurized oil. Valve timing chains are often lubricated by oil draining from the cylinder head being dumped onto the chain and gears.

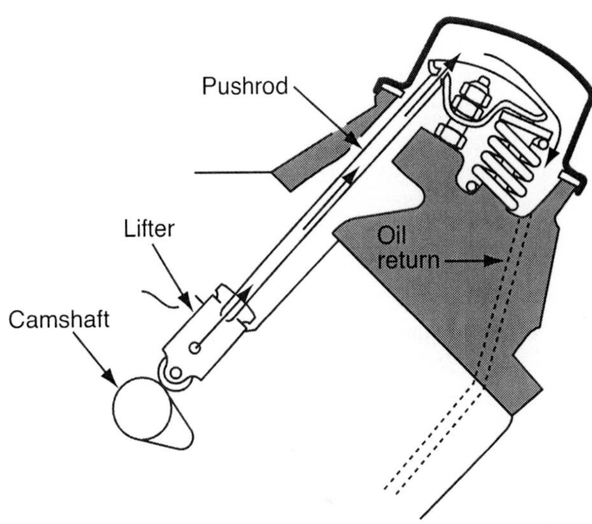


**Figure 18.** OHV engines often have hollow pushrods to supply oil from the valve lifters to the rocker arms and valve train.

## Summary

- The lubrication system provides an oil film to prevent moving parts from coming in direct contact with each other. Oil molecules work as small bearings rolling over each other to reduce friction. Another function is to act as a shock absorber between the connecting rod and crankshaft.
- Many different types of additives are used in engine oils to formulate a lubricant that will meet all of the demands in today's engines.
- Oil is rated by two organizations: the SAE and the API. The SAE has standardized oil viscosity ratings, while the API rates oil to classify its service or quality limits.
- There are two basic types of oil pumps: rotor and gear. Both types are positive displacement pumps.
- Oil filter elements have a pleated paper or fibrous material designed to filter out particles between 20 and 30 microns.

## Review Questions

1. Technician A says engine oil in the lubrication system helps to clean engine components. Technician B says engine oil helps to provide a seal between the piston rings and the cylinder walls. Who is correct?
   A. Technician A
   B. Technician B
   C. Both Technician A and Technician B
   D. Neither Technician A nor Technician B
2. Technician A says pour point depressants prevent oil from becoming too thin at high temperatures. Technician B says cohesion agents in the engine oil help the oil to adhere to friction surfaces in the engine. Who is correct?
   A. Technician A
   B. Technician B
   C. Both Technician A and Technician B
   D. Neither Technician A nor Technician B
3. Technician A says an SAE 10W-30 engine oil has the viscosity of a 10W oil at low temperatures. Technician B says at high temperature the viscosity of a 10W-30 oil increases so this oil has the same viscosity as a SAE 30 oil. Who is correct?
   A. Technician A
   B. Technician B
   C. Both Technician A and Technician B
   D. Neither Technician A nor Technician B
4. Technician A says synthetic oil is made from crude oil. Technician B says synthetic oil maintains its stability over a wide temperature range. Who is correct?
   A. Technician A
   B. Technician B
   C. Both Technician A and Technician B
   D. Neither Technician A nor Technician B
5. All of these statements about oil additives are true *except*:
   A. Pour point depressants prevent oil from becoming thicker at high engine temperatures.
   B. Corrosion inhibitors displace water from metal surfaces.
   C. Dispersants prevent carbon particles from grouping together.
   D. Viscosity index improvers prevent oil thinning.
6. Technician A says an engine oil with a SAE 50 oil rating is thinner than an oil with an SAE 20 rating. Technician B says a typical car engine oil with an SAE 30 rating is suitable for use in extremely cold weather. Who is correct?
   A. Technician A
   B. Technician B
   C. Both Technician A and Technician B
   D. Neither Technician A nor Technician B
7. The viscosity rating of a 10W-30 engine oil is determined at:
   A. 32°F.
   B. 20°F
   C. 10°F
   D. 0°F
8. All of these statements about engine oil pumps are true *except*:
   A. The oil pump may be driven off the front of the crankshaft.
   B. The oil pump may be rotor type or gear type.
   C. A calibrated spring holds the oil pressure relief valve open.
   D. The pressure relief valve prevents excessive oil pump pressure.
9. Technician A says most engine oil filters are connected so all the oil delivered by the pump flows through the filter. Technician B says if the oil filter becomes completely plugged, oil stops flowing to the engine oil system. Who is correct?
   A. Technician A
   B. Technician B
   C. Both Technician A and Technician B
   D. Neither Technician A nor Technician B
10. Technician A says in many OHV engines, oil is supplied through hollow push rods to lubricate the rocker arms. Technician B says in many OHV engines, oil is supplied to the camshaft bearings through holes drilled in the engine block. Who is correct?
    A. Technician A
    B. Technician B
    C. Both Technician A and Technician B
    D. Neither Technician A nor Technician B
11. When the engine is running, oil is forced into the oil pump by ______________ in the oil pump and ______________ ______________ above the oil level in the oil pan.
12. In many engines without a distributor, the oil pump is driven from the front of the ______________.
13. The oil pressure in the lubrication system is limited by the ______________ ______________ valve.
14. The oil filter bypass valve opens if the oil filter becomes ______________ ______________.
15. Explain the oil flow through the lubrication system in an OHV engine.
16. Explain four purposes of the engine oil.
17. List and explain seven engine oil additives.
18. Describe the design of two different types of oil pumps.

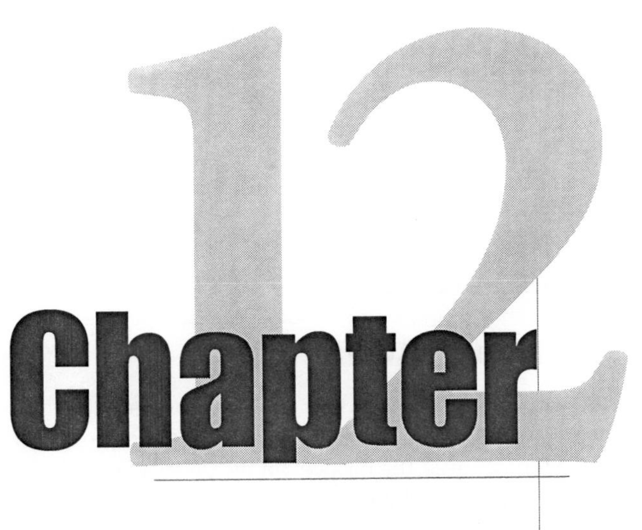


# Engine Lubrication System Maintenance, Diagnosis, and Service

## *Introduction*

Proper lubrication system maintenance and service is extremely important to provide normal engine life. If the engine oil and filter are not changed at the vehicle manufacturer's recommended intervals, contamination in the engine oil may score engine components such as cylinder walls and piston rings. Oil contamination may also cause premature crankshaft main bearing and connecting rod bearing wear.

### CHANGING ENGINE OIL AND FILTER

Lubrication system maintenance usually consists of periodic oil and filter changes. The oil pan is equipped with a drain plug to remove the oil from the crankcase. It is easier to remove all of the oil if the engine is warmed prior to draining. Contaminants are suspended in the oil when it is at operating temperature, and most of these contaminants are removed during an oil change.

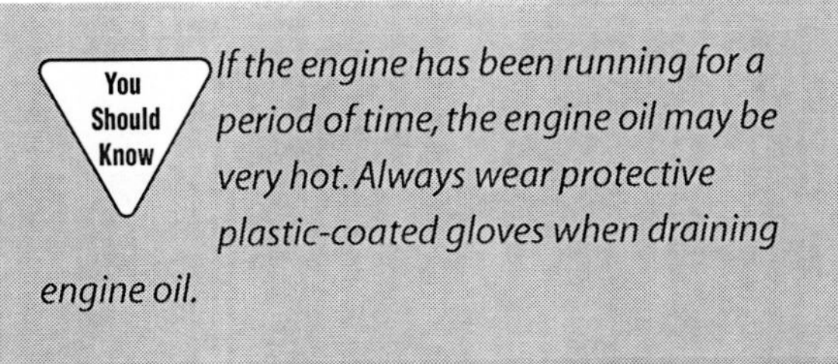


*If the engine has been running for a period of time, the engine oil may be very hot. Always wear protective plastic-coated gloves when draining engine oil.*

You Should Know

*Some customers feel they need to add additional treatments to their engine oil. A variety of aftermarket and manufacturer supplied additives are available to remove carbon, improve ring sealing, loosen stuck valves, and so forth. Before using these additives, refer to the vehicle manufacturer's warranty. Use of some of these additives may void the warranty.*

The oil filter usually threads onto the engine block or adapter. A band-type wrench or special socket is used to remove the filter from the engine **(Figure 1)**. A rubber seal

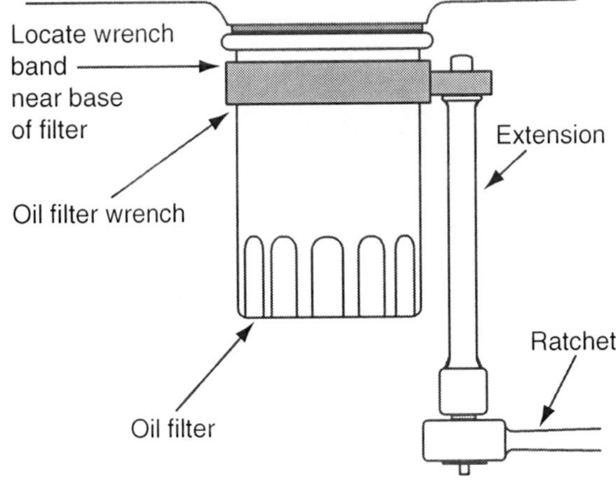


**Figure 1.** Using an oil filter wrench to remove an oil filter.

is located at the top of the oil filter to seal between the filter and block. Sometimes this seal may come off the filter and remain on the block. Be sure to remove the seal from the block if this occurs. Failure to remove the oil seal will result in oil leaks. Before installing the new oil filter, lubricate the seal with new engine oil and fill the filter with oil. Install the oil filter. Do not overtighten the filter. Turn the filter 3/4 turn after the seal makes contact. Do not use the oil filter wrench to tighten the filter. Only hand pressure is required. After the proper amount of new oil is installed in the engine, start the engine and inspect the oil filter for leaks. Be sure the oil level on the dipstick is at the full mark.

**You Should Know** *Overtightening the oil filter may cause seal damage and result in oil leaks. Oil that is drained from the oil pan may contain dirt and grit. Using this oil to lubricate the seal may cause seal damage and an oil leak.*

Used engine oil and filters are considered hazardous waste. Federal and state environmental regulations must be followed when disposing of these items. Used engine oil is usually collected and shipped to an oil recycling facility. Oil filters are usually drained, squashed, and recycled.

Some engines require preoiling before they are started after an oil change. This is especially true of engines equipped with turbochargers. The turbocharger rotates at a high rate of speed and requires lubrication to prevent damage to the turbocharger bearings. After the oil is drained and the oil filter replaced, it may take several moments before oil reaches the turbocharger. Preoiling the engine will assure all components receive oil before the engine is started. Use the following procedure to preoil the engine:

1. Make sure the oil filter and crankcase are filled with oil.
2. Disable the ignition system using the manufacturer's recommended procedure.
3. Crank the engine for 30 seconds, then allow the starter to cool, and crank the engine for another 30 seconds.
4. Repeat this procedure until oil pressure is indicated (oil pressure indicator light off or oil pressure gauge indications).
5. Enable the ignition system and start the engine.
6. Observe the oil pressure indicator. If oil pressure is not indicated, shut the engine off.
7. Recheck the oil level.

## OIL LEAK DIAGNOSIS

When performing an oil leak diagnosis, always check the oil level and condition. With the engine at normal operating temperature, the oil level on the dipstick should be at the full mark. If the crankcase is overfilled, excessive oil splash in the crankcase may cause oil leaks at the pan gasket and rear main bearing seal. If the oil is diluted with gasoline, it is excessively thin and this may aggravate oil leaks. On engines with a mechanical fuel pump, a leaking pump diaphragm may cause gasoline leaks into the crankcase and diluted oil. On a fuel-injected engine, a leaking fuel pressure regulator may cause gasoline to leak through the vacuum hose into the intake manifold resulting in diluted oil.

A thorough visual inspection of the engine often locates the source of an oil leak. When checking for oil leaks, inspect the positive crankcase ventilation (PCV) valve and hose.

A restricted PCV valve or hose causes excessive pressure buildup in the engine, and this condition may cause oil leaks from engine gaskets or seals that are in normal condition **(Figure 2)**. Another cause of excessive pressure buildup in the engine is too much blowby. A restricted PCV system or excessive blowby builds up excessive pressure in the engine, and this condition may cause oil vapors to be forced from the engine through the clean air intake hose and into the air cleaner. Some of these oil vapors enter the air intake into the engine, resulting in excessive oil consumption. Excessive **blowby** refers to leakage between the piston rings and cylinder walls caused by worn rings or scored cylinder walls.

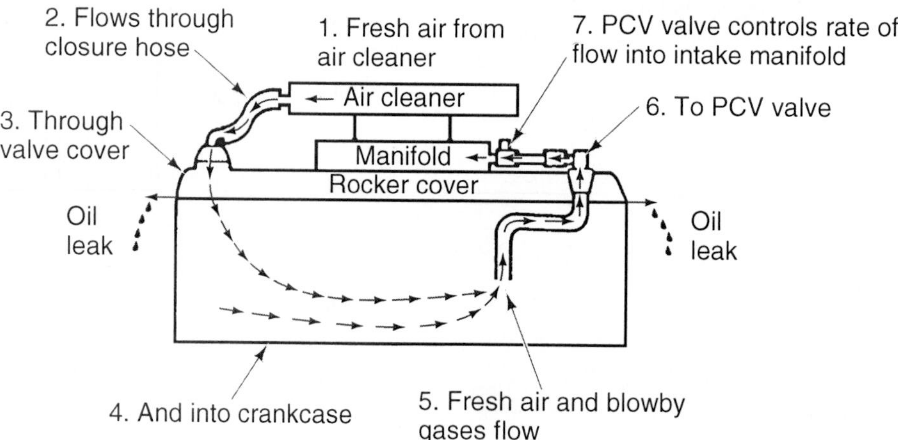


**Figure 2.** Normal PCV system operation.

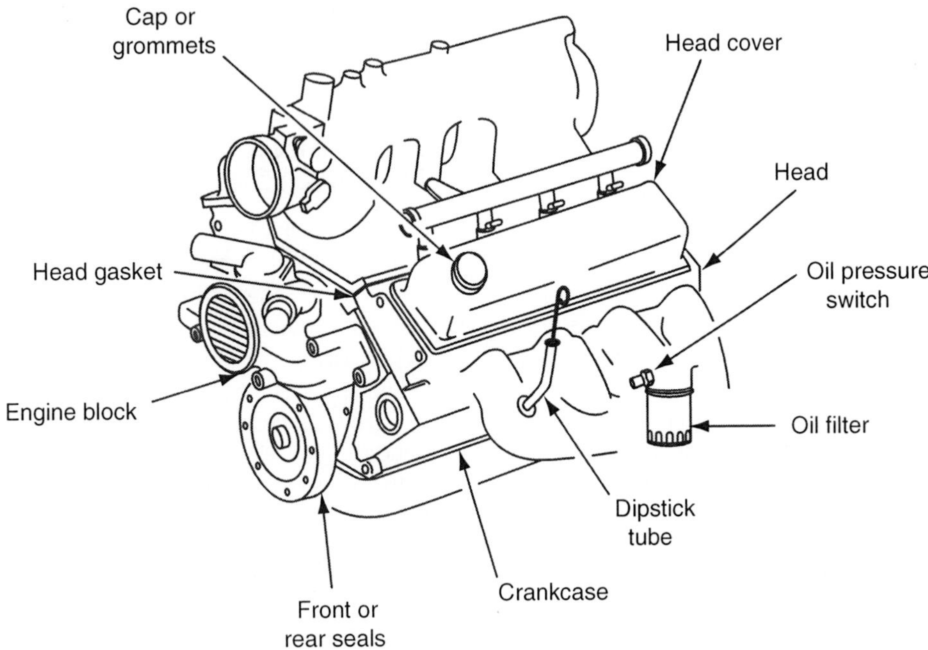


**Figure 3.** Common engine oil leak locations.

If the PCV valve is removed with the engine running and a great deal of oil vapor is escaping from the PCV valve opening, excessive blowby is indicated.

Inspect the valve cover gasket area for indications of oil leaks. In many cases, the oil may be leaking around the valve cover gaskets and running down the sides and back of the block so it drips under the vehicle. Raise the vehicle on a lift and inspect the lower side of the engine for oil leaks **(Figure 3)**. Always inspect the oil drain plug and oil filter for indications of oil leaks. Oil dripping from the oil drain plug may indicate a damaged gasket on this plug or stripped plug threads that do not allow sufficient torque on the plug. Oil leaking around the oil pan gasket area usually indicates a damaged oil pan gasket. Oil leaking from the flywheel housing usually indicates a leaking rear main bearing seal. In an OHV engine, oil leaking from the flywheel housing may indicate a leaking oil gallery plug or a leaking plug in the rear camshaft bearing opening. On some engines, oil leaks at the front of the engine may be caused by a worn timing gear cover seal or gasket or a leaking oil pan gasket. On engines with a distributor, oil leaks may be caused by a damaged gasket between the distributor housing and the block. If oil is leaking from the valve cover gaskets or oil pan gasket, be sure the retaining bolts on these components are tightened to the specified torque. Some engines, such as light-duty diesels and gasoline engines in light-duty trucks with a trailer-hauling package, have an engine oil cooler. On these vehicles, always inspect the oil cooler and lines for leaks.

## DIAGNOSING OIL PRESSURE INDICATORS

Most oil pressure warning light circuits use a normally closed switch threaded into the main oil gallery **(Figure 4)**.

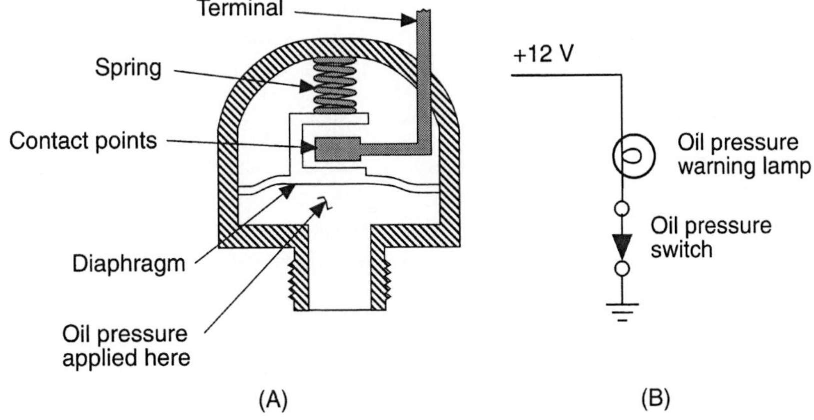


**Figure 4.** An oil pressure warning light circuit.

A diaphragm in the sending unit is exposed to the oil pressure. The switch contacts are controlled by the movement of the diaphragm. When the ignition switch is turned to the on position with the engine not running, the oil warning light turns on. Since there is no pressure to the diaphragm, the contacts remain closed and the circuit is complete to ground. When the engine is started, oil pressure builds and the diaphragm moves the contacts apart. This opens the circuit and the warning light goes off. The amount of oil pressure required to move the diaphragm is about 3 psi (20.6 kPa). If the oil warning light comes on while the engine is running, it indicates that the oil pressure has dropped below the 3 psi (20.6 kPa) limit.

When diagnosing this type of oil pressure warning light circuit, the first step is to be sure the warning light bulb is operating. Turn the ignition switch on, remove the wire from the oil pressure switch, and ground this wire to an engine ground. If the oil pressure warning light is not on, there is no power to the bulb, the bulb is burned out, or the wire from the bulb to the switch has an open circuit. If the oil pressure warning light bulb operates normally, test the engine oil pressure. When the oil pressure is within specifications but the oil pressure warning light is on with the engine running, replace the oil pressure switch.

Oil pressure gauges are usually connected to a sending unit containing a variable resistance. Engine oil pressure causes the flexible diaphragm to move in the sending unit **(Figure 5)**. The diaphragm movement is transferred to a contact arm that slides along the resistor. The position of the sliding contacts on the arm in relation to the resistance coil determines the resistance value and the amount of current flow through the gauge to ground. Always be sure the engine oil pressure is within specifications before diagnosing this system. This type of oil pressure gauge circuit may be tested by connecting a special variable resistance test tool to the oil sending unit wire. With the ignition switch on, rotate the control knob on the special test tool and observe the oil pressure gauge. If this gauge operates normally when connected to the test tool but fails to operate properly when connected to the oil pressure sending unit, replace the sending unit.

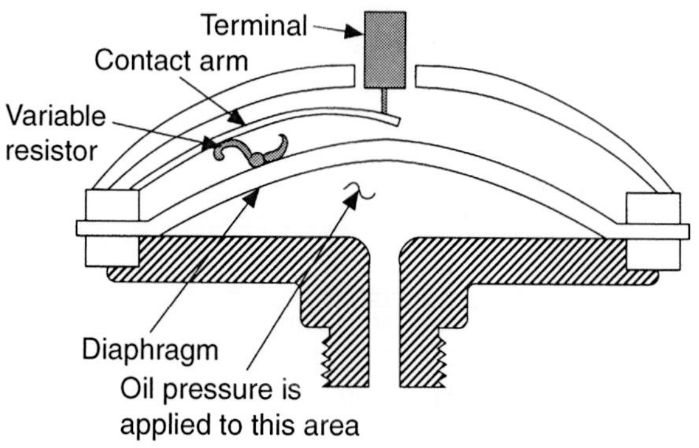


**Figure 5.** An oil pressure gauge sending unit containing a variable resistor.

## RESETTING CHANGE OIL WARNING MESSAGES

Many vehicles now have a computer-operated message center in the instrument panel that displays various warning messages. The **message center** is a digital readout in the instrument panel that provides warnings regarding abnormal operation of automotive systems, low fluid levels, or required system service.

Low engine oil pressure is indicated on the oil pressure gauge, but a LOW OIL PRESSURE warning is also displayed in the message center. When the engine oil requires changing, CHANGE OIL is displayed in the message center. The computer senses various parameters such as mileage, engine temperature, and air intake temperature. On the basis of this input information, the computer illuminates the CHANGE OIL message at the ideal oil change interval. After the oil is changed, this message must be turned off. The procedure for turning off the CHANGE OIL message varies depending on the vehicle make, but this procedure is in the service manual and owner's manual. The CHANGE OIL message is turned off on some vehicles by turning the ignition switch on and pushing the accelerator pedal to the floor three times in a 5-second interval.

## DIAGNOSIS OF EXCESSIVE OIL CONSUMPTION

When diagnosing excessive oil consumption, always check the oil level and condition. If the crankcase is overfilled, excessive oil splash in the crankcase causes too much oil on the cylinder walls which results in higher-than-normal oil consumption. Diluted oil also results in excessive oil consumption. When a customer complains about excessive oil consumption, always check the engine for oil leaks because this may be the cause of the problem.

Worn **valve guides** or valve guide seals allow excessive amounts of oil to run down between the valve guides and stems into the combustion chamber.

Worn valve guides and seals are often indicated by blue smoke from the tailpipe each time a warm engine is started. Worn piston rings and/or scored cylinder walls cause excessive blowby as explained previously. This condition also allows excessive oil to enter the combustion chamber resulting in abnormal oil consumption. If blue smoke is escaping from the tailpipe during acceleration and deceleration, worn piston rings and/or scored cylinder walls may be indicated. An engine compression test with a compression gauge or a cylinder balance test with an engine analyzer will verify this condition.

You Should Know *An engine compression test is performed with the engine warmed up, all of the spark plugs removed, and the ignition and fuel systems disabled. A pressure gauge is installed in each spark plug hole in sequence and the engine is cranked through four compression strokes on the cylinder being tested. Compression readings below specifications indicate compression leakage past piston rings, valves, or the head gasket.*

You Should Know *A cylinder balance test is performed with an engine analyzer. The analyzer prevents each cylinder from firing for a brief time period and the analyzer records the rpm drop during this time. All the cylinders should have nearly the same rpm drop. If a cylinder has low compression, it is not contributing to engine power output and that cylinder has very little rpm drop during the balance test.*

## OIL PRESSURE DIAGNOSIS

Oil pressure is dependent upon oil clearances and proper delivery. If the clearance between a journal and the bearing becomes excessive, pressure is lost. Not all low oil pressure conditions are the result of bearing wear. Other causes include improper oil level, improper oil grade, and oil pump wear. Another common cause of low oil pressure is thinning oil as a result of excessive temperatures or gas dilution. If low oil pressure is suspected, begin by checking the oil level. Too low a level will cause the oil pump to aerate and lose volume. If the oil level is too high, it may be gasoline is entering the crankcase as a result of a damaged mechanical fuel pump, improperly adjusted carburetor, sticking choke, or a leaking injector or fuel pressure regulator. If the oil level and condition are satisfactory, check oil pressure using a shop gauge.

## OIL PRESSURE TESTING

To perform an oil pressure test, remove the oil pressure sending unit from the engine **(Figure 6)**. Using the correct size adapters, connect the oil pressure gauge to the oil passage. Start the engine and observe the gauge as the engine idles. Watch the gauge as the engine warms to note any

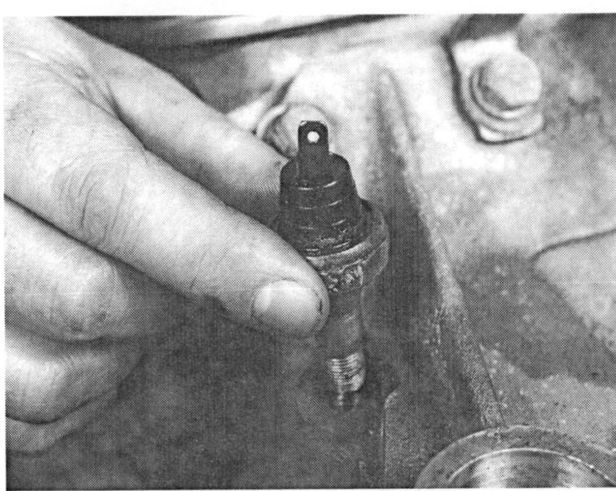

**Figure 6.** Remove the oil pressure switch and install a test pressure gauge in the switch opening to test oil pressure.

excessive drops due to temperature. Increase the engine rpm to 2,000 while observing the gauge. Compare the test results with the manufacturer's specifications. After the test is complete, reinstall the oil pressure sending unit, connect the sending unit wiring connector, start the engine, and confirm there are no leaks.

No oil pressure indicates the oil pump drive may be broken, the pickup screen is plugged, the gallery plugs are leaking, or there is a hole in the pickup tube. Lower-than-specified oil pressure indicates improper oil viscosity, a worn oil pump, a plugged oil pickup tube, a sticking or weak oil pressure relief valve, or worn bearings. An air leak in the oil pump pickup tube allows air to enter the oil pump. This lowers the oil pressure because oil is compressible. An air leak in the oil pickup tube also causes damage to engine components because of the air in the lubrication system. It is possible to have oil pressure that is too high. This can be caused by a stuck pressure relief valve or by a blockage in an oil gallery.

## OIL PUMP SERVICE

The oil pump is usually replaced whenever the engine is rebuilt or when the oil pump fails. Since the oil pump is so vital to proper engine operation, it is usually replaced along with the pickup tube and screen whenever the engine is rebuilt. The relatively low cost of an oil pump is considered cheap insurance by most technicians. However, there may be instances when the oil pump will have to be disassembled, inspected, and repaired. If the pump housing is not damaged, a pump rebuild kit containing new rotors or gears, a relief valve and spring, and seals may be available.

Since the oil is delivered to the pump before it goes through the oil filter, the pump is subject to wear and damage as contaminated oil passes through it. The particles passing through the gears or rotors of the pump wear away the surface area, resulting in a reduction of pump efficiency. If the particles are large enough, they may cause the metal of the rotor or gear surfaces to raise, resulting in pump seizure. In addition, these larger particles can form a wedge in the pump and cause it to lock up.

The pickup tube and screen should be replaced whenever the oil pump is replaced or rebuilt. If the tube attaches to the oil pump, install the new tube using light taps from a hammer. Make sure the pickup tube is properly positioned to prevent interference with the oil pan or crankshaft. Bolt-on pickup tubes may use a rubber O-ring to seal them. Do not use a sealer in place of the O-ring. Lubricate the O-ring with engine oil prior to assembly then alternately tighten the attaching bolts until the specified torque is obtained.

Prime the oil pump before installing it by submerging it into a pan of clean engine oil and rotating the gears by hand. When the pump discharges a stream of oil, the pump is primed.

If a gasket is installed between the pump and the engine block, check to make sure no holes or passages are blocked by the gasket material. Soak the gasket in oil to soften the material and allow for good compression. When installing the oil pump, make sure the drive shaft is properly seated and torque the fasteners to the specified value.

## OIL JET VALVES

Some engines use jet valves to spray oil into the underside of the piston to cool the top of the piston. These valves must be inspected and cleaned before they are returned to service. The valve consists of a check ball, spring, and nozzle **(Figure 7)**. Check the nozzle opening by inserting a small drill bit or wire into the hole. Do the same for the oil intake. With the drill bit installed in the intake hole, the check ball movement should be about 0.160 inch (4.0 millimeters). Finally, operation of the jet valve can be checked using air pressure set at 30 psi (207 kPa). At this pressure, the check ball should lift off its seat. If the nozzle is bent or damaged, replace the jet valve.

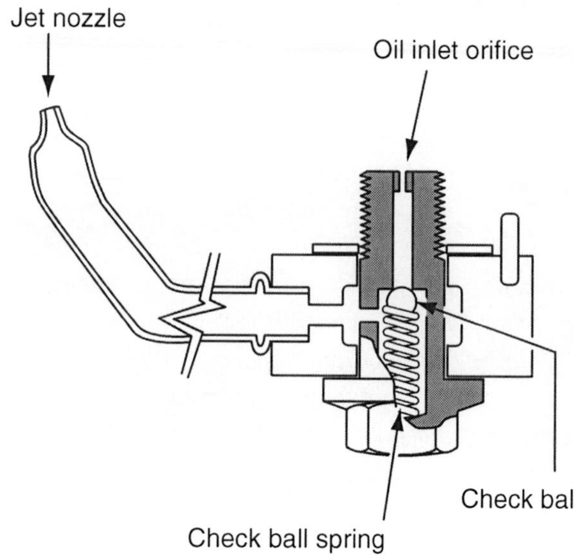


**Figure 7.** An oil jet valve.

# Summary

- The most common lubrication system maintenance is oil and filter changes.
- Used engine oil and filters must be disposed of according to hazardous material disposal regulations.
- Turbocharged engines require preoiling after an oil change or after some engine repairs.
- On fuel-injected engines, a leaking fuel pressure regulator diaphragm will cause the engine oil to be diluted with gasoline.
- A restricted PCV system or excessive engine blowby cause excessive pressure buildup in the engine and this condition may force oil vapors into the air cleaner.
- Oil leaking from the flywheel housing may be caused by a leaking rear main bearing seal, a leaking main oil gallery plug, or a leaking rear camshaft bearing plug.
- Most oil pressure warning lights are operated by an off/on switch located in the main oil gallery.
- Some CHANGE OIL messages in the message center are turned off by pushing the accelerator pedal to the floor three times in a 5-second interval with the ignition switch on.
- Excessive oil consumption may be caused by oil leaks, worn piston rings, scored cylinder walls, worn valve guides and seals, or excessive pressure buildup in the engine caused by a restricted PCV valve or excessive blowby.
- Low engine oil pressure may be caused by diluted oil, improper grade of oil, worn crankshaft or camshaft bearings, worn oil pump, or an air leak into the oil pump pickup.
- A test gauge is installed in the oil sender location to test oil pressure.
- Some engines have oil jet valves that squirt oil on the underside of the pistons to help cool these components.

# Review Questions

1. Technician A says a new oil filter should be tightened with an oil filter wrench. Technician B says used oil filters should be thrown in a dumpster. Who is correct?
   A. Technician A
   B. Technician B
   C. Both Technician A and Technician B
   D. Neither Technician A nor Technician B
2. Technician A says on fuel-injected engines, a leaking fuel pressure regulator diaphragm may cause oil dilution. Technician B says on a carbureted engine, a leaking fuel pump diaphragm may cause oil dilution. Who is correct?
   A. Technician A
   B. Technician B
   C. Both Technician A and Technician B
   D. Neither Technician A nor Technician B
3. Technician A says an air filter element saturated with oil may be caused by a restricted PCV valve. Technician B says an air filter element saturated with oil may be caused by excessive blowby in the engine. Who is correct?
   A. Technician A
   B. Technician B
   C. Both Technician A and Technician B
   D. Neither Technician A nor Technician B
4. A low oil pressure warning light is on with the engine running. With the ignition switch on and the wire to the sending unit switch grounded, this warning light is on. All of these defects could cause the problem *except*:
   A. a defective sending unit switch.
   B. excessively low engine oil pressure.
   C. an air leak in the oil pump pickup pipe.
   D. an open circuit in the wire from the oil pressure warning light to the sending unit switch.
5. All of these statements about pre-oiling an engine are true *except*:
   A. The vehicle manufacturer may recommend pre-oiling on turbocharged engines.
   B. Pre-oiling prevents damage to the oil pump.
   C. During engine pre-oiling, the ignition system is disabled.
   D. Pre-oiling is performed by cranking the engine until oil pressure is available.
6. Excessive engine blowby may cause all of these problems *except*:
   A. oil in the air cleaner.
   B. excessive crankcase pressure in the engine.
   C. excessive engine oil pressure.
   D. excessive oil vapors escaping from the PCV valve opening.
7. In an OHV engine, oil leaking from the bottom of the flywheel housing may be caused by a:
   A. leaking timing gear cover seal.
   B. leaking expansion plug in the rear of the engine block.
   C. leaking pilot bearing seal.
   D. leaking oil gallery plug.
8. An oil pressure warning light is illuminated with the engine running, but this light goes out when the wire is removed from the oil sending unit. An oil pressure test indicates the engine oil pressure is within specifications. Technician A says the wire between the ignition switch and the oil sending unit is grounded. Technician B says the oil sending unit is satisfactory. Who is correct?
   A. Technician A
   B. Technician B
   C. Both Technician A and Technician B
   D. Neither Technician A nor Technician B
9. Technician A says the CHANGE OIL message in some message centers is erased by fully depressing the accelerator pedal three times in a 5-second interval with the ignition switch on. Technician B says in some vehicles the CHANGE OIL message is erased automatically when the oil is drained from the crankcase. Who is correct?
   A. Technician A
   B. Technician B
   C. Both Technician A and Technician B
   D. Neither Technician A nor Technician B
10. Excessive oil consumption may be caused by all of these problems *except*:
    A. worn rocker arms.
    B. worn valve guides.
    C. worn piston rings.
    D. scored cylinder walls.
11. Explain the proper oil filter tightening procedure.
12. List six causes of low oil pressure.
    1. ______________________________
    2. ______________________________
    3. ______________________________
    4. ______________________________
    5. ______________________________
    6. ______________________________
13. When testing engine oil pressure, a test gauge is installed in the ______________ ______________ ______________ opening.
14. In some engines, oil jet valves squirt oil against the underside of the ______________.

15. Explain how excessive engine blowby causes the air filter to become contaminated with oil.
16. Describe the operation of a low oil pressure warning light and sending unit switch with the ignition switch on and with the engine running.
17. Describe the operation of an oil pressure sending unit containing a variable resistance.
18. Describe three data sources that an engine computer uses to determine when to illuminate the CHANGE OIL message.

# Engine Coolant and Cooling Systems

## *Introduction*

Without the cooling system, the heat created during the combustion process would quickly increase engine component temperature to a point where engine damage would occur. Engines may be cooled by air or liquid. Regardless of the type of cooling system, the purpose of this system is to disperse the heat from the engine to the atmosphere. Proper cooling system operation is extremely important to maintain engine component life! The components of a liquid cooling system include: coolant, recovery system, water pump, heater system, radiator, coolant jackets, thermostat, transmission cooler, radiator cap, hoses, cooling fan, and temperature indicator.

## ENGINE COOLANT

Water by itself does not provide the proper characteristics needed to protect an engine. It works fairly well to transfer heat, but does not protect the engine in colder climates. The boiling point of water is 212°F (100°C) and it freezes at 32°F (0°C). Temperatures around the cylinders and in the cylinder head can reach above 500°F (250°C) and the engine will often be exposed to ambient temperatures below 32°F (0°C). In addition, water reacts with metals to produce rust, corrosion, and electrolysis. Most engine manufacturers use a coolant solution of water and ethylene glycol. The proper coolant mixture will protect the engine under most conditions. Water expands as it freezes. If the coolant in the engine block freezes, the expansion could cause the block or cylinder head to crack. If the coolant boils, liquid is no longer in contact with the cylinder walls and combustion chambers. The vapors are not capable of removing the heat and the pistons may collapse or the head may warp from the excessive heat.

Ethylene glycol by itself has a boiling point of 330°F (165°C) but it does not transfer heat very well. The freezing point of ethylene glycol is –8°F (–20°C). To improve the transfer of heat and lower the freezing point, water is added in a mix of about 50/50. At a 50/50 mix, the boiling point is 226°F (108°C) and the freezing point is –34°F (–37°C). These characteristics can be altered by changing the mixture **(Figure 1)**.

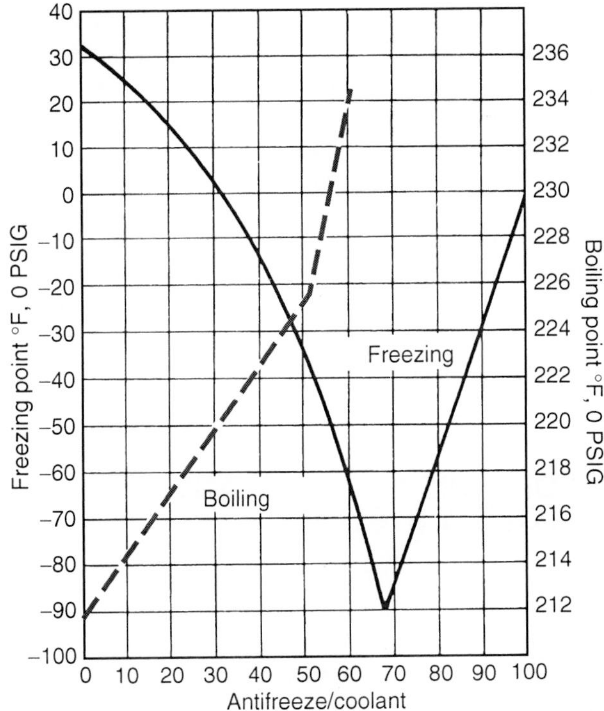


**Figure 1.** Changing the strength of the antifreeze solution changes its boiling and freezing points.

> **You Should Know** *Do not operate an engine with straight antifreeze. The poor heat transfer qualities of antifreeze can cause the engine to overheat.*

It may appear that lowering the boiling point by adding water is not desirable. In fact, this is required for proper engine cooling. Under pressure, the boiling point of the coolant mix is about 263°F (128°C).

> **You Should Know** *Every pound per square inch (psi) of cooling system pressure increase raises the coolant boiling point by approximately 2.5°.*

As was stated earlier, the temperature next to the cylinder or cylinder head may be in excess of 500°F (250°C). As coolant droplets touch the metal walls, they are turned into a gas as they boil. The superheated gas bubbles are quickly carried away into the middle of the coolant flow where they cool and condense back into a liquid. If this nucleate boiling did not occur, the coolant would not be capable of removing heat fast enough to protect the engine.

The engine coolant should always be replaced whenever an engine is rebuilt. Special agents are added to antifreezes used in aluminum engines or with aluminum radiators. Always refer to the service manual for the type and mixture recommended. In addition, proper maintenance of the cooling system is required to maintain good operation. Over time, the coolant can become slightly acidic because of the minerals and metals in the cooling system. A small electrical current may flow between metals through the acid. The electrical current has a corrosive effect on the metal used in the cooling system. Anticorrosion agents are mixed into the antifreeze; however, these agents may be depleted over time. All vehicle manufacturers provide a maintenance schedule recommending cooling system flushing and refill based on time and mileage.

## RADIATORS AND COOLANT RECOVERY SYSTEMS

The radiator is a series of tubes and fins that transfers the heat in the coolant to the air. As the coolant is circulated throughout the engine, it attracts and absorbs the heat within the engine, then flows into the radiator intake tank. The coolant then flows through the tubes to the outlet tank. As the heated coolant flows through the radiator tubes, the heat is dissipated into the air flow through the fins.

There are two basic radiator designs: downflow and crossflow. **Downflow** radiators have vertical fins that direct coolant flow from the top inlet tank to the bottom outlet tank. **Crossflow** radiators use horizontal fins to direct coolant flow across the core **(Figure 2)**.

The core of the radiator can be constructed using a tube-and-fin or a cellular-type design **(Figure 3)**. The mate-

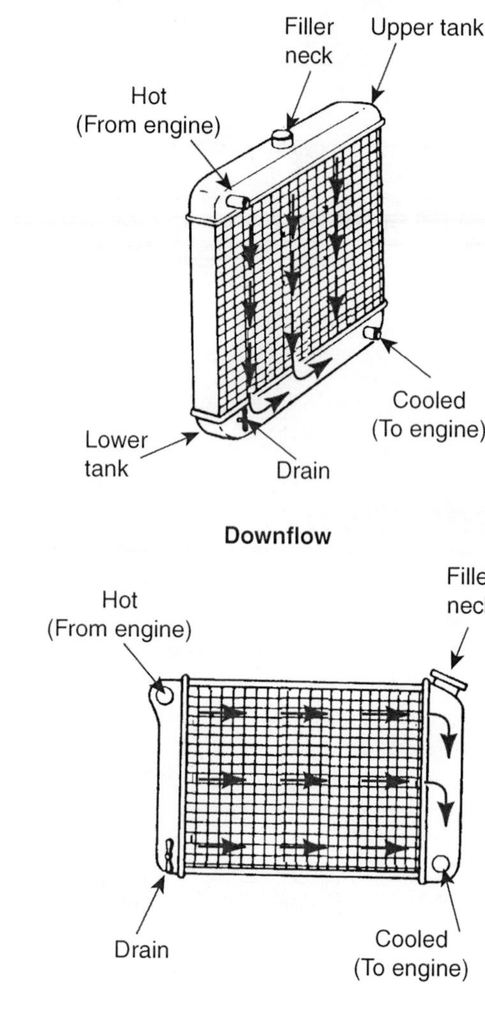


**Figure 2.** Crossflow and downflow radiators.

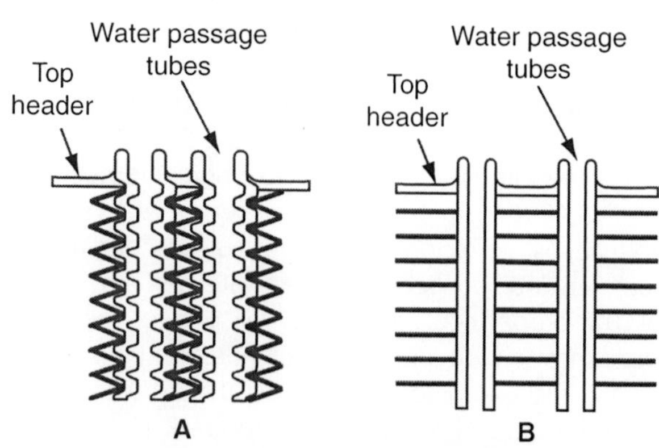


**Figure 3.** (A) Construction of a cellular radiator core. (B) Construction of a tube-and-fin radiator core.

rial used for the core is usually copper, brass, or aluminum. Aluminum-core radiators generally use nylon-constructed tanks.

For heat to be transferred to the air effectively, there must be a difference in temperature between the coolant and the air. The greater the temperature difference, the more effectively heat is transferred. The radiator cap allows for an increase of pressure within the cooling system, increasing the boiling point of the coolant. The radiator cap uses a pressure valve to pressurize the radiator to between 14 and 18 pounds **(Figure 4)**. If the pressure increases over the setting of the cap, the cap's seal will lift and release the pressure into a recovery tank **(Figure 5)**.

Vehicles with automatic transmissions usually have a transmission cooler mounted in the radiator outlet tank. As transmission fluid is circulated through this cooler, heat is transferred from the fluid to the coolant.

The coolant recovery system contains a reservoir that is connected to the radiator by a small hose. The coolant expelled from the radiator during a high-pressure condition is sent to this reservoir. When the engine cools, a vacuum is created in the radiator and the vacuum valve in the radiator cap opens **(Figure 6)**. Under this condition, atmospheric pressure on the coolant in the overflow reservoir pushes it back into the radiator. The action of the vacuum valve in the radiator cap prevents a vacuum in the cooling system as the coolant temperature decreases.

> **You Should Know** *If vacuum occurs in the cooling system, radiator hoses may be collapsed.*

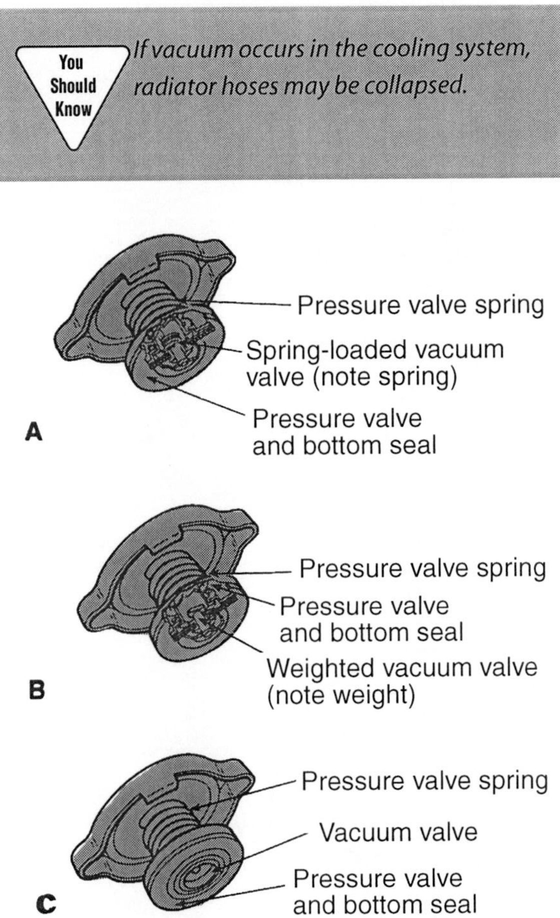


**Figure 4.** Three types of radiator caps: (A) the constant pressure type, (B) the pressure vent type, and (C) the closed system type.

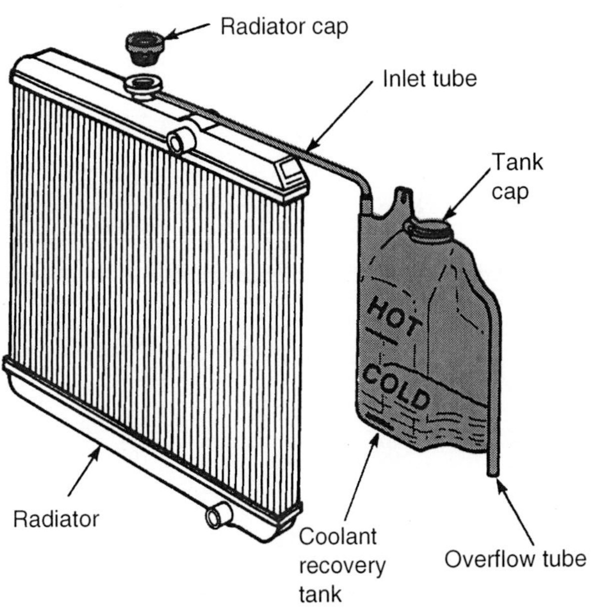


**Figure 5.** A typical coolant recovery system holds the coolant released from the radiator.

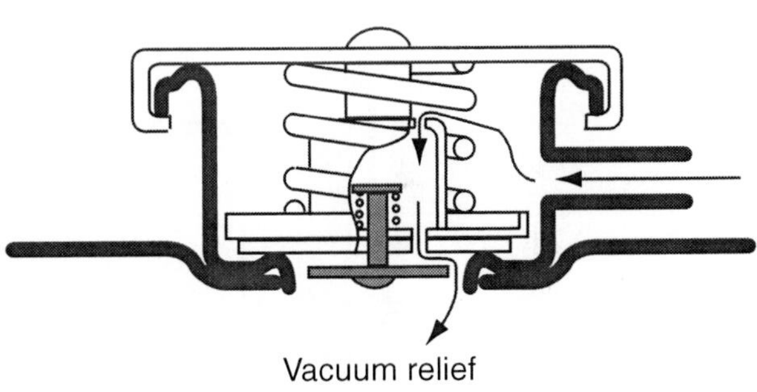


**Figure 6.** When the vacuum valve opens, coolant from the recovery reservoir enters the radiator.

Whenever an engine is rebuilt, the radiator should be thoroughly cleaned and then inspected by pressure testing it. The radiator cap should be replaced with a new one.

## HEATER CORE

The heater core is similar to a small version of the radiator. The heater core is located in a housing, usually in the

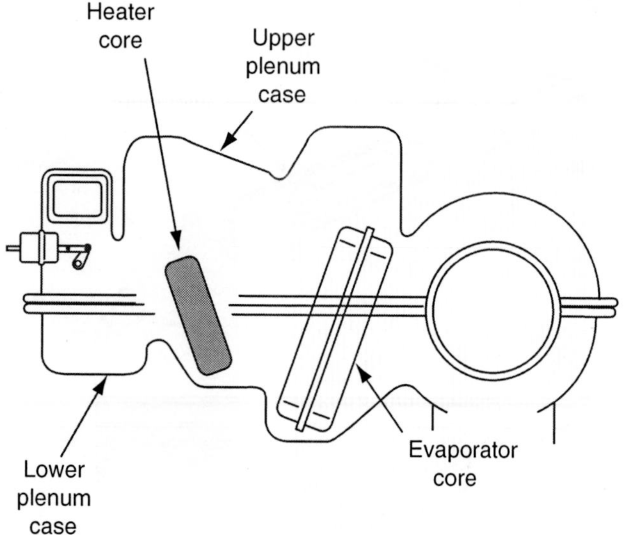


**Figure 7.** The heater core uses hot engine coolant to warm the passenger compartment.

passenger compartment of the vehicle **(Figure 7)**. Some of the hot engine coolant is routed to the heater core by hoses. The heat is dispersed to the air inside the vehicle, thus warming the passenger compartment. To aid in quicker heating of the compartment, a heater fan blows the radiated heat into the compartment.

## HOSES

Hoses are used to direct the coolant from the engine into the radiator and back to the engine. In addition, hoses are used to direct coolant from the engine into the heater core. On some engines, a bypass hose is used to bypass coolant around the thermostat when the thermostat is closed during engine warmup. This action prevents excessive coolant pressure in the engine.

Since radiator and heater hoses deteriorate from the inside out, most manufacturers recommend periodic replacement of the hoses as preventive maintenance. Whenever the engine is rebuilt, the hoses should always be replaced.

## WATER PUMP

The water pump is the heart of the engine's cooling system. It forces the coolant through the engine block and into the radiator and heater core **(Figure 8)**. The water pump may be driven by a V-belt, ribbed V-belt, timing belt,

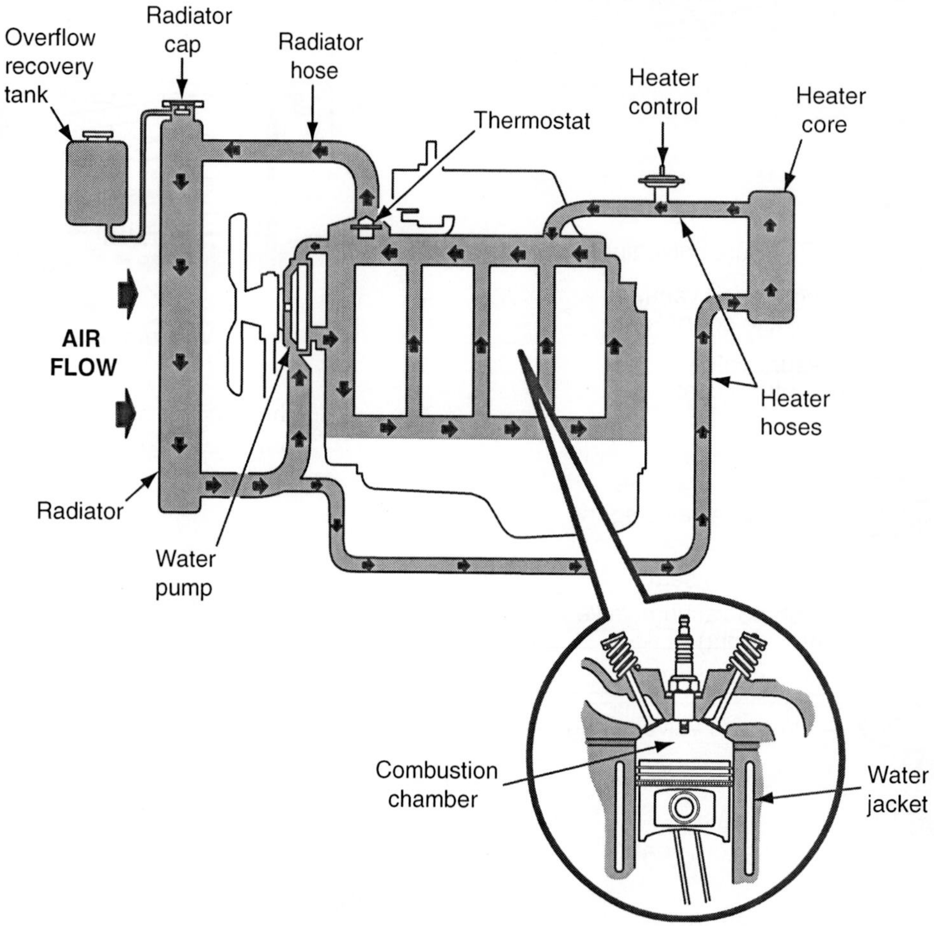


**Figure 8.** Coolant flow through the engine.

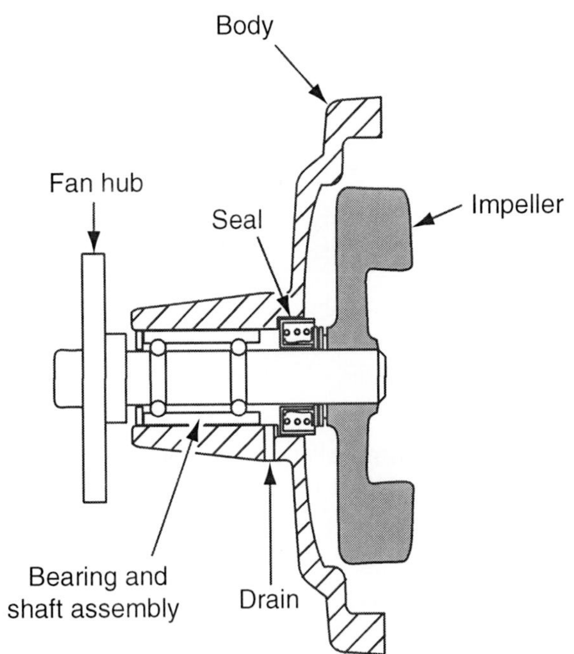


**Figure 9.** Most water pumps use an impeller to move the coolant through the system.

or directly from the camshaft. Most water pumps are centrifugal design, using a rotating impeller to move the coolant **(Figure 9)**. When the engine is running, the impeller rotates, forcing coolant from the inside of the cavity outward toward the tips by centrifugal force. Once inside the block, the coolant flows around the cylinders and into the cylinder heads, absorbing the heat from these components. If the thermostat is open, the coolant will then enter the radiator. The vacuum created by the empty impeller cavity allows the pressurized coolant to be pushed from the radiator to fill the cavity and repeat the cycle. When the thermostat is closed because coolant temperatures are too cold, the coolant is circulated through a bypass. This keeps the coolant circling through the engine block until it becomes warm enough to open the thermostat.

## WATER PUMP DRIVE BELTS

Many water pumps are driven by a V-belt that surrounds the crankshaft and water pump pulleys. This V-belt may also drive other components such as the alternator. The sides of a V-belt are the friction surfaces that contact the sides of the pulley groove and drive the water pump **(Figure 10)**. If the sides of the V-belt are worn and the lower edge of the belt is contacting the bottom of the pulley, the belt will slip. The drive pulleys on all the components driven by a V-belt must be properly aligned. Pulley misalignment causes V-belt edge wear.

A ribbed V-belt is used on many engines and this belt may be used to drive all the belt-driven components. Driving all the belt-driven components with one belt allows all these components to be placed on the same vertical plane

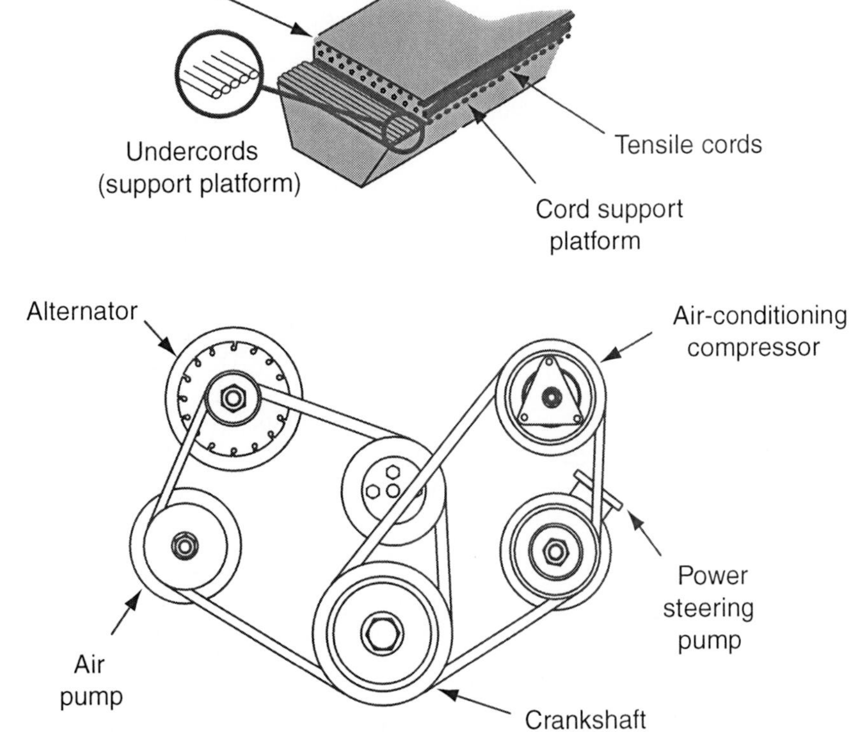


**Figure 10.** A conventional V-belt.

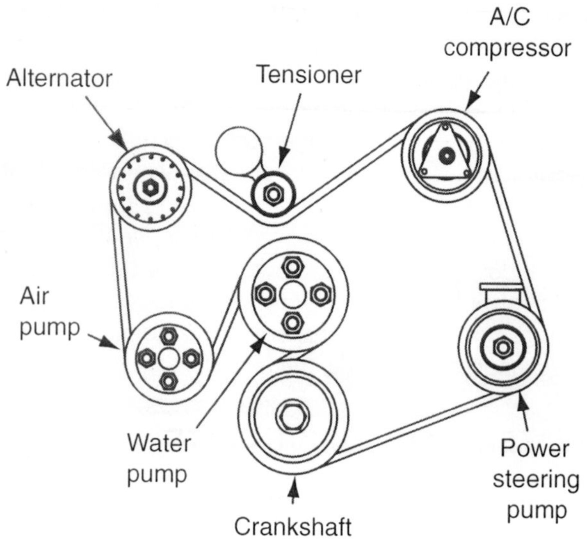


**Figure 11.** A ribbed V-belt.

which saves a considerable amount of underhood space **(Figure 11)**. The smooth backside of a ribbed V-belt may be used to drive one of the components such as the water pump. The backside of the belt also contacts the idler and tensioner pulleys. The ribbed V-belt is much wider than a conventional V-belt and the underside of this belt has a number of small ribbed grooves. Regardless of the type of belt, belt tension is critical to prevent belt slipping. Most ribbed V-belts have an automatic spring-loaded belt tensioner that eliminates periodic belt tension adjustments.

## THERMOSTAT

Control of engine temperatures is the function of the thermostat **(Figure 12)**. The thermostat is usually located at the outlet passage from the engine block to the radiator. When the coolant is below normal operating temperatures, the thermostat is closed, preventing coolant from entering the radiator. In this case, the coolant flows through a bypass passage and returns directly to the water pump.

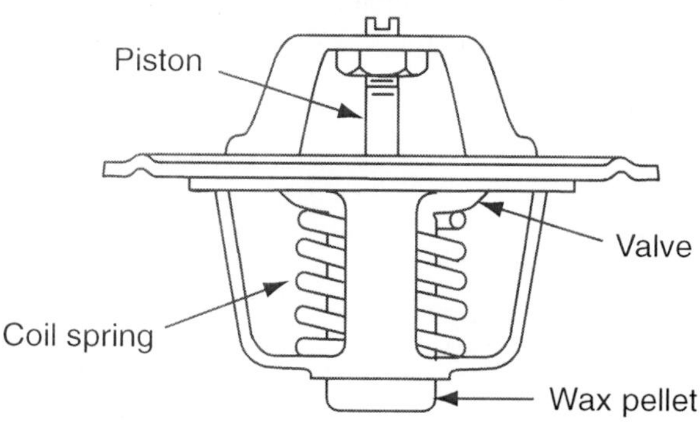


**Figure 12.** The thermostat controls the temperature of the engine.

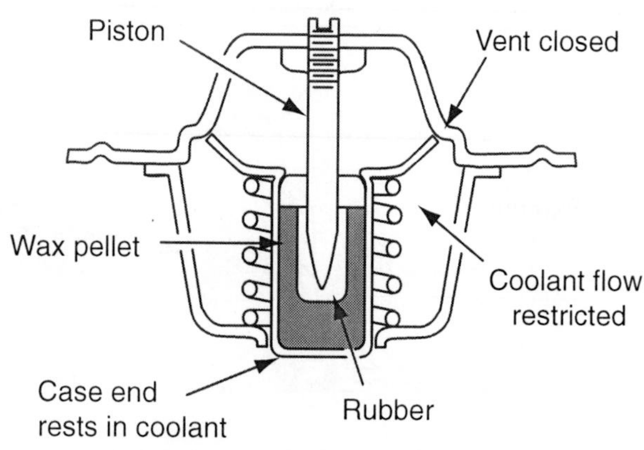


**Figure 13.** The wax pellet controls the flow of coolant through the thermostat.

The thermostat is rated at the temperature it opens in degrees Fahrenheit. If the rating is 195°F (90.5°C), this is the temperature at which the thermostat begins to open. Once the thermostat is open, it allows the coolant to enter the radiator to be cooled. The thermostat cycles open and closed to maintain proper engine temperatures.

Operation of the thermostat is controlled by a specially formulated wax and powdered metal pellet located in a heat-conducting copper cup **(Figure 13)**. When the wax pellet is exposed to heat, it begins to expand. This causes the piston to move outward, opening the valve **(Figure 14)**.

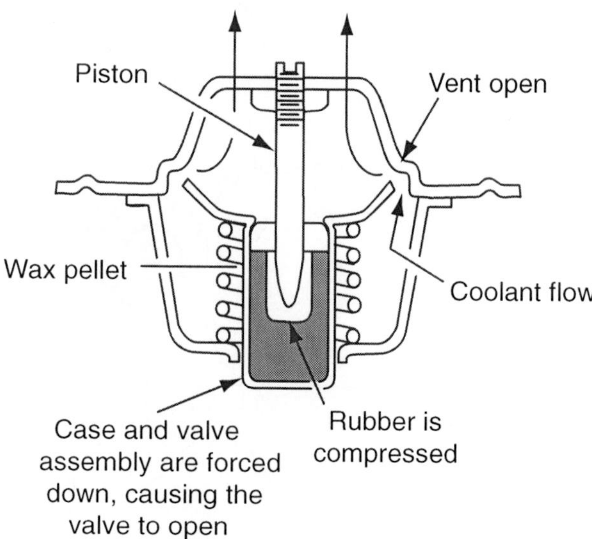


**Figure 14.** When the pellet expands, it opens the thermostat.

The thermostat should be replaced during an engine rebuild or scheduled cooling system service. Use a thermostat with a temperature rating recommended by the engine manufacturer. For proper operation, the thermostat must be installed in the correct direction. In most applications, the pellet is installed toward the engine block, but always refer to the appropriate service manual.

## COOLING FANS

To increase the efficiency of the cooling system, a fan is mounted to direct the flow of air over the radiator cores. In the past, at high vehicle speeds airflow through the grill and past the radiator core was sufficient to remove the heat. The cooling fan was really only needed for low-speed conditions when airflow was reduced. As modern cars have become more aerodynamic, airflow through the grill has declined; thus, proper operation of the cooling fan has become more critical.

Belt-driven fans are usually attached to the water pump pulley **(Figure 15)**. The cooling fan drive belt surrounds the crankshaft and water pump pulleys, and this belt may also drive other components such as the alternator. Some belt-driven fans use a viscous-drive fan clutch to enhance performance **(Figure 16)**. The clutch operates the fan in relation to engine temperature by a silicone oil **(Figure 17)**. If the engine is hot, the silicone oil in the clutch expands and locks the fan to the pump hub. The fan now rotates at the same speed as the water pump. When the engine is cold, the silicone oil contracts and the fan rotates at a reduced speed. Some fan clutches use a thermostatic coil that winds and unwinds in response to the engine temperatures **(Figure 18)**. The coil controls a piston located in a silicone-filled chamber. When it is hot, the coil unwinds and moves the piston into the chamber, which increases the pressure of the oil and locks the clutch.

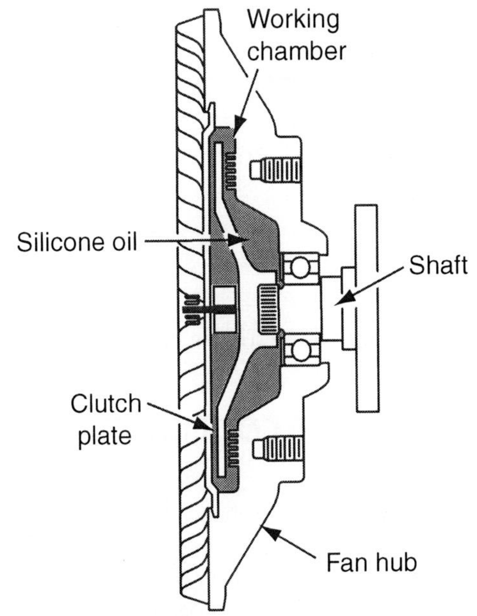


**Figure 17.** The viscous-drive clutch uses silicone oil to lock the fan to the hub.

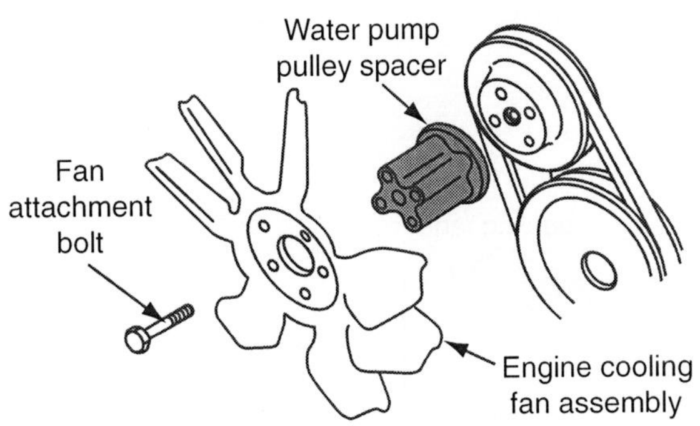


**Figure 15.** Belt-driven cooling fans are usually mounted to the water pump pulley.

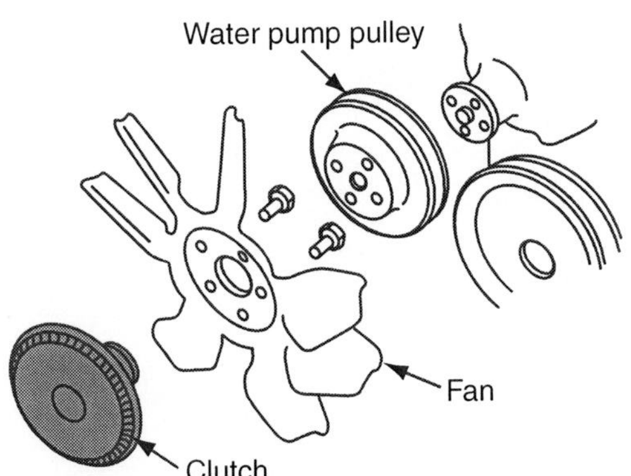


**Figure 16.** Some fans use a viscous-drive fan clutch to reduce noise and load on the engine.

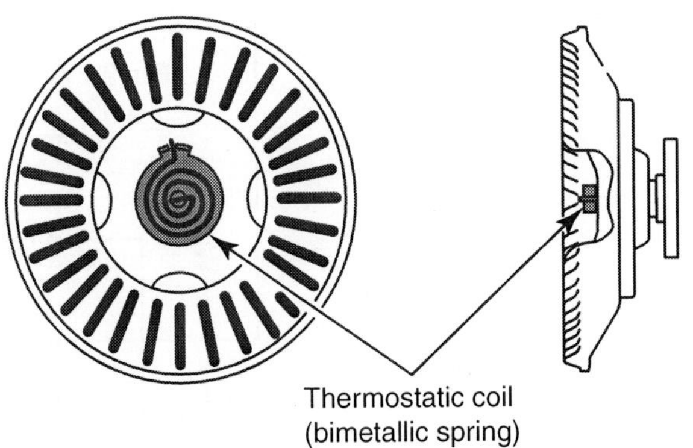


**Figure 18.** A thermostatic spring connected to a piston is another common type of clutch fan.

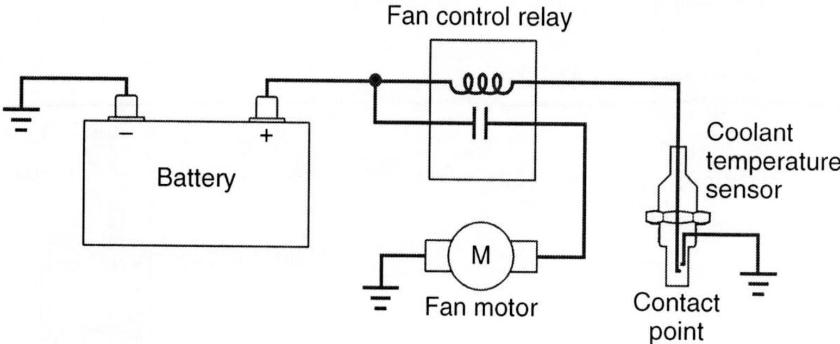


**Figure 19.** Simplified electric fan circuit.

## ELECTRIC-DRIVE COOLING FANS

Electric drive fans are common on today's vehicles because they only operate when needed, thus reducing engine loads. Some of the earlier designs of electric cooling fans use a temperature switch in the radiator or engine block that closes when the temperature of the coolant reaches a predetermined value. With the switch closed, the electrical circuit is completed for the fan motor relay control circuit and the fan turns on **(Figure 19)**. The voltage is supplied to the cooling fan motor from the battery through the relay contacts.

In recent years, most electric-drive cooling fans are operated by the PCM. The PCM receives engine coolant temperature inputs from the thermistor-type sensor. When the thermistor value indicates the temperature is hot enough to turn the fans on, the controller activates the fan control **(Figure 20)**. If the engine coolant is hot, the cooling

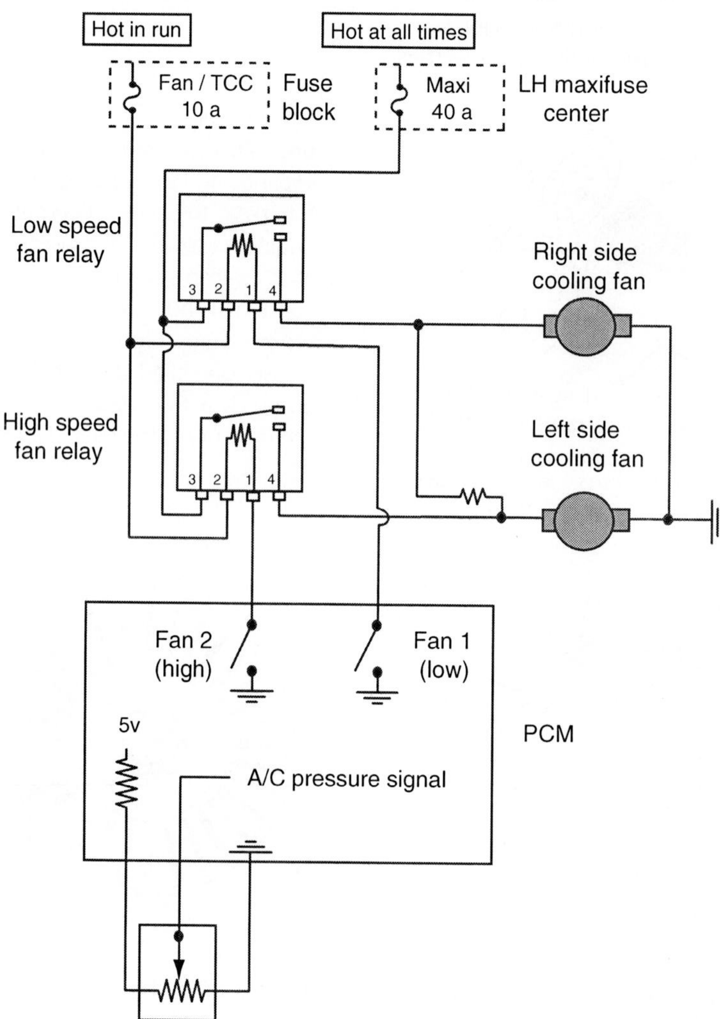


**Figure 20.** An example of the electric fan circuit using a PCM.

fan may be turned on by the controller with the ignition switch off. The controller can also turn on the fans whenever air conditioning is turned on, regardless of the engine temperature.

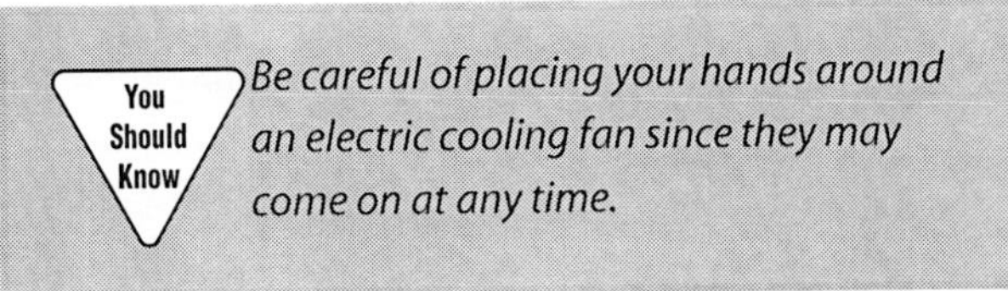


*Be careful of placing your hands around an electric cooling fan since they may come on at any time.*

## COOLING SYSTEM OPERATION

In many cooling systems, the water pump forces coolant into the engine block where it flows around the cylinders. Heat is transferred from the pistons and cylinders to the coolant **(Figure 21)**. The coolant then flows through passages in the head gasket and into the cylinder head where it flows around the valve seats and spark plug recesses. Heat is transferred from the combustion chamber, valves, valve seats, and spark plugs to the coolant. In many engines, coolant also flows through passages in the intake manifold. Coolant flows through the thermostat and the upper radiator hose into the radiator. When coolant flows through the radiator, heat is transferred from coolant to the air flowing through the air passages in the radiator core. Coolant flows from the radiator through the lower radiator hose to the water pump inlet. Many engines now have the thermostat located in a housing on the inlet side of the water pump rather than on top of the intake manifold or cylinder head.

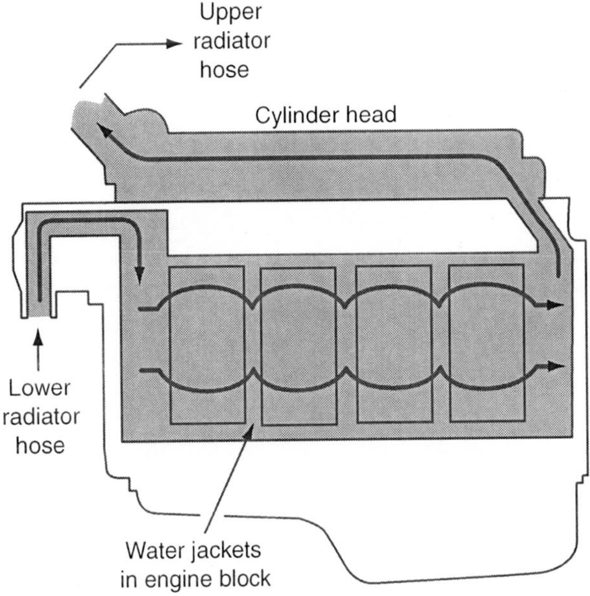


**Figure 21.** The coolant flows from the water pump around all the cylinders before it flows through the cylinder head.

Some engines have a reverse-flow cooling system. In these systems, coolant flows from the water pump into the cylinder head and then it flows through the block **(Figure 22)**.

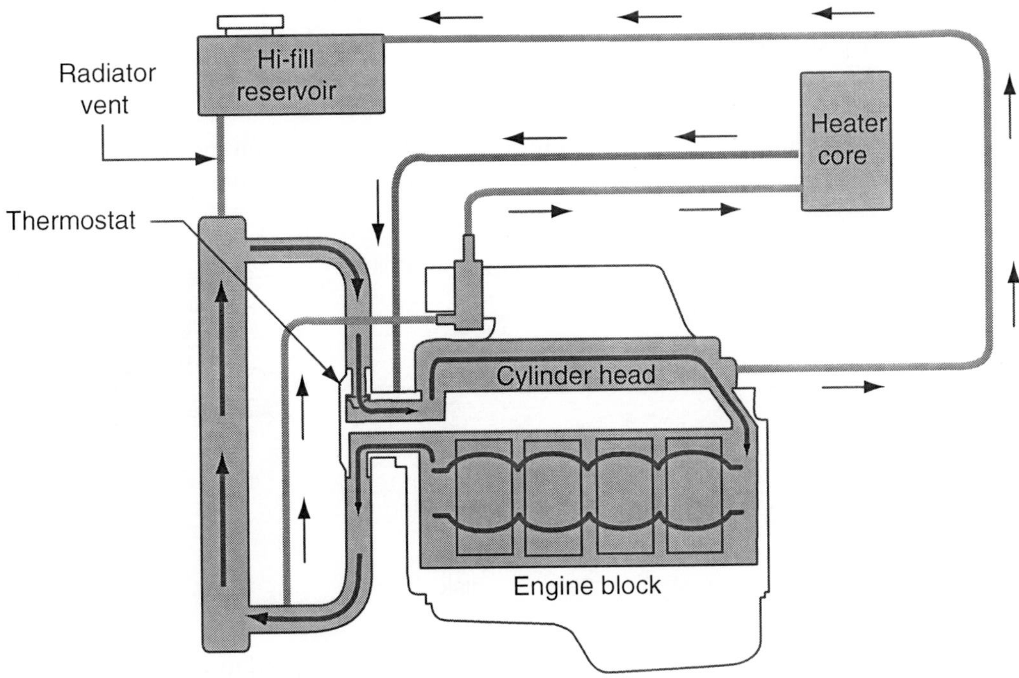


**Figure 22.** A reverse-flow cooling system.

This direction of coolant flow provides improved combustion chamber cooling. This action allows higher compression ratios without engine detonation. The engine **compression ratio** is the relationship between the combustion chamber volume with the piston at TDC and the volume with the piston at BDC. Typical compression ratios are from 8:1 to 10:1 **Detonation** occurs in a combustion chamber during the combustion process. The air-fuel mixture begins to burn normally, but glowing carbon or something ignites the remaining air-fuel mixture creating a second flame front that suddenly meets the original flame front. This action causes a sudden explosion of the remaining air-fuel mixture rather than a smooth burning action. This explosion suddenly drives the piston against the cylinder wall resulting in a rapping noise.

Corrosion or contaminants can plug the coolant passages. This results in reduced heat transfer from the cylinder to the coolant. To prevent the formation of deposits, the cooling system should be flushed and coolant changed at regular intervals as specified by the manufacturer. Any time the engine is removed and rebuilt, it is a good practice to thoroughly clean the coolant passages in the block and cylinder head.

## TEMPERATURE INDICATORS

Vehicle manufacturers provide some method of indicating cooling system problems to the driver. This is done by the use of an indicating gauge or by the illumination of a light. Some manufacturers control the operation of the gauge or light by a computer. Regardless of the type of control, sensors or switches provide the needed input.

Most coolant temperature warning light circuits use a normally open switch **(Figure 23)**. The temperature sending unit consists of a fixed contact and a contact on a bimetallic strip. As the coolant temperature increases, the bimetallic strip bends. As the strip bends, the contacts move closer to each other. Once a predetermined temperature level has been exceeded, the contacts are closed and the circuit to ground is closed. When this happens, the warning light is turned on.

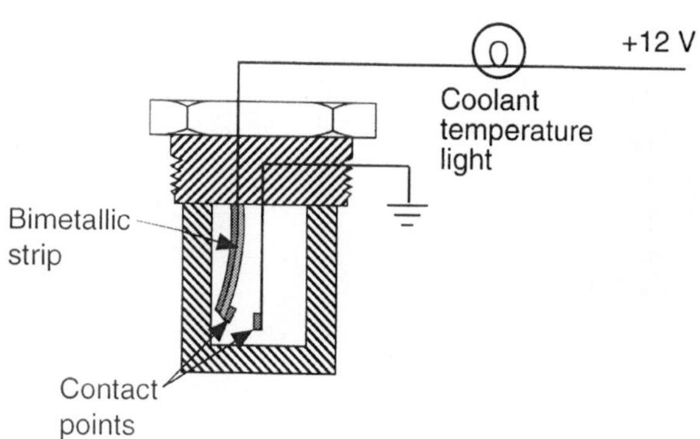


**Figure 23.** A temperature indicator light circuit.

With normally open switches, the contacts are not closed when the ignition switch is turned on. In order to perform a bulb check on normally open switches, a prove-out circuit is included that illuminates the bulb while cranking the engine **(Figure 24)**.

The most common sensor type used for monitoring the cooling system is a thermistor. In a simple coolant temperature-sensing circuit, current is sent from the gauge unit into the top terminal of the sending unit, through the variable resistor (thermistor), and to the engine block (ground). The resistance value of the thermistor changes in proportion to coolant temperature **(Figure 25)**. As the temperature rises, the resistance decreases and the current flow through the gauge increases. As the coolant temperature lowers, the resistance value increases and the current flow decreases.

Many vehicles have a computer-operated message center that displays various warning messages. In these systems, a thermistor-type engine coolant temperature (ECT) sensor is continually sending a voltage signal to the computer in relation to engine coolant temperature. The computer senses the voltage drop across the ECT sensor as this sensor changes resistance. If the coolant temperature

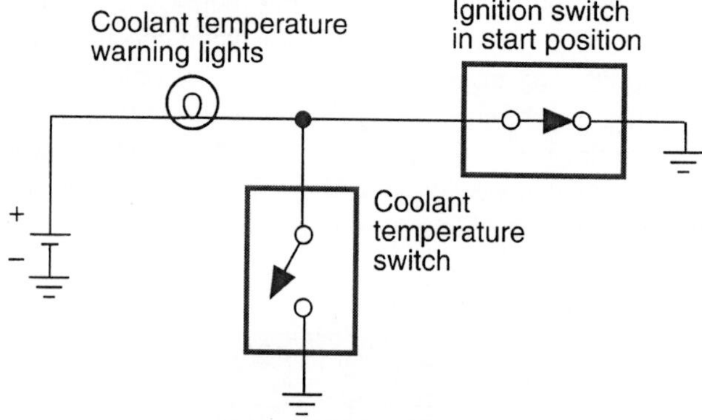


**Figure 24.** A prove-out circuit included in the temperature indicator light system.

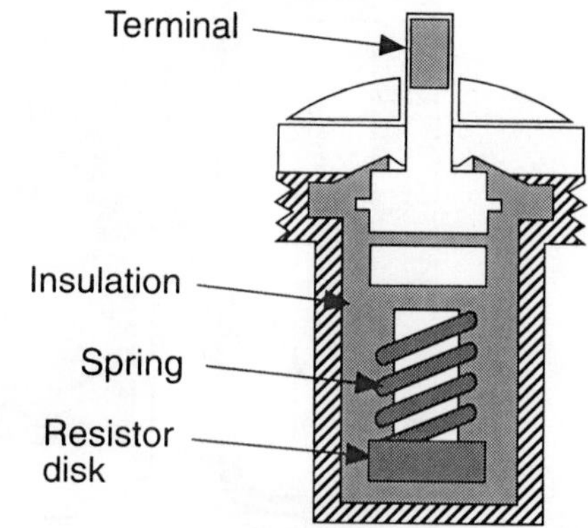


**Figure 25.** A thermistor used to sense engine temperature.

exceeds a predetermined value, the computer illuminates an ENGINE OVERHEATING message in the message center. Some message centers also display a LOW COOLANT message if the engine coolant level in the coolant recovery container is low. These systems have a coolant level sensor in the coolant recovery container.

## Summary

- Many cooling systems are filled with a 50/50 mixture of ethylene glycol and water.
- A 50/50 mixture or ethylene glycol and water freezes at −34°F (−37°C).
- Many radiator caps have a 15 psi (103 kPa) rating.
- If cooling system pressure exceeds the rating of the radiator cap, the cap pressure valve opens and releases coolant into the coolant recovery system.
- When the engine is shut off and the coolant temperature decreases, the vacuum valve in the radiator cap opens and allows coolant to flow from the recovery system into the radiator.
- The coolant is forced through the cooling system by pressure in the water pump.
- Water pump drive belts may be V type or ribbed V type, or the water pump may be driven by the timing belt.
- The thermostat controls engine temperature.
- Belt-driven cooling fans are usually mounted on the water pump pulley.
- Electric-drive cooling fans are usually operated by the engine controller.
- In many cooling systems, the coolant flows through the engine block and then into the cylinder head.
- As the coolant flows through the radiator, heat is transferred from the coolant to the air flowing through the air passages in the radiator.
- Engine temperature indicator lights have a proving circuit that proves the indicator light bulb is working each time the engine is cranked.
- In many systems, a thermistor-type coolant temperature sensor sends a voltage signal to the engine controller in relation to engine temperature. If the engine overheats, the engine controller illuminates a warning light or displays a warning message in the message center.

## Review Questions

1. Technician A says coolant boils at the same temperature regardless of the cooling system pressure. Technician B says the temperature next to the cylinder walls could be 500°F (250°C). Who is correct?
   A. Technician A
   B. Technician B
   C. Both Technician A and Technician B
   D. Neither Technician A nor Technician B
2. Technician A says some radiators have aluminum cores. Technician B says coolant is released into the recovery container if the radiator pressure exceeds the cap pressure rating. Who is correct?
   A. Technician A
   B. Technician B
   C. Both Technician A and Technician B
   D. Neither Technician A nor Technician B
3. Technician A says when the engine cools down, the vacuum valve in the radiator cap opens. Technician B says when the vacuum valve in the radiator cap opens, coolant flows from the recovery container into the radiator. Who is correct?
   A. Technician A
   B. Technician B
   C. Both Technician A and Technician B
   D. Neither Technician A nor Technician B
4. Technician A says a ribbed V-belt system allows the belt-driven components to be on the same vertical plane. Technician B says the friction surface on a V-belt is the underside of the belt. Who is correct:
   A. Technician A
   B. Technician B
   C. Both Technician A and Technician B
   D. Neither Technician A nor Technician B
5. All of these statements about ethylene glycol are true *except*:
   A. Pure ethylene glycol has a boiling point of 330°F.
   B. The freezing point of pure ethylene glycol is −20°F.
   C. The freezing point of a 50/50 ethylene glycol and water mixture is −34°F.
   D. The boiling point of a 50/50 ethylene glycol and water mixture is 226°F.
6. Every pound per square inch (psi) of cooling system pressure increases the coolant boiling point by:
   A. 1°F.
   B. 1.5°F.
   C. 2.0°F.
   D. 2.5°F

7. While discussing cooling system operation, Technician A says a small electric current may flow between the metals in the cooling system through the acid that accumulates in the cooling system. Technician B says anticorrosion agents are mixed with most brands of antifreeze. Who is correct?
   A. Technician A
   B. Technician B
   C. Both Technician A and Technician B
   D. Neither Technician A nor Technician B
8. A typical radiator cap has a pressure rating of:
   A. 8 psi.
   B. 15 psi.
   C. 24 psi.
   D. 32 psi.
9. After a hot engine is shut down and allowed to cool off, the upper radiator hose gradually collapses. The most likely cause of this problem is:
   A. a defective vacuum valve in the radiator cap.
   B. a defective pressure relief valve in the radiator cap.
   C. an engine thermostat that is stuck open.
   D. a leaking water pump seal.
10. A typical engine thermostat has a temperature rating of:
    A. 150°F.
    B. 170°F.
    C. 195°F.
    D. 225°F.
11. A thermostat is opened by a ____________ ____________ in a copper cup.
12. The engine thermostat is usually installed with the pellet facing towards the ____________.
13. Many fan clutches have a reservoir filled with ____________ ____________.
14. In a reverse-flow cooling system, the coolant flows from the water pump directly into the ____________ ____________.
15. Explain the operation of a coolant temperature warning light, including the prove-out circuit.
16. Explain the resistance change in a thermistor in relation to temperature.
17. Explain two messages that may be displayed in the message center in relation to engine temperature and discuss the cooling system conditions necessary to illuminate these warnings.
18. Describe the coolant flow through a conventional cooling system.

# Chapter 14 Engine Cooling System Maintenance, Diagnosis and Service

## Introduction

It is the function of the engine's cooling system to remove heat from the internal parts of the engine and dissipate it into the air. Proper engine operation and service life depend upon the cooling system functioning as designed.

### COOLING SYSTEM MAINTENANCE

Proper cooling system maintenance is extremely important! If cooling system maintenance is not performed, engine overheating may occur, and this condition may require a tow truck to haul the vehicle to an automotive repair shop. This experience can be time consuming and expensive. If the engine overheating is severe or prolonged, engine damage may occur.

### Checking Coolant Level

Checking the coolant level on older vehicles without a recovery system should only be done when the engine is cool. These vehicles require the radiator cap to be removed so the technician can see if the coolant level is above the radiator tubes. Regardless of the system used, removing the radiator cap on a hot engine will release the pressure in the system, causing the coolant to boil immediately. This causes coolant to be expelled from the radiator at a high rate, which can result in severe burns.

Most vehicles are now equipped with a coolant recovery system, so the radiator cap usually does not need to be removed to check or add coolant to the system. The coolant recovery system provides quick visual checks of coolant level through the translucent bottle. Simply observe that the level is between the ADD and FULL marks while the engine is idling and warmed to normal operating temperatures. If coolant needs to be added, the coolant should be added directly to the coolant recovery tank.

> **You Should Know** *On some vehicles, the radiator cap is mounted on the recovery container rather than on the radiator. The recovery container is designed to withstand cooling system pressure. Never loosen the radiator cap on this type of recovery container with the engine warm or hot.*

The use of aluminum cylinder heads, blocks, and radiator cores require the selection of proper coolant. Most manufacturers recommend antifreeze containing Alugard 340-2 or its equivalent.

### Hose Inspection

Check all the radiator and heater hoses for soft spots, cracks, bulges, deterioration, and oil contamination. Replace hoses that indicate any of these conditions. Inspect the radiator and heater hoses for contact with other components that could rub a hole in the hose. Reroute hoses away from other components if necessary. Be sure all the hose clamps are tight and in satisfactory condition.

### Drive Belt Inspection and Tension Testing

Inspect the water pump drive belt for cracks, missing chunks, fluid contamination, fraying, and bottoming in the

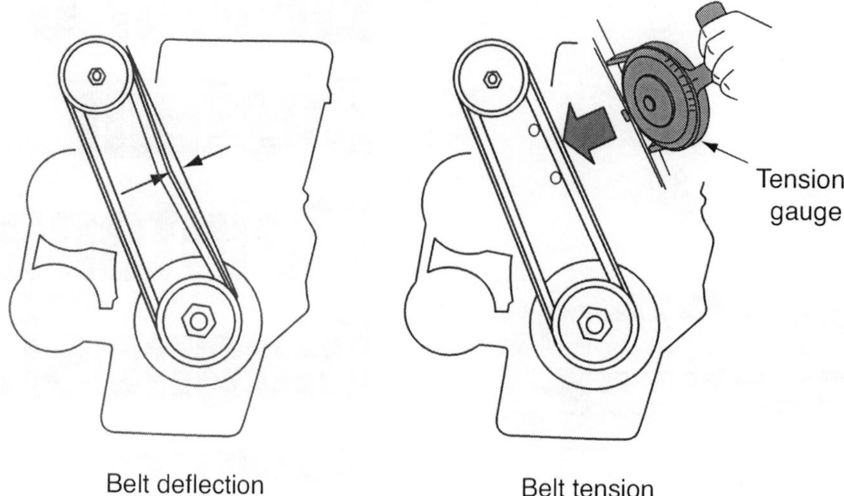


**Figure 1.** Methods of checking bolt tension.

pulley. If the belt indicates any of these conditions, belt replacement is necessary. Belt tension may be tested with the engine stopped and a belt tension gauge installed over the belt at the center of the longest belt span **(Figure 1)**. Belt tension may also be tested by measuring the specified belt deflection in the center of the longest belt span. When the belt tension is not within specifications, adjust the belt as necessary. A belt that is worn or loose will not turn the fan at a sufficient speed to draw the optimum mass of air over the radiator. Many ribbed V-belts have an automatic tensioner and do not require adjusting. Many automatic belt tensioners have a ½-inch drive opening in which a ratchet or flex-handle may be inserted to move the tensioner off the belt during belt replacement **(Figure 2)**. Some automatic belt tensioners have a wear indicator that indicates the amount of belt wear **(Figure 3)**.

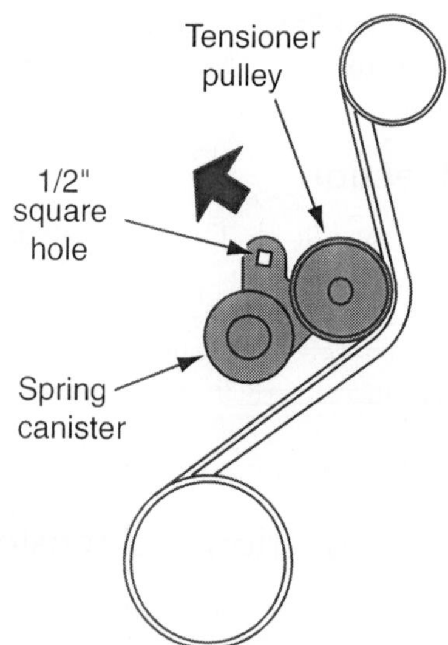


**Figure 2.** A one-half inch drive opening in the tensioner pulley.

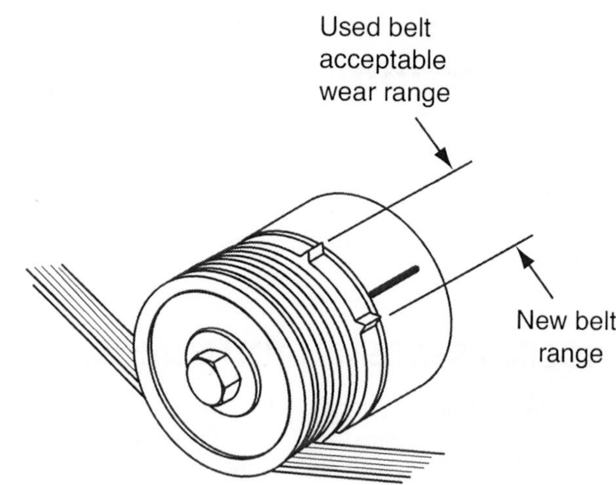


**Figure 3.** A belt tension scale.

## Coolant Testing

The antifreeze content in the coolant may be tested with a coolant hydrometer. A **coolant hydrometer** measures coolant weight or specific gravity to indicate the antifreeze content of the coolant.

The vehicle manufacturer's specified antifreeze content must be maintained in the cooling system because the antifreeze contains a rust inhibitor to protect the cooling system. Insert the hydrometer pickup tube into the coolant level in the recovery container and squeeze the hydrometer bulb to pull a coolant sample into the hydrometer to the specified level. The freezing point of the coolant is indicated on the hydrometer float.

## COOLANT CONTAMINATION

Inspect the coolant for rust, scale, corrosion, and other contaminants such as engine oil or automatic transmission fluid. If the coolant is contaminated, the cooling system must be drained and flushed. If oil is visible floating on top of the coolant, the automatic transmission cooler may be leaking. This condition also contaminates the transmission fluid with coolant. If the vehicle has an external engine oil cooler, it may also be a source of oil contamination in the coolant. These coolers may be pressure tested to determine if they are leaking. Testing or repairing radiators and internal or external oil coolers is usually done by a radiator specialty shop.

## COOLING SYSTEM DIAGNOSIS

Inspect the air passages through the radiator core and external engine oil cooler for contamination with debris or bugs. These may be removed from the radiator core with an air gun and shop air, or water pressure from a water hose.

Visually inspect the radiator and all cooling system hoses for leaks. Leaking components must be replaced or repaired. Inspect the heater core area for signs of coolant leaks. Some vehicles have the heater core mounted under the dash while other vehicles have this component under the hood. If the heater core is mounted under the dash, a leaking core may cause coolant to drip on the floor mat.

Inspect the engine for coolant leaks at locations such as the thermostat housing, core plugs, and water pump. Coolant leaks at the water pump usually appear at the water pump drain hole or behind the pulley. A leaking water pump must be replaced.

## DRAINING THE COOLING SYSTEM

Since most manufacturers recommend periodic coolant changes, a drain plug is usually provided at the bottom of the radiator **(Figure 4)**. The following is a typical procedure for draining the cooling system:

1. Allow the engine to cool. Never loosen the radiator cap or open the drain plug when the engine is hot.
2. Start the engine.
3. Move the temperature selector of the heater control panel to the full heat position.
4. Shut the engine off before it begins to warm.
5. Without removing the radiator cap, loosen the drain plug.
6. The coolant recovery tank should empty first. Then remove the radiator cap to drain the rest of the system.
7. On most engines it is necessary to drain the block separate from the radiator. Remove the drain plug(s) from the side of the engine block to drain the block.

## FILLING THE COOLING SYSTEM

The cooling system is usually filled with a mixture of 50/50 antifreeze to water. To fill the cooling system, make sure all hoses are installed and clamps are tight. Also, close the drain plug before filling. Look up the cooling capacity in the service manual. Fill the system to half of its capacity with 100 percent antifreeze through the radiator cap opening. Then complete filling with water. Since many vehicles are designed with the radiator lower than the engine, a bleed valve is opened when filling the system **(Figure 5)**. Loosen the bleed valve while the radiator is being filled. Do not allow coolant from the open bleed valve to drip on the drive belts. Close the valve when coolant begins to flow out in a steady stream without bubbles. Leave the radiator cap off, start the engine, and let the engine warm up. Continue to fill the radiator as needed as the engine warms. When the radiator is full, install the radiator cap and fill the recovery container to the cold level. Run the engine until it reaches normal operating temperature. It may be necessary to add additional antifreeze mix to the recovery tank. On some vehicles, it may take as many as four warm-up cycles

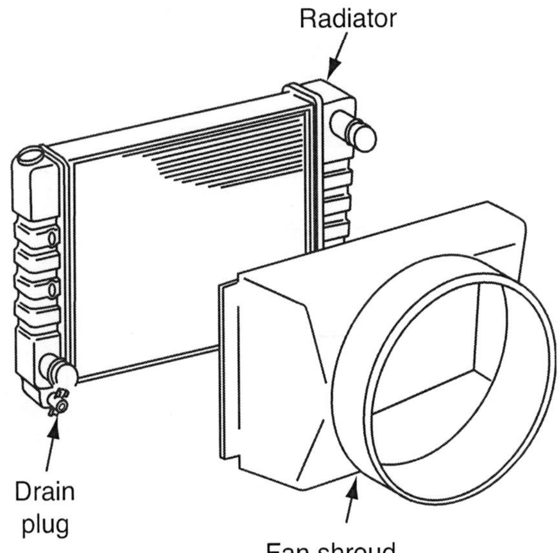


**Figure 4.** A drain plug in the lower radiator tank.

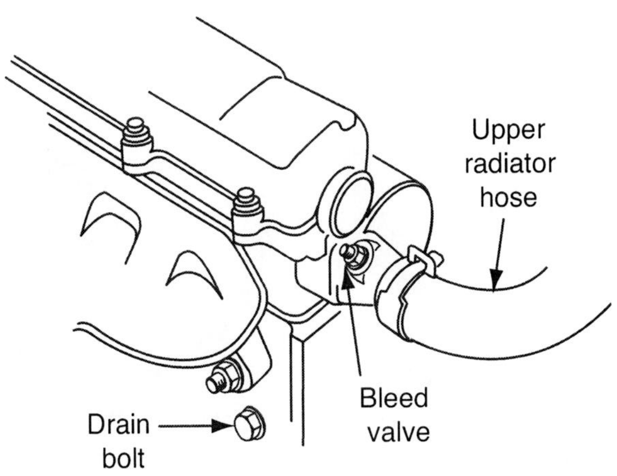


**Figure 5.** A bleed valve to remove trapped air from the cooling system.

before all of the air is removed from the system and the recovery tank equalizes.

## COOLING SYSTEM FLUSHING

The effectiveness of antifreeze and the additives mixed with it decreases over time. All manufacturers have a maintenance requirement for the cooling system. The recommended schedule for drain, flush, and refill ranges from once a year to once every five years. At the same time that the refill is performed, all cooling system components should be inspected.

Flushing of the cooling system is accomplished by using pressurized water forced through the cooling system in a reverse direction to the normal coolant flow. A special flushing gun mixes low air pressure with water. Reverse flushing causes the deposits to dislodge from the various components. The engine block and radiator should be flushed separately. To flush the radiator, drain the radiator and disconnect the upper and lower radiator hoses. A long hose may be attached to the upper hose outlet to deflect the water. Disconnect and plug any heater hoses that are attached to the radiator. Fit the flushing gun to the lower hose opening. This causes the radiator to be flushed upward **(Figure 6)**. Fill the radiator with water and turn the gun on in short bursts. Continue to flush the radiator until the water being expelled from the upper hose outlet is clean.

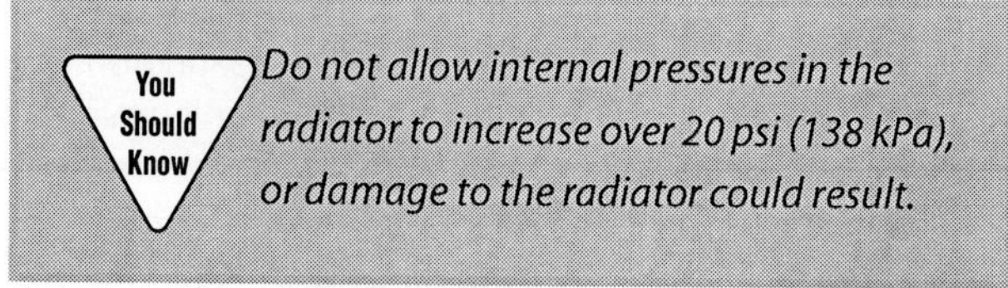


*Do not allow internal pressures in the radiator to increase over 20 psi (138 kPa), or damage to the radiator could result.*

To flush the engine block, disconnect the radiator upper and lower hoses. Also, remove the thermostat and reinstall the thermostat housing. Install the flushing gun to the thermostat housing hose **(Figure 7)**. Turn the water on until the engine is full. Then turn the air on in short bursts. Allow the engine to refill with water between blasts of air. Repeat this process until the water runs clean.

Water is usually sufficient to remove most contaminants from the cooling system. However, aluminum hydroxide deposits require a two-part cleaner to remove. The two parts are an oxalic acid and a neutralizer. The acid is usually added to the system first. Then the engine is allowed to idle for the specified length of time. The cooling system is then flushed. After the flush is completed, the neutralizer is added to prevent the acid from damaging metal components.

If chemical cleaning fails to remove the deposits, the radiator will need to be removed and cleaned out. Internal radiator cleaning is usually performed in a radiator repair shop.

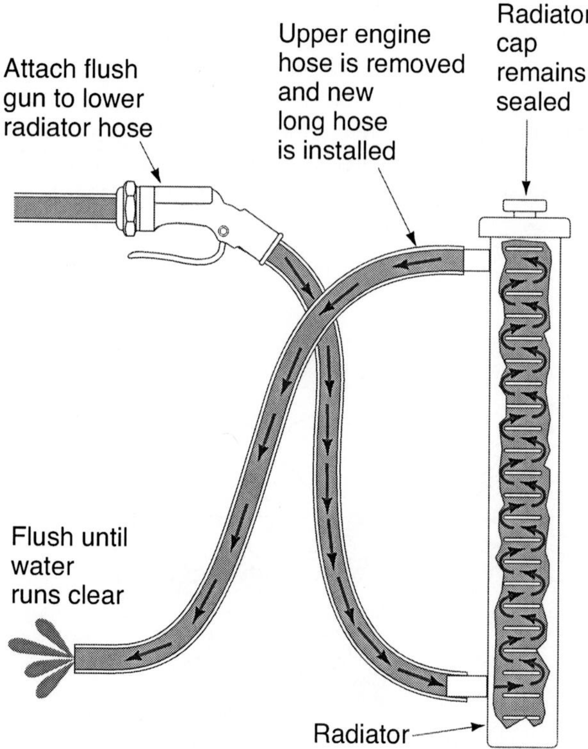


**Figure 6.** Reverse flushing the radiator.

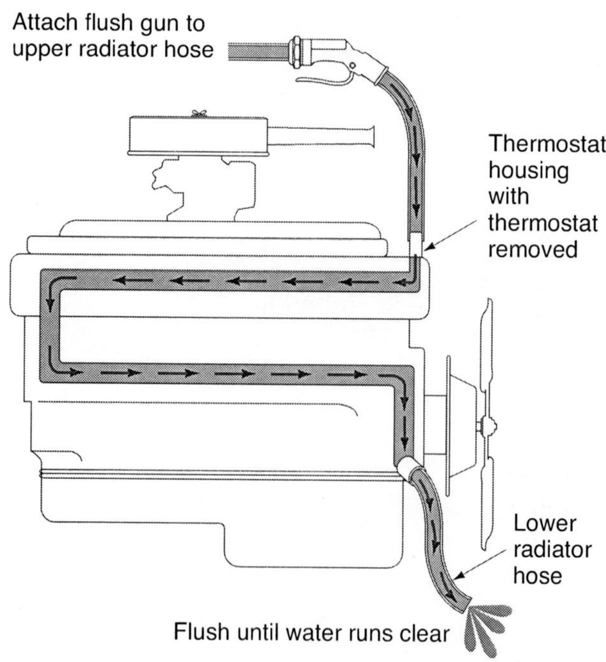


**Figure 7.** Reverse flushing the engine block.

## DIAGNOSIS OF IMPROPER OPERATING TEMPERATURE

If the engine is operating at a lower-than-normal temperature, the customer may complain about insufficient heat from the heater in cold weather. On fuel-injected vehicles with this problem, the customer may complain about reduced fuel mileage. When the engine operating temperature is below normal, the engine coolant temperature (ECT) sensor sends a cold signal to the engine controller. When this signal is received, the computer provides a richer air-fuel ratio, resulting in reduced fuel mileage. If the engine operating temperature is below normal, the most common cause is a defective thermostat. The **engine coolant temperature (ECT) sensor** is a thermistor-type sensor that is usually mounted in the engine block or cylinder head. The ECT sensor sends a voltage signal to the engine controller in relation to coolant temperature.

When the engine operating temperature is above normal, the thermostat may not be opening properly or it may be improperly installed. Most thermostats are installed in a housing on top of the cylinder head or intake manifold. These thermostats are mounted on the outlet side of the water pump and the thermostat pellet must face toward the engine block. On some late-model engines, the thermostat is mounted on the inlet side of the water pump.

You Should Know

*The 4.8L, 5.3L, and 6.0L engines in General Motors light-duty trucks have the thermostat mounted at the water pump inlet. On these engines, the water pump inlet and the thermostat must be replaced as an assembly.*

Other causes of engine overheating are: a loose water pump and fan drive belt, inoperative electric-drive cooling fan, a defective radiator pressure cap, a damaged radiator shroud, restricted air passages through the radiator, restricted coolant passages through the radiator, a defective thermostatic fan clutch, late ignition timing or insufficient spark advance on the engine, or a defective head gasket. Engine overheating causes a high coolant level in the recovery container. The technician must inspect and test these components or circuits to find the exact cause of the overheating problem.

## THERMOSTAT TESTING

If a customer brings in a vehicle with an overheating problem, it is possible the thermostat is not opening. An engine that fails to reach operating temperature may be caused by a faulty thermostat. To check thermostat operation, it must be removed from the engine. Visually inspect the thermostat for rust or other contamination. Check the temperature rating of the thermostat and confirm it is the proper one for the engine application. Also confirm it was installed in the right direction. In many engines, the thermostat pellet faces toward the engine block.

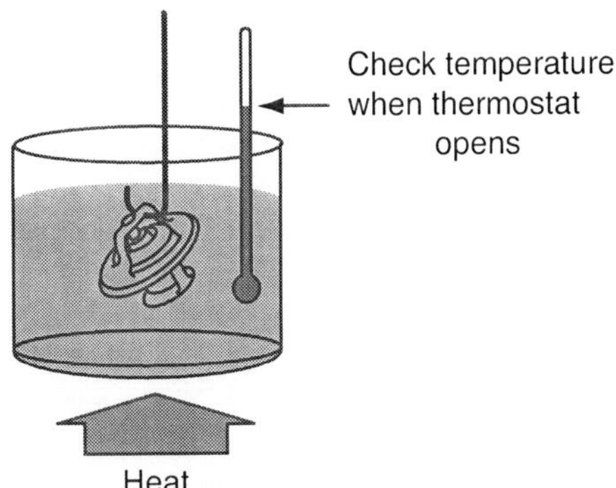


**Figure 8.** Thermostat testing.

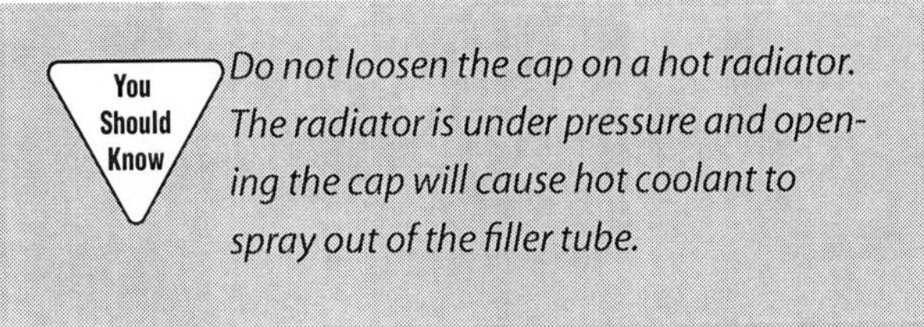
You Should Know

*Do not loosen the cap on a hot radiator. The radiator is under pressure and opening the cap will cause hot coolant to spray out of the filler tube.*

To test the thermostat, submerge it in a container of water **(Figure 8)**. Use a thermometer so the temperature can be determined when the thermostat opens. Heat the water while observing the thermostat. At the rated temperature of the thermostat, it should begin to open.

## PRESSURE CAP DIAGNOSIS AND COOLING SYSTEM LEAK DIAGNOSIS

With the engine cool, remove the radiator cap and connect the pressure tester to the radiator filler neck. If the pressure cap is located on the recovery container, install the pressure tester on the filler neck on this container **(Figure 9)**.

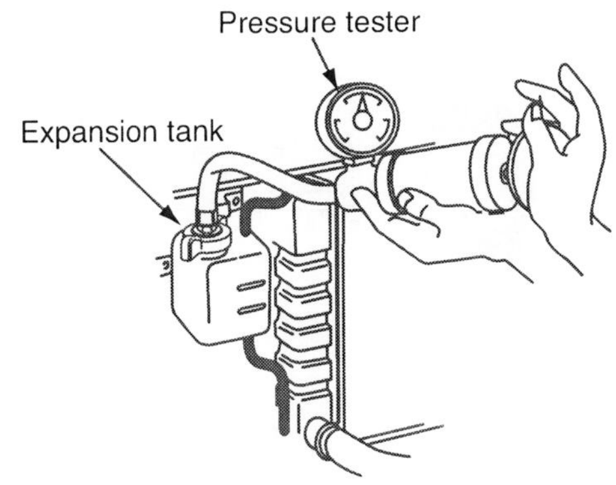


**Figure 9.** Pressure testing the cooling system.

Pump the tester handle until the gauge on the tester indicates the rated cooling system pressure stamped on the pressure cap. Leave the pressure applied to the cooling system for 5 to 15 minutes. If the pressure does not drop on the tester gauge, the cooling system is not leaking. When the reading on the tester gauge slowly drops, the cooling system is leaking. A visual inspection of the complete cooling system usually reveals the source of the coolant leak. Always remember to inspect the expansion plugs in the sides of the engine block for coolant leaks. These plugs may become rusted after a period of time. Rusted or leaking expansion plugs must be replaced.

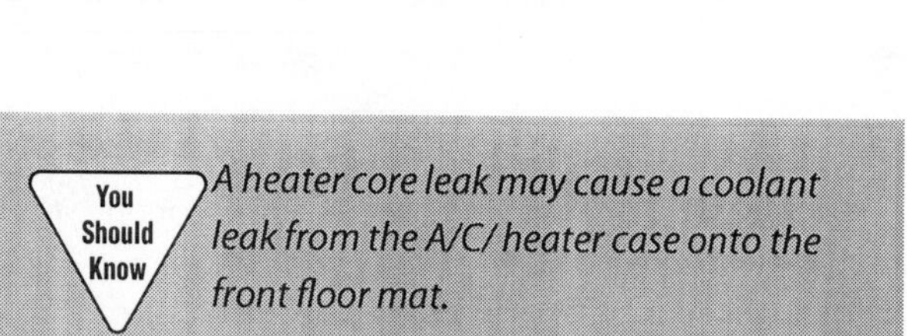


*A heater core leak may cause a coolant leak from the A/C/ heater case onto the front floor mat.*

If the pressure on the tester gauge dropped off but there is no evidence of an external leak, the leak could be in the head gasket, combustion chamber, or transmission cooler. To check head gasket and combustion chamber leaks, remove the spark plugs and check for indications of moisture in the cylinders when performing an engine compression test. The transmission cooler may be pressure tested to check it for leaks.

To test the radiator cap, use a special adaptor to connect the cap to the pressure tester **(Figure 10)**. Operate the tester pump and observe the pressure gauge. The cap should release the pressure in the tester at the specified pressure stamped on the cap. Radiator cap replacement is necessary if the cap does not release the pressure at the cap rating or fails to hold the pressure. Always check the vehicle manufacturer's specifications to be sure the radiator cap has the correct pressure rating.

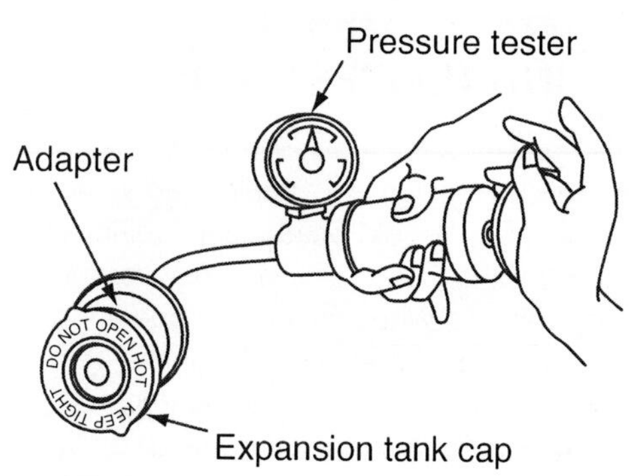


**Figure 10.** Testing the radiator pressure cap.

## COOLING SYSTEM SERVICE

The fan is driven by a drive belt from the crankshaft or operated electrically **(Figure 11)**. Regardless of the design used, inspect the fan blades for stress cracks. The fan blades are balanced to prevent damage to the water pump bearings and seals. If any of the blades are damaged, replace the fan.

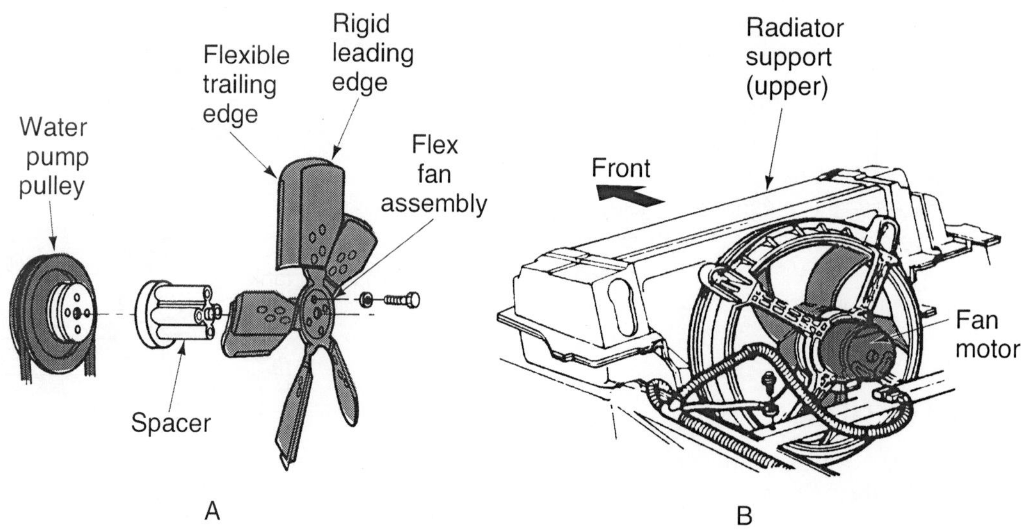


**Figure 11.** (A) Belt-driven fan. (B) An electric-driven fan.

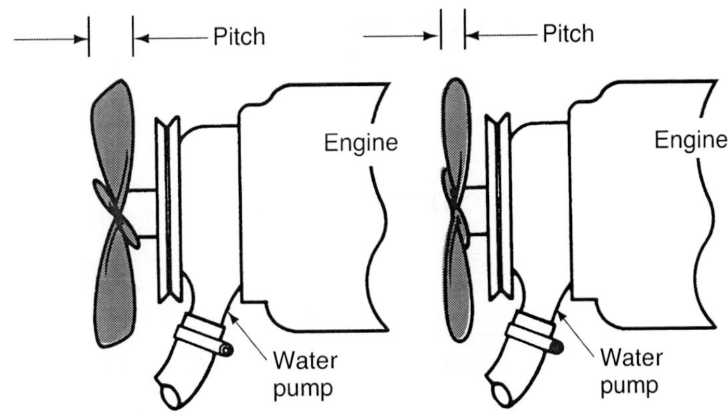


**Figure 12.** The flex fan changes its pitch in relation to engine speed.

Most modern engines with belt-driven fans use either flex fans **(Figure 12)** and/or a viscous-drive fan clutch **(Figure 13)**. The flex fan changes its pitch in relation to engine speed. A **flex fan** has blades that are designed to flex or straighten out as engine and fan speed increases. This action moves more air through the radiator at low speed and less air through the radiator at high engine speed when vehicle motion is forcing air through the radiator.

Flex fans are inspected for stress cracks as any other fan. The viscous fan clutch should be visually inspected for silicone leakage indicated by black streaks radiating outward from the center of the fan clutch. Grasp the fan blades with your hands and move these blades back and forth while checking for excessive movement in the viscous clutch. Next, observe movement of the thermostatic spring coil and shaft. If the amount of movement is out of specification, replace the clutch assembly. Also, the shaft should rotate with the coil. To check the thermostatic spring, use a screwdriver to lift one end of the spring out of its retaining slot **(Figure 14)**. Rotate the spring counterclockwise until it comes to a stop. Measure the distance from the retainer clip to the end of the spring and compare to specifications **(Figure 15)**. Return the coil spring end to its retainer slot.

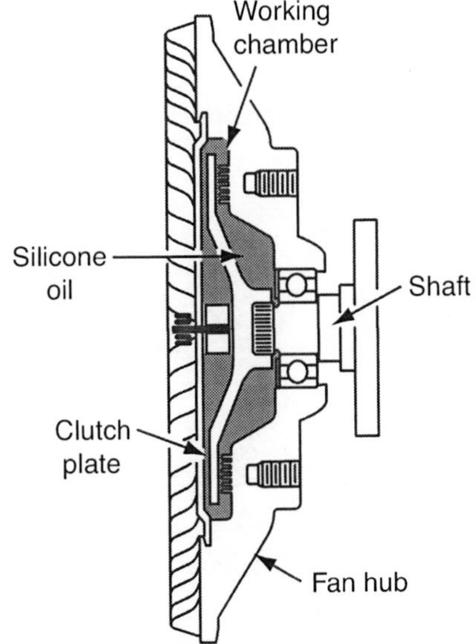


**Figure 13.** A typical viscous-drive fan clutch.

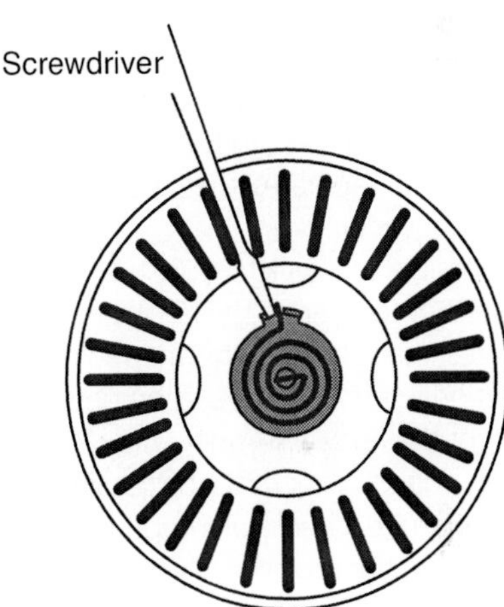


**Figure 14.** Disconnecting the end of the thermostatic spring on the viscous-drive fan clutch.

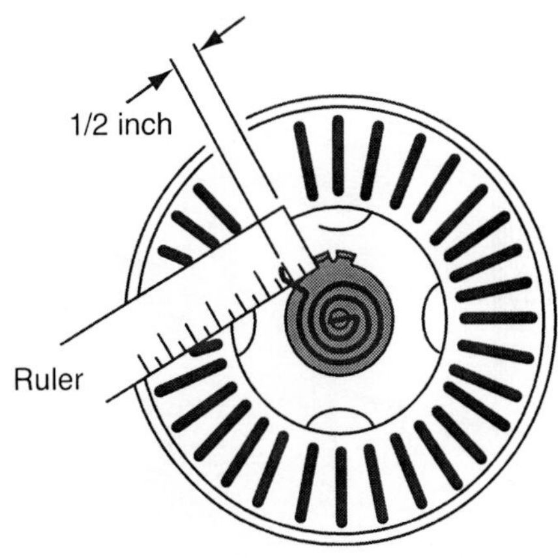


**Figure 15.** Measuring the gap betwen the spring and the clip on the viscous-drive fan clutch.

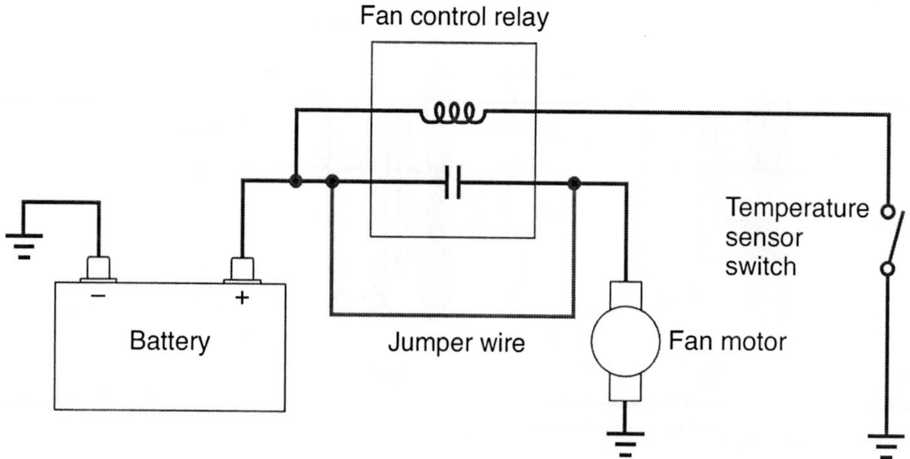


**Figure 16.** Jumping the high-current side of the relay connector to test the fan motor.

Electric fans are also inspected for damage and looseness. If the fan fails to turn on at the proper temperature, the problem could be the temperature sensor, the fan motor, the fan control relay, the circuit wires, or the controller. To isolate the cause of the malfunction, attempt to operate the fan by bypassing its control. On some computer-controlled systems, a scan tool may be used to activate the fan. If the fan operates, the problem is probably in the coolant sensor circuit.

It is also possible to check fan function by jumping the fan relay to attempt to operate the fan motor **(Figure 16)**. If the fan operates, the relay may be the faulty component; however, additional tests will have to be performed on the control circuit of the relay. A jumper wire can also be used to jump battery voltage directly to the cooling fan. If the fan motor fails to operate, check for proper ground connections before faulting the motor.

To direct airflow more efficiently, many manufacturers use a shroud **(Figure 17)**. Proper location of the fan within the shroud is also required for proper operation. Generally, the fan should be at least 50 percent inside the shroud. If the fan is outside the shroud, the engine may experience overheating due to hot under-hood air being drawn by the fan instead of the cooler outside air. If the shroud is broken, it should be repaired or replaced. Do not drive a vehicle without the shroud installed.

Inspect the idler pulley and belt tensioner, if so equipped **(Figure 18)**. Many manufacturers use idler pulleys so components such as generators, air pumps, A/C compressors, and so forth will be provided a greater area of belt contact on their pulleys **(Figure 19)**. Tensioners are used to maintain proper drive belt tension. Test the tensioner pulley for free rotation and inspect for any worn grooves or looseness. If the pulley fails inspection, it must be replaced.

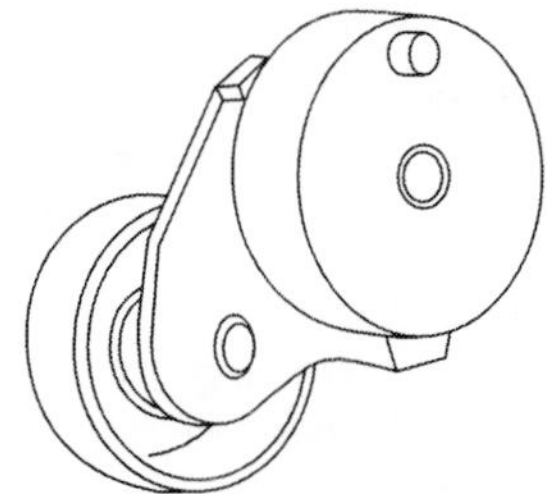

**Figure 18.** A typical belt tensioner.

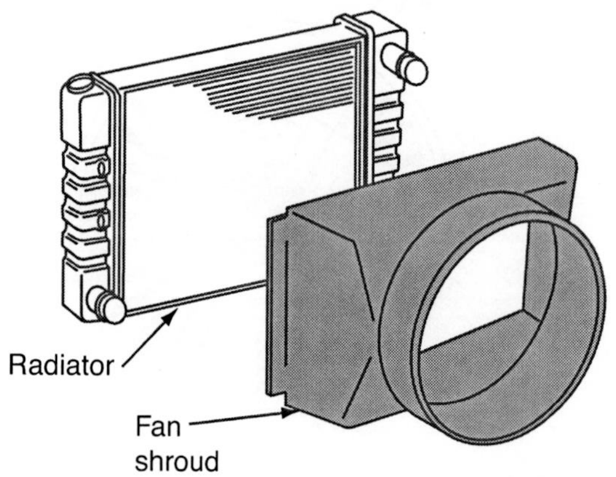


**Figure 17.** The fan shroud increases airflow through the radiator.

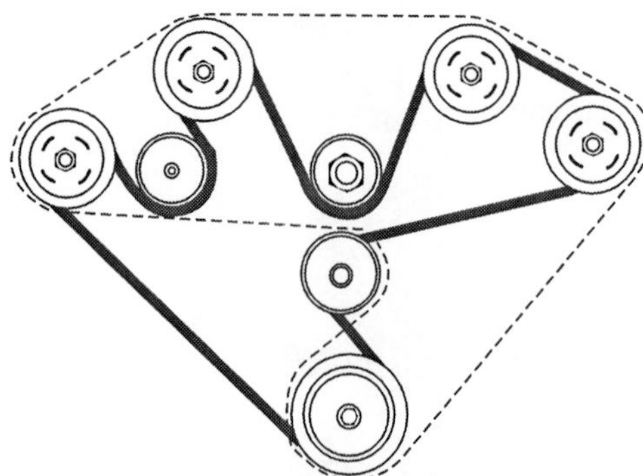

**Figure 19.** Idler pulleys provide increased belt surface area contact. The dotted line illustrates the belt routing without the idler pulleys. The increased belt surface contact prevents belt slipping.

## Summary

- In most cooling systems, the coolant level may be checked by visually inspecting the level in the coolant recovery container.
- Belt tension may be measured with a belt tension gauge or by measuring the belt deflection in the center of the longest belt span.
- Antifreeze content in the coolant is measured with a coolant hydrometer.
- The radiator pressure cap must never be loosened with the engine warm or hot.
- In many cooling systems, a bleed valve is opened to allow air to escape from the system during the filling process.
- Cooling systems may be reverse flushed to remove contaminants.
- A defective thermostat may cause engine temperature to be lower or higher than normal.
- The engine thermostat may be removed and tested in a container filled with hot water.
- A pressure tester is used to leak test the cooling system and test the radiator pressure cap.
- Fan blade assemblies must be replaced if any of the blades have stress cracks.
- A damaged fan shroud causes engine overheating.

## Review Questions

1. Technician A says the water pump drive belt tension may be tested with a belt tension gauge mounted on the shortest belt span. Technician B says belt tension may be tested by measuring the belt deflection at the center of the shortest span. Who is correct?
   A. Technician A
   B. Technician B
   C. Both Technician A and Technician B
   D. Neither Technician A nor Technician B
2. Technician A says some automatic belt tensioners have a $1/2$-inch drive opening in which a flex handle may be inserted to lift the tensioner off the belt. Technician B says some automatic belt tensioners have a wear indicator to show the amount of belt wear. Who is correct?
   A. Technician A
   B. Technician B
   C. Both Technician A and Technician B
   D. Neither Technician A nor Technician B
3. Technician A says if coolant is leaking from the water pump drain hole, replacing the water pump gasket will fix the leak. Technician B says coolant dripping on the front floor mat indicates a leaking air conditioning evaporator. Who is correct?
   A. Technician A
   B. Technician B
   C. Both Technician A and Technician B
   D. Neither Technician A nor Technician B
4. Technician A says when filling a cooling system, the bleed valve should be closed. Technician B says reverse cooling system flushing is done with water and low air pressure. Who is correct?
   A. Technician A
   B. Technician B
   C. Both Technician A and Technician B
   D. Neither Technician A nor Technician B
5. All of these statements about cooling system service are true *except*:
   A. The radiator cap should not be loosened when the engine is hot.
   B. In some cooling systems the radiator cap is mounted on the coolant recovery container.
   C. An engine with an aluminum block, cylinder heads, and/or radiator requires the selection of the proper antifreeze.
   D. When the radiator cap is mounted on the coolant recovery container, add coolant to the radiator.
6. When discussing a cooling system that is contaminated with oil, Technician A says the external engine oil cooler may be leaking. Technician B says the automatic transmission oil cooler may be leaking. Who is correct?
   A. Technician A
   B. Technician B
   C. Both Technician A and Technician B
   D. Neither Technician A nor Technician B
7. The most likely result of a loose cooling fan and water pump drive belt is:
   A. damage to the water pump bearing.
   B. engine overheating.
   C. excessive belt wear.
   D. excessive wear on the water pump pulley.
8. When discussing improper engine coolant temperature, Technician A says on a fuel-injected engine with a coolant temperature sensor, lower-than-normal coolant temperature causes a lean air-fuel ratio. Technician B says engine overheating may be caused by a defective head gasket. Who is correct?
   A. Technician A
   B. Technician B
   C. Both Technician A and Technician B
   D. Neither Technician A nor Technician B

9. During a cooling system pressure test, a pressure tester is used to pressurize the cooling system to 15 psi. After 15 minutes, the pressure decreased to 3 psi and there are no visible coolant leaks in the engine compartment or on the front floor of the vehicle. The most likely cause of the pressure decrease is:
   A. a leaking head gasket.
   B. a leaking heater core.
   C. a leaking coolant control valve.
   D. a loose water pump bearing.
10. An electric-drive cooling fan is inoperative at all engine temperatures. Technician A says on some cooling fan circuits, a scan tool may be used to activate the cooling fan. Technician B says the engine coolant temperature sensor may be defective. Who is correct?
    A. Technician A
    B. Technician B
    C. Both Technician A and Technician B
    D. Neither Technician A nor Technician B
11. List eight possible causes of engine overheating.
    1. ______________________________
    2. ______________________________
    3. ______________________________
    4. ______________________________
    5. ______________________________
    6. ______________________________
    7. ______________________________
    8. ______________________________
12. During a cooling system pressure test, the pressure gauge reading drops off and there are no external cooling system leaks. List three causes of these symptoms.
    1. ______________________________
    2. ______________________________
    3. ______________________________
13. Silicone fluid leaking from a viscous fan clutch is indicated by ______________ ______________ radiating outward from the center of the fan clutch.
14. When diagnosing the electric-drive cooling fan on some fuel-injected engines, a ______________ ______________ may be used to activate the cooling fan.
15. Explain why a lower-than-normal engine operating temperature causes reduced fuel mileage on a fuel-injected engine.
16. Explain the most likely cause of coolant dripping on the front floor mat.
17. Describe the correct installation position for a thermostat that is mounted on the outlet side of the water pump.
18. Describe the proper procedure for using a belt tension gauge to test belt tension on the water pump and cooling fan drive belt.

# Chapter 15 Intake and Exhaust Systems

## Introduction

An internal combustion engine requires airflow, fuel, and spark to provide combustion in the cylinders. The airflow is drawn into the engine by the vacuum created in the cylinders during the piston intake strokes and atmospheric pressure outside the air induction system. The air is mixed with fuel and delivered to the combustion chambers. The purpose of the air induction system is to deliver a uniform amount of air to each cylinder.

### THE AIR INDUCTION SYSTEM

The air intake system on a modern fuel-injected engine is rather complicated **(Figure 1)**. Ducts channel cool air from outside the engine compartment to the throttle body assembly. The air filter is placed below the top of the engine to allow for aerodynamic body designs. Electronic sensors measure airflow, temperature, and density. On some engines, pulse air systems provide fresh air to the exhaust stream to oxidize unburned hydrocarbons in the exhaust. These components allow the air induction system to perform the following functions: filter the air to protect the engine from wear; silence air intake noise; heat or cool the air as required; provide the air the engine needs to operate; monitor airflow, temperature, and density for more efficient combustion and a reduction of hydrocarbon (HC) and carbon monoxide (CO) emissions; and operate with the PCV system to burn the crankcase fumes in the engine.

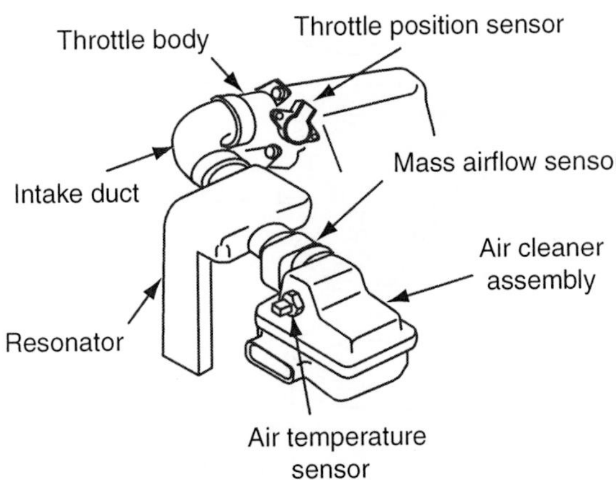


**Figure 1.** An air induction system.

### AIR INTAKE DUCTWORK

The most recent designs have remote air cleaner assemblies with a mass airflow sensor (MAF) installed in the ductwork **(Figure 2)**. The MAF sends a voltage signal to the powertrain control module (PCM) in relation to the total amount of air flowing through the intake system into the cylinders.

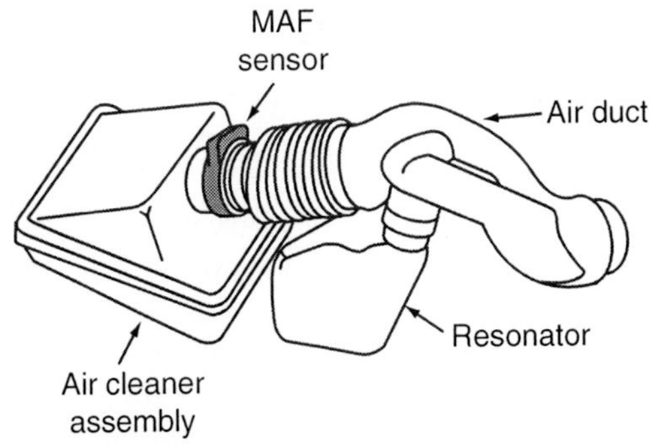


**Figure 2.** A mass airflow (MAF) sensor mounted in the air induction system.

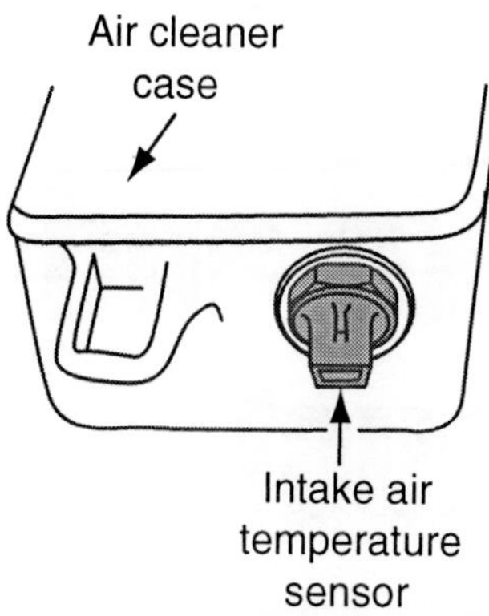


**Figure 3.** An intake air temperature (IAT) sensor mounted in the air cleaner housing.

> **You Should Know** *If the air cleaner is removed from the carburetor or the air cleaner duct is removed from the throttle body, always cover the top of the carburetor or throttle body with a clean shop towel to prevent anything from entering the intake system.*

An intake air temperature (IAT) sensor may be installed in the air cleaner assembly **(Figure 3)** or in the ductwork leading to the throttle body assembly. The air cleaner assembly also provides filtered air to the PCV system. The IAT sensor sends a voltage signal to the PCM in relation to the air intake temperature.

Be sure that the intake ductwork is properly installed and all connections are airtight, especially those between an airflow sensor or remote air cleaner and the throttle body assembly. Generally, metal or plastic air ducts are used when engine heat is not a problem. Flexible paper-metal ducts are used when they will be exposed to high engine temperatures.

## AIR CLEANER/FILTER

The primary function of the air filter is to prevent airborne contaminants and abrasives from entering into the engine. Without proper filtration, these contaminants will cause serious damage and appreciably shorten engine life. All incoming air should pass through the filter element before entering the engine.

Air filters are basically assemblies of pleated paper supported by a layer of fine mesh wire screen. The screen gives the paper some strength and also filters out large particles of dirt. A thick plastic-like gasket material normally surrounds the ends of the filter. This gasket adds strength to the filter and serves to seal the filter in its housing. If the filter does not seal well in the housing, dirt and dust can be

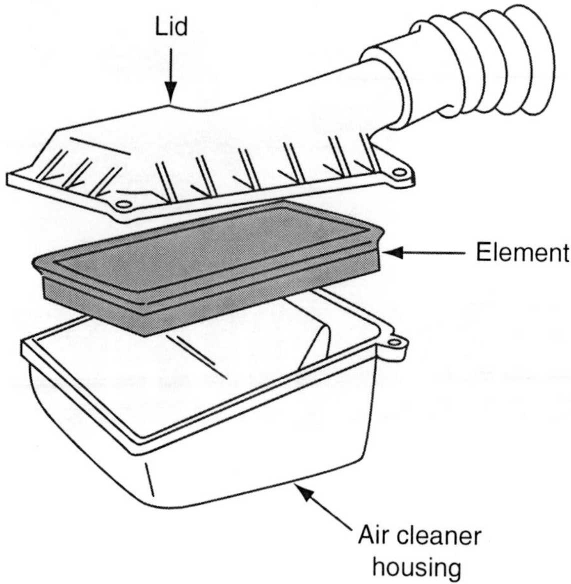


**Figure 4.** A typical flat air cleaner element.

pulled into the air stream to the cylinders. In most air filters. the air flows from the outside of the element to the inside as it enters the intake system. On some air intake systems, the air flows from the inside of the element to the outside.

The shape and size of the air filter element depend on its housing; the filter must be the correct size for the housing or dirt will be drawn into the engine. On today's engines, air filters are either flat **(Figure 4)** or round **(Figure 5)**. Air filters must be properly aligned and closed around the filter to ensure good airflow of clean air.

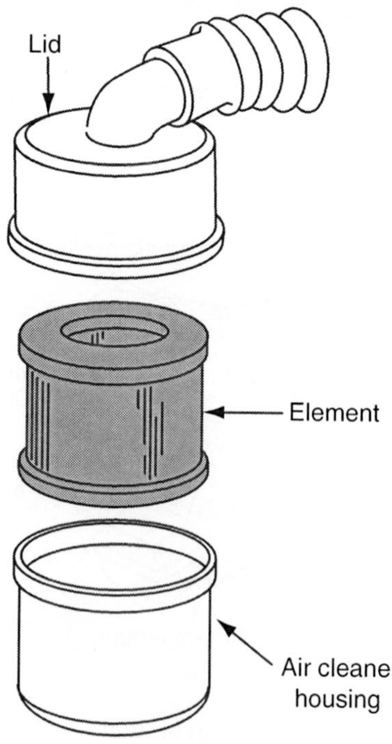


**Figure 5.** A typical round air cleaner element.

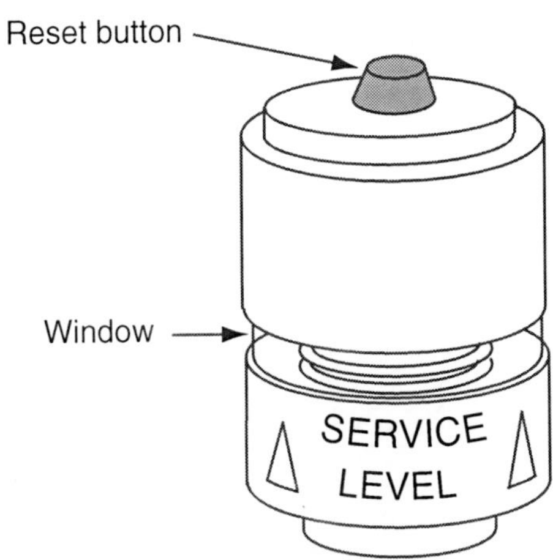


Figure 6. An airflow restriction indicator.

Many air cleaners in recent years have an airflow restriction indicator mounted in the air cleaner housing **(Figure 6)**. The **airflow restriction indicator** indicates restriction in the air cleaner element by the color of the indicator window.

If the air filter element is not restricted, a window in the side of the restriction indicator shows a green color. When the air filter element is restricted, the window in the airflow restriction indicator is orange and "Change Air Filter" appears; the air filter must be replaced. After the air filter is replaced, a reset button on top of the airflow restriction indicator must be pressed to reset the indicator so it displays green in the window.

Some air cleaners have an MAF sensor and IAT sensor mounted in the air cleaner housing. A duct is connected from the MAF sensor to the throttle body. The MAF sensor must be attached to the air cleaner so air flows through this sensor in the direction of the arrow on the sensor housing **(Figure 7)**. If the airflow is reversed through the MAF sensor, the PCM supplies a rich air-fuel ratio and increased fuel consumption. Other air cleaners contain a separate IAT sensor.

## INTAKE MANIFOLD

The intake manifold distributes the clean air or air-fuel mixture as evenly as possible to each cylinder of the engine.

Most older, carbureted engines and engines with throttle-body injection had cast-iron intake manifolds. With this type of engine, the intake manifold delivered air and fuel to the cylinders. Most early intake manifold designs have short runners **(Figure 8)**. These manifolds were either wet or dry. Wet manifolds have coolant passages cast directly in them. Dry manifolds did not have these coolant passages, but some had exhaust passages. Exhaust gases and/or coolant were used to heat up the floor of the manifold. This helped to vaporize the fuel before it arrived in the cylinders. Other dry manifold designs used some sort of electric heater unit or grid to warm up the bottom of the manifold. Heating the floor of the manifold also stopped the fuel from condensing in the manifold's plenum area. Good fuel vaporization and the prevention of condensation allowed for delivery of a more uniform air-fuel mixture to the individual cylinders, especially during engine warmup.

Modern intake manifolds for engines with port fuel injection are typically made of die-cast aluminum or plastic. These materials are used to reduce engine weight. A plastic manifold transfers less heat to the intake air and this results in a denser air-fuel mixture. Because intake manifolds for port-injected engines only deliver air to the cylinders, fuel vaporization and condensation are not design considerations. These intake manifolds deliver air to the intake ports where it is mixed with the fuel delivered by the injectors

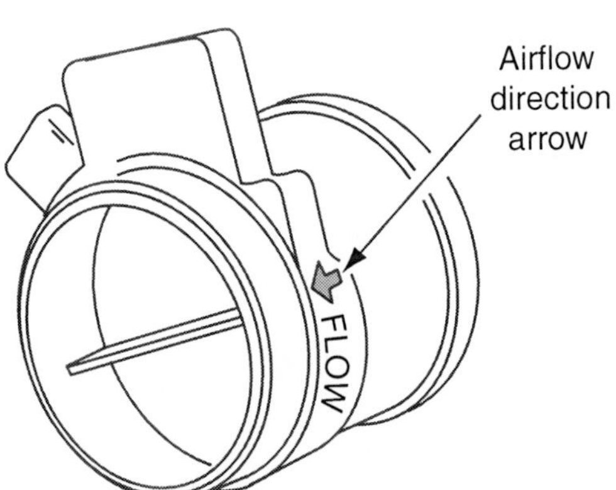


Figure 7. The mass airflow (MAF) sensor must be installed in the proper direction.

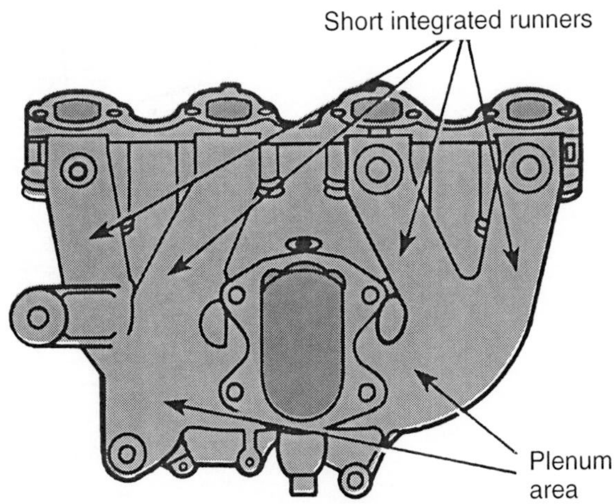


Figure 8. The intake manifold for an in-line 4-cylinder engine.

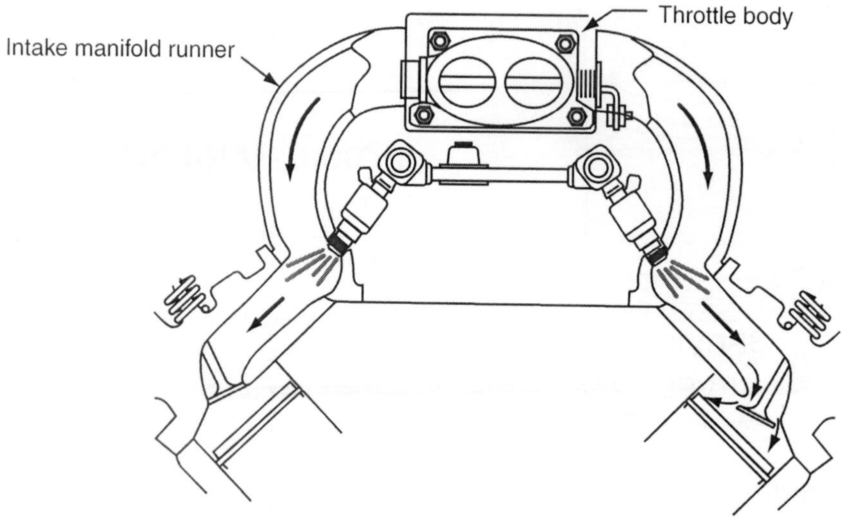


Figure 9. In a port fuel-injected engine, the intake manifold delivers air to the intake ports.

**(Figure 9)**. The primary consideration of these manifolds is the delivery of equal amounts of air to each cylinder.

> **You Should Know** *When cleaning old gasket material from the mating surfaces on aluminum or plastic intake manifolds, always use a plastic scraper. Using a steel scraper on these manifolds may scratch the mating surfaces resulting in vacuum leaks.*

Modern intake manifolds also serve as the mounting point for many intake-related accessories and sensors **(Figure 10)**. Some include a provision for mounting the thermostat and thermostat housing. In addition, connections to the intake manifold provide a vacuum source for the EGR system, automatic transmission vacuum modulators, power brakes, and/or heater and air-conditioning airflow control doors. Other devices located on or connected to the intake manifold include the manifold absolute pressure (MAP) sensor, knock sensor, various temperature sensors, and EGR passages.

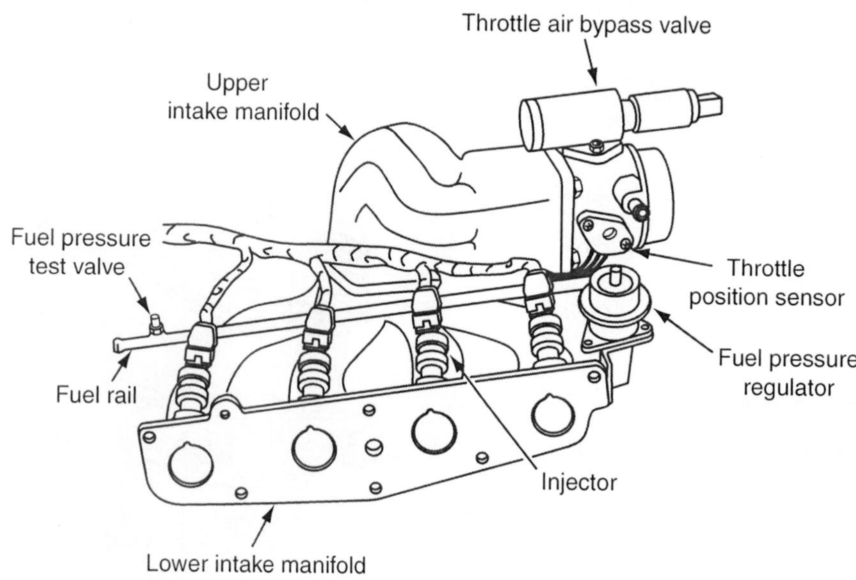


Figure 10. Various components are mounted on the intake manifold.

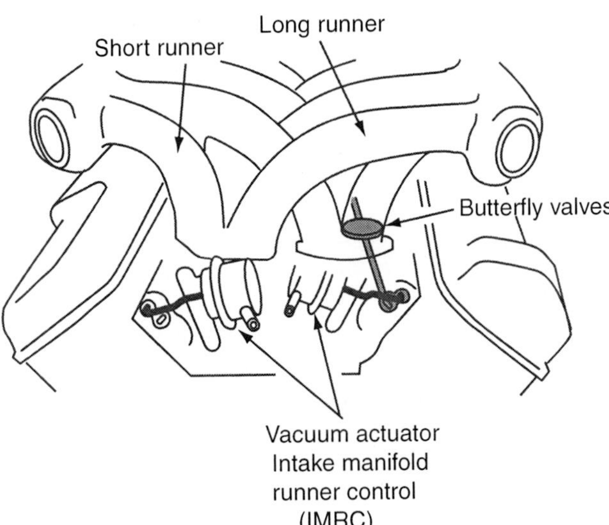


**Figure 11.** A V6 intake manifold with dual runners for each cylinder.

Some engines have two large intake manifold runners for each cylinder **(Figure 11)**. These engines have two intake ports for each cylinder. There is one intake port for each runner. Both intake runners are open when the engine is operating at high speeds. The butterfly valves in the high-speed runners are closed by the PCM when the engine is operating at lower speeds,. This action decreases airflow speed and volume at lower engine speeds and allows for greater airflow at high engine speeds. The butterfly valves in the intake manifold runners may be operated by a vacuum actuator and the PCM operates an electric/vacuum solenoid that turns the vacuum on and off to the actuator.

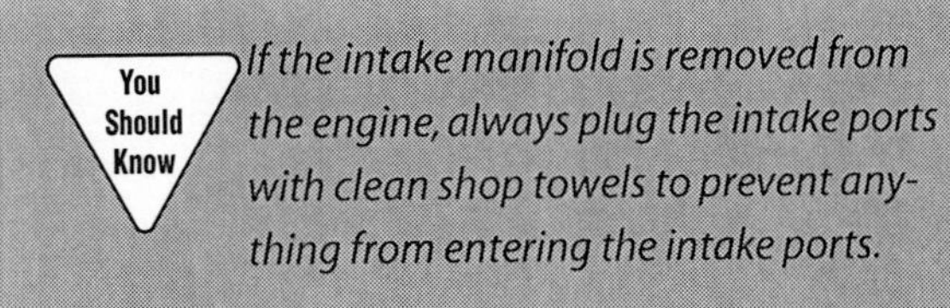


*If the intake manifold is removed from the engine, always plug the intake ports with clean shop towels to prevent anything from entering the intake ports.*

## VACUUM BASICS

Vacuum in the intake manifold is created by the downward movement of the pistons during the intake strokes. With the intake valve open and the piston moving downward, a vacuum is created within the cylinder and intake manifold. The air passing the intake valve does not move fast enough to fill the cylinder, thereby causing the lower pressure. This vacuum is continuous in a multicylinder engine since at least one cylinder is always at some stage of its intake stroke. The normal measure of vacuum is in inches of mercury (in. Hg) or kilopascals (kPa).

The amount of low pressure produced by the piston during its intake stroke depends on a number of factors. Basically it depends on the cylinder's ability to form a vacuum and the intake system's ability to fill the cylinder. When there is high vacuum (15 to 22 in. Hg (50.7 to 74.3 kPa), we know the cylinder is well sealed and not enough air is entering the cylinder to fill it. At idle, the throttle plate is almost closed and nearly all airflow to the cylinders is stopped. This is why vacuum is high during idle. Since there is a correlation between throttle position and engine load, it can be said that load directly affects engine manifold vacuum. Therefore, vacuum will be high whenever there is no or low load on the engine.

## VACUUM SYSTEM

The vacuum in the intake manifold is used to operate many systems, such as emission controls, brake boosters, parking brake releases, headlight doors, heater/air conditioners, and cruise controls. Vacuum is applied from intake manifold fittings through a system of hoses and tubes to these vacuum-operated components and systems.

## EXHAUST SYSTEM COMPONENTS

The major components of a typical exhaust system include the exhaust manifold(s), exhaust pipe, catalytic converter(s), muffler, tailpipe, resonator, and heat shields. All the parts of the system are designed to conform to the available space of the vehicle's undercarriage and yet be a safe distance above the road.

### Exhaust Manifold

The exhaust manifold **(Figure 12)** collects the burnt gases as they are expelled from the cylinders and directs them to the exhaust pipe. Exhaust manifolds for most vehicles are made of cast- or nodular-iron. Many newer vehicles have stamped, heavy-gauge sheet metal or stainless steel units.

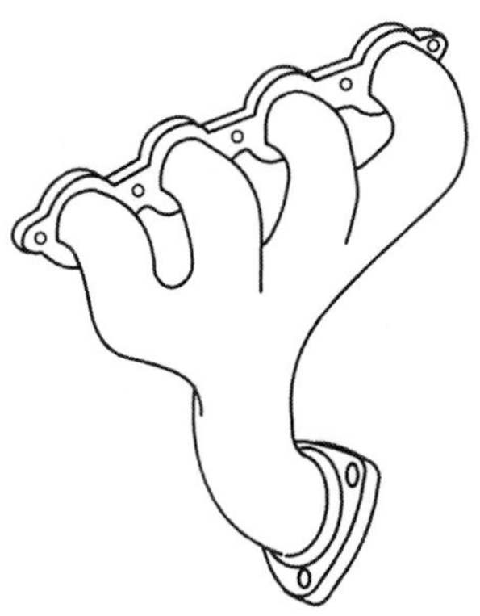

**Figure 12.** An exhaust manifold.

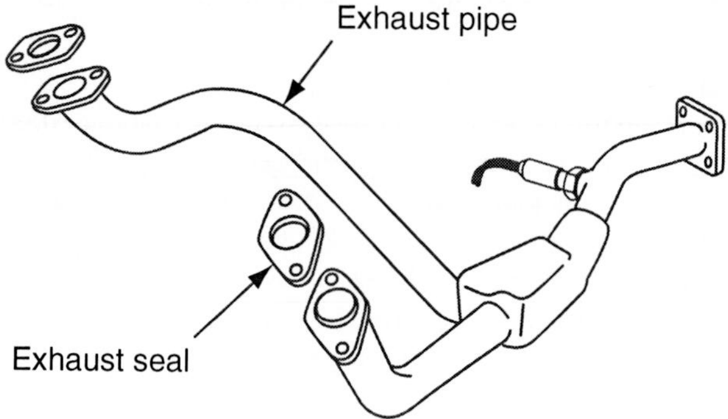


Figure 13. An exhaust pipe.

In-line engines have one exhaust manifold. V-type engines have an exhaust manifold on each side of the engine. An exhaust manifold will have either three, four, five, or six passages, depending on the type of engine. These passages blend into a single passage at the other end, which connects to an exhaust pipe. From that point, the flow of exhaust gases continues to the catalytic converter, muffler, and tailpipe, then exits at the rear of the car.

V-type engines may be equipped with a dual exhaust system that consists of two almost identical but individual systems in the same vehicle.

## Exhaust Pipe and Seal

The exhaust pipe is usually a double-walled metal pipe that runs under the vehicle between the exhaust manifold and the catalytic converter **(Figure 13)**. A seal between the exhaust pipe and the exhaust manifold prevents exhaust leaks at this location.

# CATALYTIC CONVERTERS

A catalytic converter **(Figure 14)** is part of the exhaust system and is a very important part of the emission control system. Because the converter is part of both systems, it has a role in both. As an emission control device, it is responsible for converting undesirable exhaust gases into harmless gases. As part of the exhaust system, it helps reduce the noise level of the exhaust. A catalytic converter contains a ceramic element coated with a catalyst. A catalyst is a substance that causes a chemical reaction in other elements without actually becoming part of the chemical change and without being used up or consumed in the process.

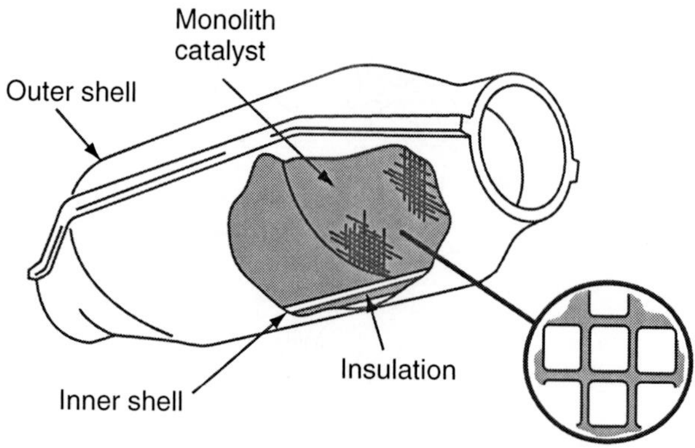


Figure 14. A monolithic-type catalytic converter.

Catalytic converters may be pellet type of monolithic type. A pellet-type converter contains a bed made from hundreds of small beads. Exhaust gases pass over this bed. In a monolithic-type converter, the exhaust gases pass through a honeycomb ceramic block. The converter beads or ceramic block are coated with a thin coating of cerium, platinum, palladium, and/or rhodium and are held in a stainless-steel container. Modern vehicles are equipped with three-way catalytic converters which means the converter reduces the three major exhaust emissions: hydrocarbons (HC), carbon monoxide (CO), and oxides of nitrogen ($NO_x$). The converter oxidizes HC and CO into water vapor and carbon dioxide ($CO_2$), and reduces $NO_x$ to oxygen and nitrogen.

Many vehicles are equipped with a mini-catalytic, or warmup, converter that is either built into the exhaust manifold or is located next to it **(Figure 15)**. These converters are used to clean the exhaust during engine warmup and are commonly called warmup converters.

Many catalytic converters have an air hose connected from the AIR system to the oxidizing catalyst. This air helps the converter work by making extra oxygen available. The air from the AIR system is not always forced into the converter; rather, it is controlled by the vehicle's PCM. Fresh air added to the exhaust at the wrong time could overheat the converter and produce $NO_x$, something the converter is trying to destroy.

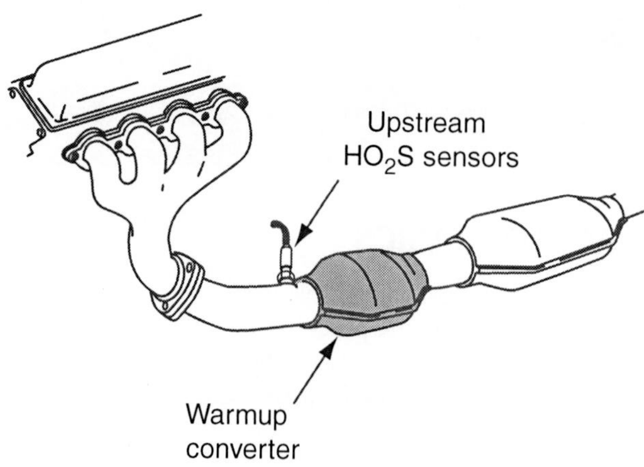


Figure 15. A warmup converter may be located between the exhaust manifold and the standard converter.

On-board diagnostic II (OBD II) regulations call for a way to inform the driver that the vehicle's converter has a problem and may be ineffective. Vehicles manufactured since 1996 are equipped with **On-Board Diagnostic II (OBD II)** systems. These systems have extensive software in the PCM that monitors various engine systems and components. If a defect occurs in the engine computer system that makes the exhaust emissions one-and-a-half times higher than the exhaust emission standards for that vehicle year, the PCM illuminates the malfunction indicator light (MIL) in the instrument panel.

The PCM monitors the activity of the converter by comparing the signals of an $HO_2S$ located at the front of the converter with the signals from an $HO_2S$ located at the rear **(Figure 16)**. If the sensors outputs are the same, the converter is not working properly and the malfunction indicator lamp (MIL) on the dash will light. The **malfunction indicator light (MIL)** is located in the instrument panel. This light is illuminated by the PCM if the PCM detects certain defects in the engine computer system a specific number of times. The MIL light may be called by various names such as the "CHECK ENGINE LIGHT."

You Should Know

*The SAE developed the J1930 standard for engine computer terminology. This standard uses universal automotive electronic terms such as the PCM and MIL. The J1930 terminology is followed by all vehicle manufacturers.*

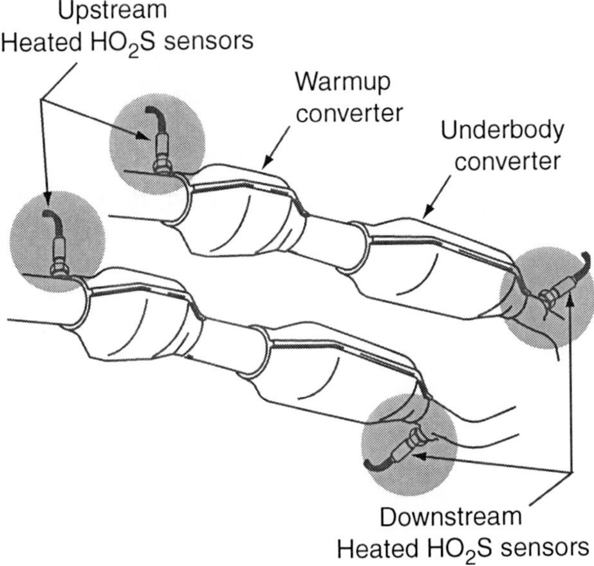


**Figure 16.** An exhaust system for an On-Board Diagnostic II (OBD II) vehicle.

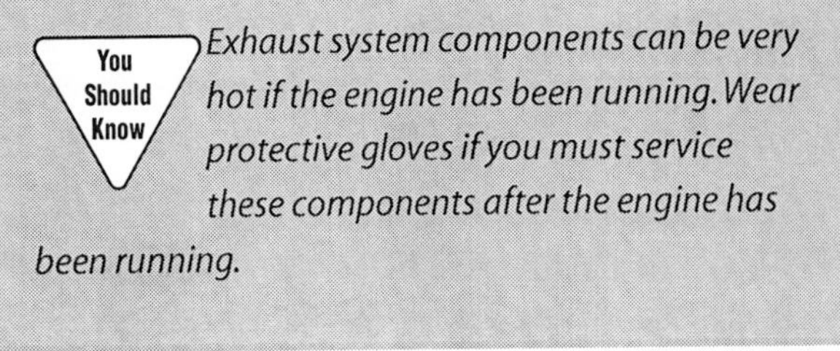

You Should Know

*Exhaust system components can be very hot if the engine has been running. Wear protective gloves if you must service these components after the engine has been running.*

## Mufflers

The muffler is a cylindrical or oval-shaped component, generally about 2 feet (0.6 meters) long, mounted in the exhaust system about midway or toward the rear of the vehicle. Inside the muffler is a series of baffles, chambers, tubes, and holes to break up, cancel out, or silence the pressure pulsations that occur each time an exhaust valve opens.

Two types of mufflers are commonly used on passenger vehicles **(Figure 17)**. Reverse-flow mufflers change the direction of the exhaust gas flow through the inside of the unit. This is the most common type of automotive muffler. Straight-through mufflers permit exhaust gases to pass through a single tube. The tube has perforations that tend to break up pressure pulsations. They are not as quiet as the reverse-flow type.

There have been several important changes in recent years in the design of mufflers. Most of these changes have been centered at reducing weight and emissions, improving fuel economy, and simplifying assembly. These changes include the following:

- New Materials—More and more mufflers are being made of aluminized and stainless steel. Using these materials reduces the weight of the units as well as extending their lives.
- Double-wall design—Retarded engine ignition timing that is used on many small cars tends to make the exhaust pulses sharper. Many cars use a double-wall

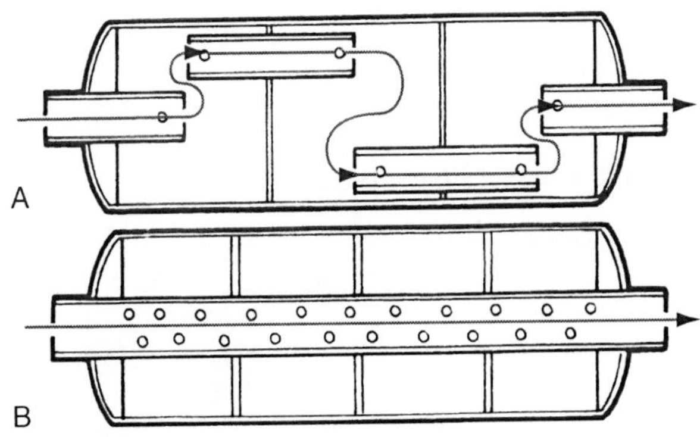


**Figure 17.** (A) A reverse-flow muffler. (B) A straight-through muffler.

exhaust pipe to better contain the sound and reduce pipe ring.

- Rear-mounted muffler—More and more often, the only space left under the car for the muffler is at the very rear. This means the muffler runs cooler than before and is more easily damaged by condensation in the exhaust system. This moisture, combined with nitrogen and sulfur oxides in the exhaust gas, forms acids that rot the muffler from the inside out. Many mufflers are being produced with drain holes drilled into them.
- Back Pressure—Even a well-designed muffler will produce some **back pressure** in the system. Back pressure reduces an engine's volumetric efficiency, or ability to breathe. Excessive back pressure caused by defects in a muffler or other exhaust system part can slow or stop the engine. However, a small amount of back pressure can be used intentionally to allow a slower passage of exhaust gases through the catalytic converter. This slower passage results in more complete conversion to less harmful gases. Also, no back pressure may allow intake gases to enter the exhaust.

## Resonator

On some older vehicles, there is an additional muffler, known as a resonator or silencer. This unit is designed to further reduce or change the sound level of the exhaust. It is located toward the end of the system and generally looks like a smaller, rounder version of a muffler.

## Tailpipe

The tailpipe is the last pipe in the exhaust system. It releases the exhaust fumes into the atmosphere beyond the back end of the vehicle.

## Heat Shields

Heat shields are used to protect other parts from the heat of the exhaust system, especially the catalytic converter **(Figure 18)**. They are usually made of pressed or perforated sheet metal. Heat shields trap the heat in the exhaust system, which has a direct effect on maintaining exhaust gas velocity.

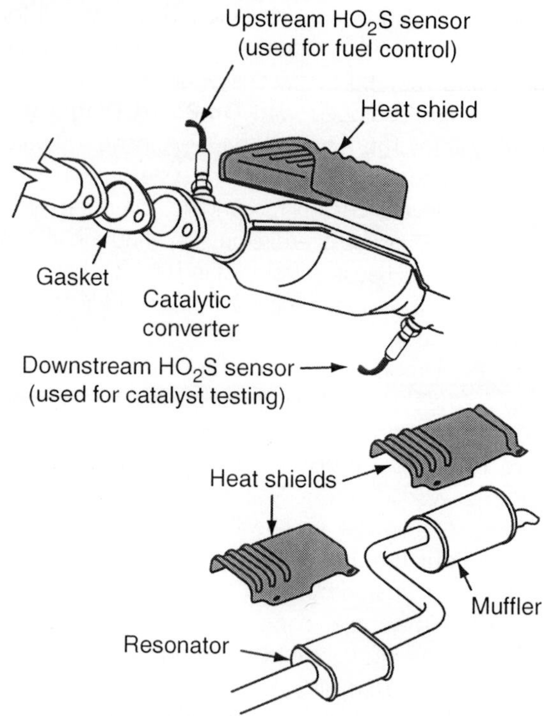


**Figure 18.** The typical location of heat shields in an exhaust system.

## Clamps, Brackets, and Hangers

Clamps, brackets, and hangers are used to properly join and support the various parts of the exhaust system. These parts also help to isolate exhaust noise by preventing its transfer through the frame **(Figure 19)** or body to the passenger compartment. Clamps help to secure exhaust system parts to one another. The pipes are formed in such a way that one slips inside the other. This design makes a close fit. A U-type clamp usually holds this connection tight **(Figure 20)**. Another important job of clamps and brackets

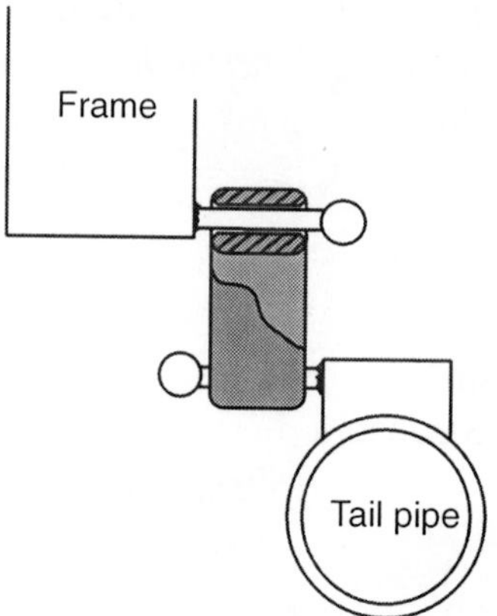


**Figure 19.** Rubber hangers are used to keep the exhaust system in place without contacting the frame.

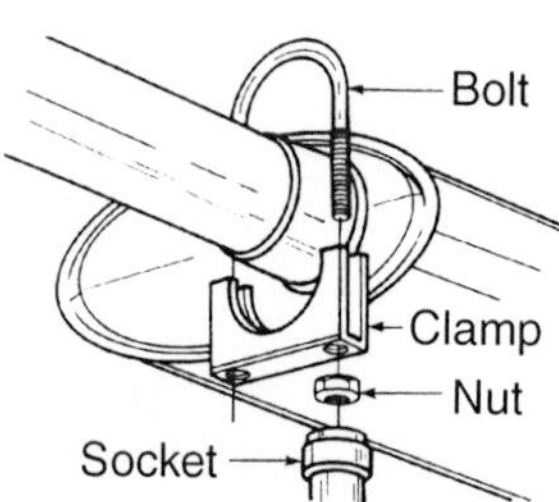


**Figure 20.** A U-clamp is often used to secure two pipes that slip together.

is to hold pipes to the bottom of the vehicle. Clamps and brackets must be designed to allow the exhaust system to vibrate without transferring the vibrations through the car.

There are many different types of flexible hangers available. Each is designed for a particular application. Some exhaust systems are supported by doughnut-shaped rubber rings between hooks on the exhaust component and on the frame or car body. Others are supported at the exhaust pipe and tailpipe connections by a combination of metal and reinforced fabric hanger. Both the doughnuts and the reinforced fabric allow the exhaust system to vibrate without breakage that could be caused by direct physical connection to the vehicle's frame.

Some exhaust systems are a single unit in which the pieces are welded together by the factory. By welding instead of clamping the assembly together, car makers save the weight of overlapping joints as well as that of clamps.

## EXHAUST SYSTEM PURPOSE AND OPERATION

The purpose of the exhaust system is to deliver the exhaust gases from the exhaust ports in the cylinder head(s) to the atmosphere with a minimum amount of noise. The exhaust system must also provide a minimum amount of restriction to the flow of exhaust gas. Thirdly, the exhaust system—and particularly the catalytic converter—plays a significant role in reducing tailpipe emissions.

Exhaust systems are designed for particular engine-chassis combinations. Exhaust system length, pipe size, and silencer size are used to tune the flow of gases within the exhaust system. Proper tuning of the exhaust manifold tubes can actually create a partial vacuum that helps draw exhaust gases out of the cylinder, improving volumetric efficiency. Separate, tuned exhaust headers **(Figure 21)** can also improve efficiency by preventing the exhaust flow of one cylinder from interfering with the exhaust flow of another cylinder. Cylinders next to one another may release exhaust gas at about the same time. When this happens, the pressure of the exhaust gas from one cylinder can interfere with the flow from the other cylinder. With separate headers, the cylinders are isolated from one another, interference is eliminated, and the engine breathes better. The problem of interference is especially common with V8 engines. However, exhaust headers tend to improve the performance of all engines.

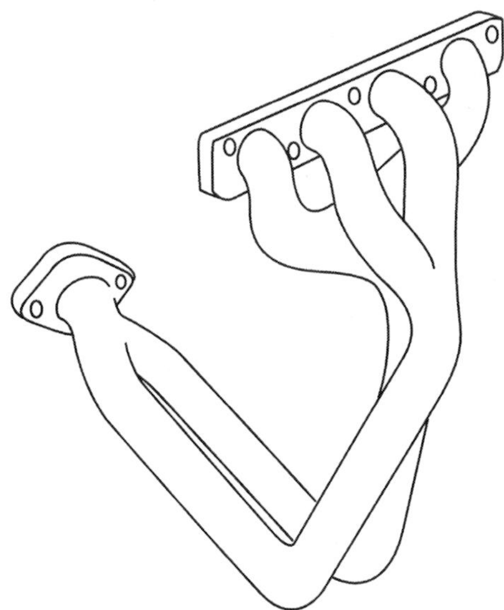

**Figure 21.** Engine efficiency can be improved with tuned exhaust headers.

Exhaust manifolds may also be the attaching point for the air injection reactor (AIR) pipe **(Figure 22)**. This pipe introduces cool air from the AIR system into the exhaust stream. Some exhaust manifolds have provisions for the EGR pipe. This pipe takes a sample of the exhaust gases and delivers it to the EGR valve. Also, some exhaust manifolds have a tapped bore that retains the oxygen sensor **(Figure 23)**.

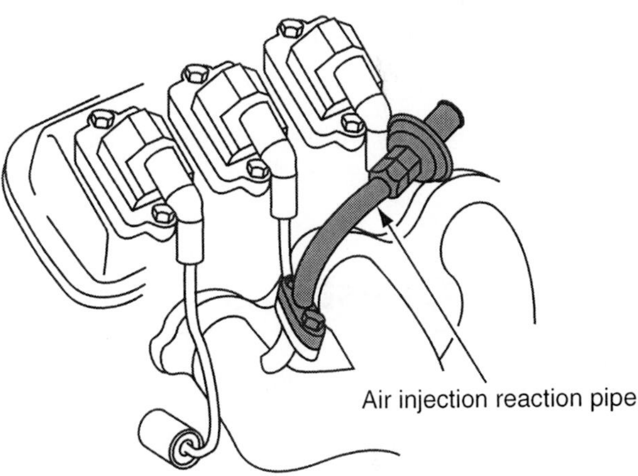


**Figure 22.** An AIR pipe mounting on an exhaust manifold.

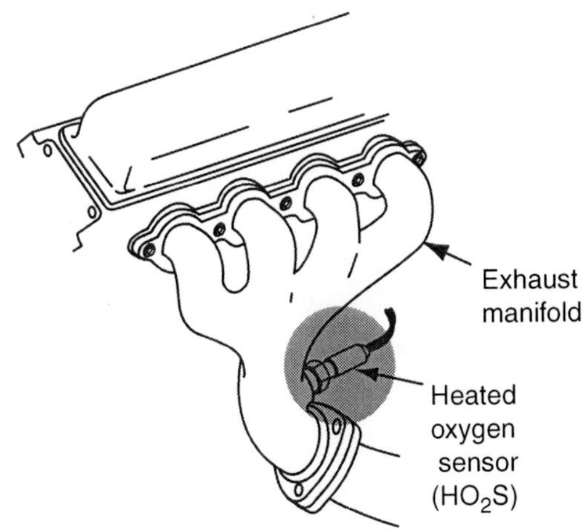


**Figure 23.** An exhaust manifold fitted with a heated oxygen sensor ($HO_2S$).

## Summary

- The air induction system allows a controlled amount of clean, filtered air to enter the engine. Cool air is drawn in through a fresh air tube. It passes through an air cleaner before entering the carburetor or throttle body.
- The intake manifold distributes the air or air-fuel mixture as evenly as possible to each cylinder, helps to prevent condensation, and assists in the vaporization of the air-fuel mixture. Intake manifolds are made of cast iron, plastic, or die-cast aluminum.
- The vacuum in the intake manifold operates many systems such as emission controls, brake boosters, heater/air conditioners, cruise controls, and more. Vacuum is applied through an elaborate system of hoses, tubes, and relays. A diagram of emission system vacuum hose routing is located on the underhood decal. Loss of vacuum can create many driveability problems.
- A vehicle's exhaust system carries away gases from the passenger compartment, cleans the exhaust emissions, and muffles the sound of the engine. Its components include the exhaust manifold, exhaust pipe, catalytic converter, muffler, resonator, tailpipe, heat shields, clamps, brackets, and hangers.
- The exhaust manifold is a bank of pipes that collects the burned gases as they are expelled from the cylinders and directs them to the exhaust pipe. Engines with all the cylinders in a row have one exhaust manifold. V-type engines have an exhaust manifold on each side of the engine. The exhaust pipe runs between the exhaust manifold and the catalytic converter.
- The catalytic converter reduces HC, CO, and $NO_x$ emissions.
- The muffler consists of a series of baffles, chambers, tubes, and holes to break up, cancel out, and silence pressure pulsations. Two types commonly used are the reverse-flow and the straight-through mufflers.
- The tailpipe is the end of the pipeline carrying exhaust fumes to the atmosphere beyond the back end of the vehicle. Heat shields protect vehicle parts from exhaust system heat. Clamps, brackets, and hangers join and support exhaust system components.
- Exhaust system components are subject to both physical and chemical damage. The exhaust can be checked by listening for leaks and by visual inspection. Most exhaust system servicing involves the replacement of parts.

## Review Questions

1. Technician A says the PCV system relieves the crankcase of unwanted pressure. Technician B says the PCV system replaces blowby gases in the crankcase with clean air. Who is correct?
   A. Technician A
   B. Technician B
   C. Both Technician A and Technician B
   D. Neither Technician A nor Technician B
2. Technician A says a vacuum leak results in less air entering the engine which causes a richer air-fuel mixture. Technician B says a vacuum leak in the intake manifold causes rough engine idle. Who is correct?
   A. Technician A
   B. Technician B
   C. Both Technician A and Technician B
   D. Neither Technician A nor Technician B
3. A restricted exhaust system can cause:
   A. stalling.
   B. backfiring.
   C. loss of power.
   D. acceleration stumbles.
4. When an airflow restriction indicator window appears orange, it is necessary to:
   A. replace the airflow restriction indicator.
   B. replace the air filter and press the reset button on the airflow restriction indicator.
   C. replace the air filter and the airflow restriction indictor.
   D. clear the diagnostic trouble codes (DTCs) from the PCM memory.
5. All of these statements about airflow restriction indicators are true *except*:
   A. A green color in the indicator window indicates an unrestricted air filter.
   B. After air filter replacement, a reset button on the indicator is used to reset the indicator.
   C. A blue color in the indicator window indicates a restricted air filter.
   D. The airflow restriction indicator may be mounted in the air cleaner housing.

6. When discussing an MAF, Technician A says the MAF sensor is mounted on the inlet to the air cleaner housing. Technician B says if the airflow through a MAF sensor is reversed, the air-fuel ratio is not affected. Who is correct?
   A. Technician A
   B. Technician B
   C. Both Technician A and Technician B
   D. Neither Technician A nor Technician B
7. All of these statements about intake manifolds are true *except*:
   A. Engine coolant circulates through a wet manifold.
   B. On engines with port-fuel injection, intake manifold temperature is critical.
   C. Intake manifolds may be manufactured from aluminum.
   D. A vacuum hose is connected from the intake manifold to the transmission vacuum modulator.
8. Many catalytic converters are designed to reduce all of these pollutants *except*:
   A. carbon monoxide (CO).
   B. oxides of nitrogen ($NO_x$).
   C. hydrocarbons (HC).
   D. carbon dioxide ($CO_2$).
9. Technician A says a bed in some catalytic converters contains a large number of small beads. Technician B says on some vehicles, airflow from the AIR system is directed into the catalytic converter at a certain engine temperature. Who is correct?
   A. Technician A
   B. Technician B
   C. Both Technician A and Technician B
   D. Neither Technician A nor Technician B
10. All of these statements about OBD II systems are true *except*:
    A. The PCM software monitors various engine systems and components.
    B. Heated oxygen sensors ($HO_2S$) are located before and after the catalytic converter.
    C. The PCM monitors catalytic converter temperature to determine converter efficiency.
    D. The MIL light is illuminated by the PCM if the emission levels are one-and-a-half times higher than the emission standards a specific number of times.
11. Without proper intake air filtration, contaminants and abrasives in the air will cause severe ________________ damage.
12. In a port fuel-injected engine, the intake manifold delivers ____________ to the cylinders.
13. A wet intake manifold has ________________ ________________ cast into the manifold.
14. When an intake manifold has dual runners, both runners are open at ______________ engine speeds.
15. Explain the operation of an airflow restriction indicator.
16. Describe the result of installing an MAF sensor backwards.
17. Explain the purposes of the intake manifold.
18. Explain why fuel vaporization and condensation are not intake manifold design considerations on port fuel-injected engines.

# Intake and Exhaust System Maintenance, Diagnosis, and Service

## Introduction

Intake and exhaust system service is extremely important! For example, if an exhaust leak is not repaired, poisonous carbon monoxide gas may enter the passenger compartment. This gas causes nausea and headaches. In high concentrations, this gas is fatal to human beings. An intake system vacuum leak may cause improper engine operation and reduced fuel mileage. Therefore, it is important to maintain proper intake and exhaust system operation.

### INTAKE SYSTEM MAINTENANCE

If an air filter is doing its job, it will become dirty. That is why filters are made of pleated paper. The paper is the actual filter. It is pleated to increase the filtering area. By increasing the area, the amount of time it will take for dirt to plug the filter becomes longer. As a filter becomes dirty, the amount of air that can flow through it is reduced. Without the proper amount of air, the engine will not be able to produce the power it should, nor will it be as fuel efficient as it should be.

Included in the preventive maintenance schedule for all vehicles is the periodic replacement of the air filter. This mileage or time interval is based on normal vehicle operation. If the vehicle is driven continuously in heavy dust, the life of the filter is shorter. Always use a replacement filter that is the same size and shape as the original.

Follow these steps for air filter service or replacement:

1. Remove the wing nut, clips, or screws that retain the air filter cover and remove the cover.
2. Remove the air filter element **(Figure 1)** from the air cleaner and be sure that no foreign material such as small stones drop into the air cleaner duct or throttle body.

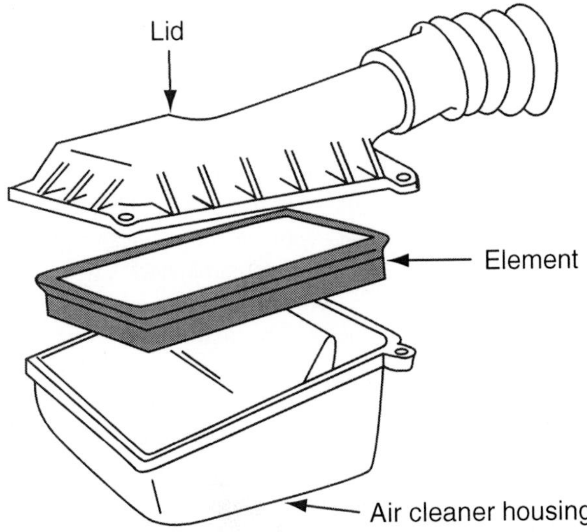


**Figure 1.** An air cleaner with air filter.

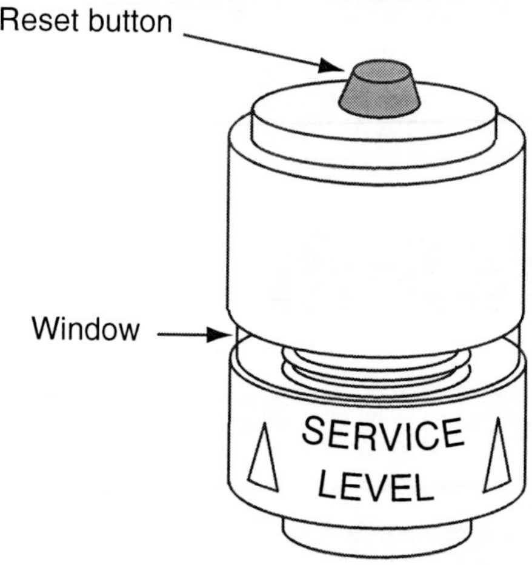


**Figure 2.** An airflow restriction indicator.

3. If the air cleaner has an airflow restriction indicator, inspect the color of the indicator window **(Figure 2)**. If the window appears orange and "Change Air Filter" appears, the air filter is restricted. When the window appears green, the air filter is not restricted. If the airflow restriction indicator window appears orange, press the reset button on top of the indicator to reset the indicator so green appears in the window.
4. Carefully clean the air cleaner housing to remove any small stones, bugs, and debris. Inspect the air cleaner housing for cracks, holes, and damage.
5. Visually inspect the air filter for pin holes in the paper element and damage to the paper element, sealing surfaces, or metal screens on both sides of the element. If the air filter is damaged or contains pin holes, replace the filter.
6. Place a shop trouble light on the inside surface of the air filter and look through the filter toward the light. The light should be clearly visible through the filter, but there must be no pin holes in the paper element. If the light is not visible over most of the filter or if the filter is contaminated with oil, the air filter must be replaced. If the air filter is contaminated with oil, the engine has excessive blowby past the piston rings or the PCV valve or hose is restricted. When either of these conditions exist, excessive crankcase pressure forces oil vapors out of the engine through the PCV clean air hose into the air cleaner. Clean air should normally flow from the air cleaner through the PCV clean air hose into the engine.
7. If there is only a small amount of dirt in the air filter, use a shop air hose and gun with a maximum of 30 psi (207 kPa) pressure to blow the dirt out of the element. Place the air gun 6 inches from the inside surface of the air filter when blowing the dirt out of the filter. After blowing out the air filter, use a shop trouble light to reinspect the filter.

**You Should Know** *In a few air cleaners, the air flows through the air filter from the inside to the outside. When blowing dirt out of these filters, the air gun must be placed 6 inches from the outside surface of the air filter. If the air intake for the air cleaner is in the center of the air filter, the air flows from the inside to the outside of the element.*

8. Clean or replace the PCV system filter in the air cleaner.
9. Install the replacement or newly cleaned air filter in the air cleaner housing. Be sure the sealing surfaces on the element fit snugly on the air cleaner housing. Install the air cleaner cover and tighten the retaining clamps, screws, or wing nut.
10. Be sure the PCV hose and any other hoses, sensors, or wiring connectors are properly connected to the air cleaner.

**You Should Know** *If shop air is directed through the air filter in the wrong direction, small dirt particles may be forced through the air filter and this creates small holes in the filter. Small holes in the air filter allow abrasives in the air to enter the engine and this action shortens engine life.*

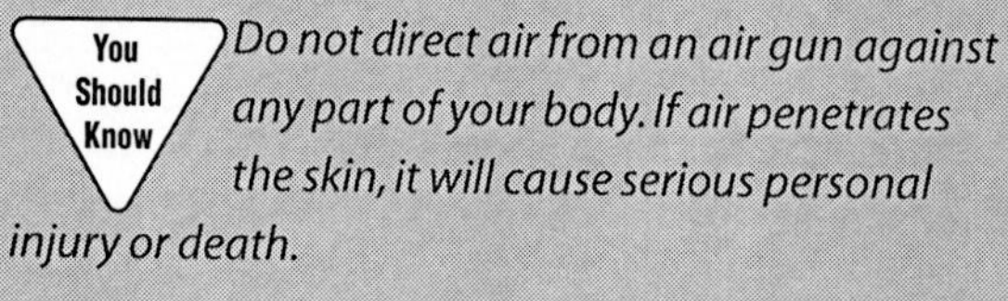

**You Should Know** *Do not direct air from an air gun against any part of your body. If air penetrates the skin, it will cause serious personal injury or death.*

## INTAKE SYSTEM DIAGNOSIS

Vacuum system problems can produce or contribute to a variety of drivability symptoms such as engine stalling, reduced fuel economy, no start or hard starting, backfiring, rough idling, acceleration stumbles, overheating, and detonation.

As a routine part of problem diagnosis, a technician who suspects a vacuum problem should first:

1. Inspect vacuum hoses for improper routing or disconnections (engine decal identifies hose routing).
2. Look for kinks, tears, or cuts in vacuum lines.

3. Check for vacuum hose routing and wear near hot spots, such as the exhaust manifold or the EGR tubes.
4. Make sure there is no evidence of oil or transmission fluid in vacuum hose connections. Transmission fluid in the vacuum hose to the transmission modulator indicates a leaking modulator diaphragm.
5. Inspect vacuum system devices for damage, such as dents in vacuum reservoirs, damaged by-pass valves, broken nipples on vacuum control valves, or broken tees in vacuum lines.

Broken or disconnected hoses allow vacuum leaks that admit more air into the intake manifold than the engine is calibrated for. The most common result is a rough-running engine due to the leaner air-fuel mixture created by the excess air.

Kinked hoses can cut off vacuum to a component, thereby disabling it. For example, if the vacuum hose to the EGR valve is kinked, vacuum cannot move the EGR valve diaphragm. Therefore, the valve will not open.

To check vacuum controls, refer to the service manual for the correct location and identification of the components. Typical locations of vacuum-controlled components are shown in **Figure 3**.

Tears and kinks in any vacuum line can affect engine operation. Any defective hoses should be replaced one at a time to avoid misrouting. OEM vacuum lines are installed in a harness consisting of 1/8-inch (3.18-millimeter) or larger outer diameter and 1/16-inch (1.59-millimeter) inner diameter nylon hose with bonded nylon or rubber connectors. Occasionally, a rubber hose might be connected to the harness. The nylon connectors have rubber inserts to provide a seal between the nylon connector and the component connection (nipple).

## Vacuum Tests

The vacuum gauge is one of the most important engine diagnostic tools used by technicians. With the gauge connected to the intake manifold and the engine warm and idling, watch the action of the gauge needle. A healthy engine will give a steady, constant vacuum reading between 17 and 22 in. Hg (432 and 559 mm/Hg). On some four- and six-cylinder engines, however, a reading of 15 inches (381 mm/Hg) is considered acceptable. With high-performance engines, a slight flicker of the needle can also be expected. Keep in mind that the gauge reading will drop

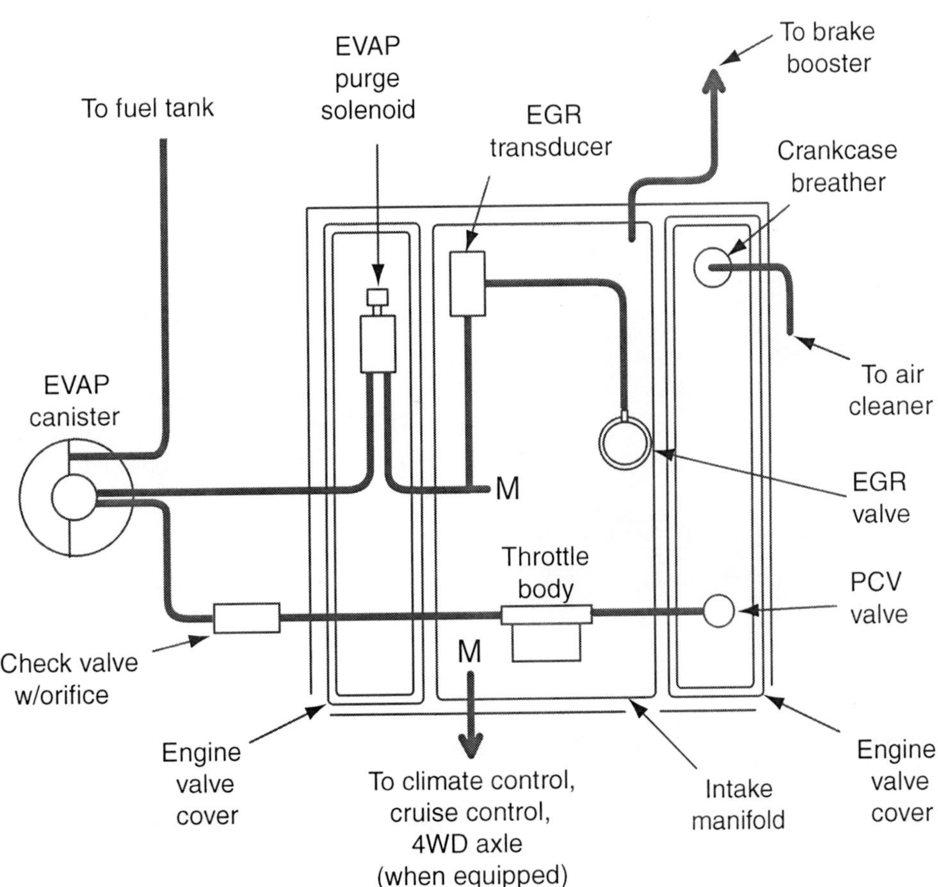


**Figure 3.** Typical vacuum devices and controls.

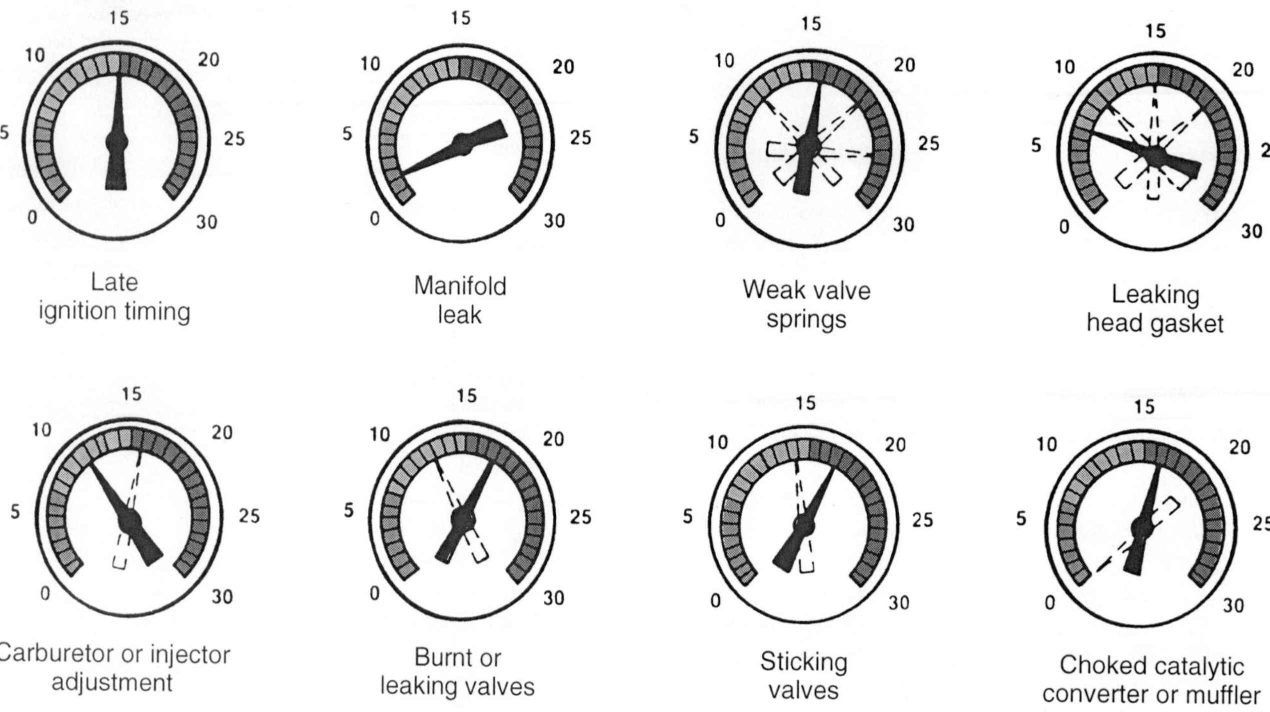


**Figure 4.** Vacuum gauge readings and the indicated engine conditions.

about 1 inch (2.54 centimeters) for each 1,000 feet (305 meters) above sea level. **Figure 4** shows some of the common readings and what engine malfunctions they indicate.

As shown in **Figure 5**, a hand-held vacuum pump/gauge is used to test vacuum-actuated valves and motors.

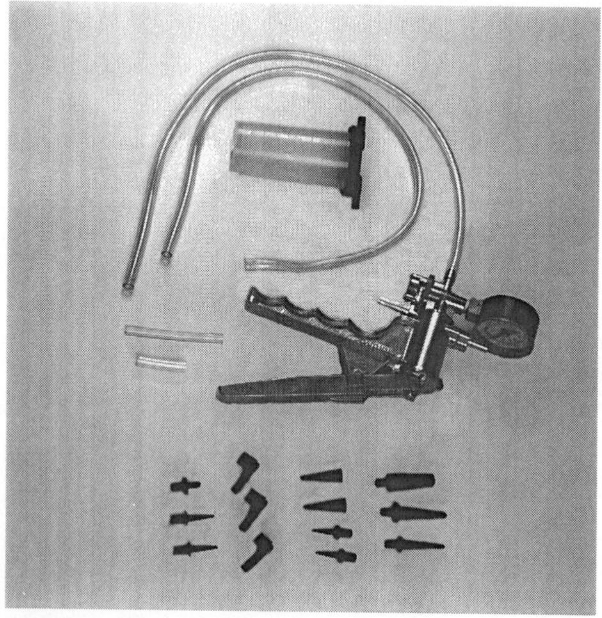

**Figure 5.** A hand-operated vacuum pump is used to test an air cleaner vacuum motor.

If the component does not operate when the proper amount of vacuum is applied, it should be serviced or replaced. An ultrasonic-type vacuum leak detector provides a beeping noise when the tester pickup is placed near a vacuum leak. Smoke-type leak detectors are available that create and blow smoke into the component with a suspected vacuum leak. The smoke may be seen escaping from the vacuum leak.

## Diagnosis of Heated Air Inlet Systems

Most carbureted fuel systems and some throttle body injection systems have a heated air inlet system. The heated air inlet system may be checked visually. When the engine is cold, vacuum should be supplied through the temperature switch in the air cleaner to the vacuum actuator on the butterfly valve in the air cleaner snorkel. Under this condition, the butterfly valve should be positioned so the cold air passage into the air cleaner is closed and warm air is moved from the exhaust manifold and flexible pipe into the air cleaner. As the engine warms up, the temperature switch in the air cleaner should gradually shut off the vacuum to the butterfly door actuator and the butterfly door slowly moves to open the cold air passage into the air cleaner.

# INTAKE SYSTEM SERVICE

There are few reasons why an intake manifold would need to be replaced. Obviously, if the manifold is cracked or the sealing surfaces are severely damaged, it should be replaced. After the intake manifold is removed, clean the mating surfaces on the intake manifold and cylinder heads with a plastic scraper. Inspect the intake manifold and cylinder head mating surfaces for nicks, scratches, cracks, and damage. Minor scratches may be polished out with crocus cloth. The intake manifold and cylinder head mating surfaces should also be checked for warpage with a straightedge and feeler gauge. Always plug the cylinder head and intake manifold intake air passages with clean shop towels when the intake manifold is removed to prevent any dirt particles or other objects from entering these passages.

- *The intake manifold, throttle body, fuel rail, and fuel injectors should be removed as an assembly when removing the intake manifold.*
- *On fuel-injected engines, the fuel pressure should be relieved before disconnecting any fuel line or fuel system component. An injector balance tester may be connected to one of the injectors and this injector may be energized briefly to relieve the fuel pressure in the system.*
- *Never disconnect any coolant hose or loosen the cooling system pressure cap if the engine coolant is warm or hot. The sudden decrease in cooling system pressure may cause the coolant to boil, resulting in personal injury.*
- *Never use a metal scraper or wire brush to clean surfaces on aluminum components. This action may scratch these components so they have to be replaced.*
- *If any dirt particles or foreign objects enter the intake manifold or cylinder head air intake passages, they will likely be pulled into the combustion chambers when the engine is started, resulting in severe engine damage.*
- *If the cylinder head or intake manifold mating surfaces are cracked or damaged, replace the necessary component(s). Inspect the intake manifold surface for cracks.*
- *Never reuse old intake manifold gaskets. This action may cause vacuum leaks and driveability problems.*

## EXHAUST SYSTEM MAINTENANCE

Exhaust system components are subject to both physical and chemical damage. Any physical damage to an exhaust system part that causes a partially restricted or blocked exhaust system usually results in loss of power. Leaks in the exhaust system caused by either physical or chemical (rust) damage could result in illness, asphyxiation, or even death from carbon monoxide in the exhaust. Remember that vehicle exhaust fumes can be very dangerous to one's health.

## EXHAUST SYSTEM DIAGNOSIS

Most parts of the exhaust system, particularly the exhaust pipe, muffler, and tailpipe, are subject to rust, corrosion, and cracking. Broken or loose clamps and hangers can allow parts to separate or hit the road as the car moves.

- *If the engine has been running, exhaust system components may be extremely hot! Wear protective gloves when working on these components.*
- *Exhaust gas contains poisonous carbon monoxide (CO) gas. This gas can cause illness and death by asphyxiation. Exhaust system leaks are dangerous for customers and technicians.*

Any exhaust system inspection should include listening for hissing or rumbling that would result from a leak in the system. An on-lift inspection should pinpoint any of the following types of damage:

- Holes, road damage, separated connections, and bulging muffler seams
- Kinks and dents
- Discoloration, rust, soft corroded metal, and so forth
- Torn, broken, or missing hangers and clamps
- Loose tailpipes or other components
- Bluish or brownish catalytic converter shell, which indicates overheating

## EXHAUST RESTRICTION DIAGNOSIS

Often leaks and rattles are the only problems looked for in an exhaust system. The exhaust system should also be tested for blockage and restrictions. Collapsed pipes or clogged converters and/or mufflers can cause these blockages.

There are many ways to check for a restricted exhaust. The most common of these is the use of a vacuum gauge.

Connect a vacuum gauge to an intake manifold vacuum source. Bring the engine speed to 2,500 rpm and hold it there for 3 minutes. Watch the vacuum gauge. If the exhaust system is not restricted, the vacuum reading will be high and will either stay at that reading or increase slightly as the engine runs at this speed. If the exhaust is restricted, the vacuum will begin to decrease after a period of time. This is caused by the cylinder's inability to purge itself of all of its exhaust gases during the exhaust stroke. The presence of exhaust in the cylinder when the intake stroke begins decreases the amount of vacuum that can be formed on that stroke.

## CONVERTER DIAGNOSIS

The **catalytic converter** is normally a trouble-free emission control device, but leaded gasoline or overheating may cause converter damage. Lead coats the catalyst and renders it useless. The difficulty of obtaining leaded gasoline has reduced this problem. Overheating is caused by raw fuel entering the exhaust because of a fouled spark plug or other problem, thereby quickly increasing the temperature of the converter. The heat can melt the ceramic honeycomb or pellets inside, causing a major restriction to the flow of exhaust.

A plugged converter or any exhaust restriction can cause damage to the exhaust valves due to excess heat, loss of power at high speeds, stalling after starting (if totally blocked), or a drop in engine vacuum as engine rpm increases.

The best way to determine if a catalytic converter is working is to check the quality of the exhaust. This is done with a four- or five-gas exhaust analyzer. The results of this test should show low emission levels if the converter is working properly.

Another way to test a converter is to use a hand-held digital pyrometer, an electronic device that measures heat. By touching the pyrometer probe to the exhaust pipe just ahead of and just behind the converter, it is possible to read an increase of at least 100°F (37.7°C) as the exhaust gases pass through the converter. If the outlet temperature is the same or lower than the inlet temperature, nothing is happening inside the converter. Do not be quick to condemn the converter. To do its job efficiently, some converters need a steady supply of oxygen from the air pump. A bad pump, a faulty diverter valve or control valve, leaky air connections, or faulty computer control over the air injection system could be preventing the needed oxygen from reaching the converter. If a catalytic converter is found to be defective, it must be replaced.

## EXHAUST SYSTEM SERVICE

Before beginning work on an exhaust system, make sure it is cool to the touch. Disconnect the negative battery cable before starting the exhaust system service to avoid short-circuiting the electrical system. Soak all rusted bolts, nuts, and other removable parts with penetrating oil. Finally, check the system for critical clearance points so they can be maintained when new components are installed.

Most exhaust work involves the replacement of parts. When replacing exhaust parts, make sure the new parts are exact replacements for the original parts. Doing this will ensure proper fit and alignment as well as ensure acceptable noise levels. Exhaust system component replacement might require the use of special tools **(Figure 6)** and welding equipment.

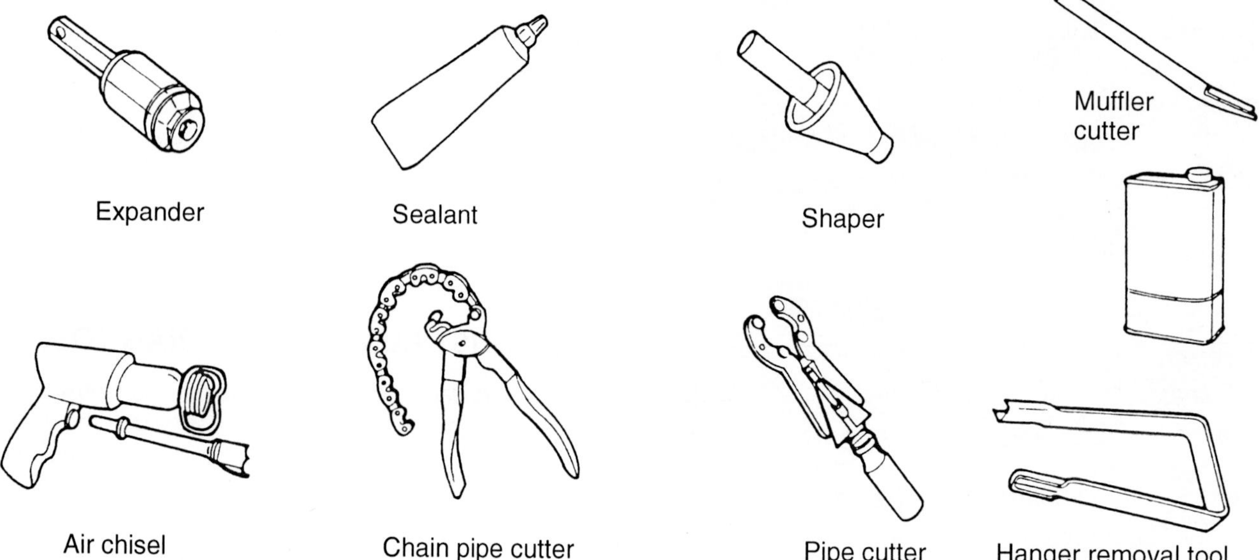


**Figure 6.** Exhaust system service tools.

## Exhaust Manifold and Exhaust Pipe Servicing

As mentioned, the manifold itself rarely causes any problems. On occasion, an exhaust manifold will warp because of excess heat. A straightedge and feeler gauge can be used to check the machined surface of the manifold.

Another problem, also the result of high temperatures generated by the engine, is a cracked manifold. This usually occurs after the car passes through a large puddle and cold water splashes on the manifold's hot surface. If the manifold is warped beyond manufacturer's specifications or is cracked, it must be replaced. Also, check the exhaust pipe for signs of collapse. If there is damage, repair it. These repairs should be done as directed in the vehicle's service manual.

## Replacing Exhaust System Gaskets and Seals

The most likely spot to find leaking gaskets and seals is between the exhaust manifold and the exhaust pipe **(Figure 7)**. When installing exhaust gaskets, carefully follow the recommendations on the gasket package label and instruction forms. Read through all installation steps before beginning. Take note of any of the OEM's recommendations in service manuals that could affect engine sealing. Manifolds warp more easily if an attempt is made to remove them while they are still hot. Remember, heat expands metal, making assembly bolts more difficult to remove and easier to break.

To replace an exhaust manifold gasket, follow the torque sequence in reverse to loosen each bolt. Repeat the process to remove the bolts. This minimizes the chance of warping the components.

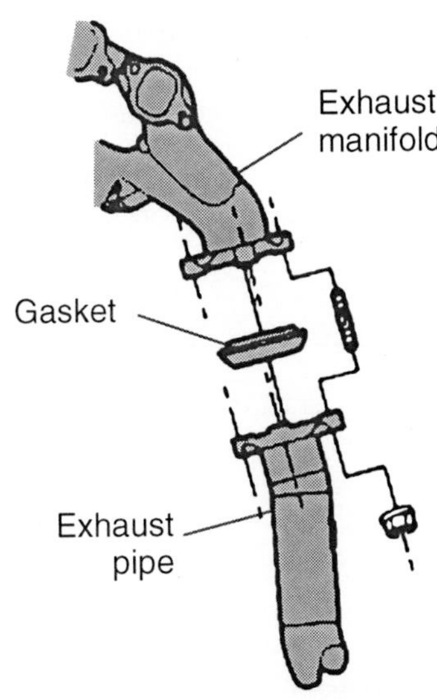


**Figure 7.** Exhaust pipe to manifold gasket.

Any debris left on the sealing surfaces increases the chance of leaks. A good gasket remover will quickly soften the old gasket debris and adhesive for quick removal. Carefully remove the softened pieces with a scraper and a wire brush. Be sure to use a nonmetallic scraper when attempting to remove gasket material from aluminum surfaces.

Inspect the manifold for irregularities that might cause leaks, such as gouges, scratches, or cracks. Replace the exhaust manifold if it is cracked or badly warped. File down any imperfections to ensure proper sealing of the manifold.

Due to high heat conditions, it is important to retap and redie all threaded bolt holes, studs, and mounting bolts. This procedure ensures tight, balanced clamping forces on the gasket. Lubricate the threads with high-temperature antiseize lubricant. Use a small amount of contact adhesive to hold the gasket in place. Align the gasket properly before the adhesive dries. Allow the adhesive to dry completely before proceeding with manifold installation.

Install the bolts finger-tight. Tighten the bolts in three steps—one-half, three-quarters, and full torque—following the torque tables in the service manual or gasket manufacturer's instructions. Torquing is usually begun in the center of the manifold, working outward in an X pattern.

To replace a damaged exhaust pipe, begin by supporting the converter to keep it from falling. Carefully remove the oxygen sensor if there is one. Unbolt the flange holding the exhaust pipe to the exhaust manifold. When removing the exhaust pipe, check to see if there is a gasket. If so, discard it and replace it with a new one. Once the joint has been taken apart, the gasket loses its effectiveness. Disconnect the pipe from the converter and pull the front exhaust pipe loose and remove it.

Although most exhaust systems use flanges or a slip joint and clamps to fasten the pipe to the muffler, a few use a welded connection. If the vehicle's system is welded, cut the pipe at the joint with a hacksaw or pipe cutter. The new pipe need not be welded to the muffler. An adapter available with the pipe can be used instead. When measuring the length for the new pipe, allow at least a 2-inch (50.8-millimeter) opening for the adapter to enter the muffler.

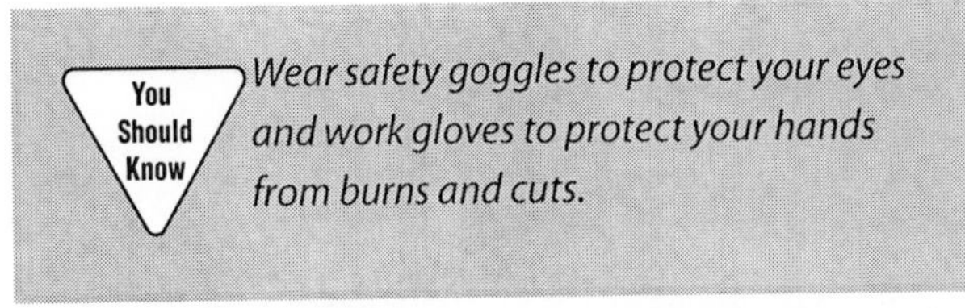


*Wear safety goggles to protect your eyes and work gloves to protect your hands from burns and cuts.*

When trying to replace a part in the exhaust system, you may run into parts that are rusted together. This is especially a problem when a pipe slips into another pipe or the muffler. If you are trying to reuse one of the parts, you

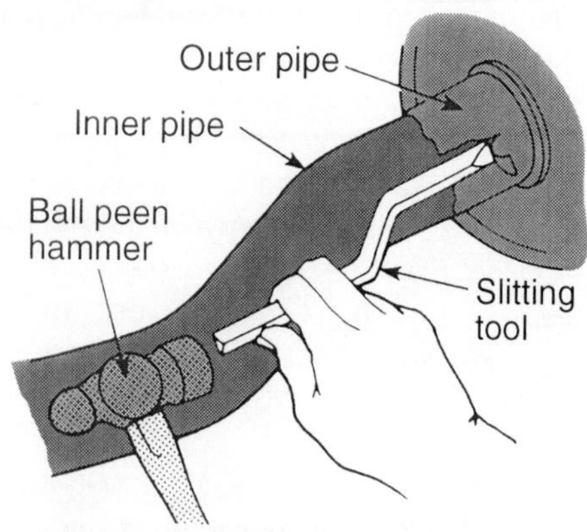


Figure 8. Using an exhaust pipe splitting tool.

should carefully use a cold chisel or slitting tool **(Figure 8)** on the outer pipe of the rusted union. You must be careful when doing this because you can easily damage the inner pipe. It must be perfectly round to form a seal with a new pipe.

Slide the new pipe over the old. Position the rest of the exhaust system so that all clearances are evident and the parts aligned, then put a U-clamp over the new outer pipe to secure the connection.

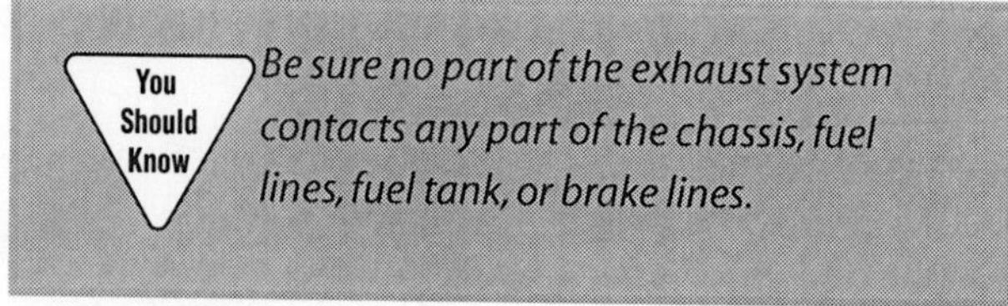


*Be sure no part of the exhaust system contacts any part of the chassis, fuel lines, fuel tank, or brake lines.*

*The Ford Focus partial zero emissions vehicle (PZEV) introduced as a 2003 mid-year model has a hydrocarbon trap mounted in the air intake duct between the air filter and the throttle body. This hydrocarbon trap prevents hydrocarbons from escaping from the air intake ducts and air filter while the engine is shut off. The hydrocarbon trap contains a screen-protected, cushioned carbon monolith in a special housing. The hydrocarbon trap is a passive device with no moving parts or electronic controls.*

## Summary

- When replacing an air cleaner element, always inspect the airflow restriction indicator. Green displayed in the indictor window indicates the air filter is not restricted, whereas an orange display indicates a restricted air filter.
- If the airflow restriction indicator window is orange, push the reset button on the indicator to reset the window to green after the new air filter is installed.
- An engine in satisfactory condition normally has an intake manifold vacuum of 17 to 22 in. Hg (432 and 559 mm/Hg) when the engine is idling.
- An intake manifold vacuum leak causes a low steady vacuum reading with the engine idling.
- An intake manifold vacuum leak may cause a cylinder misfire with the engine operating at idle and low speeds.
- The mating surfaces on aluminum intake manifolds and cylinder heads should be cleaned with a plastic scraper.
- If the exhaust system is restricted, the intake manifold vacuum will slowly decrease when the engine is operated at 2,500 rpm for several minutes.
- With the engine running, the catalytic converter outlet should be about 100°F hotter than the inlet.

## Review Questions

1. While discussing airflow restriction indicators, Technician A says if the restrictor window appears green, the air filter is restricted. Technician B says if the restrictor window appears green, the airflow restriction indicator must be replaced. Who is correct?
   A. Technician A
   B. Technician B
   C. Both Technician A and Technician B
   D. Neither Technician A nor Technician B
2. While discussing air filter service, Technician A says the air gun should be held tight against the air filter element when blowing dirt out of the filter. Technician B says when the outside of the air filter is observed with a trouble light inside the filter, small holes in the filter indicate filter replacement is required. Who is correct?
   A. Technician A
   B. Technician B
   C. Both Technician A and Technician B
   D. Neither Technician A nor Technician B

3. All of these drivability and engine problems may be caused by an intake manifold vacuum leak *except*:
   A. rough idle operation.
   B. low engine operating temperature.
   C. acceleration stumbles.
   D. engine detonation.
4. An engine has 19 in. Hg (64.2 kPa) of vacuum in the intake manifold with the engine idling. With the engine running at 2,500 rpm for 3 minutes, the vacuum slowly decreases to 10 in. Hg (33.8 kPa). This problem could be caused by:
   A. a restriction in the catalytic converter or muffler.
   B. a leak in the exhaust pipe.
   C. a leak in the intake manifold gaskets.
   D. a restricted vacuum hose connected to the brake booster.
5. All of these statements about air filter service and diagnosis are true *except*:
   A. A shop trouble light may be used to check air filter restriction.
   B. Oil contamination in an air filter may be caused by a restricted PCV valve.
   C. Oil contamination in an air filter may be caused by excessive engine blowby.
   D. Oil contamination in an air filter may be caused by a leaking rocker arm cover gasket.
6. Oil is located in the vacuum hose connected to a transmission modulator. Technician A says this problem is caused by overfilling the transmission. Technician B says the cause of this problem may be a restricted engine PCV system. Who is correct?
   A. Technician A
   B. Technician B
   C. Both Technician A and Technician B
   D. Neither Technician A nor Technician B
7. All of these statements about heated air inlet systems are true *except*:
   A. When the engine coolant is cold, the butterfly valve in the air cleaner snorkel closes the cold air passage into the air cleaner.
   B. When the engine is cold, vacuum is supplied to the vacuum actuator in the air cleaner.
   C. When the engine is at normal operating temperature, the temperature switch shuts off vacuum to the vacuum actuator.
   D. When the engine coolant is cold, the butterfly valve closes the air passage from the exhaust manifold stove into the air cleaner.
8. When diagnosing a catalytic converter with the engine at normal operating temperature, the converter inlet and outlet temperatures are the same. The most likely cause of this problem is:
   A. low fuel pump pressure.
   B. a lean air-fuel ratio.
   C. an inoperative EGR valve.
   D. a defective catalytic converter.
9. When replacing a catalytic converter on an engine with port fuel injection, the monolith material inside the old converter is severely melted. Technician A says the air-fuel ratio may be excessively rich. Technician B says the fuel pump pressure may be lower than specified. Who is correct?
   A. Technician A
   B. Technician B
   C. Both Technician A and Technician B
   D. Neither Technician A nor Technician B
10. A throttle body injected engine has a cylinder misfire at idle speed. When the engine is accelerated gradually to 2,500 rpm, the misfire disappears. The most likely cause of this problem is:
    A. lower-than-specified compression on one cylinder.
    B. an exhaust manifold leak.
    C. an intake manifold vacuum leak.
    D. excessive fuel pump pressure.
11. Never use a ______________ ______________ or a ______________ ______________ when cleaning the gasket surface on an aluminum cylinder head.
12. Late ignition timing causes a ______________ ______________ intake manifold vacuum with the engine idling.
13. When installing exhaust system components, pipes that slide over or fit inside other pipes should have a(n) ______________-inch overlap.
14. When an air cleaner element is not restricted, the window in the airflow restrictor should appear ______________ in color.
15. Describe a normal vacuum gauge reading with the engine idling.
16. Explain the results of a restricted PCV valve.
17. Describe the vacuum gauge reading caused by sticking valves.
18. Explain how an intake manifold vacuum leak may cause a cylinder misfire with the engine operating at idle and low speeds.

## Electrical Systems

*Approximately 95 percent of the vehicles retired from service each year are processed for recycling and at least 75 percent of the material content of these vehicles is recycled. Recycled materials from retired vehicles include: automotive shredder residue—plastic, rubber, and glass; non-ferrous metals—aluminum, copper, lead, zinc, and magnesium; ferrous metals—steel and iron; batteries and fluids—engine oil, coolant, and refrigerants.*

## SECTION OBJECTIVES

After you have read, studied, and practiced the contents of this section, you should be able to:

- Describe the parts of an atom.
- Explain the importance of the number of electrons on the valance ring of an atom.
- Describe how electrons flow through a conductor.
- Define conductor, semiconductor, and insulator.
- Explain voltage, amperes, and ohms in an electric circuit.
- Describe the main features of a series circuit.
- Explain the factors that determine the strength of an electromagnet.
- Describe the design of single- and dual-filament light bulbs.
- Describe sealed beam design.
- Describe the operation of halogen and high-intensity discharge headlights.
- Describe the headlight switch operation in the PARK position.
- Explain the headlight switch operation in the HEADLIGHT position.
- Describe the current flow through the headlights during high beam operation.
- Explain the operation of a circuit breaker.
- Describe the operation of the signal light switch and related circuit during a right turn.
- Explain the operation of the interior light circuit with the left-front door open.
- Explain voltage drop in a headlight circuit.
- Describe the precautions to be observed when servicing halogen bulbs.
- Explain how to test bulbs and fuses.
- Explain the operation of oil pressure indicator and charge indicator lights.
- Describe the operation of a bimetallic gauge.
- Explain the purpose and operation of an instrument voltage limiter.
- Describe the proper ammeter and voltmeter connections in an electrical circuit.
- Diagnose oil pressure indicator light problems.
- Diagnose charge indicator light problems.
- Test thermistor-type gauge sending units.

# Chapter 17 Basic Electricity and Electric Circuits

## Introduction

An understanding of basic electricity is absolutely essential before a study of more complex electrical/electronic systems. It is very difficult to understand electrical/electronic systems if you do not have a clear understanding of basic electricity. Conversely, if you are familiar with the principles of basic electricity, these principles are used in the operation of electrical/electronic system and the understanding of these systems becomes much easier.

### ATOMIC STRUCTURE

An **atom** may be defined as the smallest particle of an element in which all the chemical characteristics of the element are present. Atoms are very small particles that cannot be seen even with a powerful electron microscope that magnifies millions of times. Atoms contain even smaller particles called **protons**, **neutrons**, and **electrons**. Protons and neutrons are located at the center or nucleus of each atom. Protons contain a positive electrical charge and neutrons add weight to the atom. Heavier elements contain more neutrons.

Electrons circle around the **nucleus** or center of the atom in various orbits **(Figure 1)**. For example, a hydrogen atom contains one proton in the nucleus and one electron orbiting around the proton. The hydrogen atom does not have any neutrons. A copper atom has four orbits or rings with two, eight, eighteen, and one electron on these orbits **(Figure 2)**. Atoms may have from one to seven orbits with different numbers of electrons on each orbit. Electrons have a negative electrical charge. The outer orbit on an atom is called the **valence ring**. Elements are listed on the **Periodic Table** or Atomic Scale according to their number of protons and electrons. For example, hydrogen is number one on this scale and copper is number twenty-nine. The nucleus does not always have the same number of protons and neutrons. A copper atom has twenty-nine protons and thirty-five neutrons.

The number of electrons on the valence ring determines the electrical characteristics of the element. If the valence ring has one, two, or three electrons, the element is

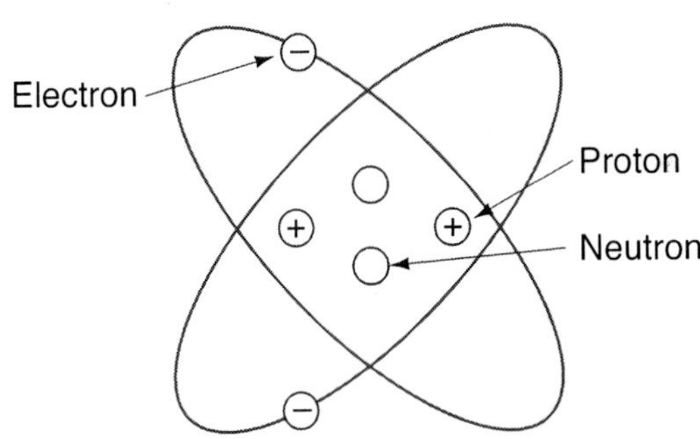


**Figure 1.** The parts of an atom.

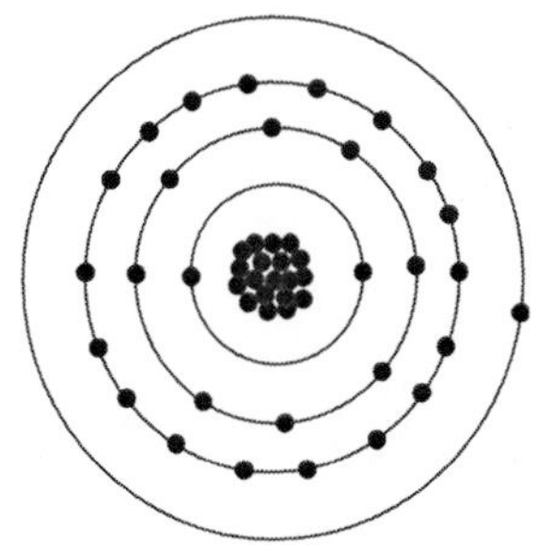

**Figure 2.** The structure of a copper atom.

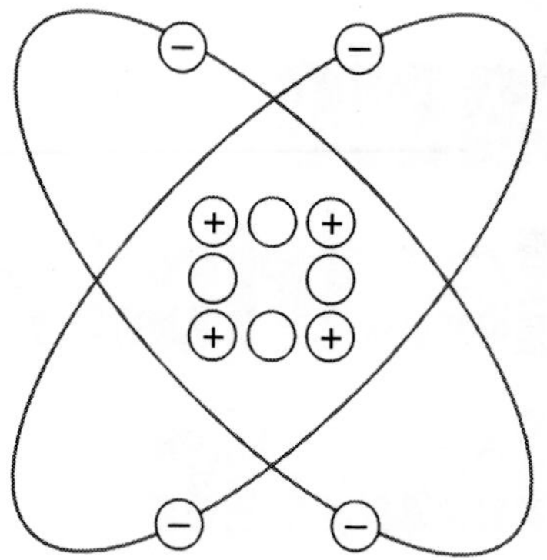

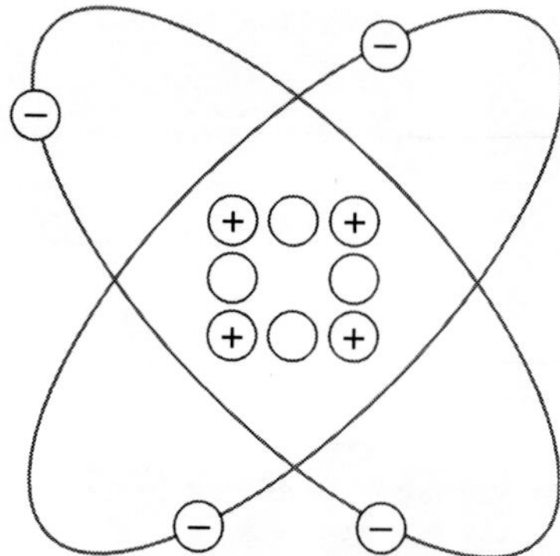

**Figure 3.** If an atom has the same number of electrons and protons, the atom is in balance.

classified as a **good conductor**, because the electron(s) on the valence ring move easily from one atom to another. When an atom has four valence electrons, it is classified as a **semiconductor**. Semiconductors have special electrical characteristics and these materials are used to manufacture diodes and transistors. If an element has five or more valence electrons, these electrons will not move easily and the element is classified as an **insulator**.

If an atom has the same number of protons and electrons, the atom is in balance **(Figure 3)**. Atoms always try and maintain their proper balance with the same number of electrons and protons. If an atom loses an electron, it will try and attract an electron from another atom. When an atom has more protons than electrons it is unbalanced and the positively charged protons will immediately attract other electrons from another atom.

## ELEMENTS, COMPOUNDS, AND MOLECULES

As stated previously, an **element** is a material with only one type of atom. For example, pure copper contains only copper atoms. A **compound** is a material containing two or more types of atoms. For example, water ($H_2O$) contains hydrogen and oxygen. A **molecule** is the smallest particle of a compound that retains all the characteristics of the compound.

## ELECTRIC CURRENT FLOW

A massing of electrons must occur at one point in an electric circuit and a lack of electrons must be present at another point in the circuit before electrons will move through the circuit. The massing of electrons may be referred to as electrical pressure. The electrical circuit must also be complete between the massing of electrons and the lack of electrons. When these requirements are present in a circuit, the electrons begin moving from the massing of electrons into the atoms in the conductor **(Figure 4)**. If an atom has more electrons than protons, it immediately repels an electron to another atom. When an atom lacks an electron, it attracts an electron from another atom. Current flow may be defined as the mass movement of valence electrons from atom to atom in a conductor.

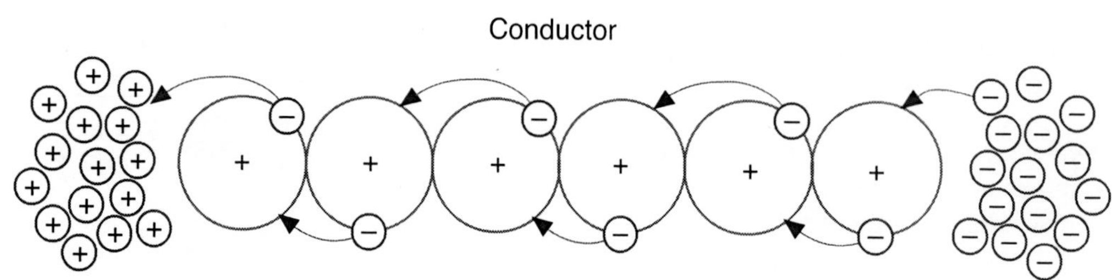


**Figure 4.** Electron flow from atom to atom in a conductor.

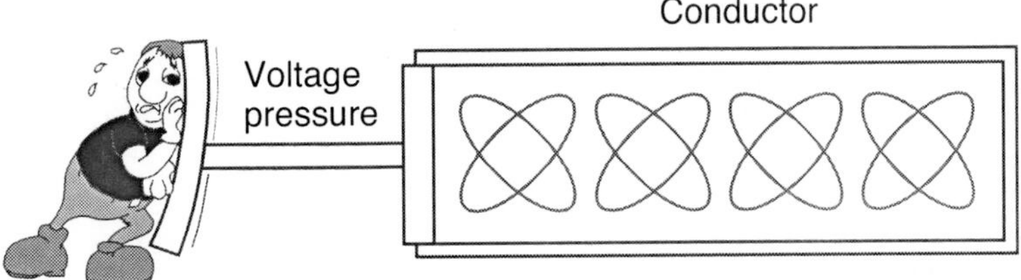


**Figure 5.** Voltage is the pressure that moves electrons through a circuit.

## ELECTRIC CIRCUIT MEASUREMENTS

**Voltage** is an electrical pressure caused by a massing of electrons at one point in an electrical circuit. Voltage causes the electrons to move through a circuit **(Figure 5)**. Voltage is a measurement for electrical pressure difference. When electric current flows through a circuit, there must be a high voltage (massing of electrons) at one point in the circuit and a low voltage (lack of electrons) at another point in the circuit.

If a voltmeter is connected across the terminals of an automotive battery, the voltmeter may indicate 12.6 volts if the battery is fully charged. This reading indicates there are 12.6 volts at one battery terminal in relation to 0 volts at the opposite battery terminal.

### Amperes

**Amperes** is a measurement for the rate of electron flow or the amount of current flowing through a circuit **(Figure 6)**.

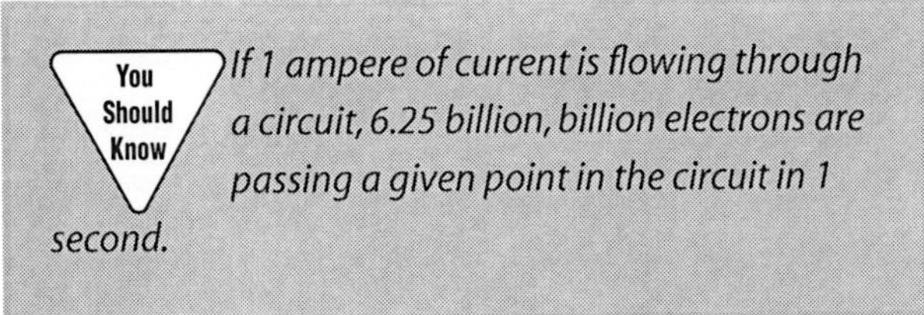


*If 1 ampere of current is flowing through a circuit, 6.25 billion, billion electrons are passing a given point in the circuit in 1 second.*

Amperes will continue to flow through a circuit as long as the massing of electrons and lack of electrons are maintained in the circuit. There are two theories regarding the direction of electron movement or current flow. The **electron theory** says that electrons are negatively charged and electrons move from a massing of electrons (negative charges) to a lack of electrons (positive charges). The **conventional theory** says that for illustration purposes in the automotive industry we assume that current flows from positive to negative through a circuit.

**Direct current (DC)** flows in only one direction. Nearly all automotive circuits operate on direct current. **Alternating current (AC)** flows alternately in one direction and then in the opposite direction. The windings in the alternator stator have AC flowing in them, but this AC is rectified to DC by the diodes in the alternator. Therefore, DC is delivered from the alternator to the battery and electrical components on the vehicle.

Electrical resistance may be considered the opposition to electron movement in a circuit and this resistance is measured in **ohms.** The size, length, type, and temperature of a conductor determines its resistance. For example, a smaller diameter wire has a higher resistance to electron movement compared to a large diameter wire.

**Watts** is an electrical measurement that is calculated by multiplying the amperes by the volts. Although watts is not measured when servicing vehicles, it plays an important role in electrical system design. For example, if the total amperage load of all the electrical accessories on a vehicle is 250 amperes in a 12-volt electrical system, the total watts of energy required are 12 × 250 = 3,000 watts. Therefore, the charging circuit on the vehicle must be capable of supplying over 3,000 watts.

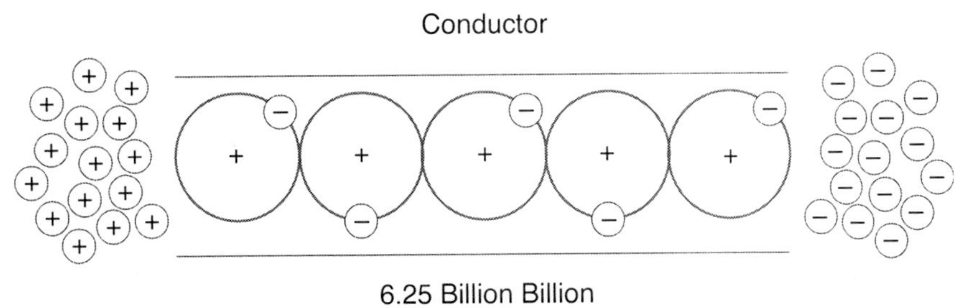


**Figure 6.** Amperes is a measurement for electron movement through a circuit.

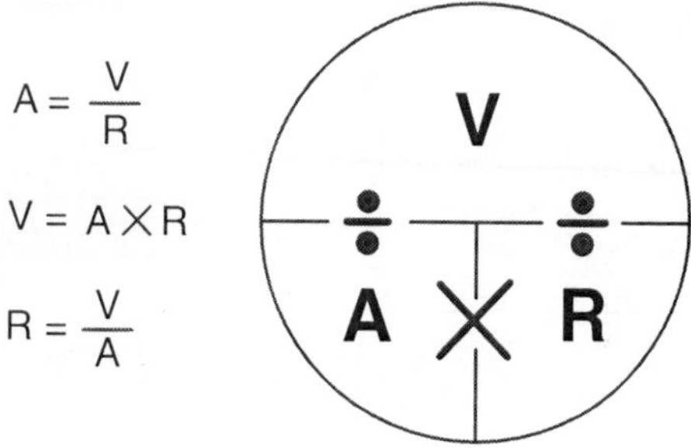


Figure 7. Ohm's Law formula.

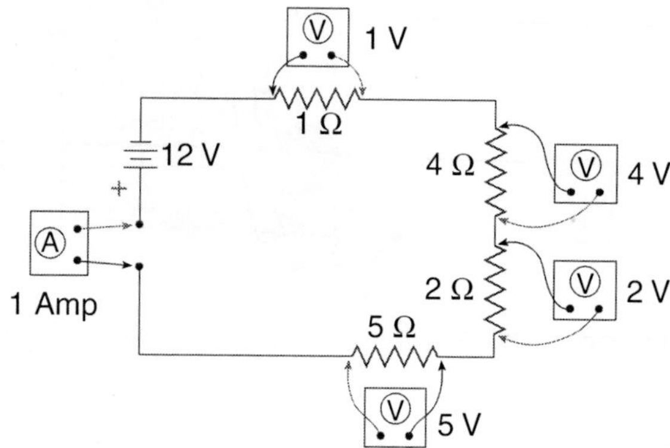


Figure 8. A series circuit.

## OHM'S LAW

Ohm's Law states that the current flow in a circuit is directly proportional to the voltage and inversely proportional to the resistance. This indicates that an increase in voltage causes a corresponding increase in current flow, whereas a decrease in voltage reduces current flow. Conversely, an increase in resistance decreases current flow and a decrease in resistance increases current flow. Ohm's Law may be used to calculate a value in a circuit if the other two values are known. In this formula, voltage is indicated by V, amperes is represented by A, and R indicates resistance **(Figure 7)**. If 12 volts are supplied to a circuit and the resistance in the circuit is 2 ohms, the current flow is 12 ÷ 2 = 6 amperes. When a circuit has 4 amperes and 3 ohms resistance, the voltage is 4 × 3 = 12 volts. If a circuit has 12 volts and a current flow of 1.5 amperes, the resistance is 12 ÷ 1.5 = 8 ohms.

## VOLTAGE DROP

Voltage drop is the difference in voltage across a resistance when current flows through the resistance. There is always some voltage drop when current flows through a resistance. For example, a 0.05-volt drop across some switch contacts is normal. However, higher-than-normal resistance causes high voltage drop in a circuit. High resistance and high voltage drop result in reduced current flow in a circuit. When a set of switch contacts becomes pitted and corroded, they may have a 2-volt drop, which reduces current flow in the circuit.

## SERIES CIRCUIT

In a series circuit, the same current flows through all the resistances in the circuit **(Figure 8)**. A series circuit has these features:

- The same current flows through all the resistances in the circuit because there is only one path for current flow.
- The total resistance is the sum of all resistances in the circuit.
- The voltage drop across each resistance depends on the ohm value of that resistance.
- The sum of the voltage drops across each resistance equals the source voltage.

The heater blower circuit is an example of a series circuit. The resistors in the blower circuit that control blower speed are switched in series with the blower motor.

## PARALLEL CIRCUIT

In a parallel circuit, each resistance is a separate path for current flow **(Figure 9)**. These facts may be summarized regarding parallel circuits:

- Each resistance is a separate path for current flow.
- The amount of current flow through each resistance depends on the amount of resistance in that part of the circuit.
- Equal full source voltage is supplied to each resistance and equal full source voltage is dropped across each resistance.
- The total resistance is always less than the ohm value of the lowest resistor in the circuit.

The light circuit on a vehicle is an example of a parallel circuit. Each of the lights is connected in parallel to the battery.

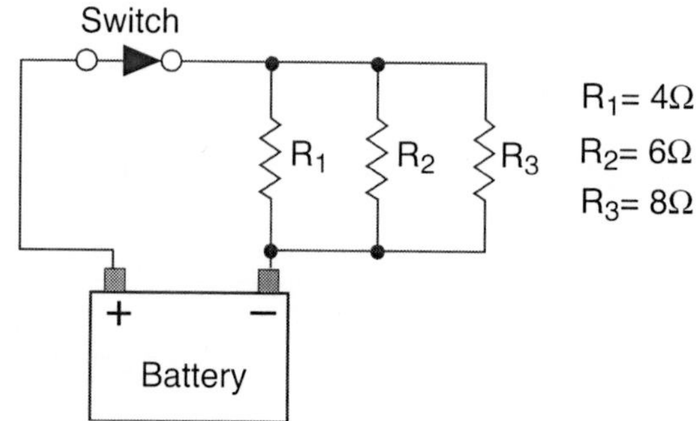


Figure 9. A parallel circuit.

## SERIES-PARALLEL CIRCUIT

In a series-parallel circuit, a component is connected in series with the parallel circuit. Many electrical automotive electrical circuits are series-parallel circuits. For example, the instrument panel lights are connected parallel to the battery, but the variable resistor that dims the instrument panel lights is connected in series with these lights.

## ELECTROMAGNETS

A permanent magnet has an invisible magnetic field surrounding the magnet. These lines of force move from the north pole of the magnet to the south pole. One of the basic magnetic principles is that like magnetic poles repel each other and unlike magnetic poles have an attracting force **(Figure 10)**. This principle is used in a starting motor.

An electromagnet is manufactured by winding a coil of wire around a metal core. When current flows through a straight wire, a magnetic field surrounds the wire. This magnetic field is concentric to the wire and the same strength at any point along the wire. In the coil of wire on an electromagnet, the magnetic strength of each loop of wire adds to the strength of the other loops and the magnetic field surrounds the entire coil. An iron core placed in the center of the coil helps to concentrate the magnetic field because iron is a better conductor for magnetic lines of force compared to air **(Figure 11)**. The strength of an electromagnet is determined by the number of turns on the coil and the amount of current flow through the coil. Increasing either of these factors provides a stronger magnetic field around the electromagnet **(Figure 12)**. The magnetic strength is calculated by multiplying the amperes by the number of turns. For example, 1 ampere × 1,000 turns = 1,000 ampere turns. Electromagnets are used in relays and solenoids. If the current flow through the coil is reversed, the polarity of the coil is also reversed.

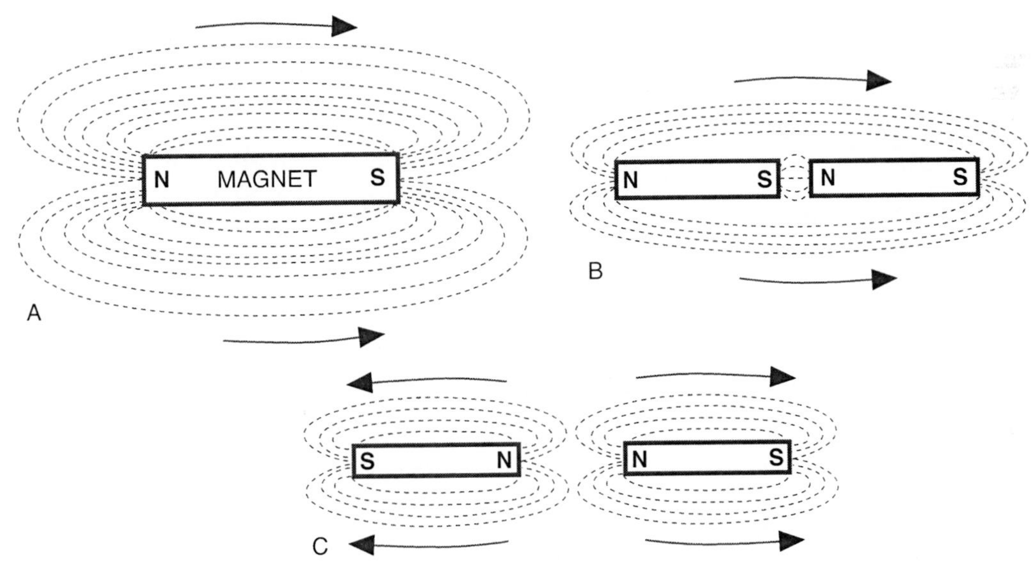


**Figure 10.** Magnetic principles: (A) All magnets have a north pole and a south pole, (B) unlike poles attract, and (C) like poles repel.

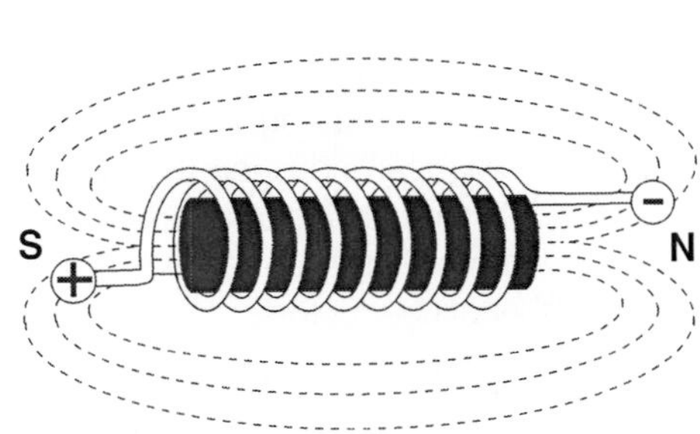


**Figure 11.** A metal core placed in the center of an electromagnet concentrates the lines of force.

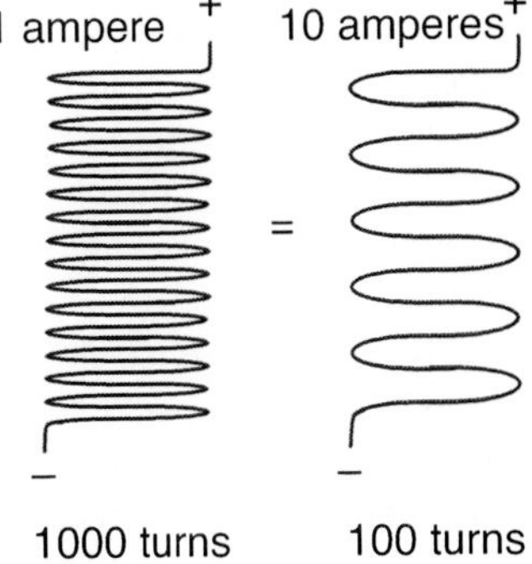


**Figure 12.** The strength of an electromagnet is determined by the number of coils and the amperes flowing through the winding.

## ELECTROMAGNETIC INDUCTION

When a conductor is moved through a magnetic field, a voltage is induced in the conductor **(Figure 13)**. This process is called **electromagnetic induction**. If the conductor is connected to a complete circuit, current flows through the circuit.

> You Should Know
>
> *During electromagnetic induction, it does not matter if the conductor is moved through a magnetic field or the magnetic field is moved across the conductor; voltage is still induced in the conductor.*

> You Should Know
>
> *After magnetic lines or force pass through a piece of hard steel, the steel retains magnetism. After magnetic lines of force pass through a piece of soft iron, it becomes de-magnetized immediately.*

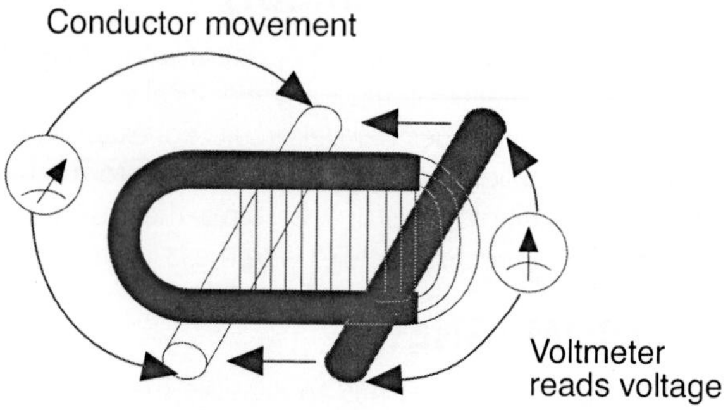

Conductor movement

Voltmeter reads voltage

**Figure 13.** When a conductor is moved through a magnetic field or vice versa, a voltage is indicated in the conductor.

During electromagnetic induction, the amount of voltage induced in a conductor is determined by the number of conductors, the strength of the magnetic field, and the speed of motion. In an alternator, the revolving magnetic field of the rotor cuts across the stator windings and induces voltage in these windings. This voltage supplies current to the battery and electrical accessories on the vehicle. In an ignition coil, the windings are stationary and the magnetic field moves across these windings to induce voltage in the windings. This voltage is used to fire the spark plugs.

## *Summary*

- Protons and electrons are located at the center or nucleus of an atom. The protons have a positive electrical charge and the neutrons do not have any electrical charge, but they add weight to the atom.
- Electrons have a negative electrical charge and they move in orbits around the nucleus of an atom.
- An element is a liquid, solid, or a gas with only one type of atom.
- A compound contains two or more different types of atoms.
- The outer orbit on an atom is called the valence ring and the number of electrons on this ring determines the electrical characteristics of the element.
- A good conductor is a material with one, two, or three valence electrons.
- A semiconductor has four valence electrons.
- An insulator has five or more valence electrons.
- When current flows through a conductor, the valence electrons move from atom to atom in the conductor.
- Voltage is a measurement for electrical pressure difference.
- Amperes is a measurement for the amount of current flow in a circuit.
- Ohms are a measurement for the opposition to current flow in a circuit.
- In a series circuit, the same current flows through all the resistances in the circuit.
- In a parallel circuit, each part of the circuit is a separate path for current flow.
- In a series-parallel circuit, a component is connected in series with the parallel components in the circuit.
- The number of coils and the current flow through the winding determines the strength of an electromagnet.
- During electromagnetic induction, the amount of voltage inducted in the conductor is determined by the strength of the magnetic field, the speed of motion, and the number of conductors.

## Review Questions

1. Technician A says protons have a positive electrical charge. Technician B says electrons are located in the nucleus of an atom. Who is correct?
   A. Technician A
   B. Technician B
   C. Both Technician A and Technician B
   D. Neither Technician A nor Technician B
2. Technician A says if the atoms in a material have four valence electrons, the material is a good conductor. Technician B says if the atoms in a material have five valence electrons, the material is a semiconductor. Who is correct?
   A. Technician A
   B. Technician B
   C. Both Technician A and Technician B
   D. Neither Technician A nor Technician B
3. Technician A says if an atom has the same number of protons and electrons, it is considered to be in balance. Technician B says if an atom loses an electron, it will attract an electron from another atom. Who is correct?
   A. Technician A
   B. Technician B
   C. Both Technician A and Technician B
   D. Neither Technician A nor Technician B
4. Technician A says an element has two or more types of atoms. Technician B says a molecule is the smallest particle of a compound. Who is correct?
   A. Technician A
   B. Technician B
   C. Both Technician A and Technician B
   D. Neither Technician A nor Technician B
5. All of these statements about atoms are true *except*:
   A. Protons are located in the nucleus of an atom.
   B. Neutrons have a positive electrical charge.
   C. Electrons move in orbits around the nucleus.
   D. Electrons have a negative electrical charge.
6. Technician A says the outer ring on an atom is called the valence ring. Technician B says the number of electrons on the valence ring of each atom determines the electrical characteristics of an element. Who is correct?
   A. Technician A
   B. Technician B
   C. Both Technician A and Technician B
   D. Neither Technician A nor Technician B
7. All of these statements about insulators, conductors, and semiconductors are true *except*:
   A. A good conductor has one valence electron on each atom.
   B. An element with four valence electrons is a semiconductor.
   C. Semiconductors are used to manufacture transistors.
   D. An element with three valence electrons is an insulator.
8. Technician A says voltage is a measurement for the amount of energy in an electric circuit. Technician B says amperes is a measurement for resistance in an electrical circuit. Who is correct?
   A. Technician A
   B. Technician B
   C. Both Technician A and Technician B
   D. Neither Technician A nor Technician B
9. All of these statements about electrical circuits and Ohm's Law are true *except*:
   A. A voltage increase causes an increase in current flow.
   B. An increase in resistance causes a decrease in current flow.
   C. If a circuit has 5 amperes of current and 2.5 ohms of resistance, the voltage is 12.5.
   D. If a circuit has 14 volts and 7 ohms of resistance, the current flow is 2.5 amperes.
10. In an electric circuit with three unequal resistances connected in series:
    A. the total resistance is the sum of all three resistances.
    B. the voltage drop across each resistance is the same.
    C. the current flow varies in each resistance.
    D. the sum of the voltage drops across each resistance is higher than the source voltage.
11. Voltage is a measurement for ______________ ______________.
12. The rate of electron flow is measured in ______________.
13. The opposition to current flow in a circuit is measured in ______________.
14. When current flows in one direction only, it is called ______________ current.
15. Explain the result of increasing the voltage in an electrical circuit.
16. Describe the result of decreasing the resistance in an electrical circuit.
17. List the factors that determine the strength of an electromagnet.
18. List the factors that determine the amount of voltage induced during electromagnetic induction.

# Chapter 18 Light Circuits

## Introduction

Technicians must understand basic light circuits to be able to maintain, diagnose, and service these systems. When these circuits are understood, diagnosing becomes much easier and faster.

### LAMPS

An automotive light bulb usually contains one or two filaments. In a single filament bulb, the terminal is connected to one side of the filament and the opposite end of the filament is usually connected to the bulb case **(Figure 1)**. Voltage is supplied to the bulb terminal and current flows through the filament to the bulb case. The circuit is completed from the bulb case through the vehicle ground back to the battery. The indexing pins on the sides of the case retain the bulb in the socket. Many automotive bulbs have two filaments and two terminals that supply voltage to the filaments. These dual filament bulbs serve two purposes such as stop and tail lights. The indexing pins position the bulb terminals properly in the socket. A variety of different bulbs are used in a typical vehicle **(Figure 2)**.

When current flows through a bulb filament, it becomes very hot. The electrical energy in the filament is changed to heat energy and this action is so intense that the filament glows and gives off light. This process of changing electrical energy to heat energy that produces light is called **incandescence**. The filament is surrounded by a vacuum that prevents overheating and destruction of the filament. When a bulb is manufactured, a vacuum is sealed inside the glass envelope surrounding the bulb.

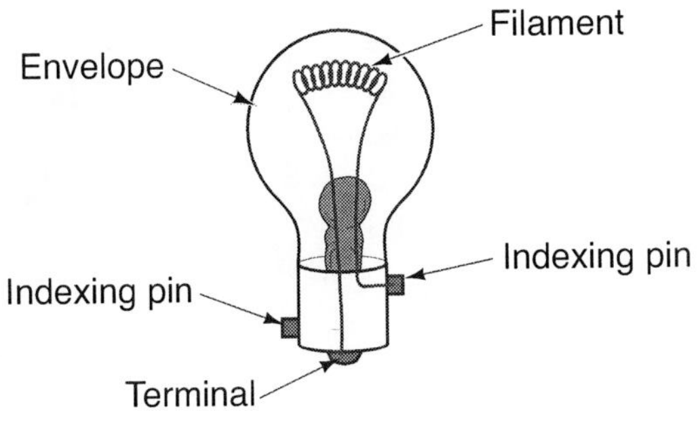


**Figure 1.** A single filament bulb.

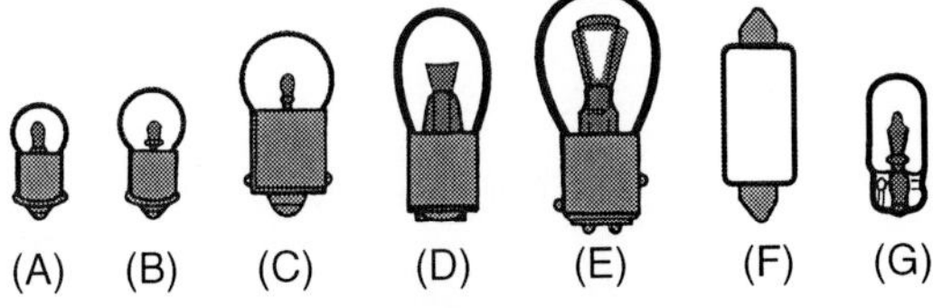


A,B Miniature bayonet for indicator and instrument lights
C — Single contact bayonet for license and courtesy lights
D — Double contact bayonet for trunk and underhood lights
E — Double contact bayonet with staggered indexing lugs for stop, turn signals, and brake lights
F — Cartridge type for dome lights
G — Wedge base for instrument lights

**Figure 2.** Various types of automotive bulbs.

**You Should Know** *When replacing automotive bulbs, be sure the replacement bulb is the same as the original bulb, including the position of the indexing pins. Most bulbs have the part number stamped on the bulb case.*

## SEALED BEAM HEADLIGHTS

Sealed beam headlights may be round or rectangular-shaped. Sealed beam headlights have a parabolic reflector sprayed with vaporized aluminum in the rear of the sealed beam. This reflector is fused to a glass lens in the manufacturing process **(Figure 3)**. All the oxygen is removed from the sealed beam and then it is filled with argon gas. If oxygen were allowed to remain in the sealed beam, the filament would become oxidized and burn out quickly. Sealed beams may contain one or two filaments. If the sealed beam operates on both high and low beam, it has two filaments and three terminals. Some sealed beams that operate only on high beam contain a single filament and two terminals.

The light from the filament in a sealed beam is reflected from the reflector through concave prisms in the lens **(Figure 4)**. The prisms in the lens direct the light beam downward in a flat, horizontal pattern **(Figure 5)**. The filaments are precisely located in the reflector to properly direct the light. If a sealed beam has two filaments, the lower filament is for high beam and the upper filament is for low beam **(Figure 6)**.

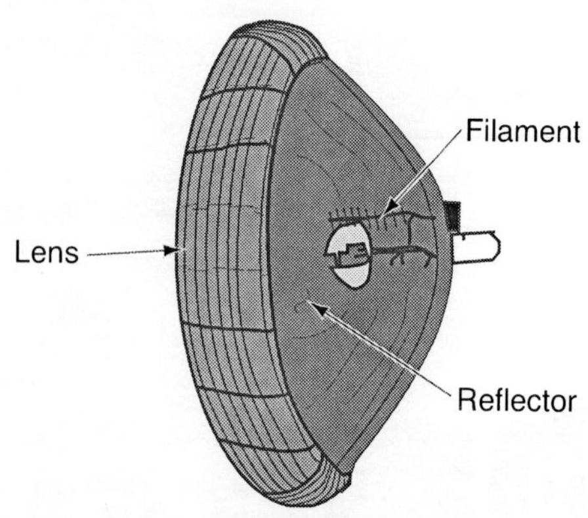


**Figure 3.** Sealed beam headlight design.

**Figure 4.** The lens in a sealed beam uses prisms to redirect the light.

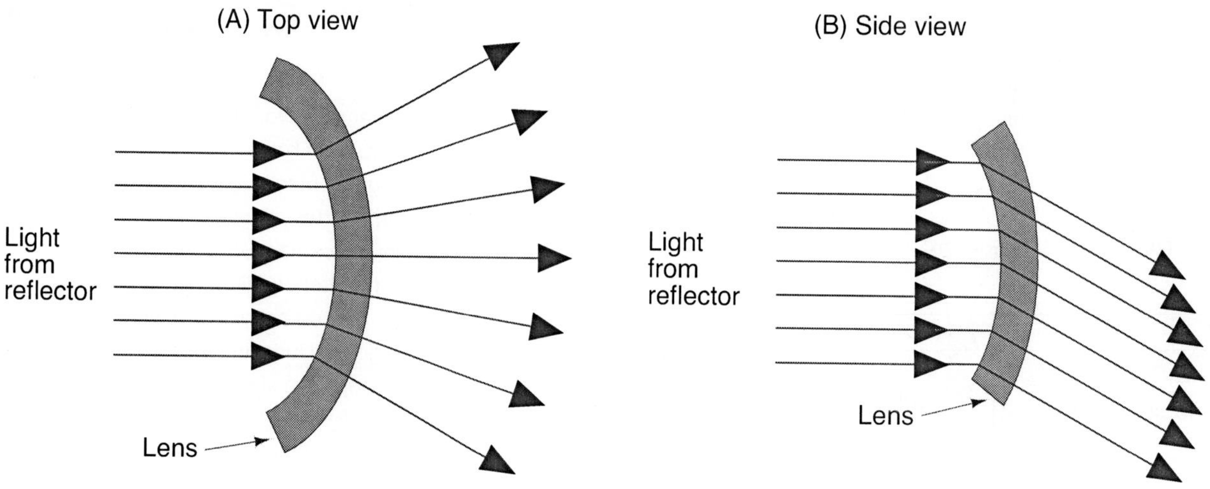


**Figure 5.** The prism directs the light beam into (A) a flat, horizontal pattern, and (B) downward.

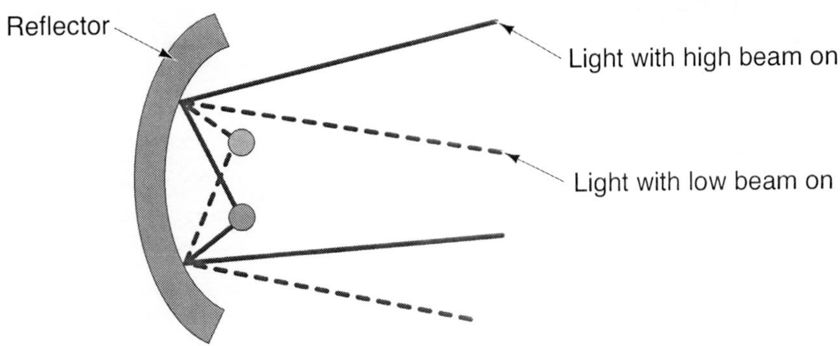


**Figure 6.** Filament location in a sealed beam controls the light beam projection.

## HALOGEN HEADLIGHTS

Many newer vehicles have halogen headlights. This type of headlight contains a small bulb filled with iodine vapor. The bulb has a glass or plastic envelope surrounding a tungsten filament. The bulb is installed in a sealed glass housing **(Figure 7)**. **Halogen** is a term for a group of chemically related nonmetallic elements including chlorine, fluorine, and iodine.

The tungsten filament can withstand higher temperatures and burn brighter because of the halogen added to the bulb. Halogen headlights produce approximately 25 percent more light compared to sealed beam headlights.

You Should Know *Because the bulb in a halogen headlight is self contained, a cracked lens does not prevent headlight operation. However, a cracked lens should be replaced because it results in poor light quality.*

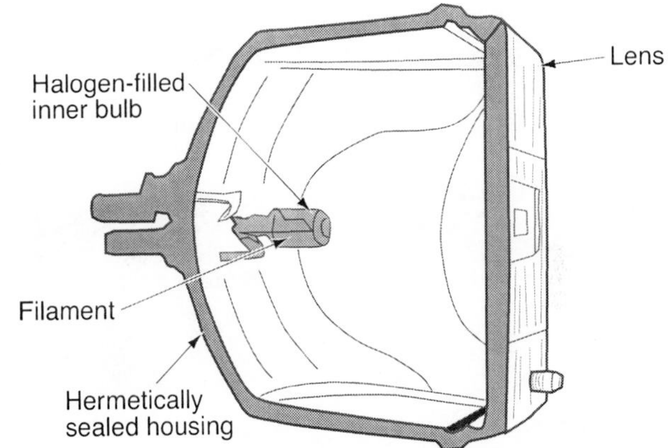


**Figure 7.** A halogen headlight.

Many vehicles are presently equipped with composite headlights and replaceable halogen bulbs **(Figure 8)**. The composite headlights allow the vehicle manufacturers to design the headlights in various shapes to conform to more aerodynamic body styling. For example, some composite headlights wrap around the front corner of the vehicle.

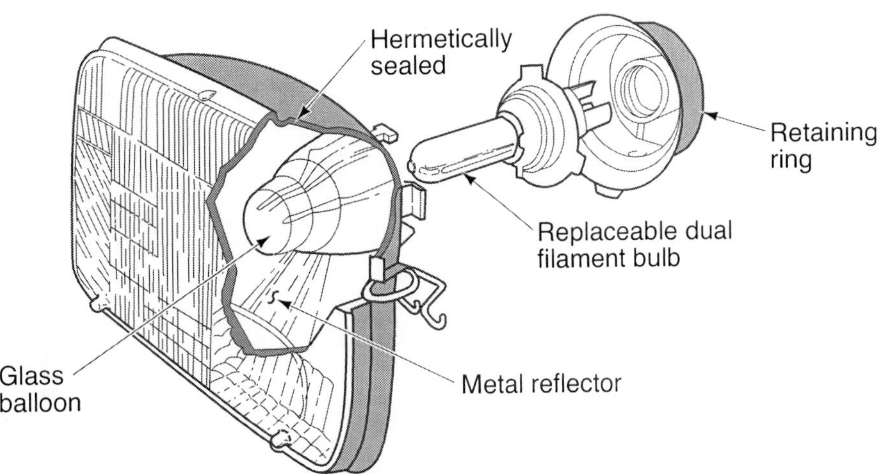


**Figure 8.** A composite headlight with a replaceable halogen bulb.

## HIGH INTENSITY DISCHARGE (HID) HEADLIGHTS

In recent years, some vehicles have been equipped with high intensity discharge (HID) headlights. In HID headlights, the light is produced from high voltage arcs across an air gap between two electrodes. An inert gas in the headlight amplifies the light provided by the high voltage arcing. Approximately 15,000 to 25,000 volts are required to initially force current across the air gap between the electrodes. Once this gap is bridged with a stream of electrons, about 80 volts are required to maintain the current flow across the gap. A voltage booster and controller are required to provide this higher voltage. Compared to halogen headlights, the HID lights provide approximately three times more light, draw less current, and last twice as long. The improved light output from HID lights allows these lights to be smaller and this makes it possible for the vehicle manufacturers to be more flexible in front-end body styling **(Figure 9)**.

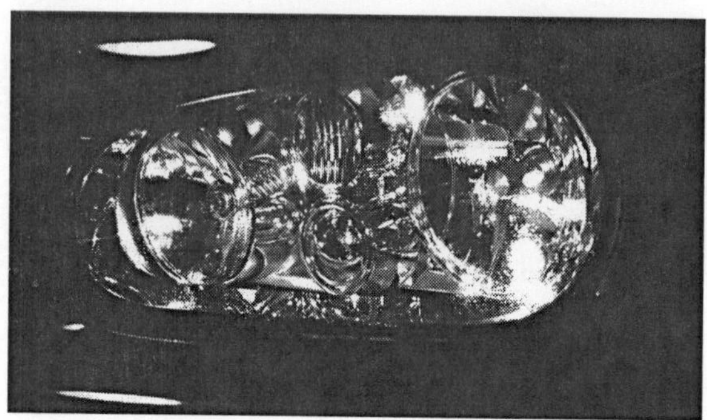

Figure 9. A high intensity discharge (HID) headlight assembly.

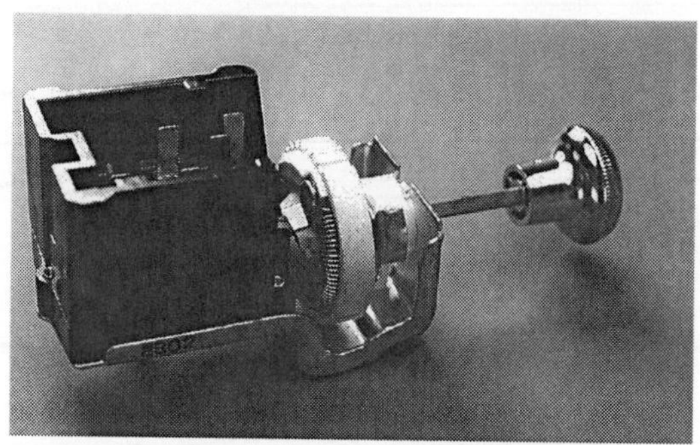

Figure 10. A push-pull, dash-mounted headlight switch.

## HEADLIGHT AND DIMMER SWITCHES

The headlight switch is usually mounted on the front of the instrument panel **(Figure 10)**. Some headlight switches are pulled outward to turn on the lights, whereas other switches require a rotary or pushbutton action to turn on the lights. Most headlight switches have two positions. In the first (PARK) position, the switch turns on the park, tail, sidemarker, and instrument panel lights **(Figure 11)** and in the second position the headlights are also turned on **(Figure 12)**. Many push-pull-type headlight switches contain a variable resistor that is connected in the instrument panel light circuit. The headlight switch knob is rotated to operate the variable resistor and vary the voltage supplied to the instrument panel lights. The action of the variable resistor allows the driver to dim or brighten the instrument panel lights as desired. If the vehicle has a rotary-type headlight switch, the variable resistor is mounted separately beside the headlight switch. A thumb wheel in the instrument panel operates this type of variable resistor.

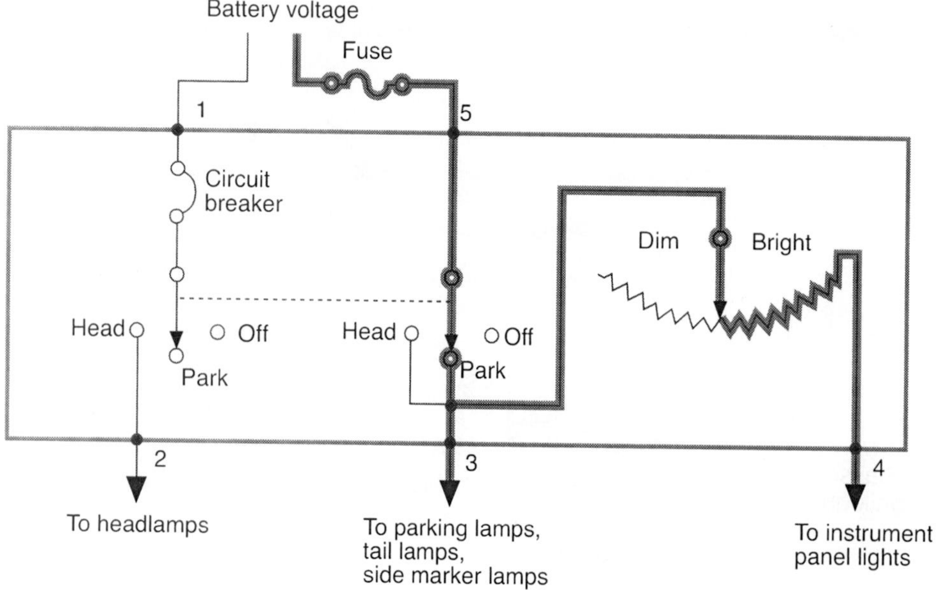


Figure 11. A headlight switch in the PARK position.

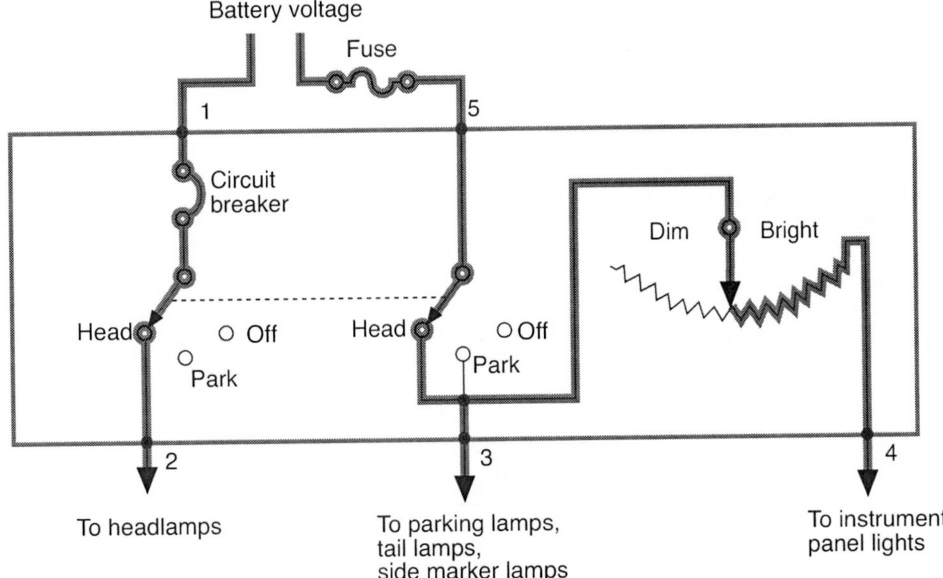


**Figure 12.** A headlight switch in the HEADLIGHT position.

In older vehicles, the dimmer switch is mounted on the floor pan and the driver's left foot was used to operate the dimmer switch. On most of these vehicles, the dimmer switch was mounted in a compartment so it was not exposed to road splash. However, in this location the dimmer switch was subject to moisture and corrosion. In recent years, the dimmer switch has been mounted in the steering column and operated by pulling the signal light lever upward toward the driver **(Figure 13)**. On some vehicles, the signal light lever operates the signal lights, dimmer switch, and headlights. The end of the signal light lever is rotated to operate the park lights and headlights. On these vehicles, another lever on the opposite side of the steering column operates the wipe/wash switch. On other vehicles, the headlight switch is mounted in the instrument panel and the signal light lever operates the signal light switch, the dimmer switch, and the wipe/wash switch. To operate the dimmer switch the signal light lever is pulled upward until a distinct click is heard.

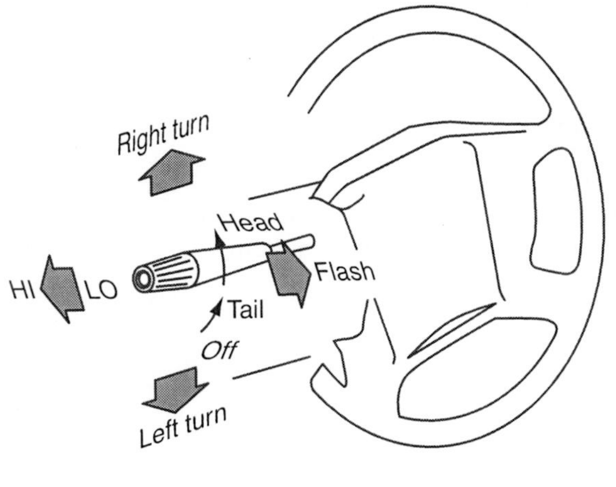

**Figure 13.** A signal light switch, dimmer switch, and headlight switch combined.

**You Should Know** *Many dimmer switches mounted in the steering column are combined with other switches, such as the signal light switch and wipe/wash switch. This combined switch may be called a smart switch or a multifunction switch.*

Many dimmer switches have three terminals. One terminal is a voltage input terminal from the light switch and the other terminals are output terminals to the high or low beam headlight circuits.

## HEADLIGHT AND PARK LIGHT CIRCUITS

Two terminals on the headlight switch are supplied with voltage directly from the battery. The circuit that supplies voltage to the headlight circuit is not fused externally from the switch. However, a circuit breaker inside the headlight switch is connected in the headlight circuit. An external fuse is connected in the circuit to the headlight switch that supplies voltage to the park, tail, instrument panel, and sidemarker lights. Depending on the vehicle, several fuses may be connected in the light circuits.

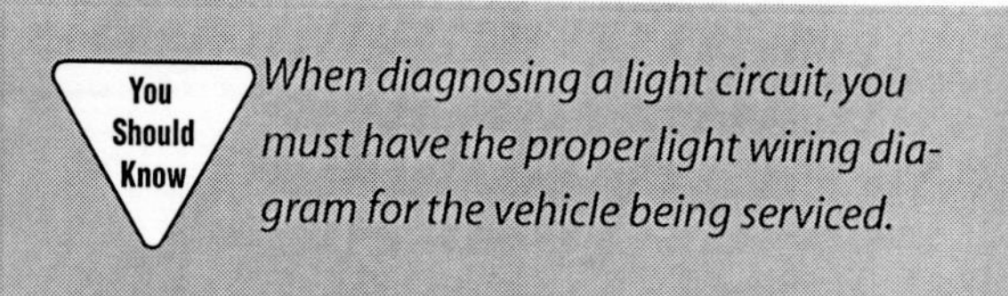


*When diagnosing a light circuit, you must have the proper light wiring diagram for the vehicle being serviced.*

A **circuit breaker** is a protection device that opens an electric circuit if excessive current flows through the circuit. This action prevents damage to wiring and circuit components. After a circuit breaker cools, it will reset itself and close the electrical circuit. A **fuse** is a protection device containing a fusible strip that burns out and opens an electric circuit if excessive current flows through the circuit. This action protects circuit wiring and components. Burned-out fuses must be replaced because they have no reset action.

Fuses, relays, and flashers are usually located in an underhood fuse and relay center **(Figure 14)**. Each component in this center is identified on the center cover. On older vehicles the fuse and relay center is located under the instrument panel.

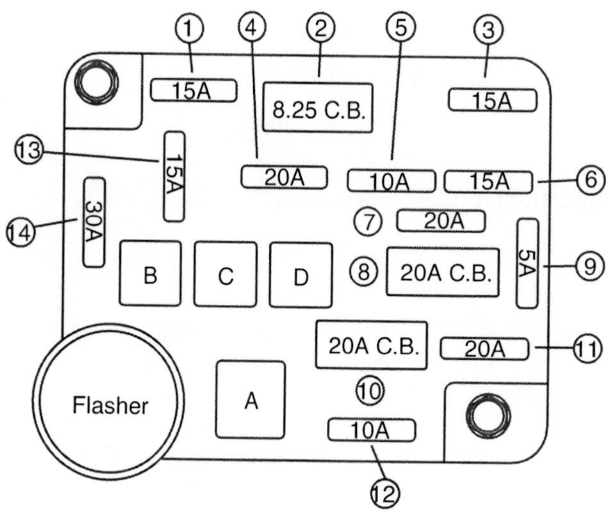


**Fuses and Circuit Breakers**

1. Stop lights, emergency warning system, speed control module, cornering light relays, and trailer tow relays
2. Windshield wiper/washer
3. Tail, park, license, coach, cluster illumination, side marker lights, and trailer tow relay
4. Trunk lid release, cornering lights, speed control, chime, heated backlite, and control A/C clutch
5. Electric heated mirror
6. Courtesy lights, clock feed, trunk light, miles-to-empty, ignition key warning chime, garage door opener, autolamp module, keyless entry, illuminated entry system, visor mirror light, and electric mirrors
7. Radio, power antenna, CB radio
8. Power seats, power door locks, keyless entry system, and door cigar lighters
9. Instrument panel lights and illuminated outside mirror
10. Power windows, sun roof relay, and power window relay
11. Horns and front and instrument panel cigar lighters
12. Warning lights, seat belt chimes, throttle solenoid positioner, autolamps system, and low fuel module
13. Turn signal lights, back up lights, trailer tow relays, keyless entry module, illuminated entry module, and cornering lights
14. ATC blower motor

**Relays**

A. Engine main relay
B. MFI main relay
C. Fan relay
D. Horn relay

**Figure 14.** A fuse and relay center.

> **You Should Know** *Some vehicles have more than one fuse and relay center. The location of these centers is shown in the service manual. Some vehicles have the flasher(s) mounted separately from the fuse and relay center.*

When the headlight switch is moved to the PARK position, voltage is supplied through the fuse and the headlight switch contacts to the park, tail, and sidemarker lights. Voltage is also supplied through the variable resistor to the instrument panel lights. Each instrument panel bulb is grounded to the instrument panel. The variable resistor controls the brilliance of the instrument panel lights by varying the voltage supplied to these lights.

When the headlight switch is moved to the second position, voltage is also supplied through the headlight switch contacts to the headlights. This voltage is supplied from the headlight switch to the dimmer switch. If the dimmer switch is on the low beam position, current flows through the low beam contacts in the dimmer switch to the low beams in both headlights to ground **(Figure 15)**. When the dimmer switch is in the high beam position, current flows through the high beam contacts in the dimmer switch to the high beam filaments in both headlights

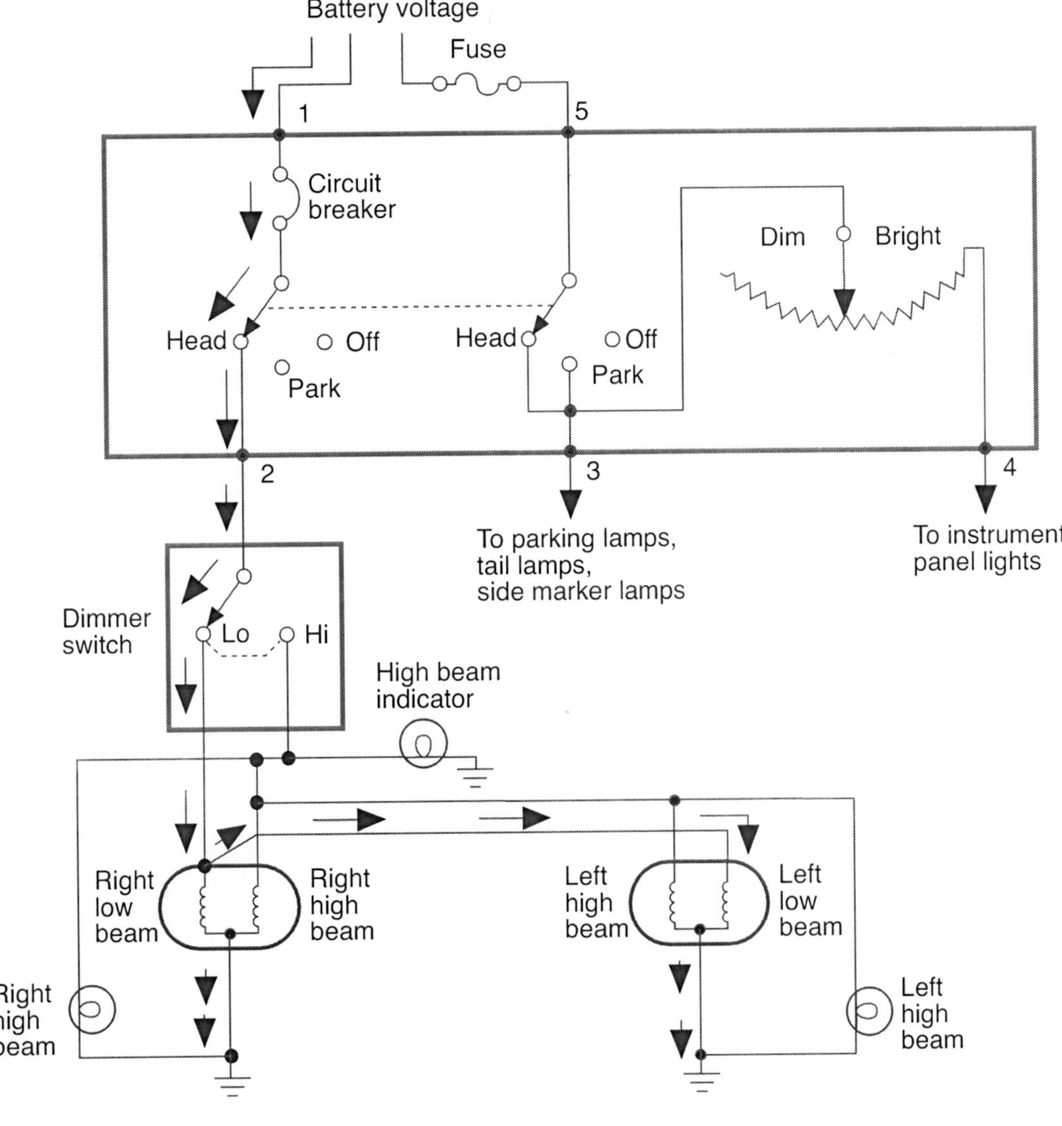


**Figure 15.** A low beam headlight circuit.

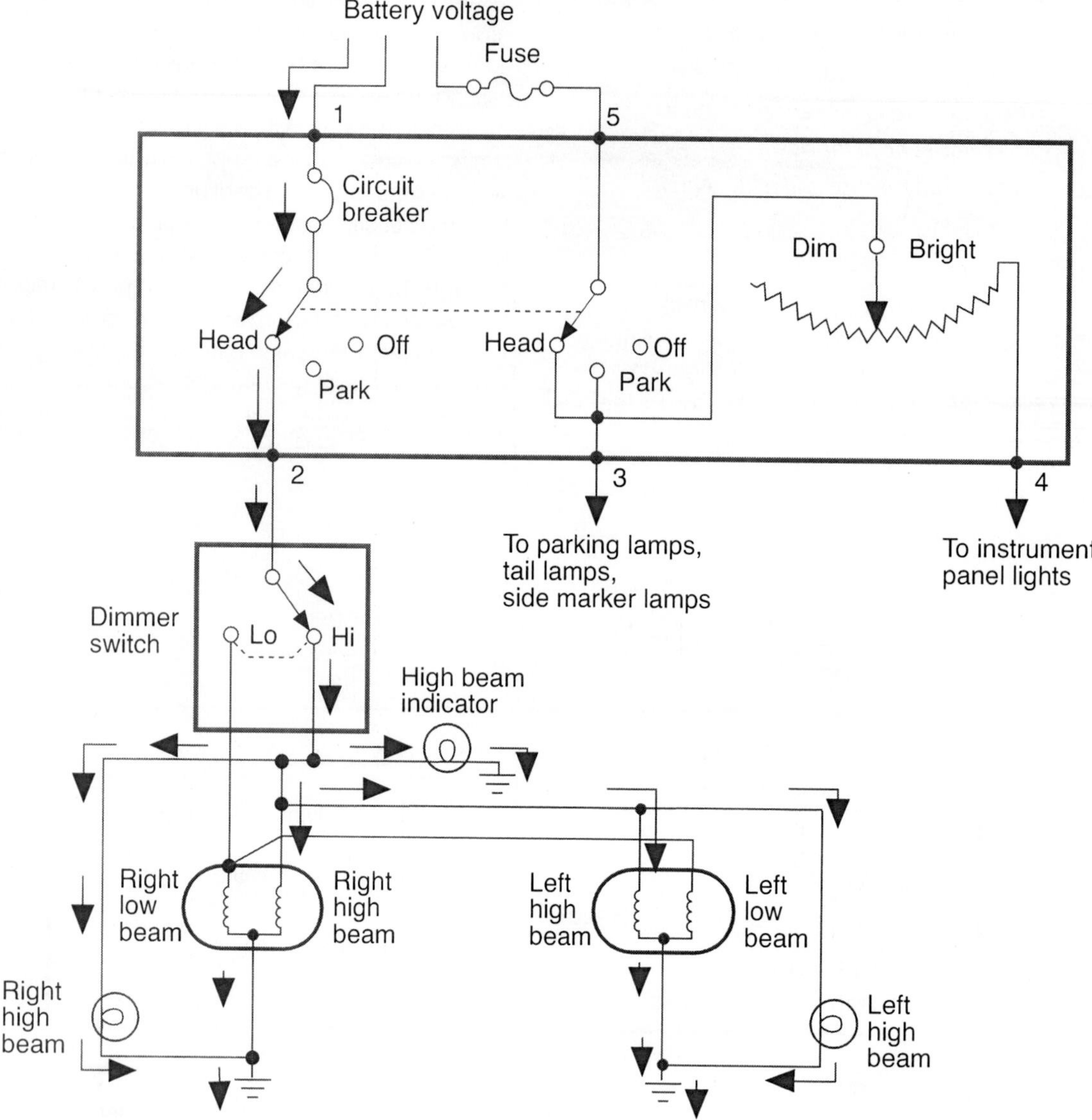


Figure 16. A high beam headlight circuit.

(**Figure 16**). Current also flows from the high beam circuit to the high beam indicator bulb in the instrument panel. The illumination of this bulb reminds the driver that the highlights are on high beam.

Many headlight systems have a **flash-to-pass feature**. During daylight hours when a driver wants to pass a vehicle in front, this feature may be used to turn the headlights on and signal the driver in front regarding the intended pass. To operate the flash-to-pass feature, the signal light lever is pulled upward, but not so far that it clicks. Regardless of the dimmer switch position, the headlights are on high beam when activated by the flash-to-pass feature, and the high beam indicator light in the instrument panel is illuminated. The headlights remain on as long as the signal light lever is held upward. Release the signal light lever to turn the headlights off.

Most vehicles have a light buzzer or beeper. If the lights are on, the ignition switch is off, and one of the front doors is opened, the buzzer sounds to remind the driver to shut off the lights. Some buzzers are activated if the lights are on and the ignition switch is turned off. The buzzer is often mounted on or near the fuse panel.

Some vehicles have automatic headlight dimmers. These vehicles have a light sensor mounted behind the grill. When the lights of an approaching vehicle supply a specific amount of light to the light sensor, the headlights dim automatically.

**You Should Know** *Many newer vehicles have a light sensor mounted on top of the instrument panel. This sensor turns on the front and rear lights automatically when ambient light decreases to a specific level. This is very useful when driving through tunnels because the lights come on automatically without the driver trying to suddenly turn on the light switch.*

## CONCEALED HEADLIGHT SYSTEMS

Some vehicles have the headlights concealed behind headlight doors. When the headlights are turned on, the doors open to expose the lights. Concealed headlight doors may be vacuum or electrically operated. On vacuum-operated systems, the doors are held closed by vacuum. When the headlights are turned on, the vacuum is bled off in the concealed door system. If there is a vacuum leak in these systems, one or both headlight doors may slowly open with the headlights shut off. An electric motor operates the headlight doors in electrically-operated systems. If the electric headlight door system is inoperative, a manual override knob under the hood may be rotated to open the headlight doors.

## TAIL LIGHT, STOP LIGHT, SIGNAL LIGHT, AND HAZARD WARNING LIGHT CIRCUITS

Many vehicles have dual stop and tail light filaments in all or some of the rear lights. The same bulb filaments used for stop lights are also used for signal lights. If the headlight switch is moved to the PARK or HEADLIGHT position, voltage is supplied from the headlight switch to the tail light bulbs **(Figure 17)**. The signal light switch contains three triangular-shaped contacts. If the signal light switch is in the center position so the signal lights are not operating, the center contact in the signal light switch completes the circuit between the brake light switch input terminal and the rear stop light terminals. If the driver depresses the brake pedal, current flows through these signal light switch contacts to the rear stop lights and the center high-mounted stop light (CHMSL) bulbs.

When the signal light switch is moved to the right turn position, the contacts in the signal light switch are moved to the left. This contact movement shifts the right switch contact so it completes the circuit between the signal light flasher and the terminal to right rear signal lights. The right switch contact also completes the circuit between the flasher and the right dash turn indicator light and right front signal light **(Figure 18)**. When the circuit is completed

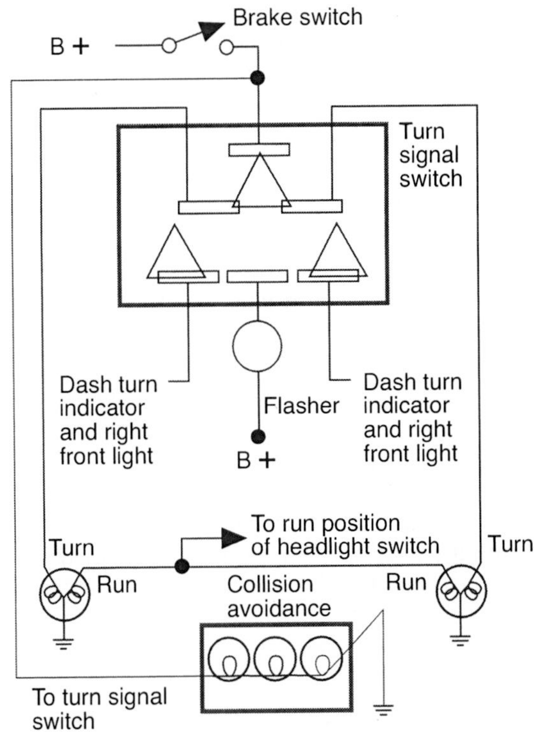


**Figure 17.** A signal light, a stop light, and a turn signal circuit.

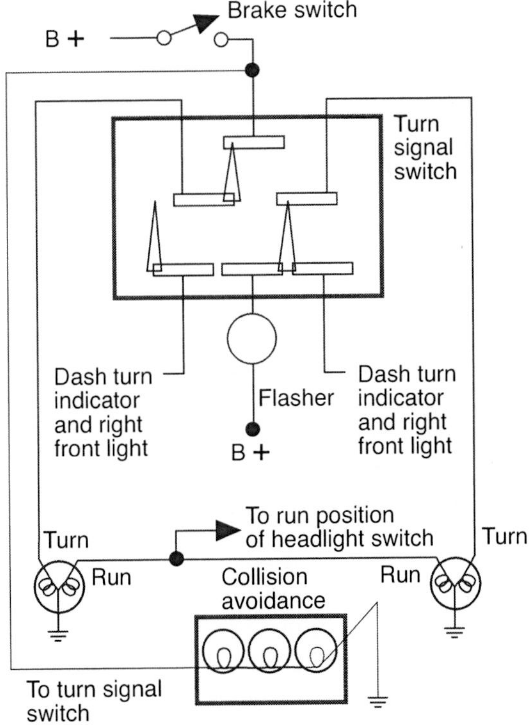


**Figure 18.** The position of the signal light switch contacts during a right turn.

through the flasher, it begins opening and closing the circuit to turn the right signal lights on and off. Many flashers contain a set of contacts mounted on a bimetallic strip that bends when heated. A heating coil is mounted on the bimetallic strip. Current flow through the heating coil heats the bimetallic strip and opens the contacts. This strip immediately cools and closes the contacts. The heating and cooling of this bimetallic strip causes the flasher contacts to open and close at a specific rate. Some vehicles have a solid state electronic flasher. This type of flasher provides light flashes at the same speed regardless of the current flow through the signal lights.

In the right turn position, the center switch contact maintains the circuit between the brake light switch and the left rear stop light. Therefore, if the driver depresses the brake pedal, current flows through this circuit to the left rear stop light and the CHMSL. The brake light switch is a mechanical switch operated by movement of the brake pedal. In the right turn position, the left switch contact is not completing any circuit.

When the signal light switch is moved to the left turn position, the contacts in the signal light switch are moved to the right. This contact movement shifts the left switch contact so it completes the circuit between the signal light flasher and the terminal to left rear signal lights. The right switch contact also completes the circuit between the flasher and the dash turn indicator light and left front signal light **(Figure 19)**. When the circuit is completed through the flasher, it begins opening and closing the circuit to turn the left signal lights on and off. In the left turn position, the center switch contact maintains the circuit between the brake light switch and the right rear stop light. Therefore, if the driver depresses the brake pedal, current flows through this circuit to the right rear stop light and the CHMSL.

When the steering wheel is rotated back to the straight ahead position after a turn, a cam mounted on the steering shaft returns the signal light switch to the off position to cancel the signal light operation. Signal light switches have a special feature that allows the driver to move the signal light switch lever enough to operate the signal lights on either side without latching the signal light switch into position.

When this special feature is used, the driver moves the signal light switch lever a small amount to flash the signal lights several times on the desired side to indicate an impending lane change. When the driver releases the signal light switch lever, the switch returns to the off position without the operation of the canceling mechanism. During a lane change, the steering wheel is not turned enough to operate the canceling mechanism.

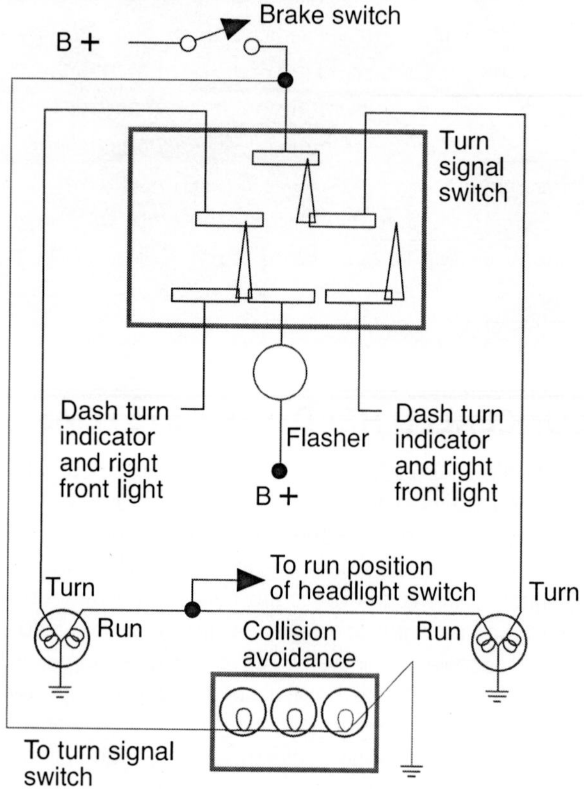


**Figure 19.** The position of the signal light switch contacts during a left turn.

The hazard warning switch is usually mounted on the steering column. This switch is connected into the signal light circuit **(Figure 20)**. When the hazard warning switch is pressed, all front and rear signal lights flash together. The hazard warning switch may be used when a vehicle is parked in an emergency situation. The signal lights continue flashing as long as the hazard warning switch is pressed.

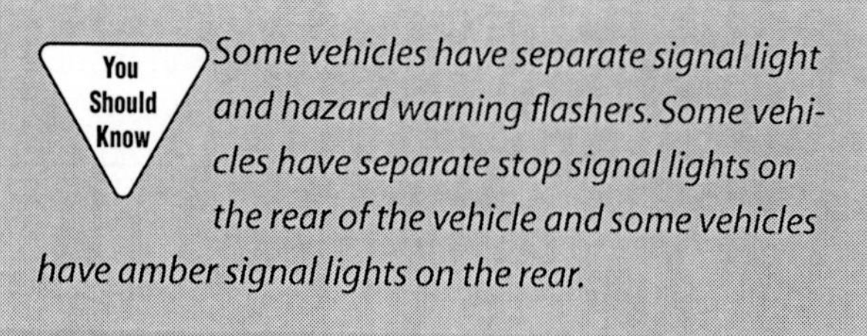


*Some vehicles have separate signal light and hazard warning flashers. Some vehicles have separate stop signal lights on the rear of the vehicle and some vehicles have amber signal lights on the rear.*

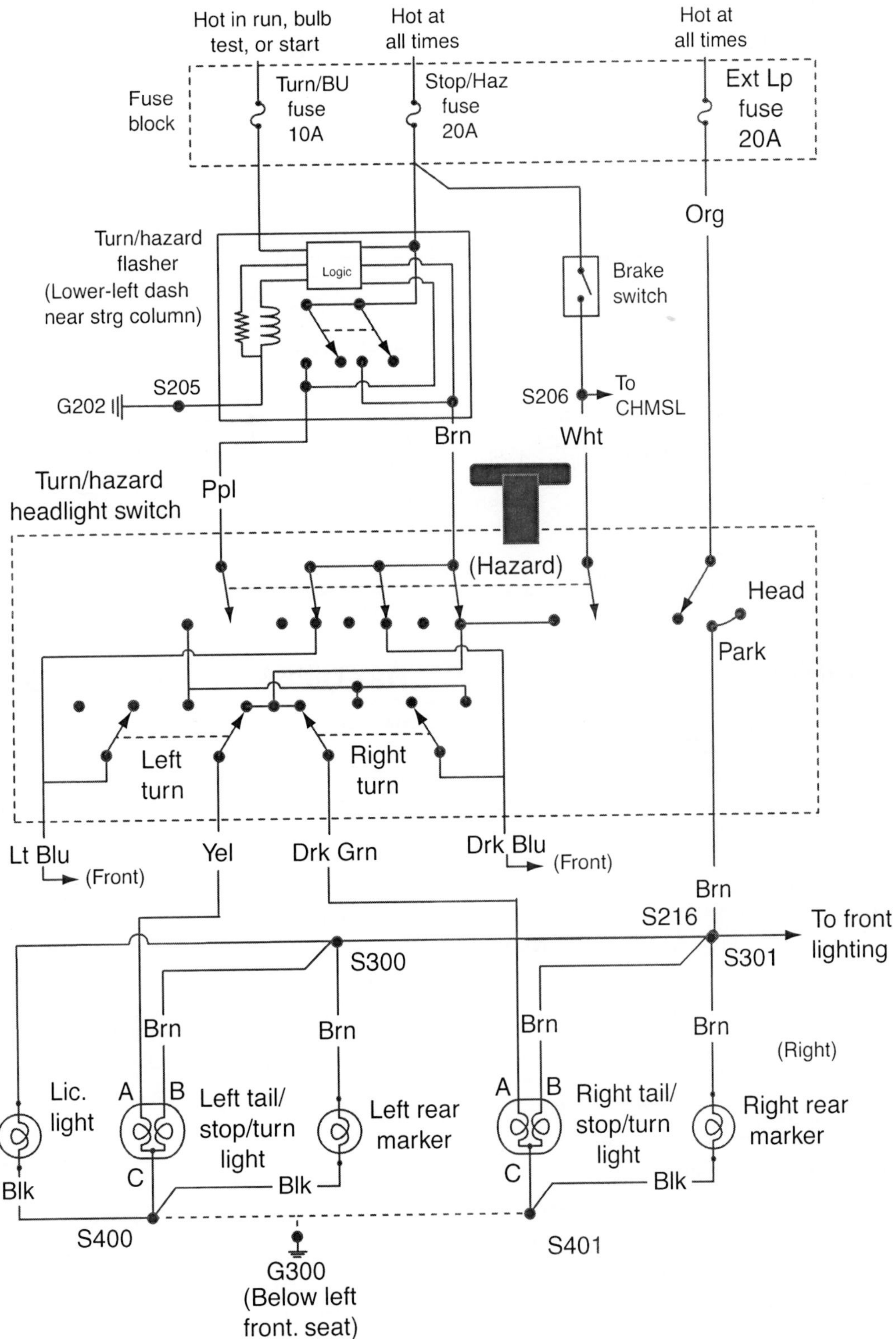


**Figure 20.** A hazard warning switch circuit.

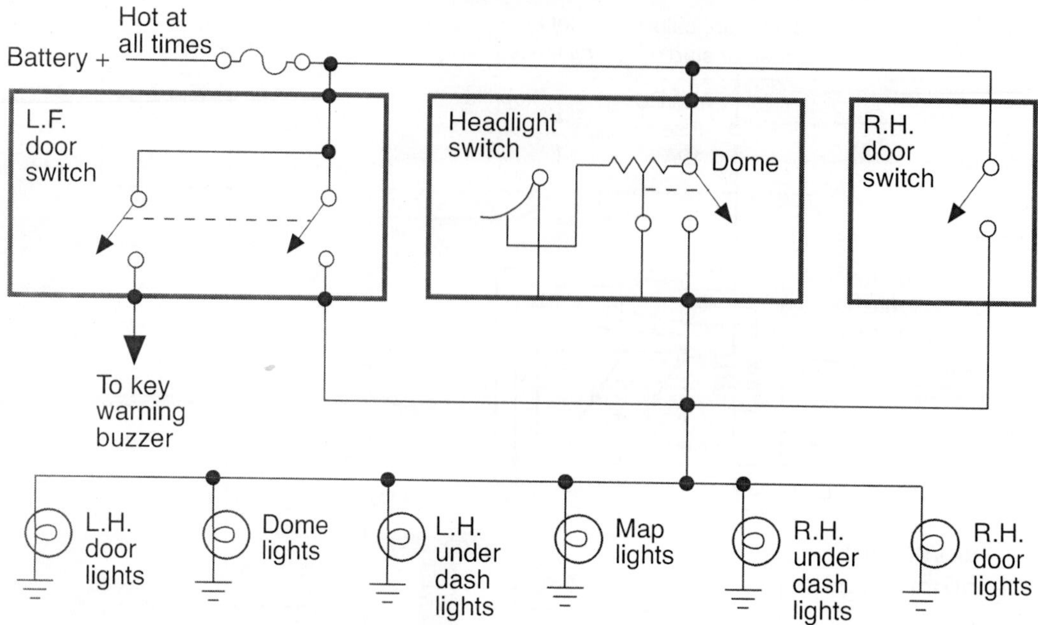


**Figure 21.** An interior light circuit.

## INTERIOR LIGHTS

Interior lights usually include under-dash lights, door lights, map lights, and a dome light. Door jamb switches are connected in the interior light circuit. Some vehicles have door jamb switches on the front doors, whereas other vehicles have these switches on the front and rear doors. The door jamb switches are normally open. When a door is opened, the door jam switch closes and supplies voltage to the interior lights. These lights are grounded to the vehicle chassis so they are illuminated when voltage is supplied to them **(Figure 21)**. On some vehicles the interior lights are turned on when the headlight switch knob is rotated fully counterclockwise. Other vehicles have a separate switch mounted near the dome light to turn on the dome light and interior lights.

**You Should Know** *On some vehicles, the interior light bulbs are insulated and the ground side of these bulbs is connected to the door jamb switches. When a door is opened, the door jamb switch completes the circuit to ground. In these circuits, the door jamb switches have a single wire and the interior lights have two wires.*

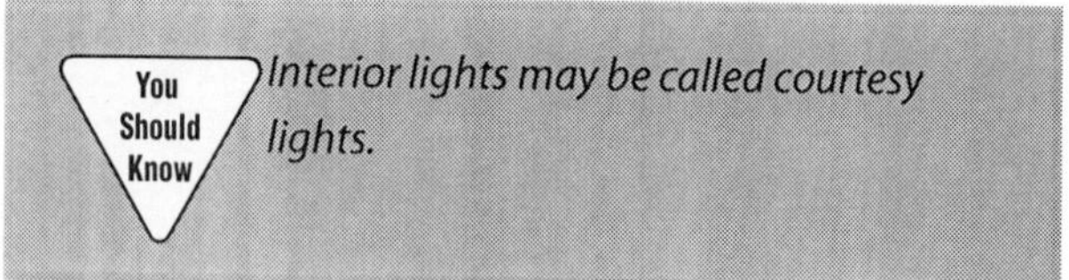


*Interior lights may be called courtesy lights.*

## DAYTIME RUNNING LIGHTS

On some vehicles, daytime running lights (DRLs) are separate lights mounted on the front of the vehicle. These lights have replaceable bulbs that provide more light than a stoplight but less than a headlight **(Figure 22)**. The DRLs are on when driving during the day. When the headlights are turned on the daytime running lights are turned off. DRLs are intended to warn drivers that a vehicle is approaching. On some vehicles, the DRLs are not illuminated until the transmission selector is placed in drive or reverse. This allows service personnel to work on the vehicle in the shop with the DRLs off. Other daytime running lights are illuminated when the engine is started during the day, but these lights are turned off if the parking brake is pressed.

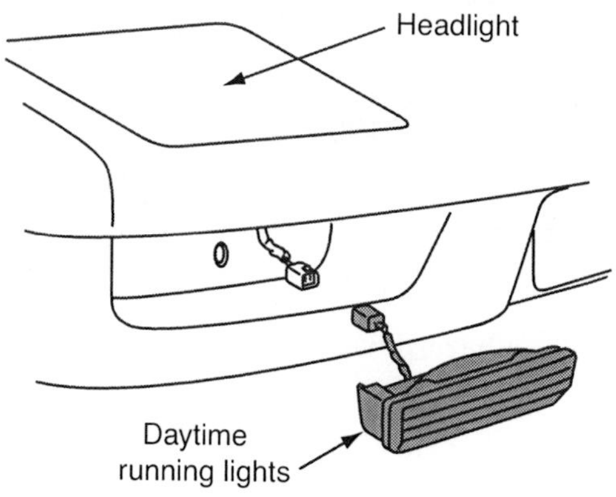


**Figure 22.** Daytime running lights.

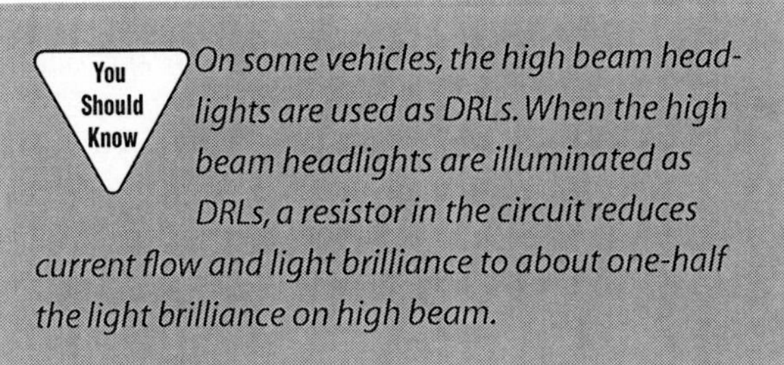


*On some vehicles, the high beam headlights are used as DRLs. When the high beam headlights are illuminated as DRLs, a resistor in the circuit reduces current flow and light brilliance to about one-half the light brilliance on high beam.*

## BACKUP LIGHTS

Backup lights have clear lenses and these lights are mounted on the rear of the vehicle. When the ignition switch is on and the transmission selector is placed in reverse, the backup lights are turned on **(Figure 23)**. The backup light switch is mounted on the transmission selector linkage or in the transmission. The backup lights provide some illumination behind the vehicle when backing up at night. These lights also warn anyone behind the vehicle about the backup action.

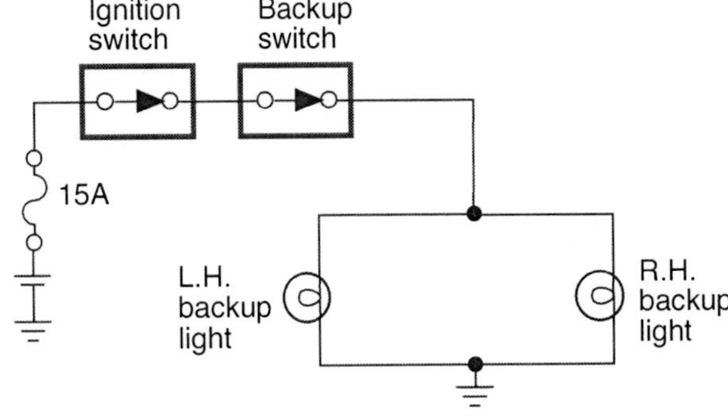


**Figure 23.** A backup light circuit.

# Summary

- Automotive bulbs may have single or dual filaments.
- Automotive bulbs contain a vacuum to prevent overheating the filament.
- During the manufacturing process, oxygen is removed from sealed beams and they are filled with argon gas.
- Sealed beams may contain only high beam filaments or both high and low beam filaments.
- Many halogen headlights have a replaceable bulb.
- Halogen headlights provide more light than conventional sealed beams.
- HID headlights produce light when high voltage arcs across an air gap. These lights contain an inert gas that amplifies the light.
- HID headlights provide more light, draw less current, and last longer than other types of headlights.
- A driver-operated variable resistor controls instrument panel light brilliance by changing the voltage supplied to these lights.
- The dimmer switch directs voltage to the high or low beam headlights depending on the dimmer switch position.
- Many headlight switches contain a circuit breaker that is connected in the headlight circuit.
- Doors for concealed headlights may be vacuum or electrically operated.
- The signal light switch connects the flasher to the signal lights on one side of the vehicle depending on the signal light switch position.
- Many flashers contain a bimetallic strip and a heating coil.
- Some vehicles have separate signal light and hazard warning flashers.
- When the hazard warning switch is pressed, all signal lights begin flashing.
- The interior lights are switched on and off by the door jamb switches.

# Review Questions

1. Technician A says in a dual filament bulb, one side of both light filaments is connected to the bulb case. Technician B says the dual filaments in a rear bulb may be used for stop and tail lights. Who is correct?
   A. Technician A
   B. Technician B
   C. Both Technician A and Technician B
   D. Neither Technician A nor Technician B

2. Technician A says a tail light bulb contains a vacuum. Technician B says if the glass envelope leaks on a bulb, the filament will burn out quickly. Who is correct?
   A. Technician A
   B. Technician B
   C. Both Technician A and Technician B
   D. Neither Technician A nor Technician B

3. Technician A says composite headlights have replaceable halogen bulbs. Technician B says a halogen headlight provides reduced light compared to a conventional sealed beam. Who is correct?
   A. Technician A
   B. Technician B
   C. Both Technician A and Technician B
   D. Neither Technician A nor Technician B
4. Technician A says a HID headlight has a halogen bulb. Technician B says a HID headlight uses more current than a conventional sealed beam. Who is correct?
   A. Technician A
   B. Technician B
   C. Both Technician A and Technician B
   D. Neither Technician A nor Technician B
5. All of these statements about sealed beam headlights are true *except*:
   A. Sealed beam headlights are filled with argon gas.
   B. Sealed beam headlights may contain one or two filaments.
   C. A dual-filament sealed beam headlight has two terminals.
   D. In the dual filament sealed beam headlight, the upper filament is for the low beam.
6. All of these statements about halogen headlights are true *except*:
   A. The tungsten filament in a halogen headlight can withstand higher temperatures than a sealed beam filament.
   B. Halogen headlights have a replaceable bulb.
   C. A cracked headlight lens results in poor light quality.
   D. A cracked headlight lens causes rapid headlight failure.
7. While discussing HID headlights, Technician A says HID headlights have a long-life replaceable bulb. Technician B says an HID light contains an inert gas that magnifies light. Who is correct?
   A. Technician A
   B. Technician B
   C. Both Technician A and Technician B
   D. Neither Technician A nor Technician B
8. When a push-pull-type headlight switch is pulled outward to the first position:
   A. the tail, park, and instrument panel lights are turned on.
   B. the high beam headlights are turned on.
   C. the low beam headlights are turned on.
   D. voltage is supplied to the dimmer switch.
9. In a headlight circuit with a circuit breaker, the wire connected between the dimmer switch and one of the high-beam headlights is shorted to ground. Technician A says this condition will cause the circuit breaker to open the headlight circuit. Technician B says this problem will burn out the high-beam headlights. Who is correct?
   A. Technician A
   B. Technician B
   C. Both Technician A and Technician B
   D. Neither Technician A nor Technician B
10. All of these statements about concealed headlights are true *except*:
    A. Concealed headlight doors may be operated electrically or by vacuum.
    B. Vacuum-operated headlight doors are held open by vacuum.
    C. A vacuum leak in a headlight door system may cause the doors to open.
    D. A manual override knob may be rotated to force electrically-operated headlight doors open.
11. To supply the necessary voltage to HID headlights, a voltage ______________ and ______________ are required.
12. On modern vehicles, the dimmer switch is usually operated by the ______________ ______________ ______________.
13. If the headlight switch is in the PARK position, voltage is supplied through this switch to the ______________ ______________ ______________ and ______________ lights.
14. Many headlight switches have an internal ______________ ______________ connected in the headlight circuit.
15 Describe the current flow through the headlight circuit when the headlight switch is on and the dimmer switch is on high beam.
16. Describe the operation of the signal light switch and related circuit during a left turn.
17. Explain the operation of the interior lights when a front door is opened.
18. Explain the operation of the hazard warning lights.

# Chapter 19 Light Circuit Maintenance, Diagnosis, and Service

## *Introduction*

It is extremely important for all the lights on a vehicle to be operating properly. For example, if the tail lights on a vehicle are not working, a driver approaching the vehicle from the rear after dark may not see the vehicle soon enough to avoid an accident. Inoperative brake lights may cause a vehicle to be hit from the rear because the driver of the vehicle approaching from the rear did not realize the vehicle in front was going to stop. If one headlight is not working or the headlights are aimed too low, the driver has reduced visibility at night and this may result in a collision. Vehicle owners may not be aware that some of the lights on their vehicle are not working. Therefore, the lights should be checked when the vehicle is brought into the shop for service.

### HEADLIGHT MAINTENANCE, DIAGNOSIS, AND SERVICE

Headlight maintenance involves checking the headlights each time a vehicle is brought to the shop for service. The headlight wiring should be inspected for worn insulation, corrosion, and contact with other components. Be sure to check the headlights on both low and high beam.

### Headlight Diagnosis

Be sure the battery is fully charged and the battery terminals are clean and tight before any light diagnosis. When the engine is running, charging circuit voltage is supplied to all the lights that are turned on. Therefore, improper charging circuit operation affects the light circuits. For example, if the headlights become considerably brighter when the engine rpm is increased, the charging voltage may be excessive. Normal charging voltage is between 14.2 volts and 14.8 volts with the engine at normal operating temperature. Excessively high charging circuit voltage may also cause repeated failure of headlights and other lights. Excessively high charging system voltage also results in too much gassing of water from the battery. High resistance in the headlight circuit reduces current flow and causes dim headlights. If the high resistance is in the ground of one headlight, only that light is dim. When the high resistance is between the dimmer switch and the headlights, all the headlights are dim. A voltage drop test with the headlights on high beam measures high resistance in the headlight circuit. **Voltage drop** is the difference in voltage between two points in a circuit.

Select a low scale on the voltmeter and connect the voltmeter across the part of the circuit in which high resistance is suspected **(Figure 1)**. For example, to measure the voltage drop across the headlight ground, connect the voltmeter positive lead to the headlight ground and the negative lead to the negative battery terminal. Voltage drop should not exceed the vehicle manufacturer's specifications. In many parts of the headlight circuit, except the headlight itself, voltage drop over 0.2 volt is excessive. Headlight diagnosis is provided in **Figure 2**).

*Some composite headlights are not vented and some moisture formation in these lights is normal.*

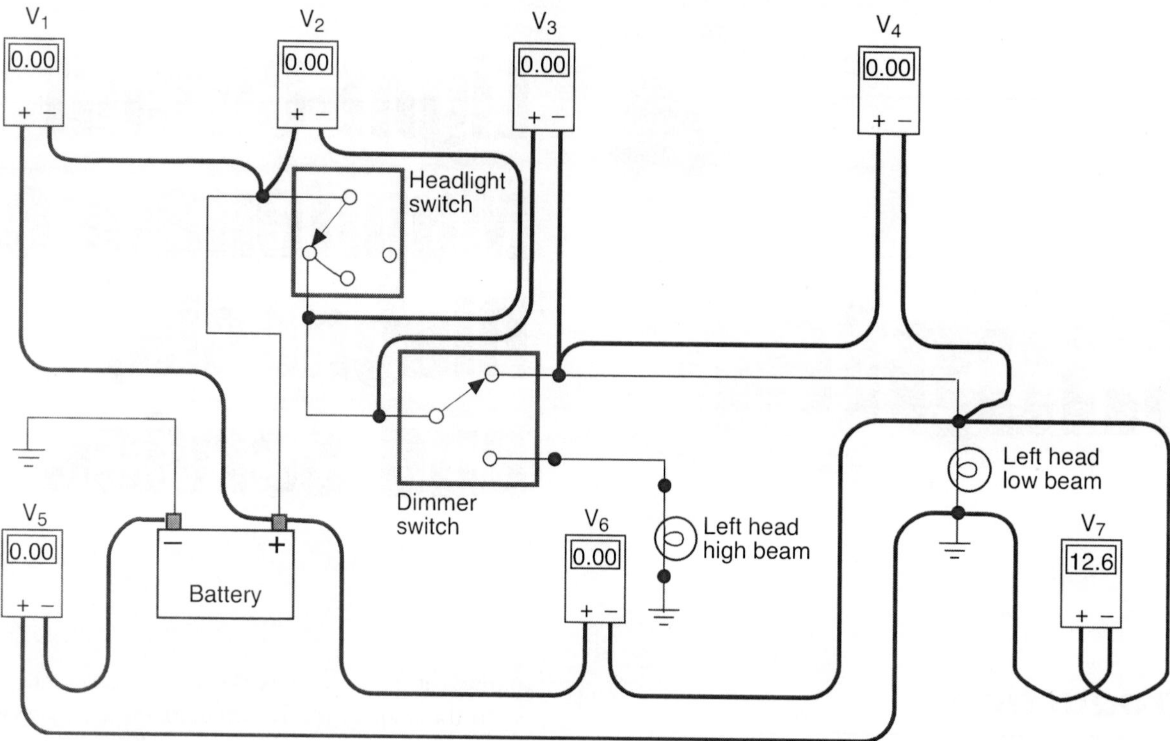


Figure 1. Measuring voltage drop in a headlight circuit.

| Symptom | Possible Cause | Remedy |
|---|---|---|
| Headlights do not light. | Loose wiring connections | Check and secure connections at headlight switch and dash panel connector. |
| | Open circuit in wiring | Check power to and from headlight switch. Repair as necessary. |
| | Worn or damaged headlight switch | Verify condition. Replace headlight switch if necessary. |
| One headlight does not work. | Loose wiring connections | Secure connections to headlight and ground. |
| | Sealed beam bulb burned out | Replace bulb. |
| | Corroded socket | Repair or replace, as required. |
| All headlights out; park and tail lights are okay. | Loose wiring connections | Check and secure connections at dimmer switch and headlight switch. |
| | Worn or damaged dimmer switch | Check dimmer switch operation. Inspect for corroded connector. Replace, if required. |
| | Worn or damaged headlight switch | Verify condition. Replace headlight switch as necessary. |
| | Open circuit in wiring or poor ground | Repair if required. |
| Both low beam or both high beam headlights do not work. | Loose wiring connections | Check and secure connection at dimmer switch and headlight switch. |
| | | Check dimmer switch operation. Inspect for corroded connector. Replace if required. |

Figure 2. Headlight diagnosis.

## Headlight Service

The most common headlight service is changing and aiming headlights. The headlight replacement procedure varies depending on the vehicle and the type of headlights. Be sure the headlights are turned off before a headlight replacement procedure. When these lights are turned off, there is no voltage at the lights; therefore, the battery disconnecting is not required. The following is a typical sealed beam headlight replacement procedure:

1. Place fender covers over the vehicle in the work area.
2. Remove the bezel retaining screws and remove the bezel surrounding the headlight.
3. Remove the four sealed beam retaining ring screws but do not turn the headlight adjusting screws **(Figure 3)**.
4. Remove the sealed beam retaining ring and the sealed beam.
5. Disconnect the wiring connector from the back of the sealed beam.
6. Inspect the terminals in the wiring connector for corrosion and clean these terminals as required.
7. Place dielectric grease on the connector terminals and on the new sealed beam terminals to prevent corrosion.
8. Install the wiring connector on the sealed beam terminals and be sure this connector is fully seated.
9. Install the new sealed beam with the embossed number at the top.
10. Install the sealed beam retaining ring and screws.
11. Turn the headlight switch on and be sure the sealed beam operates on both low and high beam.
12. Install the headlight bezel and retaining screws.

## Halogen Bulb Replacement in Composite Headlights

On some composite headlights, the bulb wiring terminals are not accessible behind the headlights. On most of these lights, two steel pins on top of the headlight assembly may be removed to allow the complete assembly to come forward out of the chassis. This action allows access to the wiring connectors and bulbs.

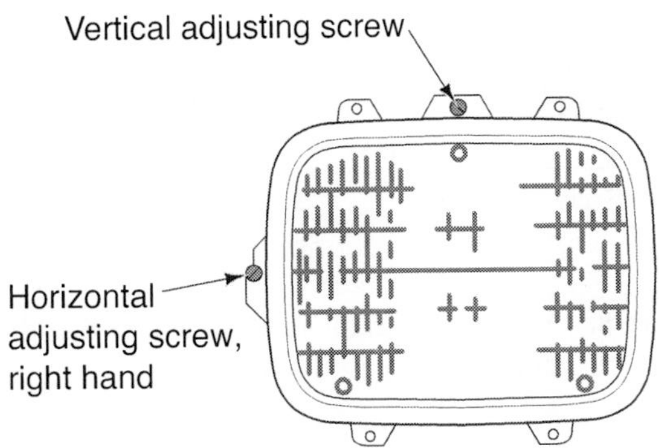


**Figure 3.** Headlight adjusting screws.

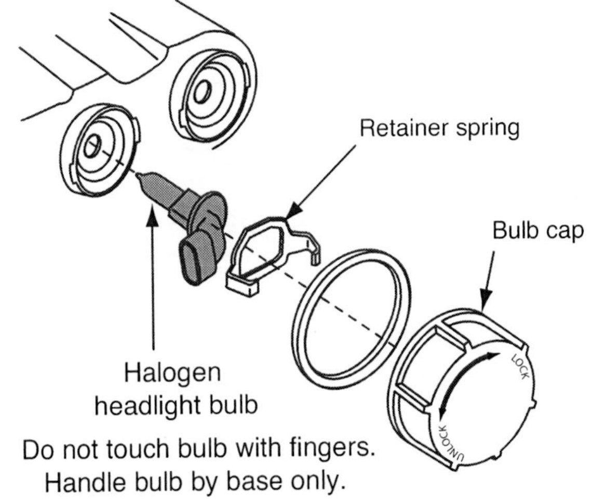


**Figure 4.** Removing a halogen headlight bulb.

The following is a typical replacement procedure for a halogen bulb in a composite headlight:

1. Be sure the headlight switch is off.
2. Place fender covers over the vehicle body in the work area.
3. Remove the cap over the rear of the bulb **(Figure 4)**.
4. Remove the wiring connector from the bulb and remove the retaining spring.

> **You Should Know** *The retaining ring must be rotated about $1/8$ of a turn counterclockwise to remove the retaining ring and bulb on some composite headlights.*

5. Remove the bulb from the composite headlight.
6. Inspect the wiring connector for corrosion and clean as required.
7. Install the new bulb in the headlight and install the retaining ring.
8. Install the wiring connector and the bulb cap.
9. Turn the headlight switch on and be sure the headlight operates on low and high beam.

> **You Should Know** *When halogen headlight bulbs are illuminated, they become very hot. Do not touch halogen bulbs or attaching hardware immediately after the headlights have been turned off. Do not attempt to clean or replace halogen bulbs when the lights are on. Do not touch the glass envelope on a halogen bulb with your fingers. When oil from your skin is deposited on this envelope, bulb life is shortened.*

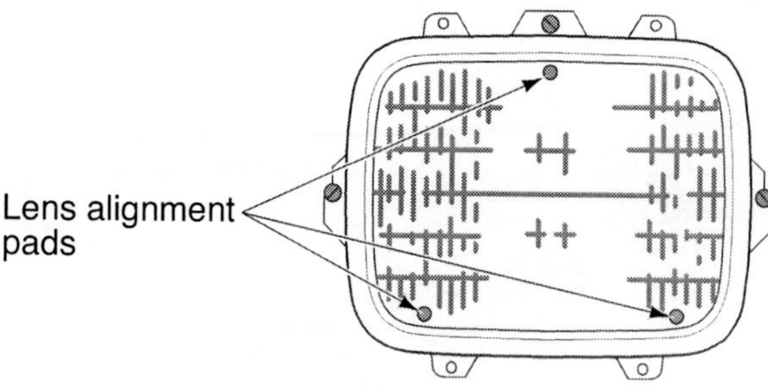


**Figure 5.** Headlight aiming pads.

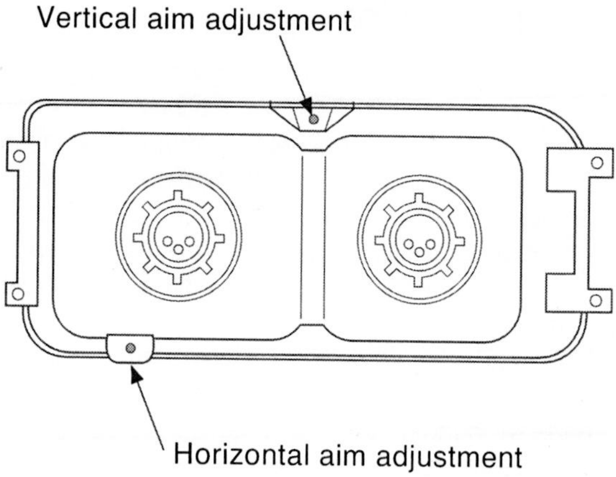


**Figure 7.** Headlight horizontal and vertical adjustment screws.

## Headlight Aiming

Headlight aim is very important! A headlight that is misaimed by one degree downward reduces the driver's vision distance by 156 feet. If the headlight aim is too high, the headlights tend to reduce the vision of oncoming drivers.

The following is a typical procedure for aligning sealed beam headlights:

1. Park the vehicle on a level floor, apply the parking brake, and place the automatic transmission in PARK or manual transmission in REVERSE. Be sure the vehicle has a normal load and the specified tire inflation.
2. Install fender covers over the vehicle body in the work area.
3. Select the proper headlight aimer adapter for the headlights being aligned. These adapters must fit over the aiming pads on the headlights **(Figure 5)**. **Aiming pads** are small projections on the front of a headlight lens.
4. Be sure the sealed beam outer surface is clean. Attach the headlight aimers to the sealed beams using the suction cups on the aimers. Be sure the aimer sight openings face each other **(Figure 6)**.
5. Zero the horizontal adjustment dial on one of the aimers.
6. Be sure the split image target lines are visible in the aimer viewport.
7. Turn the headlight horizontal adjustment screw **(Figure 7)** until the target lines in the view port are aligned **(Figure 8)**.
8. Repeat steps 5 through 7 on the opposite headlight.
9. Set the vertical adjustment dial to zero on one of the aimers and then turn the vertical aim adjustment screw until the bubble is centered in the aimer.
10. Repeat step 9 on the opposite headlight.

Many composite headlights have curved outer surfaces and special adapters are required to mount the headlight aimers properly on these headlights. Each com-

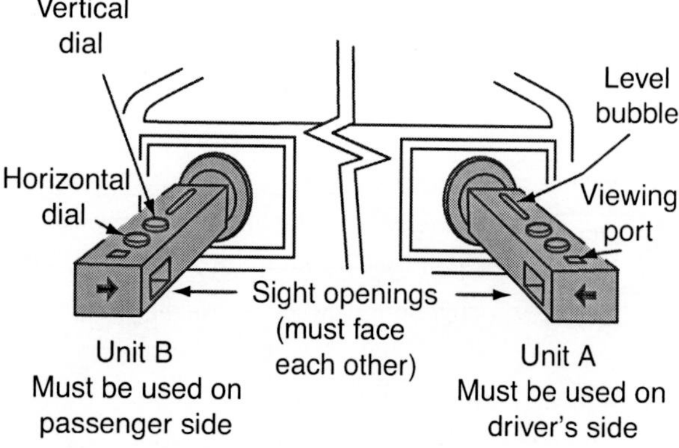


**Figure 6.** Headlight aimers installed on headlights.

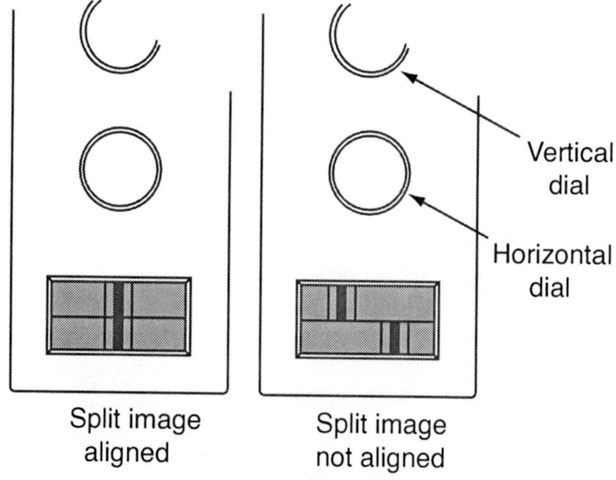


**Figure 8.** Target lines in the headlight aimers indicate horizontal headlight alignment.

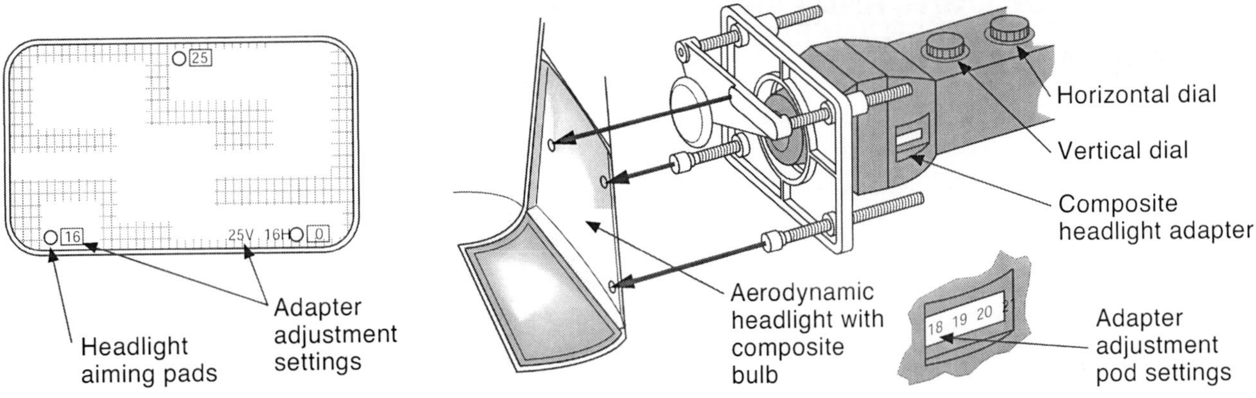


**Figure 9.** Composite headlight aiming pads and related numbers.

posite headlight assembly has numbers molded into the headlight beside each aiming pad **(Figure 9)**. Each adjustment rod setting on the aimer adapter must match the number on the composite headlight aiming pad on which the adjustment rod will be placed. The adjustment rod setting must be locked in this position. After the adjustment rod setting is completed, the aimers are attached to the composite headlights and the headlight adjustment is performed using the same procedure for sealed beams.

Some vehicles do not have aiming pads on the headlights. These headlights may be aligned with the vehicle parked on a level floor 25 feet back from a blank wall. The vehicle must have a normal load and the specified tire inflation. The centerline of the vehicle and the vertical and horizontal centerlines of each headlight must be marked on the wall and the vehicle must be perpendicular to the wall **(Figure 10)**. With the headlights on high beam, the center of each light beam should be at the specified location on the wall. The amount of headlight beam drop may be specified by state or federal regulations. Rotate the headlight aiming screws as necessary to place the light beams at the specified location.

Some late-model vehicles have levels built into the headlight assemblies. This spirit level allows the driver to adjust the headlights in relation to the load in the vehicle. If a heavy load is placed in the trunk, the headlight beams become too high. Under this condition the driver may adjust the vertical headlight aiming screws to level the bubbles in each spirit level. After the load is removed from the trunk, the headlight beams will be too low and the driver may turn the vertical aiming screws to re-center the bubbles in the spirit levels.

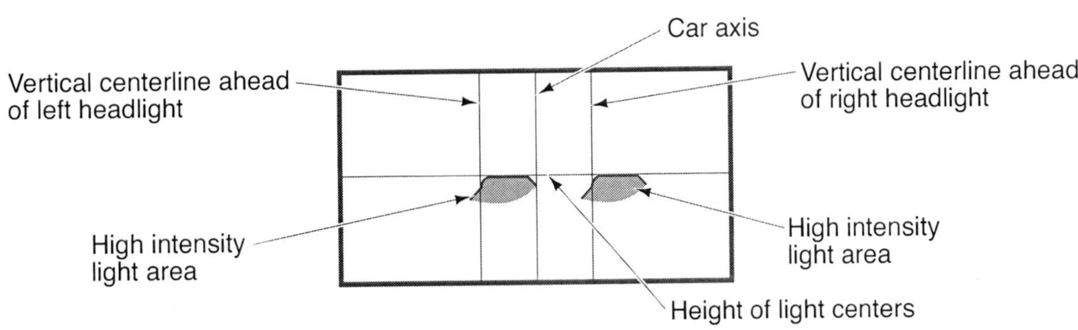


**Figure 10.** Headlight alignment using light beam projection.

## TAIL LIGHT, STOP LIGHT, AND PARK LIGHT CIRCUIT MAINTENANCE, DIAGNOSIS, AND SERVICE

The maintenance on any light circuit is basically the same. This maintenance includes regular inspections to determine if all the lights are working properly. Light circuit inspections should also include checking the wiring harness for frayed insulation, broken harness retainers, and improper contact between the harness and other components.

Diagnosis of tail, stop, and park lights is provided in **Figure 11**.

Follow these steps to remove a park light or front turn signal bulb:

| Symptom | Possible Cause | Remedy |
|---|---|---|
| One taillight out. | Bulb burned out | Replace bulb. |
| | Open wiring or poor ground | Repair as necessary. |
| | Corroded bulb socket | Repair or replace socket. |
| All taillights and maker lamps out; headlights okay. | Loose wiring connections | Secure wiring connections where accessible. |
| | Open wiring or poor ground | Check operation of front park and marker lamps. Repair as necessary. |
| | Blown fuse | Replace fuse. |
| | Damaged headlight switch | Verify condition. Replace headlight switch if necessary. |
| Stop lights do not work. | Fuse or circuit breaker burned out | Replace fuse or circuit breaker. If fuse or circuit breaker blows again, check for short circuit. |
| | Worn or damaged turn signal circuit | Check turn signal operation. Repair as necessary. |
| | Loose wiring connections | Secure connection at stop-light switch. |
| | Worn or damaged stop-light switch | Replace switch. |
| | Open circuit in wiring | Repair as required. |
| Stop lights stay on continuously. | Damaged stop-light switch | Disconnect wiring connector from switch. If lamp goes out, replace switch. |
| | Switch out of adjustment | Adjust switch. |
| | Internal short circuit in wiring | If lamp stays on, check for internal short circuit. Repair as necessary. |
| One parking lamp out. | Bulb burned out | Replace bulb. |
| | Open wiring or poor ground | Repair as necessary. |
| | Corroded bulb socket | Repair or replace socket. |
| All parking lamps out. | Loose wiring connections | Secure wiring connections. |
| | Open wiring or poor ground | Repair as necessary. |
| | Bad switch | Replace switch. |

**Figure 11.** Tail, stop, and park light diagnosis.

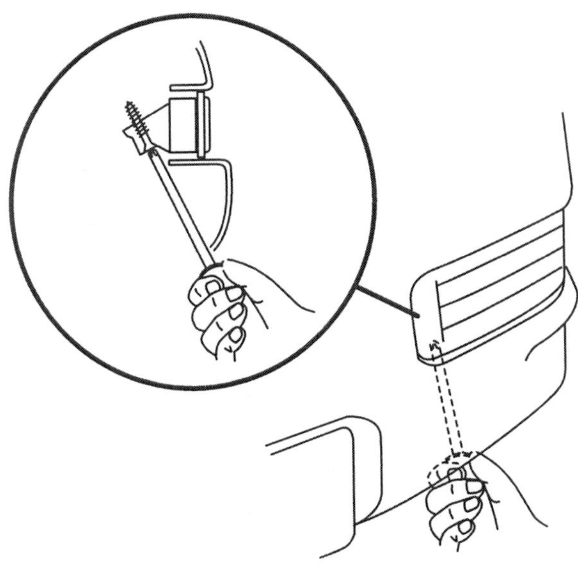

**Figure 12.** Removing front park and signal light assembly.

1. Be sure the lights are turned off.
2. Remove the screw that retains the front signal and park light assembly to the chassis **(Figure 12)**.
3. Pull the front signal and park light assembly forward as far as the wiring harness will allow.
4. Turn the bulb socket 45 degrees counterclockwise to remove the socket from the housing **(Figure 13)**.
5. Pull the bulb from the socket.

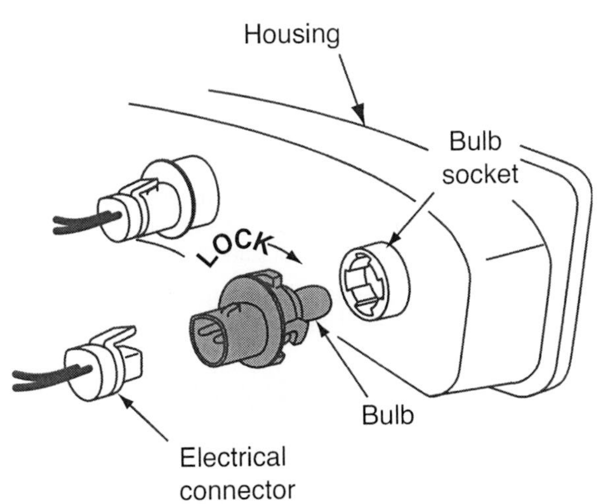


**Figure 13.** Removing signal and park light sockets and bulbs.

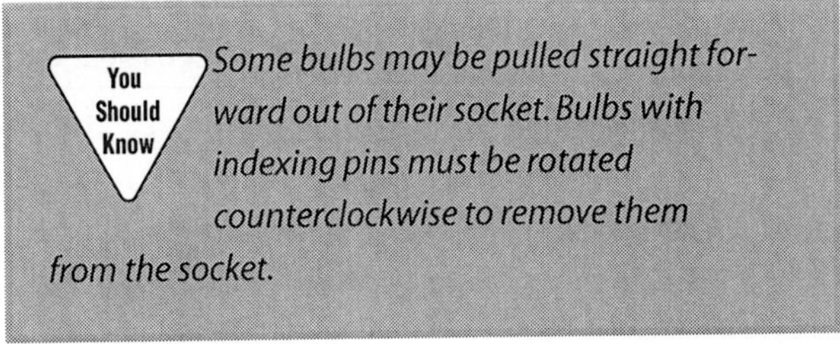

You Should Know
*Some bulbs may be pulled straight forward out of their socket. Bulbs with indexing pins must be rotated counterclockwise to remove them from the socket.*


6. Inspect the light socket for corrosion on the socket and terminals. Clean the socket and terminals as necessary.

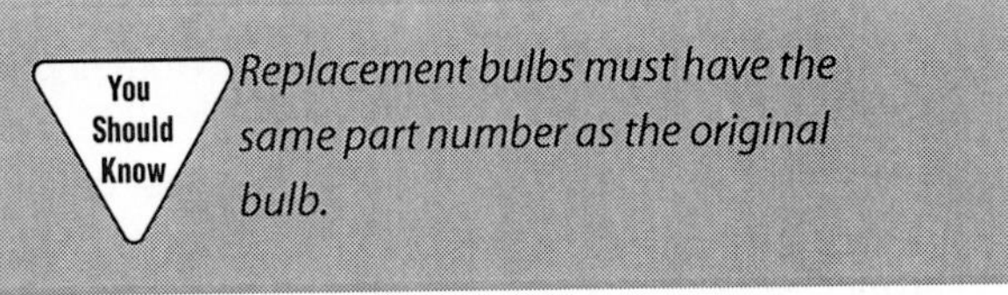

You Should Know
*Replacement bulbs must have the same part number as the original bulb.*


7. Install the new bulb in the socket, re-install the socket in the light assembly, and then re-install the light assembly in the chassis.
8. Install and tighten the light assembly retaining screw.
9. Be sure all the lights operate properly after the light assembly is installed.

## SIGNAL LIGHT AND HAZARD WARNING LIGHT CIRCUIT MAINTENANCE, DIAGNOSIS, AND SERVICE

Maintenance of signal lights and hazard warning lights is similar to the maintenance of other light circuits. When inspecting signal and hazard warning lights, always be sure they flash at the proper speed. If the signal lights flash normally on one side of the vehicle but they flash faster or slower on the opposite side, a problem is indicated in the circuit. The cause of the different flashing rates on the two sides of the vehicle may be a burned out signal light bulb or resistance in the circuit.

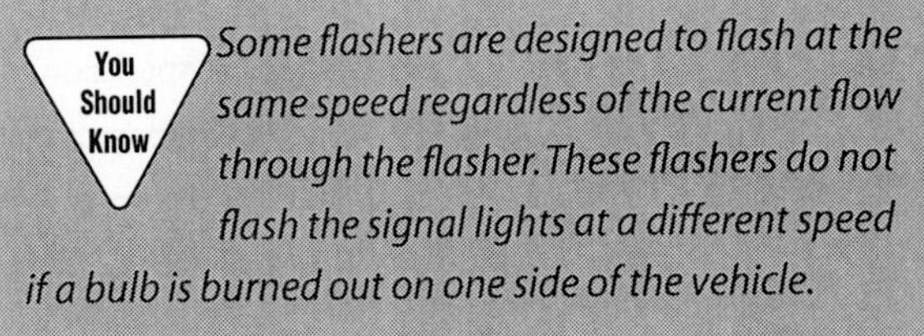

You Should Know
*Some flashers are designed to flash at the same speed regardless of the current flow through the flasher. These flashers do not flash the signal lights at a different speed if a bulb is burned out on one side of the vehicle.*

| Symptom | Possible Cause | Remedy |
|---|---|---|
| Turn signal lamps do not light. | Fuse or circuit breaker burned out | Replace fuse or circuit breaker. If fuse or circuit breaker blows again, check for short circuit. |
| | Worn or damaged turn signal flasher | Substitute a known good flasher. Replace if required. |
| | Loose wiring connections | Secure connections where accessible. |
| | Open circuit in wiring or poor ground | Repair as required. |
| | Damaged turn signal switch | Check continuity of switch assembly. Replace turn signal switch and wiring assembly as necessary. |
| Turn signal lamps light but do not flash. | Worn or damaged turn signal flasher | Substitute a known good flasher. Replace if required. |
| | Poor ground | Repair ground. |
| Front turn signal lamps do not light. | Loose wiring connector or open circuit | Repair wiring as required. |
| Rear turn signal lamps do not light. | Loose wiring connector or open circuit | Repair wiring as required. |
| One turn signal lamp does not light. | Bulb burned out | Replace bulb. |
| | Open circuit in wiring or poor ground | Repair as required. |

**Figure 14.** Signal light diagnosis.

Signal light diagnosis is provided in **Figure 14** and hazard warning light diagnosis is listed in **Figure 15**.

A typical procedure for changing a rear signal light bulb is the following:

1. Open the trunk lid and remove the rear trim panel **(Figure 16)**.
2. Turn the light socket 45 degrees counterclockwise on the bulb to be replaced **(Figure 17)**.

| Symptom | Possible Cause | Remedy |
|---|---|---|
| Hazard flasher lamps do not flash. | Fuse or circuit breaker burned out | Replace fuse or circuit breaker. If fuse or circuit breaker blows again, check for short circuit. |
| | Worn or damaged hazard flasher | Substitute a known good flasher. Replace flasher if damaged. |
| | Worn or damaged turn signal operation | Repair turn signal system. |
| | Open circuit in wiring | Repair as required. |
| | Worn or damaged hazard flasher switch | Repair or replace turn signal switch and wiring assembly which includes hazard flasher |

**Figure 15.** Hazard warning light diagnosis.

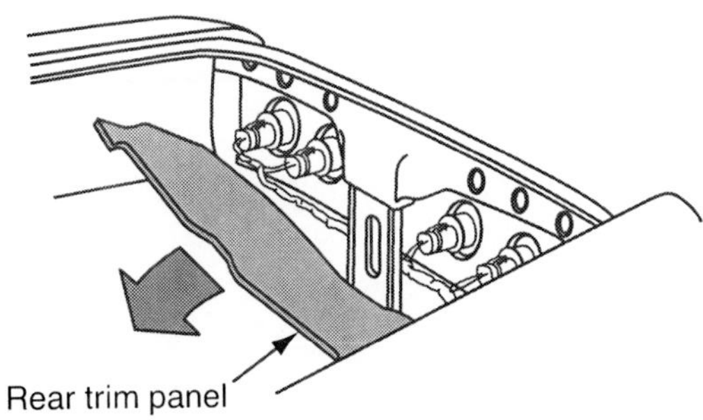


**Figure 16.** Removing the rear trim panel.

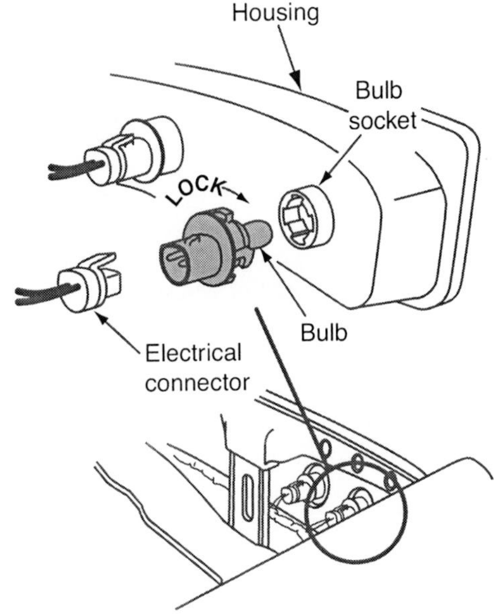


**Figure 17.** Removing the rear sockets and bulbs.

3. Remove the bulb from the socket.
4. Inspect the light socket and terminals for corrosion and clean as necessary.
5. Install the new bulb in the socket and install the socket and trim panel.

## TESTING BULBS AND FUSES

Bulbs and fuses may be tested with an ohmmeter. When testing a dual filament bulb, connect the digital ohmmeter leads from one of the bulb terminals to the case and then connect the meter leads from the other bulb terminal to the case. An infinite reading with either meter connection indicates a burned out bulb filament.

When testing a fuse, remove the fuse from the fuse panel and connect the ohmmeter leads across the fuse terminals. A zero ohm reading indicates a satisfactory fuse and a high or infinite reading indicates an open fuse. Fuses and circuit breakers may be tested using the same procedure and the test results are also the same.

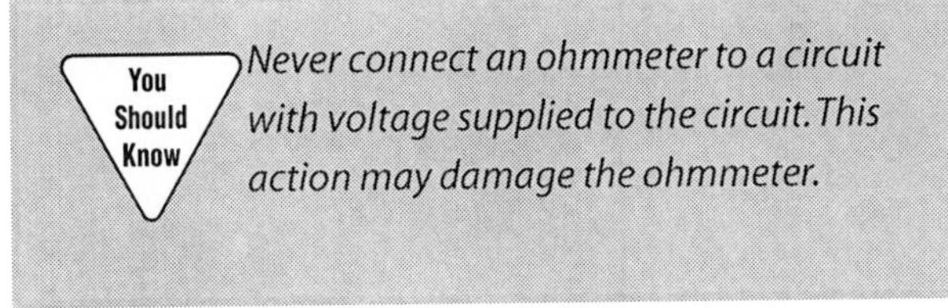

You Should Know
*Never connect an ohmmeter to a circuit with voltage supplied to the circuit. This action may damage the ohmmeter.*


## Summary

- High charging voltage may cause the headlights to become excessively bright when engine rpm is increased.
- High resistance in a headlight circuit reduces current flow and causes dim lights.
- If only one headlight is dim, the ground circuit on that light likely has high resistance.
- Measure the voltage drop in the headlight circuit to check the circuit resistance.
- Some moisture formation in non-vented composite headlights is normal.
- The glass envelope on halogen light bulbs should not be touched with your hands.
- Many light sockets are removed by rotating them 45 degrees counterclockwise.
- Headlights should be properly aimed vertically and horizontally.
- Digital ohmmeters do not require calibration.
- An ohmmeter may be damaged if it is connected to a circuit with voltage supplied to the circuit.
- When an ohmmeter is connected to fuse terminals, a low reading indicates a satisfactory fuse.
- When an ohmmeter is connected across a filament in a bulb, an infinite ohmmeter reading indicates an open filament.

## Review Questions

1. Technician A says if all the headlights are dim, the alternator may be inoperative. Technician B says if all the headlights are dim, there may be high resistance in one headlight ground connection. Who is correct?
   A. Technician A
   B. Technician B
   C. Both Technician A and Technician B
   D. Neither Technician A nor Technician B
2. Technician A says when installing halogen bulbs, hold the glass envelope with your fingers. Technician B says some moisture formation in composite headlights is normal. Who is correct?
   A. Technician A
   B. Technician B
   C. Both Technician A and Technician B
   D. Neither Technician A nor Technician B
3. Technician A says when aiming sealed beam headlights, the aimer adapters must fit over the aiming pads on the sealed beams. Technician B says suction cups hold the aimers onto the sealed beams. Who is correct?
   A. Technician A
   B. Technician B
   C. Both Technician A and Technician B
   D. Neither Technician A nor Technician B
4. Technician A says headlights may be aimed with the headlight beams projected against a wall and the vehicle parked 50 feet back from the wall. Technician B says vehicle load and tire inflation should be normal before aiming the headlights. Who is correct?
   A. Technician A
   B. Technician B
   C. Both Technician A and Technician B
   D. Neither Technician A nor Technician B
5. A vehicle with two headlights has one headlight that is dim on low and high beam and one headlight that has normal brilliance. The most likely cause of this problem is:
   A. low charging system voltage.
   B. high resistance in the battery ground cable.
   C. high resistance in the dimmer switch.
   D. high resistance in the dim headlight ground circuit.
6. When measuring the voltage drop across the ground circuit from a headlight to the battery negative terminal, an acceptable voltage drop reading is:
   A. 0.2 volt.
   B. 0.8 volt.
   C. 1.0 volts.
   D. 1.2 volts.
7. All of these statements about servicing halogen headlight bulbs are true *except*:
   A. When illuminated, halogen bulbs become very hot.
   B. If a halogen bulb is burned out, the complete headlight assembly must be replaced.
   C. Do not attempt to clean or replace a halogen bulb with the headlights on.
   D. Do not touch the glass envelope in a halogen bulb with your fingers.
8. The signal lights on the left side of a vehicle flash faster than the signal lights on the right side of the vehicle. The most likely cause of this problem is:
   A. a burned out signal light bulb on the left side of the vehicle.
   B. a defective signal light flasher.
   C. an open circuit in the signal light fuse.
   D. high resistance between the signal light fuse and the fuse holder terminals.
9. When using an ohmmeter to test the fuse in the stop light circuit, Technician A says the brake lights should be on. Technician B says an open fuse provides a very low reading on the ohmmeter. Who is correct?
   A. Technician A
   B. Technician B
   C. Both Technician A and Technician B
   D. Neither Technician A nor Technician B
10. After the headlights are turned on high beam, in a headlight circuit with a circuit breaker, both headlights keep going off and on. The most likely cause of this problem is:
    A. a defective circuit breaker.
    B. a grounded wire between the dimmer switch high beam terminal and the headlights.
    C. an open ground wire on one of the headlights.
    D. a loose high beam wiring connection on one headlight.
11. When the ohmmeter leads are connected to the fuse terminals, a very low ohmmeter reading indicates the fuse is ________________.
12. When testing bulb filaments, the ohmmeter leads are connected from one of the bulb terminals to the ________________.
13. When an ohmmeter is connected across a bulb filament, an infinite reading indicates the filament is ________________.
14. If the ohmmeter leads are connected across the terminals on a circuit breaker, a high reading indicates the breaker is ________________.
15. Explain the importance of proper headlight aim.
16. Explain the cause of excessive headlight brilliance when the engine rpm is increased.
17. Describe the proper test procedure for resistance in a headlight circuit.
18. Describe the ohmmeter test procedure for a circuit breaker.

# Indicator Lights and Gauges

## *Introduction*

When technicians understand normal indicator light and gauge operation, they are able to diagnose problems in these circuits. Technicians must be able to quickly and accurately diagnose whether a defect is in the indicator light or gauge itself or in the system that the indicator light or gauge is monitoring.

### OIL PRESSURE INDICATOR LIGHTS

The sending unit for the oil pressure indicator light is an on/off switch with a set of normally closed contacts. This sending unit is usually threaded into an opening in the main oil gallery of the engine block **(Figure 1)**. **Normally closed** contacts are closed with no pressure supplied to the unit and opened when pressure is supplied to the unit.

Full pressure from the lubrication system is supplied to this sending unit. If less than 3-psi (20.6-kPa) oil pressure is supplied to the oil sending unit, the contacts in this unit are closed. Under this condition, current flows through the oil pressure indicator light and the sending unit contacts to ground and the light is on. When the engine is started, oil pressure is supplied to the diaphragm in the oil sending unit. If the oil pressure exceeds 3 psi (20.6 kPa), the sending unit contacts are forced open and the oil indicator light goes out **(Figure 2)**.

Oil filter
Engine oil pressure switch

**Figure 1.** An oil pressure switch.

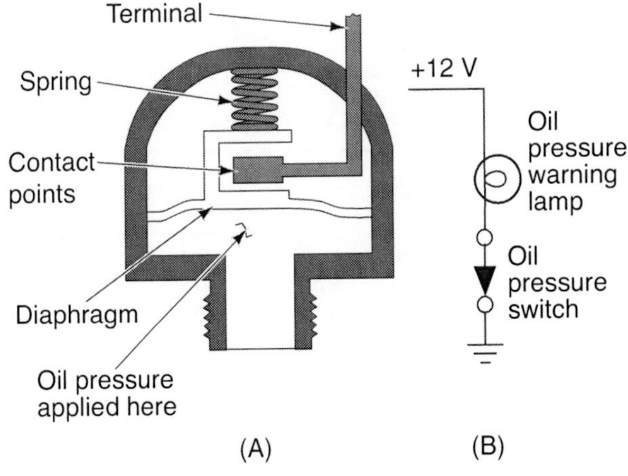


**Figure 2.** Oil pressure switch internal design.

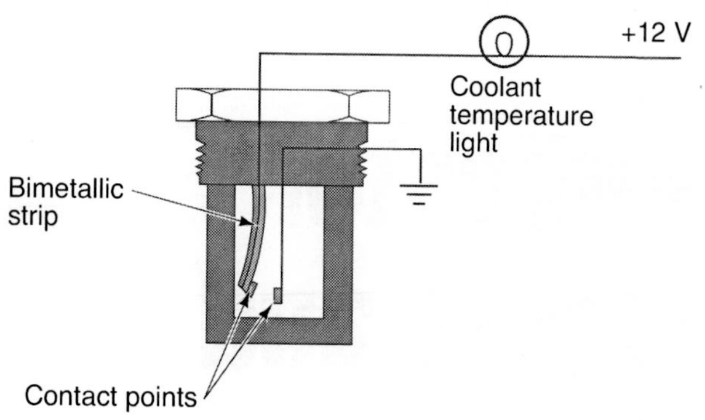


**Figure 3.** A temperature sending switch and related circuit.

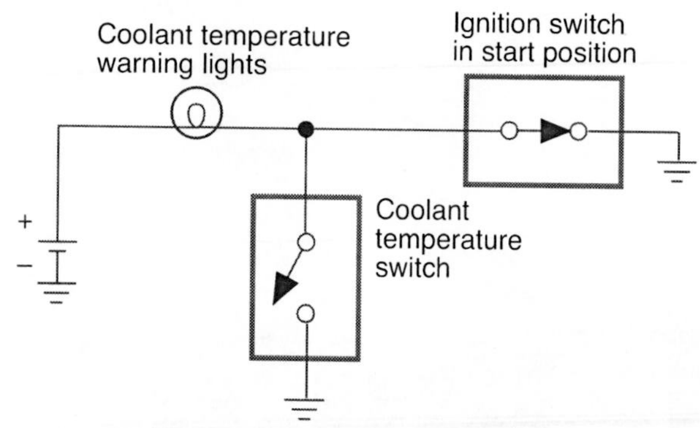


**Figure 4.** A coolant temperature switch proving circuit.

## ENGINE TEMPERATURE WARNING LIGHTS

The sending unit for the engine temperature warning light contains a bimetallic strip and a set of normally open contacts. **Normally open** contacts remain in the open position until they are acted upon by temperature or pressure.

When the engine temperature is below a specific temperature, the contacts in the temperature sending unit remain open and the temperature warning light remains off **(Figure 3)**. If the coolant temperature increases to a specific overheated condition, the bimetallic strip bends and closes the contacts in the sending unit. As long as the overheated condition is present, the sending unit contacts remain closed and the temperature warning light is on.

Many temperature warning lights have a proving circuit in the ignition switch. When the ignition switch is in the start position, the proving circuit contacts in the ignition switch ground the temperature warning light **(Figure 4)**. This illuminates the temperature warning light while cranking the engine and proves the temperature warning light bulb is operating.

## CHARGE INDICATOR LIGHTS

Some charge indicator lights have a resistor connected in parallel with the bulb. When the ignition switch is turned on, current flows through the charge indicator bulb and parallel resistor to one of the alternator terminals. This current flows through the alternator number 1 terminal to the alternator field coil and electronic voltage regulator to ground and the charge indicator light is on **(Figure 5)**. Once the engine starts, approximately 14.2 volts are supplied from the alternator battery terminal to the battery and electrical system. This same voltage is also supplied from the alternator stator windings through the diode trio to the field coil and to the number 1 alternator terminal. Because equal voltage is supplied to both sides of the charge indicator bulb, this light remains off.

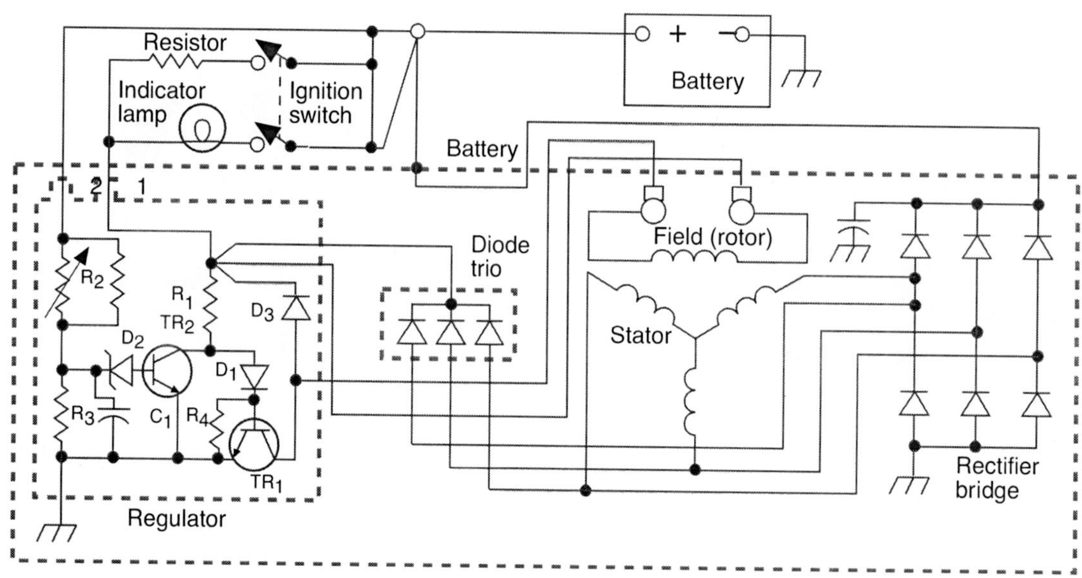


**Figure 5.** A charge indicator light circuit.

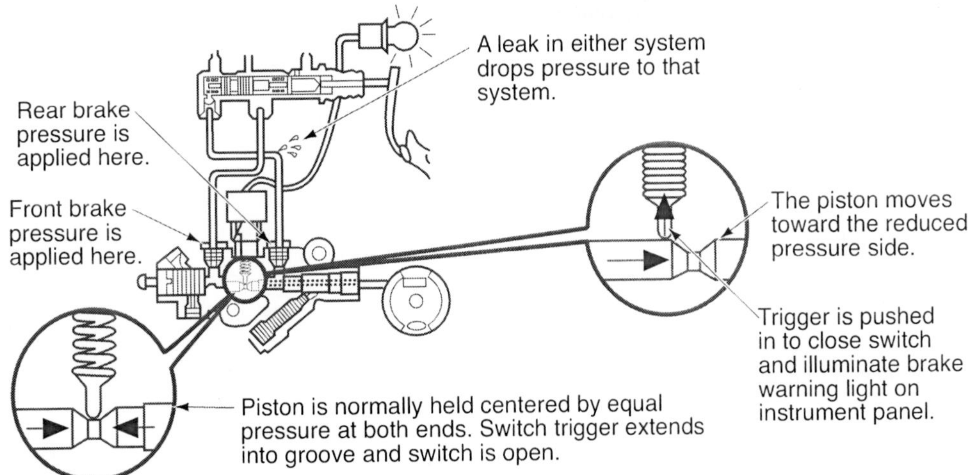
A leak in either system drops pressure to that system.
Rear brake pressure is applied here.
Front brake pressure is applied here.
The piston moves toward the reduced pressure side.
Trigger is pushed in to close switch and illuminate brake warning light on instrument panel.
Piston is normally held centered by equal pressure at both ends. Switch trigger extends into groove and switch is open.

**Figure 6.** A brake warning light circuit.

## BRAKE WARNING LIGHT

The red brake warning light in the instrument panel is connected to a switch in the combination brake valve. The combination brake valve also contains two hydraulic valves, the metering valve and the proportioning valve. These hydraulic valves are discussed in chapter 44. All vehicles manufactured since 1967 have dual master cylinders. One piston in the master cylinder supplies fluid pressure to half the brake system and the other piston supplies pressure to the opposite half. On front-wheel drive vehicles, the brake system is usually designed so one master cylinder piston supplies pressure to the left front and right rear wheels and the other master cylinder piston supplies pressure to the right front and left rear wheels. Brake systems on rear-wheel drive vehicles are divided so one master cylinder piston supplies the front brakes and the other master cylinder piston supplies the rear brakes.

Pressure from each master cylinder piston is supplied to opposite ends of the combination valve piston that operates the brake warning light. If the master cylinder fluid level is satisfactory and both master cylinder pistons supply the same pressure, the piston in the brake warning light circuit remains centered. Under this condition, the brake warning switch is open and the warning light is off. If a fluid leak occurs and pressure from one master cylinder piston is low, the brake warning light piston moves toward the low pressure side of the piston. This action grounds the brake warning light bulb through the switch and the bulb is illuminated **(Figure 6)**. This circuit also has a prove-out circuit for bulb check when cranking.

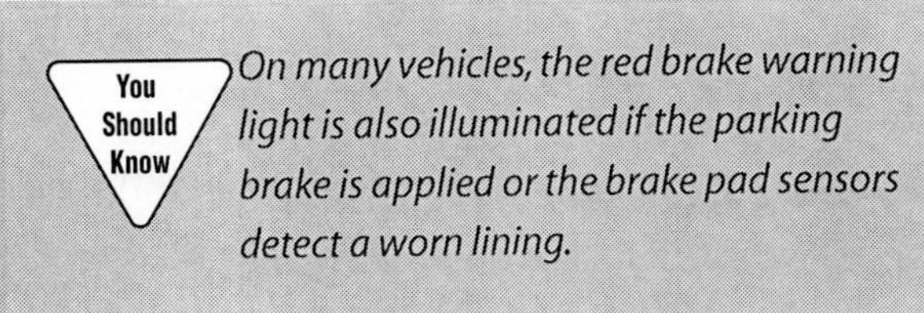
You Should Know

*On many vehicles, the red brake warning light is also illuminated if the parking brake is applied or the brake pad sensors detect a worn lining.*

## TYPES OF GAUGES

Various types of gauges are used in instrument panels. Each type of gauge has different operating principles and technicians must understand gauge operation to accurately diagnose and service each type of gauge and its related circuit.

### Bimetallic Gauges

Many vehicles are equipped with bimetallic gauges. In this type of gauge, the needle is linked to a bimetallic strip. A **bimetallic strip** contains two different metals fused together. As a bimetallic strip is heated, these metals expand at different rates and cause the strip to bend.

When this strip is heated, it pushes the needle across the gauge scale. A heating coil surrounds the bimetallic strip and the amount of heat supplied to this strip depends on the current flow through the heating coil **(Figure 7)**. The

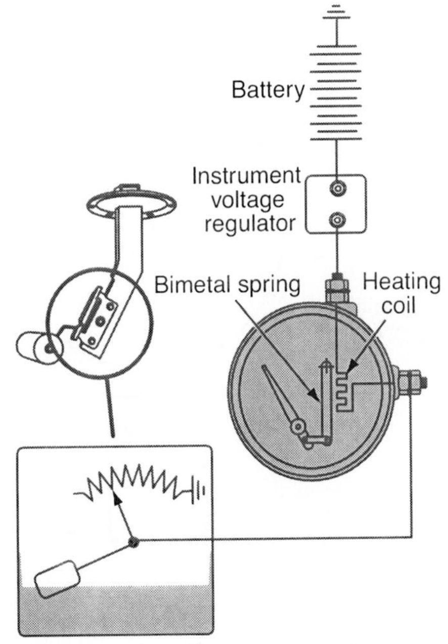
Battery
Instrument voltage regulator
Bimetal spring
Heating coil

**Figure 7.** Bimetallic gauge design.

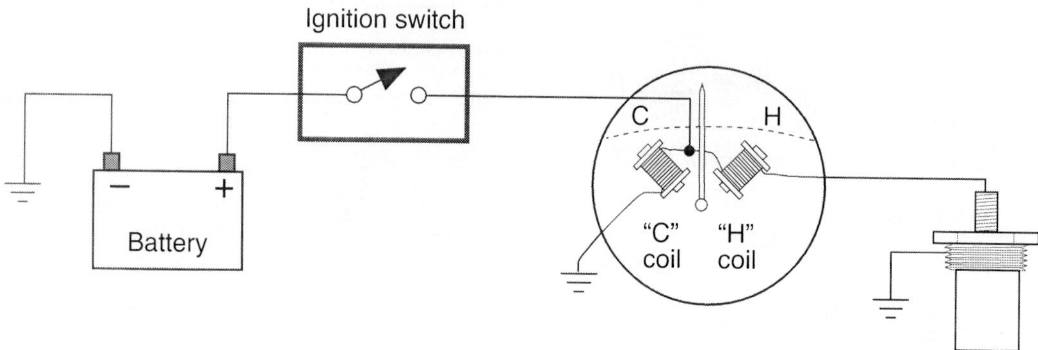


Figure 8. A balancing coil temperature gauge.

gauge sending unit contains a variable resistor that controls the current flow through the heating coil.

## Balancing Coil Gauge

Some gauges contain a low coil and a high coil. The gauge needle is pivoted between the two coils and a permanent magnet is mounted on the needle. When the ignition switch is turned on, voltage is supplied between the two coils. The low coil is grounded and the high coil is connected to the sending unit **(Figure 8)**. If the sending unit has high resistance, most of the current flows through the low coil to ground because this coil has lower resistance than the high coil and the sending unit. The magnetic field of the low coil attracts the magnet on the needle and moves the needle to the low position. When the sending unit has low resistance, most of the current flows through the high coil and the sending unit. Under this condition, the magnetic field of the high coil attracts the pointer magnet so the pointer moves to the high position.

In some balancing coil fuel gauges, voltage is supplied to the low or empty coil and the sending unit is connected between the empty coil and the full coil. The full coil is grounded **(Figure 9)**. When the fuel level is low in the tank, the sending unit has low resistance. Under this condition, current flows through the empty coil and the sending unit to ground and the magnetic field of the empty coil attracts the needle near the empty position. If the fuel level in the tank is high, the sending unit has high resistance. Under this condition, the current flows through the empty coil and the full coil to ground. Since the full coil has more coils of wire, it develops the stronger magnetic field and attracts the needle near the full position.

## Oil Pressure Gauges

The sender for an oil pressure gauge contains a diaphragm and a variable resistor. Engine oil pressure is supplied to the sending unit diaphragm. As the engine oil pressure increases, the sending unit diaphragm moves upward and the contact arm slides along the resistor **(Figure 10)**. An increase in oil pressure reduces sending unit resistance and a decrease in oil pressure increases sending unit resistance. When the oil sending unit is connected to a bimetallic gauge, the reduced sending unit resistance with higher oil pressure increases current flow through the gauge heating coil and the sending unit. This increased cur-

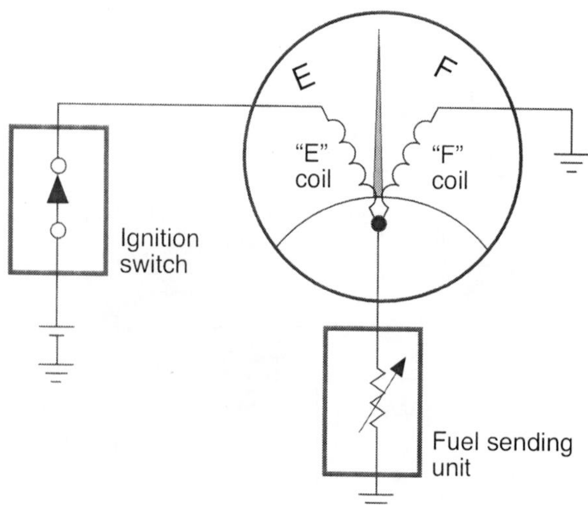


Figure 9. A balancing coil fuel gauge.

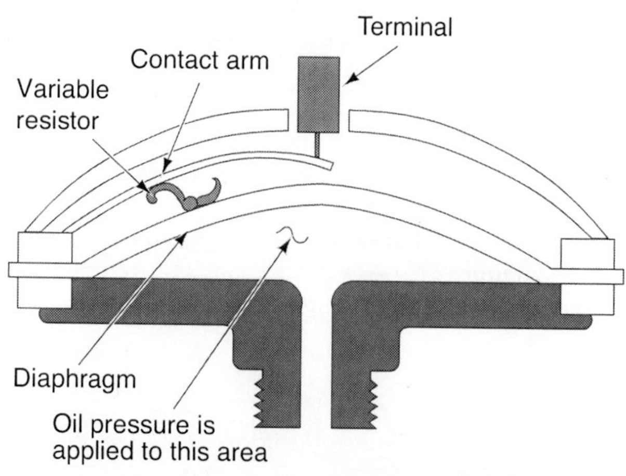


Figure 10. An oil pressure sending unit with a variable resistor.

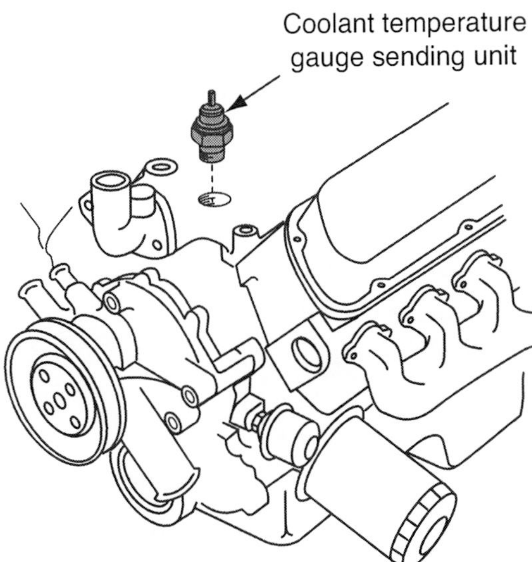


**Figure 11.** Coolant temperature gauge sending unit location.

rent flow heats the bimetallic strip. Under this condition, the bimetallic strip bends and pushes the gauge needle to the high position.

## Engine Temperature Gauge

The temperature sending unit is threaded into an opening in the cooling system **(Figure 11)**. This sending unit is often mounted in the top of the intake manifold or in the cylinder head. The lower end of the sending unit is in contact with engine coolant. The sending unit for most temperature gauges contains a resistor disc called a thermistor **(Figure 12)**. A **thermistor** is a special resistor that changes resistance in relation to temperature.

At low temperatures, the thermistor has high resistance and as the coolant temperature increases, the resistance decreases. If the temperature sending unit is connected to a bimetallic gauge, as the coolant temperature increases, the reduced sending unit resistance increases the current flow through the gauge heating coil. This action heats the bimetallic strip and the bending of this strip pushes the gauge needle to a higher gauge reading.

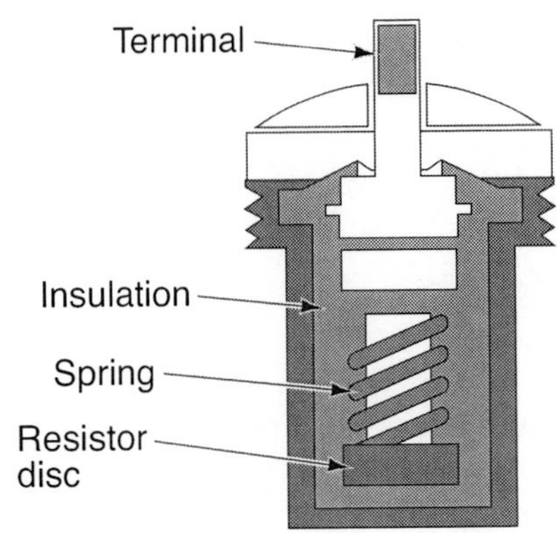


**Figure 12.** Coolant temperature sending unit design.

## Fuel Level Gauge

The fuel level gauge contains a float mounted on an arm attached to the gauge. The float moves up and down with the fuel level in the tank. As the float moves up and down, it moves a sliding contact on a variable resistor in the sending unit. High fuel level in the tank results in low sending unit resistance and a low fuel level increases sending unit resistance **(Figure 13)**. If the fuel sending unit is connected to a bimetallic fuel gauge, high fuel level and low sending unit resistance increases current flow through the gauge heating coil. Under this condition, the bimetallic strip bends and pushes the gauge needle near the full position.

**You Should Know** *The fuel sending unit that is connected between the empty and full coils on a balancing coil fuel gauge operates the opposite way compared to the sending unit for a bimetallic gauge. The sending unit for this type of balancing coil gauge has high resistance when the fuel level in the tank is high.*

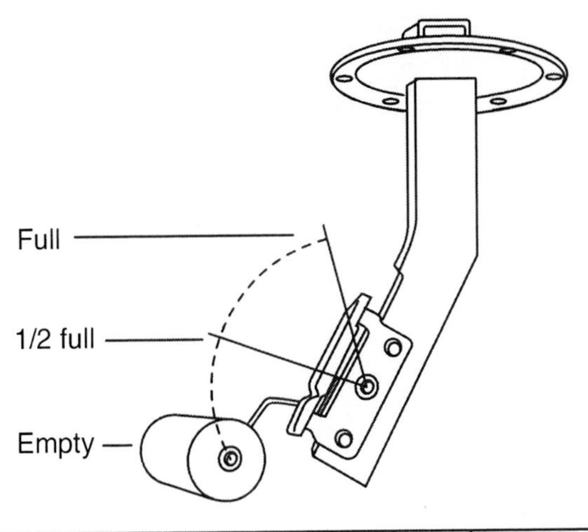


| Float position | F | 1/2 | E |
|---|---|---|---|
| Resistance (Ω) | 2 - 5 | 25.5 - 39.5 | 105 - 110 |

**Figure 13.** A fuel gauge sending unit.

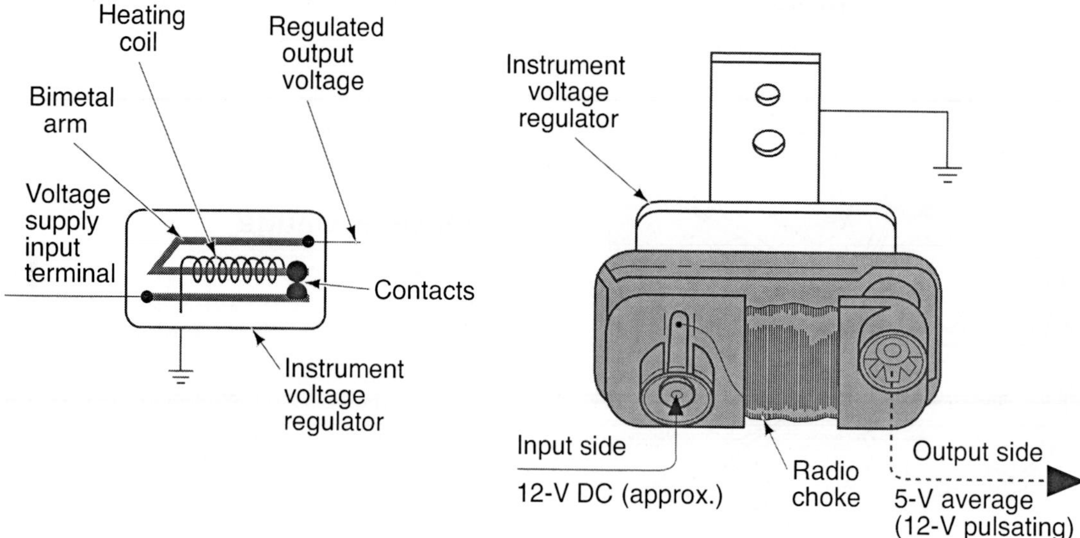


**Figure 14.** An instrument voltage limiter.

## INSTRUMENT VOLTAGE LIMITERS

A instrument voltage limiter may be connected to bimetallic gauges. This limiter contains a set of contacts mounted on a bimetallic strip. The voltage supply to the gauges is connected through the limiter contacts. A heating coil surrounds the bimetallic strip and this heating coil is connected to ground on the limiter.

> **You Should Know** *The instrument voltage limiter must be grounded on the instrument panel and this panel must have a satisfactory ground connection to the battery. If the instrument voltage limiter does not have a satisfactory ground, the limiter contacts remain closed and this action supplies 12 volts to the gauges. This voltage will damage the gauges very quickly.*

When the ignition switch is turned on, voltage is supplied through the limiter contacts to the gauges. Current also flows through the heating coil to ground. The heating coil heats the bimetallic strip very quickly and the limiter contacts open the circuit to the gauges and also to the heating coil. The bimetallic strip cools quickly and the contacts close. The voltage limiter supplies a pulsating 5 volts to the gauges regardless of the input voltage and this provides more stable gauge operation **(Figure 14)**.

## SPEEDOMETERS AND ODOMETERS

A mechanical speedometer has a cable drive from the transmission output shaft. A gear on the speedometer cable drive is meshed with a gear on the transmission output shaft **(Figure 15)**. Therefore, output shaft rotation turns the speedometer cable. This cable is connected to a permanent magnet surrounded by a metal drum in the speedometer. The speedometer needle is attached to the drum, but there is no direct mechanical connec-

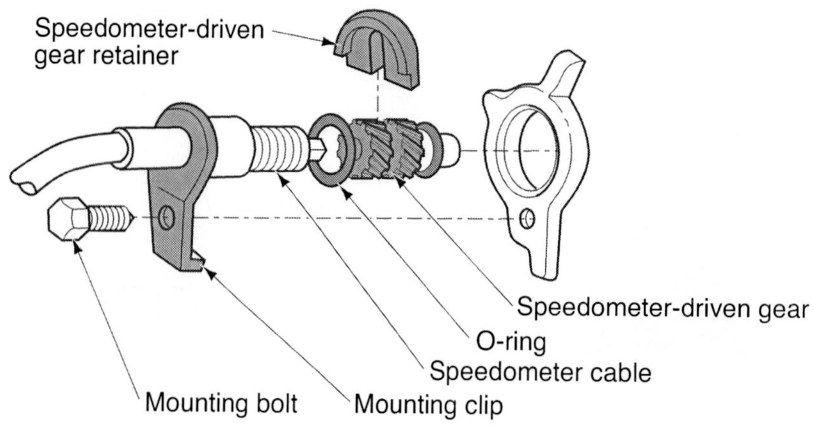


**Figure 15.** A speedometer cable drive.

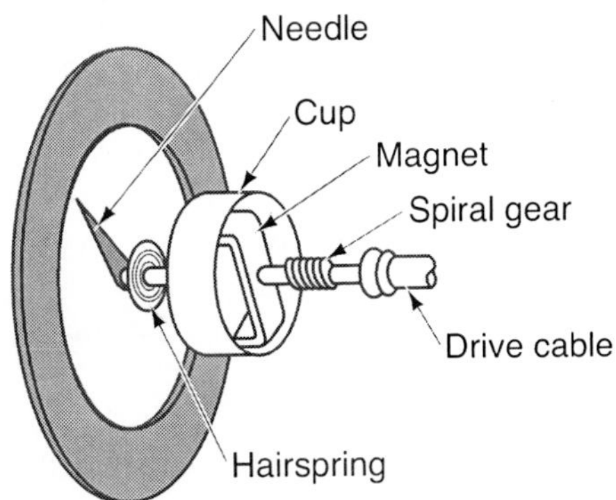


**Figure 16.** Speedometer design.

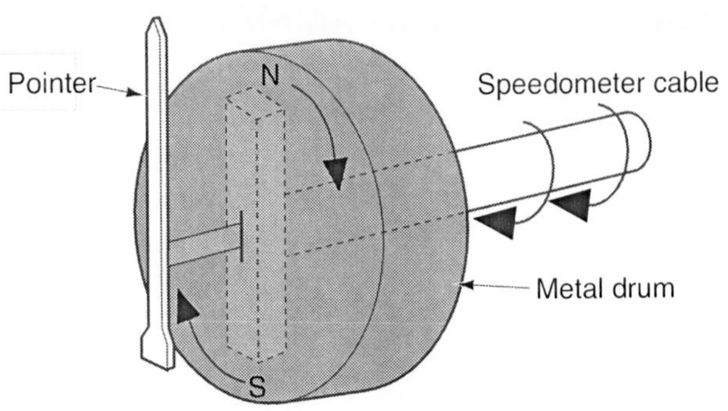


**Figure 17.** Speedometer operation.

tion between the cable and the speedometer needle **(Figure 16)**. When the speedometer cable rotates the permanent magnet, a rotating magnetic field is created around the drum. This magnetic field pulls the drum and speedometer needle in a rotary motion to provide an accurate speedometer reading **(Figure 17)**.

An odometer is driven by a worm gear drive from the speedometer cable **(Figure 18)**. Some odometers have six wheels and numbers from zero to nine are stamped on the outer surface around the wheels. These wheels are designed so when the right wheel makes one revolution, the wheel on the left moves one position. Odometers with seven wheels indicate mileage over 100,000.

> You Should Know — *Many modern vehicles are equipped with electronic speedometers. These speedometers may be analog or digital and they do not require a cable. The vehicle speed sensor (VSS) mounted in the transmission sends a voltage signal to the computer and the computer controls the speedometer reading.*

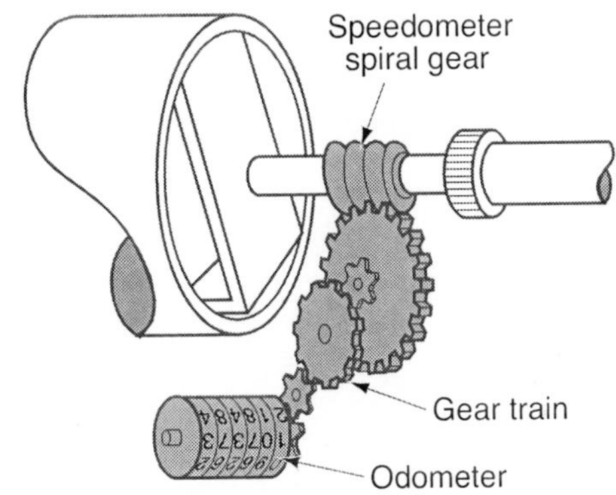


**Figure 18.** An odometer and drive mechanism.

## TACHOMETERS

Some vehicles have tachometers that read engine rpm. The tachometer usually receives voltage input signals from the primary ignition system **(Figure 19)**. These input voltage signals are in direct proportion to engine speed and the tachometer changes these signals to a rpm reading.

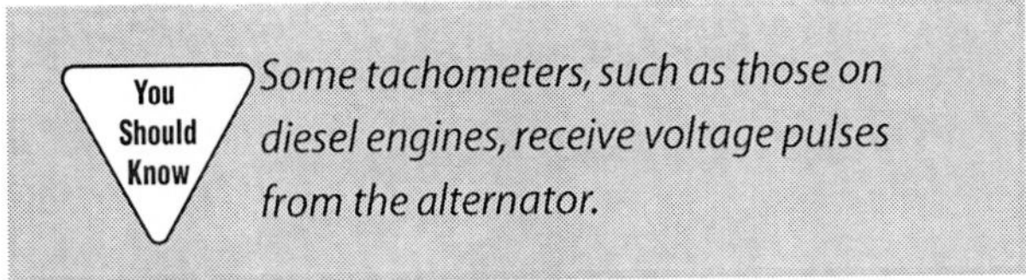

You Should Know
Some tachometers, such as those on diesel engines, receive voltage pulses from the alternator.


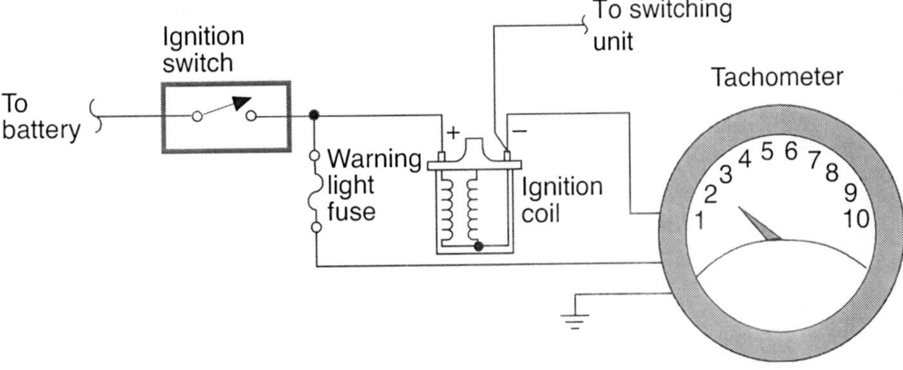


**Figure 19.** A tachometer circuit.

## VOLTMETERS AND AMMETERS

Some vehicles have a voltmeter to indicate the charging system condition. The voltmeter has high internal resistance and it must be connected in parallel. Because a voltmeter has high internal resistance, it draws a very low current. The voltmeter leads are connected from the battery positive circuit to ground. On some vehicles, the voltmeter is connected to the battery positive circuit at the ignition switch. On these systems, there is no voltage supplied to the voltmeter when the ignition switch is turned off **(Figure 20)**. With the engine running, the voltmeter should indicate 13.2 volts to 15.2 volts. If the voltage is below 13.2 volts, the alternator voltage is too low and the battery will become discharged. When the voltage is above 15.2 volts, the alternator voltage is too high and the battery will be overcharged. Excessive charging voltage may also burn out electrical accessories on the vehicle.

Some vehicles have an ammeter to indicate charging circuit and electrical system operation. The ammeter is connected in series in the circuit between the alternator battery terminal and the positive battery terminal. The ammeter reads the amount of current flow from the alternator into the battery with the engine running. The alternator side of the ammeter is also connected to a junction block and this block connects the ammeter to the ignition switch and the vehicle accessories **(Figure 21)**. Therefore, the ammeter will also indicate the current flow out of the battery and through the accessories with the engine stopped or with the engine running and the alternator inoperative.

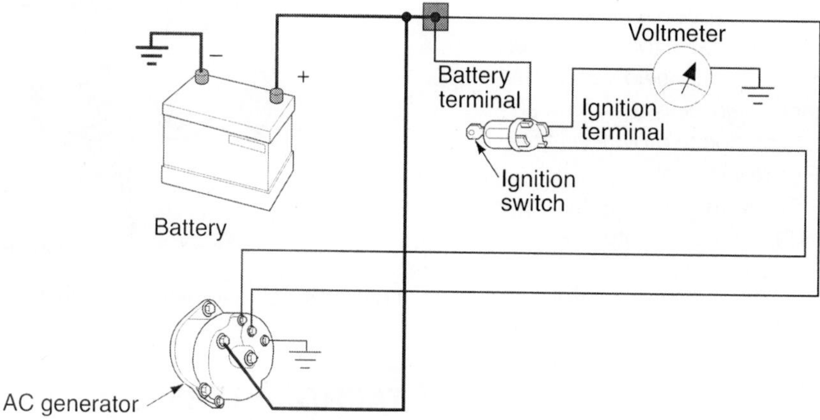


**Figure 20.** Voltmeter connections.

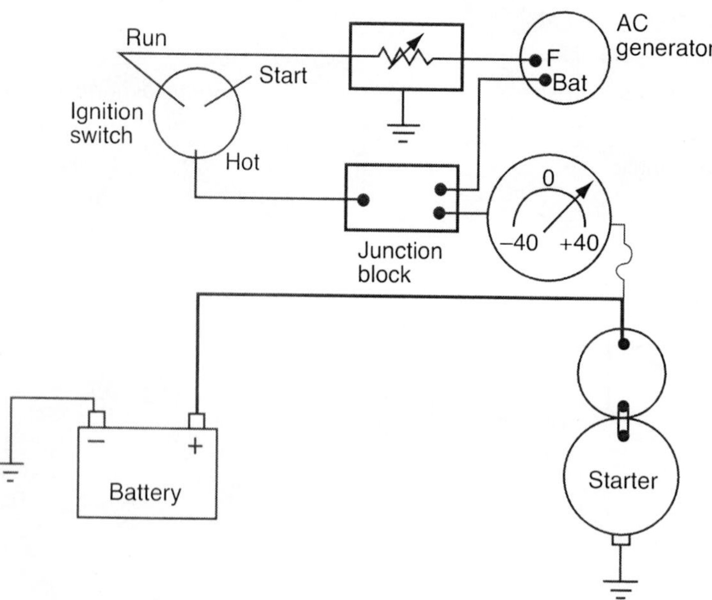


**Figure 21.** Ammeter connections.

## Summary


- The oil pressure switch opens and turns the oil pressure indicator light off when the oil pressure reaches 3 psi. (20.6 kPa).
- The engine temperature switch contains a bimetallic strip and a set of normally open contacts.
- When the engine is running, equal voltage on each side of the charge indicator light keeps the light off.
- If one-half of the master cylinder is low on brake fluid, the brake switch in the combination valve closes and this action turns on the red brake warning light.
- Bimetallic gauges contain a bimetallic strip wrapped with a heating coil and are linked to the gauge needle.
- A balancing coil gauge contains two coils and the magnetism of these coils determines the gauge needle position by attracting a permanent magnet on the gauge needle.
- A fuel gauge sending unit contains a float linked to a variable resistor. As the float moves up and down with the fuel level in the tank, the resistance changes in the variable resistor.
- An instrument voltage limiter limits voltage to the gauges to 5 volts regardless of the input voltage.
- An instrument voltage limiter contains a bimetallic strip surrounded by a heating coil and a set of normally closed contacts.
- An instrument voltage limiter must be grounded to the instrument panel.
- A mechanical speedometer has a cable driven by a set of gears on the transmission output shaft.
- The odometer in a mechanical speedometer is driven through a gear set from the speedometer cable.
- Many tachometers use a voltage input from the primary ignition circuit.
- A voltmeter contains high internal resistance and it is connected in parallel to the circuit.
- An ammeter contains low internal resistance and it is connected in series in the circuit.


## Review Questions

1. Technician A says an oil pressure switch contains a set of normally open contacts. Technician B says the oil pressure warning light should be on if the engine is running and the oil pressure is 8 psi (55kPa). Who is correct?
   A. Technician A
   B. Technician B
   C. Both Technician A and Technician B
   D. Neither Technician A nor Technician B
2. Technician A says the charge indicator light is grounded through the alternator field circuit if the ignition switch is on and the engine is not running. Technician B says when the engine starts, equal voltage on both sides of the charge indicator bulb turns this light off. Who is correct?
   A. Technician A
   B. Technician B
   C. Both Technician A and Technician B
   D. Neither Technician A nor Technician B
3. Technician A says the red brake warning light is on if the parking brake is applied. Technician B says the red brake warning light is on if both halves of the master cylinder are overfilled with brake fluid. Who is correct?
   A. Technician A
   B. Technician B
   C. Both Technician A and Technician B
   D. Neither Technician A nor Technician B
4. Technician A says in a bimetallic-type temperature gauge, current through the gauge and sending unit increases as the temperature decreases. Technician B says in a bimetallic-type temperature gauge, the resistance of the sending unit decreases as the coolant temperature decreases. Who is correct?
   A. Technician A
   B. Technician B
   C. Both Technician A and Technician B
   D. Neither Technician A nor Technician B
5. In a typical oil pressure indicator light circuit with the engine idling at normal operating temperature, the oil pressure indicator light is on if the oil pressure is below:
   A. 18 psi.
   B. 15 psi.
   C. 9 psi.
   D. 3 psi.
6. All of these statements about engine temperature warning light circuits are true *except*:
   A. A proving circuit illuminates the warning light when cranking the engine.
   B. The sending unit contains a set of normally closed contacts.
   C. The sending unit contains a bimetallic strip.
   D. The warning light is illuminated when the sending unit contacts close.

7. When discussing charge indicator light circuits, Technician A says with the engine running, the charge indicator light may be turned off by unequal voltage on each side of the light. Technician B says when the engine starts, 12.6 volts are supplied to one side of the light and 14. 2 volts are supplied to the other side of the light. Who is correct?
   A. Technician A
   B. Technician B
   C. Both Technician A and Technician B
   D. Neither Technician A nor Technician B
8. All of these statements about red brake warning lights are true *except*:
   A. The red brake warning light is on if the parking brake is applied.
   B. The red brake warning light is on if there is no fluid in one-half of the master cylinder.
   C. The switch in the combination valve is closed each time the brakes are applied.
   D. The red brake warning light is connected to the switch in the combination valve.
9. A vehicle has a bimetallic-type fuel gauge that indicates FULL regardless of the amount of fuel in the tank. The most likely cause of this problem is:
   A. an open wire between the fuel gauge and the fuel tank.
   B. a grounded wire between the ignition switch and the fuel gauge.
   C. an open circuit in the gauge's fuse.
   D. a grounded wire between the fuel gauge and the fuel tank.
10. A vehicle has a bimetallic-type oil pressure gauge and an oil pressure sending unit containing a diaphragm and a variable resistor. Technician A says the resistance of the variable resistor increases as the oil pressure decreases. Technician B says the gauge pointer moves to the HIGH position as the resistance of the variable resistor decreases. Who is correct?
    A. Technician A
    B. Technician B
    C. Both Technician A and Technician B
    D. Neither Technician A nor Technician B
11. In a fuel gauge sending unit used with a bimetallic gauge, the resistance in the sending unit ____________________ as the tank is filled with fuel.
12. A temperature sending unit used with a bimetallic gauge contains a ____________________.
13. A fuel level sending unit contains a ____________ linked to a ____________ ____________.
14. When the contacts open in an instrument voltage limiter, the current flow through the heating coil on the bimetallic strip is ____________.
15. Explain the basic operation of a mechanical speedometer.
16. Describe the operation of a mechanical odometer.
17. Explain the proper voltmeter connection in an electrical circuit.
18. Explain the operation of an ammeter in the instrument panel.

# Indicator Light and Gauge Maintenance, Diagnosis, and Service

## *Introduction*

Technicians must understand normal indicator light and gauge operation to detect and diagnose improper operation of these components. When normal indicator light and gauge operation is understood, technicians can often detect problems in the circuits monitored by these indicator lights and gauges. Detecting these problems before they become serious usually saves the customer money and inconvenience.

## INDICATOR LIGHT MAINTENANCE, DIAGNOSIS, AND SERVICE

Indicator lights should be inspected for proper operation when a vehicle is brought into the shop for minor service. Some shops have a policy of inspecting indicator light operation each time a vehicle is brought into the shop for lubrication service. A customer may not be aware that an indicator light is not working. For example, if the oil pressure indicator light bulb is burned out, the oil pressure warning light never comes on even with the ignition switch on and the engine not running. If a customer does not notice this condition, the engine could be ruined from low oil pressure and the customer would not be aware of the problem until the engine is severely damaged. Therefore, during routine lubrication service it is a good shop policy to inspect all the warning lights for proper operation. When one or more indicator lights are not working, the first step in diagnosing the problem is to visually inspect the indicator light wiring for damage, worn insulation, and contact with other components. Inspect the indicator light electrical connections for corrosion and looseness.

### Oil Indicator Light Diagnosis

If the oil pressure indicator light does not come on with the ignition switch on and the engine not running, remove the wire from the oil sender switch and use a jumper wire to connect this wire to ground. If the indicator light does not come on under this condition, check the bulb, the voltage input to the bulb, and the wire from the light to the oil sender switch. When the oil indicator light is on with the oil sender wire grounded and the ignition switch on, the bulb and connecting wires are satisfactory. Therefore the problem must be in the oil sender switch contacts. With the oil sender switch wire disconnected and the engine not running, connect an ohmmeter from this switch terminal to an engine ground near the terminal. A high or infinite reading indicates a defective oil sender switch and a very low ohmmeter reading indicates the switch contacts are grounding the circuit.

If the oil pressure indicator light is on with the engine running, the engine oil pressure could be very low. Under this condition, shut the engine off and check the oil level. Inspect the oil for contamination with gasoline or coolant. If the oil level is very low but the oil does not appear to be contaminated, the proper grade and viscosity of oil should be added to the crankcase until the oil level is at the FULL mark on the dipstick. If the oil indicator light now goes out with the engine running and there is no evidence of engine oil leaks, the vehicle may be driven. However, the engine should be checked as soon as possible for the cause of the excessive oil consumption.

When the oil indicator light is on with the engine running and the oil level and condition are satisfactory, the vehicle should be towed to a service facility. Under this condition, the engine oil pressure may be extremely low or

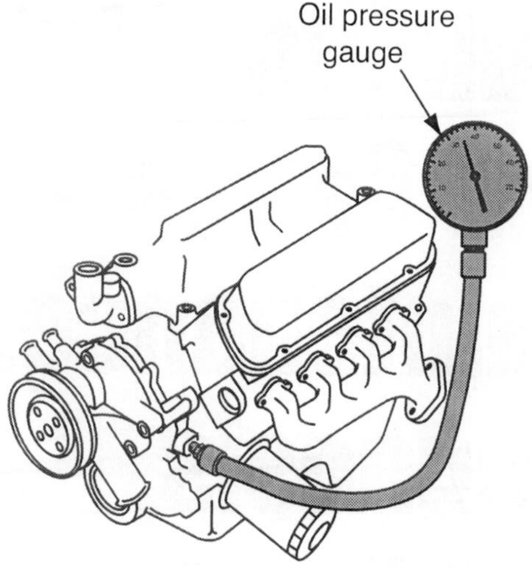


Figure 1. Oil pressure testing.

zero and severe engine damage may result from running the engine. When this condition is present, the oil sending switch should be removed and a test gauge installed in place of this switch **(Figure 1)**. If the engine oil pressure meets or exceeds the vehicle manufacturer's specifications, replace the oil sending switch. If the engine oil pressure is below the vehicle manufacturer's specifications, the oil pump and pump pickup, crankshaft, and camshaft bearings should be inspected and measured.

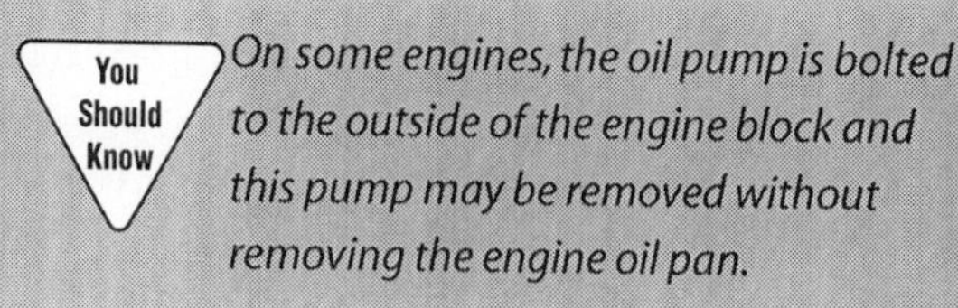


*On some engines, the oil pump is bolted to the outside of the engine block and this pump may be removed without removing the engine oil pan.*

## Oil Indicator Light Service

If it is necessary to change the oil sending switch, special oil sender sockets are available to fit the serrations on the sender. Use a ratchet or ratchet and extension to turn the oil sender socket. When installing the new sending switch, coat the sender threads with pipe thread lubricant and always torque this switch to the specified torque.

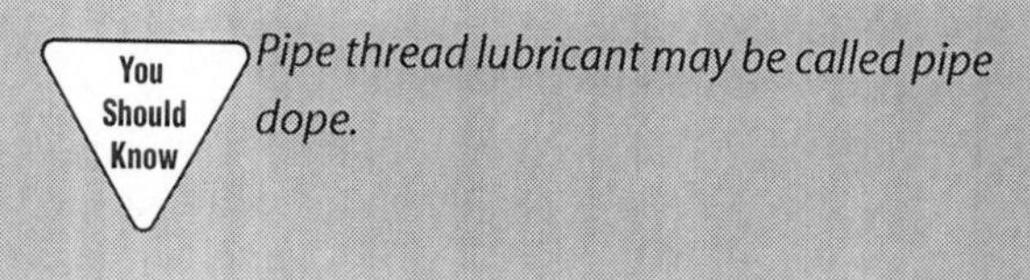


*Pipe thread lubricant may be called pipe dope.*

If the bulb in the oil pressure indicator light is burned out, it may be necessary to remove the retaining screws and pull the instrument panel partially out of the dash to gain access to the charge indicator bulb. Always disconnect the battery negative cable and wait the specified time period (2 to 10 minutes) before removing the instrument panel.

You Should Know

*Disconnecting the battery prevents accidental grounding of any circuit during instrument panel removal. This grounding could damage expensive electronic components. Waiting for the specified time period allows the backup power supply to power down in the air bag system. This power supply provides voltage to deploy the air bags if the battery is disconnected in a collision. If this procedure is not followed, the air bags could be deployed accidentally while servicing the instrument panel.*

You Should Know

*Disconnecting the battery erases the station programming in the vehicle radio. Before disconnecting the battery, it is a good idea to write down the station programming selections on AM and FM. After the battery is reconnected, the original radio station programming may be restored. Disconnecting the battery also erases the engine computer memory. After the battery is reconnected, the engine operation may be somewhat erratic until the engine computer relearns the input sensor data. These problems with erasing computer memories when a battery is disconnected may be eliminated by connecting an appropriate power supply to the vehicle's cigarette lighter before disconnecting the battery.*

Many light bulbs are retained in the back of the instrument panel with a plastic retainer **(Figure 2)**. Rotate the retainer counterclockwise to release the bulb. Install the new bulb and turn the retainer clockwise to hold the bulb in place. Most instrument panels have a printed circuit board. A **printed circuit board** is made from a thin insulating material and electrical solder or metal tracks are imbedded into the insulating material. The tracks in the printed circuit board connect the bulbs and gauges to the external circuits. A wiring connector plugged into the printed circuit board connects the external circuits to this board.

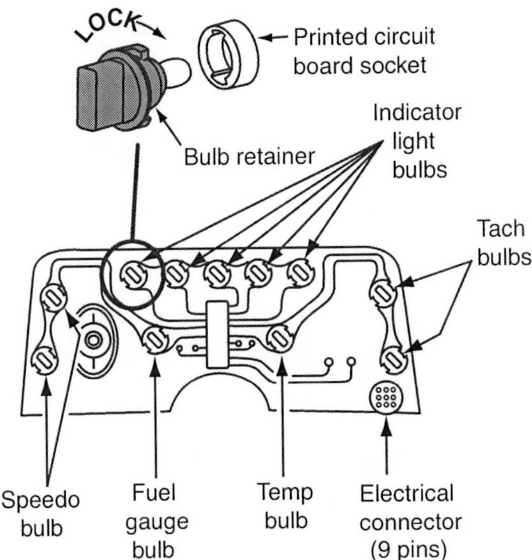


**Figure 2.** Indicator bulbs in the instrument panel.

> **You Should Know** *Do not reconnect the negative battery cable and turn the ignition switch on to test the indicator bulb with the instrument panel pulled forward. If the vehicle has a instrument voltage limiter, this limiter is not grounded under this condition. When the battery is reconnected and the ignition switch is turned on, the instrument voltage limiter supplies 12 volts to the gauges, resulting in gauge damage. Always reinstall the instrument panel and tighten all the retaining screws before reconnecting the battery and turning the ignition switch on.*

# CHARGE INDICATOR LIGHT MAINTENANCE, DIAGNOSIS, AND SERVICE

The charge indicator light should be illuminated when the ignition switch is on and the engine is not running. When the engine starts, the charge indicator light should go out and it should remain off as long as the engine is running. Each time minor service is performed on a vehicle, the charge indicator light operation should be checked. The customer may not notice the charge indicator light is on with the engine running. If this condition is present and the alternator is not working, the battery may become discharged and fail to start the vehicle. This could result in an expensive service call for the customer. During a routine check of the charge indicator light operation, inspect the alternator wiring for worn insulation and loose or corroded terminals.

## Charge Indicator Light Diagnosis

If the charge indicator light does not come on with the ignition switch on and the engine not running, remove the connector from the number 1 and 2 alternator terminals and use a jumper wire to connect the number 1 terminal to ground **(Figure 3)**. If the indicator light does not come on under this condition, check the bulb, the voltage input to the bulb, and the wire from the light to the number 1 alternator terminal. When the charge indicator light is on with the number 1 terminal grounded and the ignition switch on, the bulb and connecting wires are satisfactory. Therefore, if the charge indicator bulb is not on with the ignition switch on and the engine not running, there must be an open circuit in the alternator field circuit. The most likely location for an open field circuit is at the slip rings and brushes.

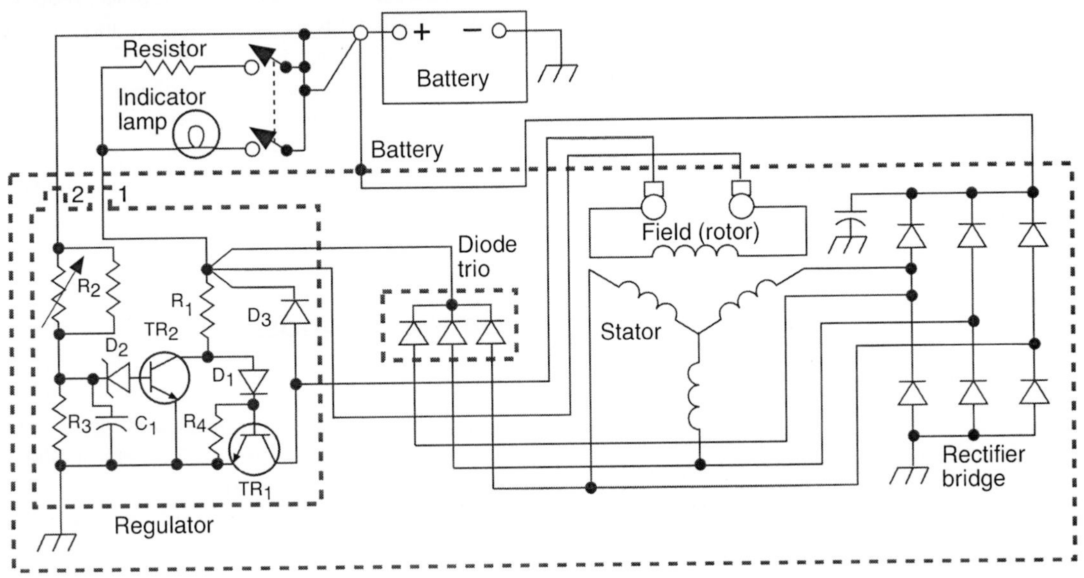


**Figure 3.** Charge indicator light and related circuit.

When the charge indicator light is on with the engine running, stop the engine and use a belt tension gauge to measure the belt tension. If the belt tension is satisfactory, test the alternator output to prove the alternator is operating properly. If the alternator output is less than specified, the alternator must be repaired or replaced. Alternator diagnosis is explained in chapter 24.

If the charge indicator light is illuminated dimly with the engine running, remember this light is kept off by equal voltage on each side of the light with the engine running. High resistance between the alternator battery terminal and the junction block or high resistance between the alternator number 1 terminal and the charge indicator light causes low voltage on one side of the charge indicator bulb. Unequal voltage across the charge indicator light causes a low current through this light. Measure the voltage drop across the circuits mentioned above to locate the high resistance.

## Charge Indicator Light Service

If the charge indicator bulb must be replaced, follow the same procedure and precautions that were explained previously regarding oil pressure indicator light service.

# BRAKE WARNING LIGHT MAINTENANCE, DIAGNOSIS, AND SERVICE

The red brake warning light should be checked for proper operation when performing minor service on a vehicle. With the parking brake applied, the red brake warning light should be on. If this light is not on, remove the wiring connector from the parking brake switch, turn the ignition switch on, and ground the wire from the red brake warning light. If this light is not on, check the bulb, voltage supply to the bulb, and the wire from the bulb to the parking brake switch.

If the red brake warning light is on with the engine running, remove the wire from the brake switch in the combination valve. If the brake warning light goes out under this condition, there is unequal pressure between the two halves of the master cylinder. Check the fluid levels in both halves of the master cylinder. Fill both halves of the master cylinder to the proper level with the brake fluid specified by the vehicle manufacturer.

The **combination valve** contains a proportioning valve, a metering valve, and a switch that grounds the red brake warning light if there is unequal pressure in the two halves of the master cylinder. The combination valve is usually mounted near the master cylinder.

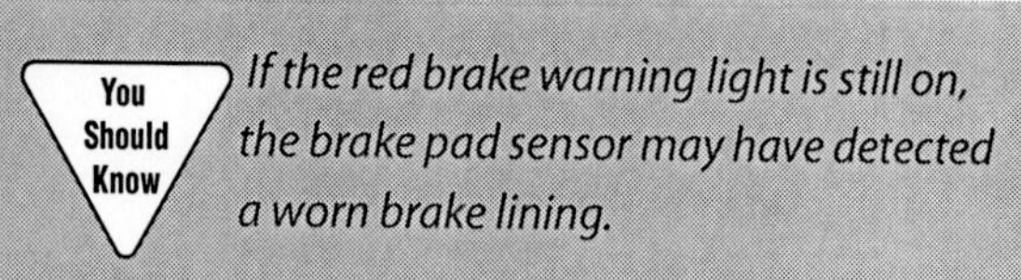

You Should Know
*If the red brake warning light is still on, the brake pad sensor may have detected a worn brake lining.*


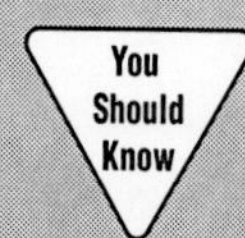


*If one half of the master cylinder was low on brake fluid, it may be necessary to bleed the air from the brake system. This procedure is provided in chapter 45.*

You Should Know *Many ABSs have an amber and a red brake warning light. The amber ABS warning light is illuminated if the ABS computer senses a defect in the ABS electronic system. Specific ABS problems cause the ABS computer to illuminate both the amber and red brake warning lights. The red brake warning light is also illuminated by unequal pressure between the two halves of the master cylinder or a parking brake application.*

If the brake warning light bulb must be replaced, follow the same procedure and precautions that were explained previously regarding oil pressure indicator light service.

# GAUGES AND RELATED CIRCUIT MAINTENANCE, DIAGNOSIS, AND SERVICE

Gauges should be inspected for proper operation when a vehicle is in the shop for minor service. If all the gauges are inoperative or erratic, the instrument voltage limiter should be tested. When only one gauge is inoperative or erratic, concentrate the diagnosis on that gauge and its related circuit. Prior to gauge diagnosis, always inspect the wiring harness and electrical connections on the gauge-sending units. Repair any damaged wiring or corroded, loose connections.

## Gauge Diagnosis

When all the gauges are inoperative or erratic, use a voltmeter to test the instrument voltage limiter. Connect a voltmeter from the battery side of the limiter to ground. With the ignition switch on, the input voltage to the limiter should be 12 volts or more. If this voltage is zero, test the gauge fuse. When the input voltage to the limiter is satisfactory, connect the voltmeter from the gauge side of the

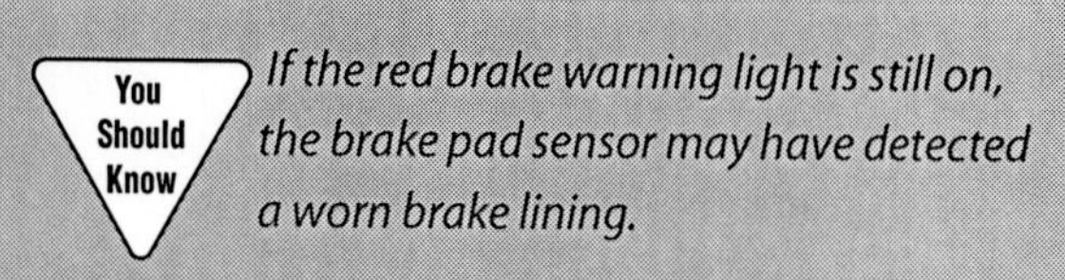

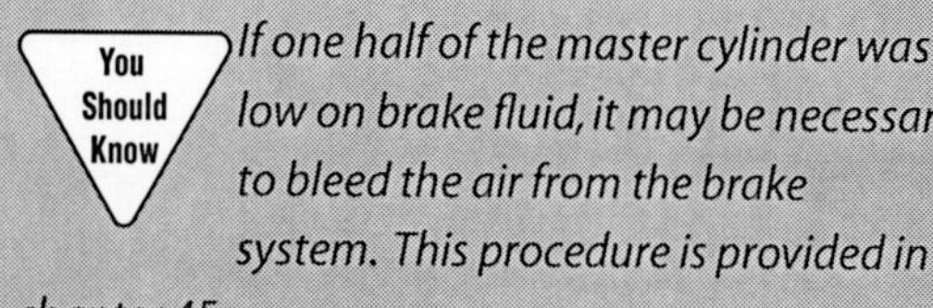

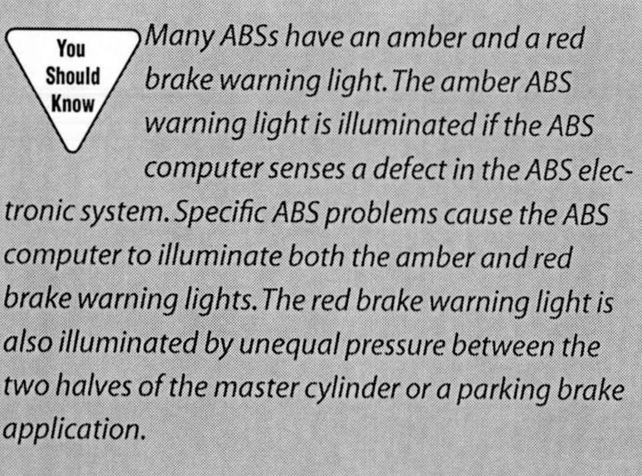

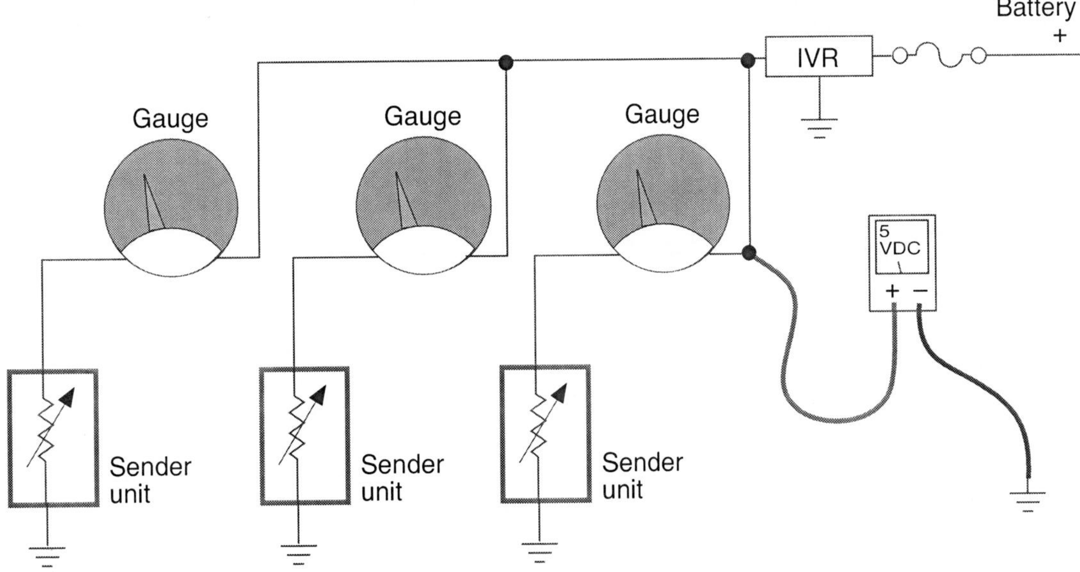


**Figure 4.** Testing the instrument voltage limiter.

limiter to ground **(Figure 4)**. With the ignition switch on, the voltage supplied by the limiter should be a pulsating 5 volts. Replace the limiter if it does not supply the specified voltage.

To test individual gauges, disconnect the wire from the sending unit and connect the gauge tester from the sending unit wire to ground (**Figure 5**). Always connect and use the gauge tester as specified by the equipment manufacturer. With the gauge tester connected from the sending unit wire to ground, turn the ignition switch on and rotate the specified control knob on the tester while observing the gauge operation. The gauge tester contains a variable resistor(s) operated by the control knobs. As the control knob is rotated, the resistance in the tester is varied and the gauge should move from low to high. If the gauge does not move from low to high, test the wire from the gauge to the sending unit. When this wire is satisfactory, replace the gauge.

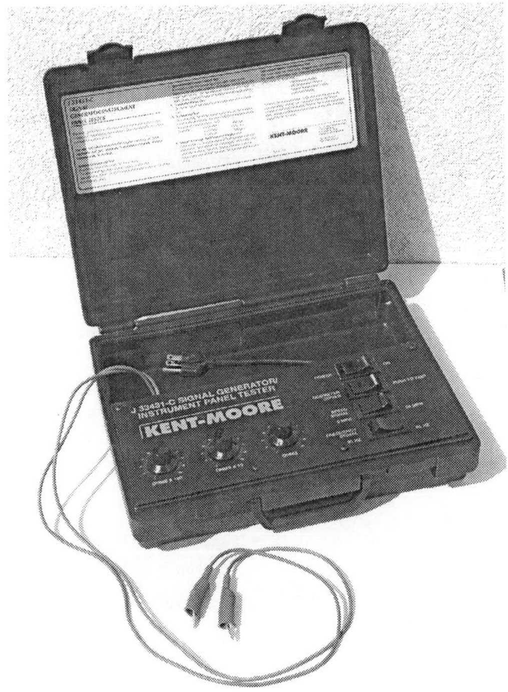

**Figure 5.** A gauge tester.

## Sending Unit Diagnosis

If a sending unit contains a variable resistance, such as a fuel gauge sending unit, connect an ohmmeter from the sending unit terminal to ground on the sending unit case. When the float arm on the fuel gauge sending unit is moved from the low fuel to the high fuel position, the ohmmeter should read the specified ohms resistance. If the sending unit does not provide the specified resistance, replace the sending unit.

**You Should Know** *Most fuel tank sending units have a wire connected from the sending unit case to a ground on the chassis. If this ground wire is damaged, corroded, or had an open circuit, the fuel gauge may be inoperative or erratic. With the ignition switch off, use an ohmmeter to test this wire. It should have very low resistance.*

You Should Know A grounded or open wire between the fuel gauge and the sending unit causes different gauge readings depending on the type of fuel gauge. With a bimetallic-type fuel gauge, a grounded sending unit wire provides a continually high gauge reading and an open circuit in this wire provides a continually low gauge reading. Some balancing coil fuel gauges provide the same reaction to a grounded or open sending unit wire. Other balancing coil fuel gauges provide a high reading with an open sending unit wire and a low reading with a grounded sending unit wire.

Thermistor-type sending units may be tested with an ohmmeter. With the ohmmeter leads connected from the sending unit terminal to the case, the ohmmeter should indicate the specified resistance in relation to the sending unit temperature **(Figure 6)**.

If the temperature gauge sending unit must be replaced, always drain some of the coolant from the cooling system before loosening this unit. If the engine is at normal operating temperature, use caution because the cooling system is pressurized. If a gauge must be replaced, follow the same procedure and precautions that were explained previously regarding oil pressure indicator light service.

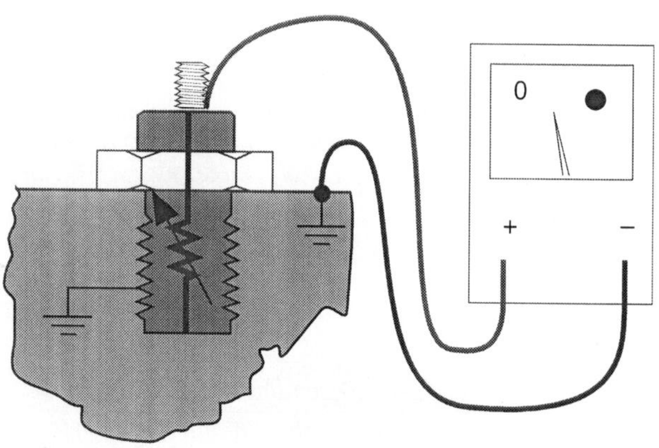


**Figure 6.** Testing a thermistor-type sending unit.

## VOLTMETER AND AMMETER MAINTENANCE, DIAGNOSIS, AND SERVICE

Voltmeters and ammeters usually do not require any maintenance or service. However, the technician should inspect the ammeter or voltmeter operation to detect problems in other circuits. For example, if the technician has just tested the battery and found it to be fully charged but the ammeter indicates a very high charging rate, the alternator voltage may be higher than specified. This condition results in excessive gassing of water from the battery and possibly burned out lights and electrical accessories. This problem should be verified by testing the charging system voltage. If this voltage is higher than specified, the customer should be advised before proceeding with the repairs. When this problem is detected before the battery and electrical accessories are damaged, the customer probably saves some money and possible inconvenience.

## SPEEDOMETER AND ODOMETER MAINTENANCE, DIAGNOSIS, AND SERVICE

Speedometers and odometers require a minimum amount of maintenance. The speedometer cable may require periodic lubrication. A dry speedometer cable provides a rasping sound, especially in cold weather. A dry cable may also cause an erratic speedometer reading. Disconnect the cable from the back of the speedometer and remove the inner cable. When the inner cable cannot be removed by pulling it from the upper end of the cable, it may be necessary to remove the cable retainer from the transmission or transaxle. Inspect the inner cable for kinks and damage. Lubricate the cable with an approved lubricant and reinstall it in the outer casing. If the speedometer and odometer are inoperative, the inner cable is likely broken. If the speedometer is working properly but the odometer is inoperative, it must be replaced.

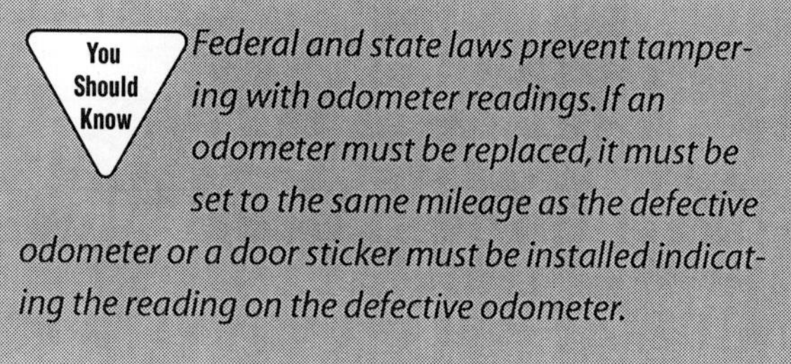

You Should Know *Federal and state laws prevent tampering with odometer readings. If an odometer must be replaced, it must be set to the same mileage as the defective odometer or a door sticker must be installed indicating the reading on the defective odometer.*

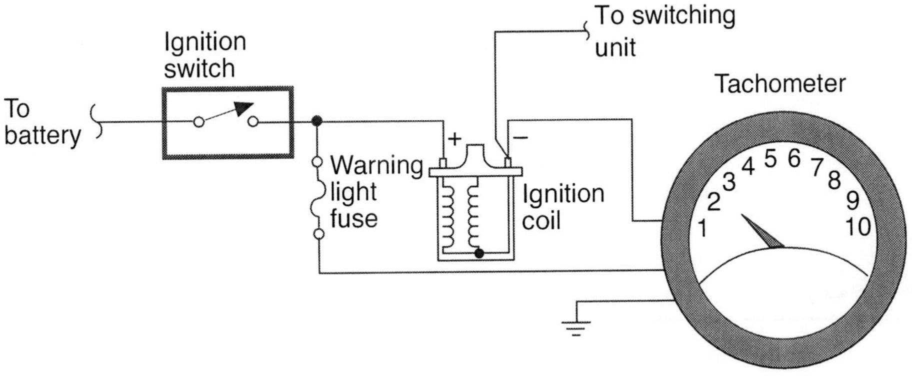


**Figure 7.** A tachometer circuit.

## TACHOMETER MAINTENANCE, DIAGNOSIS, AND SERVICE

If a tachometer is inoperative, inaccurate, or erratic, test the fuse and all the wires connected to the tachometer **(Figure 7)**. Be sure all these wires are in satisfactory condition. There is no possible service on a tachometer. If the fuse and all the wires are satisfactory, replace the tachometer.

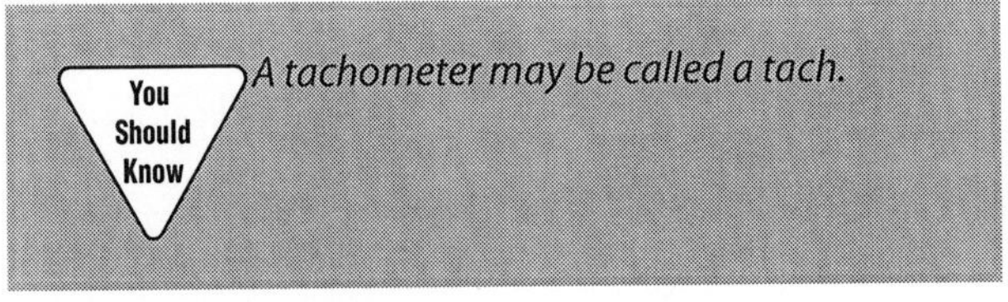


*A tachometer may be called a tach.*

*Cranfield University in the United Kingdom has developed a prototype four-seat car called the Aerocarbon. Rather than using hybrid or fuel cell technology to reduce fuel consumption and emissions, the Aerocarbon uses a very light, strong structure and a low-drag body shape. The body is designed to provide an exceptionally low 0.23 coefficient of drag (Cd). The spaceframe is manufactured from a new, very light, stiff carbon-based material called Coretex and the body panels are made from recyclable plastics. The spaceframe weighs 176 pounds and the body weight is 2,309 pounds, which is approximately half the weight of a vehicle with a steel frame and body. The Aerocarbon is powered by a 3-cylinder, 40 hp, Honda engine. Fuel mileage is 84 mpg with a top speed of 89 mph.*

## Summary

- If the oil pressure indicator light is on with the engine running, the oil level and oil pressure may be excessively low or the oil may be contaminated. The oil may also have the wrong viscosity or the oil pressure switch may be defective.
- If the oil pressure indicator light does not come on with the ignition switch on and the engine not running, the bulb may be burned out, the oil pressure switch may be defective, the electrical circuit may be open, or there is no voltage supply to the light.
- A test gauge may be installed in place of the oil pressure switch to test oil pressure.
- Before attempting to remove an instrument panel, always disconnect the battery negative cable and wait for the time period specified by the vehicle manufacturer.
- If the charge indicator light is on with the engine running, the belt tension and alternator output should be tested.
- If the charge indicator bulb is not on with the ignition on and the engine not running, the bulb may be burned out, the alternator field circuit or external wiring to the bulb may have an open circuit, or the voltage supply to bulb may be zero.

- With the engine running, a glowing charge indicator light may be caused by high resistance in the charging circuit.
- If the red brake warning light is on with the engine running, there may be unequal pressure between the halves of the master cylinder, the parking brake may be on, or the ABS computer may have illuminated this light.
- Gauges may be tested with a gauge tester containing a variable resistor.
- Sending units containing a variable resistor or a thermistor may be tested with an ohmmeter.
- A dry speedometer cable causes a rasping noise, especially when cold.

## Review Questions

1. The oil indicator light is on with the engine running. Technician A says the oil pressure switch may be defective. Technician B says the oil may be contaminated with gasoline. Who is correct?
   A. Technician A
   B. Technician B
   C. Both Technician A and Technician B
   D. Neither Technician A nor Technician B
2. Technician A says a test gauge may be installed in place of the oil pressure switch to test engine oil pressure. Technician B says before removing an instrument panel, always disconnect the battery positive cable. Who is correct?
   A. Technician A
   B. Technician B
   C. Both Technician A and Technician B
   D. Neither Technician A nor Technician B
3. The charge indicator light is not on with the ignition switch on and the engine not running. Technician A says the alternator field circuit may be open. Technician B says the fuse link in the alternator battery wire may have an open circuit. Who is correct?
   A. Technician A
   B. Technician B
   C. Both Technician A and Technician B
   D. Neither Technician A nor Technician B
4. All of the gauges read lower than normal on a vehicle with an instrument voltage limiter. Technician A says the voltage limiter may not be grounded on the instrument panel. Technician B says the wire to the fuel gauge sending unit may be grounded. Who is correct?
   A. Technician A
   B. Technician B
   C. Both Technician A and Technician B
   D. Neither Technician A nor Technician B
5. An oil indicator light does not come on when the ignition switch is turned on and the engine is not running. When the wire connected to the oil sending unit is grounded with the ignition switch on, the oil indicator light is illuminated. The most likely cause of this problem is:
   A. an open circuit in the oil sending unit.
   B. an open circuit in the wire between the oil indicator light and the oil sending unit.
   C. an open circuit between the ignition switch and the oil indicator light.
   D. an open circuit in the indicator light fuse.
6. All of these defects may be a cause of very low engine oil pressure *except*:
   A. worn main bearings.
   B. worn camshaft bearings.
   C. an air leak in the oil pump pickup.
   D. a leaking rear main bearing seal.
7. When diagnosing an instrument panel containing an instrument voltage limiter, all the gauges are tested and found to be defective. Technician A says the instrument voltage limiter may be defective. Technician B says the ignition switch may have been turned on with the instrument panel not grounded. Who is correct?
   A. Technician A
   B. Technician B
   C. Both Technician A and Technician B
   D. Neither Technician A nor Technician B
8. A charge indicator light is illuminated dimly with the engine running. The most likely cause of the problem is:
   A. excessive resistance in the charge indicator bulb.
   B. excessive resistance between the ignition switch and the charge indicator bulb.
   C. high resistance between the alternator and the battery ground.
   D. a defective voltage regulator, resulting in low alternator voltage.
9. With the ignition switch on, a typical voltage output from an instrument voltage limiter is a pulsating:
   A. 1 volt to 2 volts.
   B. 3 volts to 3.5 volts.
   C. 5 volts to 7 volts.
   D. 9 volts to 12 volts.
10. When a fuel gauge sending unit is removed from the fuel tank, it should be tested with:
    A. a voltmeter.
    B. an ohmmeter.
    C. an ammeter.
    D. a graphing voltmeter.
11. On a vehicle with ABS, the red brake warning light is illuminated with the engine running. List three causes of this problem.
    A. ________________
    B. ________________
    C. ________________
12. A gauge tester contains a ________________ ________________.
13. A fuel gauge sending unit may be tested with a(n) ________________.

14. Federal and state laws prohibit____________________ with odometer readings.
15. Explain the procedure to diagnose an oil pressure indicator light that is on with the engine running.
16. Describe the cause of a charge indicator light glowing with the engine running.
17. Explain the results of a grounded fuel gauge sending unit wire on a vehicle with a bimetallic-type fuel gauge.
18. Describe the results of excessively high charging system voltage.

# Engine Electrical Systems



## SECTION OBJECTIVES

After you have read, studied, and practiced the contents of this section, you should be able to:

- Describe battery design, including battery plate and cell groups.
- Explain battery operation during the discharge and charge cycles.
- Explain two common battery ratings.
- Describe basic starting motor armature and field coil design.
- Describe the operation of a starter solenoid.
- Explain the operation of an overrunning clutch starter drive while engaging and disengaging.
- Perform basic battery maintenance and tests.
- Perform battery charge procedures.
- Diagnose and test starting motors.
- Perform starting motor on-vehicle electrical tests.
- Inspect alternator belts and adjust belt tension.
- Diagnose charging system problems.
- Describe DI and EI ignition system operation.
- Describe briefly how the PCM controls spark advance in a DI system.
- Describe fouled spark plug conditions.
- Perform a DI ignition no-start diagnosis.
- Diagnose and test pickup coils.
- Time a distributor to an engine and adjust ignition timing.
- Test crankshaft and camshaft position sensors.

*At present there are 839,689 automotive technicians in the United States. Industry experts estimate that during the next 10 years this number will increase by 151,000. Currently there is much discussion in the automotive service industry regarding the shortage of technicians and the importance of technician training. With all the new technology being introduced to the automotive industry each year, training is extremely important. When discussing these problems, an industry spokesperson recently stated, "Increased training leads to increased retention, increased respect, and an increase in recruitment."*

# Chapter 22 Battery and Starting Systems

## Introduction

Technicians must understand battery operation to comprehend how the battery operates in the charging circuit. For example, if the battery is gassing excessively, the charging voltage must be too high. Technicians also need to understand that only water is gassed from the battery during the charging process. Therefore, only water should be added to the battery cells. Technicians also need to understand starting motor design and operation so they can accurately diagnose starting motor and system defects.

### BATTERY DESIGN

The battery case holds and protects the battery components and electrolyte **(Figure 1)**. Separating walls in the case form a separate reservoir for each cell. Some battery cases are made from translucent plastic and the electrolyte level is visible through the case which allows the electrolyte level to be checked without removing the cell vent caps. Some cases have sediment spaces in the bottom of the case which provides spaces for material shed off the plates.

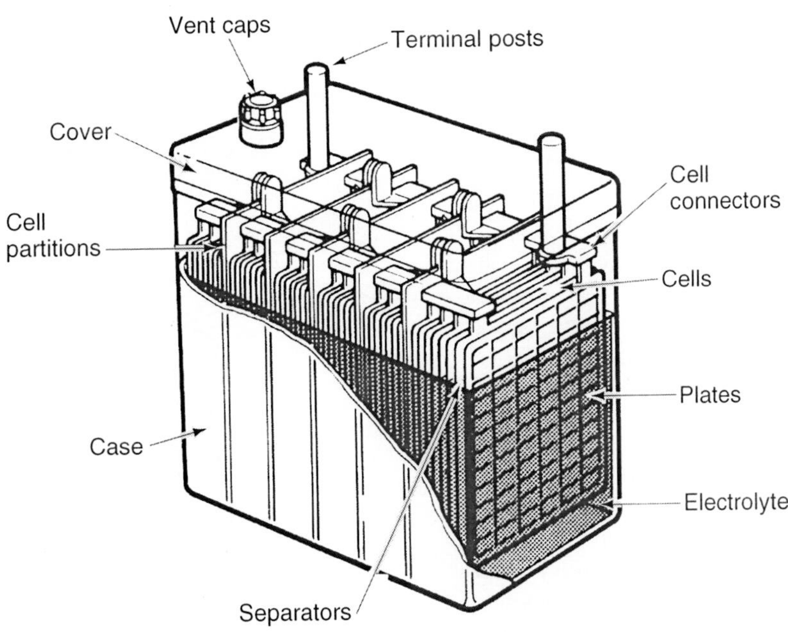


**Figure 1.** Battery internal design.

## Cover

The battery cover is permanently sealed to the top of the case. The cover has openings for the terminal posts and vent holes for the venting of gases. Some batteries have the terminals extending through the side of the case and other batteries have terminals in the sides and top. Some covers have removable vent caps that may be removed to add water or test the electrolyte. These vent caps may be individual, strip type, or box type.

## Plates

Battery plates contain a grid or coarse screen. In some batteries, the plate grids are made from lead mixed with antimony, which stiffens the lead. **Battery plates** are a coarse grid on which the active plate material is pasted.

Many batteries manufactured today have plate grids made from lead and calcium. In some plate grids, the horizontal and vertical grids are positioned at right angles to each other. Other plate grids have vertical support bars mounted at an angle to the other vertical grids. The active plate materials are pasted on the grids. The material on the positive grids is lead peroxide and the negative plate material is sponge lead. A tab on the top of each plate grid allows the plate to be connected to the cell connector.

## Cell Group

A **cell group** contains a number of alternately spaced negative and positive plates. A negative plate is positioned on the outside of each cell group; thus, each cell group has an odd number of plates. For example, a cell group may have five positive plates and six negative plates.

Thin porous separators are positioned between the negative and positive plates. Separators are positioned between each pair of battery plates to keep these plates from touching each other. **Separators** are made from fiberglass sheets or plastic envelopes that fit between or over the plates. The separators prevent the plates from touching each other and shorting together electrically.

Cell connectors are lead burned to the positive and negative plate tabs in each cell group. These cell connectors also extend through the partitions between the cells and the negative plates are connected to the positive plates in the next cell. The negative plates in one end cell and the positive plates in the opposite end cell are connected to terminal posts that extend through the top or sides of the battery.

## Terminal Posts

On top-terminal batteries, large round terminal posts extend through the top of the battery. The positive post is larger than the negative post. These terminal posts are sealed into the battery cover to prevent electrolyte leakage around the posts. POS or NEG is usually stamped on the battery cover beside the proper terminal post. The battery cable ends are clamped to the terminal posts. On side-terminal batteries, the battery terminals are threaded and the cables are bolted to the terminals.

## Electrolyte

The **electrolyte** is a mixture of approximately 64 percent water ($H_2O$) and 36 percent sulfuric acid ($H_2SO_4$). The electrolyte reacts with the plate materials to produce voltage.

The electrolyte must contain enough sulfuric acid so it does not freeze in extremely cold temperatures. If the battery contained a higher percentage of sulfuric acid, this acid would attach and deteriorate the plate grids too quickly.

**You Should Know** *Sulfuric acid is a very strong corrosive liquid. It is very damaging to human eyes or skin. Always wear face and eye protection and protective gloves and clothing when handling batteries or electrolyte. If electrolyte contacts your skin or eyes, immediately flush with clean water and obtain medical attention. Battery electrolyte is very damaging to vehicle paint surfaces, upholstery, and clothing. Never allow electrolyte to contact any of these items.*

# BATTERY OPERATION

If one negative plate and one positive plate are placed in an electrolyte solution, a chemical reaction takes place that causes about 2.1 volts difference between the plates. Because a 12-volt battery contains six cells, the total voltage of a fully charged battery is 12.6 volts. If an electrical resistance, such as a light bulb, is connected to the two battery plates, the higher voltage on one plate forces current through the bulb to the opposite plate with the lower voltage **(Figure 2)**. This action is

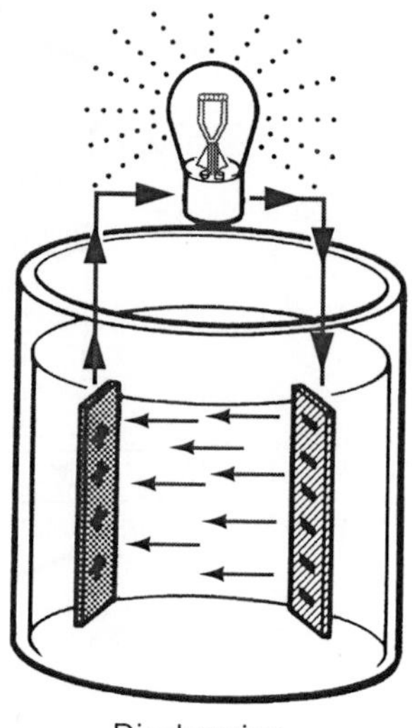


**Figure 2.** Current flows from one battery plate to the other when a conductor with some resistance is connected between the plates.

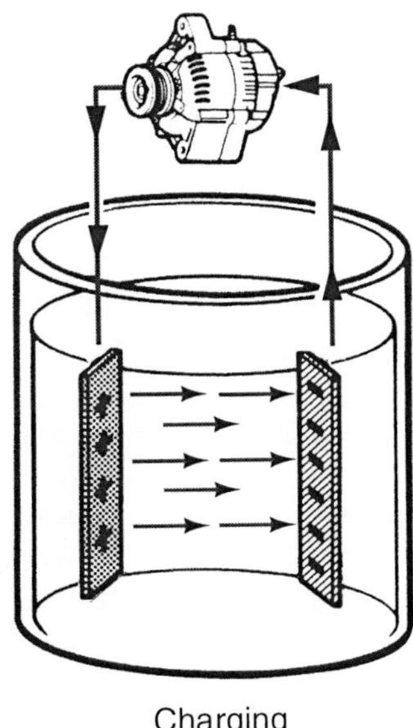


Figure 3. When the engine is running, current flows from the alternator through the battery plates.

called battery discharging. A battery cell with two plates will not supply much current flow. When more plates are added, the current capability increases.

If a voltage source, such as an alternator, is connected to the battery plates, the alternator forces current flow into one plate and out of the other battery plate **(Figure 3)**. This action is called battery charging. The voltage of the alternator must be higher than the battery voltage to allow the alternator to force current through the battery.

In a fully charged battery, the positive plate material is lead peroxide ($PbO_2$) and the negative plate material is lead (Pb) **(Figure 4)**. When an electrical load is connected across the battery terminals, the voltage difference between the plate materials forces current through the electrical load. As the battery discharges, the $H_2SO_4$ breaks up in the electrolyte and the sulfate ($SO_4$) goes to both plates, where it

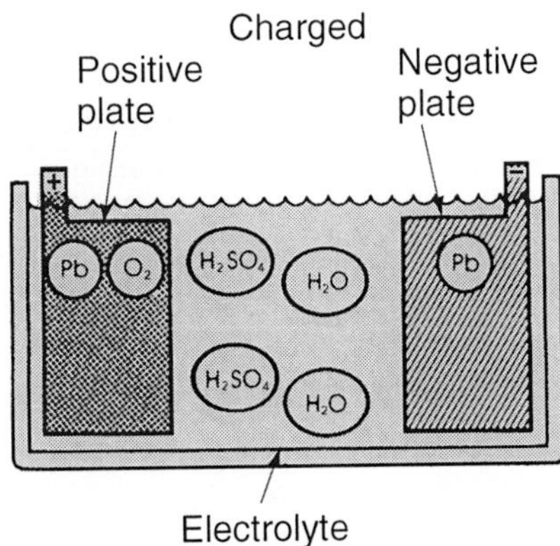


Figure 4. Plate materials and electrolyte content in a fully charged battery.

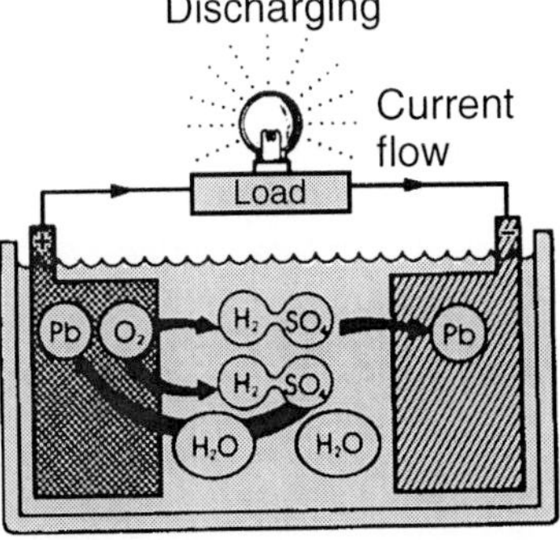


Figure 5. Chemical action in a battery while discharging.

joins with the Pb to form lead sulfate ($PbSO_4$) on both plates. The $O_2$ on the positive plates joins with the hydrogen (H) in the electrolyte to form ($H_2O$) **(Figure 5)**. As the battery discharges, the percentage of water in the electrolyte increases. In a discharged battery, both sets of plates are coated with $PbSO_4$ and the electrolyte contains a high percentage of $H_2O$ **(Figure 6)**.

When a charging source, such as the alternator, or a battery charger is charging the battery, the $SO_4$ comes off both plates and joins with the H in the electrolyte to form $H_2SO_4$ **(Figure 7)**. The $H_2O$ breaks up and the O goes to the positive plates, where it joins with the Pb to form lead peroxide ($PbO_2$).

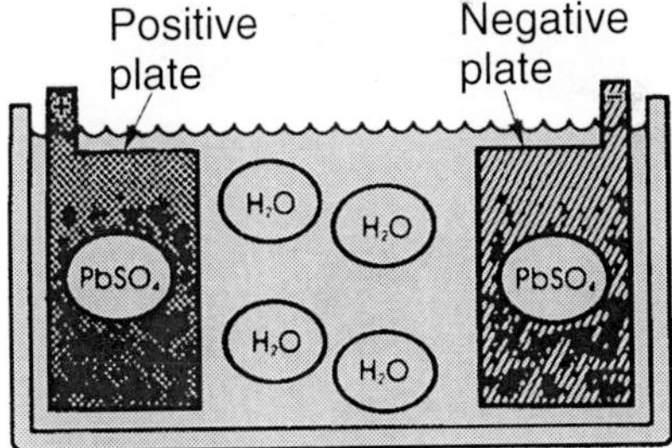


Figure 6. Plate materials and electrolyte content in a discharged battery.

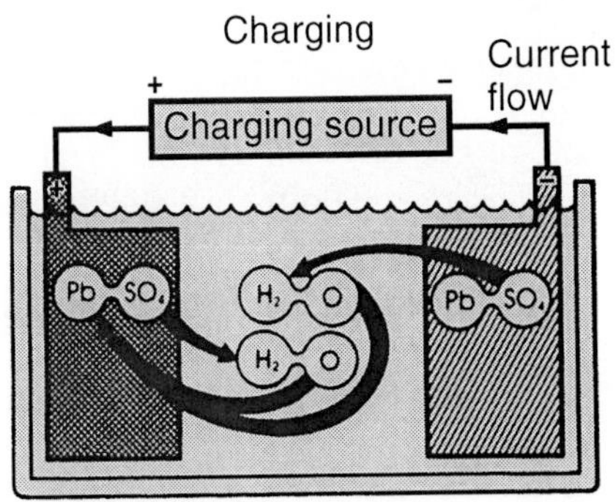


Figure 7. Chemical action in a battery while charging.

When a battery is charging, hydrogen gas escapes at the negative plates and oxygen gas is given off at the positive plates. When combined, these two gases form water ($H_2O$). The $SO_4$ is always on the plates or in the electrolyte. Because $H_2O$ is the only liquid gassed from a battery, this is the only liquid that should be added to a battery.

## LOW-MAINTENANCE BATTERIES

Low-maintenance batteries are designed to minimize heat and water loss. The filler caps may be removed to add water to the cells. In many batteries, the electrolyte level should be at a split-ring round indicator below the filler cap in each cell. The water level must be above the battery plates. The average interval for adding water to a battery is approximately 15,000 miles. If the battery requires frequent addition of water, the charging system voltage may be too high.

## MAINTENANCE-FREE BATTERIES

Maintenance-free batteries are designed to reduce internal heat and water loss. The cells in these batteries are vented but the filler caps cannot be removed and it is not possible to add water to the battery. Most maintenance-free batteries have a built-in hydrometer in the top of the battery to indicate battery state of charge. If this hydrometer indicates a green dot, the battery is sufficiently charged for test purposes. When the hydrometer appears dark, the battery charge is below 65 percent, and a clear hydrometer indicates a low electrolyte level **(Figure 8)**.

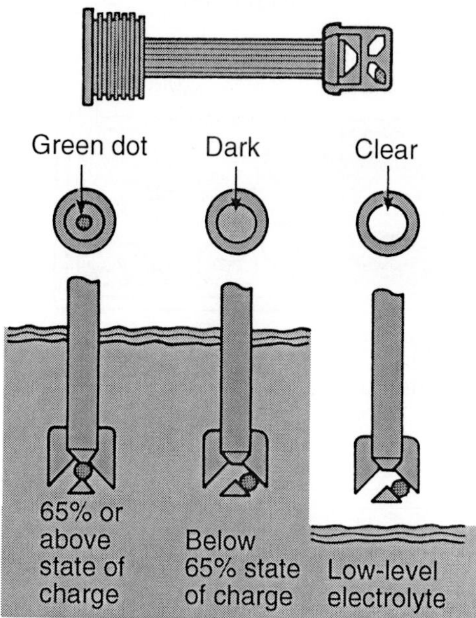


**Figure 8.** A hydrometer in the cover of a maintenance-free battery.

## BATTERY RATINGS

The cold cranking ampere (CCA) rating indicates the amperes that a battery will deliver at 0°F (−18°C) for 30 seconds while maintaining a voltage above 1.2 volts per cell or 7.2 volts for the complete battery. The CCA rating for most automotive batteries is between 350 and 800 amperes.

### Reserve Capacity Rating

The reserve capacity rating is the length of time in minutes that a fully charged battery at 80°F (27°C) will deliver 25 amperes with the voltage remaining above 1.75 volts per cell or 10.5 volts for the complete battery. Many reserve capacity ratings are between 55 and 125 minutes.

## STARTING MOTOR ELECTROMAGNETIC PRINCIPLES

Starting motors are designed so they have very low resistance and draw a very high current flow, which produces a very high torque for short time periods. The field coils in a starting motor are made of heavy copper wire and are wound on steel pole shoes that are bolted to the starter case. One end of the field windings is connected to the positive battery terminal and the other end of these windings is connected to an insulated brush in the starting motor. A copper commutator bar is soldered to each end of the armature winding. This winding is positioned between the pole shoes. The commutator bars are insulated from each other and the insulated brush contacts one commutator bar. A ground brush completes the circuit from the other commutator bar to ground. Current in the starting circuit flows from the positive battery terminal through the field coils and armature winding and returns through the ground brush to the negative battery terminal. Because the field coils are connected in series with each other and with the armature winding, the same amount of current must flow through the field coils and the armature winding. A high current flows through the field coils and armature winding because these heavy windings have very low resistance. The high current flow through the armature winding creates strong magnetic fields around the sides of this winding. Since the current flows in opposite directions in the two sides of the armature winding, the magnetic fields surrounding the sides of this winding flow in opposite directions.

The high current flow through the field coils creates a strong magnetic field between the pole shoes and the armature winding is positioned in this magnetic field

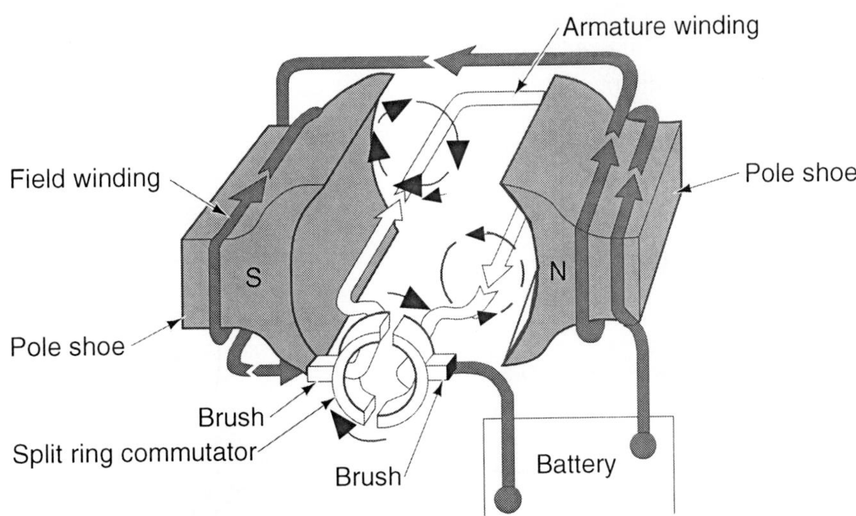


**Figure 9.** Starting motor field coils and armature winding.

**(Figure 9)**. The magnetic field around the upper side of the armature winding is moving in a counterclockwise direction. On the top side of this winding, the magnetic field around the armature winding is moving in the same direction as the field between the pole shoes. Therefore, these magnetic fields join together and a very strong magnetic force is created above this side of the armature winding **(Figure 10)**. Below the upper side of the armature winding, the magnetic field around the armature winding is moving in the opposite direction to that of the magnetic field between the pole shoes. Magnetic fields moving in opposite directions in the same space cancel each other. Therefore, a very weak magnetic force is created below the upper side of the armature winding.

The magnetic field around the lower side of the armature winding is moving in a clockwise direction. Because the magnetic field between the pole shoes is moving in the same direction as the magnetic field around the armature winding below this winding, a very strong magnetic force is created under the winding. The magnetic force above the lower side of the winding is cancelled because the magnetic lines of force in this area are moving in opposite directions. The very strong magnetic forces above and below the sides of the armature winding cause the armature winding to rotate because this winding is mounted on a steel shaft and is supported on bushings.

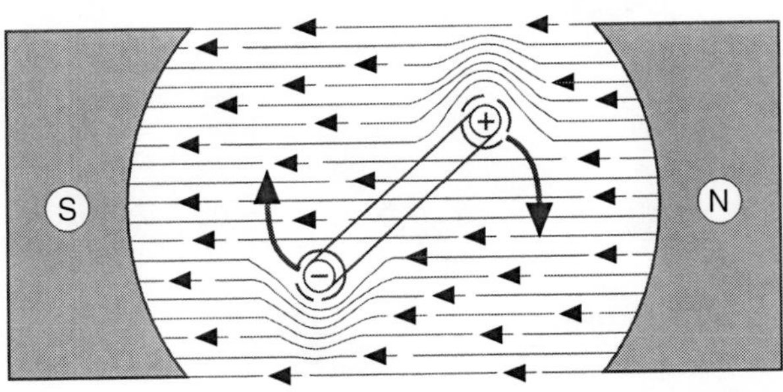


**Figure 10.** Reaction between the magnetic field between the pole shoes and the magnetic fields around the armature windings causes armature rotation.

## STARTER ARMATURE DESIGN

An armature has many windings mounted in a laminated metal core that is pressed onto the shaft **(Figure 11)**. Each winding is insulated from the other windings and from the mounting slots in the metal core. The ends of the windings are soldered to the commutator bars that are insulated from each other and from the armature shaft. Some armature windings are always in the magnetic field between the pole shoes to provide continuous armature rotation.

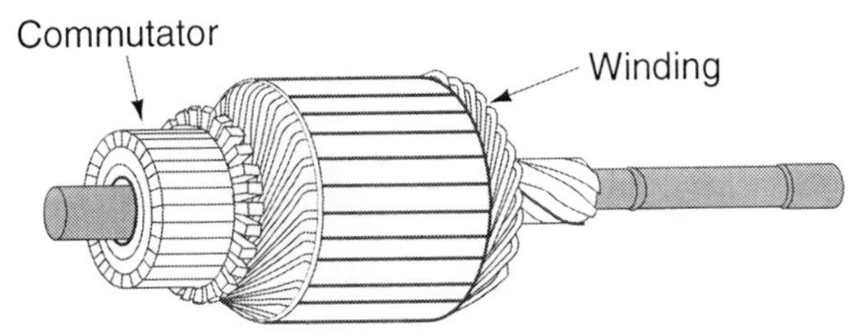


**Figure 11.** A starter armature.

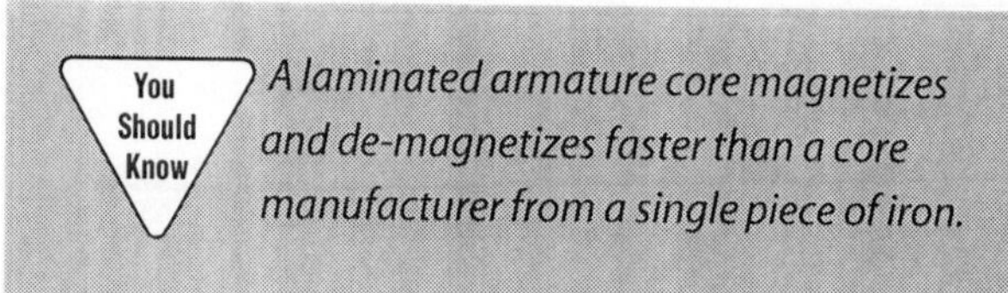

You Should Know
A laminated armature core magnetizes and de-magnetizes faster than a core manufacturer from a single piece of iron.


## STARTER FIELD COIL DESIGN

Some automotive starting motors have four field coils connected in series **(Figure 12)**. All four of these windings contain heavy copper wire with very low resistance. Other starting motors have three series field coils and one parallel or shunt field coil **(Figure 13)**. The shunt field coil has many turns of fine wire. While cranking the engine, the shunt coil has the same magnetic strength as the series field coils. The shunt coil creates a strong magnetic field from a low current flow through many turns of wire, whereas the series field coils develop their strong magnetic fields from a high current flow through a few turns of wire.

When the engine starts, the starter turns freely for a few seconds until the starter drive is pulled out of mesh. This is called an overrunning starter condition. Under this condition, the armature could rotate at a very high speed. When the starter overruns, the battery voltage increases because the high starter current load is removed from the battery. When this action occurs, the current flow through the shunt coil increases and strengthens the magnetic strength around this coil. The strong magnetic field of the shunt coil creates an induced opposing voltage in the armature windings that opposes the battery voltage and reduces the current flow through the series field coils and armature. This action reduces starter armature speed and protects the armature. If the armature rotates at excessive speed, the heavy armature windings may be thrown from their armature slots, resulting in severe armature damage.

Some starting motors have permanent magnets in place of wound field coils **(Figure 14)**. In this type of start-

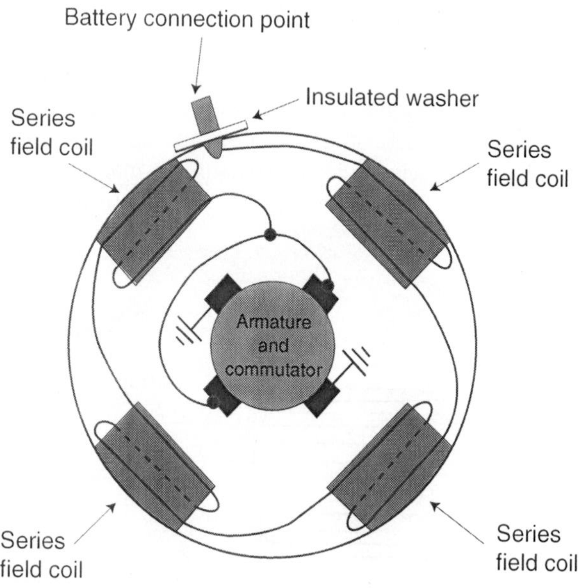


**Figure 12.** Series field coils.

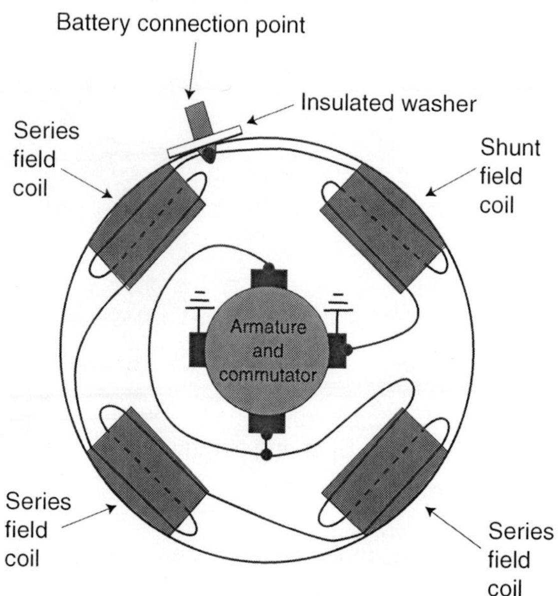


**Figure 13.** Series-shunt field coils.

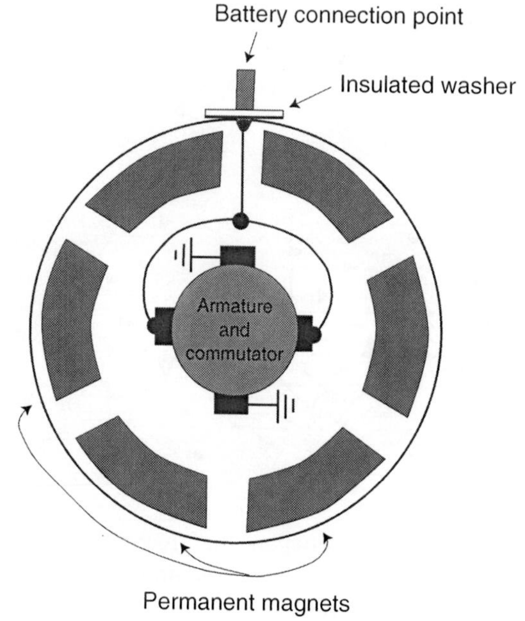


**Figure 14.** Permanent magnet field coils.

ing motor, electrical defects in the wound field coils are eliminated. A permanent magnet starting motor is also more compact compared to other starting motors. The permanent magnets are manufactured from an alloy of boron, neodymium, and iron. These permanent magnets have much greater strength than ordinary steel magnets. The permanent magnets are attached to the starting motor housing. These starters usually have more permanent magnets compared to the number of field coils in a conventional starter because no space is required for the windings. A permanent magnet motor operates in the same way as a conventional starting motor.

## SOLENOIDS

Many starting motors have a solenoid that completes the electric circuit between the battery and starting motor. The solenoid also pulls the starter drive into mesh with the flywheel ring gear. The **solenoid** is an electro-mechanical device that moves the starter drive into mesh with the flywheel ring gear and opens and closes the electrical circuit between the battery and the starting motor.

The solenoid contains a heavy pull-in coil connected from the solenoid terminal to the main starting motor terminal. The finer hold-in coil is connected from the solenoid 50 terminal to ground on the solenoid case. The solenoid terminal is connected to the start terminal on the ignition switch.

When the ignition switch is turned to the start position, current flows from this switch to terminal 50 on the starter solenoid. From this terminal, the current flows through a heavy pull winding to the main starting motor terminal C. This current flows through the series field coils and the armature windings to ground, but there is not enough current to cause armature rotation. A small amount of current also flows through the hold-in winding to ground on the solenoid case. The combined magnetic fields of the two solenoid windings attracts the solenoid plunger and moves this plunger ahead **(Figure 15)**. Forward plunger movement pulls the starter drive into mesh with the flywheel ring gear and further plunger movement pushes the solenoid disk against the two large solenoid terminals 30 and C. When this contact is completed, a very high current flows from the battery positive terminal through the battery cable and solenoid disk and terminals into the starting motor. Under this condition, the starting motor begins cranking the engine. Once the solenoid plunger is pulled forward, current no longer flows through the pull-in winding because equal voltage is supplied through the solenoid disk to the other end of the pull-in

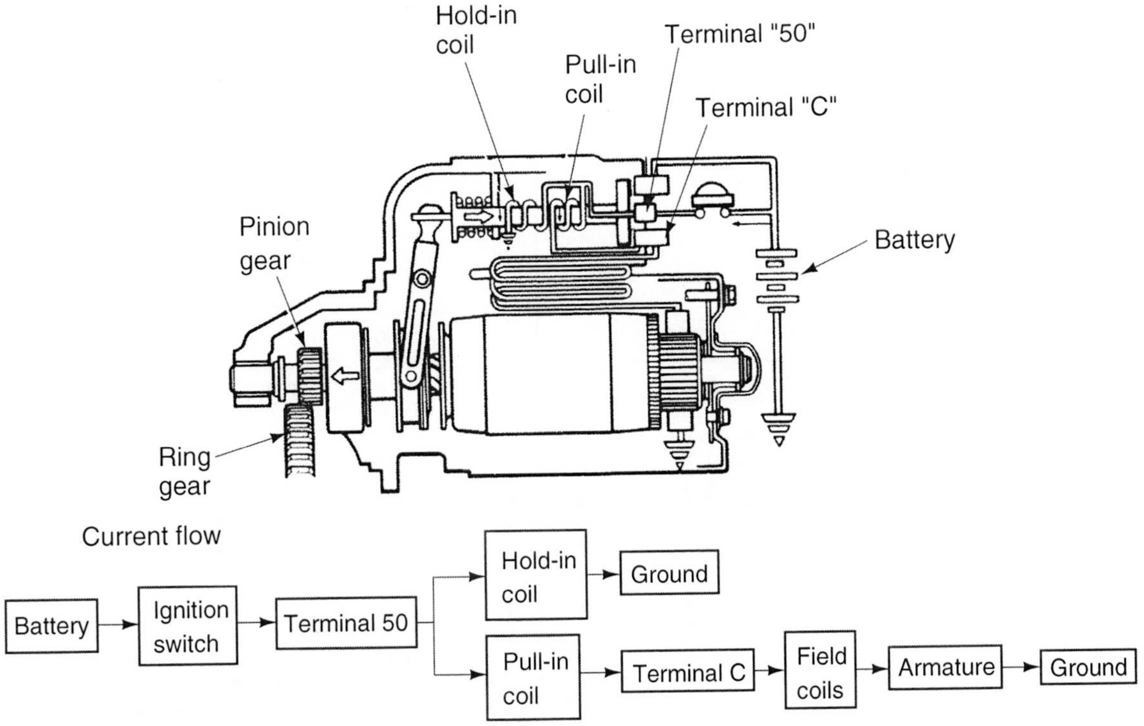


Figure 15. Solenoid current flow while engaging.

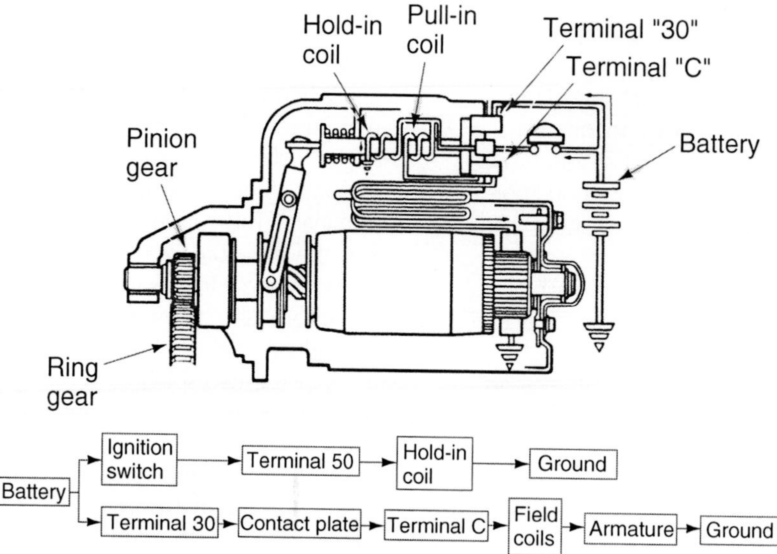


**Figure 16.** Solenoid current flow with starting motor engaged.

winding **(Figure 16)**. Under this condition, the hold-in coil magnetic field is strong enough to hold the plunger in the engaged position.

When the engine starts, the driver releases the ignition switch to the on position. Under this condition, current no longer flows through the ignition switch to the solenoid 50 terminal and there is no magnetic field around the hold-in coil. When this action occurs, a strong plunger return spring pushes the disk away from the solenoid terminals and current flow to the starting motor is stopped. The plunger return spring pulls the drive out of mesh with the flywheel ring gear and returns the plunger to its disengaged position.

## MAGNETIC SWITCHES

Some starting motors have a magnetic switch that completes the electrical circuit between the battery and the starting motor. Many of these starting motors have a moveable pole shoe that pulls the starter drive into mesh with the flywheel ring gear **(Figure 17)**. The magnetic

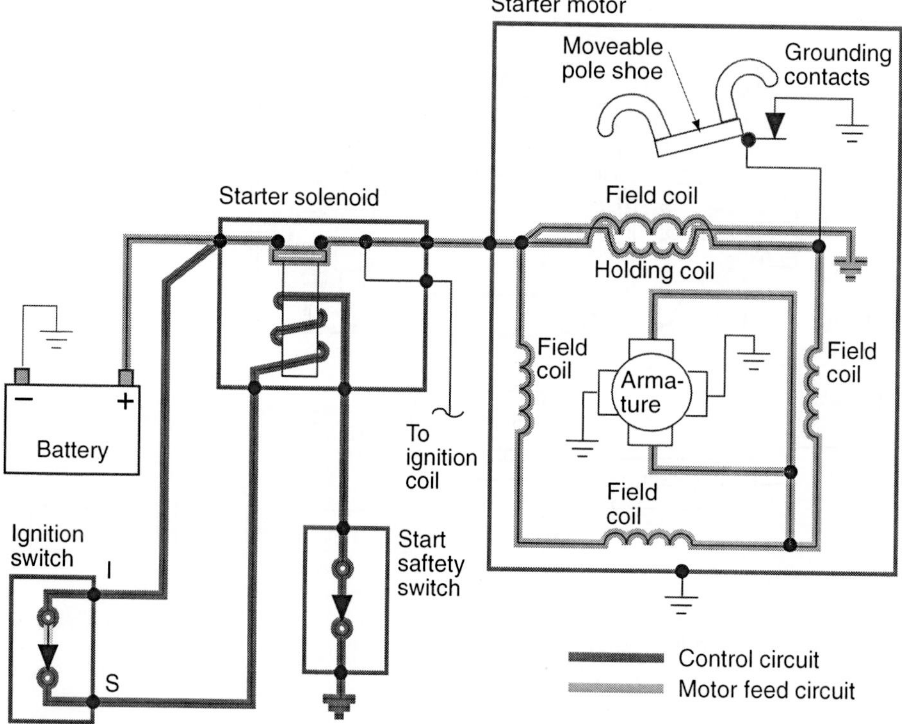


**Figure 17.** A magnetic switch circuit.

switch contains a high-resistance winding. When the ignition switch is turned to the start position, current flows through the ignition switch to the magnetic switch terminal. Current then flows from this terminal through the magnetic switch winding and the neutral safety switch to ground. This current flow creates a magnetic field around the magnetic switch winding, which attracts the switch plunger. Plunger movement forces the disk against the magnetic switch terminals connected to the positive battery cable and the starting motor. A very high current now flows through the magnetic switch disk to the starting motor.

A shunt field coil and a holding coil surround the moveable pole shoe. The shunt field coil contains many turns of fine, high-resistance wire. The holding coil is grounded through a set of contacts on top of the starting motor and this coil is also connected in series with one of the other field coils.

When the magnetic switch closes, most of the current flows through this winding and the contacts to ground because the holding coil is the path of least resistance. This high current flow creates a very strong magnetic field and pulls the moveable pole shoe downward. A fork on the back of this pole shoe moves the starter drive into mesh with the flywheel ring gear. When the pole shoe is fully downward, it pushes the contacts open. Under this condition, current cannot flow through the holding coil and the contacts to ground. Current now flows through the series field coils and armature and the starter begins to crank the engine. The magnetic field around the shunt coil is strong enough to hold the moveable pole shoe downward. When the driver releases the ignition switch from the start to the on position, current flow no longer flows through the magnetic switch winding. A spring pushes the disk away from the magnetic switch terminals and a spring on the moveable pole shoe pushes the pole shoe upward and pulls the starter drive out of mesh with the flywheel ring gear.

The neutral safety switch is mounted on the transmission linkage or in the transmission. The neutral safety switch is closed when the gear selector is in PARK or NEUTRAL, but this switch is open in other gear selector positions. Therefore, the gear selector must be in PARK or NEUTRAL to provide a ground on the magnetic switch winding and allow the starting motor to operate.

**You Should Know** *Some neutral safety switches are mounted on top of the steering column under the dash. Vehicles with a manual transmission usually have a clutch switch in place of the neutral safety switch.*

## STARTER DRIVES

The overrunning clutch is the most common type of starter drive. The starter drive housing is mounted on armature shaft splines. The **starter drive** connects and disconnects the armature shaft and the flywheel ring gear.

The starter drive housing contains four spring-loaded rollers in tapered grooves. The drive gear has a machined shoulder on the rear part of the gear. This machined shoulder is in the center of the rollers. When the drive is engaged with the flywheel ring gear and the armature begins to rotate, the rollers are turned so they move to the narrow end of the tapered grooves. Under this condition, rollers are jammed between the tapered grooves and the machined shoulder on the gear so the drive housing and the gear must rotate as a unit **(Figure 18)**. Therefore, the armature shaft rotates the solenoid housing and gear to crank the engine.

When the engine starts, the flywheel ring gear momentarily rotates the drive gear faster than the drive housing. This action rotates the rollers so they move into

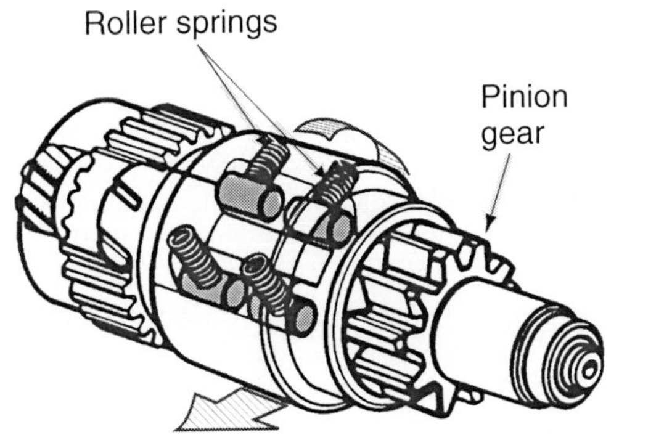


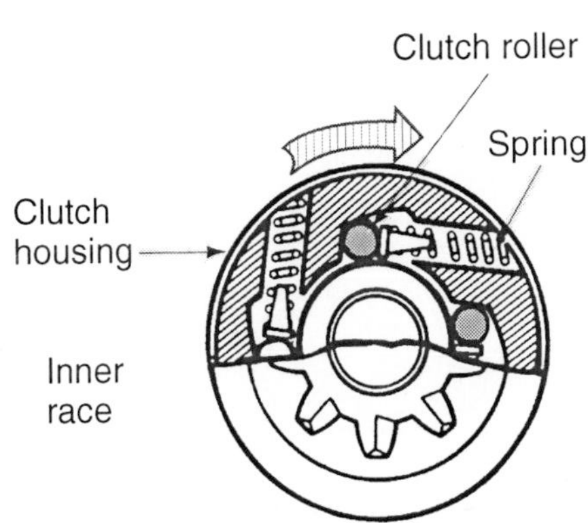


**Figure 18.** Starter drive operation while cranking the engine.

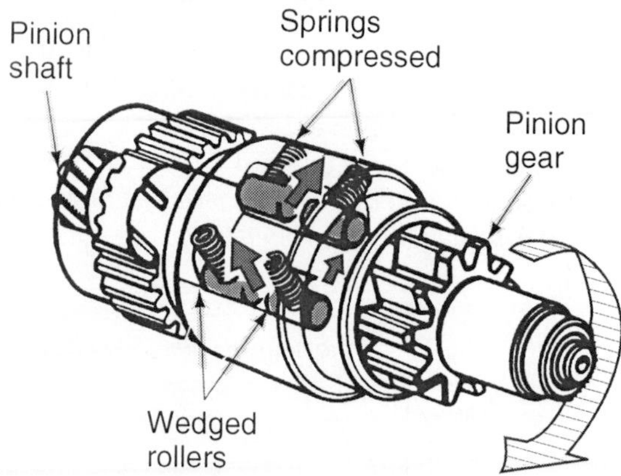


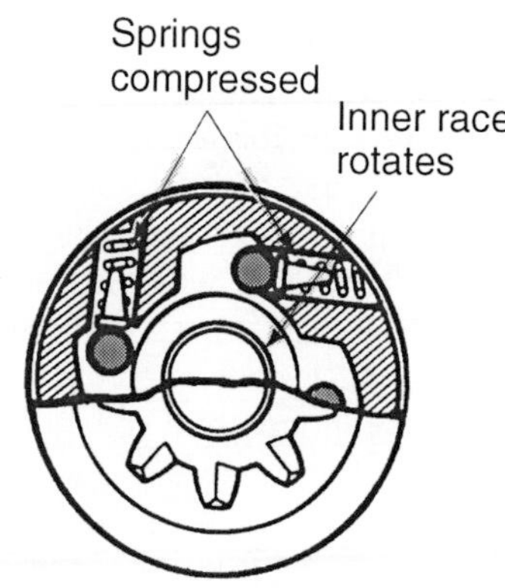


**Figure 19.** Starter drive operation while overrunning.

the wide end of the tapered grooves. Under this condition, the drive gear free-wheels so it does not turn the armature at high speed and damage the armature **(Figure 19)**.

You Should Know *The ratio between the starter drive gear and the flywheel ring gear is usually between 15:1 and 20:1. If this ratio is 20:1 and the drive is allowed to turn the armature momentarily when the engine starts, an engine speed of 1,000 rpm would rotate the armature at 20,000 rpm. This speed would throw the armature windings from their slots and destroy the starting motor.*

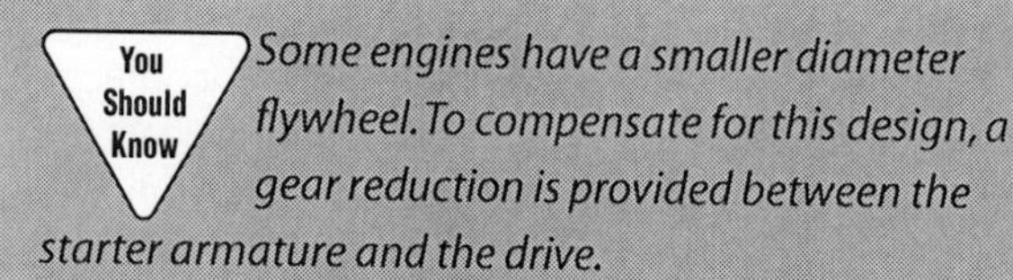
You Should Know *Some engines have a smaller diameter flywheel. To compensate for this design, a gear reduction is provided between the starter armature and the drive.*

## Summary

- The battery case contains a separate compartment for each battery cell.
- Battery plate grids are manufactured from lead and antimony or lead and calcium.
- The active material on positive battery plates is lead peroxide ($PbO_2$).
- The active material on negative battery plates is sponge lead (Pb).
- A battery cell always contains an odd number of plates.
- Separators made from fiberglass are positioned between each pair of battery plates.
- The battery cells are connected in series so the voltage of the cells adds together.
- Battery electrolyte contains about 64 percent water and 36 percent sulfuric acid.
- When a battery is charging, hydrogen is given off at the negative plates and oxygen is given off at the positive plates.
- The most common battery ratings are the cold cranking ampere (CCA) rating and the reserve capacity rating.
- A starting motor develops torque from the interaction of the magnetic field between the pole shoes and the magnetic fields around the armature windings.
- A starting motor has very low resistance and draws a very high current flow.
- A starting motor is designed to develop a very high torque for a short time period.

- Some starting motors have permanent magnet fields in place of field coil windings.
- A starter solenoid moves the starter drive into mesh with the flywheel ring gear and completes the electrical circuit between the battery and the starting motor.
- A magnetic switch is used in some starting circuits to complete the circuit between the battery and the starting motor.
- A starter drive connects and disconnects the armature with the flywheel ring gear.

## Review Questions

1. Technician A says in a fully charged battery, the material on the positive plates is lead peroxide. Technician B says in a discharged battery, the material on the negative plates is sponge lead. Who is correct?
   A. Technician A
   B. Technician B
   C. Both Technician A and Technician B
   D. Neither Technician A nor Technician B
2. Technician A says the positive plates in one cell are connected to the positive plates in the next cell. Technician B says in a discharged battery the electrolyte contains 36 percent sulfuric acid. Who is correct?
   A. Technician A
   B. Technician B
   C. Both Technician A and Technician B
   D. Neither Technician A nor Technician B
3. Technician A says during battery discharge the oxygen ($O_2$) comes off the positive plates and joins with the hydrogen (H) in the electrolyte to form water ($H_2O$). Technician B says during battery discharge, both negative and positive plate materials are changed to lead sulfate ($PbSO_4$). Who is correct?
   A. Technician A
   B. Technician B
   C. Both Technician A and Technician B
   D. Neither Technician A nor Technician B
4. Technician A says the need to add water frequently to a battery is caused by low charging system voltage. Technician B says water may be added to a maintenance-free battery. Who is correct?
   A. Technician A
   B. Technician B
   C. Both Technician A and Technician B
   D. Neither Technician A nor Technician B
5. All of these statements about a lead-acid battery are true *except*:
   A. The active material on the negative plates is sponge lead.
   B. The active material on the positive plates is lead peroxide.
   C. Each battery cell group has an even number of plates.
   D. The electrolyte solution contains 64 percent water,
6. Separators in a battery cell:
   A. keep the plates from touching the bottom of the case.
   B. keep the plates from touching each other.
   C. prevent the electrolyte from touching the plates.
   D. prevent the cell connectors from touching each other.
7. The voltage of a fully charged lead-acid battery cell is:
   A. 1.5 volts.
   B. 1.8 volts.
   C. 2.0 volts.
   D. 2.1 volts.
8. While discussing lead-acid battery operation, Technician A says when a battery is discharged, both plates are coated with lead sulfate. Technician B says when a battery is discharged, the electrolyte has a high sulfuric acid content. Who is correct?
   A. Technician A
   B. Technician B
   C. Both Technician A and Technician B
   D. Neither Technician A nor Technician B
9. A lead-acid battery may be explosive when exposed to sparks of a flame because:
   A. petroleum products are used in the manufacture of the battery case.
   B. sulfuric acid is vented from the battery during the discharge process.
   C. hydrogen gas is vented from the battery during the charging process.
   D. lead peroxide gas slowly escapes from the battery when the engine is not running.
10. All of these statements about starting motor design are true *except*:
    A. The armature windings are insulated from the commutator bars.
    B. The armature windings are insulated from the armature core.
    C. The commutator bars are insulated from each other.
    D. Typically, four brushes contact the commutator bars.
11. If a built-in battery hydrometer contains a green dot, the battery is ____________.

12. The CCA rating indicates the amperes that a battery will deliver at 0°F for 30 seconds with the battery voltage remaining above ______________________ volts.
13. In a starting motor, a ______________________ ______________________ is soldered to each end of the armature windings.
14. The shunt coil in a starting motor limits armature speed when the starter is ______________________.
15. Explain the operation of a starter solenoid while engaging and after engagement.
16. Describe the operation of a magnetic switch used with a moveable pole shoe starting motor.
17. Explain the operation of a starter drive while the starter is engaging and disengaging.
18. Explain battery operation while charging.

# Battery and Starting System Maintenance, Diagnosis, and Service

## Introduction

The battery is the heart of the vehicle's electrical system and the condition of the battery affects the operation of the entire electrical system. When the engine is not running, the battery supplies voltage and current to operate any electrical accessories that are turned on. The battery supplies voltage to maintain on-board computer memories when the engine is not running. The battery must also supply the high current required by the starting motor to crank the engine. Therefore, battery and cable condition are very important to provide proper electrical system operation.

Starting motor maintenance, diagnosis, and service are also extremely important. If the starting motor is defective, the engine may be prevented from starting, resulting in an expensive service call and/or tow bill.

### BATTERY MAINTENANCE

The battery should be kept clean and dry and the battery terminals and cable ends should be clean and tight. Moisture on top of a battery and corrosion on the battery cable ends causes a battery to slowly self discharge because a very low current leaks through the corrosion and moisture from one battery terminal to the other. A battery may be removed and cleaned with a baking soda and water solution. The baking soda helps to neutralize the sulfuric acid.

**You Should Know** *When servicing a battery, always wear a face shield and protective gloves. Sulfuric acid is very damaging to human skin and eyes. The sulfuric acid solution in a battery is very damaging to clothing and automotive paint or upholstery.*

When disconnecting the battery cables, always disconnect the negative cable first.

**You Should Know** *If you disconnect the positive cable first and the wrench slips and makes contact between the positive cable and ground, a very high current flows through the wrench. This action may heat the wrench and burn your hand and possibly cause the battery to explode. Never create sparks near a battery. Hydrogen gas given off while charging a battery is explosive. Even though the battery has not been charged for a period of time, some hydrogen gas may still be present in the battery cell vents.*

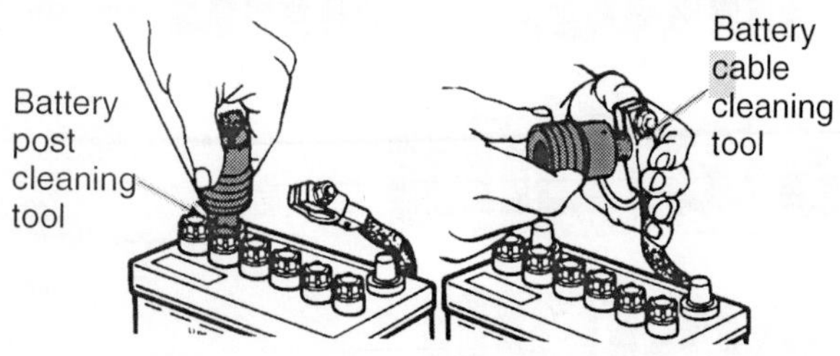


**Figure 1.** Cleaning battery terminals and cable ends.

Always reconnect the positive cable first, followed by the negative cable. The battery terminals and cable ends may be cleaned with a battery terminal cleaner **(Figure 1)**. Inspect the battery case for damage and leaks. If these conditions are present, battery replacement is necessary. Be sure the battery holddown is in satisfactory condition **(Figure 2)**. A badly eroded and corroded battery tray should be replaced. If the battery holddown does not hold the battery securely, the battery may bounce out of place while driving. This action may allow contact between the battery and the cooling fan, resulting in damage to the battery case. Always use a battery carrying strap to lift and carry the battery. Check the electrolyte level by removing the filler caps or observing the level through the translucent case. If the filler caps are removable, add water to the cells as necessary. When the electrolyte level is below the top of the plates in a maintenance-free battery, replace the battery. If the water loss from a battery is excessive and the case is not leaking, check the charging circuit for high voltage.

## BATTERY DIAGNOSIS AND SERVICE

Battery diagnosis and service is extremely important to supply adequate battery life and avoid time consuming and frustrating experiences. For example, if a battery is being undercharged by the charging system and this problem is not diagnosed and corrected, a no-start condition will likely occur. This experience could be time consuming, frustrating, and expensive.

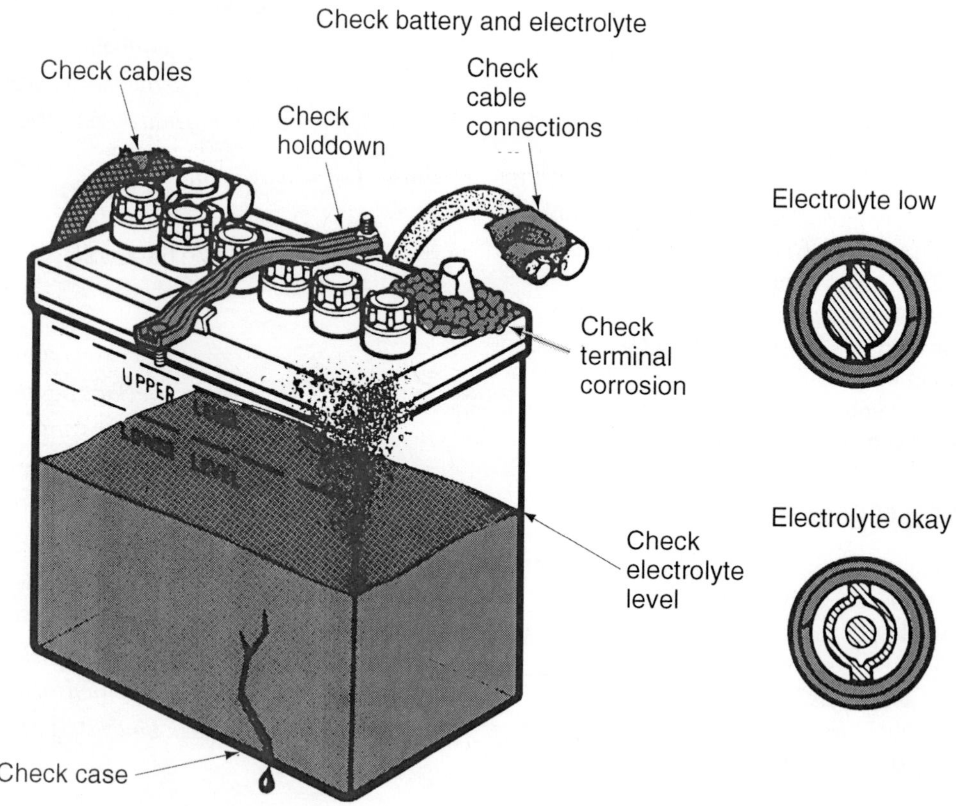


**Figure 2.** Battery inspection and maintenance.

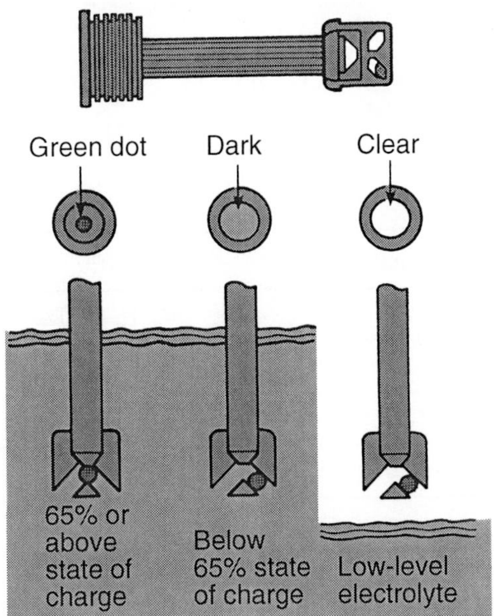


**Figure 3.** A built-in battery hydrometer.

## Hydrometer Test

If the battery has a built-in hydrometer, observe the indicator color **(Figure 3)**. A green indicator means the battery is over 65 percent charged, and may be tested in its present condition. If the hydrometer indicator is dark, the battery is less than 65 percent charged. Under this condition, the battery should be charged and load tested. A clear hydrometer indicator means the battery is low on electrolyte. If this condition is present on a battery with removable filler caps, add water to the cell(s) as required. When a clear hydrometer is present in a maintenance-free battery, replace the battery.

A portable hydrometer may be used to test the state of charge in a battery with removable filler caps **(Figure 4)**. A **hydrometer** is designed to test the specific gravity of a liquid.

Place the hydrometer pickup into the electrolyte in a battery cell. Squeeze the hydrometer bulb and draw enough electrolyte into the hydrometer until the float moves upward and floats freely. Do not allow the top of the float to contact the top of the hydrometer because this causes an inaccurate reading. Read the specific gravity of the electrolyte at the level of the electrolyte on the float. A specific gravity scale is provided on the upper float extension **(Figure 5)**. Repeat the hydrometer test on all the cells. If the specific gravity is 1.265, the battery is fully charged, whereas a specific gravity of 1.190 indicates the battery is 50 percent charged **(Figure 6)**. The variation in specific gravity readings between the battery cells should not exceed 0.050 specific gravity points. The **specific gravity** of the electrolyte is the weight of the electrolyte in relation to an equal volume of water. For example, if battery electrolyte has 1.265 specific gravity, a quart of electrolyte is 1.265 times heavier than a quart of water.

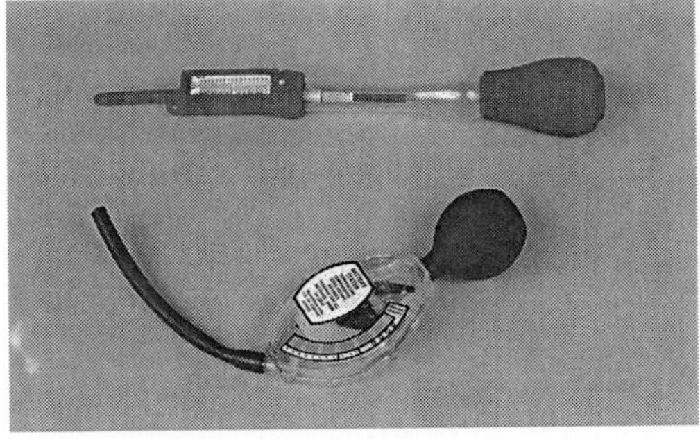

**Figure 4.** Portable hydrometers.

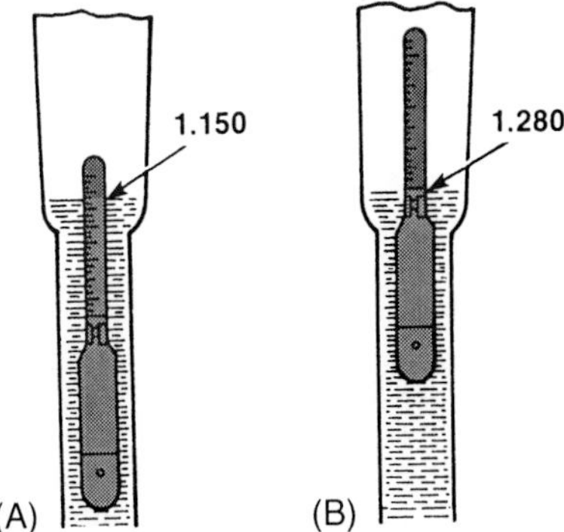


**Figure 5.** Specific gravity scale on hydrometer float.

**You Should Know** *A hydrometer test indicates the chemical condition of the battery, but this test does not always indicate the electrical capability of the battery. For example, if one of the cell connectors inside the battery is not making electrical contact, the specific gravity may be satisfactory but the battery will not deliver any current.*

| STATE OF CHARGE | SPECIFIC GRAVITY |
|---|---|
| 100% | 1.265 |
| 75% | 1.225 |
| 50% | 1.190 |
| 25% | 1.155 |
| DEAD | 1.120 |

**Figure 6.** Battery state of charge in relation to specific gravity.

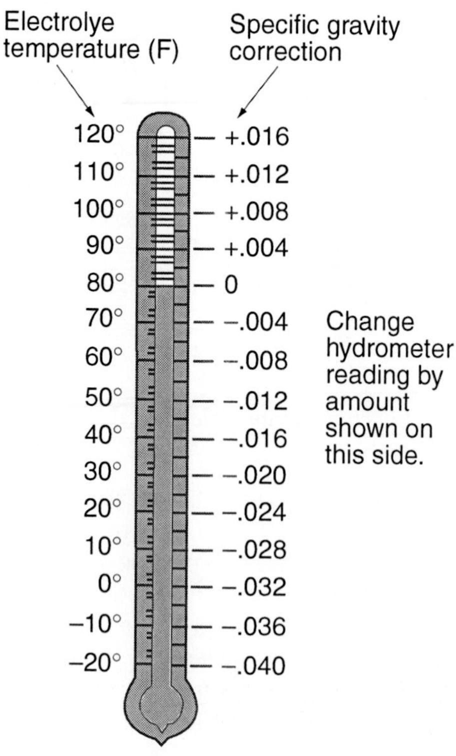


**Figure 7.** Specific gravity correction in relation to electrolyte temperature.

Some hydrometers have a built-in thermometer and temperature correction scale. If the battery is very cold, the electrolyte is heavier and more dense. Therefore, the electrolyte moves the float higher in the hydrometer. As indicated on the temperature correction scale, for every 10 degrees of electrolyte temperature below 80°F, 4 specific gravity points must be subtracted from the hydrometer reading **(Figure 7)**. Conversely, if the battery is hot, the electrolyte is thin and less dense. Therefore, the electrolyte does not move the electrolyte as high in the hydrometer. If the electrolyte is hot, for every 10° of electrolyte temperature above 80°F, add 4 specific gravity points to the hydrometer reading.

## Open Circuit Voltage Test

If the battery has recently been charged, perform a load test to stabilize the battery voltage before completing an open circuit voltage test. **Open circuit voltage** is the voltage available when a circuit is open and there is no current flow in the circuit.

**BATTERY OPEN CIRCUIT VOLTAGE AS AN INDICATOR OF STATE OF CHARGE**

| Open Circuit Voltage | State of Charge |
|---|---|
| 12.6 or greater | 100% |
| 12.4 to 12.6 | 75–100% |
| 12.2 to 12.4 | 50–75% |
| 12.0 to 12.2 | 25–50% |
| 11.7 to 12.0 | 0–25% |
| 11.7 or less | 0% |

**Figure 8.** Open circuit voltage in relation to battery state of charge.

Connect the voltmeter leads to the battery terminals with the correct polarity and read the open circuit voltage on the voltmeter. Compare the battery voltage reading to specifications. If the battery is fully charged, the open circuit voltage should be 12.6 volts. When the battery is partially charged, the voltage reading should be as specified in **Figure 8**.

## Load Test

The load test indicates the capability of the battery to deliver a specific current flow with the battery voltage remaining above a certain value. Therefore, the load test indicates the electrical condition of the battery.

**You Should Know** *To obtain accurate load test results, a hydrometer test should indicate the battery is 65 percent charge and the battery temperature must be above 0°F (−18°C). Do not load test maintenance-free batteries below this temperature.*

Follow these steps to perform a load test:

1. Connect the load tester cables to the battery terminals with the proper polarity.

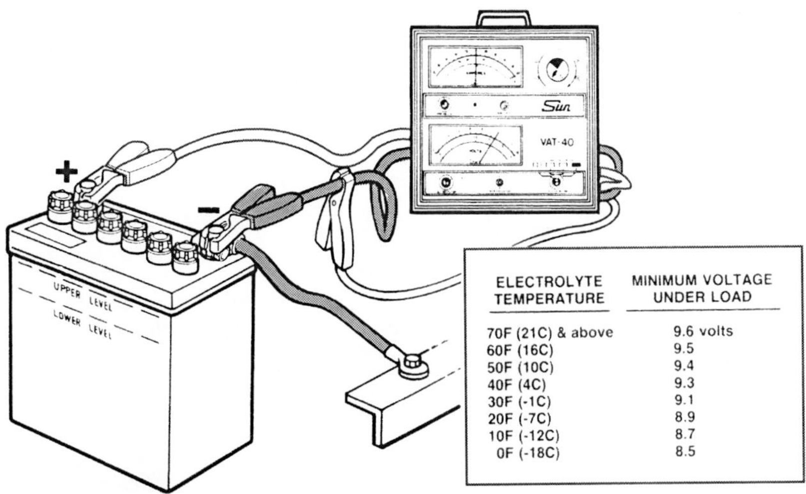


| ELECTROLYTE TEMPERATURE | MINIMUM VOLTAGE UNDER LOAD |
|---|---|
| 70F (21C) & above | 9.6 volts |
| 60F (16C) | 9.5 |
| 50F (10C) | 9.4 |
| 40F (4C) | 9.3 |
| 30F (-1C) | 9.1 |
| 20F (-7C) | 8.9 |
| 10F (-12C) | 8.7 |
| 0F (-18C) | 8.5 |

**Figure 9.** Battery load test connections and specifications.

2. Connect the inductive ammeter clamp over the negative load tester cable **(Figure 9)**.
3. Rotate the load tester control knob clockwise until the ammeter reading is one-half the CCA rating of the battery.
4. Maintain this load on the battery for 15 seconds.
5. At the end of the 15 seconds, read the voltmeter reading and then rotate the tester control knob fully counterclockwise to the off position.
6. Disconnect the load tester cables and compare the voltage reading obtained in step 5 to specifications.

A satisfactory battery with the electrolyte temperature at 70°F (21°C) has a load test voltage of 9.6 volts or above. If the battery is fully charged and fails the load test, replace the battery.

## Battery Drain Test

If the customer complains about the battery becoming discharged after the engine has not been started for a period of several days, there may be an electrical drain on the battery through one of the vehicle electrical systems. The battery drain test allows the technician to locate the cause of battery electrical drain.

**You Should Know** *Nearly all computers have a very low electrical drain to maintain the computer memory. Modern vehicles have a significant number of on-board computers and the electrical drain through these computers adds up. The technician must know how to measure battery drain and determine if this drain is satisfactory or excessive.*

**You Should Know** *On vehicles without on-board computers, some technicians used to connect a test light in series between the negative battery terminal and the negative battery cable to test battery drain. This method is not accurate on computer-equipped vehicles because the test light does not indicate the exact drain in milliamperes.*

Follow this procedure to perform a battery drain test:

1. Be sure all the electrical accessories are turned off and the vehicle's doors are closed. If the vehicle has an underhood light, remove the bulb from this light.
2. Disconnect the negative battery cable and connect the battery drain test switch to the negative battery terminal.
3. Connect the negative battery cable to the outer end of the test switch **(Figure 10)**.

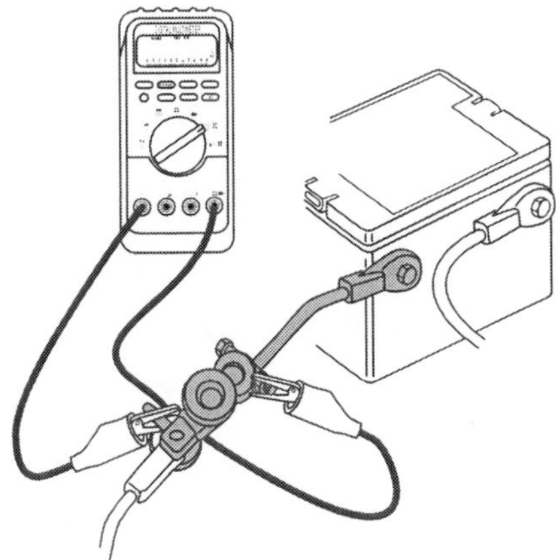

**Figure 10.** Battery drain test connections.

4. Connect a digital multimeter to the test switch terminals and select the highest ammeter scale on the multimeter.
5. Be sure the test switch is closed and start the engine. With the test switch closed, the starter current flows through this switch.
6. Operate the vehicle until the engine is at normal operating temperature and turn on all the electrical accessories.
7. Turn off all the electrical accessories and close the vehicle doors with the driver's window down. Turn the ignition switch off and remove the ignition key.
8. Turn the test switch to the open position. Current now flows from the battery through the ammeter to the electrical system.
9. Note the ammeter reading. Many computers gradually decrease their current draw as they enter a "SLEEP" mode. Therefore, the ammeter reading may gradually decrease.

**You Should Know** *A slow battery drain may be called a parasitic drain. Always follow the vehicle manufacturer's recommended procedure for testing battery drain. On some vehicles, it is necessary to connect a scan tool to the DLC under the dash and perform a power down procedure on the body computer module (BCM) during this procedure.*

10. Wait 3 minutes to allow the computers to enter the "SLEEP" mode. Switch the ammeter to a lower scale and read the battery drain in milliamperes. Compare the battery drain to specifications. If specifications are not available, 50 milliamperes drain or less is the average acceptable drain.
11. Disconnect all the fuses and circuit breakers one at a time and observe the ammeter. When a fuse is disconnected and the milliampere drain decreases, the circuit connected to that fuse is causing the battery drain.
12. Close the test switch and disconnect the multimeter leads. Disconnect the test switch and reconnect the negative battery cable.

## Battery Charging

Battery slow charging has two advantages. This method of battery charging brings a battery to a fully charged condition and it reduces the possibility of overcharging and/or overheating the battery. A slow charger usually supplies a charging rate of 3 amperes to 10 amperes. Several batteries may be connected in series on a shop slow charger. When batteries are connected in series, the negative terminal on one battery is connected to the positive terminal on the next battery. The slow charging process should be continued until the battery specific gravity reaches 1.265.

**You Should Know** *Disconnecting a battery cable(s) erases the memories in on-board computers, including the radio station programming. If the memory in the PCM (engine computer) is erased, the engine may have a rough idle problem and other adverse operating conditions until the computer relearns the sensor data. These problems can be avoided by connecting a 9 volts to 12 volts power supply to the cigarette lighter before disconnecting the battery cables.*

**You Should Know** *When fast charging a battery in a vehicle, disconnect the battery cables and connect the charger to the battery terminals. This action prevents the charger voltage from being applied to the on-board computers.*

An average fast charger has a maximum output of 50 to 100 amperes. Fast charging a battery is much faster than slow charging, but fast charging has some disadvantages. Fast charging may overheat and damage a battery and a battery should not be fully charged at a high rate on a fast charger. When the specific gravity reaches 1.225, reduce the charging rate to less than 10 amperes to avoid gassing excessive water from the battery. When fast charging a battery, never allow the electrolyte temperature to exceed 125°F (52°C). Damage to battery plate grids and active plate materials occurs above this temperature.

# STARTING MOTOR MAINTENANCE

The most important starting motor maintenance involves making sure the battery and starting motor cable connections are clean and tight. Inspect starting motor cables for worn insulation caused by improper contact with other components. Replace and reroute these cables as required. Be sure the starter mounting bolts are tight. Some starting motors have a ground cable from one of the starting motor mounting bolts to a bolt on the engine block. Be sure the bolts securing this cable are tight.

## STARTING MOTOR DIAGNOSIS

Each time a vehicle is in the shop for lubrication or minor service, always listen to the starting motor as it starts the engine. If the starting motor cranks the engine slowly, the battery may be partly discharged, the cables may have high resistance, or the starting motor may be defective. Test the battery as explained previously and test the starter circuit and starting motor as described in following pages.

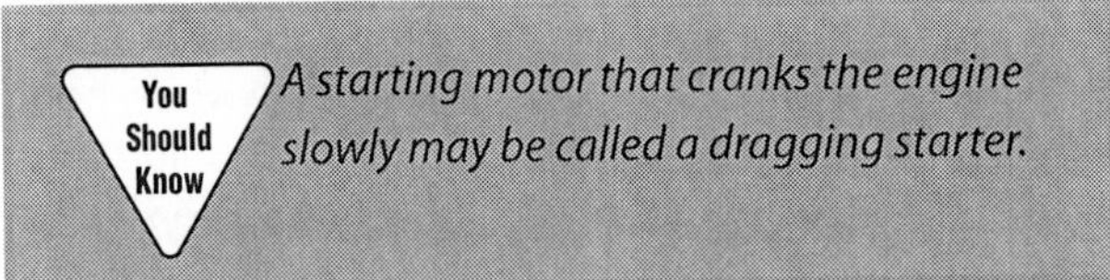

You Should Know
A starting motor that cranks the engine slowly may be called a dragging starter.


If the starting motor provides a grinding noise when cranking the engine, the drive gear and/or flywheel ring gear may be worn. Worn starting motor bushings also cause a noisy starting motor. When the starter armature spins but the engine fails to crank, the starter drive may be slipping or sticking on the armature shaft. If a clicking noise is heard when the ignition switch is turned to the START position, the battery cables may not be making proper electrical contact on the battery terminals. A defective solenoid disk and terminals may also cause a clicking noise when the ignition switch is turned to the START position.

## STARTING MOTOR AND SOLENOID DIAGNOSIS AND TESTING

When the starting motor slowly cranks the engine, test the battery and be sure it is in satisfactory condition and fully charged. When the engine cranks slowly, the problem may be caused by high resistance in the battery and/or starting motor cables. To test these cables, measure the voltage drop across each cable. Before measuring the voltage drop, disconnect the voltage supply to the primary ignition system to disable the ignition system and prevent the engine from starting. If the engine is fuel injected, disconnect the wiring connector from the fuel pump relay to prevent the injectors from discharging fuel into the intake manifold while cranking the engine.

### Starting Circuit Resistance Tests

Connect the voltmeter leads across the cable to be tested and crank the engine. For example, to test resistance and voltage drop across the cable from the positive battery terminal to the solenoid terminal, connect the voltmeter leads across to the ends of this cable. **Voltage drop** is the voltage difference between two points in an electrical circuit.

While cranking the engine, the voltage drop across this cable should not exceed 0.5 volt **(Figure 11)**. If the voltage drop across the cable exceeds specifications, the cable should be replaced. Voltage drop from the starting motor

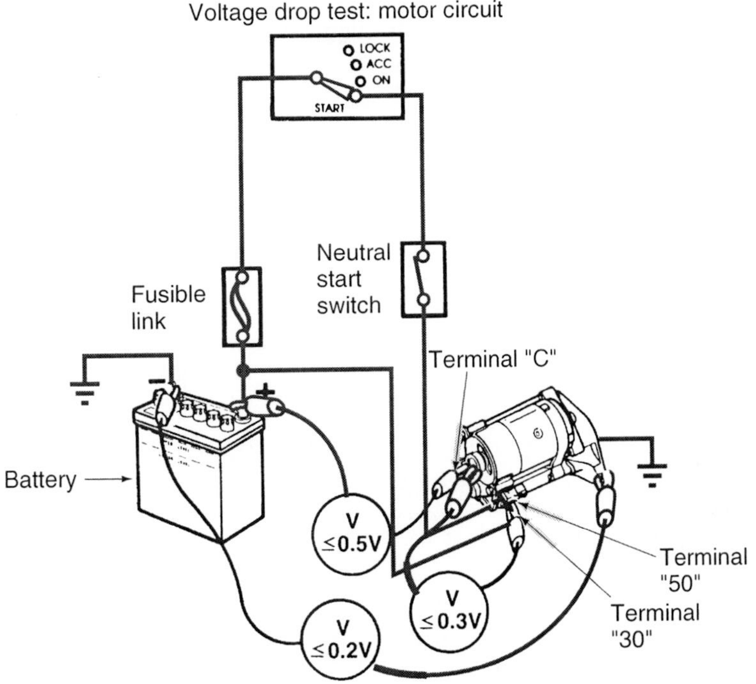


**Figure 11.** Starter circuit resistance tests.

ground to the battery ground terminal should not exceed 0.2 volt. If the battery cables are satisfactory and the starter cranks slowly, measure the starter current draw.

## Starter Current Draw Test

To measure the starter current draw, connect the battery starter tester cables to the battery terminals with the correct polarity. Place the inductive ammeter clamp over the negative battery cable **(Figure 12)**. Place the test selector switch in the STARTING TEST position. Disable the ignition and crank the engine for 10 to 15 seconds and observe the ammeter and voltmeter reading. If the starting motor is satisfactory, the voltage remains above 9.6 volts and the starter current draw is within the vehicle manufacturer's specifications. An average V8 engine has a starter draw of 200 amperes, whereas the normal starter draw on a V6 engine is 150 amperes, and a 4-cylinder engine may have a starter draw of 125 amperes. If the starter current draw is excessive, remove and replace or test and repair the starter.

**You Should Know** *Some starting motors have a shim(s) between the starter housing and the flywheel housing to provide proper starter alignment. When removing a starting motor always check to see if these shims are present and note the position of the shims, because they must be installed in their original position.*

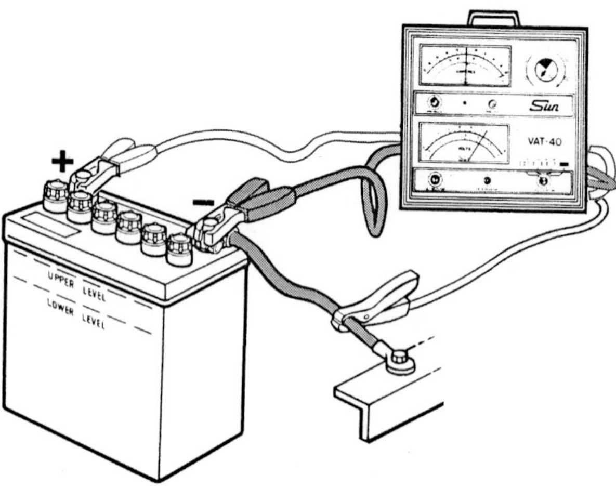


**Figure 12.** Starter current draw test connections.

## STARTING MOTOR SERVICE AND INSPECTION

If the starting motor has excessive current draw or any of the other problems described previously, the common practice in the automotive service industry is to replace the starting motor rather than to perform disassembly, testing, and repair procedures. Always disconnect the battery negative cable before removing a starting motor and reconnect this battery cable after the starting motor is reinstalled. When removing a starting motor, check for shims between the starting motor and the flywheel housing. If shims are present, the same number of shims must be reinstalled with the starting motor. Always be sure the starting motor mounting surface on the flywheel housing is clean.

## Solenoid Tests

Inspect the solenoid disk and terminals for a burned condition. If the solenoid disk and terminals are burned, replace the solenoid or install a solenoid repair kit. To test the hold-in winding, connect a pair of ohmmeter leads from the solenoid S terminal to ground on the solenoid case **(Figure 13)**. A low ohmmeter reading indicates the hold-in winding is not open and an infinite ohmmeter reading indicates an open hold-in winding.

Connect a pair of ohmmeter leads from the solenoid S terminal to the M terminal. A low ohmmeter reading indicates the pull-in winding is not open and the pull-in winding is open if the ohmmeter reading is infinite.

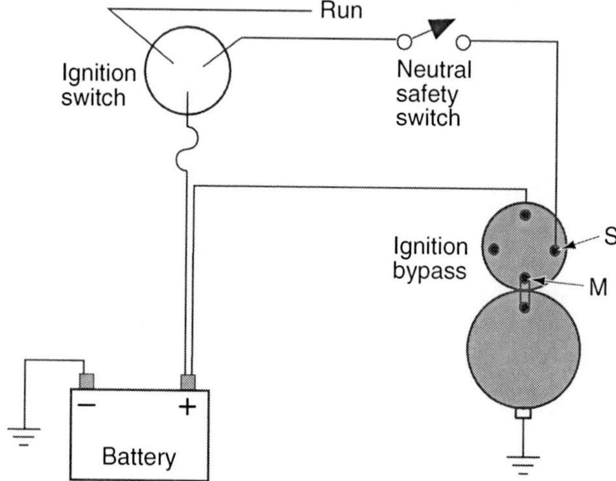


**Figure 13.** Solenoid terminal identification.

## Summary

- One important aspect of battery maintenance is keeping the battery and the battery terminals clean and dry.
- When disconnecting and reconnecting battery terminals, disconnect the negative terminal first and reconnect this terminal last.
- A hydrometer test indicates the chemical condition of a battery.
- The specific gravity of a fully charged battery should be 1.265.
- When a battery is fully charged, it should have an open circuit voltage of 12.6 volts.
- A battery load test indicates the capability of the battery to deliver high current flow.
- A battery drain test measures the milliamperes of current flow out of a battery with the ignition switch and all electrical accessories turned off.
- When fast charging a battery, the battery temperature should not exceed 125°F (52°C).
- If a starting motor cranks the engine slowly, the battery may be discharged or defective, the battery and starting motor cables may have high resistance, or the starting motor may be defective.
- When the armature spins but the engine fails to crank, the starter drive is slipping or sticking on the armature shaft.
- Starting circuit resistance may be measured by testing the voltage drop across the cables and components in the circuit.

## Review Questions

1. Technician A says when disconnecting battery cables, always disconnect the negative battery cable first. Technician B says when reconnecting battery cables, always connect the negative battery cable first. Who is correct?
   A. Technician A
   B. Technician B
   C. Both Technician A and Technician B
   D. Neither Technician A nor Technician B
2. Technician A says the specific gravity of the electrolyte in a fully charged battery is 1.225. Technician B says if the specific gravity is above 1.225, a battery may be charged at a high ampere rate on a fast charger. Who is correct?
   A. Technician A
   B. Technician B
   C. Both Technician A and Technician B
   D. Neither Technician A nor Technician B
3. Technician A says an open circuit voltage of 12.6 volts indicates a fully charged battery. Technician B says a load test voltage of 9 volts on a fully charged battery at 70°F (21°C) indicates a satisfactory battery. Who is correct?
   A. Technician A
   B. Technician B
   C. Both Technician A and Technician B
   D. Neither Technician A nor Technician B
4. Technician A says during a load test, the battery should be discharged at one half the CCA rating. Technician B says during a load test, the load should be applied to the battery for 15 seconds. Who is correct?
   A. Technician A
   B. Technician B
   C. Both Technician A and Technician B
   D. Neither Technician A nor Technician B
5. All of these statements about battery service are true *except*:
   A. Sulfuric acid in battery electrolyte is damaging to human skin and eyes.
   B. Sulfuric acid in battery electrolyte is damaging to paint and upholstery.
   C. Moisture on a battery top has no effect on battery operation.
   D. The battery top may be washed with a baking soda and water solution.
6. If the battery case is not leaking, the most likely cause of excessive electrolyte loss from the battery is:
   A. a slipping alternator belt.
   B. excessively high charging system voltage.
   C. high resistance in the battery terminals.
   D. excessive resistance at the alternator battery terminal.

7. A battery has just been disconnected from a fast charger. The specific gravity of the electrolyte is 1.220 and the battery temperature is 120°F (48.8°C). The actual battery specific gravity is:
   A. 1.236.
   B. 1.242.
   C. 1.248.
   D. 1.250.
8. The maximum variation in specific gravity readings between the cells in a battery is:
   A. 0.050 points.
   B. 0.075 points.
   C. 0.100 points.
   D. 0.120 points.
9. The specific gravity of the electrolyte in a battery is 1.170 on all the cells with the battery temperature at 70°F. When the battery is load tested, the battery voltage is 8.6 volts at the end of the load test. Technician A says the battery should be charged and retested. Technician B says the battery is defective and should be replaced. Who is correct?
   A. Technician A
   B. Technician B
   C. Both Technician A and Technician B
   D. Neither Technician A nor Technician B
10. When a drain test is performed on a battery, an acceptable battery drain is:
    A. 450 milliamperes.
    B. 375 milliamperes.
    C. 250 milliamperes.
    D. 50 milliamperes.
11. Excessive battery drain should be tested using a(n) ____________________.
12. Immediately after the test switch is opened during a battery drain test, the current drain ____________________.
13. Several batteries may be connected in ____________________ on a shop slow charger.
14. Starting circuit resistance may be tested by measuring the ____________________ ____________________ in the circuit.
15. Explain the resulting starting motor operation if the starter drive is slipping.
16. Explain the ohmmeter connection for testing a solenoid pull-in winding for an open circuit.
17. Describe the procedure for testing voltage drop across the starting motor ground circuit.
18. Describe the ohmmeter connections for testing a solenoid hold-in coil.

# Chapter 24 Charging Systems

## Introduction

The purpose of the alternator is to supply current to recharge the battery and power the electrical accessories on the vehicle. Alternator operation is extremely important to provide proper operation of the electrical accessories and maintain the battery in a fully charged condition. If the alternator voltage is too high, it forces excessive current flow through the electrical accessories and the battery, resulting in damaged electrical components and excessive battery gassing. When the alternator voltage is too low, it does not supply enough current to the electrical accessories and the battery. This may lead to improper operation of some electrical accessories and a discharged battery.

### ALTERNATOR DESIGN

The alternator contains a rotor and stator mounted between two end housings. Bearings in the end housings support the rotor shaft and allow rotor rotation **(Figure 1)**. The drive end of the rotor shaft extends through the drive end frame and a pulley is bolted or pressed onto this end of the rotor shaft. A cooling fan may be part of the pulley or it may be a separate component that is mounted between the pulley and the drive-end bearing. In some alternators, the cooling fins are attached to the rotor pole piece inside the alternator. The cooling fan rotates with the pulley to move air through the alternator and cool the alternator diodes **(Figure 2)**. A drive belt surrounding the alternator pulley and the crankshaft pulley rotate the rotor. The drive belt may also drive other components such as a power steering pump or air-conditioning compressor. The stator is mounted in a frame and is bolted between the two end frames so that it is stationary. The stator contains three insulated windings mounted in stator frame slots. There is a very small clearance between the rotor poles and the inside diameter of the stator frame. A rectifier plate is mounted in the slip ring end frame.

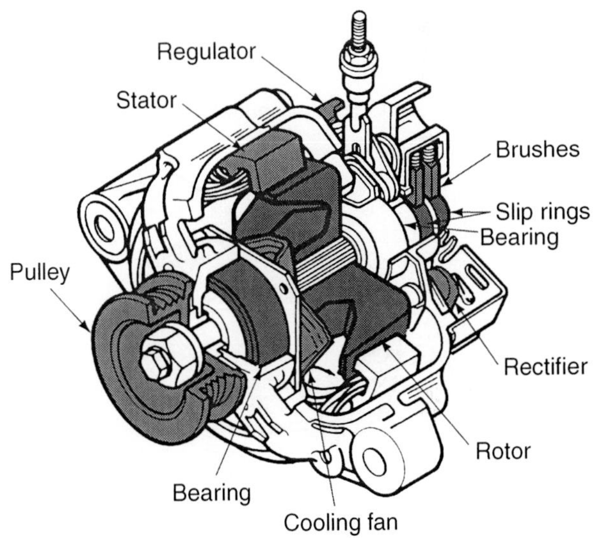


**Figure 1.** Alternator components.

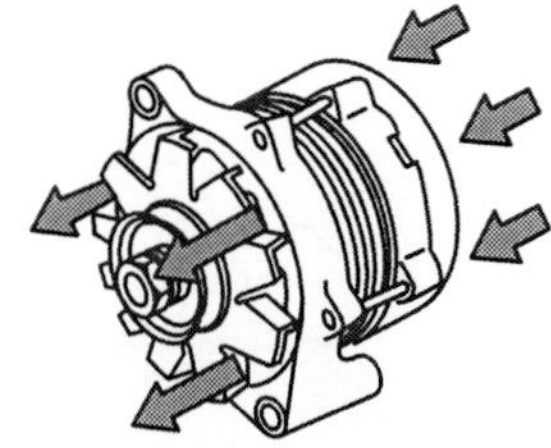

**Figure 2.** The cooling fan moves air through the alternator to cool the diodes.

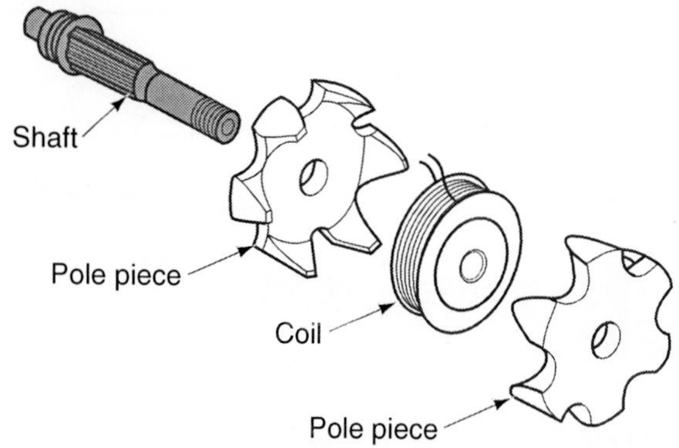


Figure 3. Rotor design.

## ROTOR

An insulated field winding is mounted on a spool that is pressed onto the rotor shaft. Two metal poles are pressed onto the rotor shaft on each side of the winding and these poles have interlacing fingers positioned above the winding **(Figure 3)**. Two insulated copper slip rings are mounted on the end of the rotor shaft and the ends of the field winding are connected to these slip rings. In some alternators, both brushes are insulated and connected to the two alternator field terminals. In other alternators, one brush is grounded and the other brush is connected to a field terminal.

## STATOR

The stator assembly contains three insulated windings. These windings are mounted in insulated stator frame slots. Some stators have wye-connected windings and in other stators the windings are delta connected. Wye-connected stator windings have three ends of these windings connected together in a wye connection **(Figure 4)** and the other three ends of these windings are connected to the diodes. In a delta-connected stator, the end of one stator winding is connected to the next stator winding **(Figure 5)**. In this type of stator winding, the junction where each pair of stator windings is connected is attached to the diodes.

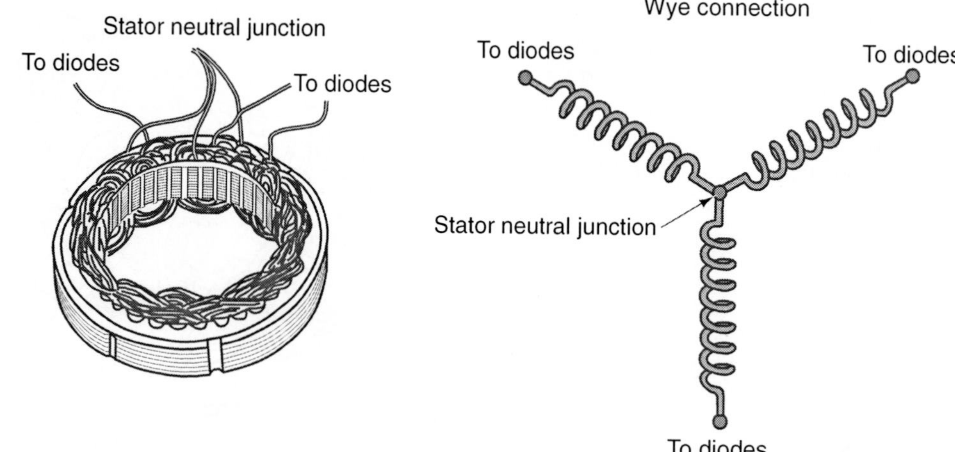


Figure 4. A wye-connected stators.

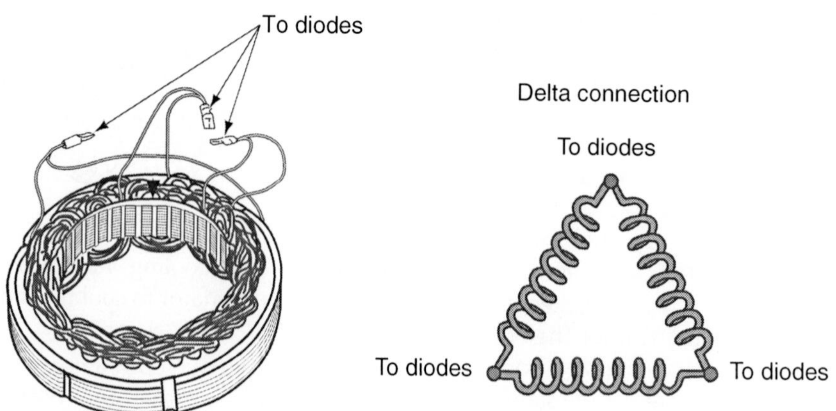


Figure 5. A delta-connected stator.

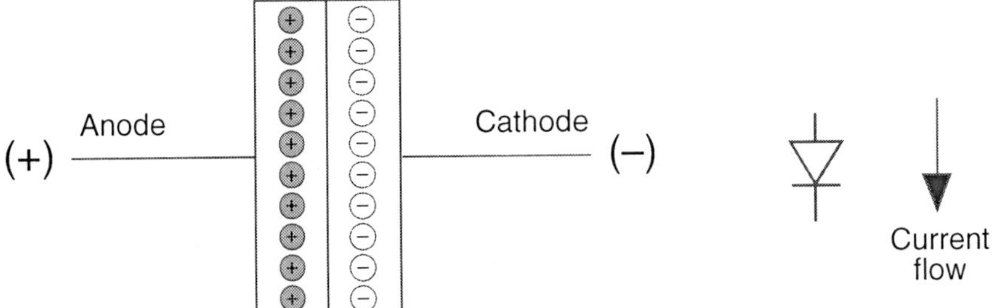


**Figure 6.** A diode and its symbol.

## DIODES

A diode is made of two semiconductor materials joined together. A semiconductor material is usually manufactured from silicon. In the manufacturing process, the silicon is mixed with other elements such as boron and phosphorus. When a precise quantity of boron is melted into a silicon wafer, a positive material is created with a lack of electrons. When phosphorus is melted into the other side of the silicon wafer, a negative material is created with an excess of electrons. These excess electrons in the negative material want to cross the junction area in the diode to the positive material with a lack of electrons **(Figure 6)**. However, these electrons will not move until they are acted upon by an external voltage source.

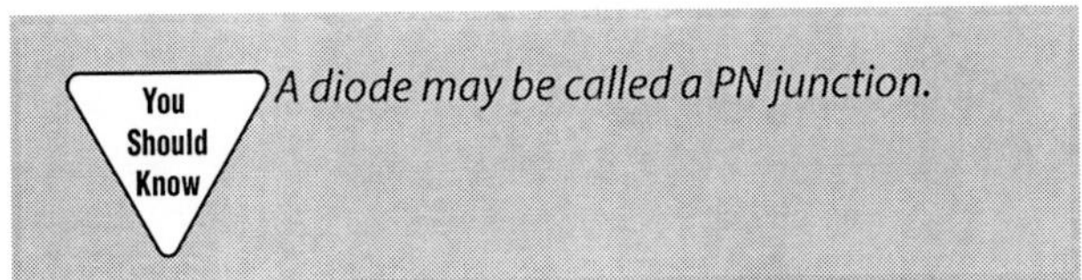


*A diode may be called a PN junction.*

When a voltage source is connected to a diode with positive polarity connected to the positive side of the diode and negative polarity connected to the negative side of the diode, the excess electrons in the negative material are repelled across the junction area and the diode conducts current **(Figure 7)**. This condition may be referred to as **forward bias**. If a voltage source is connected to a diode with positive polarity connected to the negative side of the diode and negative polarity connected to the positive side of the diode, the diode blocks current flow **(Figure 8)**. This condition is called **reverse bias**.

A rectifier assembly containing six diodes is mounted on a plate in the slip ring end of the alternator. Three of these diodes are mounted on an insulated plate that is connected to the alternator battery terminal. The other

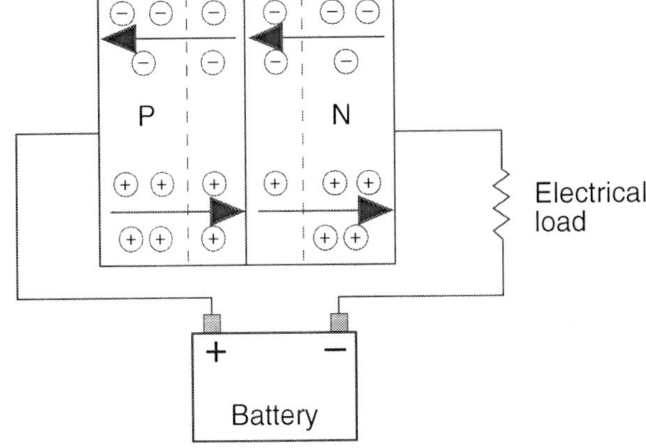


**Figure 7.** A forward bias connection to a diode result in current flow.

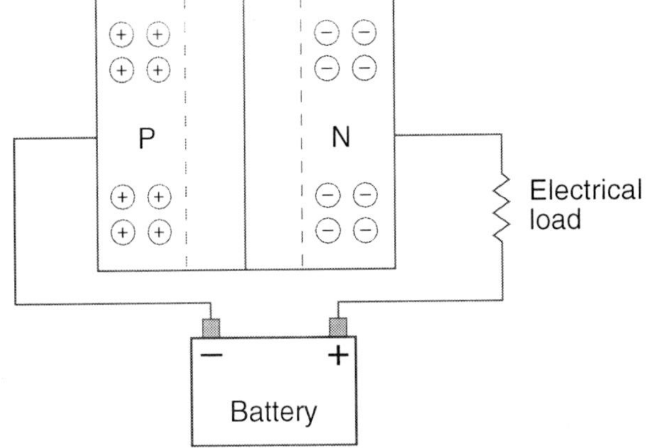


**Figure 8.** A reverse-bias connection to a diode prevents current flow.

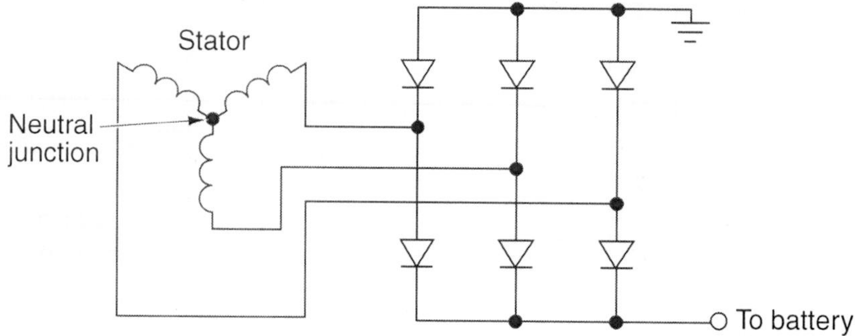


Figure 9. Stator windings connected to the diodes.

three diodes are mounted on a plate that is grounded to the end frame. The ends of the stator windings are connected to the diodes **(Figure 9)**.

## ALTERNATOR OPERATION

The alternator battery terminal is connected through a 12-gauge wire to the positive battery terminal **(Figure 10)**. Therefore, battery voltage is available at the alternator battery terminal with the ignition switch off. A fusible link is usually connected in the wire attached to the alternator battery terminal. This fusible link protects the battery wire and wiring harness if this wire is accidentally shorted to ground.

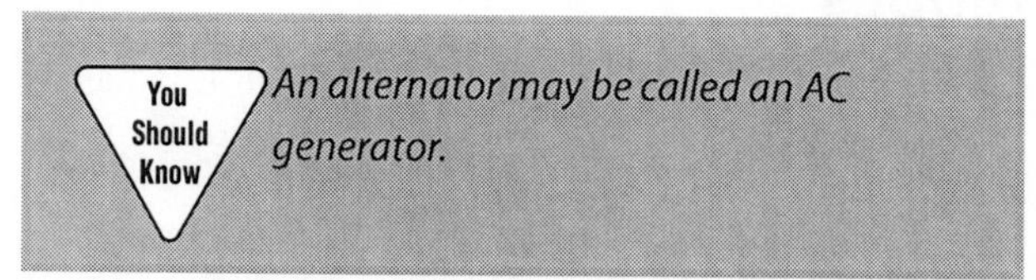


*An alternator may be called an AC generator.*

In many alternator circuits, voltage is supplied from the ignition switch to the alternator field terminal when the ignition switch is turned on **(Figure 11)**. Current then flows through the insulated brush, slip ring, field winding, and the other slip ring and brush to ground. This current flow through the field winding creates a magnetic field around the rotor. The interlacing fingers on one side of the rotor

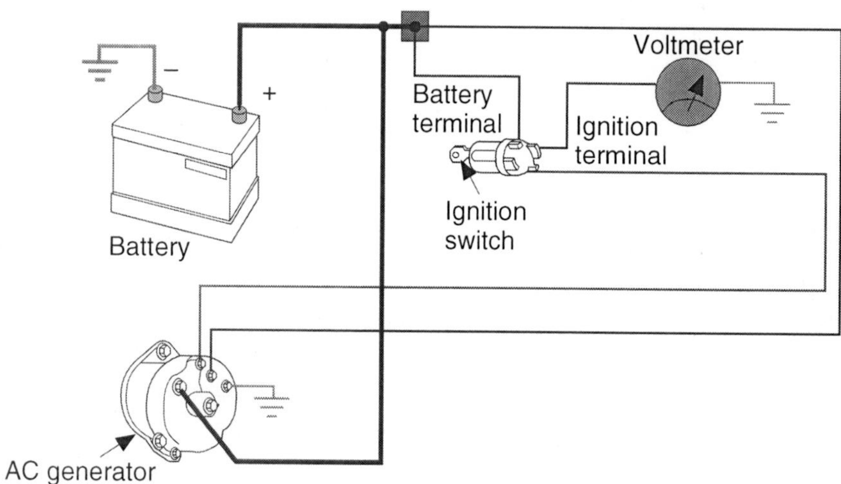


Figure 10. The alternator battery terminal is connected to the positive battery terminal in the charging circuit.

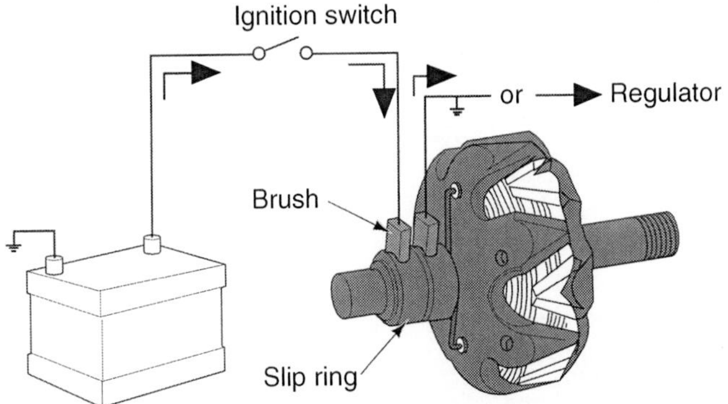


**Figure 11.** When the ignition switch is turned on, current flows through the field circuit.

become north poles and the interlacing fingers on the opposite side of the rotor become south poles. The magnetic lines of force travel from the north to the south poles on the interlacing fingers **(Figure 12)**.

When the engine starts, the rotor revolves inside the stator and the rotor magnetic field cuts across the stator windings. Notice the rotor poles around the circumference of the rotor are alternately north and south poles. Therefore, each stator winding is influenced by a north and a south pole followed by a south and a north pole. When the alternate magnetic poles on the rotor cut across a stator winding, an AC is produced in the winding **(Figure 13)**. Since the battery and electrical accessories on the vehicle must be supplied with DC, the AC in the stator windings must be rectified to DC **(Figure 14)**. This rectification is accomplished by the alternator diodes. The diodes change the AC current in the stator windings to a flow of DC current through the battery and electrical accessories.

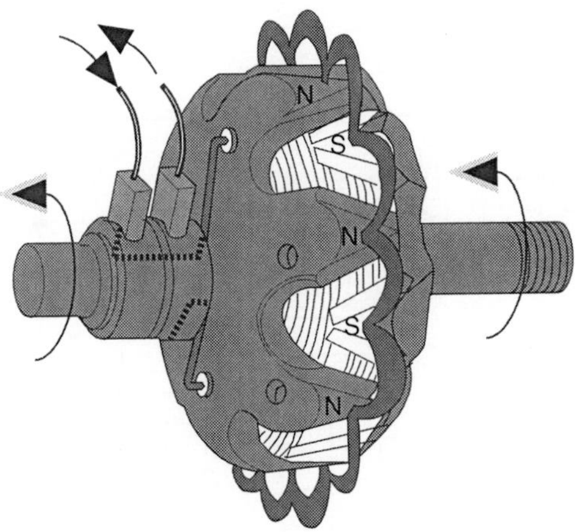


**Figure 12.** When current flows through the field coil, a magnetic field is created between the rotor poles.

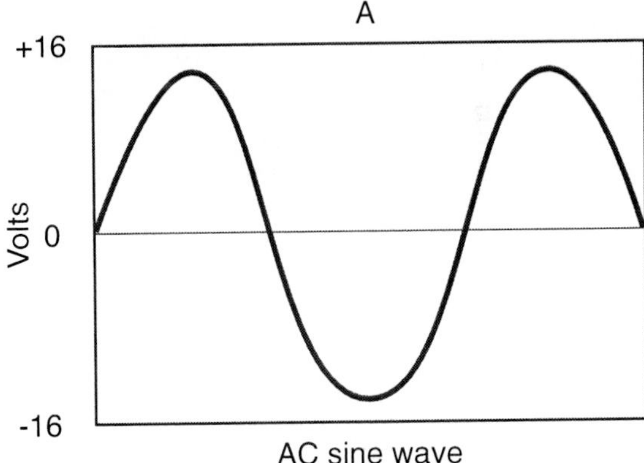


**Figure 13.** AC is produced in the stator windings.

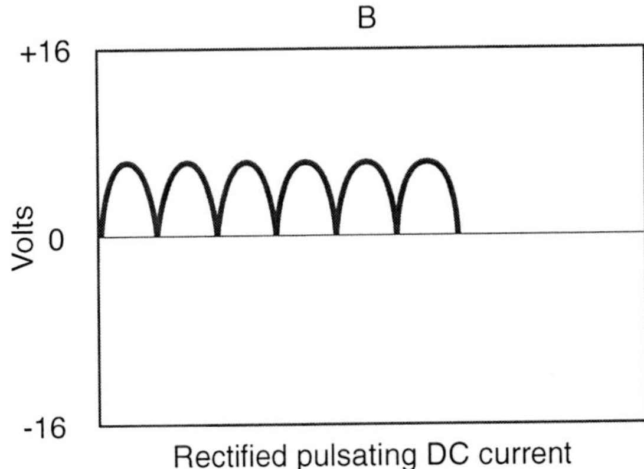


**Figure 14.** The AC in the stator windings is changed to DC by the diodes.

## VOLTAGE REGULATOR OPERATION

Modern vehicles are equipped with electronic voltage regulators. These regulators are often mounted inside the alternator or on the outside of the alternator end frame. Electronic regulators are non-adjustable. Some older vehicles have electronic regulators mounted externally from the alternator.

> **You Should Know** *DaimlerChrysler vehicles manufactured since 1985 have the voltage regulator designed into the PCM. If the voltage regulator is defective, the PCM must be replaced.*

> **You Should Know** *The term voltage regulator is somewhat misleading. If something is regulated, it is prevented from going too high or too low. The voltage regulator only limits the alternator voltage and prevents it from becoming too high.*

Electronic voltage regulator circuits vary depending on the vehicle. Always use the charging circuit wiring diagram for the vehicle you are working on. We will discuss the operation of a typical electronic charging circuit. When the ignition switch is turned on, current flows through the ignition switch, charge indicator bulb, and parallel resistor to the alternator number 1 terminal. From this location, current flows through the integral electronic voltage regulator, slip rings, brushes, and field coil to transistor TR1 in the regulator **(Figure 15)**. Current flows through this transistor to ground. This current flow creates a magnetic field around the rotor.

When the engine starts, the rotor magnetic field induces voltage in the stator windings and current begins to flow from the stator windings to the battery positive terminal. Current also flows from the stator windings through the diode trio to the alternator number 1 terminal. Under this condition, equal voltage is supplied to both sides of the charge indicator bulb and this bulb goes off. Current also flows through the diode trio and the slip rings, brushes, and field coil to TR1. Since TR1 is turned on, current flows through this transistor to ground **(Figure 16)**. Under this condition, field current and rotor magnetic strength are higher and the voltage in the stator windings increases. The **diode trio** is a small assembly containing three diodes. These diodes are connected from the stator terminals to the number 1 terminal. The purpose of the diode trio is to supply DC to the voltage regulator and field coil when the engine is running.

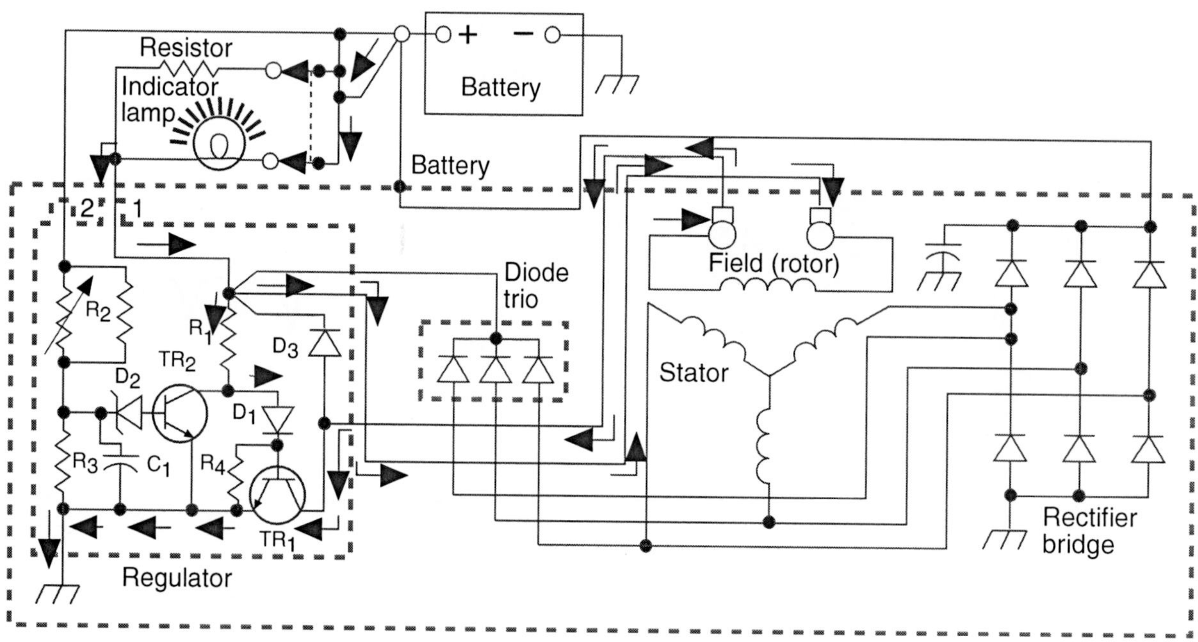


**Figure 15.** Current flow through the field circuit with the ignition switch in the ON position.

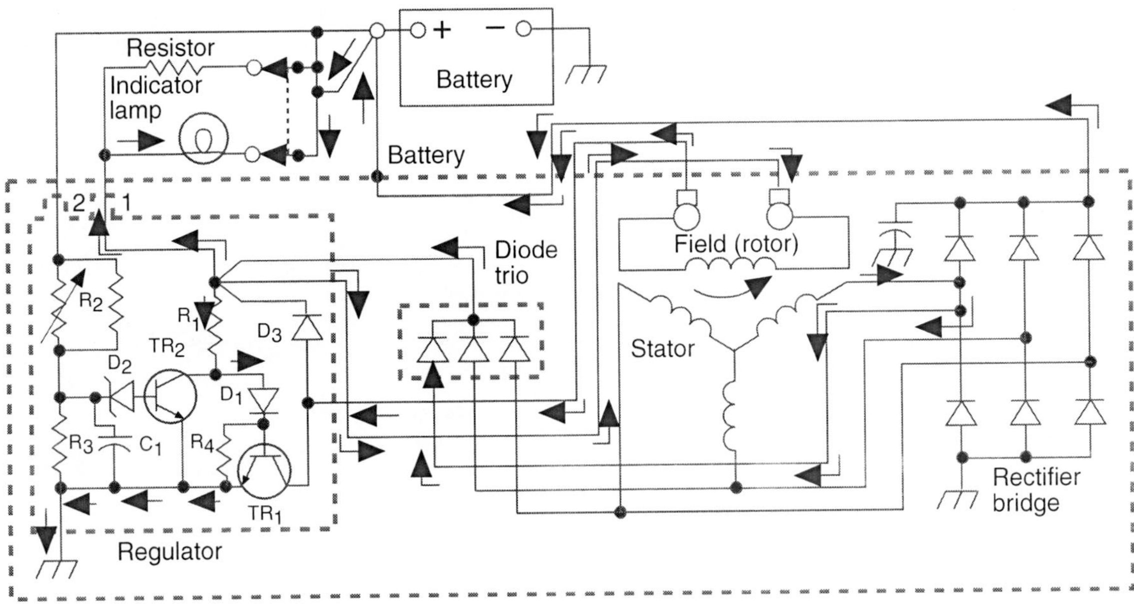


**Figure 16.** Current flow through the field coil and voltage regulator with the engine running and the field current turned on by the regulator.

> **You Should Know** *The resistance of the field coil limits the alternator field current. Because this coil typically has 3 to 4 ohms of resistance, the maximum field current is 3 to 4 amperes.*

The alternator number 2 terminal is connected to the positive battery terminal. Therefore, alternator voltage is sensed at this terminal when the engine is running. When the alternator voltage reaches a predetermined value, TR2 in the voltage regulator is turned on and TR1 is turned off **(Figure 17)**. When TR1 is turned off, the field current through the field coil, slip rings, and brushes is stopped.

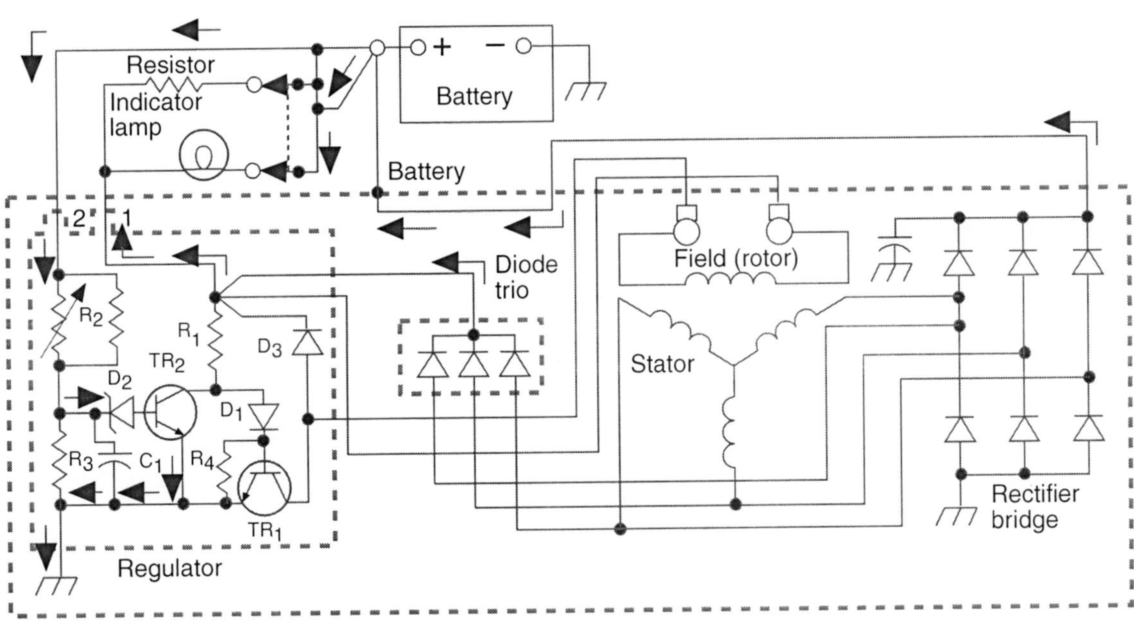


**Figure 17.** Current through the regulator when the regulator turns the current flow OFF through the field coil.

Under this condition, the rotor magnetic strength decreases rapidly and the alternator voltage also decreases. When this action occurs, the regulator switches back to the condition where TR2 is off and TR1 is on and the field current through the field coil and TR1 is restored. The regulator cycles the field current on and off very quickly to provide the necessary field current and magnetic strength around the rotor and limit the voltage induced in the stator windings. The regulator typically limits the alternator voltage to between 13.8 volts to 14.8 volts.

Resistor R2 in the voltage regulator is a thermistor with a parallel resistor. The thermistor allows the regulator to provide a higher alternator voltage when the atmospheric temperature is cold. This action compensates for additional resistance in the battery when it is cold and maintains the charging rate from the alternator through a cold battery. A **thermistor** is a special resistor that changes resistance in relation to temperature. When the thermistor is cold, its resistance increases.

**You Should Know** *Older vehicles are equipped with point-type voltage regulators. These voltage regulators have a voltage relay with a dual set of vibrating contacts. One set of vibrating contacts switches a resistor in and out of the field circuit to control rotor magnetic strength and the voltage induced in the stator windings. The second set of contacts intermittently grounds the field circuit through the resistor to further reduce the field current, weaken the rotor magnetic field, and limit the voltage in the stator.*

## Summary

- The cooling fan moves air through the alternator to cool the diodes.
- The ends of the field winding are connected to the slip rings.
- The field winding and the slip rings are insulated from the rotor shaft and poles.
- Three positive diodes in the rectifier assembly are mounted in an insulated heat sink and are connected to the alternator battery terminal.
- Three negative diodes in the rectifier assembly are grounded to the alternator end frame.
- The revolving rotor magnetic field induces an AC in the stator windings.
- The diodes change the AC in the stator windings to a DC for battery and electrical accessories.
- The electronic voltage regulator cycles the field current on and off to control field current and magnetic strength around the rotor. This action limits the voltage induced in the stator windings.
- The electronic voltage regulator provides a higher alternator voltage in cold weather to compensate for additional resistance in the battery under this condition.

## Review Questions

1. Technician A says a typical alternator stator contains four windings. Technician B says in a wye-connected stator, one end of each stator winding is connected to the diodes. Who is correct?
   A. Technician A
   B. Technician B
   C. Both Technician A and Technician B
   D. Neither Technician A nor Technician B

2. Technician A says a diode conducts current when it is connected to a voltage source with forward bias. Technician B says a forward bias connection occurs when the positive side of a voltage source is connected to the positive side of a diode and the negative side of the voltage source is connected to the negative side of the diode. Who is correct?
   A. Technician A
   B. Technician B
   C. Both Technician A and Technician B
   D. Neither Technician A nor Technician B

3. Technician A says a diode contains three semiconductor materials mounted together. Technician B says if a voltage source is connected to a diode with reverse bias, the diode will be damaged. Who is correct?
   A. Technician A
   B. Technician B
   C. Both Technician A and Technician B
   D. Neither Technician A nor Technician B
4. Technician A says the wire from the alternator battery terminal to the positive battery terminal is protected by a fuse link. Technician B says if the ignition switch is turned off, the voltage at the alternator battery terminal is zero. Who is correct?
   A. Technician A
   B. Technician B
   C. Both Technician A and Technician B
   D. Neither Technician A nor Technician B
5. All these statements about alternator rotor and stator design are true *except*:
   A. The ends of the field winding are connected to the rotor slip rings.
   B. The two rotor slip rings are insulated from each other.
   C. The wye connection in a stator is connected to the stator frame.
   D. In a wye-connected stator, three ends of the stator windings are connected to the diodes.
6. All of these statements about alternator diodes are true *except*:
   A. The rectifier assembly in an alternator usually contains four positive diodes and two negative diodes.
   B. The diodes in an alternator change the AC voltage and current in the stator windings to DC voltage and current.
   C. A diode contains two semiconductor materials joined together.
   D. When phosphorus is melted into a silicon wafer, a negative material is created with an excess of electrons.
7. The magnetic field surrounding the alternator rotor:
   A. is created by current flow through the stator windings.
   B. induces a DC voltage in the stator windings.
   C. is present when the ignition switch is in the off position.
   D. travels from the north to the south poles on the interlacing rotor fingers.
8. When discussing electronic alternator voltage regulators, Technician A says most electronic voltage regulators are adjustable. Technician B says the electronic voltage regulator is connected in the circuit from the alternator battery terminal to the positive battery terminal. Who is correct?
   A. Technician A
   B. Technician B
   C. Both Technician A and Technician B
   D. Neither Technician A nor Technician B
9. An electronic alternator voltage regulator increases the alternator voltage by:
   A. weakening the magnetic field surrounding the rotor.
   B. increasing the field current.
   C. increasing the resistance in the regulator ground circuit.
   D. decreasing the resistance in the stator circuit.
10. A vehicle frequently experiences burned out light bulbs. Technician A says to check the alternator voltage regulator setting. Technician B says to measure the resistance between the alternator battery terminal and the positive battery terminal. Who is correct?
   A. Technician A
   B. Technician B
   C. Both Technician A and Technician B
   D. Neither Technician A nor Technician B
11. The two ends of the field coil in the rotor are connected to the ________________ ________________.
12. In a wye-connected stator, three ends of the stator windings are connected to the diodes and the other three ends of these windings are connected ________________.
13. The diodes change ________________ current in the stator windings to ________________ current for the battery and accessories.
14. The voltage regulator limits the alternator voltage by controlling the rotor ________________ ________________.
15. Describe the design of a delta-connected stator.
16. Describe the design of an alternator rotor.
17. Explain how an electronic regulator limits alternator voltage.
18. Explain the purpose of a thermistor in an electronic regulator.

# Chapter 25 Charging System Maintenance, Diagnosis, and Service

## Introduction

Technicians must understand charging system design and operation to perform accurate diagnosis of these systems. Technicians must also understand the proper charging system diagnostic procedures to provide accurate diagnosis of these systems without damaging the charging system components and other electrical/electronic systems. Charging system maintenance, diagnosis, and service are explained in this chapter, including the newer systems in which the PCM controls the alternator voltage. The diagnosis of these late-model charging systems using DTCs is also explained.

## ALTERNATOR AND VOLTAGE REGULATOR MAINTENANCE

Alternator belt condition and tension are extremely important for satisfactory alternator operation. A loose belt causes low alternator output, which may result in a discharged battery. A loose, dry, or worn belt may cause squealing and chirping noises, especially during engine acceleration.

### Checking Alternator Belt Condition and Tension

The alternator belt should be inspected for cracks, oil-soaking, worn or glazed edges, tears, and splits **(Figure 1)**. If any of these conditions are present, the belt should be replaced.

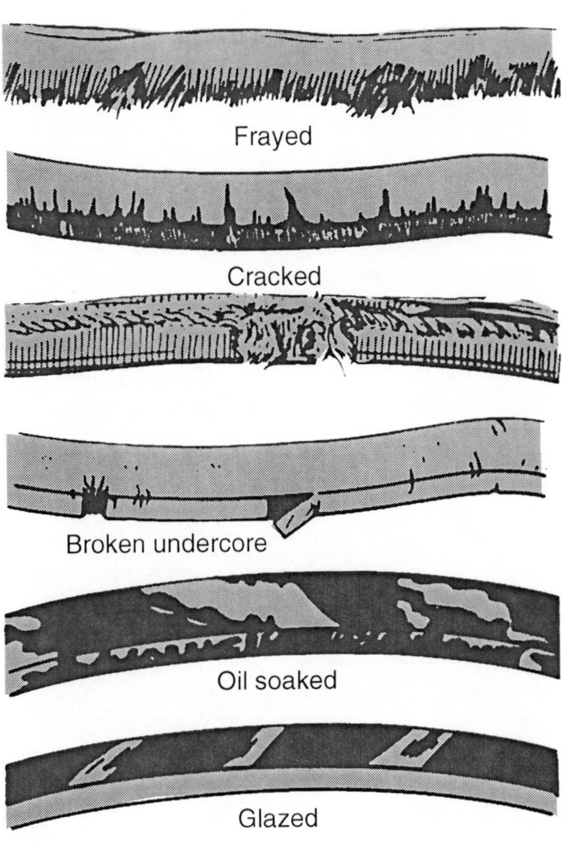


**Figure 1.** Defective alternator belt conditions.

Since the friction surfaces are on the sides of a V-belt, wear occurs in this area. If the belt edges are worn, the belt may be rubbing on the bottom of the pulley. This condition causes belt slipping and belt replacement is necessary to correct this problem.

With the engine stopped, belt deflection may be measured to test belt tension. The maximum deflection should be 0.5 inch per foot of free span. The belt tension may be measured with a belt tension gauge placed over the center of the longest belt span **(Figure 2)**. The tension indicated on the gauge should equal the vehicle manufacturer's specifications.

If the belt requires tightening, follow this procedure:

1. With the engine stopped, loosen the alternator tension adjusting bolt in the alternator bracket.
2. Loosen the alternator mounting bolts.
3. Check the alternator bracket and mounting bolt for excessive wear. If these bolts or bolt openings are worn, replacement is necessary.
4. Pry against the alternator housing with a prybar to tighten the alternator belt.
5. Hold the alternator in the position obtained in step 4 and tighten the tension-adjusting bolt in the alternator bracket.
6. Retest the belt tension with the tension gauge. If the belt does not have the specified tension, repeat step 1 through step 5.
7. Tighten the tension-adjusting bolt and the alternator mounting bolt to the specified torque.

Some alternators have a ribbed V-belt. Many ribbed V-belts have an automatic tensioning pulley; therefore, a belt tension adjustment is not required. The ribbed V-belt should be inspected to be sure it is properly installed on each pulley in the belt drive system **(Figure 3)**. The tension of a ribbed V-belt may be measured with a belt tension gauge in the same way as the tension of a V-belt.

Many ribbed V-belts have a spring-loaded tensioner pulley that automatically maintains belt tension. As the belt wears or stretches, the spring moves the tensioner pulley to maintain the belt tension. Some of these tensioners have a belt length scale that indicates new belt range and used belt range **(Figure 4)**. If the indicator is out of the used belt length range, belt replacement is required. Many belt tensioners have a 0.5-inch drive opening in which a ratchet or flex handle may be installed to move the tensioner pulley off the belt during belt replacement **(Figure 5)**.

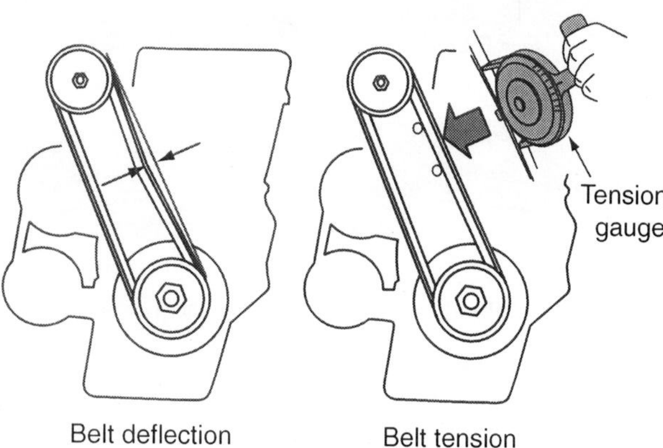


**Figure 2.** Methods of testing belt tension.

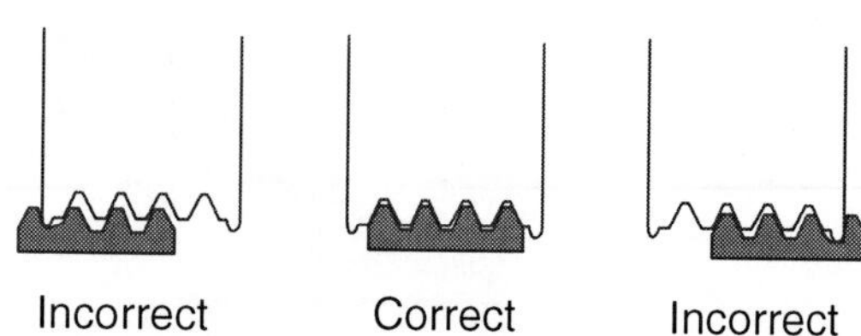


**Figure 3.** Proper and improper ribbed V-belt installation.

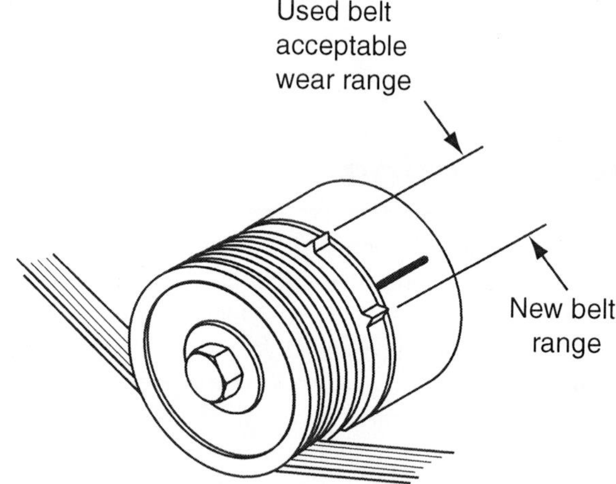


**Figure 4.** A belt tension scale.

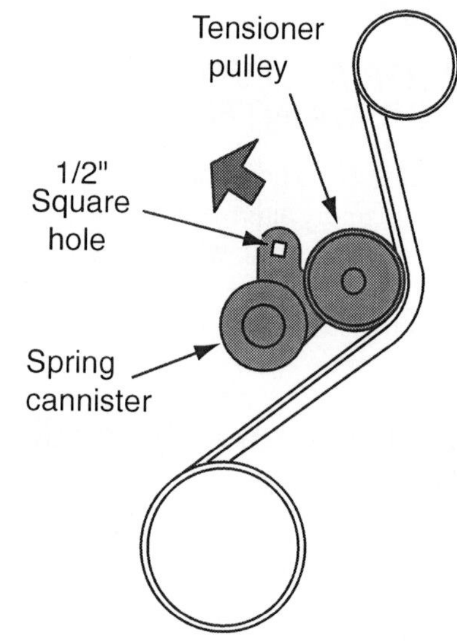


**Figure 5.** A one-half inch drive opening in the tensioner pulley.

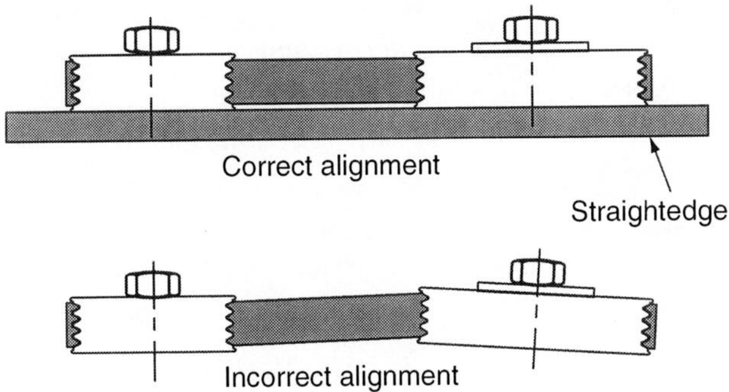


**Figure 6.** Checking pulley alignment.

Belt pulleys must be properly aligned to minimize belt wear. The edges of the pulleys must be in line when a straightedge is placed on the pulleys **(Figure 6)**. A misaligned alternator pulley may cause repeated belt failure.

## General Charging System Maintenance

When a vehicle is in the shop for minor service, the operation of the charge indicator light should be checked. If the charge indicator light is on with the engine running, further alternator diagnosis is required. Inspect the alternator wiring for worn insulation and loose or corroded connections. Be sure the insulating boot is in place over the alternator battery terminal. The electronic voltage regulator is usually mounted integrally in the alternator or on the back of the alternator housing. This type of regulator is non-serviceable and non-adjustable. If the regulator is mounted on the alternator end frame, be sure the regulator mounting bolts are tight and check any wiring terminals on the regulator for looseness and corrosion.

With the engine running, listen for any unusual alternator noises. Some diode or stator defects cause a whining noise from the alternator and a growling noise may be caused by defective alternator bearings.

Inspect the battery electrolyte level. If this level is excessively low, the alternator voltage may be too high. When the engine is accelerated, observe the brilliance of the headlights. If these lights become considerably brighter, the alternator voltage may be too high.

## CHARGING SYSTEM DIAGNOSIS

When diagnosing a charging system, one of the most important tests is the alternator output test. During the output test, the alternator is forced to produce full output. If the alternator passes this test, the alternator is satisfactory and this proves the problem must be in some other part of the charging system.

### Alternator Output Test

If the battery is discharged, the alternator may not be producing the specified voltage and current. An alternator output test may be performed to determine if the alternator output in amperes is satisfactory. Follow this procedure to perform an alternator output test:

1. Inspect the alternator belt condition and measure the belt tension. Adjust or replace the alternator belt as necessary.
2. Connect the volt-amp tester to the battery terminals with the correct polarity **(Figure 7)**.
3. Install the ammeter inductive clamp over the negative battery cable.
4. Turn off all the electrical accessories on the vehicle.
5. Start the engine and increase the engine rpm to 2000.
6. Turn the load control clockwise on the volt-amp tester until the voltmeter indicates 13 volts, and read the ammeter. The ammeter reading should be within 10 amperes of the rated alternator output if the alternator output is satisfactory. If the alternator output is not

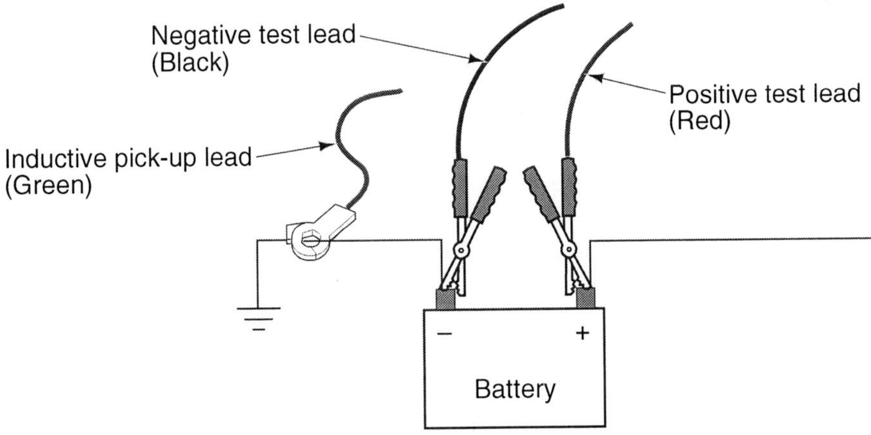


**Figure 7.** Meter test connections for an alternator output test.

satisfactory, the alternator should be disassembled and individual alternator components should be tested.

7. Turn the tester load control fully counterclockwise and allow the engine to idle. Shut the engine off and disconnect the tester cables.

You Should Know

*When performing an alternator output test, do not lower the charging system voltage below 12.6 volts. This action causes excessive current flow from the alternator and battery through the carbon pile load in the volt-amp tester. Under this condition, the carbon pile load and other tester components may be damaged.*

You Should Know

*On some older vehicles, the manufacturers recommended testing alternator output by full fielding the alternator. Vehicle manufacturers have discontinued this procedure because of the possibility of high voltage damaging electronic systems and components on the vehicle.*

**Full fielding** the alternator involved connecting a jumper wire or making a connection to the alternator field circuit that bypassed the voltage regulator and allowed maximum current to flow through the field coil in the rotor. This resulted in maximum voltage and current output in the stator windings. A carbon pile load connected across the battery terminals limited the voltage to a safe value, such as 15 volt, during this test.

## Charging System Voltage Test

In most charging systems, the voltage regulator limits the charging system voltage to 13.8 volts to 14.8 volts. If the charging system voltage is continually below 13.8 volts, the battery may become discharged. A charging system voltage above 14.8 volts may overcharge the battery, resulting in excessive battery gassing. A charging system voltage above 14.8 volts may also damage electrical/electronic components on the vehicle.

You Should Know

*Electronic and mechanical voltage regulators are temperature compensated, which allows the regulator to provide a higher charging system voltage when the atmospheric temperature and the regulator are cold. This higher voltage compensates for the higher resistance that naturally occurs in a cold battery. This higher charging system voltage maintains some charging rate through a cold battery.*

Follow this procedure to test the charging system voltage.

1. Connect the amp-volt tester to the battery terminals as described for the alternator output test.
2. Start the engine and turn all the electrical accessories off.
3. Increase the engine speed to 1,500 to 2,000 rpm. Observe the ammeter reading.
4. The charging rate on the ammeter should be less than 10 amperes. If the charging rate is high, the charging system voltage may not be at peak. Allow the engine to run until the charging rate is below 10 amperes.
5. Observe the charging voltage on the voltmeter. If the charging voltage is above or below the vehicle manufacturer's specifications, replace the voltage regulator.

## SCAN TOOL DIAGNOSIS OF CHARGING SYSTEMS

On many vehicles, DTCs are set in the PCM memory if there is a defect(s) in the charging system. DTCs are set in the PCM memory when charging system defects occur on systems with the voltage regulator in the PCM and also on some systems with the voltage regulator in the alternator. On some systems with the voltage regulator in the alternator, the PCM senses the charging system voltage and commands the regulator to turn the alternator field on and off. If the PCM senses a charging system defect the required number of times and sets a DTC, the PCM illuminates the MIL in the instrument panel. A scan tool is connected to the

**Figure 8.** A scan tool.

DLC under the left side of the dash to read the DTCs **(Figure 8)**. On 1996 and newer vehicles with OBD II systems, the DLC is a 16-terminal connector **(Figure 9)**.

**You Should Know** *When connecting or disconnecting a scan tool, the ignition switch must be off. If the ignition switch is on during this connection or disconnection, electronic systems on the vehicle or the scan tool may be damaged.*

After the scan tool is connected, turn the ignition switch on and complete the initial entries on the scan tool. These entries may include vehicle make, model year, and engine code. After the initial entries, select READ CODES on the scan tool. DTCs vary depending on the vehicle make and model year. Always use the DTCs and diagnostic procedure in the vehicle manufacturer's service manual. Typical DTCs related to charging system faults are the following:

1. P0562 Battery voltage low—The PCM sets this DTC if the battery-sensed voltage is 1 volt below the specified charging voltage range for 13.47 seconds. The PCM senses the battery voltage and then turns off the alternator field circuit and senses the voltage again. When the two sensed voltages are the same, the PCM sets the DTC. The PCM sets this DTC the first time the fault is sensed.
2. P0563 Battery voltage high—The PCM sets this DTC if the battery voltage is 1 volt higher than the specified charging system voltage range. The PCM sets this DTC the first time the fault is sensed.
3. P0622 Alternator field control circuit—The PCM sets this DTC if the PCM tries to regulate the field current with no result. The PCM sets this DTC the first time the fault is sensed.
4. P2503 Charging system voltage low—The conditions required to set this DTC are the same as the conditions for P0562.

The DTCs in step 1 through step 4 apply to 2002 Intrepid, Concorde, LHS, and 300M vehicles.

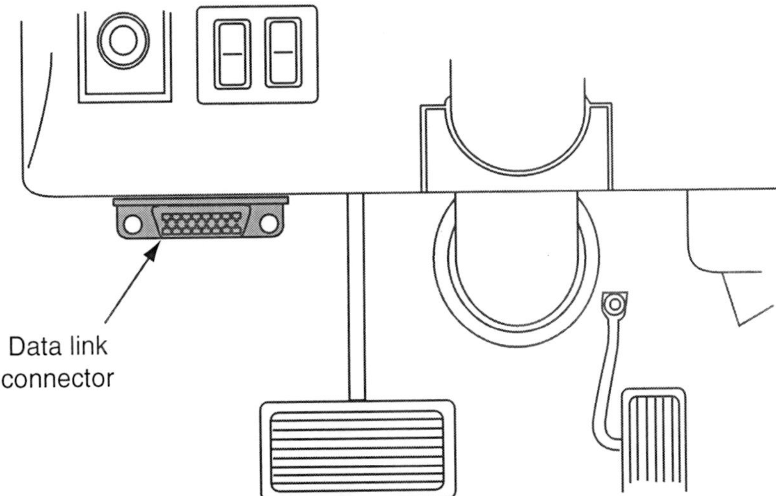


**Figure 9.** A data link connector (DLC) in an on-board diagnostic II (OBD II) system.

You Should Know

*There are some standards, such as SAE J2012, established for the DTCs on OBD II vehicles. The P in the above codes indicates they are related to powertrain systems. If the second digit is a zero or two in the powertrain DTCs, the DTC is mandated by the SAE. When the second digit is a one or a three, the DTC is established by the vehicle manufacturer. The third digit in the DTC indicates the subgroup to which the DTC belongs. For example, if the third digit is a five, the DTC is in the vehicle speed, idle control, and auxiliary inputs subgroup. When the third digit is a six, the DTC is in the computer and auxiliary outputs subgroup.*

## CHARGING CIRCUIT VOLTAGE DROP TESTING

Excessive resistance in the charging circuit between the alternator battery terminal and the positive battery terminal reduces the voltage supplied to the battery. For example, a higher-than-normal resistance in this circuit may cause a 1-volt drop across the circuit. Under this condition, if the alternator voltage is 14.5 volts, the voltage supplied to the battery is 13.5 volts. When this condition is present, the current flow through the battery is reduced and the battery may become discharged.

To measure the voltage drop across the charging circuit, connect the positive voltmeter lead to the alternator battery terminal and connect the negative meter lead to the positive battery terminal.

Select the lowest voltmeter scale. If necessary, connect a carbon pile load across the battery terminals and adjust the load to obtain 10 to 15 amperes charging rate with the engine running. A voltage drop above 0.5 volt indicates excessive charging circuit resistance. To measure voltage drop across the alternator ground circuit, connect the negative voltmeter lead to the negative battery terminal and attach the positive meter lead to the alternator case. With the engine running and a 10- to 15-ampere charging rate, the maximum voltage drop should be 0.2 volt.

## ALTERNATOR SERVICE AND INSPECTION

If an alternator output test indicates the alternator is defective, the common procedure in the automotive service industry is to replace the alternator. Because of high labor costs, it is usually more economically feasible to replace rather than repair an alternator. Always disconnect the negative battery cable before attempting to remove the alternator. Some replacement alternators are supplied without a pulley, so the pulley must be removed from the old alternator and installed on the replacement alternator. If the alternator has a press-on pulley, the proper pulling and installation tools must be used to remove and replace the pulley. On many alternators, the pulley is retained on the shaft with a nut and lock washer. These alternators usually have a hex opening in the center of the rotor shaft. An Allen wrench may be installed in this opening to hold the shaft while the pulley nut is loosened. After the pulley nut and lock washer are installed, be sure the nut is tightened to the specified torque. If the mounting holes in the alternator bracket are worn, replace the alternator bracket.

# *Summary*

- The alternator belt must have the specified tension to provide proper alternator output.
- The friction surfaces are on the sides of a V-belt.
- Ribbed V-belts usually have an automatic tensioner pulley.
- Some tensioner pulleys have a belt length scale that indicates used belt range and new belt range.
- Defects in the alternator diodes or stator may cause a whining noise from the alternator with the engine running.
- The alternator output may be tested by lowering the charging system voltage to 13 volts with a carbon pile load connected across the battery.
- The specified voltage regulator operating range is 13.8 volts to 14.8 volts.
- On many late-model charging systems, the PCM will store DTCs when a defect occurs in the charging system.
- A scan tool is connected to the DLC to obtain the DTCs.
- The resistance in the charging system is measured by testing the voltage drop in the system.

## Review Questions

1. Technician A says a loose alternator belt may cause a discharged battery. Technician B says a V-belt that is contacting the bottom of the alternator pulley may cause a discharged battery. Who is correct?
   A. Technician A
   B. Technician B
   C. Both Technician A and Technician B
   D. Neither Technician A nor Technician B
2. An alternator has a whining noise with the engine running. Technician A says the alternator may have an open field circuit. Technician B says one of the alternator diodes may be defective. Who is correct?
   A. Technician A
   B. Technician B
   C. Both Technician A and Technician B
   D. Neither Technician A nor Technician B
3. Technician A says during an alternator output test, the engine should be running at 2,000 rpm. Technician B says during an alternator output test, the charging system voltage should be lowered to 13 volts. Who is correct?
   A. Technician A
   B. Technician B
   C. Both Technician A and Technician B
   D. Neither Technician A nor Technician B
4. A vehicle has a charging system voltage of 15.5 volts with the engine at normal operating temperature. Technician A says this voltage may cause a discharged battery. Technician B says this voltage may damage electrical/electronic components on the vehicle. Who is correct?
   A. Technician A
   B. Technician B
   C. Both Technician A and Technician B
   D. Neither Technician A nor Technician B
5. All of these statements about testing alternator output are true *except*:
   A. A slipping alternator belt may cause low alternator output.
   B. Low alternator output may be caused by worn brushes and slip rings.
   C. Low alternator output may cause a discharged battery.
   D. Low alternator output may cause damaged electronic equipment on the vehicle.
6. An alternator produces a squealing noise when the vehicle is accelerated. When the alternator field wire is disconnected, the noise is not present. The most likely cause of this problem is:
   A. an open alternator field winding.
   B. a defective alternator diode.
   C. a loose alternator drive belt.
   D. a defective alternator voltage regulator.
7. When testing alternator output, a carbon pile load may be used to lower the alternator voltage to:
   A. 14.2 volts.
   B. 13.8 volts.
   C. 13.6 volts.
   D. 13.0 volts.
8. When discussing charging circuit diagnosis, Technician A says some charging circuits may be diagnosed with a scan tool. Technician B says when disconnecting a scan tool from the DLC, the ignition switch should be in the ON position. Who is correct?
   A. Technician A
   B. Technician B
   C. Both Technician A and Technician B
   D. Neither Technician A nor Technician B
9. With a 10-ampere charging rate, the maximum voltage drop between the alternator battery terminal and the positive battery terminal is:
   A 0.1 volt.
   B. 0.2 volt.
   C. 0.5 volt.
   D. 1.2 volts.
10. The most likely result of high resistance in the wire between the alternator battery terminal and the positive battery terminal is:
    A. excessive battery gassing.
    B. an undercharged battery.
    C. damage to electronic components.
    D. damage to the electronic voltage regulator.
11. In a charging system that has the voltage regulator in the PCM, DTCs related to charging system defects may be retrieved with a(n) ________________ ________________.
12. When testing the resistance between the alternator battery terminal and the positive battery terminal, connect the positive voltmeter lead to the ________________ ________________ ________________.
13. When testing the resistance in the alternator ground circuit, connect one voltmeter lead to the negative battery terminal and connect the other voltmeter lead to the ________________ ________________.
14. Describe one precaution to be observed when connecting and disconnecting a scan tool from the DLC.
15. Describe the results of higher-than-specified charging system voltage.
16. Describe the result of higher-than-specified resistance in the charging system between the alternator battery terminal and the positive battery terminal.
17. If the PCM senses charging system defects, list four DTCs that may be stored in the PCM memory.
18. Explain the reason why vehicle manufacturers no longer recommend full fielding an alternator during an output test.

# Chapter 26 Ignition Systems

## Introduction

Proper ignition system operation is extremely important to obtain satisfactory engine performance and fuel economy. Ignition defects, such as misfiring or reduced spark advance, reduce engine performance and economy. The ignition operating principles in this chapter provide the necessary background information so the ignition diagnostic procedures described in Chapter 27 can be easily understood.

### SPARK PLUGS

In SI engines, spark plugs provide a critical air gap for the spark to jump across. The purpose of the ignition system is to create a spark across the spark plug electrodes at the correct instant in relation to piston movement. The spark at the plug electrodes ignites the air-fuel mixture in the combustion chamber.

A spark plug is contained in a metal shell and the lower end of this shell has threads that match the threads in the cylinder head spark plug opening. A metal center electrode is positioned in the center of the spark plug and a ceramic insulator surrounds this electrode. The center electrode and insulator assembly is mounted in the metal shell. The upper end of the shell is crimped over the insulator. A sealing powder is positioned between the insulator and the shell to prevent cylinder pressure from leaking up through the spark plug. A ground electrode is attached to the lower edge of the shell and the outer end of the ground electrode is positioned over the top of the center electrode. The spark plug gap is the distance between the center and ground electrodes. This gap may vary from 0.035 inch (.889 millimeter) to 0.080 inch (2.032 millimeter) depending on the engine and ignition system design. Most spark plugs have a resistor in the center electrode to reduce radio static. An electric terminal is threaded into the top of the insulator. This terminal contacts the center electrode **(Figure 1)**. The spark plug wire contacts the terminal on top of the spark plug.

The spark plug **reach** is the length of the threaded area. The spark plug reach usually varies from 3/8 inch (9.52 millimeters) to 3/4 inch (19.050 millimeters) depending on cylinder head and combustion chamber design.

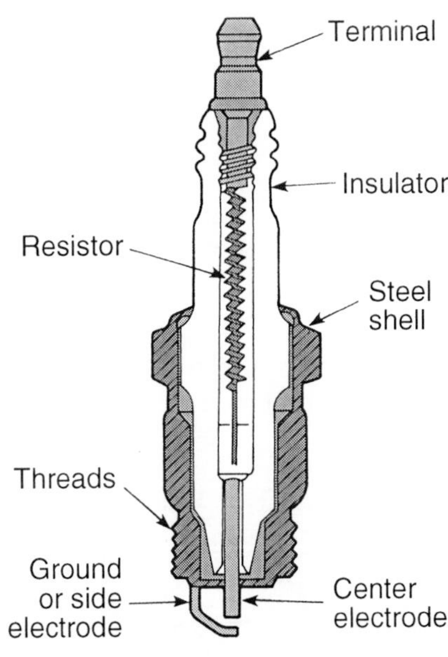


**Figure 1.** A typical spark plug design.

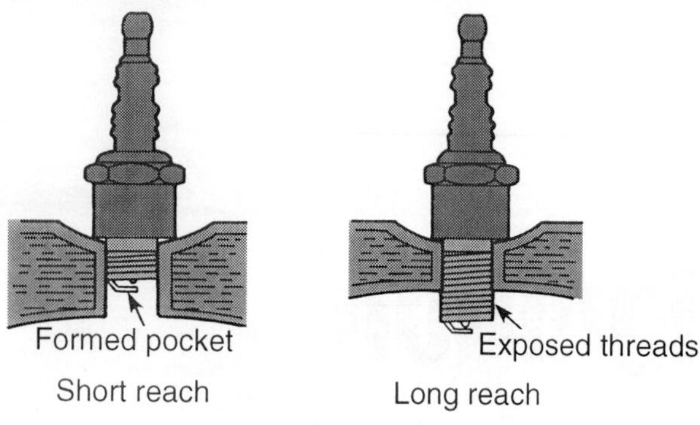


Figure 2. Spark plug reach.

Replacement spark plugs must have reach specified by the vehicle manufacturer. If the spark plug reach is longer than specified, the spark plug extends too far into the combustion chamber **(Figure 2)**. Under this condition, the piston may strike the spark plug electrodes, resulting in severe piston damage. If the spark plug reach is too short, the electrodes are not positioned properly in the combustion chamber. This may result in improper combustion. The spark plug shell is sealed to the cylinder head opening by a metal gasket or a tapered seat on the shell.

The spark plug **heat range** indicates the ability of the spark plug to dissipate heat from the center electrode through the spark plug insulator to coolant circulated through a cylinder head passage surrounding the spark plug.

Hot spark plugs are designed with a long heat path so the heat has to travel further up through the insulator before it is dissipated through the shell into the cylinder head and coolant. In a spark plug with a colder heat range, the distance is shorter from the center electrode to the point where the insulator contacts the shell. This design allows faster heat dissipation through the spark plug so the electrode temperature is reduced **(Figure 3)**. On many spark plugs, the number on the spark plug indicates the heat range; a higher number represents a hotter range spark plug.

In recent years, many engines have platinum-tipped electrodes **(Figure 4)**. The platinum coating on the electrodes resists chemical erosion and corrosion to provide longer spark plug life. Some manufacturers specify platinum-tipped spark plug replacement at 100,000 miles (160,000 kilometers).

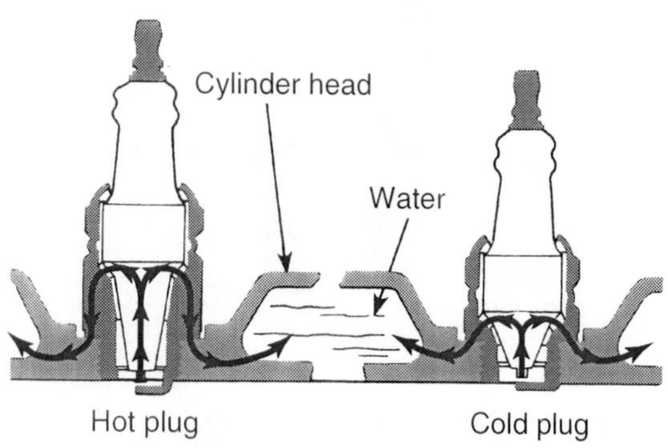


Figure 3. Spark plug heat range.

Figure 4. A platinum-tipped spark plug.

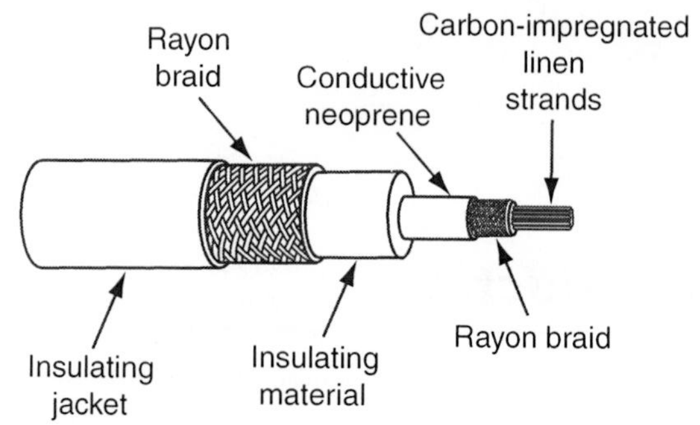


Figure 5. A suppression-type spark plug wire.

Most vehicles are equipped with spark plug cables that contain a resistance to provide television/radio suppression (TVRS). These wires have a conductor in the center that contains carbon-impregnated linen strands **(Figure 5)**. The center conductor is surrounded by a double braid and insulating material. Because spark plug wires conduct high voltage, they must have heavy insulation made from hypalon or silicone. Metal ends on the spark plug cable connect the center conductor to the spark plug terminal and the distributor cap or coil. Many spark plug cables have an outer diameter of approximately 0.312 inch (8 millimeters).

## DISTRIBUTOR IGNITION (DI) SYSTEMS

As the name suggests, distributor ignition systems contain a distributor. One purpose of the distributor is to distribute the spark from the ignition coil to the spark plugs. This is accomplished by the distributor cap and rotor. The distributor components also open and close the primary ignition circuit at the right time. Modern distributors have a pickup coil and an ignition module to provide this function. In some ignition systems, the module is mounted externally from the distributor. Distributors may have mechanical and vacuum advance mechanisms to provide spark advance or the PCM may perform this function. Regardless of the distributor design, this component plays a very important role in providing proper ignition operation and maintaining engine performance, economy, and emission levels.

## Ignition Coils

A conventional ignition coil has a laminated metal core in the center of the coil. This core contains many strips of metal. This type of core magnetizes and de-magnetizes faster than a solid metal core. A secondary winding containing thousands of turns of very fine wire is wound around the core. One end of the secondary winding is connected to a terminal in the coil tower and the other end of this winding is usually connected to one of the primary terminals in the tower. Insulation is applied to the wire in the secondary winding so the turns of wire do not contact each other and insulating paper is positioned between each layer of secondary turns. The primary winding contains a few hundred turns of heavier wire and this winding is wound on top of the secondary winding. The ends of the primary winding are connected to the primary terminals in the coil tower **(Figure 6)**. Some metal sheathing is placed around the outside of the primary winding and the winding assembly is installed in a metal can. The metal sheathing concentrates the magnetic field on the outside of the coil windings. In the manufacturing process, the coil is filled with oil through the screw opening in the coil tower. Then the screw is installed in the tower. The oil helps to cool the coil and also prevents air space in the coil that would allow moisture formation because of temperature change. If moisture forms in a coil, arcing will occur because of the high voltage developed in the coil.

**You Should Know** *An ignition coil is a step-up transformer. Battery or charging system voltage is supplied to the primary winding and the ignition coil steps this voltage up to a very high voltage to fire the spark plugs in the secondary circuit.*

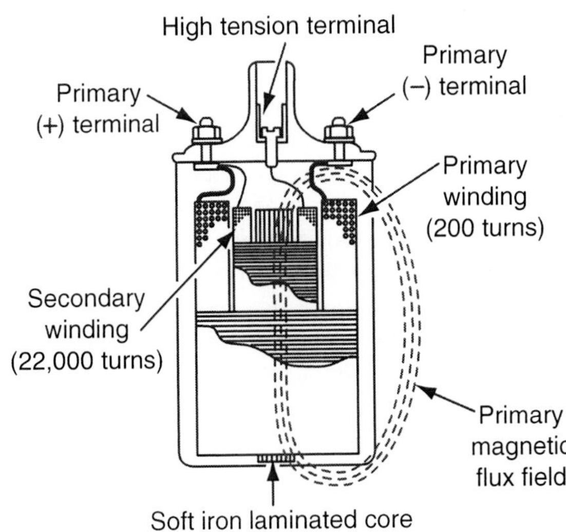


**Figure 6.** Ignition coil design.

## Primary and Secondary Ignition Systems

The primary ignition system contains the ignition switch, primary coil winding, ignition module, and pickup assembly. Some primary ignition circuits on older vehicles have a resistor connected in series between the ignition switch and the coil primary winding. The wire from the ignition switch is connected to the positive primary coil terminal and the negative primary coil terminal is connected to the ignition module, which is often mounted in the distributor. The primary circuit operates on battery or charging system voltage. The secondary ignition circuit contains the secondary coil winding, coil secondary wire, distributor cap and rotor, spark plug wires, and spark plugs **(Figure 7)**. The secondary circuit operates on a very high voltage produced in the coil secondary winding.

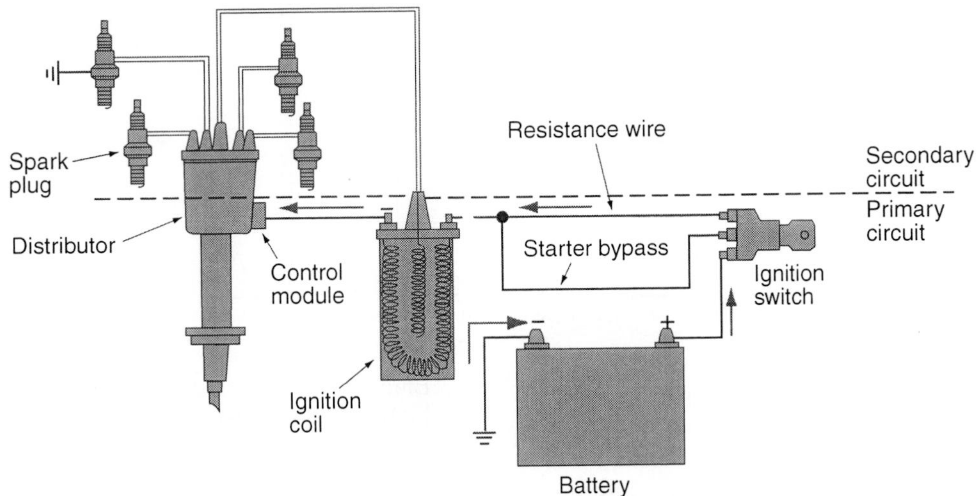


**Figure 7.** Primary and secondary ignition circuits.

## Pickup Assembly

The pickup assembly is usually mounted in the distributor. Many pickup assemblies contain a coil of wire surrounding a magnet. The ends of the pickup winding are connected to the ignition module. A reluctor attached to the distributor shaft has a high point for each engine cylinder. A small, specified clearance is provided between the reluctor high points and the pickup coil. Each time a reluctor high point rotates past the pickup coil, a voltage is inducted in the pickup coil **(Figure 8)**. This induced voltage in the pickup coil signals the module to open the primary circuit.

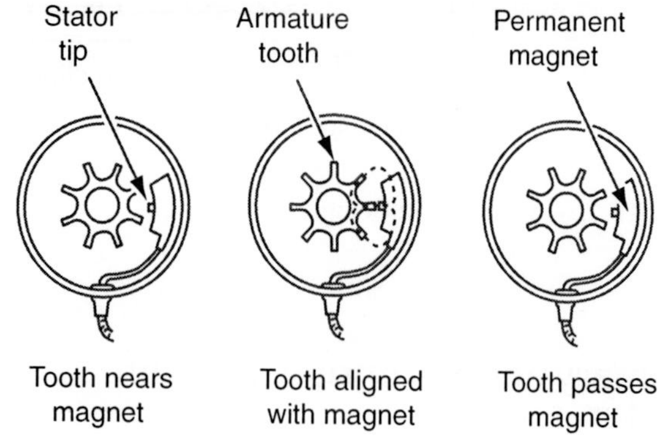


**Figure 8.** Pickup coil operation.

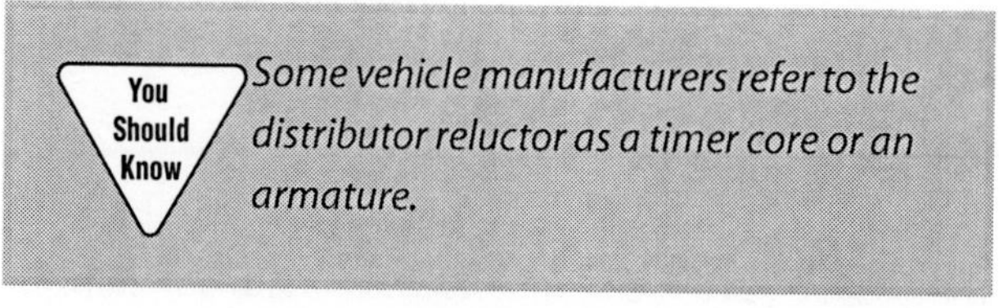


*Some vehicle manufacturers refer to the distributor reluctor as a timer core or an armature.*

Some distributors contain a pickup assembly with a Hall effect switch. In this type of pickup, a metal blade assembly attached to the underside of the rotor has an individual blade for each engine cylinder. As these blades rotate through the pickup, the Hall element signal goes from low voltage to high voltage **(Figure 9)**. The Hall element is connected to the ignition module.

## Ignition Modules

The ignition module contains many electronic components such as diodes, transistors, capacitors, and resistors. The ignition module is non-serviceable. Some ignition modules are mounted inside the distributor where they are bolted to the distributor housing, whereas other modules are bolted on the outside of the distributor. Some modules are mounted externally from the

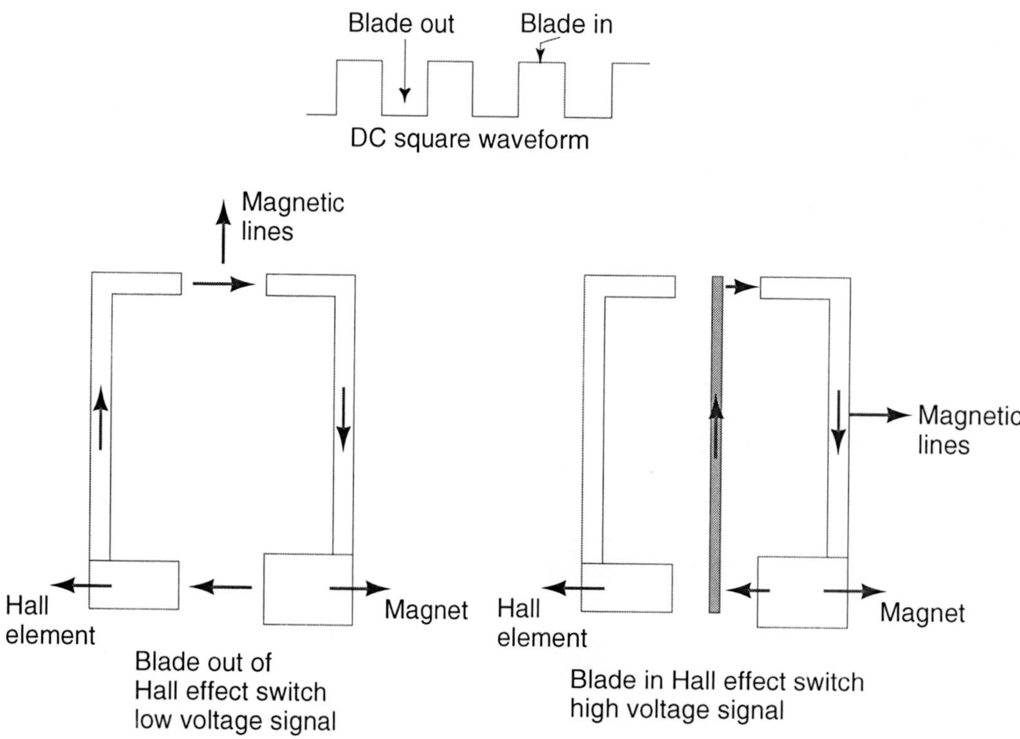


**Figure 9.** Operation of a distributor pickup containing a Hall effect switch.

**Figure 10.** An ignition module.

distributor **(Figure 10)**. If the ignition module is mounted in or on the distributor, the module is grounded to the distributor housing. Most of these modules require a heat-dissipating grease on the module surface that is in contact with the distributor housing to prevent module overheating. Externally mounted modules usually have a ground wire connected from the module to the vehicle ground. The purpose of the ignition module is to open and close the primary circuit at the right instant.

## Ignition Operation

When the ignition switch is turned on, current flows from the battery through the ignition switch, coil primary winding, and the ignition module to ground. This current flow through the primary winding creates a strong magnetic field around both coil windings.

> **You Should Know** *The ignition module must turn the primary ignition circuit on long enough to allow the magnetic field to build up in the ignition coil. This buildup time is called dwell time. The dwell time is determined by the module design and it is not adjustable on DI systems. If the module does not supply enough dwell time, the magnetic field will not have time to build up in the coil. This condition causes a weak magnetic field and reduced secondary voltage to fire the spark plug, resulting in spark plug misfiring.*

When the engine begins cranking and a reluctor high point rotates past the pickup coil, a voltage signal is sent from the pickup coil to the module. When the module receives this voltage signal, the module opens the primary circuit, causing the magnetic field in the coil to collapse rapidly across both coil windings. As the magnetic field collapses across the thousands of turns in the secondary winding, a very high voltage is induced in the secondary winding. This high voltage in the secondary winding forces current to flow through the coil secondary wire, distributor cap and rotor, and spark plug wire to the appropriate spark plug. As the secondary current arcs across the spark plug gap, it ignites the air-fuel mixture in the combustion chamber and the engine starts. When the magnetic field collapses in the coil, a low voltage is induced in the primary winding. This voltage is absorbed by protective circuitry in the module.

Each time a reluctor high point rotates past the pickup coil, the module opens the primary circuit and the magnetic field collapses in the coil, resulting in high induced voltage in the secondary winding and spark plug firing. The distributor cap and rotor distribute the secondary current to the appropriate spark plug.

The maximum secondary coil voltage may be 35,000 volts on a typical coil. However, if the spark plug gaps are set to specifications and the engine is operating at idle, the average voltage required to fire the spark plugs is 10,000 volts. The difference between the voltage required to fire the spark plugs and the maximum available coil voltage is called reserve coil voltage. Reserve secondary coil voltage is necessary to compensate for extra resistance in the secondary circuit as the spark plug gaps erode and become wider. Reserve secondary coil voltage is also necessary to compensate for high cylinder pressures when the engine is operating under heavy load. For example, if the engine is operating at wide-open throttle, resulting in high cylinder pressures, the voltage required to fire the spark plugs may be 18,000 volts.

## DISTRIBUTOR ADVANCES

When the distributor is installed in the engine, the engine must be positioned with number 1 piston at TDC on the compression stroke and the timing marks on the crankshaft pulley or flywheel in the 0 degree position. The distributor is then installed with the rotor under number 1 spark plug wire terminal in the distributor cap and one of the reluctor high points lined up with the pickup coil. The other spark plug wires must be installed in the distributor cap in the direction of distributor shaft rotation and in the proper engine firing order **(Figure 11)**.

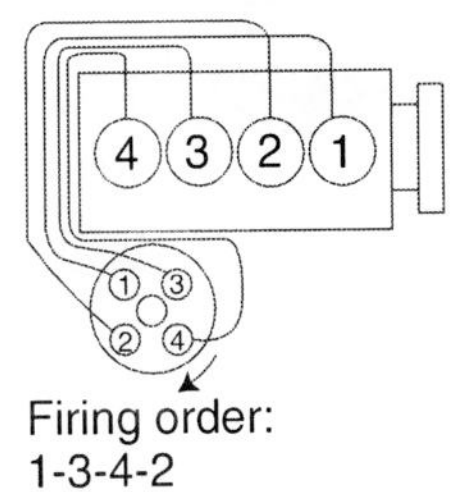


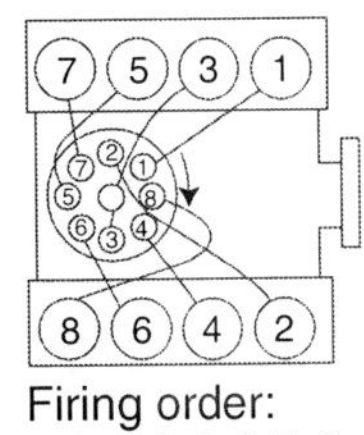


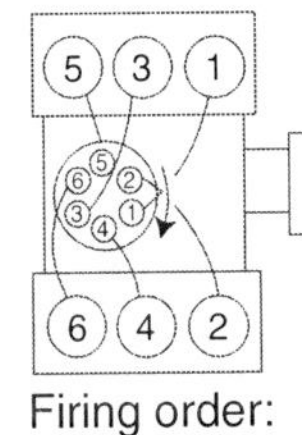


**Figure 11.** Proper spark plug wire installation in relation to firing order and distributor shaft rotation.

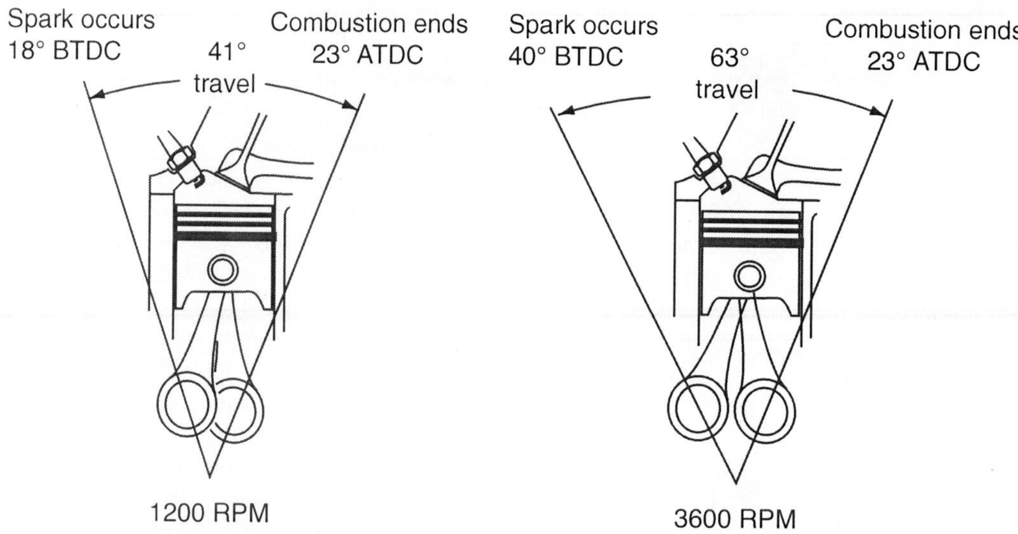


**Figure 12.** Spark timing in relation to engine speed.

With the engine operating at idle speed, the distributor is rotated to provide the specified timing. This specified timing is usually between 2 degrees and 12 degrees BTDC depending on the engine. When the engine speed increases, the pistons move up and down much faster and the ignition system must advance the spark timing, so the air-fuel mixture has just started to burn when the piston is at TDC **(Figure 12)**. This action maintains maximum downward force on the piston from the combustion process. If the ignition system does not advance the spark timing in relation to engine speed, the piston will be part way down in the power stroke by the time the air-fuel mixture begins to burn. Under this condition, engine power is greatly reduced.

Some distributors have a centrifugal and a vacuum advance to control spark timing. The centrifugal advance contains pivoted weights that fly outward against a spring tension as the distributor shaft speed increases. This outward weight movement rotates the reluctor ahead of the distributor shaft to provide spark advance in relation to engine speed **(Figure 13)**. The vacuum advance contains a diaphragm in a sealed chamber. Manifold vacuum or ported vacuum from above the throttle plate is supplied through a vacuum hose to the vacuum advance chamber. The other side of the vacuum advance chamber is open to atmospheric pressure and a rod on this side of the diaphragm is connected to the distributor pickup plate.

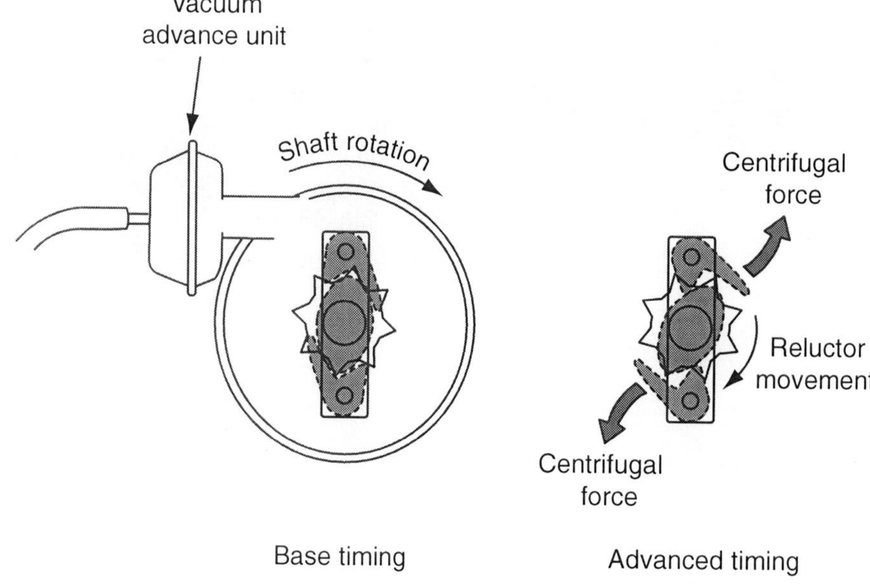


**Figure 13.** Centrifugal advance operation.

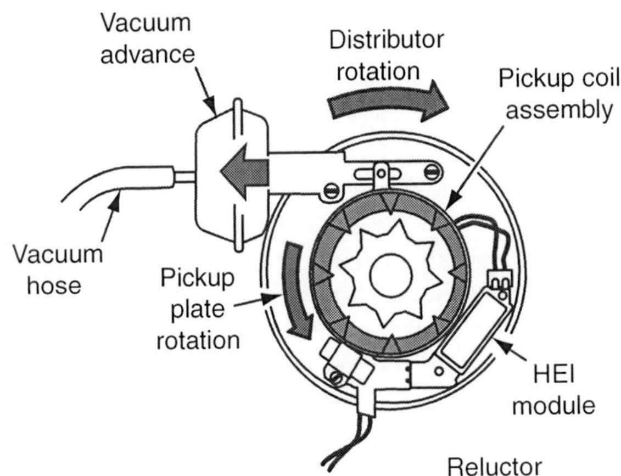


**Figure 14.** Vacuum advance operation.

When manifold vacuum is high during light throttle operation, the high manifold vacuum pulls the vacuum advance diaphragm against the spring tension. As the diaphragm moves, it rotates the pickup plate in the opposite direction to distributor shaft rotation. This action causes the pickup coil to be aligned with the reluctor high points sooner. This results in additional spark advance **(Figure 14)**. When the throttle is 75 percent open or more, manifold vacuum decreases. Under this condition, the spring on the vacuum advance diaphragm moves the diaphragm back to the retarded position. The vacuum advance controls spark advance in relation to engine load. During part throttle, light load operation, the vacuum advance provides additional spark advance to improve fuel economy and engine performance. If the engine is operating at wide throttle under heavy load conditions, the vacuum advance retards to prevent engine detonation.

## COMPUTER-CONTROLLED SPARK ADVANCE

In many ignition systems, the spark advance is controlled by the PCM and the distributor advances are not required. The ignition module is connected to the PCM and the negative coil primary terminal is connected to the ignition module **(Figure 15)**. On the basis of the pickup coil signals and signals from other sensors, such as the ECT sensor, the PCM determines the precise spark advance required by the engine. The PCM signals the ignition module to open the primary ignition circuit and fire each spark plug at the correct instant to provide the necessary spark advance.

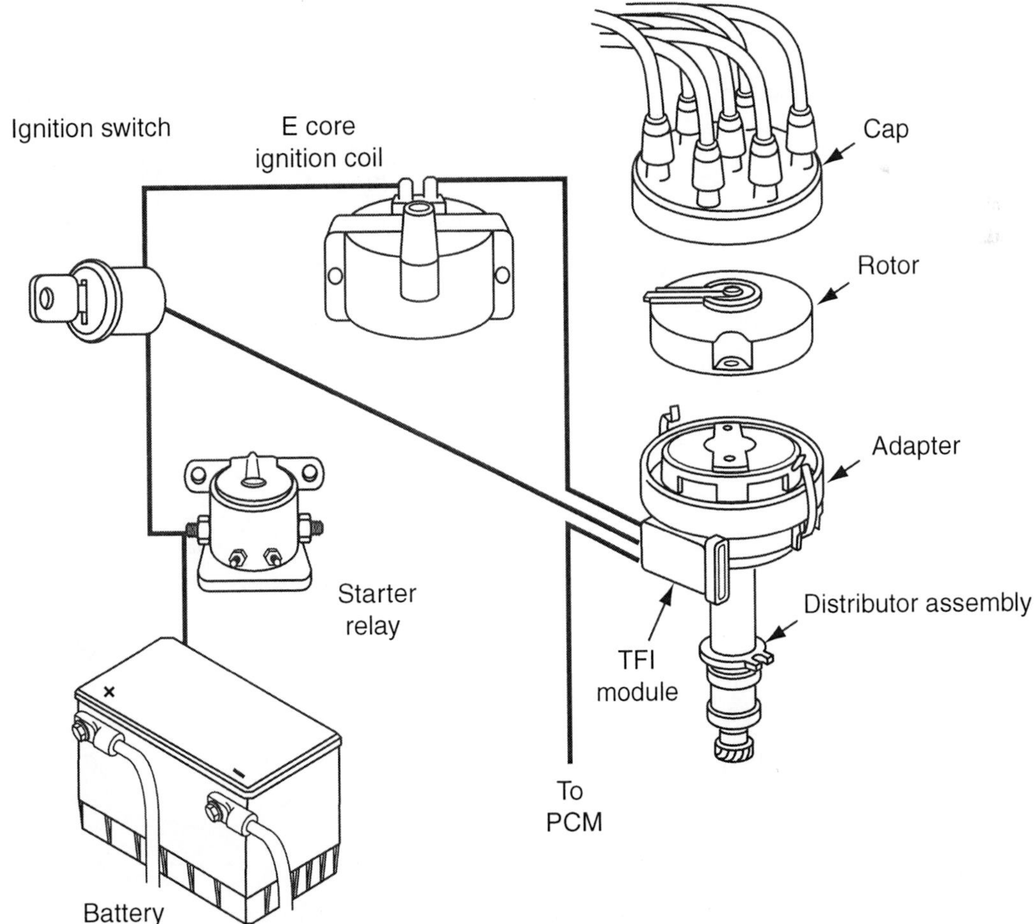


**Figure 15.** A distributor ignition (DI) with computer-controlled spark advance.

## ELECTRONIC IGNITION (EI) SYSTEMS

The SAE Standard J1930 is an attempt to standardize automotive powertrain terminology. This standard was developed in 1991 and has been revised since that time. In the J1930, the DI refers to ignition systems with a distributor and EI refers to distributorless ignition systems. In the J1930 terminology, the term for an engine computer is PCM. Most vehicle manufacturers use the J1930 terminology in their service publications.

### Crankshaft Position Sensor

A distributor is not required in the EI system and maximum secondary coil voltage is 20 percent higher compared to previous ignition systems. On some vehicles, the crankshaft position sensor is mounted in an opening in the transaxle bell housing **(Figure 16)**. The inner end of this sensor is positioned near a series of notches and slots that are integral with the transaxle flexplate **(Figure 17)**. A group of four slots is located on the transaxle drive plate for each pair of engine cylinders; thus a total of twelve slots are positioned around the drive plate on a V6 engine. The slots in each group are positioned 20 degrees apart. When the slots on the transaxle drive plate rotate past the crankshaft position sensor, the voltage signal from the sensor changes from 0 volt to 5 volts. This varying voltage signal informs the PCM regarding crankshaft position and speed and the PCM calculates spark advance from this signal. The PCM also uses the crankshaft position sensor signal along with other inputs to determine the air-fuel ratio. Base timing is determined by the signal from the last slot in each group of slots and base timing adjustment is not possible.

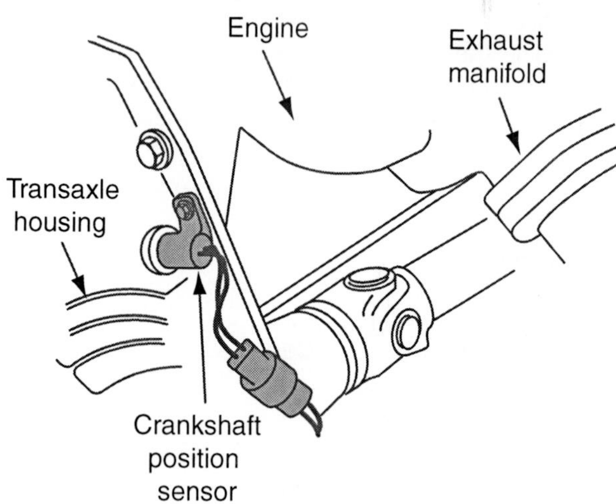


**Figure 16.** A crankshaft position sensor.

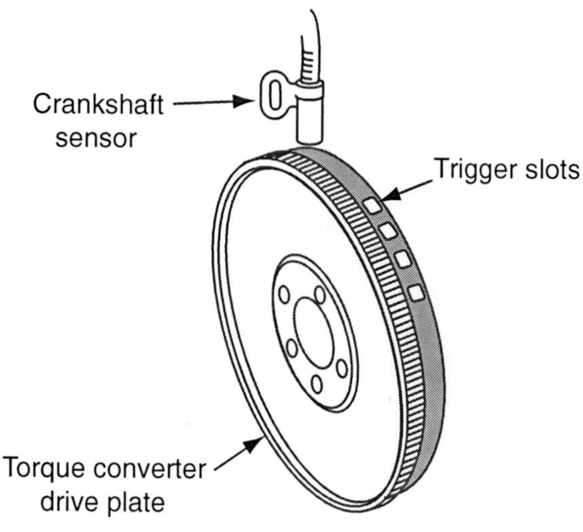


**Figure 17.** A crankshaft position sensor with transaxle drive plate.

### Camshaft Reference Sensor

The camshaft reference sensor is mounted in the top of the timing gear cover **(Figure 18)**. A notched ring on the camshaft gear rotates past the end of the camshaft reference sensor. This ring contains two single slots, two double slots, and a triple slot **(Figure 19)**.

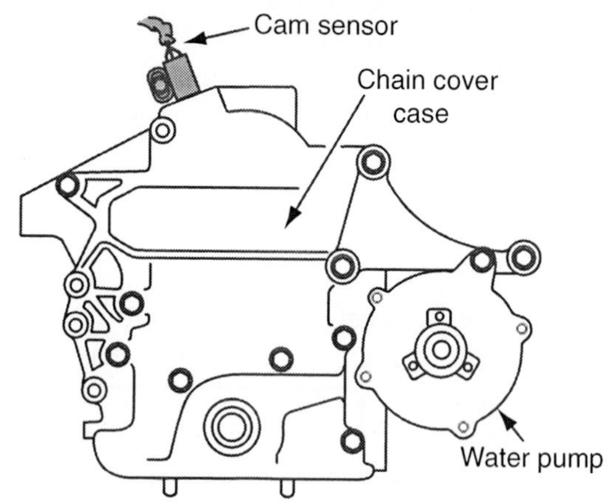


**Figure 18.** A camshaft reference sensor.

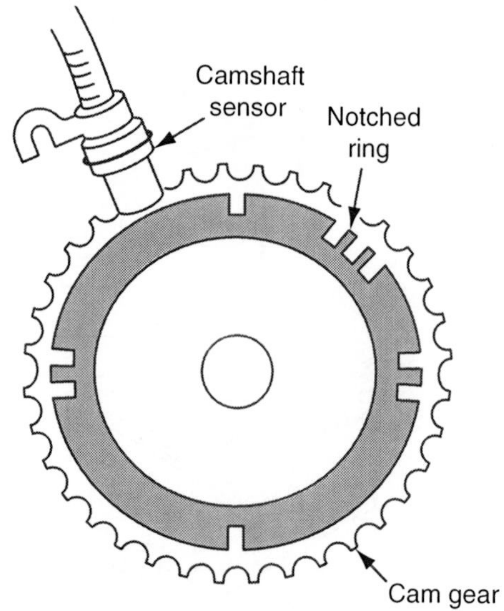


**Figure 19.** A notched ring on a camshaft gear.

When a camshaft gear notch rotates past the camshaft reference sensor, the signal from the sensor changes from 0 volt to 5 volts. The single, double, and triple notches provide different voltage signals from the camshaft reference sensor. These voltage signals are sent to the PCM. The ignition module is contained in the PCM so an external module is not required. The PCM determines the exact camshaft and crankshaft position from the camshaft reference sensor signals. The PCM uses these signals to sequence the coil primary windings and each pair of injectors at the correct instant. The PCM supplies 9.2 volts to 9.4 volts from terminal 7 through an orange wire to both the crankshaft position sensor and the camshaft reference sensor. A black ground wire is connected from both of these sensors to PCM terminal 4. The camshaft reference sensor signal is connected to PCM terminal 44, whereas the crankshaft position sensor signal is sent to PCM terminal 24.

## Coil Assembly

The coil assembly contains three ignition coils on a V6 engine **(Figure 20)**. The ends of each secondary winding are connected to the dual secondary terminals on each coil. Two spark plug wires are connected from the secondary terminals on each coil to the spark plugs. When the ignition switch is turned on, 12 volts are supplied through the ignition switch to the positive primary terminal on each ignition coil. The negative primary terminal on each coil is connected to the PCM.

**Figure 20.** An EI system coil assembly.

## Ignition System Operation

When the engine starts cranking, the spark plugs fire and the injectors discharge fuel within one crankshaft revolution. The spark plug wires from coil number 1 are connected to cylinders 1 and 4 and the spark plug wires from coil number 2 go to cylinders 2 and 5. The spark plug wires from coil number 3 are connected to cylinders 3 and 6. The firing order of some V6 engines is 1-2-3-4-5-6. When the single camshaft reference sensor slot rotates past the camshaft reference sensor, one high- and one low-digital voltage signal are received by the PCM. After these signals are received, the PCM receives the digital signals from the crankshaft position sensor as cylinder number 2 is approaching TDC on the compression stroke **(Figure 21)**.

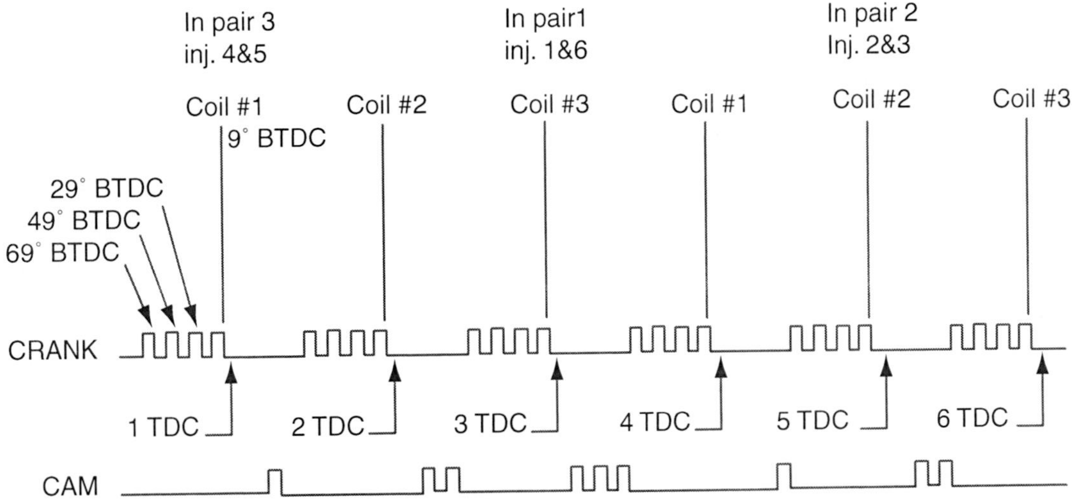


**Figure 21.** An EI system coil firing and injector sequencing.

When the engine is cranking and the last decreasing voltage signal is received from the crankshaft position sensor at 9 degrees BTDC, the PCM opens the primary circuit on the number 2 coil and fires spark plugs 2 and 5. Cylinder number 5 is on the exhaust stroke and firing the spark plug in this cylinder has no effect at this time. The coils in many EI systems operate on a **waste spark** principle of firing two spark plugs simultaneously with one spark plug being in a cylinder on the compression stroke and the other spark plug located in a cylinder on the exhaust stroke **(Figure 22)**.

When a coil fires two spark plugs in a waste spark system, the secondary current flows down through one pair of spark plug electrodes and up through the other pair of spark plug electrodes.

You Should Know

*In the EI system being explained, the ignition module is in the PCM. Some other EI systems have an ignition module that is mounted externally to the PCM, but the operating principle of both systems is similar.*

When the engine is cranking, the PCM fires all cylinders at 9 degrees BTDC. When the PCM fires spark plugs 2 and 5, it also sequences injectors 2 and 3 in this multiport fuel injection (MFI) system. In a **multiport fuel injection** system, the PCM opens two or more injectors simultaneously.

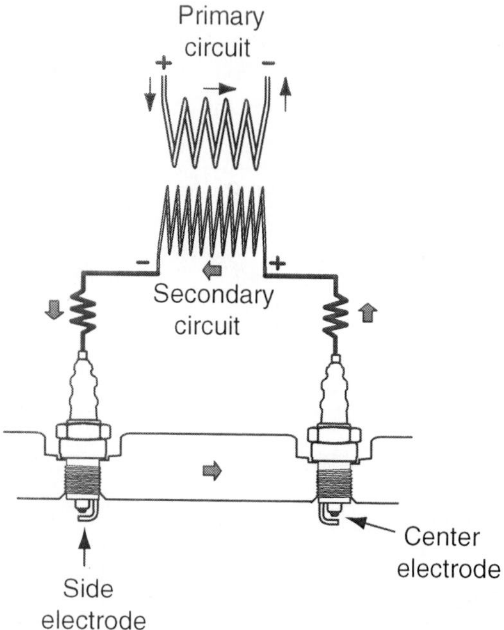


**Figure 22.** Spark plugs firing in an EI system.

After the double notch on the camshaft gear rotates past the camshaft reference sensor, the PCM fires coil 3, which is connected to spark plugs 3 and 6. The PCM also sequences injectors 1 and 6 when this camshaft reference sensor signal is received. After the triple notch on the camshaft gear rotates past the camshaft reference sensor, the PCM fires coil 1, which is connected to spark plugs 1 and 4 and the PCM also sequences injectors 4 and 5. If the engine is running, the PCM fires each coil to provide the precise spark advance required by the engine. The PCM receives crankshaft position sensor signals at 69 degrees, 49 degrees, 29 degrees, and 9 degrees BTDC on each cylinder. By counting the time interval between these crankshaft position sensor signals, the PCM can fire any cylinder at the precise spark advance required by the engine. The PCM determines this precise spark advance requirement from all the input data it receives from other input sensors. Some engines have a detonation sensor located in the cowl side of the engine block. If the engine detonates, this sensor signals the PCM to reduce the spark advance.

Many engines are now equipped with **sequential fuel injection (SFI)**, in which the PCM opens each injector individually, but the operating principle we have just described remains basically unchanged.

## COIL-ON-PLUG AND COIL-NEAR-PLUG IGNITION SYSTEMS

Many engines are presently equipped with coil-on-plug ignition systems. On these systems, the coil secondary terminals are connected directly to the spark plugs and secondary plug wires are not required **(Figure 23)**. This design eliminates the possibility of high voltage leakage

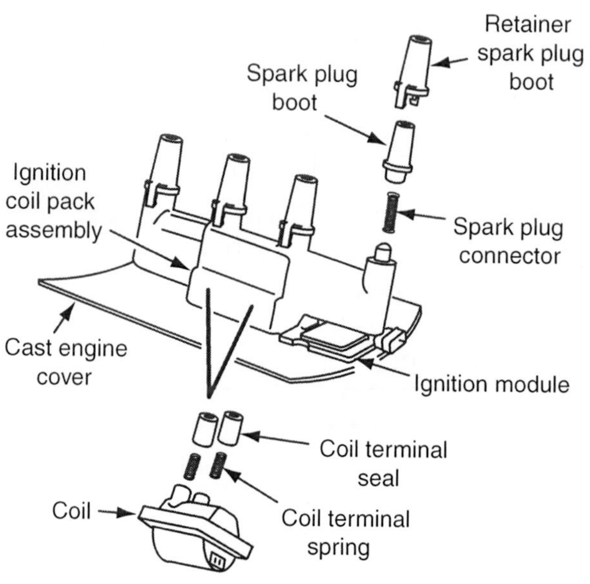


**Figure 23.** A coil-on-plug ignition system.

from secondary spark plug wires. A coil-on-plug ignition system may have two spark plugs connected to each coil as described in the previous EI system. Some coil-on-plug ignition systems have individual coils connected to each spark plug. In these systems, 12 volts are supplied to each positive primary terminal when the ignition switch is turned on and each negative primary terminal is connected to the ignition module. The crankshaft position and camshaft reference sensors are similar to those on the EI system described previously.

Coil-near-plug ignition systems have individual coils for each spark plug and a short spark plug wire is connected from each coil secondary terminal to the spark plug **(Figure 24)**.

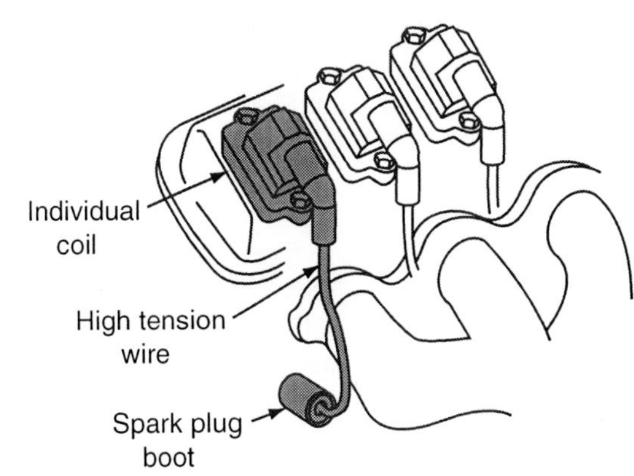


**Figure 24.** A coil-near-plug ignition system.

## Summary

- Spark plug reach is the distance from the lower edge of the plug shell to the shoulder on the shell that fits against the cylinder head.
- Spark plug heat range is determined by the length of the heat path from the center electrode to the location where the insulator contacts the plug shell.
- The recommended replacement interval for platinum-tipped spark plugs is 100,000 miles (160,000 kilometers).
- The primary winding in an ignition coil contains a few hundred turns of heavy wire and the secondary winding contains thousands of turns of very fine wire.
- The ignition switch, primary coil winding, ignition module, and pickup coil are in the primary ignition circuit.
- The secondary coil winding, coil wire, distributor cap and rotor, spark plug wires, and spark plugs are in the secondary ignition circuit.
- A voltage is produced in the distributor pickup coil each time a reluctor high point rotates past the pickup coil.
- The ignition module opens the primary ignition circuit each time it receives a voltage signal from the pickup coil.
- The ignition module must turn the primary circuit on long enough to allow the magnetic field to buildup in the ignition coil. This buildup time is referred to as dwell time.
- The centrifugal advance controls spark advance in relation to engine speed.
- The vacuum advance controls spark advance in relation to engine load.
- In many DI systems, the PCM controls the spark advance and the centrifugal and vacuum advances are not required.
- In an EI system, the distributor is eliminated and a coil is provided for each pair of cylinders.
- In an EI system, each coil fires two spark plugs simultaneously. One of these spark plugs is in a cylinder on the compression stroke and the other spark plug is in a cylinder on the exhaust stroke.
- In an EI system, the ignition module may be located in the PCM or may be mounted externally to the PCM.
- In a coil-on-plug ignition system, the spark plug wires are eliminated and the coil secondary terminals are connected directly to the spark plugs.
- A coil-near-plug ignition system has an individual coil mounted near each spark plug and a short secondary wire is connected from the coil secondary terminal to the spark plug.

## Review Questions

1. Technician A says that spark plugs with a hotter heat range have a higher number compared to spark plugs with a colder heat range. Technician B says spark plugs with a hotter heat range have a longer heat path compared to spark plugs with a colder heat range. Who is correct?
   A. Technician A
   B. Technician B
   C. Both Technician A and Technician B
   D. Neither Technician A nor Technician B
2. Technician A says the secondary coil winding is connected to the ignition module. Technician B says the primary coil winding contains many turns of very fine wire. Who is correct?
   A. Technician A
   B. Technician B
   C. Both Technician A and Technician B
   D. Neither Technician A nor Technician B

3. Technician A says in some EI systems, each coil fires two spark plugs simultaneously. Technician B says in some EI systems, the camshaft reference sensor is mounted in the transaxle bell housing. Who is correct?
   A. Technician A
   B. Technician B
   C. Both Technician A and Technician B
   D. Neither Technician A nor Technician B
4. Technician A says in an MFI system, the PCM fires each injector individually. Technician B says in an EI system, the PCM uses the camshaft reference sensor signal to fire the coils in the proper order. Who is correct?
   A. Technician A
   B. Technician B
   C. Both Technician A and Technician B
   D. Neither Technician A nor Technician B
5. A typical service interval on platinum-tipped spark plugs is:
   A. 35,000 miles.
   B. 50,000 miles.
   C. 80,000 miles.
   D. 100,000 miles.
6. All of these statements about DI systems are true *except*:
   A. The distributor pickup coil is part of the secondary circuit.
   B. The pickup coil leads are connected to the ignition module.
   C. The ignition module opens and closes the primary circuit.
   D. The ignition module may be mounted externally from the distributor.
7. In a DI system, the high voltage in the secondary winding is produced:
   A. when the reluctor high points are not aligned with the pickup coil.
   B. when the magnetic field begins to build up in the coil.
   C. when there is no voltage signal from the pickup coil to the module.
   D. when the magnetic field collapses in the coil.
8. Technician A says secondary reserve voltage in an ignition system is necessary to compensate for high combustion chamber pressures at heavy engine loads. Technician B says secondary reserve voltage in an ignition system is necessary to compensate for wear at spark plug electrodes. Who is correct?
   A. Technician A
   B. Technician B
   C. Both Technician A and Technician B
   D. Neither Technician A nor Technician B
9. When discussing DI systems, Technician A says the spark plug wires must be installed in the distributor cap in the opposite direction to distributor shaft rotation. Technician B says the spark plug wires must be installed in the distributor cap in the engine firing order. Who is correct?
   A. Technician A
   B. Technician B
   C. Both Technician A and Technician B
   D. Neither Technician A nor Technician B
10. All of these statements about EI systems are true *except*:
   A. The crankshaft position sensor informs the PCM regarding crankshaft position and speed.
   B. On many EI systems, the PCM uses the camshaft position sensor signals to properly sequence the injectors and ignition coils.
   C. On many EI systems, each coil fires two spark plugs at the same time.
   D. On some EI systems, each coil fires one spark plug on the compression stroke and another spark plug on the intake stroke.
11. The vacuum advance controls ignition spark advance in relation to engine ________________.
12. When the centrifugal advance weights move outward, they move the ________________ in relation to the distributor shaft.
13. In a DI system, the pickup coil leads are connected to the ________________ ________________.
14. In a primary ignition circuit, the dwell time is the time that the module keeps the primary circuit ________________ ________________.
15. Explain the importance of reserve voltage in the secondary ignition system.
16. Explain the waste spark principle in an EI system.
17. Describe the difference between MFI and SFI systems.
18. Explain the importance of dwell time in the primary ignition circuit.

# Chapter 27 Ignition System Maintenance, Diagnosis, and Service

## Introduction

Proper ignition system operation is extremely important to obtain satisfactory engine performance and fuel economy. For example, a defective spark plug or spark plug wire results in cylinder misfiring that greatly reduces engine performance and fuel economy. A defective ignition coil causes cylinder misfiring on some or all cylinders depending on the ignition system. This misfiring is most noticeable during hard engine acceleration and this problem also results in a significant reduction in engine power and fuel economy. Proper ignition system operation is maintained by performing accurate ignition system maintenance, diagnosis, and service.

## SPARK PLUG MAINTENANCE, DIAGNOSIS, AND SERVICE

In a four-cylinder engine running at 3,000 rpm, each spark plug is firing twenty-five times per second. A considerable amount of intermittent spark plug misfiring goes undetected because of the very rapid spark plug firing. However, this intermittent misfiring causes increased exhaust emissions and reduced fuel economy. Therefore, spark plug maintenance, diagnosis, and service are extremely important to provide the intended fuel economy and emission levels designed into the vehicle by the manufacturer. Spark plug conditions indicate what is happening in the cylinder. In some cases, spark plug conditions indicate a cylinder problem. If this problem is corrected when it is first detected, it may prevent a more expensive repair bill and/or tow bill later on.

### Spark Plug Removal

Spark plugs should be removed at the vehicle manufacturer's recommended service interval. This service interval may be from 20,000 to 100,000 miles (32,000 to 160,000 kilometers) depending on the type of spark plug and the operating conditions. For example, many engines are presently equipped with platinum-tipped spark plugs with a specified replacement interval of 100,000 miles (160,000 kilometers). When removing the spark plugs, grasp the spark plug wire and twist it back and forth on the spark plug before pulling the wire off the spark plug. Applying excessive pulling force to the spark plug wire may damage the wire. When loosening and removing spark plugs, use a special spark plug socket and keep the ratchet or breaker bar as close as possible to a 90-degree angle to the socket. This procedure helps to prevent the socket from cocking to one side and breaking the spark plug insulator. A spark plug socket has an internal rubber grommet that reduces the possibility of cracking the spark plug insulator. After spark plugs are loosened, use a shop air gun to blow any debris out of the spark plug recesses in the cylinder head. This action prevents debris from entering the cylinder when the spark plug is removed. When the spark plugs are removed, keep them in order so you can identify the spark plug removed from each cylinder. Cylinder operating conditions may be diagnosed by inspecting the spark plug conditions.

### Spark Plug Maintenance and Diagnosis

If normal combustion is occurring in all the cylinders, the lower end of each spark plug insulator should have a minimum amount of light tan or gray carbon deposits.

Inspect the spark plug electrodes. The current arcing across the electrodes slowly erodes the electrodes, but this electrode erosion should not exceed 0.001 inch (.0254 millimeter) for every 10,000 miles (16,000 kilometers).

**You Should Know** *In DI systems, the center spark plug electrodes erode faster than the ground electrodes because current is flowing from the center electrode to the ground electrode. In some EI systems, each coil fires two spark plugs, with current flowing down through one set of spark plug electrodes and up through the other set of electrodes. On the spark plugs with current flowing up through the electrodes, the ground electrode erodes faster than the center electrode. When center spark plug electrodes become eroded and rounded instead of having a flat surface, a higher secondary voltage is required to fire the spark plugs.*

Excessive electrode burning may be caused by too hot a spark plug heat range, improper spark plug torque and cooling, a lean air-fuel mixture, or engine detonation. Severely damaged spark plug electrodes indicate **preignition** damage **(Figure 1)**. Inspect the upper and lower ends of the insulator for cracks. Spark plugs with burned electrodes or cracked insulators must be replaced.

**Detonation** occurs in a cylinder when the compressed air-fuel mixture explodes violently instead of burning smoothly. This action is usually caused by a second flame front located away from the spark plug in the combustion chamber.

**Preignition** is the ignition of the air-fuel mixture in the combustion chamber by something other than the spark at the spark plug electrodes. This action is usually caused by hot carbon deposits or spark plugs with a heat range that is too hot.

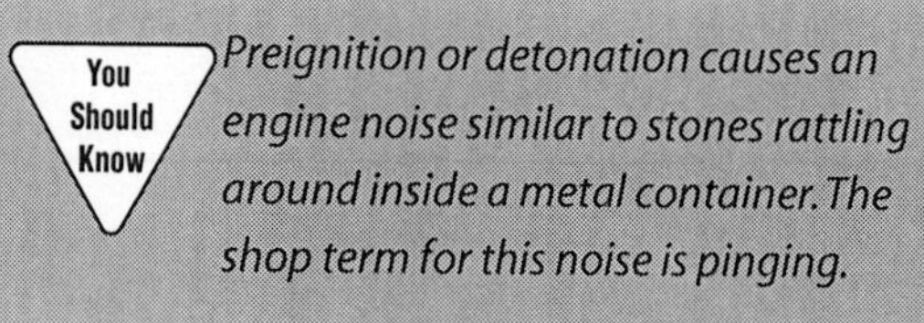

*Preignition or detonation causes an engine noise similar to stones rattling around inside a metal container. The shop term for this noise is pinging.*

Carbon-fouled spark plugs have dry, black carbon deposits. Any type of carbon deposits on the spark plug insulator causes the high secondary ignition voltage to arc from the center electrode across the carbon deposits to the spark plug shell. This action results in spark plug and cylinder misfiring and the air-fuel mixture in the combustion chamber is not ignited. Therefore, the cylinder does not contribute to engine power. Carbon-fouled spark plugs may be caused by a rich air-fuel mixture, secondary ignition misfiring on that spark plug, a colder-than-specified spark plug heat range, or low cylinder compression.

Oil-fouled spark plugs are recognized by wet oil deposits on the spark plug insulator and electrodes. Oil-fouled spark plugs have the same effect as carbon-fouled spark plugs. Oil-fouled spark plugs are caused by oil entering the combustion chamber. The causes of this problem may be worn piston rings or cylinder walls, worn valve guide seals or valve guides, leaking intake manifold gaskets on some V8 or V6 engines, a leaking transmission modulator, or a completely restricted PCV system.

Splash-fouled spark plugs have hard carbon deposits splashed on the spark plug insulators (refer to Figure 1). A splash-fouled spark plug is usually caused by combustion chamber deposits that loosen and stick to the spark plug insulator. This condition may occur after a cylinder has been misfiring, theregby allowing some carbon deposits to occur in the combustion chamber. When the cause of the misfiring is corrected, these carbon deposits may loosen and cause a splash-fouled spark plug.

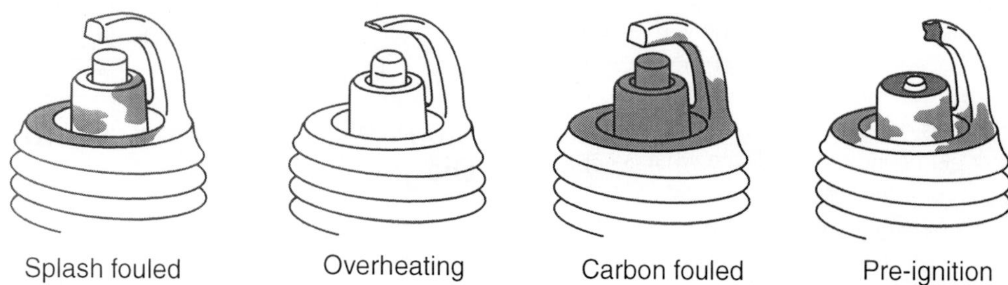


**Figure 1.** Defective spark plug conditions.

> **You Should Know** *Fouled spark plugs may not cause misfiring at idle speed because secondary ignition voltage is low under this condition. During hard acceleration, the secondary ignition voltage must increase because of the increase in combustion chamber pressure. This higher voltage increases the possibility of arcing across the carbon, oil, or splash-fouling spark plug deposits.*

Occasionally hot combustion chamber deposits loosen and become lodged between the spark plug electrodes. This spark plug condition is called gap bridging.

Spark plugs that show any of these defective conditions must be replaced. The cause of the defective spark plug condition must also be corrected. For example, if spark plugs indicate oil fouling, the cause of this problem must be corrected and new spark plugs installed.

## Spark Plug Service and Installation

When installing a new set of spark plugs, always use the spark plugs with the number specified by the vehicle manufacturer. For example, in an R46TS spark plug the R indicates a resistor-type spark plug, 4 represents 14-millimeter thread size, 6 indicates the heat range, T represents tapered seat, and S indicates extended electrodes. Spark plug electrode gaps must be set to the vehicle manufacturer's specifications using a round wire feeler gauge or a gapping tool **(Figure 2)**.

> **You Should Know** *Several years ago it was a recommended practice to clean spark plugs in a sand blaster. In modern engines, this procedure is not practical because spark plugs are inexpensive and the labor cost is high to remove and replace them on many engines. Attempting to clean platinum-tipped spark plugs in a sand blast-type cleaner will destroy the platinum coating on the electrodes.*

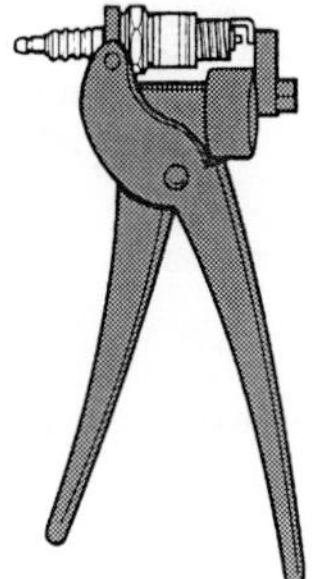

**Figure 2.** A spark plug gapping tool.

If spark plugs require a metal gasket, always be sure this gasket is in place before installing the plugs. Place a small amount of anti-seize compound on the first two spark plug threads. This is especially important on aluminum cylinder heads to prevent electrolytic action between the steel spark plug shell and the aluminum cylinder head.

Always start the spark plug into the cylinder head threads by hand to avoid damaging these threads. Turn the spark plug into the cylinder head threads by hand until the gasket or tapered seat contacts the cylinder head. Follow the vehicle manufacturer's recommended procedure and tighten the spark plugs to the specified torque.

> **You Should Know** *When spark plug torque is less than specified, the spark plug electrodes are not properly cooled. Under this condition, the excessive electrode temperature may cause engine detonation and severe engine damage.*

# DI SYSTEM MAINTENANCE, DIAGNOSIS, AND SERVICE

If DI system maintenance, diagnosis, and service is ignored or is not provided at the vehicle manufacturer's recommended interval, the result may be an expensive tow bill and a considerable amount of inconvenience. When maintenance, diagnosis, and service is ignored, minor ignition defects may become major problems that cause the engine to quit running on a busy highway or perhaps in an area where there is very little traffic and no service facilities nearby. Ignoring DI system maintenance, diagnosis, and service is usually more expensive in the long term than having this vehicle service completed at the vehicle manufacturer's recommended service interval.

## DI System Maintenance

At the vehicle manufacturer's recommended service interval, the components in the DI system should be inspected, diagnosed, and serviced. Spark plug wires should be inspected for corroded terminals on both ends. These wires should also be inspected for insulation cracks and heat damage. Test the spark plug wires with an ohmmeter on the X1,000 scale **(Figure 3)**. Most TVRS spark plug wires

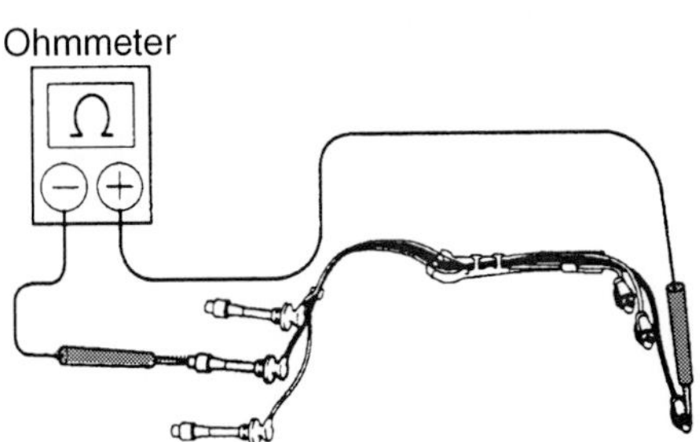


**Figure 3.** Testing spark plug wires with an ohmmeter.

should have a maximum resistance of 10,000 ohms per foot. Replace the spark plug wires as required. Spark plug wires should be placed in retaining clips so they are positioned away from exhaust manifolds and engine drive belts. When spark plug wires are placed in retaining clips, the wires from cylinders that fire one after the other should not be positioned beside each other. Spark plug wires should be routed away from computers and computer wiring harnesses. Be sure the spark plug wires are installed completely into the distributor cap terminals.

**You Should Know** *If spark plug wires from cylinders that fire one after the other are placed beside each other, the magnetic field from current flow through one wire may build up and collapse across the adjacent wire. This action induces a voltage in the adjacent wire that fires the spark plug connected to this wire when the piston in this cylinder is approaching TDC on the compression stroke. This action may cause preignition, resulting in spark plug electrode and possible engine damage.*

When performing basic ignition system maintenance, the distributor should be inspected. Inspect the distributor cap for corroded or worn terminals and cracks or carbon tracking **(Figure 4)**. Cracked distributor caps must be replaced. Corroded spark plug wire terminals in the cap may be cleaned with a special round wire brush, but do not remove any metal from these terminals. Terminals inside the cap may be cleaned with a small, flat file. Inspect the rotor for cracks and corroded or worn terminals **(Figure 5)**. If the distributor has advances, inspect the centrifugal advance for free movement and wear on the weight pivots. Use a vacuum hand pump to apply 20 in. hg. to the vacuum advance diaphragm. If this diaphragm is leaking, the vacuum will slowly decrease. When this vacuum is applied to the diaphragm, be sure the pickup plate rotates freely. If the vacuum advance diaphragm is leaking, replacement is necessary.

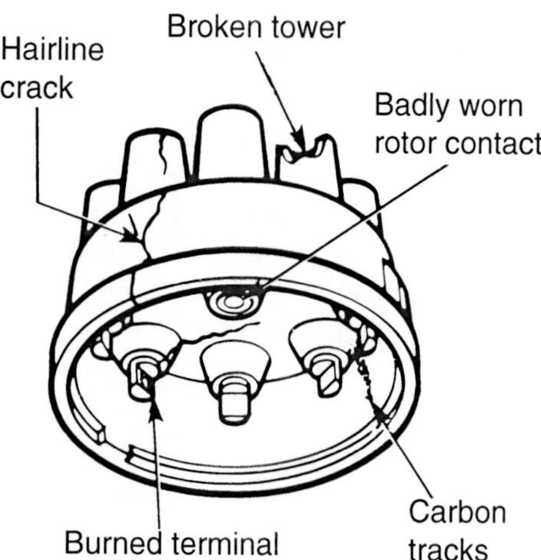


**Figure 4.** Distributor cap inspection.

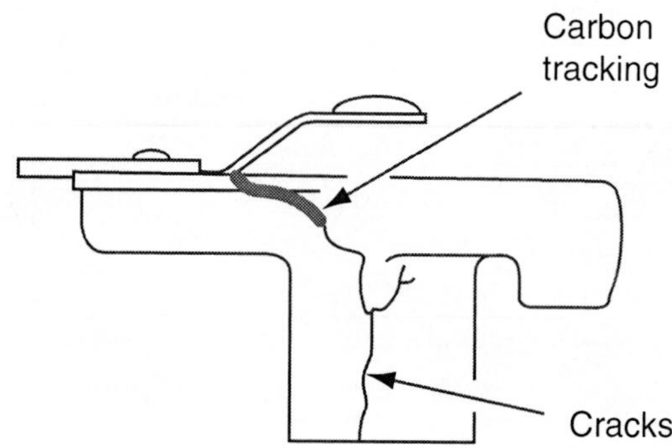


**Figure 5.** Rotor inspection.

Place the distributor housing in a soft-jawed vice and grasp the top of the shaft assembly. Move the shaft sideways and check the amount of shaft movement. Worn distributor shaft bushings result in excessive sideways shaft movement. If this movement is excessive, most vehicle manufacturers do not supply replacement bushings, and so the distributor replacement is required. Inspect the distributor drive gear for worn or damaged teeth. Inspect the pickup coil and module leads for worn insulation and loose terminals **(Figure 6)**.

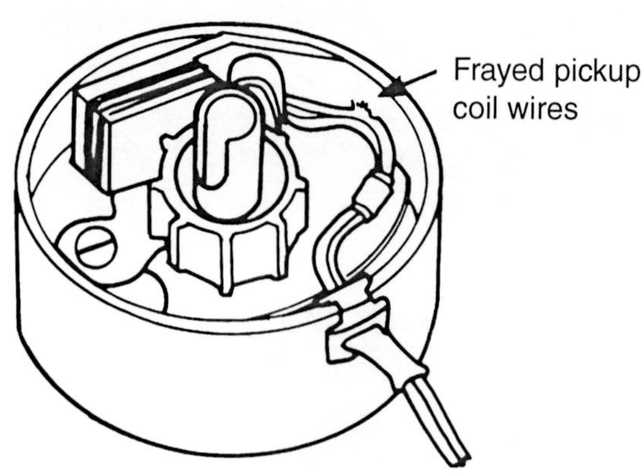


**Figure 6.** Inspecting pickup coil leads.

## DI System No-Start Diagnosis

If the engine fails to start, use a voltmeter to measure the voltage from the positive primary ignition coil terminal to ground with the ignition switch on. On most ignition systems, 12 volts should be available at this terminal. Some older ignition systems have a resistor in the primary circuit that reduced the voltage at the positive primary terminal to approximately 6 volts. When the specified voltage is not available at the positive primary ignition coil terminal, test the ignition switch and the wire from this switch to the coil.

If the specified voltage is available at the positive primary coil terminal, connect a 12-volt test light from the negative primary terminal to ground and crank the engine. If the test light flashes on and off, the primary circuit is being triggered on and off by the module, indicating the pickup coil and module are satisfactory. When the test light does not flutter, one of these components is defective and individual component testing is necessary to determine the defective component. If the test light flashes on and off, connect the proper test spark plug from the coil secondary lead to ground and crank the engine **(Figure 7)**. If the test spark plug does not fire, the ignition coil is defective. When the test spark plug fires normally, the coil is satisfactory. Connect the test spark plug from several spark plug wires to ground and crank the engine If the test spark plug does not fire when connected to the spark plug wires, the secondary voltage is leaking in the distributor cap and rotor.

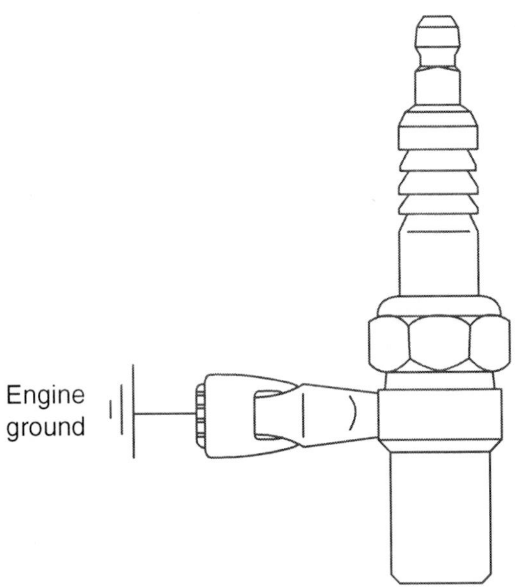


**Figure 7.** A test spark plug.

## Pickup Coil Diagnosis

An electromagnetic-type pickup coil may be tested with an ohmmeter on the X100 scale. Connect the ohmmeter leads to the pickup lead terminals **(Figure 8)**. A satisfac-

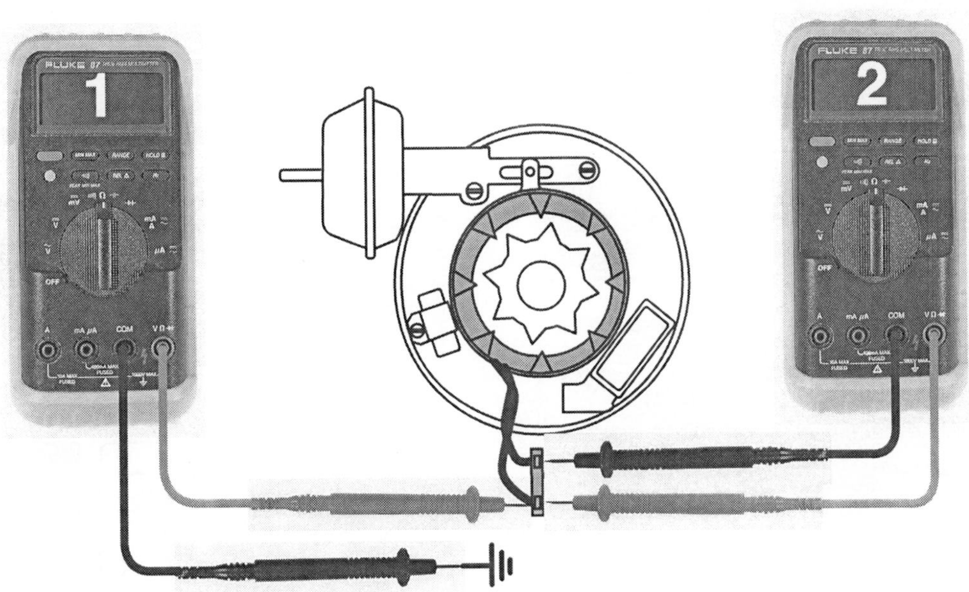


**Figure 8.** Pickup coil tests.

tory pickup coil indicates the specified resistance. This resistance is usually between 150 and 900 ohms. An infinite ohmmeter reading indicates an open pickup coil, whereas a reading below the specified value indicates a shorted pickup coil.

An **electromagnetic pickup coil** contains a permanent magnet surrounded by a coil of wire. An **infinite** ohmmeter reading is an ohm reading beyond measurement, causing an "OL" display on a digital ohmmeter. An **open pickup coil** has an unwanted break in the winding. In a **shorted pickup coil,** the turns of wire are touching each other, thereby reducing the effective number of turns and the coil resistance.

Connect the ohmmeter leads from one of the pickup coil leads to ground. An infinite reading indicates the pickup coil is not grounded, whereas a low reading indicates a grounded pickup coil. If either of the pickup coil tests are unsatisfactory, the pickup coil must be replaced. A **grounded pickup coil** has an unwanted connection between the pickup coil winding or lead wires and ground on the distributor housing.

The gap between the head on the pickup coil and the reluctor high points may be measured with a non-magnetic feeler gauge. On some pickup coils, the mounting screws may be loosened and the pickup coil moved to adjust the pickup coil gap. Always check for excessive side-to-side distributor shaft movement, which indicates worn distributor bushings.

You Should Know

*A non-magnetic feeler gauge must be used to measure the pickup coil gap because a metal gauge sticks to the pickup magnet and provides an inaccurate reading. Some pickup coils are riveted to the breaker plate and the gap on these pickup coils is not adjustable.*

You Should Know

*A variety of testers is available to test ignition modules. These testers usually check the ability of the module to switch the primary ignition circuit on and off.*

## Testing Ignition Coil Windings

Place an ohmmeter on the lowest scale and connect the meter leads to the primary coil terminals **(Figure 9)**. A satisfactory primary winding provides the specified resistance, which is usually between 0.5 and 2.0 ohms. If the ohmmeter reading is less than specified, the primary winding is shorted. An infinite ohmmeter reading indicates an open primary winding.

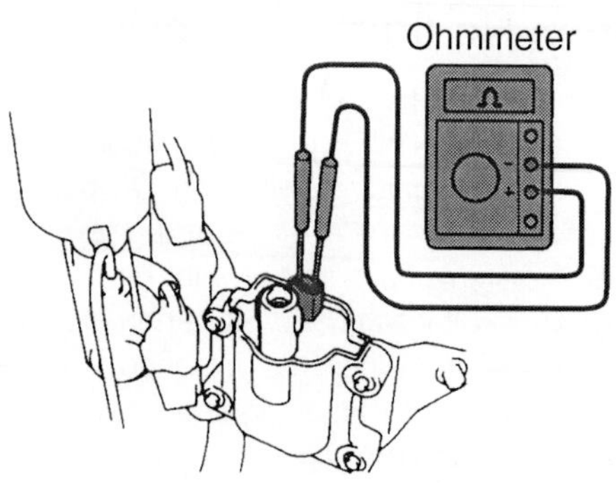


**Figure 9.** Testing the ignition coil primary winding.

You Should Know

*A shorted primary winding causes excessive primary current, which may damage other components in the primary circuit, such as the ignition module.*

Switch the ohmmeter to a higher scale and connect the ohmmeter leads from the secondary coil terminal in the coil tower to one of the primary terminals **(Figure 10)**. A satisfactory secondary winding provides the specified ohmmeter reading, which is usually 8,000 to 20,000 ohms. An infinite reading indicates an open secondary winding and a lower-than-specified reading indicates a shorted secondary winding.

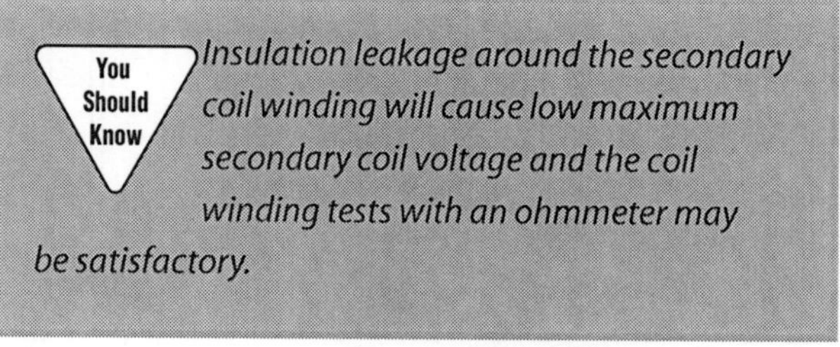
You Should Know

*Insulation leakage around the secondary coil winding will cause low maximum secondary coil voltage and the coil winding tests with an ohmmeter may be satisfactory.*

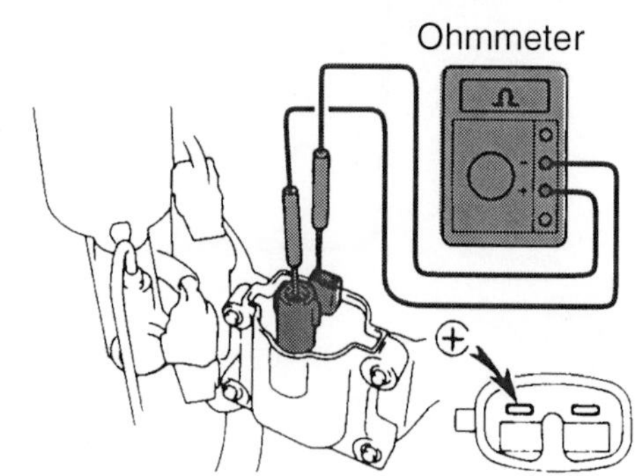


**Figure 10.** Testing the secondary ignition coil winding.

## Timing the Distributor to the Engine

If the distributor is removed, it must be properly timed to the engine when it is re-installed. Follow this procedure to time the distributor to the engine:

1. Remove the spark plug from number 1 cylinder and place a compression gauge hose fitting in the spark plug hole.
2. Crank the engine a small amount at a time until compression pressure appears on the gauge.
3. Crank the engine a very small amount at a time until the 0 degree position on the timing marks is aligned with the timing indicator.
4. Locate the number 1 spark plug wire position in the distributor cap. The wire terminals in some caps are marked (the manufacturer's service manual provides this information).
5. Install the distributor in the block with the rotor positioned under the number 1 plug wire position in the distributor cap and the vacuum advance in the original position. The distributor drive gear easily meshes with the camshaft gear, but many distributors also drive the oil pump with a hex-shaped or slotted drive in the lower end of the distributor gear or shaft. It may be necessary to hold down on the distributor housing and crank the engine to get the distributor shaft into mesh with the oil pump drive. When this action is required, repeat step 2 and step 3 and be sure the rotor is under number 1 plug wire terminal in the distributor cap with the timing marks aligned.
6. Rotate the distributor a small amount until a high point on the reluctor is aligned with the head on the pickup coil.
7. Install the distributor holddown clamp and bolt, leaving this bolt slightly loose.
8. Install the spark plug wires in the distributor cap in the direction of distributor shaft rotation and in the cylinder firing order **(Figure 11)**.

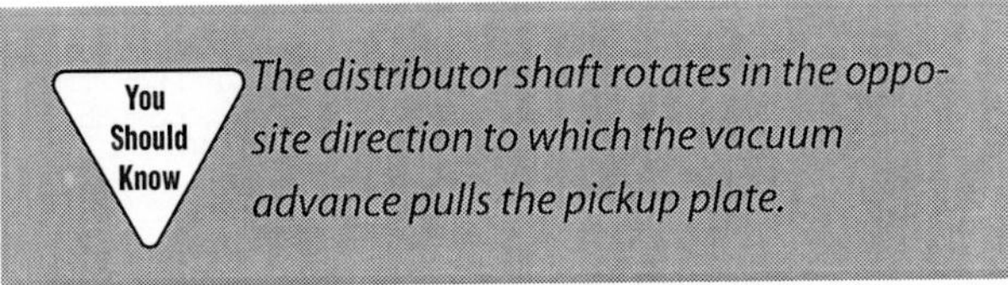


*The distributor shaft rotates in the opposite direction to which the vacuum advance pulls the pickup plate.*

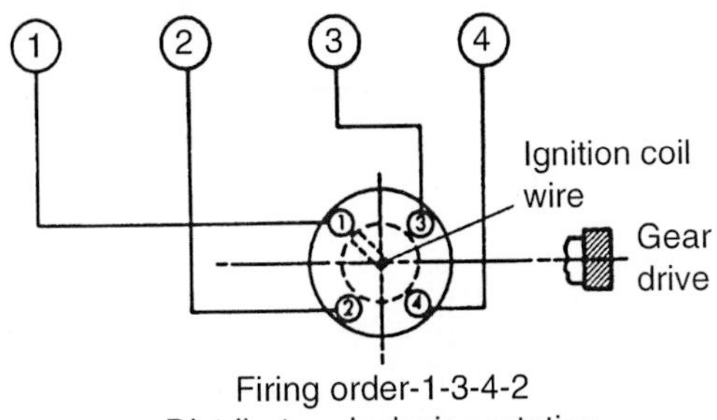


**Figure 11.** Installation of spark plug wires in the firing order and direction of distributor shaft rotation.

9. Connect the distributor wiring connectors. The vacuum advance hose is usually left disconnected until the ignition timing is set with the engine running.

## Checking and Adjusting Ignition Timing

The ignition timing procedure varies depending on the make and year of vehicle and the type of ignition system. Ignition timing specifications and instructions are included on the underhood emissions label. More detailed instructions are provided in the vehicle manufacturer's service manual. The ignition timing procedure recommended by the vehicle manufacturer must be followed. On distributors with advance mechanisms, manufacturers usually recommend disconnecting and plugging the vacuum advance hose while checking the ignition timing. On carbureted engines, the manufacturer usually specifies a certain engine rpm while checking ignition timing. The timing light pickup is connected to number 1 spark plug wire and the power supply wires on the light are connected to the battery terminals with the correct polarity. Follow these steps for ignition timing adjustment:

1. If the spark advance is controlled by the PCM, disconnect the in-line timing connector. The underhood emissions label usually provides the location of this connector. When the timing connector is disconnected, the PCM cannot provide any spark advance.
2. Connect the timing light and start the engine.
3. The engine must be idling at the specified rpm and all other timing procedures must be followed.
4. Aim the timing light at the timing indicator and observe the timing marks **(Figure 12)**.

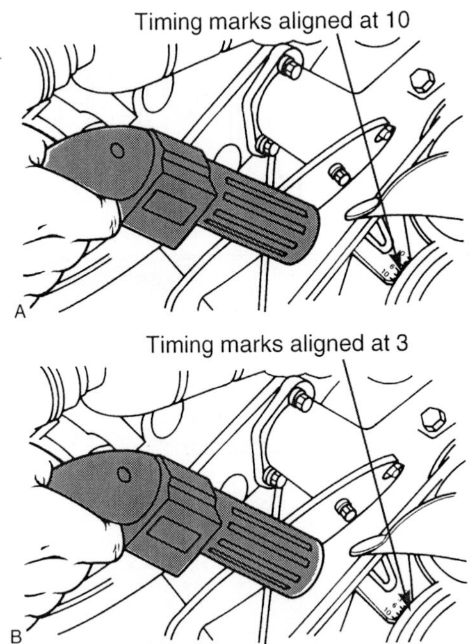


**Figure 12.** Checking ignition timing.

5. If the timing mark is not at the specified position, rotate the distributor a small amount until this mark is at the specified location.
6. Tighten the distributor holddown bolt to the specified torque and recheck the timing mark position.
7. Connect the vacuum advance hose, the in-line timing connector, and any other connectors, hoses, or components that were disconnected for the timing procedure.

## EI MAINTENANCE, DIAGNOSIS, AND SERVICE

Spark plug and spark plug wire inspection, diagnosis, and service is basically the same on DI and EI systems. Since EI systems do not have a distributor, servicing of this component is eliminated.

On many EI systems, if the crankshaft position sensor or camshaft reference sensor are defective, the engine will not start. On some vehicles, if the camshaft reference sensor becomes defective when the engine is running, the engine continues to run but after the engine is shut off it will not restart. On other EI systems, if the camshaft reference sensor is defective, the PCM will allow the engine to start because it estimates the camshaft position after the engine cranks a few times.

If the crankshaft position sensor or camshaft reference sensor is defective, a DTC is stored in the PCM memory. These DTCs may be read by connecting a scan tool **(Figure 13)** to the data link connector, which is usually located under the dash **(Figure 14)**. A DTC indicates a problem in a specific area. For example, a DTC representing a crankshaft position sensor fault indicates a defective crankshaft position sensor, faulty wires between the sensor and the PCM, or a problem with the PCM receiving this signal.

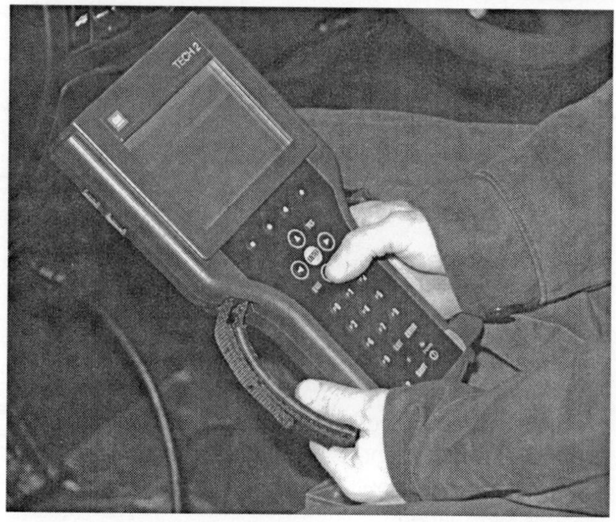


**Figure 13.** A scan tool.

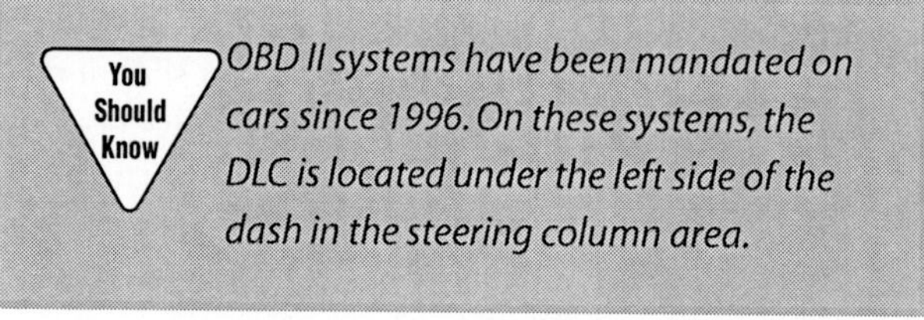


*OBD II systems have been mandated on cars since 1996. On these systems, the DLC is located under the left side of the dash in the steering column area.*

The crankshaft position sensor and camshaft reference sensor may be tested using the procedure in the vehicle manufacturer's service manual. If the EI system has electromagnetic crankshaft and camshaft sensors, they may be

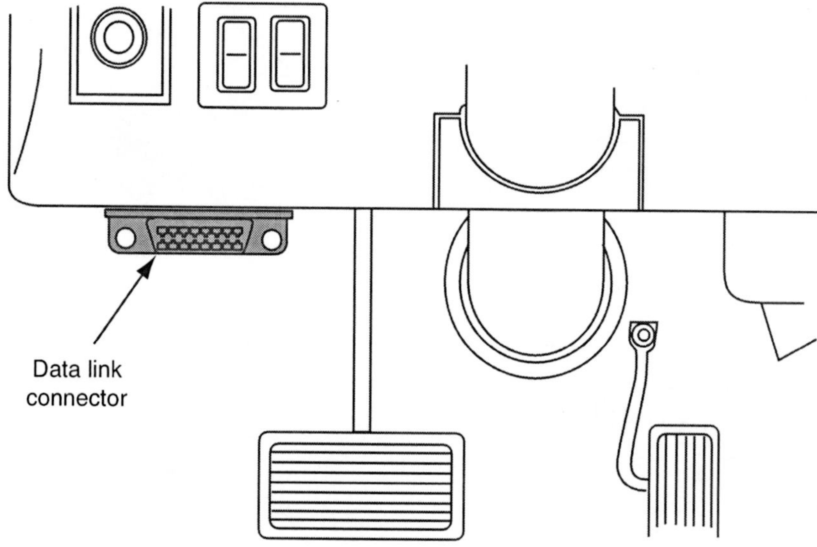


**Figure 14.** A data link connector (DLC).

tested with an ohmmeter using the same procedure as described previously in this chapter for testing electromagnetic pickup coils. When the EI system has Hall effect crankshaft and camshaft sensors, these sensors may be tested with a digital voltmeter, graphing voltmeter, or lab scope. Hall effect sensors have three wires. These wires include a voltage supply wire, a signal wire, and a ground wire. When testing this type of sensor, always be sure the specified voltage is available on the voltage supply wire with the ignition switch on. This voltage is usually between 5 volts and 12 volts depending on the system. Connect a digital voltmeter to the Hall effect sensor signal wire and crank the engine. The signal voltage should vary from a very low voltage to a high voltage if the sensor is satisfactory. On a graphing voltmeter or lab scope, a satisfactory Hall Effect switch should provide a square waveform as indicated in **Figure 15**. If the Hall effect crankshaft and camshaft sensor signals are satisfactory, it may be necessary to test the wires between these sensors and the PCM to locate the cause of the DTC representing one of these sensors.

A **graphing voltmeter** or **lab scope** provides waveforms or voltage traces representing the voltages in electronic circuits. However, the lab scope scans voltage signal much faster, actually in billionths of a second. For this reason, the lab scope waveform will indicate momentary defects in electronic components that other types of test equipment may fail to display.

> **You Should Know** *In some EI systems, the crankshaft position sensor is mounted at the front of the crankshaft and a notched ring on the crankshaft pulley rotates near the tip of this sensor. On some EI systems, a gap adjustment is necessary between the crankshaft position sensor and the notched ring. The crankshaft sensor position may be moved to correct this gap.*

A test spark plug may be used to test for secondary voltage in EI systems. If the spark plugs are not firing, always be sure 12 volts are supplied to the primary winding of each coil with the ignition switch on.

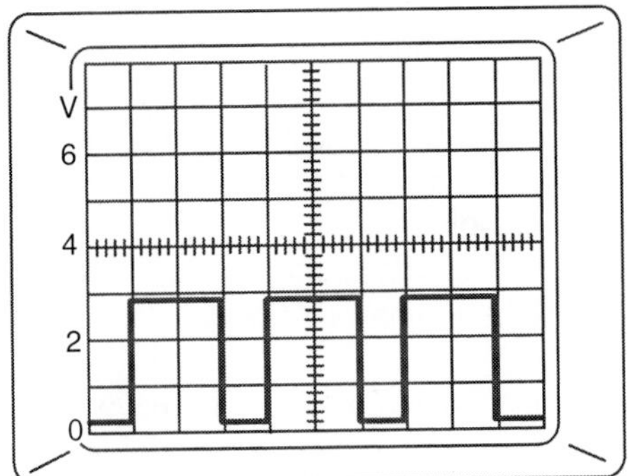


**Figure 15.** A voltage waveform from a Hall effect-type sensor.

> **You Should Know** *On most DaimlerChrysler vehicles, voltage is supplied to the primary coil windings through an automatic shutdown (ASD) relay. On most systems, this relay also supplies voltage to the fuel injectors and the oxygen sensor heaters. If the ASD relay is defective, the primary ignition circuit and the fuel injectors are inoperative.*

Ignition coils in these systems may be tested with an ohmmeter as discussed previously in this chapter. Timing adjustments are not possible on EI systems.

## COIL-ON-PLUG AND COIL-NEAR-PLUG IGNITION SYSTEM MAINTENANCE, DIAGNOSIS, AND SERVICE

Spark plug and spark plug wire maintenance, diagnosis, and service on coil-on-plug and coil-near-plug ignition systems is basically the same as the service on these components in DI systems described previously in this chapter. Since coil-on-plug ignition systems do not have spark plug wires, servicing of these items is not required. Distributor service and timing procedures are not required on these systems because the distributor is eliminated.

The crankshaft position sensor and camshaft reference sensor on coil-on-plug and coil-near-plug systems may be diagnosed using the same procedure as described previously on the EI systems. An ohmmeter may be used to test the coil windings as explained previously on DI systems.

*At the Paris Motor Show in 2003, Jaguar displayed a polished (instead of painted) new XJ model with an aluminum body. During the design process, the target weight reduction for the new XJ's body was 40 percent less than a comparable steel body. This weight reduction was achieved and the aluminum body weight is 485 pounds. The advantages of reducing vehicle weight are improved fuel economy and performance with reduced emissions. David Scholes, Chief Program Engineer for Jaguar's new XJ model stated, "I can see within 25 years more aluminum cars being produced than steel. We need to find increasingly efficient ways of producing cars, and I believe that if predictions about fuel and taxation trends come to fruition, aluminum could be used very widely, including for the high-volume end of the market."*

## Summary

- Platinum-tipped spark plugs usually have a service interval of 100,000 miles (160,000 kilometer).
- Spark plug conditions are an indicator of cylinder operating conditions.
- Carbon-fouled spark plugs may be caused by a rich air-fuel mixture, too cold a spark plug heat range, cylinder misfiring, or low cylinder compression.
- Oil-fouled spark plugs may be caused by worn piston rings or cylinders, worn valve guides or seals, or a defective transmission modulator.
- Splash-fouled spark plugs or spark plugs with bridged gaps are caused by loosened combustion chamber deposits that adhere to the spark plug insulator or electrodes.
- Proper spark plug torque is very important to provide adequate spark plug cooling.
- A no-start diagnosis may be performed with a test spark plug and a 12-volt test light.
- If a 12-volt test light connected from the negative primary coil terminal to ground does not flash when cranking the engine, the pickup coil or ignition module is defective.
- An electromagnetic-type pickup coil may be tested for grounds by connecting an ohmmeter from one of the pickup coil leads to ground.
- The ignition coil windings may be tested for opens and shorts with an ohmmeter.
- When timing a distributor to the engine, number 1 cylinder must be located at TDC on the compression stroke, the rotor must be under number 1 plug wire terminal in the distributor cap, and one of the reluctor high points must be aligned with the pickup coil.
- On many DI systems with PCM-controlled spark advance, an in-line timing connector must be disconnected when checking ignition timing.
- Hall effect-type sensors may be tested with a digital voltmeter, a graphing voltmeter, or a scope.
- Timing adjustments are not possible on EI, coil-on-plug, or coil-near-plug systems.

## Review Questions

1. While diagnosing an EI system with a scan tool, a DTC is obtained representing the crankshaft position sensor. Technician A says this DTC proves the crankshaft position sensor is defective. Technician B says the wires from the crankshaft sensor to the PCM may be defective. Who is correct?
   A. Technician A
   B. Technician B
   C. Both Technician A and Technician B
   D. Neither Technician A nor Technician B
2. Technician A says a Hall effect-type sensor may be tested with a lab scope. Technician B says this type of sensor may be tested with a digital voltmeter. Who is correct?
   A. Technician A
   B. Technician B
   C. Both Technician A and Technician B
   D. Neither Technician A nor Technician B
3. Technician A says a defective crankshaft position sensor causes a no-start condition on an EI system. Technician B says if the spark plugs are not firing on an EI system, check the voltage supply to the positive primary terminal on each ignition coil. Who is correct?
   A. Technician A
   B. Technician B
   C. Both Technician A and Technician B
   D. Neither Technician A nor Technician B
4. Technician A says when installing spark plug wires in a distributor cap, they should be installed in the cylinder firing order and in the direction of distributor shaft rotation. Technician B says in a DI system, the distributor shaft rotates in the same direction as the vacuum advance pulls the pickup plate. Who is correct?
   A. Technician A
   B. Technician B
   C. Both Technician A and Technician B
   D. Neither Technician A nor Technician B
5. When the spark plug is removed from number 3 cylinder, the spark plug is wet with oil. The most likely cause of this problem is:
   A. excessive valve clearance.
   B. excessive low engine operating temperature.
   C. worn piston rings in number 3 cylinder.
   D. a rich air-fuel ratio.
6. Spark plug misfiring is most likely to occur:
   A. when the engine is at normal operating temperature and idle speed.
   B. when it is at or near the wide-open position.
   C. when the engine is decelerating and the throttle is closed.
   D. at a constant cruising speed of 60 mph.

7. All of these conditions must be present to properly time a distributor to an engine *except*:
   A. The pickup coil must be aligned with the mark on the distributor housing.
   B. The piston in number 1 cylinder must be positioned at TDC on the compression stroke with the timing marks in the 0 degree position.
   C. The distributor rotor must be under number 1 spark plug wire terminal in the distributor cap.
   D. One of the reluctor high points must be aligned with the pickup coil.
8. When diagnosing a no-start complaint on a DI system, a test spark plug fires when connected from the coil secondary lead wire to ground and the engine is cranked, but the test spark plug does not fire when connected from several spark plug wires to ground and the engine is cranked. Technician A says the ignition coil is defective. Technician B says the pickup coil is defective. Who is correct?
   A. Technician A
   B. Technician B
   C. Both Technician A and Technician B
   D. Neither Technician A nor Technician B
9. All of these statements about crankshaft position sensors and camshaft reference sensors in EI systems are true *except*:
   A. A defective crankshaft position (CP) sensor causes a no-start condition.
   B. A Hall effect CP sensor has three wires.
   C. When tested with a lab scope, a satisfactory Hall effect CP sensor produces a square-wave voltage signal.
   D. When tested with a lab scope, a satisfactory electromagnetic CP sensor produces a digital voltage signal.
10. An engine with a coil-on-plug ignition system has a no-start condition and the spark plugs are not firing. Technician A says to test the voltage supply to each coil positive primary terminal with the ignition switch on. Technician B says to test the CP sensor signal. Who is correct?
    A. Technician A
    B. Technician B
    C. Both Technician A and Technician B
    D. Neither Technician A nor Technician B
11. Carbon-fouled spark plugs have ____________ ____________ carbon deposits.
12. Normal spark plug carbon deposits should be ____________ or ____________ colored.
13. TVRS spark plug wires should have a maximum resistance of ____________ ohms per foot.
14. When testing an electromagnetic-type pickup coil, the ohmmeter leads are connected to the pickup leads and an infinite meter reading is obtained. This reading indicates the pickup coil is ____________.
15. List the causes of oil-fouled spark plugs.
16. Define preignition and detonation in the combustion chambers and explain the causes of these problems.
17. Describe a no-start diagnosis on a DI system using a 12-volt test light and a test spark plug.
18. Explain the proper procedures for diagnosing and testing crankshaft position and camshaft reference sensors.

# Section 7

## Engine Control Systems



## SECTION OBJECTIVES

After you have read, studied, and practiced the contents of this section, you should be able to:

- Describe an analog voltage signal.
- Recognize a digital voltage signal.
- Explain binary coding.
- Describe the purpose of a microprocessor chip in a computer.
- Explain how a computer controls the system outputs.
- Describe adaptive strategy.
- Describe the operation of the fuel pump circuit.
- Describe the operation of a zirconium-type oxygen sensor.
- Explain open loop and closed loop computer operation.
- Define a speed density fuel injection system.
- Describe the operation of the operation of various input sensors.
- Perform input sensor tests.
- Explain the design of TBI, MFI, and SFI, and CPI systems.
- Define trip and drive cycle in an OBD II system.
- Perform fuel pump pressure and injector tests and interpet.
- Obtain DTCs on OBD I and OBD II systems.
- Remove and replace fuel pumps and fuel injectors.

*One of the latest concept cars is the Hy-Wire introduced by General Motors. This car is powered by fuel cells that produce electricity for the front- and rear-wheel drive motors. The fuel cells use hydrogen fuel to produce electricity. The fuel cell stack, hydrogen storage tanks, and drive motors are all housed in a chassis that is 11 inches (280 millimeters) thick in the center and 7 inches (178 millimeters) thick at the ends. When the body is removed, the chassis resembles a giant skateboard. The only byproducts from the fuel cells are warm air and water. Different bodies can be fitted to the same chassis. Mechanical linkages, such as the brake pedal, steering, and throttle, are eliminated. The driver uses two handgrips to send all handling instructions to the vehicle electronics. Byron McCormick, head of GM's fuel cell research says, "Because hydrogen can be made from renewable sources, in the future we believe that fuel cells and hydrogen will allow us to have sustainable mobility." In other words, the environmental impact of the automobile will be reduced to almost zero.*

# Engine Control Computers and Output Controls

## Introduction

It is extremely important for technicians to understand the function of engine computers and output controls. When the understanding of these components is inadequate, fast and accurate diagnosis of component malfunctions becomes more difficult. If the technician understands the function of these components, it is much easier to recognize and diagnose improper component operation.

### ANALOG VOLTAGE SIGNALS

When a rheostat is used to control a 5-volt bulb if the rheostat voltage is low, a small amount of current flows through the bulb to produce a dim light from the bulb **(Figure 1)**. If the rheostat voltage is 5 volts, higher current flow produces increased light brilliance. As the rheostat voltage is decreased, the light becomes dimmer. The rheostat voltage may be anywhere between 0 volts and 5 volts. This is an example of analog voltage. Many of the input sensors in an engine computer system produce analog voltages. **Analog voltage** is continuously variable within a specific range.

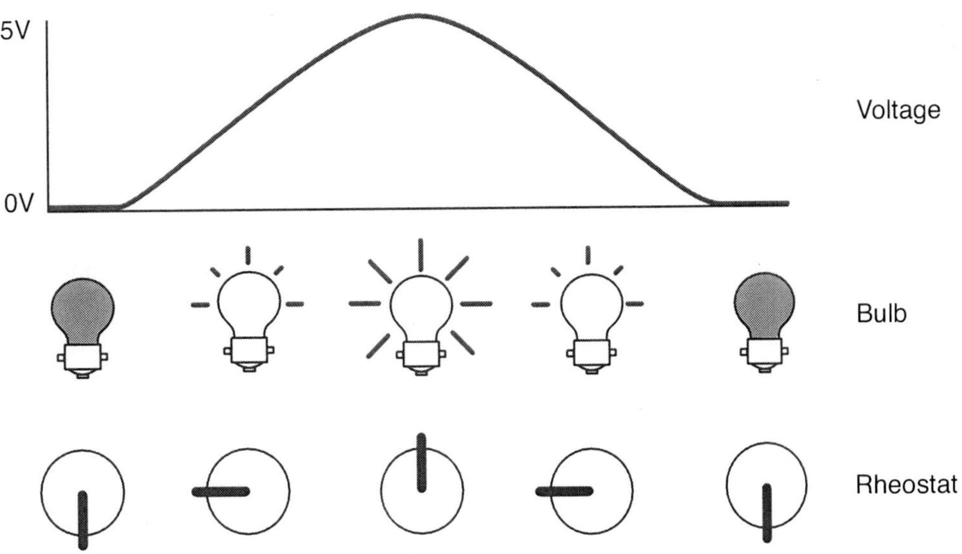


**Figure 1.** An analog voltage signal.

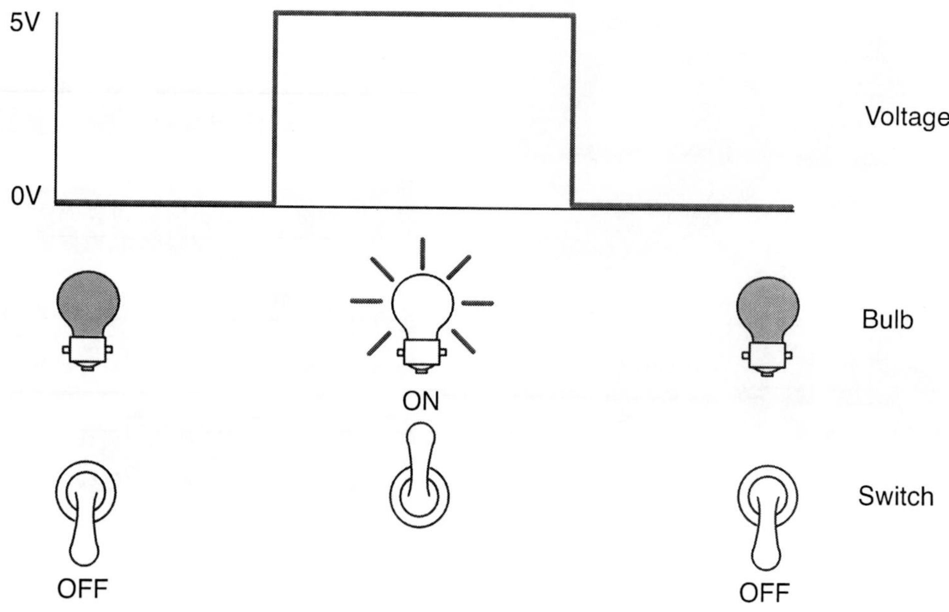


**Figure 2.** A digital voltage signal.

## DIGITAL VOLTAGE SIGNALS

If a conventional on/off switch is connected to a 5-volt bulb and the switch is off, 0 volts are available at the bulb. When the switch is turned on, a 5-volt signal is sent from the switch to the bulb and the bulb is illuminated to full brilliance **(Figure 2)**. If the switch is turned off, the voltage at the bulb returns to 0 volts and the bulb goes out. The voltage signal from the switch is either 0 volts or 5 volts, or we could say the voltage signal is either high or low. This type of signal is called a digital voltage signal. If the switch is turned on and off rapidly, a square wave digital voltage signal is applied from the switch. Many computers have the capability to vary the on time of digital signals. A **digital voltage signal** is either high or low.

## BINARY CODE

A numeric value may be assigned to digital signals. For example, a low digital signal may be given a value of 0, and a high digital signal may be given a value of 1 **(Figure 3)**. This assignment of numeric values to digital signals is called **binary coding**. The word binary means two values and in the binary coding system, the two values are 0 and 1 (Figure 3). In an automotive computer, information is transmitted in binary codes. Conditions, numbers, and letters can be represented by a series of zeroes and ones.

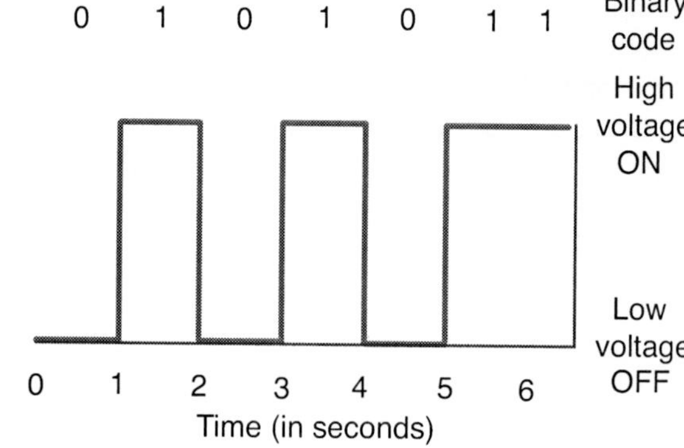


**Figure 3.** Binary coding.

You Should Know

*In computer language, each zero and one represents a bit of information. Eight bits make a byte, which may be called a word. Electronic information is exchanged in bytes.*

Many computer input sensors operate in the 0 volts to 5 volts range. The TPS may produce the following voltages: at closed throttle 0.6 volt, at part open throttle 2.5 volts, and at wide open throttle 4.8 volts.

A numeric value may be assigned to each of these voltages by the engine computer.

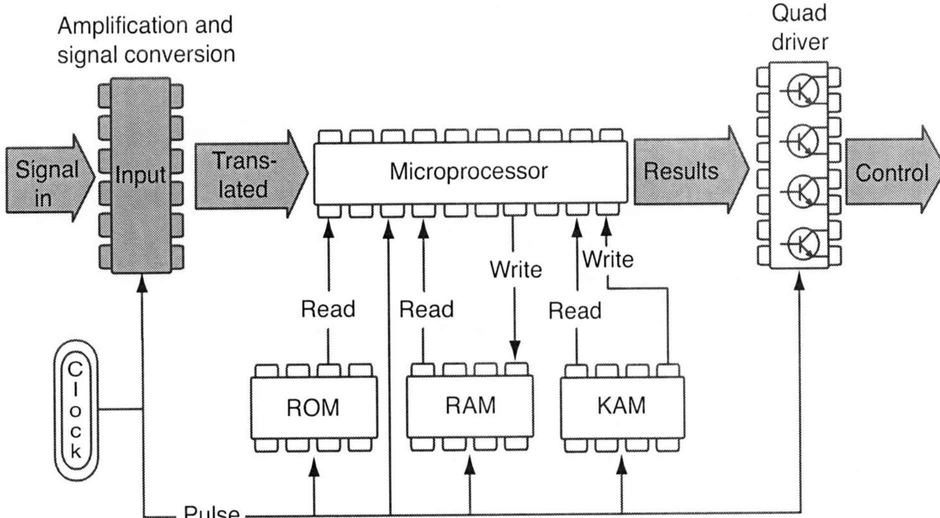


**Figure 4.** A computer with an input conditioning chip, a microprocessor, and memory chips.

## INPUT SIGNAL CONDITIONING

Since many input sensors produce analog voltage signals and the microprocessor in the computer operates on digital signals, something must change these analog signals to digital signals. This job is performed by the input amplification and signal conversion chip in the computer **(Figure 4)**. The input amplification and signal conversion chip may be called an analog/digital (A/D) converter. This chip continually scans the input sensor signals, assigns numeric values to the signal voltages, and then translates the numeric values to a binary code.

Some input sensors such as the $O_2$ sensor produce a very low voltage signal with a low current flow. This type of signal must be amplified or increased before it is sent to the microprocessor. The input amplification and signal conditioning chip also provides the necessary signal amplification.

## MICROPROCESSORS

The **microprocessor** is the calculating and decision-making chip in a computer. Thousands of miniature transistors and diodes are contained in the microprocessor. These transistors act as electronic switches that are either on or off. The components in the microprocessor are etched on an integrated circuit (IC) that is small enough to fit on a fingertip. The silicon chip containing the IC is mounted in a flat, rectangular protective box. Metal connecting pins extend from each side of the microprocessor container. These pins connect the microprocessor to the circuit board in the computer.

## COMPUTER MEMORY CHIPS

The microprocessor is supported by various memory chips which store information and help the microprocessor in making decisions. These memory chips are similar in appearance to the microprocessor chip. Computer memory chips are called by various names, including random access memory (RAM), read only memory (ROM), programmable read only memory (PROM), electronically erasable programmable read only memory (EEPROM), and keep alive memory (KAM).

The memory chips store information regarding the ideal operating conditions of various components and systems and the input sensors inform the computer about the engine and vehicle operating conditions. The microprocessor reads the ideal operating conditions in the memory chip programs and compares this information with the sensor inputs. After this comparison, the microprocessor makes the necessary decisions and operates the various components to meet the ideal operating conditions in the computer program.

You Should Know

*Many vehicle manufacturers now supply new computers that are not programmed. These computers must be programmed when they are installed or the engine will not start. This computer programming is done with the appropriate scan tool, personal computer (PC), or the vehicle manufacturer's recommended test equipment such as DaimlerChrysler's Mopar Diagnostic System (MDS). This computer programming may be called* ***flash programming.*** *This programming also applies to other new computers, such as ABS computers.*

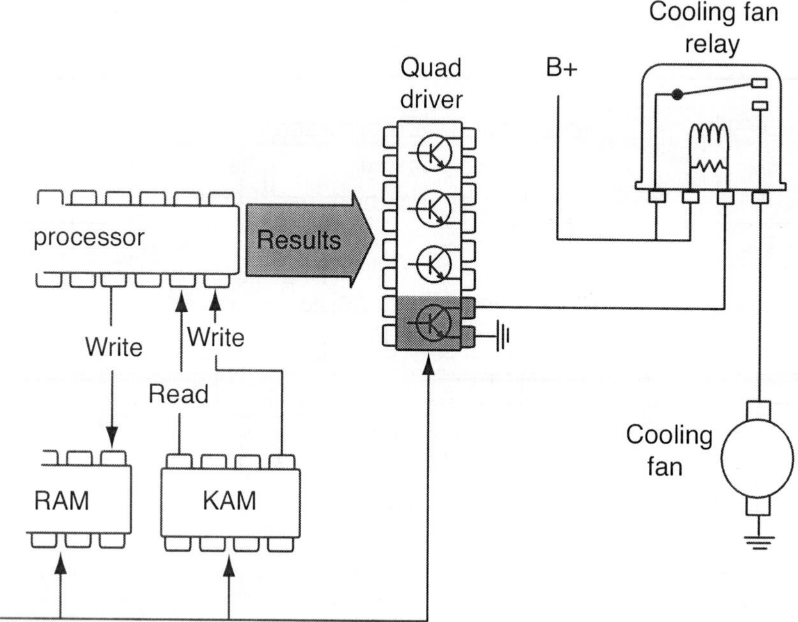


**Figure 5.** Computer output controls.

## COMPUTER OUTPUT DRIVERS

The computer operates many different output controls, such as relays and solenoids. The computer contains a number of **quad drivers**, or single transistors, that switch the output controls on and off. The microprocessor commands the quad drivers to operate the output controls. For example, if the ECT sensor indicates the engine temperature is high enough to require cooling fan operation, the microprocessor commands the appropriate quad driver to ground the cooling fan relay winding. This action closes the relay contacts that supply voltage to the cooling fan motor **(Figure 5)**.

## ADAPTIVE STRATEGY

Many automotive computers have an adaptive strategy that allows the computer to compensate for various changes in some of the inputs or outputs. For example, if the injector orifices become partially restricted, the injectors deliver less fuel and the air-fuel mixture becomes lean. Under this condition, the $O_2$ sensor voltage signal is continually low. The computer senses this condition and automatically provides a slight increase in injector pulse width to allow the injector fuel delivery to return to normal. **Injector pulse** width is the length of time in milliseconds that the computer keeps the injectors open.

If these injectors are replaced, the computer still supplies the increased injector pulse width and the air-fuel mixture tends to be rich. Under this conditionm the engine may experience rough idle operation. After about 5 minutes of driving with the engine at normal operating temperature, the computer learns about the injector replacement and returns the injector pulse width to normal.

The adaptive strategy in a computer is erased if the battery voltage is disconnected from the computer. If this condition occurs, the engine operation may be erratic until the computer re-learns the system and restores the adaptive strategy.

## FUEL PUMP ELECTRIC CIRCUIT

Fuel-injected vehicles have an electric fuel pump in the fuel tank. The PCM operates a relay that supplies voltage to the fuel pump. When the ignition switch is turned on, the PCM supplies voltage to the fuel pump relay winding **(Figure 6)**. Under this condition, current flows from the PCM through the fuel pump relay winding to ground and the relay contacts close. Under this condition, voltage is supplied through these relay contacts to the in-tank fuel pump. If the engine is not cranked or started within 2 seconds, the PCM shuts off the current flow to the fuel pump relay winding and the fuel pump stops operating. While the engine is cranking or running, the PCM maintains the voltage supply to the fuel pump relay winding and the fuel pump relay contacts remain closed to maintain fuel pump operation.

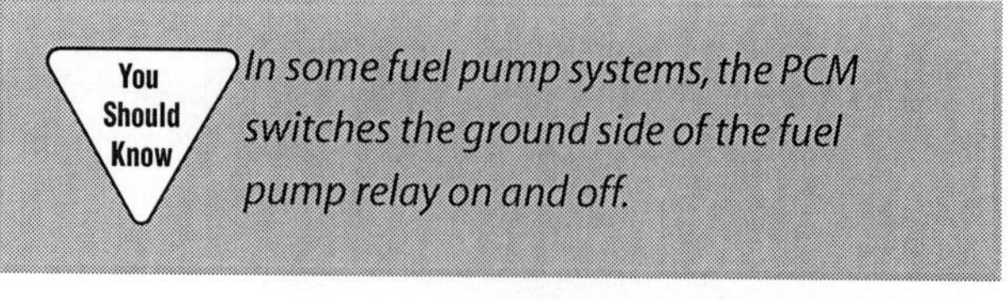


*In some fuel pump systems, the PCM switches the ground side of the fuel pump relay on and off.*

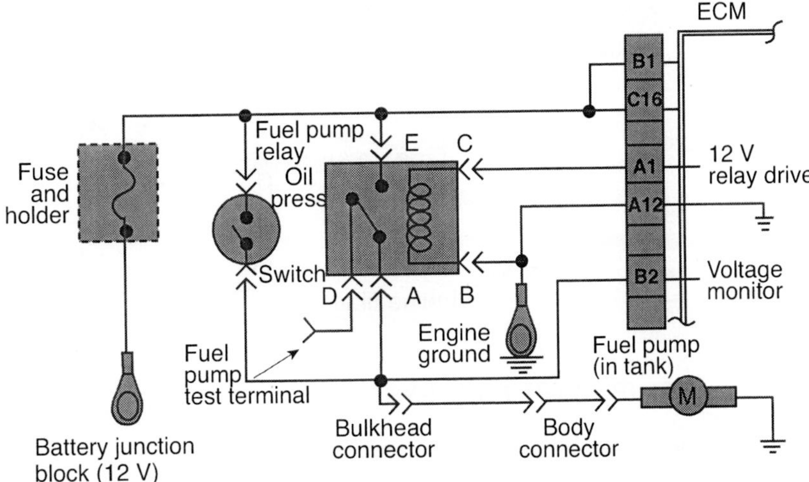


**Figure 6.** A fuel pump electrical circuit.

The 2-second shutoff feature on the fuel pump turns the pump off if the vehicle is in a collision and the engine stalls while the ignition switch is left in the on position. Many fuel pump circuits on Ford vehicles have an inertia switch in the circuit between the relay and the fuel pump **(Figure 7)**. If the vehicle is in even a moderate collision, the inertia switch opens the fuel pump circuit and turns the pump off. A reset button on top of the inertia switch may be pressed to restore the switch operation. On cars the

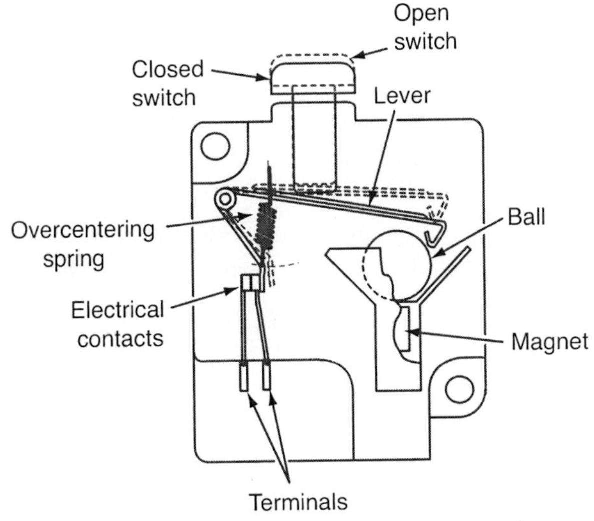


**Figure 7.** A fuel pump inertia switch.

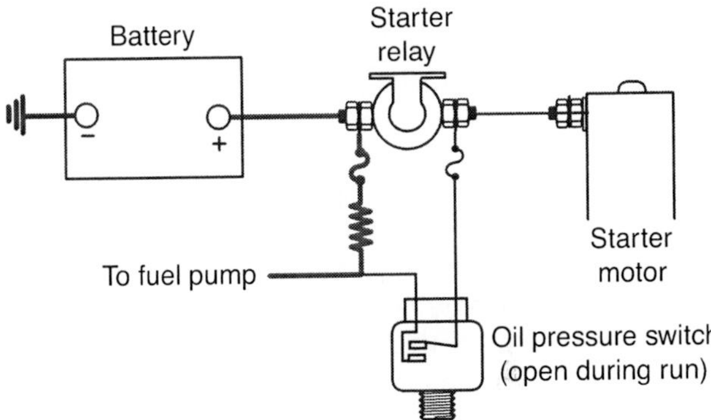


**Figure 8.** A fuel pump oil pressure switch.

inertia switch is in the trunk area and on light-duty trucks the inertia switch is under the dash.

On some General Motors vehicles, the fuel pump circuit has an oil pressure switch connected in the fuel pump circuit **(Figure 8)**. If the fuel pump relay fails, current is supplied through the oil pressure switch to maintain fuel pump operation.

## FUEL PUMP FILTER AND PRESSURE REGULATOR

When the fuel pump is running, it forces fuel through the fuel line filter and pressure regulator to the fuel rail and injectors. The fuel filter may be located in the fuel line under

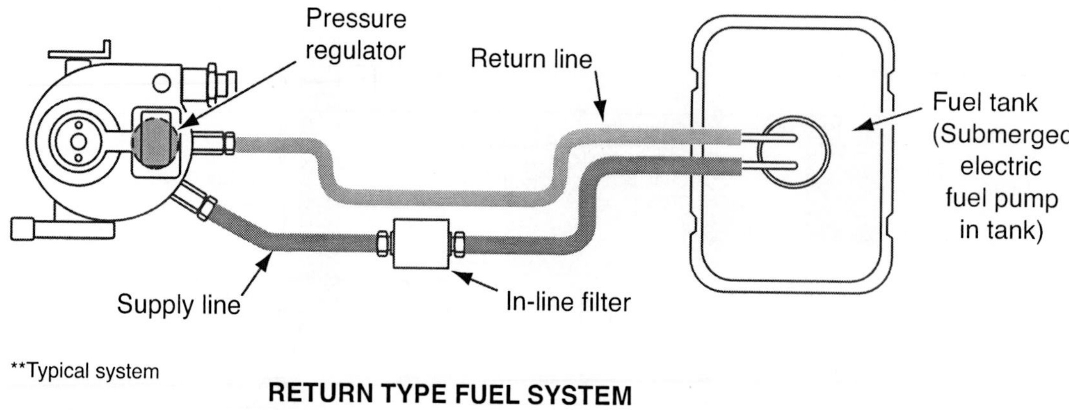


**Figure 9.** A fuel filter, a throttle body assembly, and a return fuel line.

the vehicle or in the engine compartment **(Figure 9)**. The fuel filter has a pleated paper filter element in a metal container. The fuel filter must be installed in the proper direction. The inlet and outlet fittings are identified on some filters or an arrow on the filter housing indicates the direction of fuel flow.

On throttle body injection (TBI) systems, a fuel line is connected from the fuel filter to the throttle body assembly. On port fuel injection (PFI) systems, the fuel line is connected from the filter to the fuel rail **(Figure 10)**. Fuel pump pressure is supplied to the injectors and to the pressure regulator in the TBI assembly or fuel rail. At the specified fuel pressure, the valve opens in the pressure regulator and some fuel is allowed to return through the return line to the fuel tank **(Figure 11)**. When the fuel pressure drops slightly, the pressure regulator valve closes and fuel pressure increases to open this valve again. The pressure regulator valve opens and closes to keep fuel pressure at the specified value. On TBI systems, the average fuel pressure is 10 to 25 psi (68.9 to 172 kPa). The average fuel pressure on PFI systems is 35 to 65 psi (241 to 448 kPa).

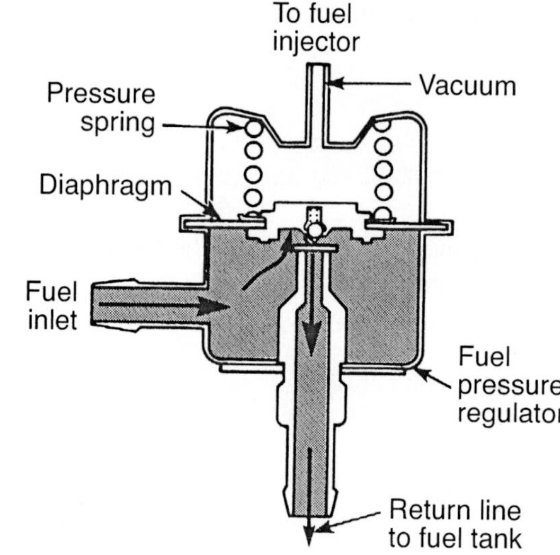


**Figure 11.** Fuel pressure regulator operation.

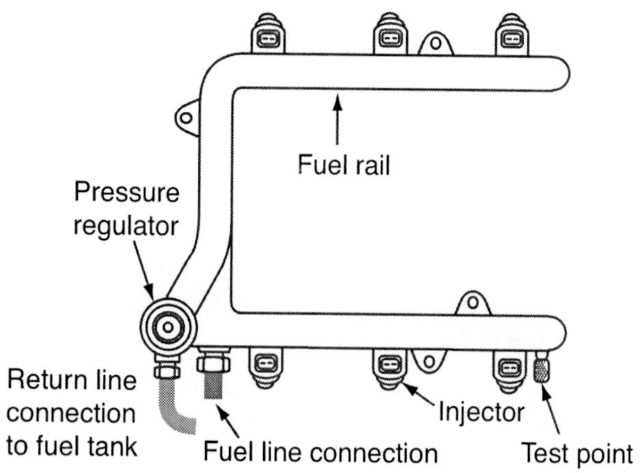


**Figure 10.** A pressure regulator on a fuel rail.

> **You Should Know** *Fuel pressure is extremely important to maintain proper fuel system operation. If the fuel pressure is less than specified, the injectors deliver a reduced amount of fuel and the air-fuel mixture is lean. Excessive fuel pressure causes the injectors to deliver too much fuel and the air-fuel mixture is rich.*

On most PFI systems, a vacuum hose is connected from the intake manifold to the pressure regulator. When the engine is running, this hose supplies manifold vacuum to the upper side of the regulator diaphragm. If the engine is idling or operating at moderate throttle opening, mani-

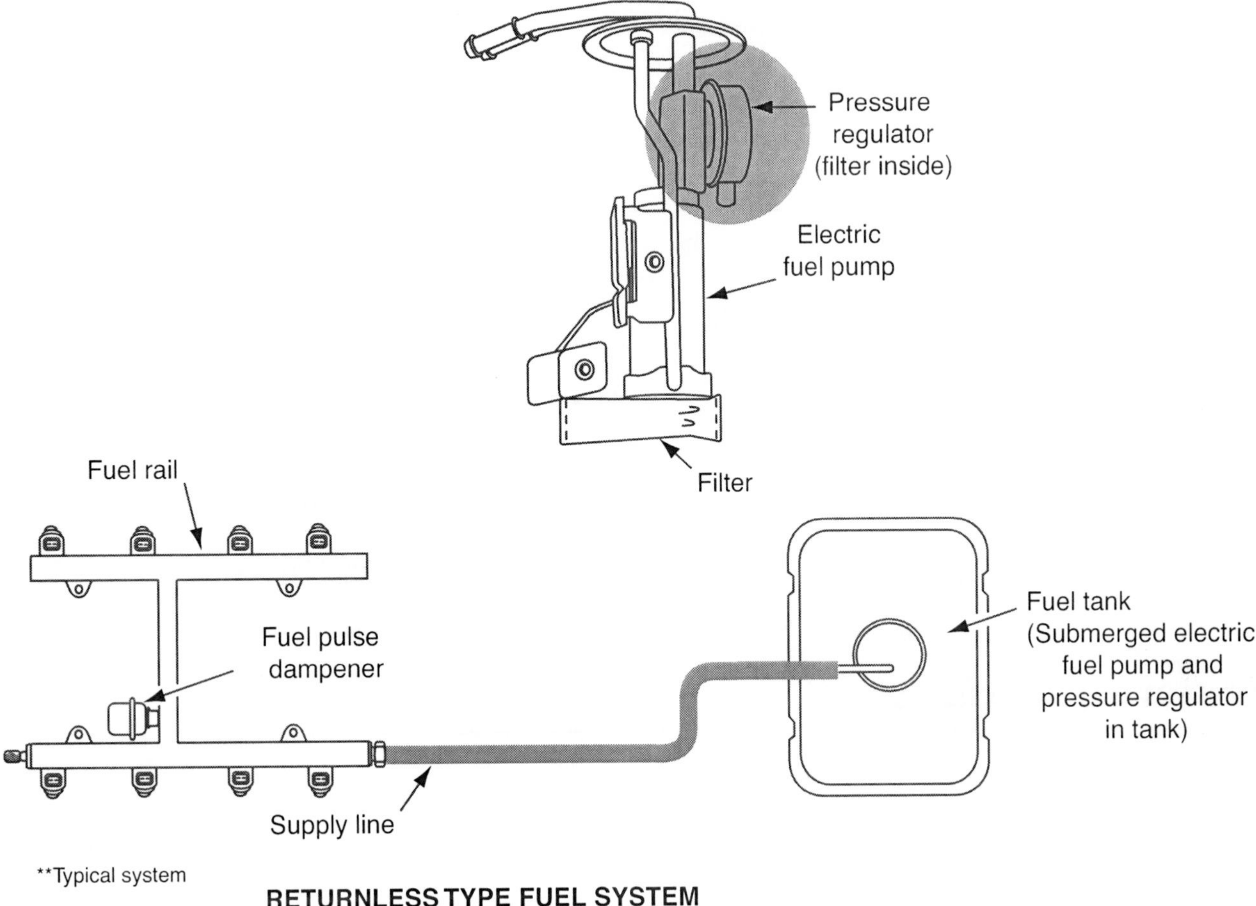


**Figure 12.** A returnless fuel system.

fold vacuum is high and this vacuum above the regulator diaphragm allows the diaphragm to move upward at a lower pressure and open the regulator valve. When the engine is operating at wide open throttle, manifold vacuum is low and a higher fuel pressure is required to move the regulator diaphragm upward and open the regulator valve. Therefore, fuel pressure is higher when operating at wide open throttle. On a PFI system, the injectors in each intake port deliver fuel into intake manifold at the intake ports. At wide open throttle, the manifold vacuum is low or closer to atmospheric pressure compared to idle operation. This pressure increase at the injector tips tends to decrease the amount of fuel delivered by the injectors. However, this tendency is offset by the higher fuel pressure supplied by the pressure regulator at wide open throttle.

Some vehicles have a returnless fuel system. In these systems, the pressure regulator is mounted on top of the fuel tank or inside the top of the fuel tank **(Figure 12)**. Fuel is returned from the pressure regulator directly into the fuel tank. On these systems, there is no return line.

You Should Know *Some vehicles have a fuel pressure sensor in the fuel rail. These vehicles do not have a pressure regulator. In response to the fuel pressure sensor signal, the PCM pulses the fuel pump on and off to maintain the specified fuel pressure.*

## POWERTRAIN CONTROL MODULE (PCM) OUTPUTS

The PCM controls a variety of solenoids, relays, and motors to operate the output devices. In most cases, the PCM provides a ground to operate these output controls. Many computers operate the output controls by opening and closing the ground circuit on the relay or solenoid. When switching the ground side of the circuit, a lower current flows through the computer compared to switching the positive side of the circuit on and off. Computer outputs include the fuel injectors, EGR solenoid, purge solenoid in the EVAP emission system, solenoids in the AIR system, and AIR pump.

## Summary

- Analog voltage signals are variable within a specific range.
- Digital voltage signals are either on or off, high or low.
- Binary coding is the assignment of numbers to digital signals.
- The microprocessor is the decisionmaking and calculating chip in a computer.
- Computers usually operate outputs by providing a ground.
- A quad driver is a group of four transistors used to operate computer outputs.
- Adaptive strategy allows a computer to compensate for some variations that may occur in the inputs or outputs.
- The PCM operates a relay that supplies voltage to the electric in-tank fuel pump.
- If the ignition switch is on and the engine is not cranked or started, the PCM shuts off the fuel pump relay and fuel pump in 2 seconds.
- The fuel pressure regulator regulates fuel pressure.
- If the fuel pressure is higher or lower than specified, the air-fuel mixture is affected.

## Review Questions

1. Technician A says a rich air-fuel ratio may be caused by higher-than-normal fuel pressure. Technician B says higher-than-normal fuel pressure may be caused by the pressure regulator valve sticking open. Who is correct?
   A. Technician A
   B. Technician B
   C. Both Technician A and Technician B
   D. Neither Technician A nor Technician B
2. Technician A says in a TBI system, the injectors discharge fuel below the throttles. Technician B says in a PFI system, the injectors are mounted in the intake manifold near the intake ports. Who is correct?
   A. Technician A
   B. Technician B
   C. Both Technician A and Technician B
   D. Neither Technician A nor Technician B
3. Technician A says the fuel pump in a PFI system will be running if the ignition switch is on for 5 minutes and the engine is not cranked or started. Technician B says many fuel injection systems return excess fuel from the pressure regulator to the fuel tank. Who is correct?
   A. Technician A
   B. Technician B
   C. Both Technician A and Technician B
   D. Neither Technician A nor Technician B
4. Technician A says an inertia switch opens the fuel pump circuit if this circuit overheats. Technician B says an oil pressure switch in the fuel pump circuit supplies voltage to the fuel pump if the relay fails. Who is correct?
   A. Technician A
   B. Technician B
   C. Both Technician A and Technician B
   D. Neither Technician A nor Technician B
5. All of these statements about digital signals are true *except*:
   A. A digital signal varies continuously within a specific voltage range.
   B. A digital signal is either high or low.
   C. A digital signal may be called a square-wave signal.
   D. A conventional on/off light switch produces a digital voltage signal.
6. The input signal conditioning chip in a computer may perform all of these functions *except*:
   A. change analog signals to digital signals.
   B. amplify input voltage signals.
   C. assign numeric values to input voltage signals.
   D. reduce input voltage signals.

7. Technician A says the adaptive strategy is erased if the battery voltage is disconnected from the PCM. Technician B says after 5 minutes of driving the vehicle, the adaptive strategy relearns about component replacements in the PCM system. Who is correct?
   A. Technician A
   B. Technician B
   C. Both Technician A and Technician B
   D. Neither Technician A nor Technician B
8. The process of flash programming an engine computer:
   A. installs new information only in the RAM.
   B. may be necessary when a computer is replaced.
   C. may be done with a lab scope and a digital voltmeter.
   D. is required before replacing engine computer system components.
9. Technician A says the adaptive strategy in an engine computer is erased if the computer is disconnected from the battery. Technician B says the adaptive strategy in an engine computer allows the computer to adapt to minor system defects. Who is correct?
   A. Technician A
   B. Technician B
   C. Both Technician A and Technician B
   D. Neither Technician A nor Technician B
10. All of these statements about a typical electric fuel pump circuit in a port fuel injection system are true *except*:
    A. The fuel pump operates for 2 seconds if the ignition switch is turned on and the engine is not cranked.
    B. Voltage is supplied through the fuel pump relay contacts to the fuel pump.
    C. The inertia switch opens the fuel pump circuit if the vehicle is in a collision.
    D. The fuel pump is usually mounted on the frame rail under the vehicle.
11. An analog voltage signal is continuously ____________ in a specific voltage range.
12. A digital voltage signal may be considered ____________ or ____________.
13. The input amplification and conditioning chip in a computer changes ____________ signals to ____________ signals.
14. The amount of fuel discharged from an injector is determined by the ____________ ____________ ____________.
15. Explain adaptive strategy in a computer.
16. Describe the operation of the fuel pump circuit if the ignition is left in the on position and the engine is not started.
17. Explain how the computer operates many of its outputs.
18. Explain the effect of low fuel pump pressure on the air-fuel ratio.

# Chapter 29 Input Sensors for Engine Control Systems

## Introduction

A thorough understanding of input sensors provides the basis for fast, accurate diagnosis of these sensors. If technicians do not understand input sensor operation and typical input sensor voltage signals, diagnostic procedures for these sensors are often time consuming and inaccurate.

### OXYGEN SENSORS

The oxygen sensor is threaded into the exhaust manifold or exhaust pipe. OBD II systems have oxygen sensors upstream and downstream from the catalytic converter. An oxygen-sensing element is positioned in the center of the $O_2$ sensor **(Figure 1)**. The oxygen-sensing element is coated with zirconia or titania and this element is mounted in an insulator and contained in a metal case. Zirconia and titania-type $O_2$ sensors have a similar design but different operating principles. The sensor case has external threads that fit matching threads in the exhaust system. The lower end of the sensor is mounted in the exhaust stream. A protective cap on the lower end of the sensor contains slots that allow exhaust to be applied to the outside of the sensing element. Some $O_2$ sensors have a protective rubber boot over the top of the sensor. On some $O_2$ sensors, internal air slots in the protective boot and an air port in the side of the sensor allow air to enter the inside of the sensing element. In other $O_2$ sensors, the air is sealed inside the sensing element. The sensing element is connected to the terminal on

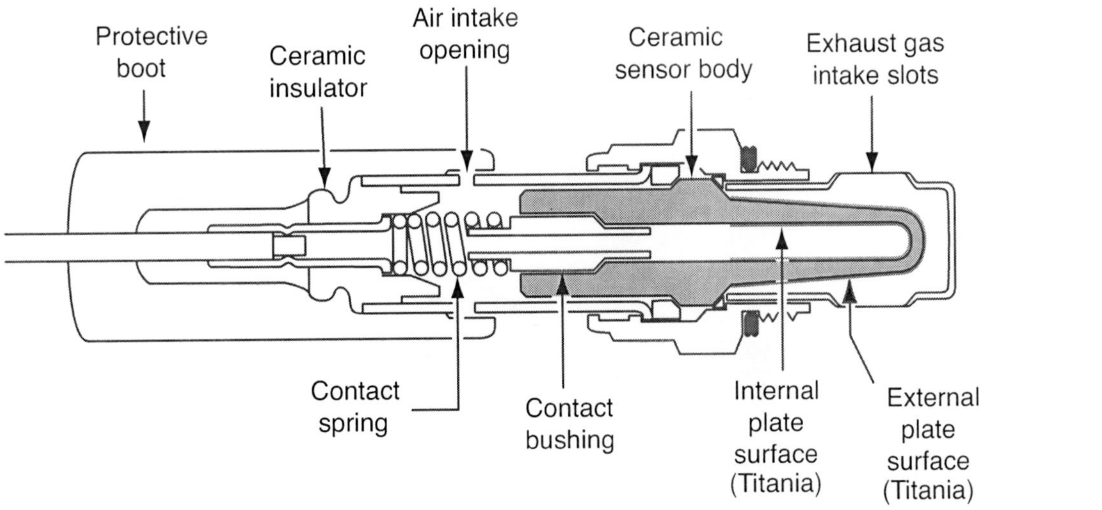


**Figure 1.** An oxygen sensor design.

top of the $O_2$ sensor and a wire is connected from the sensor terminal to the PCM. Some $O_2$ sensors have a ground wire connected from the sensor to the PCM.

On newer vehicles, the $O_2$ sensors have four wires. These sensors have an internal electric heater that has a voltage supply wire and a ground wire. These sensors are called heated oxygen sensors ($HO_2S$). Voltage may be supplied to the $HO_2S$ heater directly from the ignition switch, through a relay, or from the PCM. If the $HO_2$ heaters are powered from the ignition switch or through a relay, voltage is supplied to these heaters while the ignition switch is on. When the $HO_2S$ heaters are powered by the PCM, the PCM only supplies voltage to these heaters when necessary, such as during engine warmup and at idle and low engine speeds. If the engine is at normal operating temperature and running at higher speeds, the exhaust flow maintains $HO_2S$ sensor temperature and the PCM shuts off the sensor heaters.

## Zirconia-Type Oxygen Sensor Operation

When the engine temperature is cold, a zirconia-type $HO_2S$ does not produce a satisfactory voltage signal. When the engine approaches normal operating temperature, this sensor begins to produce a satisfactory signal. During engine's warmup, the PCM operates in **open loop**. In this mode, the PCM ignores the $HO_2S$ signal and a program in the PCM controls the air-fuel ratio. In open loop, the PCM supplies a slightly richer air-fuel ratio depending on engine temperature and throttle opening. At a specific engine temperature, the PCM enters the **closed loop** mode. In closed loop, the PCM uses the $HO_2S$ signal and other inputs to control the air-fuel ratio.

If the PCM is operating in closed loop and the air-fuel ratio is lean, nearly all of the fuel injected has been combined with air and burned in the combustion chambers and excess air is left over. Under this condition, the exhaust stream flowing past the $HO_2S$ has a high oxygen content. This causes oxygen to be supplied to both sides of the oxygen-sensing element in the $HO_2S$, resulting in a very low voltage signal from this sensor to the PCM. When this signal is received, the PCM increases the injector pulse width and provides a richer air-fuel ratio. A rich air-fuel ratio causes all the oxygen in the intake air to be mixed with fuel and excessive fuel is left over. Under this condition, the exhaust stream has very low oxygen content. This condition causes low oxygen content on the outside of the $HO_2S$ element, and high oxygen content is present inside this element. When this condition occurs, the $HO_2S$ produces a higher voltage, up to 1 volt **(Figure 2)**. In response this $HO_2S$ signal, the PCM reduces the injector pulse width and provides a leaner air-fuel ratio. The $HO_2S$ signal cycles very quickly from low voltage to high voltage and the PCM uses this signal to control the air-fuel ratio at or close to the stoichiometric ratio of 14.7:1.

## Titania Oxygen Sensor Operation

A titania-type $HO_2S$ modifies voltage whereas a zirconia-type $HO_2S$ generates voltage. The PCM supplies voltage to the zirconia-type $HO_2S$ and this voltage is lowered by a resistor in the circuit. The resistance of the titania changes as the air-fuel ratio cycles from lean to rich. If the air-fuel ratio is lean, the titania resistance is high and the sensor voltage signal is low. When the air-fuel ratio is rich, the titania resistance is low and the sensor voltage signal is high

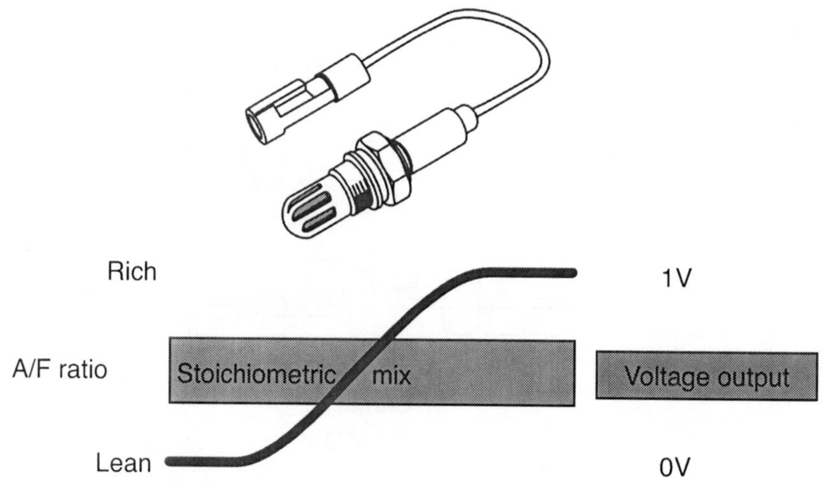


**Figure 2.** A zirconia-type oxygen sensor voltage signal.

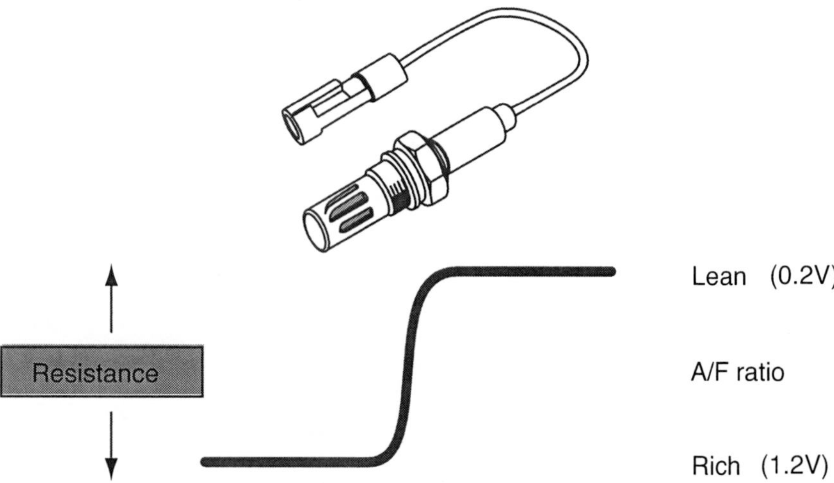


**Figure 3.** A titania-type oxygen sensor voltage signal.

**(Figure 3)**. The titania-type $HO_2S$ produces a satisfactory voltage signal almost immediately after a cold engine is started. This action provides improved air-fuel ratio control during cold engine operation. The titania $HO_2S$ can also cycle faster from low voltage to high voltage.

## ENGINE COOLANT TEMPERATURE AND INTAKE AIR TEMPERATURE SENSORS

ECT sensors and IAT sensors both contain thermistors. When a thermistor is cold it has very high resistance. This resistance decreases if the thermistor is heated. Some ECT sensors have 35,000 ohms of resistance when the engine is very cold and the same sensor has less than 1,000 ohms of resistance if the sensor is at normal operating engine temperature. Two wires are usually connected from the PCM to the ECT or IAT sensors. One of these wires is a signal wire and the other wire is a ground wire. The PCM supplies 5 volts through the signal wire to each sensor and the PCM senses the voltage drop across the sensor. When the engine is cold and sensor resistance is high, the voltage drop across the sensor may be 4.5 volts. If the engine is at normal operating temperature, the voltage drop across the ECT or IAT sensor is very low **(Figure 4** and

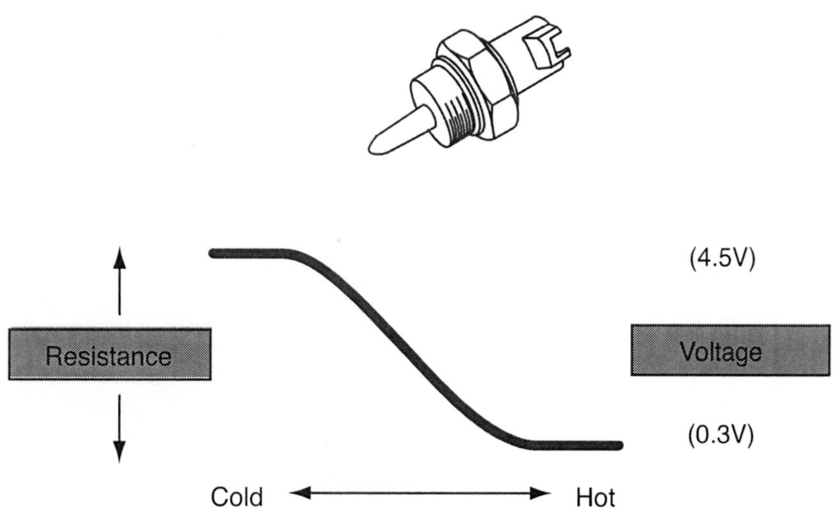


**Figure 4.** An ECT sensor voltage signal.

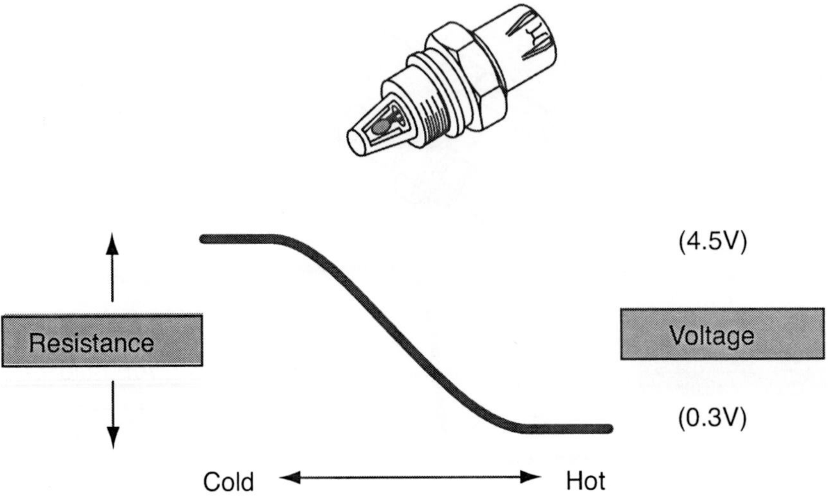


Figure 5. An IAT sensor voltage signal.

**Figure 5)**. The PCM uses the ECT and IAT sensor signals to control many of the outputs. For example, the PCM supplies a richer air-fuel ratio when the ECT sensor indicates that the engine coolant is cold. The PCM uses the ECT sensor and $HO_2S$ signals to determine when to enter closed loop.

## MANIFOLD ABSOLUTE PRESSURE SENSORS

The MAP sensor is usually mounted in the engine compartment. Some MAP sensors are threaded directly into the intake manifold. Other MAP sensors have a vacuum hose connected from the intake manifold to the sensor.

MAP sensors have three wires connected from the sensor to the PCM **(Figure 6)**. These wires are a 5-volt reference wire, a signal wire, and a ground wire connected to the PCM. The PCM supplies a constant 5 volts through the reference wire to the MAP sensor.

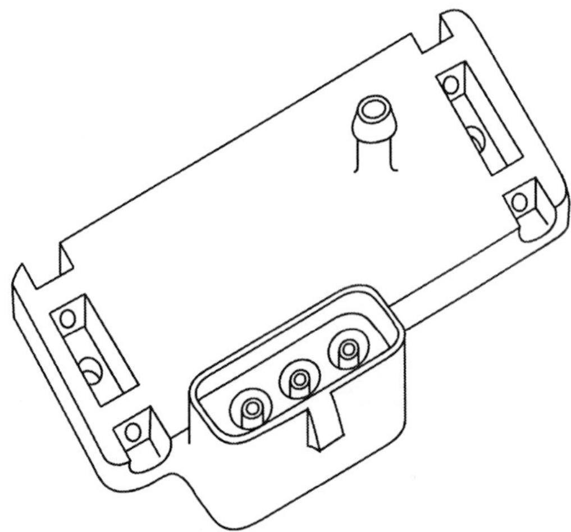

Figure 6. A MAP sensor.

In some MAP sensors, the manifold vacuum is supplied to a silicon diaphragm in the sensor. When the manifold vacuum increases, it stretches this diaphragm and this action changes the voltage signal to the PCM. A typical MAP sensor provides a 1-volt signal to the PCM with the engine idling and high vacuum supplied to this sensor. At wide open throttle, the vacuum decreases and the voltage signal from the MAP sensor is approximately 4.5 volts. A fuel injection system with a MAP sensor may be called a speed density system.

In a **speed density system,** the PCM determines the amount of air entering the engine from the engine rpm signal, and the MAP sensor signal, which indicates the density of vacuum in the intake manifold. The PCM must know the amount of air entering the engine to calculate the amount of fuel to be injected.

You Should Know

*Some MAP sensors contain a barometric pressure sensor that sends a signal to the PCM in relation to barometric pressure each time the ignition switch is turned on. Barometric pressure varies depending on atmospheric conditions and altitude. When the PCM receives this signal, it corrects the air-fuel ratio in relation to altitude.*

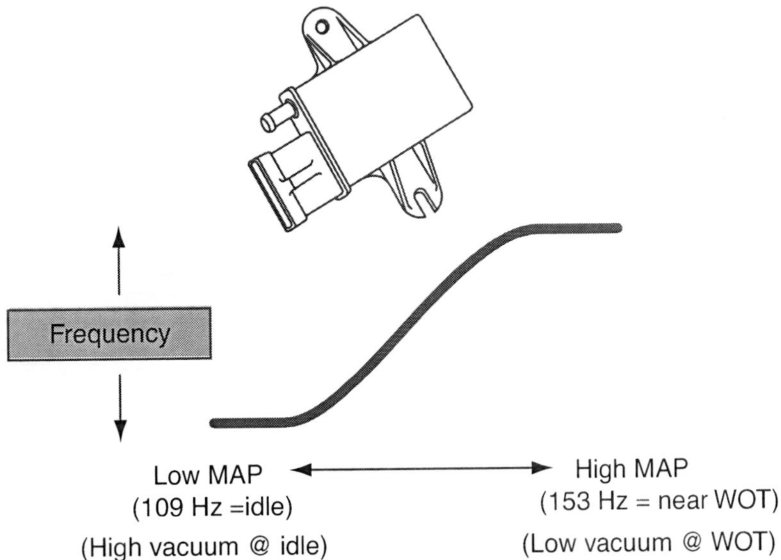


**Figure 7.** A MAP sensor with a hertz-type voltage signal.

Some MAP sensors produce a digital voltage with a continually varying frequency. This frequency increases as throttle opening and engine load increase. The signal from this type of MAP sensor is measured in cycles or hertz. The MAP sensor signal may vary from 109 hertz at idle to 153 hertz at wide open throttle **(Figure 7)**.

## MASS AIR FLOW SENSORS

The MAF sensor is mounted in the air intake hose between the air cleaner and the throttle body. An arrow on the MAF sensor housing indicates the direction in which air must flow through the sensor **(Figure 8)**. Some engines have a hot wire-type MAF sensor. In this type of sensor, a hot wire is positioned in the air stream through the sensor **(Figure 9)**. An ambient temperature-sensing wire is mounted beside the hot wire. When the ignition switch is turned on, the module in the MAF sends enough current through the hot wire to maintain the temperature of this wire at 392°F (200°C) above the temperature of the cold wire. When the engine is suddenly accelerated, the rush of air through the MAF tries to cool the hot wire. When the module senses the cooling of the hot wire, it immediately increases the current through this wire to maintain the wire temperature at 392°F (200°C). The module sends this increasing current signal to the PCM. In response to this signal, the PCM supplies the correct amount of fuel to go with the additional air entering the engine. The MAF signal informs the PCM regarding the amount of air entering the engine and the PCM provides the correct amount of fuel to mix with the air and maintain the proper air-fuel ratio.

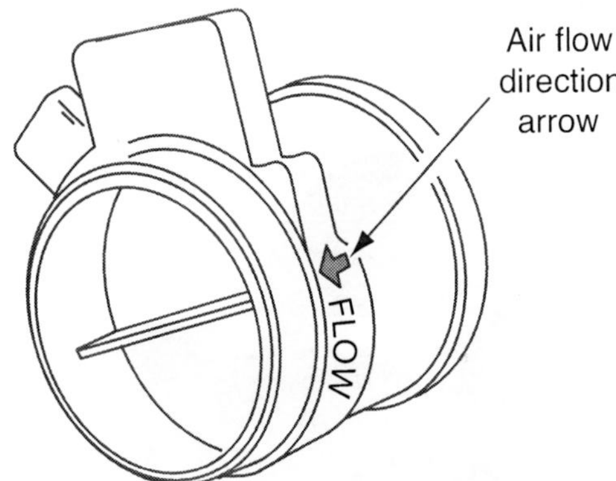


**Figure 9.** Airflow through a MAF sensor must be in the direction of the arrow on the housing.

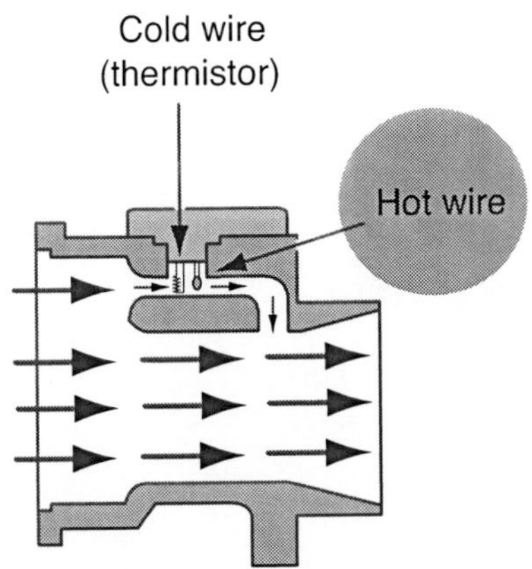


**Figure 8.** A hot wire-type MAF sensor.

> **You Should Know** *Some MAF circuits have a burn-off circuit containing a burn-off relay. When the ignition switch is turned off, the burn-off relay closes. This relay activates a burn-off circuit that provides a high current flow through the hot wire for a short time. This action burns contaminants off the wire. Even though the MAF sensor is mounted after the air cleaner, the air still contains very small particles that the air filter does not remove.*

Some MAF sensors contain a vane-type sensor. In this type of sensor, a moveable, spring-loaded vane is located in the air intake hose between the air cleaner and the throttle body **(Figure 10)**. This normally-closed vane is connected to a variable resistor in the sensor housing. The PCM supplies 5 volts to this resistor and the voltage signal from the sensor to the PCM varies as the vane moves a contact on the variable resistor **(Figure 11)**. When the engine is started, the air flow through the air intake opens the vane slightly. If the engine is accelerated, the increased air flow through the air intake opens the vane and moves the contact on the variable resistor. This action changes the voltage signal sent from the MAF to the PCM.

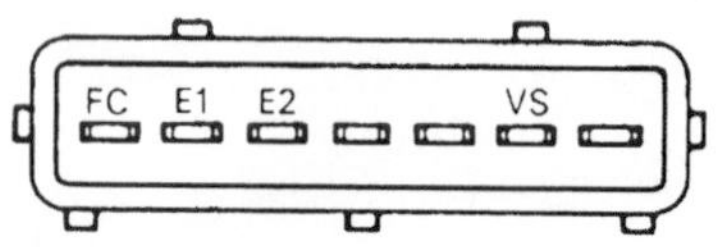


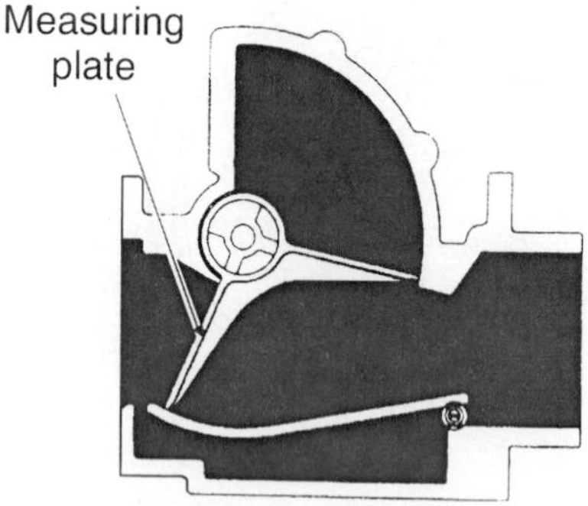


**Figure 10.** A vane-type MAF sensor.

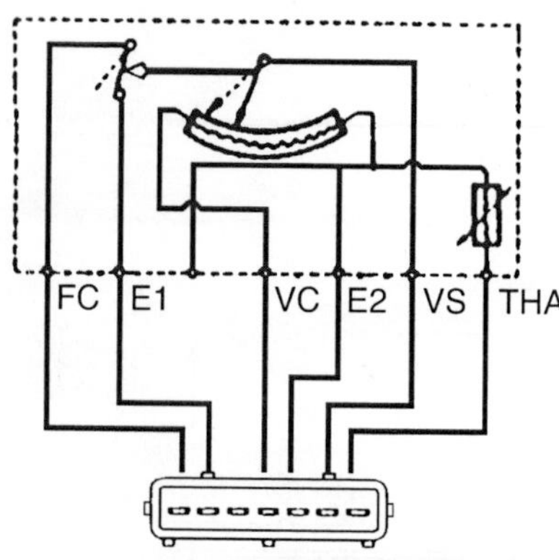


**Figure 11.** A variable resistor in a vane-type MAF sensor.

## THROTTLE POSITION SENSORS

The TPS has three wires connected from the sensor to the PCM. These wires are similar to the wires used on a MAP sensor. The same 5-volt reference is supplied to both the MAP and TPS sensors. The TPS also has a signal wire and a ground wire. The TPS contains a variable resistor that is connected to the throttle shaft. As the throttle is opened, a sliding contact moves on the variable resistor. A typical TPS voltage is 0.5 volt to 1 volt at idle and 4.5 volts at wide open throttle **(Figure 12)**. As the throttle is opened, the TPS voltage must increase smoothly. A defective variable resistor

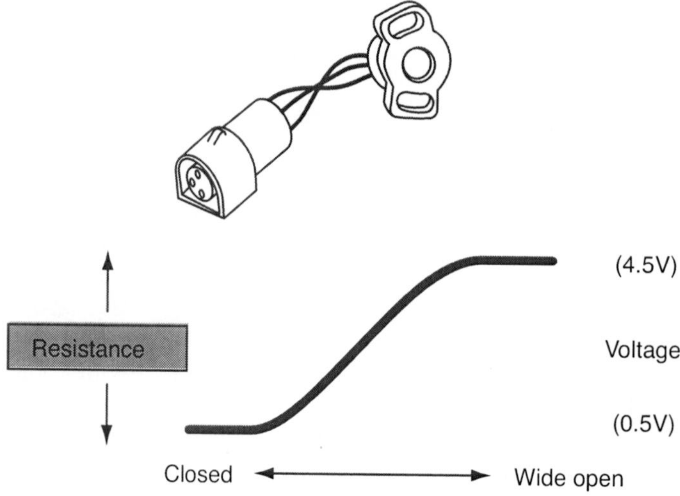


**Figure 12.** A TPS voltage signal.

may cause erratic TPS voltage and a hesitation during engine acceleration.

## KNOCK SENSORS

The knock sensor is often threaded into the engine block or cylinder head. The knock sensor contains a piezoelectric crystal **(Figure 13)**. When the engine detonates or knocks, a vibration is present in the block and cylinder head. The knock sensor changes this vibration to a voltage signal. When the PCM receives a knock sensor signal indicating the engine is detonating, the PCM reduces the spark advance to stop the detonation.

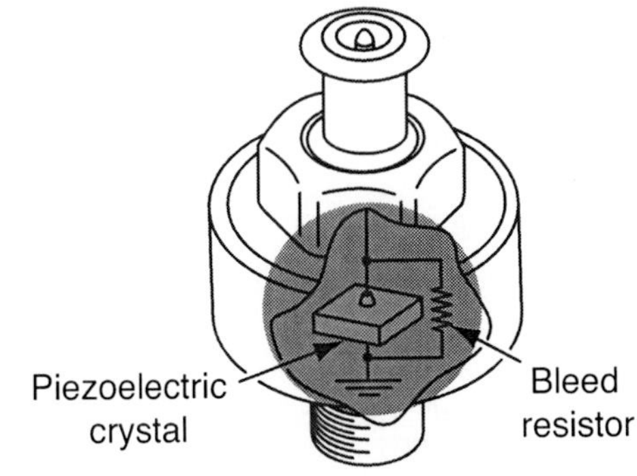


**Figure 13.** A knock sensor.

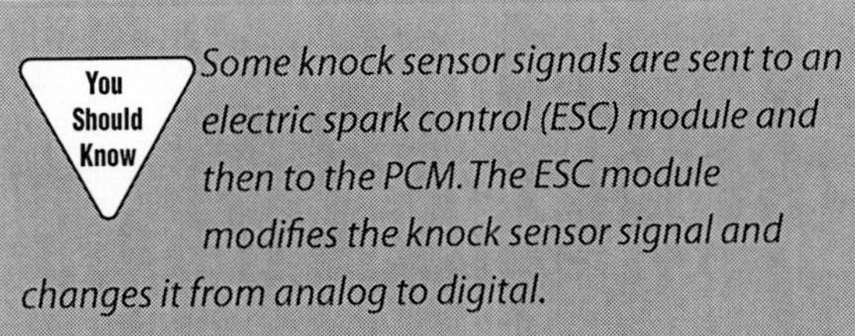


*Some knock sensor signals are sent to an electric spark control (ESC) module and then to the PCM. The ESC module modifies the knock sensor signal and changes it from analog to digital.*

# Summary

- When the air-fuel ratio is rich, the $HO_2S$ provides a high voltage signal and a lean air-fuel ratio causes a low voltage signal from this sensor.
- A typical $HO_2S$ voltage signal varies from 0.2 volt to 1 volt.
- An ECT sensor has high resistance when cold and low resistance when hot.
- The PCM senses the voltage drop across the ECT and IAT sensors.
- Some MAP sensors contain a silicon diaphragm that stretches as vacuum is applied to the diaphragm.
- A typical MAP sensor voltage signal is 1 volt at idle and 4.5 volts at wide open throttle.
- Some MAP sensors provide a hertz-type voltage signal.
- Some MAP sensors act as barometric pressure sensors when the ignition switch is first turned on.
- Some MAF sensors have a hot wire and an ambient temperature-sensing wire.
- In a hot wire-type MAF sensor, the sensor module keeps the hot wire 392°F (200°C) hotter than the ambient temperature-sensing wire.
- Some MAF sensors have a moveable vane in the air intake hose. This vane is connected to a variable resistor in the sensor.
- The TPS contains a variable resistor and a moveable contact connected to the throttle shaft slides on this resistor.
- A typical TPS voltage signal is 0.5 volt to 1 volt at idle and 4.5 volts at wide open throttle.
- The knock sensor changes a vibration to a voltage signal.

# Review Questions

1. Technician A says a rich air-fuel ratio provides an exhaust stream that has an excess amount of oxygen. Technician A says with a rich air-fuel ratio, the $HO_2S$ voltage signal is low. Who is correct?
   A. Technician A
   B. Technician B
   C. Both Technician A and Technician B
   D. Neither Technician A nor Technician B
2. Technician A says if oxygen is present on both sides of the $HO_2S$ sensing element, this sensor produces a low voltage. Technician B says a lean air-fuel ratio results in high oxygen content on both sides of the $HO_2S$ sensing element. Who is correct?
   A. Technician A
   B. Technician B
   C. Both Technician A and Technician B
   D. Neither Technician A nor Technician B

3. Technician A says when the engine coolant is cold, the ECT sensor resistance is high. Technician B says when the engine coolant is cold, the ECT sensor voltage drop is high. Who is correct?
   A. Technician A
   B. Technician B
   C. Both Technician A and Technician B
   D. Neither Technician A nor Technician B
4. Technician A says with the engine idling, a satisfactory MAP sensor voltage signal is 4.5 volts. Technician B says the MAP sensor sends a barometric pressure signal to the PCM when the ignition switch is first turned on. Who is correct?
   A. Technician A
   B. Technician B
   C. Both Technician A and Technician B
   D. Neither Technician A nor Technician B
5. All of these statements about the $HO_2S$ are true *except*:
   A. OBD II systems have the $HO_2S$ upstream and downstream from the catalytic converter.
   B. A rich air-fuel ratio causes a zirconia $HO_2S$ to produce a low voltage signal.
   C. A typical $HO_2S$ has four wires connected to the sensor.
   D. When the engine is cold, a zirconia-type $HO_2S$ does not produce a satisfactory voltage signal.
6. A typical ECT sensor:
   A. contains a variable resistor.
   B. has a higher resistance when the engine coolant is cold.
   C. has three wires connected to the sensor.
   D. provides a digital voltage signal to the PCM.
7. When discussing typical MAP sensors, Technician A says the MAP sensor voltage signal increases when the engine speed and/or load increases. Technician B says a MAP sensor may contain a barometric pressure sensor. Who is correct?
   A. Technician A
   B. Technician B
   C. Both Technician A and Technician B
   D. Neither Technician A nor Technician B
8. All of these statements about mass air flow sensors are true *except*:
   A. An arrow on the MAF housing indicates the proper direction of air flow.
   B. The MAF sensor may contain a hot wire and a cold wire.
   C. The MAF module maintains a constant hot wire temperature.
   D. Some MAF sensors have a burn-off circuit that cleans the air passage through the sensor.
9. A typical TPS voltage signal with the throttle in the wide open position is:
   A. 2.1 volts.
   B. 2.8 volts.
   C. 3.1 volts.
   D. 4.5 volts.
10. When the PCM receives a knock sensor signal, the PCM:
    A. provides a richer air-fuel ratio.
    B. reduces the spark advance.
    C. reduces the injector pulse width.
    D. ignores the $HO_2S$ signal.
11. A titania-type $HO_2S$ changes ________________ as the air-fuel ratio cycles from lean to rich.
12. The PCM uses the voltage signals from the ________________ and ________________ sensors to determine when to enter closed loop.
13. The PCM ignores the ________________ sensor signal when operating in open loop.
14. In a speed density fuel injection system, the PCM determines the amount of air entering the engine from the ________________ and ________________ signals.
15. Explain the operation of a zirconia-type $HO_2S$ in relation to air-fuel ratio.
16. Explain the design and wiring connections of a TPS.
17. Describe the operation of the TPS voltage signal in relation to throttle opening.
18. Describe the design and operation of a knock sensor.

# Input Sensor Maintenance, Diagnosis, and Service

## *Introduction*

An understanding of input sensor operation is absolutely essential to provide fast, accurate sensor diagnosis. This understanding includes satisfactory and unsatisfactory sensor readings indicated by waveforms and ohmmeter or voltmeter readings. Technicians must understand how other components can affect sensor readings or their diagnosis will often be inaccurate. For example, technicians must understand satisfactory and unsatisfactory $HO_2S$ waveforms and they must also understand the effect that ignition or fuel system defects have on the $HO_2S$ waveforms.

### INPUT SENSOR MAINTENANCE

When minor underhood service is performed, the input sensor wiring should be inspected for loose connections, worn insulation, and contact with hot components such as exhaust manifolds. Repair or replace loose sensor connections and worn insulation as required. Sensor wires must be secured away from hot components. Check the MAP sensor vacuum hose for leaks and kinks. If necessary, replace this hose and check all the other vacuum hoses for leaks. A vacuum leak in another vacuum hose decreases the manifold vacuum supplied to the MAP sensor. Inspect the engine for oil leaks that may soak input sensor wires and contribute to a shorted condition between the wires. Inspect the battery area for indications of battery acid leaks, which contaminate input sensor and other wiring harnesses. If the engine has an MAF sensor, check the hose between this sensor and the throttle body for leaks and loose clamps. Leaks in this hose will allow air to enter the air intake without flowing through the MAF sensor. This causes an incorrect MAF sensor signal that indicates there is less air entering the engine. This signal results in a lean air-fuel ratio and engine performance problems such as hesitation during acceleration.

### OXYGEN SENSOR DIAGNOSIS

When diagnosing any input sensor problem, always check the MIL in the instrument panel. If the MIL is illuminated with the engine running, the PCM has detected a defect in the system and a DTC is probably set in the PCM memory. With the ignition switch off, connect a scan tool to the DLC under the dash and retrieve any DTCs in the PCM memory. These DTCs indicate the general area where the defect is located. For example, a P0131 DTC indicates a low voltage in $HO_2S$ bank 1 sensor 1.

**You Should Know**

*In a P0131 DTC, the P indicates a powertrain code and the 0 indicates this code is mandated by the SAE standards. If the second digit is a 1, the code is a vehicle manufacturer's code. When the third digit is a 1, the code belongs to the fuel and air metering subgroup and 31 is the code representing the low $HO_2S$ voltage. The vehicle manufacturer's service information provides DTC interpretations.*

**You Should Know** *A bank 1 $HO_2S$ is located on the same side of the engine as number 1 cylinder and sensor 1 is positioned upstream from the catalytic converter. A bank 2 sensor is on the opposite side of the engine from number 1 cylinder and sensor 2 is mounted downstream from the catalytic converter.*

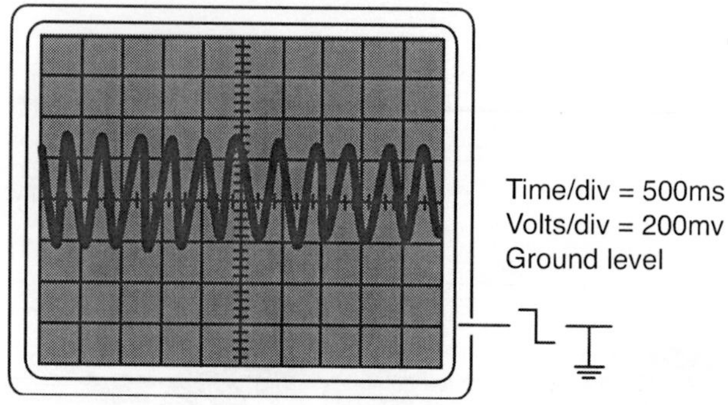


**Figure 1.** A satisfactory upstream $HO_2S$ waveform.

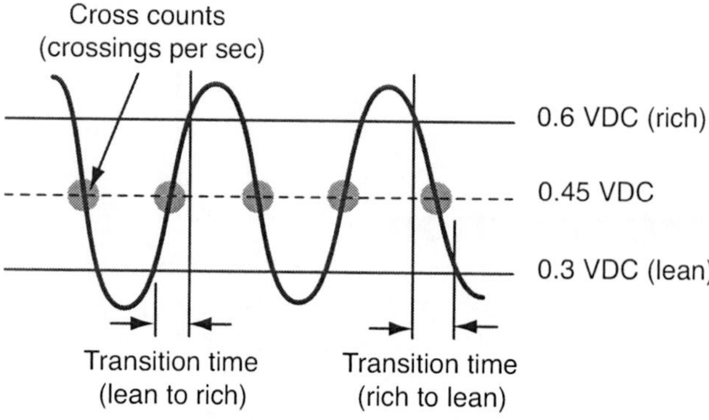


**Figure 2.** An $HO_2S$ waveform must have the proper cross counts and the correct lean-to-rich transition time.

The next step in sensor diagnosis is to inspect the sensor wiring for worn insulation and loose connections because these wiring defects may be the cause of the problem. When a MAP sensor problem is encountered, be sure the vacuum hose to this sensor is not kinked or leaking. If the sensor wiring and vacuum hose are satisfactory, further sensor diagnosis is required. With the engine operating at normal temperature, the $HO_2S$ sensor may be tested with a scan tool or a digital multimeter on the DC voltage function.

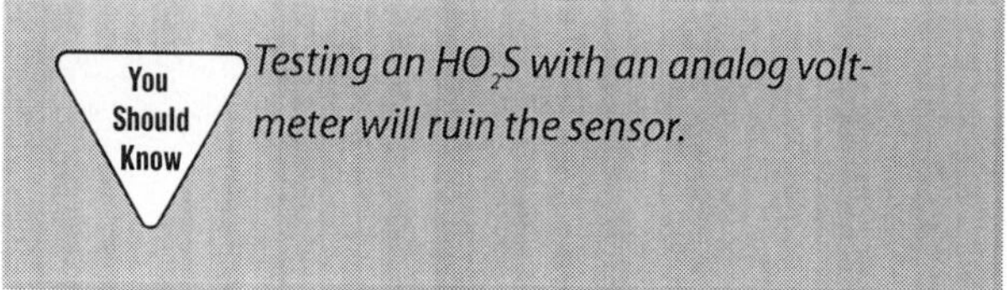

**You Should Know** *Testing an $HO_2S$ with an analog voltmeter will ruin the sensor.*

Before testing this sensor, the engine speed should be maintained at 2,000 rpm for 2 minutes to be sure the sensor is hot. Then allow the engine to idle and observe the upstream $HO_2S$ voltage displayed on the scan tool. The $HO_2S$ should provide a voltage signal that switches quickly from low voltage to high voltage. A typical $HO_2S$ voltage cycles between 0.3 volt and 0.8 volt. If the sensor voltage is not cycling quickly in the proper range, the air-fuel ratio is improper, the ignition system has a defect causing misfiring, or the $HO_2S$ sensor is defective. For example, when the fuel pressure is much higher than specified, the air-fuel ratio is rich and the $HO_2S$ voltage is always high. If the engine thermostat is stuck open and the engine temperature is always low, the air-fuel ratio is rich and again the $HO_2S$ voltage signal is always high. Therefore, when diagnosing engine computer systems and input sensors, one of the first tests is to prove that the fuel system has the specified fuel pressure. There is nothing in the computer diagnostics to indicate fuel pressure. The PCM assumes the fuel pressure is within the specified range.

An accurate $HO_2S$ test may be performed with a graphing voltmeter or a lab scope. Some scan tools display voltage graphs. A satisfactory $HO_2S$ should provide a waveform as indicated in **Figure 1**. The $HO_2S$ must have the proper number of cross counts, and display a fast transition time from lean to rich and rich to lean **(Figure 2)**. **Cross counts** are the number of times that the $HO_2S$ cycles from rich to lean in a given time period. The $HO_2S$ **transition time** is the time required for this sensor to switch from lean to rich and rich to lean.

A defective spark plug that causes cylinder misfiring provides an $HO_2S$ voltage waveform as indicated in **Figure 3**. A leaking injector that results in a rich air-fuel ratio causes

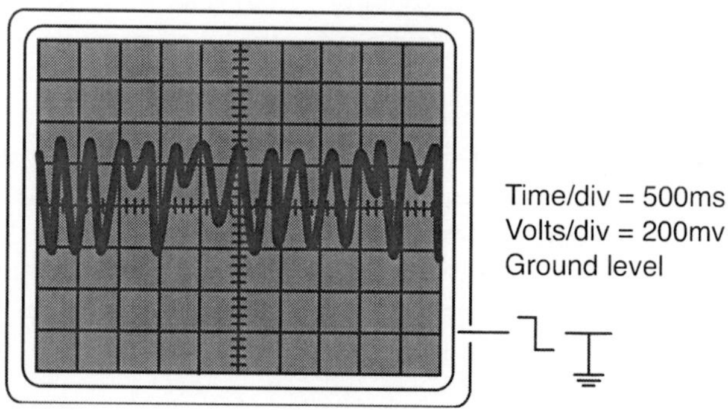


**Figure 3.** An unsatisfactory $HO_2S$ waveform caused by a defective spark plug.

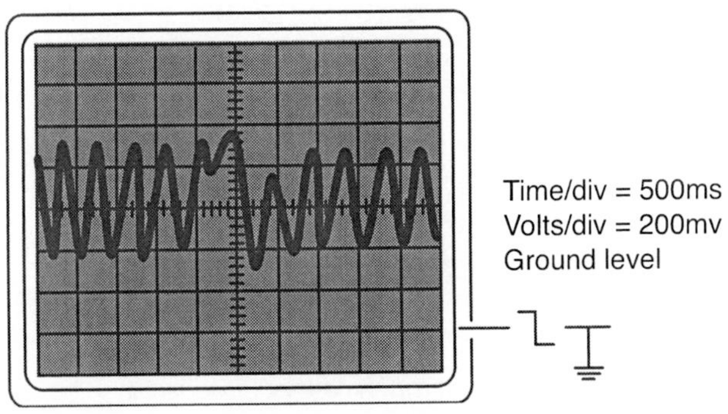


**Figure 4.** An unsatisfactory $HO_2S$ waveform caused by a leaking injector.

an $HO_2S$ voltage waveform as displayed in **Figure 4**. When diagnosing input sensors, the fuel pressure, ignition system, and the data from all the other input sensors should be tested because problems in these areas and components affect some of the input sensor signals.

**You Should Know** *Downstream $HO_2Ss$ have a lower voltage signal and cycle slower from low voltage to high voltage compared to upstream $HO_2Ss$. If the upstream and downstream $HO_2Ss$ have the same voltage waveform, the catalytic converter is defective.*

The $HO_2S$ heater may be tested by connecting an ohmmeter across the heater voltage supply and ground terminals **(Figure 5)**. The heater should have the specified

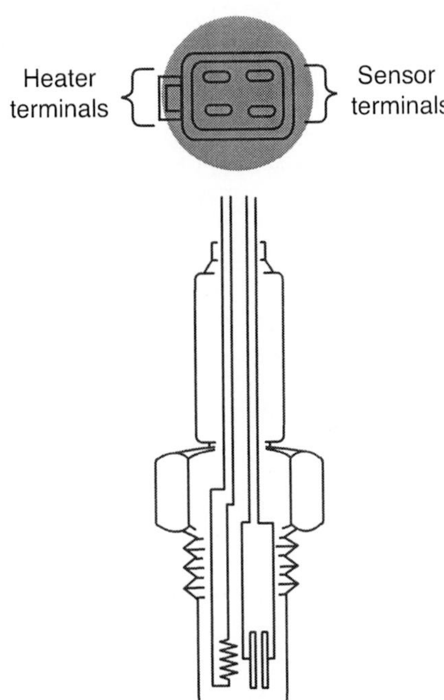


**Figure 5.** $HO_2S$ heater terminals.

resistance, which is usually about 10 ohms. If the ohmmeter reading is infinite, the heater has an open circuit and the sensor must be replaced. An inoperative $HO_2S$ heater causes slow sensor warmup and the PCM remains in open loop for a longer time. The PCM is also more likely to go back into open loop with the engine idling. This results in a richer air-fuel ratio and reduced fuel economy.

## FUEL PUMP PRESSURE TESTING

Before connecting the pressure gauge to the fuel system, depressurize the system by connecting a vacuum hand pump to the vacuum connection on the pressure regulator and supply 20 in. Hg. to the regulator. This allows any fuel pressure in the fuel rail to be released through the return line into the fuel tank. Connect the appropriate fuel pressure gauge to the Schrader valve on the fuel rail. On throttle body injection systems, connect the pressure gauge in the fuel inlet line at the throttle body. Cycle the ignition switch on and off several times and read the fuel pressure. The fuel pressure must equal the vehicle manufacturer's specifications. Low fuel pressure may be caused by a defective fuel pump, a restricted fuel filter, an air leak in a fuel line, low voltage at the fuel pump, or a high resistance in the fuel pump ground circuit. Before replacing a fuel pump, always be sure the other causes of low fuel pressure do not exist. A fuel pump flow test should also be performed to indicate the amount of fuel the pump supplies in a specific length of time.

## ENGINE COOLANT TEMPERATURE (ECT) SENSOR AND INTAKE AIR TEMPERATURE (IAT) SENSOR DIAGNOSIS

The diagnoses of the ECT sensor and IAT sensor are similar because both of these sensors contain thermistors. With the ignition switch turned off and the sensor connector disconnected, an ohmmeter may be connected to the sensor terminals to test the ECT or IAT sensor. The sensor to be tested may be removed and placed with a thermometer in a container filled with water **(Figure 6)**. When the water is heated, the sensor must have specified resistance at various

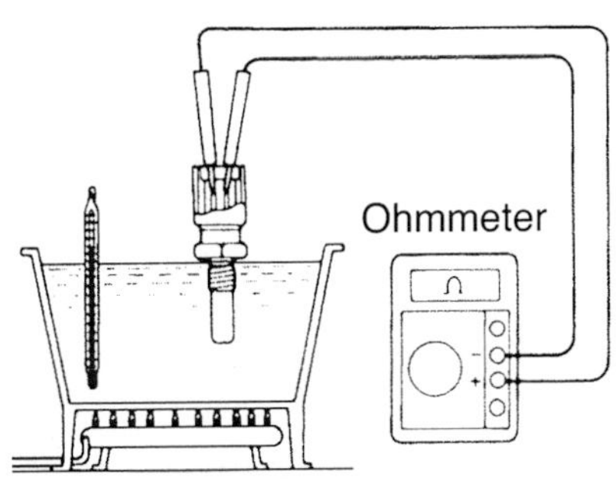


**Figure 6.** Testing an ECT sensor with an ohmmeter.

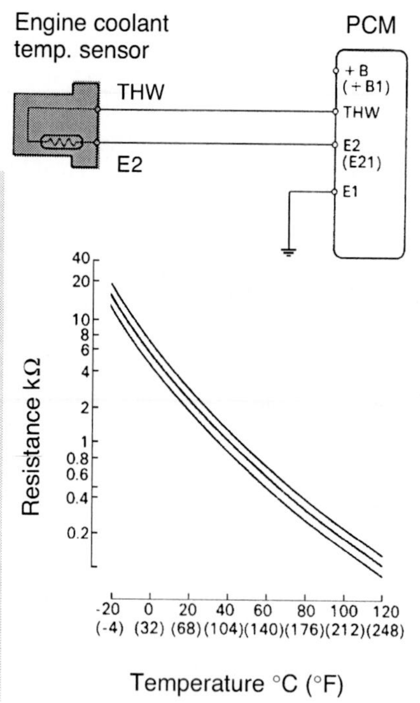


**Figure 7.** Ohm specifications for an ECT sensor at various temperatures.

temperatures **(Figure 7)**. If either sensor does not have the specified resistance, replace the sensor. The ECT or IAT sensor may also be tested with the engine running by measuring the voltage drop across the sensor terminals. These sensors must have the specified voltage drop in relation to the sensor temperature.

*In some ECT circuits, the PCM changes the resistance in the circuit by switching an internal resistor into the circuit at 120°F (48.8°C). This resistance change causes a significant change in the voltage drop across the sensor* **(Figure 8)**.

| Cold -10 K-ohm resistor | | Hot - 909-ohm resistor | |
|---|---|---|---|
| - 20 degree F | 4.7v | 110 degree F | 4.2v |
| 0 degree F | 4.4v | 130 degree F | 3.7v |
| 20 degree F | 4.1v | 150 degree F | 3.4v |
| 40 degree F | 3.6v | 170 degree F | 3.0v |
| 60 degree F | 3.0v | 180 degree F | 2.8v |
| 80 degree F | 2.4v | 200 degree F | 2.4v |
| 100 degree F | 1.8v | 220 degree F | 2.0v |
| 120 degree F | 1.25 | 240 degree F | 1.62v |

**Figure 8.** ECT sensor voltage drop specifications.

*When diagnosing input sensors, the technician must understand the effect of improper sensor signals on engine performance and other sensor signals. If the ECT sensor resistance is higher than specified, the voltage signal is also high. Under this condition, the air-fuel ratio is too rich and the PCM does not enter closed loop properly. High ECT sensor resistance may be caused by an engine thermostat that is continually open.*

## MANIFOLD ABSOLUTE PRESSURE SENSOR DIAGNOSIS

Before testing the MAP sensor, test the reference voltage to the sensor. With the ignition switch on, connect a voltmeter from the 5-volt reference wire to ground. The voltage at this terminal should be 4.8 volts to 5.2 volts. If this voltage is not present, check the battery voltage and the 5-volt reference wire from the sensor to the PCM. With the ignition switch on, connect a voltmeter from the MAP sensor ground wire to ground. The voltage drop across this wire should be less than 0.2 volt. Connect a voltmeter to the MAP sensor signal wire and ground and connect a vacuum hand pump to the sensor vacuum port **(Figure 9)**. With the ignition switch on, the sensor must have the specified voltage in relation to the vacuum supplied to the sensor. If the MAP sensor does not have the specified voltage signal, replace the sensor.

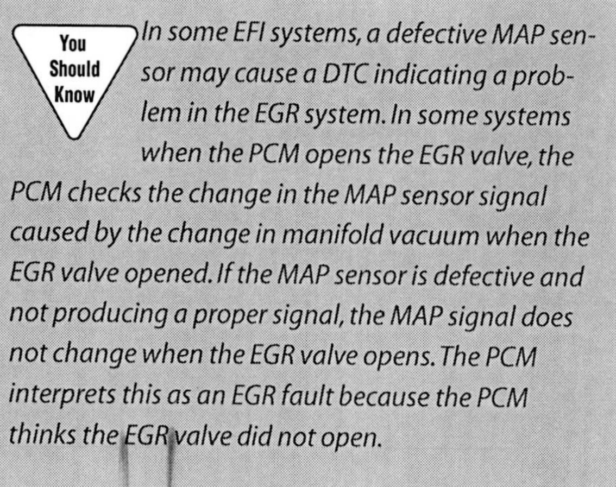
*In some EFI systems, a defective MAP sensor may cause a DTC indicating a problem in the EGR system. In some systems when the PCM opens the EGR valve, the PCM checks the change in the MAP sensor signal caused by the change in manifold vacuum when the EGR valve opened. If the MAP sensor is defective and not producing a proper signal, the MAP signal does not change when the EGR valve opens. The PCM interprets this as an EGR fault because the PCM thinks the EGR valve did not open.*

When testing a MAP sensor that provides a hertz-type voltage signal, connect a digital multimeter from the MAP sensor signal wire to ground. Switch the multimeter to the hertz scale and turn the ignition switch on. Connect a vacuum hand pump to the MAP sensor vacuum port and supply the specified vacuum to the sensor. The MAP sensor

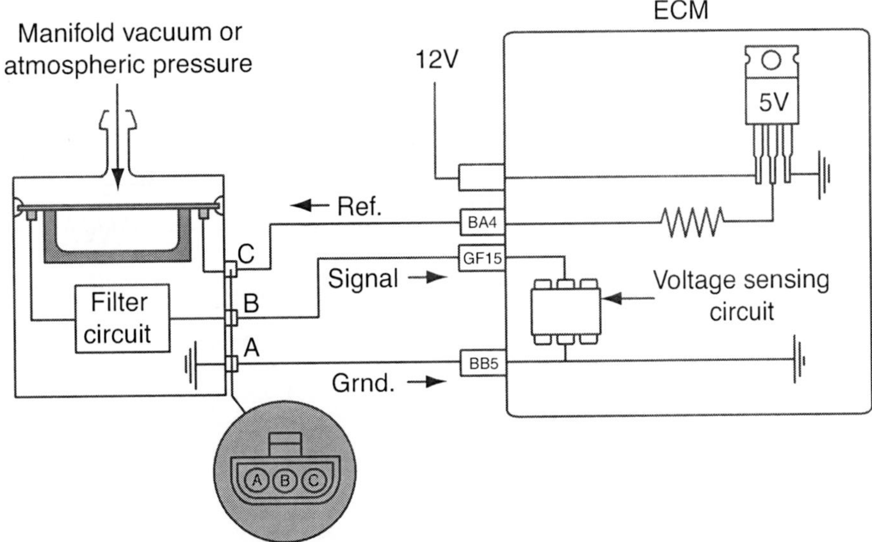


**Figure 9.** MAP sensor terminals.

must have a specified hertz signal in relation to the vacuum supplied to the sensor **(Figure 10)**. A special tester is available for hertz-type MAP sensors **(Figure 11)**. This tester changes hertz to DC volts. The tester power leads are connected to the battery terminals with the correct polarity and the test connector is installed on the MAP sensor terminals. A digital multimeter is connected to the signal wires on the tester with the correct polarity. When the specified vacuum is supplied to the MAP sensor with a vacuum hand pump, the sensor should provide the proper voltage signal.

| Vacuum Applied | Output Frequency |
|---|---|
| 0 in. Hg | 152-155 Hz |
| 5 in. Hg | 138-140 Hz |
| 10 in. Hg | 124-127 Hz |
| 15 in. Hg | 111-114 Hz |
| 20 in. Hg | 93-98 Hz |

**Figure 10.** MAP sensor hertz specifications in relation to vacuum.

**Figure 11.** A MAP sensor tester that changes hertz to DC volts.

## THROTTLE POSITION SENSOR DIAGNOSIS (TPS)

When diagnosing a TPS, test the 5-volt reference and ground wires to the sensor as explained in the MAP sensor diagnosis. Turn the ignition switch on and observe the TPS voltage on the scan tool as the accelerator pedal is slowly depressed. The TPS voltage signal should increase from 0.5 volt to 1 volt at idle to 4.5 volts at wide open throttle. Always use the specifications for the vehicle being diagnosed. If the TPS voltage does not increase smoothly and gradually, replace the sensor.

A TPS may develop a worn spot on the variable resistor, especially on vehicles driven for long periods of time at a constant speed. This worn spot on the variable resistor may cause a very momentary glitch in the TPS voltage signal that is not noticeable on a digital voltmeter. A lab scope provides a voltage sweep across the screen that is in billionths of a second. This instrument will pick out momentary glitches in TPS voltage signals. Connect the lab scope to the TPS signal wire and slowly increase the throttle opening with the ignition switch on

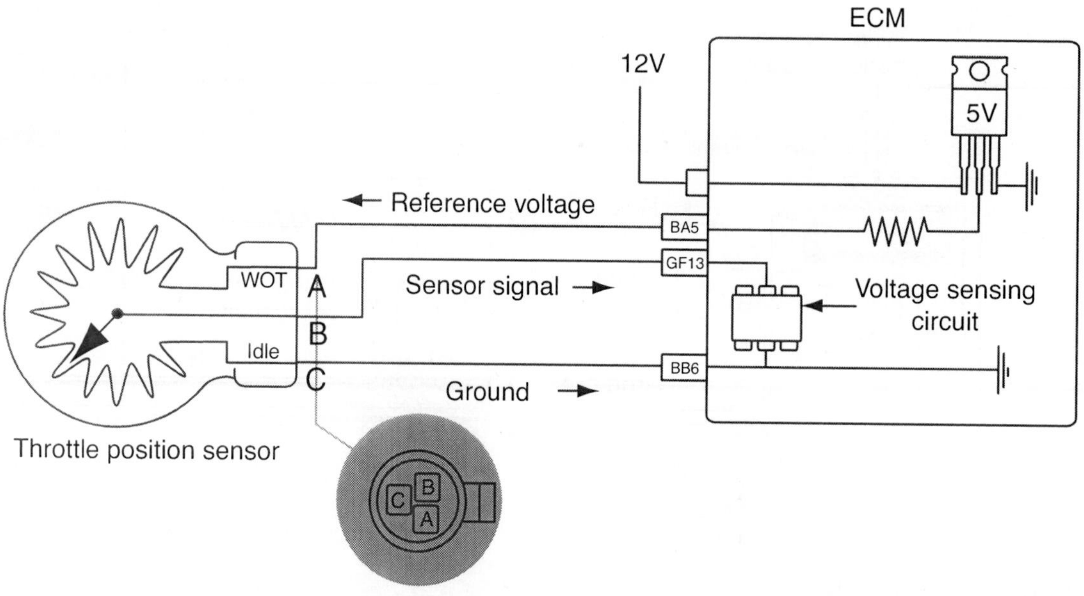


**Figure 12.** TPS terminals.

**(Figure 12)**. A satisfactory TPS voltage waveform is illustrated in **Figure 13** and a TPS waveform with a momentary glitch is pictured in **Figure 14**. A momentary glitch in a TPS voltage signal may cause acceleration stumbles.

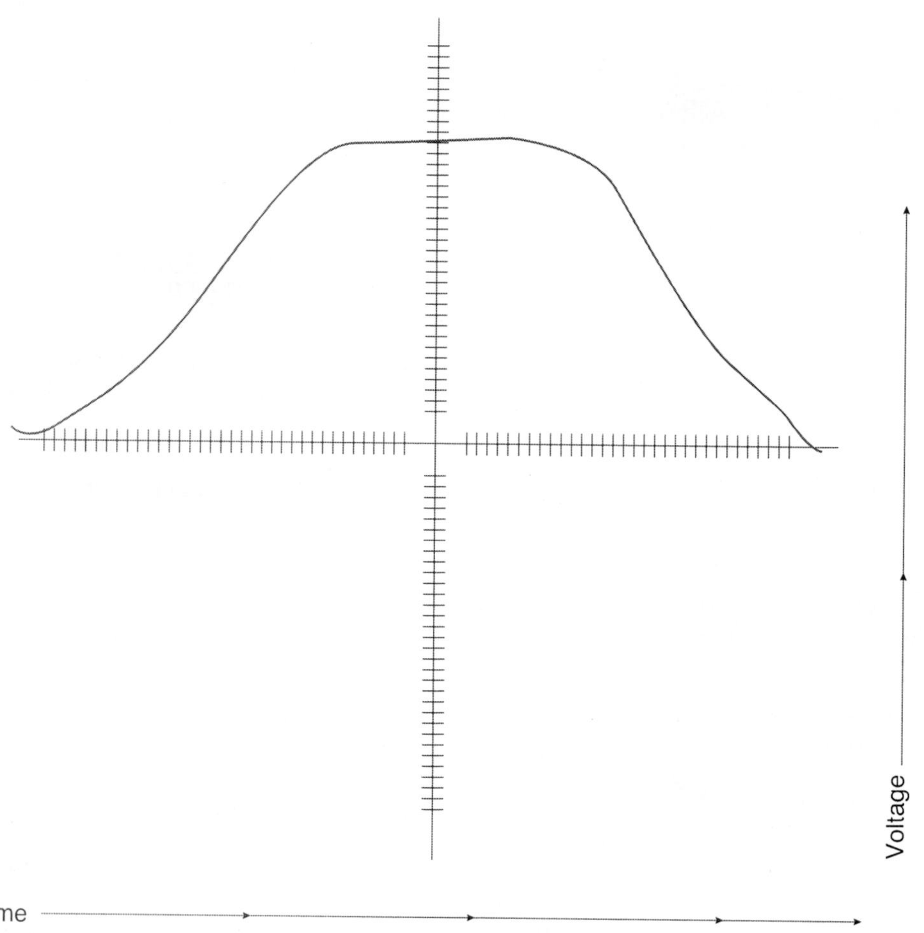


**Figure 13.** A satisfactory TPS waveform.

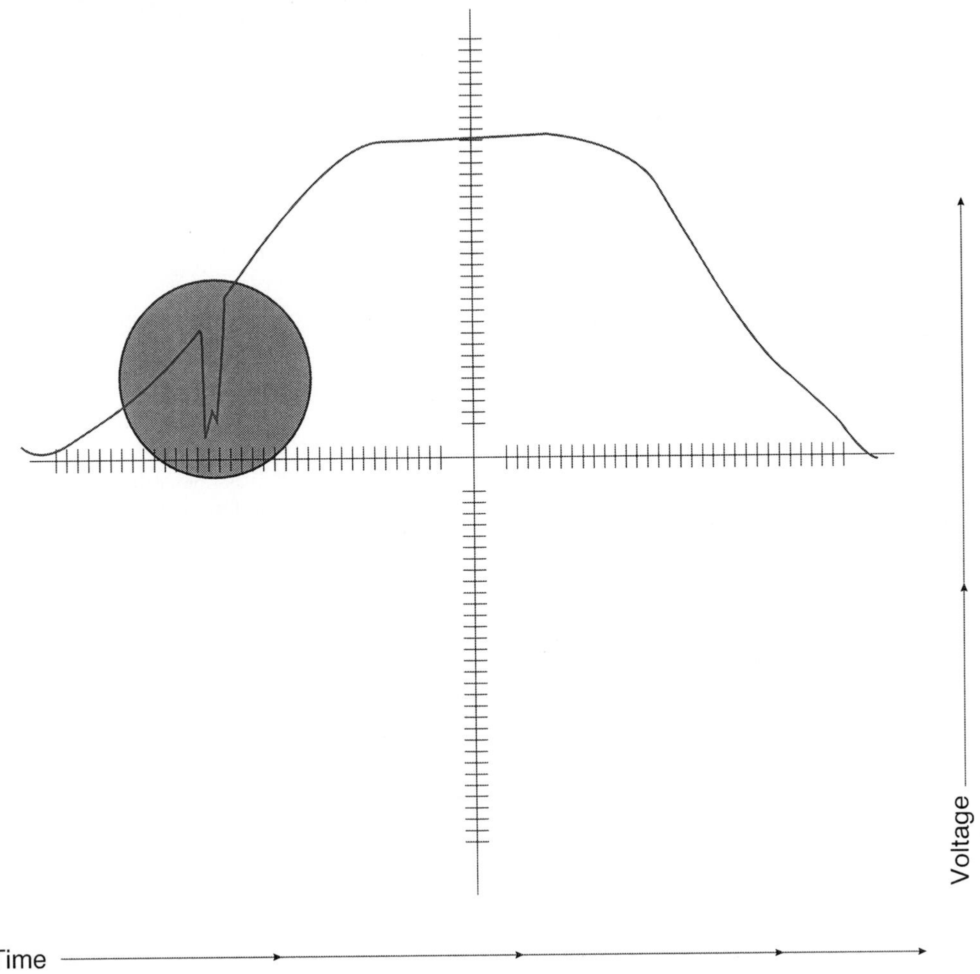


Figure 14. An unsatisfactory TPS waveform.

## MASS AIRFLOW SENSOR DIAGNOSIS

To test a hot-wire-type MAF sensor, observe the MAF grams of air per second on the scan tool. Hold the accelerator pedal steady at several different engine speeds and observe the grams per second. At a constant throttle opening, the grams-per-second reading should remain the same. If this reading fluctuates up or down, the MAF sensor hot wire is probably coated with contaminants or the sensor is defective. The result of this problem is usually rough idle and hesitation during acceleration. Some MAF sensors may be disassembled and the hot wire may be cleaned with a Q-tip dampened in throttle body cleaner. Consult the vehicle manufacturer's recommendations regarding this service procedure. If the hot wire cannot be cleaned, sensor replacement is required.

A vane-type MAF sensor may be tested with a digital multimeter connected to the MAF voltage signal wire. Switch the multimeter to DC volts. With the ignition switch on, slowly push the vane open and closed **(Figure 15)**. The variable resistor in the sensor should provide a smooth

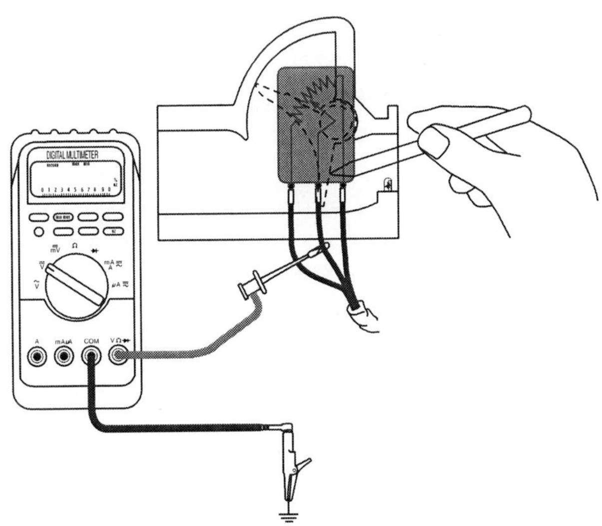

Figure 15. Testing a vane-type MAF sensor.

increase in voltage as the vane is moved to the open position. If the MAF does not provide the specified, smooth voltage signal, replace the sensor.

## KNOCK SENSOR DIAGNOSIS

The knock sensor may be tested with the engine running and the engine knock data displayed on the scan tool. Some scan tools display YES or NO for the engine knock data. During engine operation at idle or a constant throttle opening, the knock data display should indicate NO. During hard engine acceleration at normal engine operating temperature, the knock data should indicate YES momentarily.

With the ignition switch off and the knock sensor wire disconnected, an ohmmeter may be connected from the knock sensor terminal to the sensor housing. The knock sensor should have the specified resistance, which is usually 3,300 ohms to 4,500 ohms. Always use the vehicle manufacturer's specifications.

## INPUT SENSOR SERVICE

When connecting electronic test equipment to input sensors, do not probe through insulation on wires to make a meter connection. This action may cause corrosion in the wire and eventually a resistance that affects the sensor signal. A T-pin or other small probe may be used to back probe terminals on input sensors to complete a meter connection. Always be careful not to damage the weatherpack seal on the wiring connector. A **T-pin** is like a common pin used in sewing, but the T-pin is T-shaped. T-pins are available in the sewing departments of some large retailers.

When installing an $HO_2S$, always place a small amount of antiseize compound on the sensor threads. This action keeps the sensor from seizing into the exhaust system, which makes it very difficult to remove the sensor the next time it requires replacement.

# *Summary*

- $HO_2Ss$ may contain zirconia or titania.
- A rich air-fuel ratio results in a high $HO_2S$ voltage and a lean air-fuel ratio causes a low $HO_2S$ voltage.
- An $HO_2S$ may be tested with a digital voltmeter, a graphing voltmeter, or a lab scope.
- When diagnosing EFI systems, one of the first tests must be fuel pressure.
- ECT and IAT sensors may be tested with an ohmmeter, or a voltmeter may be used to measure the voltage drop across these sensors.
- A MAP sensor may be tested by measuring the MAP voltage signal in relation to the vacuum supplied to the sensor.
- A MAP sensor that produces a hertz-type signal may be tested using a multimeter with a hertz scale.
- The TPS may be tested by measuring the TPS voltage signal in relation to throttle opening.
- A hot-wire MAF sensor may be tested by measuring the grams-per-second airflow through the sensor with a scan tool.
- A vane-type MAF sensor may be tested by measuring the voltage signal from the sensor in relation to vane opening.
- A knock sensor may be tested with an ohmmeter or a scan tool.

# *Review Questions*

1. Technician A says if the $HO_2S$ voltage signal is continually high, the fuel pressure may be excessive. Technician B says if this voltage signal is always high, the fuel injectors may be restricted. Who is correct?
   A. Technician A
   B. Technician B
   C. Both Technician A and Technician B
   D. Neither Technician A nor Technician B
2. Technician A says the module in a hot-wire MAF sensor maintains the hot wire temperature at the same temperature as the cold wire. Technician B says at a constant throttle opening, the grams-per-second airflow reading on the MAF sensor should be fluctuating. Who is correct?
   A. Technician A
   B. Technician B
   C. Both Technician A and Technician B
   D. Neither Technician A nor Technician B

3. Technician A says a bank 1 sensor 2 $HO_2S$ is located upstream from the catalytic converter. Technician B says a bank 2 sensor 1 $HO_2S$ is located on the opposite side of the engine from number 1 cylinder. Who is correct?
   A. Technician A
   B. Technician B
   C. Both Technician A and Technician B
   D. Neither Technician A nor Technician B
4. Technician A says the resistance of the ECT sensor should decrease as the sensor temperature increases. Technician B says if the engine thermostat is stuck open, the air-fuel ratio will be rich. Who is correct?
   A. Technician A
   B. Technician B
   C. Both Technician A and Technician B
   D. Neither Technician A nor Technician B
5. An air leak is present in the air hose between the MAF sensor and the throttle body. The most likely result of this air leak is:
   A. excessive spark advance.
   B. a lean air-fuel ratio.
   C. low fuel pump pressure.
   D. a high voltage signal from the ECT sensor.
6. Technician A says an $HO_2S$ with a 2, 2 designation is located upstream from the catalytic converter. Technician B says an $HO_2S$ with a 2, 2 designation is located on the same side of the engine as number 1 cylinder. Who is correct?
   A. Technician A
   B. Technician B
   C. Both Technician A and Technician B
   D. Neither Technician A nor Technician B
7. At normal operating engine temperature, a typical $HO_2S$ voltage signal should cycle between:
   A. 0.3 volt and 0.8 volt.
   B. 0.1 volt and 0.3 volt.
   C. 0.2 volt and 0.5 volt.
   D. 0.5 volt and 0.9 volt.
8. The most likely result of excessive fuel pump pressure in a port injected fuel system is:
   A. reduced spark advance.
   B. engine detonation.
   C. an inoperative EGR valve.
   D. a rich air-fuel mixture.
9. An engine thermostat that is stuck in the open position may cause all of these problems *except*:
   A. an inoperative EGR valve.
   B. a rich air-fuel ratio.
   C. a higher-than-normal ECT sensor voltage signal.
   D. a lower-than-normal ECT sensor resistance.
10. A TPS has a worn spot near the center of the variable resistor. The most likely result of this problem is:
    A. hesitation during acceleration.
    B. engine surging and misfiring at high speed.
    C. high idle speed with the engine warmed up.
    D. erratic engine idle operation.
11. On a conventional MAP sensor, the voltage signal should ________________ as the vacuum supplied to the sensor increases.
12. The TPS voltage signal should be approximately ________________ volts at wide open throttle.
13. Before testing an $HO_2S$, the sensor should be ________________ by running the engine at ________________ rpm or ________________ minutes.
14. An erratic grams-per-second airflow reading on a hot-wire MAF sensor may indicate the ________________ ________________ is contaminated.
15. Explain the operation of a zirconia-type $HO_2S$ in relation to air-fuel ratio.
16. Describe the operation of a hot-wire MAF sensor in relation to engine speed.
17. Explain the cleaning procedure for the hot wire in a MAF sensor.
18. Explain four causes of low fuel pump pressure.

# Complete Engine Control Systems

## Introduction

Technicians must understand the various types of fuel injection systems to successfully diagnose these systems. For example, technicians must understand how the injectors are connected to the PCM in various injection systems and they must also understand how the PCM maintains the proper air-fuel ratio. Technicians must also understand the various monitors in OBD II systems and the requirements to illuminate the MIL in these systems. It is very important for technicians to understand OBD terminology such as trip and drive cycle.

### THROTTLE BODY INJECTION (TBI) SYSTEMS

In a TBI system, the throttle body assembly may contain one or two injectors depending on the engine displacement. Four-cylinder engines have a single injector in the throttle body assembly **(Figure 1)** and V6 or V8 engines have dual injectors. The injector body is sealed to the throttle body with O-ring seals. When the ignition switch is turned on, 12 volts are supplied through a fuse to the injector(s) **(Figure 2)**. The other terminal on each injector is connected to the PCM. The PCM supplies a ground for the injector windings to open each injector. The PCM grounds an injector each time the ignition system fires a spark plug. When the next spark plug fires in a dual throttle body, the PCM grounds the opposite injector. Grounding an injector winding allows current to flow through the winding and the resulting coil magnetism lifts the injector plunger. When the plunger tip is lifted off its seat, fuel is discharged

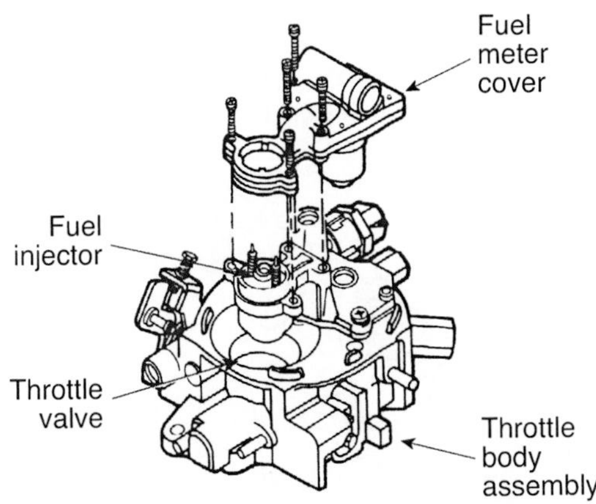


**Figure 1.** A single throttle body assembly.

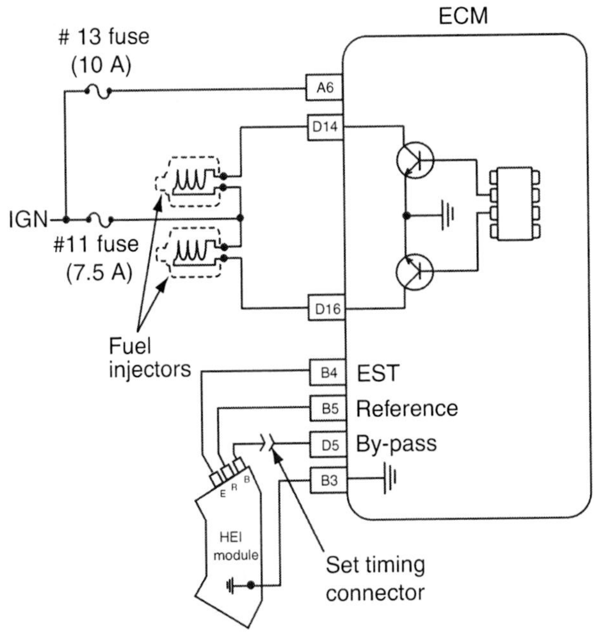


**Figure 2.** Dual throttle body injector wiring connections.

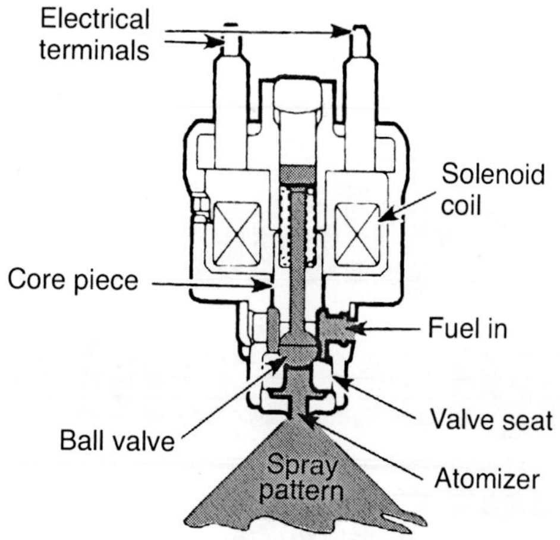


**Figure 3.** A throttle body injector.

from the injector orifice **(Figure 3)**. The amount of fuel discharged by the injector is determined by the injector pulse width. **Injector pulse width** is the length of time in milliseconds that the PCM keeps the injectors open.

At idle speed, the PCM may supply an injector pulse width of 2 milliseconds, whereas at wide-open throttle the injector pulse width may be 12 milliseconds. The PCM always provides the correct injector pulse width to maintain the stoichiometric air-fuel ratio. The **stoichiometric air-fuel ratio** is the ideal air-fuel ratio at which the engine provides the best performance and economy. In a gasoline fuel system, this ratio is 14.7 pounds of air to every 1 pound of fuel. This ratio is expressed as 14.7:1.

In a TBI system, the fuel is injected above the throttle plates and the intake manifold must be filled with fuel vapor. Therefore, on many TBI systems some type of intake manifold or intake air heating is provided to keep the fuel vapor from condensing on the cool intake manifold passages. A heated air inlet system contains a vacuum-operated air control valve in the air cleaner inlet passage and a bimetal vacuum switch mounted in the lower part of the air cleaner housing **(Figure 4)**.

When the air cleaner temperature is cold and the engine is running, manifold vacuum is supplied through the bimetal vacuum switch to the vacuum diaphragm. Under this condition, the vacuum pulls the diaphragm upward and positions the air control valve so it blocks the cold air passage into the air cleaner inlet. With the air control valve in this position, air is drawn through a flexible hose and heat stove on the exhaust manifold into the air cleaner. This warm air from the exhaust manifold heat stove increases the temperature of the air-fuel mixture in the intake manifold to prevent fuel condensation in the intake manifold. When the air cleaner temperature reaches approximately 125°F (52°C), the bimetal vacuum switch begins bleeding off the manifold vacuum, causing the air control valve to open and allow cooler air to be drawn through the air cleaner inlet into the air cleaner. If the heated air inlet system does not pull hot air through the manifold stove into the air cleaner during engine warmup, acceleration stumbles may occur.

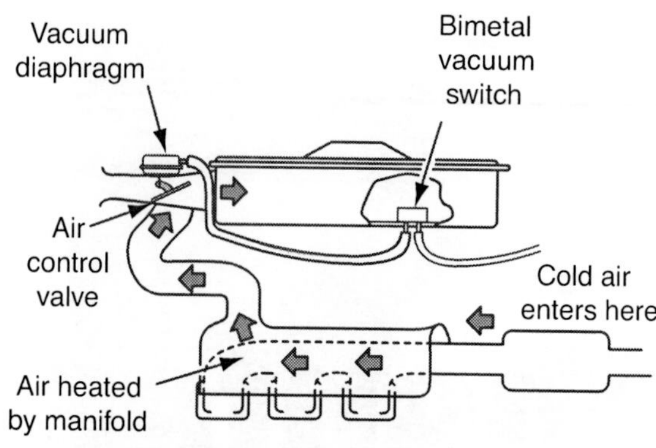


**Figure 4.** A heated air inlet system.

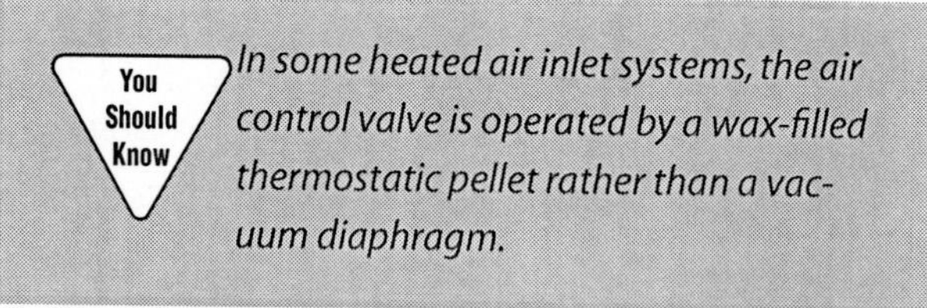


*In some heated air inlet systems, the air control valve is operated by a wax-filled thermostatic pellet rather than a vacuum diaphragm.*

Some TBI systems have an idle speed control (ISC) motor mounted on the throttle body assembly **(Figure 5)**. The ISC motor is a reversible motor that is controlled by the PCM. The stem of the ISC motor pushes against the throttle linkage. Two ISC motor windings are connected to the PCM. When one of these windings is activated, the motor stem extends and opens the throttle. Energizing the opposite winding retracts the ISC motor stem. The PCM operates the ISC motor to supply the proper idle speed at all engine temperatures. When the engine coolant is cold, the PCM operates the ISC motor to supply a faster idle speed. Some ISC motors have an idle switch in the ISC motor stem. When the throttle linkage contacts the ISC motor stem, the idle switch closes and sends a signal to the PCM. The PCM only controls the ISC motor when the throttle linkage is contacting the motor stem.

Some throttle body assemblies have an idle air control (IAC) motor mounted on the throttle body **(Figure 6)**. The IAC motor is operated by the PCM. This motor contains a plunger that opens and closes an air passage. The air flow in the IAC motor passage bypasses the throttle. The IAC motor

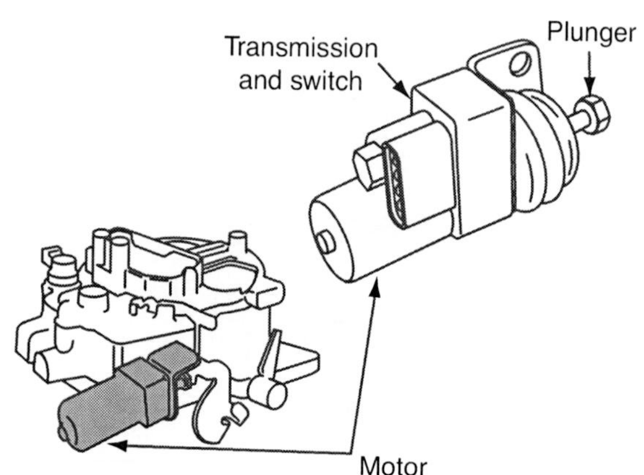


**Figure 5.** An idle speed control (ISC) motor.

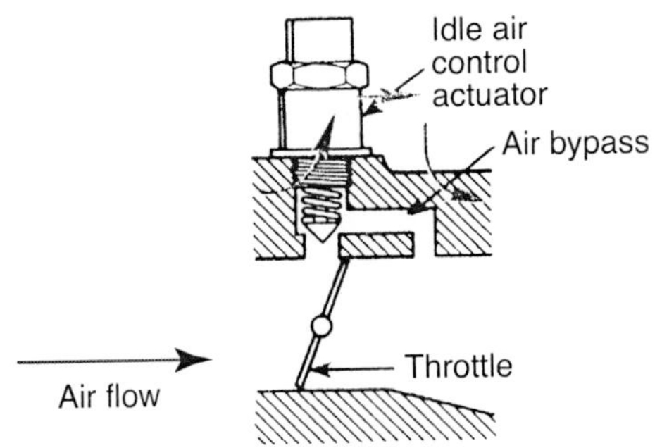


**Figure 6.** An idle air control (IAC) motor.

plunger controls idle speed by regulating the amount of air bypassing the throttle. When the engine is cold, the PCM moves the IAC motor plunger to provide more air flow through the IAC motor passage around the throttle. This action increases engine rpm. As the engine warms up, the PCM gradually moves the IAC motor plunger toward the closed position to reduce engine rpm.

## MULTIPORT FUEL INJECTION (MFI) SYSTEM

In an MFI system, the injectors are positioned in the intake manifold and the injector tips are located near the intake ports **(Figure 7)**.

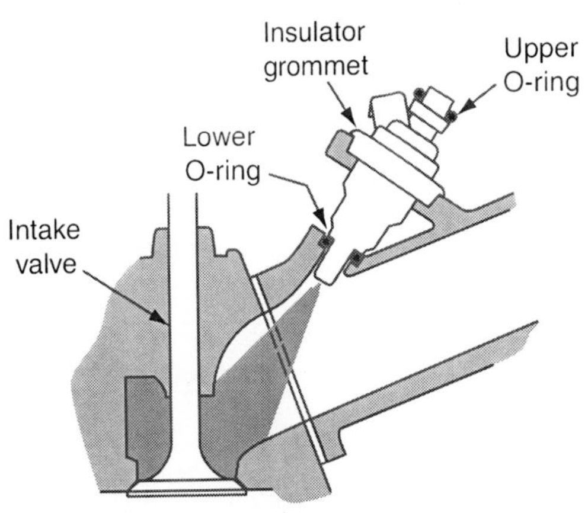


**Figure 7.** A multiport injector located in the intake port.

An O-ring on the lower end of the injector seals the injector to the intake manifold and a second O-ring on the upper end of the injector seals the injector to the fuel rail. In some MFI systems, the injectors are connected in pairs or groups to the PCM and the PCM operates two to four injectors simultaneously **(Figure 8)**. The PCM always supplies the correct injector pulse width to maintain 14.7:1 air-fuel ratio under part throttle engine operation. During cold engine operation, hard acceleration, or wide-open throttle condi-

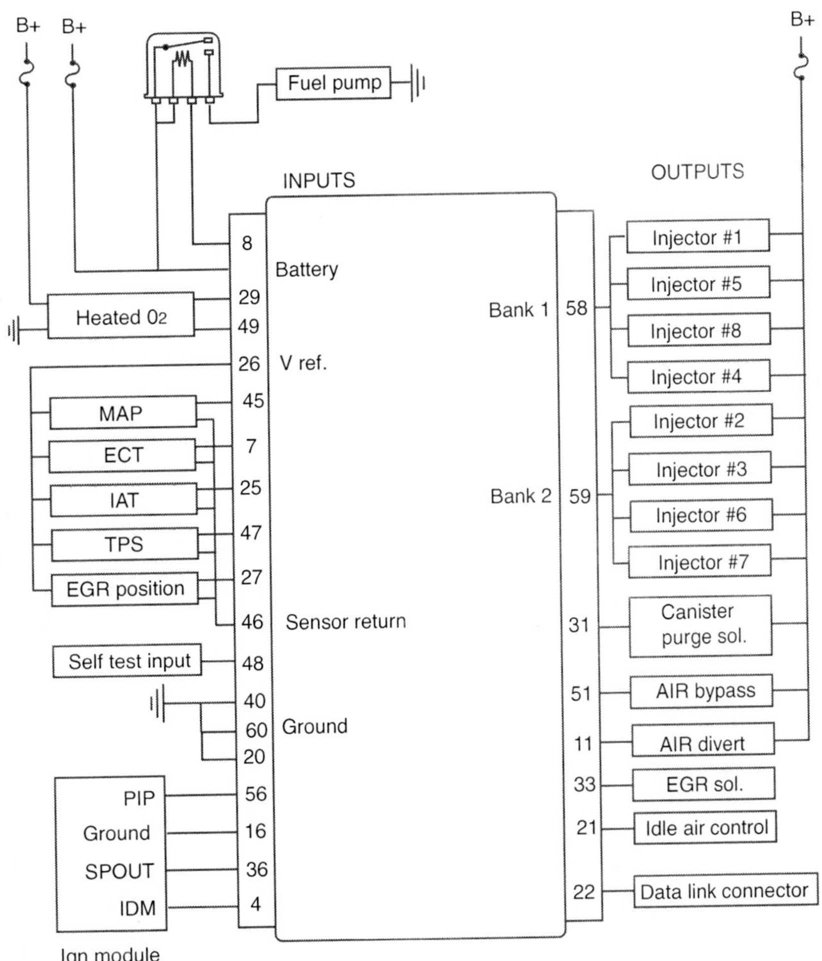


**Figure 8.** A multiport fuel injection (MFI) system.

tions, the PCM supplies a richer air-fuel ratio to maintain engine performance. When the engine is decelerating, the PCM supplies a leaner air-fuel ratio to improve fuel mileage and emissions.

In an MFI system, fuel is injected near the intake ports and the manifold passages contain air from the air cleaner and throttle body. Since the intake manifold passages contain mostly air, intake manifold or intake air heating are not required. MFI systems usually have an IAC motor to control idle speed.

## SEQUENTIAL FUEL INJECTION (SFI)

In other injection systems, the injectors are grounded individually by the PCM. This type of system is called sequential fuel injection (SFI) system **(Figure 9)**.

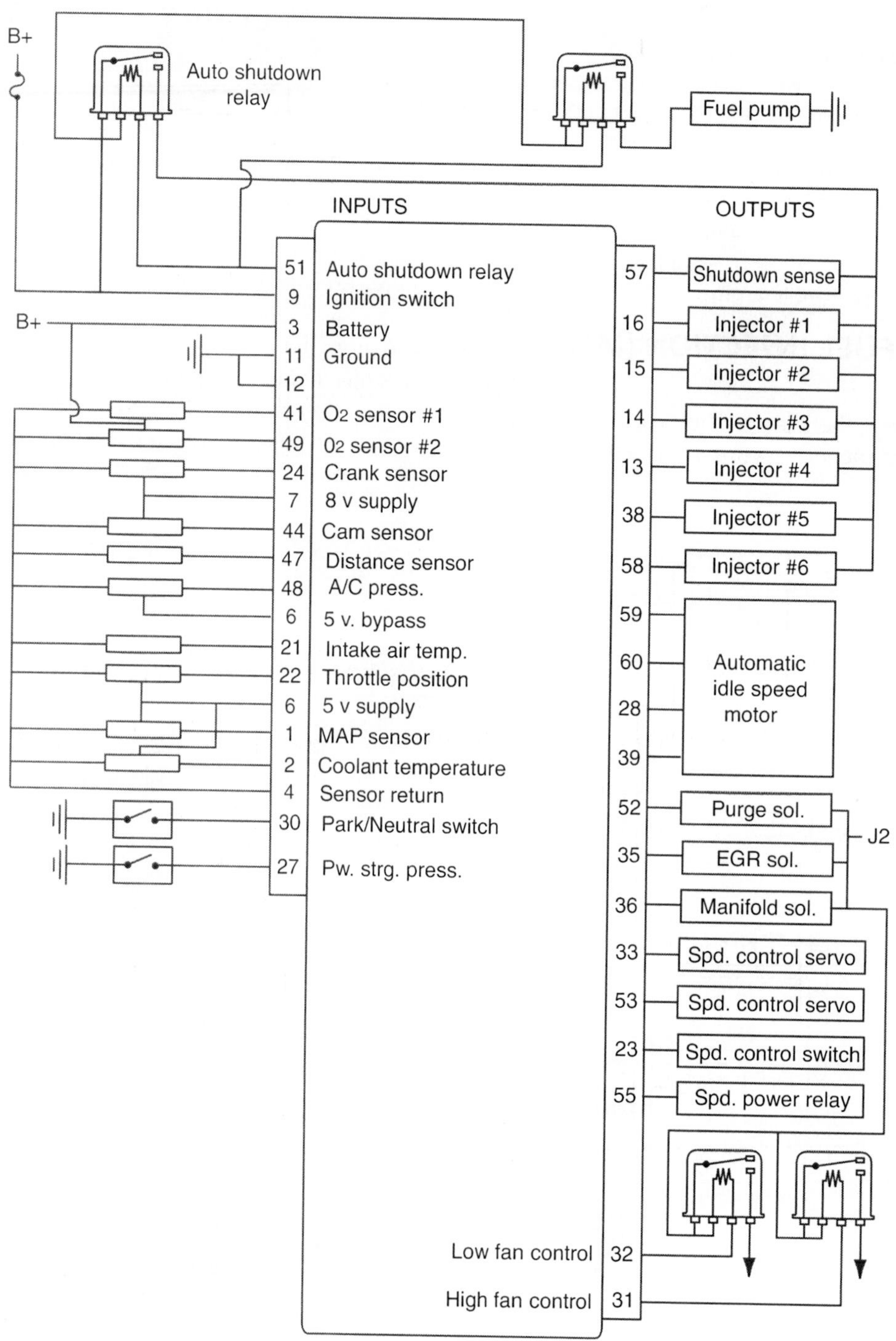


**Figure 9.** A sequential fuel injection (SFI) system.

In some SFI systems with a DI system, the number 1 blade that rotates past the distributor pickup is narrower compared to the other blades. This narrow blade provides a different signal to the PCM. When this signal is received, the PCM begins opening the injectors in the engine firing order. The PCM opens each injector a significant number of crankshaft degrees before the intake valve actually opens. This action fills the intake port with fuel vapor and ensures the proper supply of air-fuel mixture to the cylinder when the intake valve opens.

Some SFI systems have an ASD relay and a fuel pump relay. The PCM operates both relays and the ASD relay supplies voltage to the injectors. Battery voltage is supplied to the PCM on terminal 3 and ignition voltage is connected to PCM terminal 9. PCM terminals 11 and 12 are connected to a chassis ground (refer to Figure 9). The PCM must be supplied with battery voltage on terminals 3 and 9 and terminals 11 and 12 must be connected to a good chassis ground with a minimum amount of resistance. Excessive resistance in the voltage supply or ground wires may affect PCM operation and cause drivability problems.

Many PCMs have an input from the park/neutral switch. The signal from this switch is used by the PCM to help control idle rpm.

Some PCMs have an input from the power steering pressure switch. When this signal is received during a turn, the PCM opens the AC compressor clutch circuit and stops the compressor. This action helps to prevent engine stalling at low speeds from the compressor and power steering loads on the engine.

Some PCMs contain the cruise control module, so the inputs from the cruise control switches are sent to the PCM. Many PCMs operate the low fan control and high fan control relays to operate the cooling fan at low and high speed. When the PCM receives a signal from the ECT sensor indicating that cooling fan operation is required, the PCM energizes the appropriate cooling fan relay. When the relay contacts close, voltage is supplied through these contacts to the cooling fan motor.

## CENTRAL PORT INJECTION (CPI) SYSTEMS

In a **central port injection (CPI)** system, one central injector is mounted in the intake manifold. The upper half of the intake manifold is bolted to the lower half. This upper half may be removed to access the CPI components **(Figure 10)**. The pressure regulator is mounted with the central injector **(Figure 11)**. The fuel supply and return lines are connected to the central injector and the pressure regulator. These fuel lines are sealed with rubber grommets where they enter the intake manifold to prevent vacuum leaks. A port in the pressure regulator allows intake manifold vacuum to be supplied to this compo-

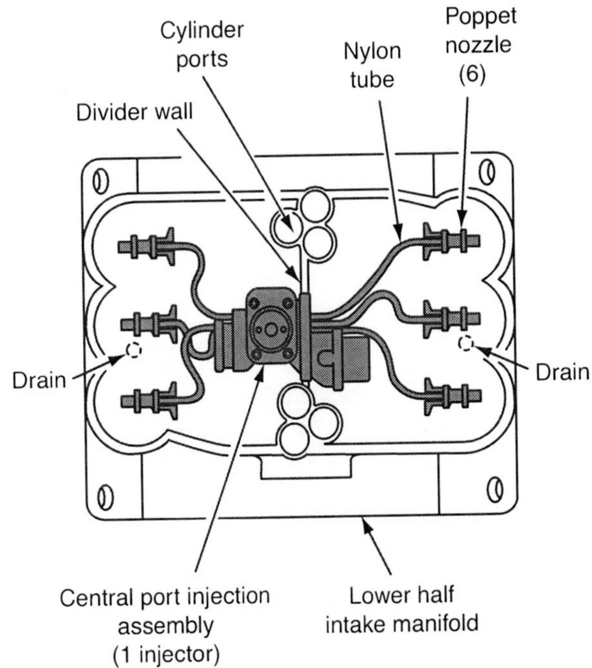


**Figure 10.** A central port injection (CPI) system.

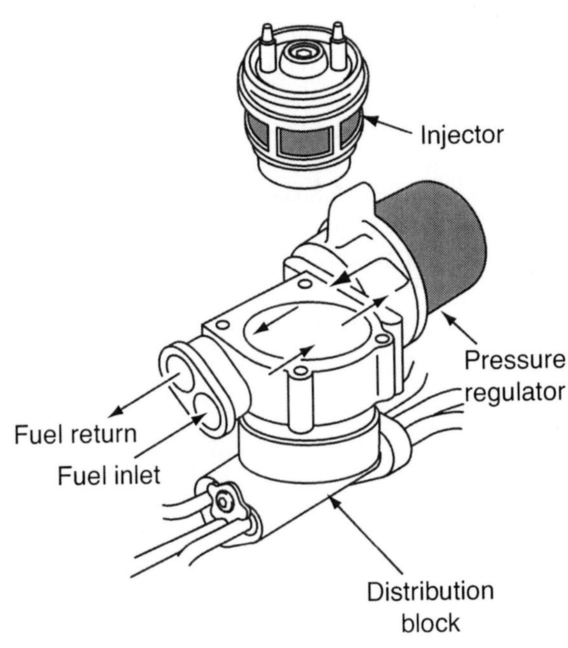


**Figure 11.** A central injector in a CPI system.

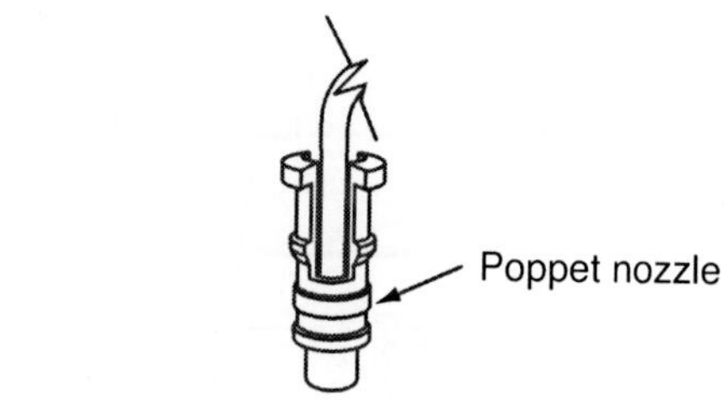


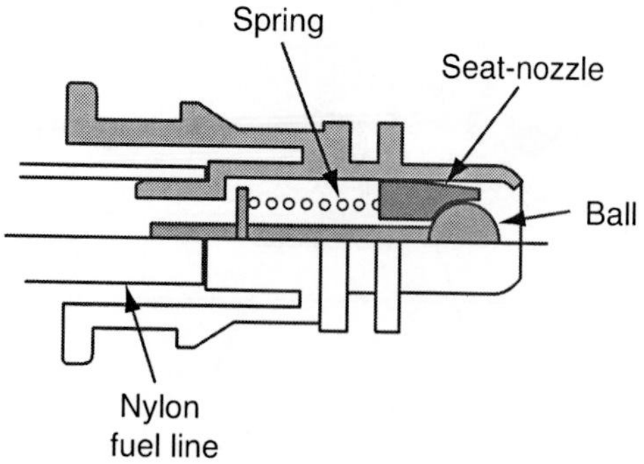


**Figure 12.** A poppet nozzle in a CPI system.

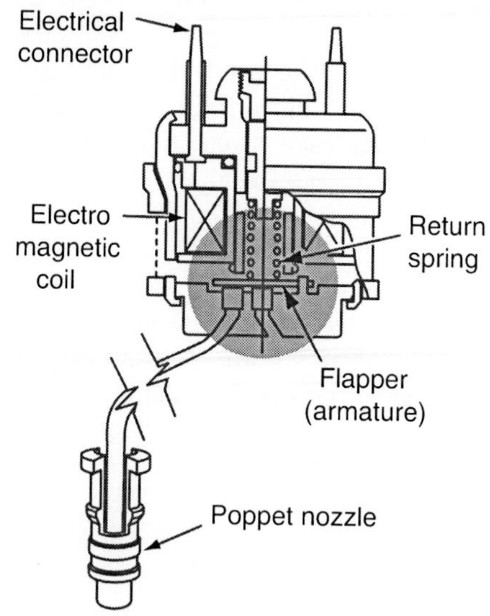


**Figure 13.** Central port injector operation.

nent. A poppet nozzle is located in each intake port and a nylon tube is connected from the central port injector to each poppet injector. The seats in the poppet nozzles are closed by spring pressure **(Figure 12)**.

Voltage is supplied to the central injector and the PCM grounds this injector. When the PCM grounds the central port injector winding, the coil magnetism lifts the injector plunger and the flapper valve on the lower end of the plunger **(Figure 13)**. Lifting this valve supplies fuel pressure to all the poppet nozzles simultaneously. When the fuel pressure at each poppet nozzle reaches 39 psi (269 kPa), the fuel pressure lifts the poppet nozzle seats and fuel is discharged from the poppet nozzles into the intake ports. The PCM supplies the proper central injector pulse width to provide the correct air-fuel ratio.

A manifold tuning valve (MTV) is positioned in the intake manifold on CPI systems. The MTV is operated by a vacuum diaphragm and vacuum is supplied to this diaphragm through a PCM-operated vacuum solenoid. At engine speeds below 3,600 rpm, there is no vacuum supplied to the MTV and this valve allows normal airflow through the intake manifold **(Figure 14)**. At speeds above

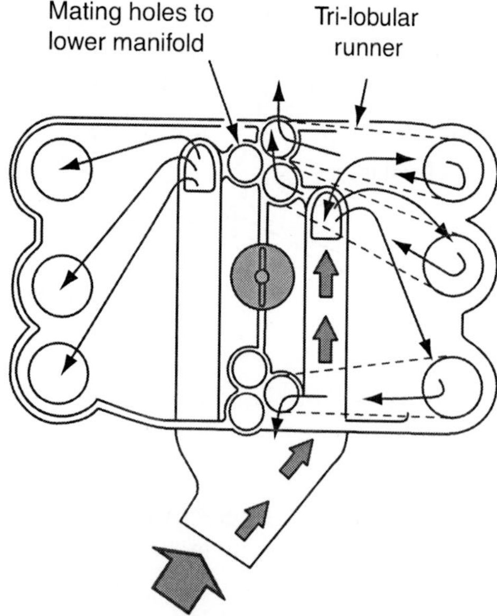


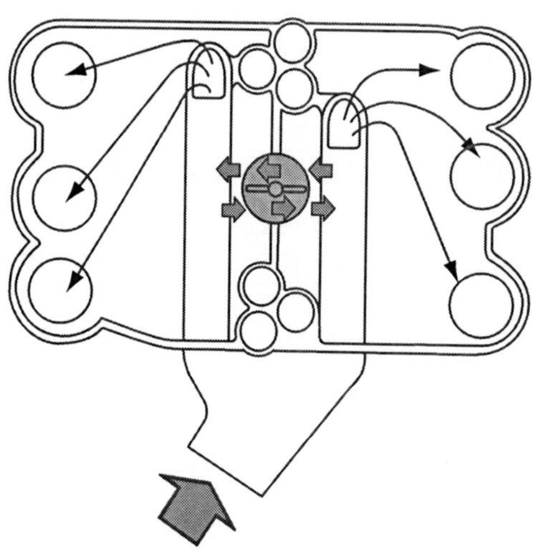


**Figure 14.** Manifold tuning valve (MTV) operation.

3,600 rpm, the PCM operates the MTV solenoid and vacuum is supplied through this solenoid to the MTV diaphragm. This action opens the MTV and this valve position provides shorter intake manifold air passages to improve engine performance.

## ON-BOARD DIAGNOSTIC I (OBD I) AND ON-BOARD DIAGNOSTIC II (OBD II) SYSTEMS

In 1988, the CARB adopted OBD I regulations for cars sold in California. These regulations required the engine computer system to monitor fuel control, EGR, emission components, and the PCM. OBD I systems also had to monitor all sensors used for fuel and emission control for opens and shorts. The PCM also had to monitor fuel trim, EGR and $HO_2S$. These systems were required to illuminate the MIL if a defect occurred and a DTC had to be set in the PCM memory.

In the late 1980s, the CARB began investigating the possibility of using OBD system diagnosis to supplement inspection maintenance (I/M) emission programs. Simple OBD diagnostic tests could be performed with much less expensive equipment compared to tailpipe emission tests. In 1990, CARB implemented the first OBD requirements combined with emission testing. However, this combination experienced some problems. One of the difficulties in using OBD diagnosis for emission testing was that the OBD systems did not have much standardization. Vehicle manufacturers were each using their own DLC. Different scan tool software was required for each manufacturer. At that time, OBD systems did not have the capability to monitor critical emission-related components such as catalytic converters, evaporative emission systems, and secondary air injection systems. The OBD systems did not even have the capability to perform a complete test of the oxygen sensor(s).

CARB and the EPA realized that a number of OBD system improvements had to be implemented if OBD systems were to be useful in I/M emission test programs. The result of these discoveries was the installation of OBD II systems on all light-duty vehicles on 1996 and later models. In 1994 and 1995, some vehicles had partial OBD II systems. Federal requirements demand that OBD II systems have many standard features including a standard DLC in the same location on all vehicles.

The EPA passed legislation requiring the implementation of OBD II diagnosis with emission I/M programs. A significant number of states have implemented OBD II diagnosis with an I/M emission program.

### OBD II Monitoring

According to EPA rules, an OBD II system must detect the following emission-related defects and alert the driver.

1. The engine misfire monitor must detect any engine misfire condition that causes exhaust emissions of one and one half times the standard for non-methane hydrocarbons (NMHC), CO, or $NO_x$. The PCM monitors the crankshaft position sensor signal in the misfire monitor. Each time a spark plug fires, the crankshaft should speed up momentarily. If a cylinder misfires, the crankshaft does not speed up, and the PCM detects this problem from the crankshaft position sensor signal. A specific number of misfires must occur in 1,000 cylinder firings before the PCM takes the necessary action such as setting a DTC.
2. The $HO_2S$ monitor must detect any $HO_2S$ defect that results in exhaust emissions of one and one half times the standard for NMHC, CO, or $NO_x$. The $HO_2S$ monitor must also detect any other system component that makes the $HO_2S$ sensor incapable of performing its intended function in the system. The PCM also monitors each $HO_2S$ heater for proper current flow. Low current flow indicates a burned-out heater or open heater wire and the PCM interprets this situation as a failed $HO_2S$ heater monitor.
3. The catalytic converter monitor must detect any catalytic converter problem that results in exhaust emissions of one and one half times the standard for NMHC using an average 4,000-mile aged converter. The PCM monitors the upstream and downstream $HO_2S$ to monitor the converter operation. The voltage cycles on the downstream $HO_2S$ should have a lower voltage and slower cycles compared to the upstream $HO_2S$. If the upstream and downstream $HO_2S$ signals are similar, the PCM takes the necessary action, such as setting a DTC.
4. The evaporative system monitor must be able to detect if there is an absence of evaporative purge air flow from the evaporative system. This monitor must also be able to detect any leak in the evaporative system equal to or greater than a leak caused by a 0.040 inch (1.016 millimeters) or a 0.020 inch (0.508 millimeter) diameter orifice, depending on the vehicle model year. This leak detection capability excludes the tubing between the purge valve and the intake manifold.
5. The exhaust gas recirculation monitor, secondary air injection monitor, and the fuel control monitor must be able to detect any malfunction or deterioration in the powertrain system or any component directly responsible for emission control that causes emissions of NMHC, CO, or $NO_x$ to be one and one half times the standard for these emissions.
6. The comprehensive monitor must be capable of detecting any defect or deterioration in the emission-related powertrain control system. This includes any defects in any PCM input sensor or output control device. Defects in these components are defined as

malfunctions that cause the component to not meet certain continuity, rationality, or functionality checks performed at specific intervals by the PCM.

7. The secondary air injection monitor checks the operation of the air pump. The PCM turns on the electric-drive air pump and checks for a leaner $HO_2S$ signal in a specific time limit. If the $HO_2S$ signal does not change, the PCM considers the AIR pump to be defective.
8. The fuel system monitor checks the long-term fuel trim to determine if this parameter is at the high or low limit. The fuel system monitor also checks the short-term fuel trim. If the PCM detects a long-term fuel trim at the high or low limit during the fuel system monitor, the PCM considers the fuel system to have failed the monitor test.
9. The thermostat monitor checks the operation of the engine thermostat. A timer circuit monitor in the PCM checks the length of time required for the engine coolant temperature to increase a specific number of degrees during engine warmup.

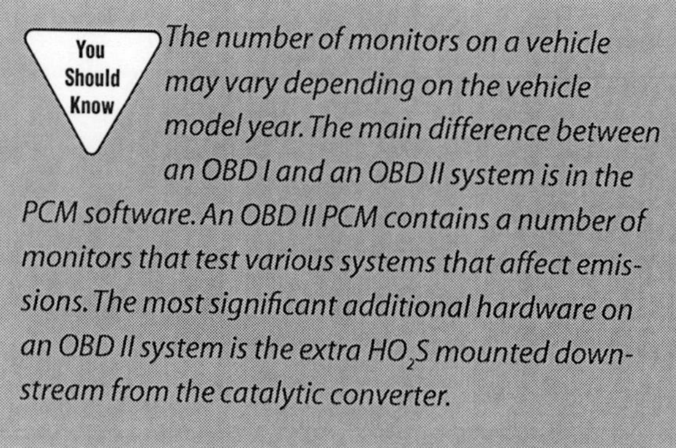


*The number of monitors on a vehicle may vary depending on the vehicle model year. The main difference between an OBD I and an OBD II system is in the PCM software. An OBD II PCM contains a number of monitors that test various systems that affect emissions. The most significant additional hardware on an OBD II system is the extra $HO_2S$ mounted downstream from the catalytic converter.*

## OBD II Definitions

A **drive cycle** may be defined as an engine startup and vehicle operation that allows the PCM to enter closed loop and allows all the monitors to complete their function. Specific driving conditions for certain lengths of time are required for all the monitors to complete their function **(Figure 15)**.

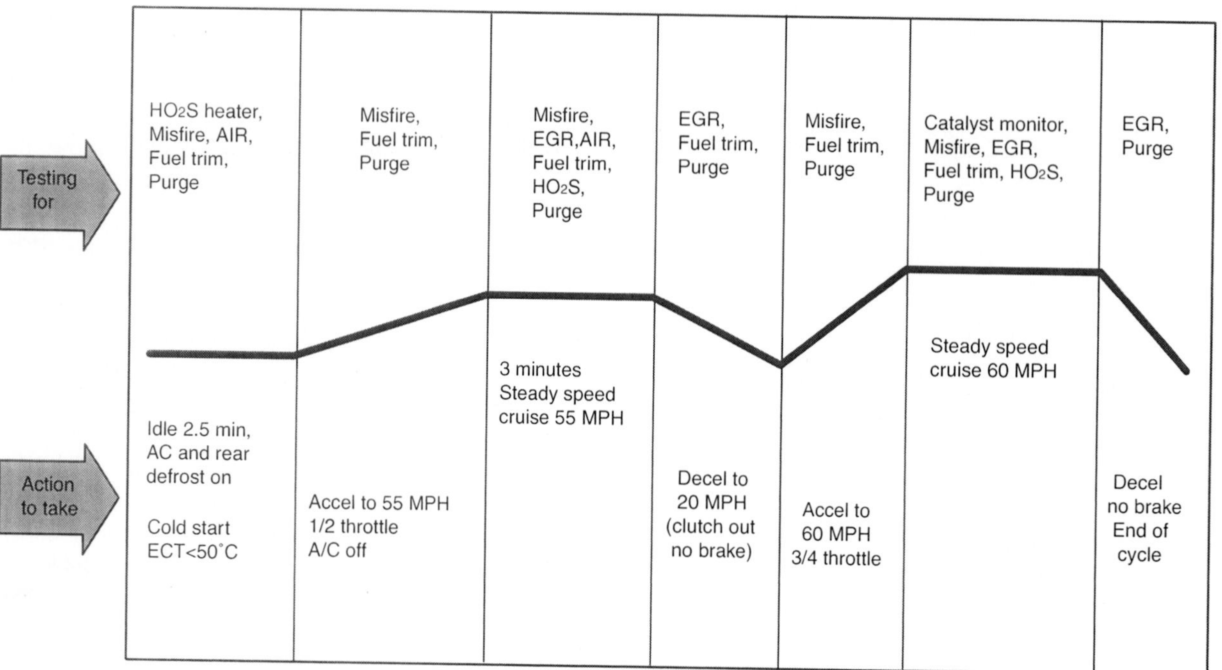


**Figure 15.** An OBD II drive cycle.

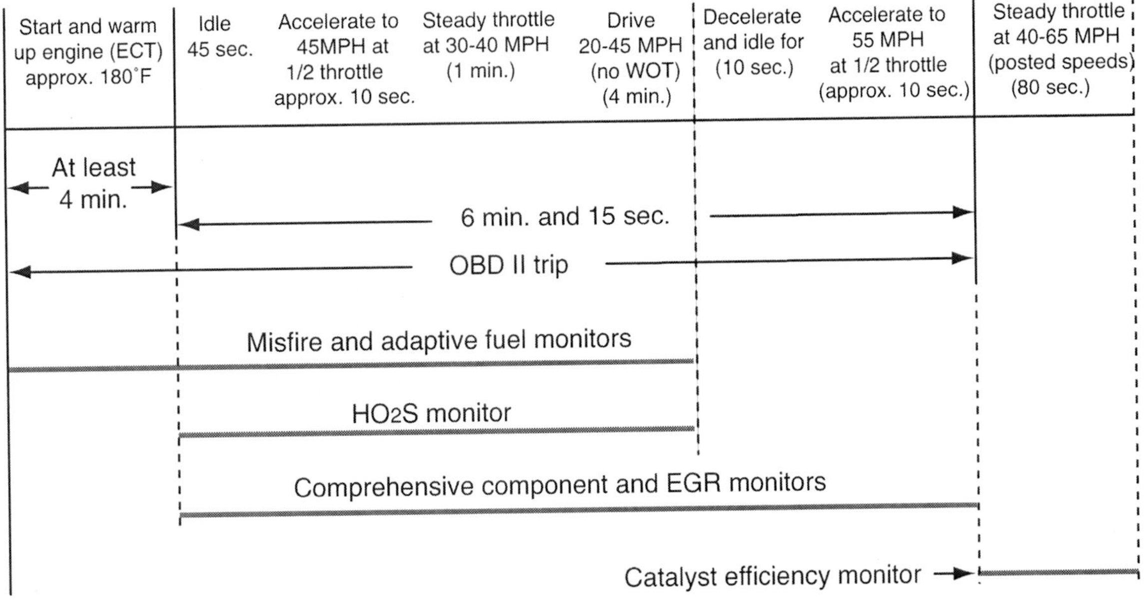


**Figure 16.** An OBD II trip.

An **engine warm-up cycle** may be defined as an engine startup and engine operation until the coolant temperature increases at least 40°F (22°C) from startup and reaches a minimum temperature of 160°F (70°C).

A **trip** may be defined as an engine startup and vehicle operation that allows all the monitors to complete their function except the catalytic converter monitor **(Figure 16)**.

## OBD II Operation

Many OBD II system defects do not cause illumination of the MIL light until the PCM senses the defect on two consecutive drive cycles. However, many defects cause a DTC to be set in the PCM memory the first time the defect is sensed. When a defect occurs and a DTC is set in the PCM memory, the PCM also stores information in the freeze frame. This information may be retrieved with a scan tool during system diagnosis. If the engine switches to any default mode that is abnormal for present driving conditions, the MIL light must be turned on. If engine misfiring occurs that may cause catalytic converter damage, the MIL light flashes once per second. Current DTCs represent defects that are continually present. Intermittent or history DTCs represent faults that were present for a short time. Pending DTCs represent faults that set a DTC in the PCM memory, but the fault did not occur the second time and turn on the MIL light.

When a defect detected by the fuel control monitor or engine misfire monitor does not reoccur in three consecutive trips under similar driving conditions, the PCM turns the MIL light off. Similar driving conditions are: engine speed within 375 rpm of the original defect occurrence, engine load within 20 percent, and engine warm-up conditions are the same and no new defects have been sensed. For defects other than those detected by the fuel control monitor and the misfire monitor, the MIL light is turned off if the defect is not sensed and no new problems are detected in three consecutive trips. History DTCs are cleared from the PCM memory after forty consecutive warm-up cycles have occurred without a malfunction.

A scan tool connected to the DLC displays system status information indicating whether each monitor has completed its monitoring function. The vehicle must be driven through the specific drive cycle conditions for all the monitors to complete their function. This system status information is useful in verifying repairs. If all the monitors have completed their function and no DTCs are set in the PCM memory, there are no defects in the OBD II system. When OBD II diagnosis is combined with I/M inspections, the system status information will be essential.

## OBD II Standardization

Vehicle manufacturers must comply with these OBD II standardization requirements:

1. A standard 16-terminal DLC must be mounted on the driver's side of the dash and the DLC must be visible (SAE standard J1962). Twelve of the 16 terminals in the DLC are dictated by regulations and the OEMs may determine the connections to the other four terminals.
2. Standard DTCs are dictated by regulations, but the OEMs can insert some other DTCs (SAE standard J2012).
3. SAE standard J1979 requires the universal use of scan tools on all vehicle makes and test modes.
4. SAE standard J1850 requires a standard communication protocol (data links).
5. SAE standard J2190 requires standard diagnostic test modes.
6. The PCM must transmit vehicle identification data to the scan tool.
7. The PCM must have the ability to erase DTCs from the PCM memory.
8. The PCM must have the ability to record and store data in snapshot form regarding conditions when a defect occurred.
9. The OBD II system must alert the driver if a defect occurs that increases emissions above a specific limit. The system must also store a DTC representing any such defect.
10. OEM terminology must conform to SAE standard J1930 regarding electronic terms, acronyms, and definitions.

## Summary


- Throttle body assemblies contain one or two injectors depending on the engine size.
- The PCM always supplies the correct injector pulse width to provide the proper air-fuel ratio.
- On a TBI system, a heated air inlet system pulls hot air through an exhaust manifold stove into the air cleaner during engine warmup.
- An ISC motor stem moves the throttle linkage to control idle speed.
- An IAC motor controls the amount of air bypassing the throttle to control idle speed.
- In an MFI system, the PCM operates the injectors in groups of two to four.
- In an SFI system, the PCM operates each injector individually.
- In a CPI system, the PCM operates a central injector that supplies fuel to poppet nozzles in each intake port.
- In an OBD II system, the PCM controls a number of monitors to detect problems in various systems and components.
- In an OBD II system, the PCM informs the driver about any problem in a system or component that causes emissions of NMHC, CO, or $NO_x$ to be one and one half times higher than the emission standards for that model year. The defect usually has to occur on two consecutive drive cycles.
- Many OBD II features are regulated by SAE standards.


## Review Questions

1. Technician A says in a TBI system, the PCM increases the injector pulse width as engine speed increases. Technician B says in a TBI system, the PCM maintains a 14.7:1 air-fuel ratio at part throttle. Who is correct?
   A. Technician A
   B. Technician B
   C. Both Technician A and Technician B
   D. Neither Technician A nor Technician B
2. Technician A says some throttle body systems require a heated air inlet system. Technician B says a heated air inlet system maintains air cleaner temperature at 160°F (71°C). Who is correct?
   A. Technician A
   B. Technician B
   C. Both Technician A and Technician B
   D. Neither Technician A nor Technician B
3. Technician A says if a signal is received from the power steering pressure switch, the PCM shuts off the AC compressor. Technician B says the power steering pressure switch signal informs the PCM to increase the spark advance. Who is correct?
   A. Technician A
   B. Technician B
   C. Both Technician A and Technician B
   D. Neither Technician A nor Technician B
4. Technician A says in a CPI system, the poppet injectors are operated by the PCM. Technician B says in a CPI system, a vacuum hose is connected from the intake manifold to the pressure regulator. Who is correct?
   A. Technician A
   B. Technician B
   C. Both Technician A and Technician B
   D. Neither Technician A nor Technician B

5. In a TBI system, the PCM supplies the proper air-fuel ratio by controlling the:
   A. fuel pump pressure.
   A. IAC motor.
   A. AIR pump.
   A. injector pulse width.
6. While discussing TBI systems with dual injectors, Technician A says the PCM alternately grounds each injector winding. Technician B says when the ignition switch is turned on, 12 volts are supplied to both injectors. Who is correct?
   A. Technician A
   B. Technician B
   C. Both Technician A and Technician B
   D. Neither Technician A nor Technician B
7. All of these statements about an IAC motor are true *except*:
   A. The IAC motor regulates the amount of air that is bypassing the throttles.
   B. If the air conditioning is turned on, the PCM moves the IAC plunger toward the open position.
   C. When the transmission is shifted from DRIVE to PARK, the PCM moves the IAC plunger toward the closed position.
   D. When the engine coolant is cold, the PCM moves the IAC plunger toward the closed position.
8. All of these statements about MFI and SFI systems are true *except*:
   A. In an MFI system, the PCM grounds two or more injectors simultaneously.
   B. In an SFI system, the PCM grounds each injector individually.
   C. In an SFI system, the PCM opens an injector when the intake valve is opening in the same cylinder.
   D. Some SFI systems have an automatic shutdown relay that supplies voltage to the injectors when the ignition switch is turned on.
9. In a CPI system:
   A. the PCM grounds each injector individually.
   B. voltage is supplied directly from the PCM to each injector.
   C. a V8 engine has eight outlets on the central port injector.
   D. the pressure regulator is mounted separately from the central port injector.
10. In a CPI system, the pressure required to open the poppet nozzles is:
    A. 10 psi.
    B. 26 psi.
    C. 32 psi.
    D. 39 psi.
11. An IAC motor controls idle speed by regulating the amount of ______________ that is bypassing the throttle.
12. Some ISC motors have an ______________ ______________ in the motor stem.
13. The PCM uses the park/neutral switch to control ______________ ______________.
14. In an OBD II system, the misfire monitor senses cylinder misfiring from the ______________ ______________ sensor signal.
15. Explain how the catalytic converter monitor senses converter condition in an OBD II system.
16. In an OBD II system, explain how the PCM monitors the air pump operation.
17. Define a drive cycle in an OBD II system.
18. Describe an engine warmup cycle in an OBD II system.

# Engine Control System Maintenance, Diagnosis, and Service

## Introduction

The engine control system affects engine performance and economy more than any other system. When an engine experiences reduced performance and/or fuel economy, the cause of the problem is usually located in the engine control system. Therefore, it is extremely important for technicians to be able to perform quick and accurate diagnoses on this system.

## ENGINE CONTROL SYSTEM MAINTENANCE

During minor underhood service, the fuel system should be inspected for fuel leaks, loose electrical connections, worn wiring insulation, and vacuum leaks. Fuel leaks reduce fuel mileage and create the possibility of a fire. Loose electrical connections or worn insulation on fuel injector wiring may cause injector misfiring that reduces engine performance and economy. Vacuum leaks may cause erratic idle operation. A vacuum leak in the fuel pressure regulator hose increases fuel pressure which causes more fuel to be injected. This situation results in reduced fuel economy.

In any fuel injection system, the fuel filter should be replaced at the vehicle manufacturer's recommended service interval. The fuel filter may be located under the vehicle or in the engine compartment. The fuel pressure should be relieved before loosening any of the fuel lines. On MFI and SFI systems, fuel pressure may be relieved by connecting a vacuum hand pump to the vacuum connection on the pressure regulator and applying 20 in. Hg of vacuum to the regulator. This action allows any pressure in the fuel system to be released back into the fuel tank. On TBI systems, disconnect the injector connector and use two jumper wires to momentarily supply power and a ground to the injector terminals.

> **You Should Know** *Energizing an injector winding for more than 5 seconds may burn out the winding because these windings are normally energized by the PCM for a few milliseconds at a time.*

After the fuel system pressure is relieved, use a clean shop towel to wipe any dust and debris from the fuel line connections to prevent this material from entering the fuel lines. Loosen and remove the line fittings from the filter and remove the filter. Install the new filter in the proper direction and tighten the fittings to the specified torque. After the filter installation, always start the engine and inspect the filter and line fittings for leaks.

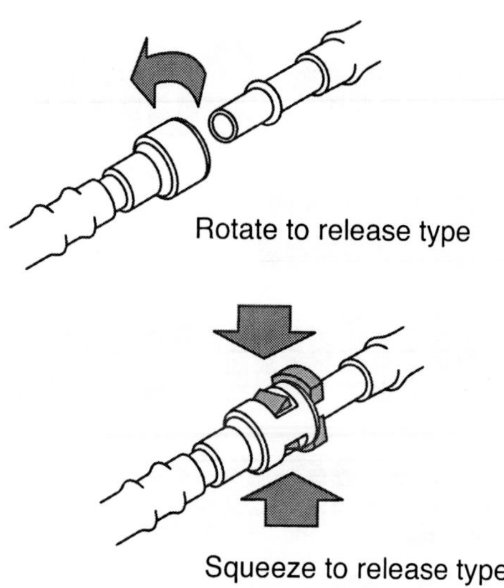


Figure 1. Quick-disconnect fuel lines.

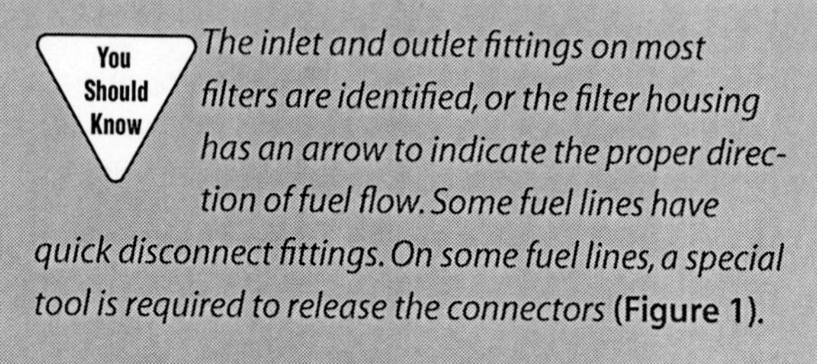


*The inlet and outlet fittings on most filters are identified, or the filter housing has an arrow to indicate the proper direction of fuel flow. Some fuel lines have quick disconnect fittings. On some fuel lines, a special tool is required to release the connectors* **(Figure 1)**.

## ENGINE CONTROL SYSTEM DIAGNOSIS

When diagnosing fuel injection systems, one of the most important tests is fuel pressure. Improper fuel pressure causes many drivability problems. For example, if the fuel pressure is higher than specified, the injectors deliver more fuel, resulting in reduced fuel economy.

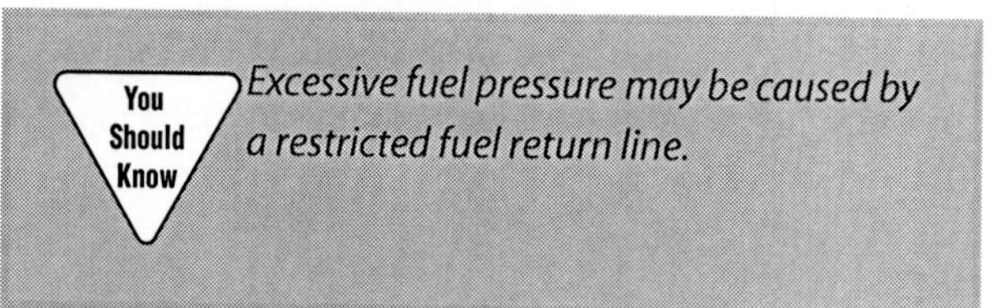


*Excessive fuel pressure may be caused by a restricted fuel return line.*

When the fuel pressure is less than specified, the injectors supply less fuel and this may cause drivability problems such as acceleration stumbles and loss of engine power or cutting out at high speed. The PCM diagnostics do not include fuel pressure. The PCM assumes the specified fuel pressure is available at the injectors.

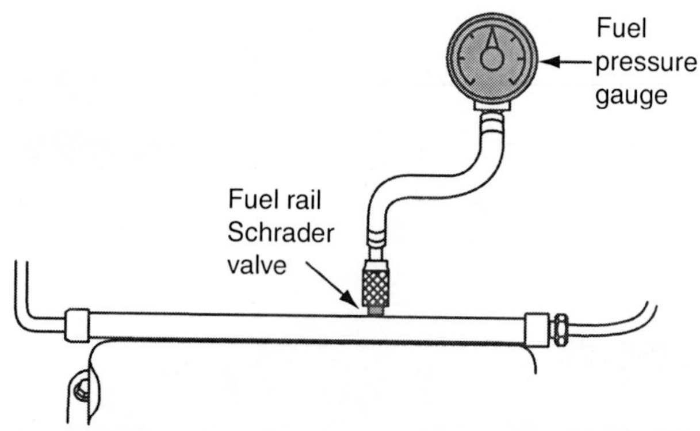


Figure 2. A fuel pressure gauge connected to the Schrader valve on the fuel rail.

### Fuel Pressure Testing

Before testing fuel pressure, always relieve the fuel pressure in the system. Be sure there is an adequate supply of fuel in the fuel tank. On MFI, SFI, and CPI systems, connect the fuel pressure gauge to the Schrader valve on the fuel rail **(Figure 2)**. On TBI systems, the fuel pressure gauge must be connected in series in the fuel supply line at the TBI assembly. Cycle the ignition switch on and off several times and read the fuel pressure on the gauge. If the fuel pressure is less than specified, the fuel filter may be severely restricted, the voltage supply or ground on the fuel pump may have high resistance, or the fuel pump may be defective. With the engine running, use a digital voltmeter to test the voltage at the fuel pump in the tank. With the engine running, the voltage at this location should be 11.5 volts or more. If this voltage is lower than specified, the wire from the fuel pump relay to the fuel pump has excessive resistance. To measure the voltage drop across the fuel pump ground, connect the voltmeter from the fuel pump ground terminal to a good chassis ground. The voltage drop across this circuit should not exceed 0.2 volt. If the fuel filter and the fuel pump voltage supply and ground are satisfactory, replace the fuel pump.

### Injector Testing

A rough idle problem may be caused by faulty injectors. Some fuel injector problems, such as an open winding or severely restricted discharge orifices, cause cylinder misfiring at idle. When the engine speed is increased, the misfiring is less noticeable. If the orifices in an injector are partially restricted, a rough idle problem results. With the air cleaner removed and the engine idling on some TBI systems, the fuel spray from the injectors may be

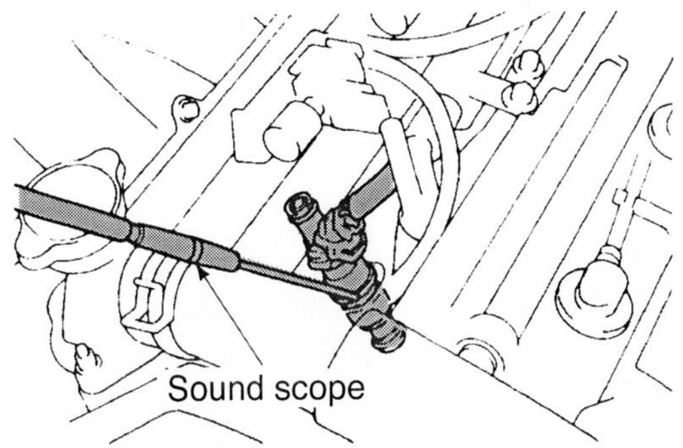


Figure 3. Injector testing with a stethoscope.

> **You Should Know** *A shorted injector winding causes high current through the injector and the PCM driver. This high current flow may damage the injector driver in the PCM. In some PCMs, the injector drivers sense high current flow and, rather than allowing injector driver damage, they shut the circuit off. After a cooldown period, the injector driver may allow injector operation.*

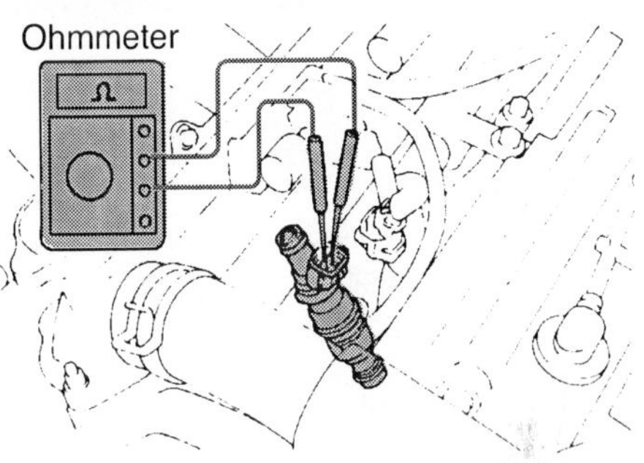


Figure 4. Testing an injector with an ohmmeter.

observed visually. If an injector is not spraying fuel, a stethoscope pickup may be placed against the injector body with the engine idling **(Figure 3)**. If a clicking noise is heard, the PCM is operating the injector and the injector winding is probably satisfactory. When the injector is not clicking with the engine idling, test the injector voltage supply and ground wire to the PCM and test the injector winding with an ohmmeter. TBI or port injectors may be tested with an ohmmeter. Turn the ignition switch off and remove the injector connector. Connect the ohmmeter leads to the injector terminals **(Figure 4)**. The injector must have the specified resistance. If the resistance is less than specified, the injector winding is shorted. An infinite ohmmeter reading indicates an open injector winding. If the injector does not have the specified resistance, replacement is necessary.

An injector balance test may be performed to test the injectors. The injector balance tester energizes each injector for a specific short time period. Each injector terminal must be disconnected and the lead from the balance tester connected to the injector terminals. The power leads on the injector balance tester must be connected to the battery terminals with the correct polarity **(Figure 5)**. During the injec-

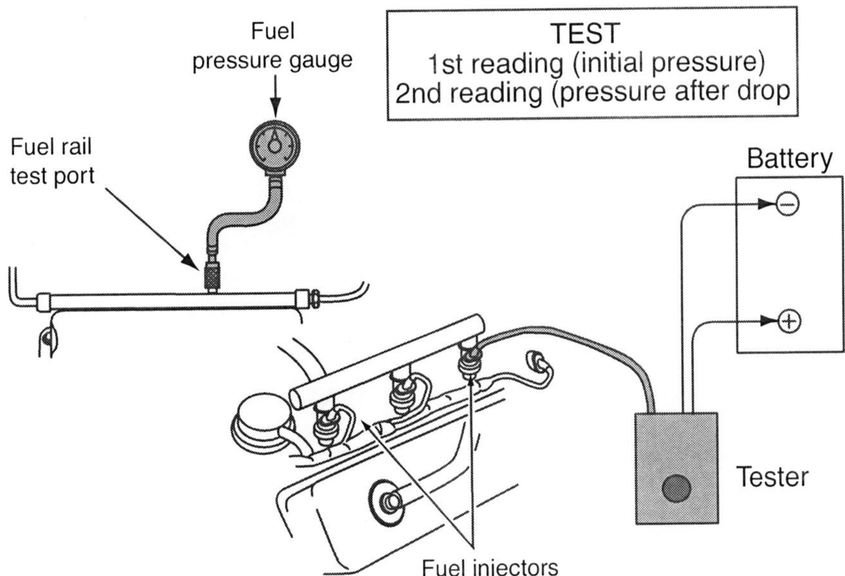


Figure 5. An injector balance test.

| CYLINDER | 1 | 2 | 3 | 4 | 5 | 6 |
|---|---|---|---|---|---|---|
| HIGH READING | 225 | 225 | 225 | 225 | 225 | 225 |
| LOW READING | 100 | 100 | 100 | 90 | 100 | 115 |
| AMOUNT OF DROP | 125 | 125 | 125 | 135 | 125 | 110 |
| RESULTS | OK | OK | OK | Faulty, rich (too much fuel drop) | OK | Faulty, lean (too little fuel drop) |

**Figure 6.** Interpeting injector balance test results.

tor balance test, the fuel pressure gauge must be connected to the Schrader valve on the fuel rail. Cycle the ignition switch several times until the specified pressure is indicated on the fuel gauge. Press the button on the balance tester to energize the injector for a specific time period. Record the fuel pressure displayed on the fuel gauge. Repeat this procedure on each injector. Obtain the maximum difference in pressure drop on each injector from the vehicle manufacturer's specifications. If an injector has considerably less pressure drop compared to the other injectors, this injector has restricted orifices. When an injector has considerably more pressure drop compared to the other injectors, the injector plunger is sticking open **(Figure 6)**.

Injectors with restricted orifices may have leaking seats that drip fuel from the injector tips into the intake manifold during idle operation and after the engine is shut down. This condition causes rough idle operation and hard starting after a hot engine has been shut down for a few minutes. When testing fuel pump pressure or injector balance, cycle the ignition switch several times until the specified fuel pressure appears on the fuel gauge. After 15 minutes, observe the fuel pressure. If the fuel pressure slowly decreases, the injectors may be leaking or the pressure regulator valve may have a leak. This decrease in fuel pressure may also be caused by a leaking one-way check valve in the fuel pump. To eliminate these leak sources and locate the problem, repeat the leak down test and use a pair of straight-jawed vise grips to close each line one at a time. If the fuel pressure still leaks down with the fuel return line and the fuel supply line blocked, the injectors are dripping fuel. If the leak down problem is eliminated when the fuel return line is closed, the pressure regulator valve is leaking. When the leak down problem is eliminated with the fuel supply line closed, the one-way check valve in the fuel pump is leaking.

**You Should Know**

*A leaking pressure regulator diaphragm on an MFI or SFI system causes fuel to be pulled through this diaphragm and the vacuum hose into the intake manifold. This results in excessive fuel consumption and erratic idle operation. A leaking nylon fuel line on a CPI system causes engine misfiring at idle because it reduces the fuel pressure so this pressure no longer opens the poppet nozzle.*

Fuel injectors may be tested with a lab scope. Connect the scope leads to the injector terminals. A satisfactory waveform from a conventional injector is shown in **Figure 7**. A satisfactory waveform from a peak-and-hold injector is

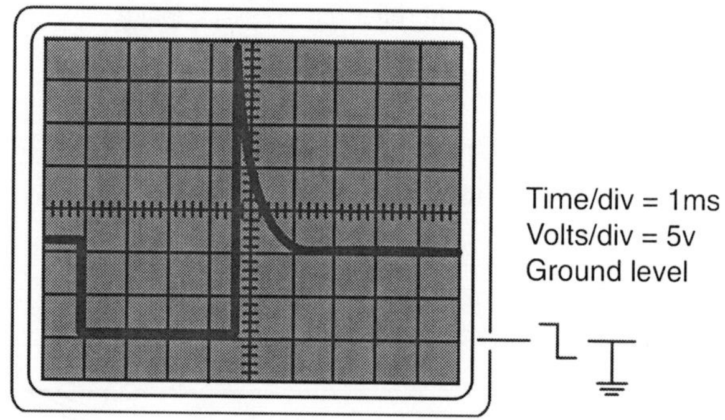


**Figure 7.** Waveform from a conventional fuel injector.

illustrated in **Figure 8**, and a waveform from a pulse-modulated injector is pictured in **Figure 9**. If an injector has an abnormal waveform, the injector has a problem.

**You Should Know**

*The PCM supplies normal current flow to a peak-and-hold injector to open the injector and then reduces the current flow to hold the injector open for a very short time period. The PCM initially supplies normal current flow for a pulse-modulated injector and then pulses the injector on and off very quickly.*

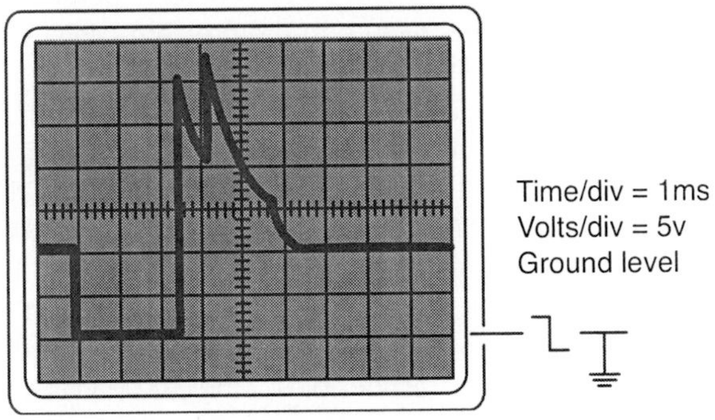

Time/div = 1ms
Volts/div = 5v
Ground level

**Figure 8.** Waveform from a peak-and-hold fuel injector.

Time/div = 1ms
Volts/div = 5v
Ground level

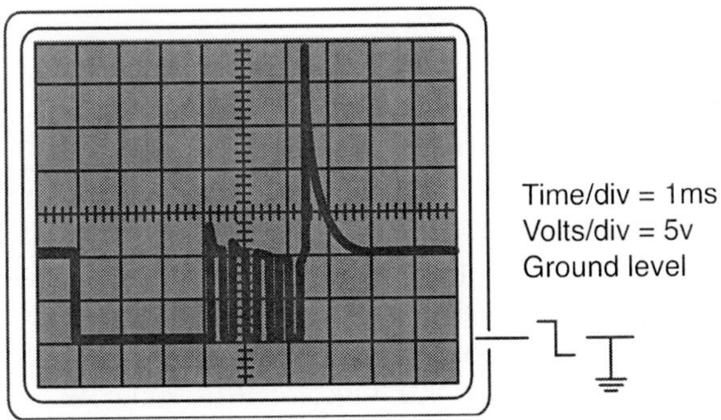

**Figure 9.** Waveform from a pulse-modulated fuel injector.

## Injector Cleaning

Injectors may be cleaned with a special injector cleaner placed in a pressurized container **(Figure 10)**. The injector cleaning solution is a mixture of cleaning solution and unleaded gasoline. After the cleaning solution is placed in the injector cleaning tool, the tester is pressurized with shop air pressure until the pressure gauge on the tester is slightly below the specified fuel pressure. At this pressure, the pressure regulator valve does not open during the injector cleaning process.

**You Should Know**

*A hand pump on some injector cleaners is used to pressurize the container. Injector cleaner is also supplied in a pressurized container so the technician does not have to pressurize the container.*

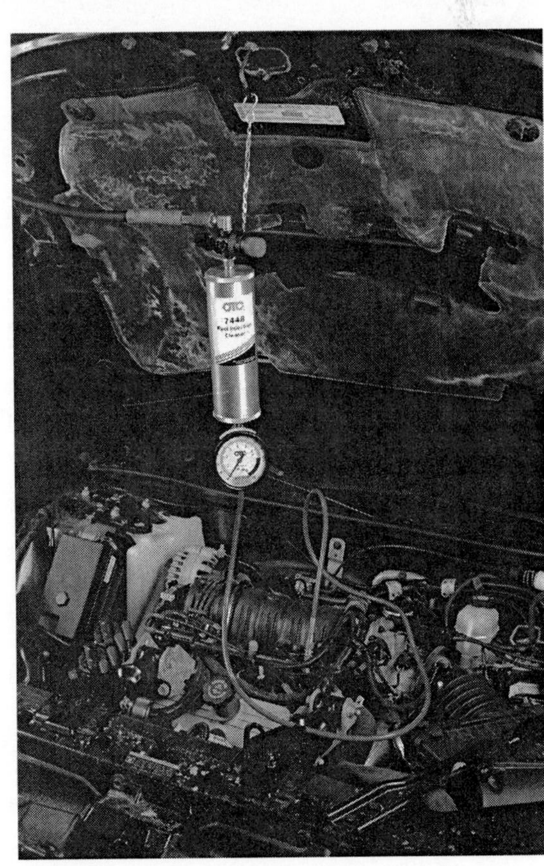

**Figure 10.** Injector cleaning equipment.

During the injector cleaning process, the fuel pump must be inoperative. Disconnect the wires from the fuel pump relay to prevent fuel pump operation. If the fuel pump circuit has an oil pressure switch, disconnect the wires from this switch. Locate a section of rubber hose in the fuel return line and use a pair of straight-jawed vise grips to close this line. This action prevents the possibility of the injector cleaning solution flowing through the fuel return line into the fuel tank. Connect the hose on the injector cleaner to the Schrader valve on the fuel rail and open the cleaner valve. Start and run the engine on the injector cleaning solution. The engine usually runs for 15 to 20 minutes on the cleaning solution. After the injectors have been cleaned, the engine may still experience a rough idle problem because the PCM has to re-learn about the cleaned injectors. With the engine at normal operating temperature, drive the vehicle for 5 minutes to allow the PCM to re-learn regarding the cleaned injectors.

You Should Know

*Some vehicle manufacturers do not recommend cleaning certain types of injectors in their vehicles. One example of this type of injector is the Multec injectors in some General Motors vehicles. These injectors have a sharp-edged orifice plate at the injector tip that provides a self-cleaning action for the injectors.*

## ENGINE CONTROL SYSTEM SCAN TOOL DIAGNOSIS AND DIAGNOSTIC TROUBLE CODES

When diagnosing fuel injection systems, observe the MIL in the instrument panel. If this light is illuminated with the engine running, the PCM has detected a problem in the system and a DTC is set in the PCM memory. In the 1980s and early 1990s, each vehicle manufacturer used a different DLC. On these systems, a jumper wire could be used to jump across two terminals in the DLC and obtain flash DTCs from the MIL. For example, if the MIL flashed quickly three times followed by a pause and three more quick flashes, DTC 33 was indicated. However, as more DTCs were programmed into the systems, it became very difficult to read flash DTCs. For example, some three-digit DTCs were as high as 999, which is difficult to read via flash DTCs. **Flash DTCs** are displayed by the flashes of the MIL.

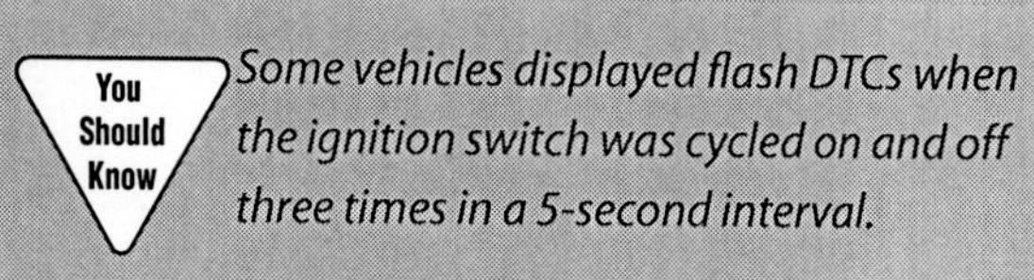

You Should Know
Some vehicles displayed flash DTCs when the ignition switch was cycled on and off three times in a 5-second interval.


A scan tool is connected to the DLC under the dash to read the DTCs and other input and output data. When performing a scan tool diagnosis, the engine should be at normal operating temperature. With the ignition switch off, connect the scan tool to the DLC. On some vehicles, the technician had to enter the vehicle make, model year, and engine code into the scan tool. On other vehicles, the PCM supplies this information to the scan tool. When "READ CODES" is selected on the scan tool, the DTCs are displayed. Some scan tools supply an interpretation for each DTC. When using other scan tools, it is necessary to obtain the DTC interpretation from the vehicle manufacturer's information. DTCs may be identified as current or history. **Current DTCs** are present at the time of testing. **History DTCs** represent intermittent faults that occurred sometime in the past, but the fault is no longer present.

A DTC indicates a problem in a certain area. For example, a DTC representing a MAF sensor may indicate a defective sensor, defective wires from the sensor to the PCM, or a PCM that is not able to receive this signal. The technician must perform specific tests, such as voltmeter or ohmmeter tests, to locate the exact cause of the problem. OBD I systems used two- or three-digit numbered DTCs.

OBD II DTCs are formatted according to SAE standard J2012. This standard requires five-digit alphanumeric DTC identification. The prefix in each DTC indicates the DTC function: P—powertrain, B—body, C—chassis, U—network communication. The first number in the DTC represents the group responsible for the code. If the first number is 0, the code is designated by SAE. When the first number is 1, the vehicle manufacturer is responsible for the DTC. Many DTCs are dictated by SAE but the vehicle manufacturer may insert some codes. The third digit in the DTC indicates the subgroup to which the code belongs as follows:

1—air-fuel control
2—air-fuel control, injectors
3—ignition system, misfire
4—auxiliary emission controls
5—idle speed control
6—PCM input and output
7—transmission/transaxle
8—transmission/transaxle

The fourth and fifth digits in the DTC indicate the area in which the defect is located. For example, in DTC P0155, P indicates it is a powertrain code, 0 indicates it is designated by SAE, 1 indicates the code is in the air-fuel control subgroup, and 55 indicates the code represents a malfunction in the $HO_2S$ heater, bank 2, sensor 1. Bank 2 indicates the bank of a V6 or V8 engine that is opposite to the bank where number 1 cylinder is located and sensor 1 is the upstream $HO_2S$ sensor located ahead of the catalytic converter. A bank 1 $HO_2S$ is located on the same side of the engine as number 1 cylinder and sensor 2 is mounted downstream from the catalytic converter.

## ENGINE CONTROL SYSTEM SERVICE

Always be sure the exact cause of the customer's complaint is diagnosed before performing any engine control system service. For example, before replacing a fuel pump, be sure no other causes of improper fuel pump operation are present. Therefore, be sure the fuel filter, fuel lines, and the fuel pump voltage supply and ground wires are satisfactory before replacing the fuel pump. Before replacing any electrical/electronic component, connect a 12-volt power supply to the cigarette lighter socket and then disconnect the negative battery terminal. Disconnecting the battery prevents accidental shorts to ground from damaging components or starting a fire when servicing vehicle components or systems. The power supply connected to the cigarette lighter socket keeps the PCM memory alive during the service procedure. This action prevents drivability problems after the vehicle is restarted.

### Fuel Pump Replacement

When replacing most electric in-tank fuel pumps, the fuel tank must be removed. On some vehicles, a fuel pump access cover is located in the trunk directly above the fuel pump in the tank. After this access door is removed, the fuel pump may be taken out of the fuel tank. If the fuel tank must be removed to gain access to the fuel pump, use an electric or hand pump to pump the gasoline out of the tank into an approved gasoline container. Be careful not to spill any fuel. If fuel is accidentally spilled, use the proper absorption material and clean up the spill immediately.

Raise the vehicle on a lift or use a floor jack to raise and lower the vehicle onto safety stands. Disconnect the fuel lines and electrical connector near the fuel pump and remove the fuel tank straps. Slowly lower the fuel tank onto the floor and remove it from under the vehicle.

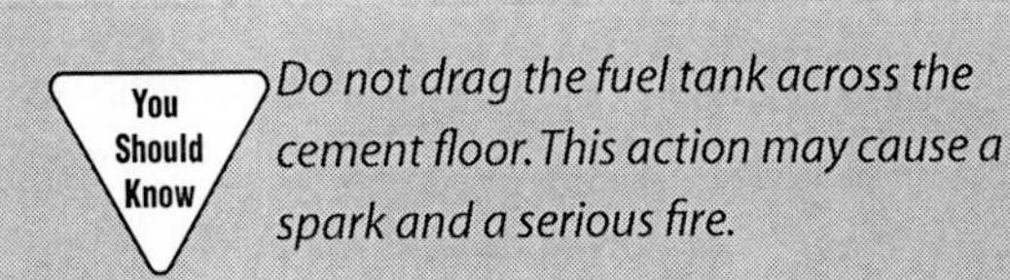

You Should Know
*Do not drag the fuel tank across the cement floor. This action may cause a spark and a serious fire.*


Use an air gun to blow debris from the top of the fuel tank and the fuel pump assembly. Remove the fuel pump retaining ring and lift the fuel pump assembly from the tank. In many vehicles the fuel pump, fuel intake filter, and fuel tank sensing unit are replaced as an assembly. In some applications, the fuel pump may be replaced separately.

Always use a new seal between the fuel pump assembly and the top of the fuel tank. Install the new fuel pump assembly and be sure the fuel pickup is very close to or touching the bottom of the tank. Install the fuel pump assembly retaining ring in the top of the fuel tank. Install the fuel tank in its proper location under the vehicle and install the holding straps. Tighten the holding strap fasteners to the specified torque. Re-connect the fuel lines and electrical connector to the fuel pump assembly. Tighten the fuel lines to the specified torque. Lower the vehicle onto the shop floor. Re-connect the negative battery cable and remove the power supply connector from the cigarette lighter. Start the engine and be sure the fuel pump has the specified pressure with no fuel leaks.

## INJECTOR REPLACEMENT

Always relieve the fuel system pressure before replacing the injectors. Connect a 12-volt power supply to the cigarette lighter and disconnect the negative battery cable. Release the lock ring and remove each injector terminal. Disconnect the fuel inlet and outlet lines on the fuel rail and the vacuum hose on the pressure regulator. Remove the fuel rail retaining bolts. Remove any other components that interfere with the fuel rail removal. Remove the fuel rail and injectors **(Figure 11)**.

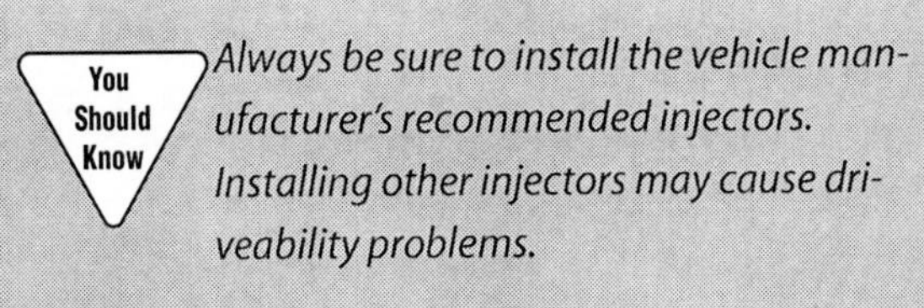

You Should Know
*Always be sure to install the vehicle manufacturer's recommended injectors. Installing other injectors may cause driveability problems.*


Always install new injector O-rings when removing and replacing injectors. Place a small amount of engine oil on each upper injector O-ring and install the injectors into the fuel rail. Use engine oil to lightly coat the lower injector

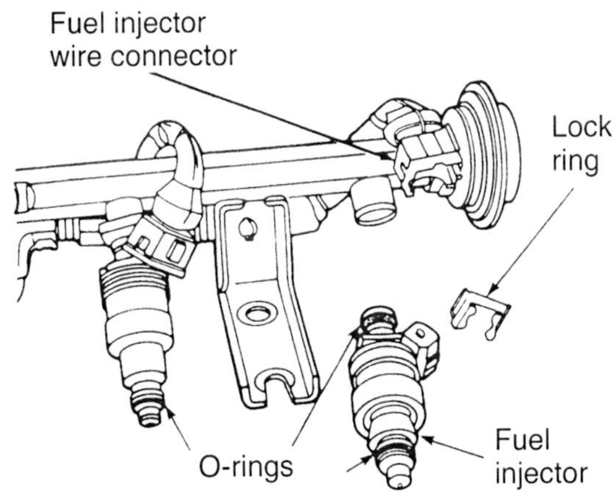


**Figure 11.** Injector removal and replacement.

O-rings and install the fuel rail and injectors in the intake manifold. Be sure the injectors are fully seated in the fuel rail and the intake manifold. Reconnect the injector wiring terminals. Install the fuel rail fasteners, and tighten these fasteners to the specified torque. Re-connect the fuel inlet and outlet lines and tighten the fittings to the specified torque. Install the vacuum hose on the pressure regulator. Reconnect the battery negative cable and disconnect the power supply from the cigarette lighter.

## Summary

- The fuel filter should be changed at the vehicle manufacturer's specified intervals.
- The fuel system pressure should be relieved before disconnecting any fuel system component.
- The PCM assumes the specified fuel pressure is available when calculating the proper amount of fuel to be injected.
- The PCM diagnostics do not provide any indication of improper fuel pressure.
- Before replacing a fuel pump, always be sure the voltage supply and ground wires to the pump are satisfactory.
- Fuel injectors may be tested by measuring the resistance in the injector windings.
- An injector balance test measures the fuel pressure drop when each injector is opened for a specific length of time.
- If the fuel system pressure gradually decreases after the engine is shut off, the one-way check valve in the fuel pump may be leaking, the pressure regulator valve may be leaking, or the injectors may be dripping fuel.
- Some injectors may be cleaned by connecting a pressurized container filled with injector cleaner to the Schrader valve on the fuel rail.
- A DTC indicates a fault in a certain area of the engine control system.
- In an OBD II DTC, the first digit indicates the area to which the DTC belongs. For example, P indicates powertrain.
- The second digit in an OBD II DTC indicates whether the DTC is established by SAE standards or introduced by the vehicle manufacturer.
- The third digit in an OBD II DTC indicates the subgroup to which the DTC belongs. For example, 5 indicates the DTC is in the idle speed control subgroup.
- The fourth and fifth digits in an OBD II DTC represent the fault in the system.
- On many vehicles, the fuel tank must be removed to access the fuel pump.
- When removing and replacing any electrical/electronic component, always connect a 12-volt power supply to the cigarette lighter socket and disconnect the negative battery cable.

## Review Questions

1. Technician A says in a P0155 DTC, the 0 indicates the DTC is developed by the vehicle manufacturer. Technician B says the 1 indicates the subgroup to which the DTC belongs. Who is correct?
   A. Technician A
   B. Technician B
   C. Both Technician A and Technician B
   D. Neither Technician A nor Technician B
2. Technician A says a bank 2, sensor 2 $HO_2S$ is located upstream from the catalytic converter. Technician B says this $HO_2S$ is located on the same side of the engine as number 1 cylinder. Who is correct?
   A. Technician A
   B. Technician B
   C. Both Technician A and Technician B
   D. Neither Technician A nor Technician B
3. Technician A says higher-than-specified fuel pump pressure causes a rich air-fuel ratio. Technician B says higher-than-specified fuel pressure may be caused by a leaking one-way check valve in the fuel pump. Who is correct?
   A. Technician A
   B. Technician B
   C. Both Technician A and Technician B
   D. Neither Technician A nor Technician B
4. Technician A says during an injector balance test, if the pressure drop on one injector is higher than on the other injectors, this injector has restricted orifices. Technician B says during an injector balance test, each injector should have nearly the same pressure drop. Who is correct?
   A. Technician A
   B. Technician B
   C. Both Technician A and Technician B
   D. Neither Technician A nor Technician B

5. In an MFI system, lower-than-specified fuel pressure may cause all of these problems *except*:
   A. a lean air-fuel ratio.
   B. engine surging and cutting out at high speed.
   C. hesitation during engine acceleration.
   D. engine backfiring during deceleration.
6. While discussing TBI system diagnosis, Technician A says the PCM has the capability to diagnose fuel pump pressure. Technician B says when testing fuel pressure with a gauge, connect the gauge hose to the Schrader valve. Who is correct?
   A. Technician A
   B. Technician B
   C. Both Technician A and Technician B
   D. Neither Technician A nor Technician B
7. All of these problems may cause low fuel pump pressure *except*:
   A. low voltage at the fuel pump.
   B. a restricted fuel filter.
   C. a restricted fuel return line.
   D. a restricted fuel supply line.
8. When diagnosing an SFI system with the fuel pressure gauge connected to the system, the ignition switch is cycled several times until the specified fuel pressure appears on a fuel pressure gauge. The fuel supply line and the return fuel line are then completely blocked. After 15 minutes, the fuel pressure has dropped from 45 psi to 10 psi and there are no indications of external fuel leaks. The most likely cause of this pressure reading is:
   A. dripping fuel injectors.
   B. a leaking pressure regulator valve.
   C. a leaking one-way check valve in the fuel pump.
   D. a leaking fuel filter.
9. When performing an injector cleaning procedure:
   A. the fuel return line must be open.
   B. the fuel pump must be in operation.
   C. the injector cleaning solution container is connected to the Schrader valve.
   D. the engine should be running for 1 hour.
10. While diagnosing an OBD II fuel system, a current P0155 DTC is obtained on a scan tool. All of these statements about this DTC are true *except*:
    A. This DTC represents a problem in the powertrain.
    B. This DTC is designated by the vehicle manufacturer.
    C. This DTC is in the air-fuel control subgroup.
    D. This DTC represents a defect that is present during the diagnosis.
11. During the injector cleaning process, the fuel pump must be ____________________.
12. While reading PCM flash DTCs, the MIL flashes quickly four times followed by a pause and one flash. This indicates DTC ________________.
13. If an OBD II DTC begins with a B, the DTC is in the ________________ category.
14. If the third digit in an OBD II DTC is 3, the DTC is related to ________________ ____________________.
15. Explain the reason for connecting a 12-volt power supply to the cigarette lighter socket before disconnecting a vehicle battery.
16. Describe the proper fuel pressure gauge connection to test fuel pump pressure on a TBI system.
17. When the fuel pump pressure is less than specified, describe the diagnostic procedure to locate the exact cause of the problem.
18. Describe the possible results of a shorted injector on the PCM.

## Emissions and Emission Systems

*An automotive industry executive recently stated, "The potential for growth in the automotive industry is tremendous: 88 percent of the world's people would like to aspire to owning an automobile; in practice today, only 12 percent can do so. We have to find ways of creating vehicles that are even more affordable than those we have and exude even more passion. We are seeing per capita incomes increase in many countries such as Brazil, Mexico, China, and India, and there is a strong relationship between vehicle ownership and per capita income."*

## SECTION OBJECTIVES

After you have read, studied, and practiced the contents of this section, you should be able to:

- List the three main automotive pollutants and explain how each is formed.
- Describe how photochemical smog is formed.
- Explain the origin of crankcase and evaporative emissions.
- Explain the difference between basic and enhanced emission test programs.
- List the items to be inspected during a visual emission pre-inspection.
- Explain the items to be checked in an emissions component inspection.
- Explain an idle test emission procedure.
- Describe an IM240 emission test procedure.
- Describe a BAR-31 emission test procedure.
- Describe an ASM emission test procedure.
- Explain the operation of the PCV system.
- Describe the operation of a typical EGR valve.
- Describe the operation of positive backpressure and negative backpressure EGR valves.
- Explain the operation of a pulsed air injection system.
- Describe the operation of belt-driven and electric-drive air pumps.
- Explain the operation of conventional and enhanced EVAP systems.
- Describe the operation of oxidation, dual-bed, and three-way catalytic converters.
- Maintain, diagnose, and service PCV systems.
- Maintain, diagnose, and service EGR systems.
- Maintain, diagnose, and service air injection systems.
- Maintain, diagnose, and service EVAP systems.
- Maintain, diagnose, and service catalytic converters.

# Vehicle Emissions and Emission Standards

## *Introduction*

Automotive emissions affect everyone! In the last 40 years, automotive emissions standards have become increasingly stringent and these emissions have been greatly reduced. This reduction in automotive emissions provides cleaner air to breathe, which, in turn, reduces health hazards such as respiratory diseases. It is very important for technicians to understand the types of vehicle pollutants and the allowable standards for these pollutants. Technicians must also understand how each automotive emission control device operates to reduce harmful emissions. Many technicians are required to diagnose and repair vehicles that have failed compulsory emission tests. The technician's understanding of emissions and vehicle emission systems is absolutely essential to perform this job quickly and accurately.

### AIR POLLUTION AND VEHICLE EMISSIONS

Vehicle emissions have been regulated since the 1960s. **Photochemical smog** is formed by the reaction of sunlight with HC and $NO_x$ in the atmosphere. Smog appears as a light-brown haze in the air. Ozone is the primary component in photochemical smog. Ozone contains three oxygen atoms ($O_3$). Ozone occurs naturally in the upper atmosphere and serves to prevent harmful ultraviolet rays given off by the sun from reaching the earth.

Hot air near the ground usually rises and is then cooled by cooler air at a higher altitude. This action usually clears smog from the air at ground level. However, in some locations such as in a valley or over a large metropolitan area, the warm air near the ground may be trapped by an inversion layer that tends to hang over the area. When this action occurs, the smog is trapped near the ground and the smog becomes more concentrated because it does not escape. Smog is known to aggravate lung conditions such as asthma.

Vehicles manufactured today emit less than 5 percent of the pollution emitted by vehicles manufactured in the 1960s because of emission control devices and improved engine design. There are many sources of air pollution other than vehicles. These sources include industry, homes, oil exploration, and airplanes.

The main regulated automotive pollutants are unburned HC, CO, and $NO_x$. Other exhaust emissions include microscopic soot and dust particles that are small enough to remain suspended in the air. Particulates are a concern with diesel engines. Other tailpipe emissions that are often measured during vehicle diagnosis include $CO_2$ and $O_2$. These emissions are not pollutants but are useful in the diagnostic process. There are three sources of emissions on a vehicle: exhaust emissions from the tailpipe, evaporative emissions from the fuel tank, and crankcase emissions. All three pollutants are present in exhaust emissions, whereas evaporative or crankcase emissions contain

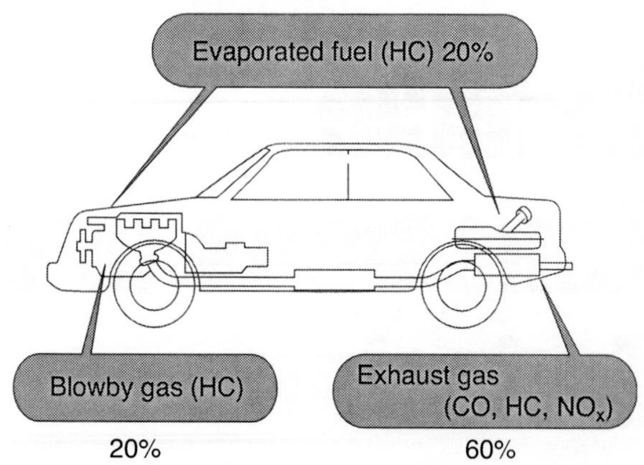


**Figure 1.** Automobile emission sources.

| Type of Gas / Composition | CO | HC | $NO_x$ |
|---|---|---|---|
| Exhaust gas | 100% | 55% | 100% |
| Blow-by gas | — | 25% | — |
| Evaporated fuel | — | 20% | — |

**Figure 2.** The percentage of emissions from automobile sources.

largely HC **(Figure 1)**. On a typical vehicle, 60 percent of the HC emissions come from the exhaust, 20 percent come from the crankcase, and 20 percent come from evaporative sources **(Figure 2)**.

## HYDROCARBONS (HC)

HC emissions are caused by unburned air-fuel mixture in the combustion chamber. At the end of each combustion event, some unburned air-fuel mixture is left on the cooler combustion chamber surfaces when the flame front is quenched near this surface **(Figure 3)**. This action results in HC emissions. A larger combustion chamber surface provides more **quench area.**

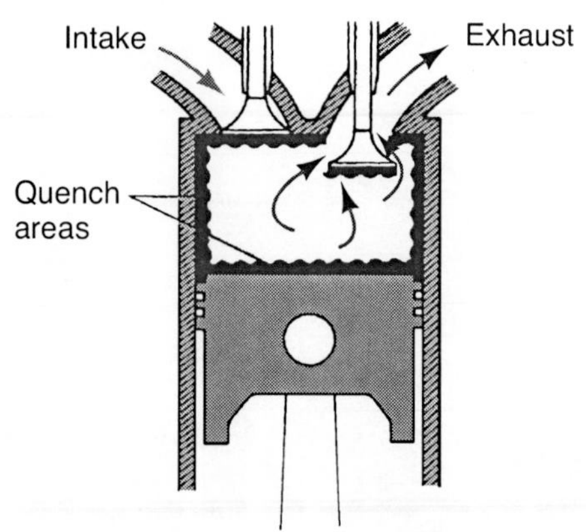


**Figure 3.** Unburned fuel left on the combustion chamber surface.

On some engines, the air-fuel ratio becomes richer during deceleration and this increases HC emissions. If a cylinder misfires, the whole air-fuel charge in the combustion chamber is forced out through the exhaust system and HC emissions increase significantly. In the United States, approximately 27 percent of the manmade HC emissions are emitted from vehicles.

## CARBON MONOXIDE (CO)

CO emissions are a byproduct of combustion. CO emissions occur when the air-fuel mixture is not completely burned. CO emissions are lowest at the ideal or stoichiometric air-fuel ratio of 14.7:1. As the air-fuel ratio becomes richer than 14.7:1, CO emissions increase in proportion to the richness of the air-fuel ratio **(Figure 4)**. In the

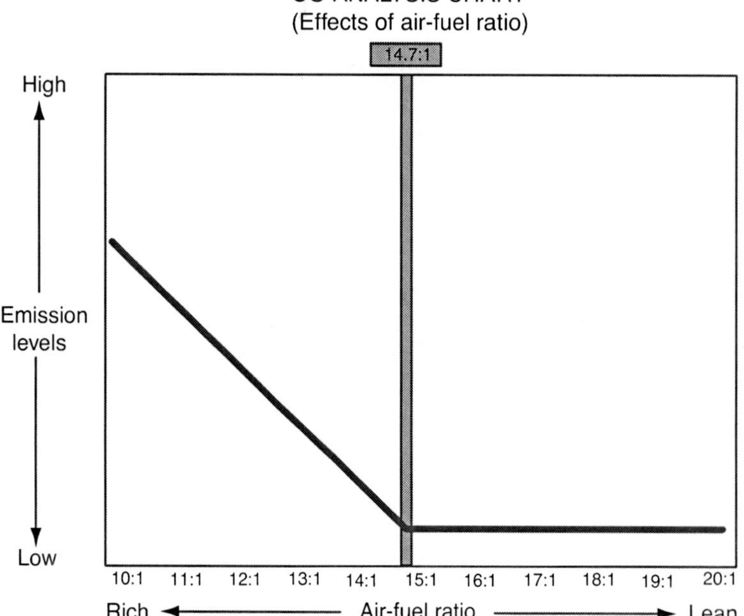


**Figure 4.** CO emissions in relation to air-fuel ratio.

United States, it is estimated that 80 percent of the CO emissions come from vehicles. In urban areas, this figure may be as high as 90 percent.

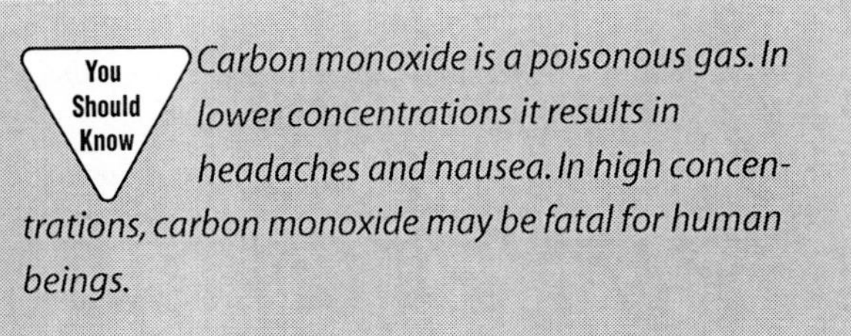


*Carbon monoxide is a poisonous gas. In lower concentrations it results in headaches and nausea. In high concentrations, carbon monoxide may be fatal for human beings.*

## OXIDES OF NITROGEN ($NO_X$)

$NO_x$ emissions are caused by oxygen and nitrogen in the air-fuel mixture uniting at combustion chamber temperatures above 2,500°F (1,371°C). $NO_x$ emissions are reduced when combustion chamber temperatures are lowered. In the United States, 32 percent of the $NO_x$ emissions come from vehicles.

## EVAPORATIVE AND CRANKCASE EMISSIONS

The source of evaporative emissions is mainly the fuel tank. Fuel vapors containing HC attempt to escape to the atmosphere. For many years, vehicles have been equipped with evaporative emission systems to reduce the escape of fuel vapors from the fuel tank to the atmosphere. The emission laws regarding evaporative emissions have become increasingly stringent. In response to these laws, evaporative systems have become more complex and efficient in reducing these emissions. Refueling is another source of evaporative emissions. In many newer vehicles, the evaporative system is designed to reduce evaporative emissions while refueling. Gasoline refueling pumps have also been designed to reduce refueling evaporative emissions. On older vehicles, the carburetor is a source of evaporative emissions.

Crankcase emissions are caused by combustion blowby past the piston rings into the crankcase. This condition causes HC emissions to enter and accumulate in the crankcase. The PCV system moves crankcase gases back into the intake manifold to greatly reduce crankcase emissions. However, excessive blowby because of worn piston rings and/or cylinders causes too much crankcase pressure, which may result in high crankcase emissions. A defective PCV system may also result in high crankcase emissions.

## EMISSION STANDARDS

In the United States, emission standards began in 1961 when all new cars sold in California were required to have a PCV system to reduce crankcase emissions. In 1966, new cars sold in California were required to have exhaust emission systems. Some of these systems involved the use of a

| | HC | CO | NOx |
|---|---|---|---|
| 1990 - U.S. | 0.41 | 3.4 | 1.0 |
| 1994 - U.S. | 0.25** | 3.4 | 0.4 |
| 1993 - Calif. | 0.25 | 3.4 | 0.4 |
| 1994 - TLEV | 0.125 | 3.4 | 0.4 |
| 1997 - LEV | 0.075 | 3.4 | 0.2 |
| 2000 - ULEV | 0.040 | 1.7 | 0.2 |
| ** non-methane HC | | | |

**Figure 5.** Emission standards.

belt-driven air pump to force air into the exhaust manifolds. The oxygen in this air allowed the air-fuel mixture left in the exhaust to continue burning in the exhaust manifolds. This action reduced HC emissions. In 1970, the United States Congress passed the Clean Air Act that established maximum emission levels for HC, CO, and $NO_x$. The Clean Air Act has been amended several times, including in 1990.

The implementation of emission standards is usually administered by federal or state organizations such as the EPA or the CARB. The EPA establishes and implements emission standards across the United States, but some states such as California have set their own more stringent emission standards. Emission standards have been established for TLEVs, LEVs, and ULEVs **(Figure 5)**. Emission standards are subject to change and sometimes the implementation date for certain emission standards is delayed.

## EMISSION TESTING

In the early 1970s, I/M emission programs were identified as an effective method for reducing emissions on in-use vehicles. By having vehicles inspected and repaired if necessary, emissions could be reduced to the level required for that model year. In 1977, The Clean Air Act was amended to require mandatory implementation of I/M programs in those areas of the United States that were not in compliance with National Ambient Air Quality Standards (NAAQS). Areas of the country that do not meet NAAQS standards are called non-attainment areas. If the states containing non-attainment areas did not implement I/M programs, they could be subjected to federal sanctions including the loss of funds for highway construction and improvement and the ability to grant permits for new industrial development.

Early I/M tests were usually basic two-speed idle tests because these tests were easy to perform and required a minimum amount of test equipment. These I/M tests could be centralized or de-centralized.

**You Should Know** *Centralized I/M programs are operated by governments or contractors working for governments. De-centralized I/M programs are operated by privately owned repair shops or franchised dealerships licensed by the state authorities.*

The EPA monitored these I/M tests and discovered the test and repair procedures did not reduce emissions as much as expected. The EPA decided that more sophisticated test procedures were necessary for the newer computer-controlled vehicle systems. The 1990 Clean Air Act amendments addressed these concerns by requiring enhanced I/M programs in all non-attainment areas identified as serious, severe, and extreme. In 1992, the EPA established the requirements for enhanced I/M programs. The purpose of enhanced I/M programs is to reduce HC, CO, and $NO_x$ emissions from in-use vehicles. Enhanced I/M programs include these features: road simulation tests run on a dynamometer for tailpipe emissions, evaporative emission system tests, more stringent program enforcement, and a $450.00 waiver for repair requirements.

In 1995, the United States Congress passed an amendment to the National Highway Systems Designation Act that allowed the state authorities considerable flexibility in designing enhanced I/M programs. As a result of this legislation, a variety of enhanced I/M programs have been developed and introduced. Most of these programs are decentralized.

## TYPES OF INSPECTION/ MAINTENANCE (I/M) PROGRAMS

Most enhanced I/M programs include a visual pre-inspection safety check. This safety check includes the following items:

1. Tires—The vehicle is rejected if any tire is excessively worn or the space-saver spare tire is mounted on one of the wheels. The tires must be safe to operate on the dynamometer during the enhanced emission test.
2. Brakes—If defective brake conditions prevent the vehicle from maintaining the drive trace during an enhanced emission test, the vehicle is rejected.
3. Exhaust system—Vehicles with leaking exhaust systems are rejected because this condition affects tailpipe emission readings and the test operators may be subjected to excessive CO.
4. Steering and suspension—The vehicle should be rejected if any suspension or steering component is worn or damaged so its normal function is adversely affected.
5. Fuel system—Vehicles with fuel system leaks are rejected because they are a fire hazard.
6. Instrument readings—Vehicles that are overheating or leaking coolant should be rejected.

An emissions component inspection is also performed. This inspection includes tampering. Included in the emissions component inspection are the fuel inlet restrictor in the fuel tank, the catalytic converter, and emission control items in the underhood area. The emissions component inspection also includes a gas tank filler cap inspection and test on applicable vehicles. The inspector tests the filler cap with a special pressurized tester.

### Basic Idle Test

The basic idle test is performed with the engine at normal operating temperature and an exhaust gas analyzer pickup inserted in the tailpipe. This analyzer must be properly calibrated. This type of analyzer uses infrared rays to measure the HC, CO, and $NO_x$ emission levels in the exhaust. The transmission selector must be in PARK for an automatic transmission or in NEUTRAL for a manual transmission. The emission levels are recorded at idle. The next step is to record the emission levels after engine operation at 2,500 rpm for a specified time period. The engine was then returned to idle and the emissions recorded again. In some jurisdictions, the pass requirement demanded that the tailpipe emissions remain below applicable emission standards at all three readings. In other areas, the pass requirement was based on the lowest of the two emission readings taken at idle. Jurisdictions establish emission test standards based on the standards for the vehicle year being tested.

### IM240 Test

The IM240 test is an enhanced emission test that requires the use of a dynamometer and a sophisticated constant volume sampling (CVS) exhaust emissions analyzer **(Figure 6)**. All the exhaust is routed through the CVS analyzer. During the IM240 test, the vehicle is safety-locked onto the dynamometer and the inspector drives the vehicle to maintain a bright dot on the IM240 drive trace dis-

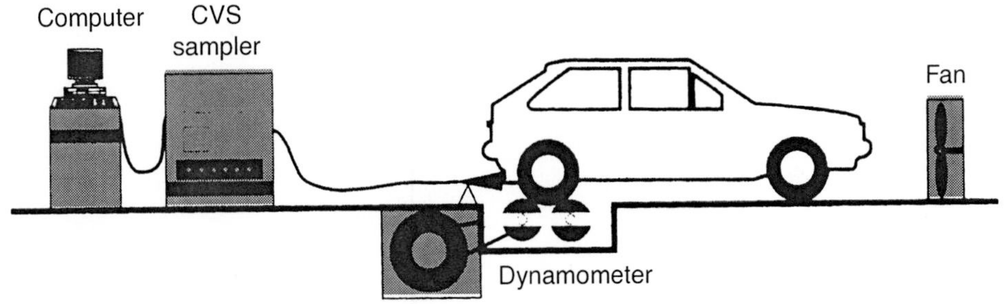


Figure 6. A typical IM240 emission test station.

played on a computer monitor **(Figure 7)**. The inspector must drive the vehicle to maintain the bright dot on the drive trace. If the bright dot goes off the drive trace to any extent, the computer aborts the test and the test procedure must be repeated. Completing the drive trace requires 240 seconds and includes a wide variation in operating conditions, including acceleration and deceleration and speeds up to 56 mph (90 kmh).

> **You Should Know** *The IM240 test procedure is taken from various parts of the Federal Test Procedures (FTP). Compared to the IM240 test, the FTP is a much longer test procedure run on a dynamometer and uses more sophisticated exhaust emission measurement equipment. The EPA uses the FTP to certify new vehicles sold in the United States.*

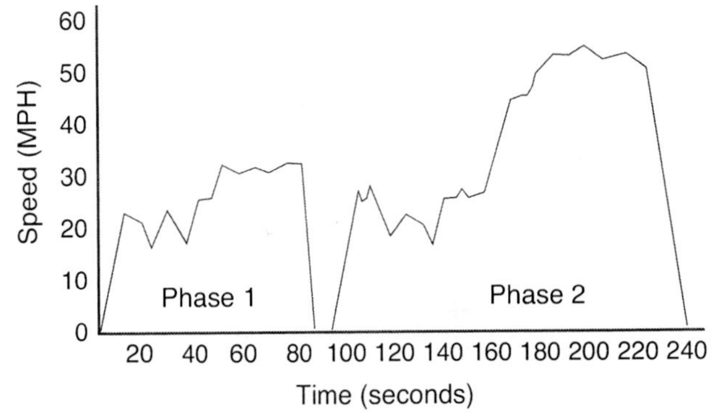


Figure 7. An IM240 emission test drive trace.

After the drive cycle is completed, the emissions analyzer prints out a trace for HC, CO, and $NO_x$. All readings are in grams per mile (GPM). These traces show the cutpoint for each pollutant and the actual emission level throughout the test **(Figure 8)**. When diagnosing the cause of high

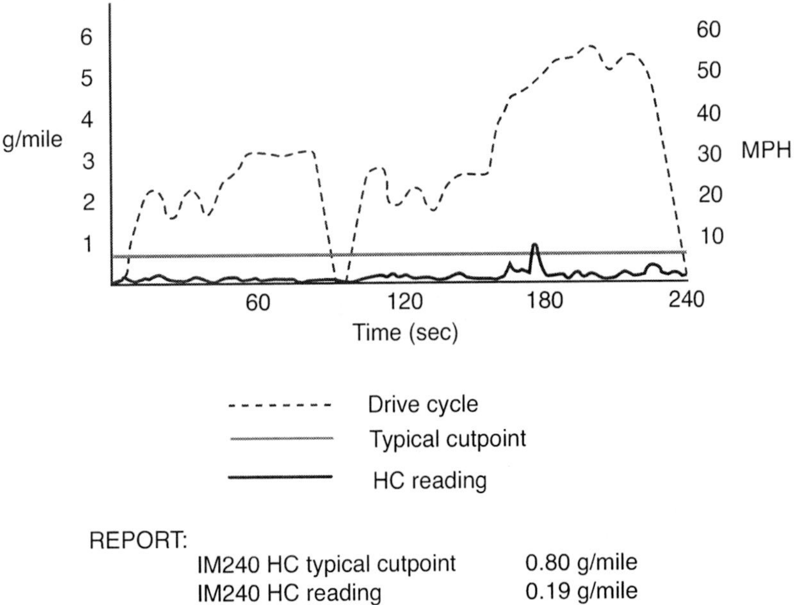


Figure 8. A hydrocarbon trace from an IM240 test.

emissions, these emission traces can be very helpful because they indicate the driving conditions that produced the high emission level above the cutpoint. The pass emission requirements are set by each jurisdiction and are based on EPA guidelines. During the IM240 test, the emissions tester measures the purge flow in the evaporative system.

## BAR-31 Test

The BAR-31 test requires the same equipment as the IM240 test and the exhaust emissions are measured in the same way. The BAR-31 test only requires 31 seconds because it has a simpler drive trace **(Figure 9)**. In most jurisdictions an emissions pass/fail decision is based on three BAR-31 tests.

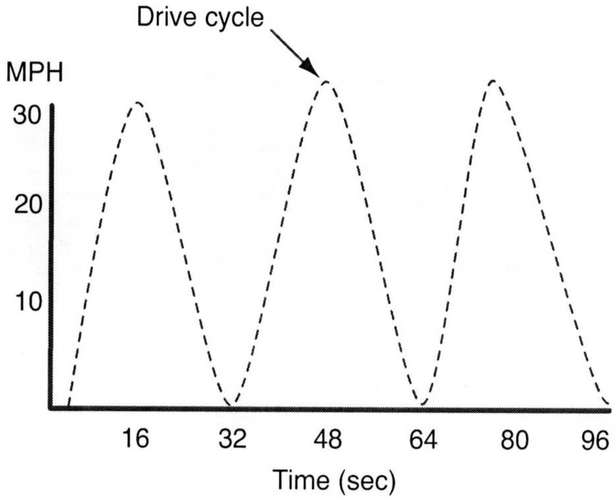


**Figure 9.** A BAR-31 emission test drive trace.

## Acceleration Simulation Mode (ASM) Test

The acceleration simulation mode (ASM) test is a steady-state test completed with a constant throttle opening. During the test, the vehicle is driven on a dynamometer at a constant speed and load setting. Two different ASM tests are used in emission testing. During the ASM 5015 test, the vehicle is driven at 15 mph (24 kmh) and at 50 percent of the maximum acceleration load during the federal test procedure cycle. The ASM 2525 test measures emissions at 25 mph (40 kmh) and at 25 percent of the maximum acceleration load during the federal test procedure cycle.

The tester used to record the emissions during an ASM test uses the same technology as the tester for the idle tests. The tester measures HC and $NO_x$ in parts per million and CO in percentage. A significant number of states use the ASM test procedure for emission testing.

You Should Know

*The ASM tester may be called a BAR-97 analyzer. Exhaust gas analyzers are usually built to specifications developed by the California Bureau of Automotive Repair (BAR). BAR specifications are labeled by the year they became effective. For example, specifications for the BAR-97 became effective in 1997.*

Many jurisdictions demand an OBD II check on applicable vehicles during an emissions test. This OBD II test usually involves checking for proper operation of the MIL with the ignition switch on and with the engine running. The OBD II test also involves connecting a scan tool to the DLC and determining that all the monitors have completed their function and there are no DTCs in the PCM memory.

# *Summary*

- Photochemical smog is formed by sunlight reacting on HC and $NO_x$ in the atmosphere.
- The main component in photochemical smog is ozone.
- The main regulated emissions are HC, CO, and $NO_x$.
- Vehicle emissions may come from the tailpipe, crankcase, or fuel tank.
- HC emissions are a result of unburned fuel after the combustion process.
- CO emissions are caused by incomplete burning of the air-fuel mixture.
- $NO_x$ emissions are caused by oxygen and nitrogen combining at combustion chamber temperatures above 2,500°F (1,371°C).
- Emission standards are implemented and administered by the EPA, but states may introduce their own, more stringent emission standards.
- A vehicle safety inspection and emissions component inspection are performed before an emission test.
- A basic idle emission test includes emissions recorded at idle speed and at 2,500 rpm.
- Enhanced emission tests are run with the vehicle on a dynamometer.
- Enhanced emission tests include the IM240 test, BAR-31 test, and ASM test.

## Review Questions

1. Technician A says that HC emissions increase as the air-fuel ratio becomes richer. Technician B says cylinder misfiring causes high HC emissions. Who is correct?
   A. Technician A
   B. Technician B
   C. Both Technician A and Technician B
   D. Neither Technician A nor Technician B
2. Technician A says $NO_x$ emissions are formed at low combustion chamber temperatures. Technician B says $NO_x$ is one of the pollutants that form smog. Who is correct?
   A. Technician A
   B. Technician B
   C. Both Technician A and Technician B
   D. Neither Technician A nor Technician B
3. Technician A says the PCV system reduces crankcase emissions. Technician B says crankcase emissions contain mainly HC. Who is correct?
   A. Technician A
   B. Technician B
   C. Both Technician A and Technician B
   D. Neither Technician A nor Technician B
4. Technician A says most evaporative emissions come from the fuel tank. Technician B says evaporative emissions contain $NO_x$ and CO. Who is correct?
   A. Technician A
   B. Technician B
   C. Both Technician A and Technician B
   D. Neither Technician A nor Technician B
5. Photochemical smog is formed by the reaction of sunlight with:
   A. CO and $CO_2$.
   B. $NO_x$ and HC.
   C. HC and CO.
   D. $NO_x$ and $CO_2$.
6. All of these statements about HC emissions are true *except*:
   A. HC tailpipe emissions are caused by unburned air-fuel mixture in the combustion chamber.
   B. When the air-fuel ratio becomes richer, the HC emissions increase.
   C. In a typical vehicle, 90 percent of the HC emissions come from the tailpipe.
   D. Approximately 27 percent of the total HC emissions come from vehicles.
7. When discussing CO emissions, Technician A says CO emissions increase when one cylinder is misfiring. Technician B says CO emissions increase when the air-fuel ratio becomes richer than 14.7:1. Who is correct?
   A. Technician A
   B. Technician B
   C. Both Technician A and Technician B
   D. Neither Technician A nor Technician B
8. $NO_x$ emissions increase considerably when combustion chamber temperatures are above:
   A. 800°F.
   B. 1200°F.
   C. 1650°F.
   D. 2500°F.
9. The most likely cause of excessive HC emissions from the crankcase is:
   A. worn piston rings.
   B. worn valve stem seals.
   C. a PCV valve that is stuck open.
   D. a lean air-fuel ratio.
10. While discussing emission test procedures, Technician A says during a BAR-31 emission test the vehicle is driven at a constant speed on a dynamometer. Technician B says during an IM240 emission test the vehicle is operated on a dynamometer at varying speed and load conditions. Who is correct?
    A. Technician A
    B. Technician B
    C. Both Technician A and Technician B
    D. Neither Technician A nor Technician B
11. Emission I/M facilities operated by a government are called ________________ facilities.
12. An idle emission test is performed at idle speed and ________________ rpm.
13. During an IM240 emission test, CO, HC, and $NO_x$ are recorded in ____________ ____________ ____________.
14. During a BAR-31 emission test, the vehicle is run on a ________________.
15. Describe the IM240 emission test procedure.
16. Describe the ASM emission test procedure.
17. List the items that are inspected during an emissions component inspection.
18. Describe a BAR-31 emission test procedure.

# Chapter 34 Emission Systems

## Introduction

Improper emission system operation may result in high emissions, drivability complaints, and reduced fuel economy. When diagnosing emission system problems, the first requirement is to understand the purpose of these systems and how they operate. When technicians do not understand the operation of various emission systems, diagnosis of these systems will likely be timeconsuming and inaccurate.

### POSITIVE CRANKCASE VENTILATION (PCV) SYSTEM DESIGN AND OPERATION

The PCV valve usually fits snugly into a rubber grommet in the rocker arm cover. A hose is connected from the PCV valve to the intake manifold. A clean air hose is connected from the rocker arm cover to the air cleaner. In a V6 or V8 engine, the clean air hose is in the opposite rocker arm cover from the PCV valve **(Figure 1)**. A filter is positioned on the air cleaner end of the clean air hose and a steel mesh is mounted in the fitting where the clean air hose enters the rocker arm cover.

The tapered, spring-loaded PCV valve is mounted so intake manifold vacuum pulls this valve toward the closed position and the spring tension pushes the valve open. The PCV system pulls clean air from the air cleaner into the engine. The PCV system removes crankcase gases containing HCs from the engine and directs them into the intake manifold. This action prevents crankcase gases containing HCs from escaping into the atmosphere.

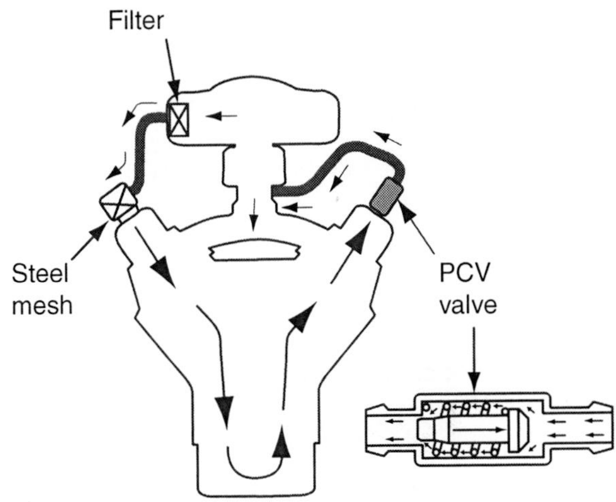


**Figure 1.** A PCV system.

When the engine is idling, the manifold vacuum pulls the PCV valve toward the closed position. At idle speed, the engine produces less blowby gas and a reduced PCV valve opening is adequate under this condition.

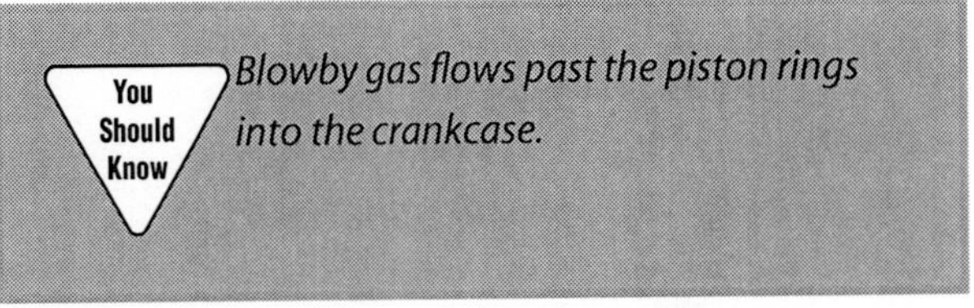


*Blowby gas flows past the piston rings into the crankcase.*

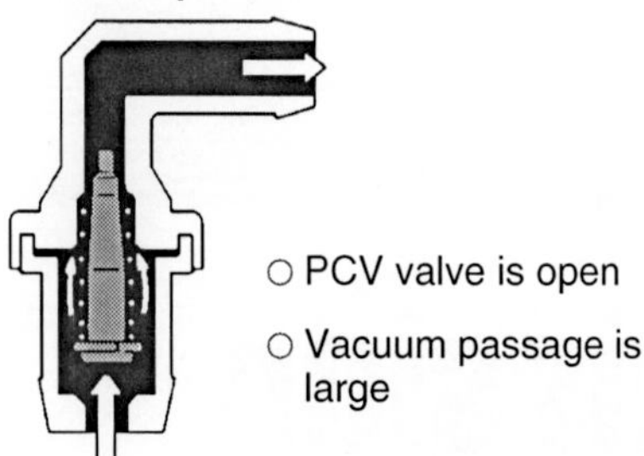


**Figure 2.** PCV valve position at part throttle.

If the engine is operating at part throttle, the manifold vacuum decreases and the spring pushes the PCV valve open to allow additional blowby gases to be pulled from the crankcase into the intake manifold **(Figure 2)**.

When the engine is operating at wide-open throttle, the manifold vacuum decreases and the spring forces the PCV valve further toward the open position **(Figure 3)**. This action allows the additional blowby gas produced under this condition to be pulled through the PCV valve into the intake manifold.

If the engine backfires into the intake manifold, the filter and steel mesh screen in the clean air hose prevent any flame front from entering the engine through the clean air hose. An engine backfire into the intake manifold seats the PCV valve against the valve housing and prevents any flame front from entering the engine through the PCV valve **(Figure 4)**.

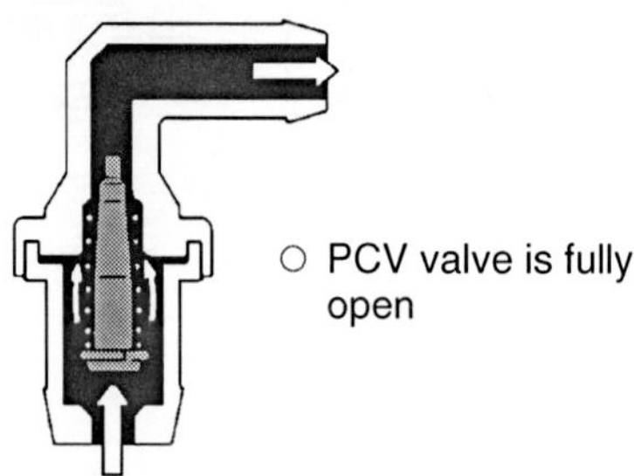


**Figure 3.** PCV valve position at wide open throttle.

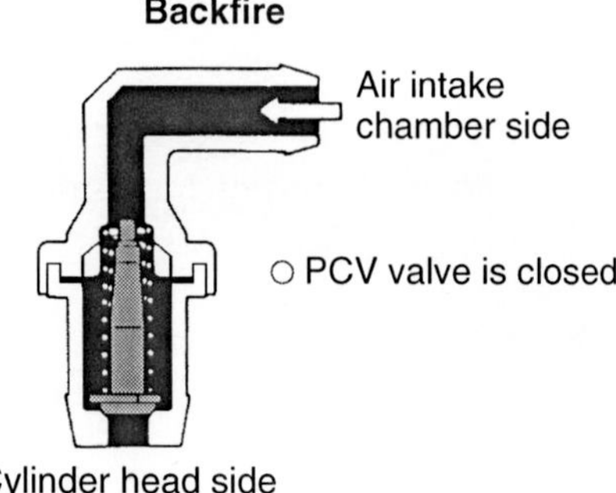


**Figure 4.** PCV valve position during engine backfire.

## EXHAUST GAS RECIRCULATION (EGR) SYSTEM DESIGN AND OPERATION

A tapered valve is mounted in the lower end of the EGR valve. This valve is linked to a diaphragm mounted in a sealed chamber on top of the valve **(Figure 5)**. Manifold or ported vacuum is usually supplied through a PCM-operated solenoid and vacuum hose to the EGR valve diaphragm **(Figure 6)**.

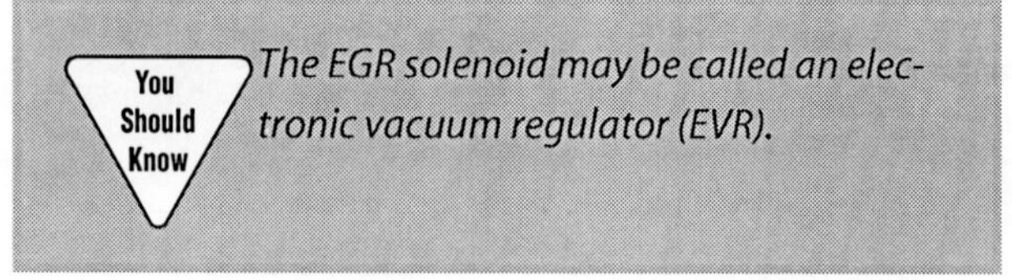


*The EGR solenoid may be called an electronic vacuum regulator (EVR).*

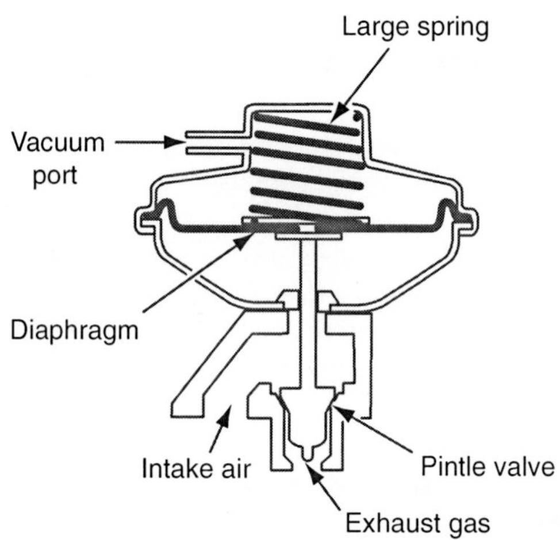


**Figure 5.** EGR valve design.

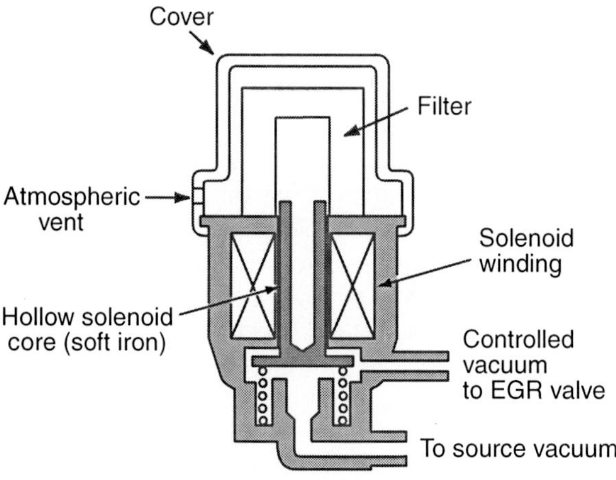


**Figure 6.** An EGR valve solenoid.

When exhaust gas recirculation is required to lower $NO_x$ emissions, the PCM energizes the EVR solenoid. Under this condition, this solenoid opens and supplies vacuum to the EGR valve. When vacuum is supplied to the EGR valve diaphragm, the EGR valve opens and allows a specific amount of exhaust gas to recirculate into the intake manifold. Since this exhaust gas is depleted of oxygen, it does not burn in the combustion chambers thereby reducing combustion chamber temperature and $NO_x$ emissions. The PCM opens the EGR valve under certain engine operating conditions. The PCM does not open the EGR valve when the engine coolant is cold because $NO_x$ emissions are not a problem on a cold engine. The EGR valve is not opened at idle because this results in rough idle and engine stalling. The PCM usually opens the EGR valve when the engine is operating at normal temperature and speeds above approximately 35 mph. The PCM does not open the EGR valve at wide-open throttle because the flow of exhaust gas through the EGR valve into the cylinders tends to reduce engine power.

On some EGR systems, the PCM pulses the EVR solenoid on and off to supply the precise exhaust flow required to lower $NO_x$ emissions without creating drivability problems. Some EGR systems have an orifice located in the exhaust stream near the EGR valve **(Figure 7)**. This orifice creates a pressure drop and a small exhaust pipe is connected from the area between the orifice and the EGR valve to the pressure feedback electronic (PFE) sensor. The PFE sensor sends a voltage signal to the PCM in relation to the amount of exhaust flow through the EGR system. If the exhaust flow through the EGR system does not match the

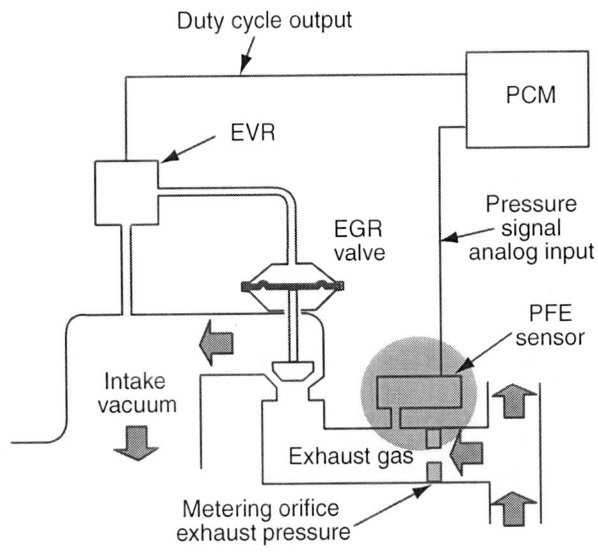


**Figure 7.** An EGR valve system with metering orifice and PFE sensor.

EGR flow demanded by the PCM input signals, the PCM makes a correction in EGR flow.

## Positive Pressure EGR Valves

Some EGR systems have a positive backpressure-type EGR valve. These EGR systems have a PCM-operated solenoid that supplies vacuum to the EGR valve, but this type of valve has an internal, normally-open control valve **(Figure 8)**.

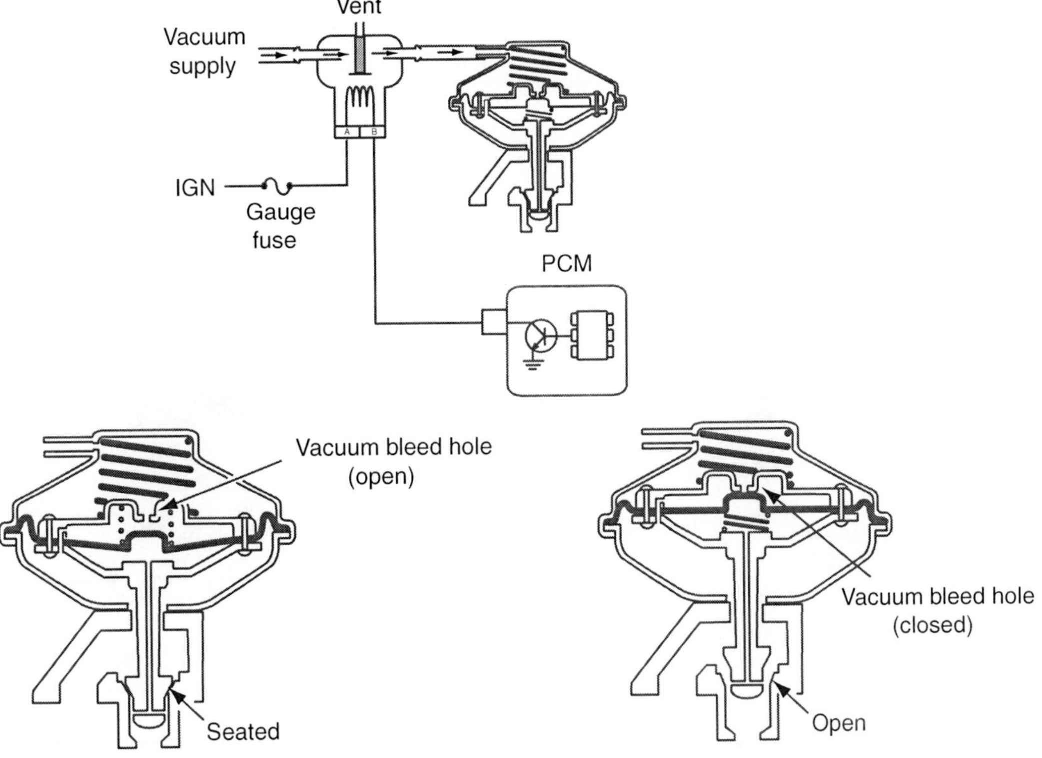


**Figure 8.** A positive backpressure EGR valve.

Exhaust gas is supplied from the lower end of the valve through a hollow valve stem to the control valve. With the engine idling, exhaust pressure is not high enough to close the control valve and this valve remains open. When the control valve is open, any vacuum in the diaphragm chamber is bled off through the valve. When the vehicle speed reaches approximately 35 mph (56 kmh), the exhaust pressure forces the control valve closed. Under this condition, vacuum supplied through the EGR valve solenoid lifts the diaphragm and opens the EGR valve.

## Negative Pressure EGR Valves

Other EGR valves have a negative backpressure-type EGR valve with an internal, normally-closed control valve **(Figure 9)**. Exhaust pressure is supplied through a hollow valve stem to the control valve. Each time a cylinder fires there is a high-pressure pulse in the exhaust, but between these high pressure pulses there is a low pressure.

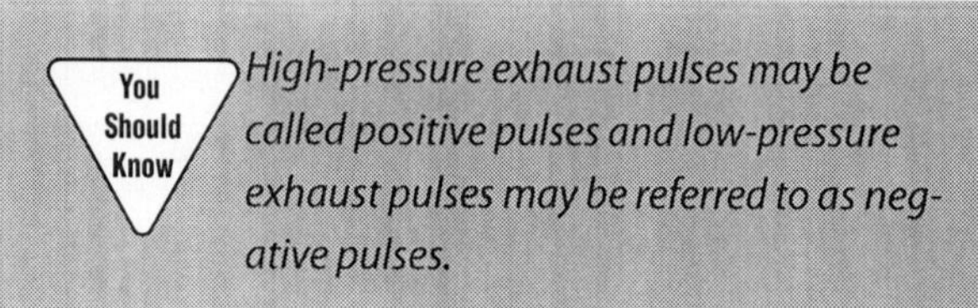


*High-pressure exhaust pulses may be called positive pulses and low-pressure exhaust pulses may be referred to as negative pulses.*

At idle speed, there is a longer time between the high-pressure pulses in the exhaust and the low-pressure pulses are more predominant. These low-pressure pulses pull the control valve open in the EGR valve. Any vacuum supplied to the diaphragm is bled off through the control valve. When the vehicle speed is approximately 35 mph (56 kmh), the high-pressure pulses in the exhaust are closer together and the negative exhaust pulses are less dominant. Under this condition, the EGR control valve closes and the vacuum supplied through the solenoid to the EGR valve diaphragm opens the EGR valve.

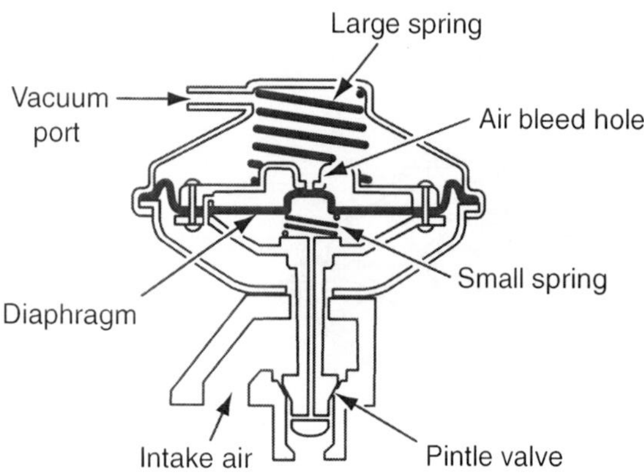


Figure 9. A negative backpressure EGR valve.

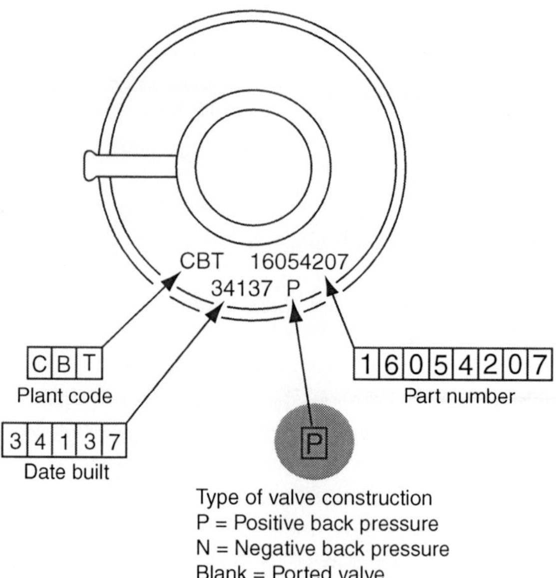


Figure 10. EGR valve identification.

Positive- or negative-type EGR valves are identified by an "N" or a "P" stamped on the top of the EGR valve housing **(Figure 10)**.

## Electronic EGR Valves

Many vehicles are now equipped with electronic EGR valves **(Figure 11)**. The PCM pulses the EGR valve winding on and off to provide the precise EGR flow required by the engine under all operating conditions.

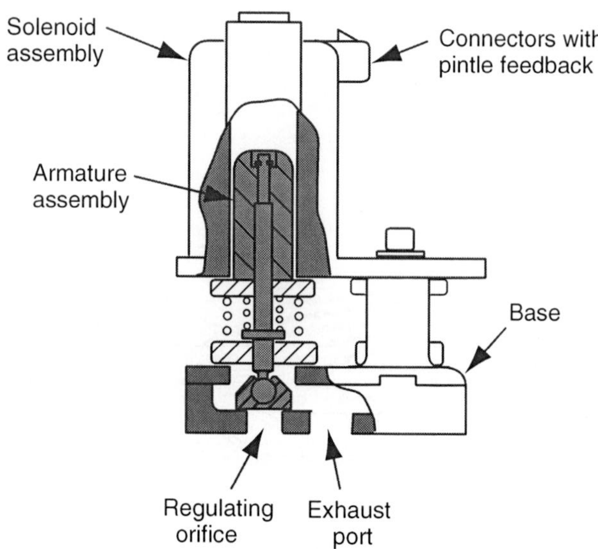


Figure 11. An electronic-type EGR valve.

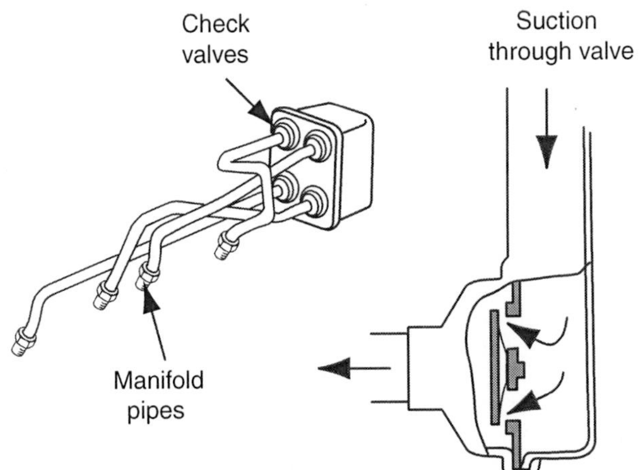


Figure 12. A pulsed air injection system.

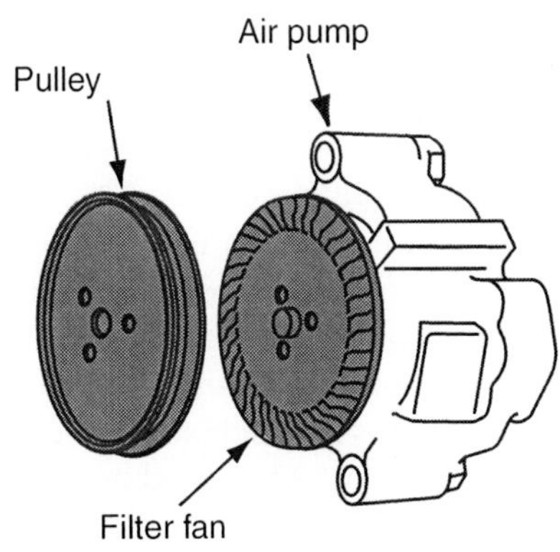


Figure 13. A belt-driven air pump.

## AIR INJECTION SYSTEM DESIGN AND OPERATION

Some older vehicles have a pulsed air injection system. In this system, a pipe with a one-way check valve is connected into each exhaust port. All the one-way check valves are attached to a common reservoir and a clean air hose is connected from this reservoir to the air cleaner **(Figure 12)**. At low engine speeds, the low-pressure pulses in the exhaust manifold open the one-way check valves and pull air into the exhaust manifold. The oxygen in the air mixes with the HC emissions in the exhaust manifold. This mixture ignites and burns in the exhaust manifold to reduce HC emissions from the tailpipe. When engine speed reaches a specific rpm, the pulsed air injection system becomes ineffective because the low-pressure pulses in the exhaust are greatly reduced.

Some vehicles have a belt-driven air pump. In some of these pumps, air enters the pump through a centrifugal filter behind the pump pulley **(Figure 13)**. Air is delivered from the pump to a bypass valve and a diverter valve **(Figure 14)**. These valves are opened and closed by vacuum supplied through vacuum/electric solenoids operated by the PCM. When the engine coolant is cold, the PCM does not energize either of the vacuum/electric solenoids. Under this condition, air from the pump is bypassed to the atmosphere at the bypass valve. When the engine coolant reaches a specific temperature, the PCM energizes the

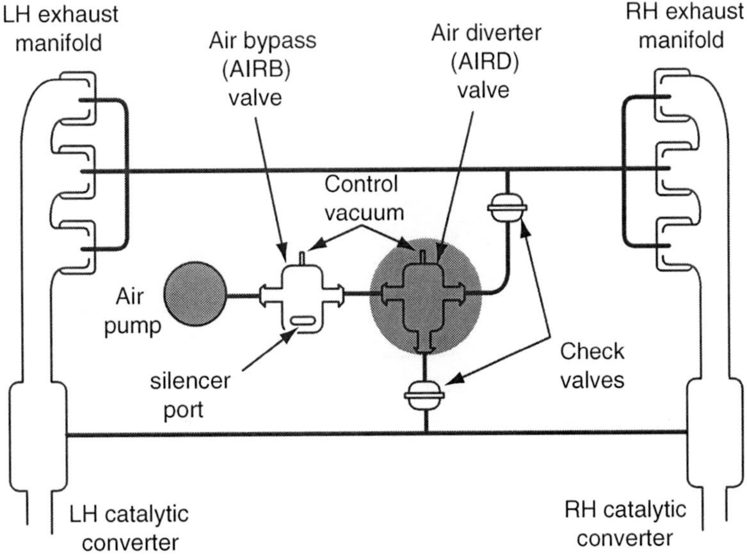


Figure 14. A belt-driven air pump system.

bypass valve solenoid, which supplies vacuum to the bypass valve. This action opens the bypass valve and allows air from the pump to flow through the bypass valve to the diverter valve. The PCM also energizes the diverter valve solenoid, which supplies vacuum to the diverter valve. Under this condition, the diverter valve is positioned so it delivers air from the pump to the exhaust ports. This airflow into the exhaust ports provides oxygen to mix with unburned hydrocarbons in the exhaust manifold. When this mixing occurs, the hydrocarbons are hot enough to ignite and burn. This action reduces hydrocarbon emissions during engine warmup. One-way check valves in the hoses connected to the exhaust manifold pipes prevent exhaust gas from blowing back into the air pump system when the air pump is not blowing air into the exhaust manifolds. If exhaust gas entered the air pump system, the pipes and hoses would be severely burned and damaged.

When the engine reaches normal operating temperature, the PCM de-energizes the diverter valve solenoid and shuts off vacuum to the diverter valve. Under this condition, the diverter valve is positioned so the airflow from the pump is directed to the catalytic converter. This airflow into the catalytic converter is essential for efficient converter operation.

Many newer vehicles have an electric-drive air pump **(Figure 15)**. Air is delivered from the pump to an air bypass valve and a diverter valve. These valves are controlled by vacuum supplied through a solenoid operated by the PCM. The air bypass and diverter valves operate in basically the same way as in the belt-driven air pump system. During engine warmup, the PCM energizes the diverter valve solenoid, which supplies vacuum to the diverter valve. Under this condition, the diverter valve is positioned to deliver air pump air to the exhaust ports. When the engine coolant reaches normal operating temperature, the PCM shuts off the solenoid and vacuum supply to the diverter valve. Under this condition, the diverter valve is positioned so it delivers air from the pump to the catalytic converter **(Figure 16)**. The operation of the air pump may vary depending on the engine application.

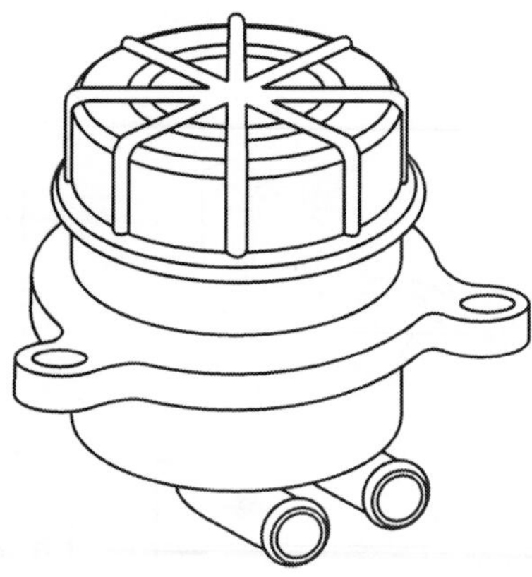

**Figure 15.** An electric-drive air pump.

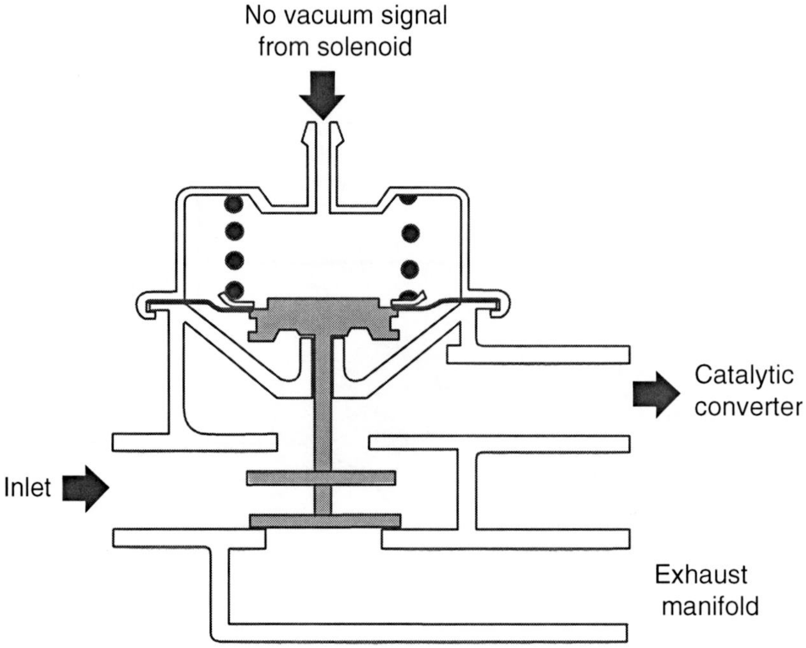


**Figure 16.** A diverter valve for an electric-drive air pump.

## EVAPORATIVE (EVAP) EMISSION SYSTEM DESIGN AND OPERATION

The EVAP system reduces the amount of fuel vapors containing HC that escape to the atmosphere. In a typical EVAP system, a hose is connected from the top of the fuel tank to a charcoal canister that is usually located in the engine compartment. The fuel tank has a domed top to allow fuel vapors to collect in the top of the tank. The tank pressure control valve (TPCV) is located in the hose between the canister and the fuel tank. With the engine running, intake manifold vacuum is supplied to the TPCV. Under this condition, the valve opens and allows fuel vapors to flow from the fuel tank into the canister. With the engine stopped and no vacuum supplied to the TPCV, this valve is closed. If the tank pressure exceeds a specific value, the pressure pushes the TPCV open and allows vapors to flow into the canister. The fuel tank is equipped with a pressure/vacuum valve **(Figure 17)**. This valve is similar to a radiator cap, but it operates at a much lower pressure or vacuum.

A normally-closed purge solenoid is mounted in the hose between the canister and the intake manifold **(Figure 18)**. When the purge solenoid is closed, fuel vapors from the tank flow into the canister. The charcoal in the canister holds the fuel vapors. The PCM energizes and opens the purge solenoid when the engine is at or near normal operating temperature and the vehicle speed and engine speed are above specific values. When the purge solenoid is open, manifold vacuum is supplied to the canister. This vacuum purges fuel vapors out of the canister into the intake manifold. As the canister is purged, air flows through the air separator into the canister.

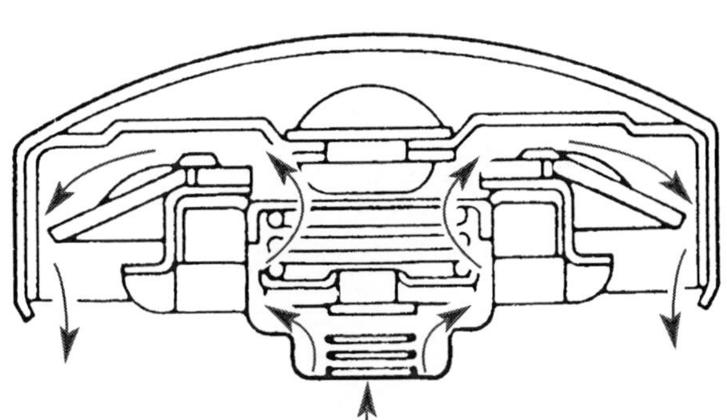


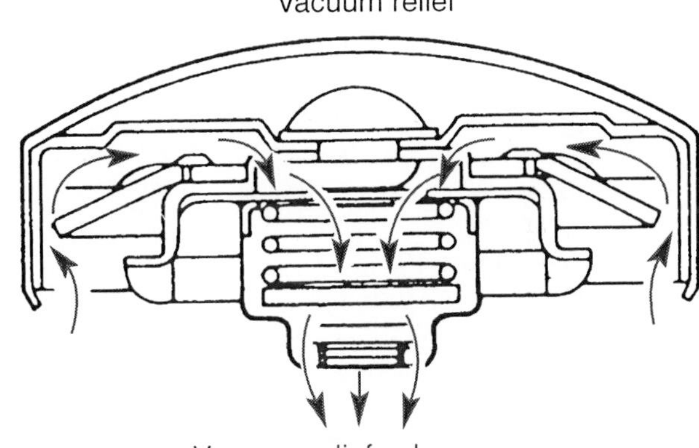


**Figure 17.** A fuel tank filler cap with pressure relief and vacuum valves.

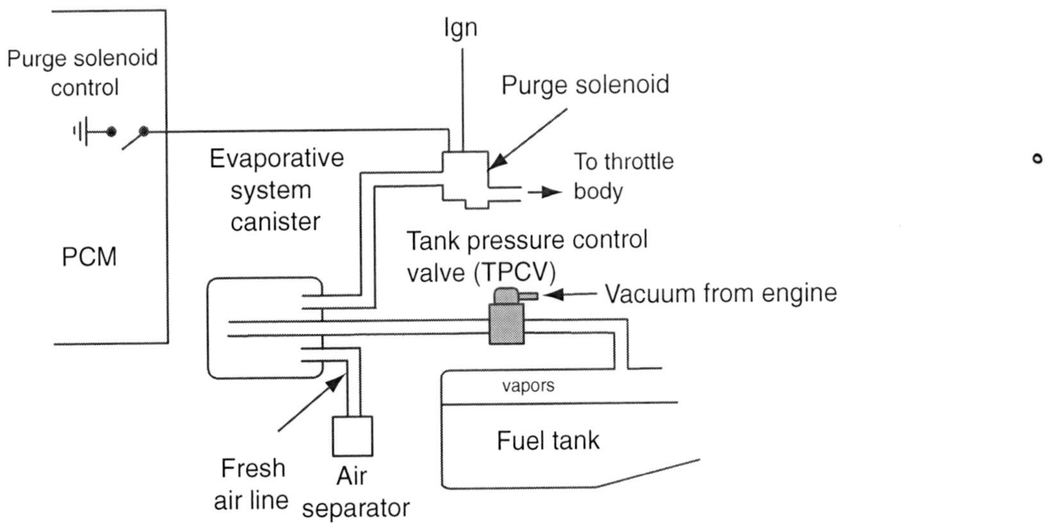


**Figure 18.** A typical evaporative (EVAP) system.

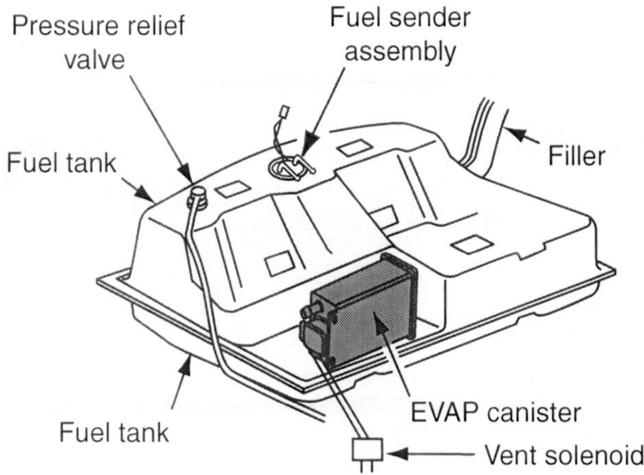


**Figure 19.** An enhanced evaporative (EVAP) system.

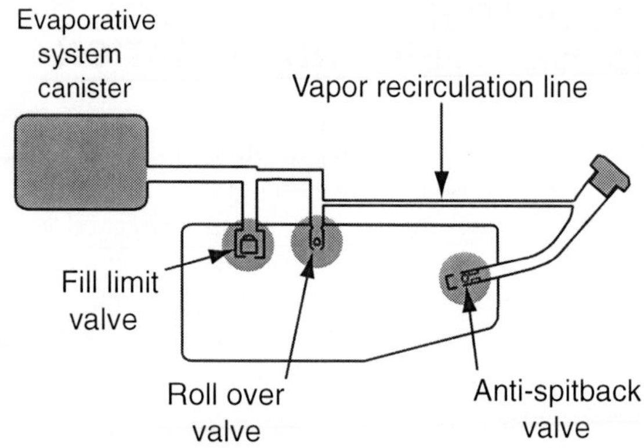


**Figure 20.** An on-board refueling vapor recovery (ORVR) system.

Newer vehicles must meet more stringent evaporative standards and these vehicles have enhanced EVAP systems. Some enhanced EVAP systems have the canister mounted near the fuel tank and a vent valve positioned in the air inlet to the canister **(Figure 19)**.

An enhanced EVAP system has the capability to check for leaks in the system. These systems are used on OBD II systems. A fuel tank pressure sensor is mounted in the top of the fuel tank and is connected electrically to the PCM. This sensor sends a voltage signal to the PCM in relation to fuel tank pressure. The fuel tank vent valve replaces the previous vent valve in the filler cap. A conventional purge solenoid is mounted in the engine compartment. The PCM operates the purge solenoid with a **pulse width modulated (PWM)** signal to control the vapor flow from the canister to the intake manifold. The PCM operates the purge solenoid and the vent solenoid to maintain the specified fuel tank pressure. A PWM signal is a digital signal with a variable on time. Periodically, the PCM tests the evaporative system for leaks. To test the system, the PCM de-energizes and closes the purge solenoid and the energizes the vent solenoid to seal the system. The PCM then scans the fuel tank pressure sensor signal for any decrease in fuel tank pressure that indicates a leak in the evaporative system. If a leak is detected on two occasions, the PCM sets a DTC in memory. The PCM must be able to detect a leak equivalent to a 0.040-inch (1.016-millimeter) or 0.020-inch (0.508-millimeter) diameter opening in the EVAP system depending on the year of vehicle.

Some EVAP systems on OBD II vehicles have an on-board refueling vapor recovery (ORVR) system. The ORVR system reduces the escape of fuel vapors during refueling. The fuel tank filler neck has a smaller diameter neck that provides a liquid seal in the filler neck to prevent fuel vapors from escaping while refueling. A one-way check valve in the end of the filler pipe allows fuel to flow into the tank, but this valve prevents fuel from spitting back out of the filler neck. While refueling, fuel vapors flow through the fill limiter vent valve (FLVV) on top of the fuel tank into the canister **(Figure 20)**. A larger canister is used on a vehicle with an ORVR system. The FLVV closes and prevents fuel from escaping in a vehicle roll-over situation.

## CATALYTIC CONVERTER DESIGN AND OPERATION

HC, CO, and $NO_x$ emissions can be converted to harmless gases at temperatures above 1832°F (1000°C). However, it is impossible to maintain this exhaust gas temperature. A catalyst in the exhaust system containing platinum (Pt), palladium (Pd), and/or rhodium (Rh) allows conversion of HC, CO, and $NO_x$ to harmless gases at temperatures of 572° to 1652°F (300° to 900°C). This temperature range can be achieved in the exhaust system. A **catalyst** is a material that accelerates a chemical reaction without being changed itself.

All types of catalytic converters must reach a temperature of approximately 600°F (315°C) to light off and start the chemical reaction that oxidizes and reduces pollutants.

Catalytic converters may be pellet type or monolith type. A **pellet-type** catalytic converter contains somewhere between 100,000 and 200,000 small pellets. The design of a **monolith-type** catalytic converter is similar to a honeycomb.

A monolith catalytic converter has a surface area about the size of ten football fields or 500,000 square feet. Metal monoliths are manufactured from an iron, chrome, and aluminum alloy. A material called corierite is used in ceramic monoliths. An aluminum oxide called alumina is sprayed on the ceramic or metal monoliths. The alumina is also sprayed on the pellets in a pellet-type converter. The alumina is highly porous and contains many small openings called micropores. Spaces left between the micropores are referred to as macropores. Base metals such as cerium (Ce) or iron (Fe) are dispersed in the alumina to help prevent shrinkage. The monolith or pellets are coated with the noble metals platinum, palladium, and/or rhodium. These noble metals are dispersed into the micropores and macro-

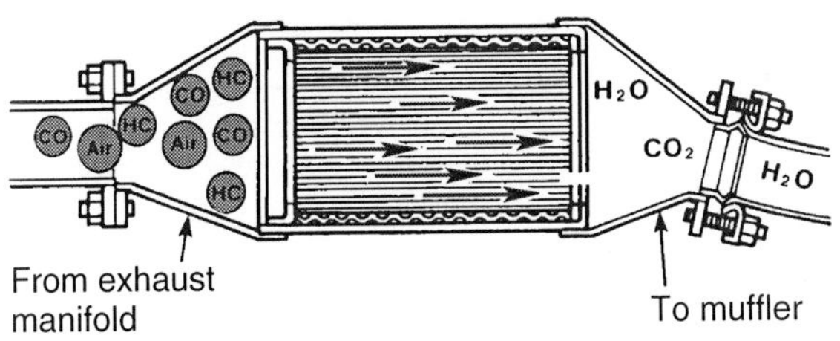


**Figure 21.** An oxidization-type catalytic converter.

pores in the alumina. The base metals in the alumina can chemically absorb $O_2$ and CO.

## Oxidation Converter Operation

The first catalytic converters installed on vehicles in the 1970s were oxidation converters. This type of converter oxidizes HC and CO into $CO_2$ and $H_2O$ **(Figure 21)**. The oxidation converter does not reduce $NO_x$ emissions, so these engines are equipped with an EGR valve to reduce this pollutant. On engines with an oxidation catalytic converter, the fuel system is calibrated to deliver an air-fuel ratio that is leaner than the 14.7:1 stoichiometric air-fuel ratio. This lean air-fuel ratio provided additional oxygen for the converter.

## Dual-Bed Converter Operation

Beginning in the mid 1980s, many engines have a dual-bed converter. This type of converter has a dual bed containing two monoliths or two pellet beds. The first bed in a dual-bed converter contains rhodium and palladium. This bed reduces $NO_x$ to $N_2$ and $CO_2$ **(Figure 22)**. The second bed in a dual-bed converter contains platinum and palladium.

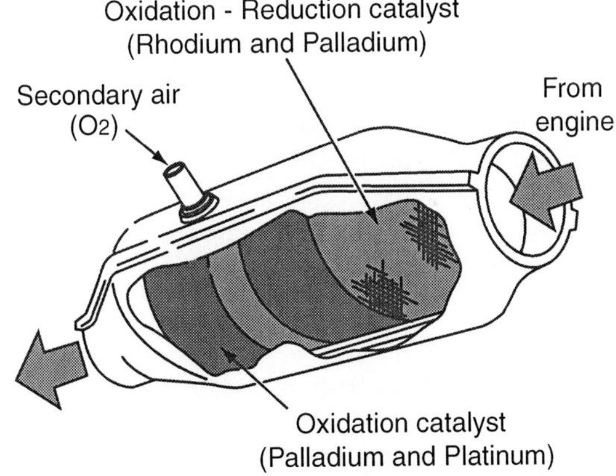


| From engine | | Between chamber | Exhaust |
|---|---|---|---|
| CO | 0.1 - 1.0% | 0 - 0.5% | 0 |
| HC | 50 - 200 ppm | 0 - 50 | 0 |
| CO2 | 13.6 - 14.3% | 14.7 -15.5% | 9.6 - 12% |
| O2 | 0.3 - 0.7% | 0 - 0.5% | 2.5 - 5.5% |
| NOx | 1200 ppm | 120 | 120 |
| | | | Water vapor |
| | | | Nitrogen |

**Figure 22.** A dual-bed catalytic converter.

This bed oxidizes HC and CO to $H_2O$ and $CO_2$. Most engines with a dual-bed converter have an air pump that delivers air to the converter between the two beds. This airflow provides additional oxygen to help oxidize the HC and CO in the rear bed. On an engine with a dual-bed converter, the fuel system is calibrated to deliver an air-fuel ratio of 14.6:1, which is slightly richer than stiochiometric. A typical dual-bed converter reduces HC, CO, and $NO_x$ emissions by 80 percent.

## Three-Way Converter Operation

Beginning in the 1990s, many vehicles were equipped with a three-way catalytic converter. These converters contain platinum and/or palladium and rhodium in a single bed. The three-way converter operates in a similar manner to the dual-bed converter. On engines with a three-way converter, the fuel system must deliver an air-fuel ratio that is at or very close to the stoichiometric air-fuel ratio of 14.7:1. When this air-fuel ratio is available, the converter operates at peak efficiency in oxidizing HC and CO and reducing $NO_x$ **(Figure 23)**. If the air-fuel ratio is leaner than stoichiometric, combustion temperature increases and the converter does not reduce all the $NO_x$. When the air-fuel ratio is richer than stoichiometric, the converter does not oxidize all the HC and CO. If the engine is operating at a 14.7:1 air-fuel ratio, a three-way converter provides a 90 percent reduction in HC, CO, and $NO_x$ emissions.

When the air-fuel ratio cycles slightly to the lean side of 14.7:1, there is more oxygen in the exhaust stream. The base metals in the converter temporarily absorb this oxygen. When the air-fuel ratio cycles slightly to the rich side of 14.7:1, there is less oxygen in the exhaust stream and the base metals in the converter release the stored oxygen. The oxygen storage capacity of the converter must be matched to the engine cubic inch displacement (CID) and the fuel system calibration.

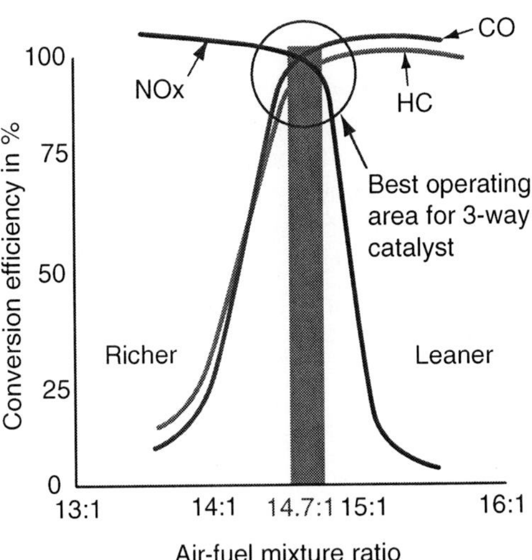


**Figure 23.** A three-way catalytic converter requires a 14:7:1 air-fuel ratio for efficient converter operation.

## Summary

- The PCV system prevents crankcase gases from escaping to the atmosphere.
- The EGR system recirculates exhaust gas into the intake manifold to reduce $NO_x$ emissions.
- The EGR valve is open in the cruising speed range with the engine at normal operating temperature.
- EGR valves may be conventional type, positive backpressure type, negative backpressure type, or electronically operated.
- A pulsed air injection system is operated by low-pressure pulses in the exhaust system at low engine speeds.
- A belt-driven or electric-drive air pump delivers air to the exhaust ports during engine warmup to reduce HC emissions.
- A belt-driven or electric-drive air pump typically delivers air to the catalytic converter with the engine at normal operating temperature.
- EVAP systems reduce the amount of fuel vapors that escape from the fuel tank to the atmosphere.
- In an enhanced EVAP system, the PCM has the capability to test for leaks in the system.
- Catalytic converters may be oxidation type, dual-bed, or three-way type.
- An oxidation converter oxidizes HC and CO into $H_2O$ and $CO_2$.
- A dual-bed or three-way catalytic converter performs the same function as an oxidation converter, but these converters also reduce $NO_x$ to $N_2$ and $CO_2$.
- A three-way catalytic converter requires an air-fuel ratio of 14.7:1.

## Review Questions

1. Technician A says the PCV valve reduces $NO_x$ emissions. Technician B says the PCV valve opening increases as the throttle opening becomes wider. Who is correct?
   A. Technician A
   B. Technician B
   C. Both Technician A and Technician B
   D. Neither Technician A nor Technician B
2. Technician A says the EGR valve should be closed at idle. Technician B says a solenoid operated by the PCM supplies vacuum to the EGR valve. Who is correct?
   A. Technician A
   B. Technician B
   C. Both Technician A and Technician B
   D. Neither Technician A nor Technician B
3. Technician A says in a positive backpressure-type EGR valve, the normally-open, internal control valve is closed by exhaust pressure at a specific throttle opening. Technician B says recirculating exhaust through the EGR valve into the intake manifold reduces HC emissions. Who is correct?
   A. Technician A
   B. Technician B
   C. Both Technician A and Technician B
   D. Neither Technician A nor Technician B
4. Technician A says on some vehicles, a belt-driven air pump system delivers air to the catalytic converter during engine warmup. Technician B says a belt-driven air pump reduces HC emissions during engine warmup. Who is correct?
   A. Technician A
   B. Technician B
   C. Both Technician A and Technician B
   D. Neither Technician A nor Technician B
5. All of these statements about PCV systems are true *except*:
   A. The PCV system reduces HC crankcase emissions.
   B. Manifold vacuum pulls the PCV valve toward the closed position.
   C. When throttle opening increases, PCV valve opening decreases.
   D. The PCV valve spring moves the valve toward the open position.
6. All of these statements about EGR valves are true *except*:
   A. Vacuum is supplied to some EGR valves through a PCM-operated solenoid.
   B. The EGR valve is opened during cold engine operation.
   C. When the EGR valve is open, exhaust gas is recirculated into the intake manifold.
   D. The EGR valve reduces $NO_x$ emissions by lowering combustion temperature.

7. The PFE sensor in an EGR system:
   A. sends a voltage signal to the PCM in relation to exhaust temperature.
   B. sends a voltage signal to the PCM in relation to EGR flow.
   C. provides a higher voltage signal when the EGR valve opening is reduced.
   D. is supplied with exhaust pressure directly from the exhaust manifold.
8. When discussing various types of EGR valves, Technician A says a negative pressure EGR valve has a normally-closed, internal valve that is opened by low-pressure pulses in the exhaust system at lower engine speeds. Technician B says a positive pressure EGR valve has a normally-open, internal valve that is closed by positive pressure in the exhaust system above a certain vehicle speed. Who is correct?
   A. Technician A
   B. Technician B
   C. Both Technician A and Technician B
   D. Neither Technician A nor Technician B
9. On an OBD II vehicle, the enhanced EVAP system has all of these features *except*:
   A. a pressure sensor mounted in the fuel tank.
   B. a purge solenoid connected between the canister and the intake manifold.
   C. a vent solenoid connected on the canister air intake connection.
   D. the capability to detect a 0.005-inch diameter leak in system.
10. With the engine at normal operating temperature, a fuel injection system and a three-way catalytic converter maintain the air-fuel ratio at:
    A. 14.0:1.
    B. 14.4:1.
    C. 14.7:1.
    D. 15.2:1.
11. A typical EVAP system purges vapors from the charcoal canister when the vehicle speed is above ______________________ mph.
12. An enhanced EVAP system has a(n) ______________ ______________ ______________ sensor mounted in the fuel tank.
13. An enhanced EVAP system has the capability to detect a leak in the system equivalent to a(n) ______________-inch or ______________-inch opening, depending on the vehicle model year.
14. An oxidation catalytic converter oxidizes ______________ and ______________ into ______________ and ______________.
15. Explain the operation of a dual-bed catalytic converter.
16. Describe the air-fuel ratio requirements for a three-way catalytic converter.
17. Describe the operation of a negative backpressure-type EGR valve.
18. Explain the operation of a typical electric-drive air pump.

# Emission System Maintenance, Diagnosis, and Service

## Introduction

A restricted PCV system causes pressure buildup in the engine which results in oil vapor being forced through the clean air hose into the air cleaner. This excessive pressure in the engine may also cause oil leaks at engine gaskets. If the PCV valve is sticking in the open position, excessive air flows through this valve, resulting in a lean air-fuel ratio and erratic idle operation. Proper maintenance, diagnostics, and service of the PCV system and all other emission control systems is important to maintain satisfactory engine drivability and provide proper life of system components.

### PCV SYSTEM MAINTENANCE

When basic underhood service is performed, the PCV system should be inspected. Check the clean air hose and inspect the PCV valve hose for cracks, splits, deterioration, and internal restriction. Inspect the air cleaner and the PCV inlet air filter for oil contamination. If there is oil contamination in these components, the PCV valve or hose is restricted or the engine has excessive blowby. Replace the PCV inlet air filter if it is dirty. Be sure the PCV valve fits snugly into the rocker arm cover grommet. Remove the PCV valve and shake it beside your ear. The plunger in the valve should rattle during this action. Inspect the PCV valve for sludge deposits and replace the valve if necessary.

### PCV SYSTEM DIAGNOSIS

With the engine idling, remove the oil filler cap and watch for oil vapors escaping from the oil filler opening. Accelerate the engine and continue watching for oil vapors at this location. If an excessive amount of oil vapors are escaping from the oil filler opening, the PCV system is restricted or the engine has excessive blowby. With the engine idling, place your hand over the oil filler opening and wait for 2 minutes. When you remove your hand, you should feel a vacuum in the engine. If there was no vacuum buildup in the engine, the PCV system is restricted or some of the engine gaskets are leaking air into the engine.

**You Should Know** *If engine gaskets allow the PCV system to pull air into the engine, some dirt particles will be suspended in the air depending on the location of the vehicle. If dirt particles enter the engine from this source, engine wear is accelerated. When engine gaskets allow air to enter the engine, they will likely allow oil to leak from the engine, especially after the engine is shut off.*

With the PCV valve disconnected from the rocker arm cover grommet, start the engine and allow the engine to idle. A hissing noise from the PCV valve indicates it is not plugged. Vacuum should be felt when your finger is placed over the end of the PCV valve. If there is very little or no vacuum at the PCV valve, check the valve and hose for a plugged condition.

## PCV SYSTEM SERVICE

The PCV system does not require any adjustments. PCV system service involves replacing the PCV valve and the inlet air filter at the vehicle manufacturer's specified intervals. Be sure the PCV hose is in good condition. This hose must be fit tightly on the PCV valve and the intake manifold fitting. A large clip usually retains the inlet air filter to the air cleaner housing. After the inlet air filter is replaced, be sure this clamp is in place and check the fit of the clean air hose on the filter and the fitting in the rocker arm opening. If there is a steel mesh in the rocker arm clean air fitting, be sure this mesh is not contaminated. If necessary, wash this mesh in an approved solvent.

## EGR SYSTEM MAINTENANCE

If the EGR valve is open with the engine idling, engine idle is rough and the engine may stall. When the EGR valve is not opening under any condition, $NO_x$ emissions are high and engine detonation may occur because the combustion chamber temperature is increased. Engine detonation will likely result in reduced fuel economy.

EGR system maintenance involves inspecting all the vacuum hoses in the EGR system for cracks, deterioration, oil contamination, and looseness. Be sure the vacuum hoses are properly connected. Most vehicles have an underhood vacuum hose routing diagram to illustrate the proper vacuum hose connections. Inspect the wires on the EGR valve solenoid, PFE sensor, or electronic EGR valve for loose connections and worn insulation. If the engine has a PFE sensor, inspect the small exhaust pipe from the exhaust manifold to this sensor for leaks and kinks.

## EGR SYSTEM DIAGNOSIS

To test a conventional EGR valve, disconnect the vacuum hose from the EGR valve. Connect a vacuum hand pump to the EGR valve and start the engine. With the engine idling, operate the hand pump to supply 20 in. Hg to the EGR valve. Under this condition, the EGR valve should open and the engine speed should decrease at least 100 rpm or the engine should stall **(Figure 1)**. If the engine slows down over 100 rpm or stalls, the EGR valve is operating. When the engine does not slow down more than 100 rpm or stall, the EGR valve is defective or the EGR exhaust passages are restricted.

To test a negative backpressure EGR valve, disconnect the EGR valve vacuum hose and connect a hand vacuum pump to the EGR valve. Supply 18 in. Hg to the EGR valve with the hand vacuum pump. Under this condition, the EGR valve should open and hold the vacuum for 20 seconds. With 20 in. Hg supplied to the EGR valve, start the engine. The vacuum should quickly drop to zero and the valve should close. Replace the EGR valve if it does not operate properly.

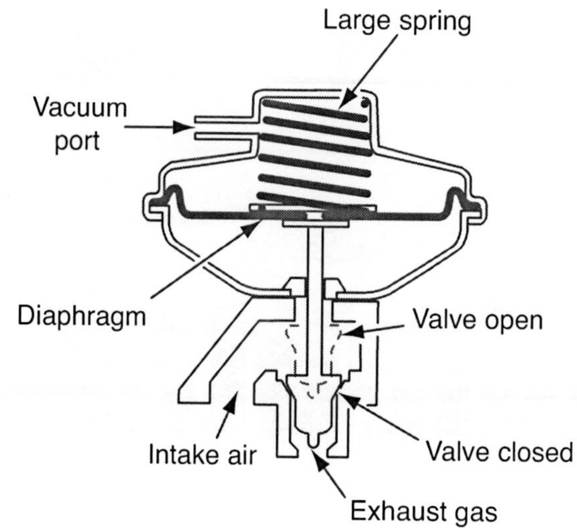


**Figure 1.** Conventional EGR valve operation.

To test a positive backpressure-type EGR valve, start the engine and be sure the engine is at normal operating temperature. Disconnect the vacuum hose from the EGR valve and connect a hand vacuum pump to the EGR valve. When the hand pump is operated, the vacuum should be bled off through the EGR valve. Accelerate the engine to 2,000 rpm and apply 18 in. Hg to the EGR valve with the vacuum hand pump. Under this condition, the EGR valve should open and the vacuum reading should be maintained on the hand pump gauge. When the engine is returned to idle speed, the vacuum should be bled off and the EGR valve should close.

### Scan Tool Diagnosis of EGR Systems

Be sure the ignition switch is off and the engine at normal operating temperature. Connect a scan tool to the DLC. Some engines have a digital EGR valve with two or three EGR solenoids that operate EGR valve plungers of different sizes. The scan tool may be used to command the PCM to energize each EGR valve plunger. Select EGR control on the scan tool followed by solenoid number 1. When the PCM energizes and opens this solenoid plunger, the engine should slow down. The engine should slow down as each EGR solenoid is selected on the scan tool. If the engine does not slow down when any EGR solenoid is energized, the solenoid winding may be open, the plunger is sticking, or the EGR passages are restricted. If the engine has a linear EGR valve, the scan tool may be used to command the PCM to provide a specific EGR valve opening. The scan tool also reads the actual percentage of EGR valve opening. If the actual EGR valve opening is not within 10 percent of the commanded opening, there is a defect in the system.

On engines with digital or linear EGR valves, check for DTCs on the scan tool. If there are DTCs displayed on the

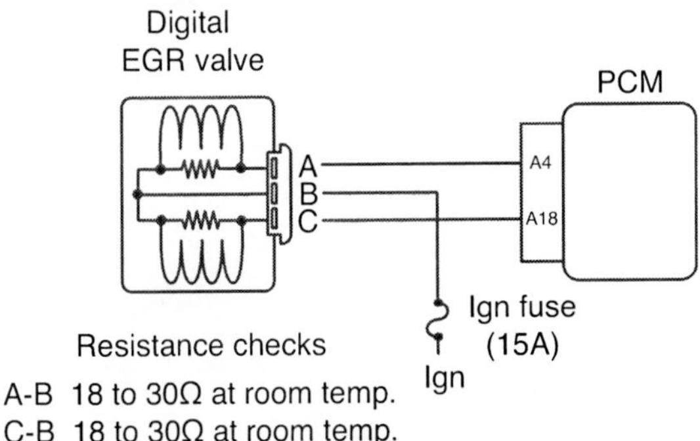


**Figure 2.** Testing digital EGR valve windings.

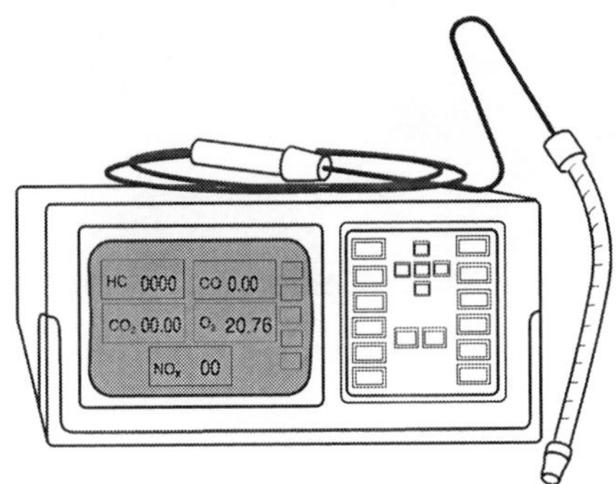

**Figure 3.** A five-gas exhaust analyzer.

scan tool representing EGR problems, the exact cause of the DTC must be diagnosed. A DTC indicates a problem in a certain area, not in a specific component. For example, if a DTC is present representing solenoid number 2 in a digital EGR valve, the solenoid winding may be defective or there may be a defect in the wires from the solenoid to the PCM. Always be sure there are 12 volts supplied to the EGR valve. The solenoid windings may be tested for opens, shorts, and grounds with an ohmmeter **(Figure 2)**. If the solenoid windings are satisfactory, test the wires between the EGR valve and the PCM.

The EGR system operation may be tested with a five-gas analyzer **(Figure 3)**. If the EGR system is not operating properly, $NO_x$ emissions are high. For example, the EGR valve may be operating normally, but restricted EGR exhaust passages may be reducing EGR flow. Under this condition, $NO_x$ emissions as indicated on the five-gas analyzer are high. Under most operating conditions, $NO_x$ emissions should be below 1,000 parts per million (ppm).

## EGR SYSTEM SERVICE

EGR systems do not require any periodic adjustments. EGR system service involves replacing system components after they have been diagnosed as being defective. EGR system service also includes removing carbon from the EGR valve exhaust passages if this condition is present. Some technicians may attempt to clean the tapered EGR valve, but this may not be practical because of high labor costs. When replacing an EGR valve, always install a new gasket under the valve.

## AIR INJECTION SYSTEM MAINTENANCE

On a pulsed air injection system, inspect the pipes for signs of burning which indicates the one-way check valves are leaking. Inspect all the pipes and hoses for leaks. If the air injection system has a belt-driven or electric-drive air pump, inspect all the air hoses and pipes for signs of burning, leaks, loose clamps, or contact with rotating components. Be sure to inspect the hose connected from the diverter valve to the catalytic converter on belt-driven air pump systems. Replace or repair the pipes and hoses as necessary. Inspect the air pump drive belt for wear, cracks, oil contamination, and proper tension. Inspect all the vacuum hoses and electric wires on the bypass and diverter valves.

## AIR INJECTION SYSTEM DIAGNOSIS

The diverter valve may be tested with the engine running at normal operating temperature and idle speed. Connect a vacuum hand pump to the diverter valve vacuum connection. With 20 in. Hg supplied to the diverter valve, the airflow from the pump should be delivered from the diverter valve outlet to the exhaust ports. When there is no vacuum supplied to the diverter valve, the airflow from the pump should be delivered from the diverter valve outlet to the catalytic converter **(Figure 4)**. If the air injection system has a

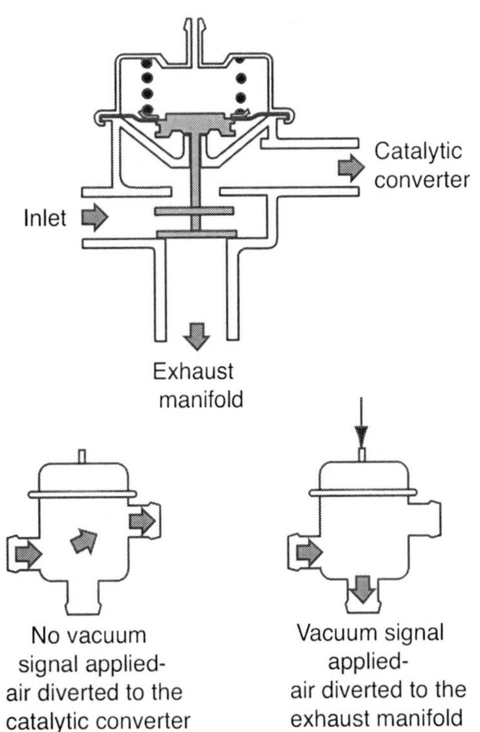


**Figure 4.** Diverter valve testing.

bypass valve, it may be tested in the same way as the diverter valve. With no vacuum supplied to the bypass valve, airflow from the pump should be bypassed from the bypass valve to the atmosphere. When vacuum is supplied to the bypass valve, airflow from the pump should be delivered from the bypass valve outlet connected to the diverter valve.

A scan tool may be used to diagnose belt-driven or electric-drive air pump systems. If there is an electrical defect in the bypass or diverter solenoids or in the electric-drive air pump, a DTC is set in the PCM memory. The scan tool indicates the status of the bypass and diverter valve solenoids. When the vehicle is driven on a road test with the scan tool connected, the bypass and diverter valve solenoids should be energized and de-energized under the appropriate vehicle operating conditions.

**You Should Know** *When a scan tool indicates a solenoid is on or off, it only indicates the PCM command to the solenoid. The PCM may provide the proper command, but an electrical defect in the solenoid may prevent the solenoid from carrying out the command. In this situation, a DTC should be set in the PCM memory.*

## AIR INJECTION SYSTEM SERVICE

The air injection system does not require any periodic adjustments except for the belt tension on a belt-driven air pump. If a belt-driven or electric-drive air pump has internal defects, replace the pump. The PCM grounds an electric-drive air pump relay winding. These relay contacts supply voltage to the electric-drive air pump. Before replacing the pump, always be sure voltage is available through the relay to the pump. The pulley and centrifugal filter may be replaced separately on a belt-driven air pump. If the bypass or diverter valve solenoids have electric defects, replace the solenoid. When a bypass or diverter valve fails to pass the tests mentioned previously, replace the valve.

## EVAPORATIVE (EVAP) SYSTEM MAINTENANCE

If the EVAP canister has a replaceable filter, this filter should be replaced at the vehicle manufacturer's specified intervals. Replace the canister if it is cracked or damaged **(Figure 5)**. Inspect all the EVAP system hoses for cracks, kinks, looseness, and oil contamination. Replace these hoses as required. Leaking purge hoses cause HC emissions to escape to the atmosphere and cause a gasoline odor. Leaking vacuum hoses may cause rough idle operation, improper idle speed, and inoperative EVAP components.

**Figure 5.** An EVAP system canister.

Inspect the wiring in the EVAP system for worn insulation, loose connections, breaks, and contact with rotating components. Repair the wiring as necessary.

## EVAP SYSTEM DIAGNOSIS

A scan tool may be connected to the DLC to diagnose the EVAP system. If there is an electrical defect in the purge solenoid, vent solenoid, or fuel tank pressure sensor, a DTC will likely be set in the PCM memory. If a DTC is displayed representing a fault in a solenoid, the solenoid windings may be tested with an ohmmeter to verify the solenoid condition **(Figure 6)**. The vehicle may be road tested with

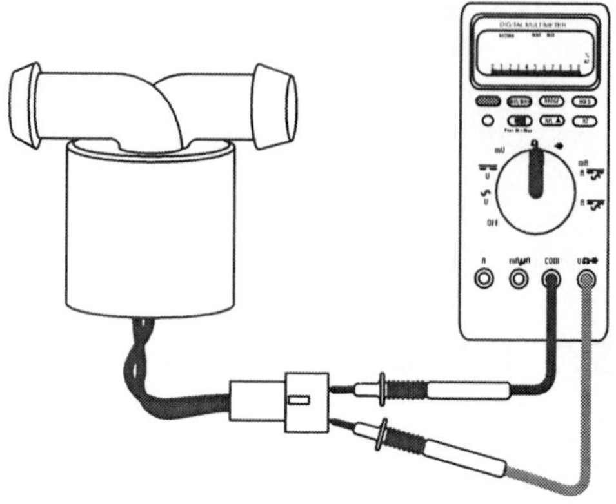

**Figure 6.** Testing the purse solenoid windings with an ohmmeter.

the scan tool connected to the DLC. The purge and/or vent solenoids should be energized under the appropriate vehicle operating conditions.

On vehicles with OBD II systems and enhanced EVAP systems, the PCM has the capability to test the EVAP system of a 0.040 inch (0.016 millimeter) or 0.020 inch (0.508 millimeter) diameter leak depending on the vehicle model year. If the PCM detects such a leak, a DTC is set in the PCM memory. The PCM performs this leak test as it monitors the EVAP system. To perform a large leak test, the PCM closes the vent valve and opens the purge solenoid to build vacuum in the system. When a specific vacuum is present in the EVAP system as indicated by the fuel tank pressure (FTP) sensor, the PCM closes the purge solenoid to close the system. The PCM then checks the rate of vacuum decrease in the EVAP system over a specific time period. If the vacuum decrease in the EVAP system is excessive, the PCM repeats the test. When the vacuum decrease is excessive on two tests, the PCM sets a DTC in memory. The PCM also performs a small leak test using the same procedure for a longer time period. A scan tool may be used to command the PCM to perform the large and small EVAP leak tests. Select SERVICE BAY TEST on the scan tool to enter this test mode.

If a leak is present in the EVAP system, some equipment manufacturers supply leak detection equipment that pressurizes the EVAP system with dry nitrogen. The leak detection tester is connected to an EVAP system service port in the engine compartment **(Figure 7)**. This service port is usually green in color. An **ultrasonic leak detector** is then positioned near any suspected EVAP system leaks. This tester produces a beep when a leak is present. When performing an EVAP system leak test, the fuel tank must be between 20 percent and 80 percent full.

You Should Know *A vacuum/pressure gauge that reads inches of water column (WC) may be connected to the EVAP system service port to check the EVAP system pressure.*

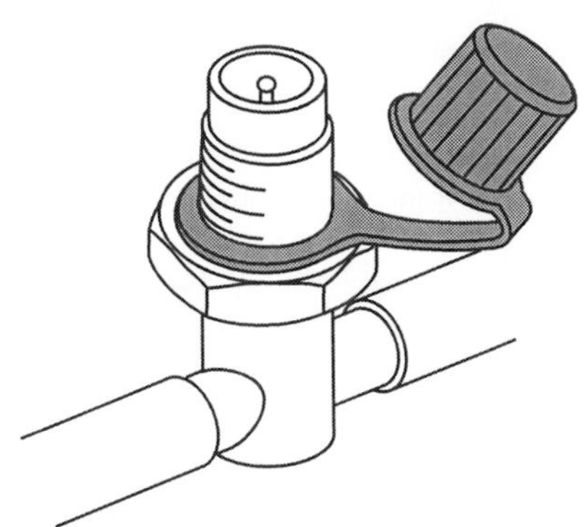

**Figure 7.** An EVAP system service port.

Other equipment manufacturers supply an **EVAP emission leak detector (EELD)** that produces and forces smoke into the EVAP system at the EVAP service port to locate system leaks. Smoke may be seen escaping at the EVAP system leak location. Some EELD testers have a portable light that produces a bright white beam to make smoke detection easier.

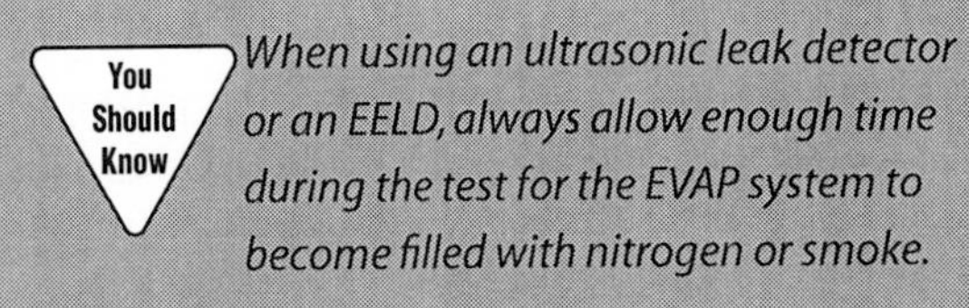

You Should Know *When using an ultrasonic leak detector or an EELD, always allow enough time during the test for the EVAP system to become filled with nitrogen or smoke.*

## EVAP SYSTEM SERVICE

If a DTC and a leak detection test indicate a leak in an EVAP system hose or component, replace the hose or component as required. When the FTP sensor voltage is not within specifications and a DTC indicates a defect in this area, always check the wires from the PCM to this sensor before replacing the sensor. In most applications, the fuel tank must be removed to access the FTP. When performing EVAP system service, always check the fuel filler cap for proper operation of the pressure relief and vacuum valves.

## CATALYTIC CONVERTER MAINTENANCE

A catalytic converter requires a minimum amount of maintenance. Inspect the outer shell of the converter for burning or damage and check the converter for exhaust leaks. If the converter has an exhaust leak or the exterior shell is severely burned or damaged, converter replacement is necessary. Be sure there are no leaks in the exhaust pipe or other exhaust system components.

You Should Know *A continual rich air-fuel ratio causes severe overheating of the catalytic converter. This condition may cause the internal components in the converter to actually melt, which makes the converter incapable of reducing emissions. The exterior of the converter may also appear overheated and burned.*

Use a rubber mallet to lightly tap the outer surface of the converter and listen for loose internal converter components that cause a rattling noise. If the converter has loose internal components, it should be replaced because these loose components will likely restrict the exhaust flow through the converter.

## CATALYTIC CONVERTER DIAGNOSIS

When testing the catalytic converter on pre-OBD II vehicles, the converter inlet and outlet temperatures may be measured to determine the converter condition. Follow these steps to measure the converter inlet and outlet temperatures:

1. Disable the air injection system. This may be done by squeezing the air pump outlet hose with a pair of straight-jawed vice grips until this hose is completely flat.
2. Run the engine until it is at normal operating temperature and then operate the engine at 2,500 rpm for 2 minutes to heat the converter and exhaust system.
3. Shut the engine off and disconnect and ground one spark plug wire in each cylinder bank on V6 or V8 engines. On a four-cylinder engine, disconnect and ground one spark plug wire.
4. Disconnect the IAC motor electrical connector.
5. Start the engine and run the engine at 1,000 rpm.
6. Use a digital multimeter with a temperature probe to measure the catalytic converter inlet and outlet temperature.

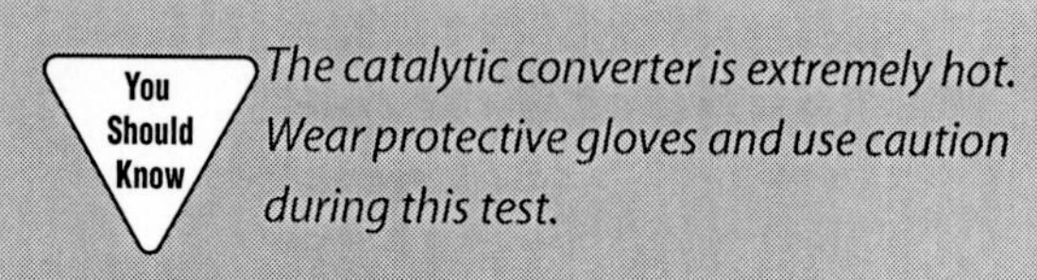


*The catalytic converter is extremely hot. Wear protective gloves and use caution during this test.*

Shut the engine off, and compare the converter inlet and outlet temperatures. The outlet temperature should be higher than the inlet temperature. When this temperature difference is less than 50°F (28°C), replace the converter. Reconnect the spark plug wires and the IAC motor connector. Restore air pump operation. Use a scan tool to clear any DTCs set during the test. When diagnosing the catalytic converter on OBD II vehicles, observe the MIL in the instrument panel. If this light is flashing, the PCM has detected a serious problem such as a very rich air-fuel ratio that may result in rapid and permanent catalytic converter damage. Connect a scan tool to the DLC under the dash and check for any DTCs indicating catalytic converter defects. Check for DTCs indicating defects in the $HO_2S$ heaters. Defects in these heaters prevent the sensors from reaching operating temperature in the required time and this affects the sensor voltages. Observe the voltage waveforms on the upstream and downstream $HO_2S$ to determine the converter condition. These voltage waveforms may be observed on a lab scope, a graphing voltmeter, or a scan tool with graphing capabilities. If the catalytic converter operation is satisfactory, the downstream $HO_2S$ voltage is considerably lower and cycling slower from low voltage to high voltage compared to the upstream $HO_2S$. When there is very little difference between the upstream and downstream $HO_2S$ voltage signals, the catalytic converter is not reducing HC, CO, and $NO_x$ emissions properly and the converter should be replaced **(Figure 8)**.

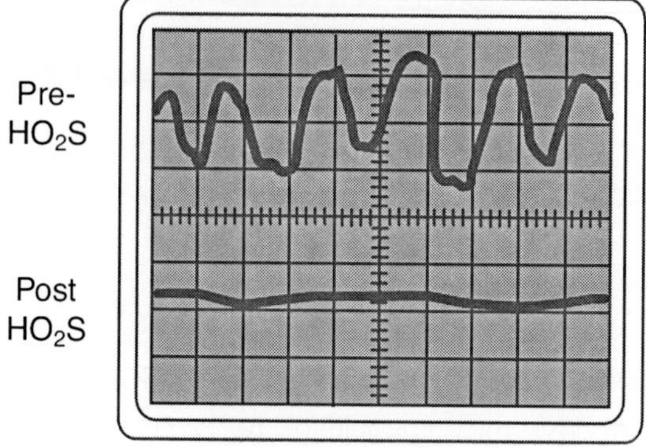


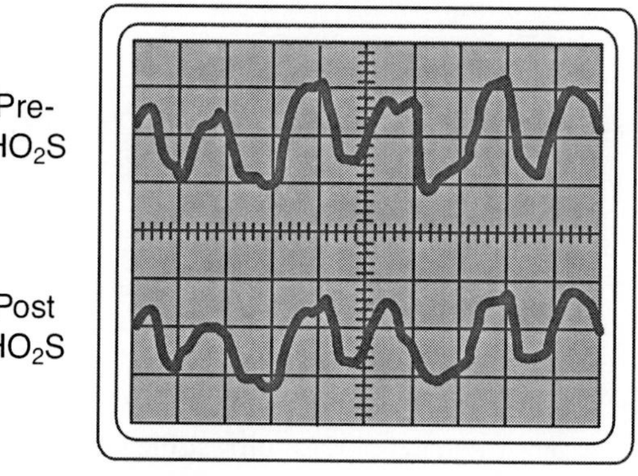


**Figure 8.** Catalytic converter diagnosis using $HO_2S$ waveforms.

## CATALYTIC CONVERTER SERVICE

When a catalytic converter is replaced, always diagnose the cause of the converter failure. If the cause of the converter failure is not diagnosed and corrected, the replacement converter will fail in a short time. For example, if the inside of the converter is melted, the air-fuel ratio may be excessively rich. This problem may be caused by excessive fuel pressure or a defective input sensor, such as the ECT sensor. Cylinder misfiring also causes excessive HC in the exhaust that overheats the converter. The cause of the rich air-fuel ratio must be diagnosed and corrected or the new converter will fail in a short time. If the inside of the converter is oil contaminated, diagnose and correct the cause of the engine oil consumption. After the converter is replaced, always be sure the voltage waveforms of the upstream and downstream $HO_2S$ are satisfactory. If there is a rich air-fuel ratio or ignition misfiring, the upstream $HO_2S$ will not be satisfactory.

The Clean Air Act allows the EPA to regulate the sale, use, and installation of aftermarket converters to ensure the replacement converters provide the same amount of emission reduction as the original converter on the vehicle. When installing an aftermarket converter, EPA regulations must be followed. According to these regulations, the replacement converter must be the same type as the original converter and it must be specified by the converter manufacturer for the vehicle being serviced. The replacement converter must be in the same location in the exhaust system. Copies of the converter invoice must be retained for 6 months and the replaced converter must be kept for 15 days.

## Summary

- The PCV system helps to prevent crankcase vapors from escaping to the atmosphere.
- When a conventional EGR valve is opened at idle, the engine should slow down to 100 rpm or stall.
- If the EGR valve is open during idle operation, rough idle will result.
- A scan tool may be used to command digital and linear EGR valves to open.
- A scan tool indicates the status of solenoids, such as the bypass and diverter valve solenoids, in an air injection system.
- A scan tool may be used to command the PCM to perform large and small EVAP leak tests.
- During an EVAP system leak test, the PCM builds vacuum in the system and seals the system. The PCM then monitors the change in EVAP system vacuum over a period of time.
- The technician may check for EVAP system leaks with equipment that pressurizes the EVAP system with dry nitrogen or forces smoke into the EVAP system.
- If the catalytic converter is operating properly, the outlet temperature should be 50°F (28°C) higher than the inlet temperature.
- If the catalytic converter is operating properly in an OBD II system, the voltage on the downstream $HO_2S$ should be lower and cycling much slower from low voltage to high voltage compared to the upstream $HO_2S$.

## Review Questions

1. Technician A says oil contamination in the air cleaner may indicate a restricted PCV system. Technician B says oil contamination in the air cleaner may indicate the engine has excessive blowby. Who is correct?
   A. Technician A
   B. Technician B
   C. Both Technician A and Technician B
   D. Neither Technician A nor Technician B
2. On an OBD II vehicle, the upstream and downstream $HO_2S$ have similar voltage waveforms. Technician A says the downstream $HO_2S$ is defective. Technician B says the catalytic converter is defective. Who is correct?
   A. Technician A
   B. Technician B
   C. Both Technician A and Technician B
   D. Neither Technician A nor Technician B
3. On a pre-OBD II vehicle, the catalytic converter outlet temperature is 100°F (55°C) hotter compared to the converter inlet temperature. Technician A says the converter is defective. Technician B says the air-fuel ratio is excessively rich. Who is correct?
   A. Technician A
   B. Technician B
   C. Both Technician A and Technician B
   D. Neither Technician A nor Technician B
4. An engine has a rough idle problem at normal operating temperature. Technician A says the EGR valve may be stuck open. Technician B says the EVAP purge solenoid may be open. Who is correct?
   A. Technician A
   B. Technician B
   C. Both Technician A and Technician B
   D. Neither Technician A nor Technician B

5. An EGR valve diaphragm is forced upward manually to open the EGR valve with the engine idling and there is no change in engine speed. The most likely cause of this problem is:
   A. a vacuum leak in the EGR valve diaphragm.
   B. blocked EGR exhaust passages.
   C. a vacuum leak in the EGR valve vacuum hose.
   D. a defective EGR solenoid.
6. When 20 in. Hg is supplied to a conventional EGR valve from a vacuum hand pump with the engine idling, the engine stalls. This action indicates:
   A. the EGR valve diaphragm has a vacuum leak.
   B. the EGR valve is operating normally.
   C. the EGR exhaust passage is restricted.
   D. the EGR valve is not opening fully.
7. All of these statements about scan tool diagnosis of digital and linear EGR valves are true *except*:
   A. A scan tool may be used to command any of the solenoids open in a digital EGR valve.
   B. A digital EGR valve has a vacuum solenoid connected between the PCM and the EGR valve.
   C. High $NO_x$ emissions may indicate the EGR valve is not opening properly.
   D. On some linear EGR valves, the scan tool indicates the commanded and the actual EGR valve opening.
8. When diagnosing a positive backpressure-type EGR valve, a vacuum hand pump should be used to open the EGR valve when the engine is running at:
   A. idle speed.
   B. 1,000 rpm.
   C. 1,200 rpm.
   D. 2,000 rpm.
9. The hoses connected to a belt-driven AIR pump are burned. The most likely cause of this problem is:
   A. a leaking one-way check valve in the AIR system.
   B. a slipping AIR pump drive belt.
   C. a defective AIR pump.
   D. a plugged AIR pump inlet filter.
10. When diagnosing an enhanced EVAP system, a DTC indicates a leak in the EVAP system. Technician A says to pressurize the EVAP system with dry nitrogen gas and use an ultrasonic leak detector to locate the leak. Technician B says to use a smoke-producing machine to force smoke into the EVAP system and locate the leak source. Who is correct?
    A. Technician A
    B. Technician B
    C. Both Technician A and Technician B
    D. Neither Technician A nor Technician B
11. When a vacuum hand pump is used to supply 20 in. Hg to a positive backpressure-type EGR valve with the engine idling, the valve should ________________ ________________.
12. When a scan tool is used to command one solenoid to open in a digital EGR valve, the engine should ________________ ________________.
13. When 20 in. Hg is supplied from a vacuum hand pump to a diverter valve, the valve should be positioned to deliver air from the pump to the ________________ ________________.
14. When performing an EVAP system leak test, the fuel tank should be between ________________ and ________________ percent full.
15. Describe the operation of a PCV valve in relation to engine speed.
16. Explain the test procedure for a negative backpressure-type EGR valve.
17. Describe two methods of leak testing an EVAP system.
18. Explain the cause of a flashing MIL light on an OBD II vehicle.

# Heating and Air Conditioning Systems



**Interesting Fact** *The popularity of environmentally friendly hybrid vehicles is increasing rapidly in some areas. These vehicles have experienced a 54 percent increase in sales since 1997. In 1997, registrations of hybrid vehicles in the United States totaled 650. From January through October 2002, hybrid vehicle registrations totaled nearly 35,000. Approximately 30 percent of these vehicles are sold in Los Angeles, San Francisco, and Sacramento, and 83 percent of the hybrid vehicles are gas-electric.*

## SECTION OBJECTIVES

After you have read, studied, and practiced the contents of this section, you should be able to:

- Explain the effects of R-12 refrigerant on the earth's ozone layer.
- Describe the harmful effects of excessive ultraviolet radiation on humans.
- Describe the refrigerant oil used with R-134a refrigerant.
- Explain the SAE standards for the purity of R-12 and R-134a recycled refrigerant.
- Describe how the air temperature entering the passenger compartment is controlled in an HVAC system.
- Explain latent heat of vaporization.
- Explain latent heat of condensation.
- Describe the changes in the refrigerant state as it flows through the refrigeration system.
- Explain the basic operation of a compressor clutch.
- Explain how the temperature door is controlled in a manual air conditioning (A/C) system.
- Describe how the temperature door is controlled in a semi-automatic A/C system.
- Explain how the blower speed is controlled in an automatic A/C system.
- Describe the results of excessive compressor clutch clearance.
- In a system with R-12 refrigerant, describe the sight glass appearance that indicates a proper refrigerant charge.
- Describe the procedure to complete a performance test on an A/C system.
- Explain the causes of excessive high-side pressure in a refrigerant system.
- Explain the procedure for refrigerant recovery.
- Describe the procedure for evacuating a refrigeration system.
- Explain the procedure for charging a refrigeration system.

# Chapter 36 Heating and Air Conditioning (HVAC) Systems

## *Introduction*

The temperature control system in a vehicle is called a heating, ventilation, and air conditioning (HVAC) system. During hot weather operation, heat in the vehicle interior comes from several sources, such as heat from the sun radiating on the vehicle glass and painted surfaces. Heat also enters the vehicle interior from the engine and exhaust system, especially the catalytic converter. Heat also comes from the outside air and passengers add to the interior vehicle heat mostly from their breath. Therefore in hot weather, the HVAC system must cool and ventilate the vehicle interior to maintain passenger comfort. In cold weather, the HVAC system must heat the vehicle interior for passenger comfort. In hot or cold weather, the HVAC system provides ventilation by forcing outside air into the vehicle interior. Air from the vehicle interior can be recirculated through the HVAC system back into the interior of the vehicle. Technicians must understand HVAC systems and the maintenance, diagnosis, and service required on these systems.

## HEATING SYSTEM DESIGN

In an HVAC system, the heating system contains a heater core mounted with the air conditioning (A/C) evaporator in the HVAC case. If the vehicle does not have A/C, the evaporator is omitted from this case. Hoses are connected from the heater core to the cooling system to provide coolant flow through the heater core. Some systems have a control valve that regulates the coolant flow through the heater core **(Figure 1)**. The coolant control valve may be operated by vacuum or a mechanical cable. Since the engine thermostat maintains coolant temperature, proper thermostat operation is very important to provide adequate heating of the vehicle interior.

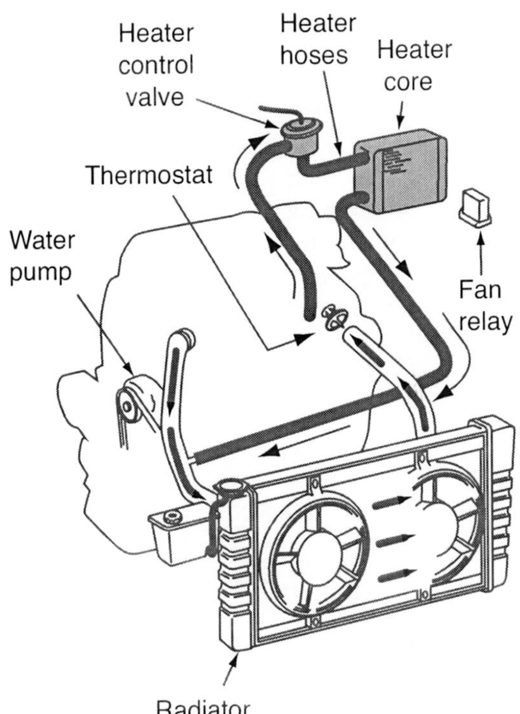


**Figure 1.** A heating system.

The blower motor moves air from the outside into the HVAC case. The position of the temperature door determines the airflow through the evaporator and the heater

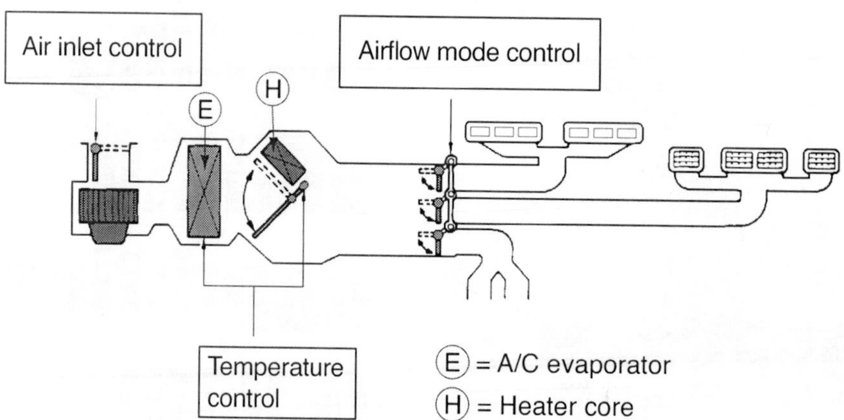


Figure 2. HVAC mode control doors.

core. In a manual HVAC system, the temperature door is operated by a cable connected to the temperature control lever in the HVAC control panel. In most automatic HVAC systems, the temperature door is operated by an electric motor. In the maximum heat position, the temperature door is positioned so all the air flows through the evaporator and then through the heater core. Under this condition, maximum interior heating is provided. In the maximum A/C mode, the temperature door is positioned so it blocks airflow through the heater core but the air flows through the evaporator. During normal HVAC operation, the air flows through the evaporator and the temperature door is positioned so a portion of the air flows through the heater core **(Figure 2)**.

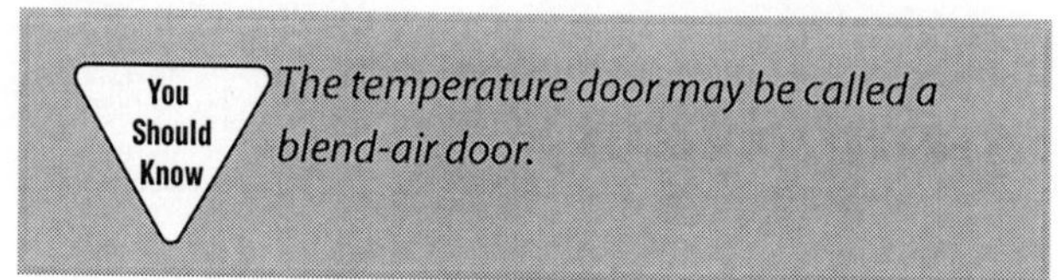


Mode control doors in the HVAC case determine whether the air is directed from this case to the A/C outlets, floor outlets, or defrost outlets. Mode doors are usually operated by vacuum actuators or electric motors. On some manually controlled systems, the mode doors are operated by cables. A recirculation (recirc) door determines whether air flows from the outside into the HVAC case or from the vehicle interior into this case **(Figure 3)**.

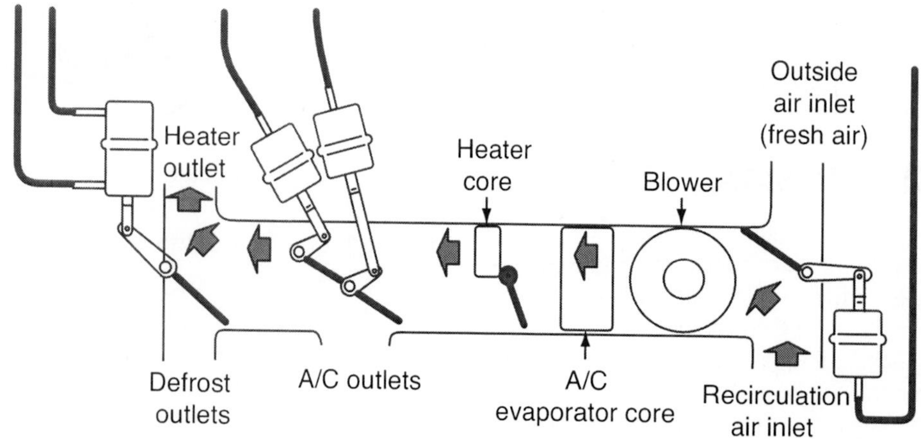


Figure 3. HVAC system recirculation door.

## HEAT ABSORPTION PRINCIPLES

Materials may be in a liquid, solid, or gaseous state. When materials change from one state to another, large amounts of heat may be transferred. If the temperature of water is reduced to 32°F (0°C), the water changes from a liquid to a solid (ice). Ice at 32°F (0°C) requires heat to change it into water, which will also be at 32°F (0°C). When the water temperature is increased to 212°F (100°C), it requires an additional 970 British thermal units (Btus) of heat per pound to make the water boil. This additional heat does not register on a thermometer. This additional heat is called latent heat of vaporization. Cooling may be defined as taking away heat. During evaporation, heat is absorbed.

A **British thermal unit (Btu)** is the amount of heat required to raise the temperature of 1 pound of water 1°F. **Latent heat of vaporization** is the amount of heat required to change a liquid to a gas after the liquid reaches the boiling point.

When a gas condenses and becomes a liquid, it releases its latent heat. When steam condenses into water, 970 Btus per pound are released. The heat released during condensation is called **latent heat of condensation**. The latent heat of vaporization and the latent heat of condensation are principles used in any air conditioning system.

## AIR CONDITIONING REFRIGERANTS

Technicians need to understand the serious effects of R-12 refrigerant on the earth's ozone layer. Technicians must also have a knowledge of alternative refrigerants and the regulations and standards regarding these refrigerants. Refrigerant system retrofitting procedures must also be familiar to service technicians

## R-12 AND THE OZONE LAYER

R-12 was introduced as a refrigerant in 1930. At that time it was considered an ideal refrigerant because it is non-poisonous, easily and cheaply manufactured, and stable. With its apparent advantages and low cost, R-12 helped to make automotive air conditioning a creature comfort in the 20th century. However, scientific tests indicate that we may have a high penalty to pay for the comfort we have enjoyed as a result of automotive air conditioning. R-12 is a chlorofluorocarbon (CFC)-based chemical and scientific tests prove that CFCs are responsible for damaging the earth's protective ozone layer located in the stratosphere 10 to 30 miles above the planet's surface. The ozone layer filters out most of the sun's harmful ultraviolet (UV) rays.

EPA sources indicate that 30 percent of all the CFCs released into the atmosphere come from mobile air conditioners. Some of these CFCs escape into the atmosphere from refrigerant system leaks, but most of the CFCs are released into the atmosphere during air conditioning service. When CFCs are released to the atmosphere, they travel high into the stratosphere where they can linger for 100 years or more.

CFCs destroy the ozone layer through chemical reaction. When influenced by sunlight, a chlorine atom is released from a CFC molecule and reacts with an oxygen atom in the ozone layer to form chlorine monoxide and free oxygen, neither of which can filter out the sun's UV rays. For each one percent of ozone reduction, one and one-half to two percent more UV radiation reaches the earth's surface.

The EPA suggests that the risks associated with exposure to excessive UV radiation include increased incidence of skin cancer, an increase in the number of eye cataracts, and damage to immune systems. EPA studies also indicate that excessive UV radiation damages vegetation, adversely affects organisms living in the oceans, and increases ground-level ozone which is a contributor to smog. Research indicates that excessive UV radiation can be harmful to crops and plankton in the oceans. Plankton is floating or weakly swimming animal and plant life, which some marine animals use for food. Loss of this source of food could have serious effects on marine life.

## R-134A REFRIGERANT

Automotive manufacturers began to install R-134a, an ozone-friendly refrigerant, in new vehicles beginning in 1992. R-134a does not contain CFCs. By 1994, the automotive manufacturers completed the transition to R-134a refrigerant on all new vehicle production. The amendments to the Clean Air Act effective January 1, 1992, required certification for technicians servicing automotive air conditioners. Certification was provided through the ASE. The Clean Air Act amendments also demanded use of approved refrigerant recycling equipment. Mandatory recovery of R-134a refrigerant was required after November 15, 1995. After this date, venting of R-134a became illegal. Although R-134a does not contain CFCs, it is a greenhouse gas. Recycling of R-134a using EPA-approved equipment became mandatory on January 29, 1998.

## ALTERNATE REFRIGERANTS

Although R-134a refrigerant was chosen by the vehicle manufacturers and regulating authorities as the replacement for R-12 refrigerant, the EPA has approved other refrigerants under the Significant New Alternatives Policy (SNAP). EPA publishes a chart listing other alternative refrigerants in place of R-12. Refrigerants listed on this chart are identified as acceptable substitutes (ASU) or unacceptable (UNA) replacements for R-12. Some refrigerants are UNA replacements for R-12 because they contain a flammable blend of hydrocarbons. Flammable refrigerants are illegal in many states.

Many of the refrigerants approved by the EPA as R-12 replacements are a blend of several different refrigerants. For example, FR-12 contains 55 percent HCFC-22, 41 percent HCFC-142b, and 4 percent isobutane. Some equipment manufacturers supply refrigerant identifiers that inform the technician regarding the type of refrigerant in the system. Because of the variety of refrigerants, including flammable products, a refrigerant identifier is a very useful piece of equipment.

## REFRIGERANT OILS

Mineral oils are used in R-12 refrigerant systems, but mineral oil is not compatible with R-134a refrigerant. Two new synthetic lubricants, polyalkylene glycol (PAG) oil and polyol ester (ester) oil, have been developed for R-134a refrigerant. A **synthetic lubricant** is one that is developed in a laboratory.

When selecting a refrigerant oil for an R-134a refrigerant system, always follow the vehicle manufacturer's or compressor manufacturer's instructions. New vehicles usually have PAG oil in the R-134a refrigerant. Installing the proper refrigerant oil has a direct bearing on system performance and compressor life.

You Should Know *PAG and ester oils are hygroscopic, which means they easily absorb moisture from the atmosphere. Always keep oil containers tightly sealed and in a dry place.*

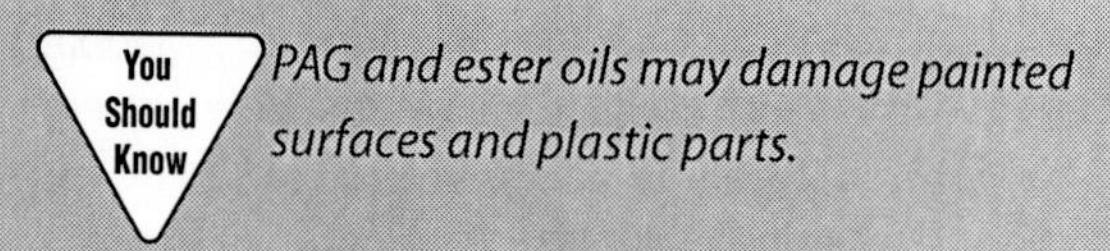

You Should Know *PAG and ester oils may damage painted surfaces and plastic parts.*

## REFRIGERANT RECOVERY AND RECYCLING STANDARDS

The SAE has drafted a number of standards that apply to the recovery and recycling of refrigerant. SAE standard J1990 applies to the recovery and recycling of R-12 and standard J2210 regulates the recovery of R-134a. These standards provide specifications for the equipment hardware such as hoses and fittings. Service hoses must have shut-off valves within 12 inches of the hose ends to prevent the unnecessary release of refrigerant. The SAE standard for purity of R-12 requires a limit of 15 parts per million (ppm) moisture by weight, 4,000 ppm refrigerant oil by weight, and 150 ppm non-condensable gases (air) by weight. These limits are the same for R-134a except the limit for refrigerant oil is 500 ppm by weight. Recovery/recycling equipment must be tested by an independent standards organization to be sure these standards are met. SAE standards are also established for R-12 and R-134a recovery-only machines. This type of machine cannot recycle refrigerant or recharge a system. The recovery equipment is used in operations such as salvage yards. Recovered refrigerant must be shipped to an off-site facility where the refrigerant is reclaimed not recycled. Reclaimed refrigerant must meet a like-new purity standard referred to as AR1700-93. This standard is established by the Air Conditioning and Refrigeration Institute (ARI).

On R-12 recovery/recycling machines, SAE standards require the low-side hose to be solid blue or black with a blue stripe. The high-side hose is solid red or black with a red stripe. The utility hose is solid yellow or white, or black with a yellow or white stripe. On recovery/recycling machines for R-134a, the low-side hose is solid blue with a black stripe, the high-side hose is solid red with a black stripe, and the utility hose is yellow with a black stripe.

To reduce the possibility of mixing R-12 and R-134a refrigerant while servicing refrigerant systems, R-12 has 7/16 inch—20 fittings for connection to manifold gauges or recovery/recycling equipment, whereas R-134a systems must have 0.5 inch—16 ACME fittings at these locations.

## REFRIGERATION SYSTEM OPERATION AND CONTROLS

When the compressor is running, it pressurizes gaseous refrigerant. A tube conducts the refrigerant to the condenser. As the refrigerant flows down through the condenser, the airflow through the condenser cools the refrigerant and reduces the pressure. This causes the refrigerant to condense into a liquid. The refrigerant transfers a large amount of heat to the air flowing through the condenser during the condensation process. The liquid refrigerant flows through the tube connected from the condenser to the receiver/drier. The receiver/drier filters and removes moisture from the refrigerant. Some refrigerant systems have an accumulator between the evaporator and the compressor in place of a receiver/drier. The refrigerant flows through a tube connected between the receiver/drier and the evaporator. A thermostatic expansion valve or a fixed orifice tube (FOT) is located in the evaporator inlet tube. The expansion valve or the FOT provides a restriction for the refrigerant flow. This restriction reduces the refrigerant pressure. The liquid refrigerant immediately boils into a vapor when the pressure is reduced. When this boiling action takes place, the refrigerant absorbs a large amount of heat from the air flowing through the evaporator into the vehicle interior and this provides cooling for the vehicle interior. A refrigeration system with a thermostatic expansion valve (TXV) is illustrated in **Figure 4**, and

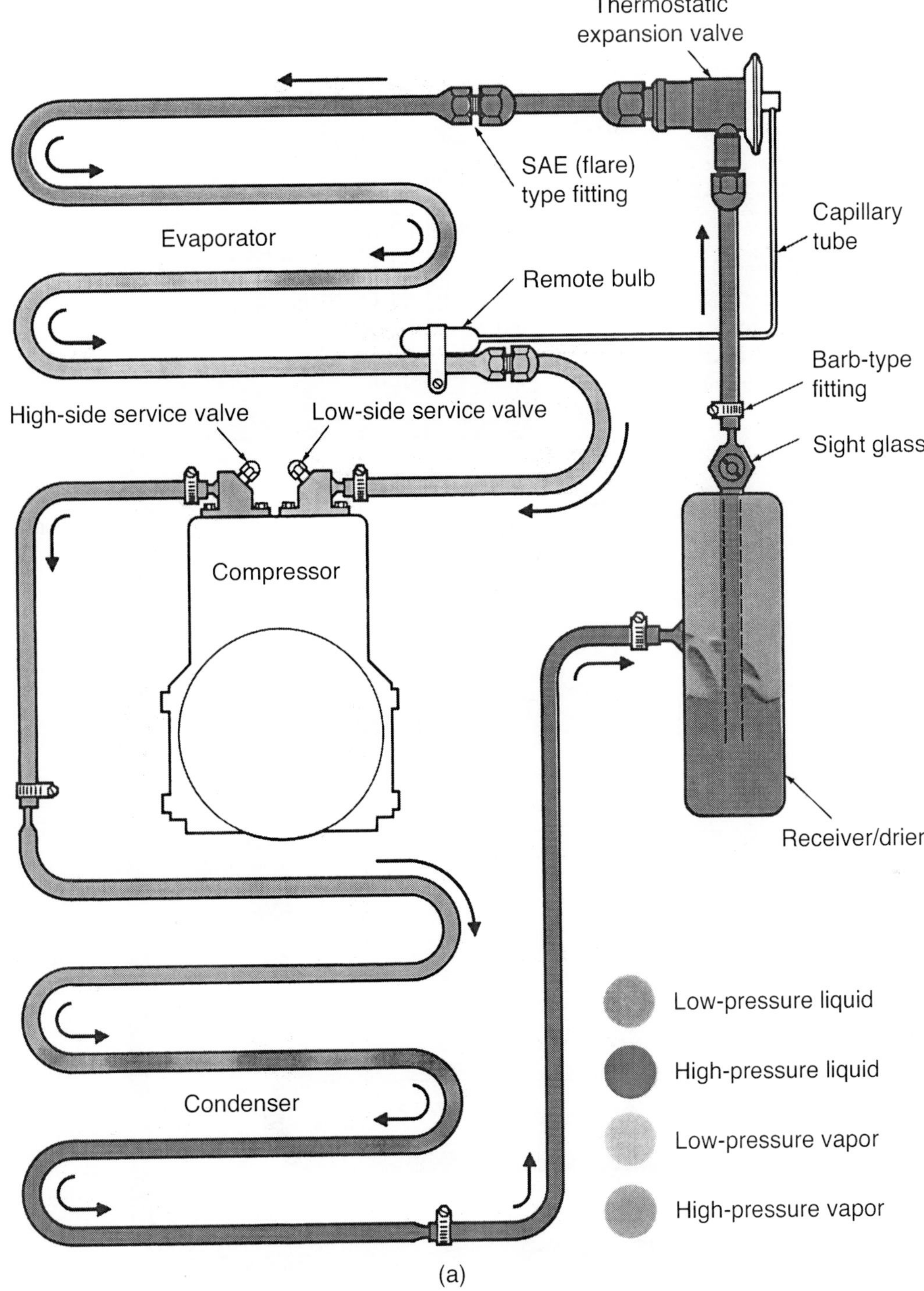


**Figure 4.** Refrigeration system with a thermostatic expansion valve.

**Figure 5** shows a refrigeration system with an orifice tube. Notice that the orifice tube system has an accumulator, whereas the TXV system has a receiver/drier. The refrigeration system is similar in any air conditioning system.

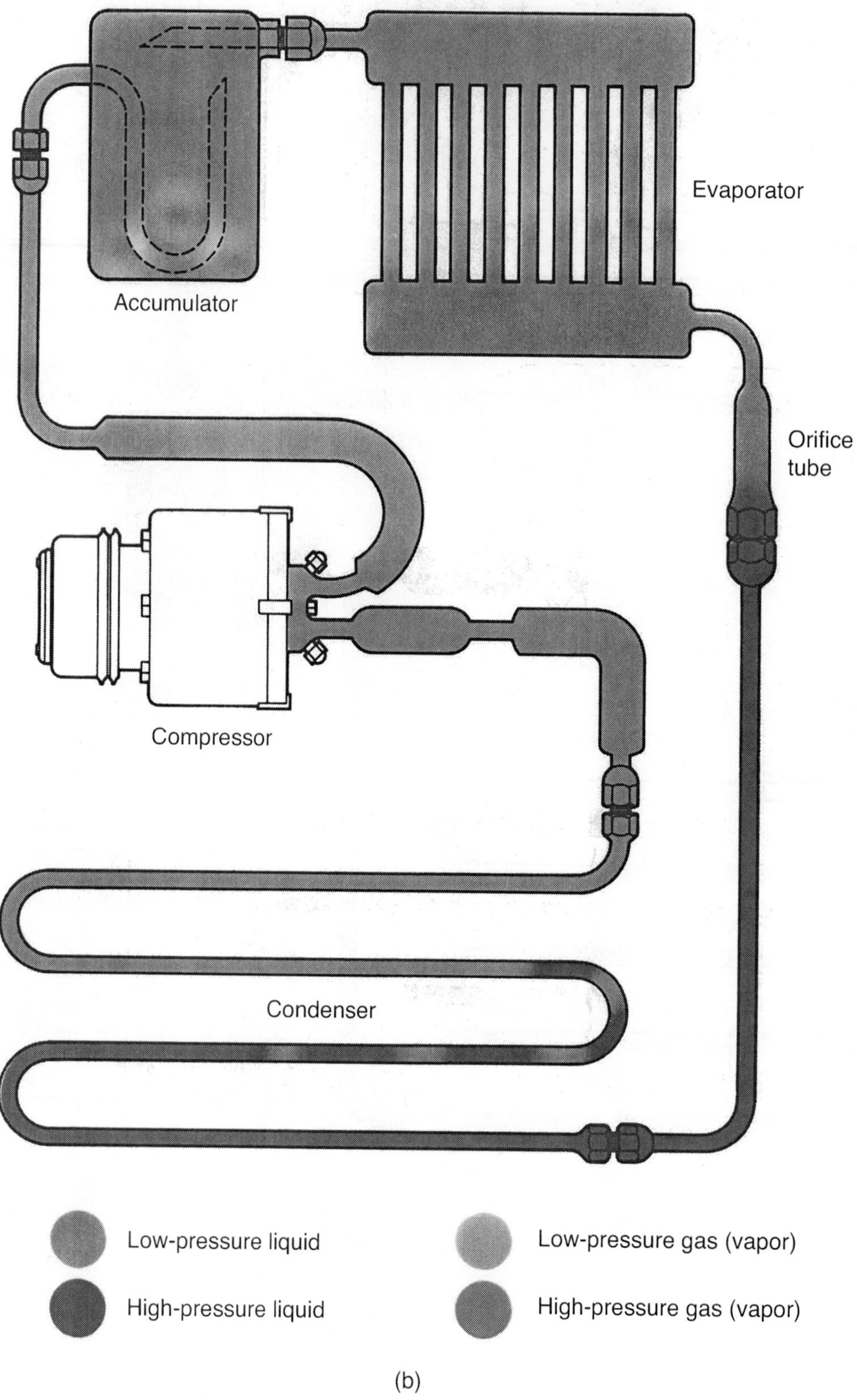


Figure 5. Refrigeration system with an orifice tube.

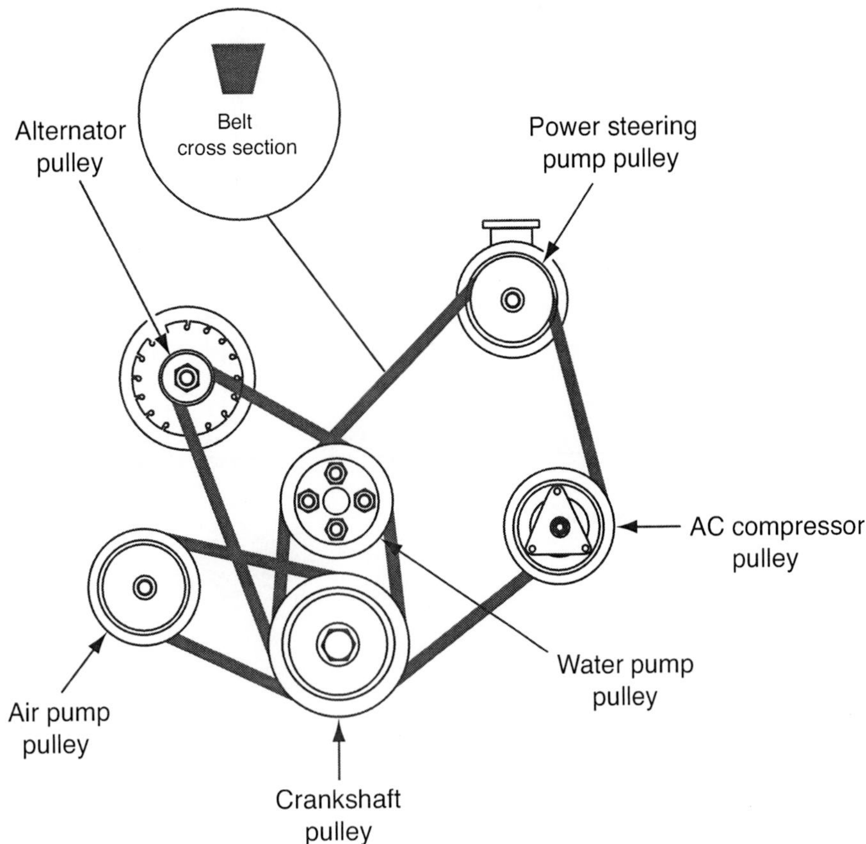


**Figure 6.** A/C compressor drive belt.

The A/C compressor is driven from the crankshaft by a V-belt or a ribbed V-belt **(Figure 6)**. Drive belt condition and tension are very important to provide proper compressor operation. There are several different types of compressors, but two common types are the piston compressor and the axial-plate compressor. In a piston compressor, a crankshaft moves a piston up and down in the compressor cylinder. Inlet and outlet reed valves are located in the cylinder head. When the piston moves down, vacuum in the cylinder moves refrigerant through the inlet valve into the cylinder **(Figure 7)**. As the piston moves upward in the cylinder, the refrigerant is pressurized and forced out

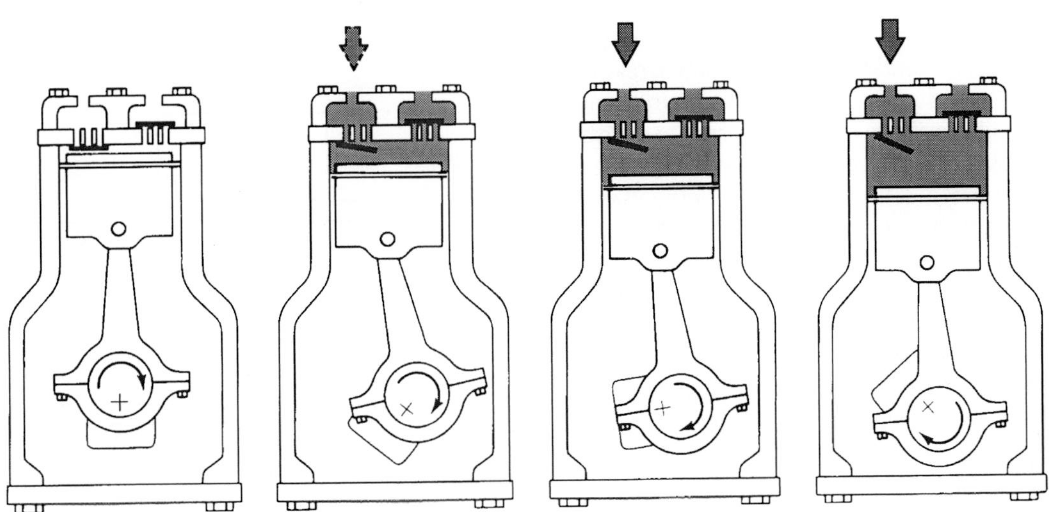

**Figure 7.** A/C compressor intake stroke.

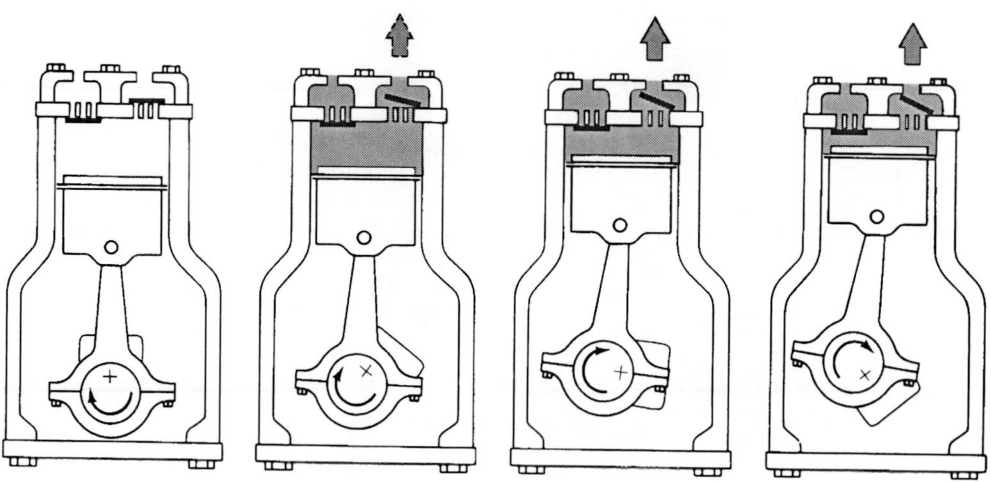

Figure 8. A/C compressor compression stroke.

through the outlet valve **(Figure 8)**. The axial-plate compressor is a different type of piston compressor. The pistons are mounted horizontally in the compressor body. These pistons have a slotted center area that fits over the axial plate attached to the compressor drive shaft. As the drive shaft rotates the axial plate, this plate moves the pistons back and forth in their horizontal cylinders **(Figure 9)**. Reed-type inlet and outlet valves are located in the ends of the cylinders.

A compressor clutch on the front of the compressor shaft is used to engage and disengage the compressor. The clutch contains a large field coil that is pressed onto the front compressor housing, and retained with a snap ring. The compressor clutch pulley is mounted on a double-row bearing that is also pressed onto the front compressor housing. The pulley is also retained with a snap ring. A clutch hub and drive plate is mounted on the compressor drive shaft and a key prevents this hub from turning on the shaft. A locknut retains the hub and drive plate to the shaft **(Figure 10)**. The drive plate on the hub is mounted on leaf springs so this plate can move toward the pulley. If the compressor is not engaged, the clutch is not energized and the drive belt rotates the pulley on the bearing. When the clutch is engaged, voltage is supplied to the clutch coil and current flow through this coil creates a strong magnetic field. This magnetic field attracts the drive plate toward the pulley surface and holds this plate firmly against the pulley. Under this condition, the drive plate, hub, and compressor drive shaft must rotate with the pulley. The clearance between the drive plate and the pulley is adjustable by removing or installing shims between the inner end of the hub and the compressor shaft.

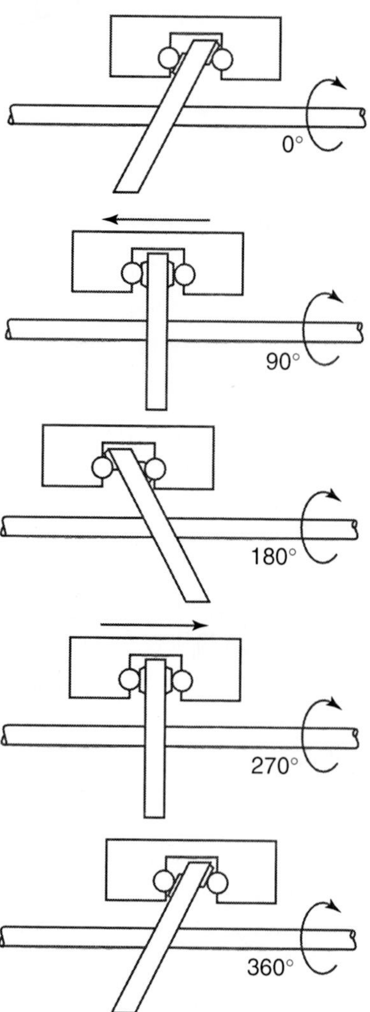


Figure 9. Piston movement in an axial-plate compressor.

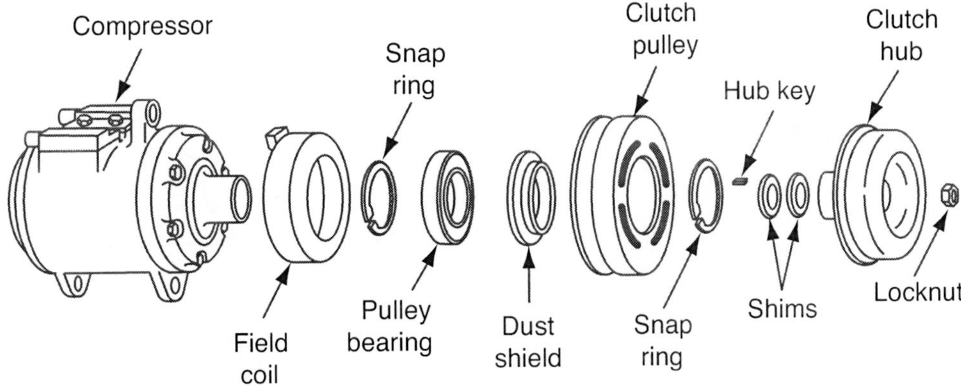


**Figure 10.** A/C compressor clutch components.

In an orifice tube system, the clutch is usually cycled on and off by a pressure switch in the accumulator **(Figure 11)**.

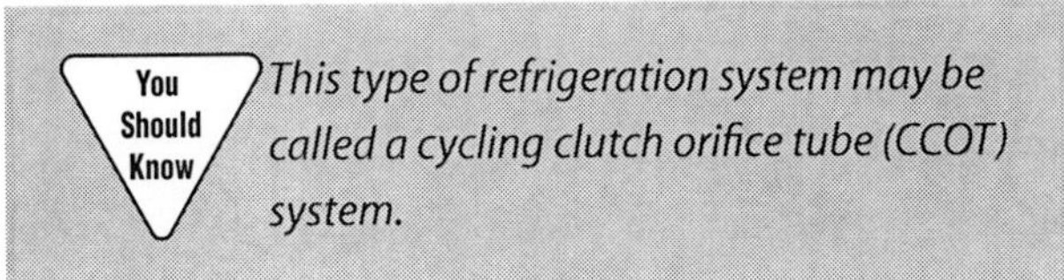


*This type of refrigeration system may be called a cycling clutch orifice tube (CCOT) system.*

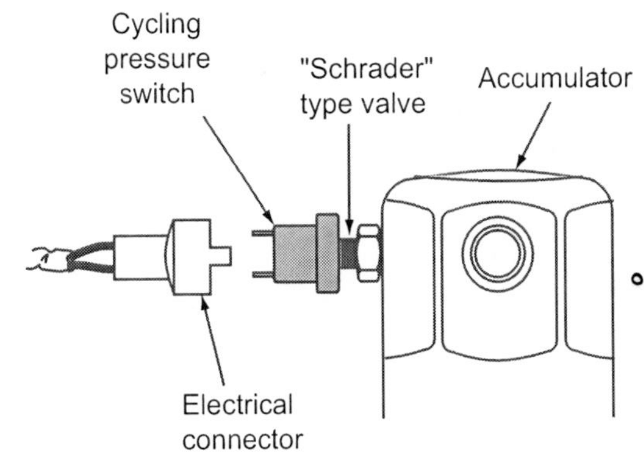


**Figure 11.** Accumulator and cycling switch.

The normally-closed cycling switch cycles the compressor on and off to provide adequate passenger compartment cooling without excessive ice formation in the evaporator. In a TXV system, the compressor clutch is cycled on and off by a thermostatic switch. A capillary tube is connected from the thermostatic switch so it contacts the suction line between the evaporator and the compressor **(Figure 12)**. The thermostatic switch cycles the compressor on and off in relation to evaporator outlet temperature. This switch performs the same function as the cycling switch in

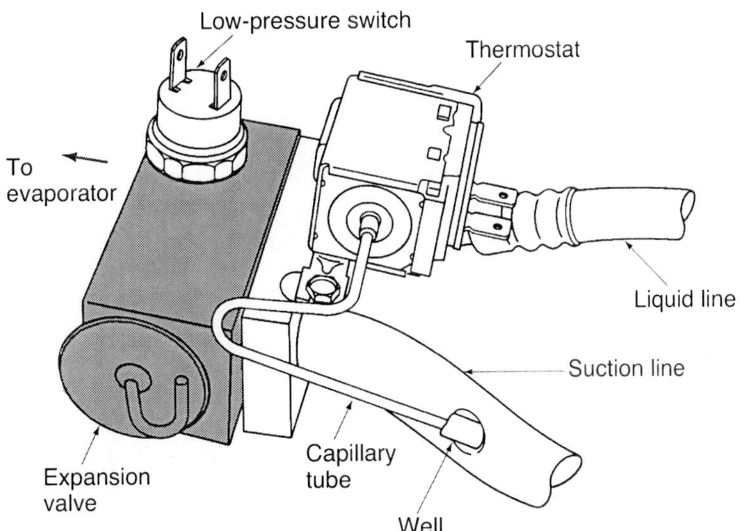


**Figure 12.** Thermostatic expansion valve with thermostatic switch.

the accumulator. A low-pressure switch in the TXV shuts the compressor off if the refrigerant pressure becomes excessively low because of a refrigerant leak.

## MANUAL A/C SYSTEMS

In a manual A/C system, a cable is connected from the temperature lever on the A/C control panel to the temperature door. In some older systems, cables are connected from the other control panel levers to the mode control doors on the A/C heater case **(Figure 13)**. In other manual A/C systems, the push buttons in the A/C control panel operate vacuum switches that shut the vacuum on and off to the mode door actuators. The blower speed control switch switches different resistors into the blower circuit to provide the desired blower speed. When high blower speed is selected, the blower switch supplies voltage to the high blower relay winding. This action closes the relay contacts and 12 volts are supplied through the relay contacts to the blower motor **(Figure 14)**.

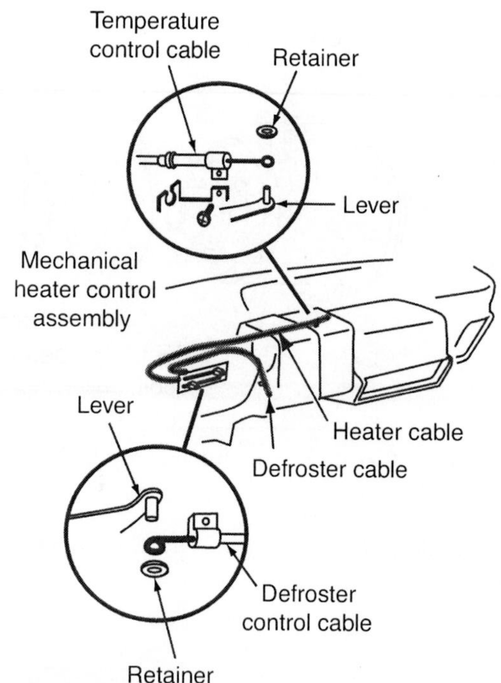


**Figure 13.** Manual A/C system control cables.

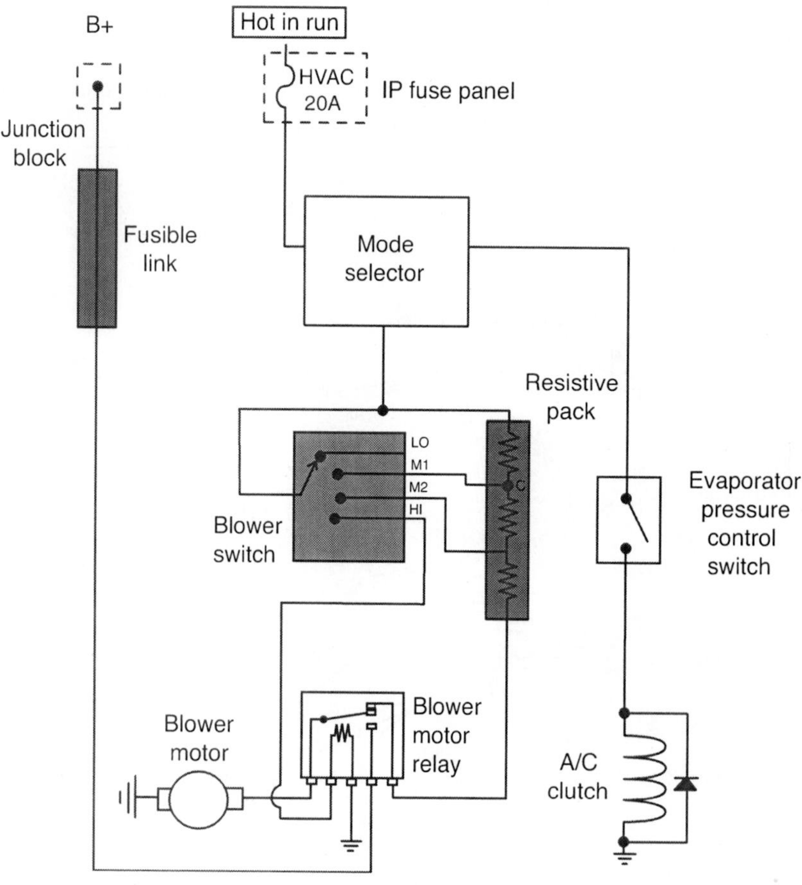


**Figure 14.** Blower motor circuit.

## SEMI-AUTOMATIC A/C SYSTEMS

In a semi-automatic A/C system, the temperature control lever operates a variable resistor. A servomotor rotates the temperature door and an electronic module is located in the servomotor. An in-car sensor is mounted in the instrument panel and an outside air temperature sensor is located on the outside surface of the A/C heater case. If the temperature lever is moved to a higher temperature, the resistance of the variable resistor increases. This variable resistor and the sensors send voltage signals to the module in the temperature door servomotor. In response to these signals, the servomotor operates the temperature door to provide the temperature selected by the driver. The other control functions in a semi-automatic A/C system are similar to those on a manual system.

## AUTOMATIC A/C SYSTEMS

In an automatic A/C systems, all the functions are operated by the A/C computer. The temperature door is operated by an electric motor that is controlled by the A/C computer. The A/C computer may be mounted in the A/C control panel **(Figure 15)**, or this computer may be mounted separately. In some automatic A/C systems, the mode doors are operated by electric motors controlled by the A/C computer. In other automatic systems, the A/C computer operates vacuum solenoids that control the vacuum supplied to the mode doors. The blower motor is usually controlled by a blower speed controller, but the A/C computer commands the blower speed controller to provide the proper blower speed. On many automatic A/C systems, the blower speed controller controls the blower speed with a pulse width modulated (PWM) signal. A **pulse width modulated signal** is an on off signal with varying on and off times.

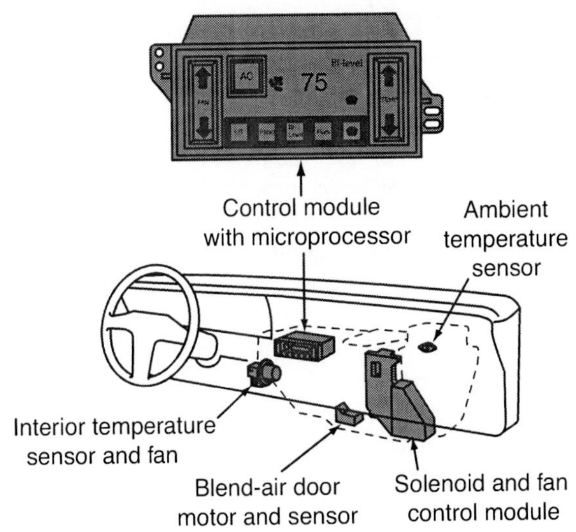


**Figure 15.** A/C control panel and computer.

The A/C compressor clutch is also operated by the A/C computer. The A/C computer receives input signals from the sunload sensor, in-car sensor, ambient sensor, engine coolant temperature sensor, driver input, and A/C pressure switch. In response to these input signals, the A/C computer controls the compressor clutch, temperature door, blower speed, and mode doors **(Figure 16)**.

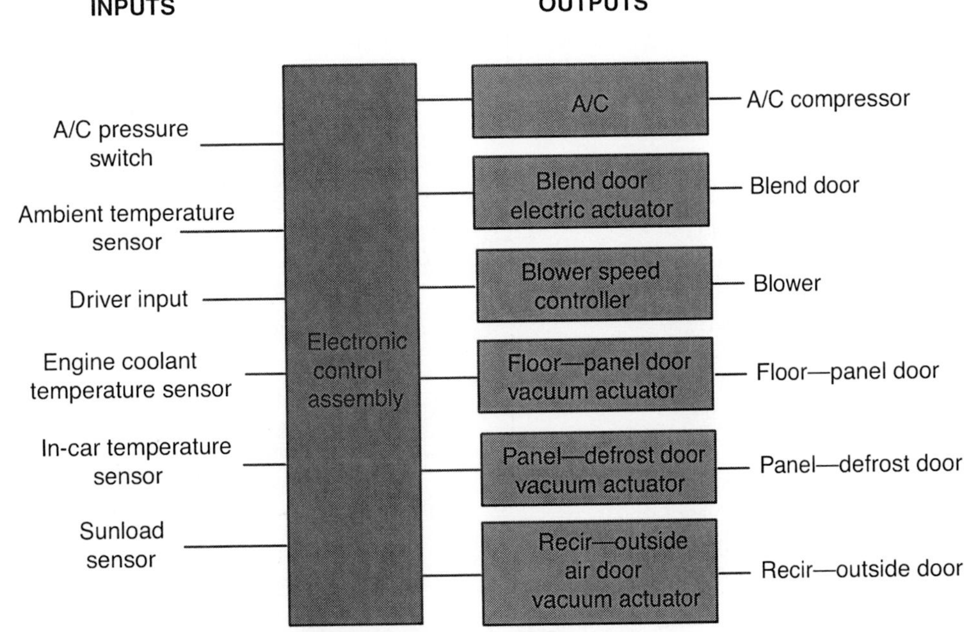


**Figure 16.** A/C computer inputs and outputs.

## Summary

- The heater coolant control valve is usually operated by a cable or by a vacuum actuator.
- The position of the temperature door determines whether inlet air flows through the heater core or the evaporator.
- The mode control doors direct the inlet air flow through the floor outlets, A/C outlets, or defrost outlets.
- The mode control doors are usually operated by vacuum actuators or electric motors.
- Latent heat of vaporization is the heat required to boil water after it reaches the boiling point.
- Latent heat of condensation is the heat released when a vapor condenses into a liquid.
- R-12 refrigerant contains chlorofluorocarbons (CFCs) that are responsible for damaging the earth's ozone layer.
- Damage to the earth's ozone layer causes excessive UV radiation to reach the earth's surface.
- Excessive UV radiation contributes to skin cancer, eye cataracts, damage to crops, and to some forms of marine life.
- R-12 refrigerant has been replaced by R-134a on all new vehicles since 1994.
- Venting of any refrigerant to the atmosphere is illegal.
- Refrigerant recovery and recycling is now mandatory.
- Refrigerant identifiers are available to inform the technician regarding the type of refrigerant in a system.
- Flammable refrigerants are illegal in many states.
- Mineral oil is used in R-12 refrigerant, but R-134a refrigerant requires PAG oil or polyol ester (ester) oil.
- The SAE has drafted standards for refrigerant recovery/recycling machines and the purity of recovered refrigerant.
- R-134a refrigerant systems have different service fittings compared to R-12 refrigerant systems.
- The compressor pressurizes the refrigerant and forces it through the system.
- The refrigerant changes from a vapor to a liquid in the condenser.
- An orifice tube or thermostatic expansion valve is mounted at the evaporator inlet.
- When refrigerant flows through the orifice tube or thermostatic expansion valve, the pressure of the refrigerant is reduced and the refrigerant immediately boils and changes from a liquid to a gas.
- The compressor clutch connects and disconnects the pulley from the compressor drive shaft.
- In a manual A/C system, the temperature door is operated by a cable connected to the temperature lever.
- In a semi-automatic A/C system, the temperature door is operated by an electric motor that is controlled by a module. This module receives inputs from the variable resistor connected to the temperature lever, the in-car sensor, and the outside air temperature sensor.
- In an automatic A/C system, the temperature door, mode doors, blower speed, and compressor clutch are operated by the A/C computer.

## Review Questions

1. Technician A says it is legal to vent R-134a to the atmosphere. Technician B says R-134a refrigerant is a greenhouse gas. Who is correct?
   A. Technician A
   B. Technician B
   C. Both Technician A and Technician B
   D. Neither Technician A nor Technician B
2. Technician A says mineral oil may be used with R-134a refrigerant. Technician B says flammable refrigerants are illegal in many states. Who is correct?
   A. Technician A
   B. Technician B
   C. Both Technician A and Technician B
   D. Neither Technician A nor Technician B
3. Technician A says the temperature door controls the airflow through the heater core and evaporator. Technician B says the temperature door may be operated by a cable. Who is correct?
   A. Technician A
   B. Technician B
   C. Both Technician A and Technician B
   D. Neither Technician A nor Technician B
4. All of these statements about A/C system operation are true *except*:
   A. In the MAX A/C mode, the temperature door blocks air flow through the heater core.
   B. In the MAX A/C mode, the temperature door allows maximum airflow through the evaporator.
   C. The coolant control valve is open in the MAX A/C mode.
   D. If the recirculation (RECIRC) door is open, air is moved from the vehicle interior into the HVAC case.
5. When refrigerant flows through the condenser:
   A. the refrigerant is cooled and the refrigerant pressure increases.
   B. the refrigerant condenses from a gas to a liquid.
   C. the refrigerant pressure decreases and the refrigerant temperature increases.
   D. the refrigerant absorbs heat from the airflow through the condenser.

6. In a refrigeration system with a TXV, the receiver/drier is connected:
   A. between the condenser and the evaporator.
   B. between the compressor and the condenser.
   C. between the evaporator and the compressor.
   D. at the evaporator inlet.
7. The clearance between the compressor pulley drive plate and the pulley may be adjusted by:
   A. loosening or tightening the pulley retaining nut.
   B. removing or installing shims between the inner end of the pulley hub and the compressor shaft.
   C. installing a pulley with a different pulley hub length.
   D. installing shims between the pulley retaining nut and the pulley.
8. In a refrigeration system with an orifice tube, the compressor clutch is cycled on and off by:
   A. a high-pressure cut-off switch in the compressor.
   B. a low-pressure cut-off switch in the compressor.
   C. a pressure cycling switch in the accumulator.
   D. a dual contact switch in the evaporator inlet.
9. In a fully automatic A/C system, a blower speed controller controls the blower speed using a:
   A. pulse width modulated voltage signal.
   B. a low voltage DC signal.
   C. a high voltage AC signal.
   D. a low digital voltage signal.
10. Technician A says when the A/C compressor clutch is engaged, the drive plate is held against the pulley surface. Technician B says current flow through the compressor clutch coil engages the compressor clutch. Who is correct?
    A. Technician A
    B. Technician B
    C. Both Technician A and Technician B
    D. Neither Technician A nor Technician B
11. In an orifice tube refrigeration system, the compressor clutch is cycled on and off by the __________ __________.
12. In a TXV refrigeration system, a capillary tube is connected from the thermostatic switch to the __________ __________.
13. If the pressure becomes excessively low in a TXV refrigeration system, the compressor is shut off by __________ __________ __________.
14. In a semi-automatic A/C system, the temperature lever is connected to a __________ __________.
15. Explain how the blower speed is controlled in an automatic A/C system.
16. Describe the temperature door position in the maximum A/C mode.
17. Explain the latent heat of vaporization.
18. Describe the changes that occur in the refrigerant as it flows through an orifice tube refrigerant system.

# Heating and Air Conditioning Maintenance, Diagnosis, and Service

## *Introduction*

Proper heating and air conditioning system maintenance, diagnosis, and service is extremely important to maintain passenger comfort. If the A/C system is properly maintained and serviced, it will provide the interior vehicle temperature desired by the driver and passengers. When the A/C system is not properly maintained and serviced, it may suddenly quit working and this causes discomfort and frustration for the driver and passengers. Technicians must understand the proper heater and A/C maintenance, diagnosis, and service procedures to properly maintain these systems and provide the system dependability expected by the customer.

### HEATING SYSTEM MAINTENANCE, DIAGNOSIS, AND SERVICE

When performing minor underhood service, the heater hoses should be inspected for unsatisfactory conditions such as cracks, soft spots, collapsed areas, loose clamps, and leaks. Loose heater hose clamps may be tightened if the hose is in satisfactory condition. If any of the other unsatisfactory hose conditions are present, replace the hoses. Inspect the area on the front floor mat directly under the HVAC case for any evidence of coolant. If the heater core is leaking, the coolant usually drips out of the HVAC case onto the front floor mat. When the heater core is leaking, some of the coolant must be drained from the cooling system. In many vehicles, the HVAC case must be removed from under the dash to access the heater core. The heater core may be repaired in a radiator repair shop or it may be replaced.

If the customer complains about lack of heat in the vehicle interior, be sure the engine thermostat is operating properly. With the engine at normal operating temperature, use a digital multimeter with a temperature probe to check the temperature of the upper radiator hose. This hose temperature should be near the temperature rating of the thermostat. If the engine coolant temperature is normal, touch each heater hose momentarily. Both of these hoses should be hot. If one heater hose is hot and the other heater hose is cool, the heater coolant control valve is closed or the heater core is partially plugged.

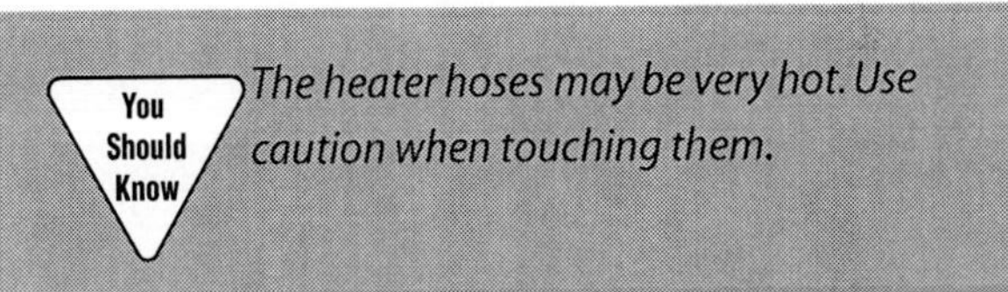


*The heater hoses may be very hot. Use caution when touching them.*

The heater core may be back-flushed in the vehicle or removed and flushed out in a radiator repair shop. To back-flush the heater core in the vehicle, drain some of the coolant from the cooling system and remove the heater hoses from the core. Connect a water hose to the heater core outlet and connect a length of hose from the inlet into a coolant drain pan. Turn the water pressure on the water hose on to back-flush the heater core. Continue back-flushing the heater core until the water flowing into the drain pan appears clear.

### AIR CONDITIONING (A/C) SYSTEM MAINTENANCE

If the compressor drive belt is slipping, the compressor does not produce sufficient refrigerant pressure and interior

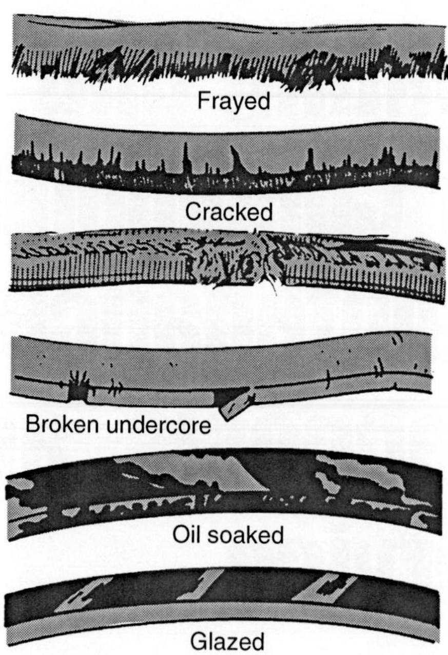


**Figure 1.** Defective compressor belt conditions.

vehicle cooling may be inadequate. Inspect the compressor drive belt for cracks, oil contamination, splits, missing chunks, and wear **(Figure 1)**. If any of these conditions are present, replace the belt.

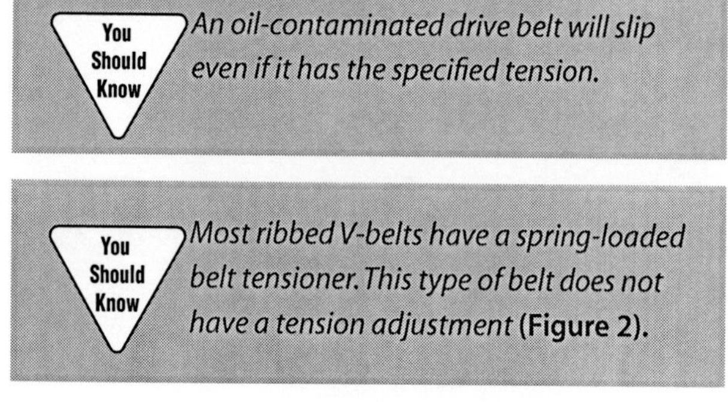


*An oil-contaminated drive belt will slip even if it has the specified tension.*

*Most ribbed V-belts have a spring-loaded belt tensioner. This type of belt does not have a tension adjustment* **(Figure 2)**.

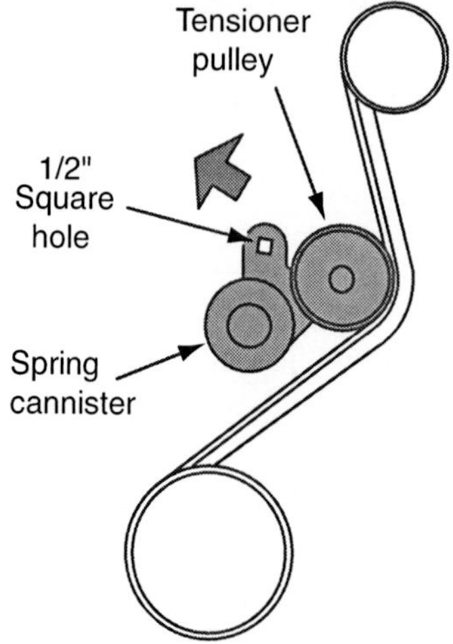


**Figure 2.** A spring-loaded belt tensioner on a ribbed V-belt.

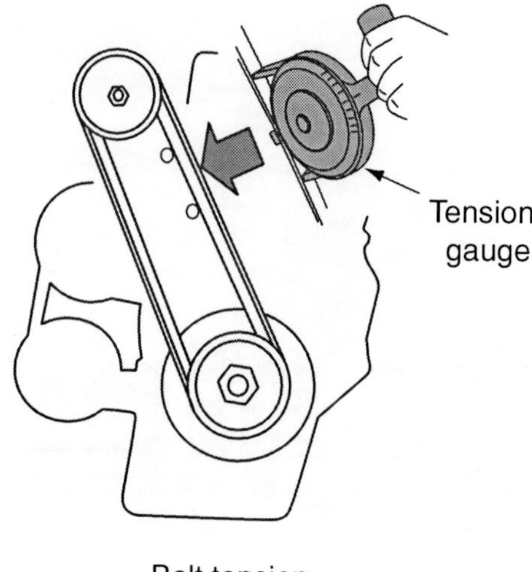


**Figure 3.** Measuring drive belt tension.

Use a belt tension gauge to measure the belt tension **(Figure 3)**. Adjust the belt tension if it is less than specified.

If the compressor clutch clearance is excessive, the clutch may slip and provide a scraping noise immediately after the clutch engages. Use a feeler gauge to measure the clearance between the clutch drive plate and the pulley surface **(Figure 4)**. If the clearance is not within specifications, shims between the clutch hub and the compressor drive shaft may be removed or installed to adjust the clutch clearance **(Figure 5)**.

Listen to the compressor with the engine running and the clutch engaged and disengaged. If a growling noise is heard with only the clutch engaged, there is a defective bearing in the compressor. When a growling noise occurs with the clutch disengaged or engaged, the pulley bearing is defective.

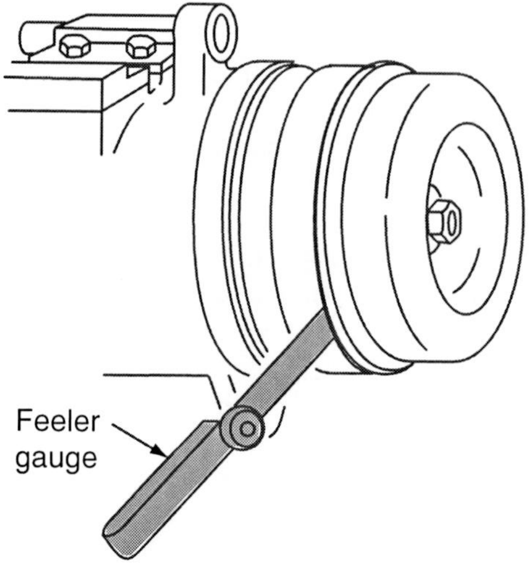


**Figure 4.** Measuring compressor clutch clearance.

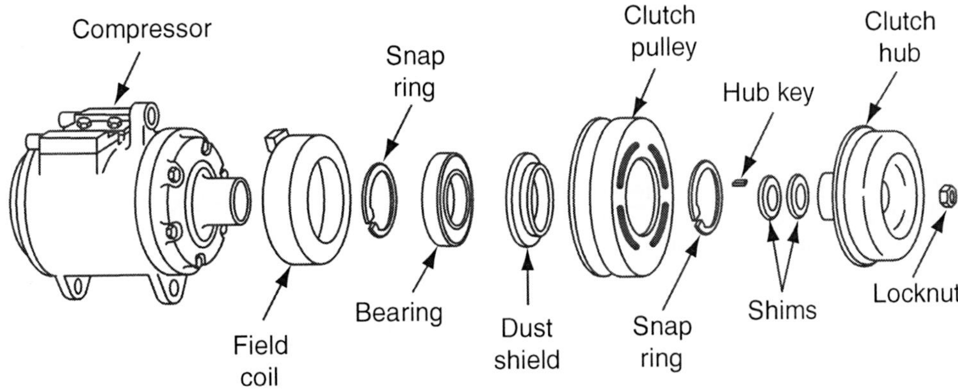


**Figure 5.** Shims between the clutch hub and the compressor drive shaft are used to adjust clutch drive plate clearance.

Inspect the refrigeration system for damaged tubing and components. Check the condenser for bugs or debris that restrict airflow through the condenser. If the condenser air passages are restricted, the refrigerant may not condense completely from a vapor to a liquid as it passes through the condenser. This reduces the cooling ability of the A/C system. A water hose may be used to wash bugs and debris from the condenser air passages. Inspect the condenser fins to be sure they are not bent so they restrict the airflow through the condenser. With the A/C system in operation, check for frosted components or lines. Frost on a refrigeration system line or component usually indicates an internal restriction.

Inspect the refrigeration system for leaks indicated by an oil smudge in the leak area. Use an electronic leak detector to check for leaks in the refrigeration system tubing and components. If all the refrigerant has escaped through a system leak, a partial refrigerant charge may be installed to locate the leak.

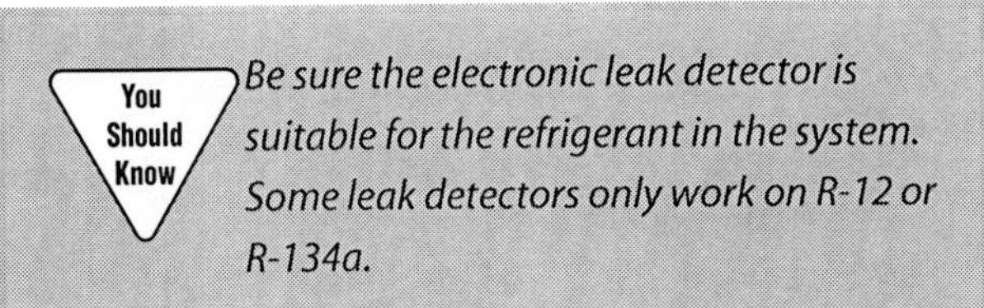


*Be sure the electronic leak detector is suitable for the refrigerant in the system. Some leak detectors only work on R-12 or R-134a.*

## AIR CONDITIONING (A/C) SYSTEM DIAGNOSIS

If the refrigeration system has a sight glass in the top of the receiver/drier, observe the sight glass with the engine at normal operating temperature and the atmospheric temperature above 70°F (21°C) **(Figure 6)**. On an R-12 system, the sight glass should appear clear. Bubbles in the sight glass may indicate the refrigerant charge in the system is low. Oil streaks in the sight glass indicate the refrigeration system is empty. If the sight glass appears cloudy, the desiccant in the receiver/drier is contaminated **(Figure 7)**.

**Figure 6.** A refrigeration system sight glass.

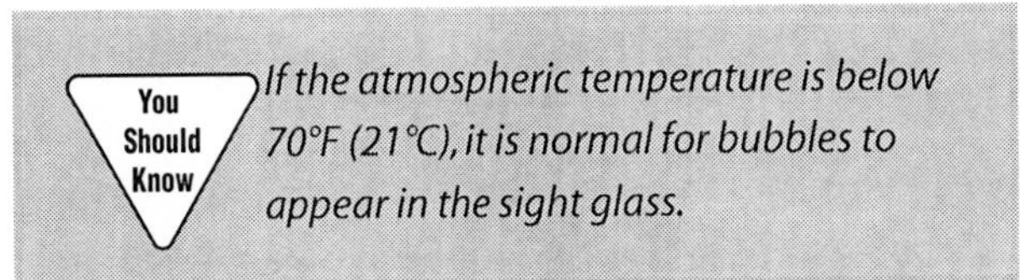


*If the atmospheric temperature is below 70°F (21°C), it is normal for bubbles to appear in the sight glass.*

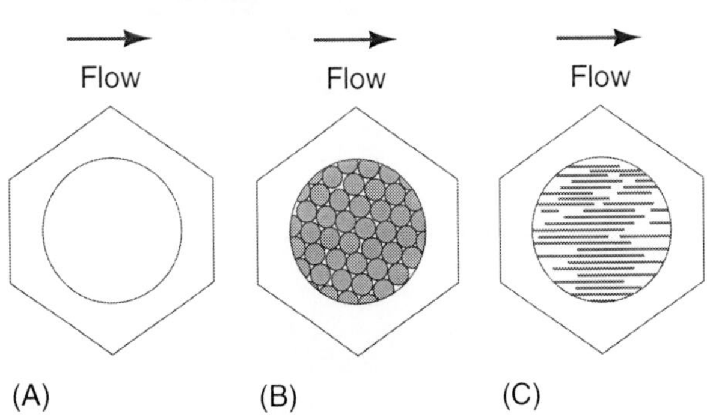


**Figure 7.** Sight glass indications in R-12 systems. (A) clear, (B) bubbles, (C) oil streaks.

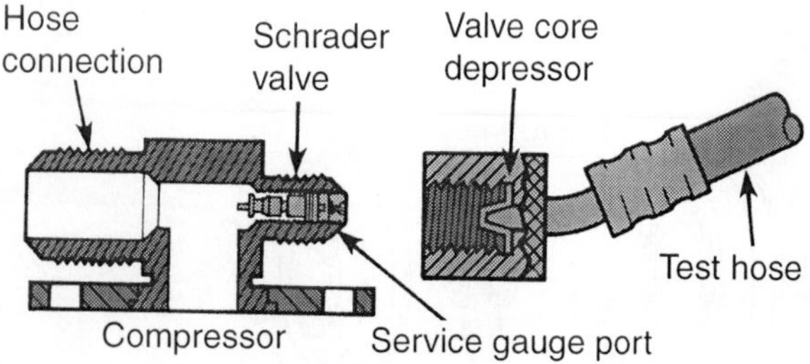


**Figure 8.** A Schrader-type service valve.

Refrigeration systems with an accumulator between the evaporator and the compressor do not have a sight glass because these systems tend to have bubbles in the liquid line from the condenser to the evaporator. R-134a refrigeration systems do not commonly have a sight glass.

A manifold gauge set is connected to the service ports in the refrigeration system to read system pressures. The pressure on the low side and high side of the refrigeration system are very useful when diagnosing system problems. Most R-12 systems have Schrader-type service valves **(Figure 8)**. Some older R-12 systems have service valves with adjustable stems **(Figure 9)**. This type of service valve is backseated during normal operation of the refrigeration system. When these service valves are midseated, refrigerant pressure is supplied to the port where the manifold gauge is connected. In the front-seated service valve position, the compressor is isolated from the rest of the system so the compressor may be removed without discharging the complete system. R-134a refrigeration systems have quick disconnect service valves.

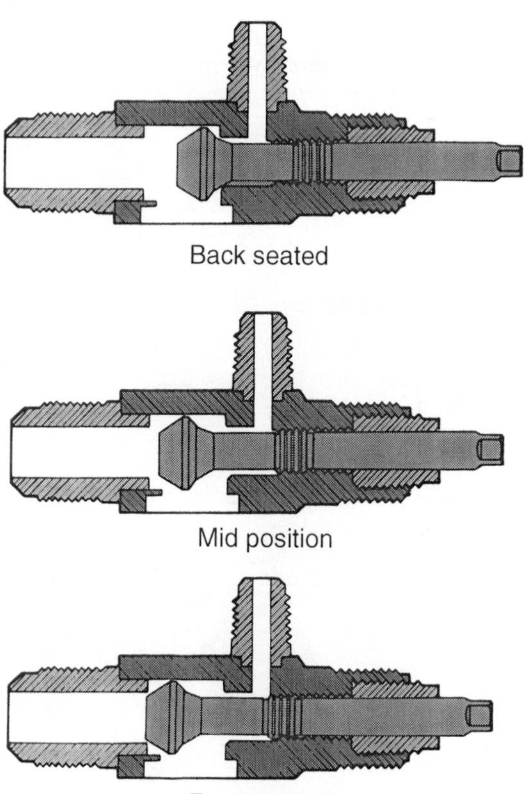


**Figure 9.** A stem-type service valve.

Remove the protective caps from the service valves. Be sure the hand valves in the manifold gauge set are closed. The valves near the end of the gauge set hoses must also be closed. Connect the high-side hose to the high-side service valve port and connect the low-side hose to the low-side service port. Open the valves in the gauge hoses and the hand valves on the manifold gauge set.

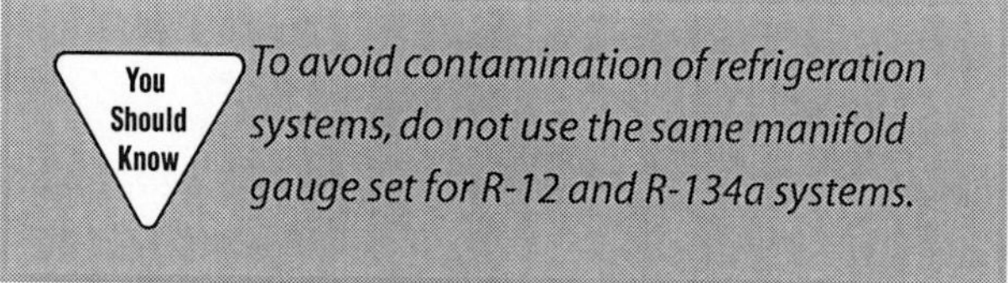


*To avoid contamination of refrigeration systems, do not use the same manifold gauge set for R-12 and R-134a systems.*

The high-side gauge only registers pressure up to 500 psi on most gauges. The high-side gauge is open to high-side pressure and the hand valve opens and closes the high-side hose to the center hose connected to the refrigerant supply when charging the system **(Figure 10)**. The low-side gauge is a **compound gauge** that reads both pressure and vacuum. The gauge is always open to the low (suction) side of the refrigeration system. The hand valve opens and closes the low-side hose from the hose in the center of the gauge set that is connected to the refrigerant supply when charging the system.

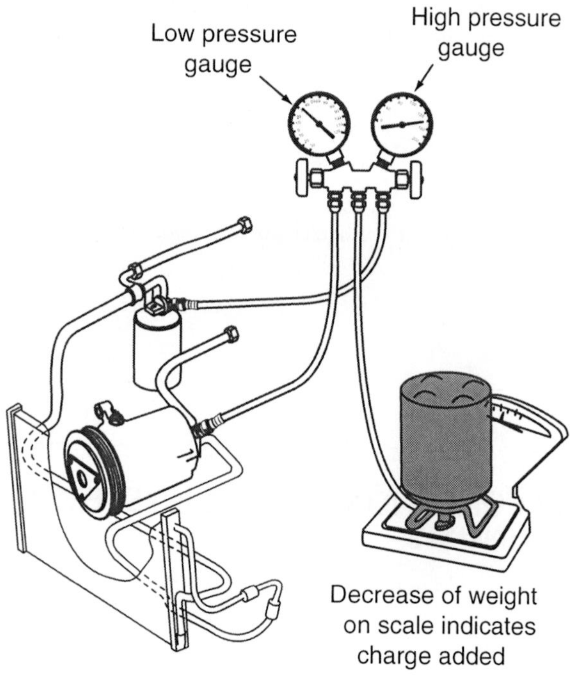


**Figure 10.** Manifold gauge set hose connections.

## Performance Test

A performance test may be completed to determine if the A/C system operation is satisfactory. Follow these steps to complete the performance test:

1. Set the temperature control in the full cold position, select MAX A/C on the A/C control panel, and move the blower switch to the high speed position.
2. Start the engine and maintain engine speed at 2,000 rpm.
3. Close the vehicle doors and windows and place a thermometer in the A/C outlet closest to the evaporator **(Figure 11)**.
4. Place an auxiliary fan in front of the condenser.
5. Continue running the engine for 5 to 10 minutes.
6. Check the temperature on the thermometer. This temperature should be 35°F to 45°F (1.6°C to 7.2°C) with an atmospheric temperature of 80°F (27°C).
7. If the temperature is too high, check the compressor cycling time and check the pressure indicated on the manifold gauge set. If the high-side pressure is excessive, the system may be contaminated with air or moisture, the system may have an overcharge of refrigerant, or the condenser air passages may be restricted.
8. Check the clutch cycling time. If the clutch cycles on and off too rapidly, the system is likely low on refrigerant. A low refrigerant charge is indicated by bubbles in the sight glass and nearly the same temperature on the compressor suction and discharge hoses. If the system is operating properly, the discharge hose should be hot and the suction hose should be cool or cold.

**You Should Know** *The compressor clutch cycling varies depending on atmospheric temperature and the load on the A/C system. The clutch cycling time is the total time required for the clutch on and clutch off cycle. A typical clutch on time is 45 seconds with the clutch on and 15 seconds with the clutch off.*

**You Should Know** *The compressor discharge hose may be very hot if the atmospheric temperature is high and the A/C system has been operating for several minutes. Use extreme caution when touching this hose.*

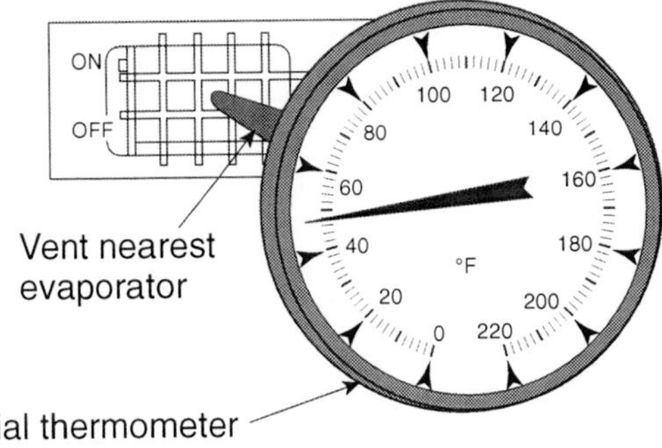


**Figure 11.** A thermometer placed in an A/C outlet.

## Refrigeration System Pressure Diagnosis

When the engine is not running, refrigeration system pressure equalizes on the high side and low side of the system. After the engine is shut off, the refrigerant can be heard hissing through the orifice tube or TXV until the pressure equalizes in both sides of the system. R-12 and R-134a systems have different refrigeration system static pressures **(Figure 12)**. The refrigeration system pressure with the engine shut off is called **static pressure.**

**Pressure Temperature Chart**

| Temperature | | Pressure | | Temperature | | Pressure | |
|---|---|---|---|---|---|---|---|
| °F | °C | HFC-134a | CFC-12 | °F | °C | HFC-134a | CFC-12 |
| –60 | –51.1 | 21.8 | 19.0 | 55 | 12.8 | 51.1 | 52.0 |
| –55 | –48.3 | 20.4 | 17.3 | 60 | 15.6 | 57.3 | 57.7 |
| –50 | –48.6 | 18.7 | 15.4 | 65 | 18.3 | 63.9 | 63.8 |
| –45 | –42.8 | 16.9 | 13.3 | 70 | 21.1 | 70.9 | 70.2 |
| –40 | –40.0 | 14.8 | 11.0 | 75 | 23.9 | 78.4 | 77.0 |
| –35 | –37.2 | 12.5 | 8.4 | 80 | 26.7 | 86.4 | 84.2 |
| –30 | –34.4 | 9.8 | 5.5 | 85 | 29.4 | 94.9 | 91.8 |
| –25 | –31.7 | 6.9 | 2.3 | 90 | 32.2 | 103.9 | 99.8 |
| –20 | –28.9 | 3.7 | 0.6 | 95 | 35.0 | 113.5 | 108.3 |
| –15 | –26.1 | 0.0 | 2.4 | 100 | 37.8 | 123.6 | 117.2 |
| –10 | –23.3 | 1.9 | 4.5 | 105 | 40.6 | 134.3 | 126.6 |
| –5 | –20.6 | 4.1 | 6.7 | 110 | 43.3 | 145.6 | 136.4 |
| 0 | –17.8 | 6.5 | 9.2 | 115 | 46.1 | 157.6 | 146.8 |
| 5 | –15.0 | 9.1 | 11.8 | 120* | 48.9 | 170.3 | 157.7 |
| 10 | –12.2 | 12.0 | 14.6 | 125 | 51.7 | 183.6 | 169.1 |
| 15 | –9.4 | 15.0 | 17.7 | 130 | 54.4 | 197.6 | 181.0 |
| 20 | –6.7 | 18.4 | 21.0 | 135 | 57.2 | 212.4 | 193.5 |
| 25 | –3.9 | 22.1 | 24.8 | 140 | 60.0 | 227.9 | 206.8 |
| 30 | –1.1 | 26.1 | 28.5 | 145 | 62.8 | 244.3 | 220.3 |
| 35 | 1.7 | 30.4 | 32.6 | 150 | 65.6 | 261.4 | 234.6 |
| 40 | 4.4 | 35.0 | 37.0 | 155 | 68.3 | 279.5 | 249.5 |
| 45 | 7.2 | 40.0 | 41.7 | 160 | 71.1 | 298.4 | 265.1 |
| 50 | 10.0 | 45.3 | 46.7 | 165 | 73.9 | 318.3 | 261.4 |

Red figures — in. Hg Vacuum
Gray figures — PSIG

*Do not heat can above 120°F

**Figure 12.** Static refrigeration system pressures.

| Ambient Temp °K | High Side PSIG, R12 | Low Side PSIG, R12 | High Side PSIG, R-134a | Low Side PSIG, R-134a |
|---|---|---|---|---|
| 60 | 120 - 150 | 5 - 15 | 120 - 170 | 7 - 15 |
| 70 | 140 - 180 | 8 - 16 | 150 - 250 | 8 - 16 |
| 80 | 160 - 250 | 10 - 18 | 190 - 280 | 10 - 20 |
| 90 | 200 - 280 | 12 - 25 | 220 - 330 | 15 - 25 |
| 100 | 220 - 300 | 15 - 30 | 250 - 350 | 20 - 30 |
| 110 | 250 - 320 | 20 - 35 | 280 - 400 | 25 - 40 |

**Figure 13.** Refrigeration system pressure in relation to atmospheric temperature.

With the engine running, the refrigeration system pressures vary depending on the atmospheric temperature. Normal pressures for R-12 and R-134a systems in relation to temperature are provided in **Figure 13**.

With the engine and the A/C system at normal operating temperature, these pressure guidelines will help you to diagnose the causes of improper refrigeration pressures.

1. When the high-side pressure is higher than normal, there may be air in the system, a refrigerant overcharge, a high-side restriction, or reduced airflow through the condenser.
2. When the high-side pressure is lower than normal, the refrigerant charge may be low or the compressor may be defective.
3. When the low-side pressure is higher than normal, there may be a refrigerant overcharge, a defective compressor, or a faulty metering device.
4. When the low-side pressure is lower than normal, the metering device may be faulty, there may be a low-side restriction, or a refrigerant undercharge.

## Compressor Clutch Diagnosis

If the compressor clutch does not engage, the system may be low on refrigerant and the low pressure shut-off switch may have opened the clutch circuit. Shut the engine off and disconnect the low pressure shut-off switch connector. Connect a jumper wire across the terminals in the low pressure shut-off switch connector and start the engine. If the clutch engages, the refrigeration system is low on refrigerant or the low pressure shut-off switch is defective.

When the compressor does not run with the jumper wire connected across the low pressure shut-off switch wiring connector terminals, shut the engine off and connect a digital voltmeter from the voltage input terminal on the compressor clutch to ground. With MAX A/C selected on the control panel and the temperature set below the atmospheric temperature, 12 volts should be supplied to the compressor clutch. If this voltage is not present, diagnose the compressor clutch circuit. The first step in this diagnosis is to check the compressor clutch fuse. In many modern A/C systems, the A/C computer energizes a compressor relay winding. When this winding is energized, the relay contacts close and supply 12 volts to the compressor clutch. If 12 volts are supplied to the compressor clutch, shut the engine off and disconnect the compressor clutch electrical connector. Connect a pair of ohmmeter leads to the compressor clutch terminals. An infinite ohmmeter reading indicates an open clutch coil and a reading lower than specified indicates a shorted clutch coil.

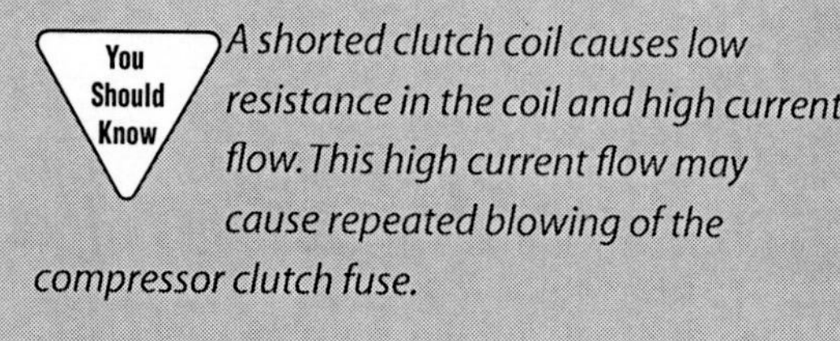


*A shorted clutch coil causes low resistance in the coil and high current flow. This high current flow may cause repeated blowing of the compressor clutch fuse.*

Connect the ohmmeter leads from one of the clutch terminals to ground. A low ohmmeter reading indicates a grounded clutch coil, whereas an infinite reading proves the coil is not grounded. Replace the clutch coil if it is grounded, shorted, or open. Connect the ohmmeter leads from the ground wire in the clutch wiring connector to a ground on the compressor. A low ohmmeter reading indicates a satisfactory ground wire, whereas an infinite reading proves the ground wire is open.

## AIR CONDITIONING (A/C) SYSTEM SERVICE

The most common A/C service procedures are refrigerant recovery, evacuation of the refrigerant system, and system charging. Most shops use a recovery/recycling machine to perform these operations.

### Refrigerant Recovery

Follow these steps to perform the refrigerant recovery operation:

1. Connect the high-side and low-side hoses on the manifold gauge set to the appropriate service fittings. If the refrigeration system has stem-type service valves, use a special wrench to rotate these valve stems so they are in the mid position.
2. Connect the center hose on the manifold gauge set to the proper fitting on the refrigerant recovery/recycling machine **(Figure 14)**.
3. Open the hose shut-off valves and the high-side and low-side hand valves on the manifold gauge set.
4. Connect the recovery/recycling machine electrical cord to a 120-volt outlet and turn the main switch on the machine on **(Figure 15)**.
5. Turn the compressor switch on the recovery machine on **(Figure 16)**.

**Figure 14.** Connect the center hose on the manifold gauge set to the inlet fitting on the recovery / recycling machine.

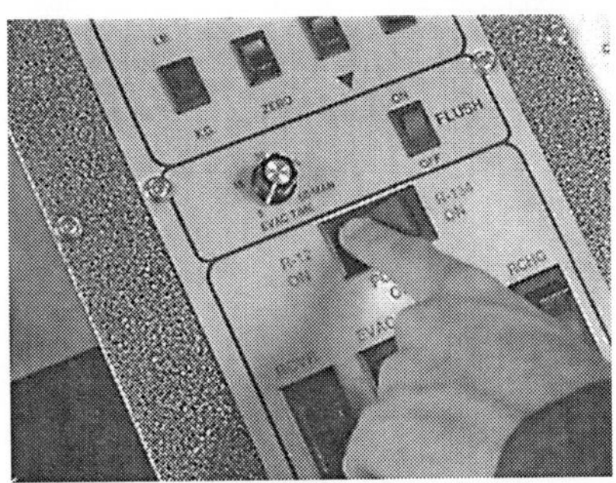


**Figure 15.** Turning on the main switch on the recovery / recycling machine.

**Figure 16.** Turning on the compressor switch on the recovery / recycling machine.

6. Operate the compressor until a vacuum is indicated on the gauge on the recovery/recycling machine **(Figure 17)**. If the machine does not have an automatic shut-off feature, turn the compressor off after achieving a system vacuum.
7. Observe the gauges on the machine for a minimum of 5 minutes. If the vacuum rises but remains at 0 psi or below, the system is leaking and must be repaired after the recovery process.

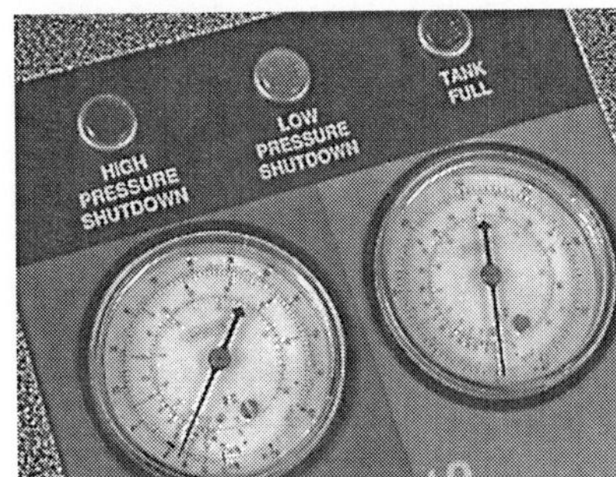


**Figure 17.** A vacuum must be indicated on the low-side gauge on the recovery / recycling machine.

8. If the vacuum reading changes to a pressure above 0 psi, the refrigerant was not completely removed from the system. Repeat steps 5 through 7.
9. After the recovery, be sure the system holds a steady vacuum for a minimum of 2 minutes.
10. Close all the manifold gauge set hand valves, service hose valves, and recovery system inlet valve.
11. Disconnect all the manifold gauge set hoses and cap all fittings.

## Refrigerant System Evacuation

When servicing a refrigerant system, it is very important to remove moisture from the system. Moisture reduces the cooling efficiency of the refrigeration system. Even a small amount of moisture may freeze in the orifice tube or TXV preventing any cooling action from the system. A vacuum pump is used to remove air from the refrigeration system and the vacuum pump also reduces the pressure in the system to the point where any moisture is boiled out of the system. Follow this procedure to evacuate a refrigeration system:

1. Be sure both hand valves on the manifold gauge set are closed and connect the manifold gauge set low-side and high-side hoses to the proper service fittings in the refrigeration system. If the refrigeration system has stem-type service valves, use a special wrench to rotate these valve stems so they are in the mid position.
2. Remove the protective caps from the inlet and exhaust fittings on the vacuum pump.
3. Connect the center hose on the manifold gauge set to the inlet fitting on the vacuum pump **(Figure 18)**.
4. Open the shut-off valves on all three service hoses.
5. Connect the pump power cord to a 120-volt outlet and turn on the pump switch.
6. Open the low-side hand valve on the manifold gauge set and observe the reading on the low-side gauge. This gauge reading should indicate a slight vacuum immediately.

**Figure 18.** Connecting the center hose on the manifold gauge set to the inlet fitting on the vacuum pump.

7. Observe both gauges after 5 minutes. The low-side gauge should indicate 20 in. Hg (33.8 kPa absolute) and the high-side gauge pointer should be slightly below zero unless the pointer movement is limited by a stop. If the high-side gauge does not read below zero, there is blockage in the refrigerant system. Discontinue the evacuation and repair the system blockage.
8. Operate the pump for 15 minutes and observe the gauges. The low-side gauge should indicate at least 26 in. Hg (13.5 kPa absolute). If this reading is not obtained, close the low-side hand valve and observe the low-side gauge reading. If the vacuum reading slowly moves toward zero on this gauge, the refrigeration system is leaking and must be repaired.
9. Be sure both hand valves on the manifold gauge set are open and operate the pump for 30 minutes. Close both hand valves, turn off the vacuum pump, and close all three service hose shut-off valves. If the vacuum pump has a shut-off valve, close this valve. Disconnect all the service hoses and install the protective caps.

## Refrigeration System Charging

After the refrigeration system is evacuated, it must be charged with refrigerant. This charging process is very important to provide proper A/C operation. A refrigerant overcharge or undercharge reduces the cooling efficiency of the refrigeration system. Follow this procedure to charge the refrigeration system:

1. Be sure all the service valves, hose shut-off valves, manifold gauge set hand valves, and the refrigerant source valve are closed. Connect the low-side and high-side hoses from the manifold gauge set to the proper service fittings.
2. Connect the center hose on the manifold gauge set to the refrigerant cylinder. Be sure you are using the correct refrigerant for the system to be charged.

*R-12 and R-134a refrigerant are not compatible and must not be mixed. R-134a containers are light blue, whereas R-12 containers are white.*

*Fluorescent dye may be added to a refrigeration system to help locate a leak source. After the dye is added to the system, a black light is used to check for leaks. The dye around a leak source appears as a luminous yellow-green color. Refrigerant containing dye may be purchased from some suppliers, or the dye may be purchased separately in a small pressurized container.*

3. Open the refrigerant cylinder hand valve. The refrigeration system is now under pressure from the refrigerant container to the service hose shut-off valve. A vacuum is present in the refrigeration system up to the low-side and high-side hose shut-off valves.
4. Open the service hose shut-off valve and open the low-side and high-side hose shut-off valves near the end of these hoses.
5. Briefly open the high-side manifold gauge set hand valve, re-close this valve, and observe the low-side gauge. If the low-side gauge does not move from a vacuum to a pressure, the refrigeration system is blocked. This blockage must be repaired before continuing with the charging procedure. When the low-side gauge moves from vacuum to a pressure, proceed with step 6.
6. Start the engine, run the engine at 1,250 rpm, and set the A/C controls to maximum cooling and high blower speed.
7. Place the refrigerant cylinder on an approved scale and note the gross weight of the cylinder and contents.
8. Open the low-side manifold gauge set hand valve and allow the refrigerant to enter the system. Do not charge the refrigerant system through the high side.
9. Closely observe the weight of the refrigerant container on the scale. When the decrease in the scale reading equals the specified amount of refrigerant for the system being charged, close the low-side manifold gauge set hand valve and the refrigerant source valve. For example, if the specified refrigerant charge is 2.5 pounds and the scale reading decreased from 28 pounds to 25.5 pounds, the system is fully charged.

You Should Know *In place of a scale that weighs the refrigerant, some shops use a clear cylinder with a graduated scale on the cylinder surface. This cylinder is connected between the refrigerant container and the center hose on the manifold gauge set. The refrigerant that is charged into the system is read on the graduated scale.*

10. Perform a refrigerant system performance test as explained previously.
11. If the performance test results are satisfactory, turn all the A/C system controls off and shut the engine off.
12. Be sure all valves are closed including the service hose valve, manifold gauge set hand valves, and low-side and high-side hose valves.
13. Disconnect the manifold gauge set hoses and install all the protective caps.

## Typical Refrigeration System Repairs

When it is necessary to change any refrigeration system component, the refrigerant must be recovered and the system should be evacuated and recharged. The receiver/drier or acccumulator should be changed at the vehicle manufacturer's specified service intervals **(Figure 19)**. If the desiccant in the receiver/drier or accumulator disintegrates, it moves through the system and usually plugs the orifice tube or TXV. If this happens, the refrigeration system requires flushing. The **desiccant** in a receiver/drier or accumulator removes moisture from the refrigeration system.

System flushing with a recovery/recycling machine was explained in Chapter 36. If debris has entered the refrigerant system, an in-line filter may be installed between the condenser and the evaporator to help remove debris from the system.

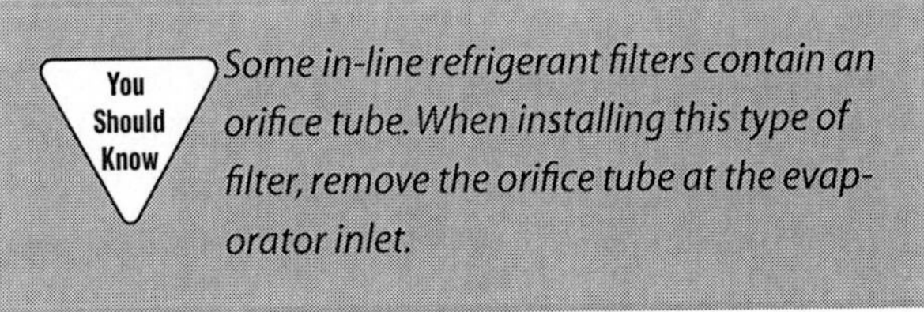


*Some in-line refrigerant filters contain an orifice tube. When installing this type of filter, remove the orifice tube at the evaporator inlet.*

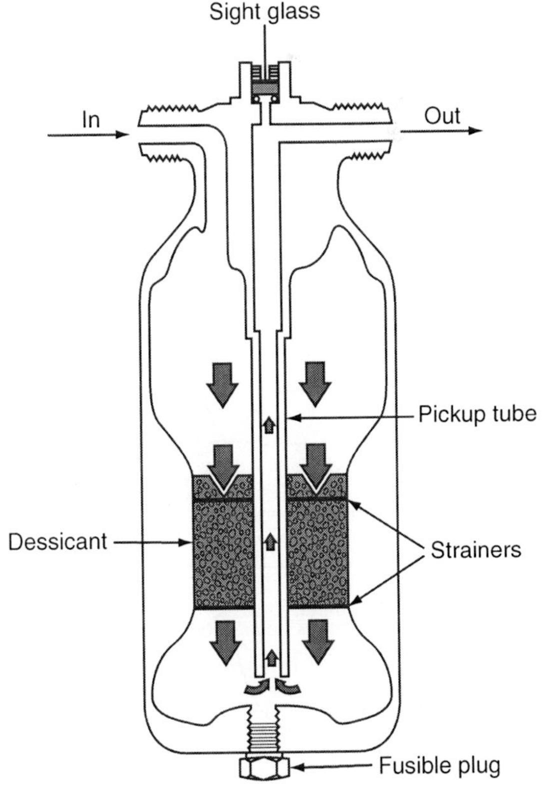


**Figure 19.** Receiver/drier internal design.

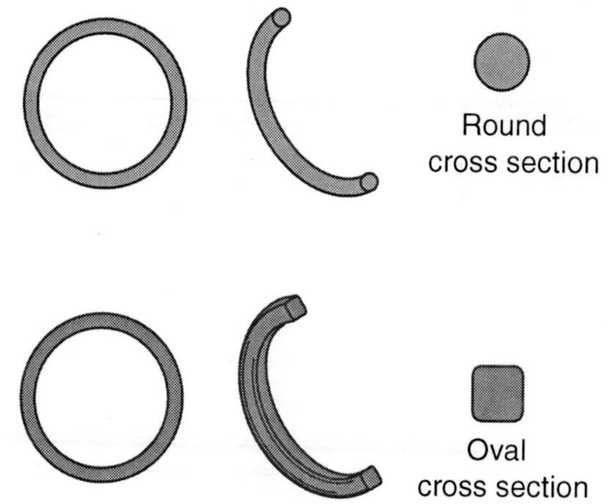


**Figure 20.** Refrigeration system O-rings.

Many fittings in the refrigerant system use O-ring seals because this allows the connection to withstand vibration better than a rigid pipe joint. Always replace these seals if a fitting is disconnected. Most refrigerant system O-ring seals are now oval-shaped **(Figure 20)**. O-rings must be compatible with the refrigerant in the system. Most newer O-rings are color coded. O-rings should always be coated with the proper refrigerant before installation.

If an orifice tube is plugged, it is better to replace this tube. After the evaporator inlet fitting is removed, a special tool is used to remove the orifice tube **(Figure 21)**. If the orifice tube breaks, a special extractor is available to remove the broken tube **(Figure 22)**.

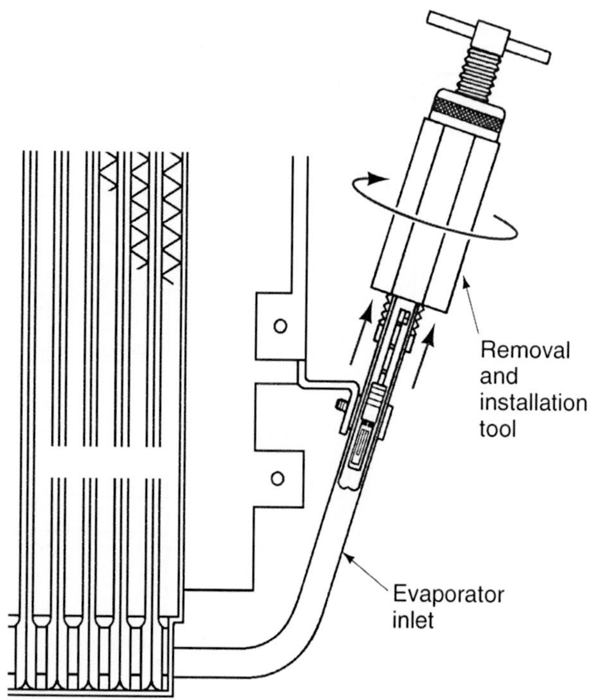


**Figure 21.** Removing an orifice tube.

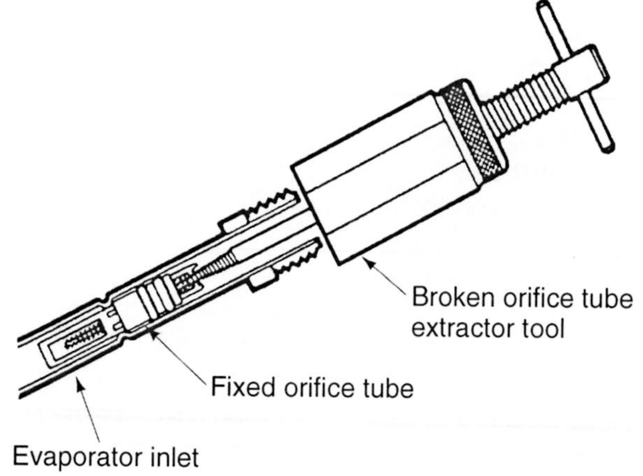


**Figure 22.** Removing a broken orifice tube.

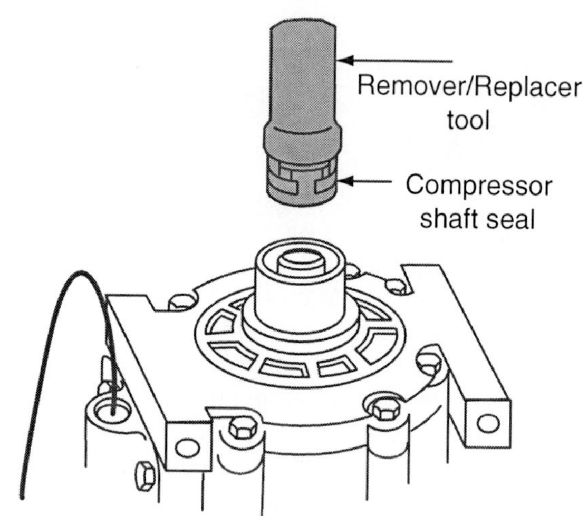


**Figure 23.** Replacing the compressor shaft seal.

If the compressor shaft seal is leaking, the compressor pulley may be removed and this seal may be replaced. On some rear-wheel drive cars, this operation is possible without removing the compressor from the engine. Special tools are available to remove and replace the compressor shaft seal **(Figure 23)**.

## RETROFITTING R-12 SYSTEMS TO R-134A SYSTEMS

R-12 systems may be retrofitted to use R-134a refrigerant or some other approved refrigerant. Prior to retrofitting an A/C system, obtain the service history from the customer. Ask the customer about the operation of the A/C system and any previous work that was done on the system. Try to determine if the customer ever had an alternate refrigerant installed in the system. A visual inspection of the A/C system is the next step in an R-12-to-R-134a retrofit. Inspect all the hoses, tubing, service ports, condenser, receiver/drier or accumulator, and compressor for indications of refrigerant leaks and damage.

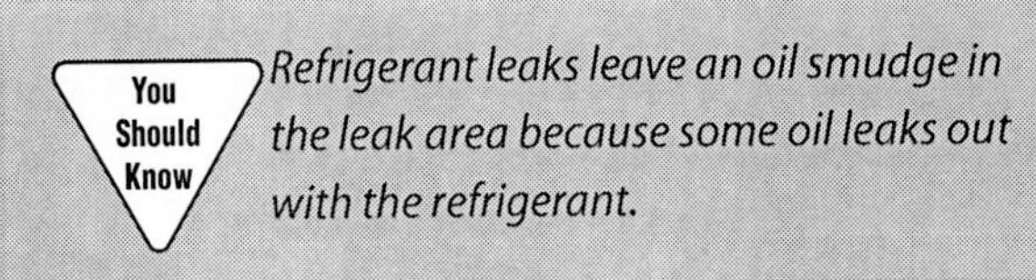


*Refrigerant leaks leave an oil smudge in the leak area because some oil leaks out with the refrigerant.*

Inspect the compressor drive belt for wear, cracks, and oil contamination. Look under the hood for the original equipment manufacturer's label or aftermarket label indicating the type of refrigerant **(Figure 24)**. If you are in doubt about the type of refrigerant, use a refrigerant identifier to provide this information. Damaged or failed components must be replaced.

Performance check the refrigerant system by connecting a manifold gauge set to the refrigerant system and installing a thermometer in the A/C outlet closest to the evaporator **(Figure 25)**. Run the engine for 15 minutes and be sure the system has the specified low-side and high-side pressure and the correct compressor clutch cycling time. Be sure the A/C system provides adequate passenger compartment cooling as indicated on the thermometer. Shut the engine off and use an electronic leak tester to check the refrigerant system for leaks. If any A/C system problems are indicated, the system must be repaired before installing the R-134a refrigerant.

Use a recovery/recycling machine to recover the R-12 refrigerant **(Figure 26)**. Remove as much of the mineral oil from the system as possible. Mineral oil left in the system may reduce the cooling efficiency of the system. On compressors with a sump, the mineral oil may be drained from the compressor. Some recovery/recycling machines have a flush cycle that flushes the refrigerant system using the original type of refrigerant and a closed loop flushing procedure. These machines filter the flushing refrigerant before it is returned to the recovery tank. If the purpose of the system flushing is to remove mineral oil, follow the closed-loop flushing procedure with the refrigerant system intact. If the purpose of the flushing procedure is to remove contaminants and debris, each refrigerant system component must be flushed separately. A **closed-loop flushing procedure** flushes the

Figure 24. An air conditioning system label indicating the type of refrigerant.

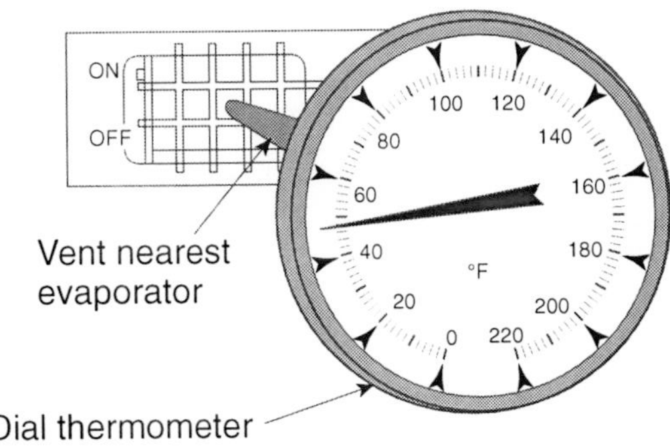


Figure 25. Checking outlet air temperature.

Figure 26. Refrigerant recovery/recycling machine.

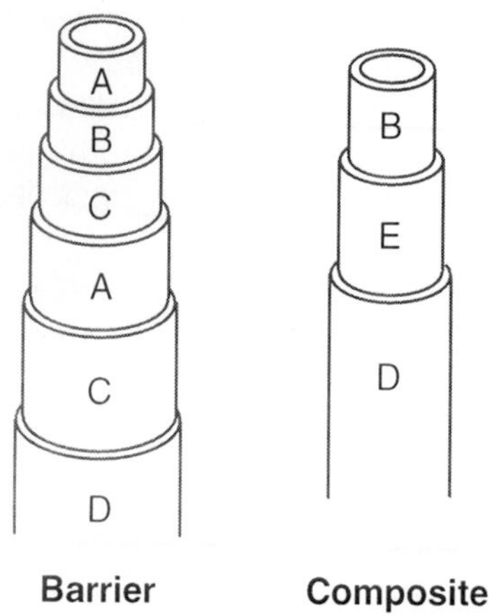


A Nitrile tube
B Nylon barrier
C Fabric reinforced (spiral or braided)
D Butyl covered
E Fabric yarn (spiral or braided)

**Figure 27.** Conventional and barrier-type A/C system hoses.

**Figure 28.** A manifold gauge set for R-134a refrigerant.

refrigerant system with the system intact. Under this condition, no refrigerant escapes to the atmosphere.

Install a new receiver/drier or accumulator depending on the refrigerant system. Repair any refrigerant system leaks as indicated in the previous leak test. Remove the R-12 service fittings and install the R-134a quick disconnect service fittings.

Install new barrier hoses if these hoses are recommended by the EPA for the refrigerant being installed **(Figure 27)**. Some refrigerants have smaller molecules than R-12 and they require the use of barrier hoses that are more leak resistant.

If the refrigerant system does not have a high-pressure compressor shut-off switch, install one of these switches in the system. This switch shuts the compressor off if refrigerant system pressure approaches the point where the refrigerant may be vented to the atmosphere through a pressure relief valve.

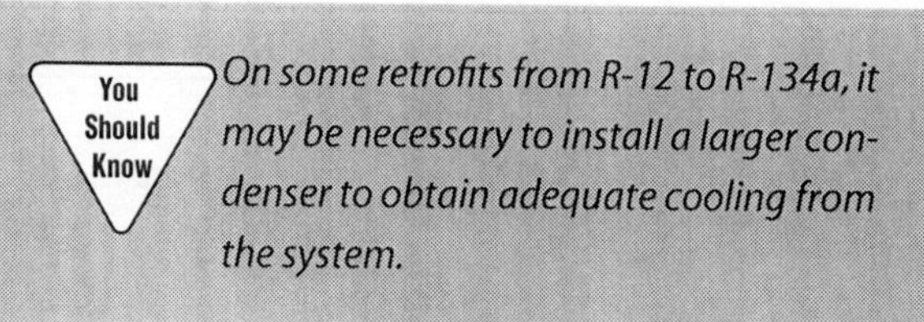


*On some retrofits from R-12 to R-134a, it may be necessary to install a larger condenser to obtain adequate cooling from the system.*

Connect a manifold gauge set to the refrigerant system **(Figure 28)**. Evacuate the refrigerant system with the recovery/recycling machine or a vacuum pump. Be sure the system holds the specified amount of vacuum for at least two minutes indicating the absence of leaks.

Use the recovery/recycling machine to charge the system with R-134a. Be sure the specified amount and type of refrigerant oil is installed in the system. The recovery/recycling machine performs this function. The proper charge of R-134a refrigerant is very important for efficient system operation. An R-134a charge of 80 percent to 90 percent of the original specified R-12 charge is usually satisfactory.

After the system is charged with R-134a refrigerant, performance check the system as mentioned previously. Be sure the system pressures are normal. Use an R-134a leak detector to check for refrigerant system leaks.

Install the retrofit label in the underhood area. The R-134a retrofit label is blue and contains the name and address of the shop and the name of the technician. This label also contains the date, type of refrigerant, amount of refrigerant, refrigerant manufacturer, and the type and amount of lubricant installed.

## Summary

- A slipping A/C compressor drive belt reduces the cooling efficiency of the A/C system.
- Excessive compressor clutch clearance may cause clutch slipping and a scraping noise when the clutch engages.
- The compressor clutch clearance may be measured with a feeler gauge.
- The compressor clutch clearance is adjusted by removing or installing shims between the clutch hub and the compressor shaft.
- Restricted airflow through the condenser reduces the cooling efficiency of the A/C system.
- On an R-12 refrigeration system, bubbles in the sight glass indicate a low refrigerant charge.
- During an A/C system performance test, the temperature of the air at the A/C outlet closest to the evaporator should be 35°F to 45°F with an atmospheric temperature of 80°F.
- If the compressor clutch cycles too fast, the refrigeration system charge may be low.
- If the refrigeration system is low on refrigerant, the low pressure shut-off switch opens the compressor clutch circuit.
- The compressor clutch coil may be tested with an ohmmeter.
- When removing the refrigerant from a refrigeration system, the refrigerant should be recovered with a recovery/recycling machine.
- The refrigeration system is evacuated with a vacuum pump to remove air and water. The vacuum pump lowers the pressure so any moisture is boiled out of the system.
- During the evacuation process, the refrigeration system is tested for leaks.
- During the charging process, the amount of refrigerant dispensed into the system may be measured by weight.
- Fluorescent dye may be added to a refrigeration system to help locate leaks.
- R-12 systems may be retrofitted to R-134a refrigerant.
- After an R-12-to-R-134a retrofit, a retrofit label must be installed in the underhood area.

## Review Questions

1. A vehicle has a lack of interior heat complaint. One heater hose feels hot while the other heater hose is cool. Technician A says the heater coolant control valve may be stuck closed. Technician B says the heater core may be plugged. Who is correct?
   A. Technician A
   B. Technician B
   C. Both Technician A and Technician B
   D. Neither Technician A nor Technician B
2. Technician A says a scraping noise when the compressor clutch engages may be caused by excessive clutch plate clearance. Technician B says the clutch plate clearance is adjusted by the torque on the compressor pulley retaining nut. Who is correct?
   A. Technician A
   B. Technician B
   C. Both Technician A and Technician B
   D. Neither Technician A nor Technician B
3. Technician A says an oil smudge around a refrigeration system fitting indicates excessive oil in the refrigeration system.. Technician B says if all of the refrigerant has escaped from a refrigeration system, a partial charge may be installed to locate the source of the leak. Who is correct?
   A. Technician A
   B. Technician B
   C. Both Technician A and Technician B
   D. Neither Technician A nor Technician B
4. Technician A says a cloudy appearance in the sight glass on an R-12 system indicates the refrigeration system is empty. Technician B says the sight glass indications are accurate at 60°F (15.5°C). Who is correct?
   A. Technician A
   B. Technician B
   C. Both Technician A and Technician B
   D. Neither Technician A nor Technician B

5. When diagnosing a lack of heat in the vehicle interior, one heater hose is hot and the other heater hose is cool with the engine at normal operating temperature. The most likely cause of this problem is:
   A. an engine thermostat that is stuck open.
   B. a coolant control valve that is stuck closed.
   C. partially restricted coolant passages in the radiator core.
   D. partially restricted air passages in the heater core.
6. All of these A/C system problems may cause insufficient vehicle interior cooling *except*:
   A. a low refrigerant charge.
   B. moisture in the refrigeration system.
   C. partially restricted air passages in the condenser.
   D. excessive compressor drive belt tension.
7. When an A/C system is operating, frost forms on the receiver/drier. Technician A says the refrigeration system is overcharged. Technician B says the receiver/drier is restricting the flow of refrigerant. Who is correct?
   A. Technician A
   B. Technician B
   C. Both Technician A and Technician B
   D. Neither Technician A nor Technician B
8. An R-12 refrigeration system has bubbles in the sight glass, and inadequate interior vehicle cooling. The atmospheric temperature is 80°F. The most likely cause of this problem is:
   A. a partially restricted evaporator core.
   B. a defective compressor.
   C. a refrigerant leak in the condenser.
   D. a desiccant deterioration in the receiver/drier.
9. After a refrigeration system performance test is performed at an atmospheric temperature of 80°F, if the system is operating normally, the temperature on the thermometer should be:
   A. 28°F to 32°F.
   B. 35°F to 45°F.
   C. 55°F to 65°F.
   D. 70°F to 75°F.
10. After a refrigeration system is evacuated for 15 minutes, the low-side gauge indicates 27 in. Hg. When the vacuum pump is shut off and the low-side hand valve is closed, the low-side gauge reading decreases to 18 in. Hg in 5 minutes. Technician A says there is refrigeration oil left in the system and it must be flushed out with dry nitrogen. Technician B says the refrigeration system has a leak and a partial charge should be installed to locate the leak. Who is correct?
    A. Technician A
    B. Technician B
    C. Both Technician A and Technician B
    D. Neither Technician A nor Technician B
11. The low-side gauge in a manifold gauge set is a compound gauge that reads ________________ and ________________.
12. Air or moisture in the refrigeration system causes high pressure in the ____________ ______________ of the refrigeration system.
13. A defective compressor may cause ____________ high-side pressure and ______________ low-side pressure.
14. Moisture is removed from a refrigeration system by creating a vacuum in the system which causes the moisture to ________________.
15. Explain how a refrigeration system leak is indicated while evacuating the system.
16. Describe two methods of measuring the refrigerant entering the system during the charging process.
17. Explain the necessary service procedures if debris has entered the refrigeration system.
18. Describe normal test results during an A/C system performance test.

# Section 10

## Tires and Wheels



## SECTION OBJECTIVES

After you have read, studied, and practiced the contents of this section, you should be able to:

- Describe general tire purposes.
- Describe three types of tire ply and belt designs.
- Explain the purpose of the tire performance criteria (TPC) rating.
- Define tire contact area, free tire diameter, and rolling tire diameter.
- Describe the tire motion forces while a tire and wheel assembly is rotating on a vehicle.
- Define wheel tramp and wheel shimmy.
- Diagnose steering pull problems related to tire condition.
- Rotate tires according to the vehicle manufacturer's recommended procedure.
- Demount, inspect, repair, and remount tires.
- Diagnose problems caused by excessive radial, lateral, wheel, or tire runout.
- Perform off-car static and dynamic wheel balance procedures.
- Diagnose tire wear problems caused by tire and wheel imbalance.
- Describe three different types of bearing loads.
- Explain the advantage of tapered roller bearings compared to other types of bearings.
- Clean, repack, reassemble, and adjust wheel bearings.
- Diagnose wheel bearings on the vehicle.
- Remove and replace rear axle bearings on rear-wheel drive cars.

*The Specialty Equipment Market Association (SEMA) announced at its sixth annual International Auto Salon that in the United States retail sales of accessories and other products for cars and light-duty trucks were $2.3 billion in 2002. Jim Spoonhower, vice president of Research for SEMA stated, "This market is fueled by the growing affection which younger drivers have for their vehicles. They can tune and tweak them, enhance their appearance, add mobile electronics systems, and personalize them to suit their lifestyle and sense of fashion."*

# Tires, Wheels, and Hubs

## Introduction

Although tires are often taken for granted, they contribute greatly to the ride and steering quality of a vehicle. Tires also play a significant role in vehicle safety. Improper types of tires, incorrect inflation pressure, and worn out tires create a safety hazard. When tires and wheels are out of balance, tire wear and driver fatigue are increased which may create a driving hazard. The purposes of the tires may be summarized as follows:

1. Tires cushion the vehicle ride to provide a comfortable ride for the occupants.
2. The vehicle weight must be supported firmly by the tires.
3. Tires must develop traction to drive and steer the vehicle under a wide variety of road conditions.
4. The tires contribute to directional stability of the vehicle and they must absorb all the stresses of accelerating, braking, and centrifugal force in turns.

### TIRE DESIGN

Tire construction varies depending on the manufacturer and the type of tire. A typical modern tire manufactured by Yokohama contains these components **(Figure 1)**: bead wire, bead filler, liner, steel reinforcement in the sidewall, sidewall with hard side compound, rayon carcass plies. steel belts, jointless belt cover, hard undertread compound, and hard high-grip tread compound.

The tire bead contains several turns of bronze-coated steel wire in a continuous loop. This bead is molded into the tire at the inner circumference and is wrapped in the cord plies. The bead anchors the tire to the wheel. The bead filler above the bead reinforces the sidewall and acts as a rim extender. Tire sidewalls are made from a blend of rubber that absorbs shocks and impacts from road irregularities and prevents damage to the plies. A lettering and numbering arrangement for tire identification is located on the outside of the sidewall. The sidewall material contains antioxidants and other chemicals that are gradually released to the surface of the sidewall during the life of the tire. These antioxidants help to keep the sidewall from cracking and protect it from ultraviolet radiation and ozone attack. Since the sidewall must be flexible to provide ride quality, minimum thickness of this component is essential. Tire manufacturers have reduced sidewall thickness by 40

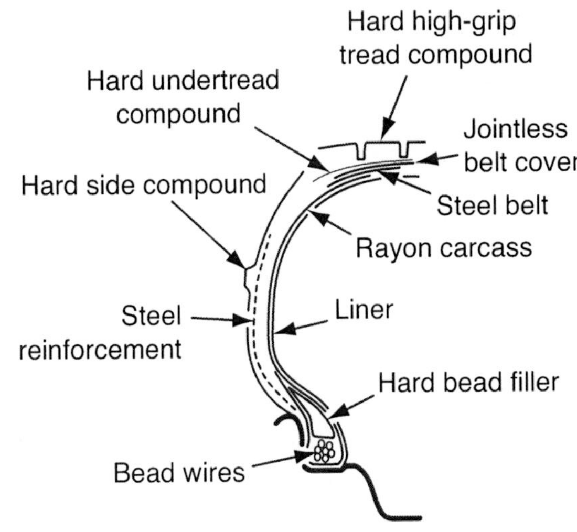


**Figure 1.** Tire design.

percent in recent years to reduce weight and heat buildup and improve ride quality.

The cord plies surround both beads and extend around the inner surface of the tire to enable the tire to carry its load. The plies are molded into the sidewalls. Each ply is a layer of rubber with parallel cords imbedded in its body. The load capacity of a tire may be increased by adding more cords in each ply or by installing additional plies. The most common materials in tire plies are polyester, rayon, and nylon. Passenger car tires usually have two cord plies, whereas many trucks and recreation vehicles are equipped with six- or eight-ply tires to carry the heavier loads of these vehicles. In general, tires with more plies have stiffer sidewalls which provide less cushioning and reduced ride quality.

Steel is the most common material in tire belts, although other belt materials such as polyester have been used to some extent. Many tires contain two steel belts. The tire belts restrict ply movement and provide tread stability and resistance to deformation. This belt action provides longer tread wear and reduces heat buildup in the tire. Steel belts expand as wheel speed and tire temperature increase. Centrifugal force and belt expansion tend to tear the tire apart at high speeds and temperatures. Therefore, high-speed tires usually have a nylon jointless belt cover. This nylon belt cover contracts as it is heated and helps to hold the tire together thereby providing longer tire life, improved stability, and better handling.

Tire treads are made from a blend of rubber compounds that are very resistant to abrasion wear. Spaces between the tire treads allow tire distortion on the road without scrubbing which accelerates wear. Modern automotive tires contain two layers of tread materials. The first tread layer is designed to provide cool operation, low rolling resistance, and durability. The outer layer, or tread, is designed for long life and maximum traction. Tread rubber is a blend of many different synthetic and natural rubbers. Tire manufacturers may use up to thirty different synthetic rubbers and eight natural rubbers in their tires. The manufacturers blend these synthetic and natural rubbers in both tread layers to provide the desired traction and durability. Tire treads must provide traction between the tire and the road surface when the vehicle is accelerating, braking, and cornering. This traction must be maintained as much as possible on a wide variety of road surfaces. For example, on wet pavement, tire treads must be designed to drain off water between the tire and the road surface. This draining action is extremely important to maintain adequate acceleration, braking, and directional control. Lines cut across the tread provide a wiping action which helps to dry the tire-road contact area.

The synthetic gum rubber liner is bonded to the inner surface of the tire to seal the tire. Nearly all passenger car and light truck tires are tubeless-type. In these tires, the tire bead must provide an airtight seal on the rim and both the tire and the wheel rim must be completely sealed. Some heavy-duty truck tires have inner tubes mounted inside the tire. On tube-type tires, the air is sealed in the inner tube and the sealing qualities of the tire and wheel rim are not important. Designing tires is a very complex engineering operation and the average all-season tire contains these components by weight:

- Synthetic rubber (thirty types)—2.49 kg
- Carbon black (eight types)—2.27 kg
- Natural rubber (eight types)—2.04 kg
- Chemicals, waxes, oils, and pigments (forty types)—1.36 kg
- Steel cord for belts—0.68 kg
- Polyester and nylon—0.45 kg
- Bead wire—0.23 kg
- Total weight 9.52 kg

Tire design varies depending on the operating conditions and the load capacity of the tire. A tire designed for improved steering and handling characteristics has a nylon bead reinforcement and a hard bead filler with a slim tapered profile **(Figure 2)**. This type of tire is suitable for

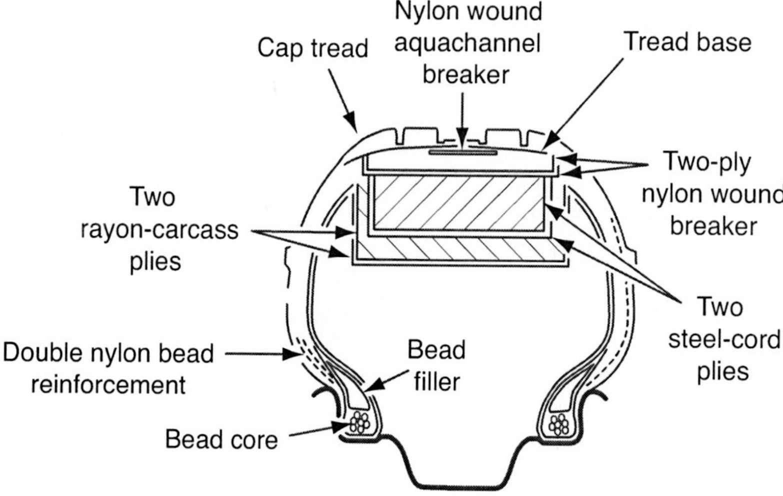


**Figure 2.** Tire designed for improved steering and handling with a nylon bead reinforcement and a hard bead filler with slim, tapered profile.

sports car operation because the design stiffens the tire and reduces tire deflection during high-speed cornering. However, this type of tire may provide slightly firmer ride quality.

## TIRE PLY AND BELT DESIGN

The most commonly used tire designs are bias, belted bias, and belted radial. In bias-ply or belted bias-ply tires, the cords criss-cross each other. These cords are usually at an angle of 25 degrees to 45 degrees to the tire center line. The belt ply cord angle is usually 5 degrees less than the cord angle in the tire casing. Two plies and two belts are most commonly used, but four plies and four belts may be used in some tires. Compared to a bias-ply tire, a belted bias-ply tire has greater tread rigidity. The belts reduce tread motion during road contact. This action provides extended tread life compared to a bias-ply tire.

In radial tires, the ply cords are arranged radially at a right angle to the tire center line **(Figure 3)**. Steel belts are the most common in radial tires, but other belt materials such as fiberglass, nylon, and rayon have been used. The steel or fiberglass cords in the belts are criss-crossed at an angle of 10 degrees to 30 degrees in relation to the tire center line. Many radial tires have two plies and two belts. Radial tires provide less rolling resistance, improved steering characteristics, and longer tread life compared to bias-ply tires.

Regardless of the type of tire construction, the tire must be uniform in diameter and width. Radial runout refers to variations in tire diameter. A tire with excessive radial runout causes a tire thumping problem as the car is driven. When a tire has excessive variations in width, this condition is called lateral runout. A tire with excessive lateral runout causes the chassis to "waddle" when the car is driven.

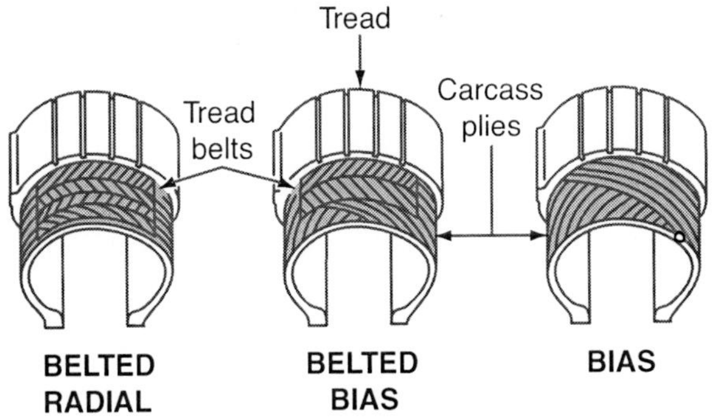


**Figure 3.** Three types of tire construction: bias-ply, belted bias ply, and belted radial-ply.

The tire plies and belts must be level across the tread area. If the plies and/or belts are not level across the tread area, the tire is cone-shaped. This condition is referred to as tire **conicity**. When a front tire has conicity, the steering may pull to one side as the car is driven straight ahead. A rear tire with conicity will not affect the steering as much as a front tire with conicity.

## TIRE RATINGS

A great deal of important information is molded into the sidewall of the average passenger car or light-truck tire. The tire rating is part of the information located on the sidewall. The tire rating is a group of letters and numbers that identify the tire type, section width, aspect ratio, construction type, rim diameter, load capacity, and speed symbol. When a tire has a P215/65R15 89H rating on the sidewall, the P indicates a passenger car tire **(Figure 4)**. The

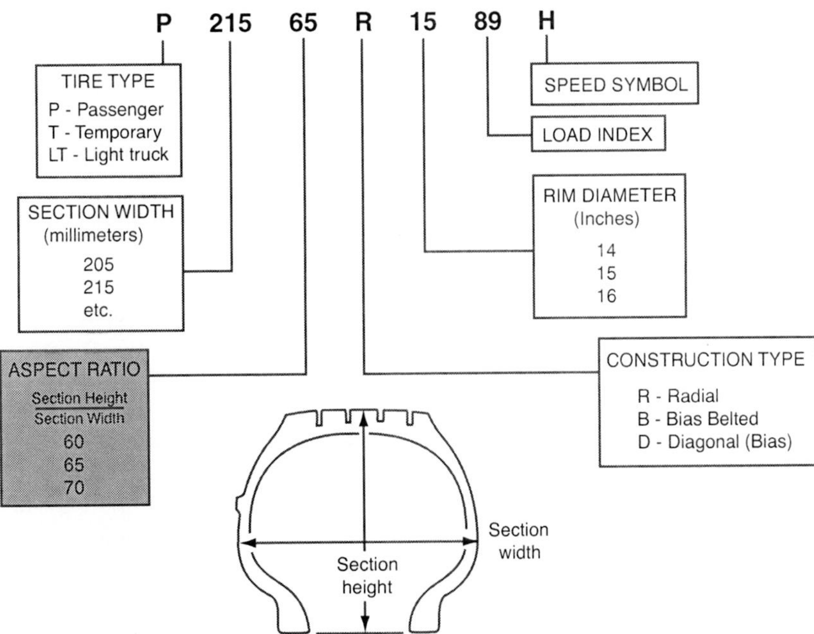


**Figure 4.** Tire sidewall information.

number 215 is the size of the tire in millimeters measured from sidewall to sidewall with the tire mounted on the recommended rim width.

The number 65 indicates the aspect ratio, which is the ratio of the height to the width. With a 65 aspect ratio, the tire's height is 65 percent of its width. The letter R indicates a radial-ply tire design. A belted bias-ply tire design is indicated by the letters A B. The letter D indicates a diagonal bias-ply tire.

The number 15 is the rim diameter in inches. The load index is represented by the number 89. This load rating indicates the tire has a load capacity of 1,279 pounds. Various numbers represent different maximum loads. Some tire manufacturers use the letters B, C, or D to indicate the load rating. The letter B indicates the lowest load rating and the letter C represents a higher load rating. A tire with a D load rating is designed for light-duty trucks. This tire will safely carry a load of 2,623 pounds when inflated to the specified pressure.

## Speed Ratings

Many tires sold in the United States are speed rated with various letters that indicate the maximum speed capabilities of the tire **(Figure 5)**. The letter designation for the speed rating is included on the sidewall markings. Although tires may be speed rated, tire manufacturers do not endorse the operation of a vehicle in an unlawful or unsafe manner. Speed ratings are based on laboratory tests and these ratings are not valid if tires are worn out, damaged, altered, underinflated, or overloaded. Tire speed ratings do not suggest that vehicles can be driven safely at the designated speed rating because many different road and weather conditions may be encountered. Also, the condition of the vehicle may affect high-speed operation.

## Tread Wear Rating

Some other tire ratings available from the manufacturers include tread wear, traction, and temperature resistance. Tread wear ratings allow consumers to compare tire life expectancies. When tires are tread-wear rated, they are installed on a vehicle and driven on a test course for 7,200 miles (11,587 kilometers). Tread wear is calculated after this test course run and the mileage to tread wearout is determined from this calculation. The projected mileage is adjusted for test condition variations and are compared to 30,000 miles (48,279 kilometers) on the test course. This calculation provides a number, or ratio, to compare various tires. For example, a tire with a 150 tread wear rating will provide 50 percent more tire wear mileage than a tire with a 100 tread wear rating. To obtain maximum tire tread life, most vehicle manufacturers recommend tire rotation at specific mileage intervals.

Q — 99 mph
S — 112 mph
T — 118 mph
U — 124 mph
V — above 130 mph without service description
V — 149 with service description
Z — above 149 mph

**Figure 5.** Tire speed ratings.

## Traction Rating

Traction ratings indicate the braking capabilities of the tire to the consumer. To determine the traction rating, ten skid tests are completed on wetted asphalt and concrete surfaces. Test conditions are carefully controlled to maintain uniformity. The results of the ten skid tests are averaged and the traction rating is designated A, B, or C, with an A rating having the best traction.

## Temperature Rating

Temperature resistance ratings indicate the tire's ability to withstand heat generated during tire operation. The National Highway Traffic Safety Administration (NHTSA) has established controlled procedures on a laboratory test wheel for temperature resistance testing of tires. The tire's temperature rating indicates how long the tire can last on the test wheel. Temperature ratings are A, B, or C. An A rating has the best temperature resistance. Tires must have a minimum letter C temperature rating to meet NHTSA standards in the United States.

## Uniform Tire Quality Grading (UTQG) and Department of Transportation (DOT) Designations

Department of Transportation (DOT) requirements specify that tires must have the UTQG system designations molded into the sidewall. The UTQG system designations include the tread wear, traction, and temperature ratings. Typical UTQG designations are Tread wear 160, Traction B, and Temperature B.

Some tire manufacturers use a DOT designation which indicates that the tire has met specific quality tests approved by the DOT. Federal law in the United States requires tire manufacturers to place these designations on tire sidewalls:

- Size
- Load range
- Maximum load
- Maximum pressure in pounds per square inch (psi)
- Number of plies under the tread and in the sidewalls
- Manufacturer's name
- Tubeless or tube construction
- Radial construction (if a radial tire)
- DOT approval number which includes manufacturer's code number, size, type, date of manufacture, and tubeless or tube type

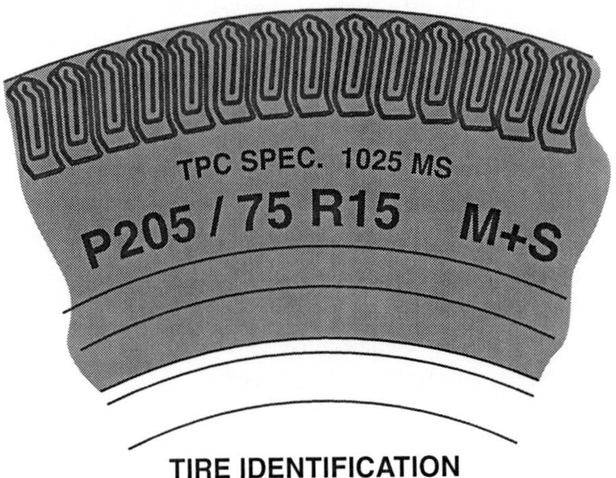


**Figure 6.** Tire performance criteria (TPC) number.

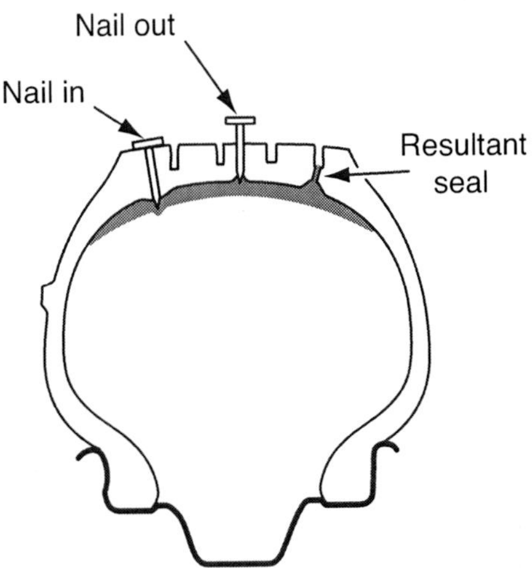


**Figure 7.** A puncture-sealing tire.

## Tire Performance Criteria (TPC) Number

A TPC specification number is molded into the sidewall near the tire rating **(Figure 6)**. The TPC number assures that the tire meets the car manufacturer's performance standards for traction, endurance, dimensions, noise, handling, and rolling resistance. Most car manufacturers assign a different TPC number to each tire size. When replacement tires are selected, these tires should have the same size, load capacity, and construction as the original tires. The replacement tires should have the same TPC number to assure that these tires meet the same performance standards as the original tires.

## ALL-SEASON AND SPECIALTY TIRES

Many tires sold at present are classified as all-season radial tires. These tires have 37 percent higher average snow traction compared to non-all-season tires. All-season tires may have slightly improved performance in areas such as wet traction, rolling resistance, tread life, and air retention. Improvements in tread design and tread compounds provide the superior qualities in all-season tires. These all-season tires are identified by an MS suffix after the TPC number.

Puncture-sealing tires are available as an option on certain car lines and some rubber companies sell these tires in the replacement tire market. These tires contain a special rubber sealing compound applied under the tread area during the manufacturing process. When a nail or other object up to $^{3}/_{16}$ inch in diameter punctures the tread area, it picks up a coating of sealant. If the object is removed, the sealant sticks to the object and is pulled into the puncture. This sealant completely fills the puncture and forms a permanent seal to maintain tire inflation pressure **(Figure 7)**. Puncture-sealing tires usually have a special warranty and these tires can be serviced with conventional tire changing and balancing equipment.

Snow and mud tires are available in various ply and belt designs. These tires provide increased traction in snow or mud compared to conventional tires. Mud and snow tires are identified with an MS suffix after the TPC number on the tire sidewall. When snow tires are installed on a vehicle, these tires should be the same size and type as the other tires on the vehicle. In areas where snow is encountered, all-season tires have replaced snow tires to a large extent. Studded tires provide improved traction on ice, but these tires are prohibited by law in many states because their use resulted in road surface damage.

## REPLACEMENT TIRES

Most tires have tread wear indicators built into the tread. When the tread wears a specific amount, the wear indicators appear as bands across the tread **(Figure 8)**.

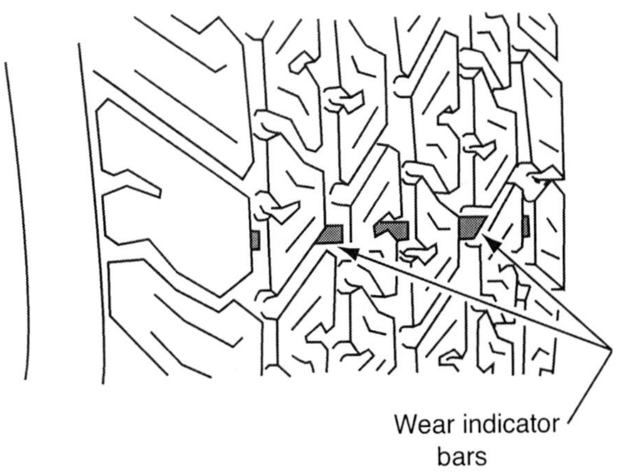


**Figure 8.** Tread wear indicators.

Some car manufacturers recommend tire replacement when the wear indicators appear in two or more tread grooves at three locations around the tire.

If replacement tires have a different size or construction type than the original tires, vehicle handling, ride quality, and speedometer/odometer calibration may be seriously affected. When replacement tires are a different size than the original tires, the vehicle ground clearance and tire-to-body clearance may be altered. Steering and braking quality may be seriously affected if different sizes or types of tires are installed on a vehicle. This does not include the compact spare tire which is intended for temporary use. Many vehicles manufactured in recent years are equipped with ABS. When different sized tires are installed on these vehicles, the ABS operation is abnormal which may result in serious braking defects.

**You Should Know** *If different sizes or types of tires are combined on the same axle or front to rear on a vehicle, handling and braking could be seriously affected. This action may result in vehicle damage and/or personal injury. If a vehicle is equipped with an ABS, installing different size tires than recommended by the vehicle manufacturer may result in braking defects and cause vehicle damage and/or personal injury. If replacement tires have any ratings of lower value than the original tires, a safety hazard may be created which could result in vehicle damage and/or personal injury.*

When selecting replacement tires, these precautions must be observed to maintain vehicle safety:

1. Replacement tires must be installed in pairs on the same axle. Never mix tire sizes or designs on the same axle. If it is necessary to replace only one tire, it should be paired with the tire having the most tread to equalize braking traction.
2. The tire load rating must be adequate for the vehicle on which the tire is installed. Light-duty trucks, station wagons, and trailer-towing vehicles are examples of vehicles that require tires with higher load ratings compared to passenger car tires.
3. Snow tires should be the same type and size as the other tires on the vehicle.
4. A four-wheel drive vehicles should have the same type and size of tires on all four wheels.
5. Do not install tires with a load rating less than the car manufacturer's recommended rating.
6. Replacement tire ratings should be equivalent to the original tire ratings in all rating designations.
7. When combining different tires front to rear, check the car manufacturer's or tire manufacturer's recommendations.

## RUN-FLAT TIRES

Some luxury and sport-type vehicles are equipped with run-flat tires. Run-flat tires eliminate the need for a spare tire and a jack on these cars. This provides a weight and space savings. Run-flat tires must minimize the difference between run-flat tires and conventional tires and provide sufficient zero-pressure durability so the vehicle can be driven a reasonable distance to a repair facility.

Some run-flat tires have stiffer sidewalls that partially support the vehicle weight without air pressure in the tire **(Figure 9)**. Other run-flat tires have a flexible rubber sup-

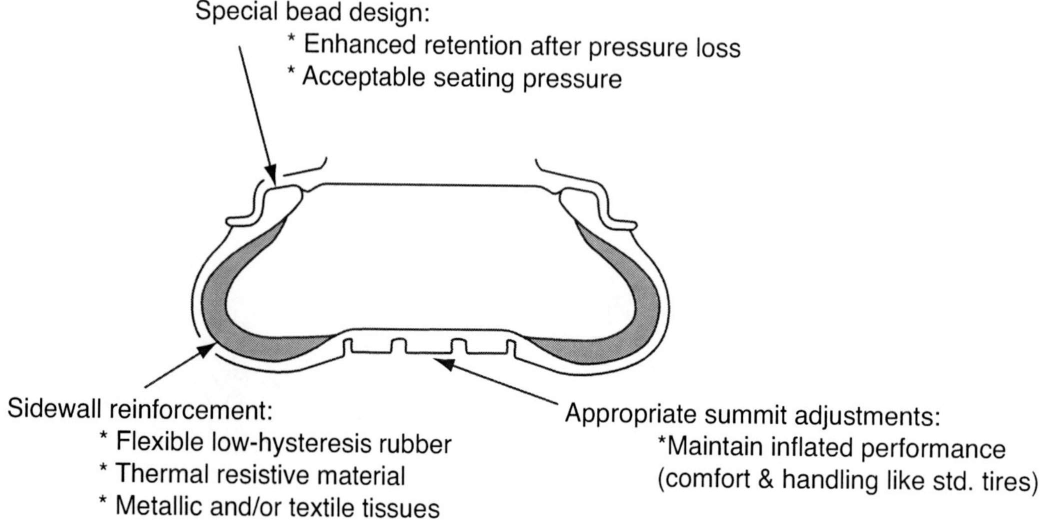


**Figure 9.** A run-flat tire with sidewall reinforcement.

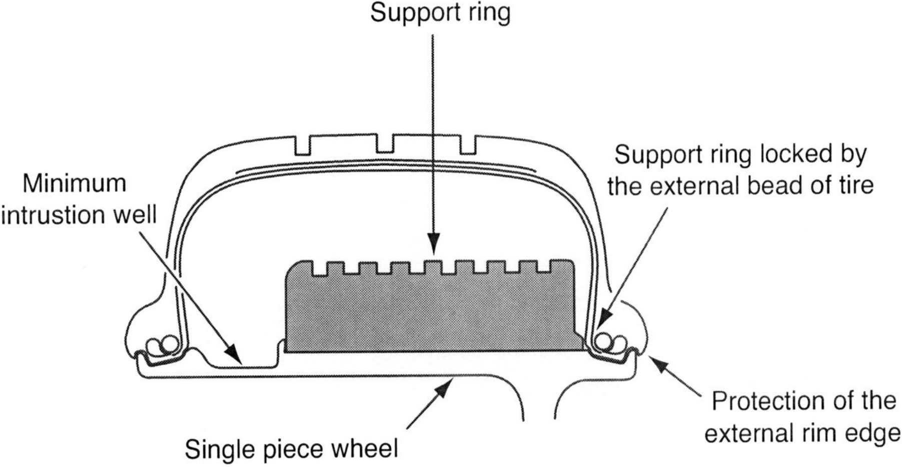


**Figure 10.** A run-flat tire with support ring.

port ring mounted on a special rim to support the vehicle weight if deflation occurs **(Figure 10)**.

## TIRE VALVES

The tire valve allows air to flow into the tire and it is also used to release air from the tire. The core in the center of the valve is spring loaded and allows air to flow inward while the tire is inflated **(Figure 11)**. Once the tire is inflated, the valve core seats and prevents airflow out of the tire. The small pin on the outer end of the valve core may be pushed to unseat the valve core and release air from the tire. An airtight cap on the outer end of the valve keeps dirt out of the valve and provides an extra seal against air leakage. A deep groove is cut around the inner end of the tire valve. When the valve assembly is pulled into the wheel opening, this groove seals the valve in the opening.

## COMPACT SPARE TIRES

Since cars have been downsized in recent years, space and weight have become major concerns for vehicle manufacturers. For this reason, many car manufacturers have marketed cars with compact spare tires to provide a weight

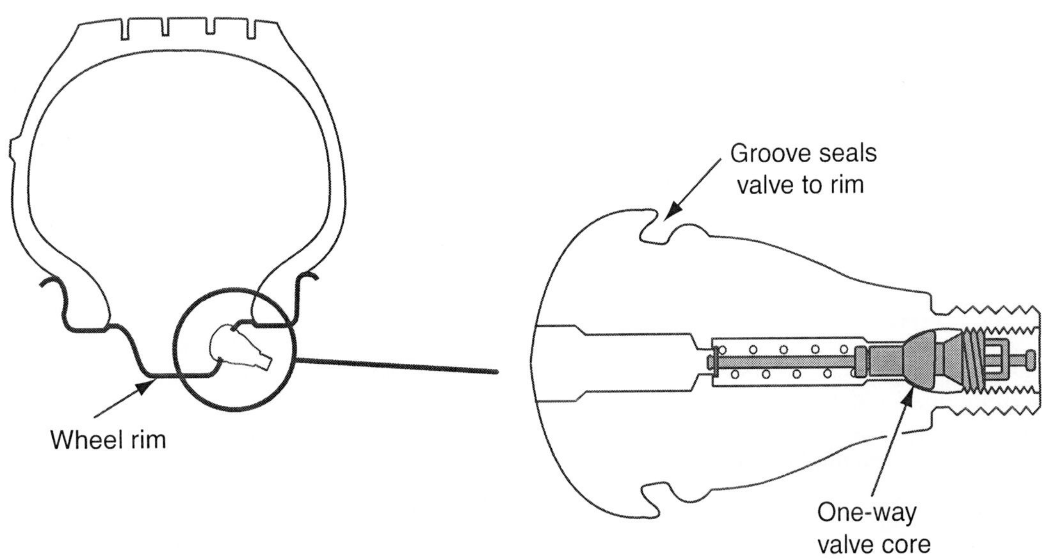


**Figure 11.** A tire valve.

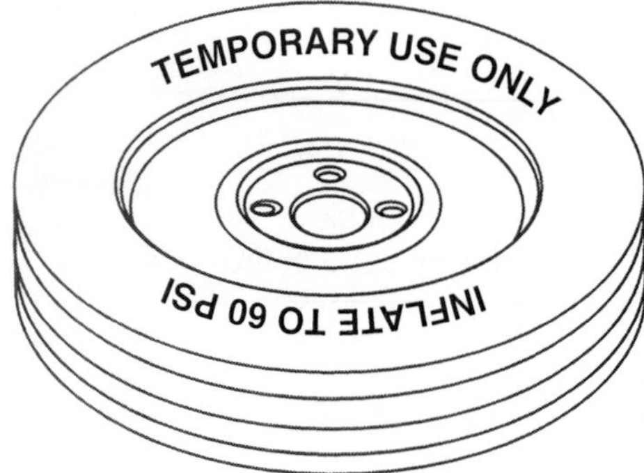


**Figure 12.** A compact, high-pressure mini spare tire.

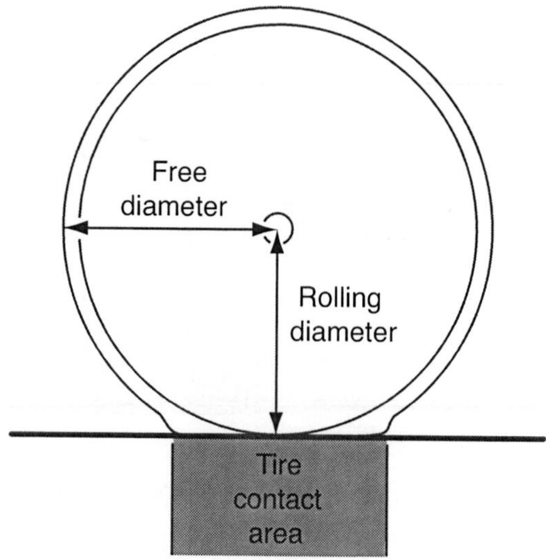


**Figure 13.** Tire rolling diameter, free diameter, and contact area.

and space savings. The high-pressure mini spare tire is the most common type of compact spare **(Figure 12)**. This compact spare rim is usually 4 inches wide, but it is 1 inch larger in diameter compared to the other rims on the vehicle. The compact spare rim should not be used with standard tires, snow tires, wheel covers, or trim rings. Any of these uses may result in damage to these items or other parts of the vehicle. The compact spare should be used only on vehicles that offered it as original equipment. Inflation pressure in the compact spare should be maintained at 60 psi (415 kPa). The compact spare tire is designed for very temporary use until the conventional tire can be repaired or replaced. Limit driving speed to 50 mph (80 kph) when the high-pressure mini spare is installed on a vehicle.

The space-saver spare tire must be inflated with a special compressor. Battery voltage is supplied to the compressor from the cigarette lighter. This type of compact spare should be inflated to 35 psi (240 kPa). After the tire is inflated, be sure there are no folds in the sidewalls.

The lightweight skin spare tire is a bias-ply tire with a reduced tread depth to provide an estimated 2,000 miles (3,200 km) of tread life. This type of spare tire is also designed for emergency use only. Driving speed must be limited to 50 mph (80 kph) when this tire is installed on a vehicle. Always inflate the lightweight skin spare to the pressure specified on the tire placard.

## TIRE CONTACT AREA

The tire contact area refers to the area of the tire that is in contact with the road surface when the tire is supporting the vehicle weight. The free diameter of a tire is the distance of a horizontal line through the center of the spindle and wheel to the outer edges of the tread. The rolling diameter of a tire is the distance of a perpendicular straight line through the center of the spindle to the outer edges of the tread when the tire is supporting the vehicle weight. The rolling diameter is always less than the free diameter. The difference between the free diameter and the rolling diameter is referred to as deflection. Tire tread grooves take up excess rubber and prevent scrubbing as the tire deflects in the contact area. The rolling diameter, free diameter, and contact area are shown in **Figure 13**.

## TIRE PLACARD AND INFLATION PRESSURE

The vehicle weight is supported by the correct air pressure exerted evenly against all the interior tire surface which produces tension in the tire carcass. Therefore, tire pressure is extremely important. Underinflation decreases the rolling diameter and increases the contact area which results in excessive sidewall flexing and tread wear. Overinflation decreases the contact area, increases the rolling diameter, and stiffens the tire. This action results in excessive wear on the center of the tread. Tire pressure should be checked when the tires are cool. Since tire pressure normally increases at high tire temperatures, air pressure should not be released from hot tires. Excessive heat buildup in a tire may be caused by underinflation. This condition may lead to severe tire damage.

On many vehicles, the tire placard is permanently attached to the rear face of the driver's door. This placard provides tire information such as maximum vehicle load,

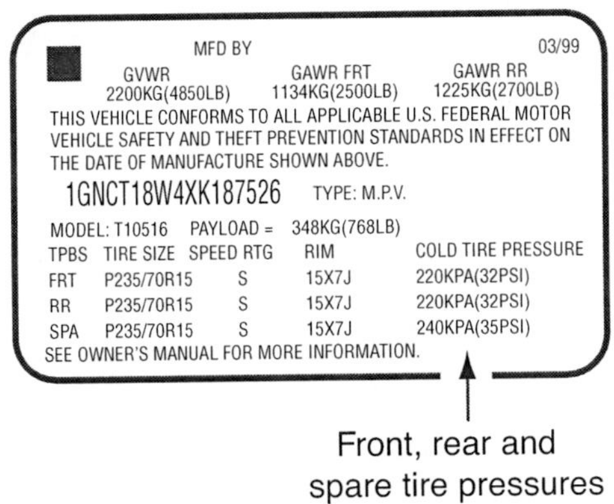


**Figure 14.** A tire placard.

tire size including spare, and cold inflation pressure including spare **(Figure 14)**.

Tire pressure is carefully calculated by the vehicle manufacturer to provide satisfactory tread life, handling, ride, and load-carrying capacity. Most vehicle manufacturers recommend that tire pressures be checked cold once a month or prior to any extended trip. The manufacturer considers the tires to be cold after the vehicle has set for three hours or when the vehicle has been driven less than one mile. The tires should be inflated to the pressure indicated on the tire placard. Tire pressures may be listed in metric or USC system values.

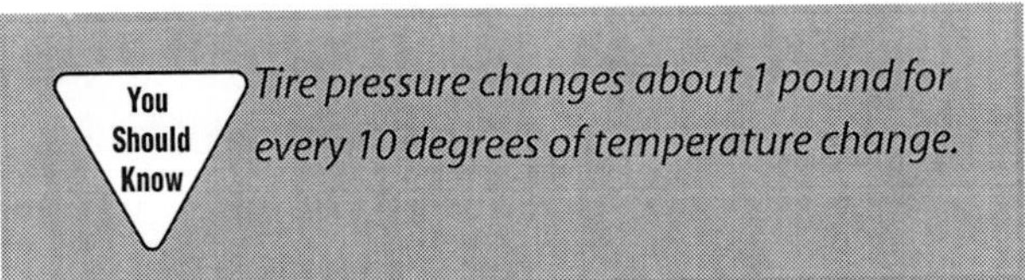

You Should Know
*Tire pressure changes about 1 pound for every 10 degrees of temperature change.*


## TIRE PRESSURE MONITORING SYSTEMS

Tire pressure monitoring systems are mandatory on all new vehicles produced in the United States starting in the 2004 model year. Some tire pressure monitoring systems have a pressure sensor strapped to the drop center in each rim **(Figure 15)**. Other systems have a pressure sensor threaded onto the end of the valve stem. The pressure sensors send radio frequency (RF) signals to the module in the tire pressure monitoring system. These RF signals change if the tire is deflated a specific amount. When the module senses a tire with low air pressure, the module illuminates a warning light in the instrument panel.

Other tire pressure monitoring systems use the wheel speed sensor signals in the ABS to monitor tire inflation pressure. When a tire is deflated to some extent, the tire diameter is smaller and wheel speed increases. Therefore, the wheel speed sensor signals may be used to indicate low tire pressure.

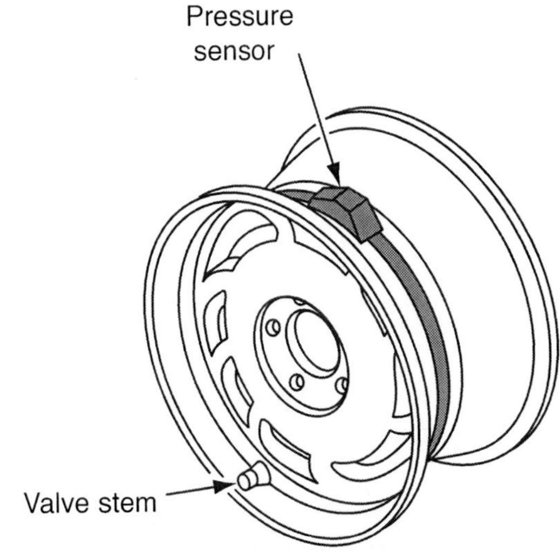


| Location | Color code |
|---|---|
| Right front | Blue |
| Left front | Green |
| Right rear | Orange |
| Left rear | Yellow |

**WARNING**: Pressure sensor inside tire. Avoid contacting sensor with tire changing equipment tools or tire bead.

Service note: Pressure sensor must be mounted directly across from valve stem.

**Figure 15.** A tire pressure sensor.

## TIRE MOTION FORCES

When a vehicle is in motion, wheel rotation subjects the tires to centrifugal force. The tires are also subjected to accelerating and decelerating forces because of their path of travel. If a vehicle is travelling at 55 miles per hour (88.5 kilometers per hour), the part of the tire exactly fore and aft of the spindle is also travelling at 55 miles per hour (88.5 kilometers per hour). At the exact top of the tire, the tire speed is 110 miles per hour (177 kilometers per hour). The tire speed actually drops to zero at the exact bottom of the tire where the arc of deceleration ends and the arc of acceleration begins. Since the tires are subjected to strong acceleration and deceleration forces, the tire construction must be uniform. For example, a soft spot in a tire will deflect further than the surrounding area. This area will be subjected to rapid wear as it strikes the road surface.

## WHEEL RIMS

Many wheel rims are manufactured from stamped or pressed steel discs that are riveted or welded together to form the circular rim. If a rim is designed with positive offset, the rim centerline is inboard of the mounting face **(Figure 16)**. A rim with negative offset has the centerline outboard of the mounting face. The rim offset affects front suspension loading and operation. The rim **offset** is the distance between the rim centerline and the mounting face of the disc.

You Should Know *If a wheel does not have the same width, diameter, offset, load capacity, and mounting configuration as the original wheel, steering quality, vehicle control, tire life, and wheel bearing life may be adversely affected. An incorrect wheel may create a safety hazard and cause vehicle damage and/or personal injury.*

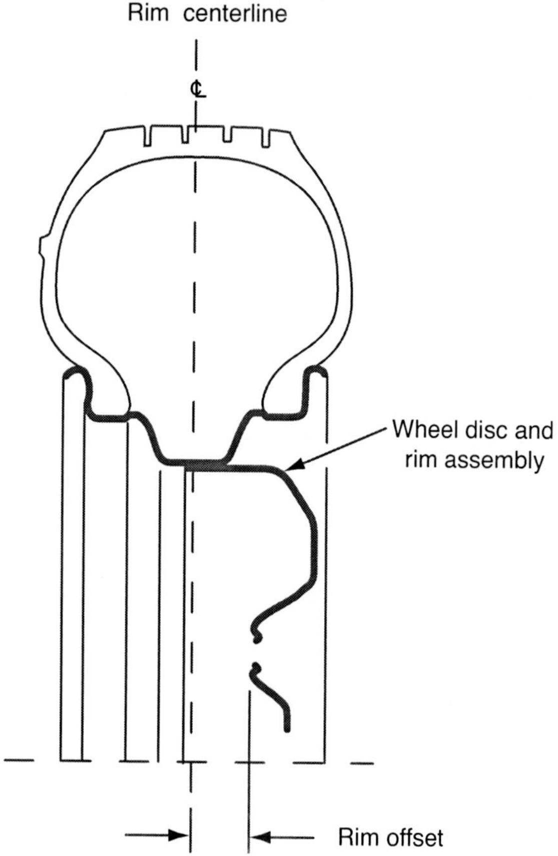


**Figure 16.** Wheel rim design.

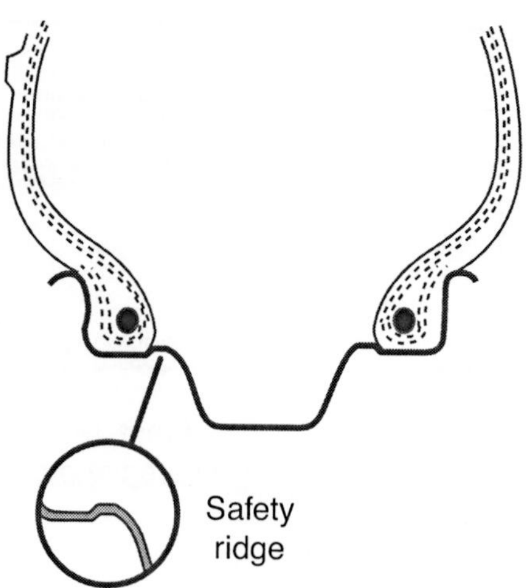


**Figure 17.** A wheel rim with drop center and safety ridges.

A large hole in the center of the rim fits over a flange on the mounting surface and the rim has a small hole for the valve stem. The wheel stud mounting holes in the rim are tapered to match the taper on the wheel nuts. The width of the wheel is measured between the rim flanges. Rim diameter is determined by measuring across the wheel from the top to the bottom. A drop center in the rim makes tire changing easier **(Figure 17)**. Rims have safety ridges behind the tire bead locations which help to prevent the beads from moving into the drop center area if the tire blows out. If the tire blows out and a bead enters the drop center area, the tire may come off the wheel. Replacement wheel rims must be the same as the original equipment wheels in load capacity, offset, width, diameter, and mounting configuration. An incorrect wheel can affect tire life, steering quality, wheel bearing life, vehicle ground clearance, tire clearance, and speedometer/odometer calibrations.

Some vehicles are equipped with cast aluminum alloy wheel rims or cast magnesium alloy wheel rims. Sometimes these wheels are referred to as "mag" wheels. These wheels are lighter and generally more accurately designed compared to pressed steel wheel rims.

## STATIC WHEEL BALANCE THEORY

When a wheel and tire has proper static balance, it has the weight equally distributed around its axis of rotation and gravity will not force it to rotate from its rest position. If a vehicle is raised off the floor and a wheel is rotated in 120-degree intervals, a statically balanced wheel will remain stationary at each interval. When wheel and tire are statically unbalanced, the tire has a heavy portion at one location. The force of gravity acting on this heavy portion will cause

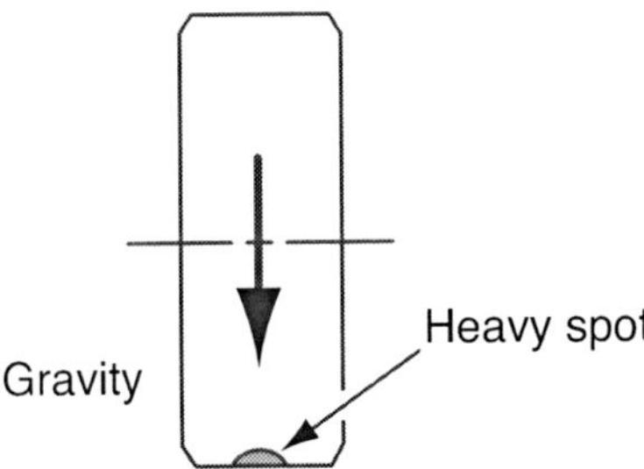


**Figure 18.** Static wheel unbalance.

the wheel to rotate when the heavy portion is located near the top of the tire **(Figure 18)**.

## Results of Static Unbalance

Centrifugal force may be defined as the force that tends to move a rotating mass away from its axis of rotation. As we have explained previously, a tire and wheel are subjected to very strong acceleration and deceleration forces when a vehicle is in motion. The heavy portion of a statically unbalanced wheel is influenced by centrifugal force. This influence attempts to move the heavy spot on a tangent line away from the wheel axis. This action tends to lift the wheel assembly off the road surface **(Figure 19)**.

The wheel lifting action caused by static unbalance may be referred to as **wheel tramp (Figure 20)**. This wheel tramp action allows the tire to slip momentarily when it is lifted vertically. When the wheel and tire move downward as the heavy spot decelerates, the tire strikes the road surface with a pounding action. This repeated slipping and pounding action causes severe tire scuffing and cupping **(Figure 21)**.

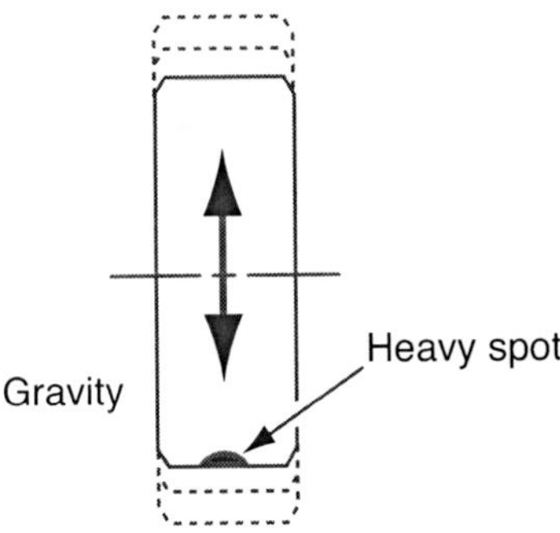


**Figure 19.** Effects of static unbalance.

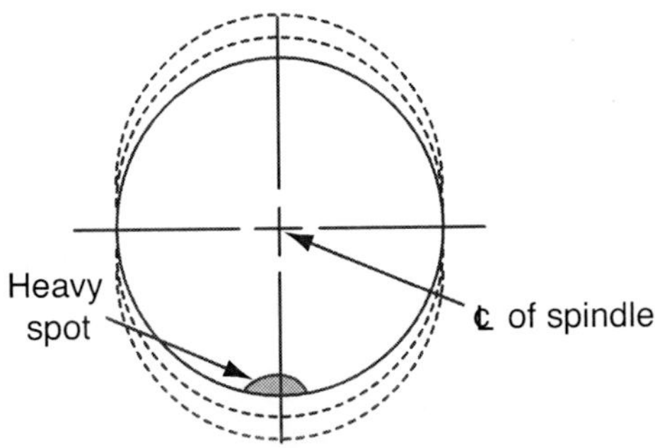


**Figure 20.** Wheel tramp.

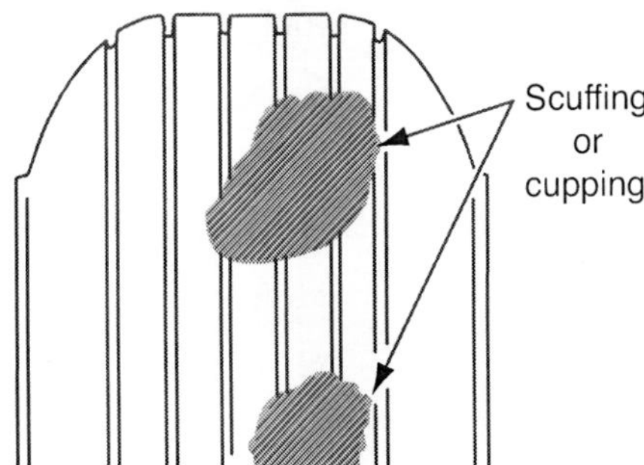


**Figure 21.** Cupping tire wear caused by static unbalance.

The vertical wheel motion from static unbalance is transferred to the suspension system and is then absorbed by the chassis and body. This action causes rapid wear on suspension and steering components. The wheel tramp action resulting from static unbalance is also transmitted to the passenger compartment which causes passenger discomfort and driver fatigue.

When a vehicle is traveling at normal highway cruising speed, the average wheel speed would be 850 revolutions per minute (rpm). A statically unbalanced tire and wheel assembly is an uncontrolled mass of weight in motion. When a vehicle is traveling at 60 miles per hour (97 kilometers per hour), if a tire has 2 ounces (57 grams) of static unbalance, the resultant pounding force is approximately 15 pounds (6.8 kilograms) against the road surface.

## DYNAMIC WHEEL BALANCE THEORY

When a wheel and tire assembly has correct dynamic balance, the weight of the assembly is distributed equally on both sides of the wheel center viewed from the front. Dynamic wheel balance may be explained by dividing the tire into four sections **(Figure 22)**.

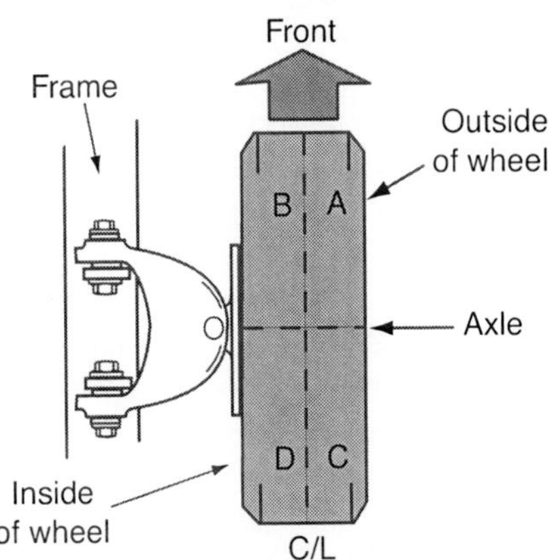


**Figure 22.** Dynamic wheel balance theory.

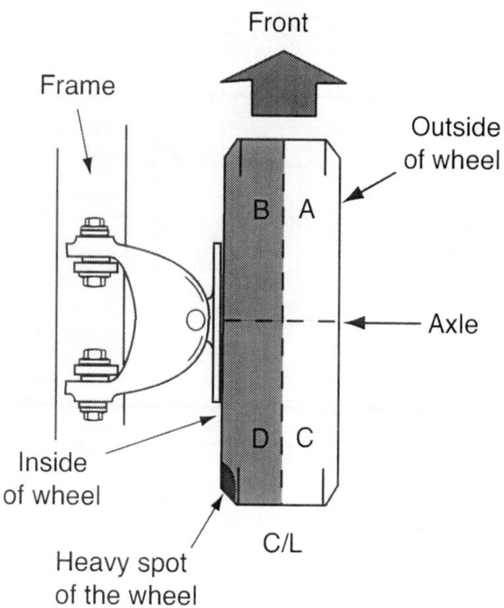


**Figure 23.** Dynamic unbalance.

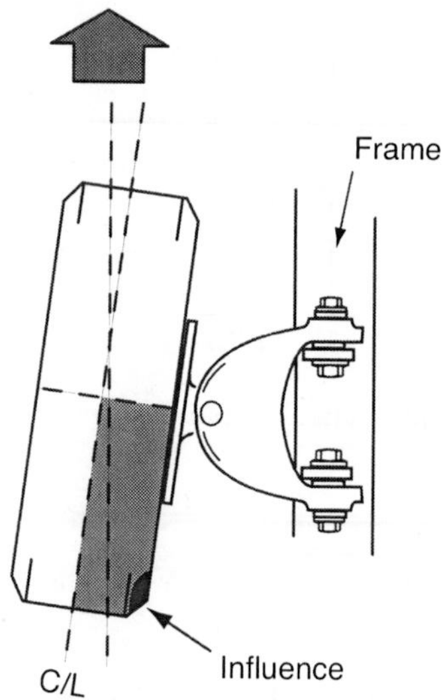


**Figure 24.** Dynamic wheel unbalance with heavy spot at the rear of the left front wheel.

In Figure 22, if sections A and C have the same weight and sections B and D also have the same weight, the tire has proper dynamic balance. If a tire has dynamic unbalance, section D may have a heavy spot, and thus sections B and D have different weights **(Figure 23)**.

From our discussion of dynamic balance, we can understand that a tire and wheel assembly may be in static balance but have dynamic unbalance. Therefore, wheels must be in balance statically and dynamically.

## Results of Dynamic Wheel Unbalance

When a dynamically unbalanced wheel is rotating, centrifugal force moves the heavy spot toward the tire center line. The center line of the heavy spot arc is at a 90 degree angle to the spindle. This action turns the true center line of the left front wheel inward when the heavy spot is at the rear of the wheel **(Figure 24)**.

When the wheel rotates until the heavy spot is at the front of the wheel, the heavy spot movement turns the left front wheel outward **(Figure 25)**.

From the these explanations we can understand that dynamic wheel unbalance causes wheel shimmy **(Figure 26)**. This action causes steering wheel oscillations at medium and high speeds with resultant driver fatigue and passenger discomfort. Wheel shimmy and steering wheel oscillations also cause unstable directional control of the vehicle. **Wheel shimmy** is the rapid, repeated lateral wheel movement.

In a previous discussion of wheel rotation in this chapter, we mentioned that a tire stops momentarily where it contacts the road surface. A wheel with dynamic unbalance is forced to pivot on the contact area which results in excessive tire scuffing and wear. Dynamic wheel unbalance

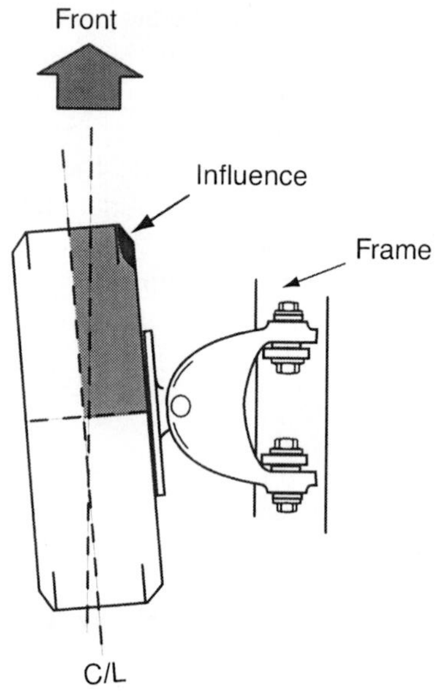


**Figure 25.** Dynamic wheel unbalance with heavy spot at the front of the left front wheel.

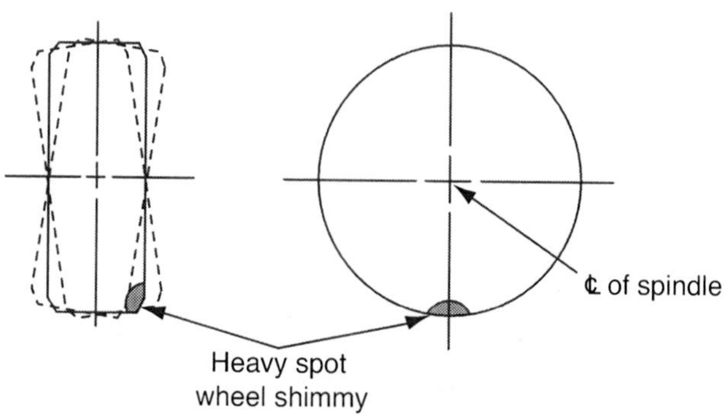


**Figure 26.** Dynamic wheel unbalance causes wheel shimmy.

causes premature wear on steering linkage and suspension components. Therefore, dynamic wheel balance is extremely important to provide normal tire life, reduce steering and suspension component wear, increase directional control, and decrease driver fatigue. The main purposes of proper wheel balance may be summarized as follows: maintains normal tire tread life, provides extended life of suspension and steering components, helps to provide directional control of the vehicle, reduces driver fatigue, increases passenger comfort, and helps to maintain the life of body and chassis components.

## WHEEL BEARINGS

Bearings are precision machined assemblies that provide smooth operation and long life. When bearings are properly installed and maintained, bearing failure is rare. When a technician understands different bearing loads, various types of wheel bearings, and the loads these bearings are designed to withstand, diagnosing wheel bearing problems becomes much easier.

### Bearing Loads

When a bearing load is applied in a vertical direction, it is called a radial load. If the vehicle weight is applied straight downward on a bearing, this weight is a radial load on the bearing. A **thrust bearing load** is applied in a horizontal direction **(Figure 27)**. For example, while a vehicle is turning a corner, horizontal force is applied to the front wheel bearings. When **angular load** is applied to a bearing, the angle of the applied load is somewhere between the horizontal and vertical positions. A **radial load** on a bearing is applied in a vertical direction.

Front and rear wheel bearings may be cylindrical ball bearings or roller bearings. Either type of bearing contains these basic parts: an inner race, or cone; a separator, also called a cage or retainer; rolling elements, balls or rollers; and an outer race, or cup.

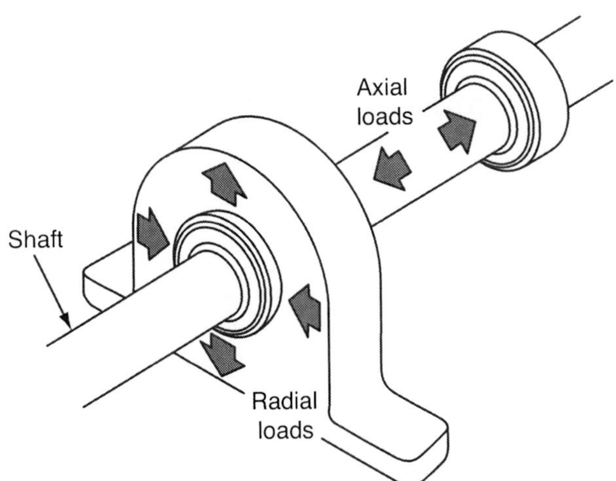


**Figure 27.** Types of bearing loads.

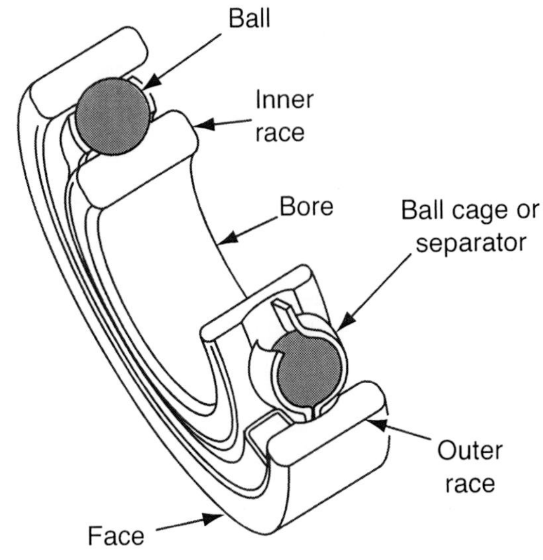


**Figure 28.** Parts of a cylindrical ball bearing.

The inner race is an accurately machined component and the inner surface of the race is mounted on the shaft with a precision fit. The rolling elements are mounted on a very smooth machined surface on the inner race. Positioned between the inner and outer races, the separator retains the rolling elements and keeps them evenly spaced. The rolling elements have precision machined surfaces. These elements are mounted between the inner and outer races. The outer race is the bearing's exterior ring and both sides of this component have precision machined surfaces. The outer surface of this race supports the bearing in the housing and the inner surface is in contact with the rolling elements.

A single-row ball bearing has a crescent-shaped machined surface in the inner and outer races in which the balls are mounted **(Figure 28)**. When a ball bearing is at rest, the load is distributed equally through the balls and races in the contact area. When one of the races and the balls begin to rotate, the bearing load causes the metal in the race to bulge out in front of the ball and flatten out behind the ball **(Figure 29)**. This action creates a certain amount of friction within the bearing and the same action

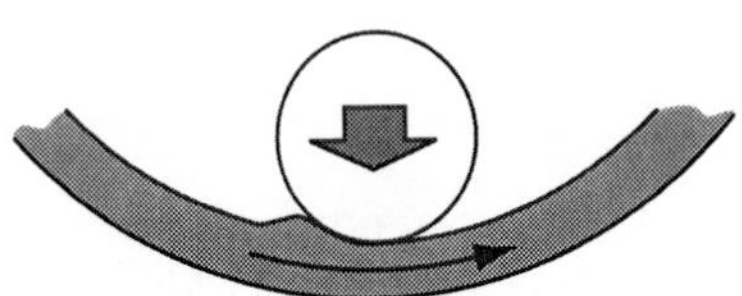

**Figure 29.** When a load is applied to a ball bearing, the metal in the race bulges out in front of the ball and flattens out behind the ball.

is repeated for each ball while the bearing is rotating. If metal-to-metal contact is allowed between the balls and races, these components would experience very fast wear. Therefore, bearing lubrication is extremely important to eliminate metal-to-metal contact in the bearing and reduce wear.

A cylindrical ball bearing is designed primarily to handle radial loads. However, this type of bearing can also withstand a considerable amount of thrust load in either direction even at high speeds. A maximum capacity ball bearing has extra balls for greater radial load-carrying capacity. Ball bearings are available in many different sizes for various applications.

Double-row ball bearings contain two rows of balls side by side. As in the single-row ball bearing, the balls in the double-row bearing are mounted in crescent-shaped grooves in the inner and outer races. The double-row ball bearing can support heavy radial loads and this type of bearing can also withstand thrust loads in either direction.

## Cylindrical Roller Bearings

A cylindrical roller bearing contains precision machined rollers that have the same diameter at both ends. These rollers are mounted in square-cut grooves in the outer and inner races **(Figure 30)**. In the cylindrical roller bearing, the races and rollers run parallel to one another. Cylindrical roller bearings are designed primarily to carry radial loads, but they can withstand some thrust load.

## Tapered Roller Bearings

In a tapered roller bearing, the inner and outer races are cone shaped. If imaginary lines extend through the inner and outer races, these lines taper and eventually meet at a point extended through the center of the bearing **(Figure 31)**. The most important advantage of the tapered roller bearing compared to other bearings is an excellent capability to carry radial, thrust, and angular loads, especially when used in pairs. In the tapered roller bearing, the rollers are mounted on cone-shaped precision surfaces in the outer and inner races. The bearing separator has an open space over each roller **(Figure 32)**. Grooves cut in the side of the separator roller openings match the curvature of the roller. This design allows the rollers to rotate evenly without interference between the rollers and the separator. Lubrication and proper end-play adjustment are critical on tapered roller bearings.

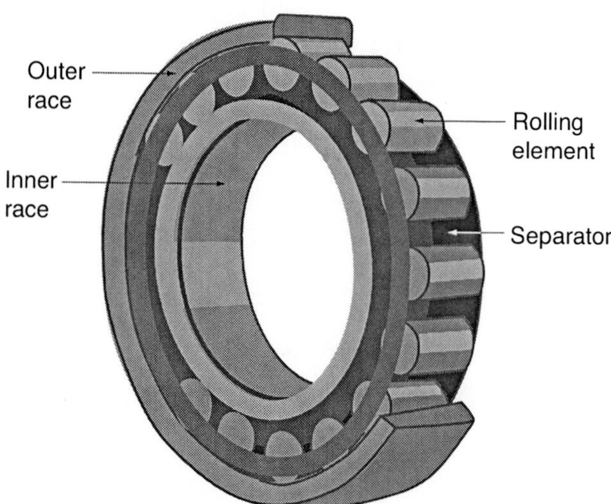


Figure 30. Parts of a cylindrical roller bearing.

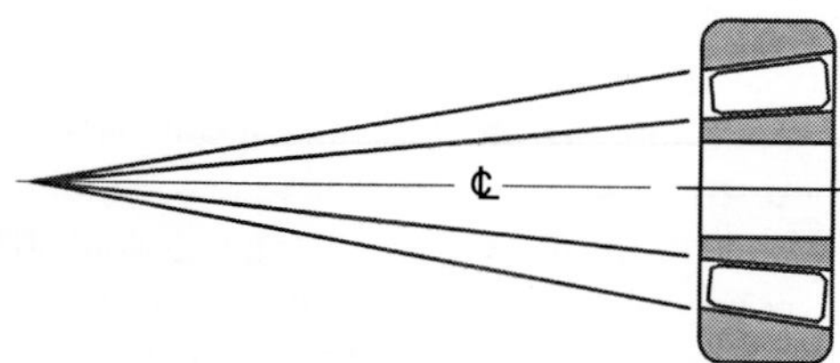


Figure 31. Imaginary lines extending from the tapered roller bearing races eventually meet at a point extending from the bearing center.

Figure 32. Tapered roller bearings.

## WHEEL BEARING SEALS

Seals are designed to keep lubricant in the bearing and prevent dirt particles and contaminants from entering the bearing. Wheel bearing seals are mounted in front and rear wheel hubs and in rear-axle housings on rear-wheel drive cars. The metal seal case has a surface coating that resists corrosion and rust and acts as a bonding agent for the seal material. Seals have many different designs including single lip, double lip, and fluted. The seal material is usually made of a synthetic rubber compound such as nitrile, silicon, polyacrylate, or a fluoroelastomer such as Viton. The actual seal material depends on the lubricant and contami-

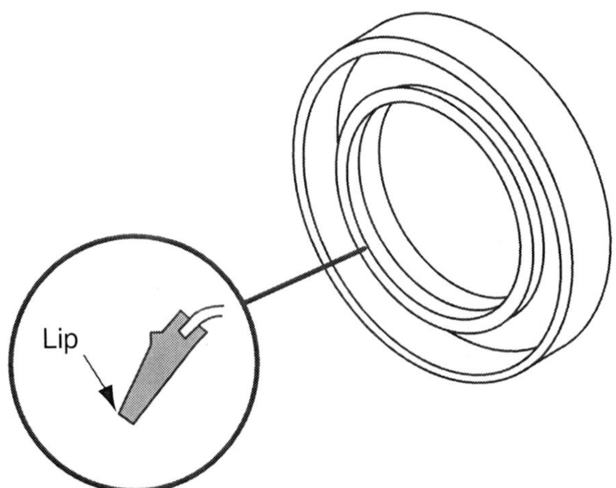


Figure 33. A springless seal.

nants that the seal encounters. All seals may be divided into two groups, springless and spring loaded. Springless seals are used in some front- or rear-wheel hubs where they seal a heavy lubricant into the hub **(Figure 33)**.

In a spring-loaded seal, the garter spring behind the seal provides additional force on the seal lip to compensate for lip wear, shaft movement, and bore eccentricity **(Figure 34)**. If a seal must direct oil back into a housing, the seal lip is fluted. This seal design provides a pumping action to redirect the oil back into the housing **(Figure 35)**.

Some seals have a sealer painted on the outside surface of the metal seal housing. When the seal is installed, this sealer prevents leaks between the seal case and the housing **(Figure 36)**. Some rear axle bearings are an example of a bearing with seals attached to each side of the bearing **(Figure 37)**.

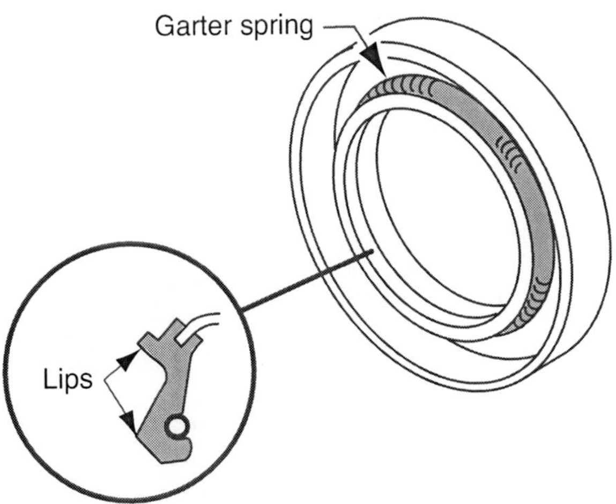


Figure 34. A spring-loaded seal.

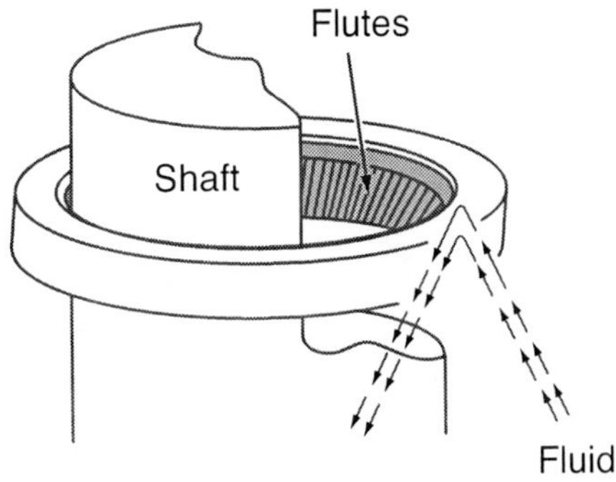


Figure 35. Fluted seal lip redirects oil back into the housing.

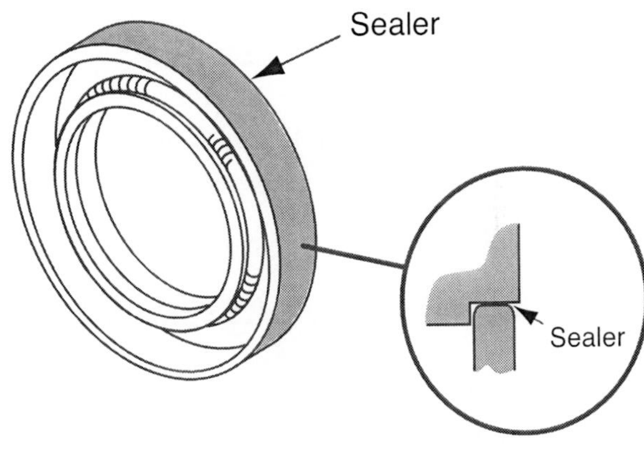


Figure 36. Sealer painted on the seal case prevents leaks between the case and the housing.

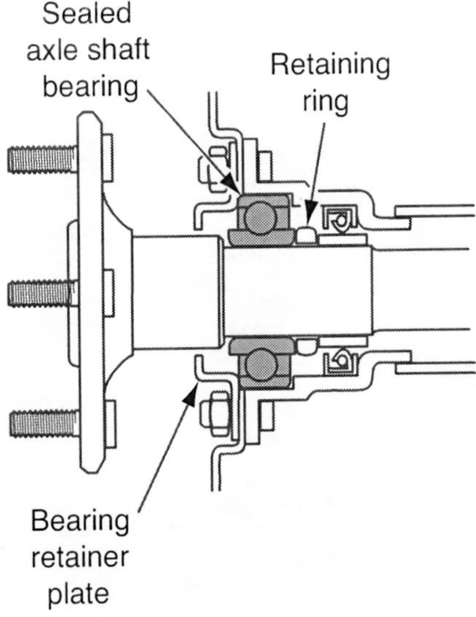


Figure 37. Some rear axle bearings are sealed on both sides and retained on the axle with a retainer ring.

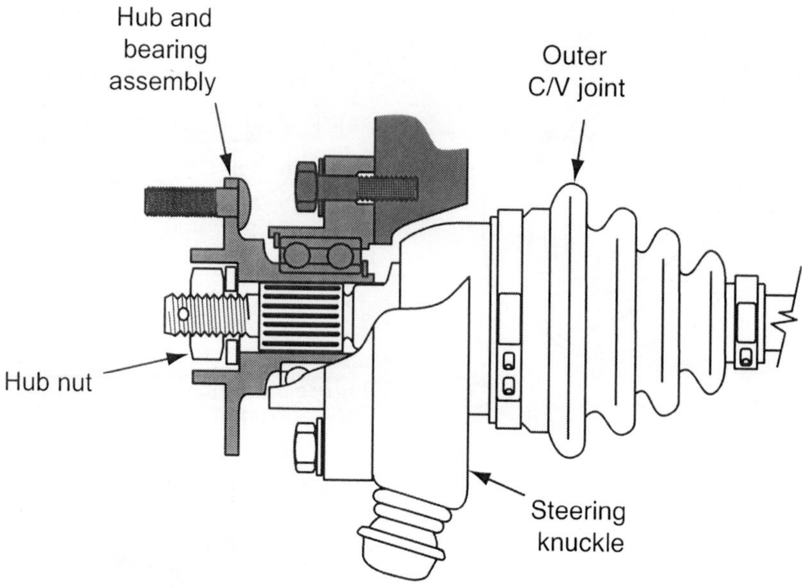


Figure 38. Wheel bearing and hub assembly.

## WHEEL BEARING HUB ASSEMBLIES

Some front-wheel drive vehicles have front-wheel bearing and hub assemblies that are bolted to the steering knuckles **(Figure 38)**. The bearings are lubricated and sealed and the complete bearing and hub assembly is replaced as a unit. The bearing and hub unit is more compact compared to other types of wheel bearings mounted in the wheel hub. This type of bearing contains two rows of ball bearings with an angular contact angle of 32 degrees **(Figure 39)**. The inner bearing assembly bore is splined and the inner ring extends to the outside to form a flange and spigot. The flange attached to the outer ring contains bolt holes. Bolts extend through these holes into the steering knuckle. This type of bearing attachment allows the bearing to become a structural member of the front suspension. Since the bearing outer ring is self supporting, the main concern in knuckle design is fatigue strength rather than stiffness. The drive axle shaft transmits torque to the inner bearing race. This shaft is not designed to hold the bearing together. This type of wheel bearing is designed for mid-sized front-wheel drive (FWD) vehicles.

Figure 39. A double-row, sealed wheel bearing hub unit.

Each front drive axle has splines that fit into matching splines inside the bearing hubs **(Figure 40)**. A hub nut secures the drive axle into the inner bearing race.

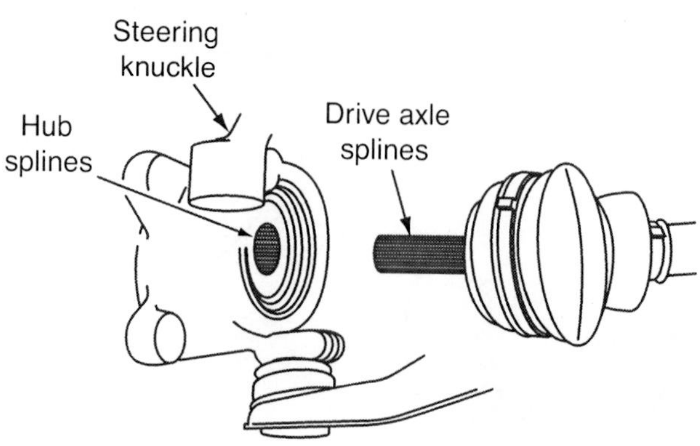


Figure 40. A front drive axle installed in a wheel bearing hub.

## Front Steering Knuckles with Two Separate Tapered Roller Bearings

Some front-wheel drive vehicles have a one-piece, dual tapered roller bearing assembly mounted in the steering knuckles **(Figure 41)**. Other front wheel drive vehicles have separate tapered roller bearings with individual races pressed into the steering knuckle and seals are located in the knuckle on outboard side of each bearing **(Figure 42)**. Correct bearing end-play adjustment is supplied by the hub nut torque. The wheel hub is pressed into the inner bearing races and the drive axle splines are meshed with matching splines in the wheel hub.

## Wheel Hubs with Two Separate Tapered Roller Bearings

Many rear-wheel drive cars have two tapered roller bearings in the front hubs that support the hubs and wheels on the spindles. This type of front wheel bearing has the bearing races pressed into the hub. A grease seal is pressed into the inner end of the hub to prevent grease leaks and to keep contaminants out of the bearings. The hub and bearing assemblies are retained on the spindle with a washer, adjusting nut, nut lock, and cotter pin. The adjusting nut must be properly adjusted to provide the correct bearing end play. A grease cap is pressed into the outer end of the hub to prevent bearing contamination.

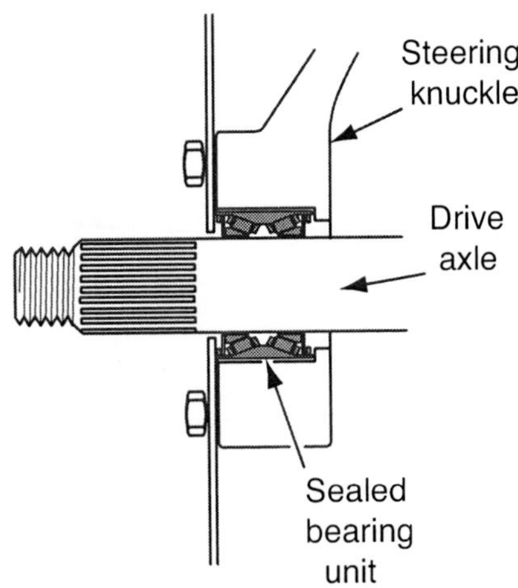


**Figure 41.** A steering knuckle with a pressed-in bearing.

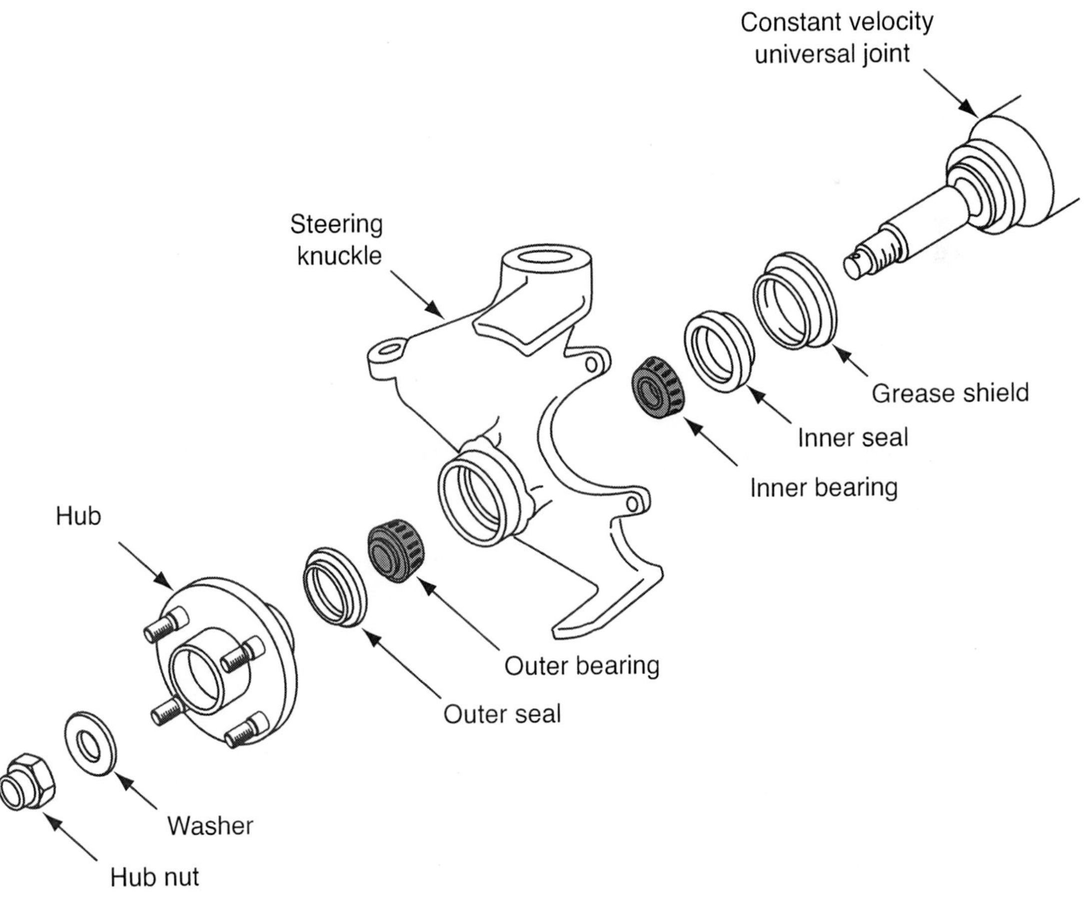


**Figure 42.** A steering knuckle with two separate tapered roller bearings.

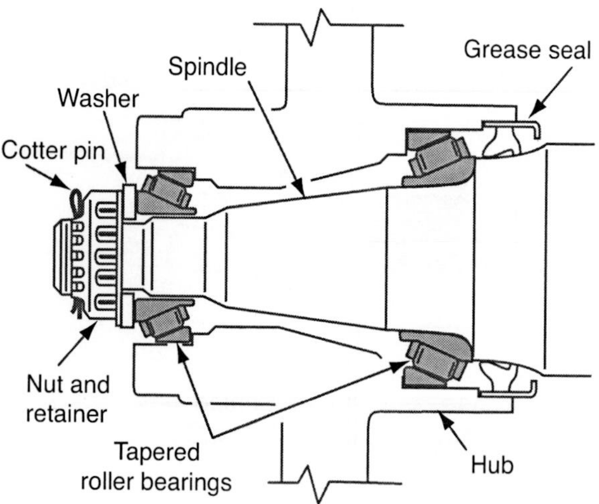


**Figure 43.** A front wheel bearing assembly, rear-wheel drive car.

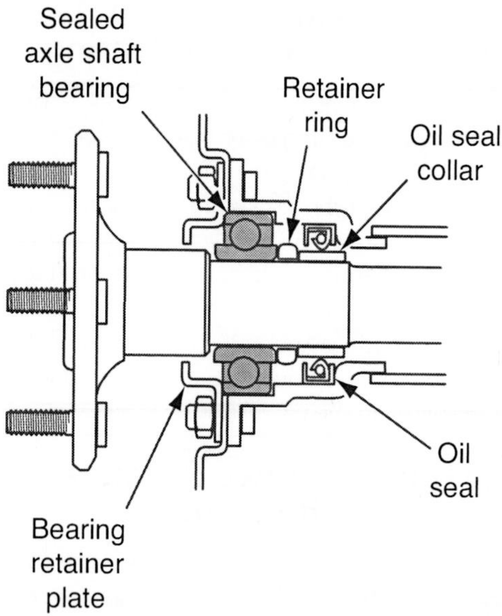


**Figure 45.** Rear axle bearing and retainer.

Some front-wheel drive cars have two tapered roller bearings in the rear-wheel hubs that are very similar to the front wheel bearings in a rear-wheel drive vehicle **(Figure 43)**.

## REAR AXLE BEARINGS

On many rear-wheel drive vehicles, the rear axles are supported by roller bearings mounted near the outer ends of the axle housing. The outer bearing race is pressed into the housing. A machined surface on the axle contacts the inner roller surface. A seal is mounted in the axle housing on the outboard side of each bearing **(Figure 44)**. This type of axle bearing is usually not sealed and lubricant in the differential and rear axle housing provides axle bearing lubrication. The seals prevent lubricant leaks from the outer ends of the axle housing and keep dirt out of the bearings.

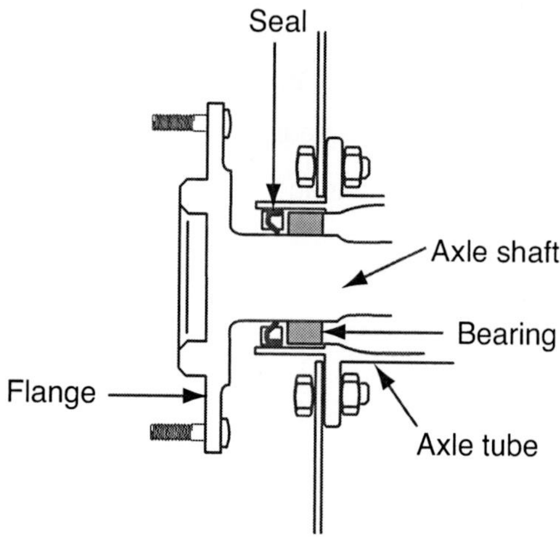


**Figure 44.** A rear axle bearing, rear-wheel drive car.

Other rear axle bearings on rear-wheel drive vehicles have sealed roller bearings pressed onto the rear axles. These axle bearings are sealed on both sides, and a retainer ring is pressed onto the axle on the inboard side of the bearing **(Figure 45)**. The outer bearing race is mounted in the rear axle housing with a light press fit. A seal is positioned in the housing on the inboard side of the bearing and adaptor ring. A retainer plate is mounted between the bearing and the outer end of the axle. This retainer plate is bolted to the axle housing to retain the axle in place.

## BEARING LUBRICATION

Proper bearing lubrication is extremely important to maintain bearing life. Bearing lubricant reduces friction and wear, dissipates heat, and protects surfaces from dirt and corrosion. Sealed or shielded bearings are lubricated during the manufacturing process. No attempt should be made to wash these bearings or pack them with grease.

Bearings that are not sealed or shielded require cleaning and repacking at intervals specified by the vehicle manufacturer. Always use the bearing grease specified by the vehicle manufacturer. Bearing lubricants may be classified as greases or oils. Many wheel bearing greases are lithium or sodium based.

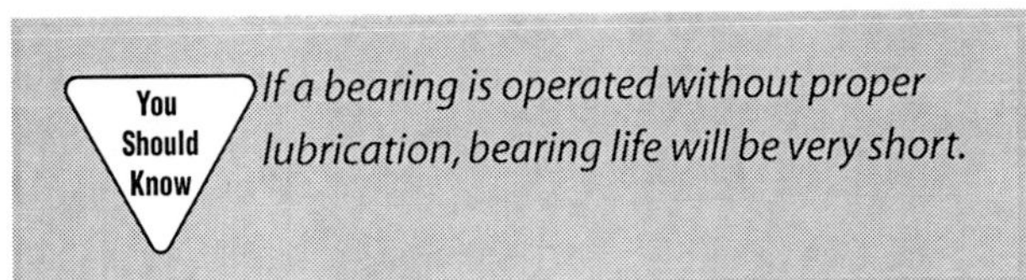

You Should Know
If a bearing is operated without proper lubrication, bearing life will be very short.


New bearings usually have a protective coating to prevent rust and corrosion. This coating should not be washed from the bearing. When rear axle bearings are lubricated from the differential housing, the type and level of oil in the housing is important.

Vehicle manufacturers usually recommend an SAE No. 75W-90 or SAE No. 140 hypoid gear oil in the differential. In very cold climates, the manufacturer may recommend an SAE No. 80 differential gear oil. The API classifies gear lubricants as GL-1, GL-2, GL-3, GL-4, and GL-5. The GL-4 lubricant is used for hypoid gears under normal conditions. The GL-5 lubricant is used in heavy-duty hypoid gears. Always use the vehicle manufacturer's specified differential gear oil.

The differential should be filled until the lubricant is level with the bottom of the filler plug opening in the differential housing. If the differential is overfilled, excessive lubricant may be present at the bearings and seals. Under this condition, the lubricant may leak past the seal. When the lubricant level is low in the differential, the lubricant may not be available in the axle housings. When this condition exists, the bearings do not receive enough lubrication and bearing life is shortened.

## Summary

- Tires are extremely important because they provide ride quality, support the vehicle weight, provide traction for the drive wheels, and contribute to steering quality and directional stability.
- Tires may be bias-ply, belted bias-ply, or radial-belted design.
- The TPC number assures that the tire meets the car manufacturer's performance standards for traction, endurance, dimensions, noise, handling, and rolling resistance.
- The UTQG designation includes treadwear, traction, and temperature ratings.
- The tire placard provides valuable information regarding the tires on the vehicle.
- Wheel rims must have the same width, diameter, offset, load capacity, and mounting configuration as the original rims to maintain vehicle safety.
- Static wheel unbalance causes wheel tramp and severe tire cupping.
- Dynamic wheel unbalance causes wheel shimmy, increased tire wear, unstable directional control, driver fatigue, and increased wear on suspension and steering components.
- A bearing reduces friction, carries a load, and guides certain components such as pivots, shafts, and wheels.
- Radial bearing loads are applied in a vertical direction.
- Thrust bearing loads are applied in a horizontal direction.
- A cylindrical ball bearing, or roller bearing, is designed primarily to withstand radial loads, but these bearings can handle a considerable thrust load.
- Bearing seals keep lubricant in the bearing and prevent dirt from entering the bearing.
- Tapered roller bearings have excellent radial, thrust, and angular load-carrying capabilities.
- Bearing hub units are compact compared to the previous bearings in the wheel hub. This compactness makes hub bearing units suitable for FWD cars.
- Some bearing hub units are bolted to the steering knuckle and other bearing hub units are pressed into the steering knuckle.
- Rear axle bearings are mounted between the drive axles and the housing on rear-wheel drive cars.

## Review Questions

1. While discussing tire design and operation, Technician A says a belted radial-ply tire provides improved steering characteristics compared to a belted bias-ply tire. Technician B says a belted radial-ply tire provides longer tread life than a belted bias-ply tire. Who is correct?
   A. Technician A
   B. Technician B
   C. Both Technician A and Technician B
   D. Neither Technician A nor Technician B

2. While discussing tires in motion, Technician A says when a vehicle is travelling at normal cruising speed, the front portion of each tire is decelerating. Technician B says the rear portion of the tire is traveling at a constant speed. Who is correct?
   A. Technician A
   B. Technician B
   C. Both Technician A and Technician B
   D. Neither Technician A nor Technician B

3. While discussing wheel balance, Technician A says static unbalance causes wear on the center of the tire tread. Technician B says static unbalance causes cupped tire wear. Who is correct?
   A. Technician A
   B. Technician B
   C. Both Technician A and Technician B
   D. Neither Technician A nor Technician B
4. While discussing wheel balance, Technician A says dynamic wheel unbalance causes lateral wheel shimmy. Technician B says dynamic wheel unbalance causes vertical wheel tramp. Who is correct?
   A. Technician A
   B. Technician B
   C. Both Technician A and Technician B
   D. Neither Technician A nor Technician B
5. All these statements about tire design are true *except*:
   A. The tire sidewalls are made from a blend of rubber that absorbs shocks and impacts from road irregularities.
   B. Antioxidants in the sidewall material help to prevent sidewall cracking.
   C. Increased sidewall thickness improves ride quality.
   D. The bead filler above the bead reinforces the tire sidewall and acts as a rim extender.
6. When manufacturing tires, a typical tire may contain:
   A. up to five different synthetic rubbers and one type of natural rubber.
   B. up to eight different synthetic rubbers and two different natural rubbers.
   C. up to fourteen different synthetic rubbers and three different natural rubbers.
   D. up to thirty different synthetic rubbers and eight different natural rubbers.
7. While discussing run-flat tires, Technician A says some run-flat tires have stiffer sidewalls. Technician B says some run-flat tires have a flexible rubber support ring mounted on special rims. Who is correct?
   A. Technician A
   B. Technician B
   C. Both Technician A and Technician B
   D. Neither Technician A nor Technician B
8. While discussing tapered roller bearings, Technician A says a tapered roller bearing can withstand high radial, thrust, and angular loads. Technician B says lubrication and proper end-play adjustment are critical on tapered roller bearings. Who is correct?
   A. Technician A
   B. Technician B
   C. Both Technician A and Technician B
   D. Neither Technician A nor Technician B
9. A front wheel bearing hub assembly
   A. is not serviceable.
   B. is more compact compared to bearings in the wheel hub.
   C. is bolted to the steering arm.
   D. may contain two rows of ball bearings.
10. All of these statements about roller bearing-type rear axle bearings in rear-wheel drive vehicles are true *except*:
    A. The outer bearing race is pressed into the rear axle housing.
    B. The bearing seal is on the inboard side of the wheel bearing.
    C. The bearing rollers contact a machined surface on the rear axle.
    D. Lubricant in the differential provides rear axle bearing lubrication.
11. To calculate the tire aspect ratio, the tire section height is divided by the ____________ .
12. Car manufacturers recommend that tire inflation pressures should be checked when the tires are ____________ .
13. The rim offset is the vertical distance between the rim centerline and the ____________ ____________ of the disc.
14. A springless seal may be used to seal a ____________ lubricant into a hub.
15. Define a radial bearing load.
16. Define a thrust bearing load and give another term for this type of load.
17. Describe the advantage of a bearing hub unit compared to the previous bearings mounted in the wheel hub.
18. Explain the purpose of the wheel rim drop center and safety ridges.

# Tire, Wheel, and Hub Maintenance, Diagnosis, and Service

## *Introduction*

Proper servicing of tires and wheels is extremely important to maintain vehicle safety and provide normal tire life. Improperly serviced and/or balanced tires and wheels cause wheel vibration and shimmy problems resulting in excessive tire tread wear, increased wear on suspension and steering components, and decreased vehicle stability and steering control.

## TIRE MAINTENANCE

Proper tire maintenance is essential to provide normal tire life and maintain vehicle safety. When tires are underinflated, tire life is shortened and worn out tires are a safety concern. A lack of tire rotation also reduces tire life.

### Tire Inflation Pressure

A tire depends on correct inflation pressure to maintain its correct shape and support the vehicle weight. Excessive inflation pressure causes excessive center tread wear **(Figure 1)**, hard ride, and damage to the tire carcass.

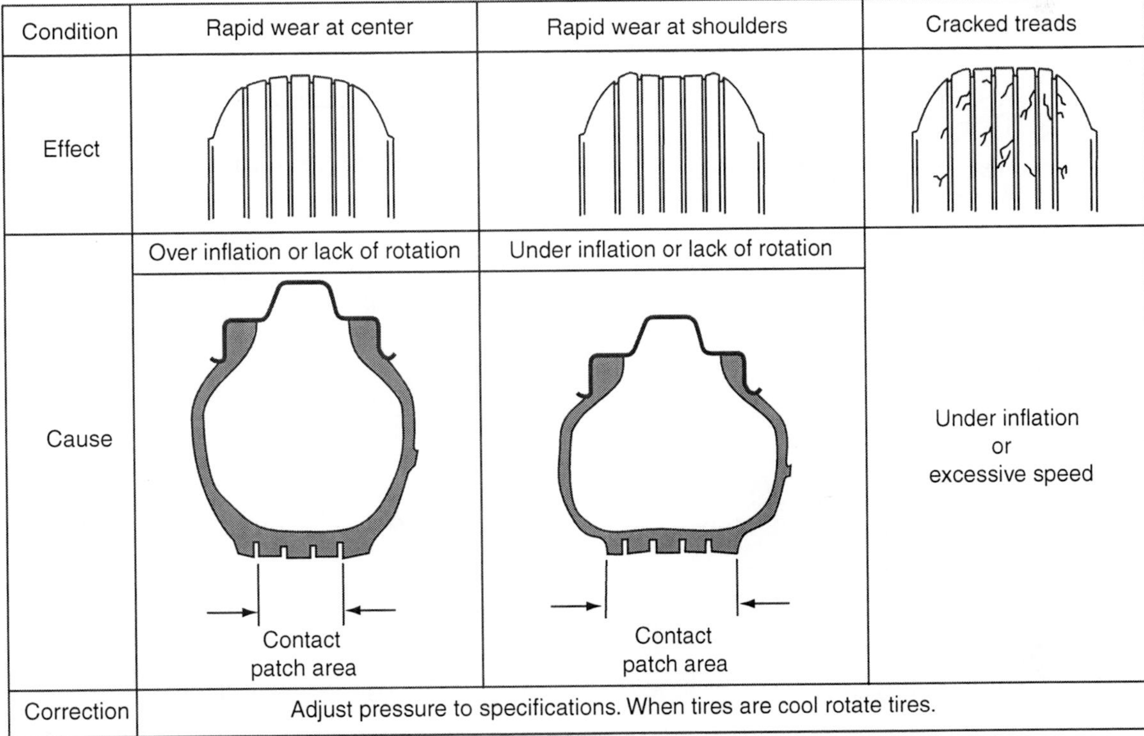

| Condition | Rapid wear at center | Rapid wear at shoulders | Cracked treads |
|---|---|---|---|
| Effect | | | |
| Cause | Over inflation or lack of rotation<br>Contact patch area | Under inflation or lack of rotation<br>Contact patch area | Under inflation or excessive speed |
| Correction | Adjust pressure to specifications. When tires are cool rotate tires. | | |

**Figure 1.** Tire tread wear caused by underinflation and overinflation.

When tires are underinflated, these problems will be evident: excessive wear on each side of the tread, hard steering, wheel damage, excessive heat buildup in the tire, and possible severe tire damage with resultant hazardous driving.

## Tread Wear Measurement

On most tires, the tread wear indicators appear as wide bands across the tread when tread depth is worn to $^1/_{16}$ inch (1.6 millimeters). Most tire manufacturers recommend tire replacement when the wear indicators appear across two or more tread grooves at three locations around the tire **(Figure 2)**. If tires do not have wear indicators, a tread depth gauge may be used to measure the tread depth **(Figure 3)**. The tread depth gauge reads in 32nds of an inch, and tires with $^2/_{32}$ tread depth or less should be replaced.

## Tire Rotation

To a large extent, driving habits determine tire life. Severe brake applications, high-speed driving, turning at high speeds, rapid acceleration and deceleration, and striking curbs are just a few driving habits which shorten tire life. Most car manufacturers recommend tire rotation at specified intervals to obtain maximum tire life. The exact tire rotation procedure depends on the model year, the type of tires, and whether the vehicle has a conventional spare or a compact spare **(Figure 4)**. Tire rotation procedures do not include the compact spare. The vehicle manufacturer provides tire rotation information in the owner's manual and service manual. Vehicle manufacturers usually recommend different tire rotation procedures for bias-ply tires compared to radial tires **(Figure 5)**.

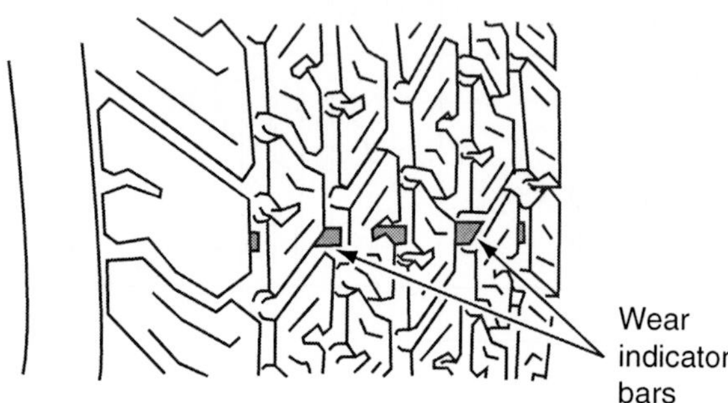


**Figure 2.** Tire tread wear indicators.

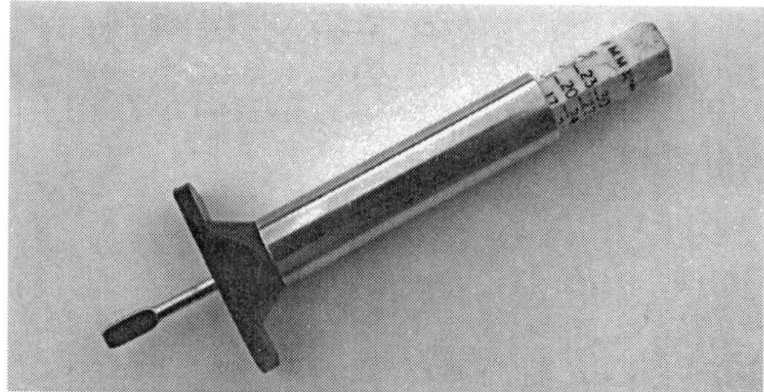

**Figure 3.** Tread depth gauge.

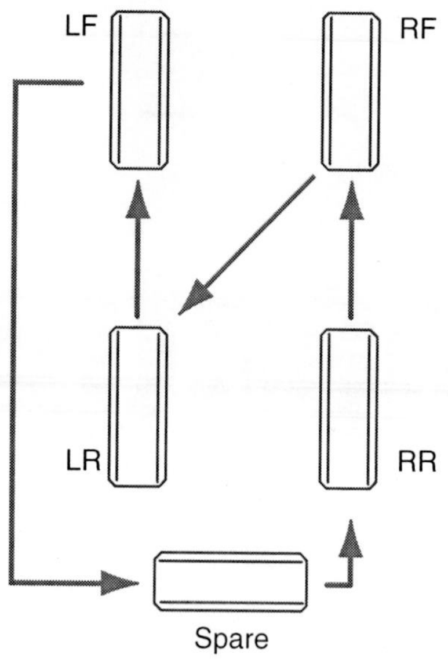


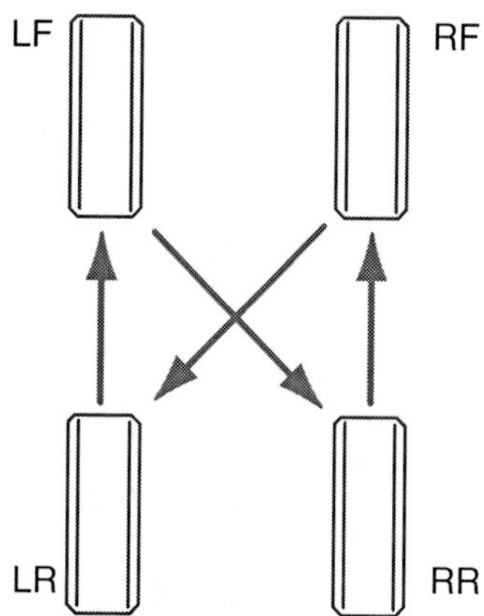


**Figure 4.** Radial tire rotation procedure.

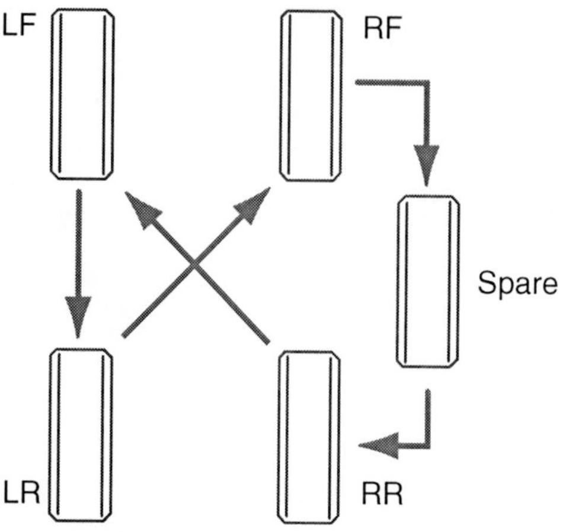


**Figure 5.** Bias-ply tire rotation procedure.

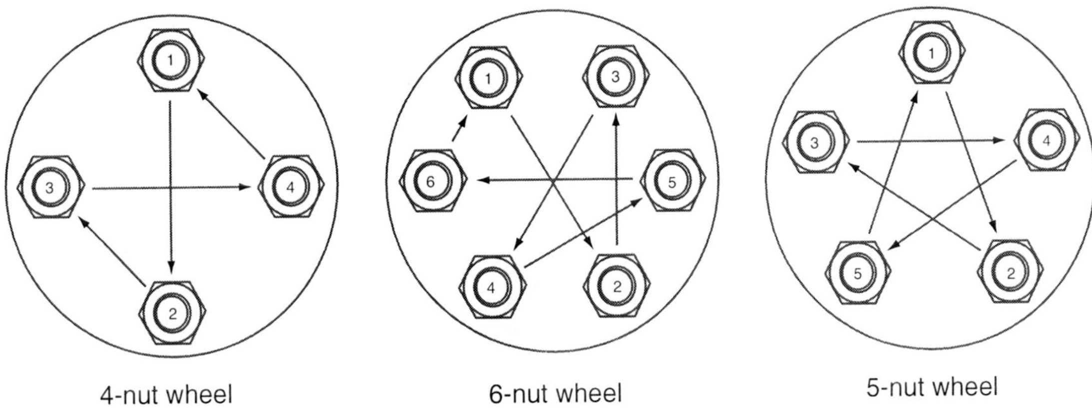


**Figure 6.** Wheel nut tightening sequence.

When tires and wheels are installed on a vehicle, it is very important that the wheel nuts are torqued to manufacturer's specifications in the proper sequence **(Figure 6)**. Do not use an impact wrench when tightening wheel nuts to the specified torque.

## TIRE DIAGNOSIS

Uneven tread surfaces may cause tire noises that seem to originate elsewhere in the vehicle. These noises may be confused with differential noise. Differential noise usually varies with acceleration and deceleration, while tire noise remains more constant in relation to these forces. Tire noise is most pronounced on smooth asphalt road surfaces at speeds of 15 to 45 miles per hour (24 to 72 kilometers per hour).

You Should Know *Tire noise varies with road surface conditions, whereas differential noise is not affected when various road surfaces are encountered.*

When tire thump and vibration is present, check for cupped tire treads; excessive tire radial runout; manufacturing defects such as heavy spots, weak spots, or tread chunking; and incorrect wheel balance.

A vehicle should maintain the straightahead forward direction on smooth, straight road surfaces without excessive steering wheel correction by the driver. If the steering gradually pulls to one side on a smooth, straight road surface, a tire, steering, or suspension defect is present. Tires of different types, sizes, designs, or inflation pressures on opposite sides of a vehicle cause steering pull. Sometimes a tire manufacturing defect occurs in which the belts are wound offcenter on the tire. This condition is referred to as conicity. A cone-shaped object rolls in the direction of its smaller diameter. Similarly, a tire with conicity tends to lead or pull to one side which causes the vehicle to follow the action of the tire **(Figure 7)**.

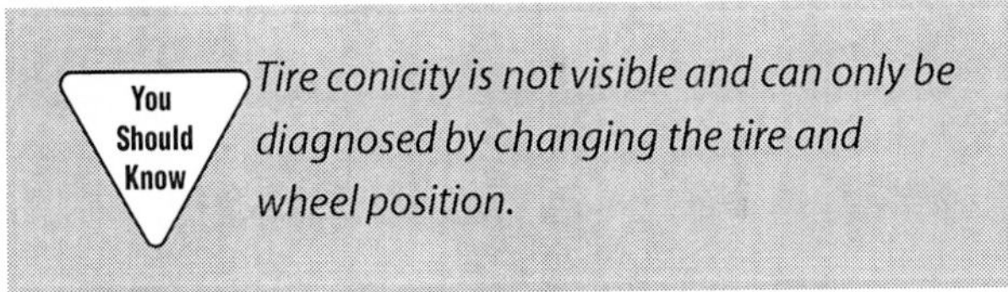
You Should Know *Tire conicity is not visible and can only be diagnosed by changing the tire and wheel position.*

**Figure 7.** Tire conicity.

Since tire conicity cannot be diagnosed by a visual inspection, it must be diagnosed by switching the two front tires and reversing the front and rear tires **(Figure 8)**. Incorrect front suspension alignment angles also cause steering pull.

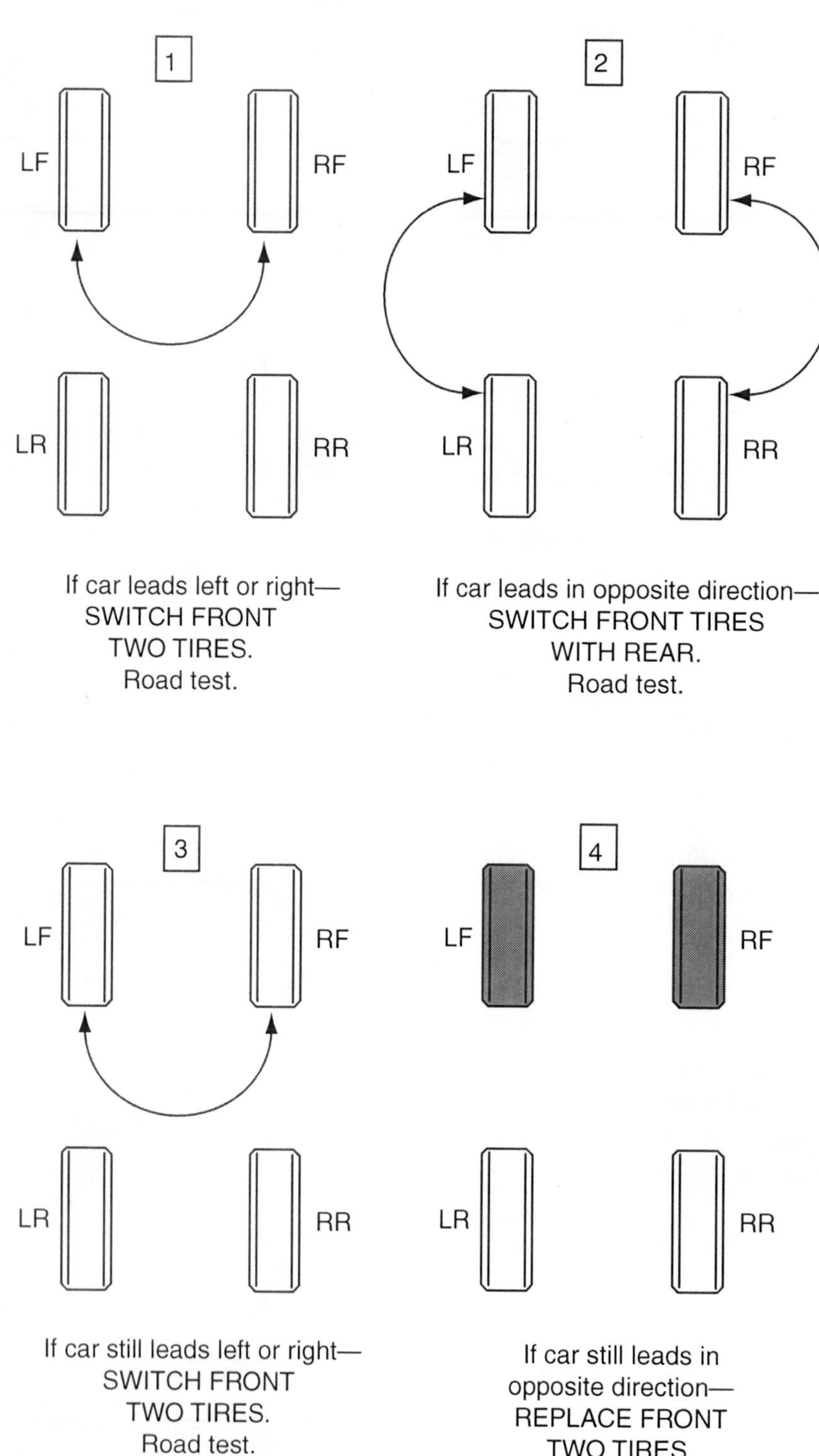


**Figure 8.** Tire conicity diagnosis.

## WHEEL AND TIRE SERVICE

Proper wheel and tire service are extremely important to maintain vehicle safety and provide normal tire life with adequate driver and passenger comfort. Improperly balanced tires and wheels greatly reduce tire life and the vibrations from this problem result in driver discomfort and fatique.

### Wheel and Tire Removal

When it is necessary to remove a wheel and tire assembly, follow these steps:

1. Remove the wheel cover. If the vehicle is equipped with anti-theft locking wheel covers, the lock bolt for each wheel cover is located behind the ornament in the center of the wheel cover. A special key wrench is supplied to the owner for ornament and lock bolt removal. If the customer's key wrench has been lost, a master key is available from the vehicle dealer.

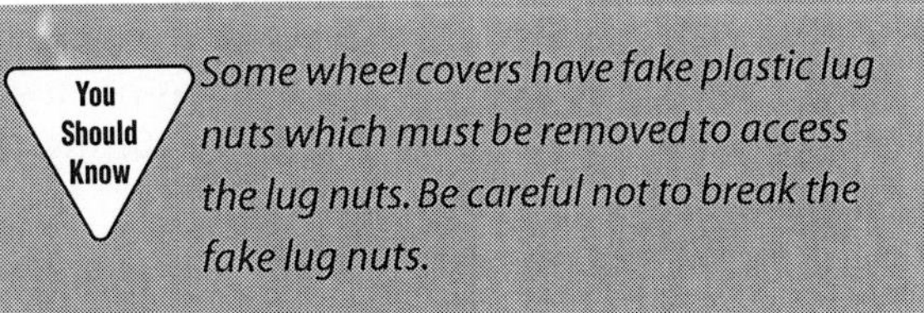

You Should Know
Some wheel covers have fake plastic lug nuts which must be removed to access the lug nuts. Be careful not to break the fake lug nuts.


2. Loosen the wheel lug nuts about one-half turn, but do not remove the wheel nuts. Some vehicles are equipped with anti-theft wheel nuts. A special lug nut key is supplied to the vehicle owner. This lug nut key has a hex nut on the outer end and a special internal projection that fits in the wheel nut opening. Install the lug nut key on the lug nuts and connect the lug nut wrench on the key hex nut to loosen the lug nuts.

You Should Know

*Before the vehicle is raised on a lift, be sure the lift is contacting the car manufacturer's recommended lift points. If the vehicle is lifted with a floor jack, place safety stands under the suspension or frame and lower the vehicle onto the safety stands. Then remove the floor jack from under the vehicle.*

3. Raise the vehicle on a lift or with a floor jack to a convenient working level.
4. Chalk mark the tire, wheel, and one of the lug nuts so the wheel and tire can be reinstalled in the same position.
5. Remove the lug nuts and remove the wheel and tire assembly. If the wheel is rusted and will not come off, hit the inside of the wheel with a large rubber mallet. Do not hit the wheel with a steel hammer because this action could damage the wheel. Do not heat the wheel.

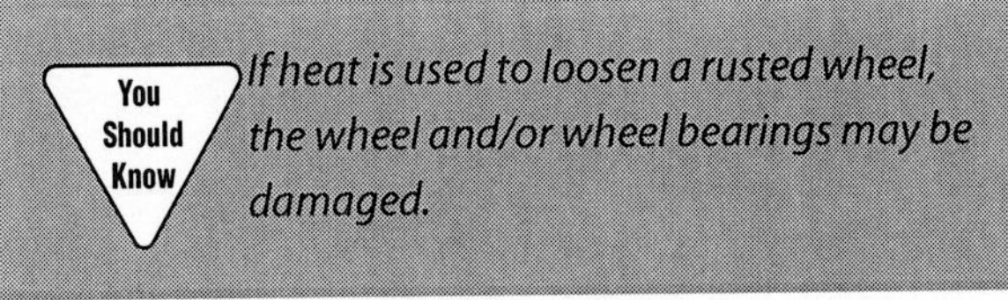

You Should Know
If heat is used to loosen a rusted wheel, the wheel and/or wheel bearings may be damaged.


### Tire and Wheel Service Precautions

There are many different types of tire-changing equipment in the automotive service industry. However, specific precautions apply to the use of any tire-changing equipment. These precautions include the following:

1. Before you operate any tire-changing equipment, always be absolutely certain that you are familiar with the operation of the equipment.
2. When operating tire-changing equipment, always follow the equipment manufacturer's recommended procedure.
3. Always deflate a tire completely before attempting to demount the tire.
4. The bead seats on the wheel rim must be clean before the tire is mounted on the wheel rim.
5. The outer surface of the tire beads should be lubricated with rubber lubricant prior to mounting the tire on the wheel rim.
6. When the tire is mounted on the wheel rim, be sure the tire is positioned evenly on the wheel rim.
7. While inflating a tire, do not stand directly over the tire. An air hose extension allows the technician to stand back from the tire during the inflation process.
8. Do not overinflate tires.
9. When mounting tires on cast aluminum alloy wheel rims or cast magnesium alloy wheel rims, always use the tire changing equipment manufacturer's recommended tools and procedures.

Figure 9. A tire changer.

## Tire Demounting

Always use a tire changer to demount tires **(Figure 9)**. Do not use hand tools or tire irons for this purpose. A typical tire demounting procedure is the following:

1. Remove the valve core and be sure the tire is completely deflated.
2. Place the wheel and tire on the tire changer with the narrow bead ledge facing upward.
3. Follow the operating procedure recommended by the manufacturer of the tire changer to force the tire bead inward and separate it from the rim on both sides.
4. Push one edge of the top bead into the drop center of the rim.

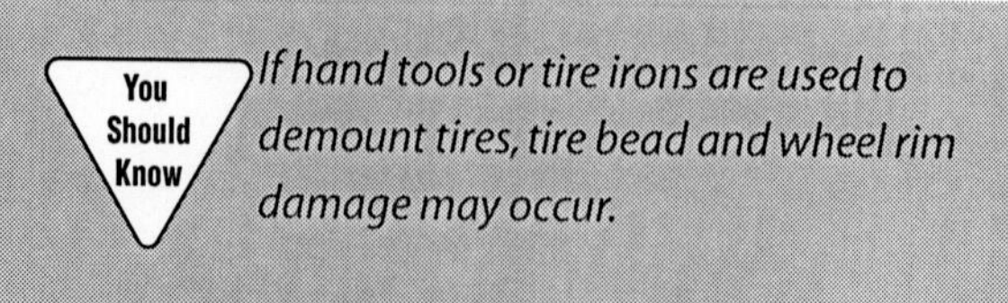


*If hand tools or tire irons are used to demount tires, tire bead and wheel rim damage may occur.*

5. Place the tire changer's bar or lever between the bead and the rim on the opposite side of the rim from where the bead is in the drop center.
6. Operate the tire changer to rotate the bar or lever and move the bead over the top of the rim.
7. Repeat steps 4, 5, and 6 to move the lower bead over the top of the rim.

## Tire Inspection and Repair

To find a leak in a tire and wheel, inflate the tire to the pressure marked on the sidewall and then submerge the tire and wheel in a tank of water. An alternate method of leak detection is to sponge soapy water on the tire and wheel. Bubbles will appear wherever the leak is located in the tire or wheel. Mark the leak location in the tire or wheel rim with a crayon and mark the tire at the valve stem location so the tire can be reinstalled in the same position on the wheel to maintain proper balance.

A puncture is the most common cause of a tire leak. Many punctures can be repaired satisfactorily. Do not attempt to repair punctures over 1/4-inch diameter. Punctures in the sidewalls or on the tire shoulders should not be repaired. The repairable area in belted bias-ply tires is approximately the width of the belts **(Figure 10)**. The belts in radial tires are wider compared to bias-ply tires and the repairable area in radial tires is also the width of the belts. Since compact spare tires have thin treads, do not attempt to repair these tires.

Inspect the tire and do not repair a tire with any of these defects, signs of damage, or excessive wear: tires with the wear indicators showing, tires worn until the fabric or belts are exposed, bulges or blisters, ply separation, broken or cracked beads, or cuts or cracks anywhere in the tire

Since most vehicles are equipped with tubeless tires, we will discuss this type of tire repair. If the cause of the puncture, such as a nail, is still in the tire, remove it from the tire. Most punctures can be repaired from inside the tire with a service plug or vulcanized patch service kit. The instructions from the tire service kit manufacturer should be followed, but we will discuss three common tire repair procedures.

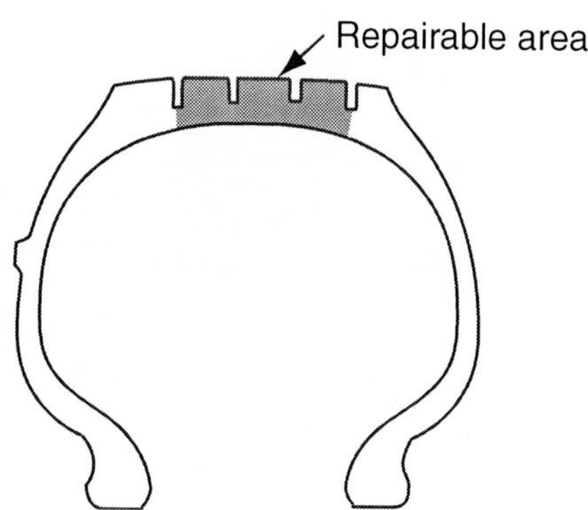


Figure 10. Repairable area on bias-ply and belted bias-ply tires.

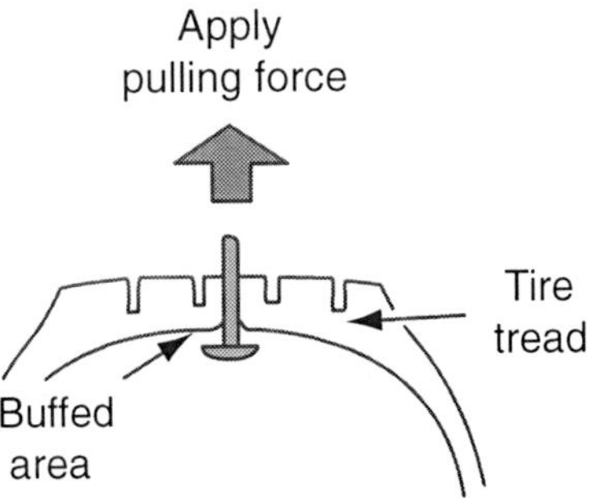


**Figure 11.** Plug installation procedure.

## Plug Installation Procedure

1. Buff the area around the puncture with a wire brush or wire buffing wheel.
2. Select a plug slightly larger than the puncture opening and insert the plug in the eye of the insertion tool.
3. Wet the plug and the insertion tool with vulcanizing fluid.
4. While holding and stretching the plug, pull the plug into the puncture from the inside of the tire **(Figure 11)**. The head of the plug should contact the inside of the tire. If the plug pulls through the tire, repeat the procedure.
5. Cut the plug off $^1/_{32}$ inch from the tread surface. Do not stretch the plug while cutting.

## Cold Patch Installation Procedure

1. Buff the area around the puncture with a wire brush or buffing wheel.
2. Apply vulcanizing fluid to the buffed area and allow it to dry until it is tacky.
3. Peel the backing from the patch and apply the patch over the puncture. Center the patch over the puncture.

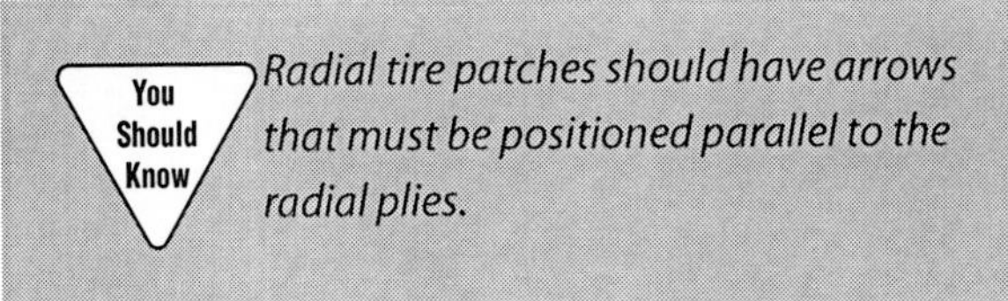


*Radial tire patches should have arrows that must be positioned parallel to the radial plies.*

4. Run a stitching tool back and forth over the patch to improve bonding.

## Hot Patch Installation Procedure

The area around the puncture must be buffed with a wire brush or buffing wheel. Many hot patches require the application of vulcanizing fluid on the buffed area. Peel the backing from the patch and install the patch so it is centered over the puncture on the inside of the tire. Many hot patches are heated with an electric heating element clamped over the patch. This element should be clamped in place for the amount of time recommended by the equipment or patch manufacturer. After the heating element is removed, allow the patch to cool for a few minutes and be sure the patch is properly bonded to the tire.

# WHEEL RIM SERVICE

Steel rims should be spray cleaned with a water hose. Aluminum or magnesium wheel rims should be cleaned with a mild soap and water solution and rinsed with clean water. The use of abrasive cleaners, alkaline-base detergents, or caustic agents may damage aluminum or magnesium wheel rims. Clean the rim bead seats on these wheel rims thoroughly with the mild soap and water solution. The rim bead seats on steel wheel rims should be cleaned with a wire brush or coarse steel wool.

You Should Know

*Steel wheel rims must not be welded, heated, or peened with a ball peen hammer. These procedures may weaken the rim and create a safety hazard. Installing an inner tube to correct leaks in a tubeless tire or wheel rim is not an approved procedure.*

Steel wheel rims should be inspected for excessive rust and corrosion, cracks, loose rivets or welds, bent or damaged bead seats, and elongated lug nut holes. Aluminum or magnesium wheel rims should be inspected for damaged bead seats, elongated lug nut holes, cracks, and porosity. If any of these conditions are present on either type of wheel rim, replace the wheel rim.

Many shops always replace the tire valve assembly when a tire is repaired or replaced. This policy helps to prevent future problems with tire valve leaks. The inner end of the valve may be cut off with a pair of diagonal pliers and then the outer end may be pulled from the rim. Coat the new valve with rubber tire lubricant and pull it into the rim opening with a special puller screwed onto the valve threads.

## Wheel Rim Leak Repair

A wheel rim leak may be repaired if the leak is not caused by excessive rust on a steel rim and the rim is in satisfactory condition. Follow these steps for wheel rim leak repair:

1. Use #80 grit sandpaper to thoroughly clean the area around the leak on the tire side of the rim.
2. Use a shop towel to remove any grit from the leak area.
3. Be sure the wheel rim is at room temperature and apply a heavy coating of silicone rubber sealer over the leak area.
4. Spread the sealer over the entire sanded area with a putty knife.
5. Allow the sealer to cure for 6 hours before remounting the tire.

## Tire Remounting Procedure

1. Be sure the wheel rim bead seats are thoroughly cleaned.
2. Coat the tire beads and the wheel rim bead seats with rubber tire lubricant.
3. Secure the wheel rim on the tire changer with the narrow bead ledge facing upward.
4. Place the tire on top of the wheel rim with the bead on lower side of the tire in the drop center of the wheel rim.
5. Use the tire changer bar or lever under the tire bead to install the tire bead over the wheel rim. Always operate the tire changer with the manufacturer's recommended procedure.
6. Repeat the procedure in steps 3 through 5 to install the upper bead over the wheel rim.
7. Rotate the tire on the wheel rim until the crayon mark is aligned with the valve stem. This mark was placed on the tire prior to demounting.

**You Should Know** *When a bead expander is installed around the tire, never exceed 10 psi (69 kPa) pressure in the tire. A higher pressure may cause the expander to break and fly off the tire causing serious personal injury or property damage. When a bead expander is not used, never exceed 40 psi (276 kPa) tire pressure to move the tire beads out tightly against the wheel rim. A higher pressure may blow the tire bead against the rim with excessive force. This action could burst the rim or tire, resulting in serious personal injury or property damage. While inflating a tire, do no stand directly over a tire. In this position, serious injury could occur if the tire or wheel rim flies apart.*

8. Follow the recommended procedure supplied by the manufacturer of the tire changer to inflate the tire. This procedure may involve the use of a bead expander installed around the center of the tire tread to expand the tire beads against the wheel rim. If a bead expander is used, inflate the tire to 10 psi (69 kPa) to move the beads out tightly against the wheel rim. Never exceed this pressure with a bead expander installed on the tire. Always observe the circular marking around the tire bead as the tire is inflated. This mark should be centered around the wheel rim. Always observe both beads while a tire is inflated. If the circular mark around the tire bead in not centered on the rim, deflate the tire and center it on the wheel rim. If a bead expander is not used, never exceed 40 psi (276 kPa) to try and move tire beads out tightly against the rim. When either tire bead will not move out tightly against the wheel rim with 40 psi (276 kPa) tire pressure, deflate the tire and center it on the wheel rim again.

## Tire and Wheel Runout Measurement

Ideally, a tire and wheel assembly should be perfectly round. However, this condition is rarely achieved and most tires have a certain amount of radial runout. **Radial runout** is the amount of diameter variation in a tire.

If the radial runout exceeds manufacturer's specifications a vibration may occur because the radial runout causes the spindle to move up and down **(Figure 12)**. A defective tire with a variation in stiffness may also cause this up and down spindle action.

A dial indicator gauge may be positioned against the center of the tire tread as the tire is rotated slowly to meas-

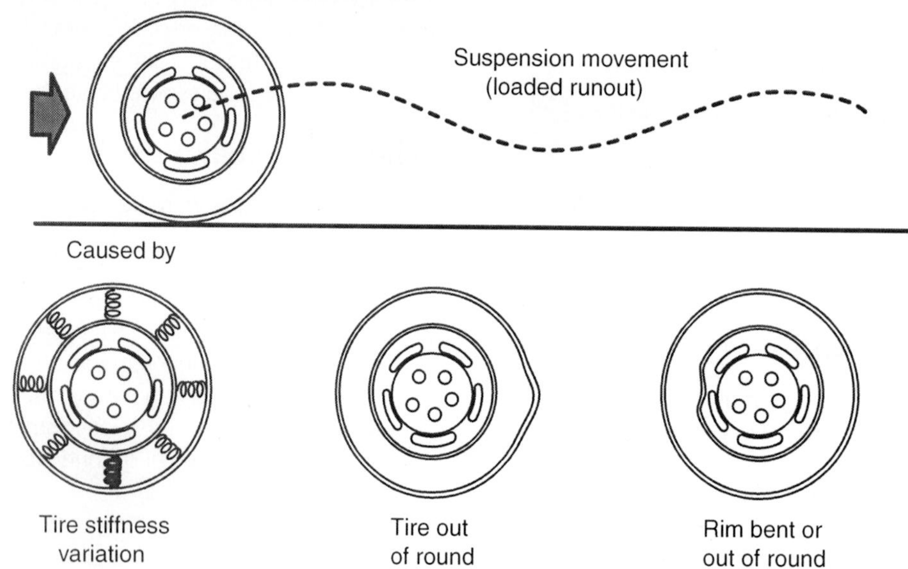


**Figure 12.** Vertical tire and wheel vibrations caused by radial tire or wheel runout or variation in tire stiffness.

Figure 13. Measuring tire radial runout.

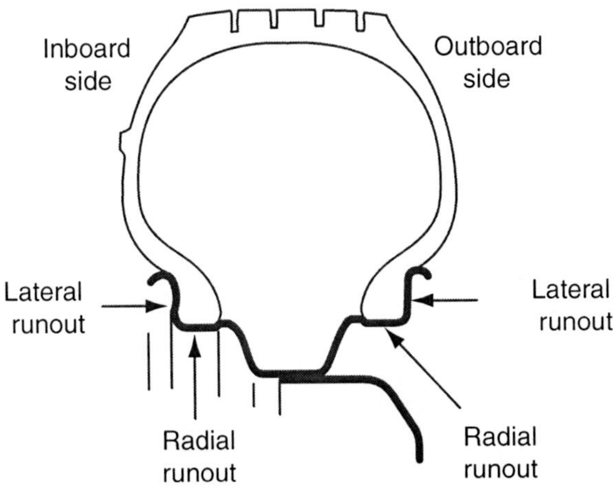


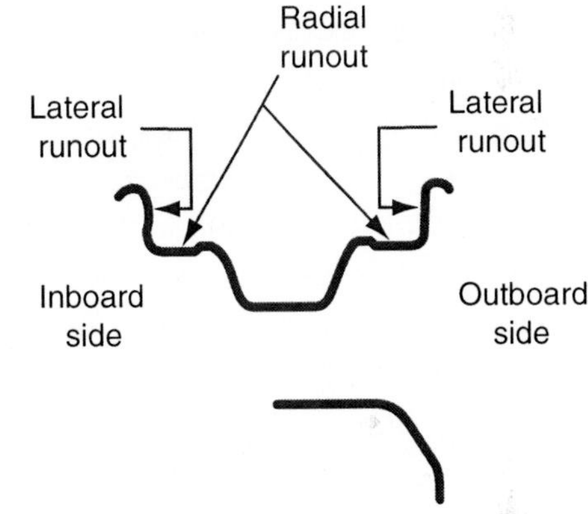


Figure 14. Measuring wheel lateral runout.

ure radial runout **(Figure 13)**. Radial runout of more than 0.060 inches (1.5 millimeters) will cause vehicle shake. If the radial runout is between 0.045 inch to 0.060 inch (1.1 millimeters to 1.5 millimeters), vehicle shake may occur. These are typical radial runout specifications. Always consult the vehicle manufacturer's specifications. Mark the highest point of radial runout on the tire with a crayon and mark the valve stem position on the tire.

If the radial tire runout is excessive, demount the tire and check the runout of the wheel rim with a dial indicator positioned against the lip of the rim while the rim is rotated **(Figure 14)**. Use a crayon to mark the highest point of radial runout on the wheel rim. Radial wheel runout should not exceed 0.035 inch (0.9 millimeter), whereas the maximum lateral wheel runout is 0.045 inch (1.1 millimeter). If the highest point of wheel radial runout coincides with the chalk mark from the highest point of maximum tire radial runout, the tire may be rotated 180 degrees on the wheel to reduce radial runout. Tires or wheels with excessive runout are usually replaced.

Lateral tire runout may be measured with a dial indicator located against the sidewall of the tire. **Lateral tire runout** is the amount of sideways wobble in a rotating tire.

Excessive lateral runout causes the tire to waddle as it turns. This waddling sensation may be transmitted to the passenger compartment **(Figure 15)**. A chassis waddling action may also be caused by a defective tire in which the belt is not straight. If the lateral runout exceeds 0.080 inch (2.0 millimeters), wheel shake problems will occur on the vehicle. Chalk mark the tire and wheel at the highest point of radial runout. When the tire runout is excessive, the tire should be removed from the wheel and the wheel lateral runout should be measured with a dial indicator positioned against the edge of the wheel as the wheel is rotated. Wheels or tires with excessive lateral runout should be replaced.

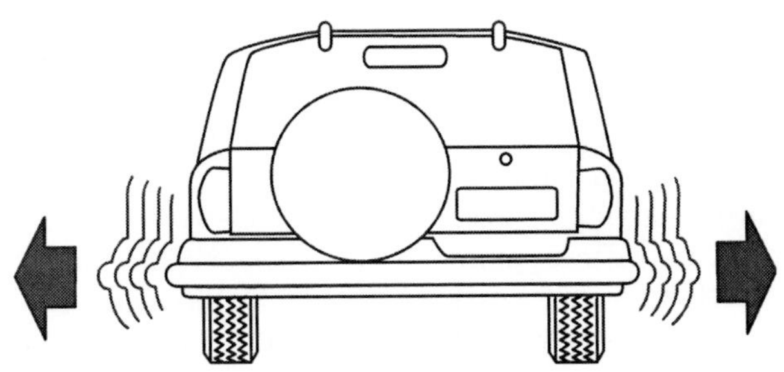

Tire waddle often caused by

- Steel belt not straight within tire
- Excessive lateral runout

Figure 15. Chassis waddle caused by lateral tire or wheel runout or a defective tire with a belt that is not straight.

## TIRE AND WHEEL BALANCING

These preliminary checks should be completed before a wheel and tire are balanced:

1. Check for objects in tire tread
2. Check for objects inside tire
3. Inspect the tread and sidewall
4. Check inflation pressure
5. Measure tire and wheel runout
6. Check wheel bearing adjustment
7. Check for mud collected on the inside of the wheel

> **You Should Know** *On many wheel balancers, the tire and wheel are spun at high speed during the dynamic balance procedure. Be sure that all wheel weights are attached securely and check for other loose objects on the tire and wheel, such as stones in the tread. If loose objects are detached from the tire or wheel at high speed, they may cause serious personal injury or property damage. On the type of wheel balancer that spins the tire and wheel at high speed during the dynamic balance procedure, always attach the tire and wheel assembly securely to the balancer. Follow the equipment manufacturer's recommended wheel mounting procedure. If the tire and wheel assembly ever becomes loose on the balancer at high speed, serious personal injury or property damage may result. Prior to spinning a tire and wheel at high speed on a wheel balancer, always lower the protection shield over the tire. This shield provides protection in case anything flies off the tire or wheel.*

All of the items on the preliminary checklist influence wheel balance or safety. Therefore, it is extremely important that the preliminary checks are completed. Since a tire and wheel assembly is rotated at high speed during the dynamic balance procedure, it is very important that objects such as stones are removed from the treads. Centrifugal force may dislodge objects from the treads and cause serious personal injury. For this reason, it is also extremely important that the old wheel weights are removed from the wheel prior to balancing and the new weights are attached securely to the wheel during the balance procedure. Follow these preliminary steps before attempting to balance a wheel and tire:

1. When most types of wheel balancers are used, the wheel and tire must be removed from the vehicle and installed on the balancer. All mud, dust, and debris must be washed from the wheel after it is removed.
2. Objects inside a tire, such as balls of rubber, make balance impossible. When the wheel and tire assembly is mounted on the balancer, be absolutely sure that the wheel is securely tightened on the balancer. As the tire is rotated slowly, listen for objects rolling inside the tire. If such objects are present, they must be removed prior to wheel balancing.
3. The tire should be inspected for tread and sidewall defects before the balance procedure. These defects create safety hazards and may influence wheel balance. For example, tread chunking makes the wheel balancing difficult.

The tires should be inflated to the car manufacturer's recommended pressure prior to the balance procedure. Loose wheel bearings allow lateral wheel shaking and simulate an imbalanced wheel condition when the vehicle is in motion. Therefore, wheel bearing adjustments should be checked when wheel balance conditions are diagnosed.

## Static Wheel Balance Procedures

Many different types of wheel balancers are available in the automotive service industry and the exact wheel balance procedure may vary depending on the type of wheel balancer.

On some types of wheel balancers, the wheel is allowed to rotate by gravity during the static balance procedure. A heavy spot rotates the tire until this spot is at the bottom. The necessary static balance weights are then added at the top of the wheel 180 degrees from the heavy spot **(Figure 16)**. When the wheel and tire assembly is balanced statically, gravity does not rotate the wheel from the at-rest position. Rotate the tire by hand and check the static balance at 120-degree intervals.

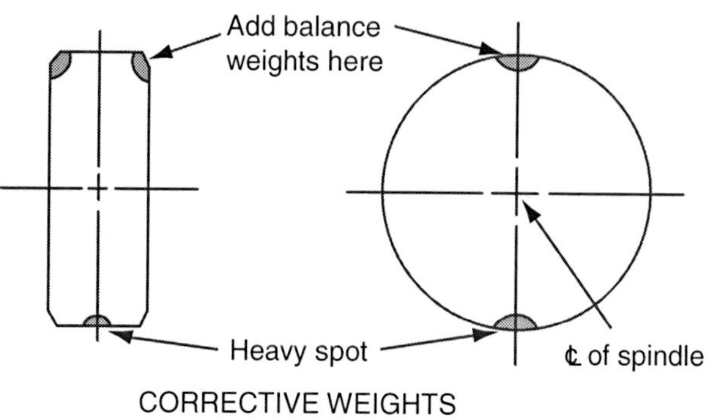


**Figure 16.** Static wheel balance procedure.

Figure 17. Electronic wheel balancer.

The electronic wheel balancer is the most common type in use at the present time **(Figure 17)**. There are many different types of electronic wheel balancers in the automotive service industry. With an electronic balancer, the operator must enter the wheel diameter, width, and offset. The balancer must have this information to perform its calculations.

## Dynamic Wheel Balance Procedure

Dynamic imbalance is corrected with lead weights. Many wheel weights have a spring clamp that holds the weight on the edge of the rim. A special pair of wheel weight pliers is used to tap the wheel weight onto the rim and remove the weight from the rim **(Figure 18)**. Magnesium or aluminum wheels may require the use of stick-on wheel weights. This type of weight must also be used on some wheels on which the wheel covers interfere with the conventional weights. Regardless of the wheel weight type, they must be attached securely to the wheel.

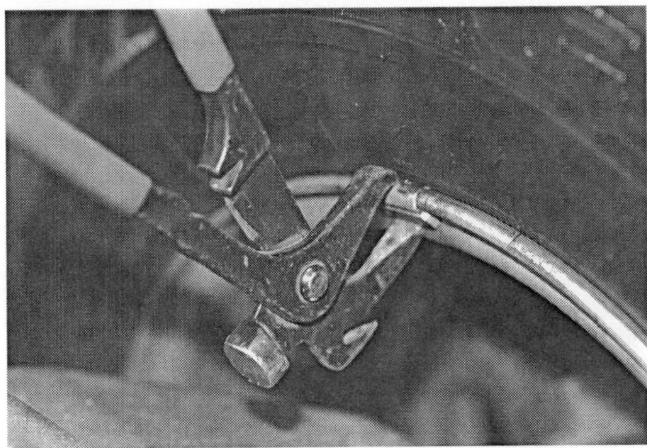

Figure 18. Wheel weight pliers are used to remove and install wheel weights.

Many electronic wheel balancers have an electric drive motor that spins the tire at high speed. On some electronic wheel balancers, the tire is spun by hand during the balance procedure. The electronic-type balancer performs static and dynamic balance calculations simultaneously and indicates the correct weight size and location to the operator. The tire and wheel assembly must be fastened securely on the balancer with the correct size adapters. Off-car electronic high-speed wheel balancers have a safety hood that must be positioned over the tire prior to dynamic balancing. Hand-spun electronic balancers do not require a safety hood. The preliminary checks mentioned previously in this chapter also apply to dynamic balancing. Wheel balancers must always be operated according to the manufacturer's instructions.

When the heavy spot is located on the inside edge of the tread, the correct size weight installed on the inside of the wheel 180 degrees from the heavy spot provides proper dynamic balance **(Figure 19)**.

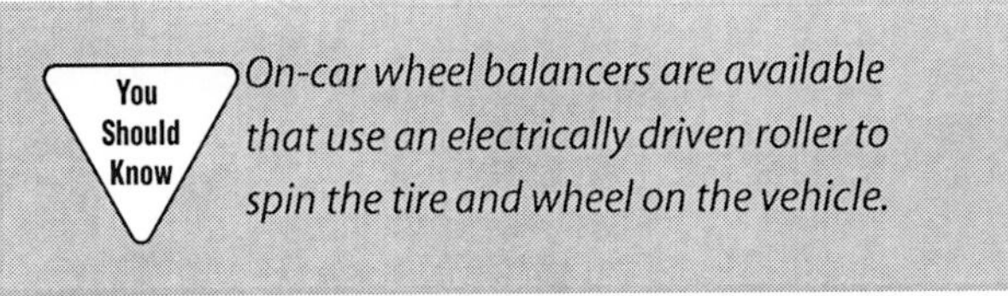


*On-car wheel balancers are available that use an electrically driven roller to spin the tire and wheel on the vehicle.*

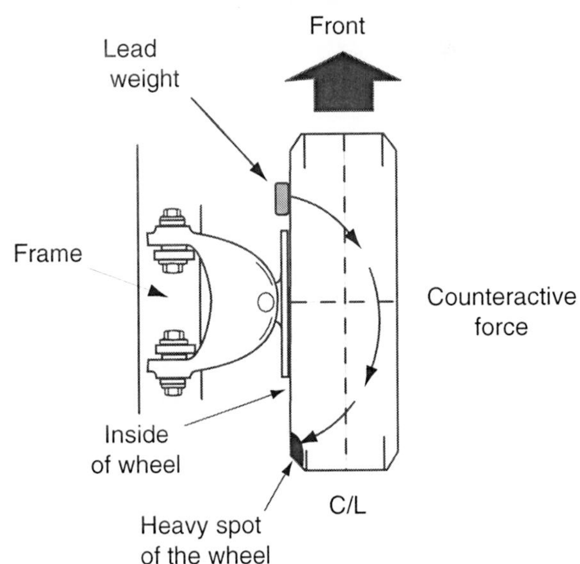


Figure 19. Dynamic wheel balance with heavy spot on the inside edge of the wheel.

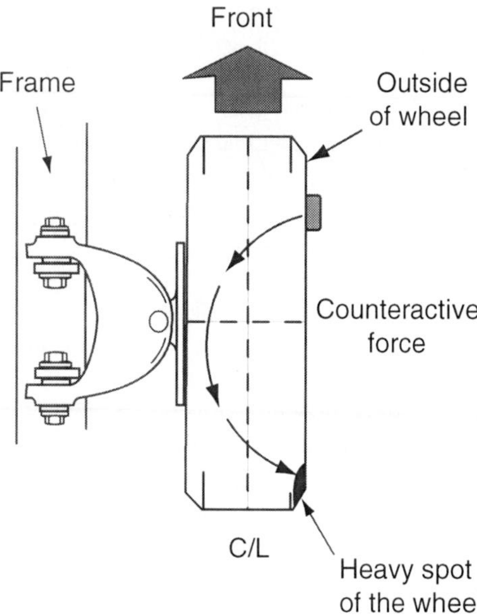


**Figure 20.** Dynamic wheel balance with heavy spot on the outside edge of the tread.

When the heavy spot is on the outside edge of the tread, the correct size wheel weight installed 180 degrees from the heavy spot on the outside of the wheel provides proper dynamic balance **(Figure 20)**.

## WHEEL BEARING MAINTENANCE

Technicians must accurately diagnosis wheel bearing problems to avoid repeat bearing failures and thus provide customer satisfaction. Accurate wheel bearing service procedures are essential to maintain vehicle safety! Improper wheel bearing service may cause brake problems, steering complaints, and premature bearing failure. Improper wheel bearing service may even cause a wheel to fly off a vehicle resulting in personal injury and vehicle damage. Technicians must also understand rear-axle bearing service procedures on rear wheel drive vehicles.

Two separate tapered roller bearings are used in the front wheel hubs of many rear-wheel drive cars and the rear wheel hubs on some front-wheel drive cars have the same type of bearings. Similar service and adjustment procedures apply to these tapered roller bearings.

These bearings should be cleaned, inspected, and packed with wheel bearing grease at the vehicle manufacturer's recommended service intervals. Remove the grease seal out of the inner hub opening with a seal puller and discard the seal. This seal should always be replaced when the bearings are serviced. Do not attempt to wash sealed bearings or bearings that are shielded on both sides. If a bearing is sealed on one side, it may be washed in solvent and repacked with grease.

**Figure 21.** Cleaning a bearing with solvent.

Bearings may be placed in a tray and lowered into a container of clean solvent. A brush may be used to remove old grease from the bearing **(Figure 21)**. The bearings may be dried with compressed air after the cleaning operation. Be sure the shop air supply is free from moisture which causes rust formation in the bearing. After all the old grease has been cleaned from the bearing, rinse the bearing in clean solvent and dry it thoroughly with compressed air.

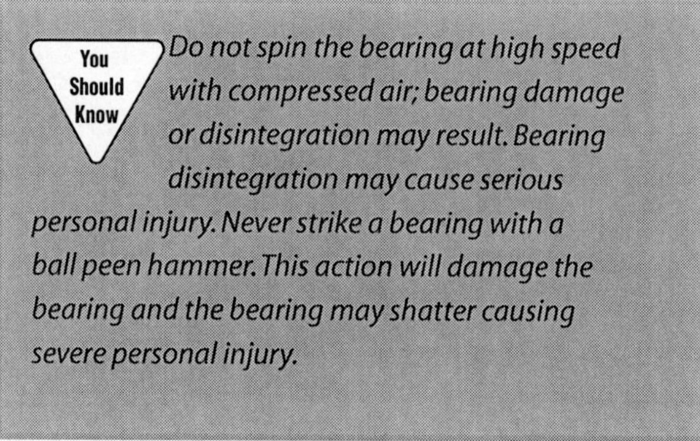


*Do not spin the bearing at high speed with compressed air; bearing damage or disintegration may result. Bearing disintegration may cause serious personal injury. Never strike a bearing with a ball peen hammer. This action will damage the bearing and the bearing may shatter causing severe personal injury.*

When bearing cleaning is completed, bearings should be inspected for the defects illustrated in **Figure 22** and

TAPERED ROLLER BEARING DIAGNOSIS

Consider the following factors when diagnosing bearing condition:

1. General condition of all parts during disassembly and inspection.
2. Classify the failure with the aid of the illustrations.
3. Determine the cause.
4. Make all repairs following recommended procedures.

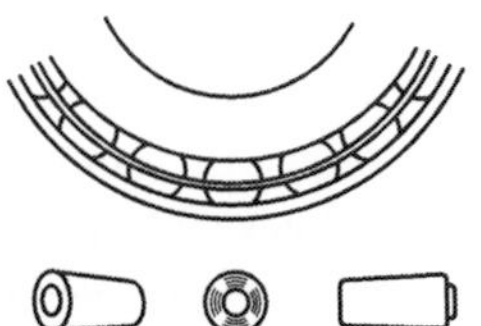

ABRASIVE STEP WEAR

Pattern on roller ends caused by fine abrasives. Clean all parts and housings, check seals and bearings, and replace if leaking, rough or noisy.

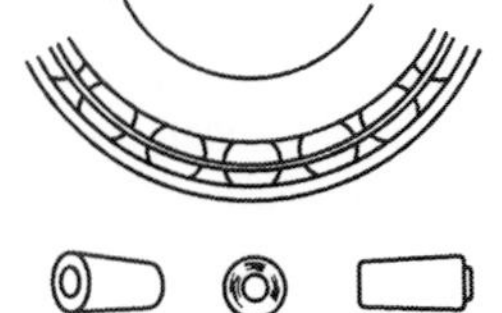

GALLING

Metal smears on roller ends due to overheating, lubricant failure, or overload. Replace bearing, check seals, and check for proper lubrication.

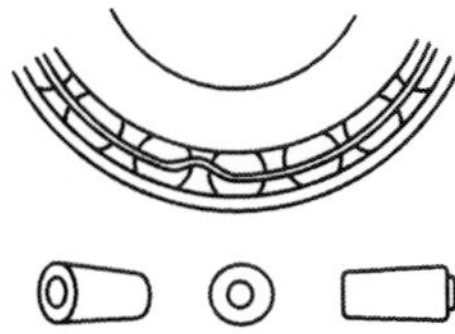

BENT CAGE

Cage damaged due to improper handling or tool usage. Replace bearing.

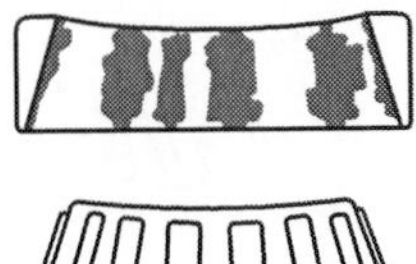

ABRASIVE ROLLER WEAR

Pattern on races and rollers caused by fine abrasives. Clean all parts and housings, check seals and bearings, and replace if leaking, rough or noisy.

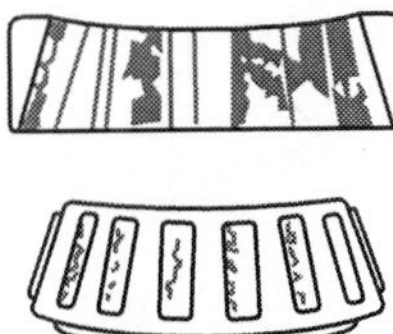

ETCHING

Bearing surfaces appear gray or grayish black in color with related etching away of material usually at roller spacing. Replace bearings, check seals, and check for proper lubrication.

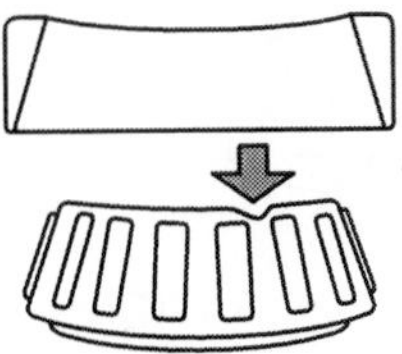

BENT CAGE

Cage damaged due to improper handling or tool usage. Replace bearing.

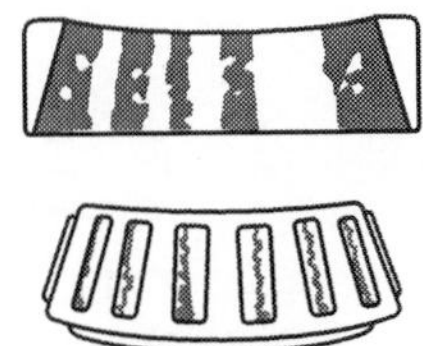

INDENTATIONS

Surface depressions on race and rollers caused by hard particles of foreign material. Clean all parts and housings. Check seals, and replace bearings if rough or noisy.

MISALIGNMENT

Outer race misalignment due to foreign object. Clean related parts and replace bearing. Make sure races are properly sealed.

**Figure 22.** Bearing failures and corrective procedures.

**Figure 23**. If any of these conditions are present on the bearing, replacement is necessary. Tapered roller bearings and their matching outer races must be replaced as a set. If the bearing installation is not done immediately, cover the

FATIGUE SPALLING

Flaking of surface metal resulting from fatigue. Replace bearing, clean all related parts.

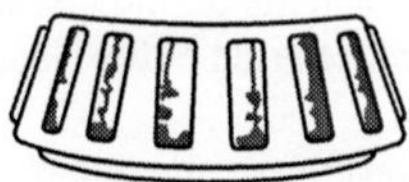

STAIN DISCOLORATION

Discoloration can range from light brown to black caused by incorrect lubricant or moisture. Re-use bearings if stains can be removed by light polishing or if no evidence of overheating is observed. Check seals and related parts for damage.

CAGE WEAR

Wear around outside diameter of cage and roller pockets caused by abrasive material and inefficient lubrication. Clean related parts and housings. Check seals and replace bearings.

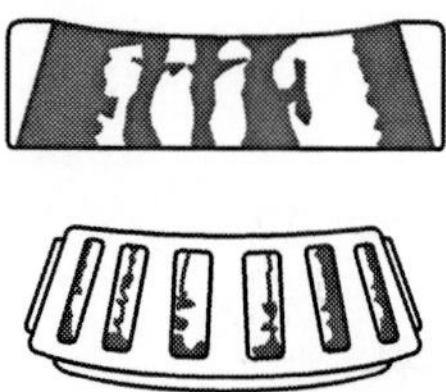

HEAT DISCOLORATION

Heat discoloration can range from faint yellow to dark blue, resulting from overload or incorrect lubricant. Excessive heat can cause softening of races or rollers. To check for loss of temper on races or rollers, a simple file test may be made. A file drawn over a tempered part will grab and cut metal, whereas, a file drawn over a hard part will glide readily with no metal cutting. Replace bearings if overheating damage is indicated. Check seals and other parts.

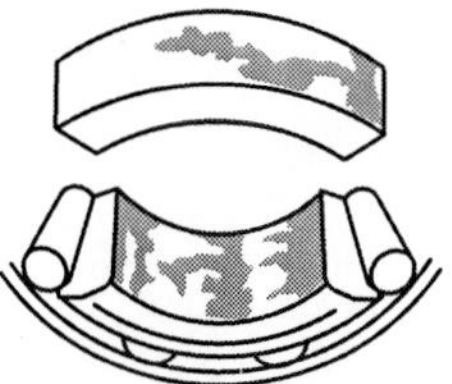

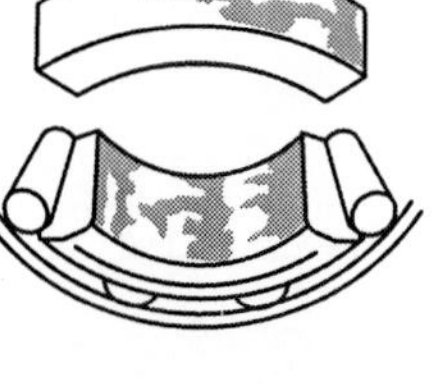

FRETTAGE

Corrosion set up by small relative movement of parts with no lubrication. Replace bearings. Clean related parts. Check seals and check for proper lubrication.

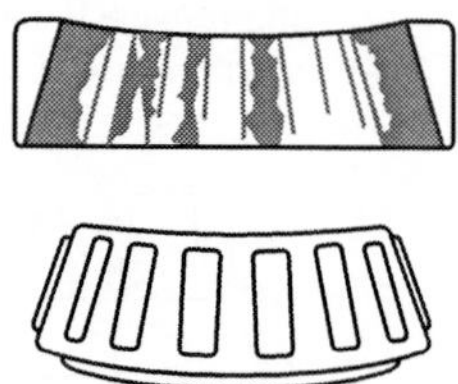

BRINELLING

Surface indentations in raceway caused by rollers either under impact loading or vibration while the bearing is not rotating. Replace bearing if rough or noisy.

SMEARS

Smearing of metal due to slippage. Slippage can be caused by poor fits, lubrication, overheating, overloads, or handling damage. Replace bearings, clean related parts and check for proper fit and lubrication.

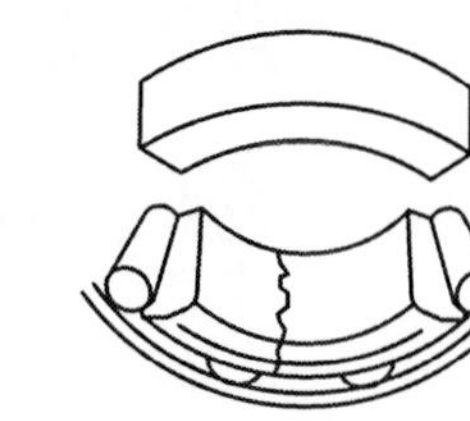

CRACKED INNER RACE

Race cracked due to improper fit, cocking, or poor bearing seats. Replace bearing and correct bearing seats.

**Figure 23.** Bearing failures and corrective procedures, continued.

**Figure 24.** Wrapping a bering in waterproof paper.

bearings with a protective lubricant and wrap them in waterproof paper **(Figure 24)**. Be sure to identify the bearings or lay them in order so you reinstall them in their original location. Do not clean bearings or races with paper towels. If you are using a shop towel for this purpose, be sure it is lint-free. Lint from shop towels or paper towels may contaminate the bearing.

Bearing races and the inner part of the wheel hub should be thoroughly cleaned with solvent and dried with compressed air. Inspect the seal mounting area in the hub for metal burrs. Remove any burrs with a fine round file. Bearing races must be replaced if they indicate any of the defects described in Figure 22 and Figure 23. The proper bearing race driving tool must be used to remove the bearing races **(Figure 25)**. If a driver is not available for the bearing races, a long punch and hammer may be used to drive the races from the hub. When a hammer and punch are used for this purpose, be careful not to damage the hub's inner surface with the punch.

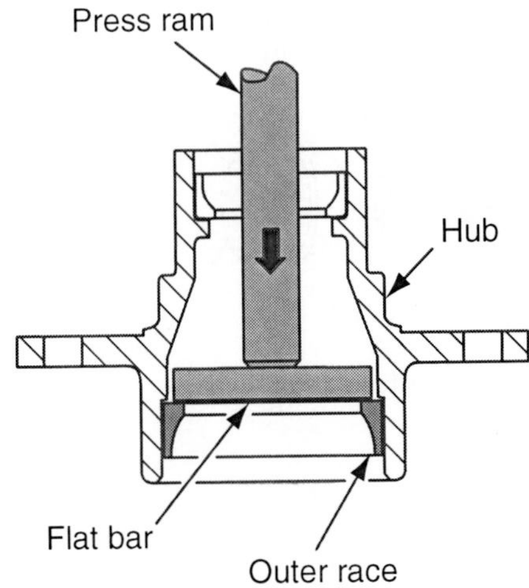


**Figure 25.** Bearing race removal.

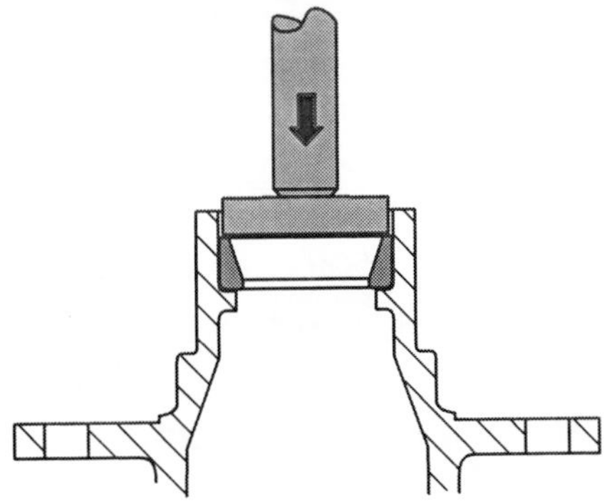

**Figure 26.** Bearing race installation.

The new bearing races should be installed in the hub with the correct bearing race driving tool **(Figure 26)**. When bearings and races are replaced, be sure they are the same as the original bearings. The part numbers should be the same on the old bearings and the replacement bearings.

Inspect the bearing and seal mounting surfaces on the spindle. Small metal burrs may be removed from the spindle with a fine-toothed file. If the spindle is severely scored in the bearing or seal mounting areas, spindle replacement is necessary.

## Bearing Lubrication and Assembly

After the bearings and races have been cleaned and inspected, the bearings should be packed with grease. Always use the vehicle manufacturer's specified wheel bearing grease. Place a lump of grease in the palm of one hand and grasp the bearing in the other hand. Force the widest edge of the bearing into the lump of grease and squeeze the grease into the bearing. Continue this process until grease is forced into the bearing around the entire bearing circumference. Place a coating of grease around the outside of the rollers and apply a light coating of grease to the races. A bearing packing tool may be used to force grease into the bearings rather than the hand method **(Figure 27)**.

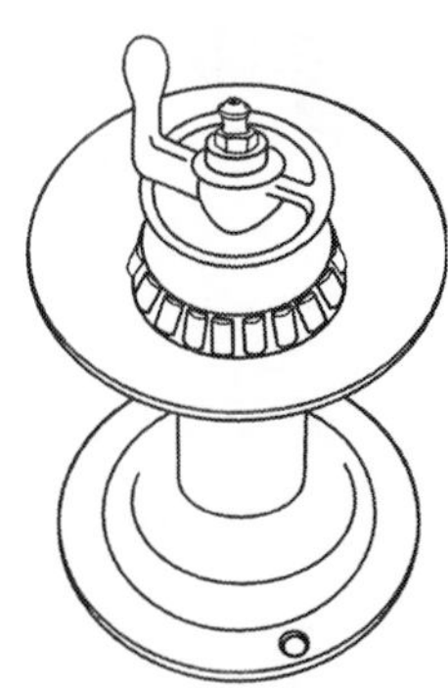

**Figure 27.** Mechanical wheel bearing packer.

**You Should Know** *Cleanliness is very important during wheel bearing service. Always maintain cleanliness of hands, tools, work area, and all related bearing components. One small piece of dirt in a bearing will cause bearing failure. Always keep grease containers covered when not in use. Uncovered grease containers are easily contaminated with dirt and moisture.*

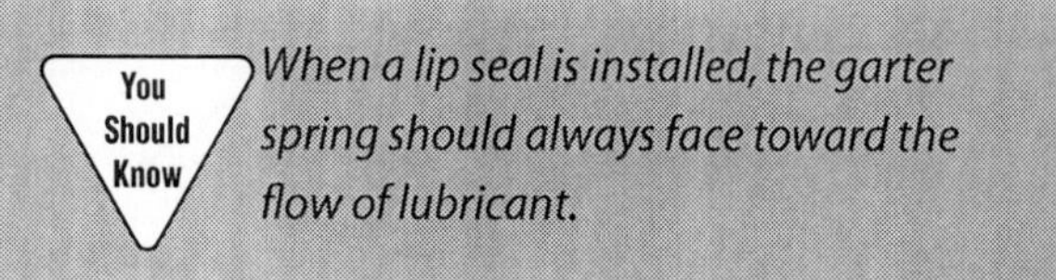
**You Should Know** *When a lip seal is installed, the garter spring should always face toward the flow of lubricant.*

Place some grease in the wheel hub cavity and position the inner bearing in the hub **(Figure 28)**. Check the fit of the new bearing seal on the spindle and in the hub. The seal lip must fit snugly on the spindle and the seal case must fit properly in the hub opening. The part number on the old seal and the replacement seal should be the same. Be sure the seal is installed in the proper direction with the garter spring and higher part of the lip toward the lubricant in the hub. The new inner bearing seal must be installed in the hub with a suitable seal driver **(Figure 29)**.

Place a light coating of wheel bearing grease on the spindle and slide the hub assembly onto the spindle. Install the outer wheel bearing and be sure there is adequate lubrication on the bearing and race. Be sure the washer and nut are clean and install these components on the spindle **(Figure 30)**. Tighten the nut until it is finger tight.

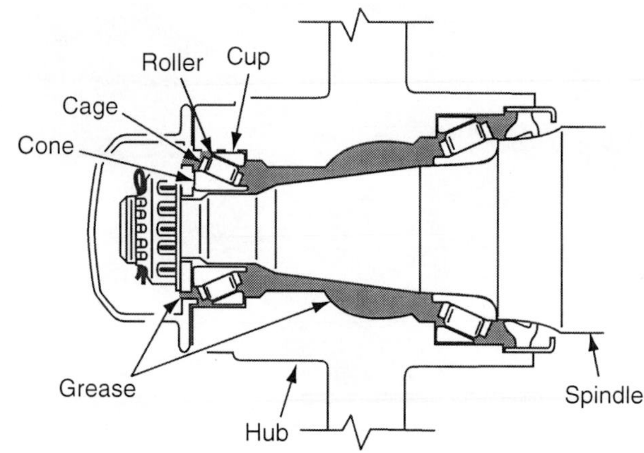


**Figure 28.** Wheel bearing lubrication.

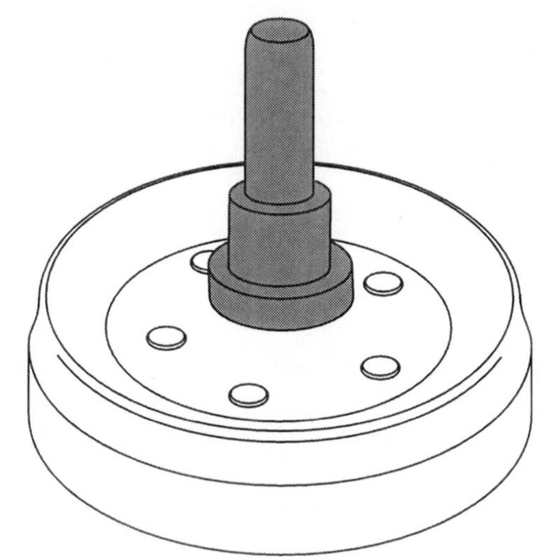

**Figure 29.** Seal installation.

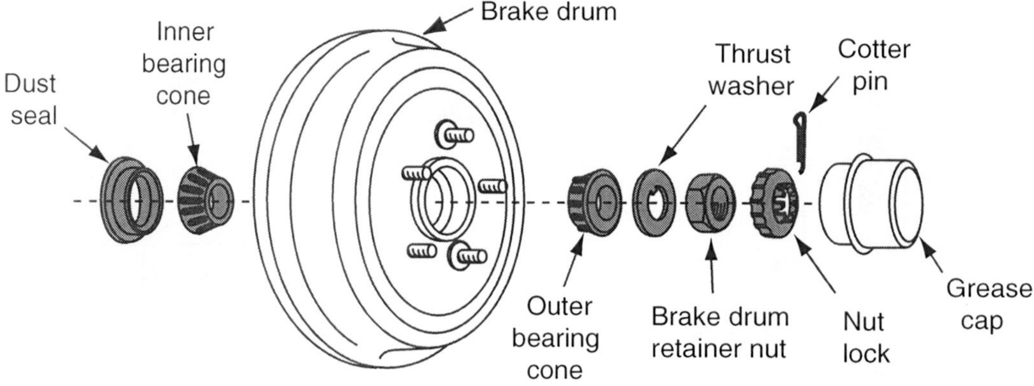


**Figure 30.** Installation of wheel bearings and related components.

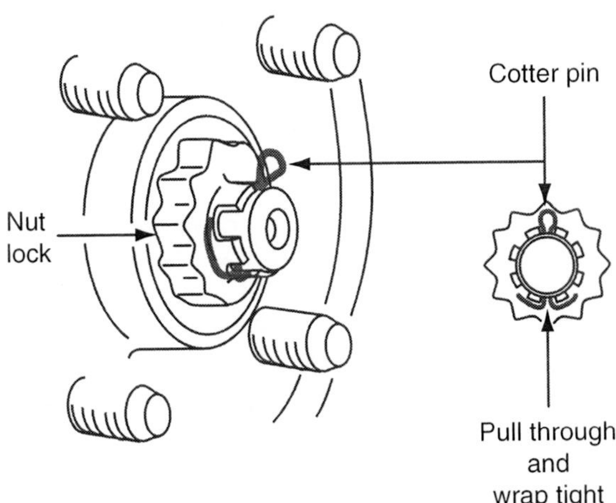


**Figure 31.** Nut lock and cotter pin installation.

## Wheel Bearing Adjustment with Two Separate Tapered Roller Bearings in the Wheel Hub

Loose front wheel bearing adjustment results in lateral front wheel movement and reduced directional stability. If the wheel bearing adjusting nut is tightened excessively, the bearings may overheat resulting in premature bearing failure. The bearing adjustment procedure may vary depending on the make of vehicle. Always follow the procedure in the vehicle manufacturer's service manual. The following is a typical bearing adjustment procedure:

1. With the hub and bearings assembled on the spindle, tighten the adjusting nut to 17 to 25 ft. lbs (23 to 34 N·m) while the hub is rotated.
2. Loosen the adjusting nut 1/2 turn and retighten it to 10 to 15 in. lbs (1.0 to 1.7 N·m). This specification varies depending on the make of vehicle. Always use manufacturer's specifications.
3. Position the adjusting nut retainer over the nut so the retainer slots are aligned with the cotter key hole in the spindle.
4. Install a new cotter pin and bend the ends around the retainer flange **(Figure 31)**.
5. Install the grease cap and make sure the hub and drum rotate freely.

## WHEEL BEARING DIAGNOSIS

Bearings are designed to provide long life, but there are many causes of premature bearing failure. If a bearing fails, the technician must decide if the bearing failure was caused by normal wear or if the bearing failed prematurely. For example, if a front wheel bearing fails on a car that is 1 year old with an original odometer reading of 15,000 miles (24,000 kilometers), experience tells us the bearing failure is premature because front wheel bearings normally last for a much longer mileage period. Always listen to the customer's complaints and obtain as much information as possible from the customer. Ask the customer specific questions about abnormal or unusual vehicle noises and operation. If a bearing fails prematurely, there must be some cause for the failure. The causes of premature bearing failure are these: lack of lubrication, improper type of lubrication, incorrect end-play adjustment (where applicable), misalignment of related components such as shafts or housings, excessive bearing load, improper installation or service procedures, excessive heat, or dirt or contamination.

When a bearing fails prematurely, the technician must correct the cause of this failure to prevent the new bearing from failing. The first indication of bearing failure is usually a howling noise while the bearing is rotating. The howling noise will likely vary depending on the bearing load. A front wheel bearing usually provides a more noticeable howl when the vehicle is turning a corner because this places additional thrust load on the bearing. A defective rear axle bearing usually provides a howling noise that is more noticeable at lower speeds. The howling noise is more noticeable when driving on a narrow street with buildings on each side because the noise resonates off the nearby buildings. A rear axle bearing noise is present during acceleration and deceleration because the vehicle weight places a load on the bearing regardless of the operating condition. The rear axle bearing noise may be somewhat more noticeable during deceleration because there is less engine noise at that time.

## WHEEL HUB UNIT DIAGNOSIS

When wheel bearings and hubs are an integral assembly, the bearing end play should be measured with a dial indicator stem mounted against the hub. If the end play exceeds 0.005 inch (0.127 millimeter) as the hub is moved in and out, the hub and bearing assembly should be replaced. This specification is typical, but the vehicle manufacturer's specifications must be used. Hub and bearing replacement is also necessary if the bearing is rough or noisy. Integral-type bearing and hub assemblies are used on the front and rear wheels on some front-wheel drive cars.

When the front wheel bearings are mounted in the steering knuckle, the wheel bearings may be checked with

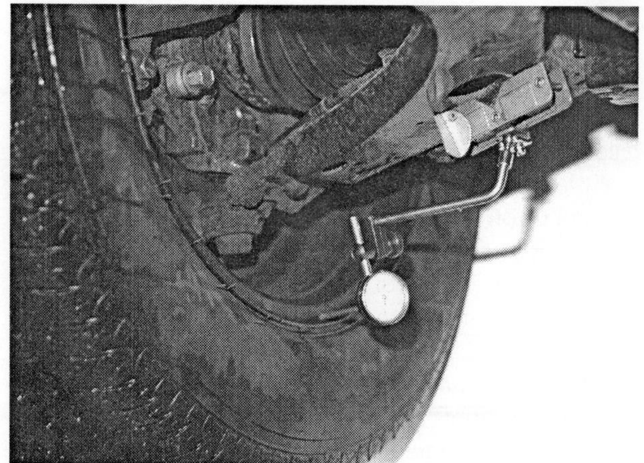

Figure 32. Wheel bearing diagnosis on vehicle.

the vehicle raised on the hoist and a dial indicator positioned against the outer wheel rim lip as pictured in **Figure 32**.

When the wheel is moved in and out, the maximum bearing movement on the dial indicator should be as follows:

0.020 inch (0.508 millimeter) for 13 inch (33 centimeter) wheels.
0.023 inch (0.584 millimeter) for 14 inch (35.5 centimeter) wheels
0.025 inch (0.635 millimeter) for 15 inch (38 centimeter) wheels

If the bearing movement is excessive, check the hub nut torque before replacing the bearing. When this torque is correct and bearing movement is excessive, the bearing should be replaced.

When a wheel is removed to service the wheel bearings, proper balance must be maintained between the wheel and tire and the hub. Therefore, the tire, wheel, and hub stud should be chalk marked prior to removal as illustrated in **Figure 33**.

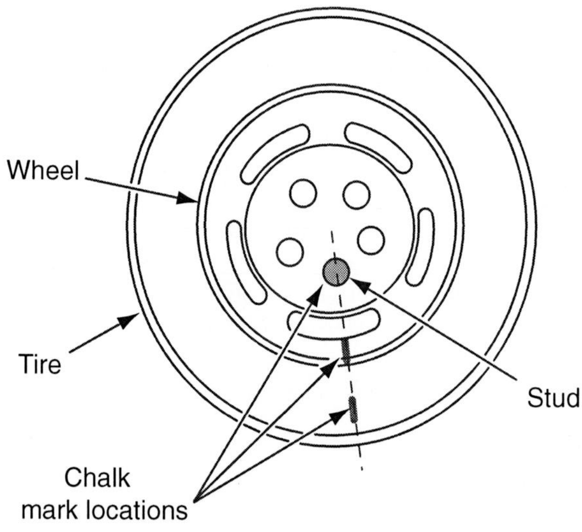


Figure 33. Chalk marking on wheel, tire, and stud.

## REAR AXLE BEARING AND SEAL SERVICE, REAR-WHEEL DRIVE VEHICLES

Rear axle bearing noise may be diagnosed with the vehicle raised on a lift. Be sure the hoist safety mechanism is engaged after the vehicle is raised on the lift. With the engine running and the transmission in drive, operate the vehicle at moderate speed, (35 to 45 miles per hour) (56 to 72 kilometers per hour) and listen with a stethoscope placed on the rear axle housing directly over the axle bearings. If grinding or clicking noises are heard, bearing replacement is necessary.

**You Should Know** *Use extreme caution when diagnosing problems with a vehicle raised on a lift and the engine running with the transmission in drive. Keep away from rotating wheels, drive shafts, or drive axles.*

Many axle shafts in rear-wheel drive cars have a roller bearing and seal at the outer end **(Figure 34)**. These axle shafts are often retained in the differential with "C" locks that must be removed before the axles.

The rear axle bearing removal and replacement procedure varies depending on the vehicle make and model year. Always follow the rear axle bearing removal and replacement procedure in the manufacturer's service manual. The following is a typical rear axle shaft removal and replacement procedure on a rear-wheel drive car with "C" lock axle retainers:

1. Loosen the rear wheel nuts and chalk mark the rear wheel position in relation to the rear axle.
2. Raise the vehicle on a lift and make sure the lift safety mechanism is in place.

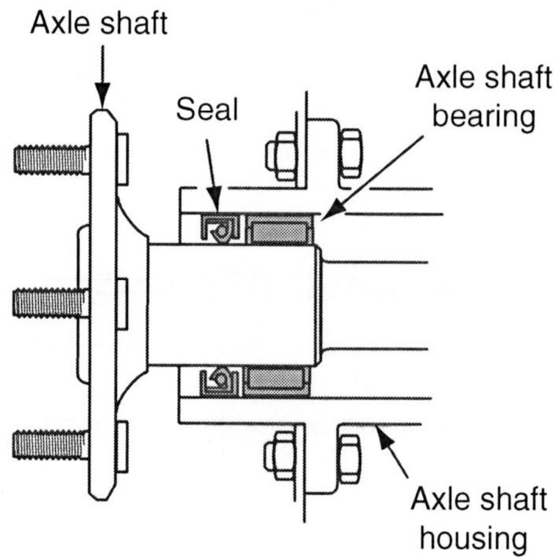


Figure 34. Rear axle roller bearing and seal, rear-wheel drive car.

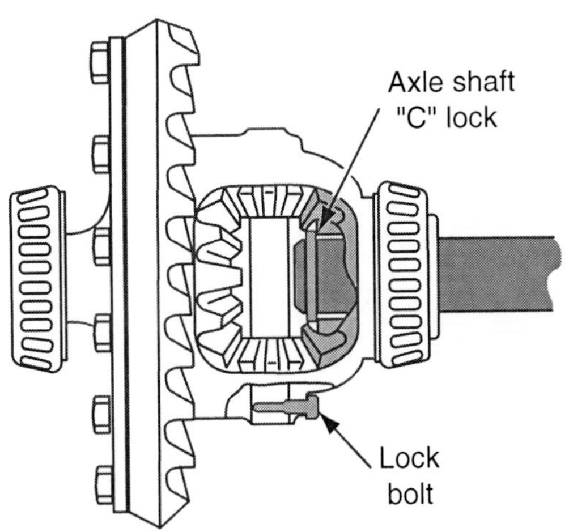


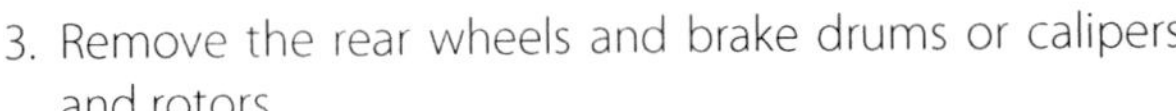
**Figure 35.** Rear axle "C" lock, lock bolt and pinion gears.

3. Remove the rear wheels and brake drums or calipers and rotors.
4. Place a drain pan under the differential and remove the differential cover. Discard the old lubricant.
4. Remove the differential lock bolt, pin, and shaft **(Figure 35)**.
5. Push the axle shaft inward and remove the axle "C" lock.
6. Pull the axle from the differential housing.

Reverse the axle removal procedure to reinstall the axle. Always use a new differential cover gasket or sealant and fill the differential to the bottom of the filler plug opening with the manufacturer's recommended lubricant. Be sure all fasteners, including the wheel nuts, are tightened to the specified torque.

The following is a typical axle bearing and seal removal procedure:

1. Remove the axle seal with a seal puller.
2. Use the proper bearing puller to remove the axle bearing **(Figure 36)**.

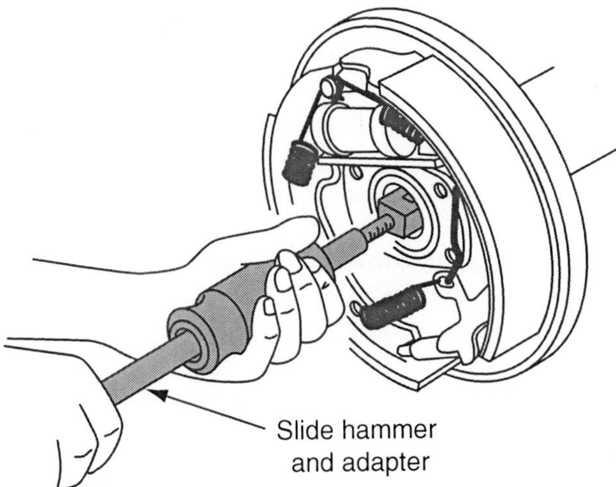


**Figure 36.** Rear axle bearing puller.

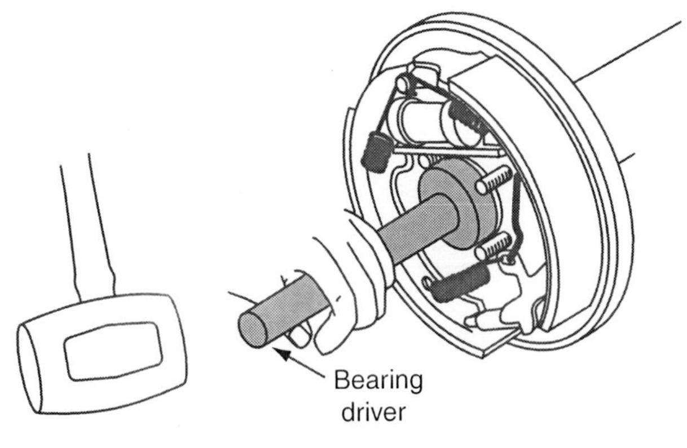


**Figure 37.** Rear axle bearing driver.

3. Clean the axle housing seal and bearing mounting area with solvent and a brush. Clean this area with compressed air.
4. Check the seal and bearing mounting area in the housing for metal burrs and scratches. Remove any burrs or irregularities with a fine-toothed round file.
5. Wash the axle shaft with solvent and blow it dry with compressed air.
6. Check the bearing contact area on the axle for roughness, pits, and scratches. If any of these conditions are present, axle replacement is necessary.
7. Be sure the new bearing fits properly on the axle and in the housing. Install the new bearing with the proper bearing driver **(Figure 37)**. The bearing driver must apply pressure to the outer race that is pressed into the housing.
8. Be sure the new seal fits properly on the axle shaft and in the housing. Make sure the garter spring on the seal faces toward the differential. Use the proper seal driver to install the new seal in the housing **(Figure 38)**.

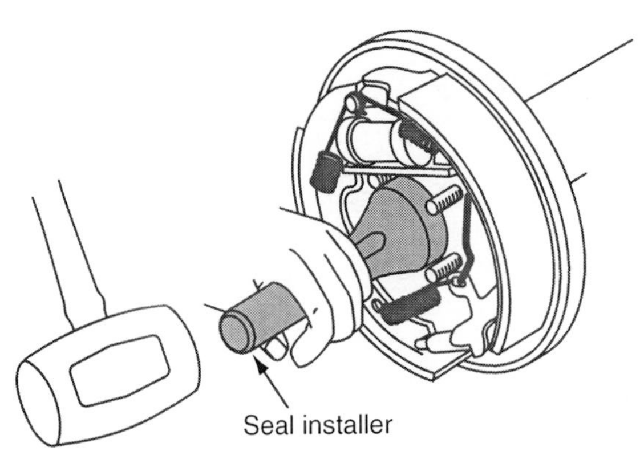


**Figure 38.** Installing rear axle seal.

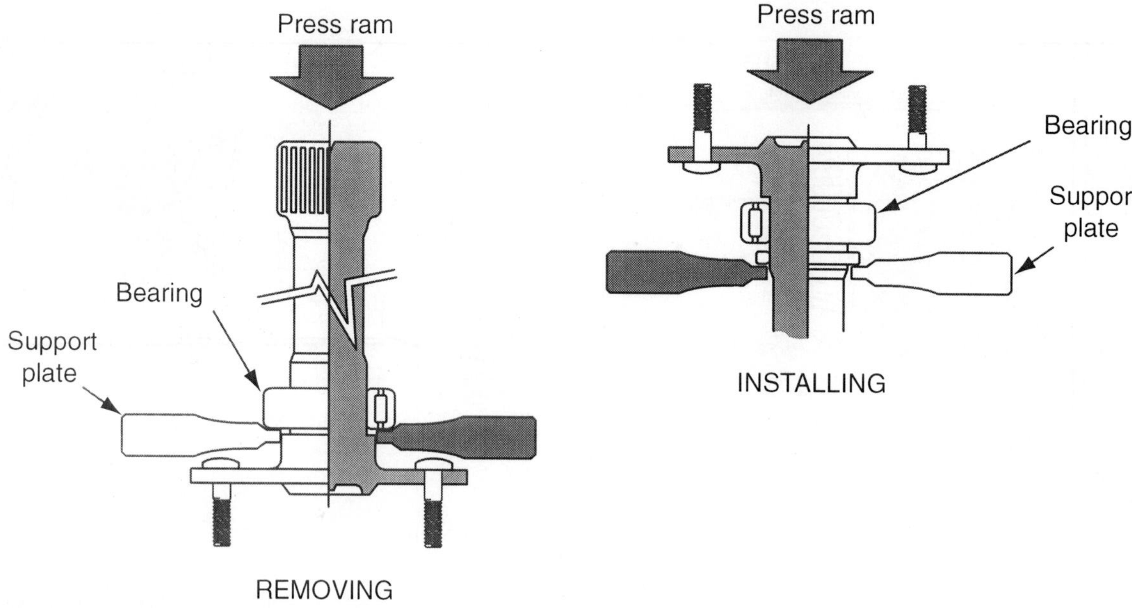


**Figure 39.** Axle bearing and retainer ring removal and replacement.

9. Lubricate the bearing, seal, and bearing surface on the axle with the manufacturer's specified differential lubricant.
10. Reverse the rear axle removal procedure to reinstall the rear axle.
11. Be sure all fasteners are tightened to the specified torque.

**You Should Know** *Never use an acetylene torch to heat axle bearings or retainer rings during the removal and replacement procedure. The heat may cause fatigue in the steel axle and the axle may break suddenly causing the rear wheel to fall off. This action will likely result in severe vehicle damage and personal injury.*

Some rear axles have a sealed bearing that is pressed onto the axle shaft and held in place with a retainer ring. These rear axles usually do not have "C" locks in the differential. A retainer plate is mounted on the axle between the bearing and the outer end of the axle. This plate is bolted to the outer end of the differential housing. After the axle retainer plate bolts are removed, a slide hammer-type puller is attached to the axle studs to remove this type of axle. When this type of axle bearing is removed, the retainer ring must be split with a hammer and chisel while the axle is held in a vise. Do not heat the retainer ring or the bearing with an acetylene torch during the removal or installation process. After the retainer ring is removed, the bearing must be pressed from the axle shaft and the bearing must not be reused. A new bearing and retainer ring must be pressed onto the axle shaft. The bearing removal and replacement procedure is illustrated in **Figure 39**.

## Summary

- Tire noises vary with road conditions, but differential noise is not affected by road conditions.
- Steering pull may be caused by defects in the suspension or steering systems and tire conicity.
- Tire conicity occurs when the belt in a tire is wound off center during the manufacturing process.
- Tire conicity cannot be diagnosed by a visual inspection.
- The tire rotation procedure varies depending on the model year, type of tires, and whether the vehicle has a compact spare or a conventional spare.
- Hand tools and tire irons should not be used to demount tires.
- Tire punctures over 1/4-inch diameter should not be repaired.
- Compact spare tires should not be repaired.

- Do not use caustic agents, alkaline-based detergents, or abrasive cleaners to clean aluminum or magnesium wheels.
- When a bead expander is used to mount a tire, do not increase tire pressure above 10 psi (69 kPa).
- When a bead expander is not used to mount a tire, do not increase tire pressure above 40 psi (276 kPa) to move the tire beads out against the wheel rim.
- While mounting a tire, be sure that the tire beads and wheel rim bead seats are coated with rubber tire lubricant.
- Radial tire and wheel runout causes tire thump.
- Lateral tire and wheel runout causes tire and chassis waddling.
- Balls of rubber or other objects inside a tire make proper balance impossible.
- Aluminum or magnesium wheels require the use of stick-on wheel weights.
- Electronic wheel balancers perform the necessary wheel weight calculations and indicate the exact locations where these weights should be installed.
- When a wheel bearing fails prematurely, the technician must determine the cause of the failure and correct this cause to prevent a second bearing failure.
- Before a tire and wheel assembly is removed from a car, the wheel and tire assembly must be chalk marked in relation to the hub or axle to maintain wheel balance when the wheel is reinstalled.
- Tapered roller bearings and races must be replaced as a matched set.
- When cleaning bearings, never spin the bearings with compressed air.
- When drying bearings with compressed air, be sure the shop air supply is free from moisture.
- Do not wipe bearings with paper towels or shop towels with lint on them.
- Always inspect bearing and seal mounting areas for metal burrs and scratches. Burrs may be removed with a fine-toothed file.
- While servicing wheel bearings, always keep hands, tools, and work area clean.
- Wheel bearing hub units must be checked for end play with a dial indicator.
- Rear axle bearings and retainer rings that are pressed onto the axle shaft should not be heated with an acetylene torch.
- On some rear-wheel drive cars, the rear axle "C" locks inside the differential must be removed prior to rear axle removal.

## Review Questions

1. While discussing a tire thumping problem, Technician A says this problem may be caused by cupped tire treads. Technician B says a heavy spot in the tire may cause this complaint. Who is correct?
   A. Technician A
   B. Technician B
   C. Both Technician A and Technician B
   D. Neither Technician A nor Technician B
2. While discussing a vehicle that pulls to one side, Technician A says that excessive radial runout on the right front tire may cause this problem. Technician B says that tire conicity may be the cause of this complaint. Who is correct?
   A. Technician A
   B. Technician B
   C. Both Technician A and Technician B
   D. Neither Technician A nor Technician B
3. While discussing tire wear, Technician A says that static imbalance causes feathered tread wear. Technician B says that dynamic imbalance causes cupped wear and bald spots on the tire tread. Who is correct?
   A. Technician A
   B. Technician B
   C. Both Technician A and Technician B
   D. Neither Technician A nor Technician B
4. While discussing tire noise, Technician A says that tire noise varies with road surface conditions. Technician B says that tire noise remains constant when the vehicle is accelerated and decelerated. Who is correct?
   A. Technician A
   B. Technician B
   C. Both Technician A and Technician B
   D. Neither Technician A nor Technician B
5. Chassis waddle may occur if the lateral tire runout exceeds:
   A. 0.010 inch.
   B. 0.025 inch.
   C. 0.050 inch.
   D. 0.080 inch.
6. Radial wheel runout should not exceed:
   A. 0.005 inch.
   B. 0.015 inch.
   C. 0.020 inch.
   D. 0.035 inch.
7. While driving a vehicle straight ahead on a smooth road surface, the steering pulls to the right. All of these defects may be the cause of the problem *except*:
   A. two tires of different sizes on the front wheels.
   B. different inflation pressures in the front tires.
   C. improper static wheel balance on one front wheel and tire assembly.
   D. different tire designs on the two front wheels.

8. The growling noise produced by a defective front wheel bearing is most noticeable when:
   A. turning a corner at low speeds.
   B. driving straight ahead at low speeds.
   C. driving at the freeway speed limit.
   D. the vehicle is just starting off after a stop.
9. On a typical sealed wheel bearing hub assembly, the maximum end play should be:
   A. 0.005 inch.
   B. 0.008 inch.
   C. 0.010 inch.
   D. 0.018 inch.
10. When servicing rear axle bearings that are pressed onto the axle shaft and held in place with a retainer ring:
   A. remove the "C" locks from the inner end of the axles.
   B. cut the retainer ring off the axle with an acetylene torch.
   C. use a hydraulic press to remove and install the bearing on the axle shaft.
   D. heat the retainer ring with an acetylene torch before installing it on the axle.
11. The maximum allowable radial tire runout is ________________ inch(es).
12. A seal should be installed with the garter spring facing toward the ____________ ____________.
13. After wheel bearings are cleaned and repacked with grease, the bearing adjusting nut should be tightened to 17 to 25 ft. lbs, loosened 1/2 turn, and retightened to ____________ to ____________ in. lbs.
14. Balls of rubber inside a tire make wheel balancing ________________.
15. Explain front wheel shimmy and describe the causes and results of this problem.
16. Describe the wheel bearing adjustment procedure on a wheel hub containing two tapered roller bearings.
17. List six causes of premature wheel bearing failure.
18. Describe the conditions when a defective front wheel bearing noise is most noticeable.

# Section 11

## Drive Shafts, Drive Axles, and Clutches

*New vehicle prices in the United States increased 53.8 percent from 1990 to 1999 and another 6.4 percent from 1999 to 2001. The main reasons for the higher prices were increased use of technology and electronic equipment.*

## SECTION OBJECTIVES

After you have read, studied, and practiced the contents of this section, you should be able to:

- Explain drive shaft and universal joint design.
- Describe drive shaft purpose.
- Describe the changes in drive shaft speed during shaft rotation.
- Describe the noises that may be produced by worn universal joints.
- Remove and replace drive shafts and universal joints.
- Explain the motion requirements for inner and outer CV joints.
- Explain how equal length drive axles are possible on some FWD vehicles.
- Describe front drive axle purpose.
- Explain the purpose and importance of CV joint boots.
- Explain how worn engine or transaxle mounts may affect CV joints.
- Describe the diagnostic procedure for inner and outer CV joints.
- Describe the necessary repair action when the grease in a CV joint is contaminated with dirt or moisture.
- Describe clutch disc, flywheel, and pressure plate design.
- Explain clutch release bearing and fork design.
- Explain clutch cables and linkages.
- Describe hydraulic clutch system design.
- Explain clutch operation during engagement and disengagement.
- Describe the clutch pedal free-play adjustment.
- Explain the results of improper clutch pedal free-play adjustment.
- Describe the indications and causes of clutch slipping.
- Describe the indications and causes of clutch chatter.
- Explain the causes of various clutch noises.
- Describe the indications and causes of clutch vibration.
- Describe the indications and causes of clutch dragging.
- Explain the causes of a pulsating clutch pedal.
- Remove and replace clutches and pressure plates.

# Chapter 40 Drive Shaft, Drive Axle, and Universal Joint Maintenance, Diagnosis, and Service

## Introduction

Technicians must understand drive shaft and universal joint design and operation to diagnose these components accurately and quickly. For example, if a technician does not understand drive shaft angles and the proper procedures for diagnosing drive shaft vibrations, it will likely be difficult and time consuming to diagnose this problem.

Technicians must also understand the design, operation, and proper diagnostic and service procedures for front drive axles on FWD vehicles. When a technician has this knowledge and the related skills, he or she is able to diagnose front-wheel drive axle problems accurately and quickly. A technician with these skills is able to repair drive axle problems correctly the first time the repair is performed.

### DRIVE SHAFT DESIGN

On rear-wheel drive vehicles with the engine in the front, the drive shaft connects the transmission output shaft to the differential. Engine torque is transferred from the transmission output shaft through the drive shaft to the pinion shaft in the differential **(Figure 1)**. A typical drive shaft has a slip yoke with internal splines that fit over the splines on the transmission output shaft. The slip yoke has a smooth, machined outer surface. A seal in the transmission extension housing rides on this surface to prevent oil leaks

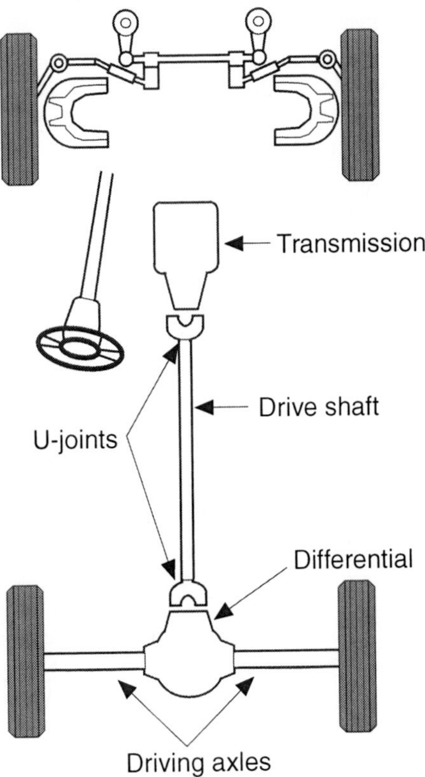


**Figure 1.** The drive shaft connects the transmission output shaft to the differential.

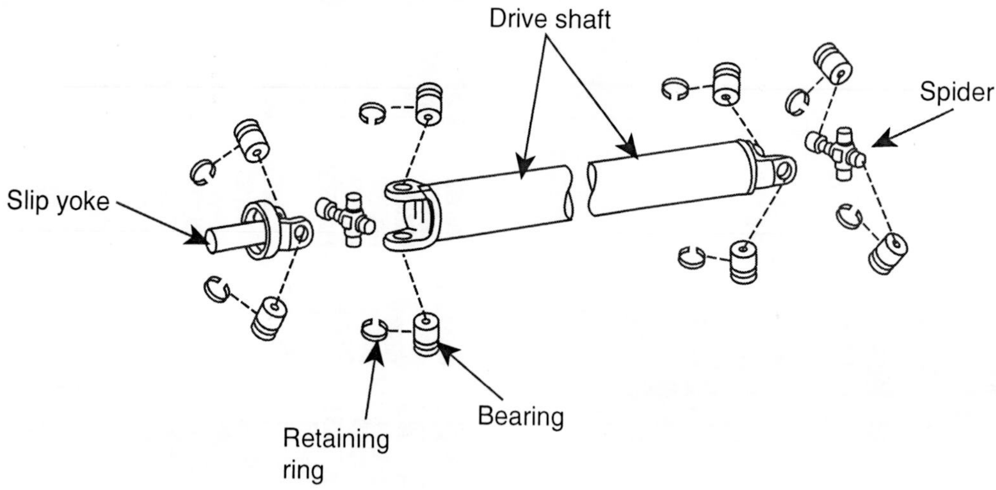


**Figure 2.** Typical drive shaft and universal joints.

at this location. A universal joint is connected between the front slip yoke and the drive shaft and another universal joint is connected between the drive shaft and the differential **(Figure 2)**.

Most drive shafts are made from seamless steel tubing with universal joint yokes welded on each end of this tubing. Some drive shafts on trucks are a two-piece unit with a center support bearing in the middle between the two shafts **(Figure 3)**. This type of drive shaft has three universal joints and a slip yoke on the rear shaft fits over splines on the rear of the front shaft.

Some drive shafts are made from aluminum or composite fiberglass. A composite drive shaft is lighter, stronger, and has improved balance characteristics compared to a steel drive shaft.

**Figure 3.** Drive shaft center bearing.

Four-wheel drive vehicles have two drive shafts, one shaft drives the rear differential and the second shaft drives the front differential **(Figure 4)**.

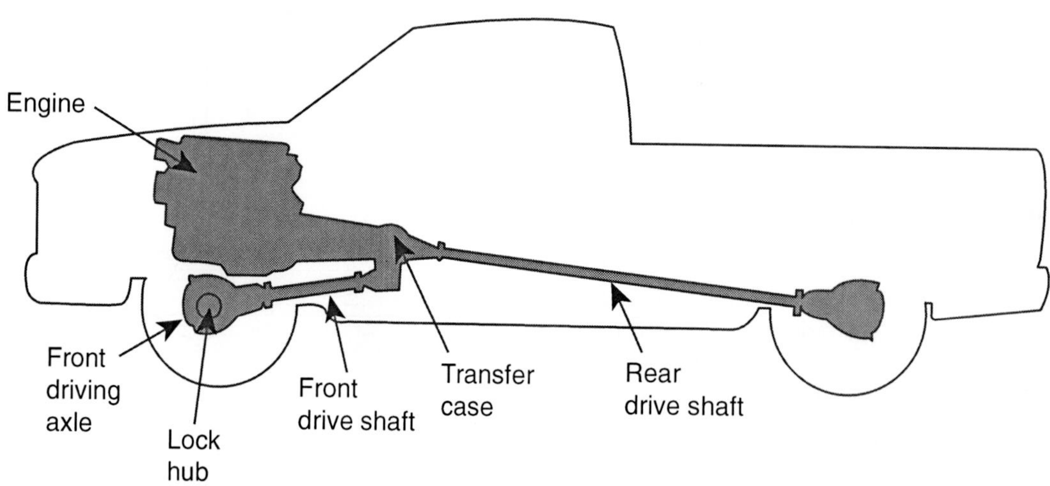


**Figure 4.** Front and rear drive shafts on a four-wheel drive vehicle.

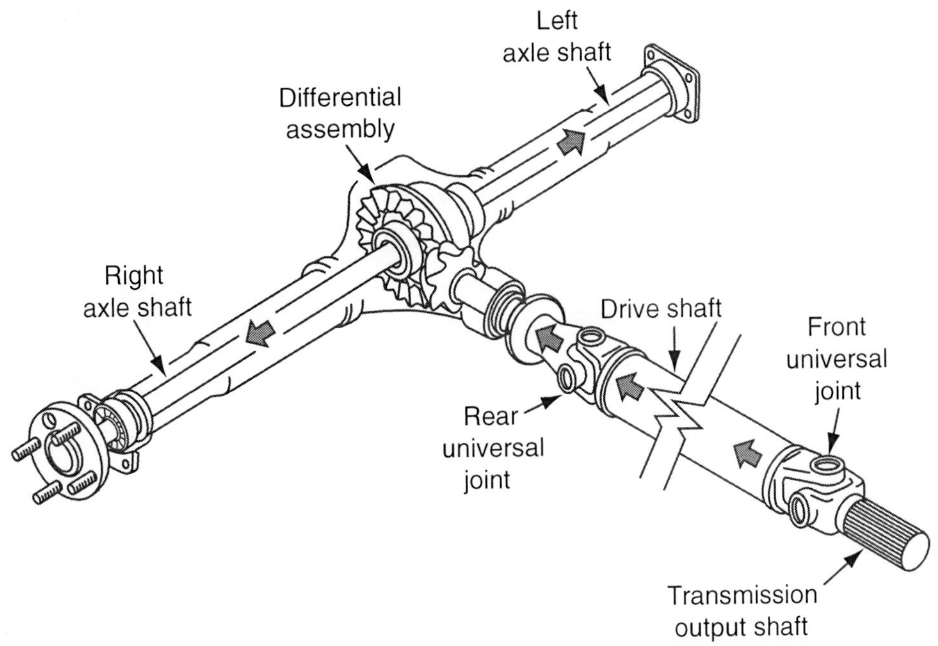


**Figure 5.** The drive shaft transmits engine torque from the transmission to the differential.

## DRIVE SHAFT PURPOSE

The purpose of the drive shaft is to transfer engine torque from the engine and transmission to the differential **(Figure 5)**. The front of the drive shaft is connected to the transmission and the transmission and engine are supported on mounts attached to the chassis. These mounts allow very limited engine and transmission movement. The differential is mounted on springs that are supported on the chassis. As the rear wheels strike road irregularities, the rear wheels and differential move upward and downward. This differential action changes the angle of the drive shaft between the differential and the transmission. It also changes the distance between these two components. The universal joints on each end of the drive shaft permit the drive shaft angle to change and the slip yoke in the front of the drive shaft can slip forward or rearward on the transmission output shaft to change the length of the drive shaft.

The drive shaft rotates three to four times faster than the vehicle drive wheels because of the differential gear ratio. Because of this high rotational speed, the drive shaft tends to vibrate at its critical speed. **Critical speed** is the rotational speed at which a component begins to vibrate. This vibration is usually caused by centrifugal force.

Drive shaft diameters are as large as possible and the drive shaft is as short as possible to keep the critical speed frequency above the normal rotational speed of the drive shaft. The most common method of drive shaft balancing is to weld balance weights to the shaft during the manufacturing process **(Figure 6)**.

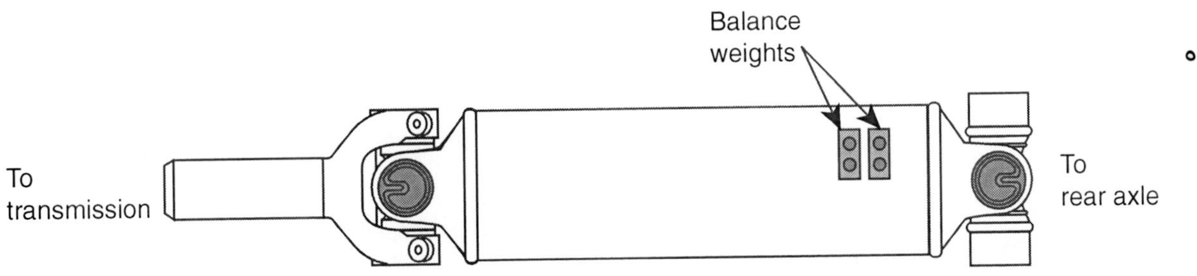


**Figure 6.** Balance weights are welded to the drive shaft.

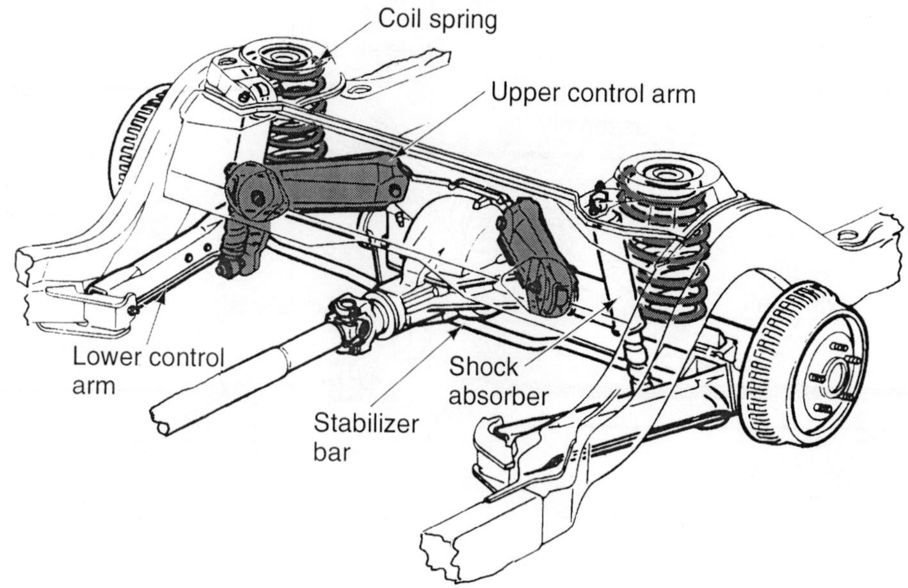


**Figure 7.** Hotchkiss drive.

## TYPES OF DRIVE SHAFTS

The most common type of drive shaft is the **Hotchkiss drive (Figure 7)** which has an open drive shaft and two universal joints on most passenger cars. On some trucks, the Hotchkiss drive has two drive shafts with a center support bearing. The center support bearing supports the drive shaft on the chassis **(Figure 8)**. The center support bearing helps to prevent vibration on the longer drive shafts required on trucks.

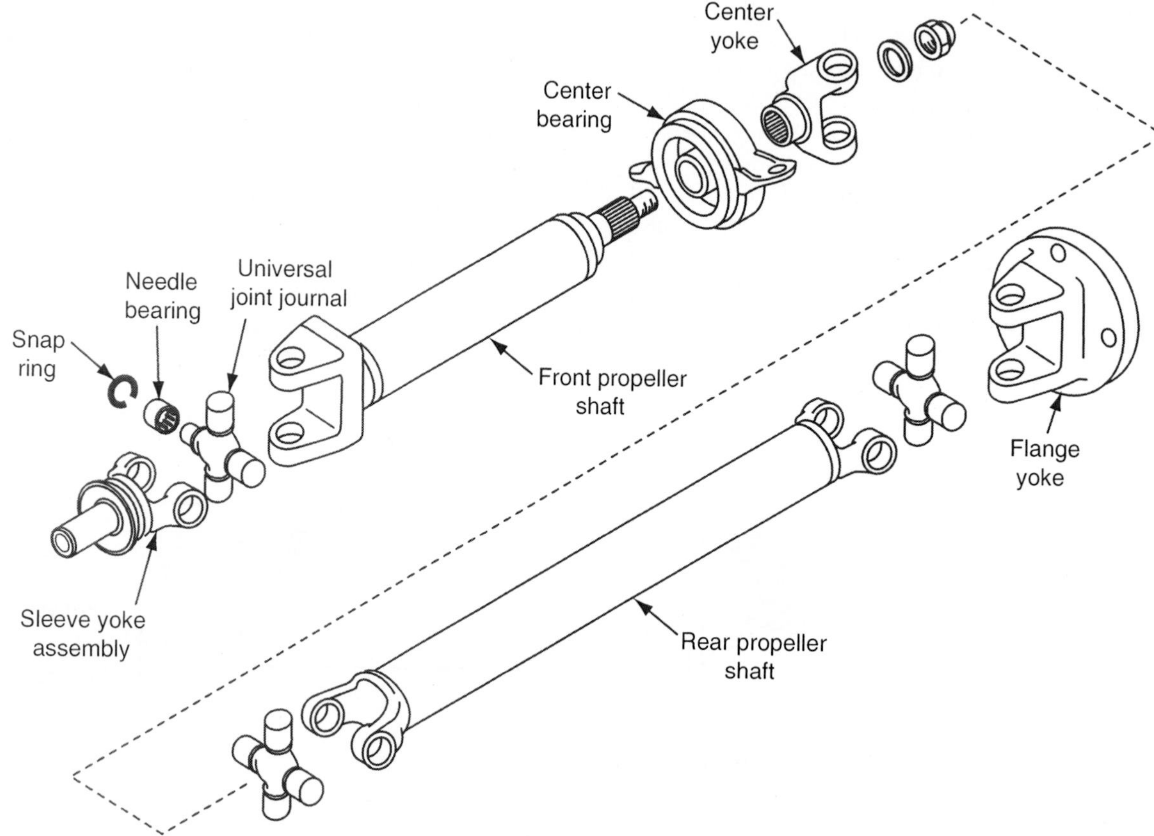


**Figure 8.** Two-piece drive shaft with a center support bearing.

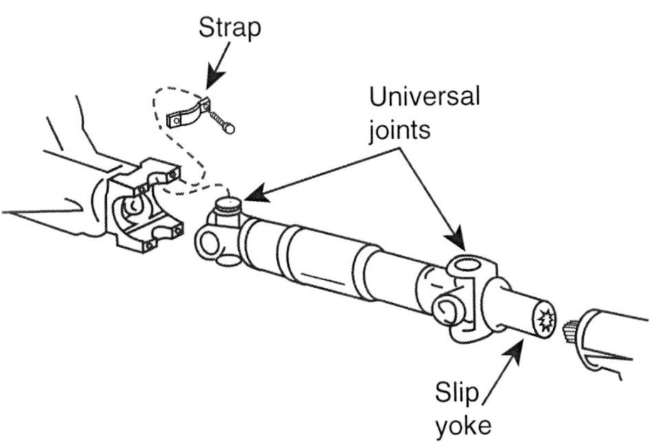


**Figure 9.** Drive shaft connections to the transmission and differential.

## UNIVERSAL JOINTS

A universal joint has two Y-shaped yokes. One of these yokes is on the drive shaft and on a front universal joint, the other yoke is on the slip joint. On the rear universal joint, the other yoke is on the differential flange **(Figure 9)**. Each yoke contains a machined opening. A spider is mounted in the center of the yokes at each end of the drive shaft.

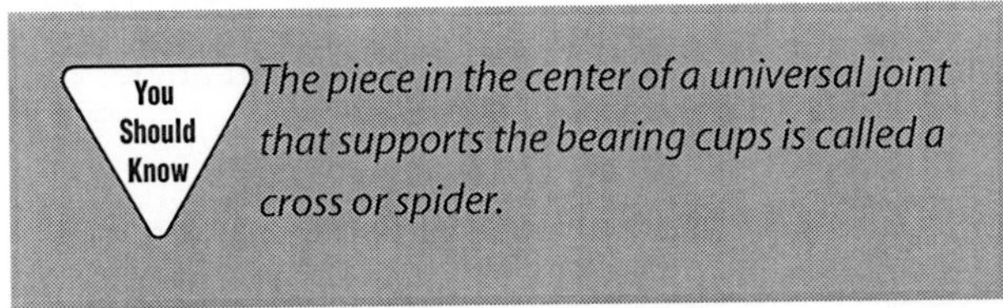


*The piece in the center of a universal joint that supports the bearing cups is called a cross or spider.*

Bearing cups containing needle bearings are mounted in each yoke opening. The needle bearings in each bearing cup rides on the machined surface on the spider **(Figure 10)**. A retaining ring fits into a groove in each bearing cup to retain this cup in the yoke. Therefore, the spider is supported in the center of the yokes and the spider and needle bearings allow the slip yoke to move sideways or vertically in relation to the drive shaft. This universal joint action allows the drive shaft to rotate smoothly and also permits

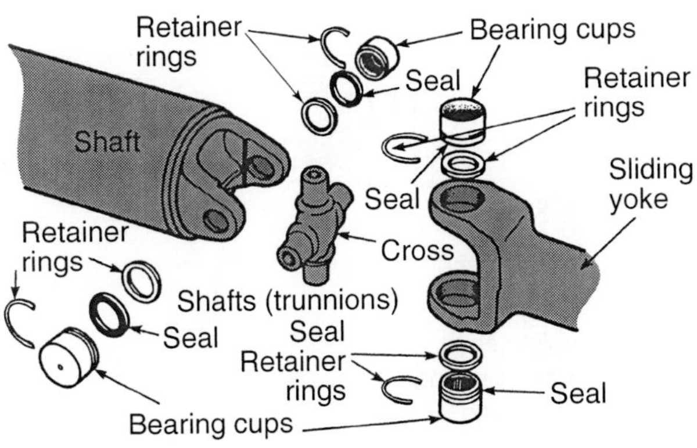


**Figure 10.** Typical Cardan universal joint.

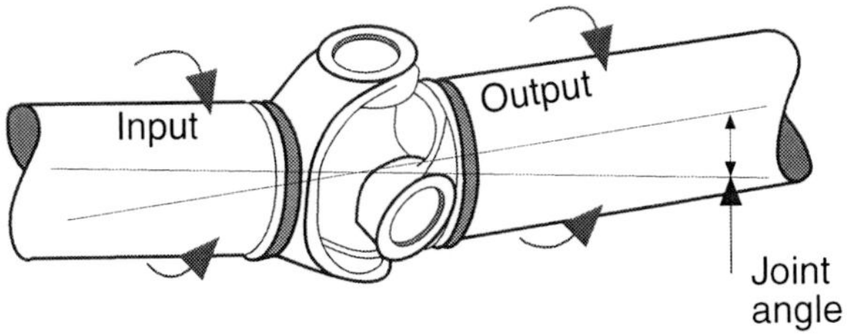


**Figure 11.** Drive shaft angles.

the drive shaft angle to change in relation to the slip yoke and transmission output shaft.

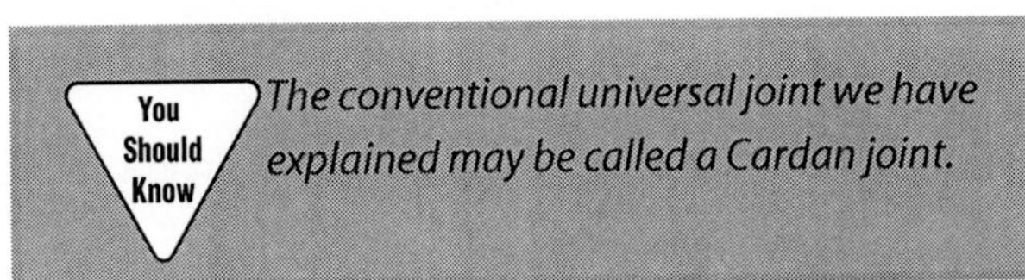


*The conventional universal joint we have explained may be called a Cardan joint.*

A universal joint transmits torque through an angle. The operating angle of a universal joint is determined by the difference between the transmission installation angle and the drive shaft installation angle **(Figure 11)**. The input shaft speed is constant, but the speed of the output shaft accelerates and decelerates twice during each revolution. The input yoke rotates at a constant speed during each revolution. However, the output yoke quadrants alternate between longer and shorter distances of travel compared to the input yoke quadrants that remain at constant distances. The rotational path of the output yoke appears like an ellipse. An **ellipse** is a compressed form of a circle.

When one point of the output yoke travels the same distance in a shorter time, it must move at a slower rate. When one point in the output yoke moves the same distance in a longer time, it must travel faster. The face of a clock may be used to illustrate the ellipse action of a drive shaft **(Figure 12)**. If the hand on the clock moves the longer distance from top to bottom on the outer circle, it must move faster compared to movement from top to bottom on the ellipse.

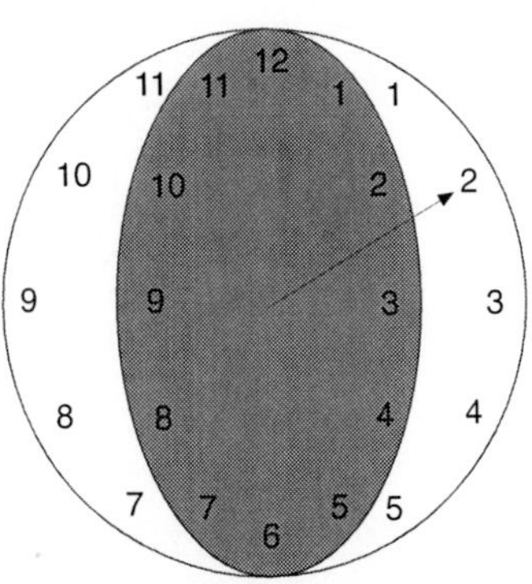


**Figure 12.** The face of a clock may be used to illustrate the elliptical movement of the drive shaft.

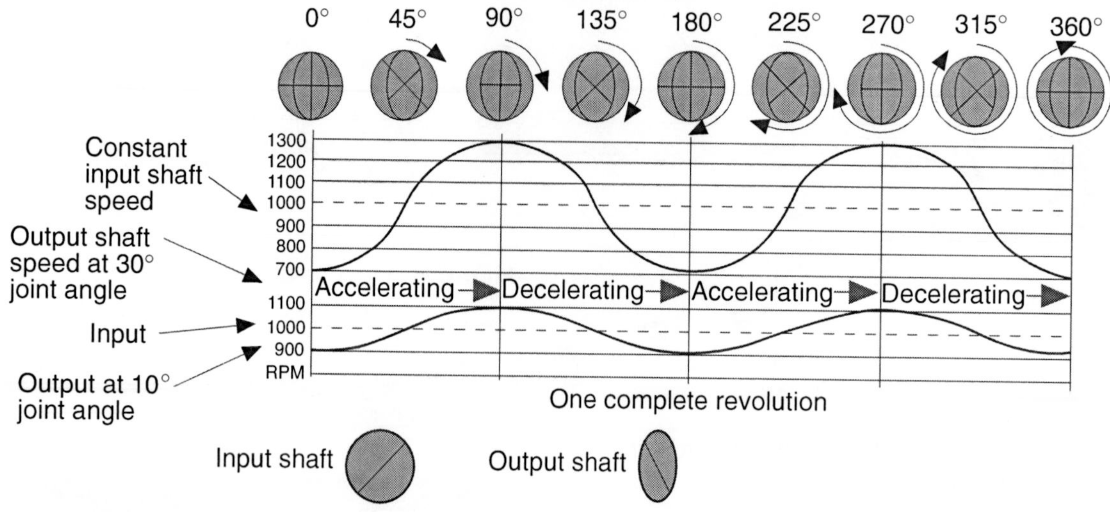


**Figure 13.** Variations in drive shaft speed in one revolution.

The output yoke in a universal joint is continually falling behind and catching up with the input yoke **(Figure 13)**. The resulting deceleration and acceleration causes fluctuating torque and torsional vibrations. A steeper drive shaft angle causes increased torsional vibrations.

Drive shaft vibration may be reduced by using canceling angles. The front and rear universal joint angle should be the same **(Figure 14)**. If this condition is present, when the front universal joint accelerates and produces a vibration, the rear universal joint decelerates and produces an equal and opposite vibration that cancels the vibration from the front universal joint. **Canceling angles** are present when the vibration from one universal joint is canceled by an equal and opposite vibration from another universal joint.

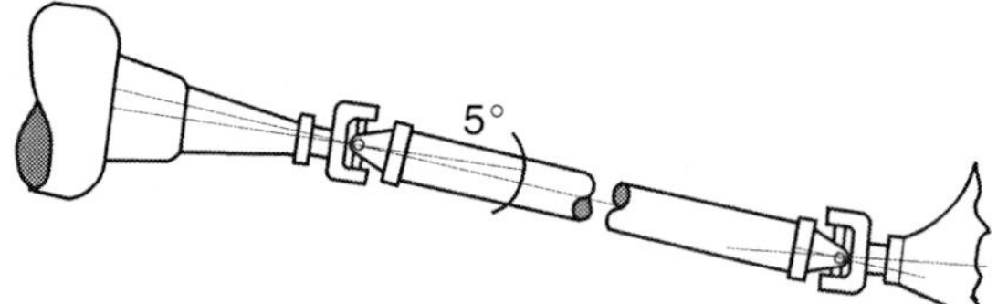


**Figure 14.** Equal front and rear drive shaft angles cancel drive shaft vibrations.

Universal joints on a drive shaft should be on the same plane **(Figure 15)**. When this condition is present, the universal joints are said to be in phase with each other. If universal joints are disassembled, the yoke should be marked in relation to the drive shaft to maintain the in-phase condition. If this action is not taken, drive shaft vibration may

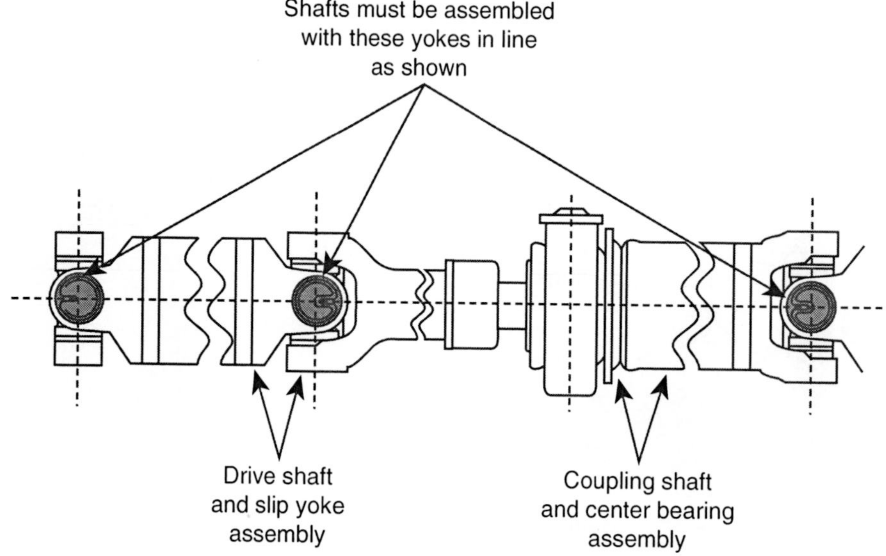


**Figure 15.** Universal joints in the same plane are in phase with each other.

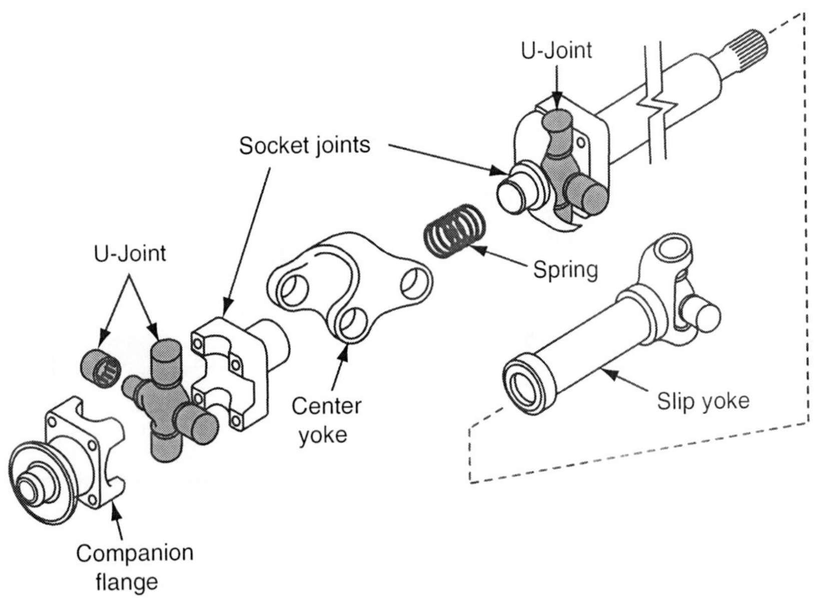


**Figure 16.** Double Cardan universal joint.

occur. Universal joints on a drive shaft are **in phase** when they are on the same plane.

Some rear-wheel drive vehicles have a double Cardan universal joint containing two individual Cardan universal joints mounted close together connected by a centering socket yoke. A round socket on the rear of the drive shaft fits into a ball socket on the rear yoke **(Figure 16)**. A short center yoke connects the two universal joints. The ball and socket of the double Cardan universal joint splits the angle of the two drive shafts equally between the two joints **(Figure 17)**.

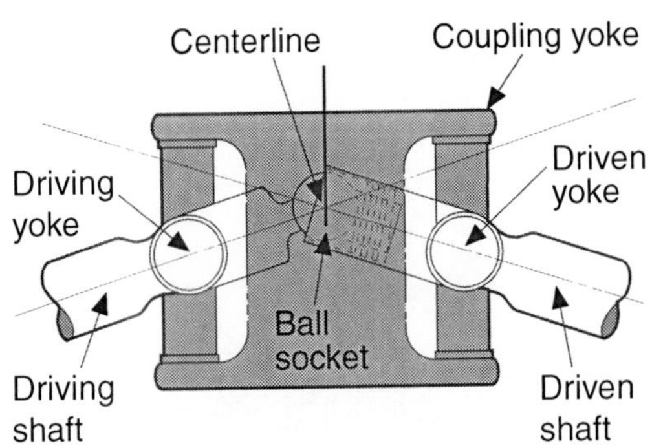


**Figure 17.** The ball and socket in a double Cardan universal joint splits the angle of the two shafts.

When these two universal joints operate at the same angle, the normal fluctuations from the acceleration and deceleration of one joint are cancelled out by the equal and opposite fluctuations at the other joint.

## DRIVE AXLE DESIGN

A typical FWD vehicle has two drive axles. Each drive axle has three main components. Inner and outer constant velocity (CV) joints are attached to the axle shaft. The axle shafts are made from solid steel or steel tubes. Each CV joint connects the axle shaft to the CV joint housing. The inner CV joint must allow axle rotation while allowing the axle to move vertically and horizontally as the front wheel moves up and down and the axle length changes. The outer CV joint must allow axle rotation and also allow vertical axle movement as the front wheel strikes road irregularities. The outer CV joint must also allow the front wheel to turn while cornering **(Figure 18)**. **Constant velocity (CV) joints** transfer a uniform torque and a constant speed while operating at a wide variety of angles.

The housing on the outer joint has a stub axle that is splined into the wheel hub. The inner joint housing has a

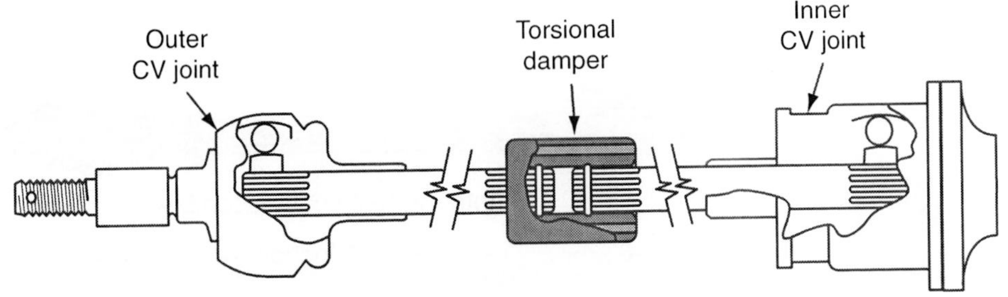


**Figure 18.** Front drive axle with inboard and outboard CV joints.

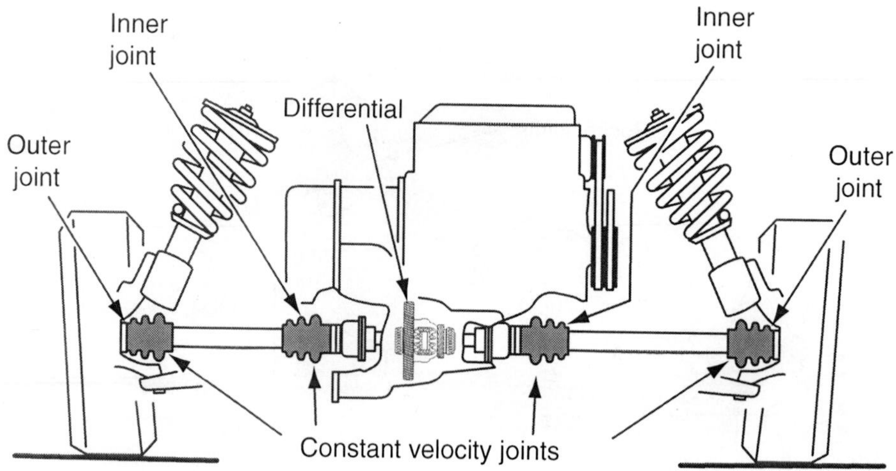


**Figure 19.** The inner axle joint housing is splined into the differential side gear.

stub axle that is splined into the differential side gear in the transaxle **(Figure 19)**. Some inner axle joint housings have a flange that is bolted to a matching flange on the transaxle. Flexible boots are clamped to the drive axle and to each joint housing to keep contaminants out of the joint and keep grease in the joint.

CV joints are used on the front wheels of many 4WD vehicles. RWD vehicles with independent rear suspension also use CV joints.

In a typical FWD vehicle, the engine is transversely mounted and the transaxle is bolted to the rear of the engine. This configuration makes it impossible to center the transaxle in the underhood area. A **transversely mounted** engine is positioned crosswise.

Because the transaxle is not centered in the vehicle, the front drive axles are unequal in length **(Figure 20)**. The CV joints on the longer drive axle usually operate at a smaller angle compared to the angle on the shorter drive axle.

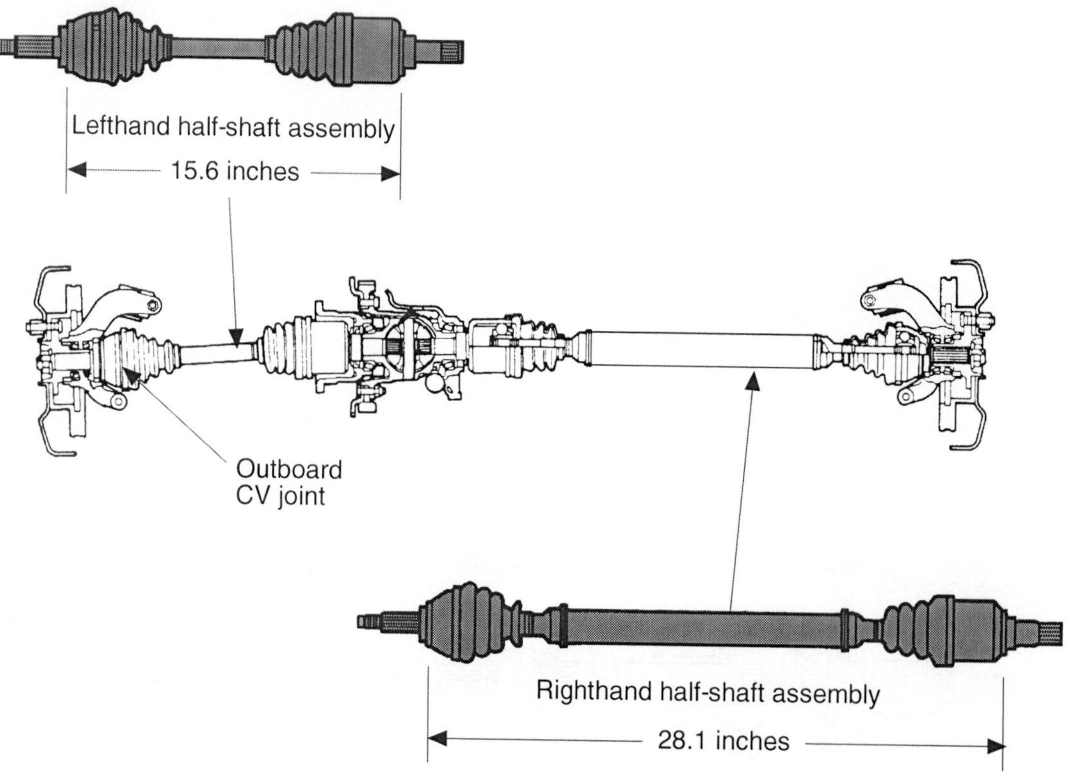


**Figure 20.** Unequal length drive axle shafts.

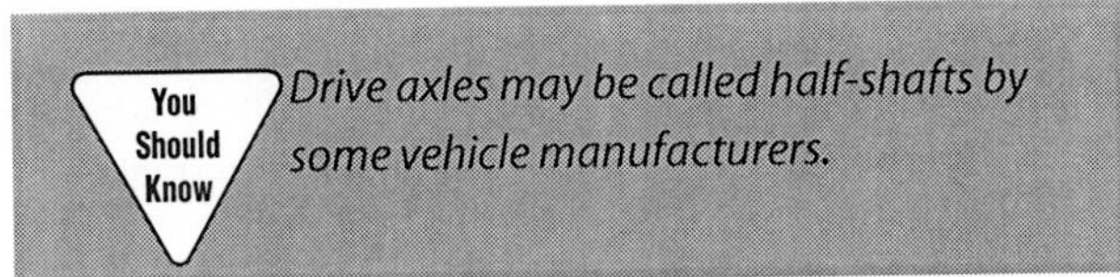

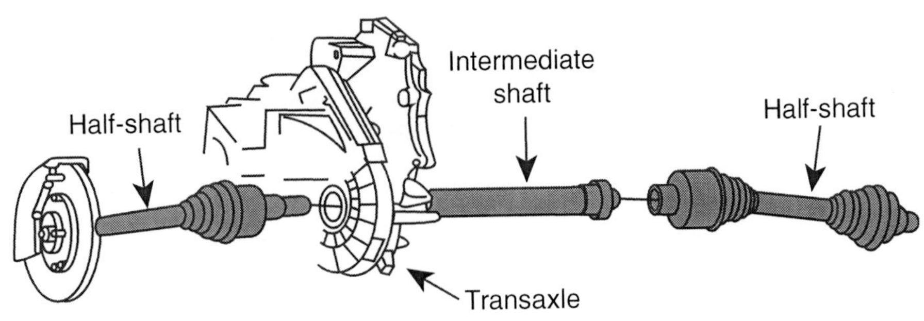


**Figure 21.** Equal length front drive axles with intermediate shaft.

When engine torque is supplied from the differential to the drive axle during hard acceleration, the longer drive axle has less resistance to turning and more engine torque is supplied to this axle and drive wheel. When more engine torque is supplied to one front drive wheel, the steering tends to pull to one side. This condition is called torque steer. **Torque steer** is a condition in which the vehicle steering pulls to one side during hard acceleration.

> You Should Know — *Drive axles may be called half-shafts by some vehicle manufacturers.*

On some FWD vehicles, the long drive axle is replaced with an intermediate drive shaft between the transaxle and the drive axle **(Figure 21)**. This allows the drive axles to be designed with equal lengths to reduce torque steer. The drive axles in FWD vehicles are connected after the differential. Therefore, these drive axles rotate much slower compared to the drive shaft in RWD vehicles. Because of the slower rotation, the balance is not critical on drive axles.

Some FWD vehicles with unequal length drive axles have a solid axle shaft on the shorter axle and a tubular portion in the longer drive axle **(Figure 22)**. This design allows the drive axles to twist equal amounts during hard acceleration and reduce torque steer.

Some FWD vehicles with unequal length drive axles have a torsional damper on the longer drive axle to dampen torsional vibrations **(Figure 23)**.

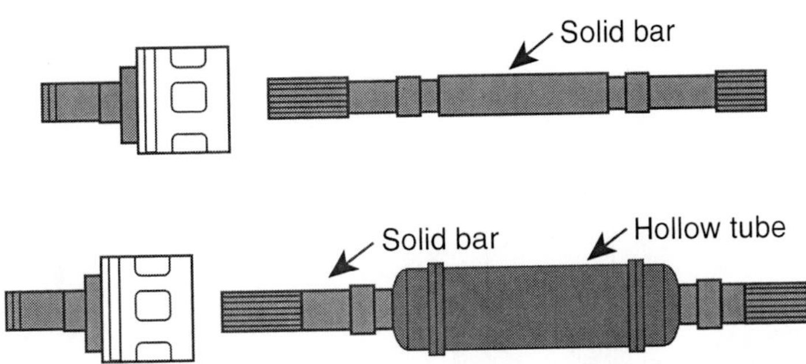


**Figure 22.** Solid and tubular drive axles.

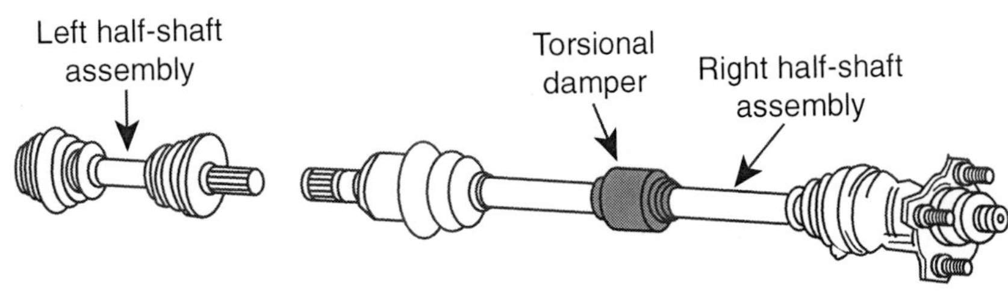


**Figure 23.** A front drive axle with torsional damper.

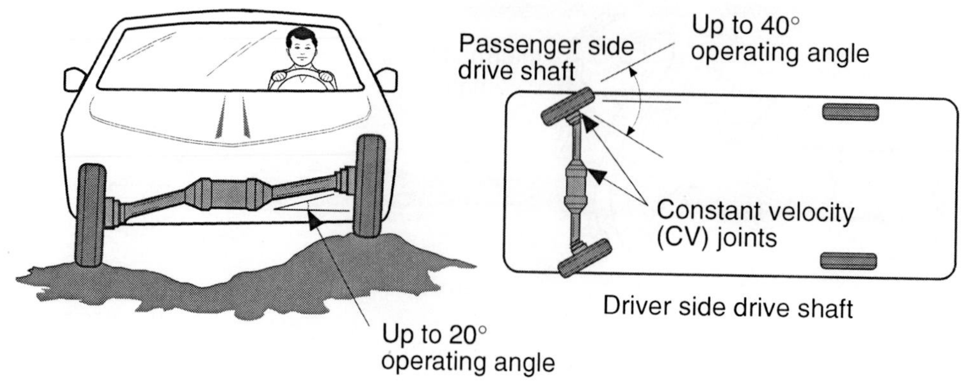


**Figure 24.** FWD drive axle angles.

## DRIVE AXLE PURPOSE

Each drive axle must transfer engine torque from the differential in the transaxle to the drive wheels. These drive axles and CV joints must transfer uniform torque at a constant speed and allow the angles of the drive axles to change considerably as the front wheels are steering and move up and down. The drive axles and CV joints must allow a drive axle angle up to 20 degrees for vertical suspension movement and up to 40 degrees for steering **(Figure 24)**. The drive axles and CV joints must also allow the drive axle length to change as the front wheels move vertically.

## TYPES OF CV JOINTS

One of the most common CV joints is the ball type. This type of joint may be called a Rzeppa joint after the original designer, A. H. Rzeppa. The ball-type CV joint is commonly used for the outboard drive axle joint. In this type of CV joint, the inner race is splined to the drive axle and a circlip retains this race on the axle **(Figure 25)**. A series of equally spaced machined grooves is located on the outer surface of the inner race. Steel balls are mounted in the inner race grooves. These balls are held in place by a ball cage. The number of balls in a CV joint may vary from three to six depending on the torque applied to

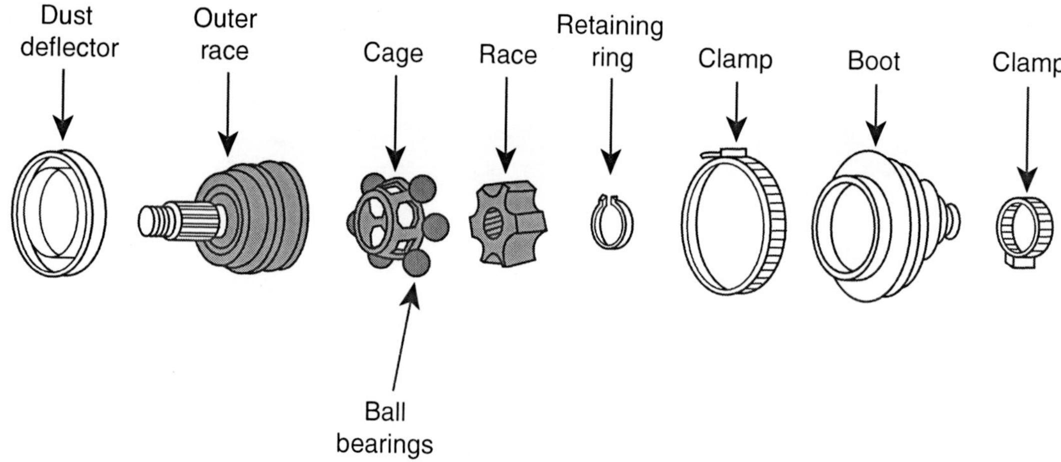


**Figure 25.** A ball-type CV joint.

the CV joint. The outer surface of the balls fits into grooves in the outer race. When engine torque is supplied to the drive shaft, the inner race and balls are forced to rotate. The balls transfer the torque to the outer race that is splined to the front wheel hub. The grooves in the inner and outer races allow the angle between the drive axle and the outer race to change as the front wheels move up and down or turn in either direction. When viewed from the side, the balls in the CV joint always bisect the angle formed by the shafts on either side of the joint regardless of the operating angle **(Figure 26)**. This action reduces the effective angle of the joint by 50 percent and eliminates vibration problems. The ball-type joint is a fixed CV joint. A **fixed CV joint** does not allow any inward and outward drive axle movement to compensate for changes in axle length.

The inner drive axle joint is commonly a tripod type. This type of CV joint has a central hub or spider with three trunnions and spherical rollers. Needle bearings are mounted between these rollers and the trunnions. A snap ring retains the spider on the drive axle shaft **(Figure 27)**. The outer surfaces of the spherical rollers are mounted in grooves in the outer race or joint housing. These grooves are long enough to allow inward and outward drive axle movement to compensate for changes in axle length as the front wheel moves upward and downward. This type of tripod joint may be called a plunging CV joint. A **plunging CV joint** allows inward and outward drive axle movement to compensate for changes in axle length as the front wheel moves upward or downward.

The outer housing or race on the inner CV joint is splined to the differential side gear. Engine torque is supplied from the differential side gear to the outer CV joint housing. From this housing, torque is transferred through the spherical rollers to the needle bearings and trunnions. Torque is then transferred from the trunnions to the drive axle. On vehicles with ABS, an exiter ring or tone wheel is pressed onto one of the CV joint outer housings **(Figure 28)**. These rings rotate past a sensor that sends a voltage signal to the ABS computer in relation to wheel speed.

Each drive axle joint is covered by a bellows-type neoprene boot. The boot is clamped to the drive axle shaft and to the joint housing. The boots are extremely important to keep moisture and dirt out of the CV joints. If a cracked boot allows moisture and dirt into a CV joint, the joint life is very short. If the drive axle boots are located close to the exhaust system, the boots are usually made from silicone or thermoplastic materials to withstand the additional heat. The appearance of the boot indicates the boot manufac-

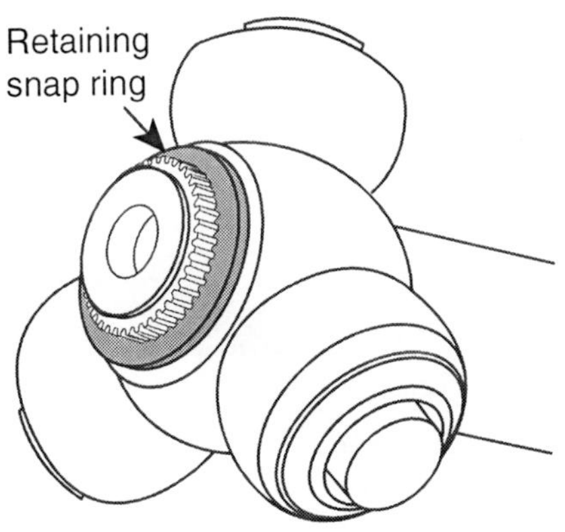


**Figure 27.** A tripod-type CV joint.

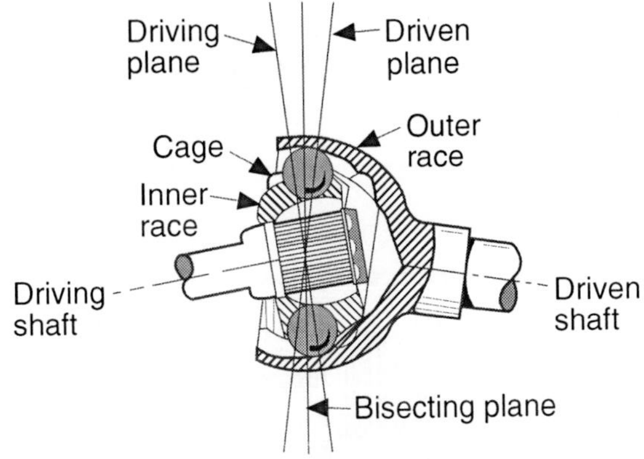


**Figure 26.** In a ball-type CV joint, the balls bisect the angle of the joint.

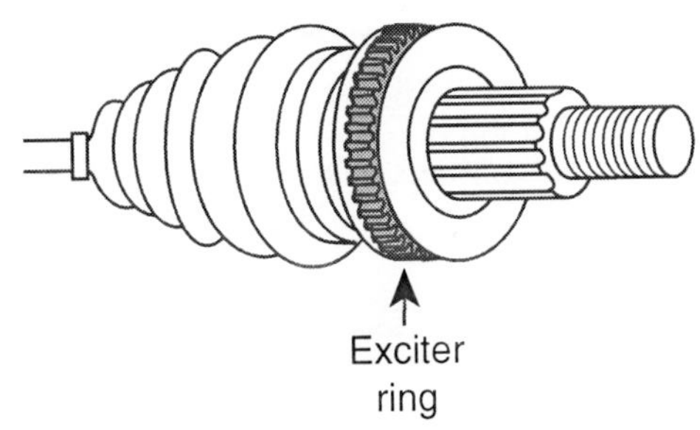


**Figure 28.** ABS exciter ring on CV joint housing.

turer **(Figure 29)**. Always use the type of boot specified by the vehicle manufacturer. Various types of boot clamps are used to retain the boots **(Figure 30)**. Always use the clamps specified by the vehicle manufacturer and follow the manufacturer's recommended procedure to install and tighten these clamps.

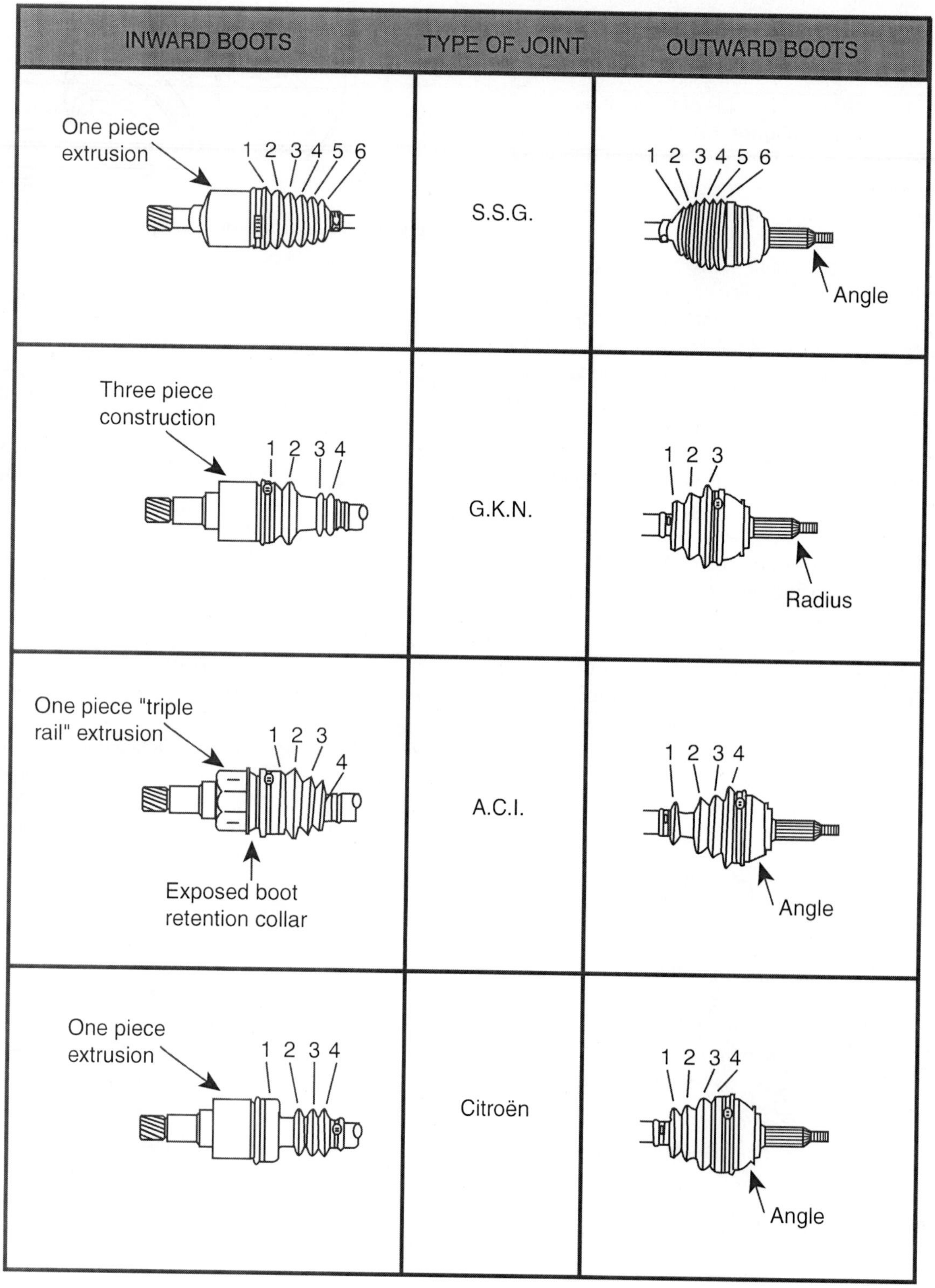

| INWARD BOOTS | TYPE OF JOINT | OUTWARD BOOTS |
| --- | --- | --- |
| One piece extrusion; 1 2 3 4 5 6 | S.S.G. | 1 2 3 4 5 6; Angle |
| Three piece construction; 1 2 3 4 | G.K.N. | 1 2 3; Radius |
| One piece "triple rail" extrusion; 1 2 3 4; Exposed boot retention collar | A.C.I. | 1 2 3 4; Angle |
| One piece extrusion; 1 2 3 4 | Citroën | 1 2 3 4; Angle |

**Figure 29.** Various types of CV joint boots.

| DESCRIPTION | APPEARANCE | TYPE |
|---|---|---|
| Large ladder clamp | | A.C.I. and G.K.N. |
| Small ladder clamp | | A.C.I. and G.K.N. |
| Small strap and buckle clamp | | Citroën |
| Large strap and buckle clamp | | Citroën |
| Large spring clamp | | A.C.I. and G.K.N. |
| Small rubber clamp | | A.C.I. and G.K.N. |

**Figure 30.** Different types of CV joint boot clamps.

## DRIVE SHAFT AND UNIVERSAL JOINT MAINTENANCE

When performing chassis lubrication, always check each universal joint to see if it has a grease fitting in the joint cross. If a grease fitting is present, use the grease gun to inject a small amount of grease into the joint. Applying excessive amounts of grease to a universal joint may damage the seal in each bearing cup. Some universal joints have a plug that may be removed to insert a grease fitting. These plugs should be removed and a grease fitting installed to allow greasing the universal joint at the vehicle manufacturer's recommended service interval.

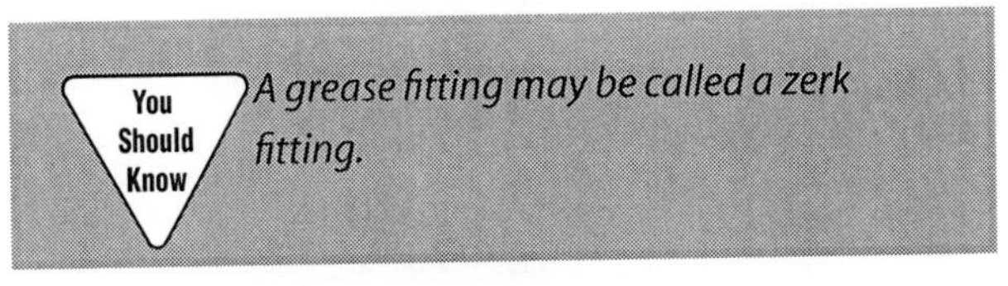


*A grease fitting may be called a zerk fitting.*

When performing an undervehicle inspection, check the universal joints for looseness. Grasp the drive shaft and determine if there is any vertical movement in the universal joint. Try and turn the drive shaft with your hand and watch for movement between the drive shaft and the yoke. If there is any vertical or rotary movement in a universal joint, it must be replaced.

Inspect the drive shaft for damage, such as dents or missing balance weights. Inspect the area around the rear of the transmission extension housing for oil leaks. An oil leak in this area usually indicates a leaking extension housing seal. If this seal is leaking, the drive shaft must be removed and the seal replaced. When the vehicle has a two-piece drive shaft with a center support bearing, inspect the center bearing for looseness.

## DRIVE SHAFT AND UNIVERSAL JOINT DIAGNOSIS

One of the most common complaints on drive shafts and universal joints is a squeaking noise whose frequency increases with vehicle speed. This noise is usually caused by a dry, worn universal joint and it is most noticeable at low vehicle speeds. Another common problem caused by a worn universal joint is a clanging noise when the transmission is shifted from park to drive or reverse. The clanging noise may also occur during initial acceleration. A severely worn universal joint may cause a clunking noise at low vehicle speeds. A universal joint that causes any of these noise complaints must be replaced. A defective center support bearing causes a steady growling noise that is more pronounced at low speed.

A vibration that increases and decreases with vehicle speed may be caused by an unbalanced drive shaft. This condition may be caused by a dented drive shaft, missing balance weights, or undercoating and other material adhered to the drive shaft. Improper drive shaft angles may cause a drive shaft vibration that is most noticeable between 30 to 35 miles per hour (48 to 56 kilometers per hour). Improper drive shaft angles may be caused by worn engine or transmission mounts. Sagged springs lower the vehicle chassis and cause improper drive shaft angles **(Figure 31)**.

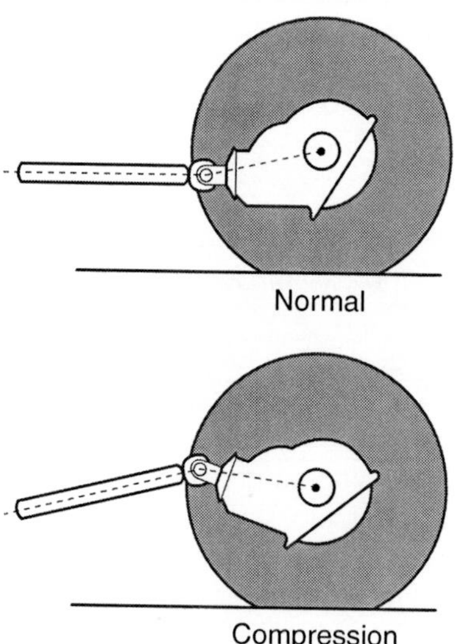


**Figure 31.** Sagged rear springs lower the chassis and cause improper drive shaft and univeral joint angles.

## DRIVE SHAFT AND UNIVERSAL JOINT SERVICE

Proper drive shaft and universal joint service is very important to prevent vibration and provide driver and passenger comfort. Vibration in the vehicle interior can greatly increase driver and passenger fatigue. Proper drive shaft and universal joint service also eliminates the annoying noises described previously.

### Drive Shaft Removal and Installation

Before removing a drive shaft, always mark the drive shaft and yoke with a chalk or paint stick so these components can be reassembled in the same position in relation to each other **(Figure 32)**. If the vehicle has a two-piece drive shaft with a center support bearing, mark each yoke in relation to the drive shaft. If this action is not taken, drive shaft vibration may occur. To remove the drive shaft, raise the vehicle on a lift. Place an oil drain pan under the rear of the transmission extension housing to catch any oil that leaks from this area when the drive shaft is removed.

Use the proper wrench to remove the bolts that retain the rear universal joint to the differential flange. It may be necessary to bend the metal locking tabs away from these retaining bolts. Slide the drive shaft forward and lower the rear end of the shaft so it is below the differential flange. Pull the drive shaft rearward so the front slip yoke comes off the transmission output shaft. Place the drive shaft on the work bench and install a plug on the transmission output shaft to keep oil from continuing to leak from this source. If the vehicle has a two-piece drive shaft, remove the center support bearing retaining bolts first and have an assistant hold the center and the front half of the drive shaft as the complete drive shaft is removed.

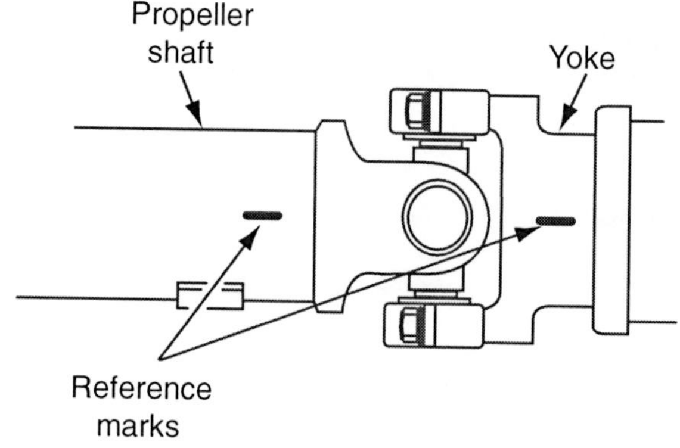


**Figure 32.** Marking the drive shaft and yokes before disassembly.

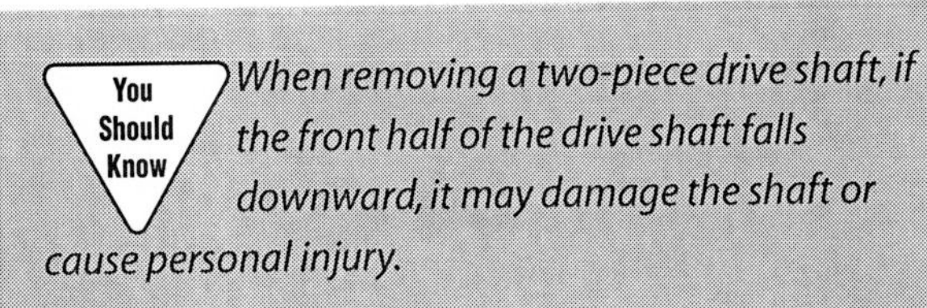


*When removing a two-piece drive shaft, if the front half of the drive shaft falls downward, it may damage the shaft or cause personal injury.*

When installing a drive shaft, slide the front slip yoke over the transmission output shaft and align the rear universal joint with the differential flange. Rotate the drive shaft so the index marks on the drive shaft and rear yoke are aligned. Position the bearing cups properly into the differential flange. Install the rear universal joint retaining bolts and be sure to install the bolt locks if applicable. Tighten these retaining bolts to the specified torque and bend the lock tabs up against the bolt heads. Use a grease gun to lubricate all the universal joints if they have grease fittings. If the vehicle has a two-piece drive shaft with a center support bearing, have an assistant guide the front slip yoke into place while you hold up the rear drive shaft and center support bearing. Install the center support bearing retaining bolts and the rear universal joint fasteners. Tighten all fasteners to the specified torque.

## Universal Joint Disassembly and Reassembly

Defective universal joints must be replaced. Do not attempt to change parts from one universal joint to another. Regardless of the type of universal joint, the disassembly and reassembly procedures are similar. Follow this procedure to disassemble a universal joint:

1. Clamp the front slip yoke in a soft-jawed vise across the bearing cup area and support the other end of the drive shaft.

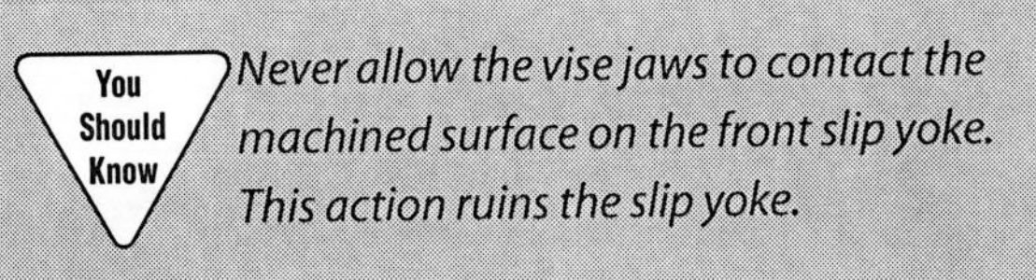


*Never allow the vise jaws to contact the machined surface on the front slip yoke. This action ruins the slip yoke.*

2. Remove the lock rings that retain the bearing cups in the drive shaft. Mark the yoke in relation to the drive shaft.
3. Select a socket that has an internal opening large enough for the bearing cup to fit into and select another socket that is small enough to fit into the bearing cup opening in the drive shaft.
4. Position the large socket over one bearing cup opening on the drive shaft and locate the smaller socket against the opposite drive shaft bearing cup.
5. Tighten the vise so the smaller socket is forced against the bearing cup and pushes the opposite bearing cup out of the drive shaft and into the large socket.
6. Turn the drive shaft over and use a brass drift and a hammer to drive the spider and the remaining bearing cup downward out of the drive shaft opening.
7. Remove the slip yoke and spider and support the yoke in a vise on the sides of the yoke. Use a brass drift and a hammer to drive the bearing cups and spider out of the yoke.

These steps may be followed to assemble a universal joint:

1. Clean the bearing cup areas, including the snap ring grooves in the front slip yoke.
2. Remove the bearing cups from the new universal joint.
3. Install the new spider in the front slip yoke and move it to one side as far as possible.
4. Install a bearing cup over the end of the spider and carefully place the assembly in the vise so the jaws are tightened against the bearing cup areas. Tighten the vise jaws so the bearing cup is partially pressed into the yoke opening.
5. Remove the yoke from the vise and move the spider toward the opposite side of the yoke. Install a second bearing cup over the end of the spider.
6. Place the slip yoke in the vise so the jaws contact the bearing cups and tighten the vise jaws to push the bearing cups into the yoke.
7. Be sure the bearing cups are fully installed into the slip yoke openings and install the bearing cup retaining rings.
8. Install the slip yoke and spider into the drive shaft with the index marks aligned and install the other two bearing cups and snap rings. Be sure all the snap rings are fully seated.

When disassembling a double Cardan universal joint, place index marks on all components so they can be reassembled in the same position. Inspect the ball and socket for wear, cracks, and scoring. On some drive shafts,

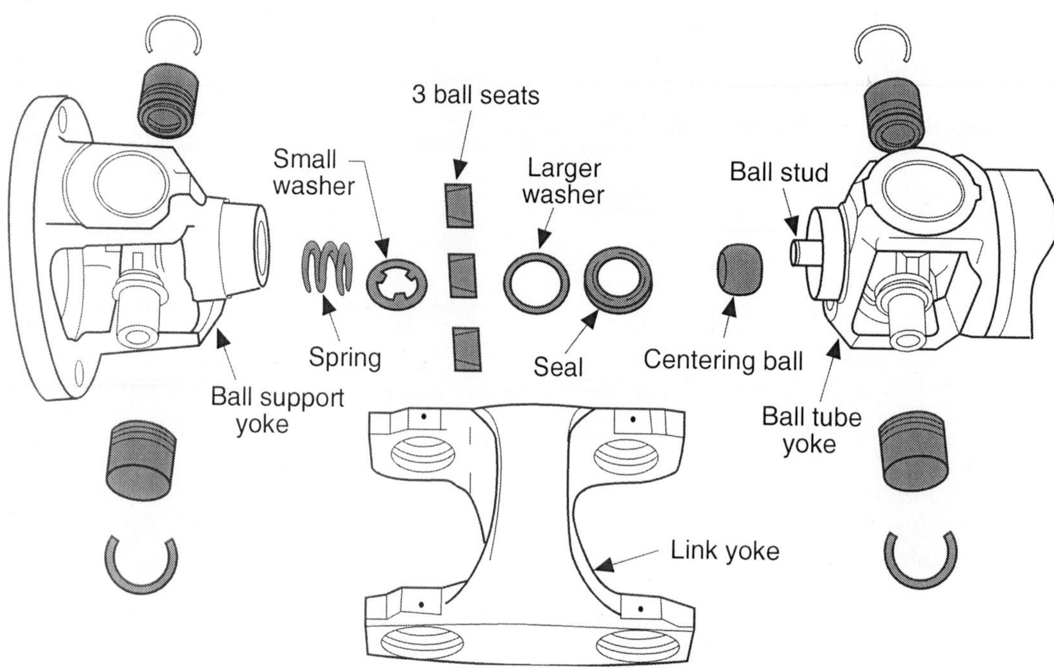


Figure 33. Double Cardan universal joint with replaceable ball and ball nuts.

the ball and socket can be replaced **(Figure 33)**. Always be sure the ball and socket are properly lubricated with the grease supplied with the replacement ball and socket.

## Measuring Drive Shaft Runout

A bent drive shaft causes vibration and noise. To measure drive shaft runout, be sure the drive shaft is clean in the center area. Mount a dial indicator to the chassis so the indicator stem is contacting the center of the drive shaft **(Figure 34)**. Zero the dial indicator and slowly rotate the drive shaft by hand. The drive shaft runout indicated on the dial indicator must not exceed manufacturer's specifications. Perform drive shaft runout readings near the front and rear of the drive shaft. Replace the drive shaft if the runout is excessive.

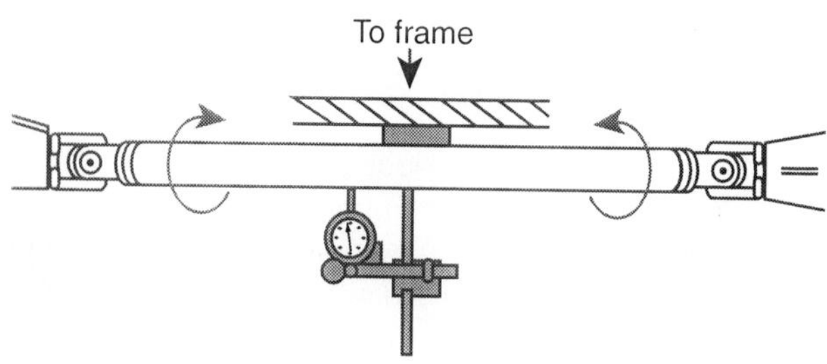


Figure 34. Measuring drive shaft runout.

## Testing Drive Shaft Balance

Drive shaft imbalance causes undesirable vibration. A strobe light and transducer may be used to check drive shaft balance. The transducer is positioned under the differential housing **(Figure 35)**.

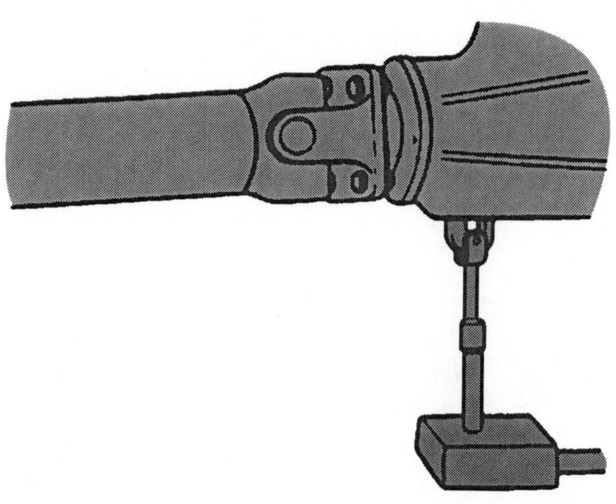

Figure 35. A transducer is used to check drive shaft vibrations.

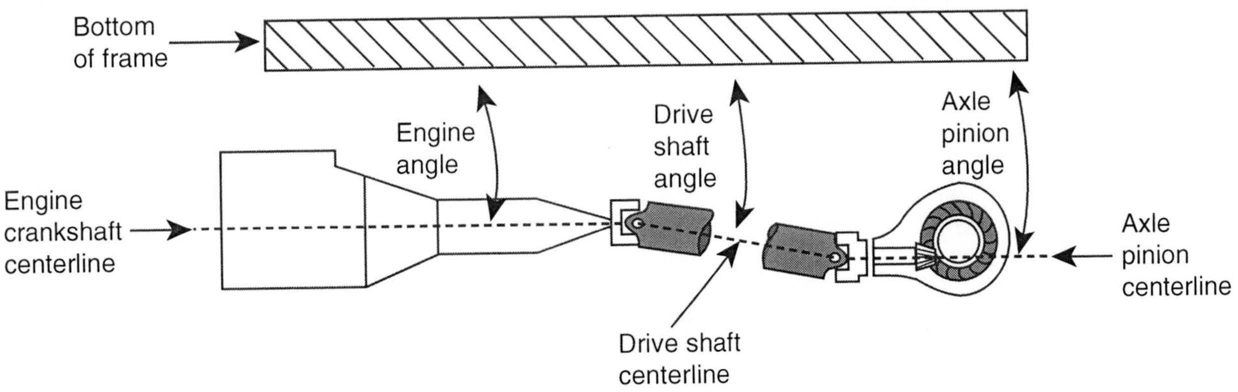


**Figure 36.** The installation angle of a drive shaft.

## Measuring Drive Shaft Angles

The angle of the drive shaft is the difference between the angle of the drive shaft and the angles of the transmission and differential **(Figure 36)**. Improper drive shaft angles may be caused by worn or sagged engine or transmission mounts or improper rear axle position. Improper drive shaft angles cause vibration. An inclinometer is used to measure drive shaft angles **(Figure 37)**.

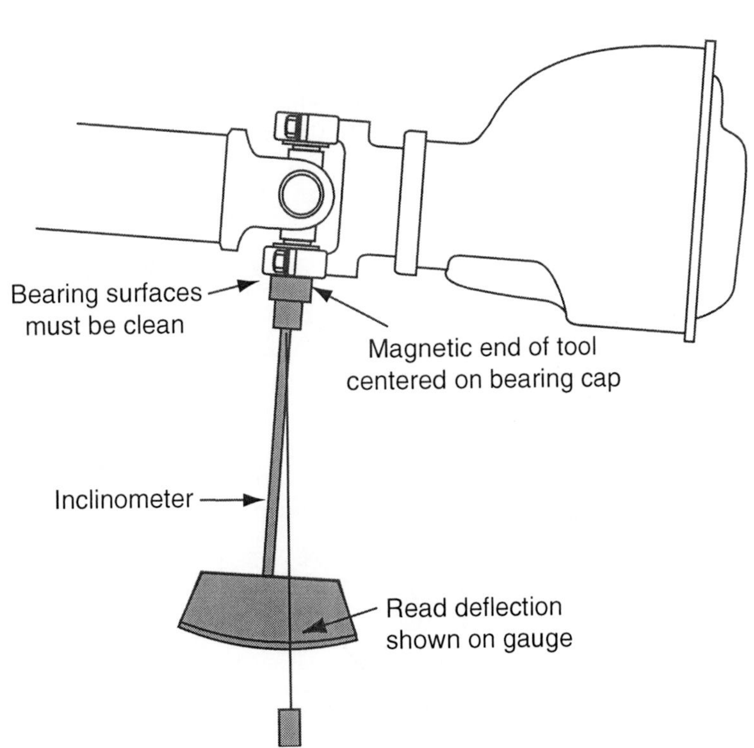


**Figure 37.** Measuring the operating angle of the rear universal joint.

## DRIVE AXLE AND CV JOINT MAINTENANCE

The CV joint boots should be inspected regularly. These boots keep lubricant in the axle joints and also prevent moisture and dirt from entering the joint. Raise the vehicle on a lift and check for cracked or torn CV joint boots. Check these boots for loose or damaged clamps. The boots must be airtight. Inspect the CV joints and the areas around the joints for evidence of grease that has been thrown out of the boot. If grease has been thrown out of a CV joint, the boot is cracked or torn or the clamps are loose. When the vehicle has an intermediate axle shaft, check the support bearing on this shaft for looseness **(Figure 38)**. If any of these boot or clamp conditions are present, the boots and clamps must be replaced. Drive axle boots may fail for any of the following reasons:

1. Improper service on a suspension component.
2. Failure of the boot clamps or improper clamp installation.
3. Deterioration caused by fluid leaks or environmental damage.
4. Ice buildup on the chassis near the boot.
5. Improper vehicle connection to a tow truck.
6. Interference from damaged sheet metal.
7. Cuts or tears from objects on the road surface.
8. Improper drive axle removal procedure.

Satisfactory drive axle boots do not ensure the axle joints are in good condition. It is possible to have satisfactory boots and worn axle joints.

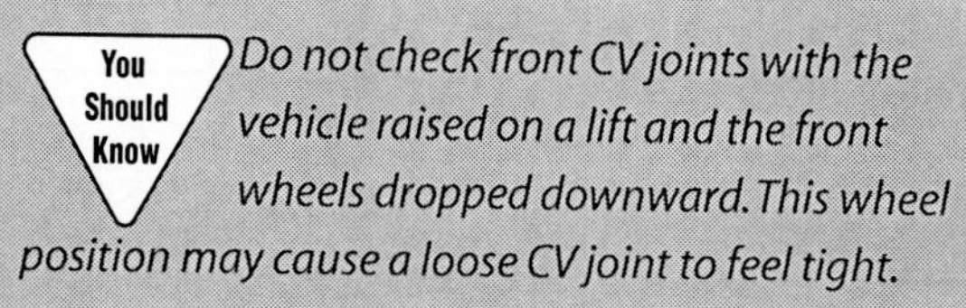


*Do not check front CV joints with the vehicle raised on a lift and the front wheels dropped downward. This wheel position may cause a loose CV joint to feel tight.*

Inspect the front drive axles for damage or a bent condition. Rotate the wheels slowly and watch for a bent drive axle. Some runout is acceptable without causing vibration problems. Always refer to the vehicle manufacturer's specifications. Grasp the inner joint near the transaxle and check for movement. Excessive movement indicates worn transaxle output shaft bushings. This condition may be evidenced by oil leaks at the output shaft seals in the transaxle case. If there is excessive movement of the inner CV joint housing in the transaxle, the output bushings and seals

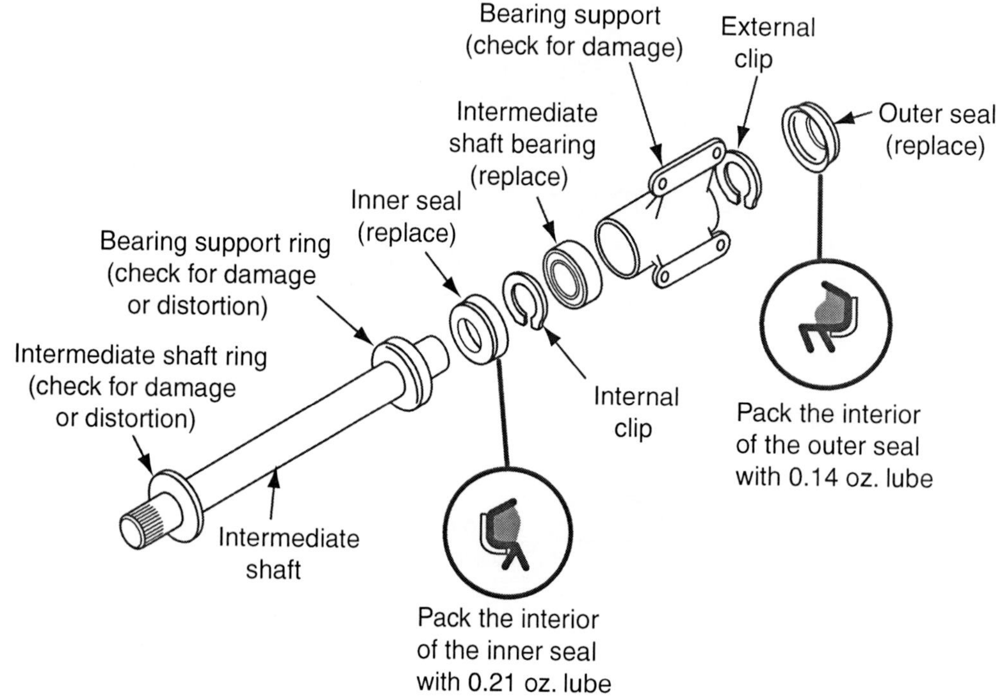


**Figure 38.** An intermediate shaft with support bearing.

must be replaced. If the inner CV joint housing is bolted to a flange on the transaxle, be sure the retaining bolts are tight in these components. Check for looseness of the outer CV joint in the front wheel hub.

During a CV joint inspection, check all other components that could affect the drive axle and CV joint operation. Inspect the engine and transaxle mounts for wear, looseness, and deterioration. Worn mounts may cause improper engine and transaxle position that may cause abnormal CV joint angles. This condition may accelerate CV joint wear, noise, or vibration. Inspect the front suspension control arms, bushings, ball joints, sway bar and bushings, and struts for wear and looseness. Inspect the steering and brake components for looseness. Inspect the tires for proper inflation pressure and excessive or unusual wear patterns.

**You Should Know** *A CV joint cannot be checked accurately for looseness by grasping the axle and/or outer joint housing and checking for movement. CV joints may have some looseness because of their design. The most accurate way to check a CV joint is to road test the vehicle and listen for unacceptable noises and observe the CV joint operation.*

## DRIVE AXLE AND CV JOINT DIAGNOSIS

The first step in drive axle and CV joint diagnosis is to talk to the customer and be sure the customer's complaint is identified. The next step is to road test the vehicle and verify the customer's complaint. Drive the vehicle in an area that is not congested with traffic. The road test area must also allow the driver to perform turns at lower speeds. Drive the vehicle slowly through several sharp turns in both directions. A rhythmic clicking, popping, or clunking noise while turning a corner is usually caused by a worn outer CV joint. Listen to this noise when the vehicle is turned to the right and left. During the turn when the clicking noise is most noticeable, the worn CV joint is on the inside of the turn **(Figure 39)**.

If there are no unusual noises while cornering, accelerate and decelerate the vehicle. A clunking noise during acceleration or deceleration may indicate a worn inner CV joint. A worn inner CV joint, loose CV joint flange bolts, or a worn drive axle damper may also cause a shudder or shaking during engine torque changes. Accelerate the vehicle to 50 to 55 miles per hour (80 to 88 kilometers per hour). If a vibration is evident at 45 miles per hour (72 kilometers per hour) and increases with vehicle speed, it is likely caused by worn output shaft bushings in the transaxle. When the vibration pulses, the problem is likely a worn inner tripod CV joint or an out-of-balance wheel. The pulsating action is caused by worn roller tracks in the tripod joint that cause

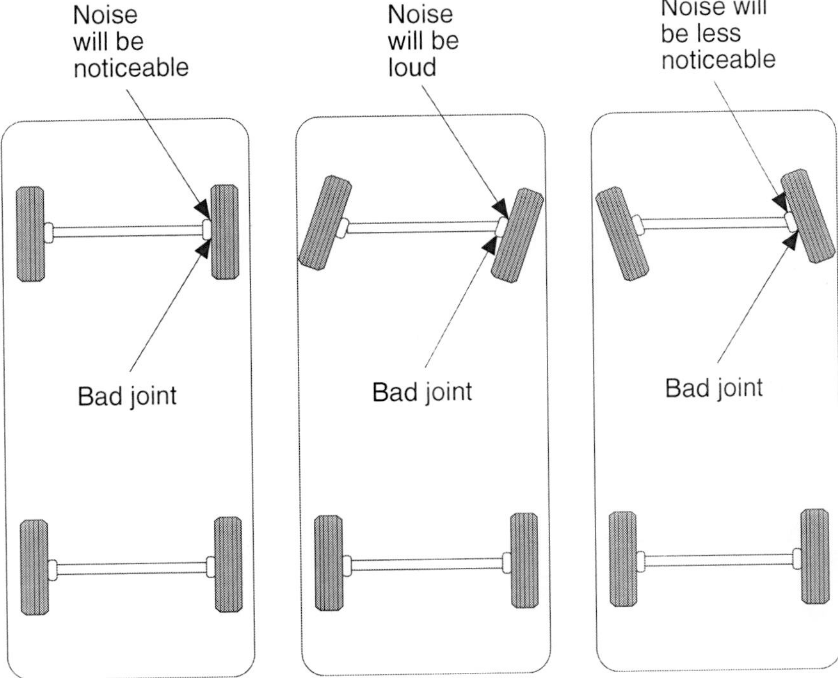


**Figure 39.** A defective outer CV joint on the inside of a turn makes a noticeable noise.

pulsating, erratic drive axle rotation. Decelerate the vehicle and coast through a sharp turn. During this driving condition, a clicking noise indicates a worn outer CV joint or a defective wheel bearing. If a shudder during the turn is accompanied by a clunking noise, a worn inner CV joint is indicated. If a shudder or vibration is intermittent while driving straight ahead, a defective drive axle damper or bent drive axle may be the cause of the problem.

## DRIVE AXLE AND CV JOINT SERVICE

Drive axle and CV joint service is very important to provide normal CV joint life and provide quiet, vibration-free drive axle operation. Drive axle and CV joint service includes drive axle removal and replacement, CV joint inspection on the bench, and CV joint replacement.

### Removing and Installing a Drive Axle

Follow this procedure to remove and reinstall a drive axle:

1. With vehicle wheels on the shop floor, remove the axle hub nut and loosen the wheel nuts.
2. Raise the vehicle on a lift and remove the tire and wheel assembly.
3. Remove the brake caliper and use a length of wire to suspend the caliper from a chassis member. Do not allow the brake caliper to hang on the end of the brake hose.
4. Remove the brake rotor.
5. Disconnect the ends of the stabilizer bar from the lower control arms.
6. Remove the pinch bolt that retains the lower ball joint to the steering knuckle and remove the ball joint stud from the knuckle **(Figure 40)**.
7. Use a slide hammer-type puller to remove the inner drive axle joint from the transaxle **(Figure 41)**.

**You Should Know** *In some early model transaxles, the inner axle joints were held in the differential with circlips. The differential cover had to be removed and these circlips pulled out to allow drive axle removal.*

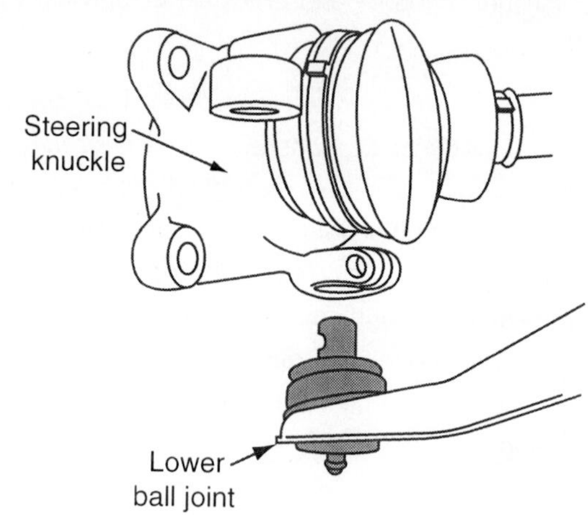


**Figure 40.** Removing the ball joint stud from the steering knuckle.

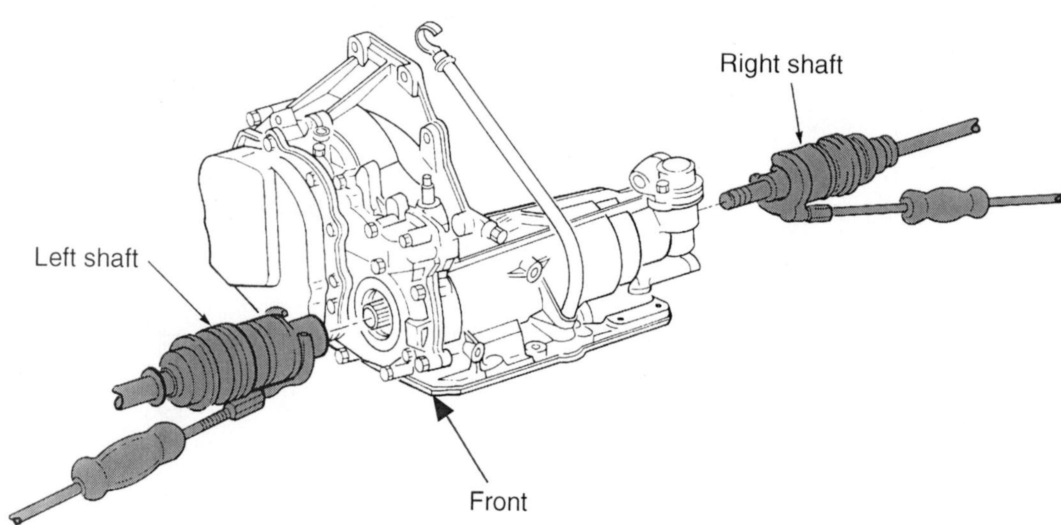


**Figure 41.** Using a slider hammer-type puller to remove the inner CV joints from the transaxle.

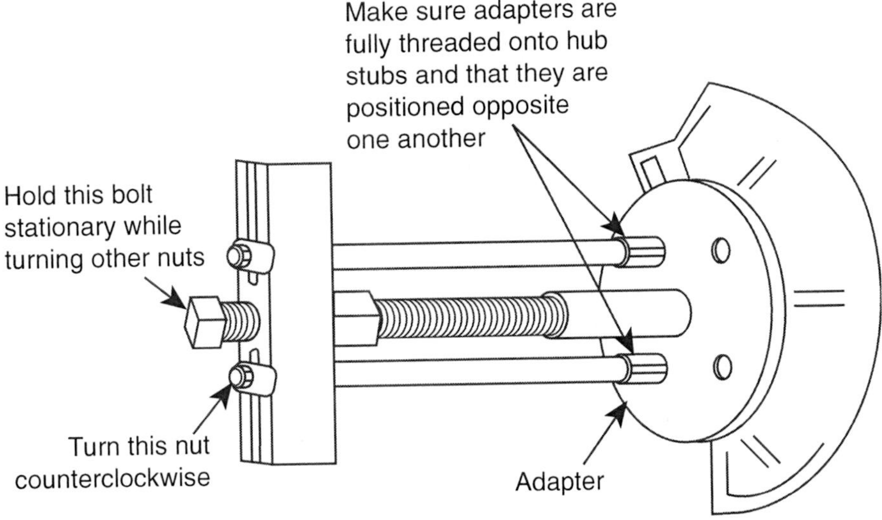
Make sure adapters are fully threaded onto hub stubs and that they are positioned opposite one another
Hold this bolt stationary while turning other nuts
Turn this nut counterclockwise
Adapter

**Figure 42.** Removing the outer end of the drive axle from the wheel hub.

8. Remove the outer end of the drive axle from the front wheel hub. On some drive axles, a special puller is required for this operation **(Figure 42)**.
9. Remove the drive axle from the chassis and place the drive axle on the workbench.
10. Install the splines on the inner drive axle joint into the differential side gear. Push or lightly tap the joint with a soft hammer until it is fully seated in the side gear. On some transaxles, the inner CV joint is bolted to a flange on the transaxle. If this type of flange is present, tighten the joint to flange retaining bolts to the specified torque.
11. Pull the spindle outward and slide the outer CV joint splines into the front wheel hub. Turn the drive axle slightly to align the splines on these two components. Pull the CV joint into the wheel hub as far as possible.

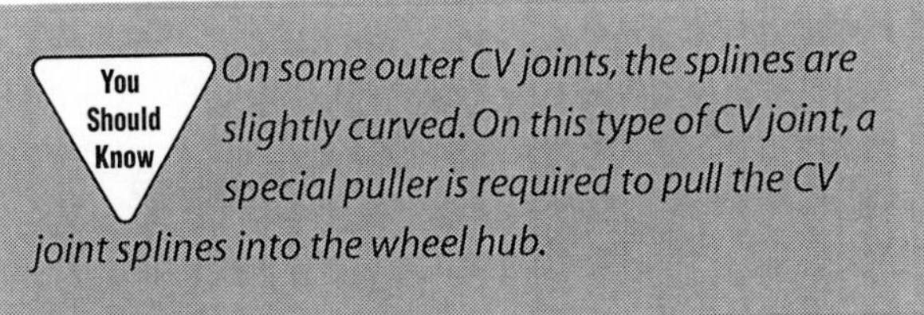
You Should Know

*On some outer CV joints, the splines are slightly curved. On this type of CV joint, a special puller is required to pull the CV joint splines into the wheel hub.*

12. Install the ball joint stud into the steering knuckle. Tighten the pinch bolt and connect the ends of the stabilizer bar to the lower control arms.
13. Install the hub nut washer and a new self-locking nut. Tighten this nut as far as possible by hand.
14. Install the brake rotor and caliper. Be sure the caliper bolts are tightened to the specified torque.
15. Install the wheel and tire assembly and the wheel nuts. Tighten the wheel nuts by hand.
16. Lower the vehicle onto the shop floor and tighten the hub nut to the specified torque. Some front drive axle nuts must be staked after they are tightened.
17. Tighten the wheel nuts to the specified torque.

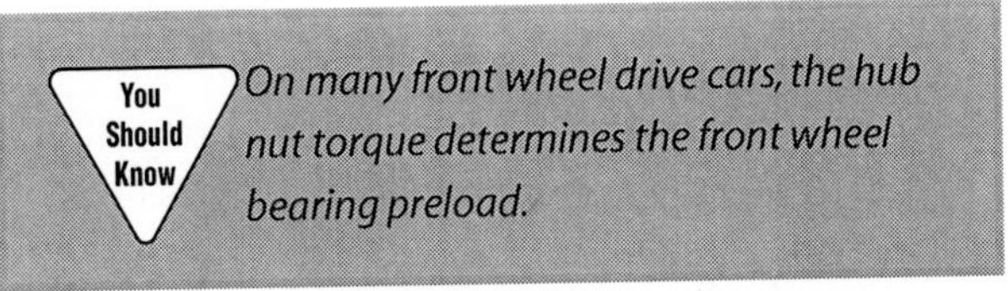
You Should Know

*On many front wheel drive cars, the hub nut torque determines the front wheel bearing preload.*

## CV Joint Inspection on the Bench

After the drive axle is removed, clamp the drive axle in a soft-jawed vice. Mark the inner end of the boot where it is positioned on the axle unless this end of the boot is mounted in a groove in the axle. Remove the boot clamps and remove the outer CV joint boot. Because the boot is always replaced, the boot may be cut from the axle shaft. Wipe the excess grease off the inner joint race and cage. Inspect the grease for grit which indicates dirt contamination. When the grease is contaminated with grit, the boot has been leaking at some time. If the grease has a milky appearance, it is contaminated with moisture. When the grease is contaminated with grit or moisture, the joint must be replaced. Inspect the balls, cage, inner race, and housing grooves for cracks, chips, pits, rust, and worn areas. Inspect the CV joint splines for wear, damage, and chips. If any of these conditions are present on these components, the joint must be replaced.

Remove the inner joint boot and check the grease in this joint for contamination. Inspect this tripod joint for

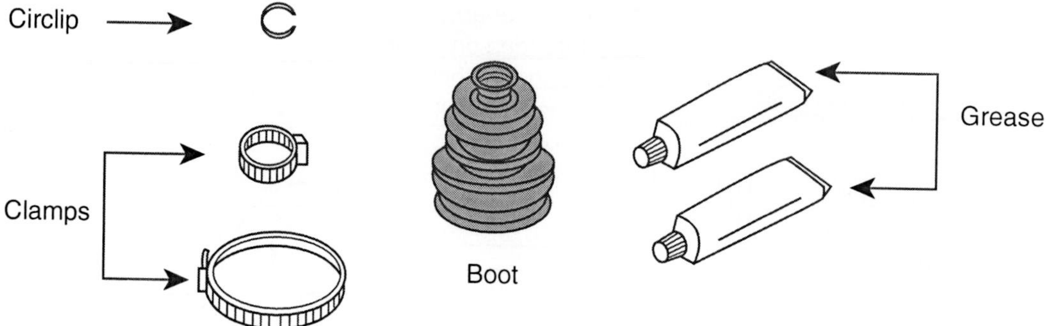


**Figure 43.** A boot kit for a CV joint.

cracks, wear, discoloration, or scoring on the spherical rollers and roller grooves in the outer housing. Check the CV joint splines for chips, damage, and wear. If any of these conditions are present, the joint must be replaced.

## Removing and Replacing an Outer CV Joint and Boot

If a CV joint is in satisfactory condition, the boot may be replaced. Boot kits are available that include the boot, clamps, circlip, and lubricant **(Figure 43)**. Whenever a CV joint is serviced, always install a new boot and clamp. Do not reuse a boot. The old boots may be cut from the drive axle. There are two methods of retaining outer CV joints to the axle shaft. Some of these CV joints have an external ring that fits into a bearing race. This clip must be expanded to remove the CV joint from the axle. Other outer CV joints have an internal retaining clip that fits into an axle shaft groove and expands into the inner bearing race. This type of CV joint can be removed from the axle shaft by tapping it with a soft hammer. After the CV joint is removed from the axle, remove the old grease from the joint. Install the new clamp and boot and all the grease supplied with the boot kit **(Figure 44)**. After the CV joint is installed on the axle, be sure the boot clamps are properly tightened.

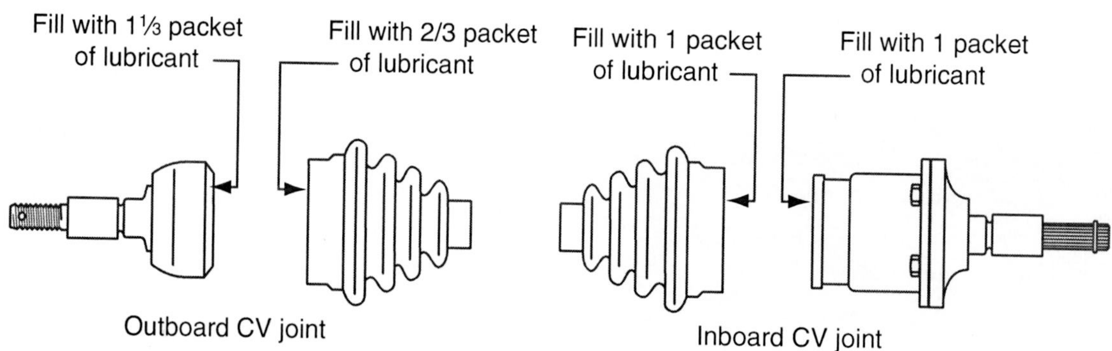


**Figure 44.** Installing the proper amount of lubricant in the CV joint housings and boots.

## Summary


- A drive shaft transmits engine torque from the transmission output shaft to the differential.
- Universal joints connect the front of the drive shaft to the front slip yoke and the rear of the drive shaft to the differential flange.
- As the drive shaft transmits torque through an angle, the drive shaft continually accelerates and decelerates.
- When universal joints in a drive shaft are on the same plane, they are in phase.
- A squeaking noise that increases with vehicle speed may be caused by a dry, worn universal joint.
- A clanging noise when the transmission is shifted from park to drive or reverse may be caused by a worn universal joint.
- An unbalanced drive shaft causes a vibration that increases and decreases with vehicle speed.
- Before removing a drive shaft, always mark the drive shaft in relation to each yoke.
- Before disassembling universal joints, always be sure the yokes are marked in relation to the drive shaft.
- Excessive drive shaft runout, improper drive shaft angles, or drive shaft imbalance may cause vibration.
- The inner CV joint allows vertical and horizontal axle movement.
- The outer CV joint allows vertical wheel movement and rotation forward and rearward as the front wheels are steered.
- Many FWD vehicles have unequal length front drive axles.
- Some FWD vehicles have equal length drive axles and an intermediate shaft on one side.
- Torque steer is the tendency of the steering to pull to one side during hard acceleration.
- Cracked or torn drive axle boots allow moisture and dirt into the CV joints resulting in very short joint life.
- When a clicking noise is heard while cornering at low speed, one of the outer CV joints may be defective.
- If a pulsating vibration or shudder is experienced while accelerating or decelerating at medium vehicle speeds, one of the inner CV joints may be defective.
- If the grease in a CV joint is contaminated, the joint must be replaced.
- When replacing a CV boot, the grease supplied with the boot must be installed in the boot and joint housing.


## Review Questions

1. Technician A says the drive shaft rotates three to four times faster than the rear wheels. Technician B says the drive shaft angle remains constant as the vehicle is driven. Who is correct?
   A. Technician A
   B. Technician B
   C. Both Technician A and Technician B
   D. Neither Technician A nor Technician B
2. Technician A says the drive shaft length remains constant as the vehicle is driven. Technician B says sagged rear springs change the drive shaft angle. Who is correct?
   A. Technician A
   B. Technician B
   C. Both Technician A and Technician B
   D. Neither Technician A nor Technician B
3. Technician A says a universal joint transmits torque through an angle. Technician B says the drive shaft speed does not change as this shaft rotates. Who is correct?
   A. Technician A
   B. Technician B
   C. Both Technician A and Technician B
   D. Neither Technician A nor Technician B
4. Technician A says universal joints that are out of phase may cause a drive shaft vibration. Technician B says universal joints are in phase when all the joints in a drive shaft are on the same plane. Who is correct?
   A. Technician A
   B. Technician B
   C. Both Technician A and Technician B
   D. Neither Technician A nor Technician B
5. All of these statements about drive shaft angles and vibration are true *except*:
   A. Drive shaft vibration may be reduced by canceling angles between the front and rear universal joints.
   B. A steeper drive shaft angle causes increased torsional vibrations.
   C. When universal joints are not on the same plane, they are in phase.
   D. When a drive shaft is disassembled, the yoke should be marked in relation to the drive shaft.

6. A rear-wheel drive vehicle with a two-piece drive shaft provides a squeaking noise when accelerating from a stop. The noise frequency increases with vehicle speed. The most likely cause of this noise is:
   A. a dry, worn universal joint.
   B. a scored outer surface on the front slip yoke.
   C. a dry, worn center support bearing.
   D. a drive shaft that is out-of-phase.
7. A rear wheel drive vehicle has a vibration between 50 and 60 mph, and this vibration changes with vehicle speed. Technician A says the drive shaft may have excessive runout. Technician B says the rear springs may be sagged resulting in an improper drive shaft angle. Who is correct?
   A. Technician A
   B. Technician B
   C. Both Technician A and Technician B
   D. Neither Technician A nor Technician B
8. All of these statements about front drive axles are true *except*:
   A. The drive axles rotate at the same speed as drive shafts in RWD vehicles.
   B. Balance is critical on front drive axles.
   C. Many vehicles have unequal length front drive axles.
   D. Some vehicles have an intermediate shaft and equal length front drive axles.
9. On a front-wheel drive vehicle, torque steer is most noticeable:
   A. during slow acceleration after the vehicle is stopped.
   B. while cruising at a steady speed.
   C. during hard acceleration at lower speed.
   D. during deceleration from high speed.
10. Technician A says the inner drive axle joint allows inward and outward axle movement. Technician B says a tripod-type joint is used for the outer drive axle joint. Who is correct?
    A. Technician A
    B. Technician B
    C. Both Technician A and Technician B
    D. Neither Technician A nor Technician B
11. A double Cardan universal joint has a ball and socket between the front ______________ ______________ and the rear ______________.
12. A dry, worn universal joint may cause a ______________ noise whose ______________ increases with vehicle speed.
13. A worn universal joint may cause a ______________ noise when the transmission is shifted from park to drive.
14. An unbalanced drive shaft may cause a vibration that ______________ and ______________ with vehicle speed changes.
15. Explain the causes of improper drive shaft angles.
16. Describe the drive shaft runout measurement procedure.
17. Describe the result of a worn outer CV joint in a front-wheel drive vehicle.
18. Describe the results of worn engine mounts on a front-wheel drive vehicle.

# Chapter 41 Clutch Maintenance, Diagnosis, and Service

## *Introduction*

On vehicles with manual transmissions, the clutch is mounted between the engine and transmission. The purpose of the clutch is to connect and disconnect engine torque from the transmission. The clutch must perform this function smoothly without grabbing, slipping, or chattering. During clutch engagement, the clutch must withstand a very high torque load from the engine and it must do this repeatedly without damaging clutch components. Technicians must understand clutch purpose and normal operation to be able to diagnose clutch problems quickly and accurately.

### CLUTCH DISC DESIGN

The clutch plate or disc contains a splined hub that is mounted on matching splines on the transmission input shaft. A steel plate is attached to the clutch hub and frictional materials are riveted or bonded to both sides of this steel plate. Asbestos was the most common material used in clutch facings until the health hazards related to breathing asbestos dust became known.

> **You Should Know** *Because of repeated engagement and disengagement, clutch facings gradually wear and dust particles from the clutch facings are distributed on clutch components and around the flywheel housing.*

Clutch facings are now made from paper-based or ceramic materials mixed with cotton, brass, and wire particles. Grooves are cut diagonally across the clutch facings to increase clutch cooling and provide a place for facing dust to accumulate **(Figure 1)**. The grooves in the clutch facings also provide smoother clutch engagement. The facings are attached to wave springs that gradually flatten out during clutch engagement. This action causes the contact pressure on the clutch facings to increase gradually. This action provides smoother clutch engagement.

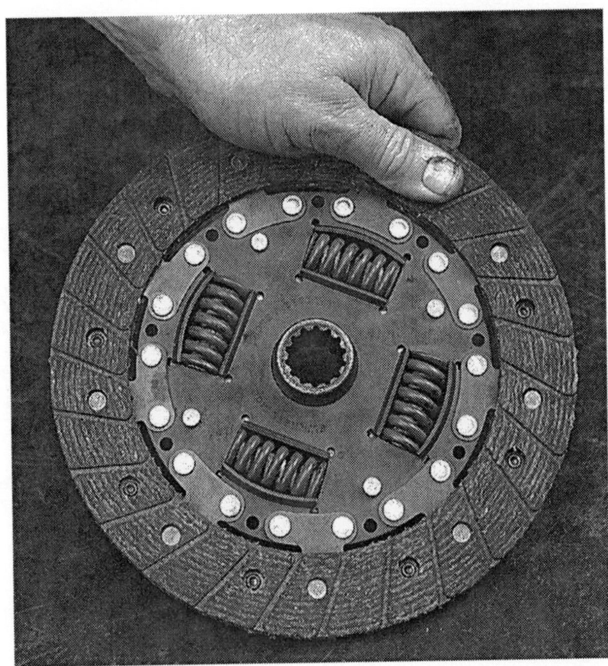

**Figure 1.** A clutch disc.

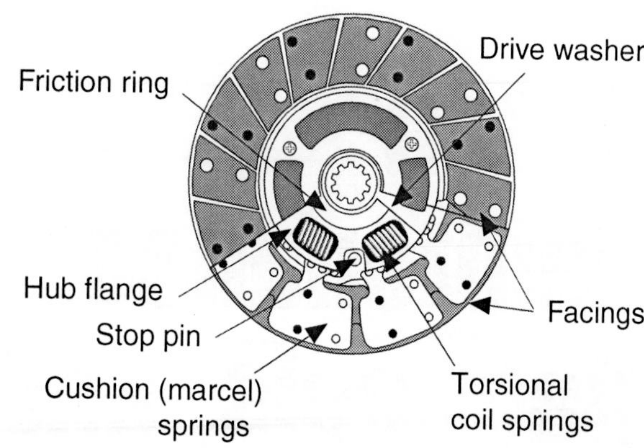


**Figure 2.** Clutch disc components.

Engines may have a flexible clutch disc or a rigid clutch disc. A flexible clutch disc has torsional springs and a friction ring between the hub and the facings. When the clutch is first engaged, engine torque is transmitted from the clutch facings through the torsional springs and friction ring to the clutch hub and the transmission input shaft. The torsional springs allow some movement between the facings and the clutch hub to cushion the clutch application **(Figure 2)**. The movement between the facings and the hub is limited by stop pins. The torsional springs also prevent engine power pulses from being transferred to the transmission input shaft. The number of torsional springs and the amount of tension they have depends on the engine torque and vehicle weight. A **flexible clutch disc** allows some movement between the facings and the hub. A **rigid clutch disc** does not allow any movement between the facings and the hub.

## FLYWHEEL DESIGN

The flywheel is a heavy, circular steel component that is bolted to the rear crankshaft flange **(Figure 3)**. The front clutch disc facings fits against a smooth, machined surface on the flywheel. A pilot bearing or bushing in the center of the rear crankshaft flange supports a machined extension on the transmission input shaft **(Figure 4)**. A ring gear is pressed onto the outside diameter of the flywheel and the teeth on this ring gear mesh with the starter drive teeth when the starting motor is engaged.

When the clutch is engaged, the clutch disc facings are jammed against the flywheel surface and engine torque is transmitted from the flywheel to the clutch facings and hub. The flywheel acts as a balancer for the engine and also dampens engine vibrations caused by cylinder firings. The flywheel provides inertia for the crankshaft between cylinder firings.

Engines with automatic transmissions have a flexplate in place of a flywheel. Because a clutch is not used with an automatic transmission, the smooth, machined surface on the flywheel is not required. The torque converter in an automatic transmission is bolted to the flexplate and the weight of the torque converter and the fluid inside it dampens engine pulses and provides inertia for the crankshaft.

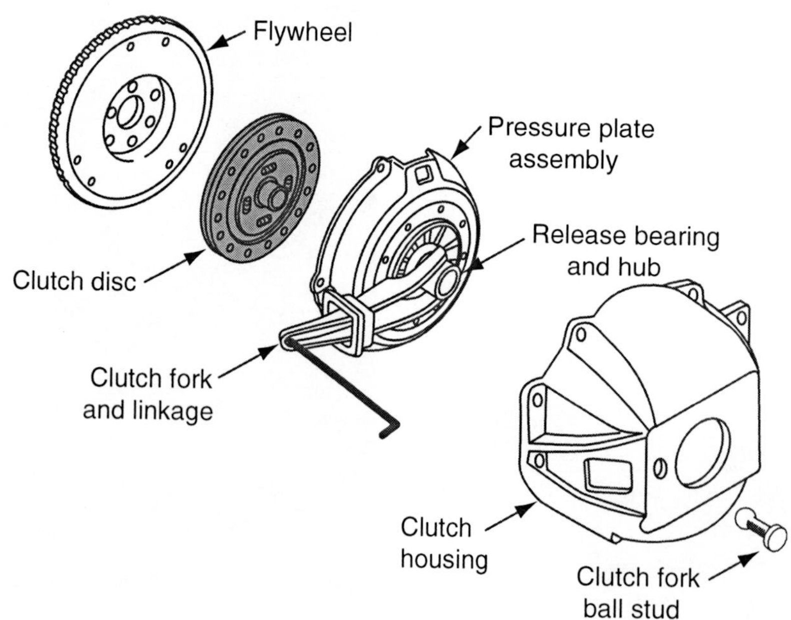


**Figure 3.** Flywheel and clutch components.

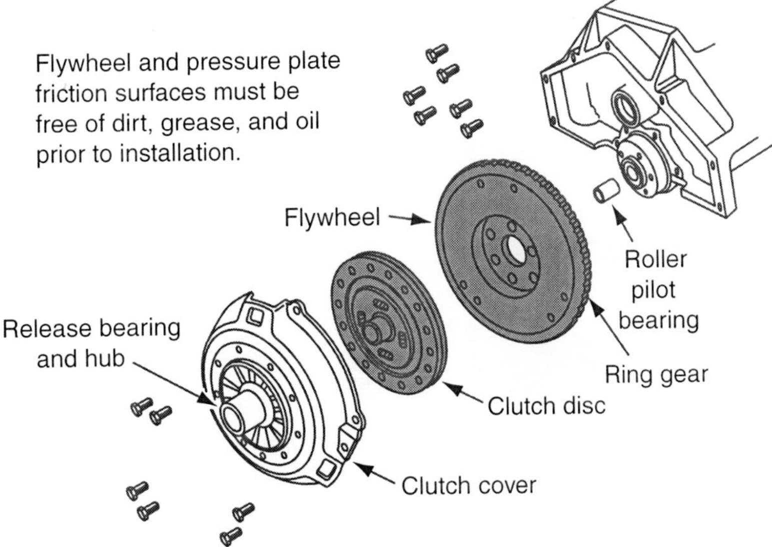


**Figure 4.** Pilot bearing and related clutch components.

Some light-duty trucks with diesel engines and some luxury or sport cars have dual-mass flywheels **(Figure 5)**. This type of flywheel reduces crankshaft oscillations before they reach the transmission. This action provides smoother transmission shifting and reduces gear noise. A dual-mass flywheel has two rotating plates connected by a spring and damper mechanism. The front rotating plate is bolted to the crankshaft flange and the pressure plate is mounted on the rear rotating plate. Engine torque is transferred from the front flywheel plate through the damper mechanism to the rear flywheel plate. The damper mechanism absorbs engine torque spikes during hard acceleration to prevent these spikes from being transferred to the transmission.

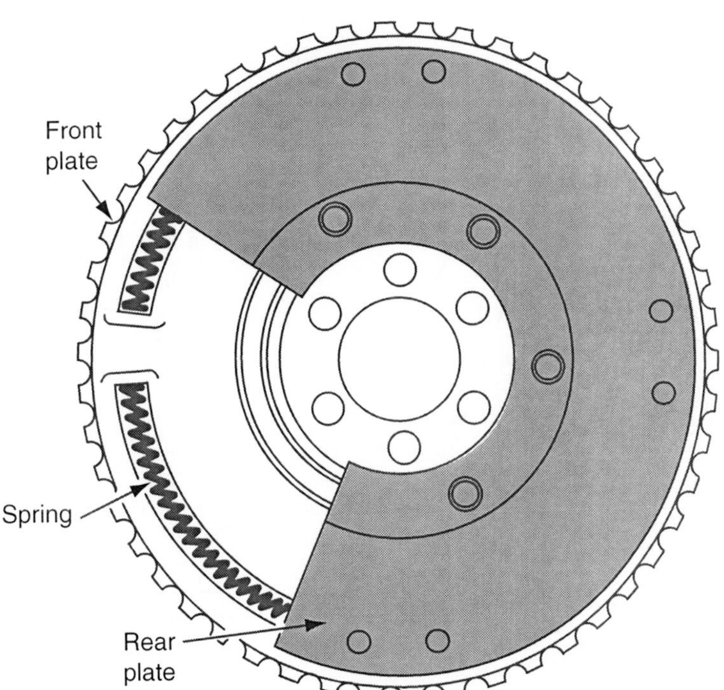


**Figure 5.** A dual-mass flywheel.

## PRESSURE PLATE ASSEMBLY DESIGN

The pressure plate has a smooth machined surface facing toward the clutch plate **(Figure 6)**. When the clutch is engaged, this machined surface squeezes the clutch plate facings against the flywheel machined surface. Under this condition, engine torque is transferred from the flywheel and pressure plate to the clutch disc and transmission input shaft. To disengage the clutch, the pressure plate is moved away from the clutch facings so these clutch facings are free to rotate. When this action takes place, engine torque is no longer transferred from the flywheel and pressure plate to the clutch disc and transmission input shaft.

Figure 6. A pressure plate.

The pressure plate assembly contains a pressure plate and cover, a diaphragm spring, spacer bolts, release levers, and rivets **(Figure 7)**. Rivets attach the diaphragm spring to the cover. The center part of the diaphragm spring is slotted to form fingers that act as release levers. The clutch release bearing is mounted near the rear surface of these release levers. When the clutch release bearing is forced against the release levers, these release levers are forced toward the flywheel. This release lever movement causes the diaphragm spring to pivot over the fulcrum ring and the outer rim on the diaphragm moves away from the flywheel. This movement on the outer rim of the diaphragm spring pulls the pressure plate away from the flywheel to release the clutch.

When the release bearing moves away from the release levers, the diaphragm spring pivots over the fulcrum ring and forces the pressure plate against the clutch facing to engage the clutch.

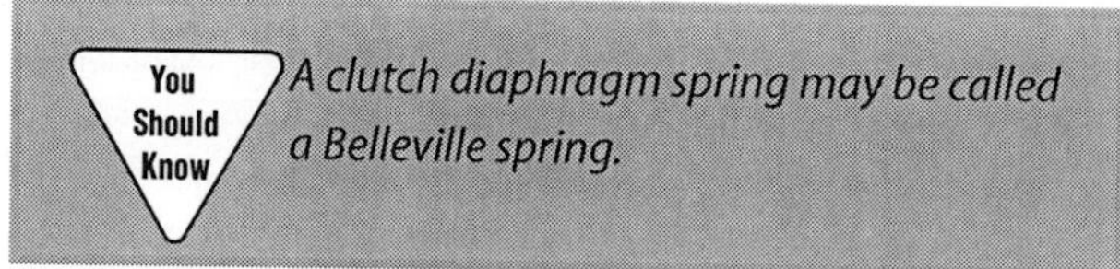


*A clutch diaphragm spring may be called a Belleville spring.*

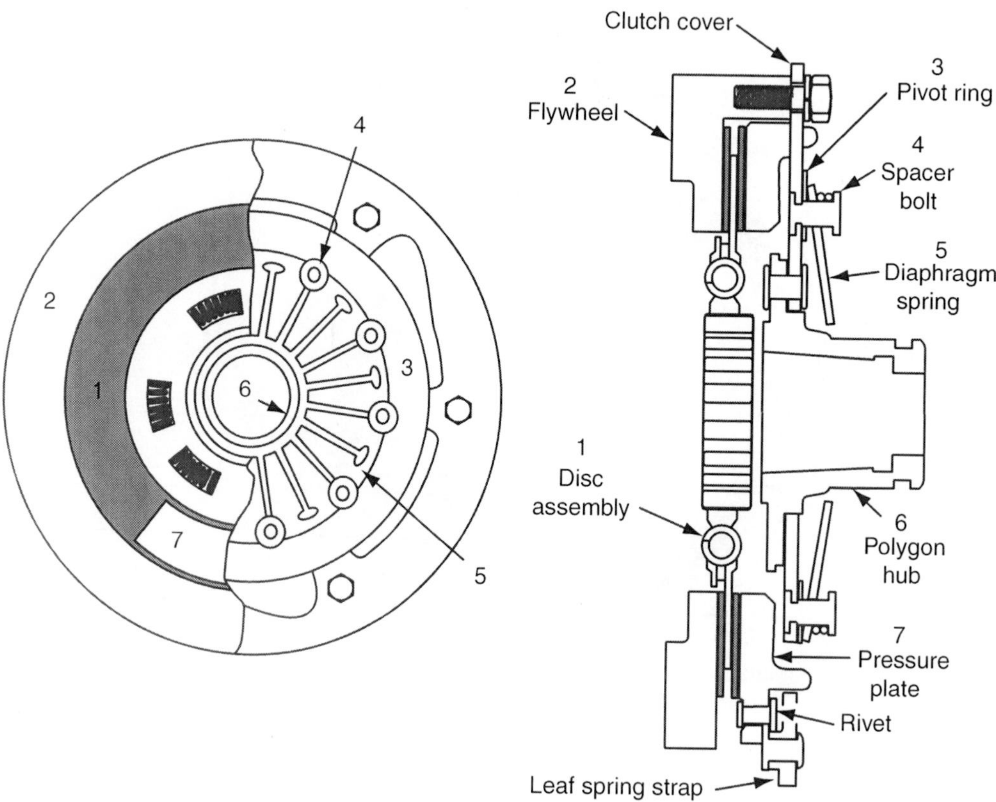


Figure 7. A pressure plate with a diaphragm spring.

A **Belleville spring** is a diaphragm spring made from thin sheet metal that is formed into a cone shape.

Some pressure plate assemblies have coil springs instead of a diaphragm spring **(Figure 8)**. This type of pressure plate has three pivoted release levers that pull the pressure plate away from the clutch facings to release the clutch. When the clutch is engaged, the coil springs force the pressure plate against the clutch facing **(Figure 9)**. Some coil spring pressure plates have a semi-centrifugal design. This type of pressure plate has pivoted weights that move outward from centrifugal force as the engine speed increases. When these weights fly outward, they increase the clamping force on the pressure plate and clutch disc **(Figure 10)**. Compared to the spring-type pressure plate,

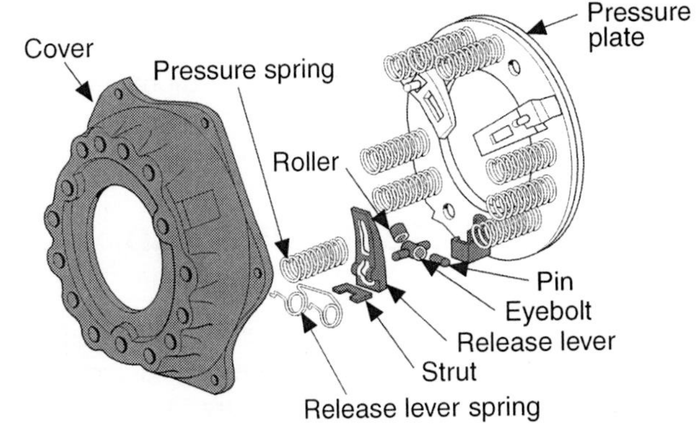


**Figure 8.** A pressure plate with coil springs.

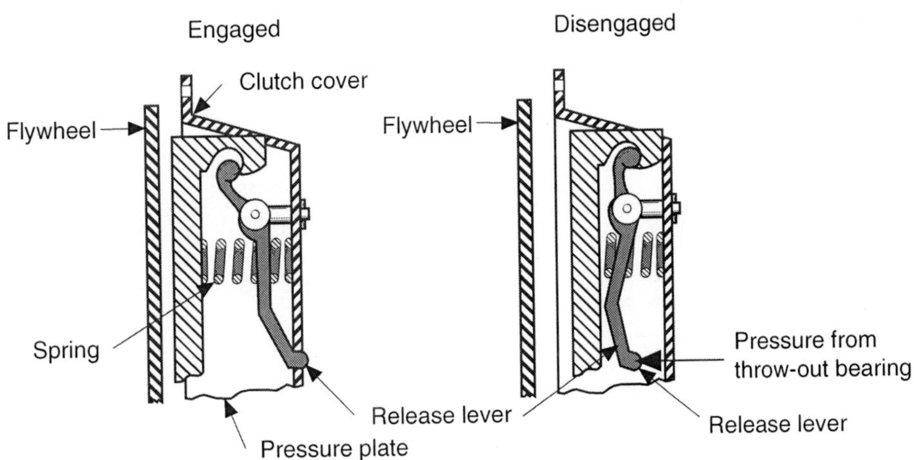


**Figure 9.** Operation of a coil spring-type pressure plate.

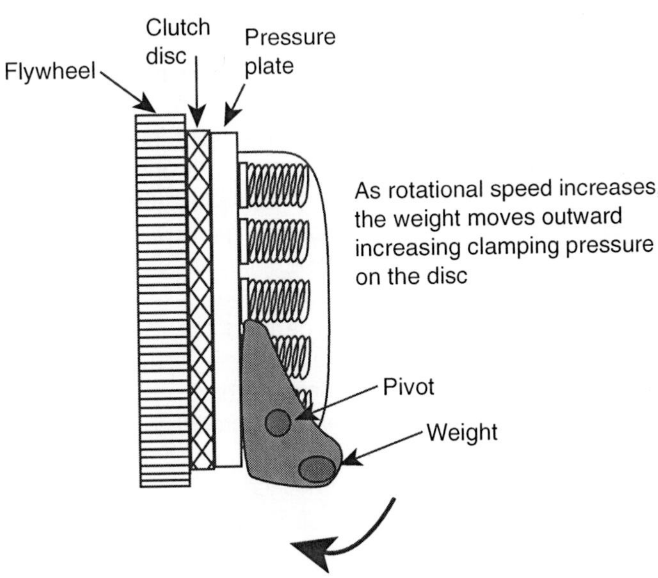


**Figure 10.** Semi-centrifugal pressure plate operation.

the diaphragm-type pressure plate is lighter, more compact, requires less pedal effort, and has fewer moving parts to wear. As the clutch facings wear on a diaphragm-type pressure plate, the pressure plate exerts more squeezing force on the clutch facings.

## CLUTCH RELEASE BEARING AND LEVER

The clutch release bearing has a ball bearing mounted on a hub and the front of the release bearing contacts the release levers on the pressure plate **(Figure 11)**. The clutch release bearing is a pre-lubricated, sealed bearing that does not require lubrication in service. A clutch release fork fits into a machined groove in the rear of the clutch release bearing. The clutch release bearing hub slides forward and rearward on a machined extension on the front bearing retainer in the transmission or transaxle **(Figure 12)**. The clutch release fork is mounted on a pivot in the flywheel housing. The outer end of this fork extends through an opening in this housing **(Figure 13)**. The flywheel housing is bolted to the rear of the engine block. This housing encloses the clutch assembly. The transmission is bolted to the rear of the flywheel housing. Some clutch release bearings must be adjusted so they are close to the pressure plate release levers, but they do not contact these levers when the clutch is engaged. This type of clutch has an adjustable clutch linkage. Many clutches have a **constant-running release bearing** that is designed to maintain light contact with the release levers when the clutch is engaged. This type of clutch has a self-adjusting clutch cable or a hydraulically-operated clutch.

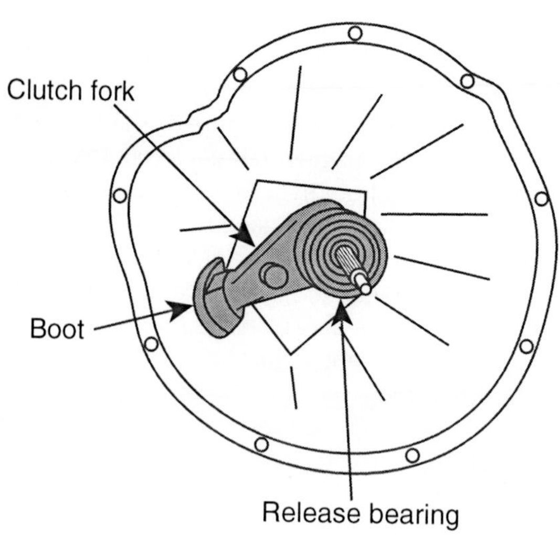


Figure 13. Clutch release fork and flywheel housing.

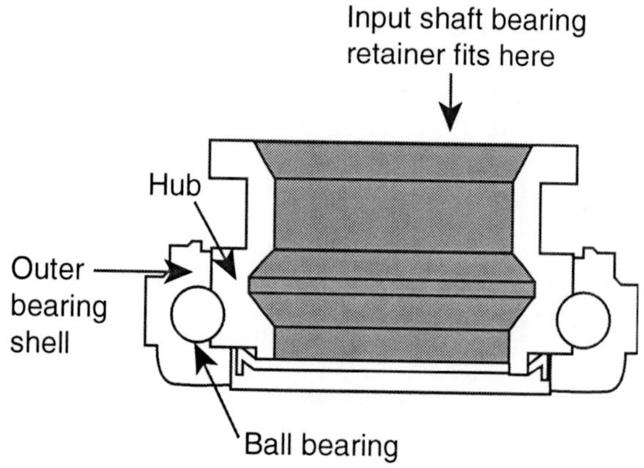


Figure 11. A clutch release bearing.

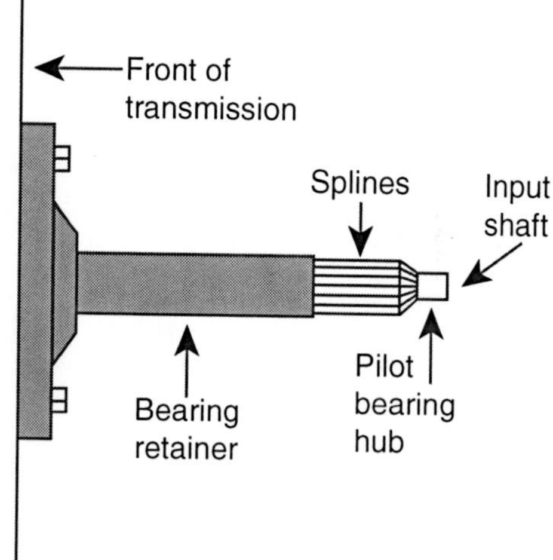


Figure 12. Transmission front bearing retainer with machined extension that supports the clutch release bearing.

## CLUTCH LINKAGES

Some clutches have a mechanical linkage that is connected from the release fork to the clutch pedal. A linkage from the clutch pedal is connected to a pivoted equalizer lever and a second linkage is connected from this lever to the release fork **(Figure 14)**. When the pedal is depressed, the linkage movement forces the release lever rearward, which in turn forces the pivoted release fork forward against the release bearing to release the clutch. When this type of clutch is engaged, a retracting spring pulls the release lever so the release bearing does not contact the pressure plate release levers. This type of clutch linkage requires proper adjustment to position the release bearing so it does not contact the release levers when the clutch is engaged.

Some clutches have a cable in place of a linkage connected between the clutch pedal and the release fork **(Figure 15)**. Some clutch cables are self-adjusting and these

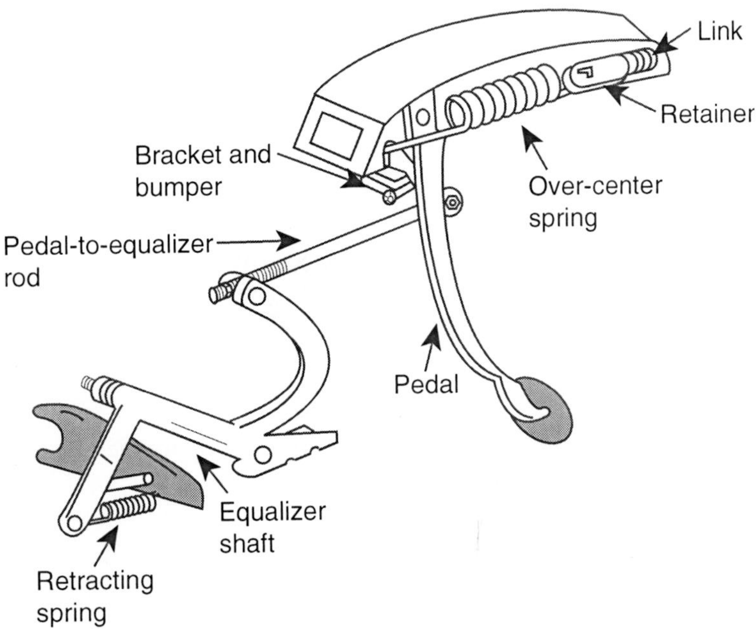


**Figure 14.** Lever-type clutch linkage.

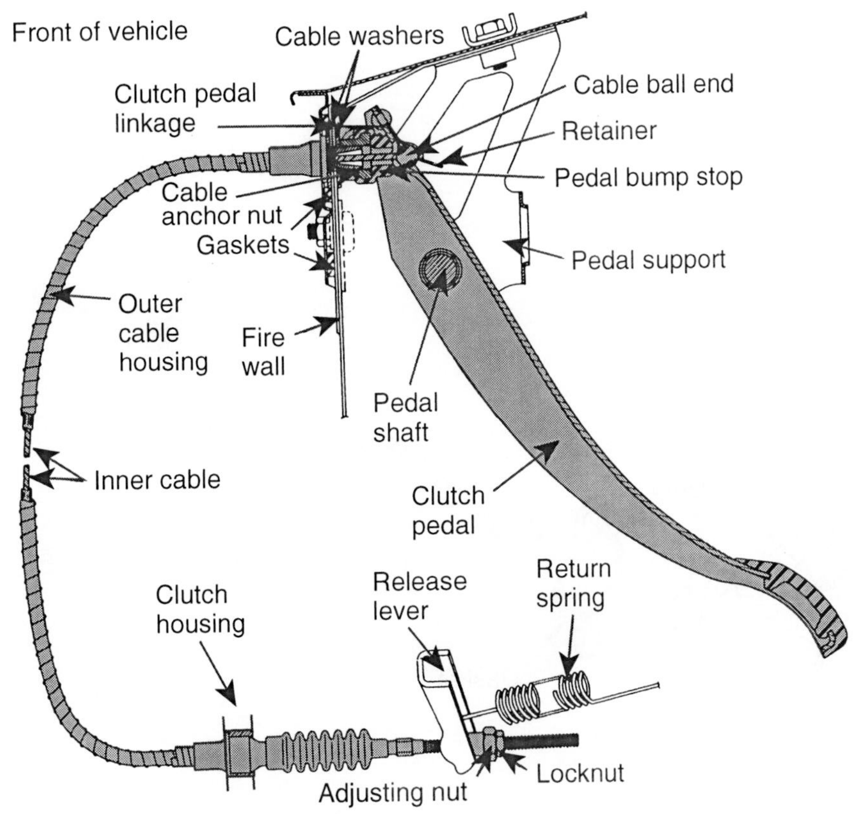


**Figure 15.** A clutch cable system.

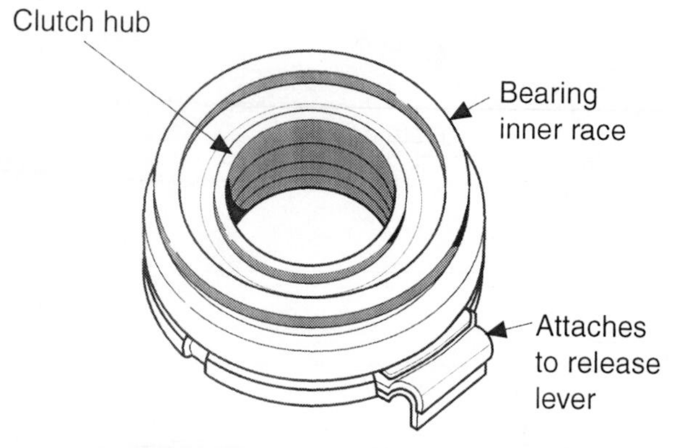


**Figure 16.** A constant-running clutch release bearing.

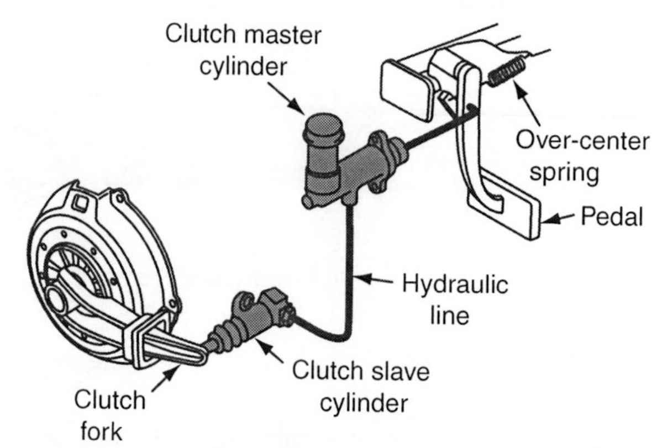


**Figure 17.** A hydraulic clutch system.

clutches have a constant-running release bearing **(Figure 16)**. Other clutch cables require adjusting and these clutches require proper adjustment to keep the release bearing away from the pressure plate release levers when the clutch is engaged.

Many vehicles in recent years are equipped with a hydraulic clutch. In these systems, a clutch master cylinder is bolted to the firewall and a slave cylinder is mounted near the outer end of the clutch release fork **(Figure 17)**. A pushrod is connected from the clutch pedal to the master cylinder and another pushrod is connected from the slave cylinder to the release fork. A hydraulic line is connected from the master cylinder to the slave cylinder. When the clutch pedal is depressed, the pushrod is forced against the piston in the master cylinder **(Figure 18)**. This action creates hydraulic pressure in the master cylinder. This pressure is supplied through the line to the slave cylinder. Hydraulic pressure in the slave cylinder moves the slave cylinder

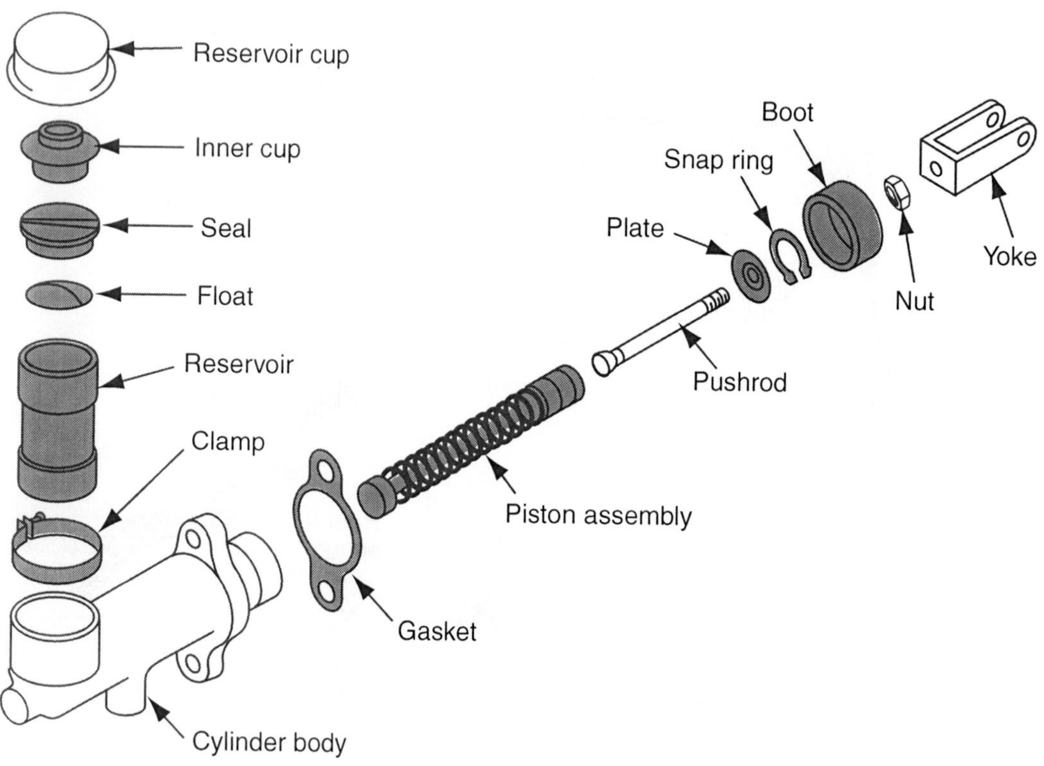


**Figure 18.** A hydraulic clutch master cylinder.

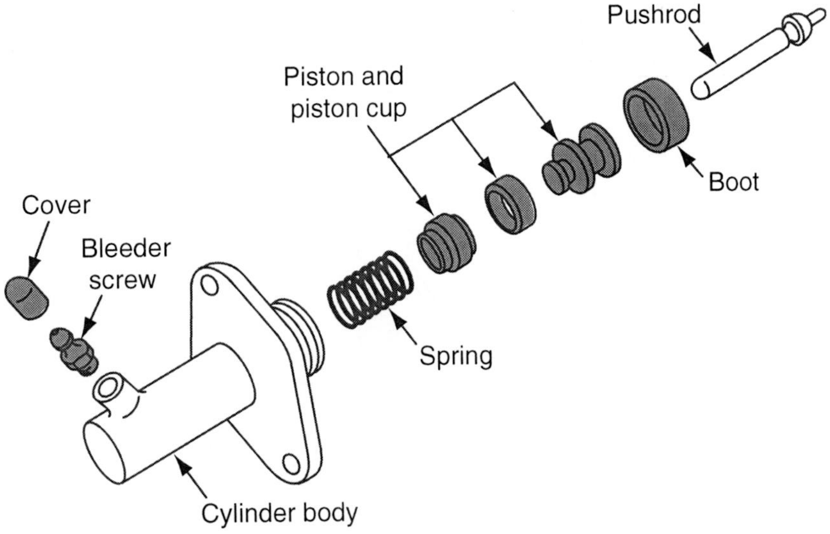


**Figure 19.** A hydraulic clutch slave cylinder.

piston, pushrod, and clutch release fork against the release bearing to disengage the clutch **(Figure 19)**. Some hydraulic clutches have a constant-running release bearing and do not require adjustment. Some hydraulic clutches have a small damper cylinder in the line between the master cylinder and the slave cylinder. This damper absorbs vibrations during clutch engagement and disengagement.

## CLUTCH OPERATION

When the driver depresses the clutch pedal, the linkage, cable, or hydraulic pressure forces the release fork rearward. Because the release fork is pivoted near the center, the inner end of this fork is moved forward. This forward movement slides the release bearing hub forward. When the release bearing hub slides forward on the front transmission bearing retainer, the release bearing is forced against the pressure plate release levers to release the clutch disc. When the clutch disc is released, the pressure plate and flywheel continue to rotate while the engine is running. Because there is no firm contact between the clutch disc facing and the pressure plate, no torque is transferred to the clutch disc and transmission input shaft.

To engage the clutch, the driver slowly allows the clutch pedal to move upward off the vehicle floor. This action allows the clutch release fork to move forward and the inner end of this release fork moves rearward. This release fork movement slowly allows the pressure plate spring(s) to squeeze the clutch disc facings between the pressure plate and the flywheel. Under this condition, engine torque is gradually transferred from the flywheel and pressure plate through the clutch disc to the transmission input shaft. On FWD vehicles, engine torque is transferred through the transaxles and drive axles to the drive wheels. As the driver allows further upward clutch pedal movement, the pressure plate exerts more pressure on the clutch disc facings. The clutch is fully engaged when the driver's foot is released from the clutch pedal.

## CLUTCH MAINTENANCE

A clutch requires a minimum amount of maintenance. If the clutch has an adjustable linkage or cable, the clutch pedal free-play should be checked at the vehicle manufacturer's recommended intervals. If the clutch pedal free-play is less than $^1/_2$ inch (13 millimeter), a free-play adjustment should be performed. **Clutch pedal free-play** is the amount of pedal movement before the release bearing contacts the pressure plate release levers.

Excessive clutch pedal free-play may not allow the clutch to release when the pedal is fully depressed. This condition may cause hard transmission shifting and/or gear clashing when shifting. Insufficient clutch pedal free-play may not allow the pressure plate to exert enough pressure against the clutch disc facings. This condition may cause clutch slipping, especially during hard acceleration.

Before performing a clutch pedal free-play adjustment, always inspect the clutch pedal linkage or cable for worn, bent, or loose linkages, and worn pivots. Observe the linkage while depressing and releasing the clutch pedal. Replace any worn, bent, or damaged linkage components before performing the free-play adjustment.

To perform the clutch free-play measurement, place a ruler beside the clutch pedal and slowly depress the clutch pedal. The free-play is the pedal movement before from the fully released position to the point where some resistance if felt when the release bearing contacts the pressure plate release levers. On many vehicles, this measurement should

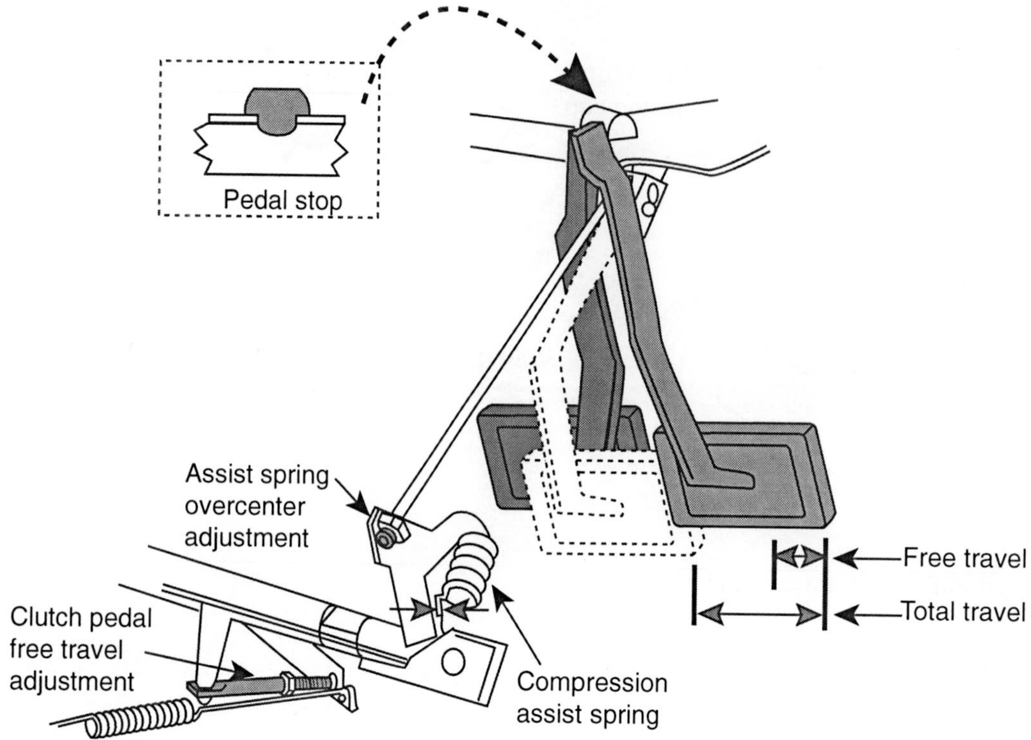


**Figure 20.** Clutch pedal free-play adjustment.

be between $\frac{1}{2}$ and $1\frac{1}{2}$ inches (13 to 44 millimeters). If the free-play is not correct, loosen the lock nut and rotate the free-play adjustment on the linkage to obtain the proper free-play **(Figure 20)**. Tighten the lock nut after the free-play adjustment is complete. When performing the free-play adjustment on a clutch with an adjustable cable, loosen the locknut and rotate the adjusting nut on the lower end of the cable **(Figure 21)**. When performing the clutch pedal free-play adjustment, always check the total pedal travel and be sure the stop for the upper end of the pedal is in good condition. On clutches with a self-adjusting cable or some hydraulic clutches, a free-play adjustment is not required. On these applications if the clutch does not engage or disengage properly, inspect the self-adjusting cable mechanism or check the hydraulic clutch system for low fluid level or air in the system. If the fluid level is low, inspect the hydraulic clutch system for leaks. If fluid is leaking from the master cylinder or slave cylinder, overhaul or replace these components. When fluid is leaking from the line between the master cylinder and slave cylinder, repair or replace the line.

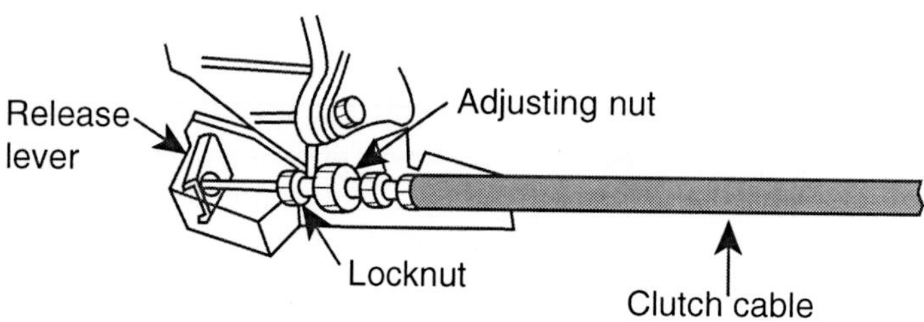


**Figure 21.** Free-play adjustment procedure on a clutch cable.

## CLUTCH DIAGNOSIS

When diagnosing clutch problems, always obtain as much information as possible from the customer. Always be sure you are aware of the exact customer complaint related to the operation of the clutch. If possible, find out the past service history of the vehicle. This information may help you to diagnose the clutch problem. A road test may be necessary to identify and verify the customer's complaint.

You Should Know *Excessive clutch facing wear and repeated clutch failure may be caused by a driver that keeps his or her foot on the clutch pedal, even when the clutch is engaged. This action may be called riding the clutch. Once the clutch is engaged, the driver's foot should be kept off the clutch pedal.*

### Clutch Slipping

When the clutch slips, the engine speed increases without the proper corresponding increase in vehicle speed. The engine rpm increases too much for the speed at which the vehicle is moving. If clutch slipping is present, check the clutch pedal free-play. If the free-play is less than specified, clutch slipping may occur. Be sure the clutch pedal is coming all the way upward. A binding linkage or some other component restricting linkage movement may prevent the pedal from coming fully upward to engage the clutch. On a hydraulic clutch, be sure the return port in the master cylinder is not restricted which prevents the clutch from fully engaging. Raise the vehicle on a lift and look for any sign of oil leaking from the bottom of the flywheel housing. If an oil leak at the rear main bearing or at the front of the transmission is contaminating the clutch facings, slipping will occur. Some vehicles have a small, removable pan on the bottom of the flywheel housing that can be removed to check for oil leaks in this area. Look for clutch facing material in the flywheel pan. If there is facing material in this area, the clutch facings are worn out. Clutch slipping may be the result of worn clutch disc facings. If this condition is present, the clutch must be replaced. The clutch must also be replaced if the facings are contaminated with oil.

If the clutch slips completely and does not transfer any engine torque to the transmission with the pedal fully released, the facings may be completely worn off the clutch disc.

You Should Know *If the clutch facings are contaminated with oil, the source of the oil leak into the flywheel housing must be corrected before the clutch is replaced. Otherwise, the new clutch will quickly become oil contaminated and the slipping problem will return.*

### Clutch Chatter

Clutch chatter is a shuddering action as the clutch engages. Once the clutch is engaged, the shuddering action stops. Road test the vehicle and check for clutch chatter each time the clutch is engaged. Clutch chatter may be caused by broken engine or transaxle mounts, worn and/or scored pressure plate and flywheel surfaces, and oil on the clutch facings and/or pressure plate and flywheel surfaces. Clutch chatter may also be caused by misalignment of the flywheel housing in relation to the engine and flywheel.

### Clutch Noises

One of the most common clutch noises is a growling noise when the clutch pedal is depressed. If the vehicle has a constant-running release bearing, the growling noise is also heard with the engine running, the transmission in neutral, and the clutch pedal released. If the vehicle has an adjustable clutch linkage or cable, the growling noise is only present when depressing or releasing the clutch pedal. This noise is caused by a defective release bearing. When a growling noise is only present with the engine running, the transmission in neutral, and the clutch pedal released, the bearing on the transaxle or transmission input shaft may be defective. A worn pilot bearing in the back of the crankshaft may cause a growling, rattling noise with the clutch pedal depressed and the clutch disengaged.

If a severe scraping, rattling noise is heard when the clutch is engaged and no torque is transferred from the engine to the transmission, the hub may be broken out of the clutch disc.

A heavy knocking or thumping noise just above idle speed may be caused by loose flywheel retaining bolts.

### Clutch Vibration

Clutch vibration may be felt in any clutch pedal position. The vibration may be felt on the clutch pedal and in the passenger compartment. If clutch vibration is present, raise the vehicle on a lift and inspect the engine and transaxle mounts. Watch for broken mounts that allow the engine or transaxle to contact the chassis. When all the mounts are in satisfactory condition, the vibration may be caused by excessive flywheel runout and improper flywheel or pressure plate balance.

**You Should Know** *Before removing a pressure plate from the flywheel, always place index marks on these components so the pressure plate can be reinstalled in the same position to maintain proper flywheel and pressure plate balance.*

## Dragging Clutch

A dragging clutch occurs when the clutch pedal is fully depressed but the clutch disc does not completely release. This condition may cause hard transmission shifting and/or gear clashing while shifting. To check for a dragging clutch, allow the engine to idle and fully depress the clutch pedal. Shift the transmission into first gear and keep the pedal depressed. Shift the transmission into neutral, wait 10 seconds, and then shift into reverse. If gear clashing is heard, the clutch is dragging. When this problem is present, be sure the clutch pedal free-play adjustment is within specifications. Excessive clutch pedal free-play causes a dragging clutch.

Raise the vehicle on a lift and inspect the clutch linkage for looseness, wear, damage, or a binding condition. Repair the linkage as required. If the vehicle has a hydraulic clutch, check the fluid level in the master cylinder and bleed the air from the hydraulic system. When the linkage or hydraulic system is satisfactory, clutch drag may be caused by a warped clutch disc or pressure plate or a defective release lever.

## Binding Clutch

If a clutch has a binding condition, it may cause erratic clutch engagement or grabbing. When this condition is present, inspect the clutch linkage or cable for wear and binding. Be sure the release fork is properly positioned on its pivot. Inspect the sleeve on which the release bearing slides. Ridges and wear on this sleeve may cause a binding clutch.

## Pulsating Clutch Pedal

Clutch pedal pulsation is a rapid upward and downward pedal movement. To check for this problem, slowly depress the clutch pedal. If noticeable pulsations are present, there is a problem in the clutch. Pedal pulsations are usually caused by uneven pressure plate release levers, a misaligned flywheel housing, a warped pressure plate, or excessive flywheel runout. The clutch must be disassembled, inspected, and various measurements performed to test for these conditions.

# CLUTCH SERVICE

Accurate clutch service is absolutely essential to provide a clutch that operates smoothly without chattering and unwanted noises.

## Clutch and Pressure Plate Removal

To remove the clutch and pressure plate, it is necessary to remove the transaxle or transmission. Before removing the transaxle or transmission, the engine must be supported so it does not drop downward and damage other components. Follow the vehicle manufacturer's recommended procedure in the service manual to support the engine prior to transaxle or transmission removal. On some FWD vehicles it is easier to remove the engine and transaxle as an assembly and then remove the transaxle from the engine. Always follow the vehicle manufacturer's recommended procedure in the service manual regarding engine and/or transaxle removal. If the engine is removed from the vehicle, bolt the engine securely onto an engine stand and then remove the transaxle from the engine. Always place index marks on the flywheel and pressure plate before removing the pressure plate from the flywheel **(Figure 22)**. After the pressure plate and clutch plate are removed from the flywheel, inspect these components and

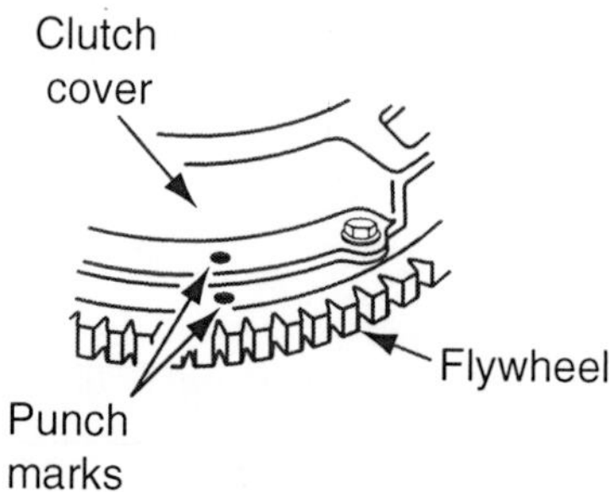


**Figure 22.** Placing index marks on the pressure plate and flywheel.

other related parts such as the release bearing, fork, and pilot bearing. Replace all worn or damaged components. Most clutch plates and pressure plates are replaced as a matched pair. When installing the clutch plate and pressure plate, insert a clutch plate alignment tool through the clutch plate into the pilot bearing to ensure proper clutch plate alignment with the pilot bearing.

## Summary

- When the clutch is engaged, the pressure plate squeezes the clutch facings between the pressure plate and the flywheel so engine torque is transferred from the flywheel and pressure plate to the clutch disc and transmission input shaft.
- If the clutch is released, the pressure plate is pulled away from the clutch disc and the flywheel and pressure plate rotate without transferring engine torque to the clutch disc.
- The torsional springs in the clutch disc prevent engine pulsations from being transferred from the engine to the transmission input shaft.
- The pressure plate may contain a diaphragm spring or coil springs.
- To disengage the clutch, the release bearing is moved forward against the pressure plate fingers.
- The pivoted release fork transfers clutch linkage or cable movement to the release bearing.
- A constant-running release bearing lightly contacts the pressure plate fingers with the clutch engaged.
- The clutch may have a mechanical linkage, a cable, or a hydraulic clutch system between the pedal and the release fork.
- Some clutch cables are self-adjusting.
- Hydraulic clutch systems have a master cylinder operated by the clutch pedal and a slave cylinder that operates the release lever.
- Clutches with mechanical linkages or adjustable cables require a free-play adjustment.
- Hydraulic clutches or clutches with self-adjusting cables do not require a free-play adjustment.
- Clutch problems include slipping, chattering, noise, vibration, dragging, binding, or pulsating.
- Flywheel runout may be measured with a dial indicator.
- Flywheel housing face and bore alignment may be measured with a dial indicator.
- When installing the clutch disc and pressure plate, a clutch disc alignment tool is used to position the clutch disc properly.

## Review Questions

1. Technician A says the flywheel provides inertia for the crankshaft between cylinder firings. Technician B says on a vehicle with an automatic transmission, the torque converter dampens engine pulses. Who is correct?
   A. Technician A
   B. Technician B
   C. Both Technician A and Technician B
   D. Neither Technician A nor Technician B
2. Technician A says a pressure plate may have a diaphragm spring or coil springs. Technician B says when the clutch is engaged, the pressure plate is held away from the clutch disc facings. Who is correct?
   A. Technician A
   B. Technician B
   C. Both Technician A and Technician B
   D. Neither Technician A nor Technician B
3. Technician A says the clutch release bearing requires periodic lubrication. Technician B says a constant-running release bearing is used with a clutch that has an adjustable cable. Who is correct?
   A. Technician A
   B. Technician B
   C. Both Technician A and Technician B
   D. Neither Technician A nor Technician B
4. Technician A says in a hydraulic clutch system, the clutch pedal operates the slave cylinder pushrod. Technician B says air in a hydraulic clutch system may cause improper clutch disengagement. Who is correct?
   A. Technician A
   B. Technician B
   C. Both Technician A and Technician B
   D. Neither Technician A nor Technician B

5. All these statements about clutch plate design are true *except*:
   A. Diagonal grooves cut in the clutch facings improve clutch cooling.
   B. Diagonal grooves cut in the clutch facings provide smoother clutch engagement.
   C. The clutch facings are attached to wave springs that gradually flatten out during clutch engagement.
   D. A rigid-type clutch disc has torsional springs and a friction ring between the facings and the hub.
6. A dual-mass flywheel has all of these design features *except:*
   A. inner and outer rotating plates connected by a damper mechanism.
   B. an inner plate that is bolted to the crankshaft flange.
   C. an outer plate on which the clutch plate facing makes contact.
   D. a pressure plate that is bolted to the inner and outer flywheel plates.
7. When discussing clutch operation, Technician A says when the clutch is engaged, the clutch plate facings are jammed between the flywheel and the pressure plate. Technician B says the clutch release bearing hub slides forward and rearward on a machined extension on the transmission front bearing retainer. Who is correct?
   A. Technician A
   B. Technician B
   C. Both Technician A and Technician B
   D. Neither Technician A nor Technician B
8. In a hydraulic clutch system:
   A. the clutch pedal pushrod is in contact with the master cylinder piston.
   B. fluid pressure from the slave cylinder operates the master cylinder.
   C. there is a specified clearance between the clutch release bearing and the pressure plate diaphragm spring.
   D. a pushrod is connected between the slave cylinder and the clutch release bearing.
9. The clutch in a vehicle with a mechanical clutch linkage does not release properly and hard gear shifting is experienced. The most likely cause of this problem is:
   A. a worn pilot bearing.
   B. a worn transmission input shaft bearing.
   C. a rough clutch release bearing.
   D. excessive clutch pedal free-play.
10. A vehicle with a hydraulic clutch has a clutch slipping problem that allows the engine rpm to increase without the proper increase in vehicle speed. The clutch master cylinder has the proper level and type of fluid. The cause of this problem could be:
    A. air in the hydraulic clutch system.
    B. oil contamination on the clutch facings.
    C. improper clutch free-play adjustment.
    D. fluid leaking past the slave cylinder piston.
11. To disengage a clutch, the release fork pushes the ____________ ________________ toward the pressure plate.
12. Excessive clutch pedal free-play may cause improper clutch ________________ .
13. Clutch slipping may be caused by ______________ clutch pedal free-play.
14. Broken engine mounts may cause clutch ________________ .
15. On a clutch with an adjustable linkage, explain the cause of a grinding noise only when the clutch pedal is depressed.
16. Describe the causes of clutch chatter.
17. Describe the causes of clutch slippage.
18. Explain the causes of clutch vibration.

# Section 12

## Manual and Automatic Transmissions and Transaxles



**Interesting Fact**

*Of the domestic new car buyers in the United States, 59.3 percent are male, 40.8 percent are female, and 5.3 percent unknown. Of the Asian car buyers in the United States, 49.1 percent are male, 45.4 percent are female, and 5.5 percent are unknown.*

## SECTION OBJECTIVES

After you have read, studied, and practiced the contents of this section, you should be able to:

- Describe two different types of transaxle gears.
- Explain gear reduction and overdrive ratios and describe how gear ratios are calculated.
- Describe synchronizer design and operation.
- Explain manual transmission operation in first, second, third, fourth, and fifth speed and reverse.
- Describe the lubricants required in manual transmissions and transaxles.
- Perform manual transmission/transaxle maintenance, including a visual inspection, leak diagnosis, and shift linkage adjustments.
- Diagnose manual transmission/transaxle problems.
- Remove and replace manual transmissions/transaxles.
- Describe the purpose of a torque converter.
- Explain the torque converter operation.
- Explain planetary gear set design.
- Explain how a planetary gear set provides reverse.
- Describe how a planetary gear set provides a 1:1 gear ratio.
- Describe the vehicle operation encountered with a slipping stator one-way clutch.
- Describe the vehicle operation encountered with a seized stator one-way clutch.
- Explain the inputs used by the PCM to provide torque converter clutch lockup.
- Perform a stator one-way clutch, interference, and end play tests.
- Explain the operation of a multiple disc clutch.
- Describe the operation of a band.
- Explain the operation of an overrunning clutch.
- List the different types of transaxle pumps and explain the operation of the pressure regulator valve.
- Explain the operation and purpose of the governor.
- Describe the operation of the throttle cable and throttle valve.
- Explain the operation of a shift valve.
- Explain the operation of an accumulator and a modulator.
- Describe basic transaxle electronic controls including shift control, pressure control, inputs, and outputs.
- Change automatic transmission or transaxle fluid and filter.
- Check automatic transmission fluid level and condition.
- Diagnosis automatic transmission and transaxle problems.
- Perform automatic transmission and transaxle service procedures.

# Manual Transmission and Transaxle Maintenance, Diagnosis, and Service

## Introduction

Manual transmissions and transaxles transfer engine torque to the differential and provide gear reductions for smooth vehicle acceleration. Depending on the transaxle, in third or fourth gear depending on the transaxle, engine torque is transmitted directly without providing a reduction or an overdrive. In fifth or sixth gear, depending on the transaxle or transmission, an overdrive gear ratio is provided to improve fuel economy at cruising speed. The transmission gears and components must be able to withstand high engine torque during hard acceleration.

Proper transmission/transaxle maintenance is essential to provide normal unit life. Accurate transmission/transaxle diagnosis is very important to determine the cause of operational problems. Professional transmission/transaxle service is critical to repair operational problems quickly and accurately.

### GEARS

Transaxle or transmission gears may be helical or straight-cut. Helical gears always have more than one tooth in mesh at once to provide additional gear strength. Helical gear teeth create a wiping action as they engage or disengage with other gear teeth. This action provides quieter operation. Helical gear teeth create axial thrust on the gear. **Helical** gears have teeth that are cut at an angle to the center line of the gear. **Straight-cut** gear teeth are parallel to the gear centerline.

Straight-cut gear teeth are noisier than helical gears, but they do not cause axial thrust. Gear teeth have a drive and a coast side. While the engine is supplying torque to the transmission, the drive side of the gear teeth transmitting torque is in contact with the teeth of another gear. During engine deceleration, the drive wheels are transmitting torque to the engine and the coast side of the gear teeth that are transmitting torque is in contact with the teeth of another gear. A typical five-speed transmission with helical and straight-cut gears is shown in **Figure 1**.

The teeth on many gears remain in constant mesh with each other. Synchronizers are used to engage or disengage constant mesh gears.

Backlash is the amount of movement between the teeth on two gears. Some backlash is required to allow proper lubrication of the gear teeth and compensate for tooth expansion when the gear temperature increases. Excessive backlash may indicate gear wear and this condition may cause gear tooth damage.

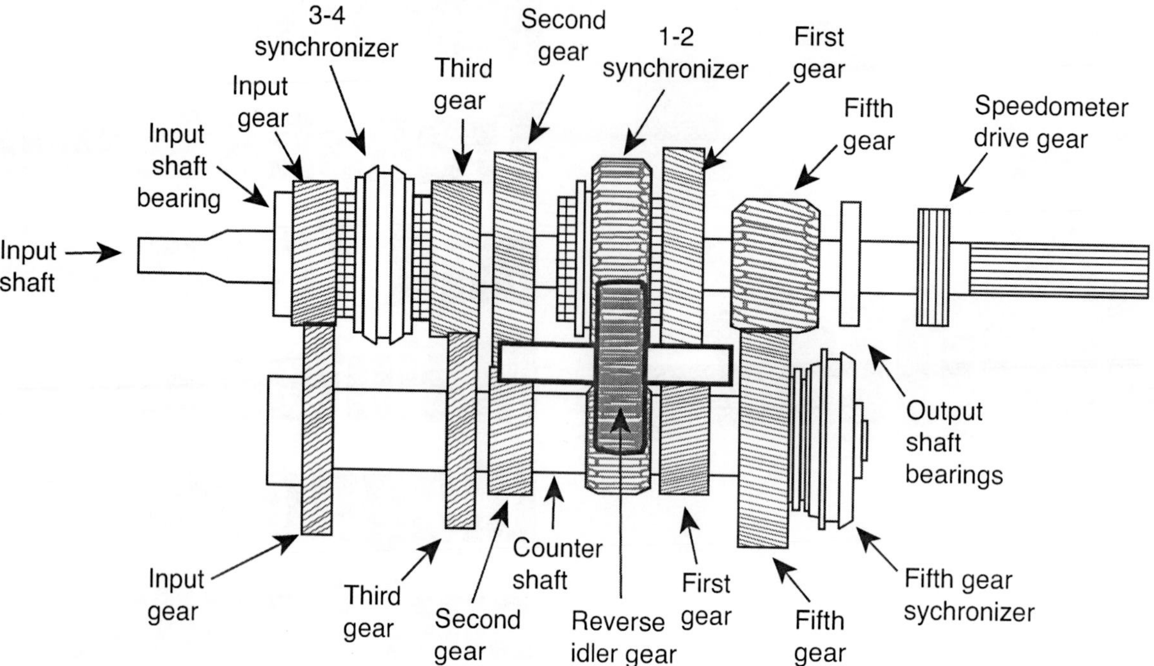


**Figure 1.** A typical five-speed transmission with helical and straight-cut gears.

## GEAR RATIOS

A gear reduction occurs when a smaller diameter gear drives a larger diameter gear. A gear reduction provides an increase in torque and a decrease in output shaft speed. When a large-diameter gear drives a smaller-diameter gear, an overdrive condition is present. An overdrive gear ratio reduces engine torque and increases output shaft speed. **Gear ratio** is the ratio between the drive and driven gears.

Gear ratio is calculated by dividing the number of drive gear teeth into the number of driven gear teeth. When the drive gear has ten teeth and the driven gear has thirty teeth, the gear ratio is 3:1. If the drive gear has ten teeth and the driven gear has eight teeth, the gear ratio is $8 \div 10 = 0.8{:}1$. In Figure 1, the fifth gear on the counter shaft is larger than the fifth driven gear on the input shaft, creating an overdrive ratio between these gears.

If torque is transmitted through two gear ratios, the total gear ratio is determined by multiplying the two ratios. For example, if the first speed gear ratio is 3.40:1 and the differential gear ratio is 3.72:1, the total gear ratio is 12.648:1.

## SYNCHRONIZERS

Synchronizers bring two gears rotating at different speeds to the same speed which provides smooth shifting. In many current transaxles and transmissions, all the gears are synchronized. Some older transmissions, especially on trucks, do not have synchronizers on some of the gears. Some transmissions and transaxles have synchronizers on all forward gears and reverse, but on other transmissions and transaxles reverse gear does not have a synchronizer.

The block-type synchronizer is presently the most common type if synchronizer. A block-type synchronizer has a hub with internal and external splines. The internal splines are mounted on matching splines on the transmission output shaft **(Figure 2)**. A synchronizer sleeve with internal splines is mounted on the external hub splines **(Figure 3)**. Therefore, the sleeve can slide forward or rearward on the hub splines. When the transmission is assembled, a shifter fork is positioned in a wide groove in the outer diameter of the sleeve. Three inserts are mounted in hub slots. A narrow raised area in each insert is mounted in an internal groove in the sleeve. A lock ring on each side of the hub holds the inserts outward against the sleeve. A brass blocking ring is mounted on each side of the synchronizer. The inserts fit into wide grooves in the blocking rings. The inside diameter of the blocking ring is a cone-shaped area with sharp grooves on this surface. This cone-shaped area matches a cone-shaped area on the gear next to each side of the synchronizer. A series of beveled dog teeth are positioned around the outer diameter of each blocking ring. The dog teeth on the blocking ring match beveled dog teeth on the gear beside the synchronizer. **Dog teeth** are a set of small gear teeth around the outer gear diameter with gaps between the teeth.

When the transmission is in neutral and the engine is idling with the clutch engaged, the input shaft and

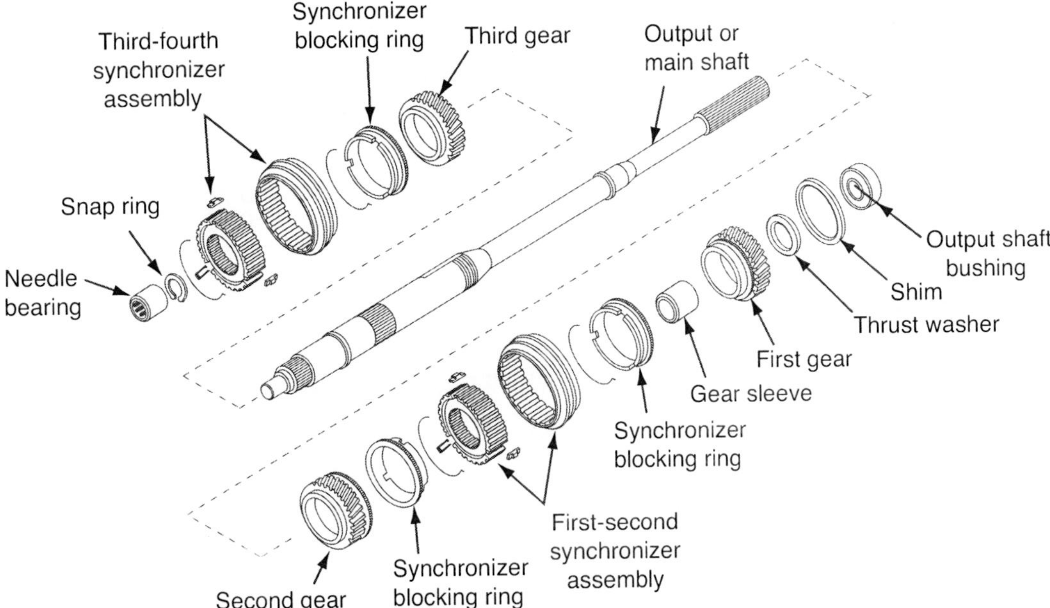


**Figure 2.** An output shaft with related gears and synchronizers.

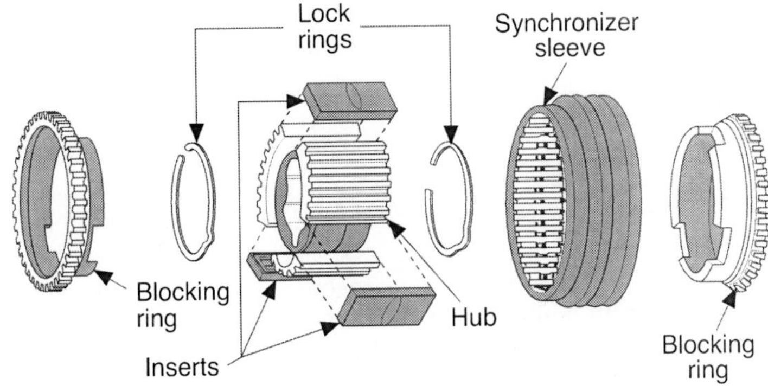


**Figure 3.** Synchronizer components.

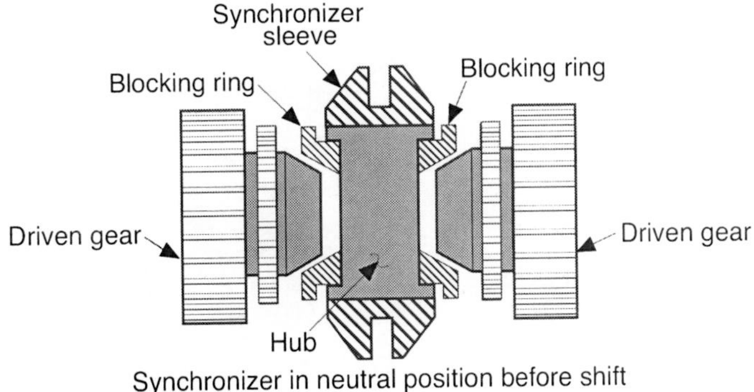


**Figure 4.** Synchronizer in the neutral position.

counter shaft are rotating. Because the gears on the counter shaft are meshed with gears on the output shaft, the gears on the output shaft also rotate. However, the synchronizers are in the neutral position and no torque is transferred from the counter shaft gears to the gears on the output shaft **(Figure 4)**. Needle bearings are positioned between the counter shaft and the counter shaft gear to reduce friction.

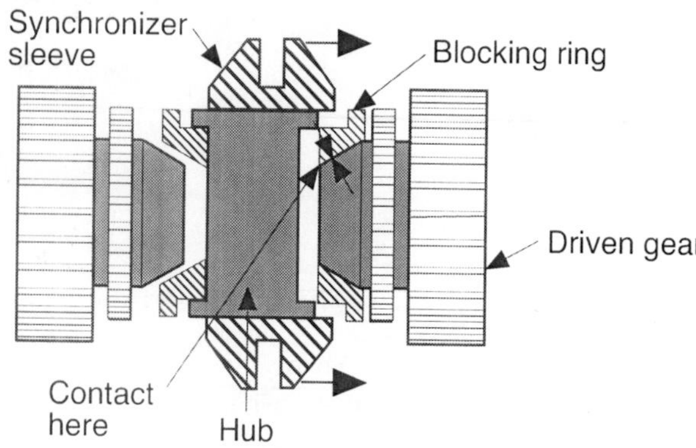


**Figure 5.** Synchronizer position during a shift.

When a shift occurs, the shifter fork moves the synchronizer sleeve toward the selected gear. The inserts move with the sleeve and these inserts push the blocking ring toward the selected gear. The grooved surface on the blocking ring cone cuts through the film of lubricant on the selected gear cone **(Figure 5)**. When the blocking ring grooves cut through the lubricant, metal-to-metal contact occurs between the blocking ring grooves and the cone on the selected gear. The friction developed by the metal-to-metal contact between these two components brings the synchronizer hub and the selected gear to the same speed. With the two components rotating at the same speed, the sleeve slides over the dog teeth on the blocking ring and the dog teeth on the selected gear **(Figure 6)**. Under this condition, engine torque is transmitted from the counter shaft gear to the output shaft gear, synchronizer hub, and output shaft.

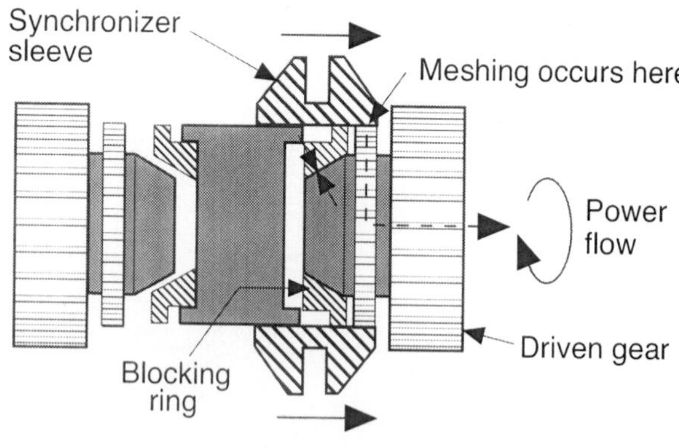


**Figure 6.** Synchronizer position when a shift is completed.

| Gear | Ratio | Overall ratio |
|---|---|---|
| First | 2.66:1 | 9.10:1 |
| Second | 1.78:1 | 6.10:1 |
| Third | 1.30:1 | 4.45:1 |
| Fourth | 1.00:1 | 3.42:1 |
| Fifth | 0.74:1 | 2.53:1 |
| Sixth | 0.50:1 | 1.71:1 |

**Figure 7.** Gear ratios in a six-speed transmission.

## TRANSMISSION TYPES

In current car production, the five-speed transmission is most commonly used. However, transmissions may be three speed, four speed, five speed, or six speed.

In a four-speed transmission, the fourth gear usually has a 1:1 ratio. The additional gear in this type of transmission provides a smaller difference in the ratios between the first-, second-, and third-speed gears.

A five-speed transmission is similar to a four-speed transmission, but a fifth overdrive speed with a synchronizer is added. In a five-speed transmission, the four-speed gear ratio is usually 1:1 and the fifth-speed gear ratio is usually between 0.70:1 and 0.90:1.

Some performance cars have a six-speed transmission. The fifth and sixth speeds have an overdrive ratio **(Figure 7)**. A sixth-speed gear ratio of 0.50:1 allows the engine to run at a lower rpm during highway cruising. In this type of transmission, all forward gears and reverse are synchronized.

## TRANSMISSION OPERATION

Shifting mechanisms and shifter forks are used to shift gears. Most shift forks have two legs fit into a groove in the synchronizer sleeve **(Figure 8)**. Each shift fork is secured to a shift rail with a tapered pin. Each shift rail has notches near the end of the rail and a spring-loaded ball or bullet rides on these notches. If the shift lever is mounted on top of the

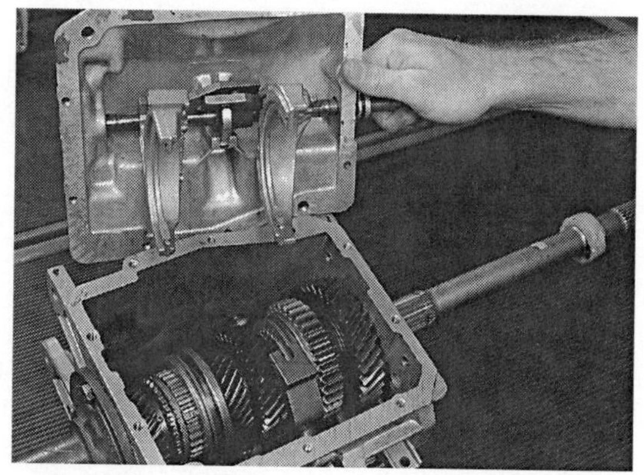

**Figure 8.** Shift forks.

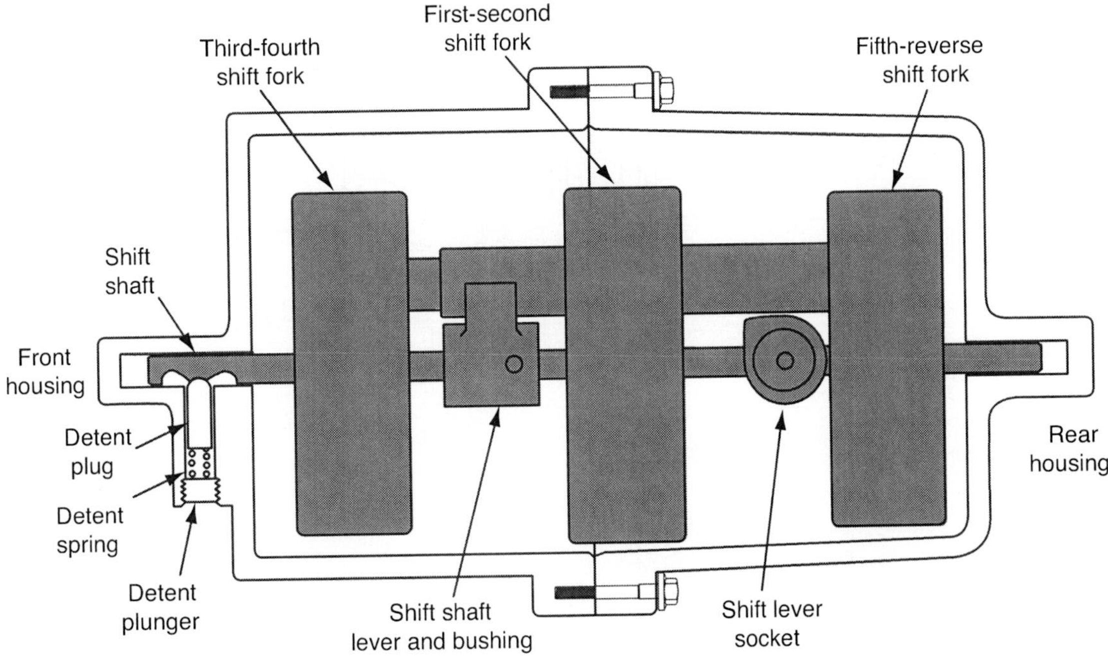


**Figure 9.** Shift forks and shift rails.

transmission, this lever is connected to a socket or slots in the shift rails **(Figure 9)**. If the gear shift lever is mounted on the rear of the transmission, linkages are connected from this lever to the shift rails **(Figure 10)**. When the gear shift lever is moved to perform a shift, this lever moves a shift rail. When a shift is completed, the spring-loaded ball fits into a shift rail notch to provide a detent feel and maintain the shift fork position. An interlock mechanism on the shift rails prevents the engagement of two gears at the same time **(Figure 11)**.

**Figure 10.** Shift lever and linkages.

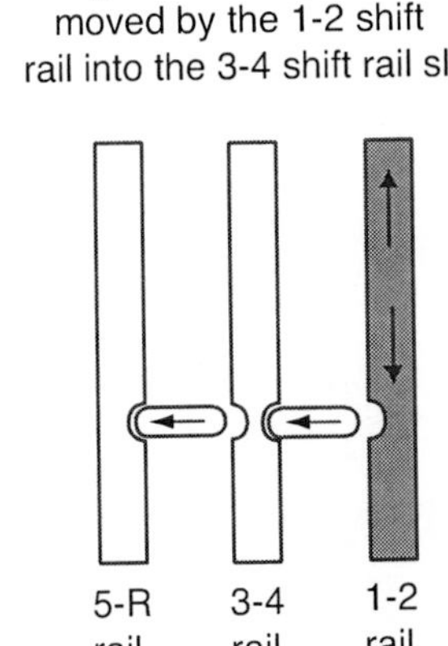


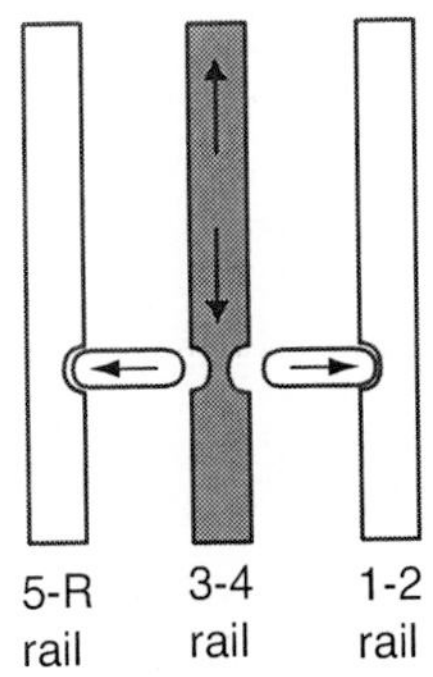


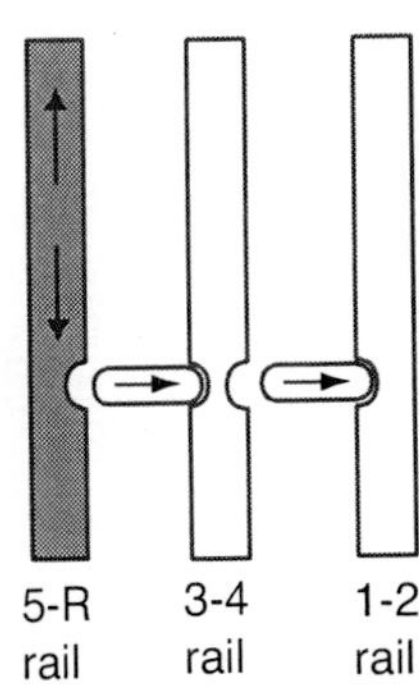


**Figure 11.** Interlock mechanism in a transmission shift assembly.

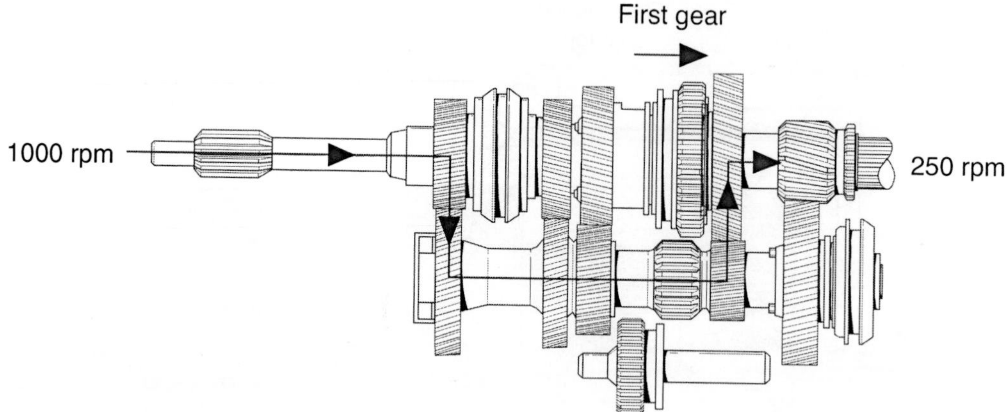


**Figure 12.** Power flow in first gear.

## First Gear Operation

In first gear, engine torque is transmitted from the input shaft and gear to the counter shaft gear in mesh with the input shaft gear. The 1–2 synchronizer sleeve is moved rearward by the shift fork. This synchronizer hub is now connected to the first-speed gear. Engine torque is now transmitted from the counter shaft gear, first-speed gear, and synchronizer hub to the output shaft **(Figure 12)**. All other gears on the output shaft rotate freely.

## Second Gear Operation

In second gear, the shift fork moves the 1–2 synchronizer sleeve forward. This sleeve connects the second-speed gear to the synchronizer hub **(Figure 13)**. Engine torque is transferred from the input shaft to the counter shaft gear in mesh with the input shaft gear. Engine torque is transferred to the second-speed gear from the counter shaft gear in mesh with the second-speed gear. From the second-speed gear, engine torque is transmitted to the sleeve and hub on the 1–2 synchronizer and the output shaft.

## Third Gear Operation

In third gear, the 1–2 synchronizer is moved to the neutral position and the 3–4 synchronizer sleeve is moved rearward by the shift fork **(Figure 14)**. Under this condition, engine torque is transmitted through the input shaft gear, counter gear third-speed gear, and synchronizer sleeve and hub to the output shaft.

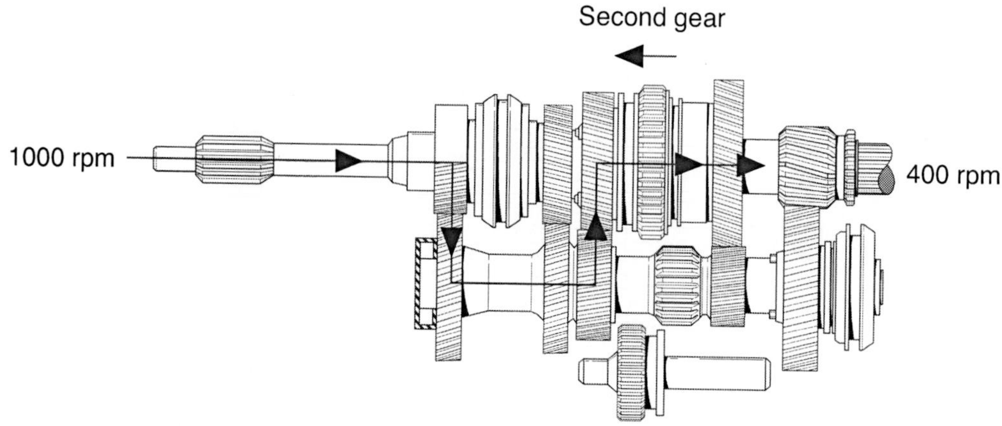


**Figure 13.** Power flow in second gear.

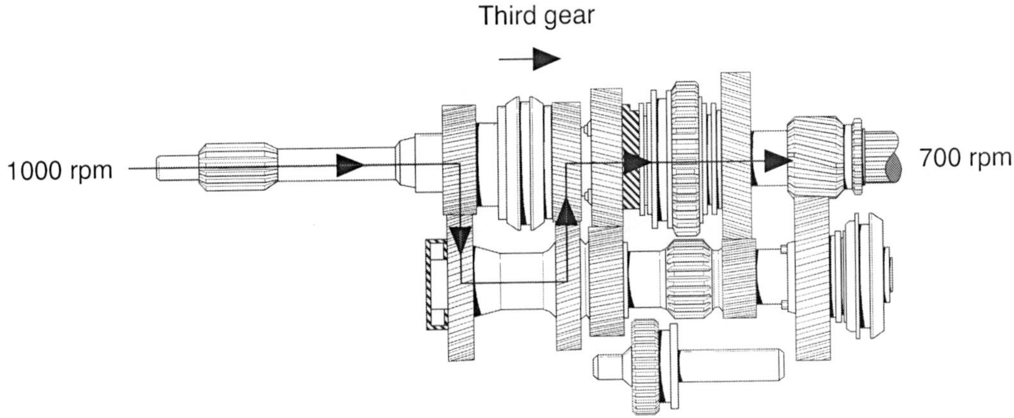


**Figure 14.** Power flow in third gear.

## Fourth Gear Operation

In fourth gear, the 3–4 synchronizer sleeve is moved forward by the shift fork. Engine torque is now transmitted from the input shaft gear, counter shaft, fourth-speed gear, and synchronizer sleeve and hub to the output shaft **(Figure 15)**.

## Fifth Gear Operation

In fifth gear, the shift fork moves the fifth-speed synchronizer toward the fifth-speed gear. Under this condition, the fifth-speed gear is connected to the counter shaft. Engine torque is now transmitted from the input shaft gear, counter shaft, fifth-speed gear, and matching gear on the output shaft to output shaft **(Figure 16)**.

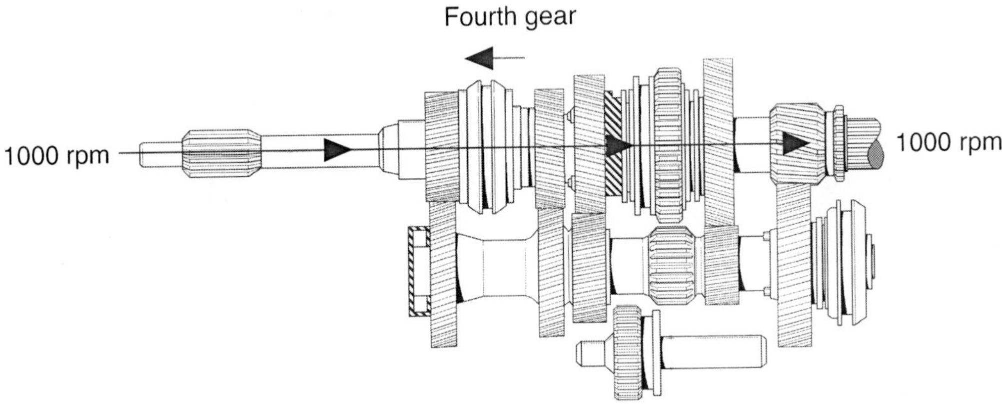


**Figure 15.** Power flow in fourth gear.

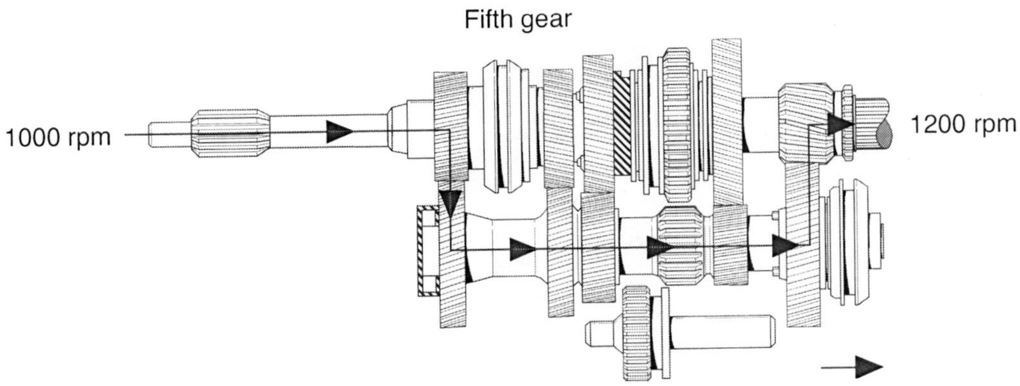


**Figure 16.** Power flow in fifth gear.

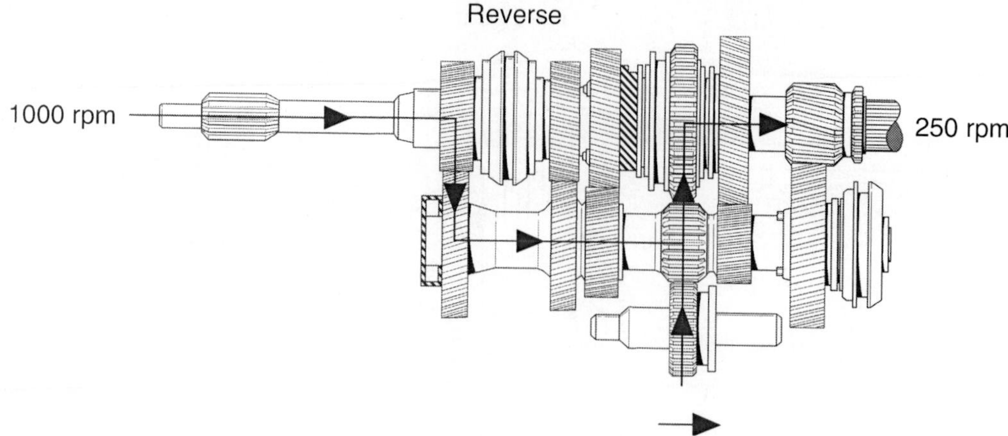


Figure 17. Power flow in reverse.

## Reverse Gear Operation

In reverse gear, all the synchronizers are shifted to the neutral position and a shift fork moves the reverse idler gear so it is in mesh with the matching gear on the counter gear and the gear on the outer diameter of the 1–2 synchronizer hub **(Figure 17)**. This movement allows engine torque to be transmitted from the input shaft gear to the matching gear on the counter shaft. Engine torque is transmitted from the counter shaft to the reverse idler gear. The input shaft rotates clockwise and the counter gear turns counterclockwise. Therefore, the reverse idler rotates clockwise. The reverse idler gear is also in mesh with the reverse gear on the outer diameter of the 1–2 synchronizer sleeve. Clockwise reverse idler rotation turns the 1–2 synchronizer sleeve, hub, and output shaft counterclockwise to drive the vehicle in reverse. For clarity, Figure 17 shows the reverse idler gear out of its normal position. The reverse idler gear would be in mesh with the reverse gear on the counter shaft and the reverse gear on the output shaft.

## TRANSAXLE DESIGN AND OPERATION

A transaxle is a combined transmission and differential that provides a compact assembly that is suitable for front-wheel drive vehicles. Manual transaxles have the same type of gears and synchronizers as manual transmissions. The shifting mechanism may have a selector rod connected from the gear shift lever to the transaxle shift forks **(Figure 18)**. A

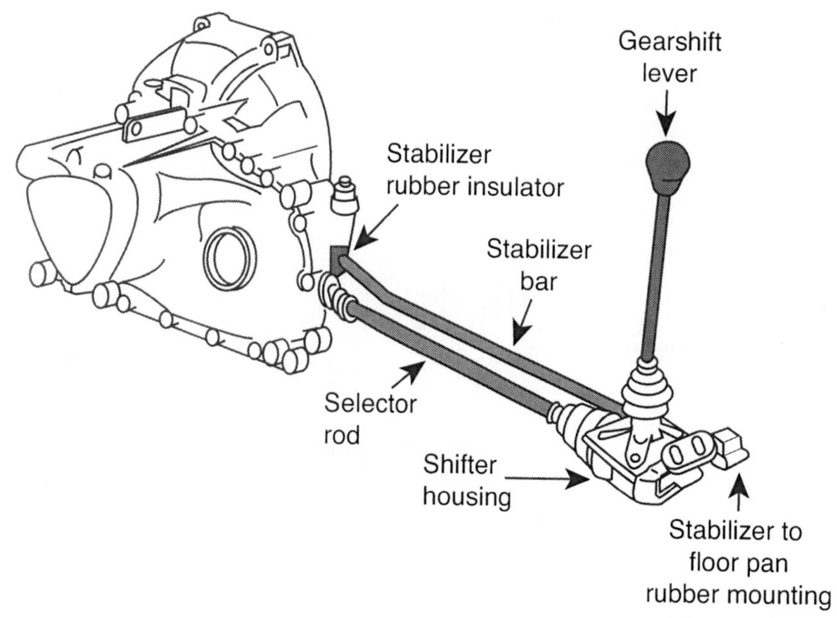


Figure 18. Transaxle gear shift linkage.

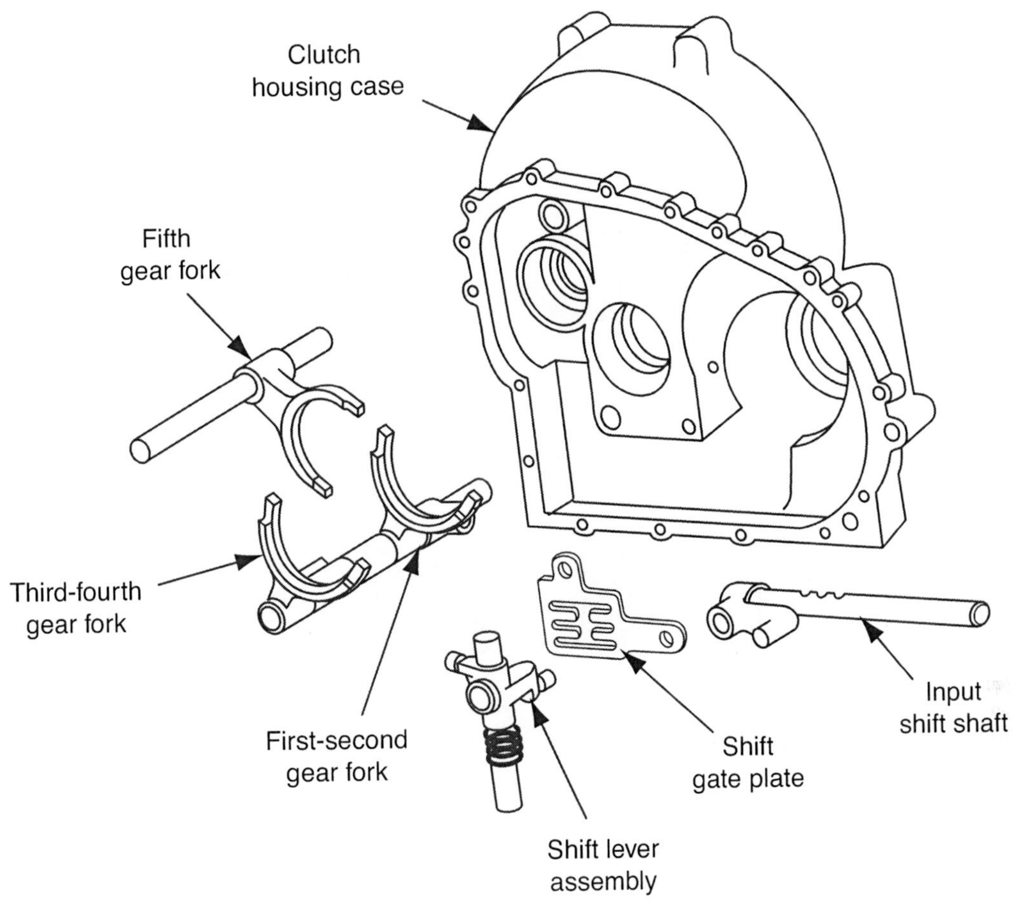


**Figure 19.** Transaxle shift forks.

stabilizer bar may be mounted between the transaxle case and the shifter housing. The shift forks in a manual transaxle are similar to those used in a manual transmission **(Figure 19)**. Some manual transaxles have two cables between the gear shift lever and the transaxle shift forks to provide proper shifting **(Figure 20)**. A manual transaxle usually has an input

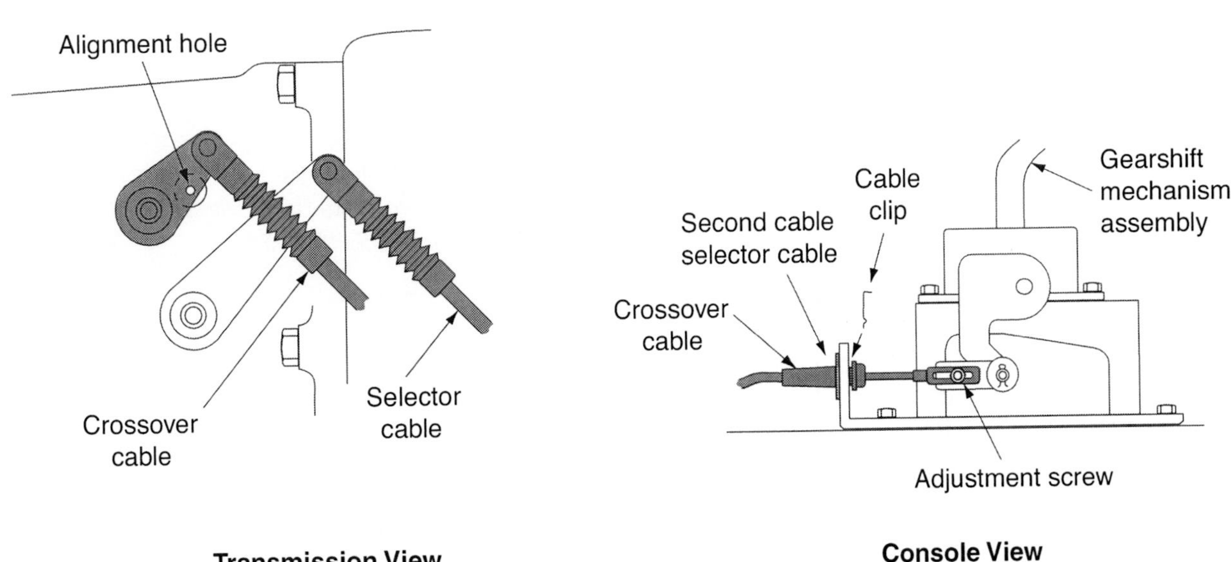


**Figure 20.** Transaxle shift mechanism with dual cables.

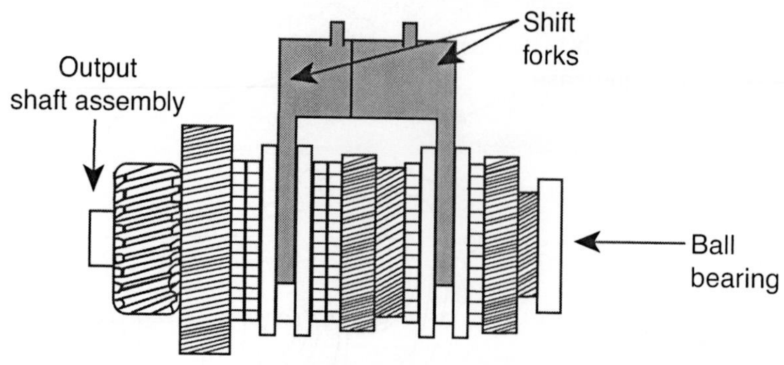


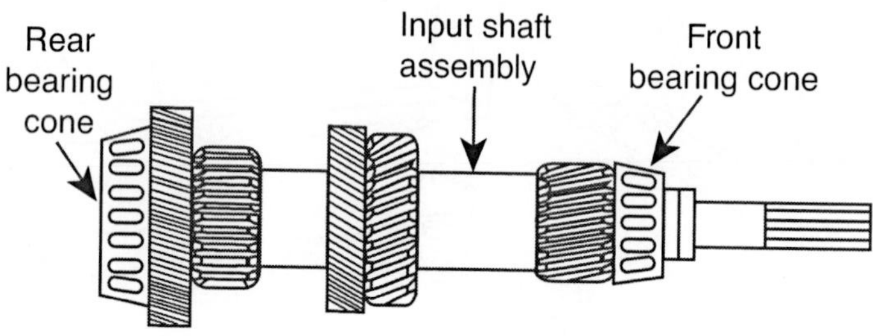


**Figure 21.** Transaxle input and output shafts.

shaft and an output shaft **(Figure 21)**. Gear shifting in various gears is similar in manual transmissions and transaxles.

## MANUAL TRANSMISSION AND TRANSAXLE LUBRICATION

Lubrication in a manual transmission or transaxle is extremely important to protect all the contacting metal parts. Bearing races and rollers must be covered with a thin film of lubricant to prevent bearing wear and overheating. Gear teeth must also be protected by a thin film of lubricant, even when high engine torque is applying extreme pressure to the gears.

Vehicle manufacturers may specify 30W engine oil, automatic transmission fluid (ATF), or gear oil for their manual transmissions or transaxles. A typical gear oil for a manual transaxle is 75W-90. Always use the lubricant specified by the vehicle manufacturer. Most vehicle manufacturers do not recommend changing the manual transmission or transaxle lubricant at specified intervals. Some vehicle manufacturers recommend changing the manual transmission or transaxle lubricant at specified intervals if the vehicle is operated under severe conditions, such as trailer towing.

## MANUAL TRANSMISSION AND TRANSAXLE MAINTENANCE

Manual transmissions or transaxles require a minimum amount of maintenance. When a chassis lubrication is performed, the fluid level in the transmission or transaxle should be checked. When the filler plug is removed, the lubricant should be level with the bottom of the filler plug opening. If necessary, add the manufacturer's specified lubricant to obtain the proper fluid level. Always use the lubricant recommended by the vehicle manufacturer.

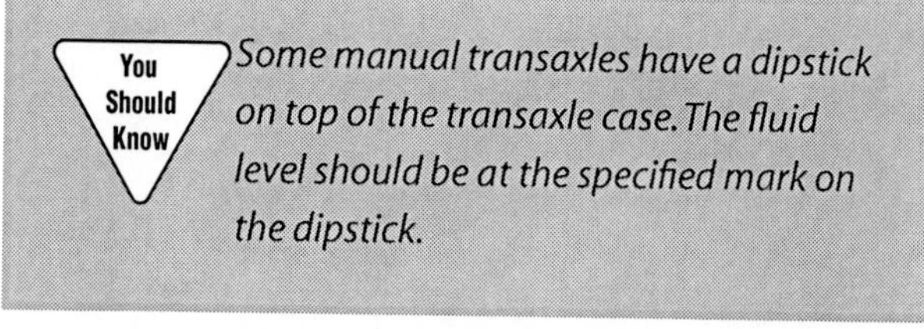


*Some manual transaxles have a dipstick on top of the transaxle case. The fluid level should be at the specified mark on the dipstick.*

Some vehicle manufacturers recommend changing the transaxle lubricant at specified intervals if the vehicle is operated under severe driving conditions such as trailer towing, continual driving in city traffic during extremely hot conditions, or continual driving in mountainous terrain.

During undercar service, inspect the transaxle for leaks at the drive axle seals. Before replacing a drive axle seal, always check the transaxle vent to be sure it is not plugged. A **transaxle vent** is located in the top of the case. When the transaxle components become hot during normal service, the vent allows air to escape from the transaxle to prevent pressure buildup in the transaxle case.

A plugged vent causes pressure buildup in the transaxle that contributes to seal leaks. To replace the drive axle seals, the drive axles must be removed. Always use the proper seal puller and driver to remove and install the seal. When replacing any seal, always inspect the seal lip contact area. If this area is scored, the component must be replaced.

Inspect the lower part of the flywheel housing for leaks. A leak in this area may be caused by a leak at the transaxle input shaft seal or at the rear main bearing seal in the engine. If the transaxle fluid is low and the engine has not been losing any oil, the leak at the flywheel housing is likely at the transaxle input shaft seal.

If the vehicle has a transmission, inspect the output shaft seal for fluid leaks. If a leak is present, always be sure the transmission vent is not restricted before replacing the seal. To replace the output shaft seal, the drive shaft must be removed. Use the proper seal puller and driver to remove and install the seal.

With the engine idling, depress the clutch pedal and check for noise and proper clutch pedal free-play. Some clutch problems may be confused with transaxle problems. For example, hard transaxle shifting may be caused by excessive clutch pedal free-play, worn synchronizers, broken engine/transaxle mounts, or improper shift cable/linkage adjustments. When diagnosing this problem, the clutch pedal free-play, engine/transaxle mounts, and cable/linkage adjustments can be inspected and measured to determine if any of these causes are the source of the problem. If these three causes are eliminated, the hard shifting is likely caused by worn synchronizers. **Synchronizers** bring two components to the same speed to allow shifting without gear clashing.

The adjustment procedure for the shift linkages or cables varies depending on whether the transaxle has shift cables or linkages. This procedure also varies depending on the vehicle make and model year. A typical shift cable adjustment procedure follows:

1. Remove the lock pin from the transaxle selector shaft housing. Reverse the lock pin and install it so the long end of the pin is downward **(Figure 22)**. This action locks the 1–2 shift fork shaft in the neutral position.
2. Remove the gear shift knob and console.
3. Loosen the selector cable and crossover cable adjusting screws and install a 3/16-inch (4.75 millimeter) drill bit in the adjusting pin openings for each cable **(Figure 23)**.

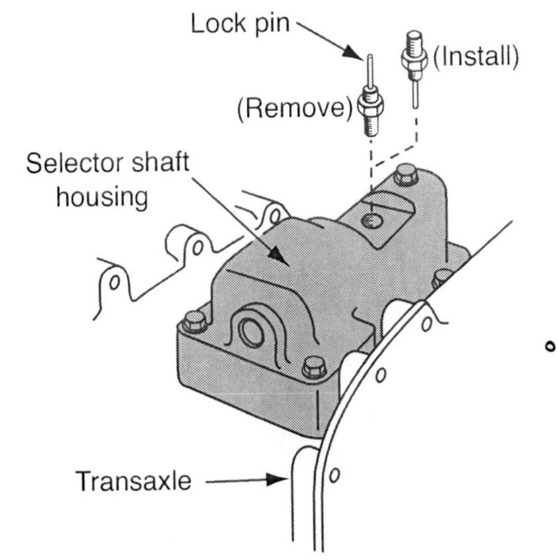


**Figure 22.** Installing a lock pin before shift cable adjustment.

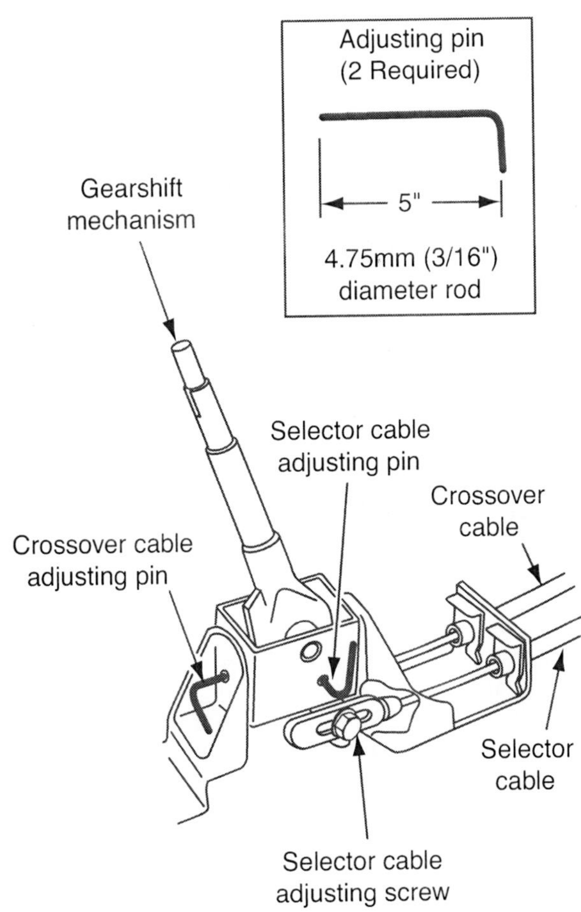


**Figure 23.** Loosening adjusting screws and installing adjusting pins.

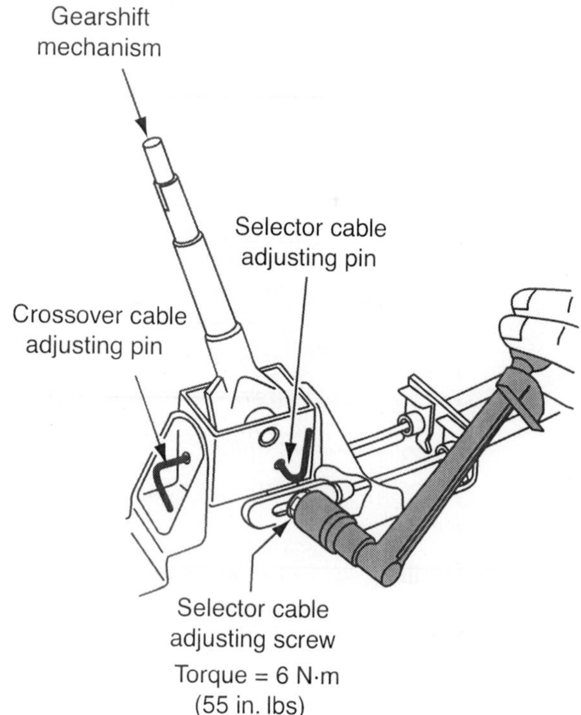


**Figure 24.** Tightening shift cable adjusting screws.

4. Tighten the selector screws on the selector cable and the crossover cable to the specified torque **(Figure 24)**.
5. Remove the adjusting pins, reinstall the lock pin, and install the console and gear shift knob.
6. Start the engine and fully depress the clutch pedal. Shift the gear selector through all the gear positions, and check for proper shifting without gear clashing.
7. Road test the vehicle and check for proper shifting without gear clashing.

## MANUAL TRANSMISSION AND TRANSAXLE DIAGNOSIS

When diagnosing transmission/transaxle problems, it is very important to determine whether the problem is caused by a defect in the transaxle or an external problem. If these problems are not diagnosed accurately, unnecessary and expensive service may be performed. For example, if a hard shifting problem is caused by excessive clutch pedal free-play and the transaxle is overhauled to correct this problem, a great deal of unnecessary, expensive service work has been performed. Before diagnosing transmission/transaxle problems, always check the transmission/transaxle fluid level and condition. Low fluid level or contaminated fluid may cause bearing or gear noises in a transmission/transaxle.

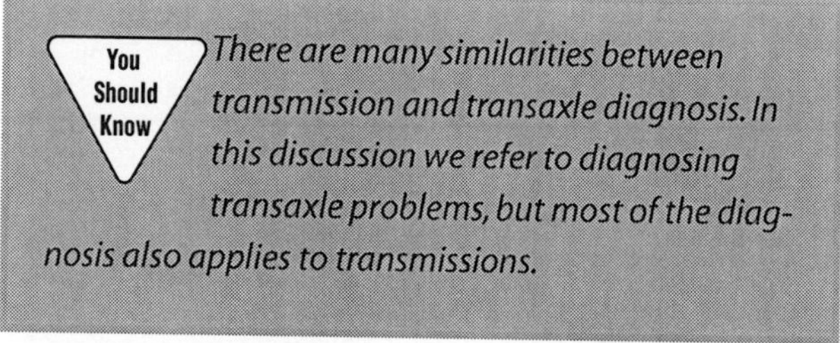


*There are many similarities between transmission and transaxle diagnosis. In this discussion we refer to diagnosing transaxle problems, but most of the diagnosis also applies to transmissions.*

### Transaxle Does Not Shift Into a Certain Gear

If the transaxle does not shift into one gear, the problem could be in the shift linkage. Be sure the shift linkage is properly adjusted and there is no interference between the shift linkage and the floor shift mechanism and the console or other components. This problem could also be caused by the shifter rails in the transaxle. When the transaxle does not shift into one gear, the synchronizer and related gear may be severely worn or damaged.

### Transaxle Jumps Out of Gear

If a transaxle jumps out of gear, the shift linkage may be the cause of the problem. An improperly adjusted or worn shift linkage may not be allowing the synchronizer to shift completely into position. Check the shift cables or linkage for proper adjustment and wear. Be sure there is no interference between any of the shift mechanisms and other components. Worn engine/transaxle mounts may cause improper transaxle position that prevents complete synchronizer movement into a certain gear. A detent, shift rail, or shift fork defect may cause the same problem. Severely worn dog teeth on a gear and a worn synchronizer may cause a transaxle to jump out of gear. Excessive end play on a gear may cause the transaxle to jump out of gear.

### The Transaxle Locks in One Gear

If a transaxle locks in one gear, the shift linkage may be binding on some other component. Internal problems, such as worn shift rails and detents, may cause the transaxle to lock in one gear. Severely worn synchronizer components, such as the sleeve and hub splines or spacers, may cause locking in one gear. This problem may also be caused by a blocking ring seizing onto the gear cone. Low fluid level in the transaxle or contaminated fluid may cause the blocking rings to seize onto the gear cones.

### Transaxle Noise

Always road test the vehicle to be sure the noise is actually in the transaxle. If the transaxle is noisy in one gear only, the problem is likely wear or broken teeth on the gears that are meshed when the noise occurs.

If a rattling noise occurs only with the clutch pedal depressed and the engine running with the transaxle in neutral, the pilot bearing or bushing is worn.

When a growling noise occurs as the clutch pedal is depressed, the clutch release bearing is defective. If the vehicle has a constant-running release bearing, a defective release bearing also causes a growling noise with the clutch pedal released and the transaxle in neutral. If the vehicle does not have a constant-running release bearing and a growling noise only occurs with the clutch pedal released and the engine running in neutral, the input shaft bearing is defective. If a rattling noise is heard under these same conditions, the input shaft gear and matching gear on the counter shaft may be worn or damaged.

**You Should Know** *Some transmissions and transaxles have a certain amount of noise with the clutch pedal released and the engine idling in neutral. This noise is aggravated by improper engine idle speed. Always be sure the engine is idling at the specified speed.*

## Excessive Vibration

Although a vibration may seem to be coming from the transaxle, the vibration is likely caused by something external to the transaxle, such as worn inner CV joints or imbalanced tires and wheels. On a rear-wheel drive vehicle with a transmission, vibration may be caused by worn universal joints or improper drive shaft angles. On either FWD or RWD vehicles, if the pressure plate is not punch-marked and reinstalled on the flywheel in the original position, vibration may occur. Vibration from internal transaxle problems is usually accompanied by noise problems which are explained in the preceding discussion.

# MANUAL TRANSMISSION AND TRANSAXLE SERVICE

After a transmission/transaxle problem is accurately diagnosed, the necessary service must be performed to correct the problem. Because of high labor rates, it is a common practice in the automotive service industry to replace rather than rebuild manual transaxles.

## Transmission/Transaxle Removal

Before attempting to remove a transmission or transaxle, connect a 12-volt supply to the cigarette lighter socket and disconnect the negative battery cable. Place a drain pan under the transaxle and remove the drain plug to drain the fluid. Reinstall the drain plug.

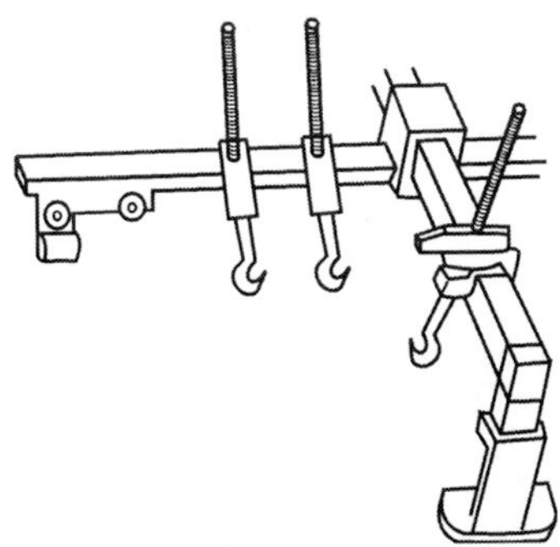

**Figure 25.** An engine support fixture.

Support the engine so it does not drop downward when the transaxle or transmission is removed. On FWD vehicles, a support fixture is placed under the hood and attached to the engine **(Figure 25)**. Raise the vehicle on a lift.

Disconnect all wiring connectors and the speedometer cable from the transaxle. Remove all the linkages from the transaxle. When removing the transmission from a RWD vehicle, place index marks on the drive shaft and differential flange and remove the drive shaft. If you are removing the transaxle from a FWD vehicle, remove the front drive axles from the transaxle. On some FWD vehicles, half of the engine cradle must be removed prior to transaxle removal. Use a transmission jack to support the transmission or transaxle, **Figure 26**. Be sure the unit is positioned securely on the transmission jack and remove the transmission or transaxle-to-engine retaining bolts. Slide the jack and transaxle to remove the input shaft from the clutch and pressure plate. Slowly lower the jack to remove the

**Figure 26.** A transmission/transaxle jack.

transaxle from the vehicle. Place the transaxle in a transaxle stand or on the workbench. Be sure the transaxle is securely bolted to the stand.

*On some FWD vehicles, it is easier to remove the engine and transaxle as an assembly rather than removing the transaxle separately.*

## Transmission/Transaxle Installation

Before installing the transaxle, always shift the transaxle through all the gears and be sure the input shaft rotates freely in all gears. Install a new clutch release bearing before installing the transaxle.

Raise the transaxle with a transmission jack and push the transaxle toward the engine so the input shaft enters the pressure plate and clutch disc. Be sure the transaxle housing is fully seated against the engine block. Do not force the transaxle into the clutch disc because this action may damage the clutch disc or transaxle input shaft or housing. If the transaxle does not slide easily into place, remove it and recheck the input shaft splines. Be sure the clutch disc is aligned with the pilot bearing opening.

After the transaxle is completely seated against the engine, install the flywheel housing-to-engine bolts and tighten these bolts to the specified torque. Remove the transmission jack. Connect all electrical wires and the speedometer cable to the transaxle. Connect the shift linkages to the transaxle. Install the front drive axles into the transaxle and be sure all fasteners are tightened to the specified torque. If part of the engine cradle was removed, install this cradle and tighten all fasteners to the specified torque.

Fill the transaxle with the specified lubricant to the bottom of the filler plug opening or to the full mark on the dipstick. Install the filler plug or dipstick. Remove the engine support fixture and reconnect the negative battery cable. Disconnect the 12-volt power supply from the cigarette lighter socket.

Road test the vehicle and be absolutely sure the original customer complaint is eliminated.

# Summary

- Transmissions and transaxles provide gear reductions to allow smooth vehicle acceleration.
- In fifth or sixth gear, a transmission provides overdrive gear ratios which supply improved fuel economy at cruising speed.
- A transaxle contains the transmission and the differential in one case.
- A gear ratio is calculated by dividing the number of gear teeth on the drive gear into the number of teeth on the driven gear.
- If torque is transmitted through two gear ratios, the total gear ratio is determined by multiplying the two ratios.
- During a shift, the synchronizer blocking ring grooves contact the cone on the selected gear to bring the synchronizer and the selected gear to the same speed.
- Vehicle manufacturers may recommend engine oil, automatic transmission fluid, or gear oil in manual transmissions and transaxles.
- Hard shifting may be caused by worn synchronizers, excessive clutch pedal free-play, broken transaxle mounts, or improper shift linkage adjustment.
- Jumping out of gear may be caused by an improperly adjusted shift linkage, worn engine or transaxle mounts, worn dog teeth on a gear, worn synchronizer, or excessive end play on a gear.
- If a growling noise occurs as the clutch pedal is depressed, the clutch release bearing may be defective.
- When a rattling noise is evident with the clutch pedal depressed, the pilot bearing or bushing may be worn.
- If a growling noise is present only with the clutch pedal released, the input shaft bearing may be defective.
- Before the transaxle or transmission is removed, the engine must be supported so it does not drop downward.
- The transaxle or transmission should be supported on a transmission jack during the removal procedure.

## Review Questions

1. Technician A says the fifth speed in a transmission has an overdrive ratio. Technician B says a gear reduction is provided when a smaller diameter gear is driving a larger diameter gear. Who is correct?
   A. Technician A
   B. Technician B
   C. Both Technician A and Technician B
   D. Neither Technician A nor Technician B
2. Technician A says a gear with straight-cut teeth does not create thrust on the gear. Technician B says backlash is the amount of movement between the teeth on two gears. Who is correct?
   A. Technician A
   B. Technician B
   C. Both Technician A and Technician B
   D. Neither Technician A nor Technician B
3. A second-speed drive gear has fourteen teeth and the second-speed driven gear has thirty-six teeth. The second-speed gear ratio is:
   A. 2.1:1.
   B. 2.5:1.
   C. 2.8:1.
   D. 3.2:1.
4. Technician A says a synchronizer sleeve is mounted on splines on the synchronizer hub. Technician B says a synchronizer hub is mounted on splines on the transmission output shaft. Who is correct?
   A. Technician A
   B. Technician B
   C. Both Technician A and Technician B
   D. Neither Technician A nor Technician B
5. A gear set has nine teeth on the drive gear and thirty-seven teeth on the driven gear. The gear ratio is:
   A. 3.6:1.
   B. 3.9:1.
   C. 4.1:1.
   D. 4.3:1.
6. All of these statements about gears and gear sets are true *except*:
   A. Helical gear teeth are cut at an angle in relation to the gear centerline.
   B. In an overdrive gear set, the driven gear is larger than the drive gear.
   C. Backlash is the amount of movement between the teeth on two gears.
   D. Helical gear teeth create axial thrust on the gear.
7. When shifting gears that have a synchronizer, Technician A says during the shift, the grooved blocking ring cone cuts through the lubricant on the selected gear cone to allow metal-to-metal contact. Technician B says the synchronizer action brings the selected gear to the same speed as the synchronizer hub. Who is correct?
   A. Technician A
   B. Technician B
   C. Both Technician A and Technician B
   D. Neither Technician A nor Technician B
8. All of these problems may cause a manual transaxle to lock in one gear *except*:
   A. a blocking ring seized onto a gear cone.
   B. worn shift rail detents.
   C. severely worn synchronizer sleeve hub splines.
   D. excessive clutch pedal free-play.
9. A vehicle with a manual transaxle has a growling noise only when the clutch pedal is released and the transmission is in neutral. The most likely cause of this problem is:
   A. a defective transaxle input shaft bearing.
   B. a rough clutch release bearing.
   C. a worn pilot bushing.
   D. worn splines on the input shaft and clutch hub.
10. A four-speed manual transmission is very hard to shift into second gear. This problem may be caused by all of these defects *except*:
    A. an improperly adjusted shift linkage.
    B. severely worn 1–2 synchronizer and second-speed gear.
    C. insufficient clutch pedal free-play.
    D. worn or damaged shifter rails.
11. An overdrive gear ratio ________________ torque and ________________ output shaft speed.
12. If torque is transmitted through two gear ratios of 3.5:1 and 4.08:1, the total gear ratio is ________________.
13. In a three-speed transaxle, the third-speed gear ratio is ________________.
14. List the possible lubricants for manual transmissions/transaxles.
    1. ________________
    2. ________________
    3. ________________
15. Explain the purpose of a synchronizer.
16. Describe the operation of a synchronizer.
17. Explain the causes of hard transaxle shifting.
18. Describe the causes of jumping out of gear.

# Chapter 43 Automatic Transmission and Transaxle Maintenance, Diagnosis, and Service

## Introduction

Transmissions and transaxles have become complex mechanical devices controlled by hydraulics and electronics. It is essential for technicians to be familiar with the mechanical principles of multiple disc clutches, bands, and planetary gear sets to gain an understanding of automatic transmissions. Technicians must also be familiar with torque converters. Technicians need to be familiar with the electronic controls in today's transaxles and transmissions. When these mechanical, hydraulic, and electronic principles are clearly understood, diagnosing and servicing transaxle and transmission problems becomes much easier.

### TORQUE CONVERTER PURPOSE AND DESIGN

A torque converter contains three main components: the impeller pump, the turbine, and the stator. The blades on the impeller and the turbine are curved in opposite directions. The converter cover is welded to the impeller pump to seal these members in a housing. The converter cover is bolted to the engine flexplate which is bolted to the crankshaft **(Figure 1)**. When the engine is running, the converter cover and impeller pump must rotate with the crankshaft. A hub on the rear of the impeller has two notches or flats that fit into matching grooves in the inner member of the transmission pump. Therefore, the converter drives the transmission pump. The pump seal lips contact a smooth machined surface on the outer diameter of the converter hub. The impeller pump hub is supported on a bushing in the transaxle pump.

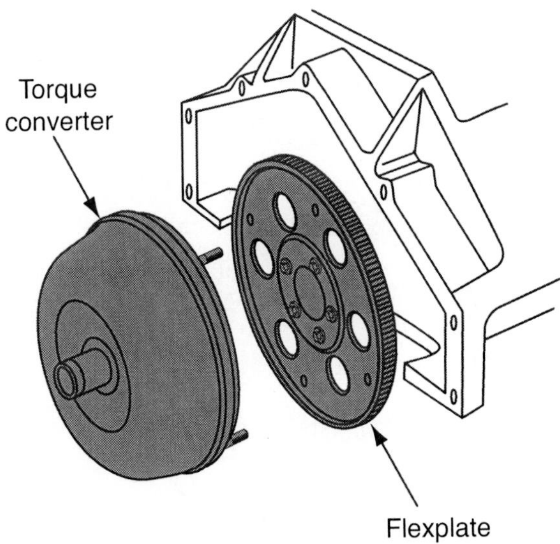


**Figure 1.** The torque converter is bolted to the flexplate.

The turbine is splined to the transaxle input shaft and the stator is connected to the reaction shaft through an overrunning clutch. The reaction shaft extends forward from the transmission pump through the converter hub into the stator. The reaction shaft is stationary and cannot

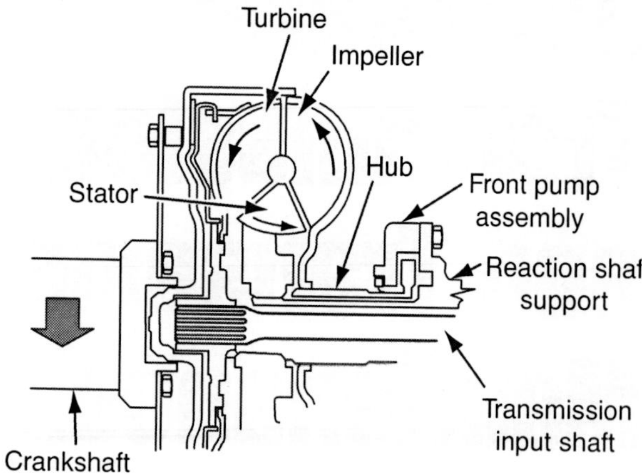


**Figure 2.** Torque converter components.

rotate **(Figure 2)**. The stator hub is splined to the reaction shaft. The stator overrunning clutch contains a series of spring-loaded rollers in tapered grooves. This overrunning clutch allows the stator to turn freely in one direction and lock up in the opposite direction **(Figure 3)**. In a typical

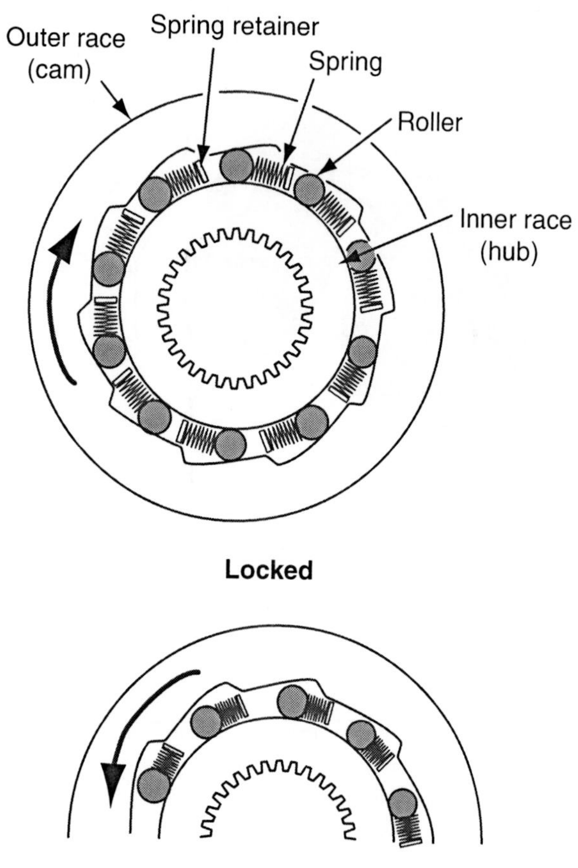


**Figure 3.** The overrunning clutch allows the stator to turn freely in one direction and lock up in the opposite direction.

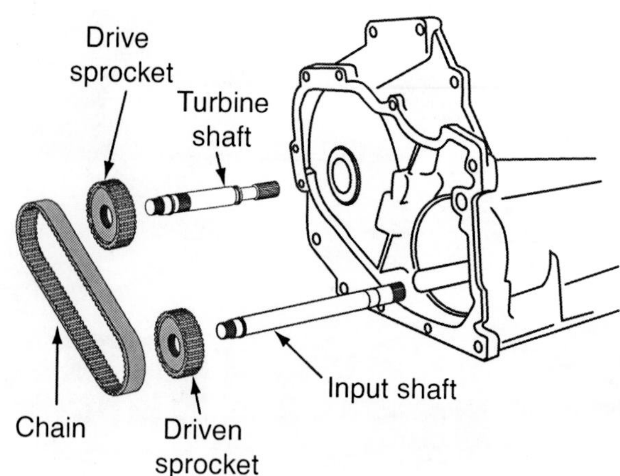


**Figure 4.** In a typical transaxle engine, torque is transmitted through the input shaft drive sprocket, chain, and driven sprocket.

transaxle, the front end of the input shaft is splined to the turbine and the rear end of this shaft is splined to a drive sprocket. The engine torque is transmitted from the input shaft and drive sprocket through a chain and driven sprocket into the transaxle **(Figure 4)**.

In recent years, most torque converters are lockup type. Many lockup converters contain a lockup disc between the turbine and the cover. The lockup disc is splined to a short extension on the front of the turbine and a friction material about 1 inch (25.4 millimeters) wide is molded onto the front of the lockup disc near the outer diameter of the disc **(Figure 5)**. The lockup disc is free to move a short distance on the turbine splines. Torsional

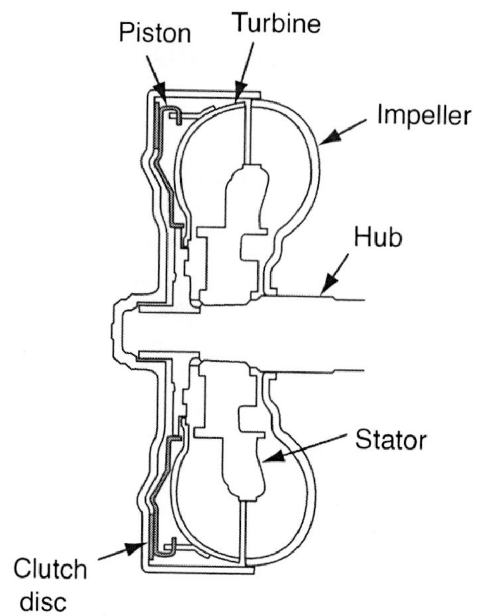


**Figure 5.** A lockup torque converter.

springs are mounted in the lockup disc between the outer part of the disc and the hub. Some lockup discs have a silicone clutch between the outer part of the disc and the hub. The silicone clutch provides smoother torque converter clutch (TCC) engagement compared to a TCC with torsional springs.

## TORQUE CONVERTER OPERATION

The torque converter contains ATF. When the engine is running, the impeller picks up ATF at its center and centrifugal force causes this fluid to be discharged between the blades at the impeller rim **(Figure 6)**. As the oil strikes the turbine blades, the turbine is forced to rotate. If the engine is idling at the specified speed, the force of the fluid striking the turbine blades is not high enough to rotate the turbine. When the engine speed is increased slightly, more rotational force is exerted on the turbine blades and the turbine begins to rotate in the same direction as the impeller pump. The impeller pump and the turbine rotate in a clockwise direction. Engine torque is now transmitted through the turbine and input shaft to the transaxle gear train, drive axles, and drive wheels.

At lower vehicle speeds, the fluid movement from the impeller pump to the turbine is redirected by the stator blades so this fluid helps to rotate the turbine **(Figure 7)**. Under this condition, the overrunning clutch prevents stator rotation. If the stator did not perform this function, this fluid movement would work against the turbine rotation

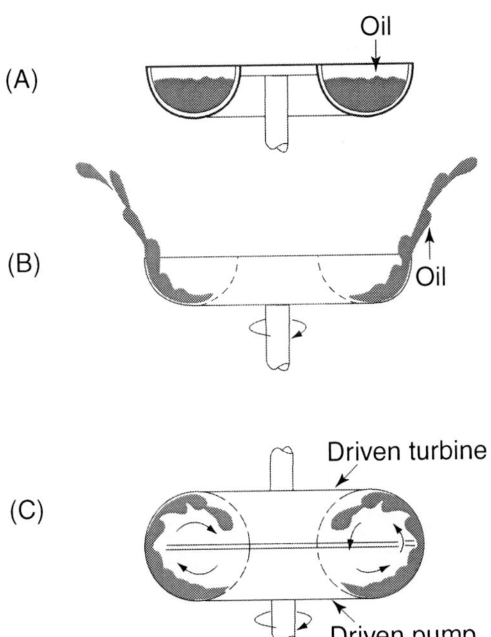


**Figure 6.** (A) Fluid in the torque converter at rest, (B) Fluid discharged from the impeller pump vanes, (C) Fluid flowing from the turbine back to the impeller pump.

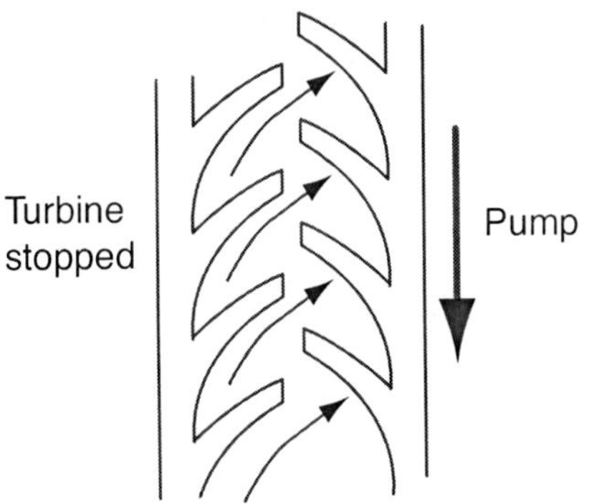


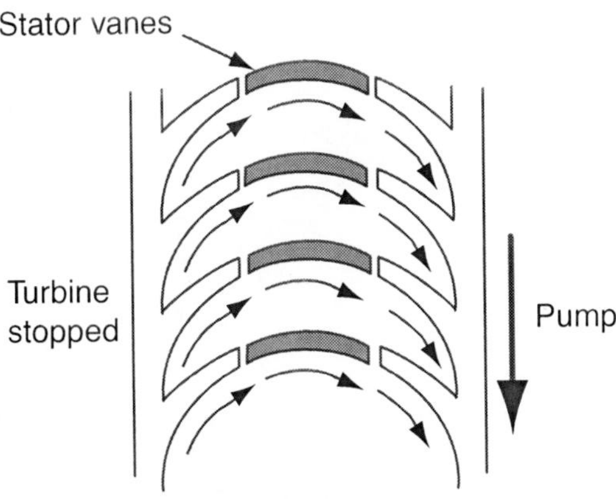


**Figure 7.** The stator redirects the fluid flowing from the turbine back into the impeller pump.

and reduce engine power and torque. The fluid follows the contour of the turbine blades so it leaves the turbine in the opposite direction to turbine rotation. Since the direction of this fluid motion is opposite the engine and impeller pump rotation, this fluid leaving the turbine and re-entering the impeller pump would work against the engine and impeller rotation to reduce engine power and torque. The term **vortex flow** is used to describe the fluid movement in the converter when the stator is redirecting the fluid movement from the impeller pump to the turbine.

When the engine and vehicle speed increase, the speed of the turbine approaches the speed of the impeller pump. This condition is called the **coupling point**. When the impeller pump and turbine reach the coupling point, the fluid returning from the turbine into the impeller pump begins striking the back of the stator blades. Under this condition, the overrunning clutch in the stator allows the stator to rotate freely with the impeller pump and turbine. The impeller pump, turbine, and stator are now rotating as a unit, the converter is no longer multiplying torque. When

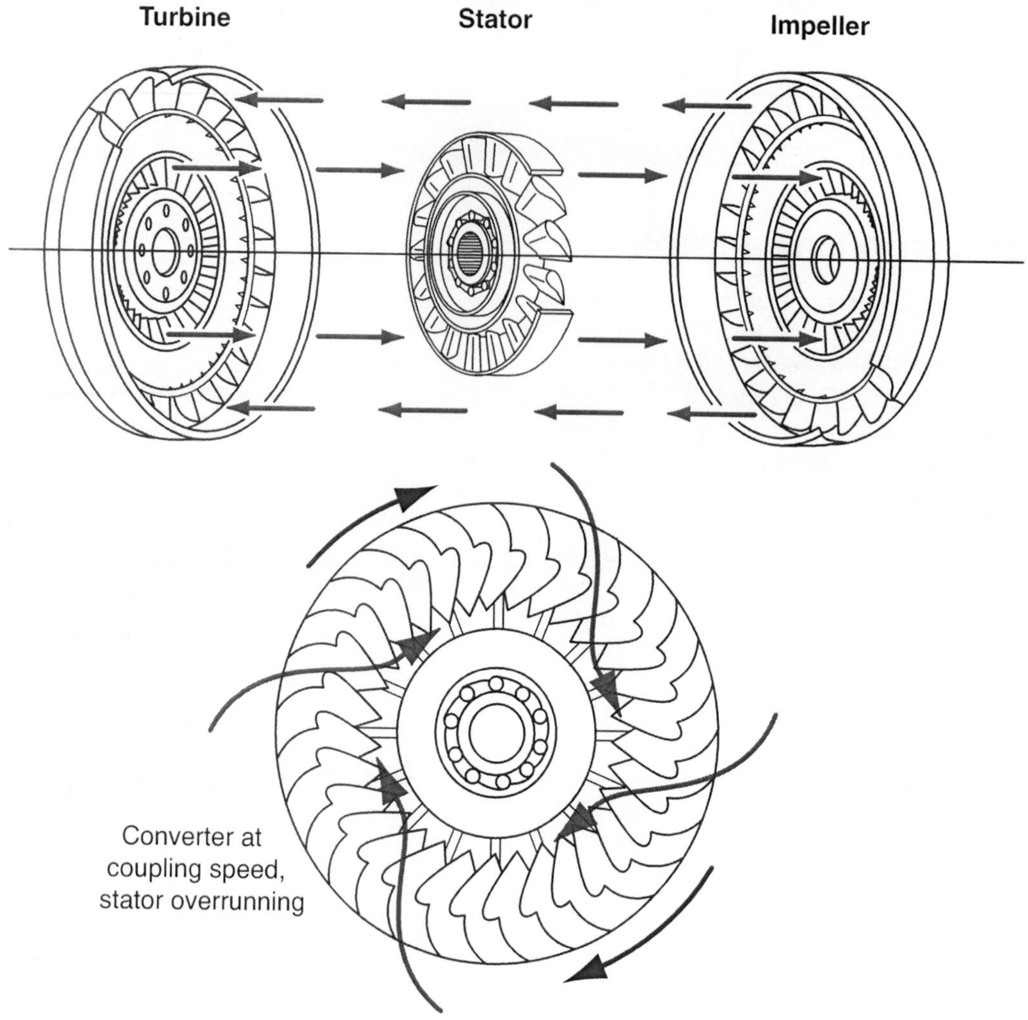


**Figure 8.** Converter fluid flow in the coupling phase with the stator overrunning.

the torque converter reaches the coupling point, the fluid movement in the converter is changed from vortex flow to **rotary flow (Figure 8).**

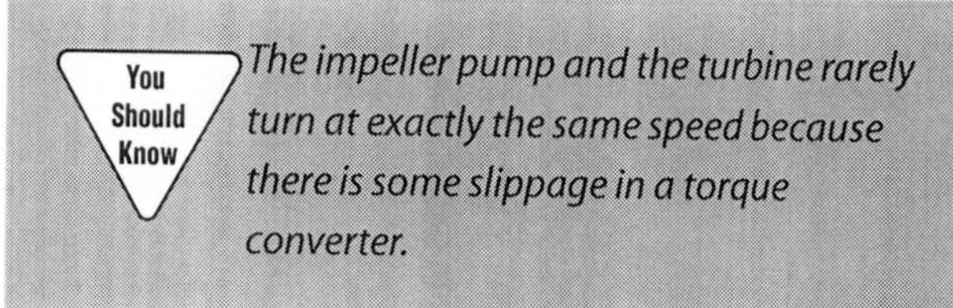


*The impeller pump and the turbine rarely turn at exactly the same speed because there is some slippage in a torque converter.*

The converter clutch lockup system is electronically operated by the PCM. When the vehicle is driven at lower speeds, the torque converter clutch is not locked up. Under this condition, fluid is directed into the converter so it flows through the hollow transmission input shaft and out in front of the lockup disc. Under this condition, the fluid forces the lockup disc away from the front of the converter so the friction material on the lockup disc does not contact the front of the converter. The torque converter clutch is always unlocked when the engine coolant is cold.

When the engine reaches the proper coolant temperature and the correct vehicle speed, the PCM operates the lockup valve in the transaxle valve body. Movement of the lockup valve and switch valve now direct fluid into rear of the torque converter hub **(Figure 9)**. Fluid is forced over the outer diameter of the turbine and in behind the lockup plate. This fluid movement forces the lockup plate against the front of the converter. This action connects the flexplate and the front of the converter directly through the lockup disc to the turbine hub and input shaft. Because the impeller pump is also bolted to the flexplate, the impeller pump and the turbine must rotate at the same speed and the converter is locked up.

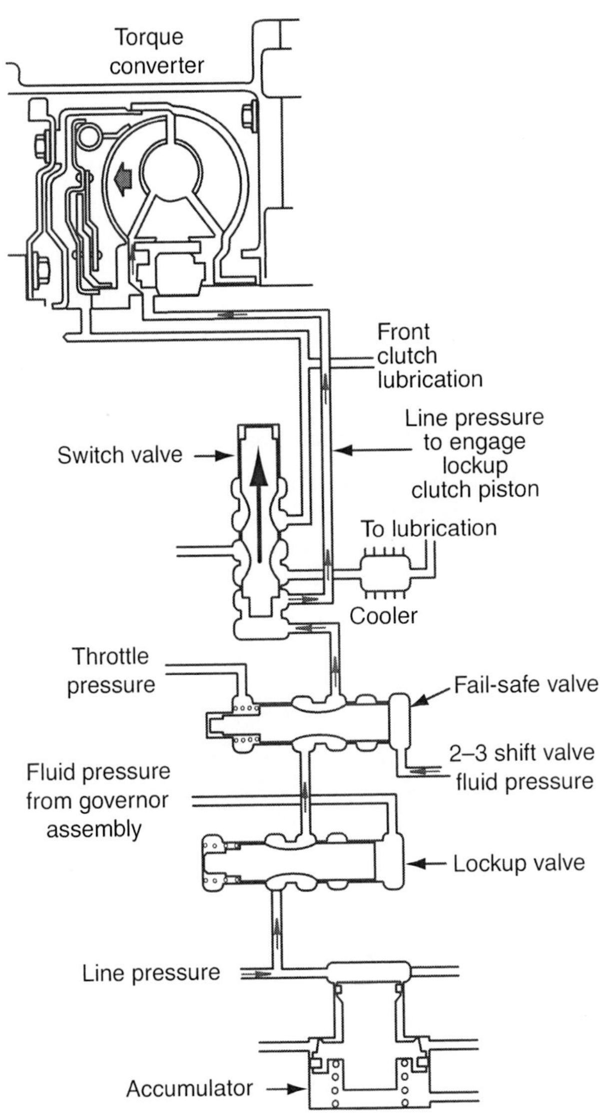


**Figure 9.** Hydraulic system for torque converter clutch lockup.

## PLANETARY GEARSET DESIGN AND OPERATION

Most automatic transmissions use planetary gearsets to provide forward gear reductions and reverse. Each planetary gearset contains an outer ring gear, a planet carrier containing planetary pinions, and a sun gear in the center **(Figure 10)**. Planetary gearsets are compact, strong, and quiet running.

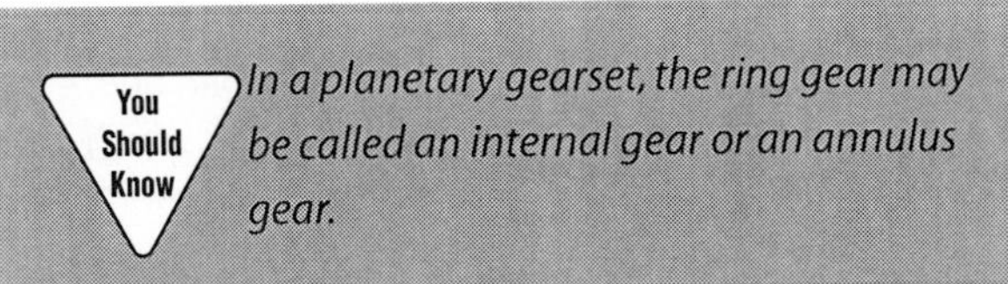


*In a planetary gearset, the ring gear may be called an internal gear or an annulus gear.*

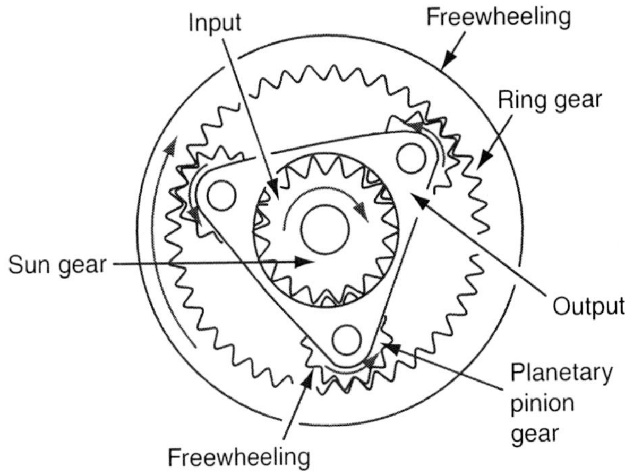


**Figure 10.** Planetary gear set in neutral.

In a planetary gearset, the planet carrier is the largest gear, the ring gear is the second largest, and the sun gear is the smallest gear. If a planetary gearset has an input but no member of the gear set is held stationary, the gear set is in neutral. Under this condition, no torque is transmitted from the input shaft. To provide gear reductions, one member or the planetary gearset must be held or locked and the input must be supplied to one of the other members. Various planetary gearset members are held by one-way clutches, groups of clutch plates, or bands.

When the sun gear is held and the ring gear is the input with the planet carrier is the output, a forward gear reduction is provided because the smaller ring gear is driving the larger planet carrier **(Figure 11)**. To calculate the gear reduction, use this formula:

$$\frac{\text{Drive gear teeth} + \text{Driven gear teeth}}{\text{Drive gear teeth}}$$

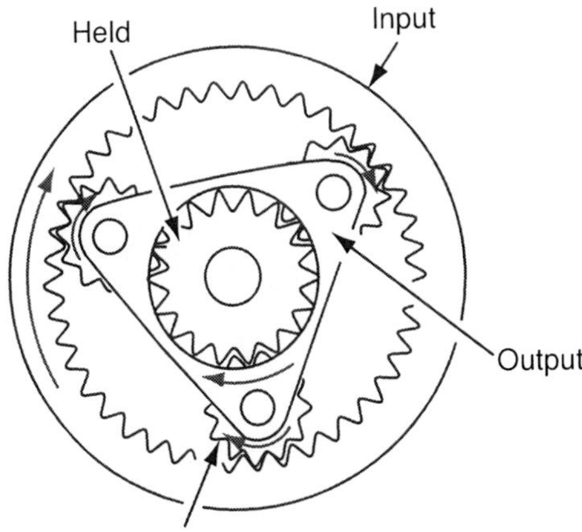


**Figure 11.** A forward gear reduction is provided when the ring gear is the input, the sun gear is held, and the carrier is the output.

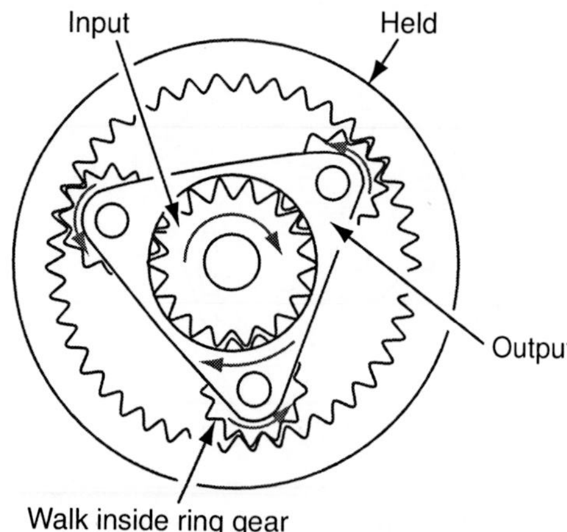


**Figure 12.** A forward gear reduction is provided when the sun gear is the input, the ring gear is held, and the carrier is the output.

Different formulas are required to calculate the planetary gear ratios in overdrive or reverse.

When the sun gear is the input, the ring gear is held, and the planet carrier is the output, a forward gear reduction is provided because the smaller sun gear is driving the much larger planet carrier **(Figure 12)**. Under this condition, a typical gear reduction is 3.3:1 which may be suitable for low gear.

If the planet carrier is the input, the sun gear is held, and the ring gear is the output, an overdrive gear ratio is provided because the larger planet carrier is driving the smaller ring gear **(Figure 13)**.

In a planetary gearset, a direct drive with a 1:1 gear ratio is provided by locking two members together. If both the ring gear and sun gears are inputs, these two members are effectively locked together and the planet carrier is the output **(Figure 14)**. The complete planetary gear set must turn together.

When the sun gear is the input, planet carrier is held, and the ring gear is the output, a reverse gear reduction is provided because the smaller sun gear is driving the larger ring gear **(Figure 15)**. Because the planet carrier is held, the ring gear must turn in the opposite direction to the sun gear rotation.

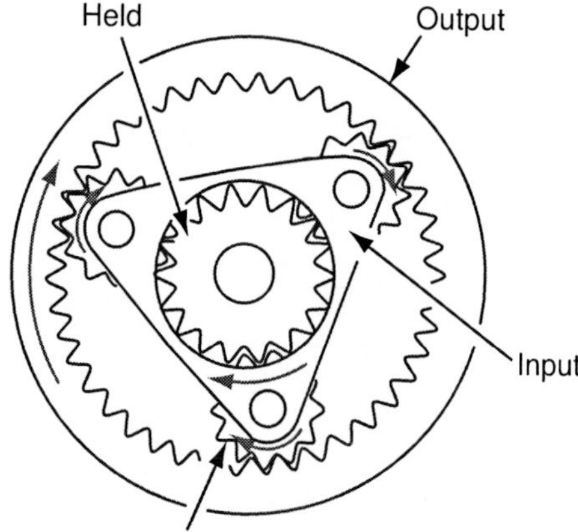


**Figure 13.** A forward overdrive gear ratio is provided if the carrier is the input, the sun gear is held, and the ring gear is the output.

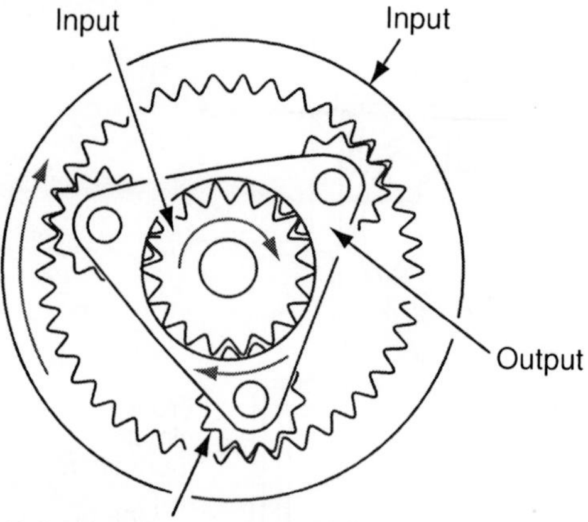


**Figure 14.** Direct drive is provided when two planetary members are locked together or two members are inputs.

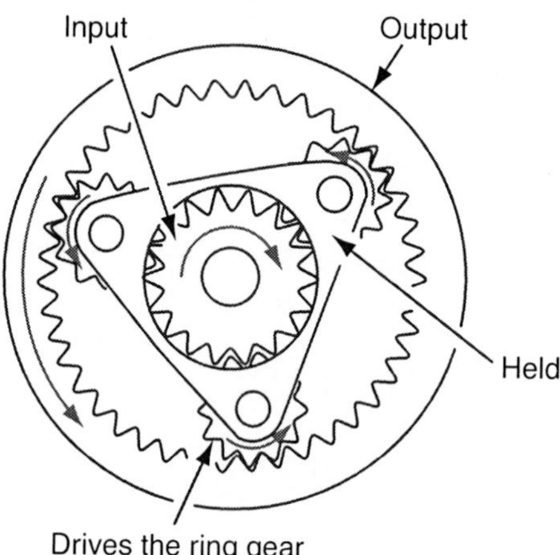


**Figure 15.** A reverse gear reduction is provided when the sun gear is the input, the carrier is held, and the ring gear is the output.

## MULTIPLE-DISC CLUTCHES

Automatic transaxles contain several multiple-disc clutches that apply and release the members in a planetary gear set. A multiple-disc clutch assembly contains a hub, driving discs, driven plates, apply piston, seals, pressure plate, release springs, and clutch assembly container. These components are held in a clutch drum by a snap ring. A friction material is bonded to the driving discs which have internal serrations mounted on the hub splines. The hub that fits inside the driving discs is from another member in the transaxle. The sides of the driven discs do not have any friction material and these plates have external tangs

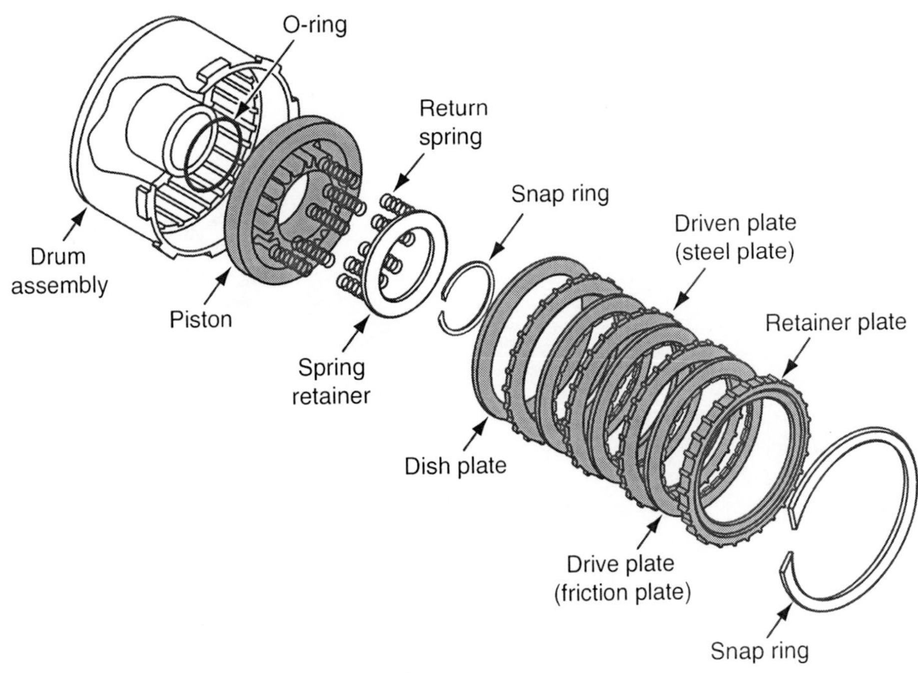


**Figure 16.** A typical multiple-disc clutch.

mounted in the clutch drum slots **(Figure 16)**. The driving and driven discs are placed alternately in the clutch drum. The spring-loaded apply piston has internal and external seals between the piston and the clutch drum hub.

When fluid pressure is not supplied to the apply piston, the drive and driven plates are free to turn within each other. If fluid pressure is applied behind the apply piston, the piston is moved against the spring pressure. This piston movement forces the clutch discs together and the hub is connected through the driving and driven clutch discs to the clutch drum.

## BANDS AND ONE-WAY CLUTCHES

Bands are used in automatic transmissions and transaxles to lock one member in a planetary gear set. The bands are made from flexible steel with friction material on the inner surface. The band is usually mounted over the outer surface of a clutch drum. One end of the band is anchored in the transaxle case and the other end is connected to a servo piston stem or a strut and apply lever **(Figure 17)**. A servo piston is in contact with the servo piston stem. The servo piston is sealed in a machined bore and

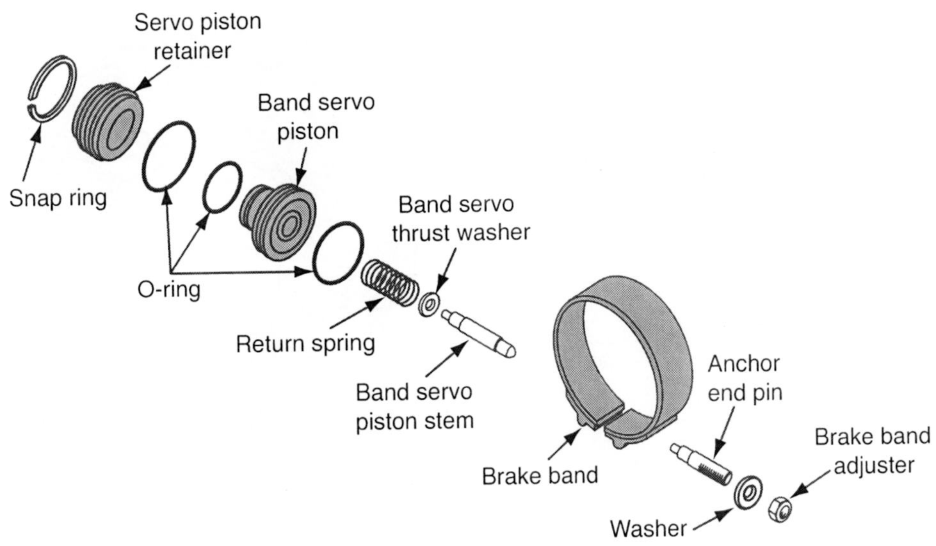


**Figure 17.** A band and servo assembly.

is retained in the bore by a retainer and snap ring. A spring is positioned between the servo piston and the end of the bore next to the band. When fluid pressure is not supplied to the servo piston, the clutch drum is free to rotate inside the band. When fluid pressure is supplied to the servo piston, this piston moves toward the band. This piston movement pushes the servo piston stem against the band and this action tightens the band on the clutch drum, effectively locking the drum in place.

Some members in an automatic transaxle are mounted on a one-way clutch. The one-way clutch acts as a locking device to prevent rotation in one direction but allows rotation in the opposite direction. The one-way clutch contains a hub and an outer race. The hub is splined to one of the transmission members. A series of tapered grooves are cut in the inner surface of the outer race. Spring-loaded steel rollers are mounted in these tapered grooves. If the hub is rotated counterclockwise, the rollers move to the narrow part of the tapered grooves **(Figure 18)**. Under this condition, the rollers jamb between the hub and the outer race. This action locks these two units and prevents hub rotation in a counterclockwise direction. If the hub is turned in a clockwise direction, the rollers move against the spring tension into the wider part of the tapered grooves. Under this condition, the hub can rotate freely.

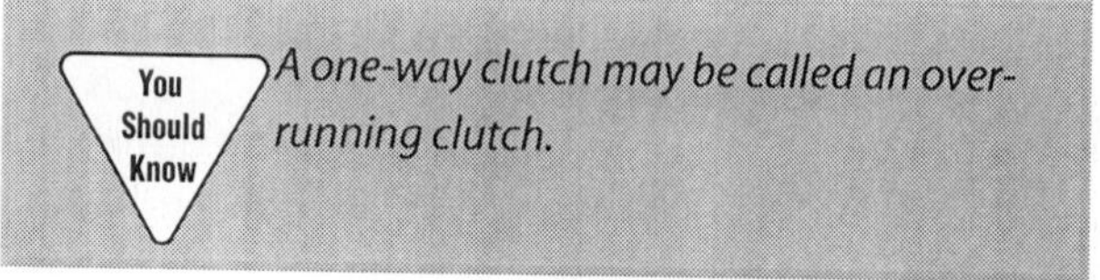


*A one-way clutch may be called an overrunning clutch.*

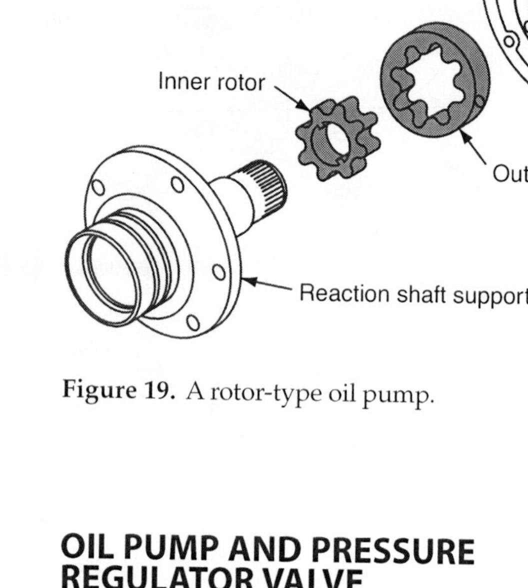

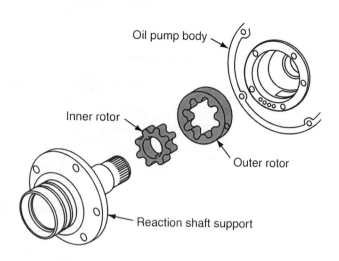

**Figure 19.** A rotor-type oil pump.

## OIL PUMP AND PRESSURE REGULATOR VALVE

Fluid pressure to operate the various clutch packs and bands in an automatic transmission or transaxle is supplied from the oil pump. On many transmissions and transaxles, the oil pump is driven by the torque converter hub. Two grooves on the converter hub fit into matching notches in the oil pump inner rotor **(Figure 19)**. When the engine starts, the oil pump builds up pressure immediately. Many transaxles have a gear-type pump **(Figure 20)**, but some transaxles have a vane-type pump **(Figure 21)**. Some

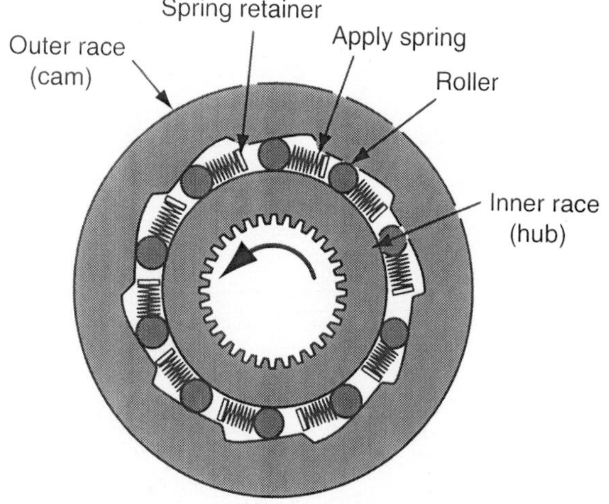


**Figure 18.** An overrunning clutch.

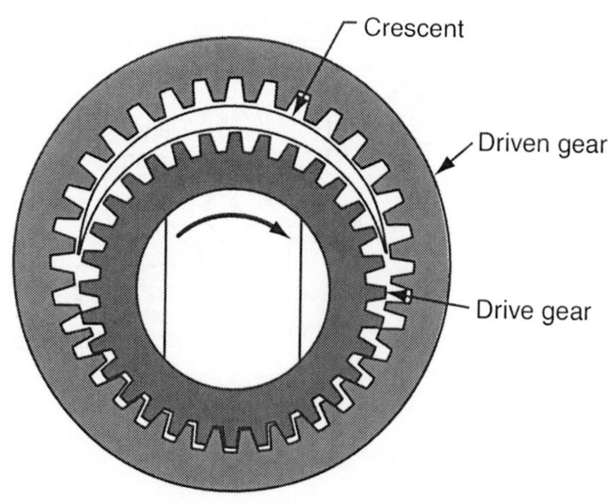


**Figure 20.** A gear-type pump.

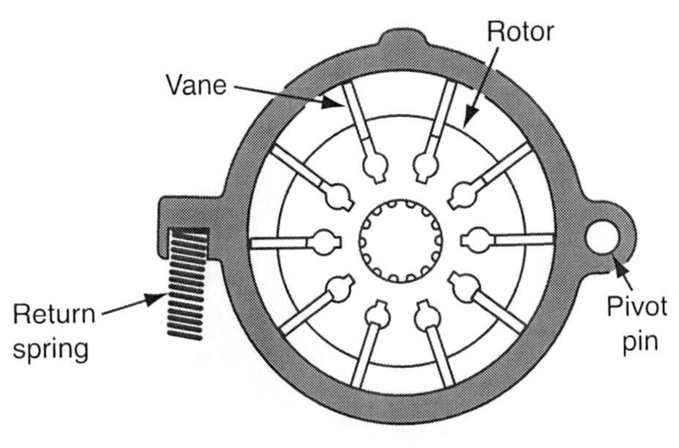


**Figure 21.** A vane-type pump.

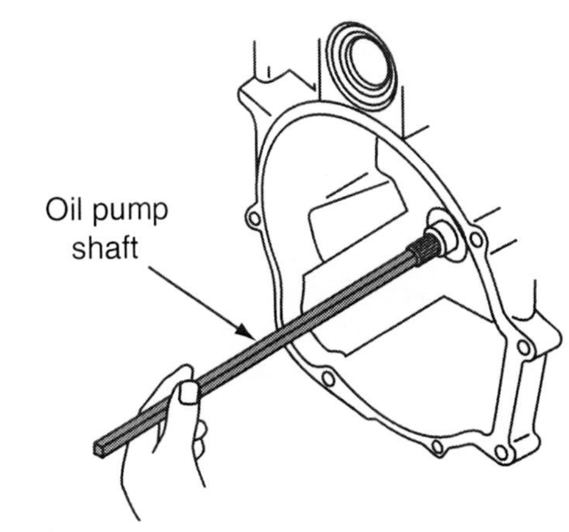


**Figure 22.** An oil pump drive shaft.

transaxles have a drive shaft extending from the front of the converter through a hollow transaxle input shaft to the oil pump hub **(Figure 22)**.

The oil pump takes in oil from the transaxle reservoir through the filter and delivers oil through the pressure regulator valve to the transaxle components **(Figure 23)**. The

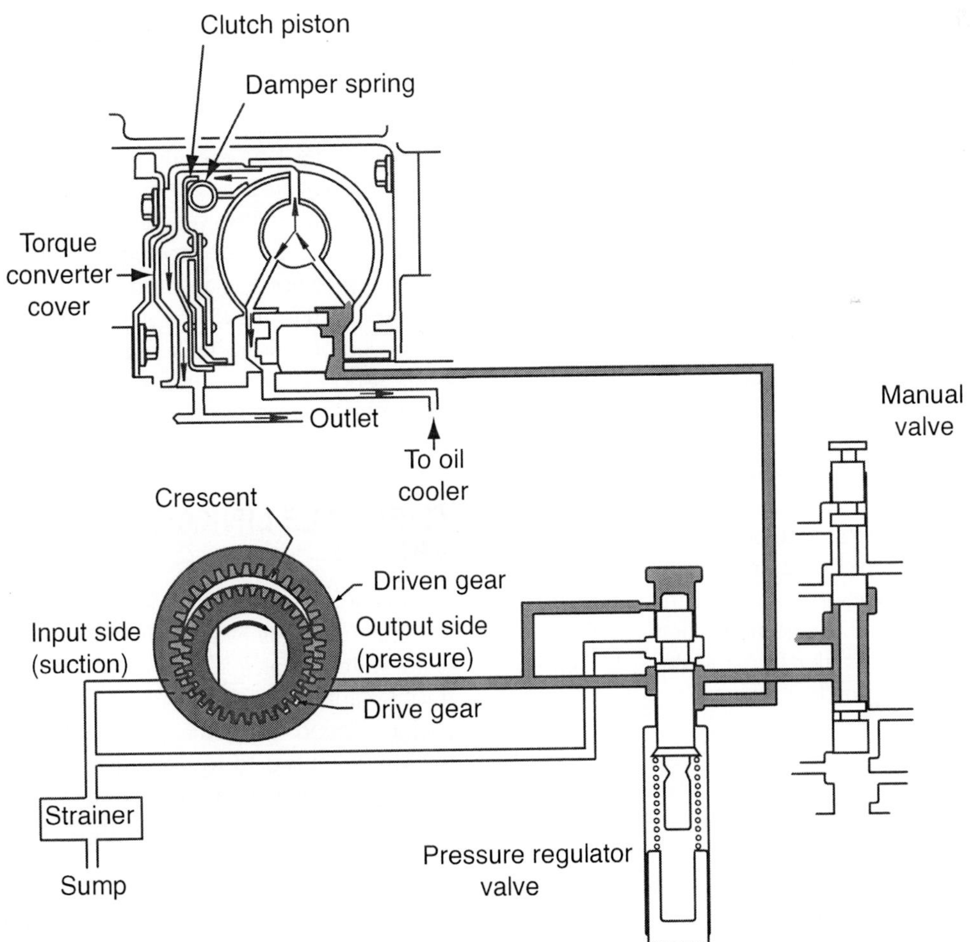


**Figure 23.** An oil pump with pressure regulator valve and manual valve.

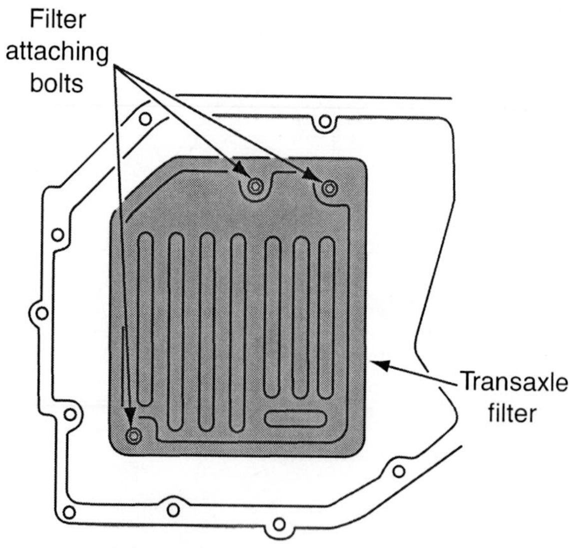


Figure 24. A transmission filter.

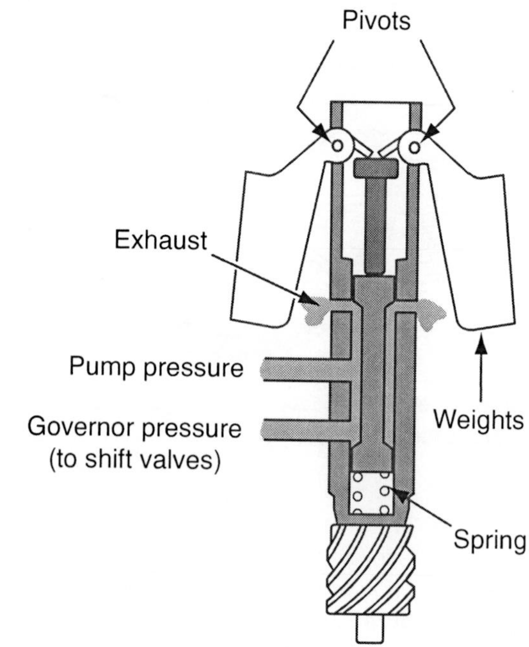


Figure 25. Governor operation.

filter is usually positioned near the bottom of the reservoir **(Figure 24)**. The pressure regulator valve regulates fluid pressure by returning some fluid to the reservoir. When the pump pressure increases, the fluid pressure forces the pressure regulator valve against the spring tension. This valve movement allows more fluid to return to the reservoir. This pressure regulator valve movement controls line pressure. The fluid pressure delivered from the pressure regulator valve is called **line pressure** or mainline pressure.

Line pressure is delivered from the pressure regulator valve to the manual valve. The manual valve is operated by the gear shift linkage. In the various gear selector positions, the manual valve delivers line pressure to the other transaxle components. Fluid pressure must be high enough to apply clutches and bands without slipping. However, the fluid pressure must be limited to prevent harsh shifting and damage to clutch pack seals. Fluid is also delivered from the pressure regulator valve to the torque converter. The return fluid from the torque converter flows through the oil cooler and the lubrication system. The lubrication system provides lubrication to many transaxle components, such as bushings and bearings.

## GOVERNORS

Governors are used in transaxles and transmissions that are not computer controlled. The governor is gear driven, usually from the output shaft. The governor has a pivoted primary and secondary weights. The primary weight flies outward at low speeds and the lighter secondary weights move outward at higher speeds. As the governor weights fly outward, they move a spool valve in the governor bore. Line pressure is supplied to the governor and governor pressure is delivered from the governor to one end of the shift valves. At low vehicle speeds, the governor valve is positioned so some of the fluid is exhausted at the governor exhaust port and governor pressure remains low **(Figure 25)**. When the vehicle speed increases, the governor weights fly outward and gradually close the exhaust port allowing governor pressure to increase. Governor pressure can reach a point where it is equal to line pressure, but it cannot exceed the line pressure.

## THROTTLE LINKAGES AND THROTTLE VALVES

Many transmissions have a throttle cable connected from the throttle linkage at the throttle body or carburetor to the throttle valve in the transaxle or transmission. The throttle cable moves the TV plunger against the spring tension in relation to throttle opening. This increased spring tension is applied to the throttle valve **(Figure 26)**. Line pressure is supplied from the pressure regulator and manual valves to the throttle valve. As the throttle is opened, the increasing pressure on the throttle valve moves this valve so the line pressure gradually flows past the valve lands into the throttle valve pressure passage to increase throttle pressure. This increasing throttle pressure is supplied to the opposite end of the shift valves to which the governor pressure is supplied.

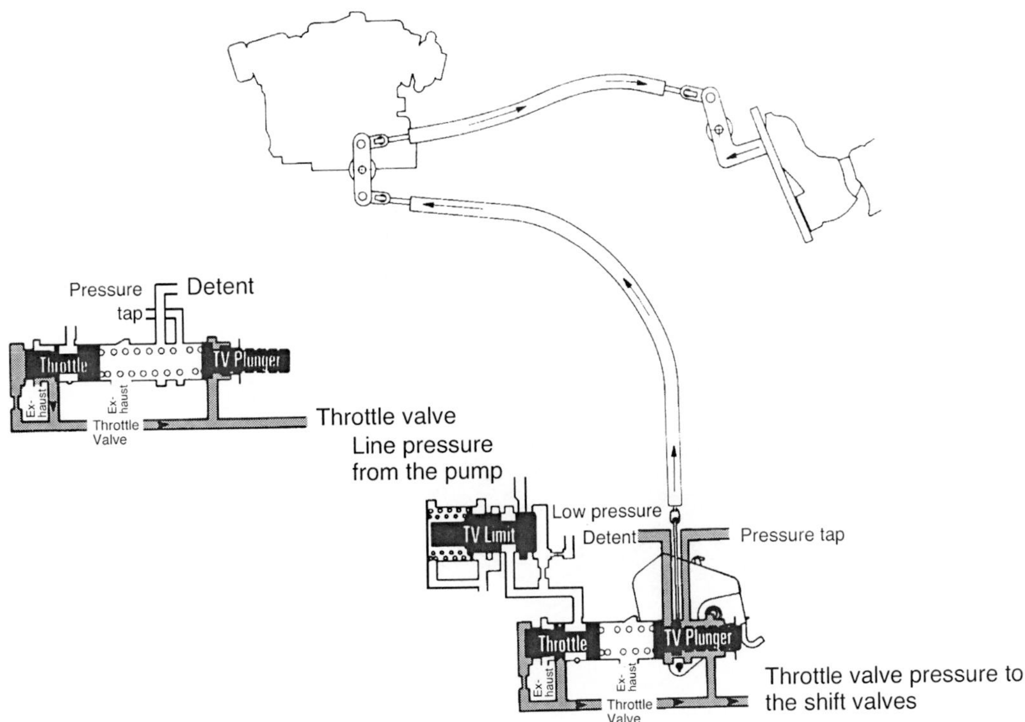


**Figure 26.** A throttle valve and cable.

## VALVE BODY AND SHIFT VALVES

The valve body is usually mounted on the bottom of the transmission or transaxle case in the reservoir. Some transaxle valve bodies are mounted on the left side of the transaxle case under the reservoir pan. The valve body contains many spool valves, poppet valves, and check balls **(Figure 27)**. A separator and separator gasket are positioned

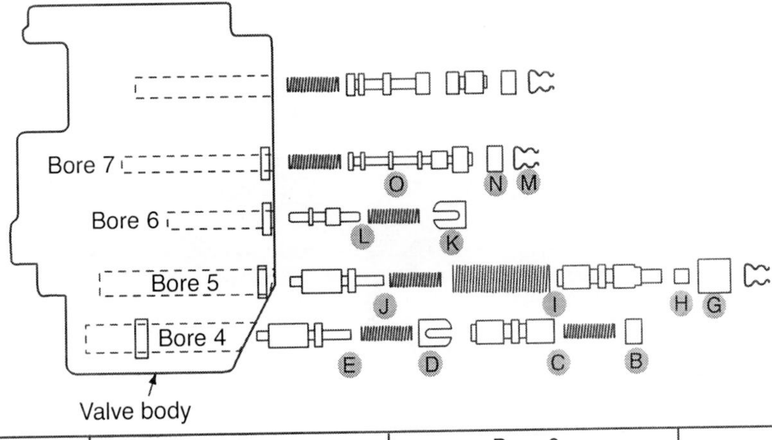


| Bore 4 | Bore 5 | Bore 6 | Bore 7 |
|---|---|---|---|
| A Spring retainer plate | F Clip | K Retainer plate | M Clip |
| B Bore plug | G Sleeve | L TV limit valve and spring | N Bore plug |
| C Orifice control valve and spring | H Plug | | O 1–2 shift valve and spring |
| D Spring retainer plate 1 | I 3–4 shift valve and spring | | |
| E 2–3 capacity modulator valve and spring | J 3–4 TV modulator valve and spring | | |

**Figure 27.** Valve body assembly.

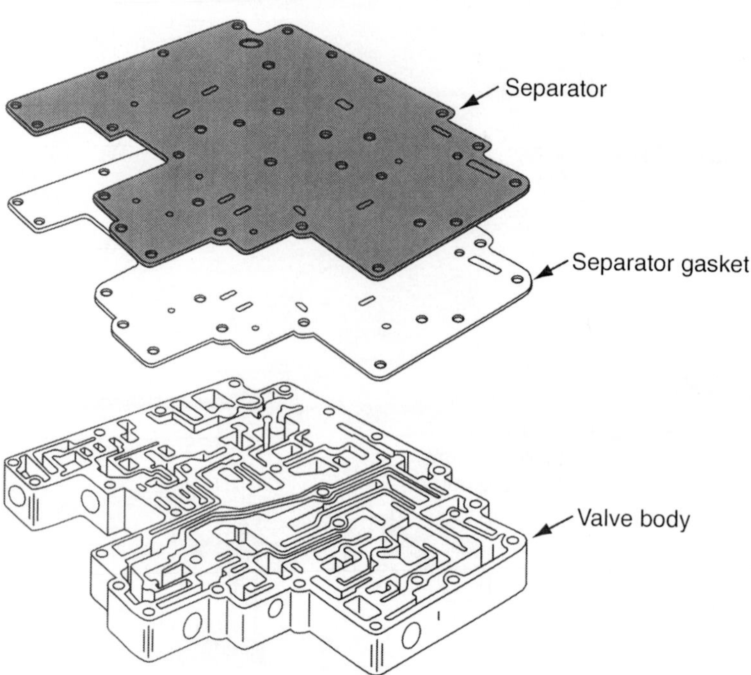


Figure 28. Valve body with separator and gasket.

between the valve body and the case **(Figure 28)**. The shift valves in the valve body control the transaxle shifting.

Governor pressure is supplied to one end of each shift valve and throttle pressure is supplied to the opposite end of the shift valves. When vehicle speed increases with a steady throttle opening, the governor pressure reaches a point where it overcomes throttle pressure on the opposite end of the 2–3 shift valve. When governor pressure overcomes throttle pressure on the 2–3 shift valve, the valve moves and allows line pressure to flow past the shift valve lands to the appropriate multiple clutch set or band **(Figure 29)**. When the clutch or band is applied, one of the

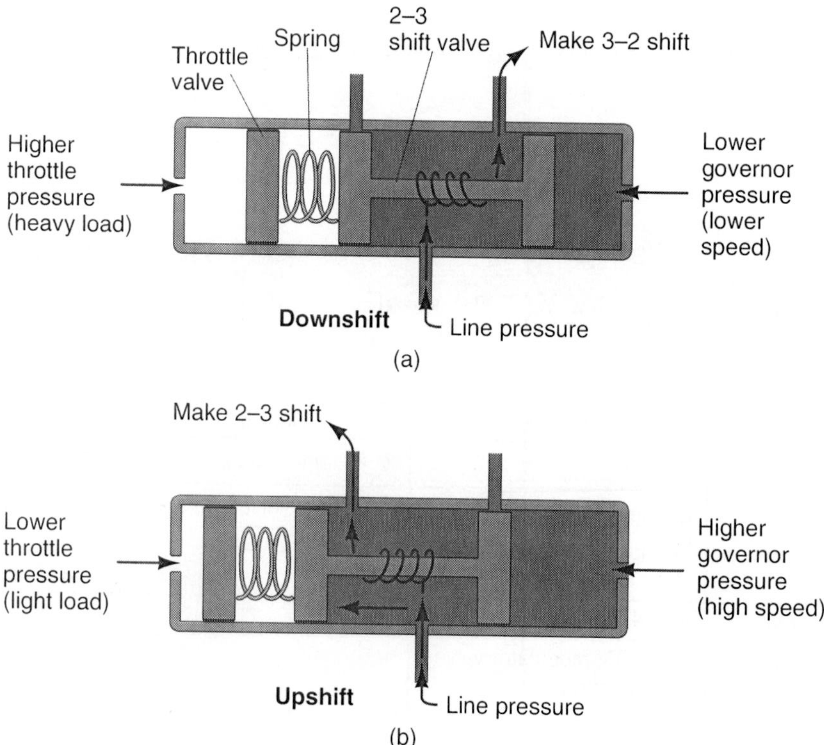


Figure 29. Shift valve operation during upshifts and downshifts.

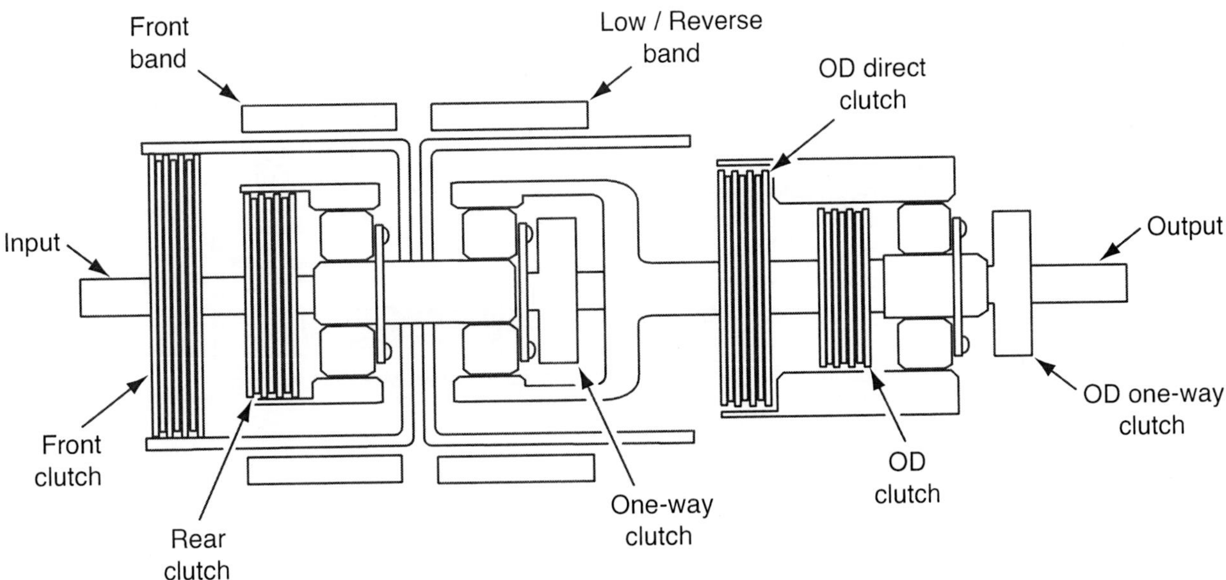


**Figure 30.** Typical gear train, clutches, and bands in a four-speed transmission or transaxle.

planetary gear members is locked while the other two members are the input and output to supply the proper gear.

When making a 1–2 shift, vehicle speed and governor pressure increase, the governor pressure overcomes the throttle pressure on the 1–2 shift valve, and this valve moves so it directs fluid to the front band and the rear clutch. Under this condition, the front band locks the sun gear **(Figure 30)**. Engine torque is transmitted through the input shaft and rear clutch to the front ring gear. Because the sun gear is locked, the smaller internal gear drives the larger carrier clockwise and provides a gear reduction. The carrier is splined to the output shaft. The overdrive (OD) direct clutch is still applied and the OD planetary gear set transmits torque with a 1:1 ratio.

If a driver suddenly accelerates the engine to pass another vehicle, throttle pressure may overcome governor pressure on the 1–2 shift valve. Under this condition, the shift valve is moved by the throttle pressure and the valve band blocks the line pressure to the clutch or band that is applied to supply third gear. This action downshifts the transaxle into second gear.

## ACCUMULATORS, MODULATORS, THRUST WASHERS, BUSHINGS, AND SEALS

Many transmissions and transaxles contain accumulators. An accumulator is a spring- loaded piston mounted in a machined bore. A seal or O-ring is positioned in a groove on the accumulator piston **(Figure 31)**. Accumulators act like a shock absorber to cushion the application of multiple disc clutches and bands. An accumulator cushions sudden increases in fluid pressure by allowing the pressure to flow into the accumulator bore against the spring-loaded piston. As the accumulator piston moves and the fluid fills the piston bore, the fluid pressure builds up gradually. This

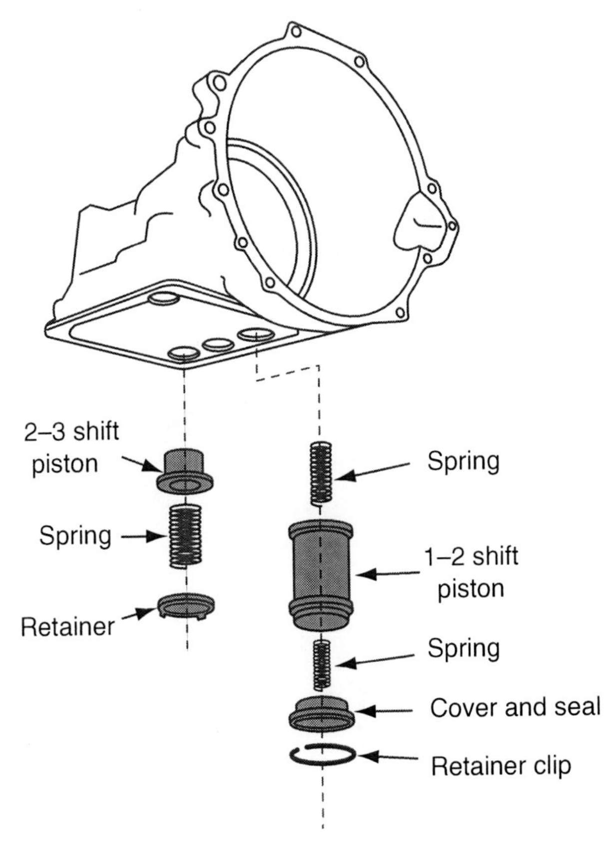


**Figure 31.** Accumulator assemblies.

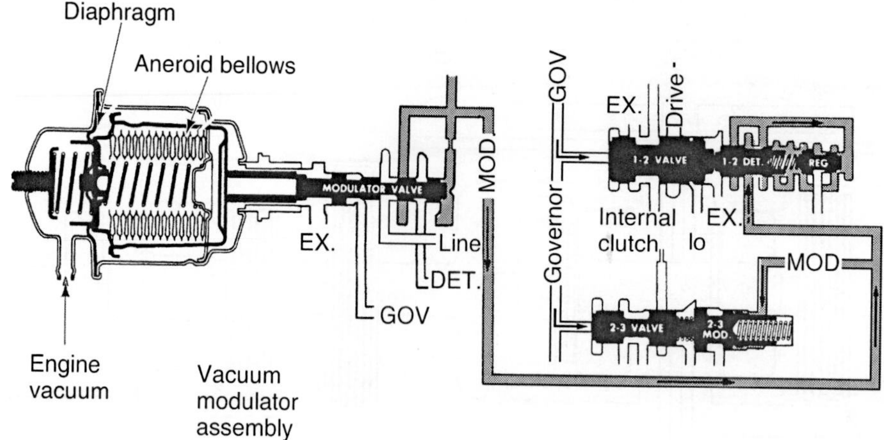


**Figure 32.** A vacuum modulator.

action delays the application of the clutch discs or band to prevent harsh shifting.

Some transmissions and transaxles have a vacuum modulator containing a diaphragm in a sealed chamber. The modulator is mounted in the side of the transaxle or transmission case. A stem on the transaxle side of the diaphragm contacts a modulator valve in the transaxle **(Figure 32)**. When the engine is running, manifold vacuum is supplied through a hose to the outer side of the modulator diaphragm and the inner side of the diaphragm is vented to the atmosphere. The diaphragm is connected to the modulator valve. At wide throttle openings, the vacuum decreases and the movement of the modulator diaphragm and valve directs modulator pressure to the spring ends of the shift valves to delay the shifts until the engine reaches a higher rpm.

Transaxles and transmissions contain numerous thrust washers, bushings, bearings, seals, and O-rings **(Figure 33)**. These items are normally replaced during an overhaul.

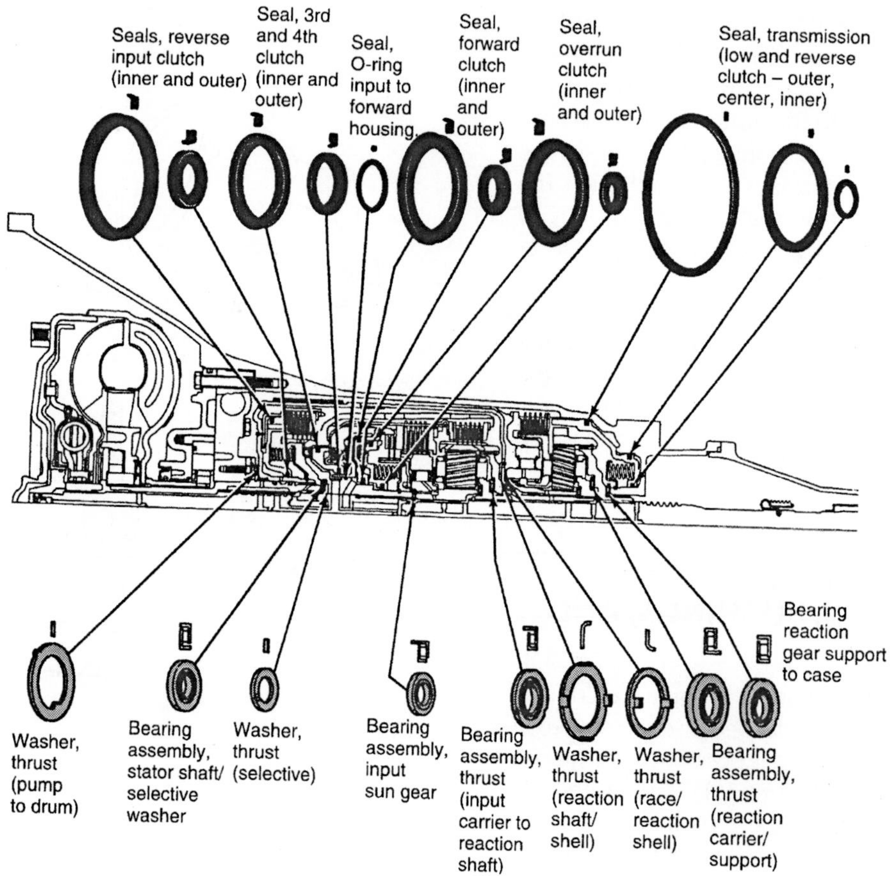


**Figure 33.** Transmission thrust washers, seals, bearings, and bushings.

## ELECTRONIC TRANSMISSION AND TRANSAXLE CONTROLS

Modern vehicles usually have computer-controlled automatic transmissions or transaxles. The computer that controls the transaxle functions may be combined in the PCM, or a separate transmission controller may be used. A governor is not required in a computer-controlled transaxle, because the computer operates two or more solenoids to supply pressure to the shift valves to control shifting. The computer also controls an electronic pressure control valve to control transaxle pressure. The electronic pressure control valve replaces the pressure regulator valve. The computer operates a TCC solenoid that supplies pressure to the TCC control valve. This valve controls the fluid supplied to the converter.

The transmission computer shares some input sensors with the engine computer. These inputs may include ECT sensor, TPS sensor, MAP sensor, VSS, and brake switch **(Figure 34)**. These input signals are usually sent to the PCM and relayed on the interconnecting data links to the transmission controller. Many computer-controlled transaxles or transmissions have a turbine speed sensor and an output speed sensor mounted in the transaxle case. A park, reverse, neutral, drive, low (PRNDL) switch is mounted on the transaxle and operated by the gear shift

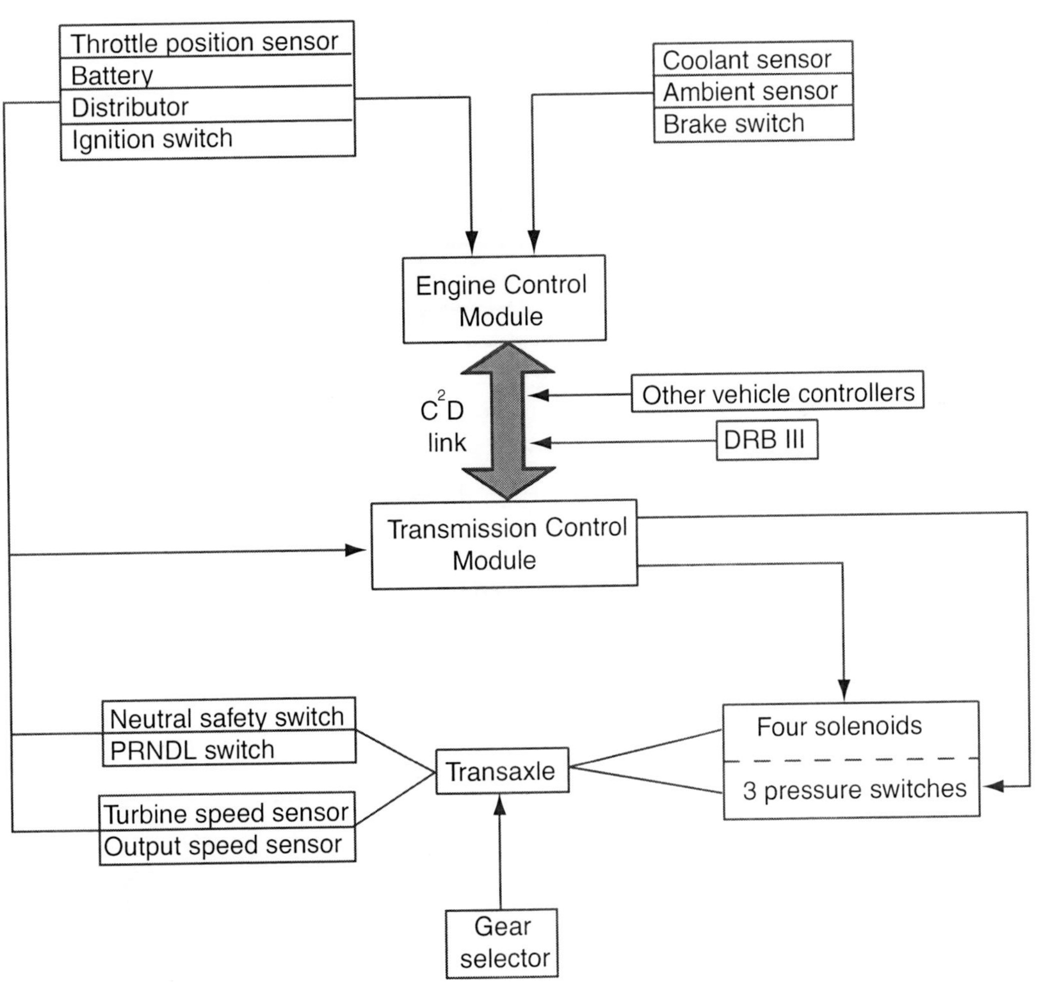


**Figure 34.** Computer-controlled transaxle inputs and outputs.

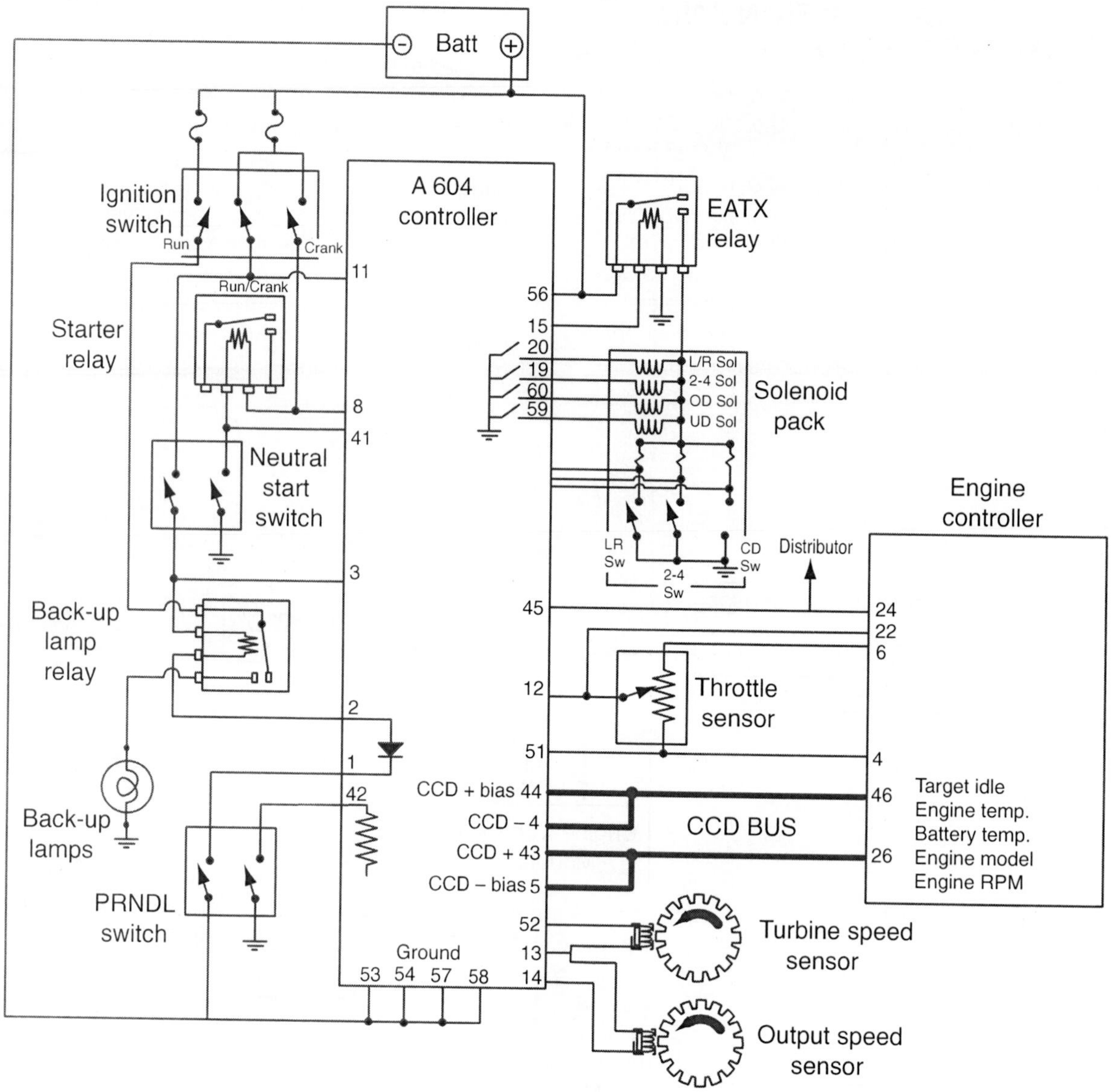

Figure 35. Transmission and engine computers with inputs and outputs.

linkage (**Figure 35**). This switch informs the transmission computer regarding the gear selection. Some transaxles have pressure switches that send feedback information to the transmission computer when a shift occurs.

## FINAL DRIVES AND DIFFERENTIALS

In RWD vehicles, engine torque is transmitted through the transmission and drive shaft to the pinion shaft and gear in the differential. The pinion gear is meshed with the differential ring gear. A small pinion shaft retains two small pinion gears inside the differential case. Two side gears are splined to the axles. These side gears are meshed with the small pinion gears (**Figure 36**). A steel pin retains the small

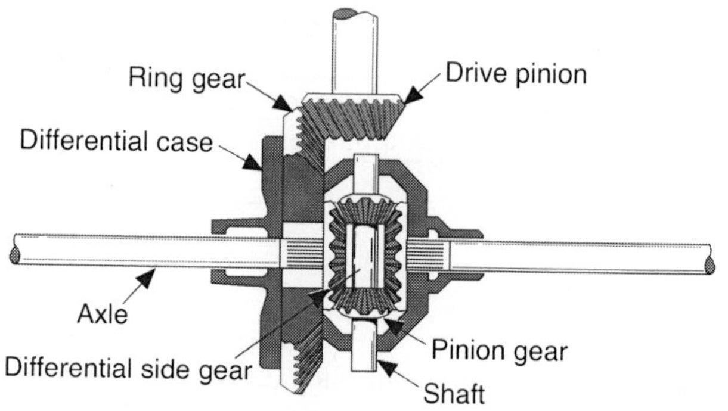


Figure 36. Basic differential.

pinion shaft in the differential case. When the vehicle is driven straight ahead, the pinion gear drives the ring gear and differential case. Because the pinion gear is much smaller than the ring gear, the differential provides a gear reduction. The differential case with the small pinion gears, side gears, and axle shafts rotates as a unit. Engine torque is transmitted from the pinion gear to the ring gear, differential case, and drive axles to the drive wheels. When the vehicle turns a corner, the side gears and small pinion gears rotate and allow the outside drive wheel to turn faster than the inside drive wheel.

In some FWD vehicles with an automatic transaxle, a gear on the transmission output shaft is meshed with a gear on the transfer shaft. A gear on the opposite end of the transfer shaft is meshed with the differential ring gear **(Figure 37)**. Engine torque is transmitted from the transmission output shaft through the transfer shaft to the differential ring gear. Some FWD vehicles with automatic transaxles have a planetary differential **(Figure 38)**. In this type of differential, the ring gear is splined to the transaxle case and the transaxle output shaft drives the differential sun gear. The differential sun gear drives the carrier which is connected to the drive axles. A gear reduction is provided when the smaller sun gear drives the larger planet carrier.

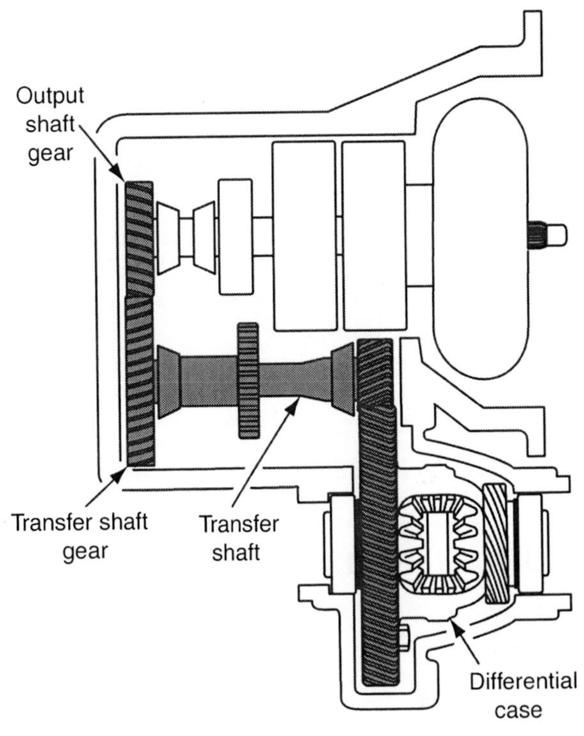


**Figure 37.** Final drive assembly in a transaxle.

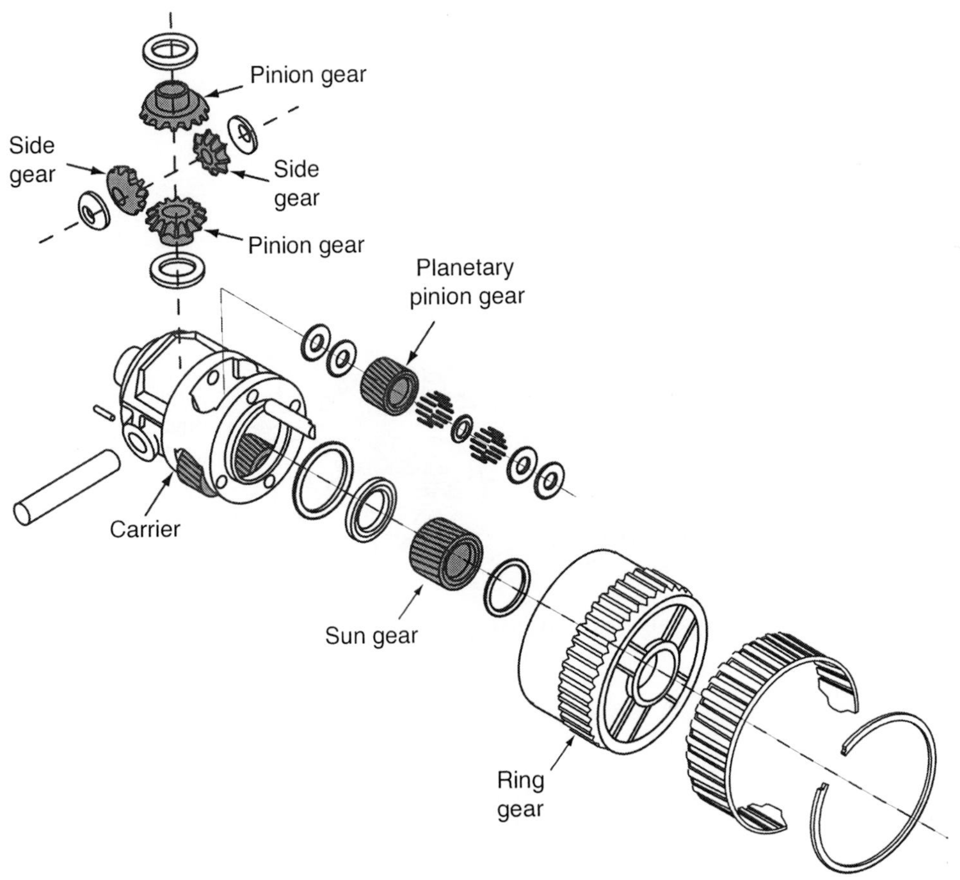


**Figure 38.** Planetary differential in a transaxle.

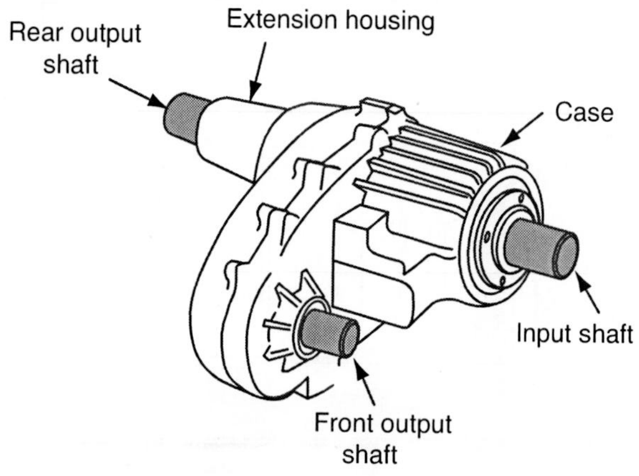


Figure 39. A typical transfer case.

## FOUR-WHEEL DRIVE (4WD)

A typical 4WD vehicle has a front, longitudinally mounted engine, a manual or automatic transmission, front and rear drive shafts, front and rear drive axle assemblies, and a transfer case. The transfer case is usually mounted behind or beside the transmission. The transfer case may be bolted to the rear of the transmission. Engine torque is transmitted from the transmission output shaft through a short drive shaft or a gear or chain drive in the transfer case. A shift lever or electric switch inside the vehicle allows the driver to select two-wheel drive or four-wheel drive. In two-wheel drive, engine torque is transmitted through the transfer case only to the drive shaft connected to the rear wheels. If the driver selects four-wheel drive, engine torque is transmitted through the transfer case to the both the front and rear drive shafts **(Figure 39)**. The driver may also shift the transfer case to four wheel low range. If this transfer case shift is completed, the transfer case provides a gear reduction which increases torque when driving in very adverse conditions such as rough terrain.

## ALL-WHEEL DRIVE

A significant number of cars and minivans are equipped with all-wheel drive. These systems are in four-wheel drive at all times and the driver cannot select between four-wheel drive and two-wheel drive. All-wheel drive systems improve traction and vehicle handling when driving on snow-covered or icy road surfaces. All-wheel drive vehicles are not intended for adverse conditions encountered in off-road operation. Many all-wheel drive systems are a FWD vehicle with a center differential designed into the transaxle to drive the rear wheels **(Figure 40)**. Some all-wheel drive systems have a viscous coupling

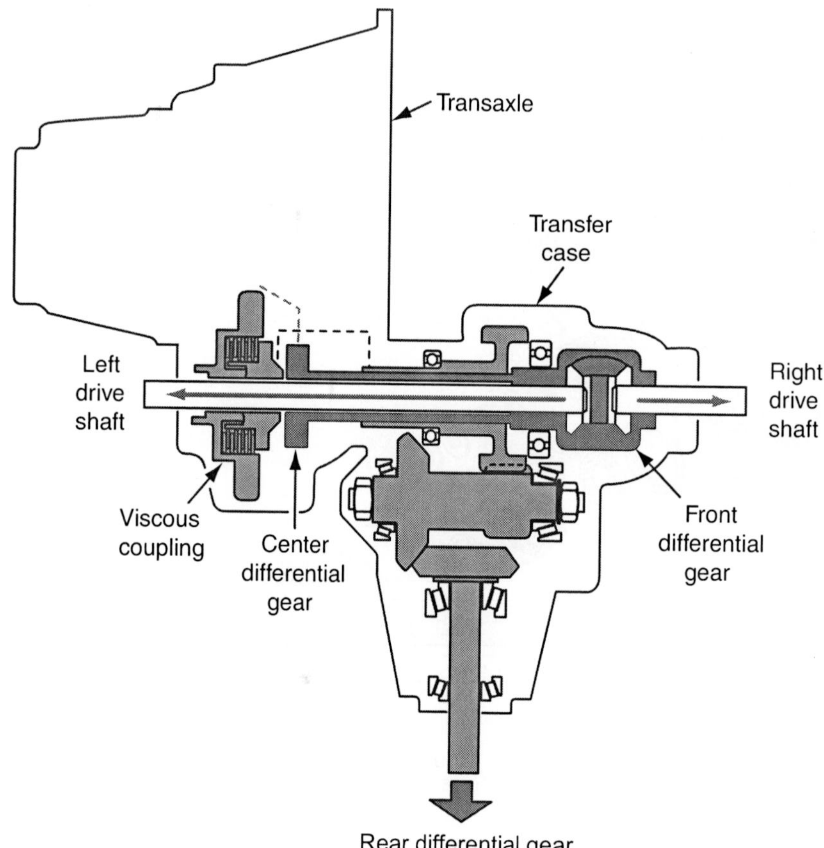


Figure 40. An all-wheel drive transaxle with rear differential drive gear and viscous coupling.

that allows some variation in front and rear axle speeds. If one wheel begins to spin on an icy road surface, the viscous coupling immediately transfers torque to the other wheels that are not spinning. A **viscous coupling** is a sealed chamber containing a thick honey-like fluid.

Planetary gears provide quiet, smooth operation in automatic transmissions and transaxles. Defective or worn planetary gears may cause excessive noise, improper shifting, and locking of the transmission. The diagnosing and servicing of planetary gears is an integral part of complete transmission and transaxle diagnosis and service.

## TORQUE CONVERTER MAINTENANCE AND DIAGNOSIS

Torque converters are a sealed assembly. When the fluid is drained from the transaxle or transmission, some of the fluid is also drained from the torque converter. On modern cars and light-duty trucks, the torque converter cannot be drained separately from the transmission or transaxle. On some vehicles, the access cover or plug may be removed from the bottom of the flywheel housing to inspect the torque converter. If ATF is leaking from the flywheel housing, the pump seal in the transmission is the most likely source of the leak. The transaxle or transmission must be removed from the vehicle to remove the torque converter.

On transmissions or transaxles with torque converter clutch lockup and/or computer-controlled shifting, visually inspect all the wires and wiring connectors entering the transmission. The next step in torque converter diagnosis is a visual inspection of all the wires going into the transaxle.

### One-Way Stator Clutch Diagnosis

If the transaxle passes the visual inspection, road test the vehicle. When the vehicle lacks power during acceleration, the exhaust system may be restricted or the one-way stator clutch in the torque converter may be slipping. To test for a restricted exhaust system, connect a vacuum gauge to the intake manifold. Operate the engine at 2,500 rpm for 3 minutes. If the vacuum reading on the gauge slowly decreases, the exhaust is restricted. This decrease in vacuum is even more pronounced during a road test. If the exhaust system is not restricted, the one-way stator clutch may be slipping. If the engine speed flares up during acceleration in DRIVE and the vehicle does not have normal acceleration, the clutches or bands in the transaxle or transmission are slipping. When the engine speed does not flare up but the engine lacks power, the one-way stator clutch is likely slipping.

A one-way stator clutch that is seized causes reduced engine performance at low speed and also at high speed. A seized one-way stator clutch may cause transmission/transaxle and engine overheating.

### Torque Converter Clutch Diagnosis

During the road test, test the operation of the TCC lockup. Typically the PCM uses inputs from the ECT sensor, VSS, TPS, and brake switch to operate the TCC lockup. The engine must be warmed up and operating at a certain speed before the PCM provides TCC lockup. Many transaxles/transmissions only allow lockup in third or fourth gear. If the brake pedal is depressed when the TCC is locked up, the brake switch sends a voltage signal to the PCM. When the PCM receives this signal, it immediately unlocks the TCC.

When TCC lockup occurs, a slight "bump" in engine operation is felt. The TCC lockup should be smooth without any shudder or vibration. If a shudder occurs during the TCC lockup, the friction material on the TCC lockup disc may be worn or the fluid pressure in the transaxle may be low. Broken damper springs in the TCC lockup disc also cause a shudder during lockup. When the shudder is only noticeable after the TCC lockup has occurred, the problem is likely in the engine or other systems in the transaxle. For example, intermittent ignition misfiring may cause a shudder after TCC lockup occurs.

### Scan Tool Diagnosis of Converter Clutch Lockup

A scan tool may be connected to the DLC under the dash to check the TCC lockup system and other transaxle electronics **(Figure 41)**. Be sure the scan tool contains the proper module for the vehicle being tested and be sure the ignition switch is off when connecting the scan tool.

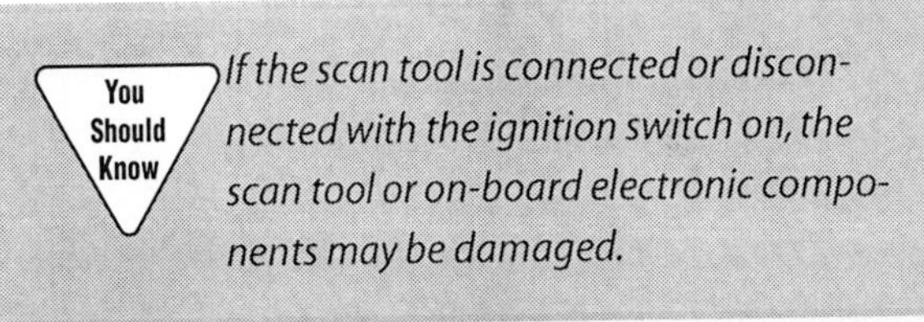


*If the scan tool is connected or disconnected with the ignition switch on, the scan tool or on-board electronic components may be damaged.*

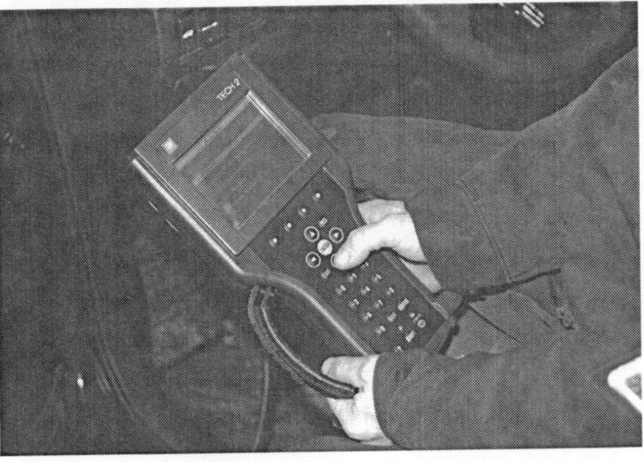

**Figure 41.** A scan tool connected to the data link connector (DLC).

With the engine and transaxle at normal operating temperature, check for DTCs on the scan tool display. If there are any defects in the TCC electrical system, such as an open wire or an open or shorted TCC solenoid winding, a DTC representing these faults is displayed on the scan tool. Check the scan tool display for other DTCs related to TCC operation. For example, if a defective ECT sensor always indicates cold engine temperature, the TCC does not lock up. A defective VSS sensor that always indicates low vehicle speed prevents TCC lockup. If a faulty VSS indicates higher-than-actual vehicle speed, the TCC locks up sooner than specified.

### Diagnosing Improper Torque Converter Clutch Release Problems

Fluid normally flows through the torque converter and then through the transaxle cooler before flowing back into the transaxle. If the fluid passages in the cooler become restricted, the fluid cannot flow out of the torque converter. This condition may cause the TCC lockup to remain applied during deceleration and idle. This condition may cause engine stalling during deceleration and idle or when the transaxle selector is moved from PARK to DRIVE or REVERSE.

To test for restricted cooler fluid passages, disconnect the cooler return line at the transmission/transaxle. Place a length of hose from the disconnected line to an empty drain pan **(Figure 42)**. Start and run the engine at idle speed for 20 seconds. Shut the engine off and measure the fluid in the drain pan. One quart of fluid should have been discharged from the cooler return line in 20 seconds. If there is less than 1 quart of fluid in the drain pan, the cooler passages are restricted. To blow out the cooler passages, disconnect the cooler inlet line and alternately supply 50 psi air pressure to the inlet and return lines.

## TORQUE CONVERTER SERVICE

After the transmission or transaxle is removed, grasp the converter and pull it from the front of the transaxle. Inspect the converter welds after it is removed. Inspect the converter hub for scoring in the pump seal contact area. If scoring is present, the converter must be replaced. Inspect the lugs on the converter hub that drive the transaxle oil pump. If these lugs are damaged or worn, replace the converter.

**You Should Know** *On many transaxles, the pump is driven by a drive shaft extending from the converter hub rather than driving the pump directly from the back of the converter hub.*

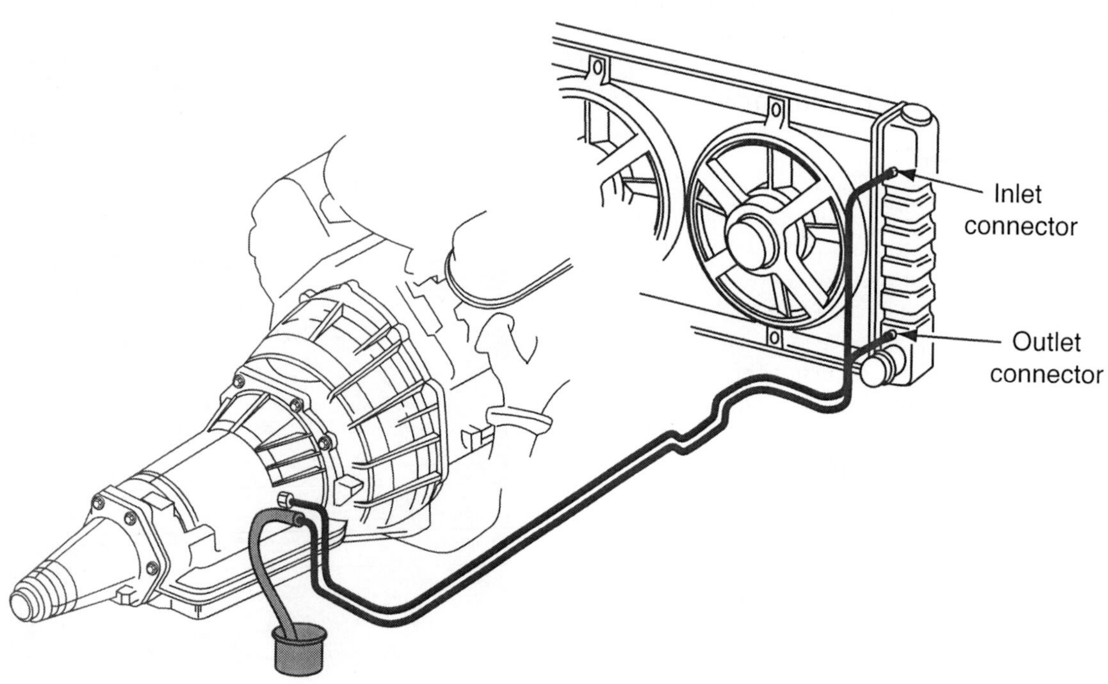


**Figure 42.** Testing fluid flow through the transmission cooler.

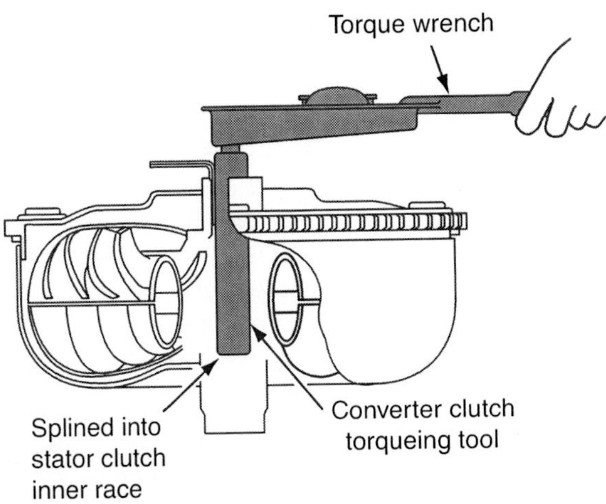


**Figure 43.** Testing the stator one-way clutch.

If the converter has an oil pump drive shaft, check the shaft support bushing in the converter for wear and inspect the drive shaft splines.

## Diagnosis of a Stator One-Way Clutch

Insert the special tool for checking the one-way stator clutch into the converter hub and connect a torque wrench to the tool **(Figure 43)**. This tool holds the inner race of the one-way clutch and allows the technician to apply torque to the outer race. While holding the inner race, rotate the torque wrench in both directions. The torque wrench should rotate easily and smoothly in one direction and lock up immediately in the opposite direction. If the one-way clutch allows rotation in both directions or if the rotation is erratic, replace the converter. When the special tool for checking the one-way stator clutch is not available, a pair of snap ring pliers with long, thin jaws may be inserted into the inner race of the one-way clutch. This method of checking the one-way clutch is not as accurate because it does not apply any torque to this clutch.

## Testing Internal Converter Interference

Wear and excessive end play in the converter may cause internal converter components to contact each other. This action may cause scraping noises in the converter. Place the converter face down on the bench and install the oil pump assembly on the back of the converter. Install the input shaft through the oil pump into the turbine hub **(Figure 44)**. Rotate the input shaft and turbine. If there is any resistance to rotation or noise, replace the converter.

Remove the input shaft and hold the pump on the back of the converter. Turn the converter and oil pump over so the oil pump is resting on the bench. Be sure the stator

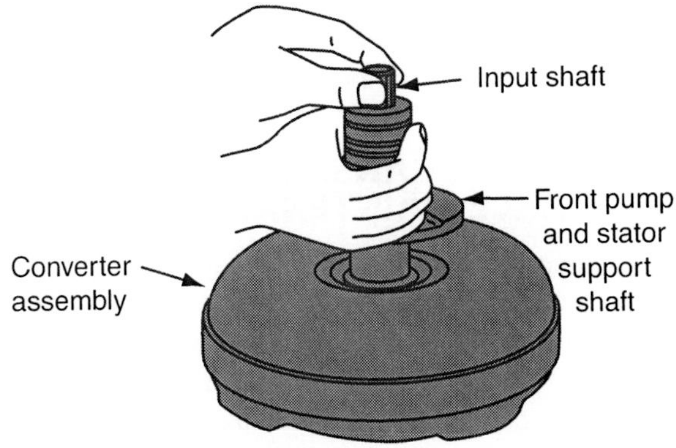


**Figure 44.** Checking the stator-to-turbine interference.

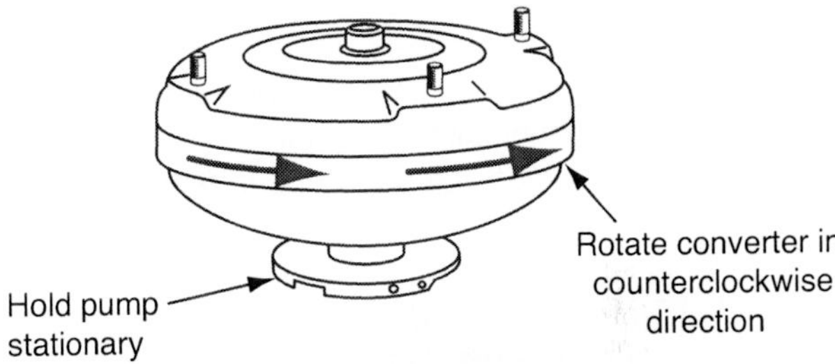


**Figure 45.** Checking the stator-to-impeller interference.

support splines are engaged in the stator. Hold the pump stationary and rotate the converter to check for interference between the impeller pump and the stator **(Figure 45)**. If the converter rotation is erratic or a scraping noise is heard, replace the converter.

## End Play Measurement

Place the torque converter face down on the bench and install a dial indicator magnetic base on the back of the converter. Insert the special tool into the turbine hub and tighten the tool in this position. Install the dial indicator stem against the end of the tool **(Figure 46)**. Lift upward on

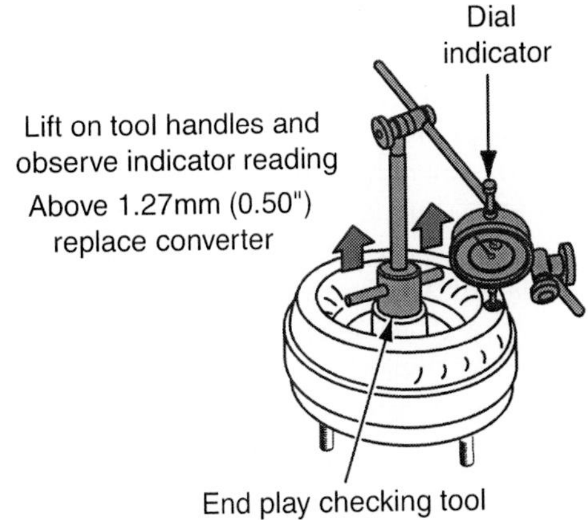


**Figure 46.** Measuring torque converter endplay.

the tool and then release the tool. Repeat this procedure and observe the turbine end play on the dial indicator. If the end play is more than specified, replace the converter.

## AUTOMATIC TRANSMISSION AND TRANSAXLE MAINTENANCE

Automatic transmissions and transaxles require a minimum amount of maintenance. Always follow the vehicle manufacturer's recommended change intervals for the transmission fluid and filter. Some vehicle manufacturers recommend different fluid and filter change intervals depending on the vehicle operating conditions. One vehicle manufacturer recommends changing transmission fluid and filter at 50,000 miles (80,000 kilometers) if the vehicle is over 8,600 gross vehicle weight rating (GVWR) or operated under one or more of the following conditions: frequent trailer towing; taxi, police, or delivery service; driving continuously in hilly or mountainous terrain; and driving continually in city traffic when the atmospheric temperature is regularly above 90°F (32°C).

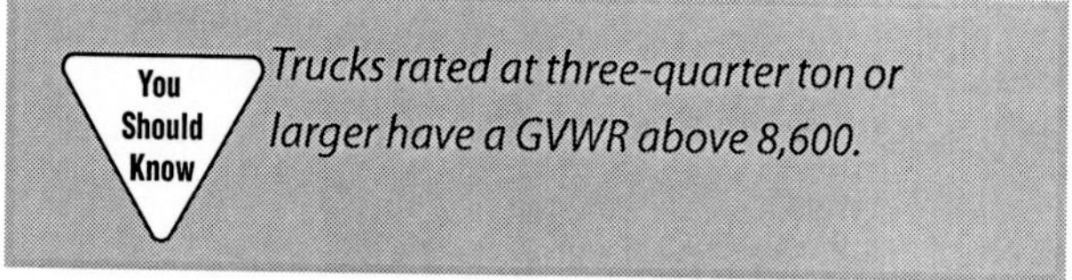


*Trucks rated at three-quarter ton or larger have a GVWR above 8,600.*

If the vehicle has a GVWR below 8,600 and is not operated under any of the above conditions, the manufacturer recommends changing the transmission fluid and filter at 100,000 miles (161,00 kilometers).

### Automatic Transmission and Transaxle Fluids

Always use the automatic transmission fluid specified by the vehicle manufacturer. For example, DaimlerChrysler recommend Mopar® ATF+4 type 9602 in many current automatic transaxles. Dexron III/Mercon ATF is recommended for many General Motors and Ford automatic transaxles and transmissions. Some vehicle manufacturers provide warnings in their service manuals indicating the use of an ATF other than the specified ATF may cause torque converter shudder or clutch slipping. Many vehicle manufacturers also have warnings in their service manuals indicating that ATF additives should never be used. The ATF contains many additives such as friction modifiers, oxidation and corrosion inhibitors, extreme pressure lubricants, anti-foaming agents, detergents and dispersants, and pour-point depressants.

### Checking Fluid Level

The fluid level in an automatic transmission or transaxle should be between the add and the full marks on the dipstick. High fluid level may cause fluid to be thrown out of the filler tube. A low fluid level may cause erratic transmission operation and damage to the clutches and bands. Before checking the fluid level, ensure that the vehicle is parked on a level surface, the engine and transmission are at normal operating temperature, and the parking brake is applied and the transmission is in PARK.

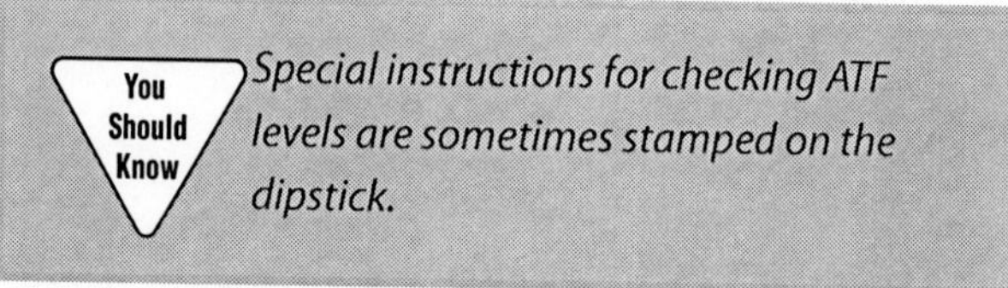


*Special instructions for checking ATF levels are sometimes stamped on the dipstick.*

### Checking Fluid Condition

After the dipstick is removed, always check the condition of the fluid on the dipstick. The fluid should be red and it should wipe off the dipstick easily. Abnormal fluid conditions are the following:

1. Milky fluid on the dipstick indicates that coolant has leaked into the transmission, usually through a leaking cooler.
2. Burned fluid is dark and has a burned smell. This condition is caused by overheating and other components in the transmission such as clutches are likely damaged. Transaxle overhaul is usually required when this condition is present.
3. If there is varnish on the dipstick, the ATF cannot be wiped easily from the dipstick. Varnish on the dipstick indicates oxidized ATF which is caused by overheating and/or lack of ATF changes. This condition usually requires a transmission overhaul.

### Visual Inspection

The automatic transmission or transaxle should be inspected for fluid leaks at the seals. If ATF is leaking from the bottom of the flywheel housing, the pump seal is likely leaking. Because the ATF is red, it should not be confused with an engine oil leak at the rear main bearing.

Inspect the transmission cooler lines for leaks and contact with other components that may wear a hole in a line. On a transaxle, inspect the drive axle seals for leaks. The rear extension housing seal in a transmission should be checked for leaks. Before replacing any seals, be sure the vent in the case is not restricted. Check for a leak at transmission or transaxle components such as the speedometer drive, vehicle speed sensor, turbine speed sensor, output speed sensor, and dipstick tube. Inspect the area around the shift linkage for a fluid leak. Check for fluid leaks around the oil pan and also around the area where the electrical connection enters the case.

If the transaxle has a vacuum modulator, be sure the vacuum hose to the modulator is not leaking and is firmly attached to the modulator. Be sure the full intake manifold vacuum is available at the modulator with the engine idling.

## AUTOMATIC TRANSMISSION AND TRANSAXLE DIAGNOSIS

The first step in transmission diagnosis is to check the fluid level and condition. Be sure the fluid level is correct and visually inspect the transmission. When the fluid level is correct and the visual inspection does not indicate any problems, road test the vehicle to verify the customer's complaint. During the road test, check the vehicle speed at which the shift points occur. If the upshifts and downshifts do not occur at the specified speed, there may be a problem with the governor, throttle cable, or modulator. When the shift points do not occur at the proper vehicle speed or some shifts are missing, perform a pressure test on the transmission.

Listen for abnormal transmission noises during the road test. A defective pump may cause a humming or buzzing noise that increases with engine speed. This noise occurs in any gear selector position. A grinding noise that increases with vehicle speed and load may be caused by a damaged planetary gear set, worn bushing, or a rough needle bearing. Refer to **Figure 47** for transmission and transaxle noise and vibration diagnosis.

| Problem | Probable Cause(s) |
|---|---|
| Ratcheting noise | The return spring for the parking pawl is damaged, weak, or misassembled |
| Engine speed-sensitive whine | Torque converter is faulty<br>A faulty pump |
| Popping noise | Pump cavitation—bubbles in the ATF<br>Damaged fluid filter or filter seal |
| Buzz or high-frequency rattle<br>Whine or growl | Cooling system problem<br>Stretched drive chain<br>Broken teeth on drive and/or driven sprockets<br>Nicked or scored drive and/or driven sprocket bearing surfaces<br>Pitted or damaged bearing surfaces |
| Final drive hum | Worn final drive gear assembly<br>Worn or pitted differential gears<br>Damaged or worn differential gear thrust washers |
| Noise in forward gears | Worn or damaged final drive gears |
| Noise in specific gears | Worn or damaged components pertaining to that gear |
| Vibration | Torque converter is out of balance<br>Torque converter is faulty<br>Misaligned transmission or engine<br>The output shaft bushing is worn or damaged<br>Input shaft is out of balance<br>The input shaft bushing is worn or damaged |

**Figure 47.** Transmission and transaxle noise and vibration diagnosis.

When the engine is decelerated, the flexplate actually flexes forward toward the engine. When this condition occurs, the pilot hub on the front of the converter moves into the bore on the rear of the crankshaft. The flexplate normally has a flex movement of approximately 0.080 to 0.100 inch (2.03 to 2.54 millimeters). A cracked flexplate causes a knocking noise that changes with engine speed and load.

## Automatic Transmission and Transaxle Pressure Testing

Prior to performing a pressure test, the transaxle must be at normal operating temperature and have the correct fluid level. Some transaxles and transmissions have several plugs in the case that may be removed to test pressure in certain gears. For example, there may be plugs in the case that can be removed to check governor pressure, low/reverse pressure, and line pressure **(Figure 48)**. Raise the vehicle on a lift and remove the desired pressure test plug from the transmission. Connect the appropriate pressure gauge to the opening from which the plug was removed. Operate the vehicle on the lift or road test the vehicle and observe the pressure in each gear **(Figure 49)**.

If the pressure is low in all gears including reverse, the pump or pressure regulator valve may be defective. When the pressure is only low in one gear, the seals on the components applied in that gear are leaking. The leaking seals are typically clutch drum seals or band servo seals.

> **You Should Know** *When reading transmission pressures during a road test on vehicles with computer-controlled transmissions or transaxles, specific vehicle speed and sensor outputs may be required. For example, a specific voltage from the TPS may be required.*

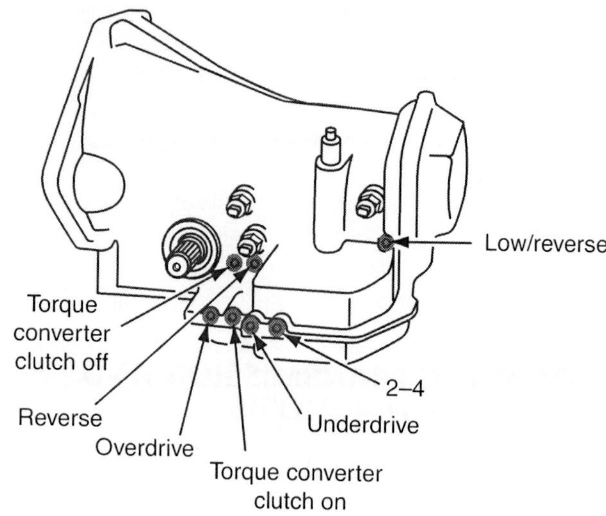


**Figure 48.** Pressure taps on a typical transaxle.

## Diagnosis of Computer-Controlled Transmissions and Transaxles

When diagnosing a computer-controlled transmission or transaxle, a scan tool may be connected to the DLC under the dash **(Figure 50)**. If faults are present in the trans-

| Transmission Pressure with TPS at 1.5 Volts and Vehicle Speed above 8 Km/h (5 mph) | | | | | |
|---|---|---|---|---|---|
| Gear | EPC Tap | Line Pressure Tap | Forward Clutch Tap | Intermediate Clutch Tap | Direct Clutch Tap |
| 1 | 276–345 kPa (40–50 Psi) | 689–814 kPa (100–118 Psi) | 620–745 kPa (90–108 Psi) | 641–779 kPa (93–113 Psi) | 0–34 kPa (0–5 Psi) |
| 2 | 310–345 kPa (45–50 Psi) | 731–869 kPa (106–126 Psi) | 662–800 kPa (96–116 Psi) | 689–827 kPa (100–120 Psi) | 655–800 kPa (95–116 Psi) |
| 3 | 341–310 kPa (35–45 Psi) | 620–758 kPa (90–110 Psi) | 0–34 kPa (0–5 Psi) | 586–724 kPa (85–105 Psi) | 551–689 kPa (80–100 Psi) |

**Figure 49.** Pressure test specifications for a computer-controlled transmission.

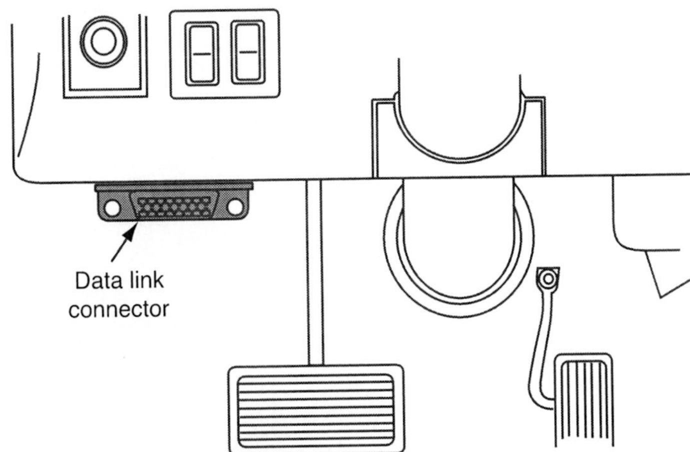


**Figure 50.** A data link connector under the dash.

mission electronic system, DTCs are stored in the transmission computer memory. These DTCs may be retrieved with the scan tool. DTCs may indicate defects in certain circuits such as a shift solenoid circuit, torque converter clutch solenoid circuit, or a pressure control valve circuit. A DTC indicates a problem in a certain circuit, but further diagnosis with a voltmeter and ohmmeter may be necessary to determine if the problem is in the solenoid winding or in the wires from the computer to the solenoid.

All 1996 and newer vehicles under 8,500 GVWR are equipped with OBD II electronic systems. In these systems, the DTCs have a standard format. For example, in a PI711 DTC, the P indicates a powertrain code and the 1 indicates this code is supplied by the manufacturer. If the second digit in the code is a 0, the code is a standard code defined by the SAE and is used by all vehicle manufacturers. When the third digit in the code is a 7, the code belongs to the transmission subgroup. The last two digits in the code indicate the exact circuit where the problem is located. In the example above, the last two digits, 11 in the P1711 code indicate a defect in the transmission oil temperature (TOT) sensor **(Figure 51)**. Data may be displayed on the scan tool

The SAE J2012 standards specify that all DTCs will have a five-digit alphanumeric numbering and lettering system. The following prefixes indicate the general area to which the DTC belongs:

1. P — powertrain
2. B — body
3. C — chassis

The first number in the DTC indicates who is responsible for the DTC definition.

1. 0 — SAE
2. 1 — manufacturer

The third digit in the DTC indicates the subgroup to which the DTC belongs. The possible subgroups are:

0 — Total system
1 — Fuel-air control
2 — Fuel-air control
3 — Ignition system misfire
4 — Auxiliary emission controls
5 — Idle speed control
6 — PCM and I/O
7 — Transmission
8 — Non-EEC power train

The fourth and fifth digits indicate the specific area where the trouble exists. Code P1711 has this interpretation:

P — Powertrain DTC
1 — Manufacturer-defined code
7 — Transmission subgroup
11 — Transmission oil temperature (TOT) sensor and related circuit

**Figure 51.** OBD II diagnostic trouble code format.

from all the sensors in the transmission including the turbine speed sensor, output speed sensor, and transmission oil temperature sensor.

## AUTOMATIC TRANSMISSION AND TRANSAXLE SERVICE

When changing the transmission or transaxle fluid, most oil pans do not have a drain plug. To change the transmission fluid, raise the vehicle on a lift and place a large diameter drain pan under the transmission pan. Place the drain pan on a stand so it is closer to the transmission oil pan to reduce oil splashing out of the pan. Loosen the oil pan retaining bolts more on one side of the pan than on the opposite side and allow the pan to drop downward on one side. Some of the fluid will begin to run into the drain pan. Remove the oil pan bolts on the side of the lower side of the oil pan and allow the pan to drop further downward to drain more fluid from the pan. Hold the oil pan and remove the remaining bolts from the pan. Lower the pan and drain the remaining fluid into the drain pan.

Check the oil pan for clutch and band material and aluminum or copper cuttings. An excessive amount of clutch and band material in the oil pan indicates the clutches and/or bands are worn and a complete overhaul is required now or in the near future. Excessive aluminum cuttings in the oil pan indicate damage to the aluminum castings, case, or torque converter. Copper cuttings in the oil pan indicate damaged bushings. Wash the oil pan in an approved solvent and use a scraper to remove the old gasket from the oil pan. Use a plastic scraper to remove old gasket material from the oil pan mating surface on the transmission case.

**You Should Know** *Using a metal scraper to remove old gasket material from an aluminum casting may scratch the casting so it is difficult or impossible for the gasket to prevent oil leaks between the two mating surfaces.*

Remove the filter retaining clips and remove the filter from the transaxle **(Figure 52)**. Lubricate the filter tube O-ring and install it on the new filter. Install the new filter and the retaining clips. Install the oil pan with a new gasket and tighten the oil pan bolts to the specified torque. Install the specified type of fluid in the transaxle until the proper fluid level is indicated on the dipstick with the engine and transaxle at normal operating temperature.

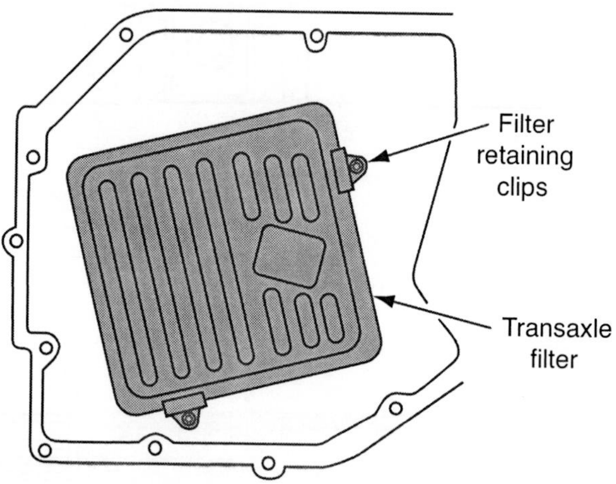


**Figure 52.** A transaxle oil filter.

### Band Adjustments

On some transmissions or transaxles, the band adjustments are on the outside of the case whereas on other vehicles, the oil pan must be removed to access these adjustments. When the band adjustments are external, locate the band adjusting nut and remove any dirt around the nut. Loosen the band adjustment locknut and back it off approximately five turns **(Figure 53)**. Tighten the band

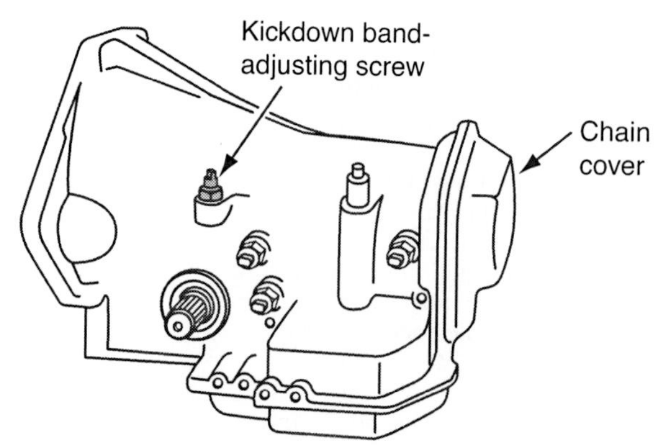


**Figure 53.** Band adjustment screw and locknut.

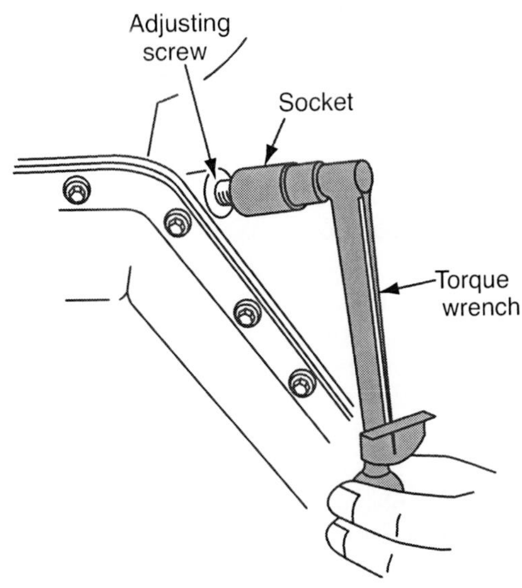


Figure 54. Adjusting a transmission or transaxle band.

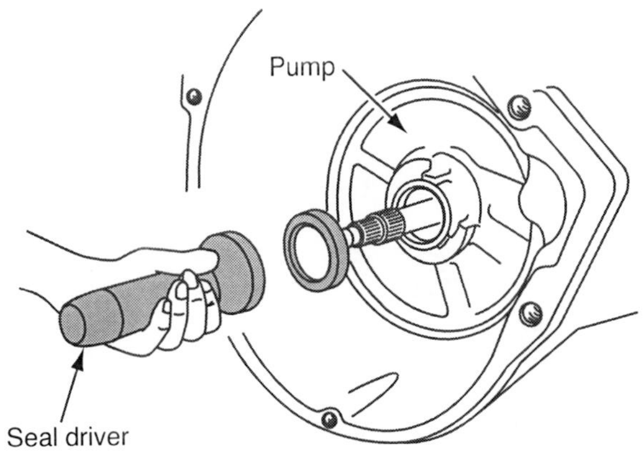


Figure 55. Installing a transmission or transaxle pump seal.

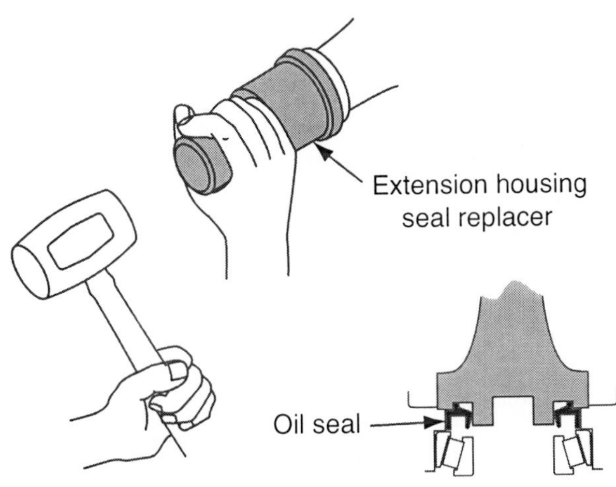


Figure 56. Extension housing seal installation.

adjustment to the specified torque with an inch-pound torque wrench **(Figure 54)**. Back off the adjustment screw the specified number of turns. Hold the band adjusting screw in this position and tighten the locknut to the specified torque.

## Seal Replacement

The converter must be removed to access the pump seal. Use the proper puller to remove the pump seal. Clean the seal mounting area in the pump and be sure there are no metal burrs in this area. Some pump seals have sealant applied to the outside diameter of the seal case. If this sealant is not present, apply sealant to the outer edge of the seal case. Use the proper driver to install the pump seal and be sure the seal is started squarely into the pump **(Figure 55)**.

The extension housing seal must be pulled from the housing with a special puller. Clean the seal mounting area and be sure there are no metal burrs in this area. Place a thin layer of sealant on the outer diameter of the seal case and use the proper driver to install the seal **(Figure 56)**.

## Linkage Adjustments

A misadjusted transmission linkage will not allow proper manual valve position resulting in improper fluid supply to some of the clutch packs or bands. Under this condition, the clutches or bands may slip and experience excessive wear. To adjust a cable-type transaxle linkage, follow this procedure:

1. Raise the vehicle on a lift and place the shift lever in PARK.
2. Loosen the clamp bolt on the shift cable bracket.
3. Be sure the preload adjustment spring engages the fork on the transaxle bracket.
4. Pull the shift lever to the front detent (PARK) position and then tighten the clamp bolt on the shift cable bracket.
5. If the cable has an adjuster, move the shift lever into the PARK position and rotate the adjuster to adjust the

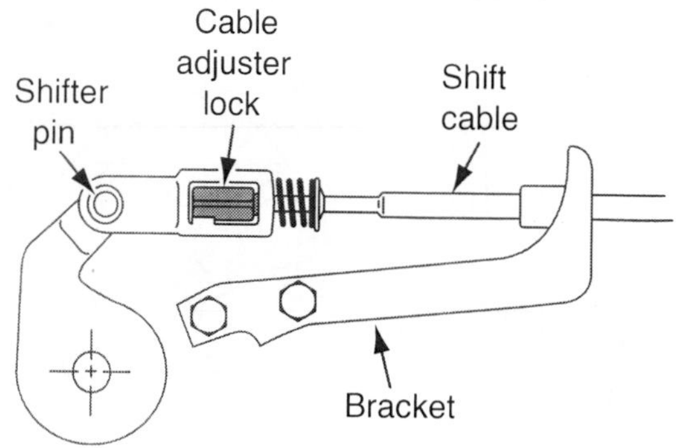


**Figure 57.** Shift cable adjuster lock assembly.

cable **(Figure 57)**. The adjuster provides a click when the lock is fully adjusted.

To adjust a rod-type transmission linkage, follow these steps:

1. Raise the vehicle on a lift and disconnect the shift rod at the shift lever bracket.
2. Place the gear selector in PARK and move the shift lever on the transmission into the PARK position detent.
3. Install and tighten the shift rod at the shift lever bracket.

After making the transmission linkage adjustment, it may be necessary to adjust the neutral safety switch. Try to start the vehicle with the gear selector in the PARK, NEUTRAL, OD, and REVERSE positions. The starter should operate only with the gear selector in PARK or NEUTRAL. If the starter operates in other gear selector positions, the neutral safety switch requires adjusting. The neutral safety switch is usually located on top of the steering column under the dash. The switch has elongated slots for the retaining screws. Loosen the neutral safety switch retaining screws slightly and slide the switch on the steering column. Continue moving the switch until the starter operates only in PARK and NEUTRAL gear selector positions.

## Throttle Cable Adjustment

If a misadjusted throttle valve (TV) cable is too long, it causes low throttle pressure that results in early transmission upshifts and clutch or band slipping. When the TV cable is too short, it causes late transmission shifts and excessive line pressure. Follow this procedure to perform a typical TV cable adjustment:

1. Be sure the engine and transmission are at normal operating temperature with the ignition switch off.
2. Depress and hold the TV cable lock tab **(Figure 58)**.
3. Pull the slider back through the fitting away from the throttle body until the slider stops against the fitting.
4. Release the lock tab.

Move the throttle linkage to fully open throttle position to automatically adjust the TV cable.

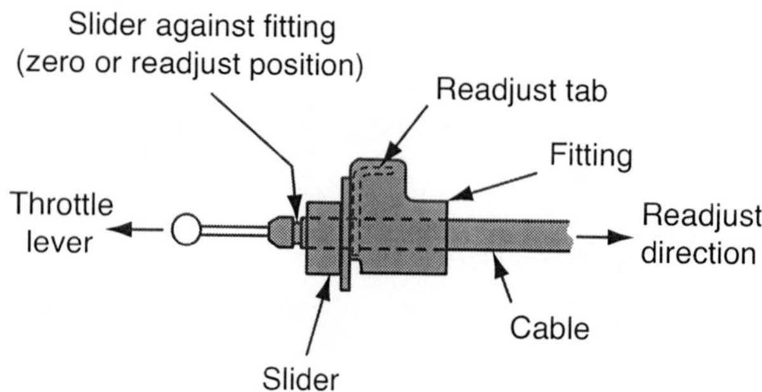


**Figure 58.** Throttle valve (TV) cable adjustment.

## Summary

- The torque converter is a fluid coupling that transfers engine torque to the transmission.
- Torque converter lockup occurs with the engine at normal operating temperature and the vehicle operating at or above a certain speed.
- Various forward reduction and overdrive gear ratios are provided by holding one member of a planetary gearset and providing input to another member with the third member as the output.
- When two members of a planetary set are locked or two members are inputs, the gearset provides a 1:1 ratio.
- A slipping one-way stator clutch causes a loss of engine power during acceleration at low speed and a seized one-way stator clutch causes a loss of engine power at low and high speeds.
- A multiple-disc clutch contains alternately placed drive and driven discs mounted in a clutch drum.
- Bands are used to lock a clutch drum or other component.
- An overrunning clutch allows the outer race to rotate freely in one direction but lock up in the opposite direction.
- The pressure regulator valve limits pump pressure.
- Fluid pressure is supplied from the throttle valve to one end of the shift valves.
- An accumulator is used to cushion a shift and prevent harsh shifting.
- In an electronically controlled transaxle, the pressure is electronically controlled.

- In an electronically controlled transaxle, the computer controls three or more solenoids to control the shifting and fluid pressure. The computer also controls another solenoid to operate the TCC lockup.
- In a four-wheel drive vehicle, the driver may select two-wheel drive or four-wheel drive operation. The driver may also select four-wheel low range operation.
- In an all-wheel drive system, if one wheel spins on a slippery road surface, the viscous coupling immediately transfers torque to the wheels that are not spinning.
- When checking the automatic transmission fluid level, the engine and transmission should be at normal operating temperature.
- Milky fluid on the dipstick indicates the transmission fluid has been contaminated with coolant, whereas dark fluid with a burned smell is usually caused by overheating and burned clutch discs.
- Transmission fluid leaking from the bottom of the flywheel housing indicates a leaking pump seal.
- A cracked flexplate causes a knocking noise that changes with engine speed and load.
- When transmission pressure is low in all gears including reverse, the pump or pressure regulator valve may be defective.
- When transmission pressure is low only in one gear, the seals on the components applied in that gear are leaking.
- A scan tool may be connected to the DLC under the dash to diagnose computer-controlled transmissions and transaxles.
- Proper transmission linkage adjustment provides the correct manual valve position.
- Proper throttle cable adjustment provides proper TV valve position and correct throttle pressure.

## Review Questions

1. Technician A says in a multiple-disc clutch, the driving discs have friction material on their surfaces. Technician B says in a multiple-disc clutch, the driving discs have internal serrations. Who is correct?
   A. Technician A
   B. Technician B
   C. Both Technician A and Technician B
   D. Neither Technician A nor Technician B
2. Technician A says the bands in an automatic transaxle may be applied by accumulators. Technician B says a band is used to lock one member in a planetary gear set. Who is correct?
   A. Technician A
   B. Technician B
   C. Both Technician A and Technician B
   D. Neither Technician A nor Technician B
3. Technician A says the oil pump in an automatic transaxle supplies fluid pressure to the pressure regulator valve. Technician B says the pressure regulator valve limits fluid pressure by reducing the oil pump speed. Who is correct?
   A. Technician A
   B. Technician B
   C. Both Technician A and Technician B
   D. Neither Technician A nor Technician B
4. Technician A says the manual valve is controlled by fluid pressure from the oil pump. Technician B says the gear shift linkage controls the pressure regulator valve. Who is correct?
   A. Technician A
   B. Technician B
   C. Both Technician A and Technician B
   D. Neither Technician A nor Technician B
5. When discussing torque converter operation, Technician A says at lower vehicle speeds the stator redirects fluid leaving the turbine so this fluid flow helps to rotate the impeller pump. Technician B says rotary fluid flow occurs in a torque converter at lower vehicle speeds. Who is correct?
   A. Technician A
   B. Technician B
   C. Both Technician A and Technician B
   D. Neither Technician A nor Technician B
6. The coupling point occurs in a torque converter when the:
   A. impeller pump is rotating faster than the turbine.
   B. turbine is rotating faster than the impeller pump.
   C. stator is rotating faster than the impeller pump.
   D. impeller pump, turbine, and stator begin to rotate together.
7. In a planetary gear set, the planet carrier is the input, the ring gear is the output, and the sun gear is held. This combination provides:
   A. a forward gear overdrive.
   B. a reverse gear overdrive.
   C. a reverse gear reduction.
   D. a forward gear reduction.
8. ATF is leaking from the bottom of the flywheel housing. The most likely cause of this problem is:
   A. a leaking torque converter weld.
   B. a leaking rear main bearing seal.
   C. a leaking transmission pump seal.
   D. a misaligned transmission.

9. When performing a pressure test on an automatic transmission, the pressure is low in all gears including reverse. Technician A says the pressure regulator valve may be sticking or defective. Technician B says the seal on the reverse clutch piston may be leaking. Who is correct?
   A. Technician A
   B. Technician B
   C. Both Technician A and Technician B
   D. Neither Technician A nor Technician B
10. When diagnosing an automatic transmission in an OBD II vehicle, the scan tool displays DTC P1711. All of these statements about DTC 1711 are true *except*:
    A. This DTC belongs to the powertrain area.
    B. This DTC is a standard code defined by SAE.
    C. This DTC belongs to the transmission subgroup.
    D. This DTC indicates a defect in the transmission oil temperature sensor or circuit.
11. The governor is driven by the transmission ______________ ______________.
12. Governor pressure is supplied to one end of the shift valves and ______________ ______________ is supplied to the opposite end of these valves.
13. To provide a transaxle downshift, ______________ pressure must overcome ______________ pressure.
14. An accumulator is used to ______________ the shifts in a transmission or transaxle.
15. Explain the power flow through a typical four-speed automatic transmission in OD, second gear.
16. Describe the operation and purpose of a modulator.
17. Explain why a governor is not required in a computer-controlled automatic transaxle.
18. Explain how an all-wheel drive vehicle transfers torque to the other drive wheels when one wheel spins on a slippery road surface.

# Conventional and Antilock Brake Systems

*Automotive industry experts predict that in the next decade, 60 percent of all vehicles sold in the world will accumulate in seven nations: China, South Korea, Thailand, Brazil, Mexico, Russia, and Poland. The largest percentage of growth in automotive sales is China where sales increased 55 percent in 2002. One industry leader predicts that China could surpass Japan as the world's second largest automotive market.*

## SECTION OBJECTIVES

After you have read, studied, and practiced the contents of this section, you should be able to:

- Explain the basic brake principle related to pressurized fluid in a confined space.
- Describe three different types of brake fluid.
- Explain master cylinder design and operation.
- Describe drum brake operation and explain brake fade.
- Describe disc brake operation, including the advantages of disc brakes.
- Explain the purpose of a metering valve, proportioning valve, and a pressure differential valve.
- Describe the operation of a vacuum brake booster.
- Explain the operation of the parking brakes on a drum brake system.
- Describe the advantages of an ABS.
- Describe the operation of a high-pressure accumulator during the antilock function.
- Explain the operation of the amber and red brake warning lights.
- Explain the difference between a three-channel and a four-channel ABS.
- Perform brake system inspections.
- Diagnose brake system problems.
- Adjust pedal free-play and drum brakes.
- Bleed brake systems.
- Service drum brake and disc brake systems.
- Observe the ABS warning light and determine the ABS condition.
- Check the master cylinder fluid level on ABS.
- Obtain flash DTCs on various ABS.
- Obtain ABS DTCs using a scan tool.
- Perform all the ABS diagnostic functions using a scan tool.
- Diagnose ABS wheel speed sensors.
- Bleed ABS using various methods, including the automated bleed procedure with a scan tool.

# Chapter 44 Brake System Design and Operation

## Introduction

The brake system is one of the most important systems on the vehicle. The brake system must slow and stop a vehicle in a short distance and it must perform this function without causing steering pull or premature wheel lockup. Technicians who service brake systems must be highly skilled experts because the work they perform can save lives. Basic brake system components and operation are explained in this chapter.

## HYDRAULIC PRINCIPLES

One of the basic principles of hydraulics is that liquids are not compressible. A second principle of hydraulics used in brake system is that pressure on a confined fluid is transmitted equally in all directions and acts with equal force on equal areas. When 10 pounds of force is applied to the brake pedal and the master cylinder piston has an area of 1 square inch, the force exerted by the fluid in the master cylinder is $10 \div 1 = 10$ psi. If the wheel cylinders also have an area of 1 square inch, the 10 psi from the master cylinder exerts a pressure of 10 psi in each wheel cylinder. Brake system principles that must be understood include the following:

1. If the master cylinder piston and bore size is decreased, the pressure exerted by the master cylinder increases for a given pressure on the brake pedal.
2. If the master cylinder piston and bore size is increased, the pressure exerted by the master cylinder decreases for a given pressure on the brake pedal.
3. A smaller diameter master cylinder requires more piston travel to displace the same amount of fluid as a larger piston.
4. The force on the brake pedal and the diameter of the master cylinder piston determine the pressure in the brake system.
5. The diameter of the wheel cylinders or calipers determines the force against the brake shoes or pads. A larger diameter wheel cylinder or caliper piston exerts more force on the brake shoes or pads.

## BRAKE FLUIDS

Brake fluid quality is extremely important to provide proper brake system operation. Brake fluid must have these qualities:

1. A controlled amount of swell. Brake fluids must provide a controlled amount of swell in cups and seals to provide adequate sealing in calipers, wheel cylinders, and master cylinders. Excessive swelling of cups and seals causes brake drag and inadequate brake response.
2. Temperature extremes. Brake fluid must operate at temperatures from −104°F (−75°C) to 500°F (260°C).
3. Compatibility with rubber. Brake fluid must be compatible with rubber in master cylinder cups, wheel cylinder cups, caliper seals, and brake hoses.
4. Lubricating ability. Brake fluid must have a satisfactory lubricating ability to provide smooth operation of brake components.
5. Resistance to evaporation. Brake fluid must resist evaporation at high temperatures.
6. Antirust and anticorrosion. Brake fluid must combat rust and corrosion in brake system components.
7. Classification standards. Brake fluid must meet classification standards established by the SAE.

Three brake fluid classifications are used at the present time:

1. A brake fluid classified as DOT 3 has a minimum dry equilibrium boiling point (ERBP) of 401°F (205°C) and a minimum wet ERBP of 284°F (140°C).
2. A brake fluid classified as DOT 4 has a minimum dry ERBP of 446°F (230°C) and a minimum wet ERBP of 356°F (180°C).
3. A brake fluid classified as DOT 5 is silicone based and this fluid has a minimum dry ERBP of 500°F (260°C) and a minimum wet ERBP of 356°F (180°C).

Always use the brake fluid specified by the vehicle manufacturer.

You Should Know

*DOT 3 and DOT 4 brake fluids are* ***hygroscopic****, which means they attract moisture from the air. Therefore, containers containing these types of brake fluid must be kept tightly closed. DOT 5 brake fluid is* ***non-hygroscopic****, which indicates it does not attract moisture from the air. DOT 5 brake fluid must not be mixed with DOT 3 or DOT 4 brake fluids.*

## MASTER CYLINDERS

The purpose of the master cylinder is to supply brake fluid pressure to the wheel cylinders or calipers during a brake application **(Figure 1)**. Dual master cylinders have been mandated since 1967. If fluid pressure is not available in one section of the master cylinder, fluid pressure in the other section is applied at two wheels to stop the vehicle. The master cylinder may be manufactured from cast iron or aluminum. Cast iron master cylinders have an integral reservoir and a removeable plastic reservoir is used on aluminum master cylinders. A cover bail is used to retain the cover and gasket on top of the master cylinder.

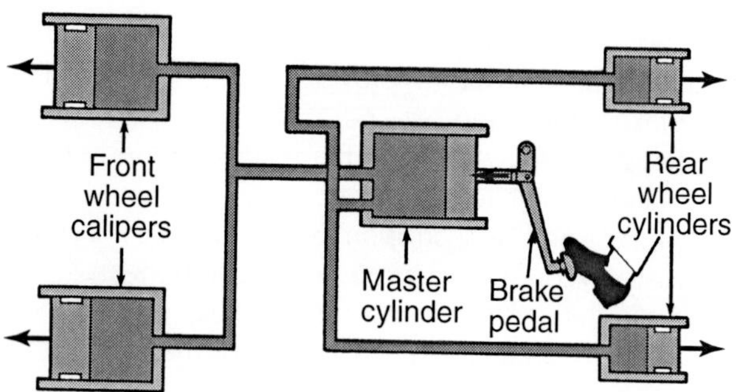


**Figure 1.** The master cylinder supplies brake fluid pressure to the calipers and wheel cylinders.

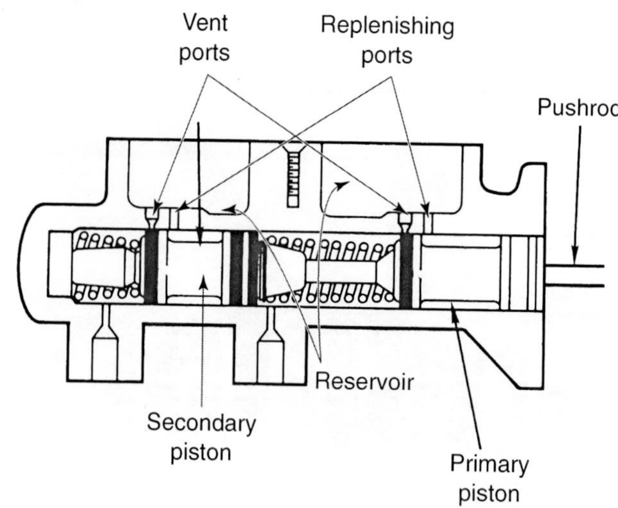


**Figure 2.** Master cylinder design.

The primary and secondary pistons are mounted in the master cylinder bore. Rubber cups are used to seal these pistons in the master cylinder bore. A spring behind each piston holds the pistons in the released position **(Figure 2)**. Steel brake lines are connected from the master cylinder to the wheel cylinders or calipers. On rear-wheel drive vehicles, the primary piston outlet is connected to the front brakes and the secondary piston outlet is connected to the rear brakes **(Figure 3)**. On many front-wheel drive vehicles, the brake system is diagonally split so the primary master cylinder piston supplies fluid to the right front and

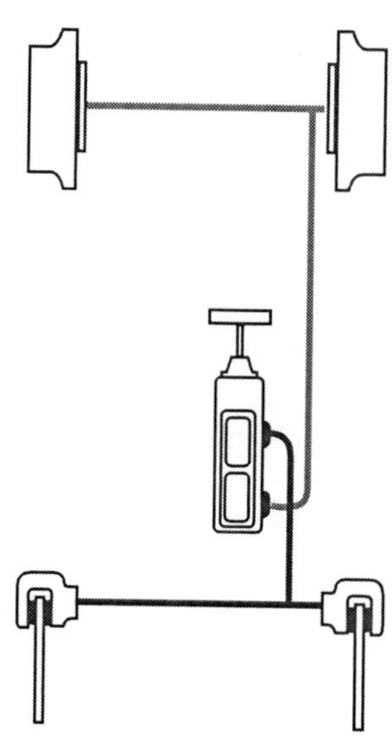

**Figure 3.** A front-to-rear split brake system.

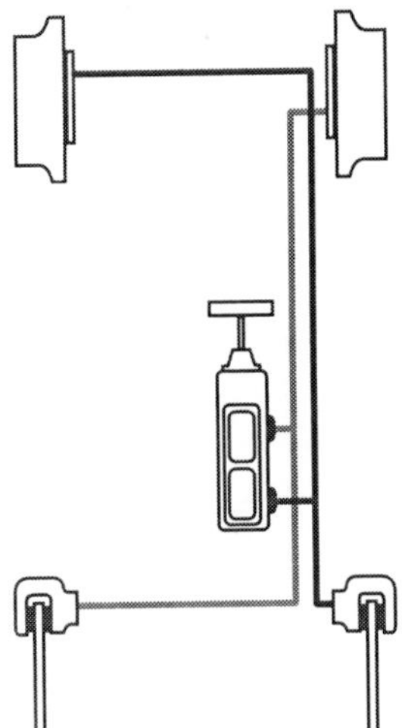

Figure 4. A diagonally-split brake system.

left rear brakes and the secondary piston supplies fluid to the left front and right rear brakes **(Figure 4).**

When the brake pedal is depressed, the primary piston moves down the master cylinder bore and the primary cup seals the vent port. This action seals the primary piston cylinder and forces fluid from the primary outlet port. Pressure in the primary bore moves the secondary piston down its bore and the vent port is sealed by the secondary piston cup. Further movement of the secondary piston forces fluid from the secondary outlet to the wheel calipers or cylinders. When fluid is forced from both outlets, the brakes are applied on all four wheels. If the brakes are applied repeatedly, the brake fluid may become very hot. Under this condition, the fluid expands and some fluid flows back through the vent ports when the brakes are released. The vent ports prevent excessive pressure buildup in the brake system.

If the brake shoe adjustment is not correct or air enters the brake system, excessive brake pedal movement is required to apply the brakes. Under this condition, the driver may pump the brake pedal rapidly to apply the brakes. This pumping action may cause a pressure drop in the master cylinder because the fluid cannot flow back from the wheel cylinders or calipers as quickly as the master cylinder pistons can move when the brakes are released. When pressure drops in the master cylinder, atmospheric pressure on top of the fluid in the reservoir forces fluid through the replenishing ports past the cup seals and through the holes in the ends of the pistons into the piston bores.

On a cast iron master cylinder, the rubber cover gasket seals the cover to the reservoir preventing moisture from entering the system. A wire cover retainer holds the cover on the master cylinder **(Figure 5)**. Atmospheric vents are located between the cover and the gasket. If the fluid level goes down in the reservoir, the bellows on the cover gasket expand into the reservoir and air flows through the vents into the area between the cover and the gasket. Some drum brake master cylinders have a residual valve at the outlet connected to the drum brakes **(Figure 6)**. This resid-

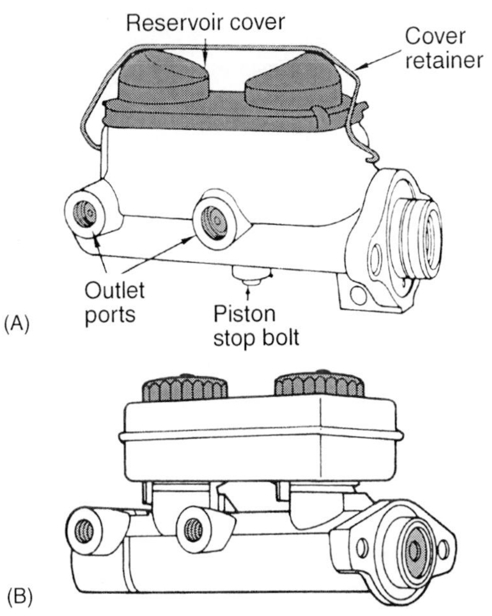


Figure 5. (A) Cast-iron master cylinder, and (B) aluminum/composite master cylinder.

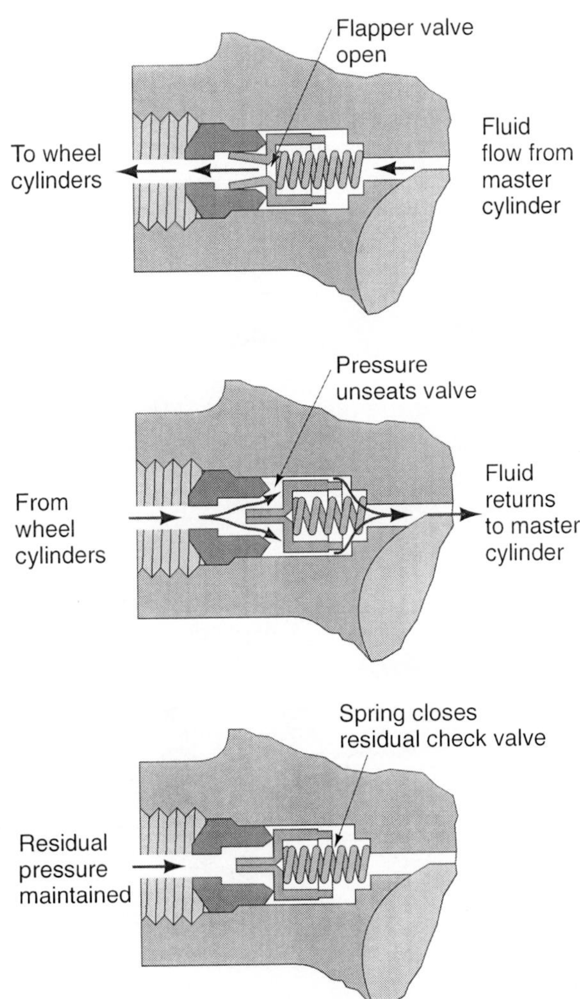


Figure 6. A residual pressure valve.

ual valve maintains a slight pressure in the brake system that keeps the wheel cylinder cups expanded outward to provide improved cup sealing. Other brake systems have wheel cylinder cup expanders in place of the residual valve.

## DRUM BRAKES

In a drum brake system, the brake shoes are forced outward against the drum by the wheel cylinder pistons. Each wheel cylinder contains two pistons and two cup seals with expanders **(Figure 7)**. The expanders prevent air from entering the system when the pistons and cups are moving. A spring is positioned between the piston cups. A rubber dust boot is mounted on each end of the wheel cylinder. Pushrods are located between the pistons and the brake shoes. The pushrods fit snugly in the dust boot openings. The dust boots keep contaminants out of the wheel cylinder. The wheel cylinder is bolted to the backing plate and a bleeder screw is mounted in the back of the wheel cylinder. The bleeder screw allows air to be bled from the wheel cylinder and brake system.

Drum brake shoes are made from stamped steel. The brake linings are riveted or bonded to the brake shoes. The curvature of the brake shoes and linings matches the contour of the brake drums. Brake linings must be able to withstand extreme heat. The brake linings may be organic, semi-metallic, metallic, or synthetic. Organic brake linings are made from non-metalic fibers bonded together to form a composite material. Organic brake linings contain friction modifiers that may include graphite and powdered metals. Fillers, binders, and curing agents are also used on organic brake linings.

You Should Know *In past years, asbestos was used in brake lining material. Because asbestos has been proven to contribute to lung cancer, asbestos has been eliminated from brake linings currently manufactured in North America.*

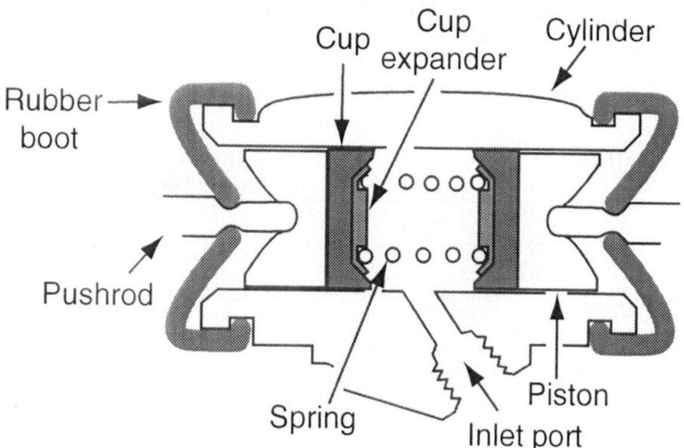


**Figure 7.** Wheel cylinder design.

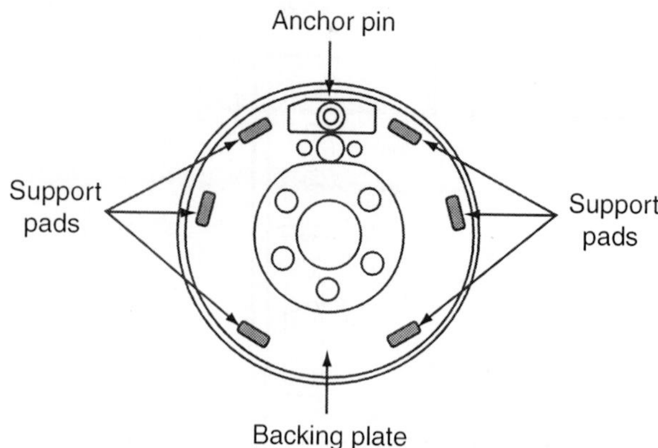


**Figure 8.** Backing plate with anchor pin and support pads.

Most drum brakes are servo-type. In this type of brake, the brake shoe return springs hold the shoes against the anchor pin mounted in the top of the backing plate. An adjusting mechanism is located between the lower ends of the brake shoes, but the lower ends of the brake shoes are not connected to the backing plate. The edges of the brake shoes contact the support pads on the backing plate **(Figure 8)**. The primary brake shoe is positioned toward the front of the vehicle and the secondary shoe is positioned toward the rear of the vehicle **(Figure 9)**. When the brakes are applied, pressure from the master cylinder forces the wheel cylinder pistons outward. As these pistons move outward, the pushrods force the brake shoes outward against the brake drums. Friction between the brake shoes and the drum provides a braking action. The brake system changes the kinetic energy of the moving vehicle into heat energy through the application of friction. **Kinetic energy** is energy in motion.

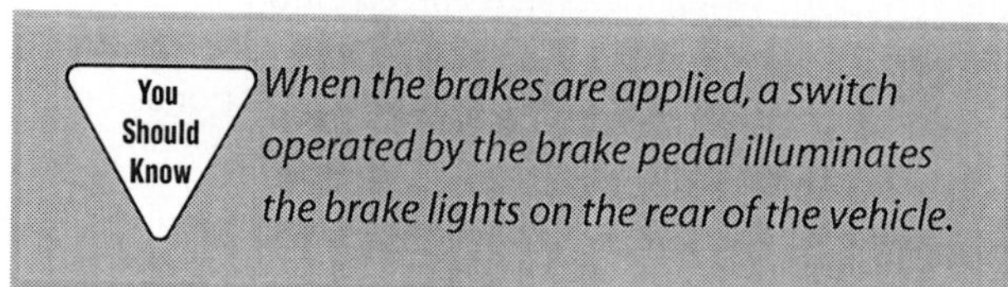

You Should Know *When the brakes are applied, a switch operated by the brake pedal illuminates the brake lights on the rear of the vehicle.*

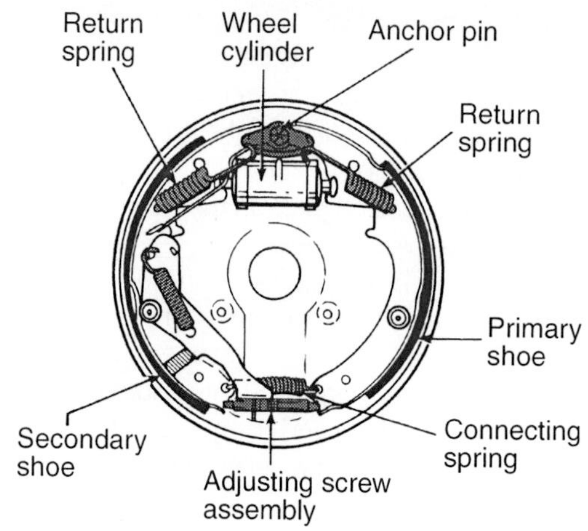


**Figure 9.** A servo-type drum brake.

This friction between the brake linings and the drums creates a great deal of heat and the drums expand as they are heated. When the drums expand, the brake shoes must move farther outward to provide braking action. This additional shoe movement causes increased brake pedal movement and brake fade. **Brake fade** occurs on drum brakes when the drums become very hot during hard, repeated braking. Under these operating conditions, the drums expand and the shoes must travel further outward to provide braking action. Increased shoe movement requires greater pedal movement.

When the primary shoe is forced against the drum surface, the wheel rotation tends to transfer movement to the secondary shoe by servo action and force this shoe against the brake drum surface **(Figure 10)**. **Servo action** occurs when the operation of the primary shoe applies mechanical force to the secondary shoe to assist in its application.

Therefore, the secondary shoe is forced against the brake drum by movement from the primary shoe and brake drum and also by fluid pressure in the wheel cylinder. In a servo-type brake, approximately 75 percent of the braking force is from the secondary shoe. Therefore, the secondary shoe has a longer lining compared to the primary shoe.

The adjusting mechanism between the lower end of the brake shoes contains a threaded star wheel with two extensions that fit over the lower ends of the brake shoes. Most drum brakes have a self-adjusting mechanism **(Figure 11)**. The self-adjusting mechanism contains a cable attached to the anchor pin. The lower end of this cable is connected to a pivoted adjusting lever mounted above the star wheel. A spring holds the adjusting lever downward against the star wheel. As the brake lining wears, the outward shoe movement increases when the brakes are applied. During a brake application in reverse, the secondary shoe moves away from the anchor pin. If the lining is worn enough to allow sufficient shoe movement, the spring pulls the adjusting lever down enough to rotate the star wheel to the next notch. This action lengthens the star wheel assembly and moves the brake shoes outward so the linings are closer to the drum surface.

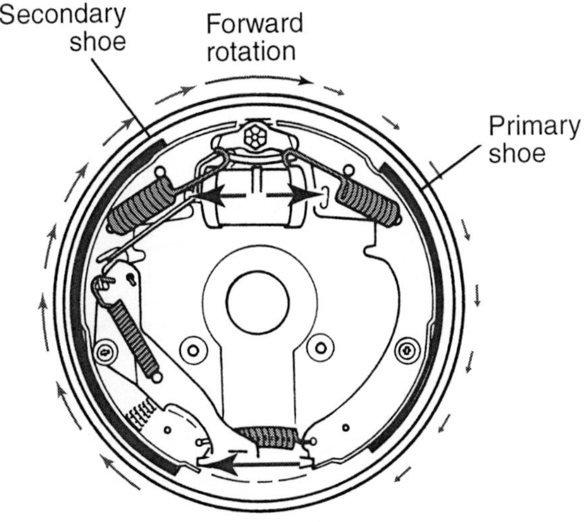


**Figure 10.** Servo-type braking forces.

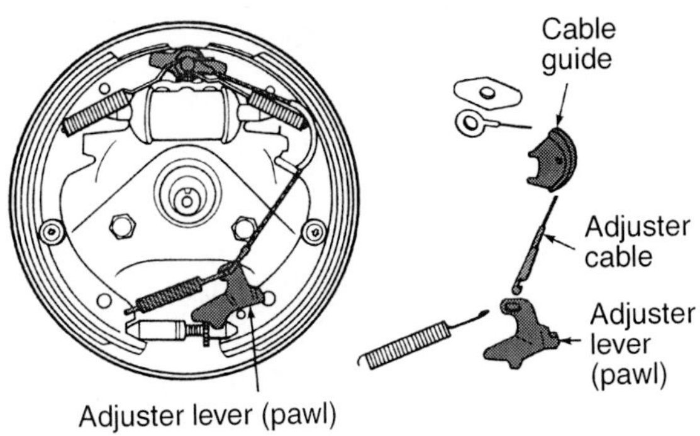


**Figure 11.** A cable-type self-adjusting mechanism.

> **You Should Know** *If a drum brake does not have a self-adjusting mechanism, a tool may be inserted through an opening in the backing plate to rotate the star wheel and adjust the brakes manually.*

> **You Should Know** *On some brake systems, the brake shoes have a backing plate anchor between the lower ends of the brake shoes* ***(Figure 12)****. This type of brake does not provide servo action and may be referred to as a leading-trailing brake.*

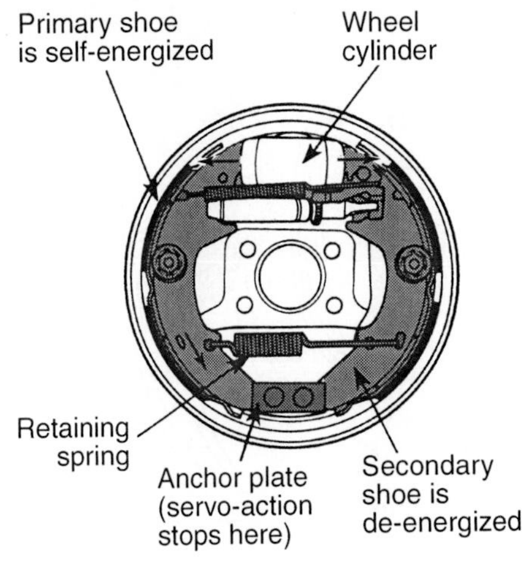


**Figure 12.** A leading-trailing brake assembly.

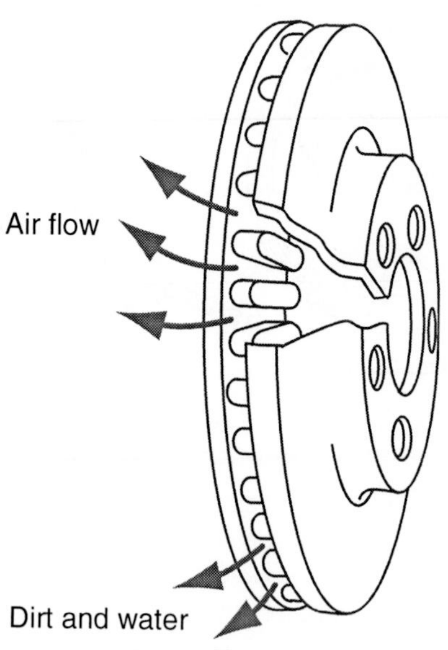


Figure 13. A ventilated brake rotor.

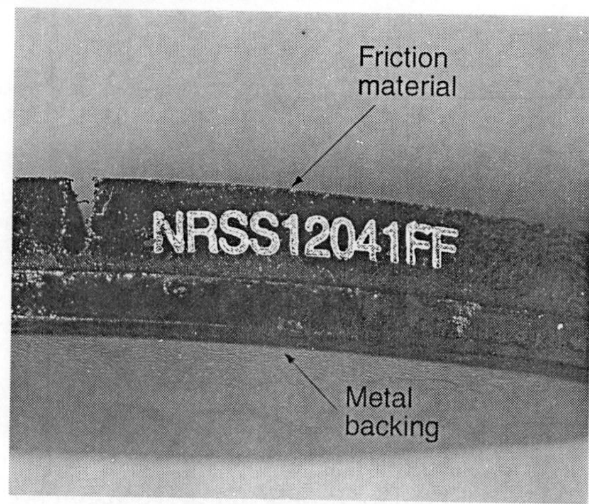


Figure 15. Brake lining edge code.

## DISC BRAKES

A disc brake has a cast iron disc or rotor mounted on the wheel hub. Both sides of the rotor have machined surfaces. During a brake application, the brake pad linings contact both sides of the rotor. Most brake rotors have ventilating slots between the two sides of the rotor to act as a fan and dissipate heat from the rotor **(Figure 13)**. The caliper and brake pad assembly is mounted over top of the rotor and the caliper is bolted to the steering knuckle **(Figure 14)**. Many calipers contain a single piston mounted in the caliper bore on the inboard side of the caliper. This type of disc brake has a floating caliper. A **floating caliper** is mounted so it is free to slide sideways on the mounting bolts.

The brake hose is threaded into the back of the caliper and a bleeder screw is also positioned in the rear of the caliper.

The brake linings are bonded or riveted to the brake pads. Brake linings on disc or drum brakes have a code on the edge of the linings to identify the lining material **(Figure 15)**. Similar materials are used in the brake linings on drum and disc brakes. Metal locating tabs on most brake pads retain the pads to the caliper. The caliper piston has a seal around the outer diameter of the piston. A dust boot is mounted between the caliper piston and the housing to keep contaminants out of the caliper bore **(Figure 16)**. When the brakes are applied, the caliper piston is forced outward. This movement pushes the inboard brake pad lining against the rotor. This action forces the floating caliper inward so the outer brake pad lining is forced against the

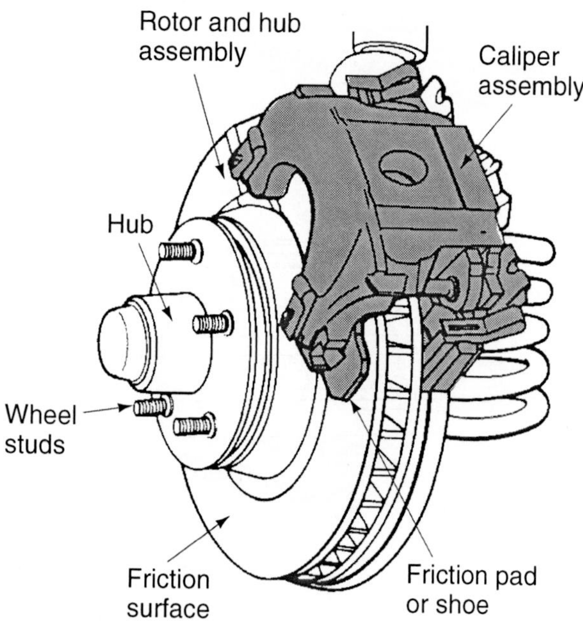


Figure 14. Disc brake assembly.

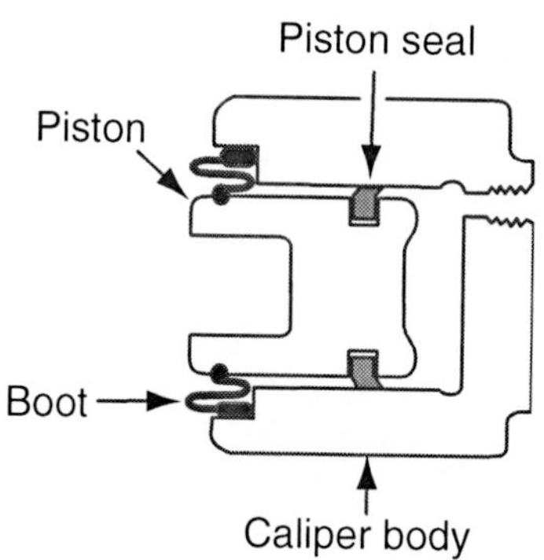


Figure 16. A caliper piston dust boot.

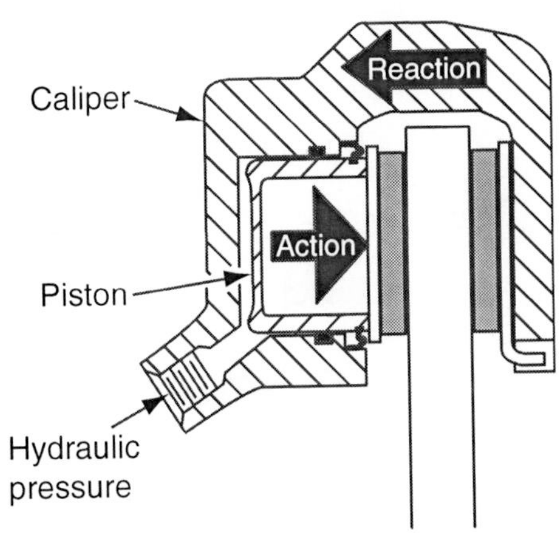


**Figure 17.** Floating caliper action during a brake application.

outside of the rotor **(Figure 17)**. When the brakes are applied, the caliper piston seal twists. After the brakes are released, the seal returns to its original shape. This seal action pulls the caliper piston and brake pad lining away from the rotor surface **(Figure 18)**. As the brake pad linings wear, the caliper pistons move outward. This eliminates the need for adjusting disc brakes. Disc brake do not experience brake fade because as the rotors are heated they expand and move slightly closer to the brake pad linings. Disc brakes do not provide any servo action. Many disc brakes have a wear indicator attached to one of the brake pads. When the brake pad lining wears a specific amount, the wear indicator contacts the rotor causing a scraping noise that alerts the driver regarding the brake problem **(Figure 19)**.

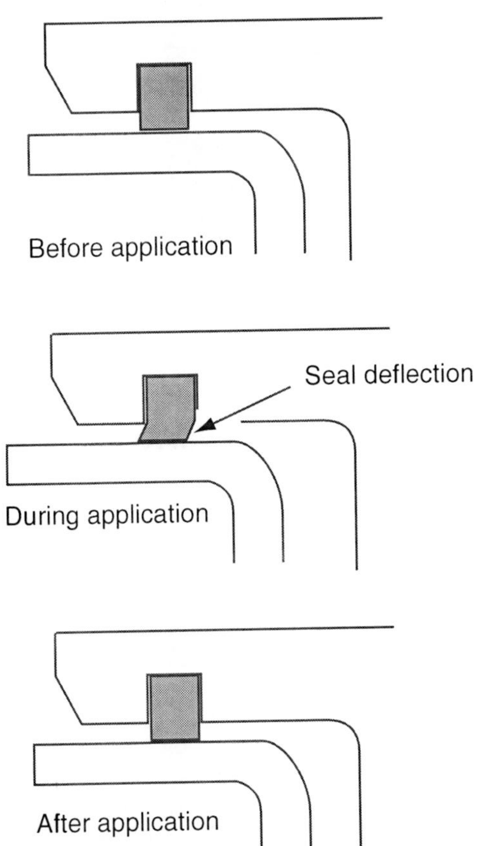


**Figure 18.** Caliper piston seal action during a brake application and brake release.

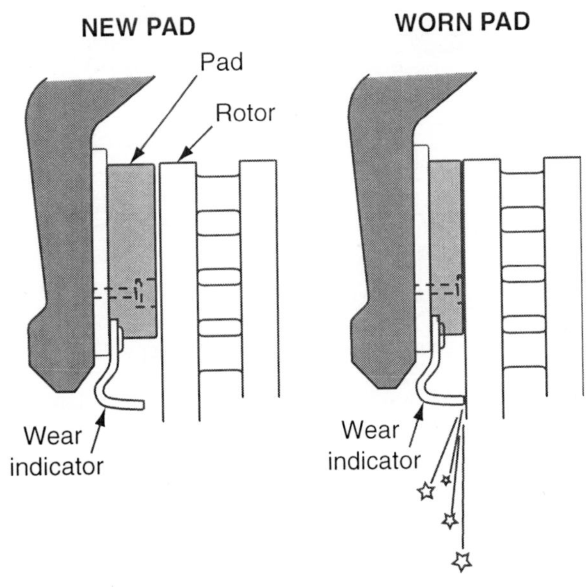


**Figure 19.** Brake lining wear indicator.

> **You Should Know** *Some calipers contain two pistons, one on each side of the rotor. Other calipers contain two pistons on each side of the rotor. Some disc brakes have an electronic sensor that illuminates a warning light in the instrument panel when the brake linings are worn a specific amount.*

## BRAKE LINES, HOSES, AND VALVES

Brake lines are made from seamless steel tubing that is coated with zinc or tin for corrosion protection. Brake lines must conform to SAE standard J1047 which requires that an 18-inch section of tubing must withstand an internal pressure of 8,000 psi. Brake lines may have an inverted double flare or an ISO flare. The inverted double flare has the end of the tubing flared out and then it is formed back onto

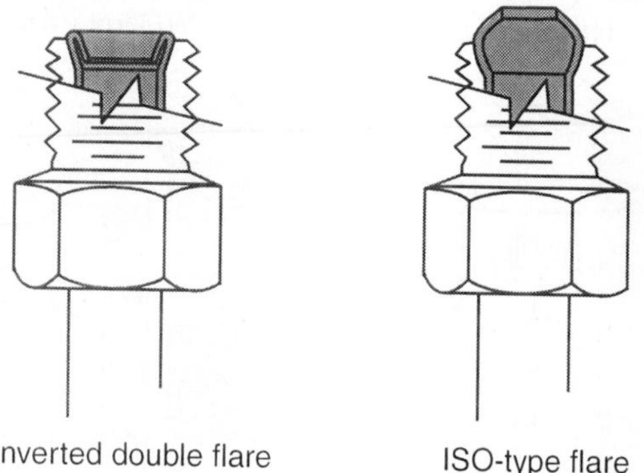


**Figure 20.** Inverted double flare and ISO flare on brake tubing.

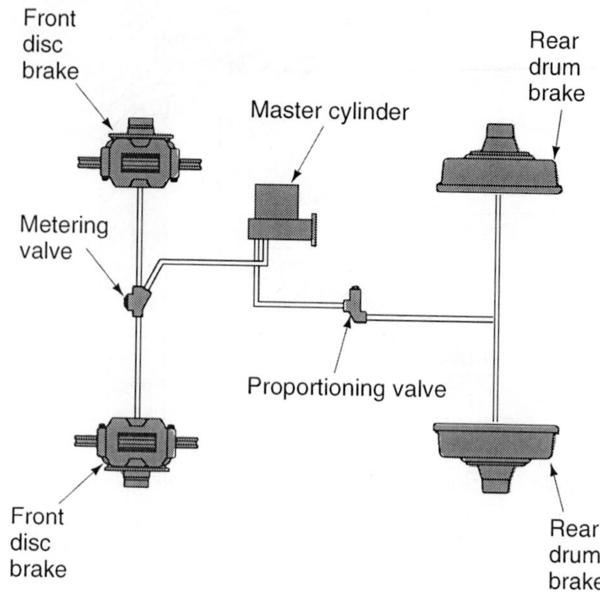


**Figure 21.** Metering valve and proportioning valve.

itself **(Figure 20)**. An ISO flare has a bubble-shaped end formed on the tubing. Different fittings are required with each type of flare and these fittings are not interchangeable. Brake tubing is specially bent to fit a specific location on the vehicle. When a brake line is leaking, the complete line must be replaced, not repaired. It is preferable to replace brake lines with the OEM pre-shaped brake line.

Brake hoses are a flexible connection between the chassis and the wheel calipers or cylinders or between the rear axle and the chassis. Brake hoses must conform to SAE J1401 standard that requires the hose to withstand 4,000 psi pressure for 2 minutes without bursting. Brake hoses contain an inner liner surrounded by metal fabric plies, a rubber separator layer, and an outer jacket. The fittings are attached to the hose during the manufacturing process.

## Metering Valve

The metering valve is connected in the brake line to the front brakes **(Figure 21)**. The metering valve is used on vehicles with front disc brakes and rear drum brakes. During a brake application, the metering valve delays the fluid pressure to the front brakes momentarily until the rear brake shoes are forced outward against the drums.

## Proportioning Valve

During a brake application, vehicle inertia and momentum tend to shift the vehicle weight forward. This weight shift is proportional to braking force and the rate of deceleration. This weight shift reduces traction between the rear tires and the road surface which may result in rear wheel lockup. To provide maximum braking, equal friction must be maintained between the front and rear tires and the road surface. During a brake application, the proportioning valve modulates pressure to the rear brakes to prevent rear wheel lockup. The proportioning valves may be integral in the master cylinder or they may be external to the master cylinder **(Figure 22)**.

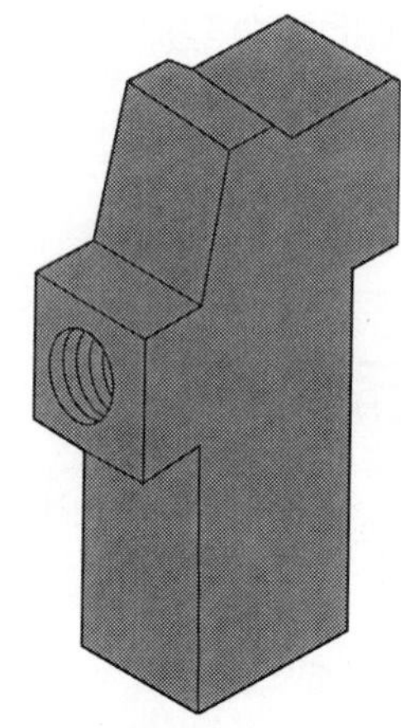

**Figure 22.** A proportioning valve.

## Pressure Differential Valve

A pressure differential valve is connected in the brake lines from the master cylinder to the front and rear brakes. A wire from the brake warning light in the instrument panel is connected to a switch in the pressure differential valve. If the brake fluid pressure is equal in both the primary and secondary sections of the master cylinder, the switch in the pressure differential valve remains open. If one section of the master cylinder is low on brake fluid and the pressure in this section decreases, the piston in the pressure differential valve moves toward the section in the pressure differential valve with the low pressure. This piston move-

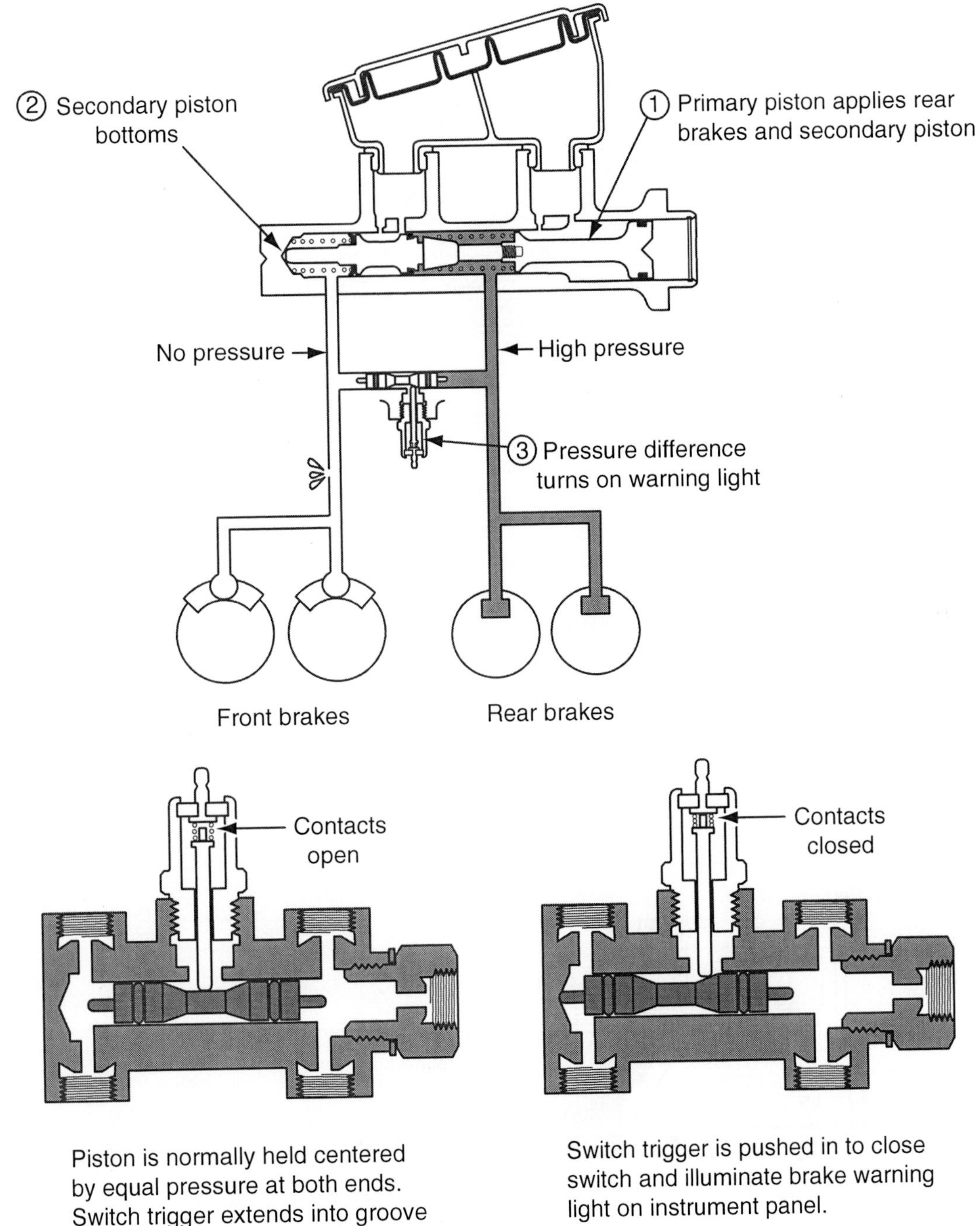

② Secondary piston bottoms
① Primary piston applies rear brakes and secondary piston
No pressure
High pressure
③ Pressure difference turns on warning light
Front brakes
Rear brakes
Contacts open
Contacts closed
Piston is normally held centered by equal pressure at both ends. Switch trigger extends into groove and switch is open.
Switch trigger is pushed in to close switch and illuminate brake warning light on instrument panel.

**Figure 23.** A pressure differential valve.

ment pushes the stem upward in the brake-warning switch and closes the switch contacts **(Figure 23)**. Under this condition, current flows through the brake warning light, the switch contacts to ground, and the brake warning light is illuminated to warn the driver that a brake problems is present. Some brake systems have the pressure differential valve combined with the metering valve and proportioning valves in a combination valve assembly. Pressure differential, metering, and proportioning valves are non-serviceable components.

## VACUUM BRAKE BOOSTERS

Most vehicles are equipped with a vacuum brake booster. The brake booster is connected between the

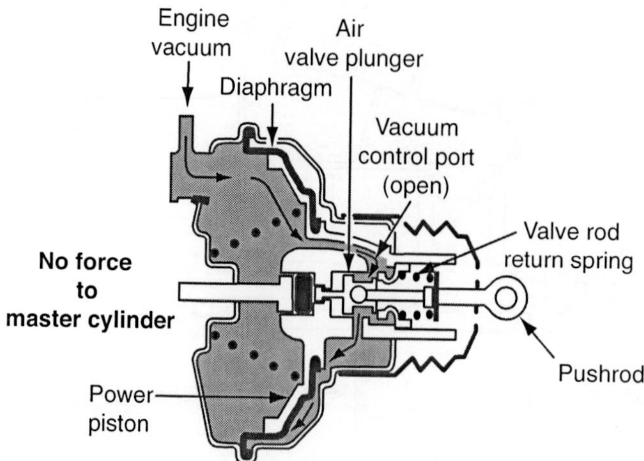


**Figure 24.** A vacuum brake booster with brakes released.

master cylinder and the brake pedal. A pushrod is connected from the booster to the brake pedal, and another pushrod is mounted between the brake booster and the master cylinder **(Figure 24)**. When the engine is running, manifold vacuum is supplied through a hose to the booster. If the brakes are released, manifold vacuum is supplied to both sides of the booster diaphragm. This diaphragm does not exert any force on the master cylinder pistons. When the brakes are applied, the brake pedal movement closes the vacuum passage to the rear of the booster diaphragm and opens an atmospheric pressure port to allow air pressure on the rear side of the diaphragm **(Figure 25)**. Because manifold vacuum is supplied to the front side of the diaphragm and atmospheric pressure is provided on the rear side of the diaphragm, the diaphragm is moved toward the master cylinder with considerable force. Under this condition, the diaphragm moves the pushrod and supplies force to the master cylinder pistons to provide brake assist. A reaction disc between the booster diaphragm and the master cylinder pushrod provides brake pedal feel to the driver.

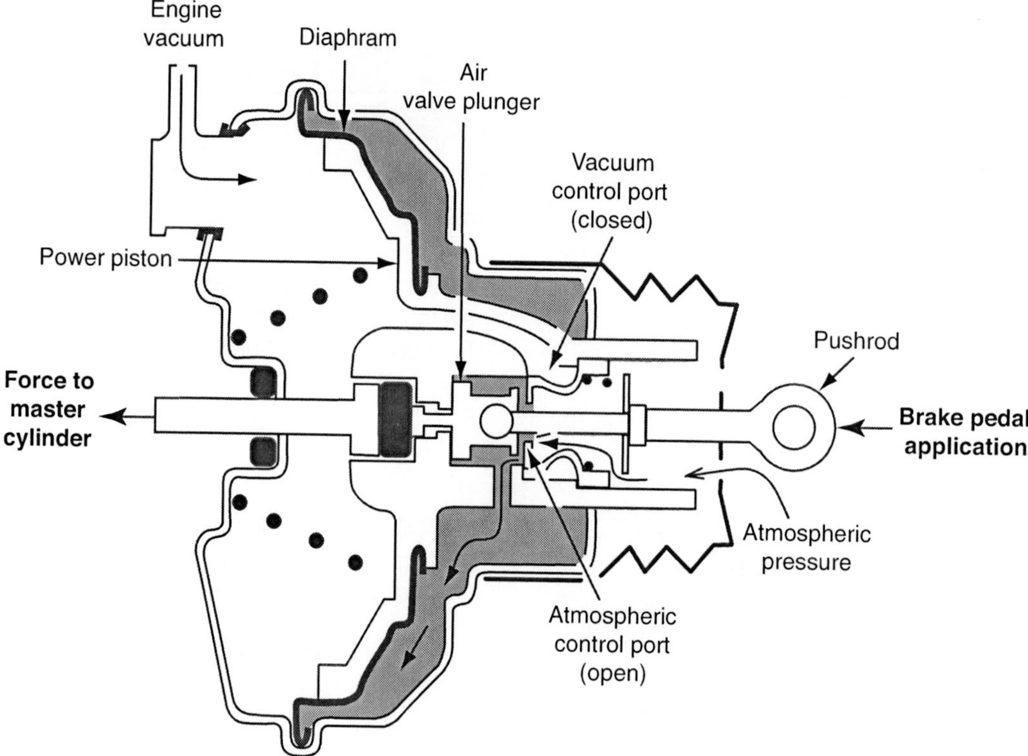


**Figure 25.** A vacuum brake booster with brakes applied.

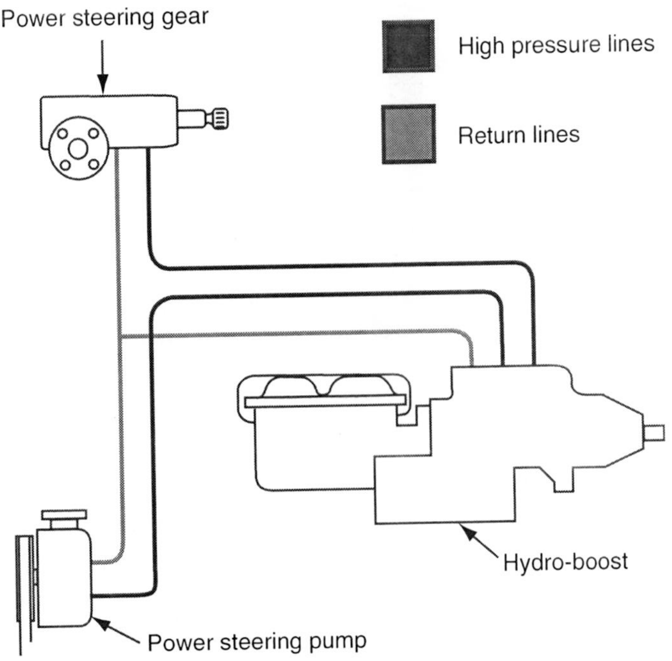


**Figure 26.** A hydro-boost power brake system.

## HYDRO-BOOST BRAKE SYSTEM

Some vehicles have a hydro-boost brake system with a hydraulic brake booster. These systems are often used on diesel engines because of the low manifold vacuum in these engines. In a hydro-boost brake system, a hydraulic brake booster is mounted between the master cylinder and the brake pedal. The power steering pump supplies fluid pressure to the hydro-boost unit and the power steering gear **(Figure 26)**. Some vehicles have a separate pump for the hydro-boost system. When the brakes are released, power steering pump pressure forces fluid through the hydro-boost unit, but this pressure is not applied to the master cylinder pistons. If the brakes are applied, brake pedal movement operates a lever in the hydro-boost unit and opens a valve that supplies power steering pump fluid pressure to the master cylinder pistons and provides brake assist.

## PARKING BRAKES

Parking brakes are applied by a foot- or hand-operated lever in the passenger compartment **(Figure 27)**. When the parking brakes are applied, a switch on the parking brake lever grounds the red brake warning light in the instrument panel. A release handle on the parking brake lever is pulled

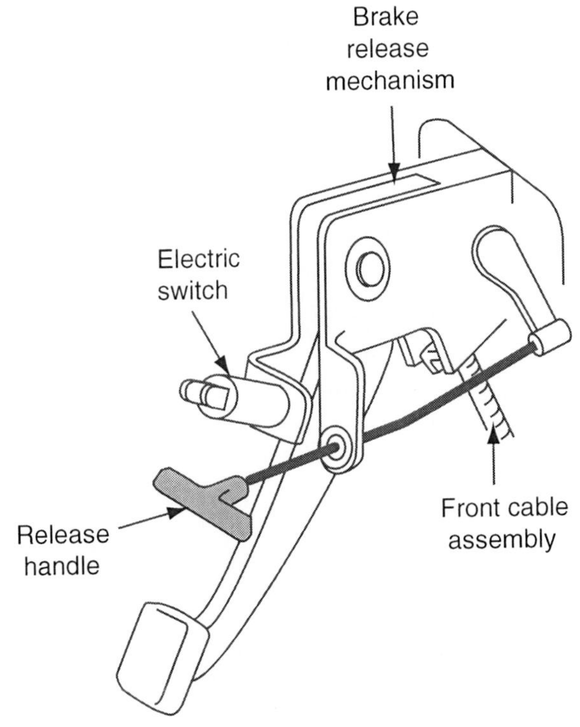


**Figure 27.** A foot-operated parking brake mechanism.

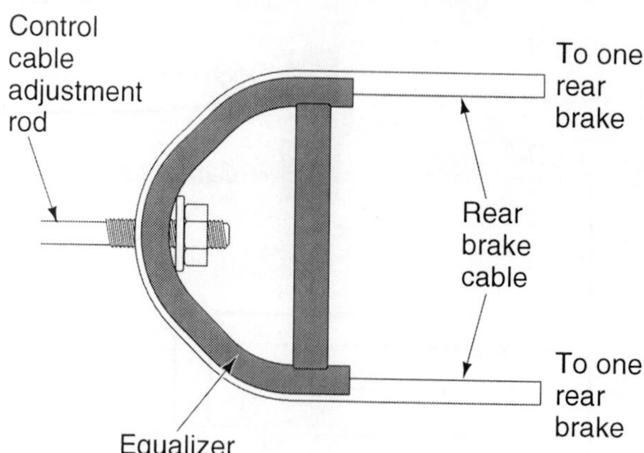


**Figure 28.** Parking brake equalizer and cables.

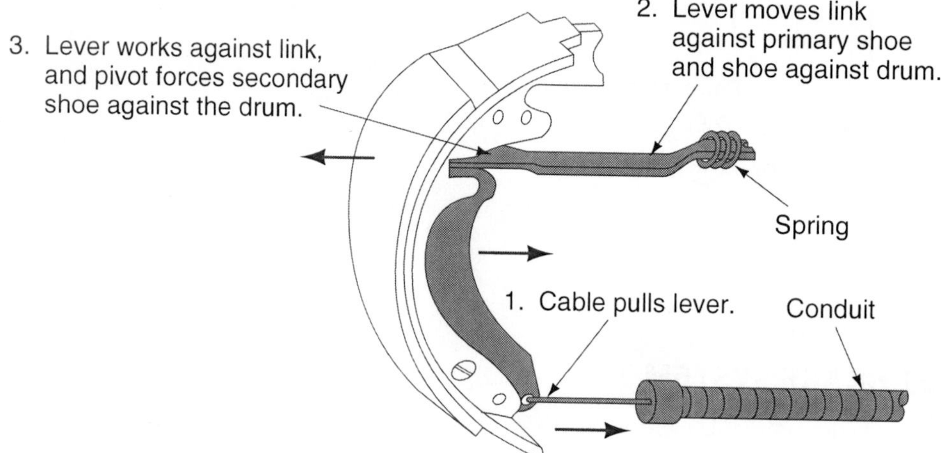


**Figure 29.** Parking brake lever and strut operation.

to release the parking brakes. Some parking brakes have a vacuum-operated release mechanism that releases the parking brake when vacuum is supplied to the vacuum diaphragm in the parking brake release actuator. This type of parking brake also has a mechanical release lever to release the parking brake if the vacuum actuator is not operating. A cable is connected from the parking brake lever to an equalizer under the vehicle. Dual cables are connected from the equalizer to the rear brakes **(Figure 28)**. When the parking brakes are applied, the cable in each rear wheel pulls a lever that forces the secondary and primary brake shoes outward against the drum to apply the brakes **(Figure 29)**. If the vehicle has rear disc brakes, the parking brake cables are connected to a lever on the rear brake calipers. When the parking brakes are applied, the cable and lever turns a threaded actuator screw inside a nut with matching threads. As the actuator screw is rotated by the parking brake cable and lever, the nut moves inward and pushes the caliper piston so it forces the brake pad lining against the rotor to apply the brakes.

## ANTILOCK BRAKE SYSTEM PRINCIPLES

If a wheel locks up during a brake application, the tire exhibits a loss of traction. Conversely, if enough braking force is supplied so the tire slips to a certain extent without wheel lockup, a traction increase is experienced. This traction increase is highest at 10 percent to 20 percent tire slip on the road surface. If there is 20 percent tire slip, the wheel is turning at 80 percent of the vehicle speed. An ABS is designed to provide approximately 20 percent tire slip without experiencing wheel lockup during the antilock brake function. This action provides improve tire traction while braking and stopping distance may be reduced depending on the road surface and other variables.

Cornering tire traction also decreases significantly if the wheel locks up during a brake application. Wheel lockup and reduced cornering traction during a brake application may result in loss of steering stability and control. If the tire slips about 20 percent without wheel lockup, cornering traction is greatly improved. Since the ABS provides about 20 percent tire slip without wheel lockup even during panic stops, steering control and vehicle stability are significantly improved. If a wheel or wheels lock up during a panic stop, the vehicle may swerve sideways and the driver usually loses steering control.

Figure 30. Integral wheel speed sensor in wheel hub.

## FOUR-WHEEL ABS WITH HIGH-PRESSURE ACCUMULATOR

Many vehicles have a four-wheel ABS system with a wheel speed sensor at each wheel. A toothed ring on the wheel hub rotates past the tip of the wheel speed sensor. On some front-wheel drive cars, the wheel speed sensors are integral with the wheel hubs and cannot be serviced separately **(Figure 30)**. Some four-wheel ABS have a high-pressure accumulator mounted on the master cylinder **(Figure 31)**. The high-pressure accumulator contains a heavy diaphragm in the center of the accumulator. A nitrogen gas charge is permanently sealed in the upper accumulator chamber above the diaphragm. A pump integral with the master cylinder pumps brake fluid into the lower accumulator chamber. The pump maintains a brake fluid pressure of 2,000 to 2,600 psi (14,000 to 16,000 kPa) in the accumulator. If the accumulator pressure drops below 2,000 psi (14,000 kPa), a pressure switch on the master cylinder signals the control module to start the pump motor and increase the pressure.

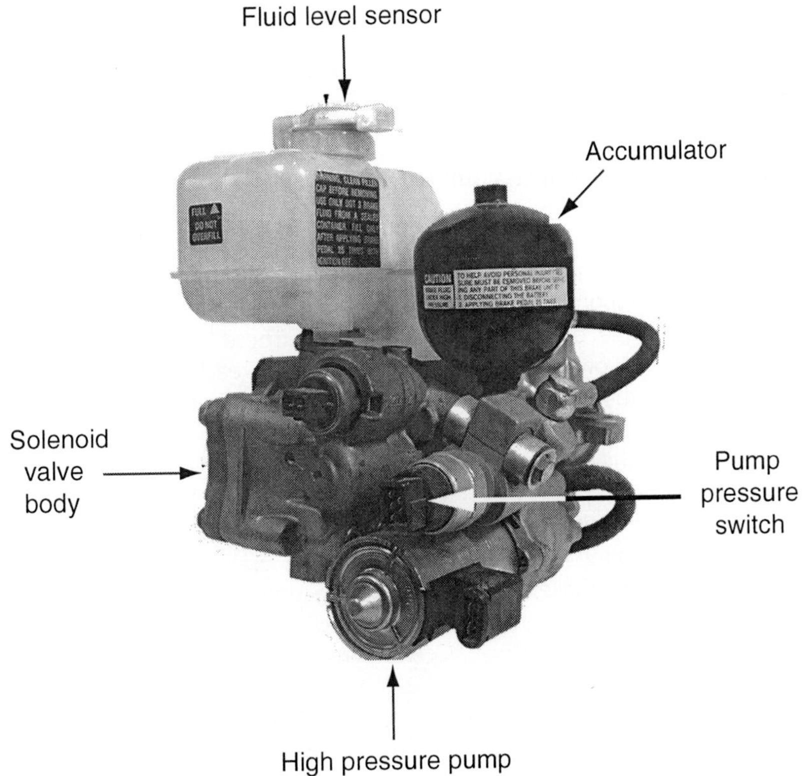


Figure 31. ABS with high-pressure accumulator.

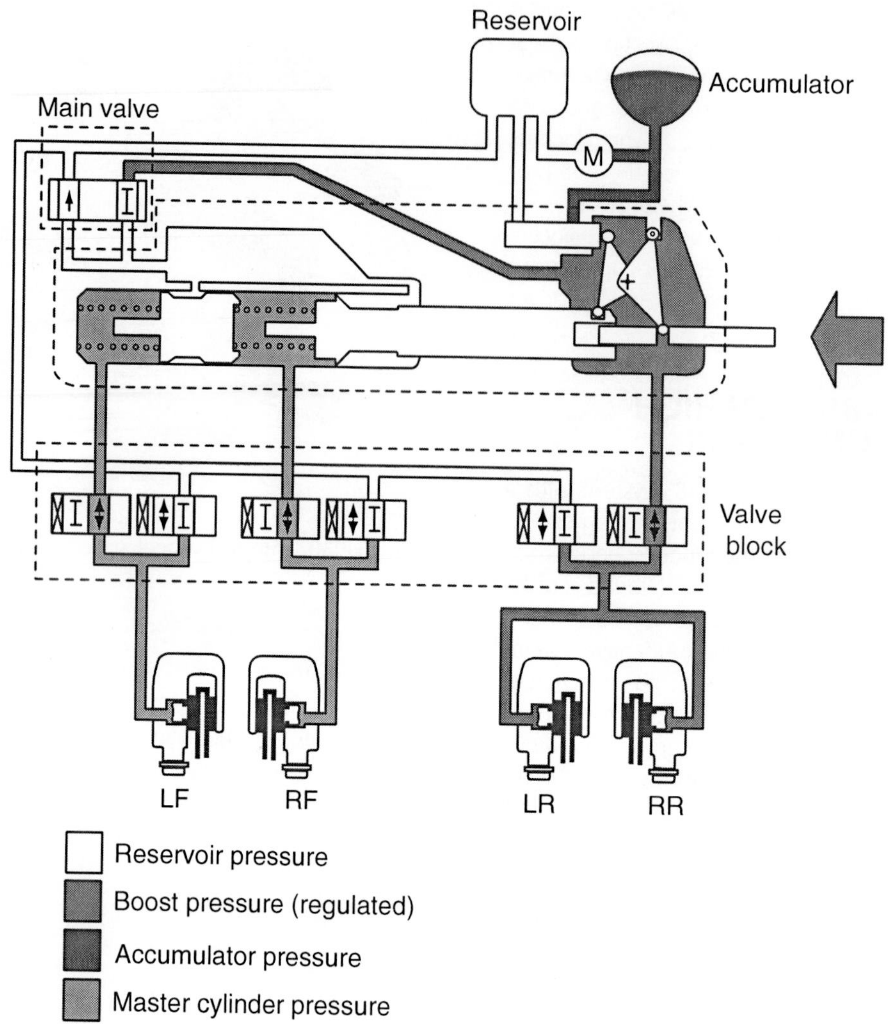


**Figure 32.** Operation of ABS with high-pressure accumulator.

When the brake pedal is depressed during a brake application, the pedal operates a precision spool-type valve in the master cylinder **(Figure 32)**. As the brake pedal moves the spool valve, brake fluid pressure flows out of the accumulator into the boost chamber in the master cylinder. The brake fluid pressure in the boost chamber pushes on the primary master cylinder piston and acts as a brake booster to help the driver apply the brakes.

Brake fluid from the boost chamber is also supplied through a normally-open solenoid valve to the rear wheel calipers or wheel cylinders to apply the rear brakes. Brake fluid is applied from the master cylinder pistons through normally-open solenoid valves to the front wheel calipers. This three-channel ABS has normally-closed solenoid valves that are connected from each front wheel caliper and both rear wheel calipers or wheel cylinders to the master cylinder reservoir. Many vehicles are equipped with four-channel ABS. All the solenoid valves are mounted in a valve block attached to the master cylinder.

A **three-channel ABS** has a pair of solenoids at each front wheel, but only one pair of solenoids for both rear wheels. These systems cannot moderate the brake fluid pressure to each rear wheel individually. A **four-channel ABS** has a pair of solenoids are each wheel. These systems can modulate the brake fluid pressure at each wheel.

During a brake application if a wheel speed sensor signal indicates to the control module that a wheel is quickly approaching a lockup condition, the module closes the normally-open solenoid connected to that wheel caliper. This action prevents any further increase in fluid pressure to the wheel caliper. In a few milliseconds of the wheel speed sensor still indicating that wheel lockup is about to occur, the module opens the normally-closed solenoid and allows some brake fluid out of the brake caliper back into the master cylinder reservoir. This action reduces pressure in the caliper and prevents wheel lockup. The module pulses the solenoids on and off to maintain maximum braking force without allowing wheel lockup. When the module pulses the solenoids on and off during the antilock function, pedal pulsations may be felt by the driver.

An amber ABS warning light is mounted in the instrument panel or in the roof console. When the ignition switch is turned on, the control module performs a check of the ABS electrical system. This check requires 3 to 4 seconds and during this time the ABS warning light is on. If the ABS warning light is on with the engine running, the module has detected a defect in the ABS.

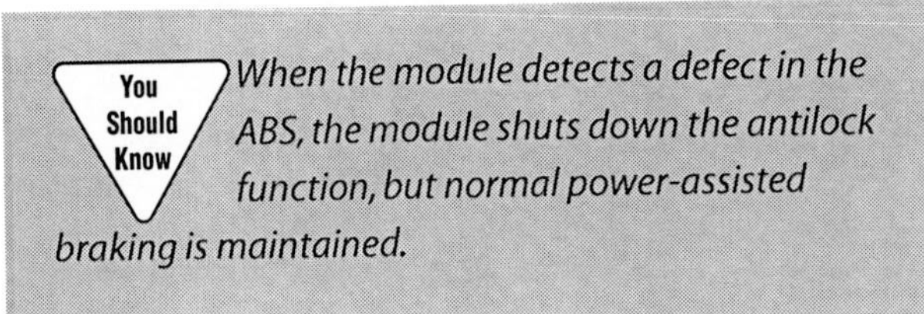


*When the module detects a defect in the ABS, the module shuts down the antilock function, but normal power-assisted braking is maintained.*

Many ABS systems also have a red brake warning light. If this light is illuminated with the engine running, the parking brake is applied or one section of the master cylinder is low on brake fluid. In some ABS, the control module illuminates both the amber and red brake warning lights when certain serious defects occur.

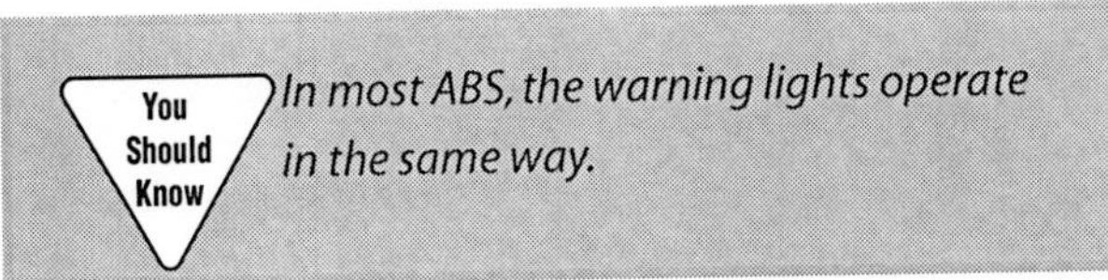


*In most ABS, the warning lights operate in the same way.*

## ABS WITH LOW-PRESSURE ACCUMULATORS

Some vehicles have an ABS with the solenoid valves, low-pressure accumulators, pressure pump, and the control module designed into one unit **(Figure 33)**. This type of ABS has a conventional vacuum brake booster to provide brake assist. The solenoids in this system are referred to as isolation and dump valves. The VSS in the transmission extension housing also acts as a rear wheel speed sensor and sends voltage signals to the ABS control module and the PCM **(Figure 34)**. One pair of solenoids controls the brake fluid pressure to both rear wheels in this three-channel system.

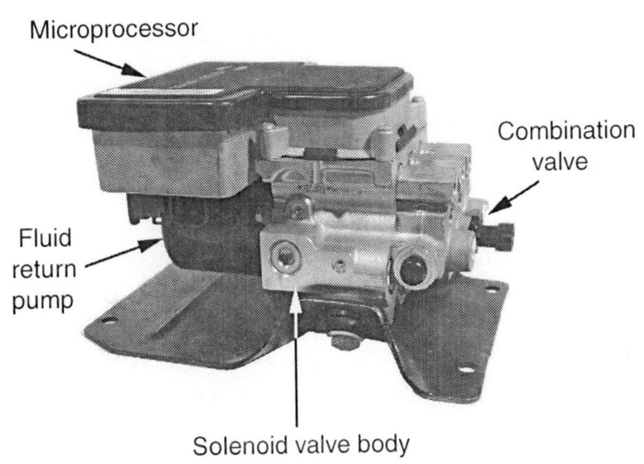


**Figure 33.** ABS with low-pressure accumulator.

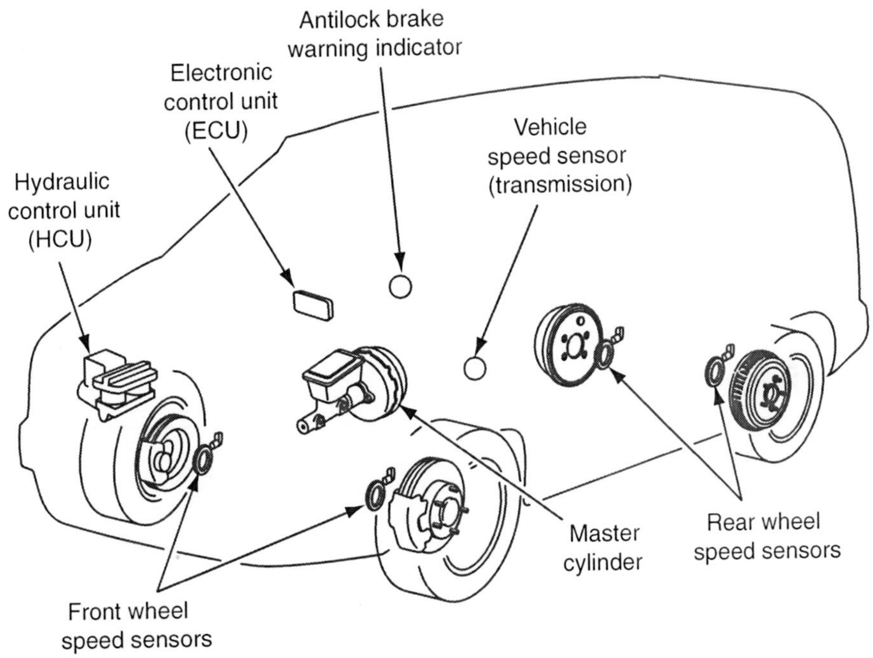


**Figure 34.** ABS with low-pressure accumulator and related components.

When a wheel speed sensor signal indicates wheel lockup is about to occur, the control module energizes the isolation valve. This valve closes the fluid passage between the master cylinder and the wheel that is about to lock up. If wheel lockup is still imminent, the control module energizes the dump valve which allows some brake fluid out of the wheel cylinder or caliper back into the low-pressure accumulator. The control module pulses the dump valve on and off very quickly to apply maximum braking force without allowing wheel lockup. During a prolonged brake application, the repeated cycling of the isolation and dump valves takes fluid out of the master cylinder and places it in the low-pressure accumulators. For this reason, the control module starts the ABS pump motor when the system enters the antilock mode. This pump forces brake fluid back against the master cylinder pistons and to the isolation valves. This action maintains brake pedal height during the antilock mode. In the antilock mode, the driver may feel pedal pulsations and a limited brake pedal fade followed by upward brake pedal movement.

*Many light-duty trucks have a rear-wheel antilock (RWAL) system which has ABS only on the rear wheels.*

## Summary

- When pressure is applied to a confined liquid, the pressure is transmitted equally in all directions and acts with equal force on equal areas.
- Brake fluids may be classified as DOT 3, DOT 4, or DOT 5.
- Dual master cylinders contain a primary and a secondary piston and two wheels on the vehicle are supplied with pressure from each piston.
- Many drum brakes are servo-type in which the operation of the primary shoe applies mechanical force to the secondary shoe.
- Many drum brakes are self-adjusting.
- Many disc brakes have floating calipers with a single piston on the inside of the caliper.
- Disc brakes do not provide a servo action. This type of brake does not experience brake fade.
- A metering valve delays fluid pressure to the front calipers until the rear brake shoes are moved outward against the drums.
- A proportioning valve reduces pressure to the rear brakes to prevent rear wheel lockup caused by weight shift to the front of the vehicle while braking.
- A pressure differential valve illuminates a brake warning light in the instrument panel if there is unequal pressure between the two sections in the master cylinder.
- Brake boosters may be vacuum operated or hydraulically operated.
- Parking brakes apply the rear brakes through a lever and cable system.
- ABS provide maximum braking force with approximately 20 percent tire slip without wheel lockup.
- A wheel speed sensor produces an AC voltage signal in relation to wheel speed.
- On some ABS, the high-pressure accumulator supplies fluid pressure to the master cylinder pistons which acts as a brake booster.
- In many ABS, a pair of solenoids operated by the ABS module control fluid pressure in the wheel caliper or cylinder during the antilock brake function.
- In a three-channel ABS, a pair of solenoids controls the brake fluid pressure to both rear wheels.
- A four-channel ABS has a pair of solenoids for each wheel.
- In place of a high-pressure accumulator, some ABS have two low-pressure accumulators and a pump motor.

## Review Questions

1. Technician A says if the master cylinder piston and bore size is decreased, the pressure exerted by the master cylinder increases for a given pedal pressure. Technician B says a smaller diameter master cylinder bore requires more piston travel to displace the same amount of fluid as a larger piston. Who is correct?
   A. Technician A
   B. Technician B
   C. Both Technician A and Technician B
   D. Neither Technician A nor Technician B
2. Technician A says a DOT 5 brake fluid is silicone-based. Technician B says a DOT 5 brake fluid has a higher dry ERBP boiling point compared to a DOT 4 brake fluid. Who is correct?
   A. Technician A
   B. Technician B
   C. Both Technician A and Technician B
   D. Neither Technician A nor Technician B

3. Technician A says the primary piston is positioned in the outer end of the master cylinder bore. Technician B says if the master cylinder has a residual valve, a slight pressure is maintained in the brake system with the brakes released. Who is correct?
   A. Technician A
   B. Technician B
   C. Both Technician A and Technician B
   D. Neither Technician A nor Technician B
4. Technician A says in a drum brake system with servo action, the primary lining has a longer lining compared to the secondary lining. Technician B says in this type of brake system, most of the braking force is applied by the primary shoe. Who is correct?
   A. Technician A
   B. Technician B
   C. Both Technician A and Technician B
   D. Neither Technician A nor Technician B
5. In a servo-type drum brake assembly:
   A. The primary shoe is positioned toward the rear of the vehicle.
   B. The lower ends of the brake shoes are connected to the backing plate.
   C. Drum heat and expansion cause brake pedal fade.
   D. The primary and secondary shoes provide equal braking force.
6. All of these statements about floating caliper disc brakes are true *except*:
   A. The brake caliper is mounted so it can slide sideways.
   B. The caliper usually contains a single piston.
   C. A piston return spring moves the lining away from the rotor surface.
   D. This type of brake does not require periodic adjustment.
7. While discussing vacuum brake boosters, Technician A says when the brakes are released, vacuum is supplied to both sides of the booster diaphragm. Technician B says when the brakes are applied, atmospheric pressure is supplied to the front side of the booster diaphragm. Who is correct?
   A. Technician A
   B. Technician B
   C. Both Technician A and Technician B
   D. Neither Technician A nor Technician B
8. All of these statements about high-pressure accumulators in ABS are true *except*:
   A. A high-pressure accumulator has an upper and lower chamber separated by a diaphragm.
   B. A high-pressure accumulator is fully charged at 800 psi.
   C. A high-pressure accumulator supplies fluid pressure that acts as a brake booster.
   D. A high-pressure accumulator supplies fluid pressure to the rear wheel calipers.
9. In an ABS with low-pressure accumulators:
   A. The module starts the motor pump when the brake pedal is depressed.
   B. The driver may feel pedal pulsations during the antilock mode.
   C. The driver should not feel any brake pedal fade in the antilock mode.
   D. The pump motor forces brake fluid back into the master cylinder reservoir.
10. During the antilock mode, the percentage of wheel slip supplied by an ABS is:
    A. 20 percent.
    B. 30 percent.
    C. 40 percent.
    D. 50 percent.
11. In a drum brake system, the self-adjusting mechanism adjusts the brakes when the brakes are applied during ______________ vehicle motion.
12. A metering valve delays fluid pressure to the ______________ brakes.
13. A proportioning valve reduces pressure to the ______________ brakes.
14. A pressure differential valve illuminates a light in the instrument panel if there is ______________ ______________ in the master cylinder sections.
15. Explain the operation of a vacuum brake booster with the brakes released and applied.
16. Describe the basic operation of a hydro-boost brake booster.
17. Describe the operation of the parking brake system on a vehicle with rear drum brakes.
18. Describe the purpose of the vent ports in the master cylinder bores.

# Brake System Maintenance, Diagnosis, and Service

## Introduction

The brake system is the most important safety system on the vehicle. Brake technicians have the customer's life in their hands. Technicians must never compromise on brake safety. When brake components are in questionable condition, they should be replaced and quality brake components must always be installed. For example, if brake drum diameter is equal to the discard diameter, replace the brake drum. Do not take a chance on a worn out brake drum that may crack and cause a brake failure. All brake components must be serviced using the vehicle manufacturer's recommended procedures.

ABS are more likely to experience basic brake problems than problems in the ABS control circuitry and components. Therefore, you must be familiar with conventional brake service because many of the same basic service procedures, such as changing brake shoes or pads, are required on ABS. Technicians must also be familiar with the maintenance, diagnostic, and service procedures required for the electronic circuits and hydraulic systems in ABS. When technicians are knowledgeable regarding these maintenance, diagnostic, and service procedures, they will be able to solve ABS problems quickly and accurately.

### BRAKE SYSTEM MAINTENANCE

During a brake inspection, one of the first items is to check the fluid level in both sections of the master cylinder. On disc brake systems, fluid level in the master cylinder becomes lower as the linings wear and the caliper pistons move outward. Fill the master cylinder to the proper level with the vehicle manufacturer's specified brake fluid. Check the brake pedal for proper free-play **(Figure 1)**. Inadequate brake pedal free-play may cause the master cylinder piston cups to cover the vent ports in the master cylinder. This action causes dragging brakes and a hard brake pedal. Excessive brake pedal free-play causes too much pedal

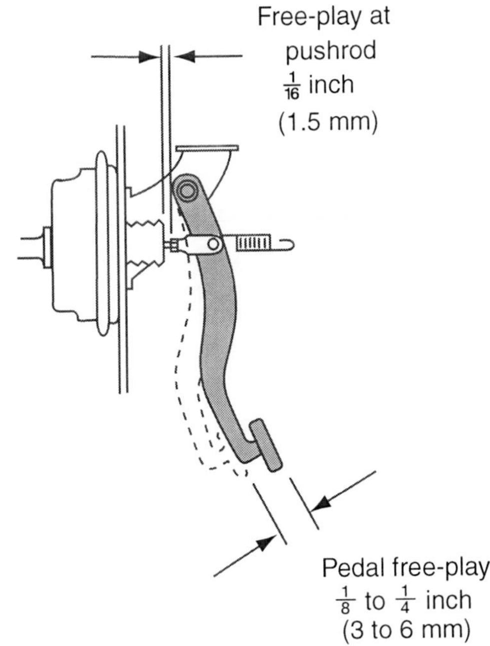


**Figure 1.** Brake pedal free-play.

movement and a low brake pedal during a brake application. The **brake pedal free-play** is the amount of pedal movement before the booster pushrod contacts the master cylinder piston.

Check all the brake lines and fittings under the hood for leaks and unwanted contact with other components. Raise the vehicle on a lift and inspect all the brake lines and hoses under the vehicle. Check the brake flexible hoses for cracks, leaks, bulges, and unwanted contact with other components. Inspect all the brake tubing for rust and corrosion, leaks, flat spots, kinks, and damage. Inspect each wheel for evidence of brake fluid or grease leaks. On many vehicles it is possible to partially inspect the condition of the rotor and brake linings on disc brakes without removing the caliper. Inspect the condition of the parking brake cables and pull on the cables to determine if they are moving freely. Check each wheel for free rotation. If one of the wheels is hard to rotate, the brakes are likely not releasing properly.

Lower the vehicle onto the shop floor and road test the vehicle. Check the brake pedal feel during a brake application. If the brake pedal feels spongy, there may be air in the brake system or the master cylinder may be low on brake fluid. When the pedal feels hard and requires excessive pedal effort, the brake linings may be oil soaked or the caliper or wheel cylinder pistons may be seized. Check the pedal for the correct amount of pedal travel during a brake application. During a brake application, check for pedal pulsations and noises in the brake system such as scraping or rattling. Check for brake grabbing, wheel lockup, and steering pull during a brake application. These problems are often the result of contaminated brake linings. To perform a complete inspection of the brake system components, the wheels and drums or calipers have to be removed. Park the vehicle and apply the parking brake. Start the engine and place the transmission in drive. Accelerate the engine slightly and determine if the parking brake holding capability is adequate.

## BRAKE SYSTEM DIAGNOSIS

The causes of various brake system problems are listed in the following diagnostic summary:

1. A low spongy pedal and excessive pedal travel with the red brake warning light on.
   (a) Low fluid level in the master cylinder.
   (b) Air in the hydraulic system.
2. Noise while braking.
   (a) Worn brake linings causing a squealing or scraping noise.
   (b) Worn bushings on floating calipers.
   (c) Interference between brake linings and splash shield or backing plate.
   (d) Chirping noise caused by lack of lubrication on backing plate supports.
3. Pedal pulsations during brake applications.
   (a) Excessive thickness variation or runout on rotors or excessive out-of-round on brake drums.
4. Grabbing brakes during a brake application.
   (a) Contaminated front brake pad linings.
   (b) Contaminated rear brake shoe or pad linings.
5. Steering pulls to one side while braking.
   (a) Contaminated brake linings.
   (b) Seized caliper piston or wheel cylinder pistons on the opposite front wheel from the direction of steering pull.
6. Excessive pedal effort on non-power brakes.
   (a) Caliper or wheel cylinder pistons seized.
   (b) Contaminated or glazed brake linings.
7. Excessive pedal effort on power brakes.
   (a) Defective brake booster.
   (b) Low manifold vacuum supplied to the booster.
   (c) Seized caliper or wheel cylinder pistons.
   (d) Glazed or contaminated brake linings.
8. The rear brakes drag.
   (a) Improper parking brake adjustment.
   (b) Seized parking brake cables.
   (c) Improper rear brake adjustment.
   (d) Weak brake shoe return springs or seized caliper or wheel cylinder pistons.
9. The brakes on all wheels drag.
   (a) Seized caliper or wheel cylinder pistons.
   (b) Improperly adjusted brake light switch preventing pedal return.
   (c) Binding pedal linkage.
   (d) Contaminated brake fluid.
   (e) Corrosion in the master cylinder that plugs the vent ports.
10. Premature rear wheel lockup.
   (a) A defective proportioning valve.
   (b) Rear brake linings contaminated.
   (c) Front caliper pistons seized.
11. Excessive pedal travel, or a low, firm pedal.
   (a) The drum brakes require adjusting.
   (b) One master cylinder section is defective.
12. The brakes release slowly and the pedal does not return fully.
   (a) Weak shoe return springs.
   (b) A defective booster or boost check valve.
   (c) Contaminated or wrong brake fluid.
   (d) The brake light switch is interfering with pedal return.
   (e) Pedal linkage binding.
13. Brake pedal fade while brakes are applied firmly.
   (a) A leaking brake line or bulging brake hose.
   (b) Leaking master cylinder piston cups.

# BRAKE SERVICE

## Brake Pedal Free-Play Adjustment

Brake pedal free-play is very important to provide proper brake system operation. If this distance is not within specifications, lengthen or shorten the pedal pushrod to obtain the proper free-play.

## Brake Shoe Adjustment

To determine if the drum brakes require adjusting, pump the brake pedal twice with the engine running. If the pedal is higher on the second pump, the drum brakes require adjustment. Raise the vehicle on a lift and remove the rubber plugs in the adjustment openings in the backing plate. Insert the end of a brake adjusting tool through the backing plate opening until it contacts the star wheel on the brake adjuster. Move the tool up and down to rotate the star wheel in the proper direction to move the brake shoes outward while rotating the tire and wheel. Continue rotating the adjuster until the wheel is hard to turn. If the vehicle has self-adjusting brakes, insert a small screwdriver through the backing plate opening to move the self-adjuster lever away from the star wheel **(Figure 2)**. Use the brake adjusting tool to rotate the star wheel to shorten the brake adjuster and move the brake shoes inward until the tire and wheel rotate freely. Reinstall the rubber plug in the backing plate opening.

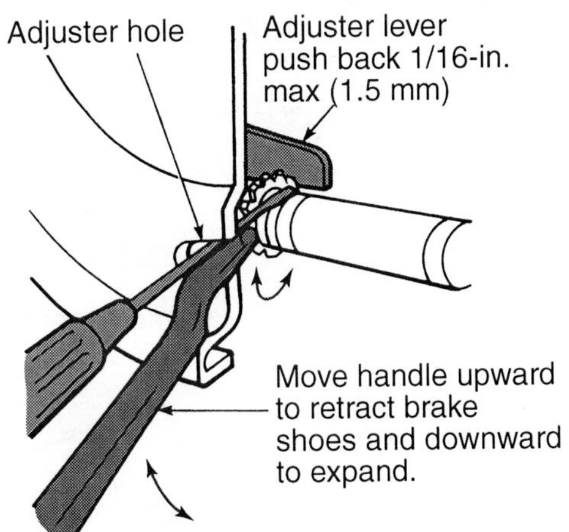


**Figure 2.** Brake shoe adjustment.

## Brake System Bleeding

Air may be bled manually from the brake system. To use the manual bleeding procedure, fill the master cylinder to the required level with the specified brake fluid and raise the vehicle on a lift. Have a coworker apply and hold the brake pedal. Connect a hose from the bleeder screw on the caliper or wheel cylinder in the right-rear wheel which is furthest from the master cylinder **(Figure 3)**. Place the other end of the hose in a glass container. Open the bleed screw and allow the brake fluid to flow into the container. Close the bleeder screw and inform your coworker to release the brake pedal. Repeat this procedure until a clear stream of brake fluid with no air bubbles is discharged when the bleeder screw is opened. Refill the master cylinder and repeat the bleeding procedure on the other wheels in this order: left rear, right front, and left front.

A pressure brake bleeder may be used to bleed the brakes. The pressure bleeder is a storage tank with two compartments separated by a diaphragm. The lower half of this tank contains brake fluid, and 15 to 20 psi shop air pressure is applied above the diaphragm. A gauge on the pressure bleeder indicates the air pressure in the bleeder tank. A hose is connected from the brake fluid reservoir in the bleeder to a special adapter installed on top of the master cylinder. A shut-off valve in the pressure bleeder hose allows the technician to open and close this hose. When using the pressure bleeder, be sure there is adequate brake fluid in the lower reservoir and close the valve in the brake fluid hose. Supply the proper air pressure to the upper reservoir

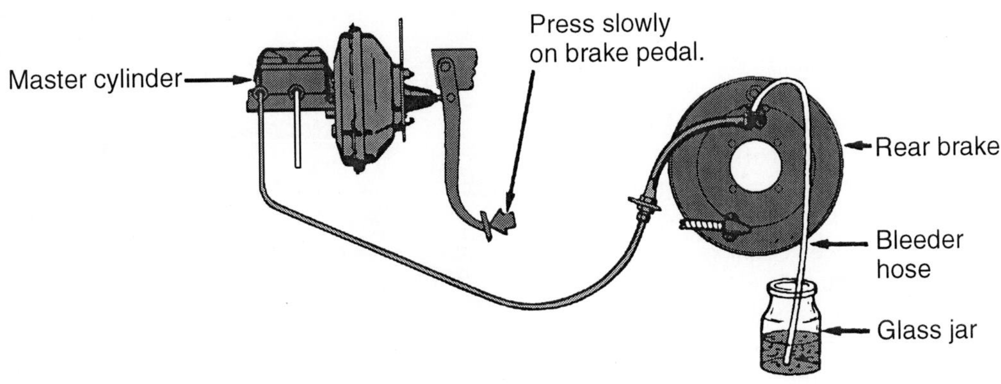


**Figure 3.** Manual brake bleeding procedure.

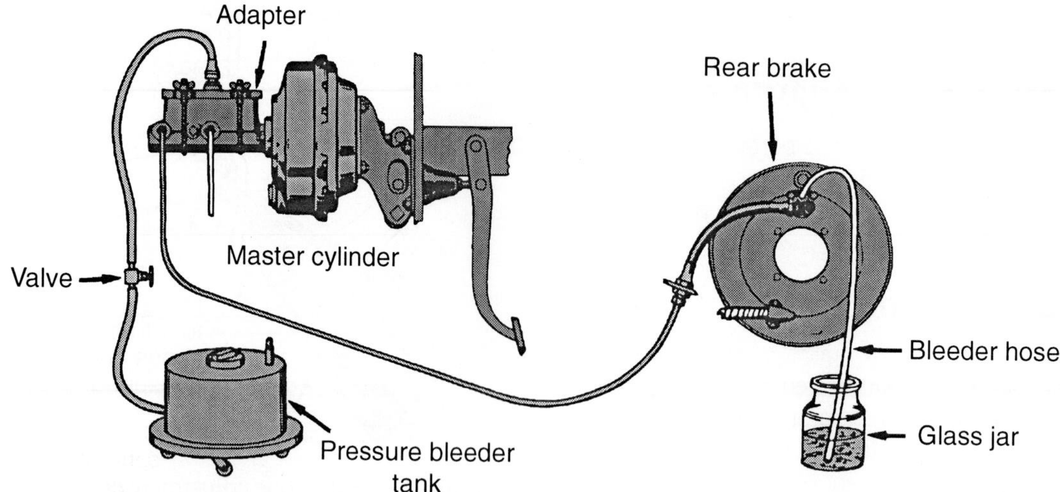


Figure 4. Pressure brake bleeding procedure.

in the bleeder. Install the adapter securely on top of the master cylinder and connect the brake fluid hose to this adapter **(Figure 4)**. Open the valve in the brake fluid hose and be sure there are no leaks between the adapter and the master cylinder. Bleed air from each wheel cylinder or caliper in the same sequence used for manual bleeding. After the bleeding procedure is completed, be sure the master cylinder has the proper brake fluid level. A hand-operated suction pump may be used to pull some brake fluid from the master cylinder to obtain the proper fluid level.

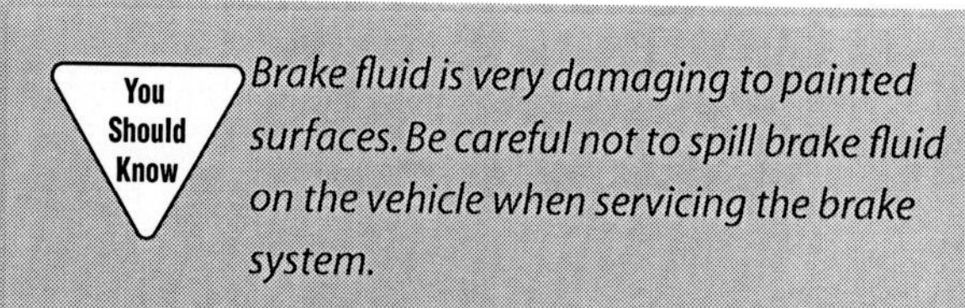


*Brake fluid is very damaging to painted surfaces. Be careful not to spill brake fluid on the vehicle when servicing the brake system.*

Figure 5. Wet brake cleaning equipment.

## Drum Brake Service

After the wheel and brake drum are removed, inspect the brake linings. A depth micrometer may be used to measure the thickness of the linings above the rivet heads. This thickness should be a minimum of 0.030 inch (0.75 millimeter). Use a low-pressure wet cleaning system to wash the brake components **(Figure 5)**. Place a retaining clip on the wheel cylinder to prevent the pistons from moving outward in this cylinder. Use a brake spring tool to remove the shoe return springs and remove the brake shoe holddown clips and springs **(Figure 6)**.

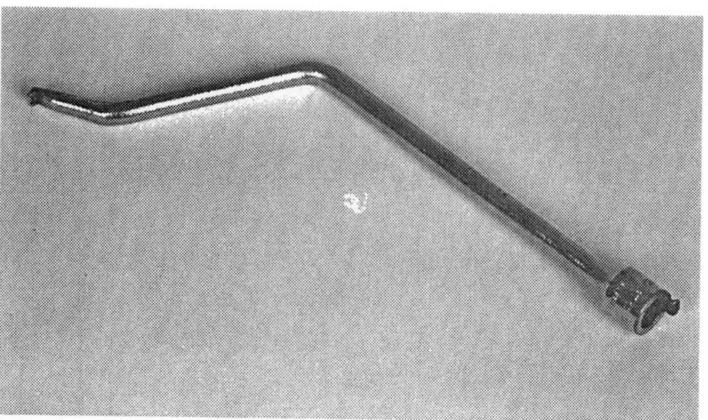

Figure 6. Brake spring tool.

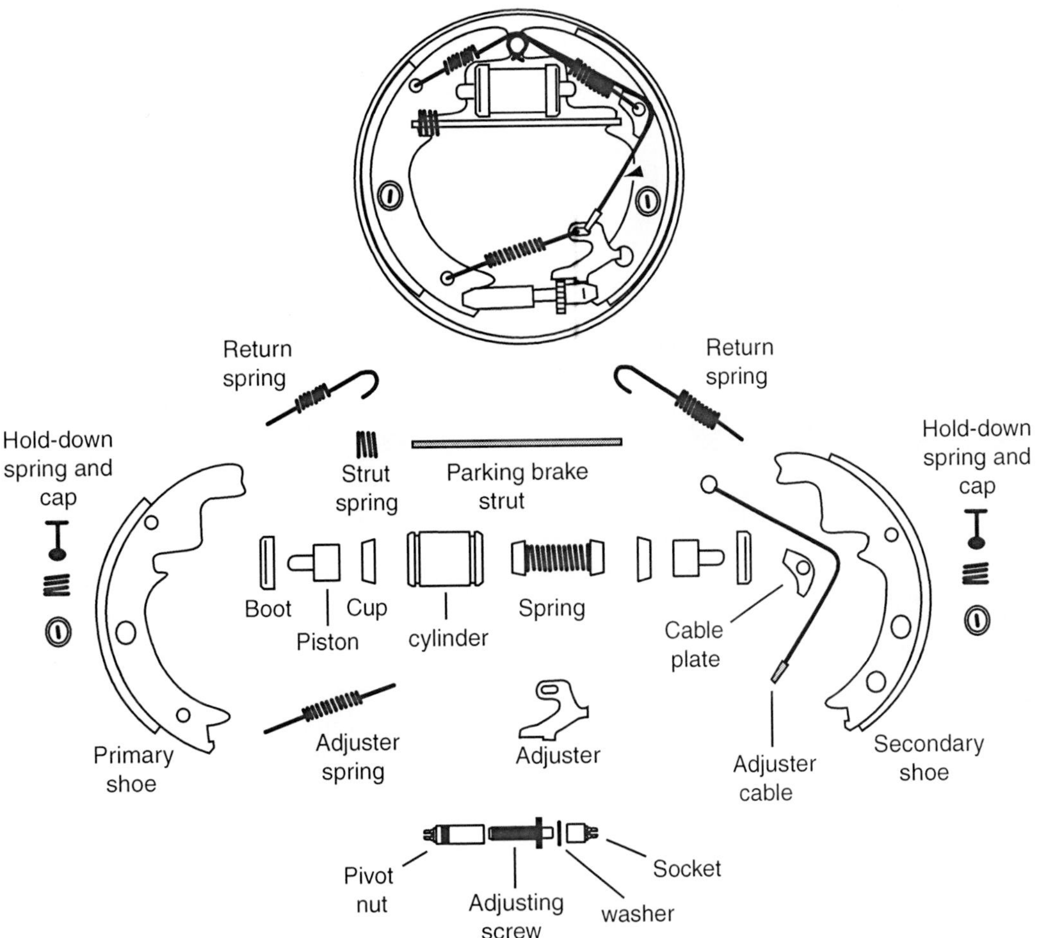


**Figure 7.** Drum brake components.

> **You Should Know** *Primary and secondary brake shoe return springs may be different. These springs must be installed in their original position.*

Remove the brake shoes and then remove the lower retracting spring and adjuster and the self-adjusting mechanism **(Figure 7)**. Replace the shoe return springs if they are damaged, distorted, bent, or corroded. Inspect the backing plates for cracks, distortion, or worn brake shoe supports and anchors. If any of these conditions are present, replace the backing plate. Peel the rubber boots back on the wheel cylinder and check for brake fluid and contaminants behind these boots. If brake fluid or contaminants are present behind the boots, wheel cylinder overhaul or replacement is required.

Inspect the brake drums for scoring, cracks, heat checks, or burned spots. Drums with heat checks, cracks, or severe burned spots must be replaced. If the burned spots are removed after turning the drum, the drum may be reused. Use a brake drum micrometer to measure the drum diameter at several locations around the drum **(Figure 8)**.

**Figure 8.** Measuring brake drum diameter with a drum micrometer.

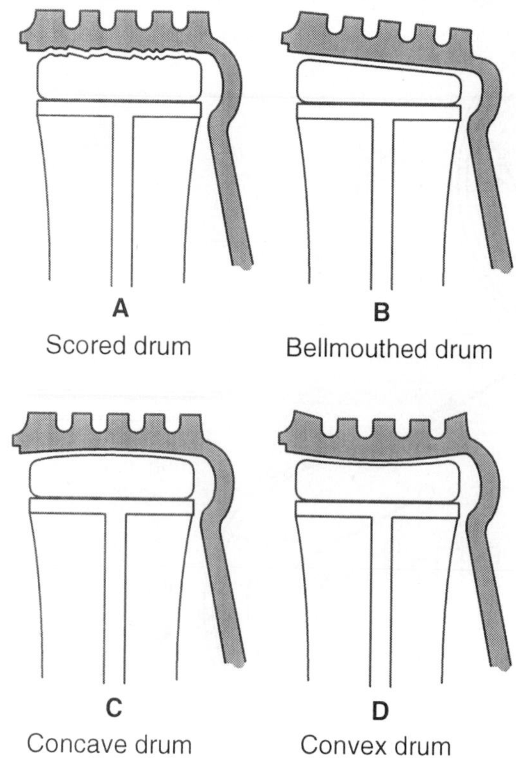


Figure 9. Defective brake drum conditions.

Measure the drum diameter at the inner and outer edges of the machined surface and also on the center of this surface. If drum taper or out-of-round exceeds 0.006 inch (0.152 millimeter), the drum must be machined or replaced. Defective drum conditions are illustrated in **Figure 9**). The drum's discard diameter is stamped on the drum **(Figure 10)**. After turning the drum, there must be 0.030 inch (0.762 millimeter) left for wear. Brake drum **out-of-round** is a variation in drum diameter at various locations.

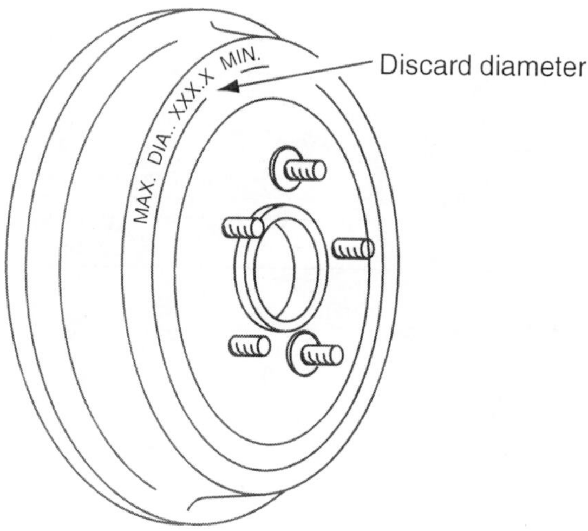


Figure 10. Brake drum discard diameter.

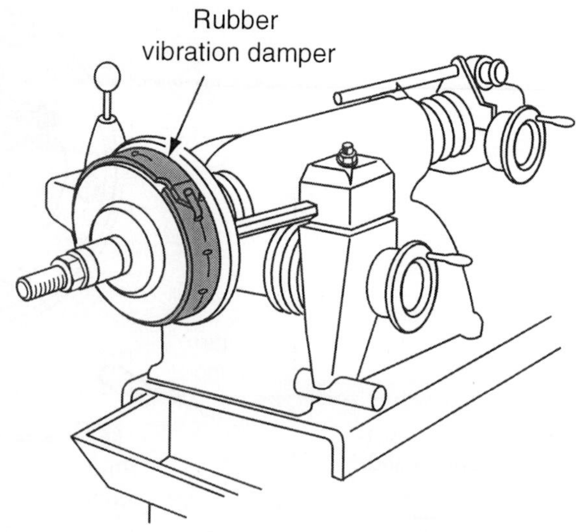


Figure 11. Brake drum with vibration damper band mounted in a drum lathe.

Mount the brake drum securely in the drum lathe and position a vibration band around the outer drum diameter **(Figure 11)**. Set the cutter on the lathe to remove a small amount of metal from the drum, usually 0.002 to 0.005 inch (0.05 to 0.15 millimeter). Follow the equipment manufacturer's recommended procedure when using the brake drum lathe. Several cuts may be required to remove scoring and other defective drum surface conditions. After turning the drum, the drum should be re-measured. If the drum is serviceable, it should be thoroughly cleaned with hot water and wiped with a lint-free shop towel to remove small metal particles from the drum.

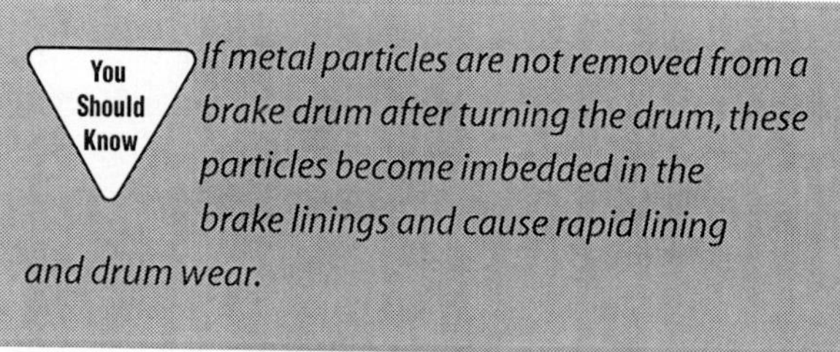


*If metal particles are not removed from a brake drum after turning the drum, these particles become imbedded in the brake linings and cause rapid lining and drum wear.*

You Should Know

*If wheel cylinders are not replaced during a brake overhaul, the wheel cylinder pistons and cups are repositioned when the new brake linings are installed. Under this condition, the wheel cylinder cups may be positioned on top of accumulated debris in the bottom of the wheel cylinder. This action results in brake fluid leaking from the wheel cylinder.*

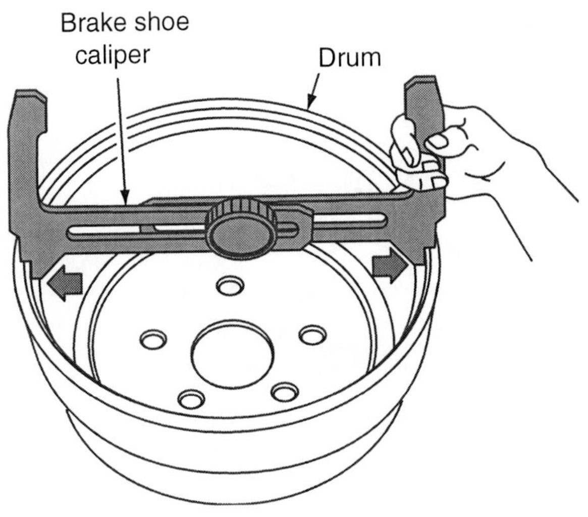


Figure 12. Brake shoe caliper installed in the brake drum.

Wheel cylinders are usually replaced rather than repaired. Install the wheel cylinder in the backing plate. Lubricate the brake shoe supports on the backing plate and install the brake shoes with the adjuster and lower retracting spring. Install the self-adjusting mechanism and brake shoe holddown clips. Use a brake spring tool to install the shoe return springs. Be sure all brake components are installed in their original position. Install one end of a brake shoe caliper in the drum and set this caliper to the drum diameter **(Figure 12)**. Install the other side of the brake shoe caliper over the brake shoes and adjust the brake shoes so they lightly contact the caliper jaws **(Figure 13)**. Using the brake shoe caliper sets the proper clearance between the brake shoes and the drum.

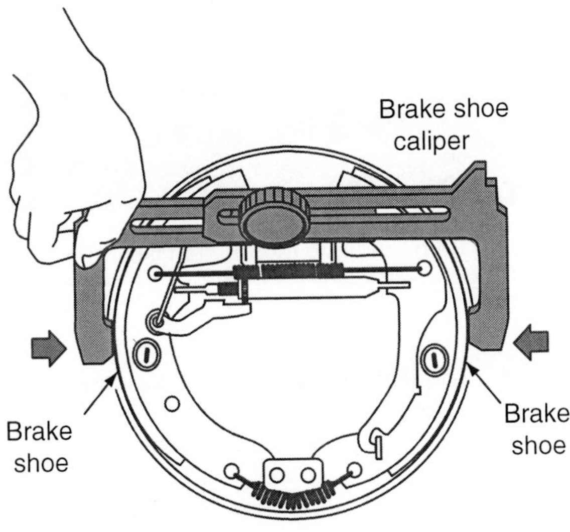


Figure 13. Brake shoe caliper installed on the brake shoes.

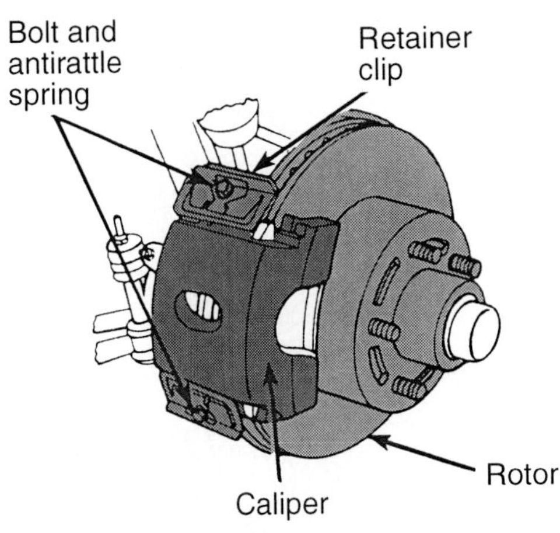


Figure 14. Sliding caliper removal.

## Disc Brake Caliper and Brake Pad Removal

The brake pad linings may be inspected through inspection holes in the caliper or through the outer ends of the caliper. If the lining thickness is equal to or less than the minimum thickness specified by the vehicle manufacturer, the brake pads must be replaced. On a sliding caliper, remove the top bolts, retainer clip, and anti-rattle spring **(Figure 14)**. If the vehicle has floating calipers, remove the two caliper pins **(Figure 15)**. On a fixed caliper, remove the bolts holding the caliper to the steering knuckle. Lift the caliper straight up off the rotor. When replacing only the brake pads, remove the brake pads from the caliper and hang the caliper from a suspension component with a length of wire. After the new brake pads and/or calipers are installed, the brake pedal should be applied several times to properly position the caliper pistons.

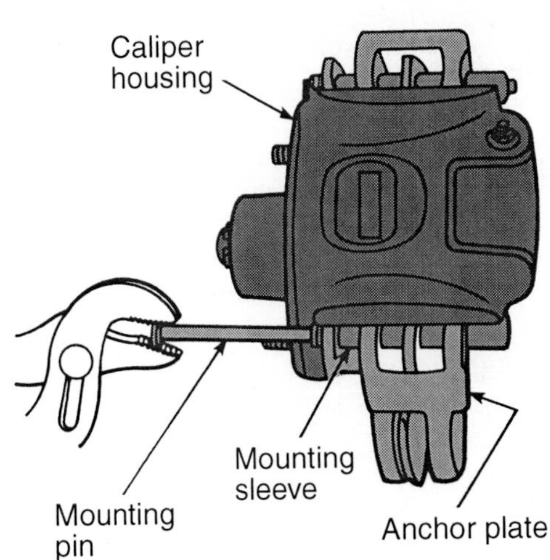


Figure 15. Floating caliper removal.

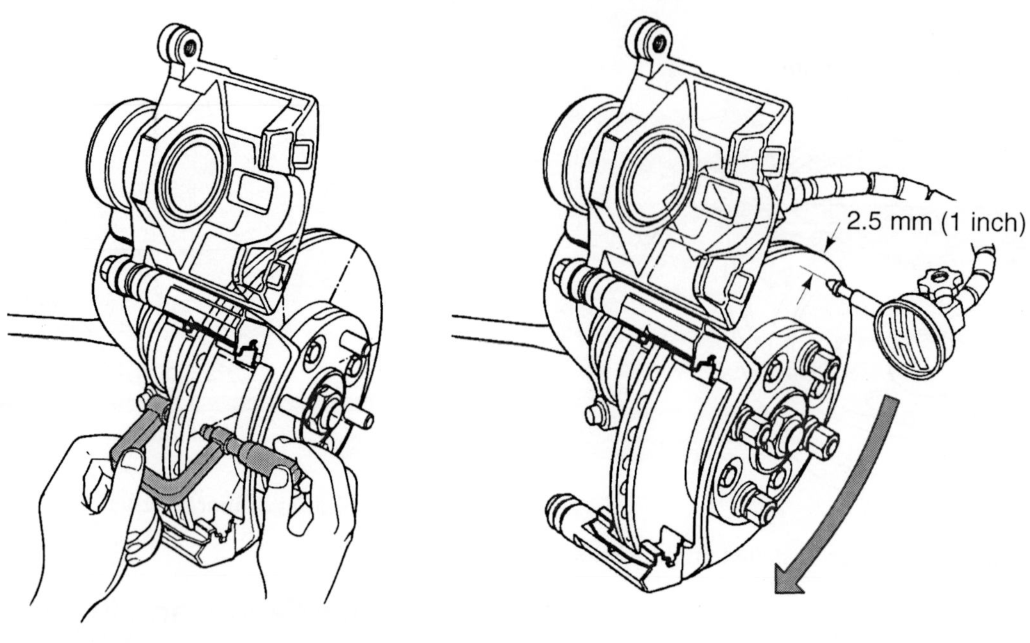
2.5 mm (1 inch)
Measure thickness at 8–12 points, equally spaced around the rotor; all about 1 inch from outer edge of rotor.

**Figure 16.** Measuring rotor thickness variation and lateral runout.

## Rotor Service

Inspect the rotor friction surfaces for scoring or grooving, cracks, broken edges, heat checking, and hard spots. Replace the rotor if it has cracks, broken edges, heat checking or excessive scoring or grooving. Use a dial indicator to measure rotor runout and measure the rotor thickness variation with a micrometer at ten locations around the rotor **(Figure 16)**. Be sure the wheel bearings are properly adjusted before measuring the rotor runout.

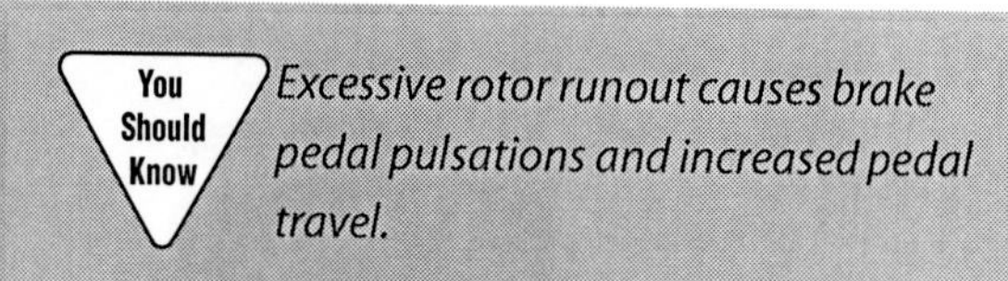
You Should Know

*Excessive rotor runout causes brake pedal pulsations and increased pedal travel.*

If the rotor thickness variation or runout exceeds the vehicle manufacturer's specifications, the rotor must be machined or replaced. The rotor discard thickness is stamped on the rotor. After machining the rotor, the rotor thickness must be 0.015 to 0.030 inch (0.40 to 0.75 millimeter) thicker than the discard dimension to allow for wear. If the rotor is to be resurfaced, remove the rotor and hub from the vehicle. Clean the hub and install the hub and rotor on the lathe using the proper cones to support the hub. Install the vibration damper ring on the outside diameter of the rotor **(Figure 17)**. Follow the equipment manufacturer's recommended procedure to machine the rotor surfaces. Adjust the cutting depth on the rotor lathe to remove a small amount of metal from the rotor surface with each cut. Several cuts are usually necessary to restore the rotor friction surfaces. After the rotor surfaces are machined, a sander may be used to apply a non-directional

**Figure 17.** Rotor and hub mounted in lathe.

**Figure 18.** Using a sander to apply a non-directional finish to the rotor surface.

finish on the rotor surfaces **(Figure 18)**. The non-directional finish on the rotor friction surfaces does not follow the arc of rotor rotation. This type of finish reduces brake noise and provides improved brake lining seating.

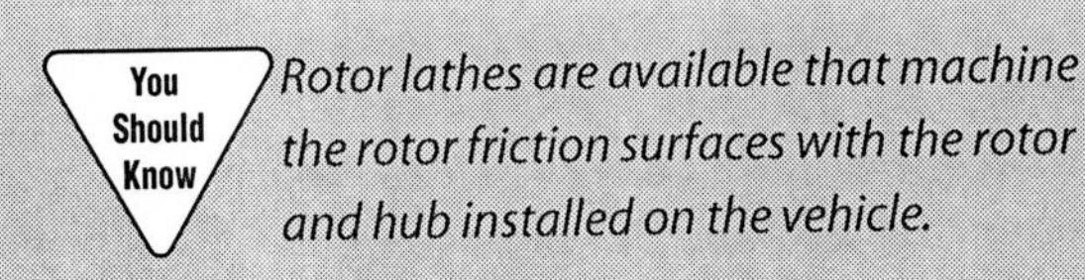

You Should Know *Rotor lathes are available that machine the rotor friction surfaces with the rotor and hub installed on the vehicle.*

After the rotor machining is completed, thoroughly clean the rotor and hub to remove any metal particles. Service, repack, or replace the wheel bearings and seals as necessary and install the hub and rotor on the steering knuckle. Install a new washer on the brake hose and install and tighten the brake hose in the caliper. Install the caliper and brake pads on the steering knuckle. On floating calipers, replace the mounting sleeves or bushings and the mounting pins. Be sure all the caliper hardware such as clips and anti-rattle springs are in good condition. Tighten all fasteners to the specified torque. Bleed the brakes as explained previously in this chapter.

## Master Cylinder Service

Master cylinders are usually replaced rather than overhauled. However, repair kits are available to rebuild master cylinders. When the master cylinder piston cups are leaking, the brake pedal gradually fades downward during steady pedal pressure. If the brake fluid is contaminated or the reservoir contains excessive corrosion, master cylinder replacement is necessary. On aluminum master cylinders, the plastic reservoir may be replaced separately from the master cylinder. To remove a master cylinder, disconnect and cap the brake lines and remove the two mounting bolts. When installing a new master cylinder, it may be bled on the bench before installation. Install plastic hoses from the master cylinder outlets into the reservoir. Use a wooden dowel to push on the primary piston. Continue stroking the primary piston until the brake fluid flowing through the plastic hoses is free from bubbles. Remove the plastic hoses and cap the fluid outlets.

## Brake Booster Service

Check the brake pedal free-play before diagnosing brake booster problems. Adjust this free-play if necessary. To check the vacuum booster operation, repeatedly pump the brake pedal with the engine not running to remove all vacuum from the booster. Hold the brake pedal down and start the engine. If the booster is working properly, the pedal should move downward slightly and then stop moving. If the pedal does not move downward, the manifold vacuum is not supplied to the booster or the booster is defective. Inspect the vacuum hose to be sure it is connected tightly to the intake manifold and to the booster. Remove the vacuum hose from the booster. The engine should stall or run very roughly. If there is not much difference in engine operation, the vacuum hose is restricted. Be sure the one-way check valve in the vacuum hose allows airflow through it in one direction only.

You Should Know *A leaking seal in the front of a vacuum booster causes brake fluid to be drawn from the master cylinder into the booster and intake manifold. If a booster has this problem, the brake fluid level goes down continually in the master cylinder with no indication of external leaks.*

## Parking Brake Service

To check the parking brake, raise the vehicle on a lift. Pull on the parking brake cable to each rear wheel while rotating the rear wheel. When the parking brake cable is pulled by hand, the rear wheel should stop. When the parking brake cable is released, the cable should move freely back to its original position. If the parking brake cables are sticking, they should be replaced. To adjust the parking brake cables, be sure the parking brake lever in the passenger compartment is released. Adjust the nut on the front parking brake cable at the equalizer until the rear wheels

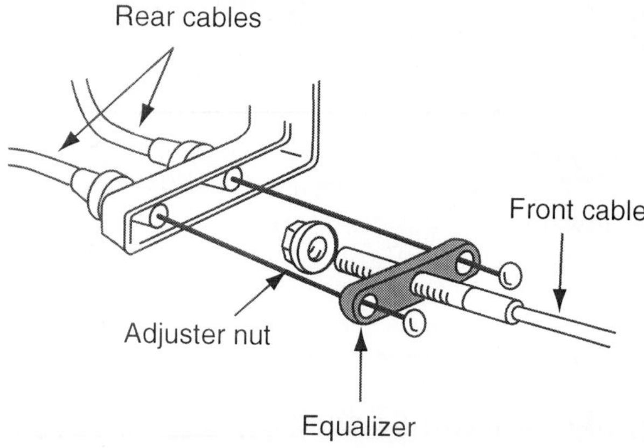


**Figure 19.** Parking brake equalizer and adjuster nut.

begin to drag when they are rotated **(Figure 19)**. Back off the adjusting nut on the front cable until the rear wheels rotate freely. Lower the vehicle onto the shop floor and apply the parking brake to be sure it operates properly.

## ANTILOCK BRAKE SYSTEM MAINTENANCE

ABS maintenance includes a complete brake system inspection. During an ABS inspection, check all the ABS wiring harness for damage, worn insulation, or corroded connections. The wiring most likely to be damaged is the wheel speed sensor wiring. Inspect all the lines, fittings, and high-pressure accumulator for leaks.

Start the engine and observe the amber ABS warning light. This light should be on when the ignition switch is turned on. It should remain on for approximately 4 seconds after the engine starts. During these few seconds, the ABS module completes a check of the ABS electrical system.

You Should Know *On some ABS while the module performs a check of the system, the module operates all the solenoids momentarily. This may cause audible clicking noises.*

If the ABS control module does not detect any electrical defects during the system check, the module turns the amber ABS warning light off. Any time the ABS module detects an electrical defect in the ABS system, the module turns on the ABS warning light with the engine running. When the ABS module turns on the ABS warning light with the engine running, the module usually cancels the ABS function. However, conventional, power-assisted braking is still available.

If the ABS has a high-pressure accumulator, follow this procedure to check the fluid level in the master cylinder.

1. Turn the ignition switch on and pump the brake pedal until the hydraulic pump motor starts.
2. Wait until the hydraulic pump motor shuts off.
3. Check the fluid level in the master cylinder. If the fluid level is below the "max" fill line on the reservoir, bring the fluid level up to this line with the vehicle manufacturer's recommended brake fluid.

You Should Know *During normal ABS operation on a system with a high-pressure accumulator, the fluid level in the master cylinder may be above the "max" fill line depending in the accumulator state of charge.*

## ABS DIAGNOSIS, OBD I VEHICLES

On OBD I vehicles prior to 1996, a variety of DLCs were used by various vehicle manufacturers for scan tool connection. These DLCs were usually located under the hood or under the dash. Many OBD I vehicles had a separate DLC for ABS diagnosis. OBD II systems were mandated in 1996 and these systems have a standard 16-terminal DLC mounted under the dash near the steering column **(Figure 20)**. On vehicles with OBD II systems, data links interconnect the various computers on the vehicle. These data links are also connected to DLC terminals. Therefore, the different computer systems on the vehicle, including the ABS computer, may be diagnosed by connecting a scan tool to the DLC.

You Should Know *OBD II systems have many standardized features including DTCs, monitors in the PCM to monitor various engine-related systems, and data links between computers on the vehicle.*

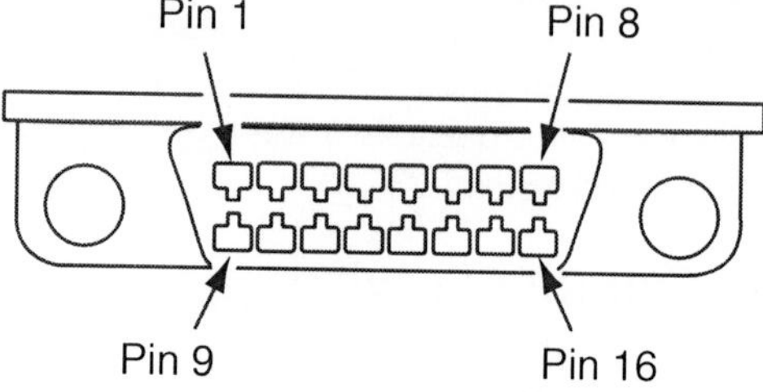


**Figure 20.** Data link connector (DLC) on an OBD II vehicle.

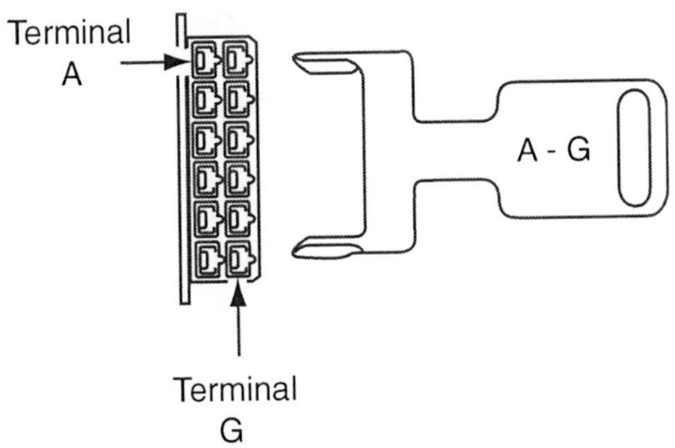
Terminal
A
A - G
Terminal
G

**Figure 21.** Connecting terminals A to G in the DLC for ABS diagnosis.

The first step in any ABS diagnosis is to check the operation of the amber ABS warning light and the red brake warning light as explained previously. When diagnosing ABS on OBD 1 vehicles, connect a jumper wire between two terminals in the DLC or ABS DLC to retrieve DTCs from the ABS computer. On General Motors vehicles with the ignition switch off, a special key may be used to connected terminals A and H or A and G in the DLC under the dash **(Figure 21)**. When the ignition switch is turned on, the ABS warning light flashes the DTCs. For example, three light flashes followed by a short pause and four light flashes indicates DTC 34. The vehicle manufacturer's service information provides the DTC interpretation. A DTC indicates an electrical problem in a certain area, such as a wheel speed sensor. Voltmeter or ohmmeter tests are usually necessary to determine if the defect is in the wheel speed sensor or the wires from this sensor to the ABS computer.

**You Should Know** *On some OBD I vehicles with ABS, the ABS provides scan tool communication from the ABS computer to the DLC. These systems display ABS DTCs or diagnostic messages on the scan tool.*

Two different DLC are used on DaimlerChrysler ABS on OBD I vehicles **(Figure 22** and **Figure 23)**. A scan tool may be connected to these DTCs to retrieve ABS DTCs and perform some other diagnostic functions.

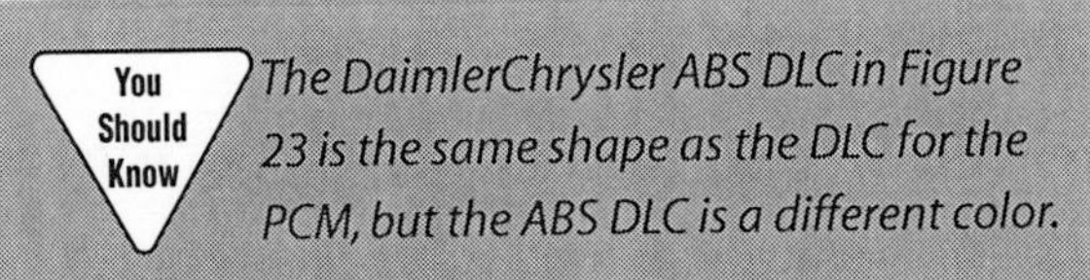
You Should Know
The DaimlerChrysler ABS DLC in Figure 23 is the same shape as the DLC for the PCM, but the ABS DLC is a different color.

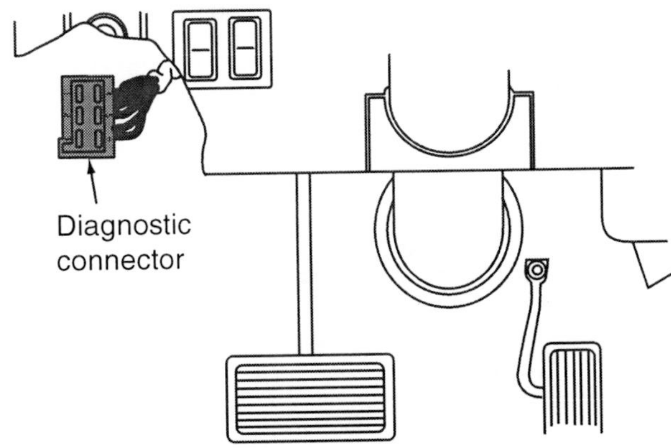
Diagnostic
connector

**Figure 22.** DaimlerChrysler DLC for body computers and ABS.

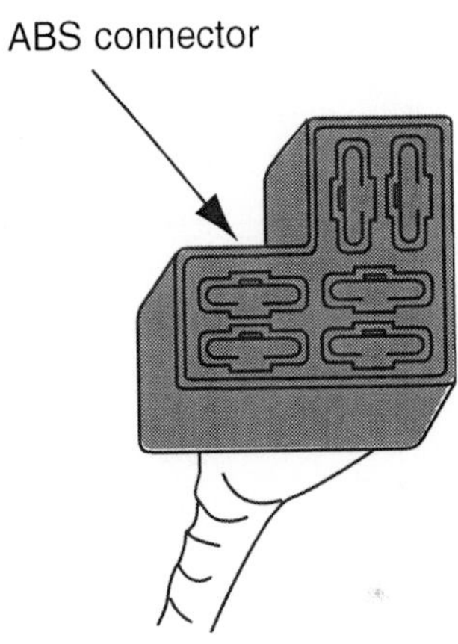
ABS connector

**Figure 23.** DaimlerChrysler ABS DLC.

**You Should Know** *ABS diagnosis varies considerably on OBD I vehicles depending on the vehicle make and model year. Always follow the vehicle manufacturer's recommended ABS diagnostic procedures in the appropriate service manual.*

On Ford OBD I vehicles with ABS, the DLC is the same shape as the DLC for the PCM, but the ABS DLC is red. The ABS DLC may be located in the trunk or under the hood. To retrieve DTCs on a these Ford ABS, turn the ignition switch off and connect a jumper wire between the trigger self-test

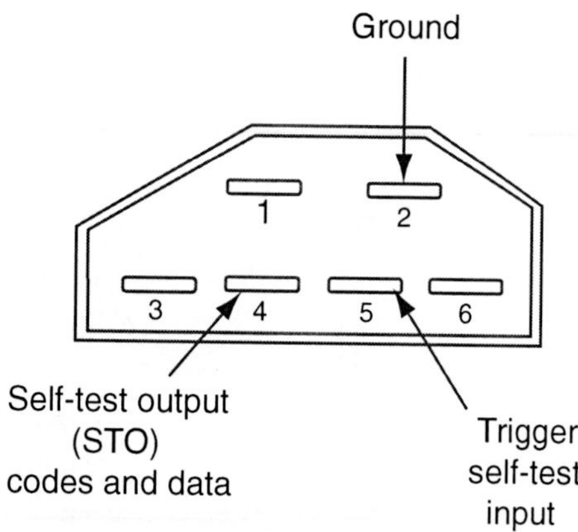


**Figure 24.** Ford ABS DLC.

input terminal and the ground terminal in the DLC **(Figure 24)**. Connect an analog voltmeter from the self-test output terminal in the DLC to ground. When the ignition switch is turned on, the voltmeter needle sweeps out the DTCs. For example, two needle sweeps followed by a pause and one needle sweep indicates DTC 21. On some Ford ABS, a scan tool may be connected to the DLC to retrieve DTCs and perform other diagnostic tests.

To erase the DTCs on many ABS on OBD I vehicles, turn the ignition switch off and disconnect the wiring harness connector from the ABS computer. Reconnect the wiring harness. On some General Motors RWAL systems, disconnect the STOP/HAZARD fuse to erase the DTCs on the RWAL computer. Always follow the vehicle manufacturer's instructions when erasing ABS DTCs because special DTC erasing procedures are required on some ABS on OBD I vehicles. On 1996 and newer vehicles, the scan tool is used to erase ABS DTCs.

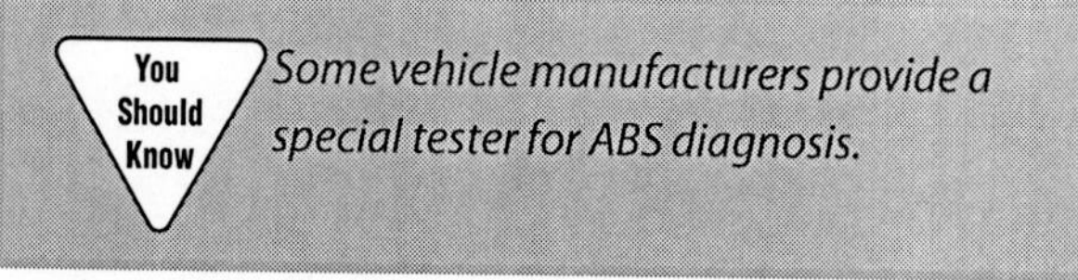


*Some vehicle manufacturers provide a special tester for ABS diagnosis.*

## ABS DIAGNOSIS, OBD II VEHICLES

To diagnose the ABS on OBD II vehicles, connect the scan tool to the DLC with the ignition switch off. Be sure the scan tool contains the proper module for the ABS diagnosis on the vehicle being tested.

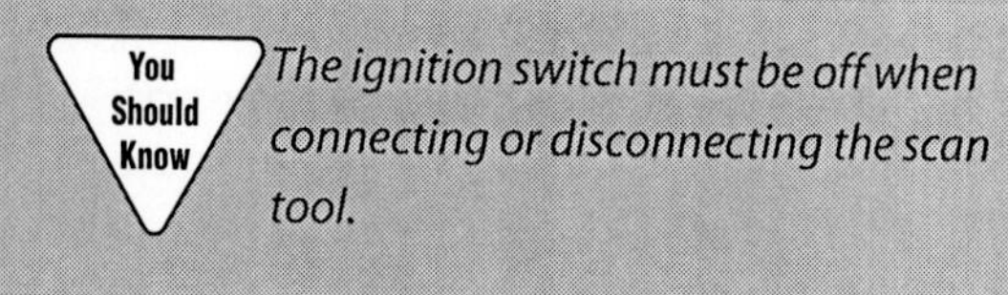


*The ignition switch must be off when connecting or disconnecting the scan tool.*

The following is a typical ABS diagnostic procedure on an OBD II vehicle:

1. Turn the ignition switch on, and select ABS diagnosis on the scan tool display. This establishes communication between the ABS computer and the scan tool.
2. Select ABS DTCs on the scan tool display. If there are any DTCs with a U prefix, there is a defect in the data links connected between the computers. This problem must be repaired before proceeding with the diagnosis.
3. Record any other DTCs displayed on the scan tool. Interpret the DTCs from the vehicle manufacturer's service information and repair the causes of the DTCs. Use the scan tool to erase the DTCs.
4. Select Functional Test on the scan tool. In this mode, the EBCM commands the ABS relay, solenoids, and pump motor on and off. If there is a defect in any of these systems, a DTC is set in the EBCM.
5. Select Automated Bleed on the scan tool. In this mode, the EBCM commands the valve solenoids and pump motor on and off in a special sequence to bleed air from the ABS system.
6. Select ABS Motor on the scan tool. This selection allows the technician to command the pump motor on and off. If a defect is present in the motor circuit, a DTC is set in the EBCM.
7. Select System Identification on the scan tool. The scan tool displays the hardware and software revision of the EBCM.
8. Select Tire Size Calibration on the scan tool. This calibration must be performed after an EBCM is replaced or when different size tires are installed on the vehicle.
9. Select Lamp Tests on the scan tool. This mode allows the technician to command the ABS amber warning light and the red brake warning light on and off.
10. Select Solenoid Tests on the scan tool. This test mode allows the technician to command a selected ABS solenoid valve on and off. The vehicle must be raised on a lift so the wheels and tires are about 1 foot off the shop floor. Command a specific solenoid on with the scan tool and depress the brake pedal. Have an assistant turn the wheel to which the solenoid is connected. The wheel should spin because the solenoid blocks fluid flow to the wheel. If the wheel does not turn, the solenoid is likely leaking. Repeat this procedure at each wheel.
11. Select ABS Relay on the scan tool. This mode allows the technician to turn the ABS relay on and off. If a malfunction is detected in the relay circuit, a DTC is set in the EBCM.

## WHEEL SPEED SENSOR DIAGNOSIS

If a DTC indicates a defect in a wheel speed sensor and related circuit, an ohmmeter may be connected to the wheel speed sensor terminals. The sensor winding should

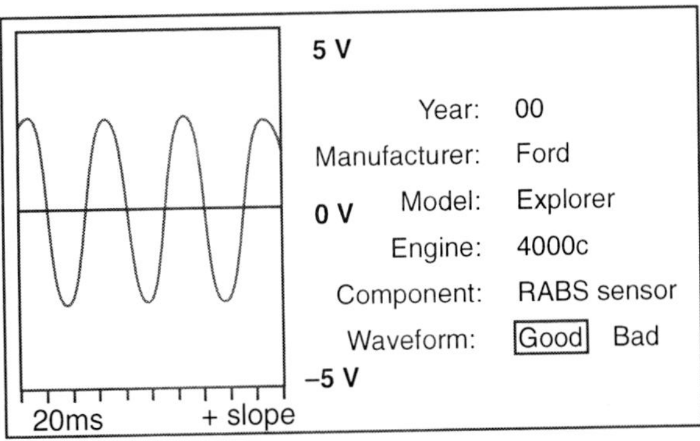


**Figure 25.** Normal wheel speed sensor voltage waveform.

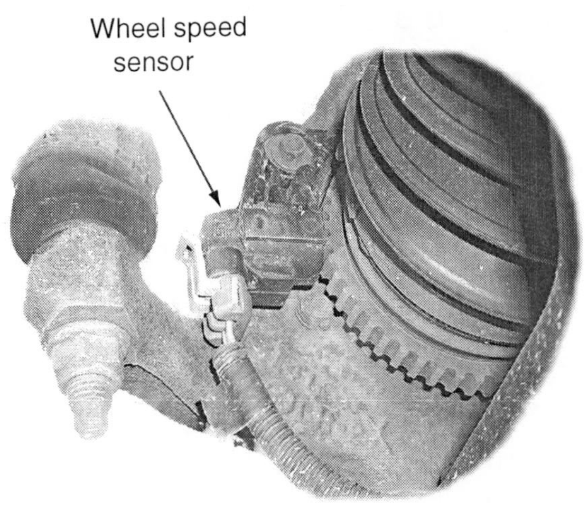


**Figure 26.** Wheel speed sensor and toothed ring.

have the specified resistance. An open winding provides an infinite ohmmeter reading and a shorted winding is indicated by an ohmmeter reading that is lower than specified. An **infinite ohmmeter reading** is a resistance reading beyond measurement that results in a full-scale meter needle position.

When the ohmmeter is connected from one of the wheel speed sensor terminals to ground, an infinite reading indicates the sensor winding is not grounded, whereas a low reading indicates a grounded sensor winding.

A lab scope or graphing voltmeter may be used to obtain a waveform from a wheel speed sensor **(Figure 25)**. If the wheel speed sensor waveform is higher, lower, or erratic compared to the one in the figure, the sensor is defective.

## ABS SERVICE

When a wheel speed sensor is removed, always check the toothed ring for damage **(Figure 26)**. If damage is present, the ring must be replaced. A paper shim must be installed on some wheel speed sensor tips before the new sensor is installed. Push the senor into place until the paper shim lightly contacts the toothed ring and tighten the sensor mounting bolt.

### ABS Bleeding

Always follow the vehicle manufacturer's recommended brake bleeding procedure. On ABS with a high-pressure accumulator, if the accumulator pressure acts as a brake booster and also supplies fluid pressure to the rear brakes, the rear brakes may be bled with a fully-charged accumulator. The front brakes may be bled manually or with a pressure bleeder. To bleed the rear brakes with a fully-charged accumulator, follow this procedure:

1. Turn the ignition switch on and pump the brake pedal until the hydraulic pump motor starts.
2. Wait until the hydraulic pump motor shuts off.
3. With the ignition switch on, have an assistant apply the brake pedal.
4. Connect a hose from the right-rear bleeder screw into a glass container and loosen the right-rear bleeder screw until the brake pedal goes down to the floor.
5. Close the bleeder screw and release the brake pedal.
6. Repeat the bleeding procedure until a clear, bubble-free stream of brake fluid flows into the glass jar when the bleeder screw is opened.
7. Fill the master cylinder reservoir as required and repeat the bleeding procedure on the left-rear wheel. ○

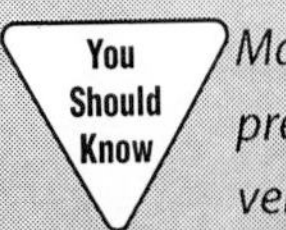


*Many ABS are bled using the manual or pressure bleeder procedures used on conventional brake systems.*

You Should Know

*On ABS with a high-pressure accumulator, always relieve the accumulator pressure before loosening any brake line or component in the hydraulic system. To relieve the accumulator pressure, turn the ignition switch off and pump the brake pedal twenty-five times. If this precaution is not followed, high-pressure brake fluid may cause personal injury or damage to vehicle paint surfaces if any brake line or component is loosened in the hydraulic system.*

## Automated Bleeding Procedure

If the brake pressure modulator valve (BPMV) has been replaced on some ABS, the vehicle manufacturer recommends a manual bleeding procedure followed by an automated bleeding procedure using a scan tool. The **brake pressure modulator valve (BPMV)** is the assembly containing the solenoid valves connected to each wheel.

Complete the manual brake bleeding procedure as explained previously in this chapter. Select ABS and AUTOMATED BLEED PROCEDURE on the scan tool. Follow any instructions provided on the scan tool. Apply the brake pedal and repeat the bleeding procedure on each wheel. Refill the master cylinder reservoir as required. Do not reuse the brake fluid that is bled from each wheel.

# *Summary*

- Brake pedal free-play is necessary for proper brake system operation.
- A low, spongy brake pedal may be caused by a low brake fluid level in the master cylinder or air in the hydraulic system.
- Brake pedal pulsations may be caused by excessive drum out-of-round or rotor lateral runout.
- Grabbing brakes may be caused by contaminated brake linings.
- Excessive brake pedal effort on a power brake system may be caused by lack of vacuum at the brake booster.
- During a brake application, brake pedal fade may be caused by leaking piston seals in the master cylinder.
- Brake system bleeding may be done manually or with a pressure bleeder.
- Brake drums must not be machined so their diameter exceeds the discard diameter.
- There must be no free-play between the caliper fingers and the brake pad flanges.
- Rotor thickness must exceed the discard thickness after machining the rotors.
- Master cylinders may be bled on the bench before installation.
- A restricted vacuum hose to the brake booster causes excessive brake pedal effort.
- The amber ABS warning light should be on for about 4 seconds after the engine starts. Then it should remain off with the engine running.
- If the ABS computer detects an electrical defect in the ABS, the computer illuminates the amber ABS warning light.
- On an ABS with a high-pressure accumulator, the brake fluid level in the master cylinder should be checked with a fully-charged accumulator.
- In OBD I vehicles prior to 1996, the various DLCs were used for ABS diagnosis and other computer system diagnosis.
- OBD II systems on 1996 and newer vehicles have a standard 16-terminal DLC and data links connected between the various computers on the vehicle. The data links are also connected to the DLC.
- On some ABS on OBD I vehicles, the ABS warning light flashes DTCs when the two terminals are connected in the diagnostic connector.
- In some ABS on OBD I vehicles, the ABS DTCs are erased by disconnecting the ABS computer wiring harness with the ignition switch off.
- On OBD II vehicles, the scan tool will display DTCs and perform many other diagnostic functions.
- Wheel speed sensor voltage waveforms may be displayed on a lab scope.
- Wheel speed sensor windings may be tested with an ohmmeter.
- On ABS with a high-pressure accumulator, the accumulator pressure must be relieved before loosening or disconnecting any brake line or component in the hydraulic system.
- The accumulator pressure may be relieved by pumping the brake pedal twenty-five times with the ignition switch off.

# *Review Questions*

1. Technician A says inadequate brake pedal free-play may cause dragging brakes. Technician B says excessive brake pedal free-play may cause a low brake pedal during a brake application. Who is correct?
   A. Technician A
   B. Technician B
   C. Both Technician A and Technician B
   D. Neither Technician A nor Technician B
2. Technician A says grabbing brakes may be caused by contaminated brake linings. Technician B says brake pedal pulsations may be caused by lateral wheel or tire runout. Who is correct?
   A. Technician A
   B. Technician B
   C. Both Technician A and Technician B
   D. Neither Technician A nor Technician B

3. Technician A says brake pedal fade during a brake application may be caused by contaminated brake linings. Technician B says a defective metering valve may cause brake pedal fade during a brake application. Who is correct?
   A. Technician A
   B. Technician B
   C. Both Technician A and Technician B
   D. Neither Technician A nor Technician B
4. Technician A says when a pressure bleeder is used to bleed the brakes, the left-front wheel should be bled first. Technician B says when using a pressure bleeder, the air pressure in the bleeder tank should be 75 psi (517 kPa). Who is correct?
   A. Technician A
   B. Technician B
   C. Both Technician A and Technician B
   D. Neither Technician A nor Technician B
5. All of these statements about the amber ABS warning light are true *except*:
   A. The amber ABS warning light should be on when the ignition switch is turned on.
   B. The amber ABS warning light should be on for about 4 seconds after the engine starts.
   C. The amber ABS warning light should remain off when the engine is running.
   D. If a serious ABS electrical defect occurs with the engine running, the amber ABS warning light begins flashing.
6. In an ABS with a high-pressure accumulator, the brake fluid level in the master cylinder should be checked with the high-pressure accumulator:
   A. discharged.
   B. charged at 500 psi.
   C. half charged.
   D. fully charged.
7. In some ABS on OBD I vehicles, the amber ABS warning light flashes DTCs when:
   A. the brake pedal is applied five times in a 10-second interval with the ignition switch on.
   B. the ABS fuse is removed and replaced.
   C. two terminals are connected in the DLC with the ignition switch on.
   D. when the ignition switch is cycled on and off three times in a 5-second interval.
8. When diagnosing an ABS, a DTC is obtained representing a wheel speed sensor. Technician A says to replace the wheel speed sensor. Technician B says to perform ohmmeter tests to determine if the problem is in the wheel speed sensor or connecting wires. Who is correct?
   A. Technician A
   B. Technician B
   C. Both Technician A and Technician B
   D. Neither Technician A nor Technician B
9. All of these statements about diagnosing typical ABS on OBD II vehicles are true *except*:
   A. A scan tool must be used to perform a Tire Size Calibration each time an ABS is diagnosed.
   B. A scan tool may be used to perform a Functional Test that commands the EBCM to turn various ABS relays, solenoids, and motors, on and off.
   C. A scan tool may be used to perform a lamp test that commands the EBCM to turn the amber and red brake warning lights on and off.
   D. A scan tool may be used to perform Solenoid Tests that allow the technician to command a specific ABS solenoid on and off.
10. With the wheel speed sensor terminals disconnected, a pair of ohmmeter leads is connected from one of the sensor terminals to ground and a low ohmmeter reading is obtained. This ohmmeter reading indicates:
    A. a shorted wheel speed sensor winding.
    B. a satisfactory wheel speed sensor winding.
    C. an open circuit in the wheel speed sensor winding.
    D. a grounded wheel speed sensor winding.
11. A convex-shaped brake drum surface has the largest diameter in the ____________ of the drum friction surface.
12. After turning a brake drum in a lathe, the drum diameter must be at least ________ to ___________ inch less than the discard diameter.
13. If the ohmmeter leads are connected to the two wheel speed sensor terminals, the sensor winding is being tested for ______________ and ______________ circuits.
14. If the ABS computer senses an electrical defect in the ABS, the computer ____________________ the ABS function.
15. Explain the manual brake bleeding procedure.
16. Describe the procedure for performing a basic test on a vacuum brake booster.
17. Describe the normal operation of the amber ABS warning light.
18. Explain the DTC erasing procedure on ABS used with OBD II systems having a 16-terminal DLC.

# Suspension Systems



## SECTION OBJECTIVES

After you have read, studied, and practiced the contents of this section, you should be able to:

- Describe shock absorber operation during wheel jounce.
- Describe the advantages of nitrogen gas-filled shock absorbers and struts.
- Explain shock absorber ratios.
- Explain basic torsion bar action as a front wheel strikes a road irregularity.
- Describe two types of load-carrying ball joints and explain the location of the control arm in each type.
- List the two steering knuckle pivot points in a MacPherson strut front suspension system.
- Describe the rear axle housing movement during vehicle acceleration.
- Explain the difference between a semi-independent and an independent rear suspension system.
- Describe how individual rear wheel movement is provided in a semi-independent rear suspension system.
- Describe the basic operation of an electronic air suspension system.
- Explain the basic operation of a continuously variable road sensing suspension system (CVRSS).
- Maintain, diagnose, and service shock absorbers.
- Maintain, diagnose, and service front and rear suspension systems.
- Measure ball joint wear.
- Diagnose electronic air suspension systems.
- Adjust front and rear trim height on electronic air suspension systems.
- Diagnose continuously variable road sensing suspension systems.

*In 1999, total vehicle registrations in the United States were 213,509,100 and there was an average of 2.2 people per vehicle. In Mexico, vehicle registrations totaled 7,850,000 and there was an average of 19.3 people per vehicle.*

# Chapter 46 Suspension Systems

## Introduction

The front and rear suspension systems must provide several extremely important functions to maintain vehicle safety and owner satisfaction. The suspension system must supply steering control for the driver under all road conditions. Vehicle owners expect the suspension system to provide a comfortable ride. The suspension, together with the frame, must maintain proper vehicle tracking and directional stability. Another important purpose of the suspension system is to provided proper wheel alignment and minimize tire wear.

Computer-controlled suspension systems now provide a soft ride during normal freeway driving and then instantly switch to a firm ride during hard cornering, braking, fast acceleration, and high-speed driving. Therefore, the computer-controlled suspension system allows the same car to meet the demands of the driver who desires a soft ride and the driver who wants a firm ride. Since computer-controlled suspension systems reduce body sway during hard cornering, these systems provide improved steering control.

Some computer-controlled suspension systems also supply a constant vehicle riding height regardless of the vehicle passenger or cargo load. This action maintains the vehicle's cosmetic appearance as the passenger and/or cargo load is changed. Maintaining a constant riding height also supplies more constant suspension alignment angles which may provide improved steering control.

### SHOCK ABSORBERS AND STRUTS

The lower half of a shock absorber is a twin tube steel unit filled with hydraulic oil and nitrogen gas **(Figure 1)**. In

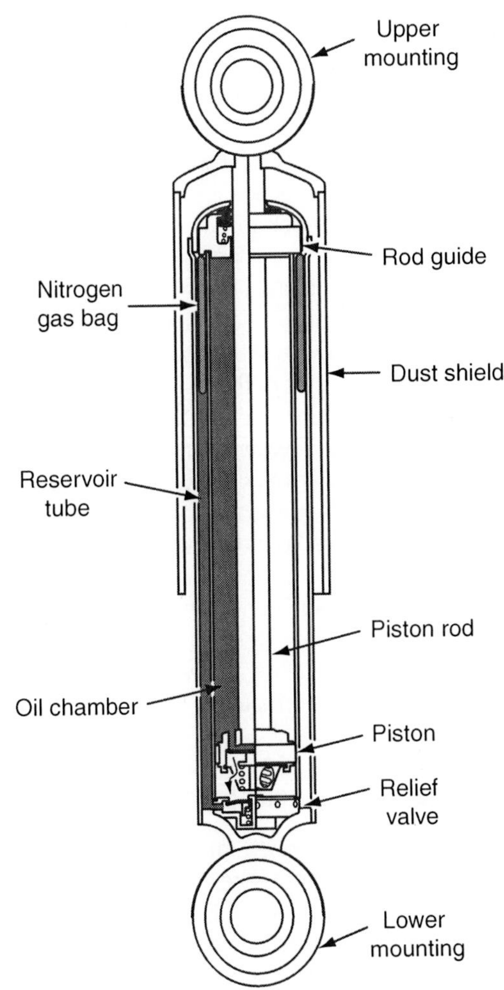


**Figure 1.** Shock absorber design.

some shock absorbers, the nitrogen gas is omitted. A relief valve is located in the bottom of the unit and a circular lower mounting is attached to the lower tube. This mounting contains a rubber isolating bushing, or grommets. A piston and rod assembly is connected to the upper half of the shock absorber. This upper portion of the shock absorber has a dust shield that surrounds the lower twin tube unit. The piston is a precision fit in the inner cylinder of the lower unit. A piston rod guide and seal are located in the top of the lower unit. A circular upper mounting with a rubber bushing is attached to the top of the shock absorber.

## Shock Absorber Operation

Shock absorbers are usually mounted between the lower control arms and the chassis. When a vehicle wheel strikes a bump, the wheel and suspension move upward in relation to the chassis. Upward wheel movement is referred to as **jounce travel**. This jounce action causes the spring to deflect or compress. Under this condition, the spring stores energy and springs back downward with all the energy absorbed when it deflected upward. This downward spring and wheel action is called **rebound travel**. If this spring action is not controlled, the wheel would strike the road with a strong downward force and the wheel jounce would occur again. Therefore, some device must be installed to control the spring action or the wheel would bounce up and down many times after it hit a bump, thereby causing passenger discomfort, directional instability, and suspension component wear. Shock absorbers are installed on suspension systems to control spring action. When a wheel strikes a bump and jounce travel occurs, the shock absorber lower tube unit is forced upward. This action forces the piston downward in the lower tube unit. Since oil cannot leak past the piston, the oil in the lower unit is forced through the piston valves to the upper oil chamber. When a wheel moves downward in rebound travel, the shock absorber piston moves downward in the lower half of shock absorber. Under this condition, oil is forced to flow through the piston valve from the lower part of the unit to the area above the piston. Shock absorber valves provide precise oil flow control and control the upward and downward action of the wheel and suspension **(Figure 2)**. Regardless of the piston orifice and valve design, the shock absorber must be precisely matched to absorb the spring's energy.

## Gas-Filled Struts and Shock Absorbers

During fast upward wheel movement on the compression stroke, excessive pressure in the lower oil chamber forces the base valve open, thus allowing oil to flow through this valve to the reservoir. The nitrogen gas provides a compensating space for the oil that is displaced into the reservoir on the compression stroke and when the oil is

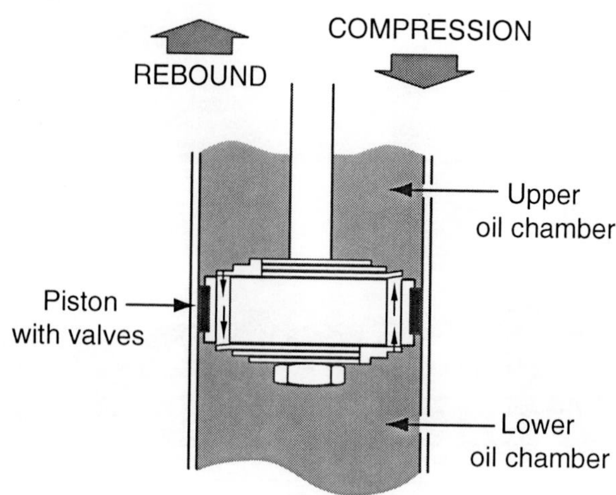


**Figure 2.** Shock absorber operation.

heated. Since the gas exerts pressure on the oil, cavitation or foaming of the oil is eliminated. When oil bubbles are eliminated in this way, the shock absorber provides continuous damping for wheel deflections as small as 0.078 inch (2.0 millimeters). A rebound rubber is located on top of the piston. If the wheel drops downward into a hole, the shock absorber may become fully extended. Under this condition, the rebound rubber provides a cushioning action.

**You Should Know** *New gas-filled shock absorbers are wired in the compressed position for shipping purposes. Exercise caution when removing the shipping wire because the shock extends when this strap is cut. After the upper attaching bolt is installed on the shock absorber, the wire may be cut to allow the unit to extend. Front gas-filled struts have an internal catch which holds them in the compressed position. This catch is released when the rod is held and the strut is rotated 45 degrees counterclockwise.*

Gas-filled units are identified with a warning label. If a gas-filled shock absorber is removed and compressed to its shortest length, it should re-extend when it is released. Failure to re-extend indicates that shock absorber or strut replacement is necessary.

## Heavy-Duty Shock Absorber Design

Some heavy-duty shock absorbers have a dividing piston in the lower oil chamber. The area below this position is pressurized with nitrogen gas to 360 psi (2482 kPa). Hydraulic oil is contained in the oil chamber above the dividing piston. The other main features of the heavy-duty shock absorber are a high-quality seal for longer life, a single tube design to prevent excessive heat buildup, and a rising rate valve that provides precise spring control under all conditions.

The operation of the heavy-duty shock absorber is similar to the conventional type **(Figure 3)**.

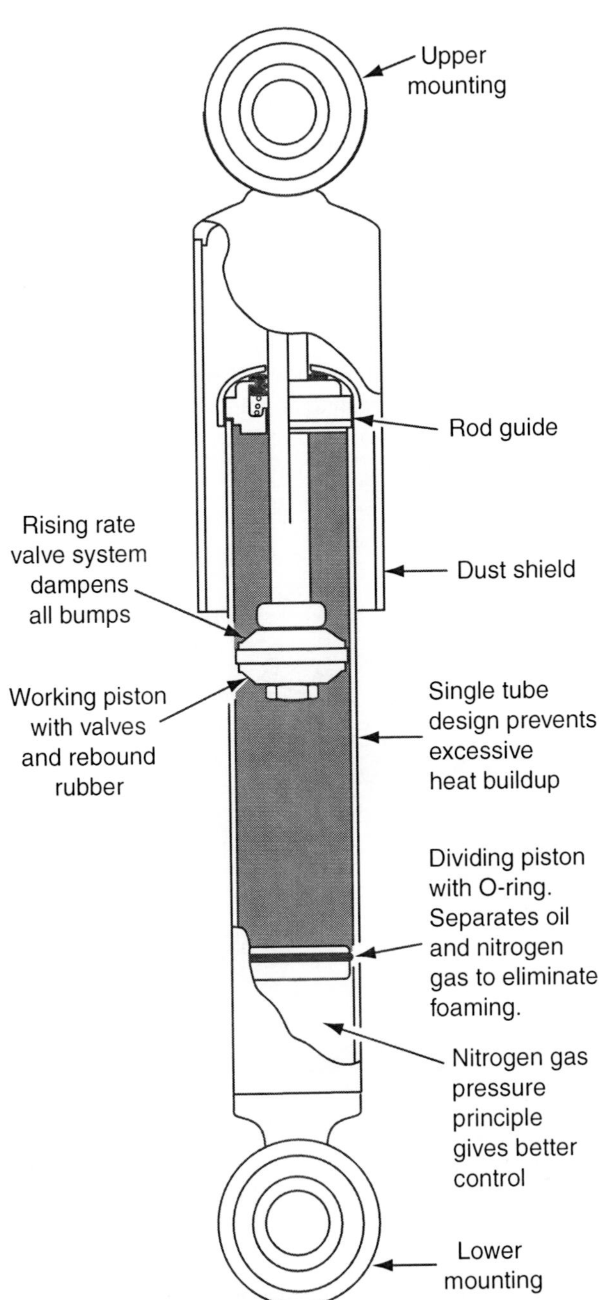


**Figure 3.** Heavy-duty shock absorber design.

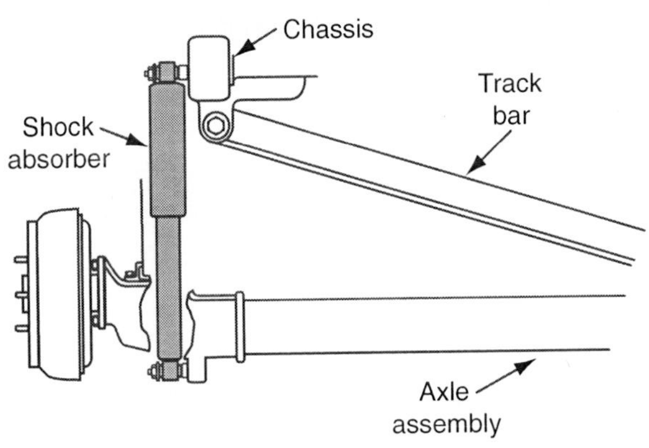


**Figure 4.** Rear shock absorber mounting on a front-wheel drive car.

## Shock Absorber Ratios

Most automotive shock absorbers are a double-acting-type that controls spring action during jounce and rebound wheel movements. The piston and valves in many shock absorbers are designed to provide more extension control than compression control. An average shock absorber may have 70 percent of the total control on the extension cycle, thus 30 percent of the total control is on the compression cycle. Shock absorbers usually have this type of design because they must control the heavier sprung body weight on the extension cycle. The lighter unsprung weight of the axle, wheel, and tire is controlled by the shock absorber on the compression cycle. **Sprung body weight** is the chassis weight supported by the springs. **Unsprung weight** is the weight of the wheel and suspension components that are not supported by the springs.

A shock absorber with this type of design is referred to as a 70/30 type. Shock absorber ratios vary from 50:50 to 80:20.

A shock absorber is mounted between the rear axle and the chassis on a front-wheel drive car **(Figure 4)**.

Mounting bolts extend through hangers on the rear axle and chassis. These bolts also pass through the isolating bushings on each of the shock absorbers. The isolating bushings are very important to prevent vibration and noise. Front shock absorbers may be mounted in a similar way between the lower control arms and the chassis.

## Strut Design

A strut-type front suspension is used on most front-wheel drive cars and some rear-wheel drive cars. Internal strut design is very similar to shock absorber design and struts perform the same functions as shock absorbers. Some struts have a replaceable cartridge. In many strut-type suspension systems, the coil spring is mounted on the strut. The coil spring is largely responsible for proper curb riding

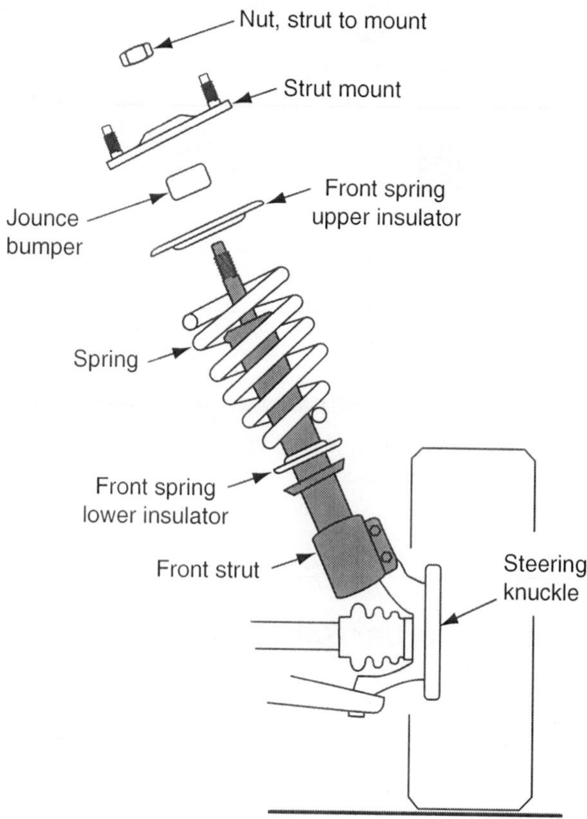


Figure 5. Front strut mounting position.

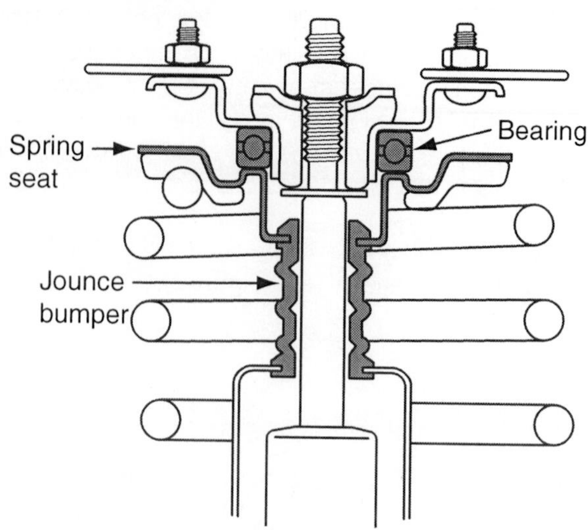


Figure 6. Upper strut mount.

height. A weak or broken coil spring reduces curb riding height and provides harsh riding. The lower end of the front suspension strut is bolted to the steering knuckle **(Figure 5)**. An upper strut mount is attached to the strut. This mount is bolted into the chassis strut tower. A lower spring seat is part of the strut assembly and a lower insulator is positioned between the coil spring and the spring seat on the strut. Another insulator is located between the coil spring and the upper strut mount. The two insulators prevent metal-to-metal contact between the spring and the strut or mount. These isolators reduce the transmission of noise and harshness from the suspension to the chassis. A rubber spring bumper is positioned around the strut piston rod. When a front wheel strikes a large road irregularity and the strut is fully compressed, the spring bumper provides a cushioning action between the top of the strut and the upper support.

The upper strut mount contains a bearing, upper spring seat, and jounce bumper **(Figure 6)**.

When the front wheels are turned, the front strut and coil spring rotate with the steering knuckle. The strut and spring assembly rotates on the upper strut mount bearing.

Some cars have a multi-link front suspension with an upper link connected from the chassis to the steering knuckle. The strut is connected from the upper link to the strut tower **(Figure 7)**. A bearing is mounted between the upper link and the steering knuckle and the wheel and knuckle turn on this bearing and the lower ball joint. Therefore, the coil spring and strut do not turn when the front wheels are turned and a bearing in the upper strut mount

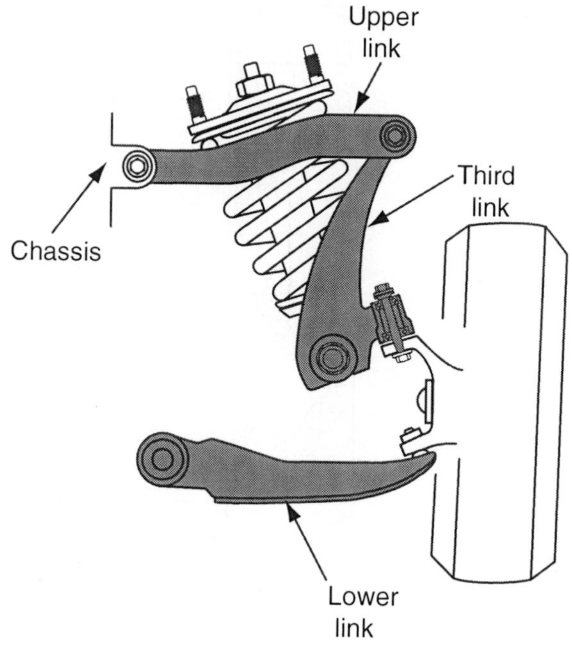


Figure 7. Multilink front suspension with strut connected between the upper link and the strut tower.

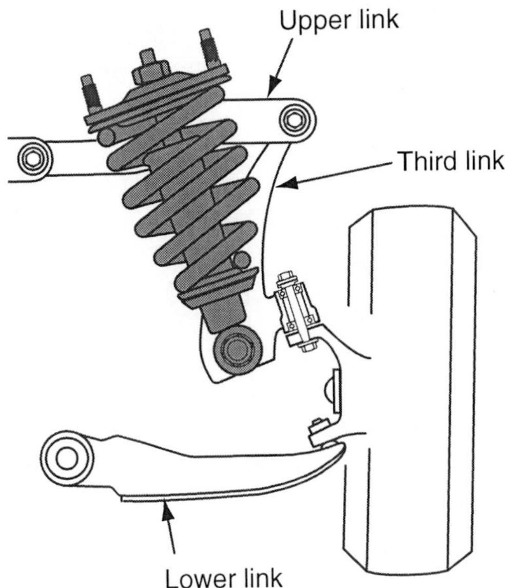


Figure 8. Multilink front suspension with knuckle and wheel pivots at the upper link bearing and lower ball joint and a non-rotating upper strut mount.

is not required **(Figure 8)**. Rear struts are similar to front struts, but rear struts do not require any turning action while cornering.

## Load-Leveling Shock Absorbers

Load-leveling rear struts (or shock absorbers) are used with an electronic height control system. An on-board air compressor pumps air into the rear shocks to raise the rear of the vehicle and an electric solenoid releases air from the shocks to lower the rear chassis. An electromagnetic height sensor may be contained in the shock absorber, or an external sensor may be used **(Figure 9)**. This sensor sends a signal to an electronic control module in relation to the rear suspension height. The module controls the air compressor and the exhaust solenoid to control air pressure in the shock absorbers. This action maintains a specific rear suspension trim height regardless of the load on the rear suspension. If a heavy package is placed in the trunk, the vehicle chassis is forced downward. However, the load-leveling shock absorbers extend to restore the original rear suspension height.

## Electronically Controlled Shock Absorbers and Struts

Many cars are now equipped with computer-controlled suspension systems. In these systems, a computer-controlled actuator is positioned in the top of each shock absorber or strut **(Figure 10)**. The shock absorber or strut actuators rotate a shaft inside the piston rod. This shaft is

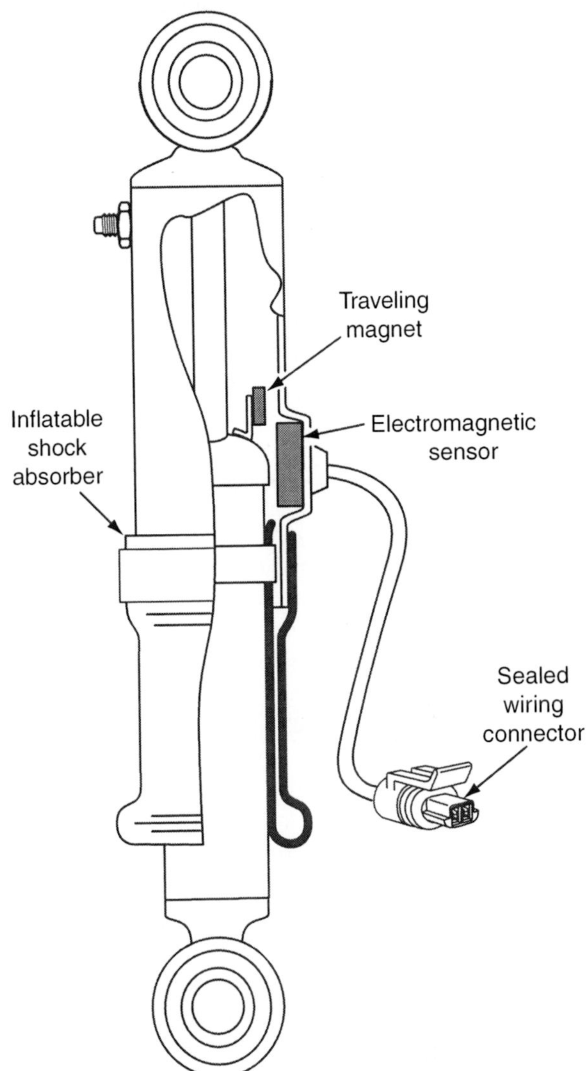


Figure 9. Load-leveling shock absorber.

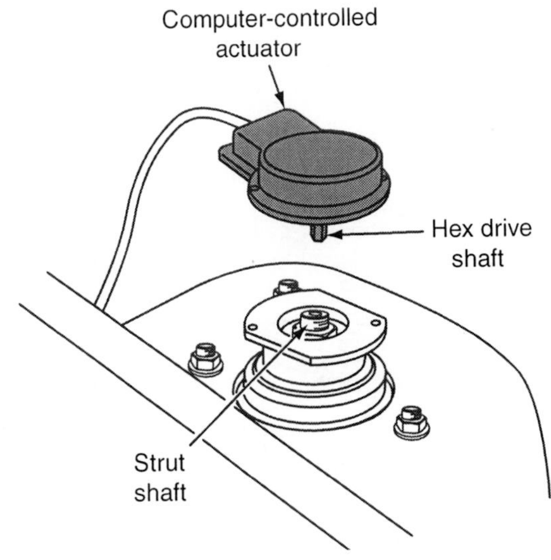


Figure 10. Computer-controlled shock absorber.

connected to the shock valve. Many of these systems have two modes, soft and firm. In the soft mode, the actuators position the shock absorber valves so there is less restriction to the movement of oil. When the computer changes the actuators to the firm mode, the actuators position the shock valves so they provide more restriction to oil movement, which provides a firmer ride.

## BALL JOINTS

Ball joints must be in satisfactory condition to provide proper steering control and vehicle safety. Worn ball joints may cause steering wander, inadequate steering control, and excessive tire tread wear.

### Ball Joint, Load Carrying

The ball joints act as pivot points that allow the front wheels and spindles (or knuckles) to turn between the upper and lower control arms. Ball joints may be grouped into two classifications, load carrying and non-load carrying. Ball joints may be manufactured with forged, stamped, cold-formed, or screw machined housings. A load-carrying ball joint supports the vehicle weight. The coil spring is seated on the control arm to which the load-carrying ball joint is attached. For example, when the coil spring is mounted between the lower control arm and the chassis, the lower ball joint is a load-carrying joint **(Figure 11)**. In a torsion bar suspension, the load-carrying ball joint is mounted on the control arm to which the torsion bar is attached.

In a load-carrying ball joint, the vehicle weight forces the ball stud into contact with the bearing surface in the joint. Load-carrying ball joints may be compression loaded or tension loaded. If the control arm is mounted above the lower end of the knuckle and rests on the knuckle, the ball joint is compression loaded. In this type of ball joint, the vehicle weight is pushing downward on the control arm

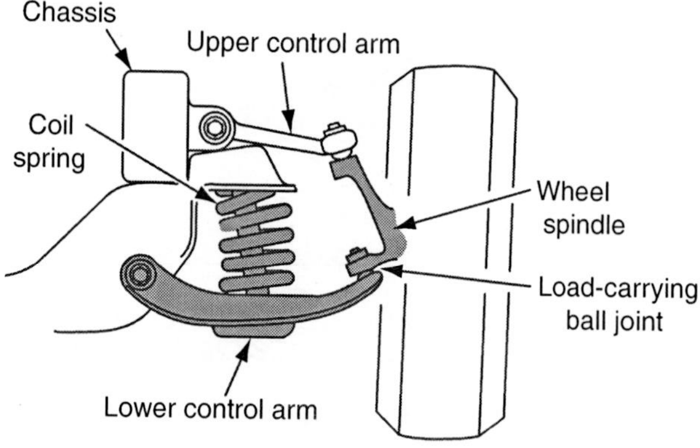


**Figure 11.** Load-carrying ball joint mounted on the control arm on which the spring is seated.

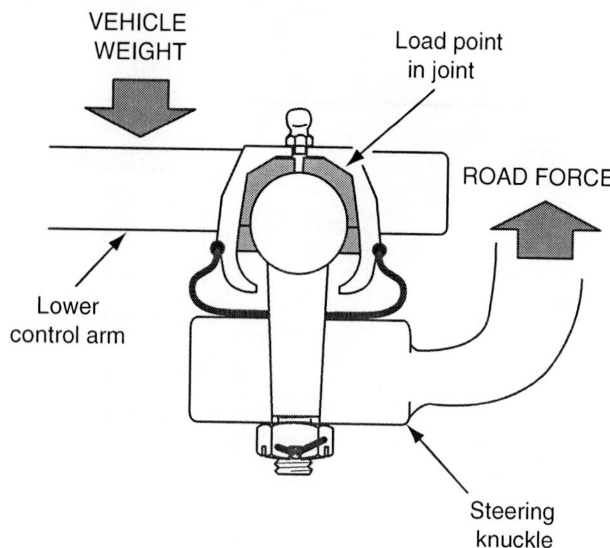


**Figure 12.** Compression-loaded ball joint.

and this weight is supported on the tire and wheel that are attached to the steering knuckle. Since the ball joint is mounted between the control arm and the steering knuckle, the vehicle weight squeezes the ball joint together **(Figure 12)**. In this type of ball joint mounting, the ball joint is mounted in the lower control arm and the ball joint stud faces downward.

When the lower control arm is positioned below the steering knuckle, the vehicle weight is pulling the ball joint away from the knuckle. This type of ball joint mounting is referred to as tension loaded. This type ball joint is mounted in the lower control arm with the ball joint stud facing upward into the knuckle **(Figure 13)**.

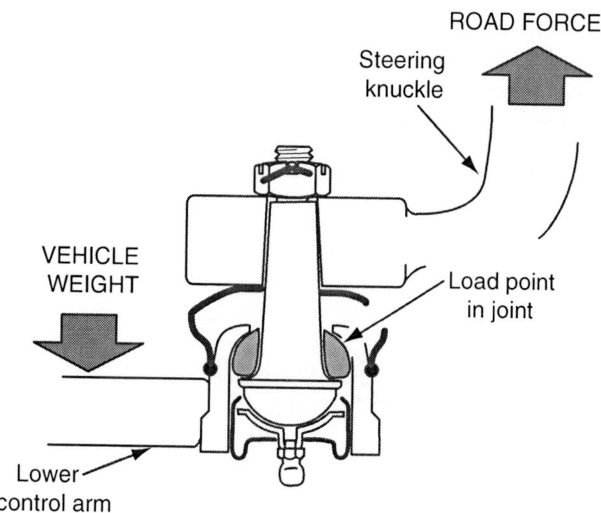


**Figure 13.** Tension-loaded ball joint.

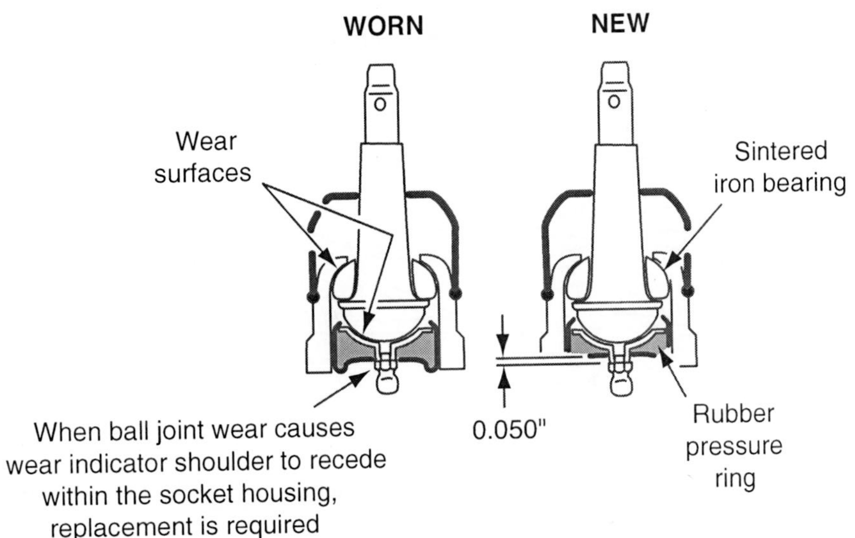


**Figure 14.** Ball joint wear indicator.

Since the load-carrying ball joint supports the vehicle weight, this ball joint wears faster compared to a non-load carrying ball joint. Many load-carrying ball joints have built-in wear indicators. These ball joints have an indicator on the grease nipple surface that recedes into the housing as the joint wears. If the ball joint is in good condition, the grease nipple shoulder extends a specified distance out of the housing. If the grease nipple shoulder is even with or inside of the ball joint housing, the ball joint is worn and replacement is necessary **(Figure 14)**.

## Ball Joint, Non-Load Carrying

A non-load carrying ball joint may be referred to as a stabilizing ball joint. A non-load carrying ball joint is designed with a preload which provides damping action **(Figure 15)**. This ball joint preload provides improved steering quality and vehicle stability.

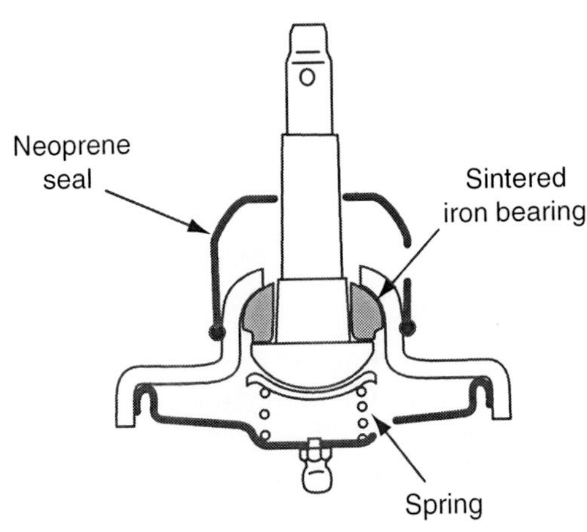


**Figure 15.** Non-load-carrying ball joint.

# SHORT, LONG-ARM (SLA) FRONT SUSPENSION SYSTEMS

An SLA front suspension system has coil springs between the lower control arms and the chassis. Since wheel jounce or rebound movement of one front wheel does not directly affect the opposite front wheel, the control arm suspension is an independent system. Many rear-wheel drive cars have SLA front suspension systems.

## Upper and Lower Control Arms

In SLA front suspension systems, the upper control arm is shorter than the lower control arm. During wheel jounce and rebound travel in this suspension system, the upper control arm moves in a shorter arc compared to the lower control arm. This action moves the top of the tire in and out slightly, but the bottom of the tire remains in a more constant position **(Figure 16)**. This SLA front suspension system provides reduced tire tread wear, improved ride quality, and better directional stability.

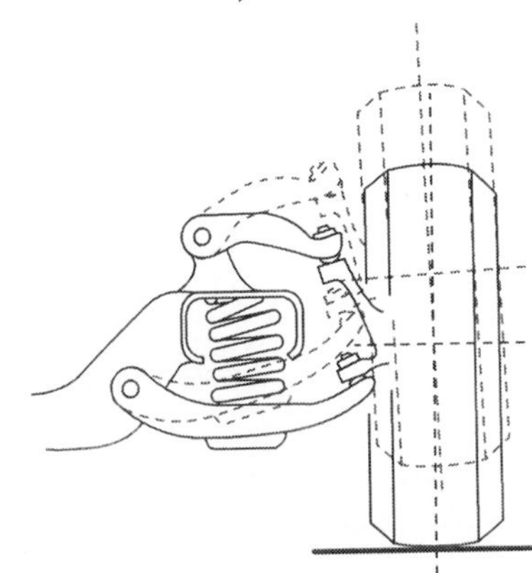

**Figure 16.** Short, long-arm front suspension system.

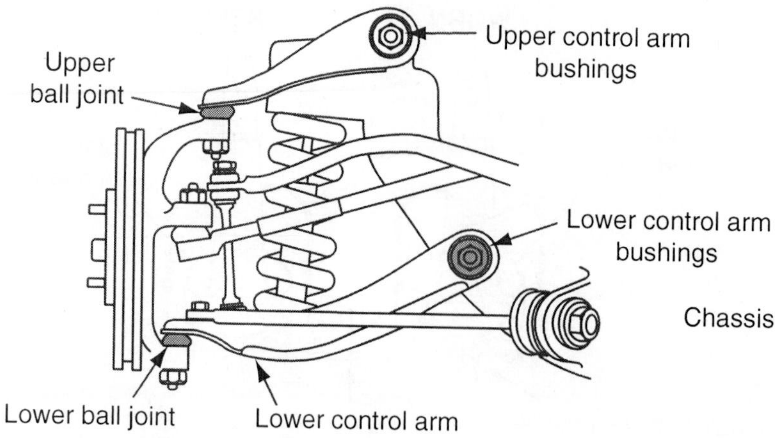


**Figure 17.** Short, long-arm front suspension system with control arm bushings and retaining bolts.

The inner end of the lower control arm contains large rubber insulating bushings and the ball joint is attached to the outer end of the control arm. The lower control arm is bolted to the front crossmember and the attaching bolts are positioned in the center of the lower control arm bushings **(Figure 17)**. The ball joint may be riveted, bolted, pressed, or threaded into the control arm. A spring seat is located in the lower control arm. An upper control arm shaft is bolted to the frame and rubber insulators are located between this shaft and the control arm.

On some SLA front suspension systems, the coil spring is positioned between the upper control arm and the chassis **(Figure 18)**. In these suspension systems, the upper ball joint is compression loaded.

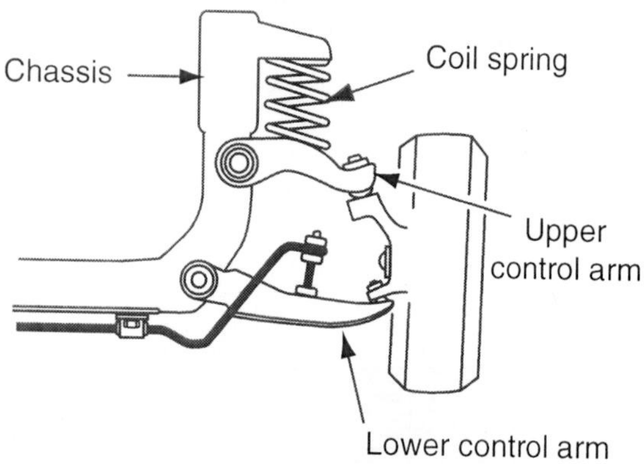


**Figure 18.** Short, long-arm front suspension with the coil spring between the upper control arm and the chassis.

## Steering Knuckle

The upper and lower ball joint studs extend through openings in the steering knuckle. Nuts are threaded onto the ball joint studs to retain the ball joints in the knuckle and the nuts are secured with cotter keys. The wheel hub and bearings are positioned on the steering knuckle extension and the wheel assembly is bolted to the wheel hub. When the steering wheel is turned, the steering gear and linkage turn the steering knuckle. During this turning action, the steering knuckle pivots on the upper and lower ball joints. The upper and lower control arms must be positioned properly to provide correct tracking and wheelbase between the front and rear wheels. The control arm bushings must be in satisfactory condition to position the control arms properly.

## Coil Spring and Shock Absorber

Coil springs are actually a coiled spring steel bar. When a vehicle wheel strikes a road irregularity, the coil spring compresses to absorb shock and then recoils back to its original installed height. Many coil springs contain a steel alloy that contains different types of steel mixed with other elements such as silicon or chromium. The coil spring is positioned between the lower control arm and the spring seat in the frame. A spring seat is located in the lower control arm and an insulator is positioned between the top of the coil spring and the spring seat in the frame. The shock absorber is usually mounted in the center of the coil spring and the lower shock absorber bushing is bolted to the lower control arm. The top of the shock absorber extends through an opening in the frame above the upper spring seat. Washers, grommets, and a nut retain the top of the shock absorber to the frame.

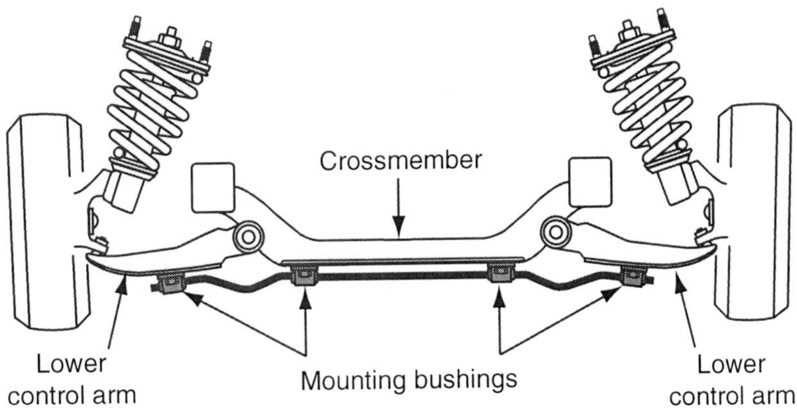


Figure 19. A stabilizer bar.

## Stabilizer Bar

The stabilizer bar is attached to the crossmember and interconnects the lower control arms. Rubber insulating bushings are used at all stabilizer bar attachment points **(Figure 19)**. When jounce and rebound wheel movements affect one front wheel, the stabilizer bar transmits part of this movement to the opposite lower control arm and wheel which reduces and stabilizes body roll. A **stabilizer bar** may be called a sway bar.

## Strut Rod

On some front suspension systems, a strut rod is connected from the lower control arm to the chassis. The strut rod is bolted to the control arm and a large rubber bushing surrounds the strut rod in the chassis opening. The outer end of the strut rod is threaded and steel washers are positioned on each side of the strut rod bushing. Two nuts tighten the strut rod into the bushing **(Figure 20)**. The strut rod prevents fore and aft movement of the lower control arm. In some suspension systems, the strut rod nut position provides proper front wheel adjustment.

# MACPHERSON STRUT-TYPE FRONT SUSPENSION SYSTEM DESIGN

When smaller front-wheel drive cars became popular, most of these cars had MacPherson strut-type front suspension systems. Since the upper control arm is not required in these suspension systems, they are more compact and therefore very suitable for smaller cars.

## Lower Control Arms and Support

On some MacPherson strut front suspension systems, a steel support is positioned longitudinally on each side of the front suspension. These supports are bolted to the unitized body. The inner ends of the lower control arms contain large insulating bushings with a bolt opening in the bushing center. The control arm retaining bolts extend through the center of these bushings and openings in the support **(Figure 21)**.

Road irregularities cause the tire and wheel to move up and down vertically. The lower control arm bushings pivot on the mounting bolts during this movement. When the vehicle is driven over road irregularities, vibration and noise

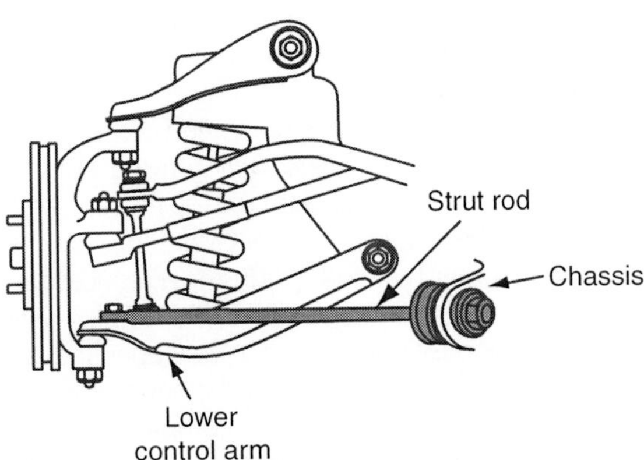


Figure 20. A strut rod.

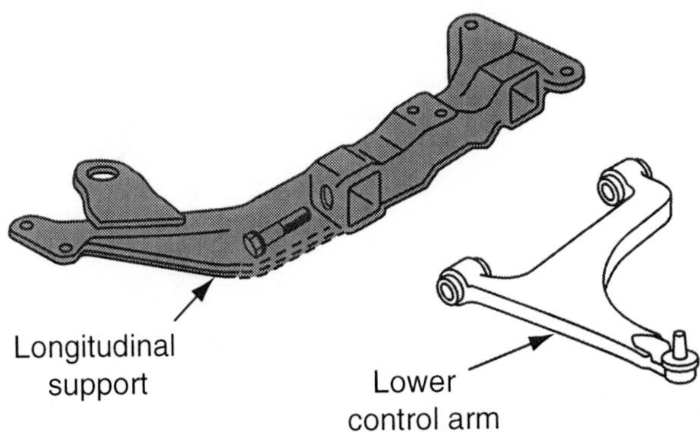


Figure 21. Lower control arm and support.

are applied to the tire and wheel. The control arm bushings help to prevent the transfer of this noise and vibration to the support, unitized body, and passenger compartment. Proper location of the support and lower control arm is important to provide correct vehicle tracking. The supports also carry the engine and transaxle weight. Large rubber mounts are positioned between the engine and transaxle and the supports. These mounts absorb engine vibration.

## Stabilizer Bar

The stabilizer bar in a MacPherson strut front suspension is similar to the stabilizer bar in a short, long-arm front suspension.

## Lower Ball Joint

The lower ball joint is attached to the outer end of the lower control arm. Methods of ball joint to control arm attachment include bolting, riveting, pressing, and threading. A threaded stud extends from the top of the lower ball joint. This stud fits snugly into a hole in the bottom of the steering knuckle. When the ball joint stud is installed in the steering knuckle opening, a nut and cotter pin retains the ball joint **(Figure 22)**.

## Steering Knuckle and Bearing Assembly

The front wheel bearing assembly is bolted to the outer end of the steering knuckle and the brake rotor and wheel rim are retained on the studs in the wheel bearing assembly. This front wheel bearing assembly is a complete non-serviceable sealed unit. The front drive shaft is splined into the center of the wheel bearing hub, thus drive axle torque is applied to the front wheel. A tie rod end connects the steering linkage from the steering gear to the steering knuckle. The top end of the steering knuckle is bolted to the lower end of the strut **(Figure 23)**.

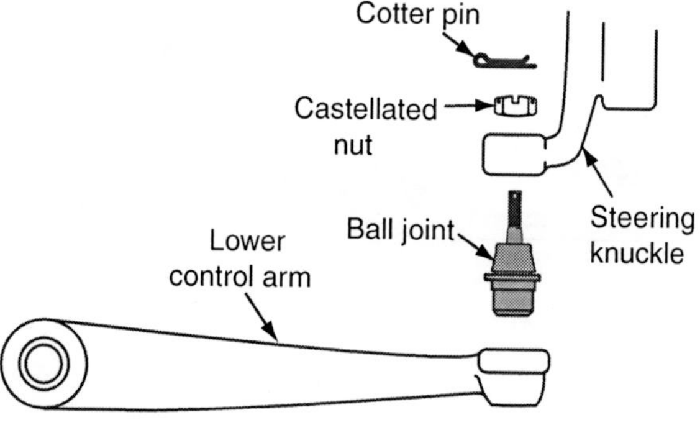


**Figure 22.** A lower ball joint.

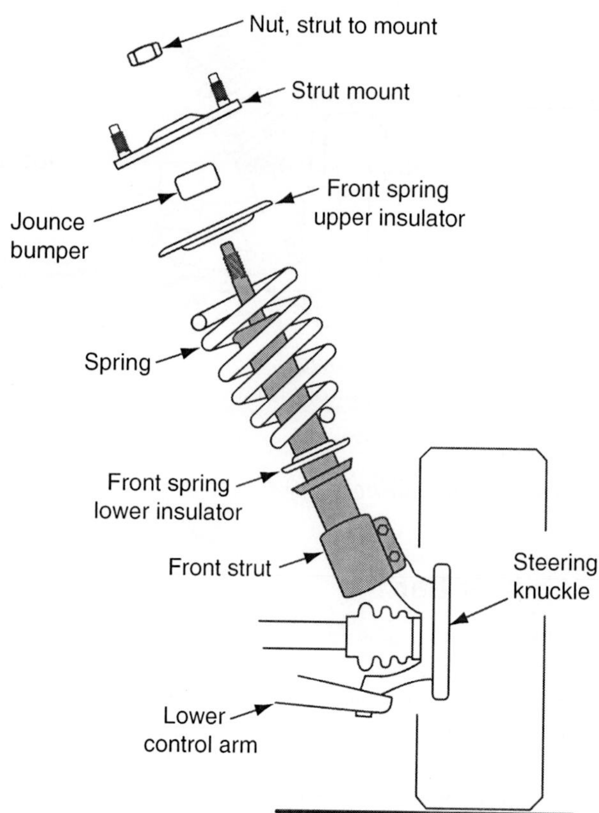


**Figure 23.** Complete MacPherson strut suspension system.

## Strut and Coil Spring Assembly

The strut is the shock absorber in the front suspension and the lower spring seat is attached near the center area of the strut. An insulator is located between the lower spring seat and the bottom of the coil spring. An upper strut mount is retained on top of the strut with a nut threaded onto the upper end of the strut rod. The upper strut mount contains a bearing and upper spring seat and an insulator is positioned between the top of the coil spring and the seat. The upper and lower insulators help to prevent the transfer of noise and vibration from the spring to the strut and body.

A bumper is located on the upper end of the strut rod. This bumper reduces harshness while driving on severe road irregularities. During upward wheel movement, the bumper strikes the lower spring seat before the coils in the spring hit each other. Therefore, this bumper reduces harshness when the wheel and suspension move fully upward. The spring tension is applied against the upper and lower spring seats and insulators. However, the nut on top of the upper mount holds the spring in the compressed position between the upper and lower spring seats. When the steering wheel is turned, the steering linkage turns the steering knuckles to the right or left. During this front wheel turning action, the strut and spring assembly pivot on the lower ball joint and the upper strut mount bearing.

All the suspension-to-chassis mounting devices, such as the lower control arm bushings, and the upper strut mount must be positioned properly and be in satisfactory condition to provide correct vehicle tracking and the same wheelbase on both sides of the vehicle.

## TORSION BAR SUSPENSION

In some front suspension systems, torsion bars replace the coil springs. During wheel jounce, the torsion bar twists. During wheel rebound, the torsion bar unwinds back to its original position. Torsion bar front suspension systems are used on some four-wheel drive light-duty trucks. Each torsion bar is anchored to the front crossmember and the lower control arm **(Figure 24)**. A pivot cushion bushing is mounted around the torsion bar. This bushing is bolted to the crossmember opposite to the torsion bar anchor. An insulating bushing is positioned on the end of the torsion bar where it is connected to the lower control arm.

Vehicle ride height is controlled by the torsion bar anchor adjusting bolts in the crossmember. Front suspension heights must be within specifications for correct wheel alignment, tire wear, satisfactory ride, and accurate bumper heights. A conventional stabilizer bar is connected between the lower control arms and the crossmember. Ball joints are located in the upper and lower control arms. Both ball joints are bolted into the steering knuckle. The shock absorbers are connected between the lower control arms and the crossmember support and the inner ends of the lower control arms are bolted to the crossmember through an insulating bushing.

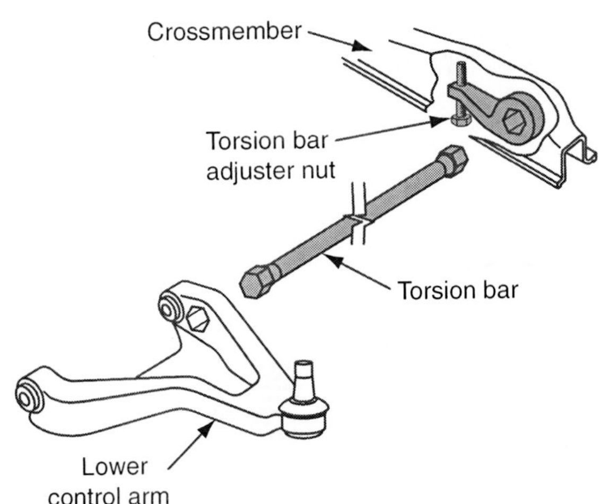


**Figure 24.** Torsion bar mounting.

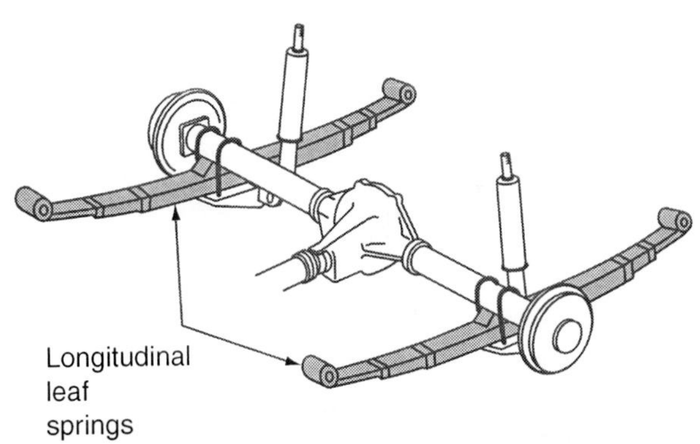


**Figure 25.** Live-axle, leaf-spring rear suspension system.

## LIVE-AXLE REAR SUSPENSION SYSTEMS

In a live-axle rear suspension with leaf springs, a leaf spring is mounted longitudinally on each side of the rear suspension on some rear-wheel drive cars and trucks **(Figure 25)**. Leaf springs may be multiple leaf or mono leaf. Multiple-leaf springs have a series of flat steel leaves of varying lengths that are clamped together. A center bolt extends through all the leaves to maintain the leaf position in the spring. The upper leaf is called the main leaf and this leaf has an eye on each end. An insulating bushing is pressed into each main leaf eye. The front bushing is attached to the frame and the rear bushing is connected through a shackle to the frame. The shackle provides fore and aft movement as the spring compresses **(Figure 26)**.

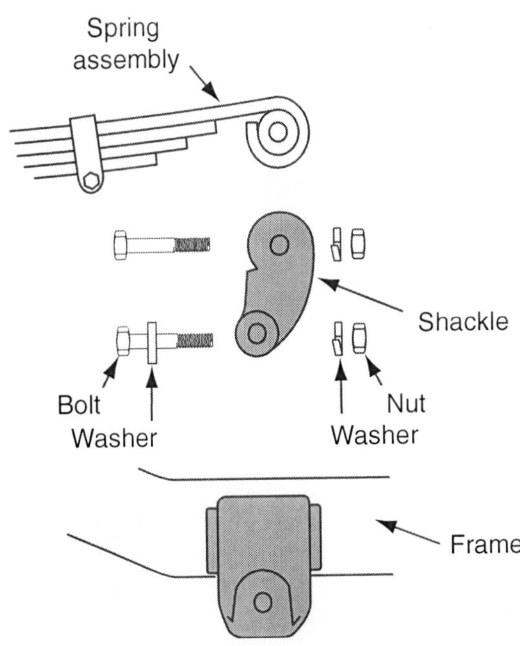


**Figure 26.** Rear leaf-spring shackle.

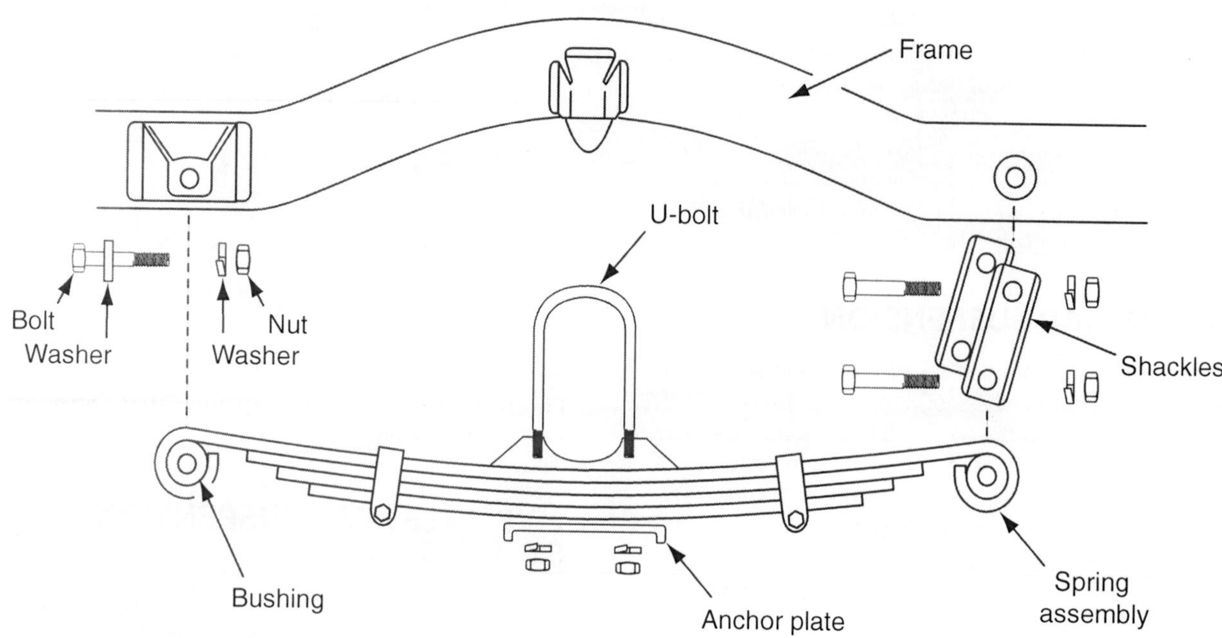


Figure 27. Rear leaf springs and U-bolts.

These relatively flat springs provide excellent lateral stability and reduce side sway which contributes to a well-controlled ride with very good handling characteristics. However, leaf spring rear suspension systems have a great deal of unsprung weight and leaf springs require a considerable amount of space. **A live-axle rear suspension** is one in which the differential housing, wheel bearings, and brakes act as a unit.

Since the differential axle housing is a one-piece unit, jounce and rebound travel of one rear wheel affects the position of the other rear wheel. This action increases tire wear and decreases ride quality and traction.

The differential axle housing is mounted above or below the springs and a spring plate with an insulating clamp and U bolts retain the springs to the rear axle housing **(Figure 27)**. The shock absorbers are mounted between the spring plates and the frame.

The vehicle weight is supported by the springs through the rear axle housing and wheels. When the vehicle is accelerated, the rear wheels are turning counterclockwise viewed from the left vehicle side. One of Newton's Laws of Motion states that for every action, there is an equal and opposite reaction. Therefore, when the wheels are turning counterclockwise viewed from the left, the rear axle housing tries to rotate clockwise. This rear axle torque action is absorbed by the rear springs and the chassis moves downward **(Figure 28)**. Engine torque supplied through the drive shaft to the differential tends to twist the differential housing and the springs. This twisting action may be referred to as axle windup. Many leaf springs have a shorter distance from the center bolt to the front of the spring compared to the distance from the center bolt to the rear of the spring. This type of leaf spring is referred to as asymetrical. The shorter distance from the center bolt to the

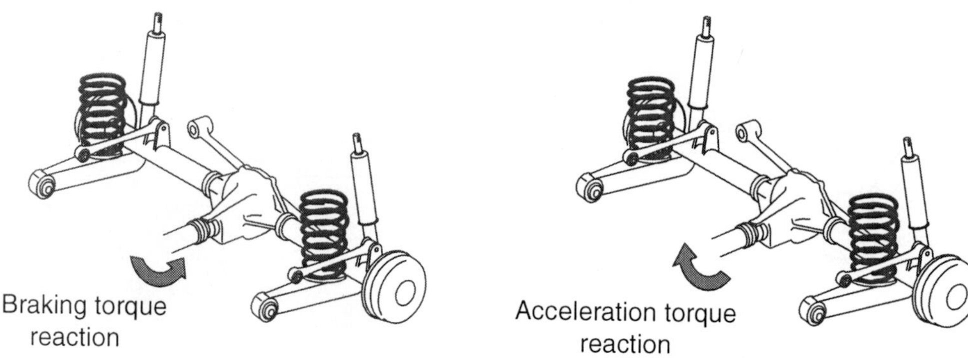


Figure 28. Rear axle torque action during acceleration and braking.

front of the spring resists axle windup. A symmetrical leaf spring has the same distance from the center bolt to the front and rear of the spring.

When braking and decelerating, the rear axle housing tries to turn counterclockwise. This rear axle torque action applied to the springs lifts the chassis.

During hard acceleration, the entire powertrain twists in the opposite direction to engine crankshaft and drive shaft rotation. The engine and transmission mounts absorb this torque. However, the twisting action of the drive shaft and differential pinion shaft tends to lift the rear wheel on the passenger's side of the vehicle. Extremely hard acceleration may cause the rear wheel on the passenger's side to lift off the road surface. Once this rear wheel slips on the road surface, engine torque is reduced and the leaf spring forces the wheel downward. When this rear tire contacts the road surface, engine torque increases and the cycle repeats. This repeated lifting of the differential housing may be called **axle tramp**. This action is provided on live-axle rear suspension systems. Axle tramp is more noticeable on live-axle leaf spring rear suspension systems in which the springs have to absorb all the differential torque. For this reason, only engines with moderate horsepower were used with this type of rear suspension. Rear suspension and axle components such as spring mounts, shock absorbers, and wheel bearings may be damaged by axle tramp. Mounting one rear shock absorber in front of the rear axle and the other rear shock behind the rear axle helps to reduce axle tramp.

Axle tramp is the repeated lifting of one rear wheel off the road surface during hard acceleration.

## Coil Spring Rear Suspension

Some rear-wheel drive vehicles have a coil spring rear suspension. Upper and lower suspension arms with insulating bushings are connected between the differential housing and the frame **(Figure 29)**. The upper arms control lateral movement and the lower trailing control arms absorb differential torque. In some rear suspension systems, the upper arms are replaced with strut rods. The front of the upper and lower arms contain large rubber bushings. When strut rods are used in place of the upper arms, both ends of these rods contain large rubber bushings to prevent noise and vibration transfer from the suspension to the chassis. The coil springs are usually mounted between the lower suspension arms and the frame, while the shock absorbers are mounted between the back of the suspension arms and the frame.

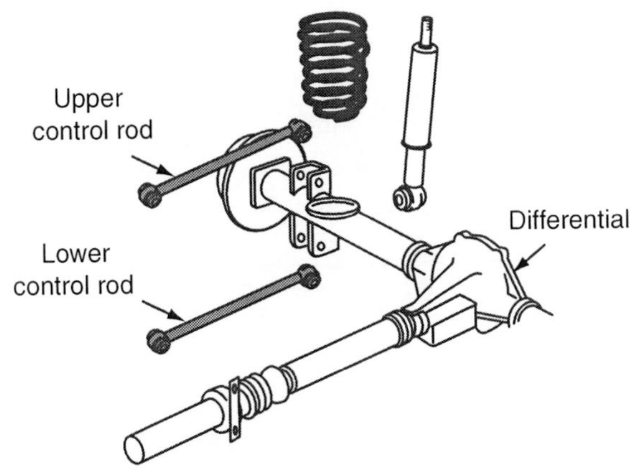


Figure 29. Coil-spring rear suspension system with upper and lower control rods.

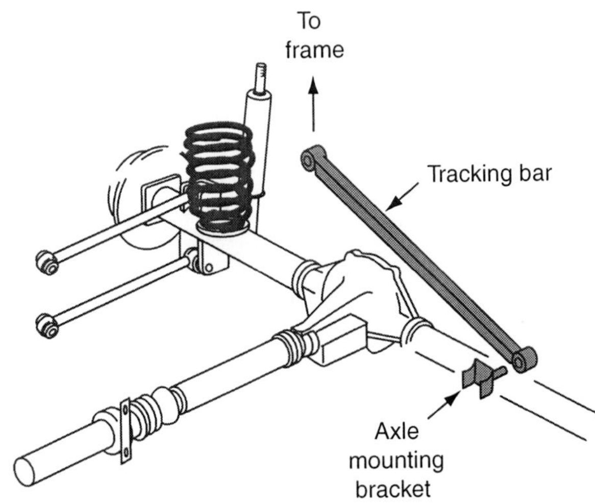


Figure 30. Tracking bar.

Some rear suspension systems have a tracking bar connected from one side of the differential housing to the chassis to prevent lateral chassis movement **(Figure 30)**. Large rubber insulating bushings are positioned in each end of the tracking bar. A **tracking bar** may be referred to as a Panhard or Watts rod.

## SEMI-INDEPENDENT REAR SUSPENSION SYSTEMS

Many front-wheel drive vehicles have a semi-independent rear suspension which has a solid axle beam connected between the rear trailing arms **(Figure 31)**. Some of

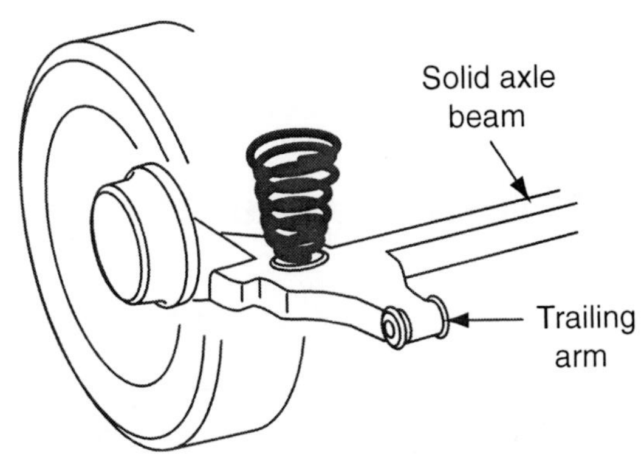


Figure 31. Semi-independent rear suspension system.

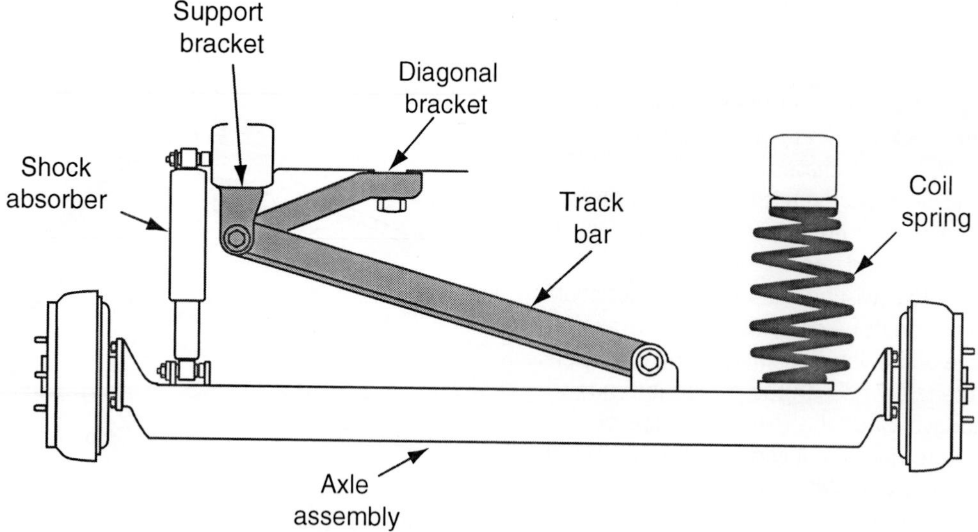


**Figure 32.** Semi-independent rear suspension with track bar and brace.

these rear axle beams are fabricated from a transverse inverted U-section channel.

In some rear suspension systems, the inverted U-section channel contains an integral tubular stabilizer bar. When one rear wheel strikes a road irregularity and the wheel moves upward, the inverted U-section channel twists which allows some independent rear wheel movement. The trailing arms are connected to chassis brackets through rubber insulating bushings. In some semi-independent rear suspension systems, the coil springs are mounted on the rear struts and the lower spring seat is located on the strut with the upper spring seat positioned on the upper strut mount.

In other semi-independent rear suspension systems, the coil springs are mounted separately from the shock absorbers. Coil spring seats are located on the trailing arms and the shock absorbers are connected from the trailing arms to the chassis. A crossmember connected between the trailing arms provides a twisting action and some independent rear wheel movement.

Some semi-independent rear suspension systems have a track bar connected from a rear axle bracket to a chassis bracket. In some applications, an extra brace is connected from this chassis bracket to the rear upper crossmember **(Figure 32)**. The track bar and the brace prevent lateral rear axle movement.

# INDEPENDENT REAR SUSPENSION SYSTEMS

In an independent rear suspension system, each rear wheel can move independently from the opposite rear wheel. Independent rear suspension systems may be found on front-wheel drive and rear-wheel drive vehicles. When rear wheel movement is independent, ride quality, tire life, steering control, and traction are improved.

## MacPherson Strut Independent Rear Suspension System

In a MacPherson strut rear suspension system, the coil springs are mounted on the rear struts. A lower spring seat is located on the strut and the upper spring seat is positioned on the upper strut mount. This upper strut mount is bolted into the inner fender reinforcement. Dual lower control arms on each side of the suspension are connected from the chassis to the lower end of the spindle **(Figure 33)**.

The lower end of each strut is bolted to the spindle. Two strut rods are connected forward from the spindles to the chassis. Rubber insulating bushings are located in both ends of the strut rods. A stabilizer bar is mounted in rubber bushings connected to the chassis and the ends of this bar are linked to the struts.

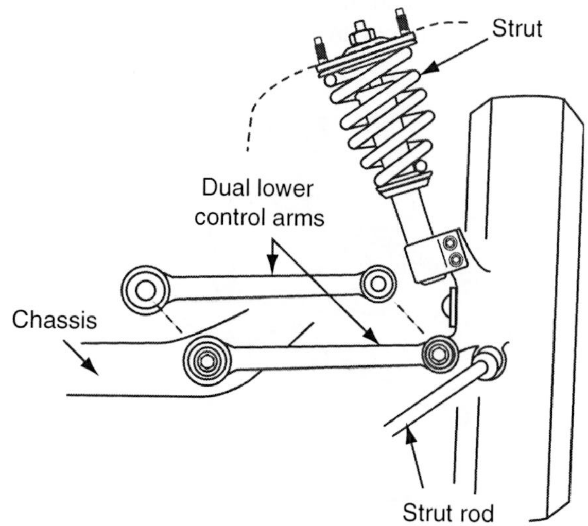


**Figure 33.** MacPherson strut independent rear suspension system.

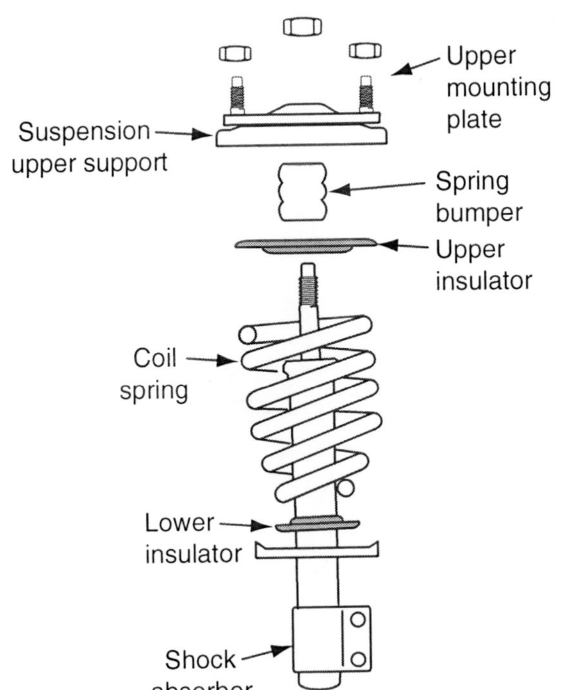


**Figure 34.** Upper and lower spring insulators, MacPherson strut independent rear suspension.

Insulators are mounted between the lower end of the coil spring and the lower spring seat and the top of the coil spring and the upper spring support **(Figure 34)**. These insulators help to prevent the transfer of spring noise and vibration to the chassis and passenger compartment.

## CURB RIDING HEIGHT

Regular inspection and proper maintenance of suspension systems is extremely important to maintain vehicle safety. The curb riding height is determined mainly by spring condition. Other suspension components, such as control arm bushings, will affect curb riding height if they are worn. Since incorrect curb riding height affects most of the other suspension angles, this measurement is critical. The curb riding height must be measured at the vehicle manufacturer's specified location which varies depending on the type of suspension system. When the vehicle is on a level floor or an alignment rack, measure the curb riding height from the floor to the manufacturer's specified location on the chassis on the front and rear suspension **(Figure 35)**.

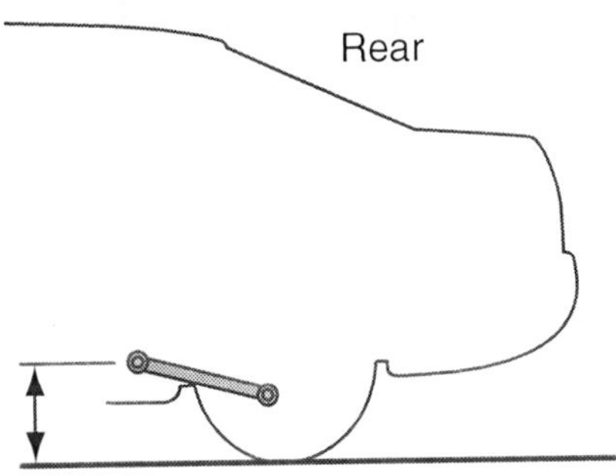


**Figure 35.** Curb riding height measurement, rear suspension.

## SPRING SAG, CURB RIDING HEIGHT, AND CASTER ANGLE

Sagged springs cause insufficient curb riding height. Therefore, the distance is reduced between the strikeout bumper and its stop. This distance reduction causes the bumper to hit the stop frequently with resulting harsh ride quality.

When both rear springs are sagged, the caster angle tilts excessively toward the rear of the vehicle. This type of angle is called positive caster **(Figure 36)**. Rear spring sag and excessive positive caster increases steering effort and causes rapid steering wheel return after a turn is completed.

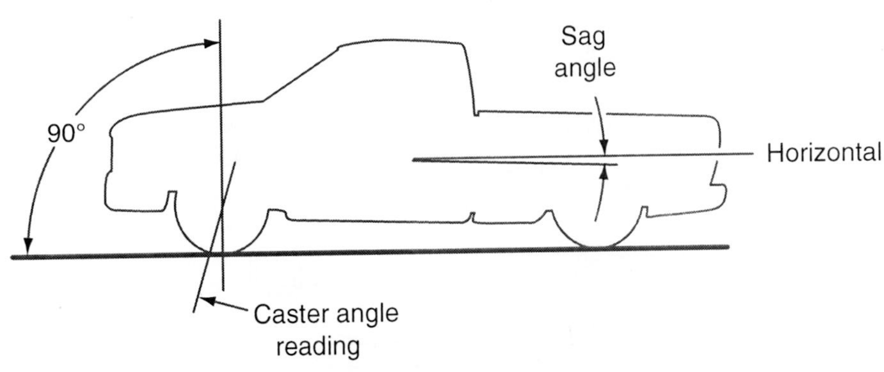


**Figure 36.** Effects of rear spring sag on caster angle.

## ELECTRONIC AIR SUSPENSION SYSTEM

In an air suspension system, the air springs replace the coil springs in a conventional suspension system. These air springs have a composite rubber and plastic membrane that is clamped to a piston located in the lower end of the spring. An end cap is clamped to the top of the membrane and an air spring valve is positioned in the end cap. The air springs are inflated or deflated to provide a constant vehicle trim height. Front air springs are mounted between the control arms and the crossmember **(Figure 37)**. The lower end of these air springs is retained in the control arm with a clip and the upper end is positioned in a crossmember spring seat.

The rear air springs are the same as the front air springs except for their mounting. The lower ends of the rear springs are bolted to the rear suspension arms and the upper ends of these springs are attached to the frame.

### Air Spring Valves

An air spring valve is mounted in the top of each air spring. These valves are an electric solenoid-type valve that is normally closed. When the valve winding is energized, plunger movement opens the air passage to the air spring. Under this condition, air may enter or be exhausted from the air spring. Two O-ring seals are located on the end of the valves to seal them into the air spring cap. The valves are installed in the air spring cap with a two-stage rotating action similar to a radiator pressure cap.

You Should Know

*Never rotate an air spring solenoid valve to the release slot in the cap fitting until all the air is released from the spring. If one of these solenoid valves are loosened with air pressure in the spring, the air pressure drives the solenoid out of the spring with extreme force. This action may result in personal injury.*

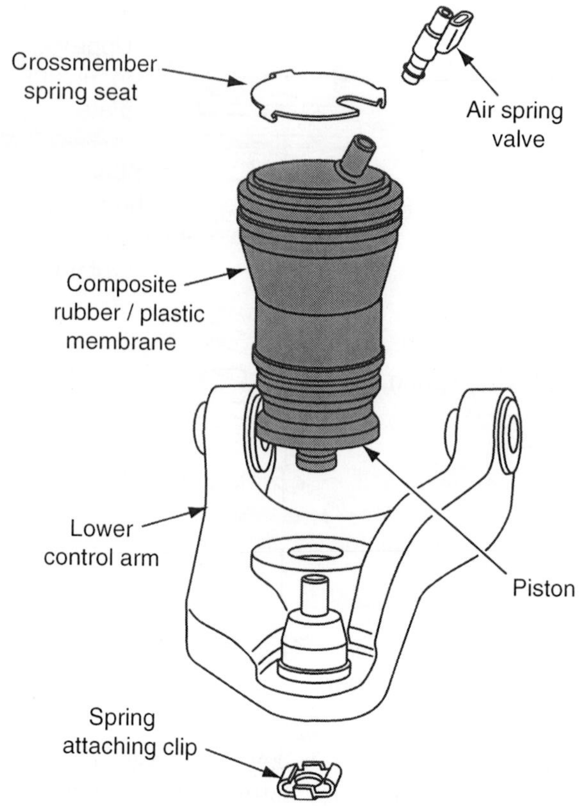


**Figure 37.** Front air spring.

### Air Compressor

A single piston in the air compressor is moved up and down in the cylinder by a crankshaft and connecting rod **(Figure 38)**. The armature is connected to the crankshaft, therefore the rotating action of the armature moves the piston up and down. Armature rotation occurs when 12 volts are supplied to the compressor input terminal. Intake and discharge valves are located in the cylinder head. An air dryer that contains a silica gel is mounted on the compressor. This silica gel removes moisture from the air as it enters the system.

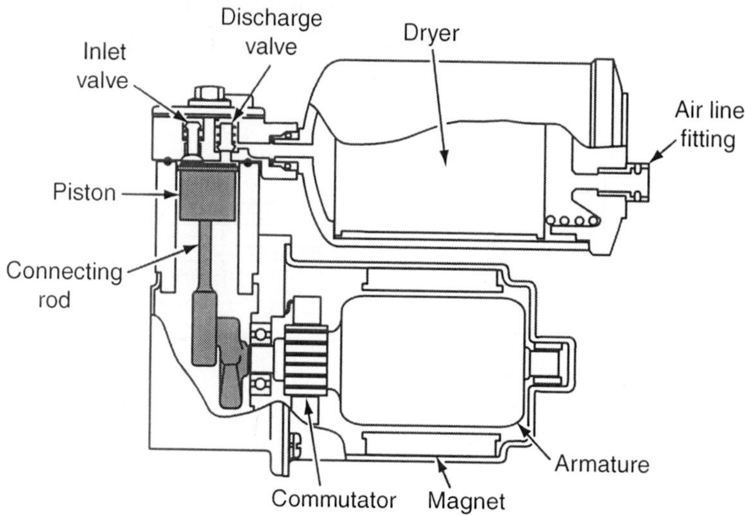


**Figure 38.** An air compressor.

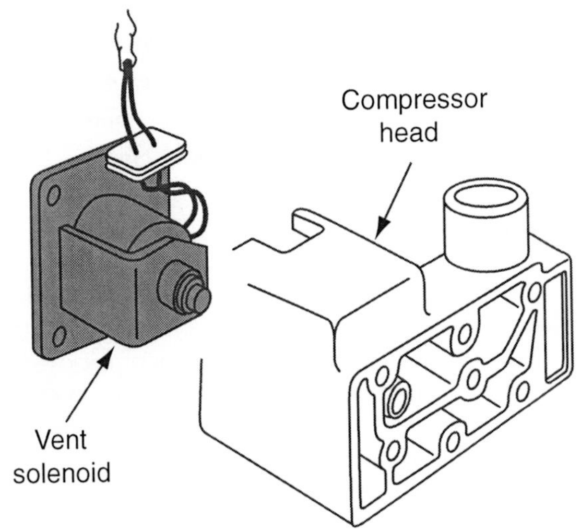


Figure 39. An air vent valve.

Nylon air lines are connected from the compressor outlets to the air spring valves. The compressor operates when it is necessary to force air into one or more air springs to restore the vehicle trim height. On non-computer-controlled suspension systems, the **trim height** may be called the curb riding height.

An air vent valve is located in the compressor cylinder head **(Figure 39)**. This normally closed electric solenoid valve allows air to be vented from the system. When it is necessary to exhaust air from an air spring, the air spring valve and vent valve must be energized at the same time with the compressor shut off. Air exhausting is necessary if the vehicle trim height is too high.

## Compressor Relay

When the compressor relay is energized, it supplies 12 volts through the relay contacts to the compressor input terminal **(Figure 40)**. The relay contacts open the circuit to the compressor if the relay is de-energized.

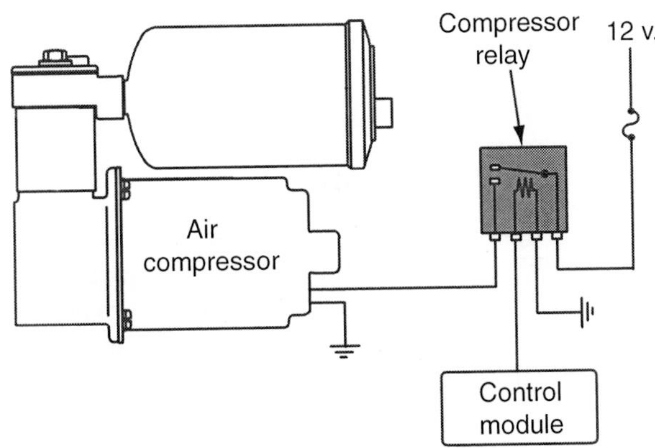


Figure 40. A compressor relay.

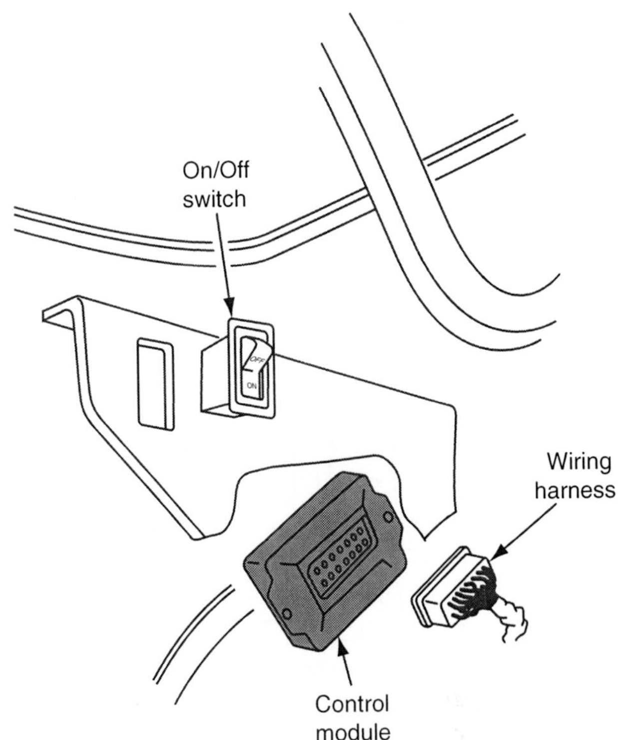


Figure 41. A control module.

## Control Module

The control module is a microprocessor that operates the compressor, vent valve, and air spring valves to control the amount of air in the air springs and maintain the trim height. The control module is located in the trunk **(Figure 41)**.

The control module turns on the service indicator light in the roof panel to alert the driver when a system defect is present. Diagnostic capabilities are also designed into the control module.

## On/Off Switch

The on/off switch opens the 12-volt supply circuit to the control module. This switch is located in the trunk near the control module. When the air suspension system is serviced or some other vehicle service is performed, the switch must be in the off position.

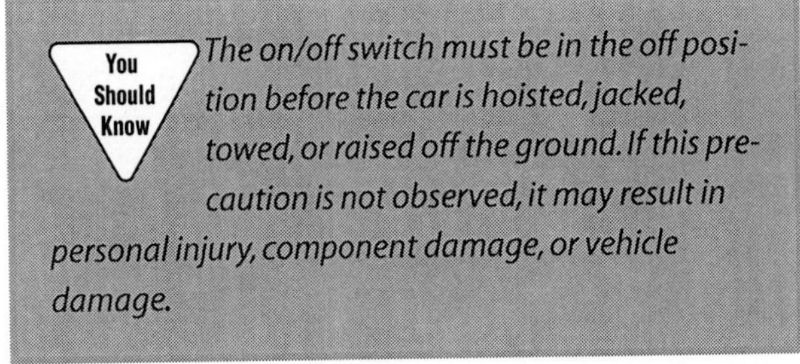


*The on/off switch must be in the off position before the car is hoisted, jacked, towed, or raised off the ground. If this precaution is not observed, it may result in personal injury, component damage, or vehicle damage.*

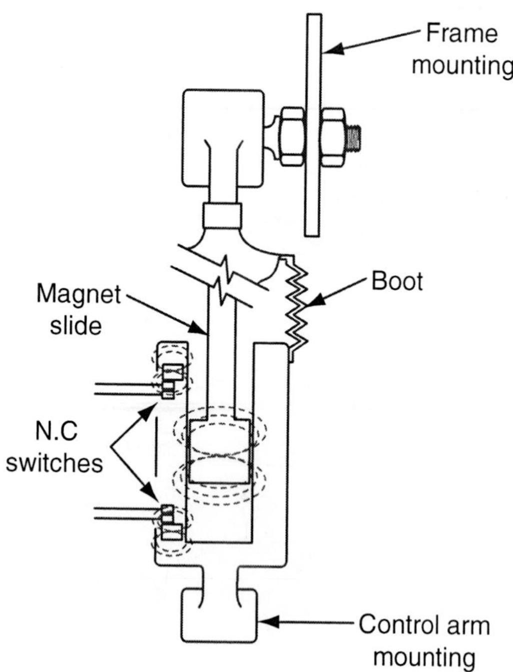


**Figure 42.** Rear height sensor mounting.

## Height Sensors

In the air suspension system, there are two front height sensors located between the lower control arms and the crossmember. A single rear height sensor is positioned between the suspension arm and the frame. Each height sensor contains a magnet slide that is attached to the upper end of the sensor. This magnet slide moves up and down in the lower sensor housing as changes in vehicle trim height occur **(Figure 42)**. The lower sensor housing contains two electronic switches that are connected through a wiring harness to the control module.

When the vehicle is at trim height, the switches remain closed and the control module receives a trim height signal. If the magnet slide moves upward, the above trim switch opens. When this signal is received by the module, it opens the appropriate air spring valve and the vent valve. This action exhausts air from the air spring and corrects the above trim height condition. Downward magnet slide movement closes the above trim switch and opens the below trim switch. When the control module receives this signal, it energizes the compressor relay and starts the compressor. The control module opens the appropriate air spring valve. This action forces air into the air spring to correct the below trim height condition. The height sensors are serviced as a unit.

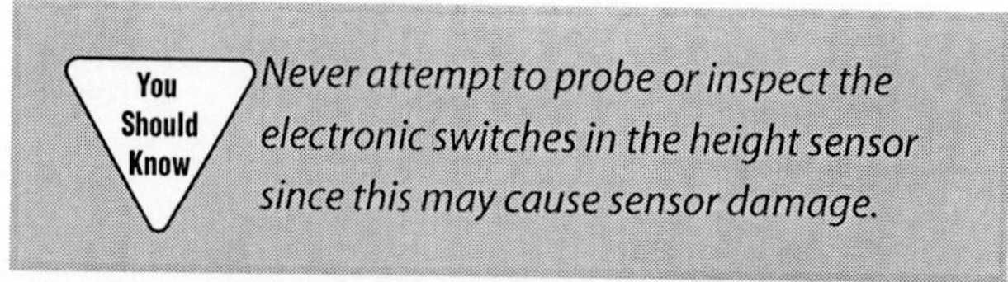


*Never attempt to probe or inspect the electronic switches in the height sensor since this may cause sensor damage.*

## General System Operation

If a door is opened with the brake pedal released, raise vehicle commands are completed immediately, but lower vehicle requests are serviced after the door is closed. This action prevents an open door from catching on curbs or other objects.

If the doors are closed and the brake pedal is released, all commands are serviced by a 45-second averaging method to prevent excessive suspension height corrections on irregular road surfaces.

If the brake is applied and a door is open, raise vehicle commands are completed immediately, but lower vehicle requests are ignored.

When the doors are closed and the brake pedal is applied, all requests are ignored by the control module. If a request to raise the rear suspension is in progress under these conditions, this request will be completed. This action prevents correction of front end jounce while braking.

## Warning Lamp

When the control module senses a system defect, the module turns the air suspension warning lamp on in the roof console which informs the driver that a problems exists. If the air suspension system is working normally, the warning lamp will be on for 1 second when the ignition switch is turned from the off to the run position. After this time, the warning lamp should remain off. This lamp does not operate with the ignition switch in the start position. The warning lamp is used during the self diagnostic procedure and the spring fill sequence.

# CONTINUOUSLY VARIABLE ROAD SENSING SUSPENSION (CVRSS) SYSTEM

The CVRSS system controls shock absorber and strut firmness very precisely to provide improved ride quality, steering control, and directional stability.

## General Description

The CVRSS system may be referred to as a real time damping (RTD) system in the on-board diagnostics. The CVRSS controls damping forces in the front struts and rear shock absorbers in response to various road and driving conditions. The CVRSS system changes shock and strut

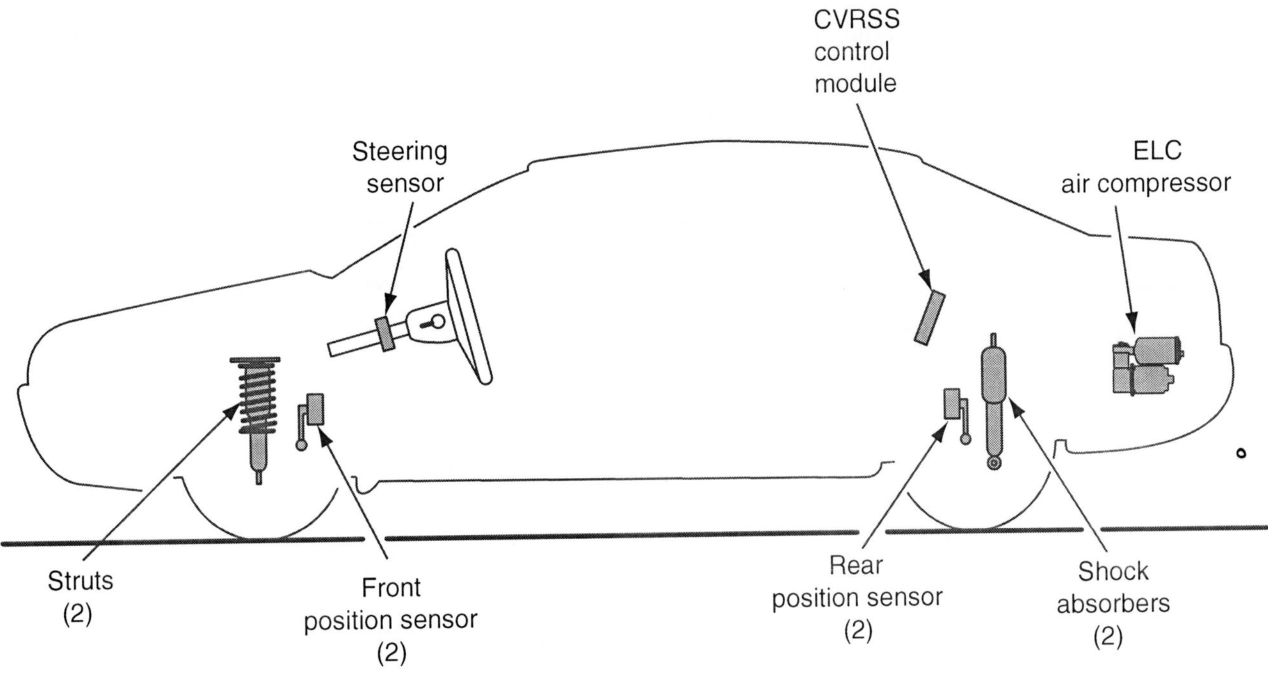


**Figure 43.** Continuously variable road sensing suspension (CVRSS) system components.

damping forces in 10 to 15 milliseconds, whereas other suspension damping systems require a much longer time interval to change damping forces. It requires about 200 milliseconds to blink your eye—this gives us some idea how quickly the CVRSS system reacts.

The CVRSS module receives inputs regarding vertical acceleration, wheel-to-body position and speed of wheel movement, vehicle speed, and lift/dive **(Figure 43)**. The CVRSS module evaluates these inputs and controls a solenoid in each shock or strut to provide suspension damping control. The solenoids in the shocks and struts can react much faster compared to the strut actuators explained previously in some systems.

## Position Sensors

A position sensor is mounted at each corner of the vehicle between a control arm and the chassis (Figure 43). These sensor inputs provide analog voltage signals to the CVRSS module regarding relative wheel-to-body movement and the velocity of wheel movement **(Figure 44)**. **Analog voltage signals** vary continually within a specific voltage range.

The rear position sensor inputs also provide rear suspension height information to the CVRSS module. This information is used by the module to control the rear suspension trim height. All four position sensors have the same design.

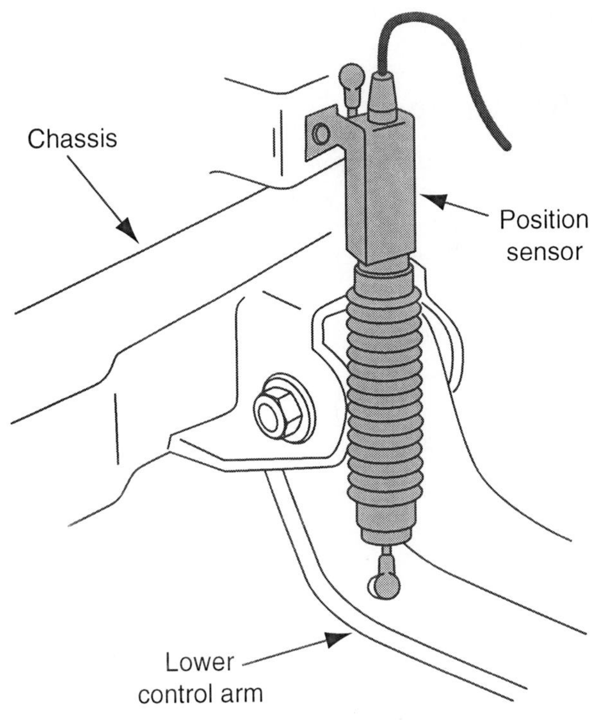


**Figure 44.** Wheel position sensor.

**You Should Know** *The CVRSS is combined with electronic level control (ELC) on the rear suspension. The ELC air compressor in the vehicle trunk is similar to the air compressor on the electronic air suspension system. However, this compressor forces air through nylon air lines to the rear air shocks to maintain rear suspension trim height. The rear suspension has conventional coil springs.*

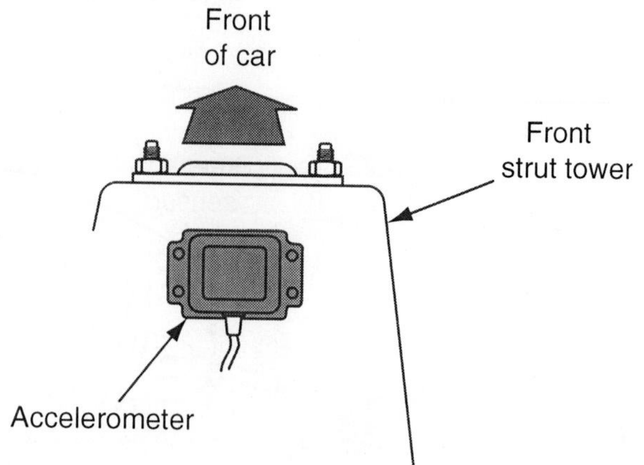


**Figure 45.** Front accelerometer mounting location.

## Accelerometer

An accelerometer is mounted on each corner of the vehicle. These inputs send information to the RSS module in relation to vertical acceleration of the body. The front accelerometers are mounted on the strut towers **(Figure 45)** and the rear accelerometers are located on the rear chassis near the rear suspension support. All four accelerometers are similar in design whereas they send analog voltage signals to the CVRSS module.

## Vehicle Speed Sensor

The VSS is mounted in the transaxle. This sensor sends a voltage signal to the PCM in relation to vehicle speed. The VSS signal is transmitted from the PCM to the CVRSS module.

## Lift/Dive Input

The lift/dive input is sent from the PCM to the CVRSS module. Suspension lift information is obtained by the PCM from the throttle position, vehicle speed, and transaxle gear input signals. The PCM calculates suspension dive information from the rate of vehicle speed change when decelerating.

## CVRSS Module

The CVRSS module is mounted on the right side of the electronics bay in the trunk. Extensive self diagnostic capabilities are programmed into the CVRSS module.

## Damper Solenoid Valves

Each strut or shock damper contains a solenoid that is controlled by the CVRSS module. Each damper solenoid provides two levels of damping, firm or soft. In the soft mode, the solenoid is switched on by the CVRSS module. This causes the oil in the shock or strut to bypass the main damper valving. Voltage is supplied through an RSS damper relay to each strut or shock damper solenoid and the RSS module energizes each damper solenoid by providing a ground for the solenoid winding **(Figure 46)**. When the system switches to

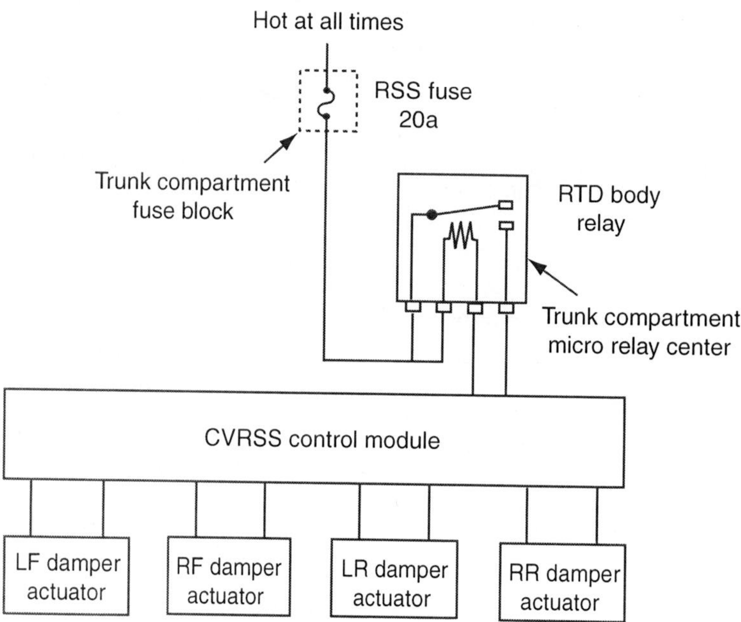


**Figure 46.** Strut damper solenoids and damper relay wiring connections.

the firm mode, the damper solenoid is switched off by the CVRSS module. This action causes the oil to flow through the main damper valving to provide a firm ride. Each strut or shock damper solenoid circuit is basically the same.

Each damper solenoid is an integral part of the damper assembly and is not serviced separately. The CVRSS system operates automatically without any driver-controlled inputs. The fast reaction time of the CVRSS system provides excellent control over ride quality and body lift or dive, which provides improved vehicle stability and handling. Since the position sensors actually sense the velocity of upward and downward wheel movements and the damper solenoid reaction time is 10 to 15 milliseconds, the RSS module can react to these position sensor inputs very quickly. For example, if a road irregularity drives a wheel upward, the CVRSS module switches the damper solenoid to the firm mode before that wheel strikes the road again during downward (rebound) movement.

## Summary

- Wheel and tire jounce travel occurs when a tire strikes a hump in the road surface and the wheel and tire move upward.
- Rebound wheel and tire travel occurs when the tire and wheel move downward after jounce travel.
- The shock absorbers control spring action and prevent excessive wheel and tire oscillations.
- Shock absorber valves are matched to the amount of energy that may be stored in the spring.
- A nitrogen gas charge is located in the oil reservoir of many shock absorbers and struts to prevent oil cavitation and foaming which provides more positive shock absorber action.
- Shock absorber ratio refers to the difference between the shock absorber control on the compression and extension cycle. Many shock absorbers provide more control on the extension cycle.
- Most front struts are connected between the steering knuckle and the upper strut mount.
- In a torsion bar suspension system, the torsion bars replace the coil springs.
- A load-carrying ball joint may be compression loaded or tension loaded.
- In an SLA suspension system, the coil springs may be mounted between the lower control arm and the frame or between the upper control arm and the chassis.
- In a live-axle leaf spring rear suspension system, the leaf springs absorb differential torque and provide lateral control.
- In a live-axle coil spring rear suspension system, the lower control arms absorb differential torque and the upper arms control lateral movement.
- A semi-independent rear suspension has a limited amount of individual rear wheel movement provided by a steel U-section channel or crossmember.
- In an independent rear suspension, each rear wheel can move individually without affecting the opposite rear wheel.
- Compared to a live-axle rear suspension system, an independent rear suspension provides improved ride quality, steering control, tire life, and traction.
- An electronic air suspension system maintains a constant vehicle trim height regardless of passenger or cargo load.
- The air spring valves are retained in the air spring caps with a two-stage rotating action, much like a radiator cap.
- An air spring valve must never be loosened until the air is exhausted from the spring.
- The on/off switch in an electronic air suspension system supplies 12 volts to the control module. This switch must be off before the car is hoisted, jacked, towed, or raised off the ground.
- In an electronic air suspension system, if the doors are closed and the brake pedal is released, all requests to the control module are serviced by a 45-second averaging method.
- If the control module in an electronic air suspension system cannot complete a request from a height sensor in 3 minutes, the control module illuminates the suspension warning lamp.
- The continuously variable road sensing suspension system changes shock and strut damping forces in 10 to 15 milliseconds.

## Review Questions

1. When discussing a torsion bar front suspension system, Technician A says the one end of the torsion bar is attached to the upper control arm. Technician B says the a suspension height adjustment is positioned on the end of the torsion bar connected to the control arm. Who is correct?
   A. Technician A
   B. Technician B
   C. Both Technician A and Technician B
   D. Neither Technician A nor Technician B
2. While discussing semi-independent rear suspensions, Technician A says some individual rear wheel movement is provided by the trailing arms. Technician B says some independent rear wheel movement is provided by the struts. Who is correct?
   A. Technician A
   B. Technician B
   C. Both Technician A and Technician B
   D. Neither Technician A nor Technician B

3. While discussing rear axle tramp, Technician A says rear axle tramp is the repeated lifting of the passenger's side tire off the road surface during hard acceleration. Technician B says rear axle tramp occurs because of the engine torque transmitted through the drive shaft. Who is correct?
   A. Technician A
   B. Technician B
   C. Both Technician A and Technician B
   D. Neither Technician A nor Technician B
4. All these statements about shock absorber and strut design and operation are true *except*:
   A. jounce wheel travel occurs when a wheel moves upward.
   B. rebound wheel travel occurs when a wheel moves downward.
   C. some shock absorbers and struts have a nitrogen gas charge in place of hydraulic oil.
   D. shock absorbers and struts control the spring action.
5. A typical shock absorber ratio is:
   A. 20:60.
   B. 40:80.
   C. 70:30.
   D. 45:90.
6. While discussing strut design, Technician A says some struts have a replaceable cartridge. Technician B says on many front suspension systems when the front wheels are turned, the strut and coil spring assembly rotate on the upper strut mount. Who is correct?
   A. Technician A
   B. Technician B
   C. Both Technician A and Technician B
   D. Neither Technician A nor Technician B
7. All these statements about ball joints are true *except*:
   A. A non-load carrying ball joint wears faster than a load carrying ball joint.
   B. If the lower control arm is mounted above the lower end of the steering knuckle, the ball joint between these components is compression loaded.
   C. A load carrying ball joint supports the vehicle weight.
   D. In a ball joint with wear-indicating capabilities, the grease fitting shoulder should be extended from the joint housing.
8. The electronic air suspension switch should be in the off position if the:
   A. vehicle is boosted with a booster battery.
   B. vehicle is diagnosed with a scan tool.
   C. vehicle is jacked up on one corner to change a tire.
   D. battery is being changed.
9. All these statements about CVRSS are true *except*:
   A. Position sensors are mounted at each corner of the vehicle.
   B. The position sensors send digital voltage signals to the CVRSS module.
   C. The position sensor signals are in relation to the amount and speed of wheel-to-body movement.
   D. An accelerometer is mounted on each corner of the vehicle.
10. In a CVRSS, the PCM sends a lift/dive signal to the CVRSS module. The PCM calculates the lift/dive signal by using information from:
    A. the engine coolant temperature sensor and engine speed signals.
    B. the mass airflow sensor and intake air temperature sensor signals.
    C. the crankshaft position sensor and camshaft position sensor signals.
    D. the throttle position sensor, vehicle speed sensor, and transaxle gear position sensor signals.
11. If the lower control arm is mounted above the steering knuckle and rests on the knuckle, the lower ball joint is ____________ ____________.
12. When one front wheel strikes a road irregularity, the stabilizer bar reduces ____________ ____________.
13. The strut rod prevents ________ and ________ lower control arm movement.
14. When the front springs are sagged, the front wheel caster becomes more ____________.
15. Explain the purpose of the nitrogen gas charge in shock absorbers and struts.
16. Explain the differential torque action during acceleration and describe how this torque is absorbed in a live-axle coil spring rear suspension.
17. List the conditions when the on/off switch in an electronic air suspension system must be turned off.
18. Explain why the control module in an electronic air suspension system services all commands by a 45-second averaging method when the doors are closed and the brake pedal is released.

# Chapter 47 Suspension System Maintenance, Diagnosis, and Service

## Introduction

Proper front and rear suspension system maintenance, diagnosis, and service are extremely important to provide adequate vehicle safety and maintain ride comfort and normal tire life. If worn or loose front suspension system components are ignored, steering control may be adversely affected, which may result in loss of steering control and an expensive collision. Defective front suspension system components, such as worn out shock absorbers and broken springs, may cause rough riding that results in driver and passenger discomfort. Other worn front suspension components, such as worn ball joints and control arm bushings, cause improper alignment angles that cause excessive front tire wear. Therefore, technicians must be familiar with front and rear suspension maintenance, diagnosis, and service.

Each year, more vehicles are equipped with computer-controlled suspension systems. These systems are becoming increasingly complex. Therefore, technicians must understand the correct procedures for diagnosing and servicing these systems. When a technician understands computer-controlled suspension systems and the proper diagnostic procedures for these systems, diagnosis becomes faster and more accurate.

### SHOCK ABSORBER MAINTENANCE

Shock absorbers should be inspected for loose mounting bolts and worn mounting bushings. If these components are loose, rattling noise is evident, and replacement of the bushings and bolts is necessary.

In some shock absorbers, the bushing is permanently mounted in the shock and the complete unit must be replaced if the bushing is worn. When the mounting bushings are worn, the shock absorber will not provide proper spring control.

Shock absorbers and struts should be inspected for oil leakage. A slight oil film on the lower oil chamber is acceptable. Any indication of oil dripping is not acceptable and replacement in pairs is necessary **(Figure 1)**.

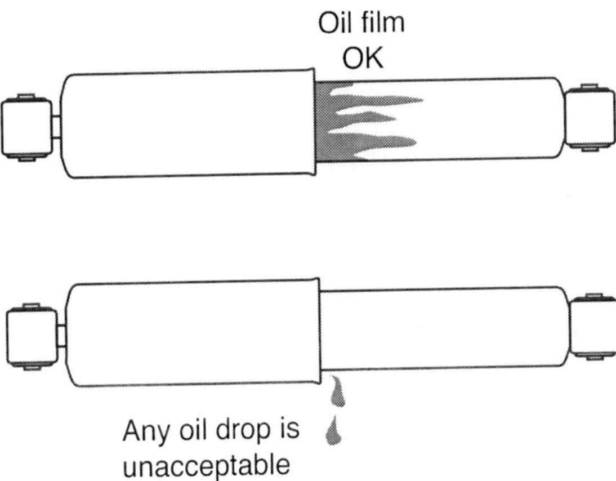


**Figure 1.** Shock absorber and strut oil leak diagnosis.

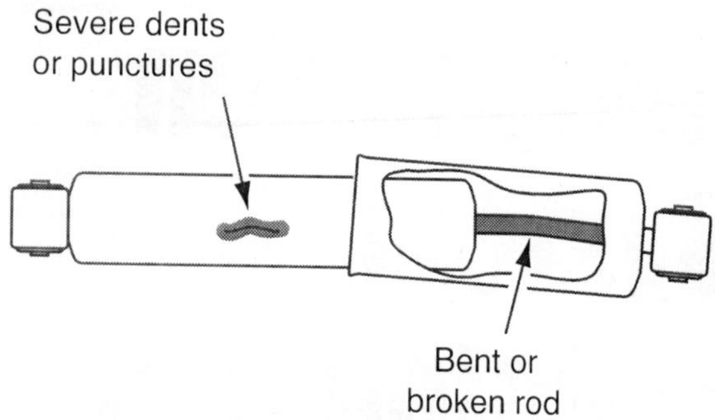


**Figure 2.** Damaged shock absorber inspection.

Struts and shock absorbers should be inspected visually for a bent condition and severe dents or punctures. When any of these conditions are present, replacement in pairs is required **(Figure 2)**.

## SHOCK ABSORBER DIAGNOSIS

A bounce test may be performed to determine shock absorber condition. When the bounce test is performed, the bumper is pushed downward with considerable weight applied on each corner of the vehicle. The bumper is released after this action and one free upward bounce should stop the vertical chassis movement if the shock absorber or strut provides proper spring control. Shock absorber replacement is required if more than one free upward bounce occurs.

A manual test may be performed on shock absorbers. When this test is performed, disconnect the lower end of the shock and move the shock up and down as rapidly as possible. A satisfactory shock absorber should offer a strong, steady resistance to movement on the entire compression and rebound strokes. The amount of resistance may be different on the compression stroke compared to the rebound stroke. If a loss of resistance is experienced during either stroke, shock replacement is essential.

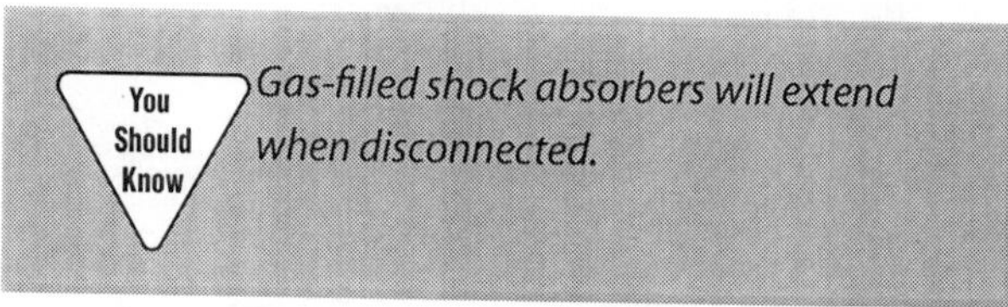


*Gas-filled shock absorbers will extend when disconnected.*

Some defective shock absorbers or struts may have internal clunking, clicking and squawking noises, or binding conditions. When these shock absorber noises or conditions are experienced, shock absorber or strut replacement is necessary. Strut chatter may be heard when the steering wheel is turned with the vehicle not moving or moving at low speed. **Strut chatter** is a repeated clicking noise when the front wheels are turned to the right or left.

To verify the location of this chattering noise, place one hand on a front coil spring while someone turns the steering wheel. If strut chatter is present, the spring binds and releases as it turns. This condition is caused by the upper spring seat binding against the strut bearing mount.

A noise that occurs on sharp turns or during front suspension jounce may be caused by interference between the upper strut rebound stop and the upper mount or strut tower, the coil spring and tower, or the coil spring and the upper mount.

On some models, these coil spring interference problems may be corrected by installing upper coil spring spacers on top of the coil spring. Spring removal from the strut is required to install these spacers.

## SHOCK ABSORBER SERVICE

When shock absorber replacement is necessary, follow this procedure:

1. Prior to rear shock absorber replacement, raise the vehicle on a lift and support the rear axle on jack stands so the shock absorbers are not fully extended.
2. When a front shock absorber is changed, lift the front end on the vehicle with a floor jack and then place jack stands under the lower control arms. Lower the vehicle onto the jack stands and remove the floor jack.
3. Disconnect the upper shock mounting nut and grommet.
4. Remove the lower shock mounting nut or bolts and remove the shock absorber.
5. Reverse steps 1 through 3 to install the new shock absorber and grommets.
6. With the full vehicle weight supported on the suspension, tighten the shock mounting nuts to the specified torque.

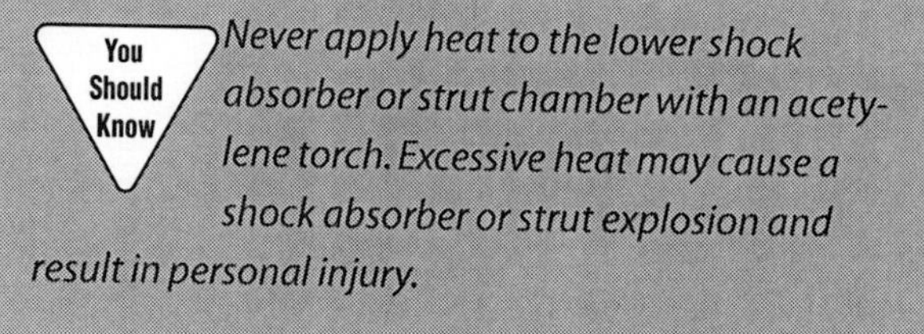


*Never apply heat to the lower shock absorber or strut chamber with an acetylene torch. Excessive heat may cause a shock absorber or strut explosion and result in personal injury.*

### Strut Removal and Replacement

Before a front strut and spring assembly is removed, the strut must be removed from the steering knuckle and

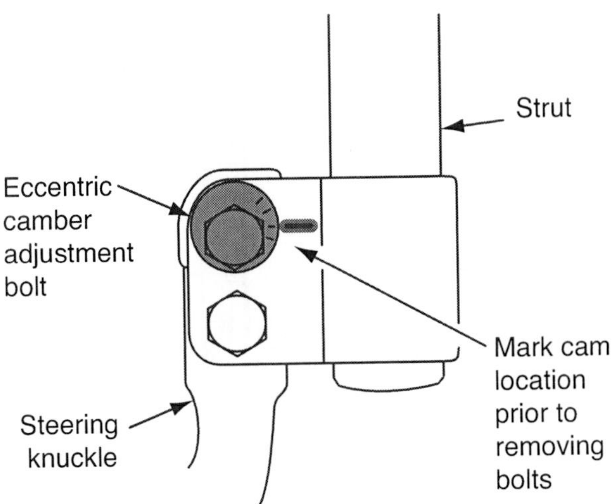


**Figure 3.** Camber bolt marking for strut removal.

top strut mount bolts must be removed from the strut tower. If an eccentric camber bolt is used to attach the strut to the knuckle, always mark the bolt head in relation to the strut and reinstall the bolt in the same position **(Figure 3)**.

Always follow the vehicle manufacturer's recommended procedure in the service manual for removal of the strut and spring assembly. The following is a typical procedure for strut and spring assembly removal:

1. Raise the vehicle on a lift or floor jack. If a floor jack is used to raise the vehicle, lower the vehicle onto jack stands placed under the chassis so the lower control arms and front wheels drop downward. Remove the floor jack from under the vehicle.
2. Remove the brake line and ABS wheel speed sensor wire from clamps on the strut **(Figure 4)**. In some cases, these clamps may have to be removed from the strut.

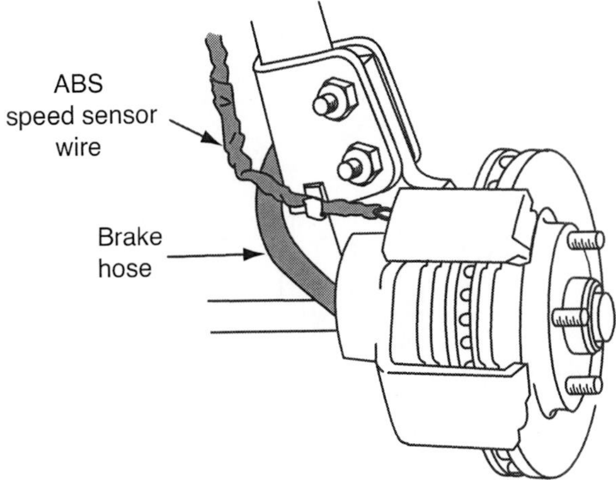


**Figure 4.** Brake hose and ABS wheel-speed sensor wire removal from the strut.

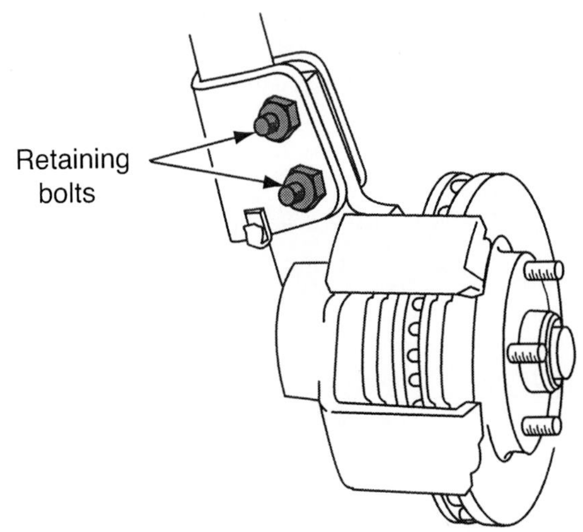


**Figure 5.** Removing strut-to-knuckle retaining bolts.

3. Remove the strut-to-steering knuckle retaining bolts and remove the strut from the knuckle **(Figure 5)**.
4. Remove the upper strut mounting bolts on top of the strut tower and remove the strut and spring assembly **(Figure 6)**.

## Removal of Strut from Coil Spring

The coil spring must be compressed with a special tool before the strut can be removed. All the tension must be removed from the upper spring seat before the upper strut piston rod nut is loosened. Many different spring compressing tools are available and they must always be used according to the manufacturer's recommended procedure.

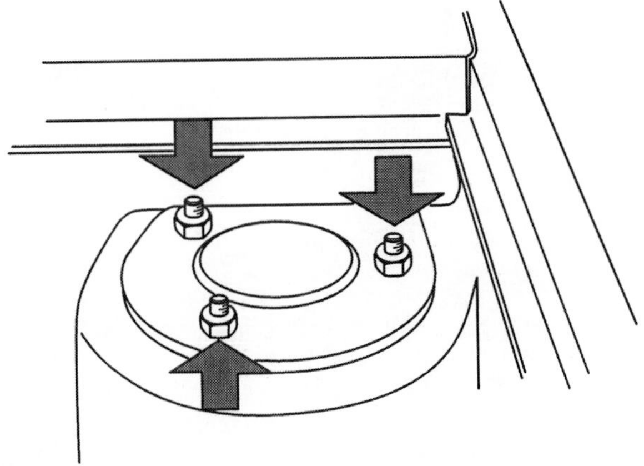

**Figure 6.** Removing upper mounting bolts on top of strut tower.

If the coil spring has an enamel-type coating, tape the spring where the compressing tool contacts the spring. The spring may break prematurely if this coating is chipped.

**You Should Know** *Always use a coil spring compressing tool according to the tool or vehicle manufacturer's recommended service procedure. Be sure the tool is properly installed on the spring. If a coil spring slips off the tool when the spring is compressed, severe personal injury or property damage may occur. Never loosen the upper strut mount retaining nut on the end of the strut rod unless the spring is compressed enough to remove all spring tension from the upper strut mount. If this nut is loosened with spring tension on the upper mount, this mount becomes a very dangerous projectile that may cause serious personal injury or property damage.*

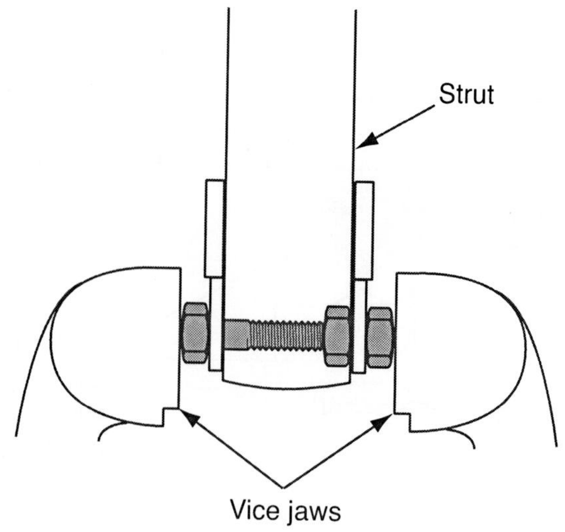


**Figure 8.** Bolt and two nuts installed in upper strut-to-knuckle mounting hole to hold strut and spring assembly in a vise.

A typical procedure for removing a strut from a coil spring is the following:

1. Install the spring compressing tool on the coil spring according to the tool or vehicle manufacturer's recommended procedure.
2. Turn the nut on top of the compressing tool until all the spring tension is removed from the upper strut mount **(Figure 7)**.

**Figure 7.** The spring compressing tool is tightened until all the spring tension is removed from the upper strut mount.

3. Install a bolt and two nuts in the upper strut-to-knuckle mounting bolt holes. Install a nut on each side of the strut flange **(Figure 8)**. Clamp this bolt securely in a vise to hold the strut and coil assembly and the compressing tool.
4. Use the bar on the spring compressing tool to hold the strut and spring assembly from turning and loosen the nut on the upper strut mount **(Figure 9)**. Be sure all the spring tension is removed from the upper strut mount before loosening this nut.
5. Remove the upper strut mount, upper insulator, coil spring, spring bumper and lower insulator **(Figure 10)**.

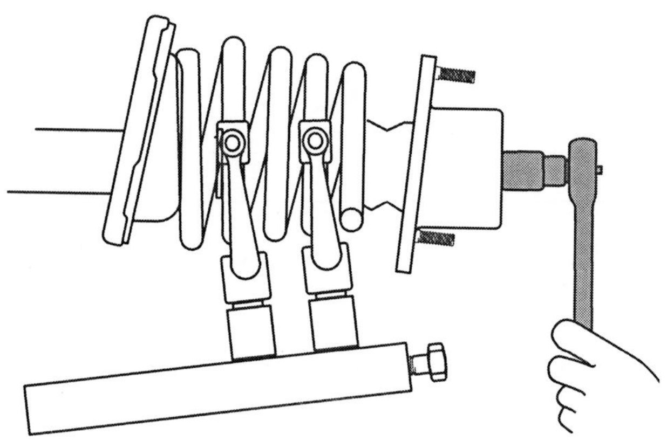

**Figure 9.** Removal of nut from strut piston rod.

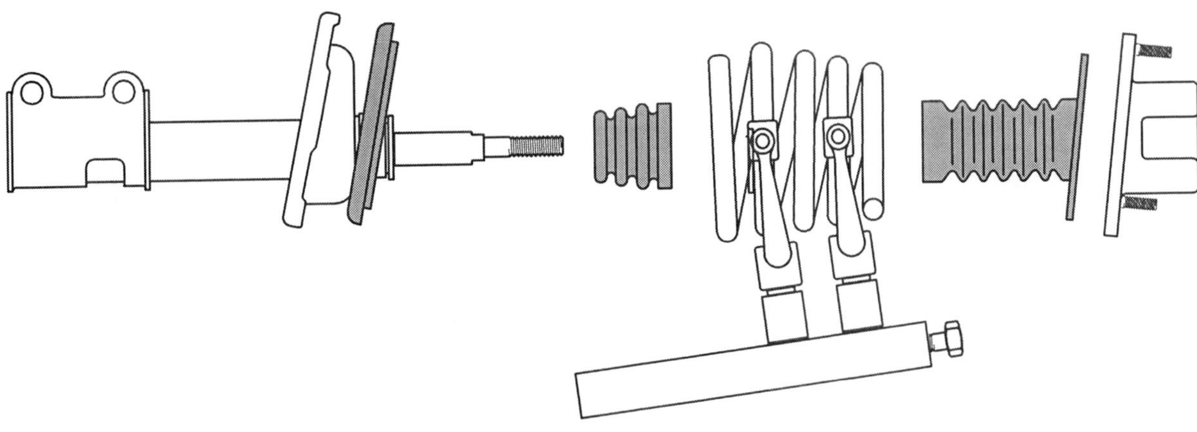

**Figure 10.** Removal of upper strut mount, upper insulator, spring bumper, and lower insulator.

You Should Know *Never clamp the lower strut or shock absorber chamber in a vise with excessive force. This action may distort the lower chamber and affect piston movement in the strut or shock absorber. Always follow the vehicle manufacturer's recommended procedure for strut disposal. Do not throw gas-filled struts or shock absorbers in a dumpster or in a fire of any kind. If the vehicle manufacturer recommends drilling the strut to release the gas charge, drill the strut at the manufacturer's recommended location.*

## Installation of Coil Spring on Strut

The following is a typical procedure for installing a coil spring on a strut:

1. Install the lower insulator on the lower strut spring seat and be sure the insulator is properly seated. **(Figure 11)**.

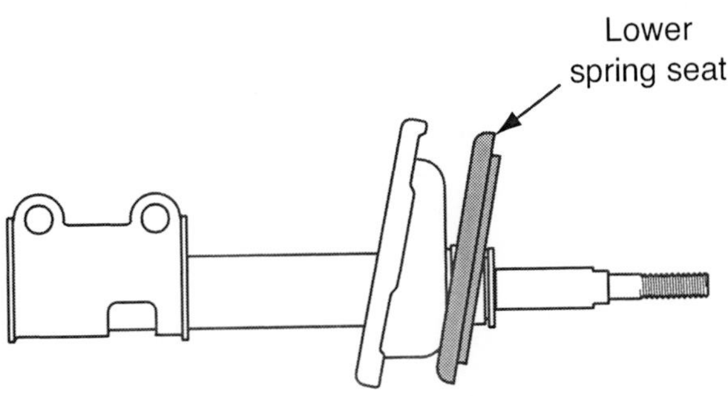


**Figure 11.** Insulator installation on lower spring seat.

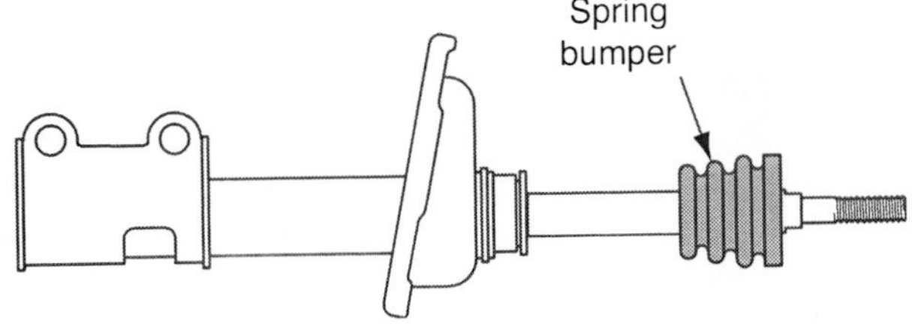


**Figure 12.** Spring bumper installation on strut piston rod.

2. Install the spring bumper on the strut rod **(Figure 12)**.
3. With the coil spring compressed in the spring compressing tool, install the spring on the strut **(Figure 13)**. Be sure the spring is properly seated on the lower insulator spring seat.
4. Be sure the strut piston rod is fully extended and install the upper insulator on top of the coil spring.

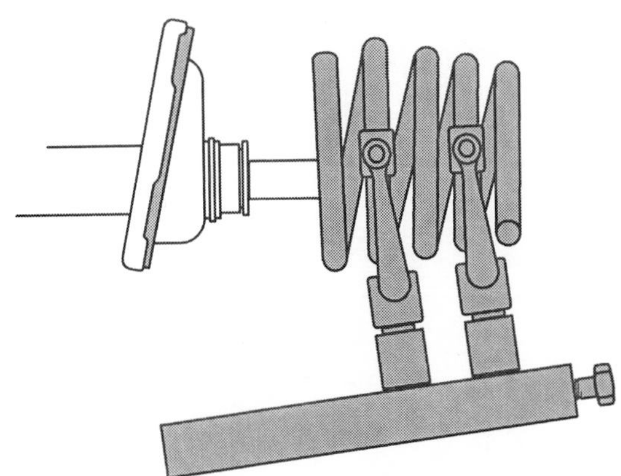

**Figure 13.** Installing compresed spring on strut with compressing tool.

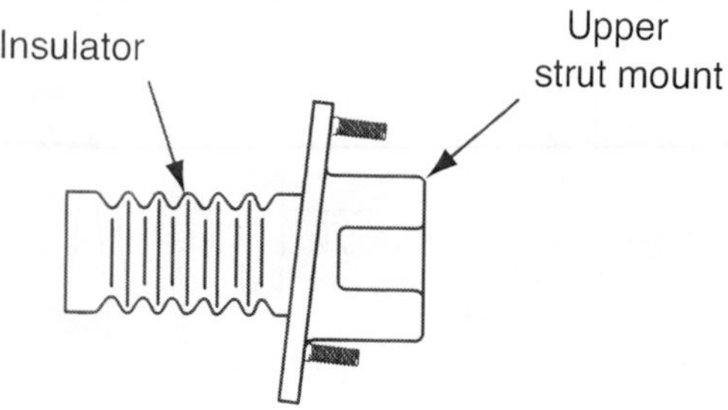


**Figure 14.** Upper insulator and upper strut mount installed on coil spring.

5. Install the upper strut mount on the upper insulator **(Figure 14)**.
6. Be sure the spring, upper insulator, and upper strut mount are properly positioned and seated on the coil spring and strut piston rod **(Figure 15)**.
7. Install a bolt in the upper strut-to-knuckle retaining bolt and clamp this bolt in a vise to hold the strut, spring, and compressing tool as in the disassembly procedure.
8. Use the compressing tool bar to hold the strut and spring from turning and tighten the strut piston rod nut to the specified torque **(Figure 16)**.

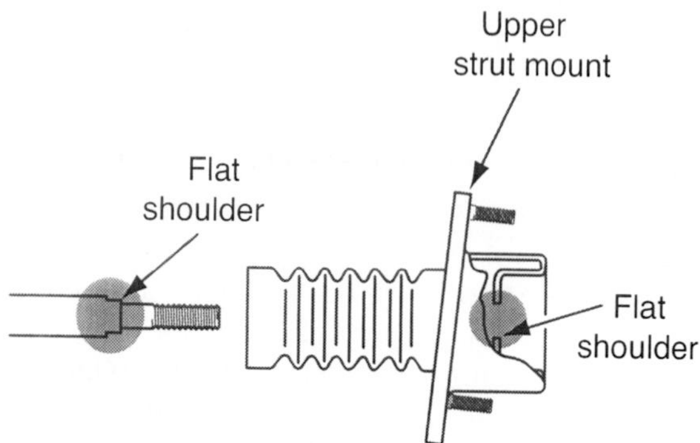


**Figure 15.** Upper strut mount properly positioned on strut piston rod.

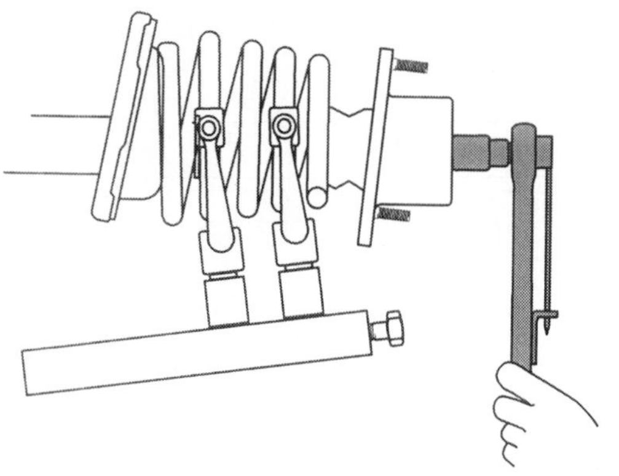

**Figure 16.** Tightening nut on strut piston rod.

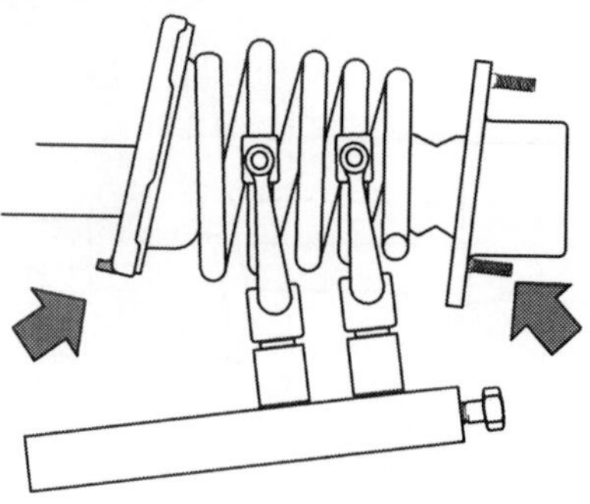

**Figure 17.** Aligning lowest bolt on upper strut mount with tab on lower spring seat.

9. Rotate the upper strut mount until the lowest bolt in this mount is aligned with the tab on the lower spring seat **(Figure 17)**.
10. Gradually loosen the nut on the compressing tool until all the spring tension is released from the tool and remove the tool from the spring.

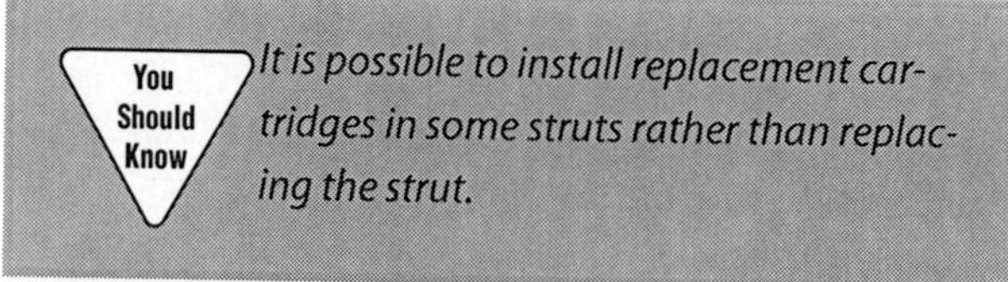


*It is possible to install replacement cartridges in some struts rather than replacing the strut.*

## FRONT AND REAR SUSPENSION SYSTEM MAINTENANCE

When performing undercar service, inspect the suspension system components. Inspect the steering knuckle for looseness where the tie rod end fits into the knuckle. There should be no looseness in the tie rod end and the tie rod end must be properly secured to the steering knuckle with a nut and cotter pin.

Upper and lower control arms should be inspected for cracks, bent conditions, or worn bushings. If the control arm bushings are worn, steering is erratic especially on irregular road surfaces. Worn control arm bushings may cause a rattling noise while driving on irregular road surfaces. Dry or worn control arm bushings may cause a squeaking noise on irregular road surfaces. Caster and camber angles on the front suspension are altered by worn upper and lower control arm bushings. Incorrect caster or camber angles may cause the vehicle to pull to one side and tire wear may be excessive when the camber angle is not within specifications.

Leaf springs should be inspected for a sagged condition which causes the curb riding height to be less than specified. Leaf springs should also be visually inspected for broken leafs, broken center bolts, and worn shackles or bushings. Weak or broken leaf springs affect front suspension alignment angles and cause excessive tire tread wear, reduced directional stability, and harsh riding. A rattling noise while driving on irregular road surfaces may be caused by worn shackles or bushings. Worn shackles and bushings lower curb riding height and cause a rattling noise on road irregularities.

Many leaf springs have plastic silencers between the spring leafs. If these silencers are worn out, creaking and squawking noises are heard when the vehicle is driven over road irregularities at low speeds.

When the silencers require checking or replacement lift the vehicle with a floor jack and support the frame on jack stands so the rear suspension moves downward. With the vehicle weight no longer applied to the springs, the spring leafs may be pried apart with a prybar to remove and replace the silencers.

Worn shackle bushings, brackets, and mounts cause excessive chassis lateral movement and rattling noises. With the normal vehicle weight resting on the springs, insert a prybar between the rear outer end of the spring and the frame. Apply downward pressure on the bar and observe the rear shackle for movement. Shackle bushings, brackets, or mounts must be replaced if there is movement in the shackle. The same procedure may be followed to check the front bushing in the main leaf. A broken spring center bolt may allow the rear axle assembly to move rearward on one side. This movement changes rear wheel tracking resulting in handling problems, tire wear, and reduced directional stability. Sagged rear springs reduce the curb riding height. Spring replacement is necessary if the springs are sagged.

Rebound bumpers are usually bolted to the lower control arm or to the chassis. Inspect the rebound bumpers for cracks, wear, and flattened conditions **(Figure 18)**. Damaged rebound bumpers may be caused by sagged springs and insufficient curb riding height or worn out shock absorbers and struts. If the rebound bumpers must be replaced, remove the mounting bolts and the bumper. Install the new bumper and tighten the mounting bolts to the specified torque.

Worn stabilizer bar mounting bushings, grommets, or mounting bolts cause a rattling noise as the vehicle is driven on irregular road surfaces. A weak stabilizer bar or worn bushings and grommets cause harsh riding and excessive body sway while driving on irregular road surfaces. Worn or very dry stabilizer bar bushings may cause a squeaking noise on irregular road surfaces. All stabilizer bar components should be visually inspected for wear.

Inspect the strut rod for a bent condition or a loose bushing. A bent strut rod or loose bushing may cause steering pull.

Rear tie rods should be inspected for worn grommets, loose mountings, or bent conditions. Loose tie rod bushings or a bent tie rod will change the rear wheel tracking and result in reduced directional stability. Worn tie rod bushings also cause a rattling noise on road irregularities. When the rear tie rod is replaced, remove the front and rear rod mounting nuts. The lower control arm or rear axle may have to be pried rearward to remove the tie rod. Check the tie rod grommets and mountings for wear and replace parts as required. When the tie rod is reinstalled, tighten the mounting bolts to specifications and check the rear wheel toe.

After replacement of front or rear suspension components such as control arms, control arm bushings, springs, and strut rods, front and rear suspension alignment should be checked.

## CURB RIDING HEIGHT MEASUREMENT

Regular inspection and proper maintenance of suspension systems is extremely important to maintain vehicle safety. The curb riding height is determined mainly by spring condition. Other suspension components, such as control arm bushings, affect curb riding height if they are worn. Since incorrect curb riding height affects most of the other suspension angles, this measurement is critical.

Reduced curb riding height on the front suspension may cause decreased directional stability. If the curb riding height is reduced on one side of the front suspension, the steering may pull to one side. Reduced rear suspension height increases steering effort and causes rapid steering wheel return after turning a corner. Harsh riding occurs when the curb riding height is less than specified. The curb riding height must be measured at the vehicle manufacturer's specified location, which varies depending on the type of suspension system.

When the vehicle is on a level floor or an alignment rack, measure the curb riding height from the floor to the center of the lower control arm mounting bolt on both sides of the front suspension **(Figure 19)**.

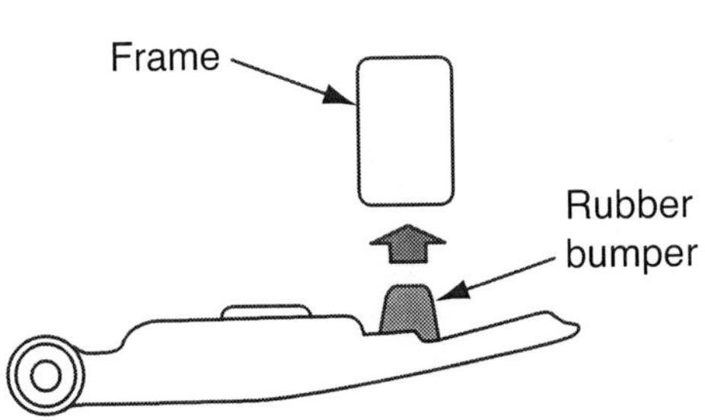


**Figure 18.** Rebound bumper.

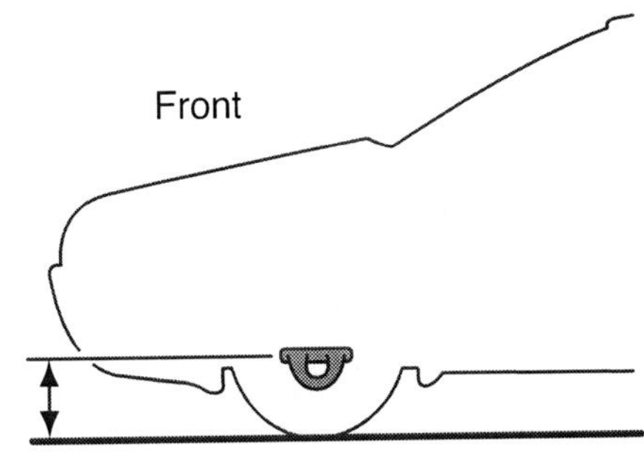


**Figure 19.** Curb riding height measurement, front suspension.

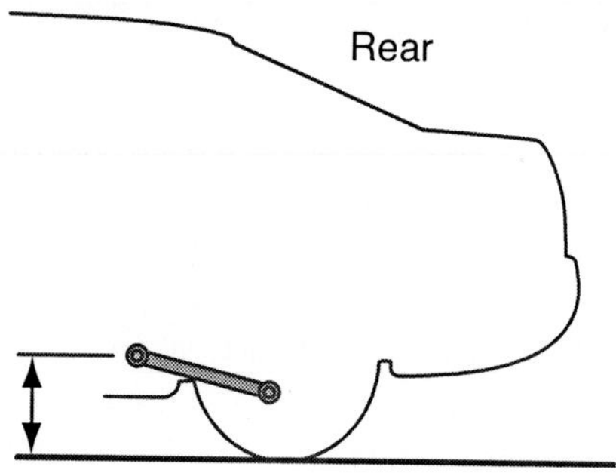


Figure 20. Curb riding height measurement, rear suspension.

On the rear suspension system, measure the curb riding height from the floor to the center of the strut rod mounting bolt **(Figure 20)**.

If the curb riding height is less than specified, the control arms and bushings should be inspected and replaced as necessary. When the control arms and bushings are in normal condition, the reduced curb riding height may be caused by sagged springs that require replacement.

## FRONT AND REAR SUSPENSION DIAGNOSIS AND SERVICE

Proper suspension system diagnosis and service is very important to maintain ride quality, steering control, passenger comfort, and vehicle safety. Accurate suspension system diagnosis is extremely critical so the recommended repair corrects the customer's complaint. Some suspension problems may be confused with brake problems. For example, a worn, loose strut rod bushing may cause steering pull. This problem can also be caused by brake problems such as contaminated brake linings. Therefore, accurate suspension system diagnosis is of utmost importance!

### Noise Diagnosis

A squeaking noise in the front or rear suspension may be caused by a suspension bushing or a defective strut or shock absorber. If a rattling noise occurs in the rear suspension, check these components:

1. Worn or missing suspension bushings, such as control arm bushings, track bar bushings, stabilizer bar bushings, trailing arm bushings, and strut rod bushings.
2. Worn strut or shock absorber bushings or mounts.
3. Defective struts or shock absorbers.
4. Broken springs or worn spring insulators.

### Rear Body Sway and Lateral Movement Diagnosis

Excessive body sway or roll on road irregularities may be caused by a weak stabilizer bar or loose stabilizer bar bushings. If lateral movement is experienced on the rear of the chassis, the track bar, or track bar bushings, may be defective.

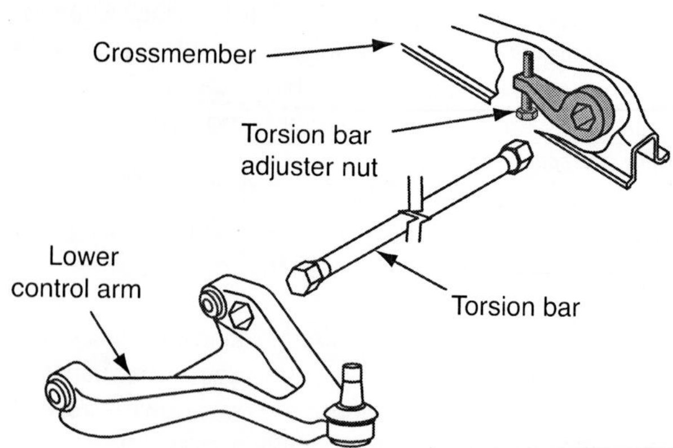


Figure 21. Curb riding height adjustment, torsion bar front suspension.

### Torsion Bar Adjustment

On torsion bar front suspension systems, the torsion bars may be adjusted to correct the curb riding height. The curb riding height must be measured at the location specified by the vehicle manufacturer. If the curb riding height is not correct on a torsion bar front suspension, the torsion bar anchor adjusting bolts must be rotated until the curb riding height equals the vehicle manufacturer's specifications **(Figure 21)**.

### Checking Ball Joints

Some ball joints have a grease fitting installed in a floating retainer. The grease fitting and retainer may be used as a ball joint wear indicator. With the vehicle weight resting on the wheels, grasp the grease fitting and check for movement **(Figure 22)**.

Some car manufacturers recommend ball joint replacement if any grease fitting movement is present. In some other ball joints, the grease fitting retainer extends a

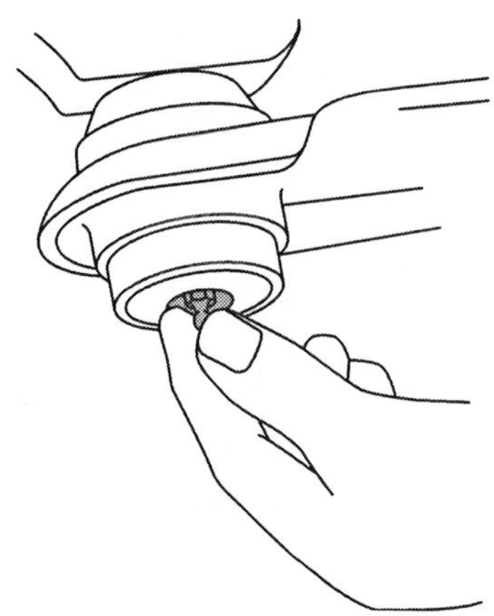

Figure 22. Ball joint grease fitting wear indicator.

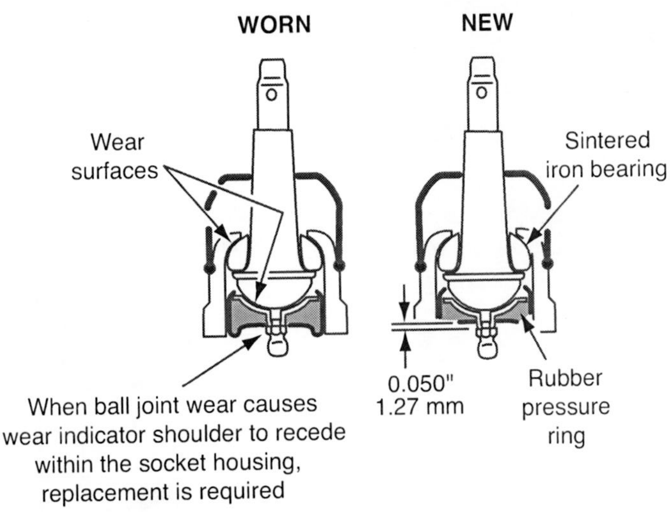


**Figure 23.** Ball joint wear indicator with grease fitting extending from ball joint surface.

short distance through the ball joint surface **(Figure 23)**. On this type of joint, replacement is necessary if the grease fitting shoulder is flush with or inside the ball joint cover.

## Ball Joint Unloading

On many suspension systems, ball joint looseness is not apparent until the weight has been removed from the joint. When the coil spring is positioned between the lower control arm and the chassis, place a floor jack near the outer end of the lower control arm and raise the tire off the floor **(Figure 24)**. Be sure the rebound bumper is not in contact with the control arm or frame.

When the coil springs are positioned between the upper control arm and the chassis, place a steel block between the upper control arm and the frame. With this block in place, raise the tire off the floor with a floor jack under the front crossmember **(Figure 25)**.

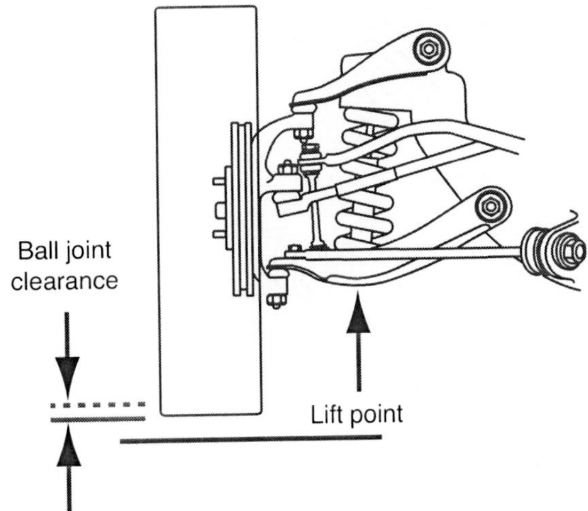


**Figure 24.** Floor jack position to check ball joint wear with spring between the lower control arm and the chassis.

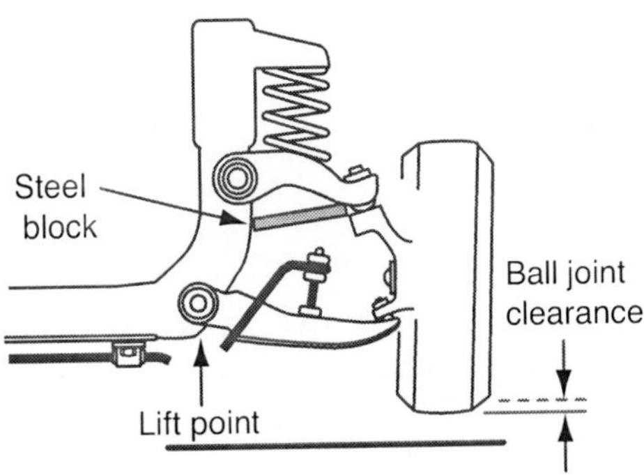


**Figure 25.** Floor jack position to check ball joint wear with spring between the upper control arm and the chassis.

## Ball Joint Axial Measurement

The vehicle manufacturer may provide ball joint axial (vertical) and radial (horizontal) tolerances. A dial indicator is one of the most accurate ball joint measuring devices **(Figure 26)**. Always install the dial indicator at the vehicle

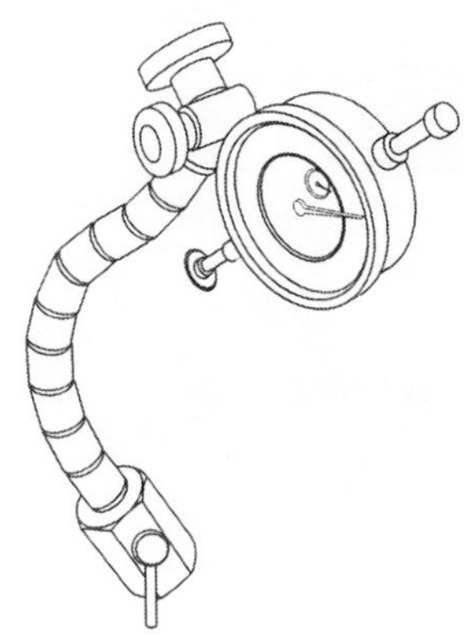

**Figure 26.** Dial indicator designed for ball joint measurement.

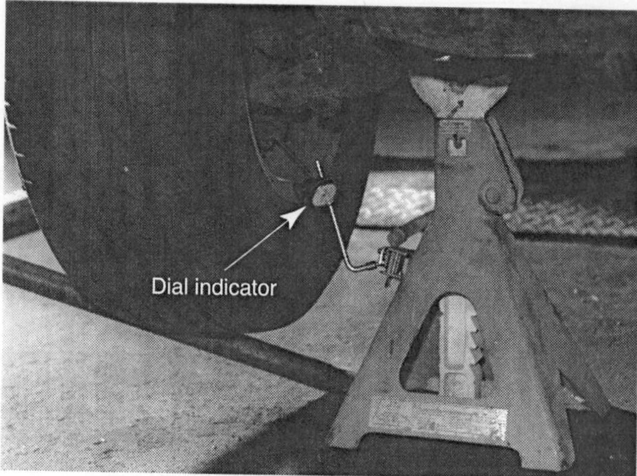
Dial indicator

Figure 27. Dial indicator installed to measure vertical ball joint movement in a compression-loaded ball joint.

manufacturer's recommended location for ball joint measurement.

Clean the top end of the lower ball joint stud and position the dial indicator stem against the top end of this stud (**Figure 27**). Depress the dial indicator plunger approximately 0.250 inch.

Lift upward with a prybar under the tire and observe the dial indicator reading. If the vertical ball joint movement exceeds manufacturer's specifications, ball joint replacement is required.

## Ball Joint Radial Measurement

Worn ball joints cause improper camber and caster angles which result in reduced directional stability and tire tread wear. Connect the dial indicator to the lower control arm of the ball joint being checked and position the dial indicator stem against the edge of the wheel rim (**Figure 28**).

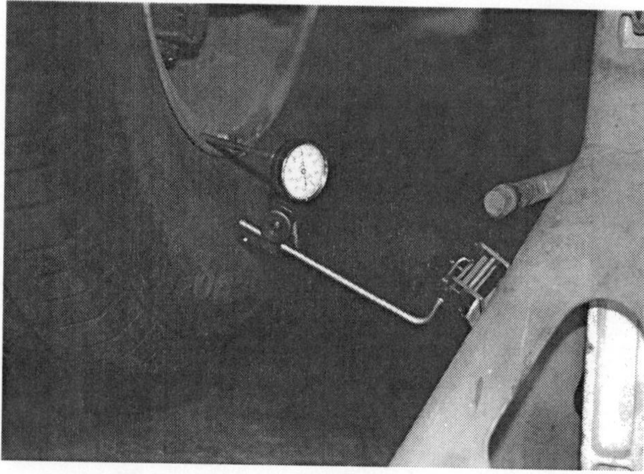

Figure 28. Dial indicator positioned to measure radial ball joint movement.

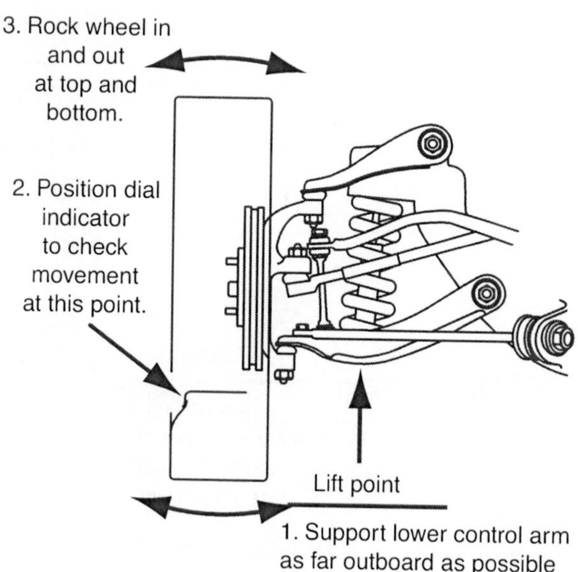
3. Rock wheel in and out at top and bottom.
2. Position dial indicator to check movement at this point.
Lift point
1. Support lower control arm as far outboard as possible

Figure 29. Measuring ball joint movement.

Be sure the front wheel bearings are adjusted properly prior to the ball joint radial measurement. While an assistant grasps the top and bottom of the raised tire and attempts to move the tire and wheel horizontally in and out, observe the reading on the dial indicator (**Figure 29**).

The lower ball joint on a MacPherson strut-type front suspension should be checked for radial movement with a dial indicator when the tire is lifted off the floor (**Figure 30**). Since the spring load is carried by the upper and lower spring seats when the tire is lifted off the floor, it is not necessary to unload this type of ball joint.

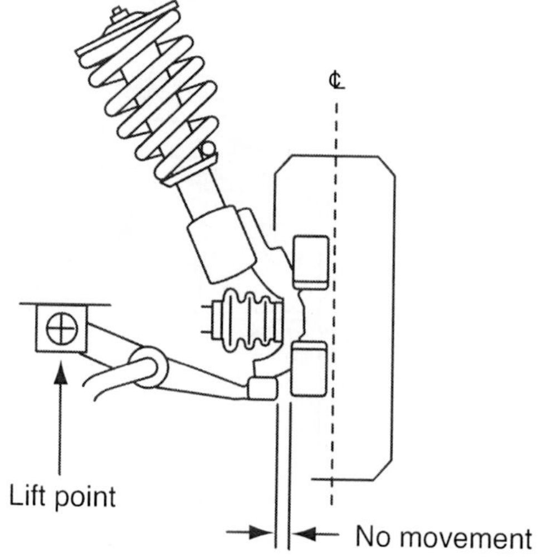
℄
Lift point
No movement

Figure 30. Ball joint wear measurement on MacPherson strut front suspension.

## ELECTRONIC AIR SUSPENSION SYSTEM MAINTENANCE

An electronic air suspension system requires a minimum amount of maintenance. When performing undercar service, inspect all the air lines for kinks, breaks, or contact with other components that would damage or wear the lines. Inspect all the wires in the system, including the module and switch wiring in the trunk, for worn insulation and unwanted contact with other components. The air springs should be inspected for worn spots on the membranes and proper retention to the chassis. Be sure the compressor mounting is secure. Most compressors have a rubber mount to reduce noise transmission to the passenger compartment. Inspect the height sensors for damage and loose mounting bushings.

## ELECTRONIC AIR SUSPENSION DIAGNOSIS

If the air suspension warning lamp is illuminated with the engine running, the control module has detected a defect in the electronic air suspension system. The electronic air suspension diagnostic and service procedures vary depending on the vehicle. Always follow the vehicle manufacturer's recommended procedures in the service manual. When the air suspension warning lamp indicates a system defect, the diagnostic procedure may be entered as follows:

1. Be sure the air suspension system switch is turned on.
2. Turn the ignition switch on for 5 seconds and then turn it off. Leave the driver's door open and the other doors closed.
3. Ground the diagnostic lead located near the control module and close the driver's door with the window down.
4. Turn the ignition switch on. The warning lamp should blink continuously at 1.8 times per second to indicate that the system is in the diagnostic mode.

There are ten tests in the diagnostic procedure. The control module switches from one test to the next when the driver's door is opened and closed. The first three systems tested in the diagnostic procedure are the rear suspension, right front suspension, and left front suspension.

During these three tests, each suspension location should be raised for 30 seconds, lowered for 30 seconds, and raised for 30 seconds. For example, in test 2 this procedure should be followed on the right front suspension. If the expected signal or an illegal signal is received during the test procedure, the test stops and the air suspension warning lamp is illuminated. If all the signals and commands are normal during the first three tests, the warning lamp continues to flash at 1.8 times per second.

While tests 4 through 10 are being performed, the air suspension warning lamp flashes the test number. For example, during test 4 the warning lamp flashes four times followed by a pause and four more flashes. This flash sequence continues while test 4 is completed. The driver's door must be opened and closed to move to the next test. During tests 4 through 10, the technician must listen to or observe various components to detect abnormal operation. The warning lamp only indicates the test number during these tests. Actions performed by the control module during tests 4 through 10 are as follows:

Test 4. The compressor is cycled on and off at 0.25 cycles per second. This action is limited to 50 cycles.

Test 5. The vent solenoid is opened and closed at 1 cycle per second.

Test 6. The left front air valve is opened and closed at 1 cycle per second and the vent solenoid is opened. When this occurs, the left front corner of the vehicle should drop slowly.

Test 7. The right front air valve is opened and closed at 1 cycle per second and the vent solenoid is opened. This action causes the right front corner of the vehicle to drop slowly.

Test 8. During this test, the right rear air valve is opened and closed at 1 cycle per second and the vent valve is opened. This action should cause the right rear corner of the vehicle to drop slowly.

Test 9. The left rear solenoid is opened and closed at 1 cycle per second and the vent valve is opened, which should cause the left rear corner of the vehicle to drop slowly.

Test 10. Return the module from the diagnostic mode to normal operation by disconnecting the diagnostic lead from ground. This mode change also occurs if the ignition switch is turned off or when the brake pedal is depressed.

If defects are found during the test sequence, specific electrical tests may be performed on air valve or vent valve windings and connecting wires to locate the problem.

## ELECTRONIC AIR SUSPENSION SYSTEM SERVICE

Many components in an electronic air suspension system such as control arms, shock absorbers, and stabilizer bars, are diagnosed and serviced in the same way as the components in a conventional suspension system. However, the air spring service procedures are different compared to coil spring service procedures on a conventional suspension system.

*The system control switch must be in the off position when system components are serviced.*

You Should Know *The system control switch must be turned off prior to hoisting, jacking, or towing the vehicle. If the front of the chassis is lifted with a bumper jack, the rear suspension moves downward. The electronic air suspension system will attempt to restore the rear trim height to normal. This action may cause the front of the chassis to fall off the bumper jack, resulting in personal injury or vehicle damage.*

You Should Know *When air spring valves are being removed, always rotate the valve to the first stage until all the air escapes from the air spring. Never turn the valve to the second (release) stage until all the air is released from the spring.*

## Trim Height Adjustment

The trim height should be measured on the front and rear suspension at the locations specified by the vehicle manufacturer **(Figure 31)**.

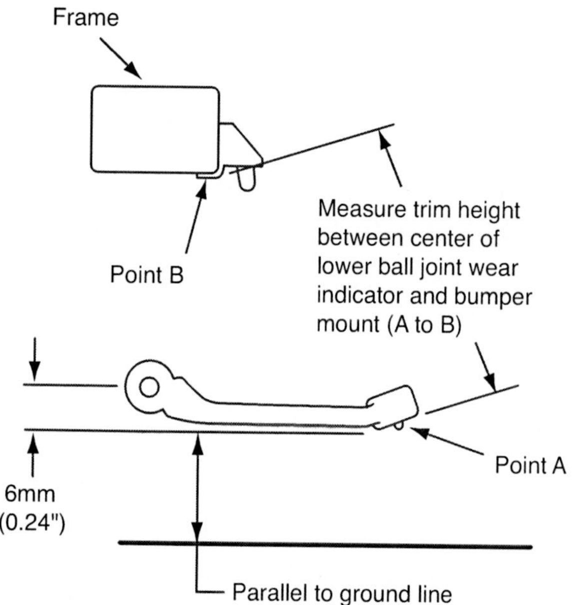


**Figure 31.** Trim height measurement location.

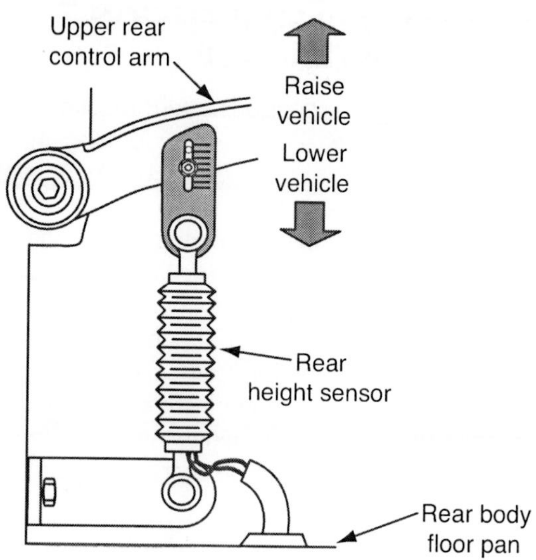


**Figure 32.** Rear trim height adjustment.

If the rear suspension trim height is not within specifications, it may be adjusted by loosening the attaching bolt on the top height sensor bracket **(Figure 32)**. When the bracket is moved one index mark up or down, the ride height is lowered or raised, 0.25 inch (6.35 millimeters).

The front suspension trim height may be adjusted by loosening the lower height sensor attaching bolt. Three adjustment positions are located in the lower front height sensor bracket **(Figure 33)**. If the height sensor attaching bolt is moved one position up or down, the front suspension height is lowered or raised, 0.5 inch (12.7 millimeters).

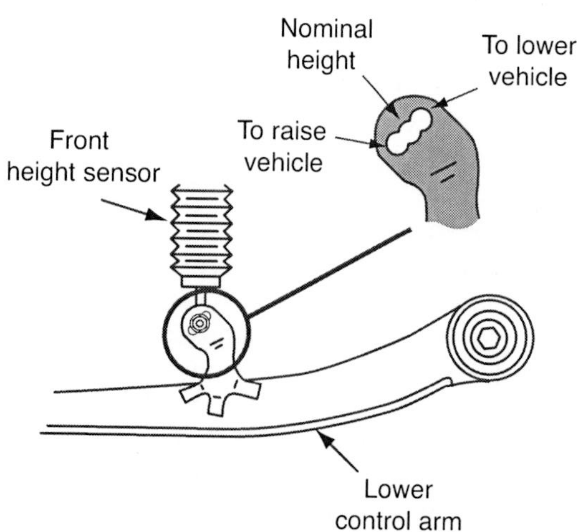


**Figure 33.** Front trim height adjustment.

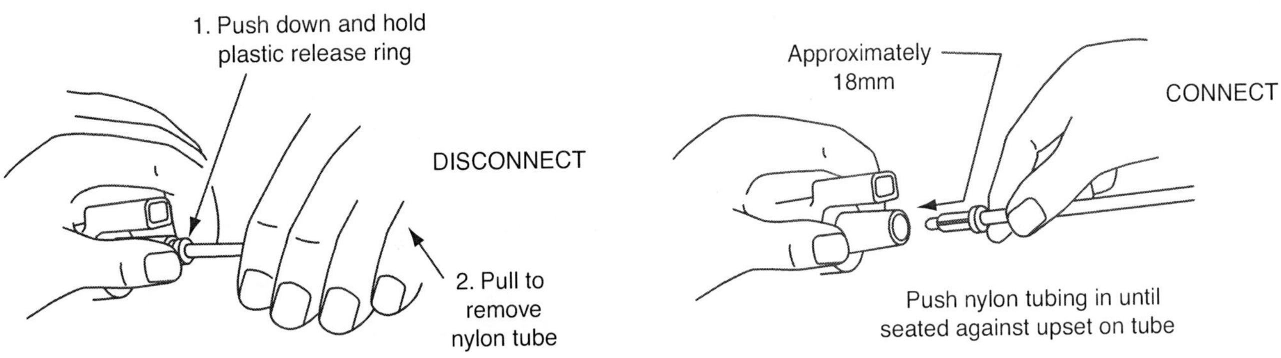


**Figure 34.** Air line removal from air spring valve or compressor outlet.

## Line Service

Nylon lines on the electronic air suspension system have quick disconnect fittings. These fittings should be released by pushing downward and holding the plastic release ring and then pulling outward on the nylon line **(Figure 34)**. Simply push the nylon line into the fitting until it seats to reconnect an air line.

If a line fitting is damaged, it may be removed by looping the line around your fingers and pulling on the line without pushing on the release ring **(Figure 35)**. When a new collet and release ring is installed, the O-ring under the collet must be replaced. If a leak occurs in a nylon line, a sharp knife may be used to cut the defective area out of the line. A service fitting containing a collet fitting in each end is available for line repairs. After the defective area is cut out of the line, push the two ends of the line into the service fitting.

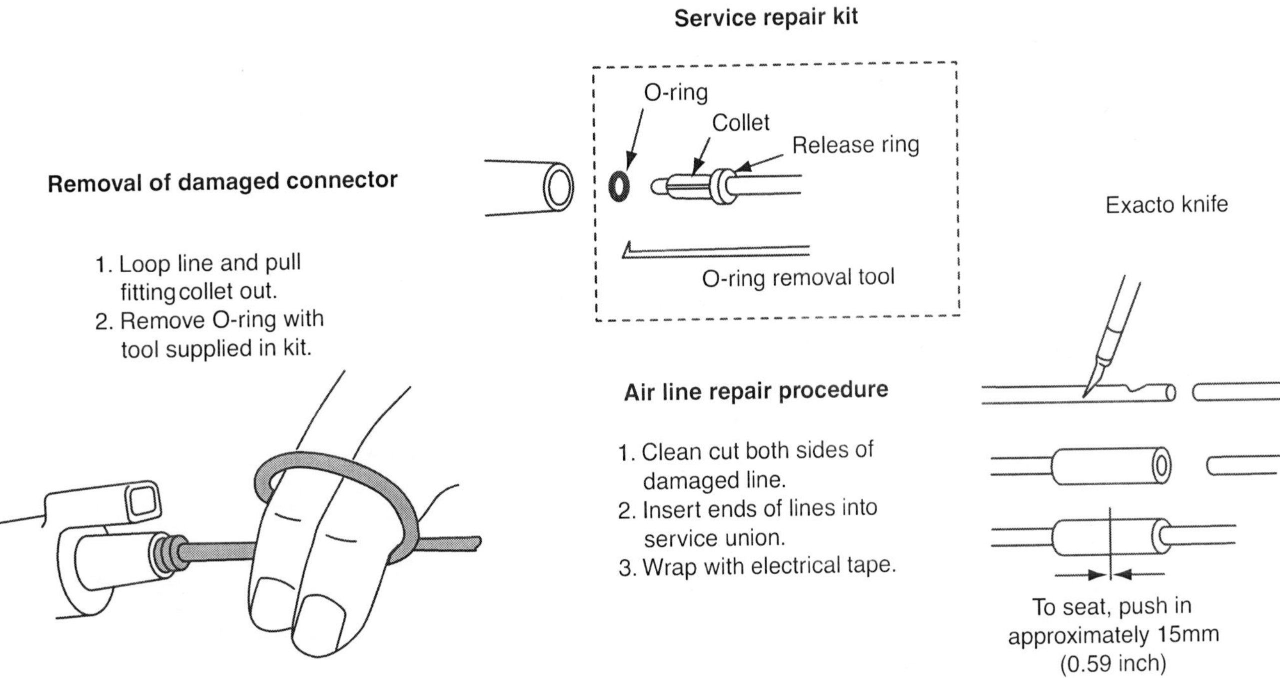


**Figure 35.** Removal of quick disconnect fittings and air line repair procedures.

## CONTINUOUSLY VARIABLE ROAD SENSING SUSPENSION SYSTEM (CVRSS) MAINTENANCE

When performing undercar service, the CVRSS should be inspected. Check all the system wiring, especially the wiring connectors on the shock absorbers or struts. Be sure all the wiring connectors are tight and check the wiring for worn insulation or unwanted contact with other components. Always inspect the module wiring in the trunk for damage. Inspect the wheel position sensors for loose mountings and wiring connectors. Check the accelerometers for secure mounting and loose or damaged wiring connectors.

## CONTINUOUSLY VARIABLE ROAD SENSING SUSPENSION SYSTEM (CVRSS) DIAGNOSIS AND SERVICE

Proper CVRSS diagnosis and service is very important. A malfunctioning CVRSS may reduce ride quality, passenger comfort, and steering control.

### Trouble Code Display

On some vehicles with CVRSS, a SERVICE RIDE CONTROL message is displayed in the message center if a defect occurs in the CVRSS.

To enter the diagnostic mode, turn the ignition switch on and press the off and warmer buttons simultaneously on the climate control center (CCC). Hold these buttons until the segment display occurs in the IPC. The segment display verifies all the segments are operational in the IPC. The turn signal indicators are not illuminated during this check. Do not proceed with the diagnostics unless all IPC segments are illuminated during the segment display. If any of the IPC segments are not illuminated, erroneous diagnosis may occur. The IPC must be replaced if any segments do not illuminate.

**You Should Know**

*When removing, replacing, or servicing an electronic component on a vehicle, always disconnect the negative battery cable before starting the service procedure. If the vehicle is equipped with an air bag or bags, wait for the time period specified by the vehicle manufacturer after the battery negative cable is removed to prevent accidental air bag deployment. Many air bag computers have a backup power supply that is capable of deploying the air bag for a specific length of time after the battery is disconnected.*

Once the diagnostics are entered, the HI and LO buttons in the CCC may be used to select or reject test displays. Pressing the HI button may be compared to a yes input, while pressing the LO button may be considered as a no input. After the diagnostics are entered, the technician may select diagnostic code displays from these computers:

- Powertrain control module (PCM)
- Instrument panel cluster (IPC)
- Air conditioning programmer (ACP)
- Supplemental inflatable restraint (SIR)
- Traction control system (TCS)
- Real time damping (RTD)

The abbreviation for each computer appears in the instrument panel display. The technician presses the HI button to select fault codes from the computer display. If the technician does not want fault codes from a computer, the LO button is pressed and the display moves to the next computer abbreviation **(Figure 36)**.

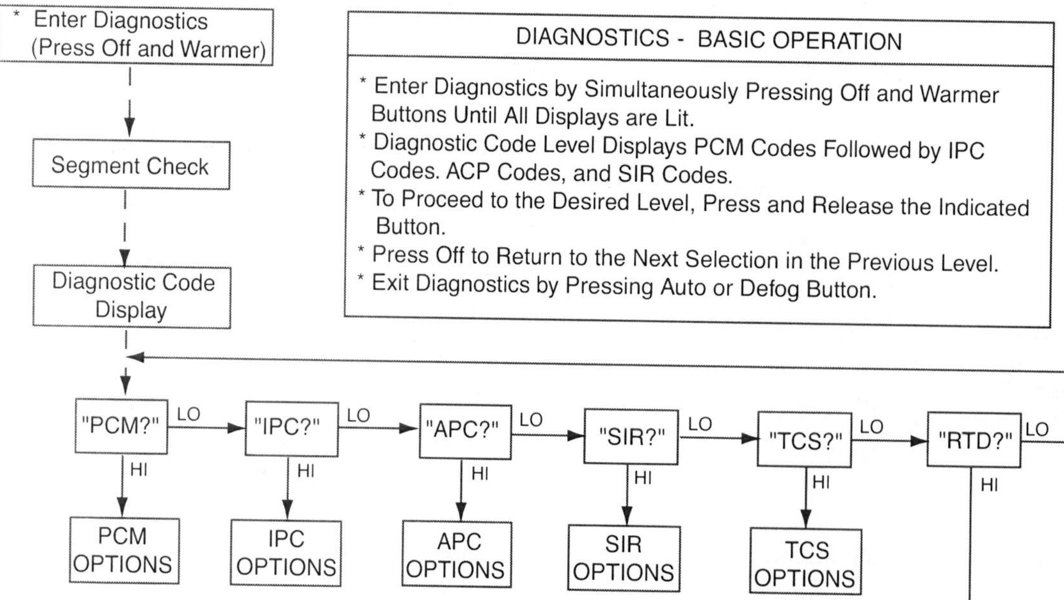


**Figure 36.** Selecting fault code displays from various computers during the diagnostic procedure.

The fault codes from any computer are three-digit numbers with a one-letter prefix and a one-letter suffix. The fault codes have these prefixes:

- PCM - P
- IPC - I
- APC - A
- SIR - R
- TCS - T
- RTD - S

Therefore, any fault codes in the RTD module are prefixed by the letter S. The suffix is a C or an H. When a code has a C suffix, it is a current code that is present at the time of diagnosis. An H suffix indicates the fault code is a history or intermittent code. The fault codes are displayed in numerical order.

If there are no fault codes in the RTD module, NO S CODES appears in the IPC display. If NO S DATA is displayed, the RTD module cannot communicate with the IPC display. When the AUTO button is pressed any time during the fault code display, the system changes from the diagnostic mode to normal operation.

*Never remove or install the wiring connector on a computer or computer system component with the ignition switch on. This action may result in computer damage. Do not supply voltage to or ground any circuit or component in a computer system unless instructed to do so in the vehicle manufacturer's service manual. This action may damage computer system components.*

## Optional Diagnostic Modes

After the fault code display, the technician may select these options: RTD data, RTD inputs, RTD outputs, and RTD clear codes.

When one of these options is displayed in the IPC, the technician selects the option by pressing the HI button. The technician rejects the displayed option by pressing the LO button and the display moves to the next option.

If RTD data is selected, the data from a number of inputs may be displayed in numerical order beginning with number SD01, RF accelerometer **(Figure 37)**. Then the car

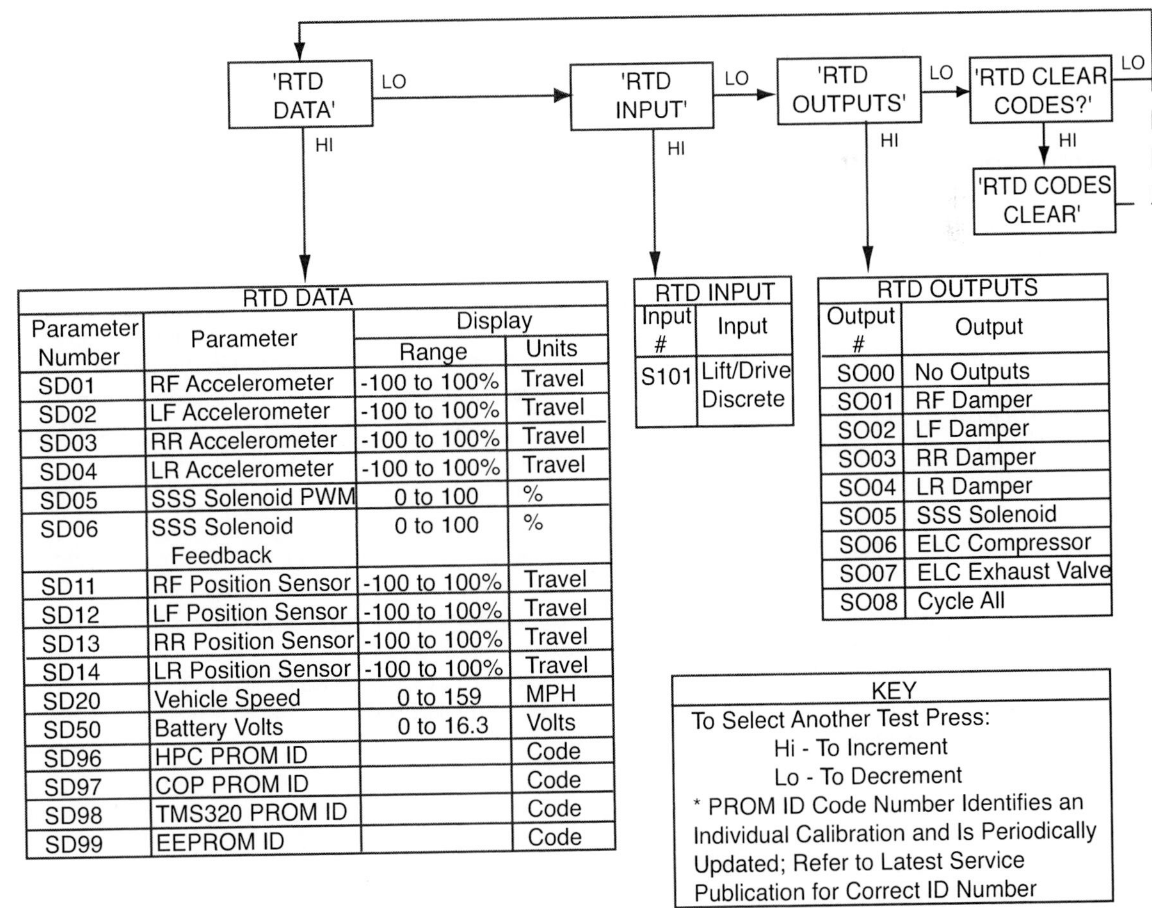


| RTD DATA | | | |
|---|---|---|---|
| Parameter Number | Parameter | Display Range | Display Units |
| SD01 | RF Accelerometer | -100 to 100% | Travel |
| SD02 | LF Accelerometer | -100 to 100% | Travel |
| SD03 | RR Accelerometer | -100 to 100% | Travel |
| SD04 | LR Accelerometer | -100 to 100% | Travel |
| SD05 | SSS Solenoid PWM | 0 to 100 | % |
| SD06 | SSS Solenoid Feedback | 0 to 100 | % |
| SD11 | RF Position Sensor | -100 to 100% | Travel |
| SD12 | LF Position Sensor | -100 to 100% | Travel |
| SD13 | RR Position Sensor | -100 to 100% | Travel |
| SD14 | LR Position Sensor | -100 to 100% | Travel |
| SD20 | Vehicle Speed | 0 to 159 | MPH |
| SD50 | Battery Volts | 0 to 16.3 | Volts |
| SD96 | HPC PROM ID | | Code |
| SD97 | COP PROM ID | | Code |
| SD98 | TMS320 PROM ID | | Code |
| SD99 | EEPROM ID | | Code |

| RTD INPUT | |
|---|---|
| Input # | Input |
| S101 | Lift/Drive Discrete |

| RTD OUTPUTS | |
|---|---|
| Output # | Output |
| SO00 | No Outputs |
| SO01 | RF Damper |
| SO02 | LF Damper |
| SO03 | RR Damper |
| SO04 | LR Damper |
| SO05 | SSS Solenoid |
| SO06 | ELC Compressor |
| SO07 | ELC Exhaust Valve |
| SO08 | Cycle All |

KEY
To Select Another Test Press:
Hi - To Increment
Lo - To Decrement
* PROM ID Code Number Identifies an Individual Calibration and Is Periodically Updated; Refer to Latest Service Publication for Correct ID Number

**Figure 37.** Optional test modes including RTD data, inputs, outputs, and clear codes.

must be driven to obtain a variable input from this sensor. This sensor input appears as a numerical range from −100 to 100 percent.

As the vehicle is accelerated and decelerated, this sensor input should change within the specified range. Always consult the car manufacturer's service manual for more detailed diagnosis and specifications. The next test in the data mode is selected by pressing the HI button and the display goes back to the previous test parameter when the LO button is pressed.

If the technician selects the RTD inputs, the lift/dive discrete input may be checked. When the technician selects RTD outputs, the RTD module cycles the displayed output. The next output in the test sequence is selected by pressing the HI button.

When CLEAR CODES is selected, the codes are erased and RTD CODES CLEAR is displayed. The AUTO or DEFOG button may be pressed to exit the diagnostics.

*During computer system diagnosis, use only the test equipment recommended in the vehicle manufacturer's service manual to prevent damage to computer system components.*

# Summary

- Shock absorber condition may be determined by performing a visual inspection, bounce test, or manual test.
- Front strut chatter may be caused by upper strut mount binding.
- Front strut noise on sharp turns or during suspension jounce may be caused by interference between the coil spring and the strut tower or interference between the coil spring and the upper mount.
- If one of the front strut-to-steering knuckle bolts has an eccentric cam, the cam position should be marked on the strut prior to bolt removal.
- When a coil spring compressing tool is used, it is extremely important to follow the tool manufacturer's and vehicle manufacturer's recommended spring compressing procedure. Never loosen the strut piston rod nut until the spring is compressed so all the tension is removed from the upper strut mount.
- The curb riding height is critical because it affects most other front suspension alignment angles.
- Bent control arms or worn control arm bushings affect curb riding height and front suspension alignment angles.
- A weak stabilizer bar or worn stabilizer bar bushings cause harsh riding, excessive body sway, and suspension noise.
- Excessive wear on strikeout bumpers may be caused by improper curb riding height or worn out shock absorbers.
- Bent strut rods or worn strut rod bushings cause improper front suspension alignment and reduced directional stability.
- Excessive body sway or roll when one wheel strikes a road irregularity may be caused by a defective stabilizer bar or bushings.
- Excessive lateral movement of the rear chassis may be caused by a defective track bar or bushings.
- Sagged rear springs cause excessive steering effort and rapid steering wheel return after a turn.
- If a coil spring has a vinyl coating, the spring should be taped in the compressing tool contact areas to prevent chipping the coating.
- A broken leaf spring center bolt may allow the rear axle to move, resulting in improper tracking and reduced directional stability.
- When servicing, hoisting, jacking, towing, hoisting, or lifting a vehicle equipped with an electronic air suspension system, always turn the air suspension switch off in the trunk.
- If a defect occurs in an electronic air suspension system, the control module illuminates the suspension warning lamp with the engine running.
- There are ten steps in the electronic air suspension diagnostic procedure. During the first three tests, the suspension warning lamp flashes at 1.8 times per second if the system is normal. While tests 4 through 10 are performed, the suspension warning lamp flashes the test number. The driver's door is opened and closed to move to the next test in the sequence.
- In an electronic air suspension system, the front and rear height sensors may be adjusted to obtain the proper trim height.
- A service fitting is available to splice damaged nylon air lines.
- To enter the diagnostics on the road sensing suspension system, the off and warmer buttons in the climate control center are pressed simultaneously with the ignition switch on.
- When diagnosing the road sensing suspension system; the HI and LO buttons in the climate control center are used as yes and no inputs to select fault code displays from various on-board computers.
- When diagnosing the road sensing suspension system, the HI and LO buttons are used to select various test options and move ahead or back up within the parameters in a specific test.

# Review Questions

1. While discussing curb riding height, Technician A says worn control arm bushings reduce curb riding height. Technician B says incorrect curb riding height affects most other front suspension angles. Who is correct?
   A. Technician A
   B. Technician B
   C. Both Technician A and Technician B
   D. Neither Technician A nor Technician B
2. While discussing ball joint radial measurement, Technician A says the dial indicator should be positioned against the ball joint housing. Technician B says the front wheel bearing adjustment does not affect the ball joint radial measurement. Who is correct?
   A. Technician A
   B. Technician B
   C. Both Technician A and Technician B
   D. Neither Technician A nor Technician B
3. While discussing rear wheel tracking, Technician A says a bent tie rod could result in improper rear wheel tracking. Technician B says improper tracking occurs when both rear springs are sagged the same amount. Who is correct?
   A. Technician A
   B. Technician B
   C. Both Technician A and Technician B
   D. Neither Technician A nor Technician B
4. A vehicle with a MacPherson strut front suspension has a strut chatter problem when the front wheels are turned to the right or left. The most likely cause of this problem is:
   A. a defective lower ball joint.
   B. a defective upper strut mount bearing.
   C. worn lower control arm bushings.
   D. a binding outer tie rod end.
5. When discussing strut removal from a coil spring, Technician A says to tighten the coil spring compressor tool until all the spring tension is removed from the upper strut mount. Technician B says to loosen the strut rod nut before the coil spring and strut assembly is removed from the vehicle. Who is correct?
   A. Technician A
   B. Technician B
   C. Both Technician A and Technician B
   D. Neither Technician A nor Technician B
6. A vehicle with a MacPherson strut front suspension has severely damaged rebound bumpers on both sides of the front suspension. All of these defects may be the cause of this problem *except*:
   A. Worn out front struts.
   B. Sagged front coil springs.
   C. Worn lower control arm bushings.
   D. Defective upper strut mount bearings.
7. When measuring ball joint movement on the right-hand side of a short arm, long arm front suspension with the coil spring mounted between the lower control arm and the chassis, Technician A says to position a floor jack under the lower control arm and raise the right front tire off the shop floor. Technician B says to lift the right-front tire with a prybar under the tire and measure the ball joint vertical movement with a dial indicator. Who is correct?
   A. Technician A
   B. Technician B
   C. Both Technician A and Technician B
   D. Neither Technician A nor Technician B
8. Worn ball joints may cause all of these problems *except*:
   A. Harsh ride quality.
   B. Excessive tire tread wear.
   C. Steering wander.
   D. Reduced directional stability on irregular road surfaces.
9. On a vehicle with a CVRSS, the diagnostic mode is entered by pressing these buttons simultaneously in the CCC with the ignition switch on:
   A. defrost and warmer.
   B. off and warmer.
   C. on and defrost.
   D. rear defog and off.
10. All these statements about CVRSS diagnosis are true *except*:
    A. CVRSS trouble codes have an S prefix.
    B. Pressing the AUTO button during diagnosis erases the trouble codes.
    C. A trouble code with a C suffix is a current code that is present at the time of testing.
    D. A trouble code with an H suffix is a history code representing an intermittent fault.
11. When the coil spring is positioned between the lower control arm and the chassis, place the floor jack under the ____________ ____________ ____________ to unload the ball joint prior to ball joint wear measurement.
12. A false measurement of ball joint radial movement may be caused by a loose ____________ ____________ adjustment.
13. To adjust the rear suspension trim height on an electronic air suspension system, the top ____________ ____________ ____________ must be moved up or down.
14. If an electrical defect occurs in a CVRSS, a ____________ ____________ message is displayed in the message center.

15. Explain the most likely cause of rear body lateral movement.
16. Describe the results of excessive front ball joint wear.
17. Explain the difference between history and current fault codes.
18. When diagnosing a CVRSS system, explain the procedure for selecting diagnostic code displays from the various computers.

## Steering Systems

*Twenty-five percent of all injuries from automobile accidents are from side impacts. Side impacts account for more than thirty percent of the serious injuries and fatalities in vehicle accidents. At least one-half of the injuries from side impacts are head injuries. Therefore, side air bags are a very important safety feature in modern vehicles.*

## SECTION OBJECTIVES

After you have read, studied, and practiced the contents of this section, you should be able to:

- Define the purpose of a clock spring electrical connector.
- Describe air bag sensor operation required to deploy an air bag.
- Describe the normal operation of an air bag system warning light.
- Describe the clock spring centering procedure.
- Describe two methods of steering column movement to protect the driver in a frontal collision.
- Explain the purpose of the pitman arm and the idler arm.
- Explain the rack-and-pinion steering linkage design.
- Diagnose steering columns.
- Inspect collapsible steering columns for damage.
- Diagnose and service steering linkage mechanisms.
- Check power steering fluid level.
- Drain, flush, and bleed air from the power steering system.
- Perform power steering pump pressure test.
- Remove and replace power steering pump pulleys.
- Describe the purpose of the wormshaft preload adjustment.
- Define gear ratio.
- Explain the purpose of the rack bearing and adjuster plug.
- Describe the fluid movement in a rack-and-pinion steering gear during a right turn.
- Explain the operation of the spool valve and rotary valve.
- Diagnose manual and power recirculating ball steering gear problems.
- Adjust manual and power steering gears.
- Diagnose manual and power steering systems.
- Diagnose oil leaks in manual and power steering gears.

# Supplemental Restraint Systems

## Introduction

All vehicles manufactured and sold in the United States must have passive restraints. Passive restraints may be air bags or automatic seat belts. Most vehicles with air bags also have active restraints. **Passive restraints** operate automatically with no action required by the driver. **Active restraints** require action by the driver or passengers before the restraints provide any protection.

Vehicle safety is one of the most important considerations of the average vehicle buyer today. In simple terms, safety sells vehicles! Therefore, vehicle manufacturers have spent a large amount of money engineering improved safety systems. Passive restraints at the present time include the driver's air bag, passenger-side air bag, side-impact air bags, air bag curtains, and seat belt pretensioners. The seat belt and air bag systems are intended to work together to protect the driver and passengers. On an air bag-equipped vehicle, conventional seat belts perform these functions:

1. Hold the occupants in proper position when air bags inflate.
2. Reduce the risk of injury in a less severe collision in which the air bag(s) do not deploy.
3. Reduce the risk of occupant ejection from the vehicle and thus reduce the possibility of injury.

### PASSIVE SEAT BELT RESTRAINTS

A passive seat belt system uses electric motors to automatically move the shoulder belts across the driver and front seat passenger. The upper ends of the belts are attached to a carrier mounted in a track just above the top of the door frame. The other end of each shoulder belt is secured by an inertia lock retractor that is mounted in the center console.

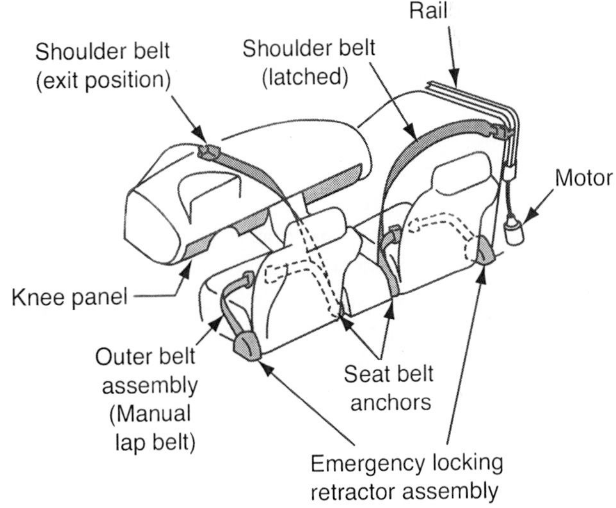


**Figure 1.** A passive sealt belt restraint system.

When a front door is opened, the outer end of the shoulder belts move forward in the door tracks to allow easy entry and exit from the vehicle **(Figure 1)**. When the door is closed and the ignition switch turned on, the shoulder belts move rearward in the door tracks to secure the front seat occupants. The active lap belt must be buckled by the driver or passenger and must be worn with the shoulder belt.

### AIR BAG SYSTEM COMPONENTS

Understanding air bag system components is essential to comprehend the complete system operation. A knowledge of air bag system operation is absolutely necessary to maintain, diagnose, and service air bag systems quickly and accurately.

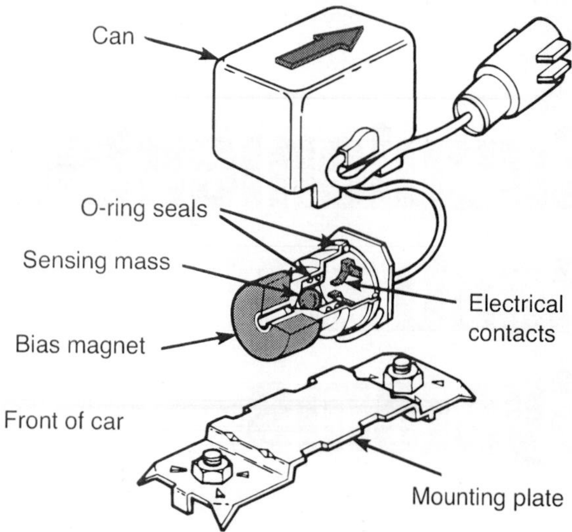


Figure 2. A mass-type air bag system sensor.

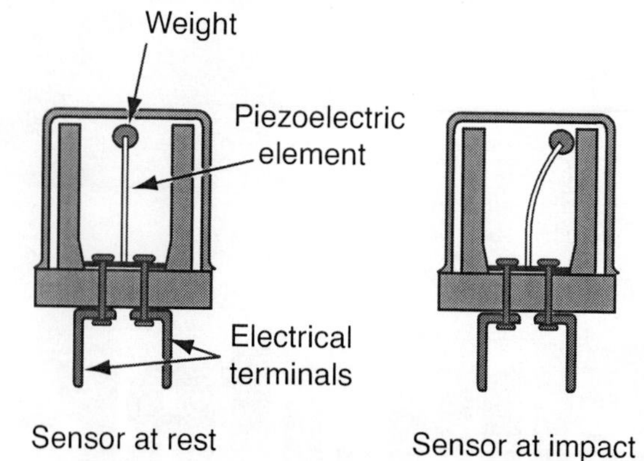


Figure 4. An accelerometer-type air bag system sensor.

## Sensors

Some air bag system sensors contain a set of normally open, gold-plated contacts and a gold-plated ball that acts as a sensing mass **(Figure 2)**. This ball is mounted in a stainless steel-lined cylinder. A magnet holds the ball about 1/8 inch away from the contacts. When the vehicle is involved in a collision of sufficient force, the ball moves away from the magnet and closes the switch contacts. These contacts remain closed for 3 milliseconds before the magnet pulls the ball away from the contacts. The sensor is completely sealed in epoxy to prevent contaminants and moisture from entering the sensor. Sensors must be mounted with the forward marking on the sensor facing toward the front of the vehicle. To operate properly, sensors must be mounted in their original mounting position and sensor brackets must not be distorted.

Some air bag sensors contain a roller on a ramp. This roller is held against a stop by small, retractable springs on each side of the roller. If the vehicle is involved in a collision of sufficient force, the roller moves up the ramp and strikes a spring contact completing the electrical circuit between the contact and the ramp **(Figure 3)**.

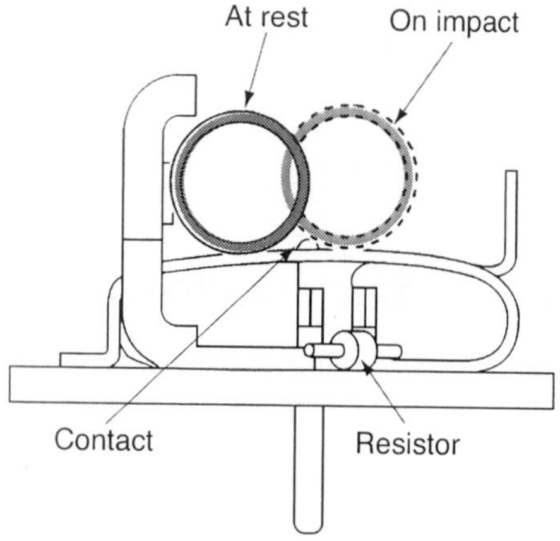


Figure 3. A roller-type air bag system sensor.

Some air bag sensors contain an accelerometer that contains a piezoelectric element **(Figure 4)**. If the vehicle is involved in a collision, this element is distorted. The voltage signal from the sensor to the air bag system module depends on the force of the collision and the amount of element distortion.

## Inflator Module

The inflator module contains the air bag, air bag container and base plate, inflator, and trim cover. The retainer and base plate are made from stainless steel and are riveted to the inflator module. The air bag is made from porous nylon and some air bags have a neoprene coating. The purpose of the inflator module is to inflate the air bag in a few milliseconds when the vehicle is involved in a collision. A typical driver's side air bag has a volume of 2.3 cubic feet. The driver's side air bag inflator module is usually retained with four bolts in the top of the steering wheel **(Figure 5)**.

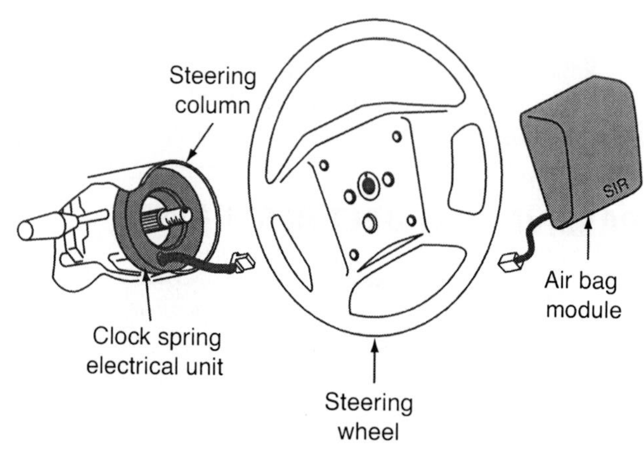


Figure 5. A driver's side air bag inflator module.

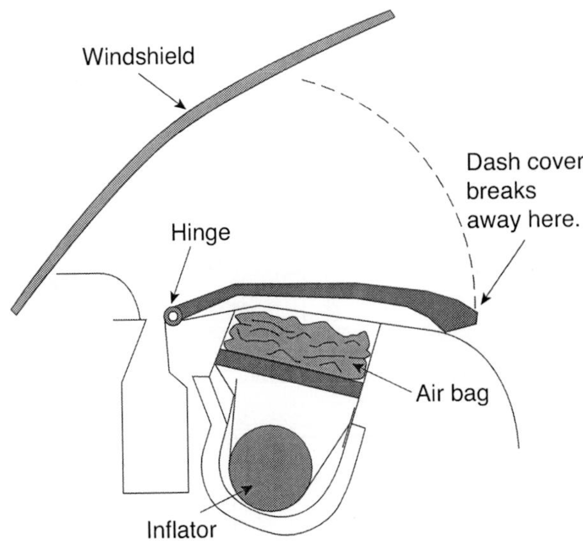


Figure 6. A passenger's side air bag inflator module.

The passenger's side inflator module is mounted in the passenger's side of the instrument panel **(Figure 6)**. Because there is a greater distance between the passenger and the dash compared to the distance between the driver and the steering wheel, the passenger's side air bag's average volume is 7.4 cubic feet.

## Clock Spring Electrical Connector

The clock spring electrical connector is mounted in the steering column directly below the steering wheel **(Figure 7)**. The clock spring electrical connector maintains electrical contact between the air bag inflator module and the air bag electrical system. The clock spring electrical connector contains a conductive ribbon that winds and unwinds as the steering wheel is turned. One end of the conductive ribbon is connected to the air bag electrical system and the opposite end of this ribbon is connected to the inflator module.

## Air Bag System Diagnostic Module

The air bag system diagnostic module (ASDM) is often mounted under the instrument panel. Other mounting locations for the ASDM include the center console or under the front passenger's seat. Some ASDMs contain an air bag sensor(s). Many ASDMs contain a backup power supply that deploys the air bag(s) if battery voltage is disconnected from the ASDM during a collision **(Figure 8)**. Some air bag systems have a backup power supply that is external from the ASDM. A typical ASDM performs these functions:

1. Controls the air bag system warning light in the instrument panel.
2. Continuously monitors the complete air bag electrical system.
3. Controls air bag system diagnostic functions.

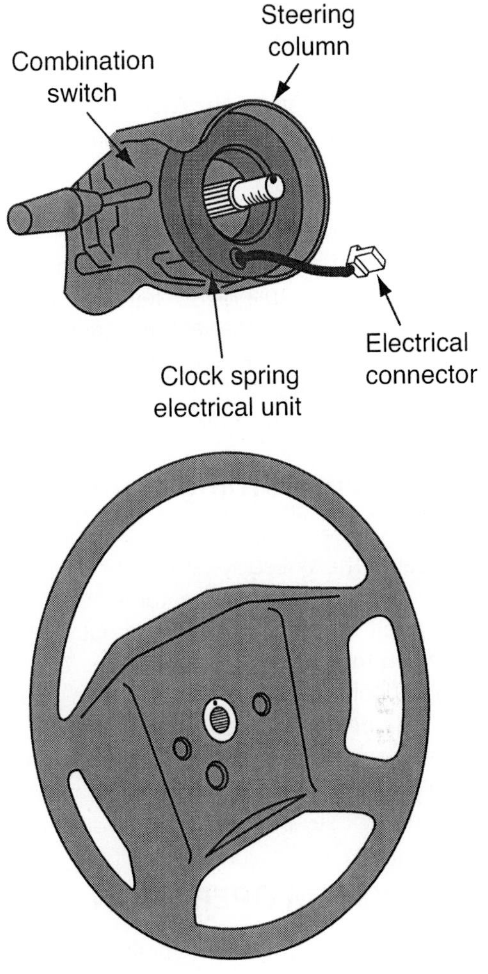


Figure 7. A clock spring electrical connector.

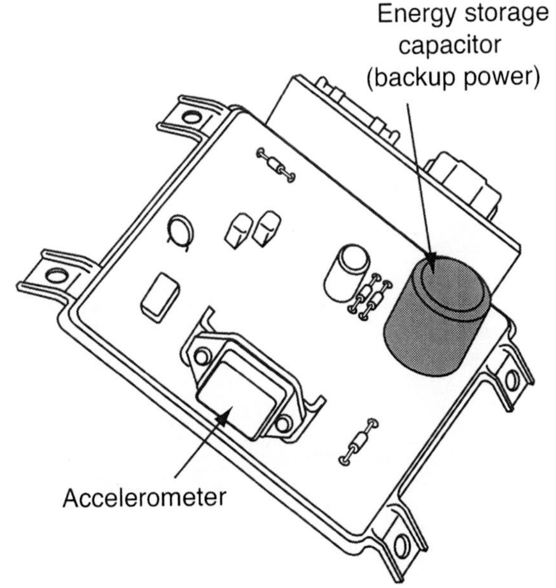


Figure 8. An air bag system diagnostic module (ASDM).

4. Provides a backup power supply to deploy the air bag(s) if battery voltage is disconnected from the air bag system during a collision.
5. On some air bag systems, the ASDM is responsible for deploying the air bag(s) when appropriate signals are received from the sensors.

If the ASDM detects an electrical defect in the air bag system, the ASDM illuminates the air bag system warning light in the instrument panel with the engine running. Some ASDMs have the capability to detect a collision and record collision information. The vehicle manufacturer uses special test equipment to download the information from the ASDM. This information indicates if the air bag system was operating properly at the time of the collision.

## Air Bag System Warning Light

The air bag system warning light is mounted in the instrument panel. The air bag system warning light should come on when the ignition switch is turned on. If the ignition switch is left in the on position without starting the engine or when the engine is started, the air bag system warning light should flash a few times and then go out. This light action indicates the air bag system is satisfactory. If the air bag system warning light is on with the engine running, there is an electrical defect in the air bag system and the air bag system may not be operational in a collision.

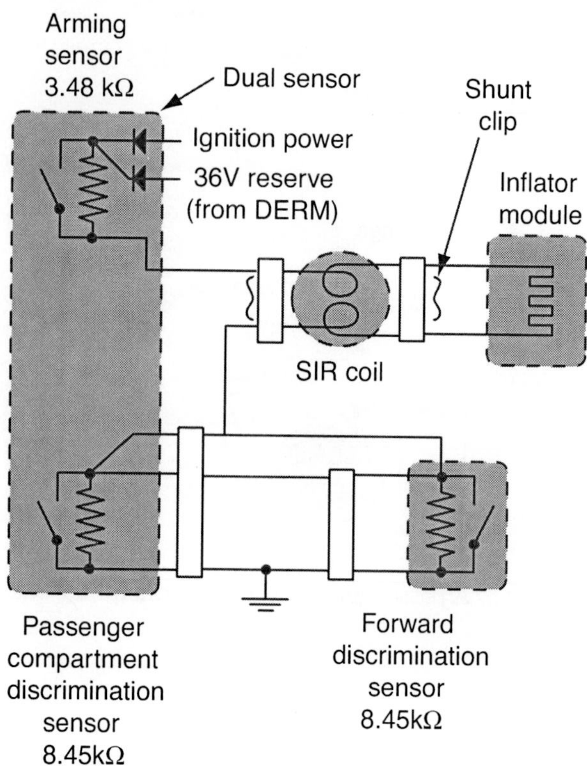


**Figure 9.** Typical squib and sensor wiring.

## AIR BAG SYSTEM OPERATION

Some air bag inflator modules contain use pyrotechnology (explosives) to inflate the air bag(s). The air bag inflator module contains a squib or igniter in the center of the module. Many air bag systems require the closing of two sensors to deploy the air bag(s). An arming sensor supplies voltage to the squib in the inflator module if this sensor closes during a collision. A passenger compartment discriminating sensor and a forward discriminating sensor are connected on the ground side of the squib **(Figure 9)**. One of these sensors must close to complete the circuit from the squib to ground.

To close an arming sensor and a discriminating sensor and deploy the driver's side and passenger's side air bags requires the vehicle to be in a frontal collision where the maximum collision force is within 30 degrees on either side of the vehicle centerline. When the arming sensor and one of the discriminating sensors close, approximately 1.75 amperes flow through the squib and the sensor circuit. This current flow heats the squib and ignites the igniter charge next to the squib. The burning igniter charge ignites the generant in the inflator module. This generant explodes very quickly, producing large quantities of hot nitrogen gas which flows through the inflator module filter into the air bag in about 30 milliseconds **(Figure 10)**. The inflated air

**You Should Know** *Many vehicles use a unique color, such as yellow, on all the wiring connectors in the air bag system so the system components are easily recognized. The wiring connectors on many air bag system components contain shorting bars that connect the wiring terminals together when the wiring connector is disconnected. These shorting bars help to prevent accidental air bag deployment from improper service procedures.*

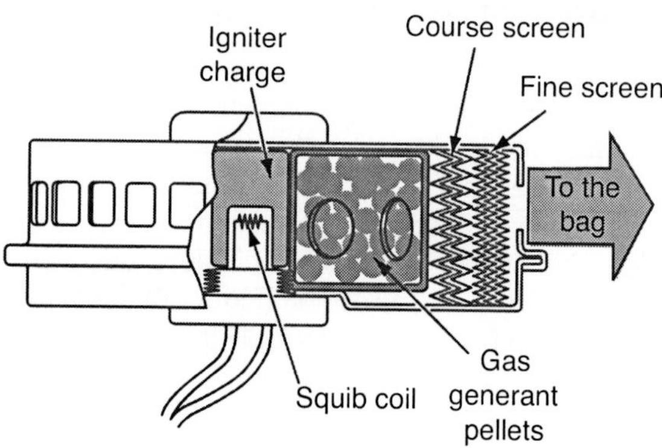


**Figure 10.** Inflator module using pyrotechnology.

bag opens the tear seams in the inflator module cover and prevents the driver's or passenger's head and chest from contacting the windshield, steering wheel, or dash.

Large openings under the air bag allow the bag to deflate in about 1 second so it does not block the driver's view if the vehicle is still in motion. This deflating action also prevents the driver or passenger from smothering in the air bag. The combustion temperature in the air bag may reach 2,500°F, but the air bag temperature is slightly above room temperature.

> **You Should Know** *During an air bag deployment, a small amount of sodium hydroxide is formed, which is an irritating caustic. Technicians should always wear eye and hand protection when servicing deployed air bags.*

Some hybrid inflator modules contain a squib (initiator) and a small amount of propellant. If the squib is energized during a collision, the exploding propellant pierces the larger argon gas container. The propellant heats the argon gas which quickly escapes into the air bag **(Figure 11)**. A pressure sensor is mounted in the opposite end of the hybrid inflator module from the squib informs the ASDM if the pressure decreases in the gas chamber during normal vehicle operation.

Some light-duty trucks have a switch in the instrument panel that allows the driver to insert the ignition key and turn the passenger's side air bag on or off **(Figure 12)**.

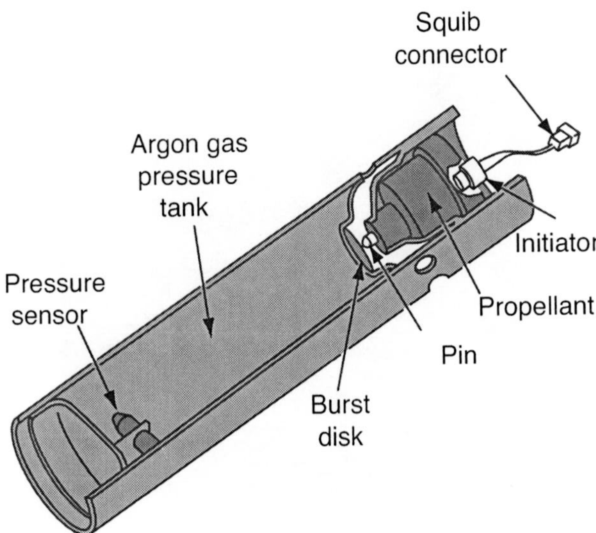


**Figure 11.** Inflator module containing pressurized argon gas.

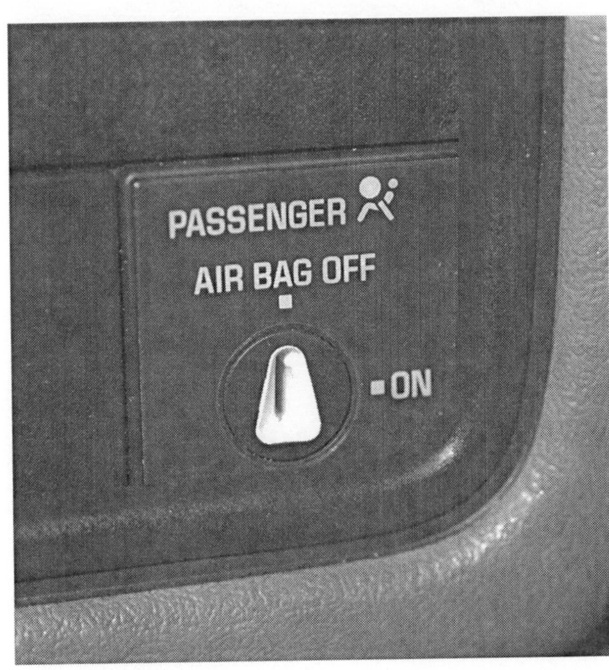


**Figure 12.** Passenger's side air bag on/off switch.

## MULTI-STAGE AIR BAG DEPLOYMENT

Many vehicles are now equipped with multi-stage air bags on both the driver's side and passenger's side. The inflator module in a multi-stage inflator module contains two squibs **(Figure 13)**. If the vehicle is involved in a collision at lower speeds that requires air bag deployment with reduced force, only one squib is ignited which inflates the air bag with less force. The second squib may be ignited 160 milliseconds later to use up the other igniter charge, but the air bag has already deployed, so this action does not affect air bag deployment. When the vehicle is involved in a severe collision at higher speeds, both squibs are fired close together so the air bag deploys faster with more force.

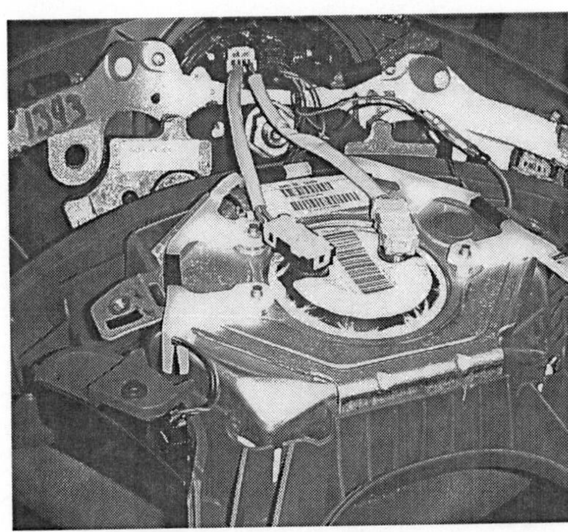

**Figure 13.** A multi-stage inflator module with dual squibs.

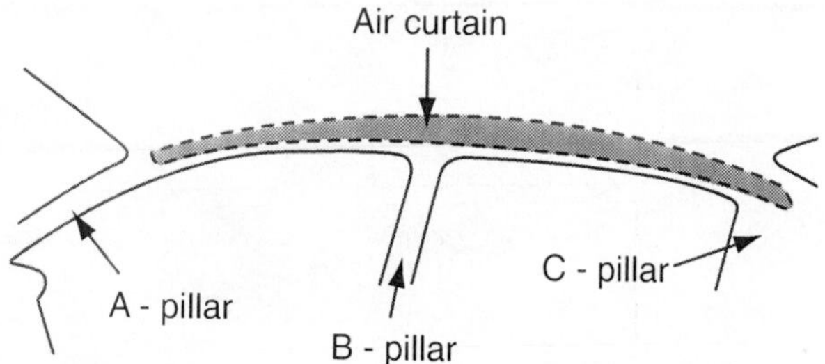


**Figure 14.** Side impact air curtains.

## SIDE IMPACT AIR BAGS

Some vehicles are now equipped with side impact air bags that protect the driver or passengers during a side collision. The side impact air bag systems are separate from the driver's side and passenger's side air bag systems. Side impact air bag systems have their own sensors and ASDM. Separate ASDMs are usually installed for each side impact air bag. The ASDMs for the side impact air bags may be under the front seats or behind the B-pillar panels.

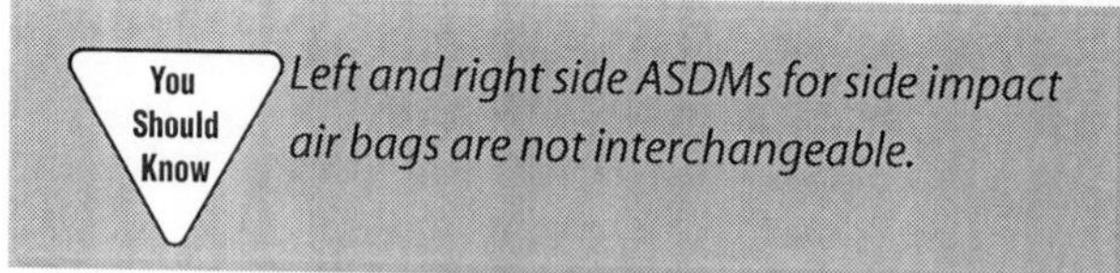


*Left and right side ASDMs for side impact air bags are not interchangeable.*

Some vehicles have a side impact air bag that deploys out of the door paneling. In other systems, the side air bags are mounted in the side of the seat back near the top. Some vehicles have a side air bag curtain that deploys out of the headliner just above the doors **(Figure 14)**. This type of air bag protects the front and rear seat occupants from head injury.

## SMART AIR BAG SYSTEMS

Some air bag systems have a switch in the passenger's side of the front seat. This switch informs the ASDM if anyone is sitting in the front passenger seat or not. If no one is sitting in this seat, the passenger's side air bag does not deploy if the vehicle is involved in a collision. Some smart air bag systems can detect from the passenger's weight. If a child below a certain weight is occupying this seat, the air bag does not deploy. In some air bag systems, the weight of the person in the passenger's seat also determines the force with which the passenger's side air bag deploys. On some smart air bag systems, if a rearward facing child's seat is placed in the front passenger's seat, the passenger's side air bag does not deploy.

You Should Know

*If a child in a rearward facing child's seat is placed in the front passengers seat and the air bag deploys, the deployed air bag may force the seat and the child against the vertical part of the front seat causing the child to suffocate.*

Some vehicles have a small knee air bag that deploys out of the dash in front of the driver's knees. This air bag protects the driver from knee injury and also keeps the driver from sliding under the seat belt during an accident. This action maintains the driver in a better position to be protected by the driver's side air bag.

## SEAT BELT PRETENSIONERS

Some vehicles have seat belt pretensioners on the seat belts. The pretensioners contain materials similar to a single-stage air bag inflator module. The pretensioners may be mounted on the buckle side of the seat belt **(Figure 15)**. If the front air bags are deployed, the ASDM also fires the pretensioners. A thin cable is connected between the buckle and a small piston in the pretensioner. When a pretensioner is fired, the piston moves up the cylinder and the cable pulls the buckle tight. This action holds the occupant tightly against the seat and helps to prevent injury.

Some pretensioners are on the retractor side of the seat belt **(Figure 16)**. When the pretensioner is fired, balls shoot out of the pretensioner against a fan wheel causing the retractor to rotate and tighten the seat belt **(Figure 17)**. The last ball to be shot out of the pretensioner is larger and sticks in the fan wheel to lock the seat belt.

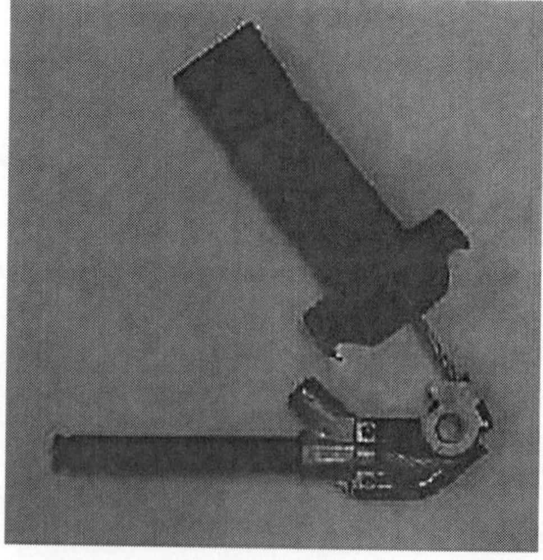

**Figure 15.** A buckle-mounted seat belt pretensioner.

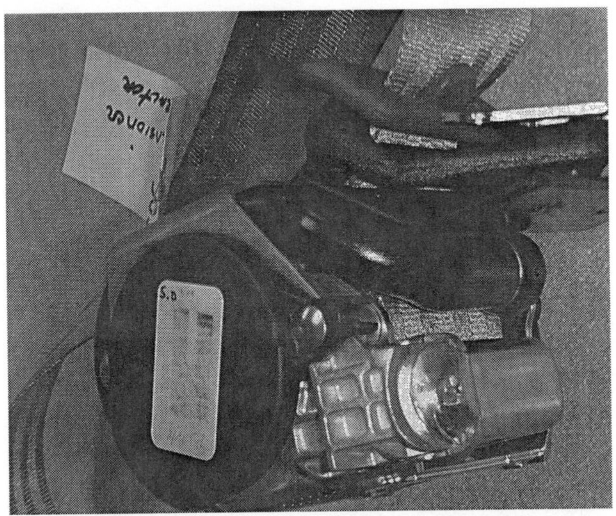

Figure 16. A retractor-mounted seat belt pretensioner.

Figure 17. Internal design of a retractor-mounted seat belt pretensioner.

## SUPPLEMENTAL RESTRAINT SYSTEM MAINTENANCE

Periodically, the complete air bag system wiring should be inspected for worn insulation, damaged wiring, loose connectors, loose sensor mounting bolts, or distorted sensor mounting areas.

**You Should Know** *Loose sensor mounting bolts or distorted sensor brackets and mountings cause improper sensor operation and inaccurate air bag deployment.*

During any air bag system diagnostic check, the air bag warning light in the instrument panel should be observed for proper operation. As mentioned previously, this warning light should be illuminated when the ignition switch is turned on. It should flash a few times and go out when the engine starts. Some air bag warning lights remain on for a few seconds after the engine starts rather than flashing. If the air bag warning light operates properly, the air bag system is operational.

Inspect the seat belts for webbing damage. Fully extend each seat belt from the retractor and check the webbing for cuts, broken or pulled threads, cut loops at the belt edge, and bowed conditions **(Figure 18)**. If any of these conditions are present, replace the seat belt assembly.

Check the seat belt buckles for damage and proper latching and unlatching. Check the seat belt retractors for proper operation and proper retention on the vehicle chassis. If the vehicle has passive seat belts, inspect the belts for free movement in the tracks. The electric motor for the passive seat belts is usually mounted behind the rear seat trim panel. This motor operates the plastic tape connected to the seat belts. Be sure the seat belt moves through its complete travel without sticking when the door is opened and closed.

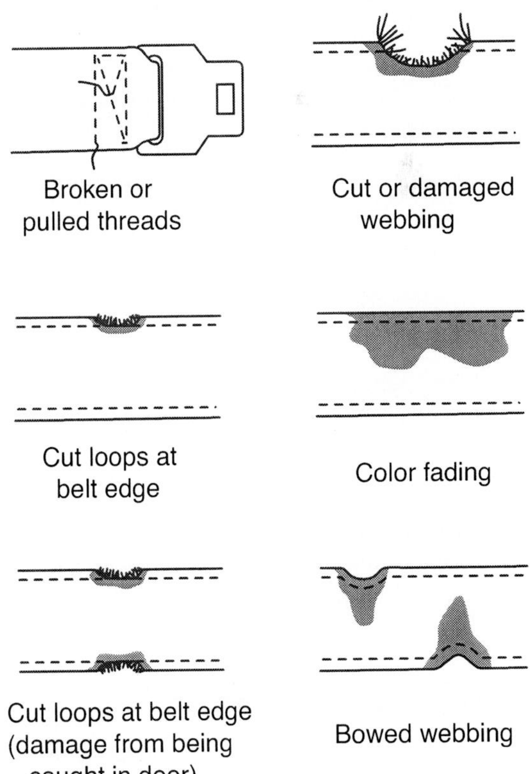


Figure 18. Seat belt webbing defects.

## SUPPLEMENTAL RESTRAINT SYSTEM DIAGNOSIS

Most air bag systems are diagnosed with a scan tool. The scan tool and the scan tool module must be compatible with the vehicle and the air bag system. Follow this procedure for typical scan tool diagnosis of an air bag system:

1. Roll the driver's window down.
2. Turn the ignition switch off and disconnect the negative battery cable. Wait for the time period specified by the vehicle manufacturer.
3. Connect the scan tool to the DLC under the dash.
4. Move the scan tool through the driver's window opening and stand outside the vehicle.
5. Reach into the vehicle and turn the ignition switch on. Be sure there is no one in the vehicle.
6. Reconnect the negative battery cable.
7. Select Air Bag System on the scan tool and read and record any air bag DTCs.
8. Disconnect the negative battery cable and wait for the time period specified by the vehicle manufacturer.
9. Turn the ignition switch off and disconnect the scan tool.
10. Reconnect the negative battery cable.

You Should Know

- *On some of the first-generation air bag systems, the air bag warning light began flashing DTCs when a defect occurred in the air bag system.*
- *On some General Motors first-generation air bag systems, flash codes could be obtained from the air bag system warning light by connecting terminals A and K in the DLC.*

You Should Know *When servicing or diagnosing air bag systems, use only the vehicle manufacturer's recommended tools. Use of other tools, such as 12-volt test lights or self-powered test lights, may cause accidental air bag deployment.*

## AIR BAG SYSTEM SERVICE

When servicing air bag system components, some vehicle manufacturers recommend disabling the air bag system to prevent accidental air bag deployment. A typical air bag system disabling procedure on a vehicle with driver's side, passenger's side, and side impact air bags follows:

1. Turn the steering wheel until the front wheels are in the straight-ahead position.
2. Turn the ignition switch off and remove the ignition key from the switch.
3. Remove the air bag fuse from the fuse block.

You Should Know *Some air bag system wiring connectors are retained with a connector position assurance (CPA) pin. This pin must be removed before the connector can be disconnected.*

A **connector position assurance (CPA) pin** holds two wiring connectors together so they cannot be separated until the pin is removed.

4. Disconnect the following connectors.
   (a) Driver's side air bag two-wire air bag system connector at the base of the steering column.
   (b) Passenger's side air bag two-wire connector behind the passenger's side air bag inflator module.
   (c) Driver's side impact inflator module two-wire connector under the driver's seat.
   (d) Passenger's side impact inflator module two-wire connector under the passenger's seat.
5. Perform the necessary service work on the air bag system.
6. Connect all the disconnected air bag system connectors and install the CPA pins in each connector.
7. Install the air bag system fuse.
8. Turn the ignition switch on and check the air bag system warning light for proper system operation.

You Should Know *The air bag system disabling procedure varies depending on the vehicle make and model year. Always use the disabling procedure in the vehicle manufacturer's service manual. When servicing an air bag system, follow the instructions and precautions provided on the vehicle air bag system warning decals.*

## Clock Spring Centering

Before installing a clock spring, it must be centered and installed with the front wheels straight ahead. To center the clock spring, turn the clock spring fully clockwise. Turn the clock spring fully counterclockwise counting the total number of turns. From the fully counterclockwise position, rotate the clock spring one-half the total number of turns clockwise. The clock spring is now in the centered position. Install the centered clock spring with the front wheels straight ahead.

You Should Know *If a clock spring is not installed in the centered position with the front wheels straight ahead, the conductive ribbon in the clock spring will be broken when the steering wheel is rotated.*

*Some clock springs have a mark that indicates the centered position.*

# Summary

- Seat belts must be used with passive air bag systems.
- Many air bag system sensors complete an electrical circuit during a collision.
- Air bag system inflator modules use pyrotechnology or pressurized argon gas to inflate the air bag.
- The air bag system warning light indicates the readiness of the air bag system.
- Many vehicles now have multi-stage air bags.
- Side impact air bags have separate sensors and inflator modules.
- Smart air bag systems can sense the presence and weight of a person in the passenger's seat.
- During a collision, seat belt pretensioners tighten the seat belts using pyrotechnology.
- A scan tool is used to diagnose the air bag system.
- Air bag systems must be disabled when servicing system components.
- The clock spring electrical connector must be installed in the centered position with the front wheels straight ahead.

# Review Questions

1. Technician A says a mass-type air bag system sensor may be installed facing in either direction. Technician B says on a mass-type air bag system sensor, distortion of the sensor mounting area may affect sensor operation. Who is correct?
   A. Technician A
   B. Technician B
   C. Both Technician A and Technician B
   D. Neither Technician A nor Technician B
2. Technician A says the clock spring electrical connector contains spring-loaded copper contacts. Technician B says the clock spring electrical connector is mounted directly below the inflator module. Who is correct?
   A. Technician A
   B. Technician B
   C. Both Technician A and Technician B
   D. Neither Technician A nor Technician B
3. Technician A says the inflator module deploys the air bag in a few seconds. Technician B says the passenger's side inflator module is mounted in the passenger's side of the instrument panel. Who is correct?
   A. Technician A
   B. Technician B
   C. Both Technician A and Technician B
   D. Neither Technician A nor Technician B
4. Technician A says multi-stage air bag inflator modules have dual squibs. Technician B says smart air bag systems can sense the presence of a person in the passenger's seat. Who is correct?
   A. Technician A
   B. Technician B
   C. Both Technician A and Technician B
   D. Neither Technician A nor Technician B

5. All these statements about passive seat belt systems are true *except*:
   A. When a door is opened, the outer end of the shoulder belts move forward in the door tracks.
   B. When a door is closed and the ignition switch is turned on, the shoulder belts move rearward in the door tracks.
   C. Electric motors move the shoulder belts across the driver and passenger.
   D. The lap belts are connected without any action from the driver or passenger.
6. All these statements about mass-type air bag system sensors are true *except*:
   A. The contacts and ball are gold plated.
   B. A magnet holds the ball about 1/2 inch away from the contacts.
   C. The sensor is completely sealed in epoxy to eliminate contaminants.
   D. An arrow on the sensor housing must face toward the front of the vehicle.
7. In an air bag system:
   A. The passenger's side air bag inflator module is larger than the driver's side inflator module.
   B. The air bags are manufactured from porous plastic.
   C. The retainer and base plate are made from lead.
   D. The retainer and base plate are bolted to the inflator module.
8. When discussing air bag system operation, Technician A says the ASDM controls the air bag system warning light. Technician B says the ASDM continuously monitors the air bag electrical system. Who is correct?
   A. Technician A
   B. Technician B
   C. Both Technician A and Technician B
   D. Neither Technician A nor Technician B
9. In a vehicle with typical driver's side and passenger's side air bags, to deploy the air bags the vehicle must be involved in a frontal collision of sufficient force and the collision force must be within:
   A. 10 degrees on either side of the vehicle centerline.
   B. 20 degrees on either side of the vehicle centerline.
   C. 30 degrees on either side of the vehicle centerline.
   D. 54 degrees on either side of the vehicle centerline.
10. In a smart air bag system:
    A. The passenger's side air bag deploys when there is no one sitting in the front passenger's seat and the vehicle is in a severe frontal collision.
    B. The passenger's side air bag deploys regardless of the weight of the front passenger's seat occupant if the vehicle is in a severe frontal collision.
    C. On some smart air bag systems, the passenger's side air bag does not deploy if a rearward facing child's seat is in the front passenger's seat.
    D. The driver's side air bag does not deploy if the driver's weight is below 110 pounds and the vehicle is involved in a severe frontal collision.
11. Side impact air bag systems have separate ________________ and ________________.
12. If a rearward facing child's seat is placed in the front passenger's seat, a smart air bag system will not ________________ the passenger's side air bag.
13. Seat belt pretensioners ________________ the seat belts if the vehicle is involved in a collision of sufficient force.
14. Some air bag system disabling procedures require the removal of the air bag ________________ before disconnecting any system connectors.
15. Explain the clock spring centering procedure.
16. Describe the sensor requirements to deploy an air bag.
17. Describe the purpose of the backup power supply in the inflator module.
18. Explain the operation of an inflator module containing pressurized argon gas during an air bag deployment.

# Chapter 49 Steering Columns, Linkages, and Power Steering Pumps Maintenance, Diagnosis, and Service

## Introduction

Steering columns play a significant part in steering control, safety, and driver convenience. The steering column connects the steering wheel to the steering gear. Collapsible steering columns provide some driver protection in a collision.

Steering linkage mechanisms are used to connect the steering gear to the front wheels. Steering linkages help to position the front wheels and tires properly to provide proper steering control and minimize front tire wear.

Power steering systems have contributed to reduced driver fatigue and have made driving a more pleasant experience. Nearly all power steering systems at the present time use fluid pressure to assist the driver in turning the front wheels. Since driver effort required to turn the front wheels is reduced, driver fatigue is decreased. The advantages of power steering have been made available on many vehicles and safety has been maintained in these systems.

### STEERING COLUMN DESIGN

Steering columns may be classified as non-tilting, tilting, and tilting/telescoping. Tilt steering columns facilitate driver entry and exit to and from the front seat. These columns also allow the driver to position the steering wheel to suit individual comfort requirements **(Figure 1)**. A tilting/telescoping steering column allows the driver to tilt

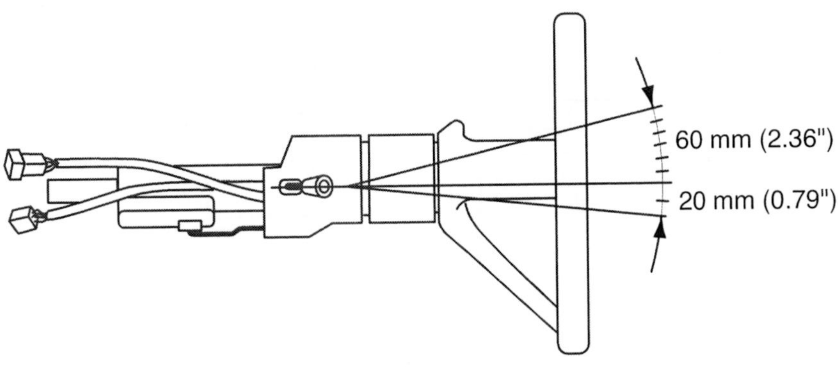


**Figure 1.** A tilt steering column.

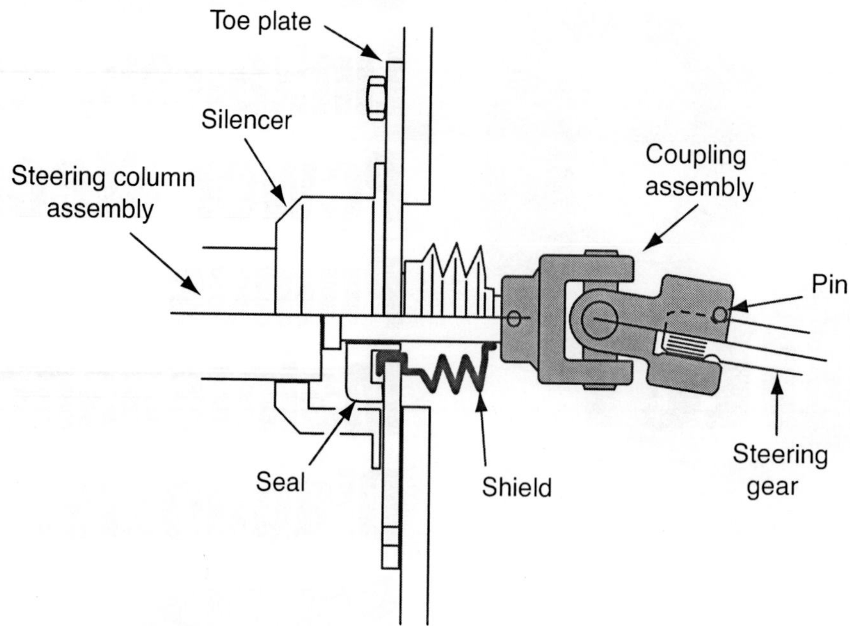


**Figure 2.** Toe plate, seal, and silencer surrounding the lower end of the steering column.

and extend or retract the steering wheel. In this type of steering column, the driver has more steering wheel position choices.

Many steering columns contain a two-piece steering shaft connected by two universal joints. A jacket and shroud surround the steering shaft and the upper shaft is supported by two bearings in the jacket. A toe plate, seal, and silencer surround the lower steering shaft and cover the opening where the shaft extends through the floor **(Figure 2)**. The lower steering shaft is surrounded by a shield underneath the toe plate.

The lower universal joint couples the lower shaft to the stub shaft in the steering gear. In some steering columns, a flexible coupling is used in place of the lower universal joint **(Figure 3)**.

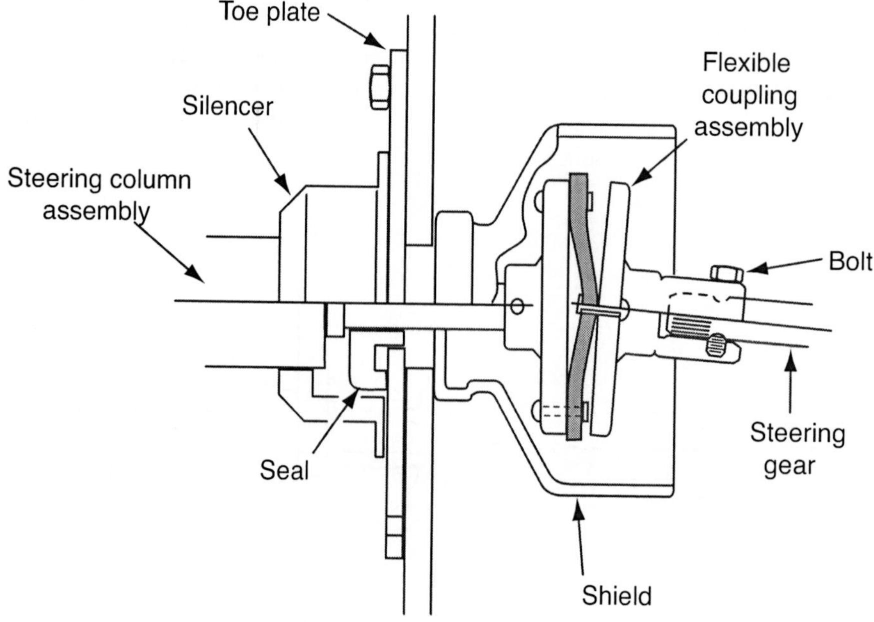


**Figure 3.** A flexible coupling.

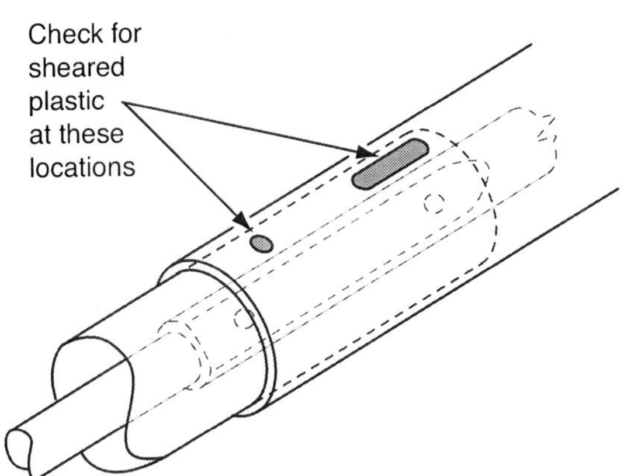


**Figure 4.** Injection plastic in a collapsible outer steering column jacket.

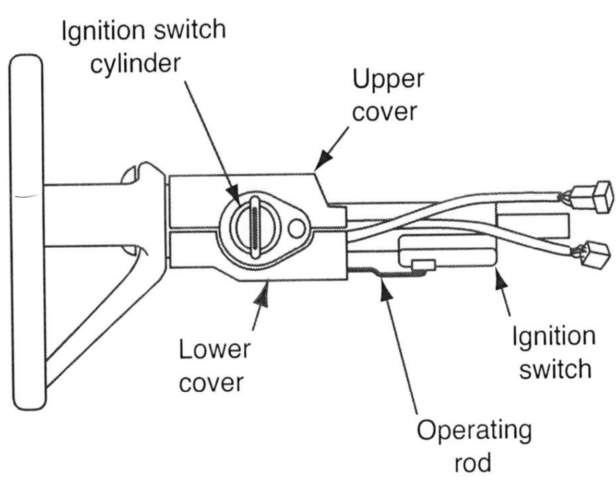


**Figure 6.** Ignition switch and ignition switch cylinder mounted in steering column.

Studs and nuts retain the steering column bracket to the instrument panel support bracket. The steering column is designed to protect the driver if the vehicle is involved in a frontal collision. An energy-absorbing lower bracket and lower plastic adapter are used to connect the steering column to the instrument panel mounting bracket. This bracket allows the column to slide forward if the driver is thrown forward into the wheel in a frontal collision. The mounting bracket is also designed to prevent rearward movement toward the driver in a collision.

In some steering columns, the outer column jacket is a two-piece unit retained with plastic pins **(Figure 4)**. In this type of column the lower steering shaft is also a two-piece sliding unit retained with plastic pins **(Figure 5)**. When the driver is thrown against the steering wheel in a frontal collision, the plastic pins shear off in the lower steering shaft and outer column jacket. The shearing action of the plastic pins allows the steering column to collapse away from the driver which reduces the impact as the driver hits the steering wheel.

The steering wheel splines fit on matching splines on the top of the upper steering shaft. A nut retains the wheel on the shaft. Most steering wheels and shafts have matching alignment marks which must be aligned when the steering wheel is installed.

An ignition switch cylinder is usually mounted in the upper right side of the column housing. The ignition switch is bolted on the lower side of the housing **(Figure 6)**. An operating rod connects the ignition switch cylinder to the ignition switch. Some ignition switches are integral with the lock cylinder in other steering columns.

The turn signal switch and hazard warning switch are mounted on top of the steering column under the steering wheel. Lugs on the bottom of the steering wheel are used to cancel the signal lights after a turn is completed. On many vehicles, the signal light lever also operates the wipe/wash switch and the dimmer switch **(Figure 7)**.

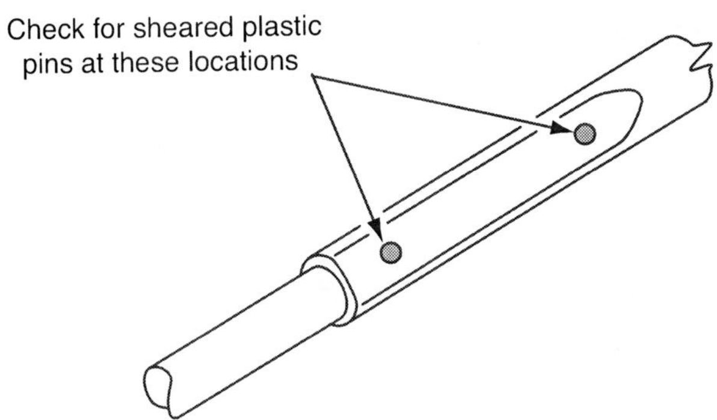


**Figure 5.** Injection plastic in a collapsible lower steering shaft.

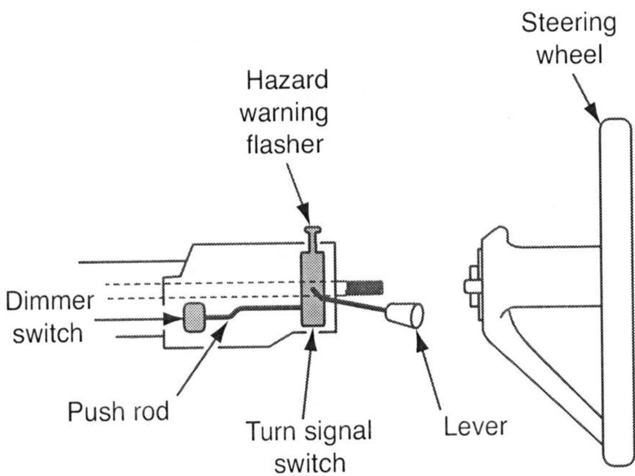


**Figure 7.** Turn signal switch, hazard warning switch, dimmer switch, and wipe/wash switch mounted in steering column.

You Should Know *The signal light switch, wipe/wash switch, dimmer switch, and hazard warning switch may be called a combination switch, multi-function switch, or smart switch.*

If the gear shift is mounted in the steering column, a tube extends from the gear shift housing to the shift lever at the lower end of the steering column. This shift lever is connected through a linkage to the transaxle or transmission shift lever. A lock plate is attached to the upper steering shaft and a lever engages with slots in this plate to lock the steering wheel and gear shift when the gear shift is in park and the ignition switch is in the lock position **(Figure 8)**.

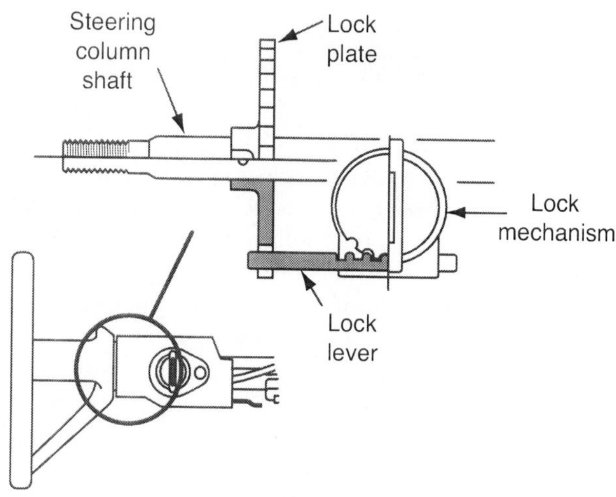


**Figure 8.** Upper steering column with locking plate and lever.

## PARALLELOGRAM STEERING LINKAGES

Parallelogram steering linkages may be mounted behind the front suspension **(Figure 9)** or ahead of the front suspension **(Figure 10)**. The parallelogram steering linkage must not interfere with the engine oil pan or chassis components.

Regardless of the parallelogram steering linkage mounting position, this type of steering linkage contains the same components. The main components in this steering linkage mechanism are the pitman arm, center link, idler arm, tie-rods with sockets, and tie-rod ends.

Parallelogram steering linkages are found on independent front suspension systems. In a parallelogram steering linkage, the tie-rods are connected parallel to the lower control arms. Road vibration and shock are transmitted from the tires and wheels to the steering linkage. These forces tend to wear the linkages and cause steering looseness. If the steering linkage components are worn, steering control is reduced. Since loose steering linkage components cause intermittent toe changes, this problem increases tire wear. The wear points in a parallelogram steering linkage are the tie-rod sockets and ends, idler arm, and center link end.

You Should Know *Always remember that a customer's life may depend on the condition of the steering linkages on his or her vehicle. During undercar service, always make a quick check of steering linkage condition.*

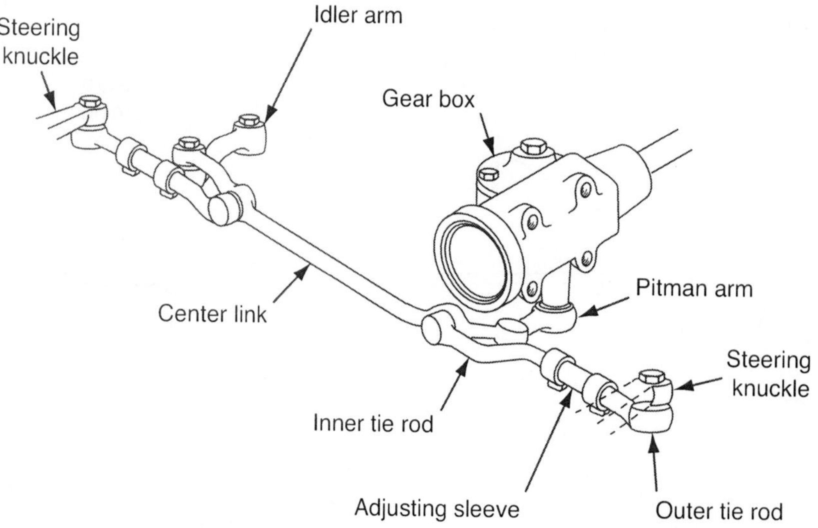


**Figure 9.** Parallelogram steering linkage behind the front suspension.

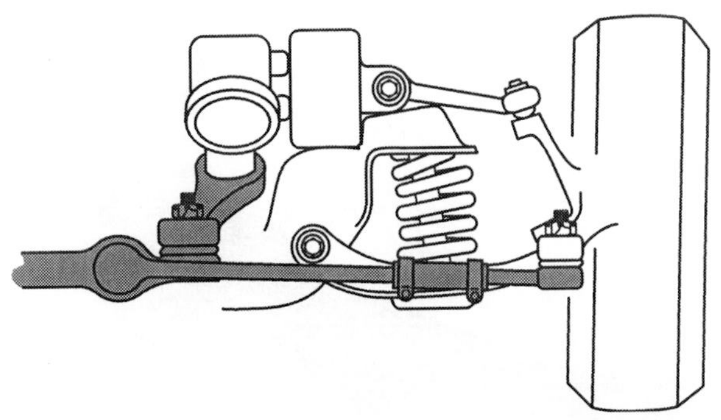

Figure 10. Parallelogram steering linkage positioned in front of front suspension.

## Tie Rods

The tie-rod assemblies connect the center link to the steering arms that are attached to the front steering knuckles. In some front suspensions, the steering arms are part of the steering knuckle while in other front suspension systems, the steering arm is bolted to the knuckle. A ball socket is mounted on the inner end to each tie-rod and a tapered stud on this socket is mounted in a center link opening. A castellated nut and cotter pin retains the tie-rods to the center link. A threaded sleeve is mounted on the outer end of each tie-rod and a tie-rod end is threaded into the outer end of this sleeve **(Figure 11)**. Some tie-rod ends have a hardened steel upper bearing and a high-strength polymer lower bearing for increased durability **(Figure 12)**.

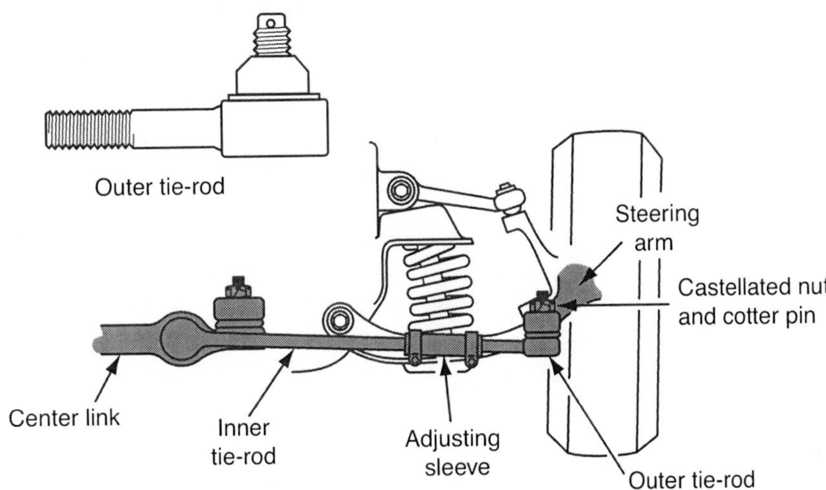


Figure 11. Tie-rod design.

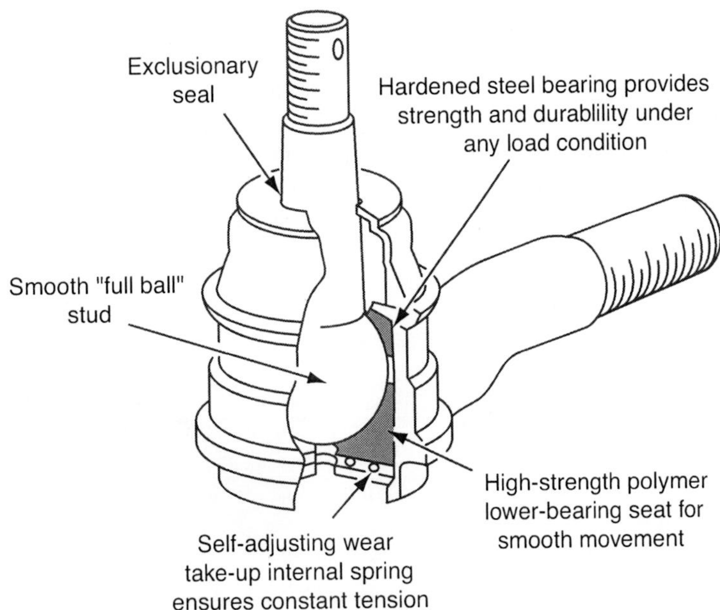


Figure 12. Outer tie-rod end with hardened steel upper bearing and high-strength polymer lower bearing.

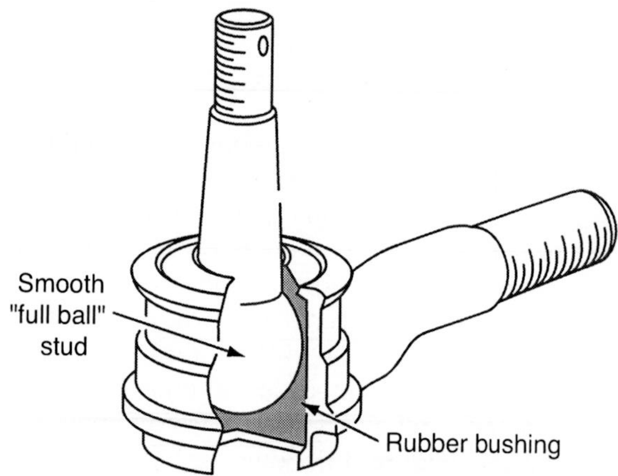


**Figure 13.** Outer tie-rod end with rubber-encapsulated ball stud.

Other tie-rod ends have a rubber-encapsulated ball stud **(Figure 13)**.

Each tie-rod sleeve contains a left-hand and a right-hand thread where they are threaded onto the tie-rod end and the tie-rod. Therefore, sleeve rotation changes the tie-rod length and provides a toe adjustment. Clamps are used to tighten the tie-rod sleeves. The clamp opening must be positioned away from the slot in the tie-rod sleeve. The design of the steering linkage mechanism allows multi-axial movement since the front suspension moves vertically and horizontally. Ball and socket-type pivots are used on the tie-rod assemblies and center link.

If the front wheels hit a bump, the wheels move up and down and the control arms move through their respective arcs. Since the tie-rods are connected to the steering arms, these rods must move upward with the wheel. Under this condition, the inner end of the tie-rod acts as a pivot and the tie-rod also moves through an arc. This arc is almost the same as the lower control arm arc because the tie-rod is parallel to the lower control arm. Maintaining the same arc between the lower control arm and the tie-rod minimizes toe change on the front wheels during upward and downward wheel movement. This action improves the directional stability of the vehicle and reduces tread wear on the front tires.

## Pitman Arm

The pitman arm connects the steering gear to the center link. This arm also supports the left side of the center link. Motion from the steering wheel and steering gear is transmitted to the pitman arm. This arm transfers the movement to the steering linkage. This pitman arm movement forces the steering linkage to move to the right or left and the linkage moves the front wheels in the desired direction. The pitman arm also positions the center link at the proper height to maintain the parallel relationship between the tie-rods and the lower control arms.

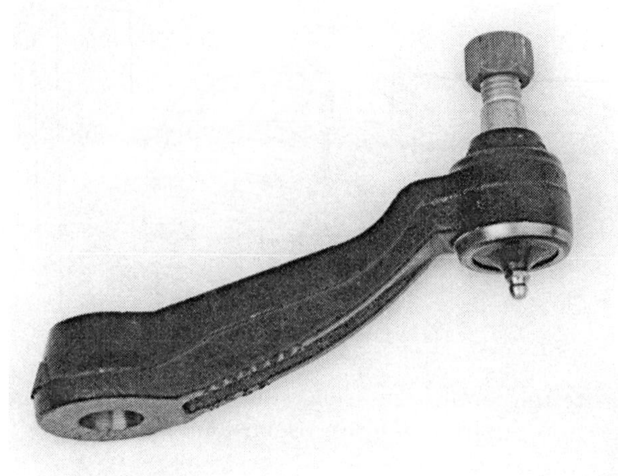

**Figure 14.** Pitman arm design.

Wear-type pitman arms have ball sockets and studs at the outer end. This stud fits into the center link opening **(Figure 14)**. The ball stud and socket are subject to wear and pitman arm replacement is necessary if the ball stud is loose. A non-wear pitman arm has a tapered opening in the outer end. A ball stud in the center link fits into this opening. The non-wear pitman arm only needs replacing if the arm is damaged, bent, or in a collision. The opening in the inner end of both types of pitman arms have serrations which fit over matching serrations on the steering gear shaft. A nut and lock washer retain the pitman arm to the steering gear shaft.

## Idler Arm

An idler arm support is bolted to the frame or chassis on the opposite end of the center link from the pitman arm. The idler arm is connected from the support bracket to the center link. Two bolts retain the idler arm bracket to the frame or chassis. In some idler arms, a ball stud on the outer end of the arm fits into a tapered opening in the center link, whereas in other idler arms a ball stud in the center link fits into a tapered opening in the idler arm **(Figure 15)**.

The idler arm supports the right side of the center link and helps to maintain the parallel relationship between the

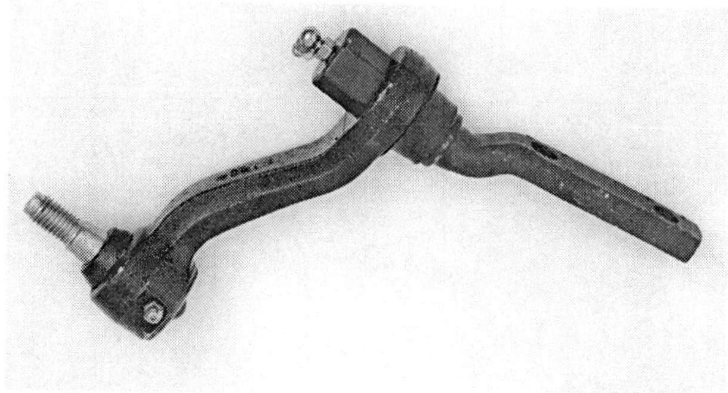

**Figure 15.** Idler arm design.

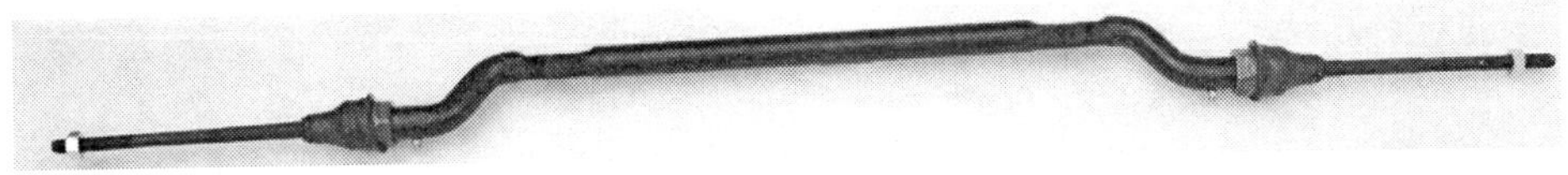

**Figure 16.** Center link design.

tie rods and the lower control arms. The outer end of the idler arm is designed to swivel on the idler arm bracket. This swivel is subject to wear. A worn idler arm swivel causes excessive vertical steering linkage movement and erratic toe. This action results in excessive steering wheel freeplay with reduced steering control and front tire wear.

## Center Links

The center link controls the sideways steering linkage movement. The center link, together with the pitman arm and idler arm, provide the proper height for the tie-rods which is very important to minimize toe change on road irregularities. Some center links have tapered openings in each end and the studs on the pitman arm and idler arm fit into these openings. This type of center link may be called a taper end or non-wear link. Other wear-type center links have ball sockets in each end with tapered studs extending from the sockets **(Figure 16)**. These tapered studs fit into openings in the pitman arm and idler arm and they are retained with castellated nuts and cotter pins.

# RACK-AND-PINION STEERING LINKAGES

The rack-and-pinion steering linkage is used with rack-and-pinion steering gears. In this type of steering gear, the rack is a rod with teeth on one side. This rack slides horizontally on bushings inside the gear housing. The rack teeth are meshed with teeth on a pinion gear and this pinion gear is connected to the steering column. When the steering wheel is turned, the pinion rotation moves the rack sideways. Tie-rods are connected directly from the ends of the rack to the steering arms. The tie-rods are similar to those found on parallelogram steering systems. An inner tie-rod end connects each tie-rod to the rack. Bellows boots are clamped to the gear housing and the tie-rods keep dirt out of these joints **(Figure 17)**. The inner tie-rod end contains a spring-loaded ball socket and the outer tie-rod ends connected to the steering arms are basically the same as those in parallelogram steering linkages **(Figure 18)**. Some inner tie-rod ends contain a bolt and bushing. These tie-rod ends are threaded onto the rack. Since the rack is connected directly to the tie-rods, the rack replaces the center link in a parallelogram steering linkage.

The rack-and-pinion steering gear may be mounted on the front crossmember **(Figure 19)** or attached to the

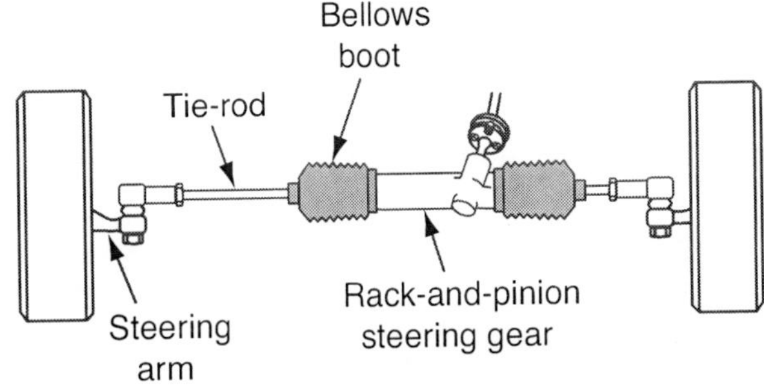


**Figure 17.** Rack-and-pinion steering gear.

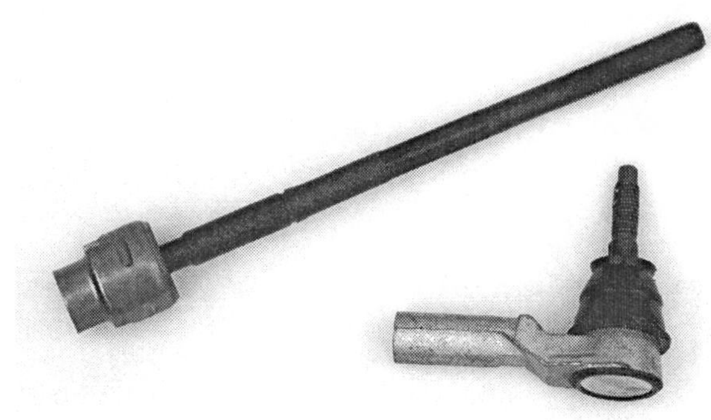

**Figure 18.** Inner tie-rod and outer tie-rod end, rack-and-pinion steering.

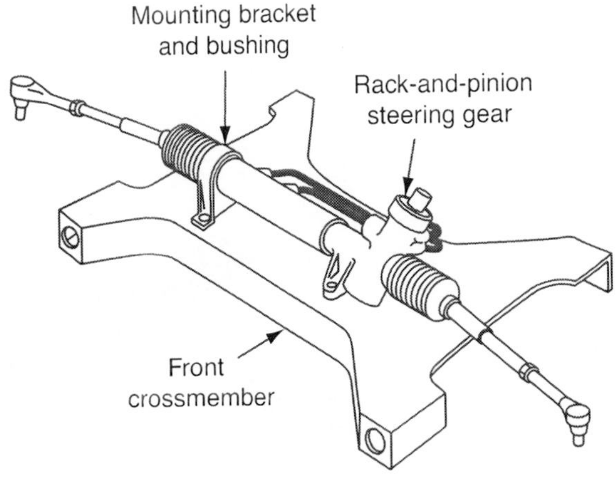


**Figure 19.** Rack-and-pinion steering gear mounting on front crossmember.

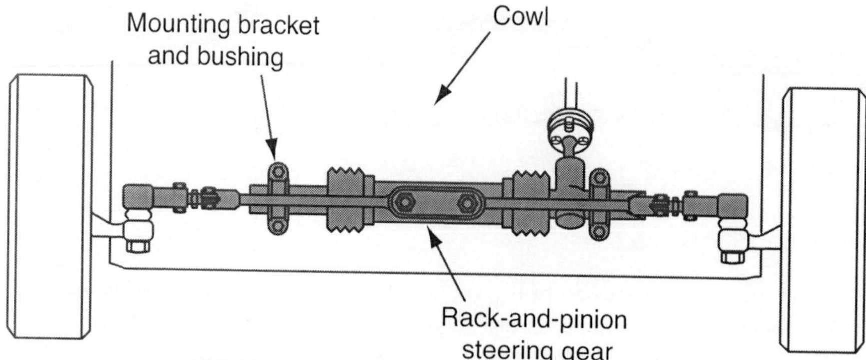


**Figure 20.** Rack-and-pinion steering gear mounted on cowl.

cowl behind the engine **(Figure 20)**. Rubber insulating bushings surround the steering gear. These bushings are clamped to the crossmember or cowl. The rack-and-pinion steering gear is mounted at the proper height to position the tie-rods and lower control arms parallel to each other. The number of friction points is reduced in a rack-and-pinion steering system. This system is light and compact. Most FWD unibody vehicles have rack-and-pinion steering. Since the rack is linked directly to the steering arms, this type of steering linkage provides good road feel.

## STEERING DAMPER

A steering damper or stabilizer may be found on some parallelogram steering linkages. The steering damper is similar to a shock absorber. This component is connected from one of the steering links to the chassis or frame **(Figure 21)**. When a front wheel strikes a road irregularity, a shock is transferred from the front wheel to the steering linkage, steering gear, and steering wheel. The steering damper helps to absorb this road shock and prevents it from reaching the steering wheel. Heavy-duty steering dampers are available for severe road conditions, such as those sometimes encountered by 4WD vehicles.

## POWER STEERING PUMP DESIGN

The power steering pump is driven by a V-belt or ribbed V-belt. These belts were explained previously in Chapter 37. Various types of power steering pumps have been used by vehicle manufacturers. Many vane-type power steering pumps have flat vanes which seal the

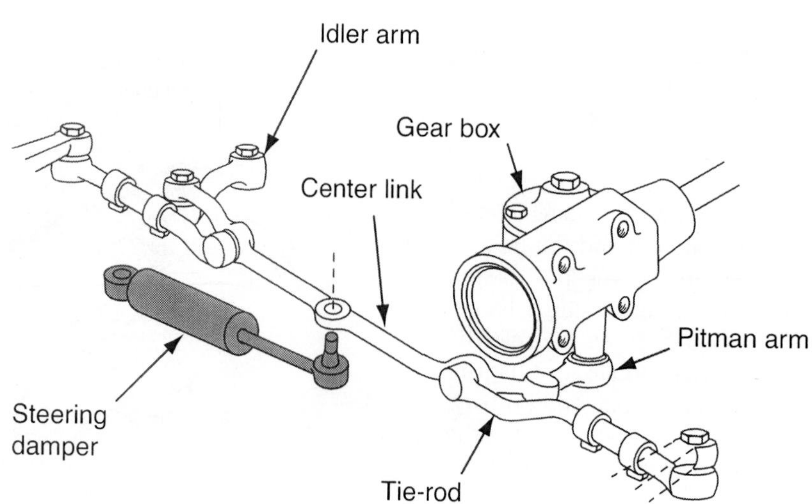


**Figure 21.** Steering damper and linkage.

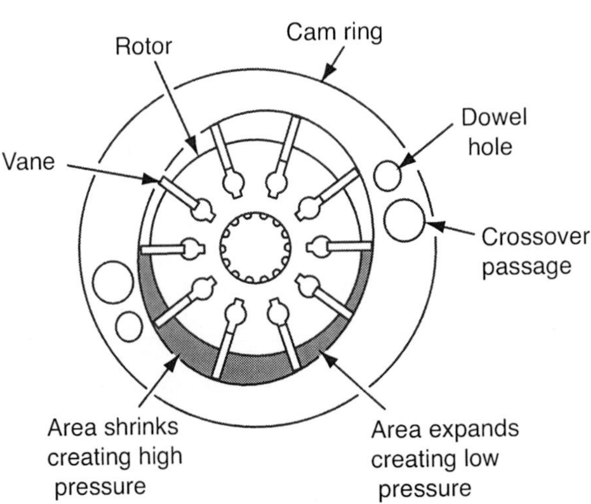


**Figure 22.** Power steering pump rotor and vanes.

pump rotor to the elliptical pump cam ring **(Figure 22)**. Other vane-type power steering pumps have rollers to seal the rotor to the cam ring. In some pumps, inverted, U-shaped slippers are used for this purpose. The major difference in these pumps is in the rotor design and the method of sealing the pump rotor in the elliptical pump ring. The operating principles of all three types of pumps are similar.

A balanced pulley is pressed on the steering pump drive shaft. This pulley and shaft are belt driven by the engine. The oblong pump reservoir is made from steel or plastic. A large O-ring seals the front of the reservoir to the pump housing **(Figure 23)**. Smaller O-rings seal the bolt fittings on the back of the reservoir. The combination cap and dipstick keep the fluid reserve in the pump and vents the reservoir to the atmosphere.

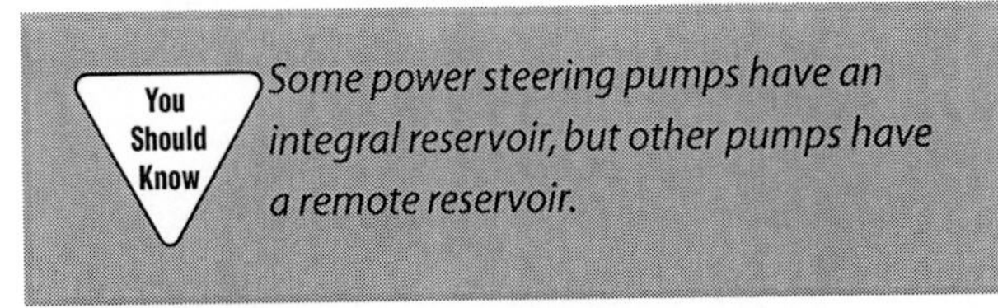


*Some power steering pumps have an integral reservoir, but other pumps have a remote reservoir.*

The rotating components inside the pump housing include the shaft and rotor with the vanes mounted in the rotor slots. As the pulley drives the pump shaft, the vanes rotate inside an elliptical-shaped opening in the cam ring. The cam ring remains in a fixed position inside the pump housing. As the vanes rotate and move toward the narrow-

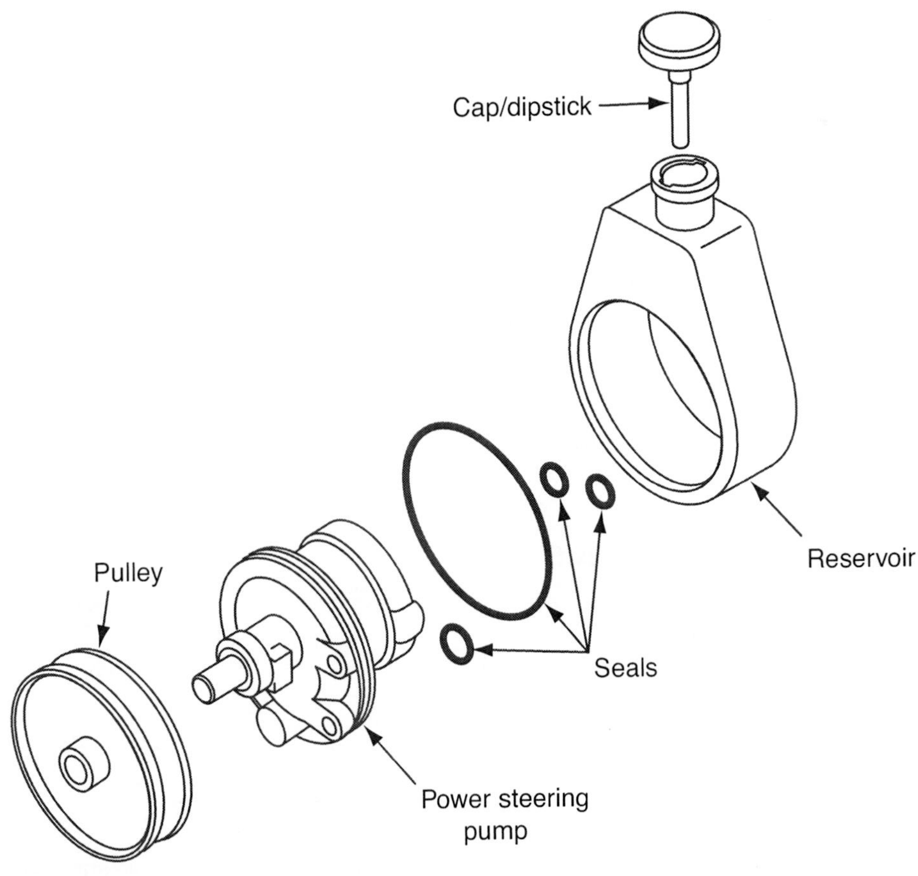


**Figure 23.** Power steering pump housing and reservoir.

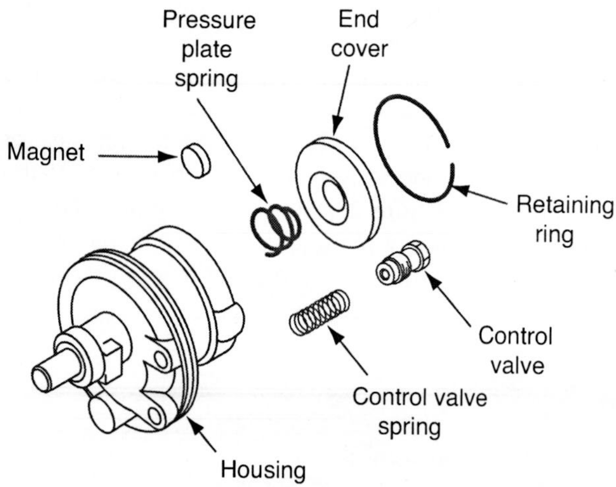


Figure 24. Power steering pump housing assembly with end cover, flow control valve, and magnet.

est part of the cam ring, the space between the vanes becomes smaller. This action pressurizes the fluid between the vanes. When the rotating vanes move into the wider part of the cam ring, the fluid pressure decreases and fluid flows from the pump reservoir into the area between the vanes. A seal between the drive shaft and the housing prevents oil leaks around the shaft.

The flow control valve is a precision-fit valve controlled by spring pressure and fluid pressure. Any dirt or roughness on the valve results in erratic pump pressure. The flow control valve contains a pressure relief ball **(Figure 24)**. High-pressure fluid is forced past the control valve to the outlet fitting. The flow control valve controls pump pressure to provide adequate pressure for steering assist and also protect system hoses from excessive pressure. A high-pressure hose connects the outlet fitting to the inlet fitting on the steering gear, while a low-pressure hose returns the fluid from the steering gear to the inlet fitting in the pump reservoir.

## STEERING COLUMN MAINTENANCE AND DIAGNOSIS

Proper steering column maintenance and diagnosis is very important to locate minor problems in the column before they become a dangerous safety concern. For example, when a rattling noise in the steering column is caused by a worn flexible coupling and this noise is ignored for a period of time, the flexible coupling may break completely and result in a loss of steering control. This loss of steering control may cause a collision resulting in personal injury and vehicle damage.

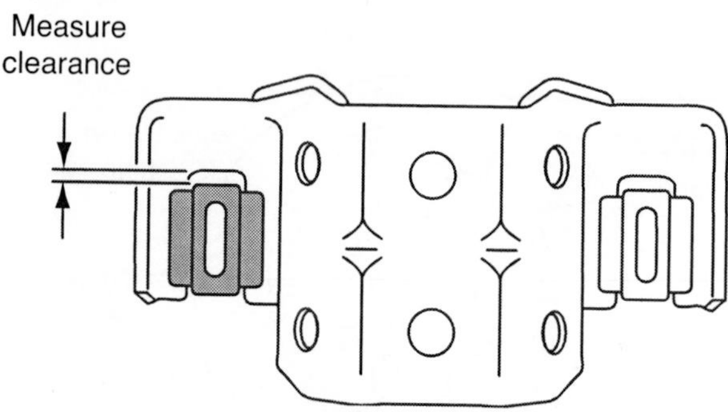


Figure 25. Capsules in steering column bracket.

### Collapsible Steering Column Inspection

Since steering column design varies depending on the vehicle, the collapsible steering column inspection procedure should be followed in the vehicle manufacturer's service manual. Measure the clearance between the capsules and the slots in the steering column bracket **(Figure 25)**. If this measurement is not within specifications, replace the bracket. Check the contact between the bolt head and the bracket **(Figure 26)**. If the bolt head contacts the bracket, the shear load is too high and the bracket must be replaced. Check the steering column jacket for sheared injected plastic in the openings on the side of the jacket **(Figure 27)**. If sheared plastic is present, the column is collapsed. Measure

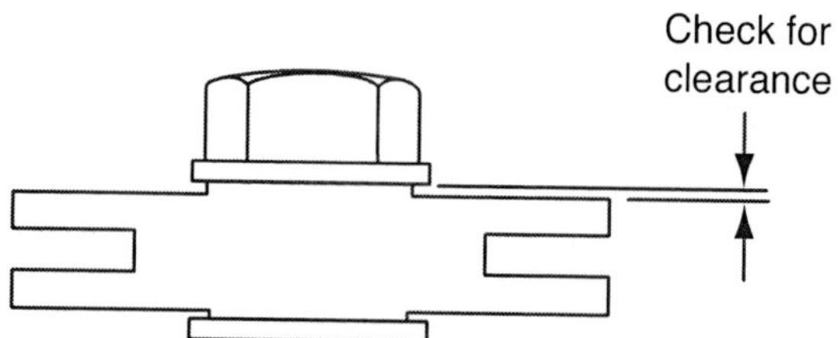


Figure 26. Bolt head to bracket clearance.

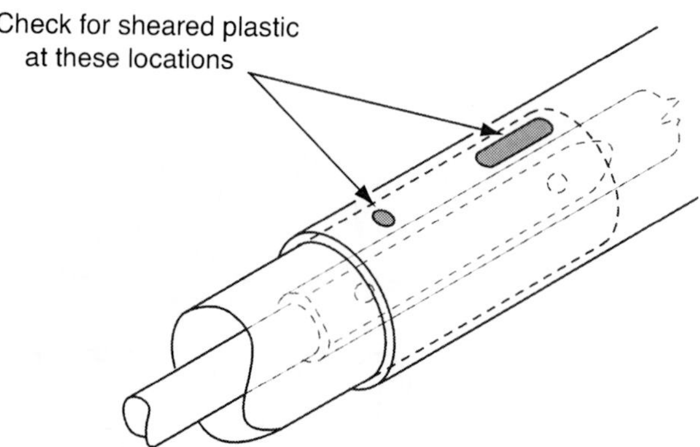


Figure 27. Inspecting for sheared plastic in jacket openings.

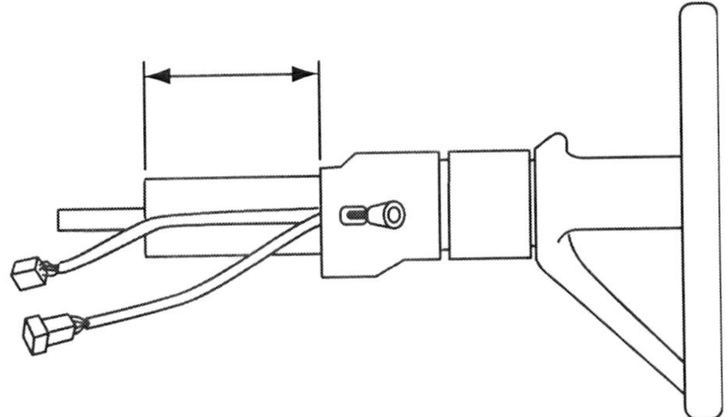

Figure 28. Measuring distance from the end of the bearing assembly to the upper steering column jacket.

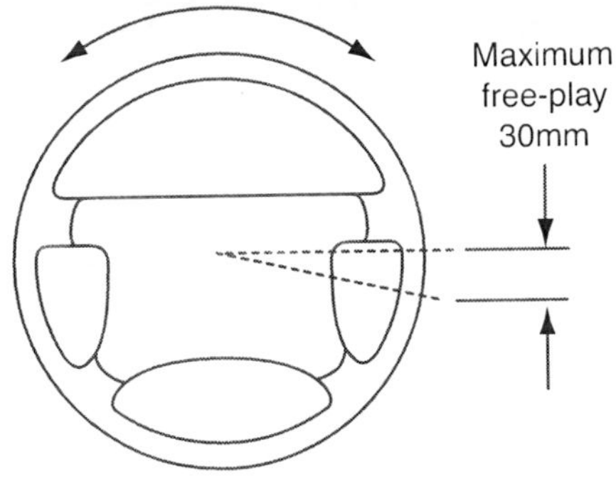


Figure 29. Measuring steering wheel free-play.

the distance from the end of the bearing assembly to the lower edge of the upper steering column jacket **(Figure 28)**. If this distance is not within the vehicle manufacturer's specification, a new jacket must be installed.

## Steering Column Noise Diagnosis

If a binding condition is present while turning the steering wheel, the problem may be in the steering column, the steering linkage, or suspension components. To determine the source of the binding condition, disconnect the flexible coupling or lower universal joint in the steering shaft. If the binding condition is still present, the steering column is the source of the problem. A worn upper universal joint or spherical bearing on a tilt column may cause a binding condition in the steering column. A binding problem may also be caused by interference between the lower steering shaft and the toe plate or silencer. A worn flexible coupling or loose universal joints may cause a rattling noise in the steering column.

## Checking Steering Wheel Free-Play

With the engine stopped and the front wheels in the straight ahead position, move the steering wheel in each direction with light finger pressure. Measure the amount of steering wheel movement before the front wheels begin to turn **(Figure 29)**. This movement is referred to as steering free-play. On some vehicles, this measurement should not exceed 1.18 inches (30 millimeters). Always refer to the vehicle manufacturer's specifications. Excessive steering free-play may be caused by worn steering shaft universal joints or a worn flexible coupling. Other causes of excessive steering wheel free-play include worn steering linkage mechanisms or a worn or out-of-adjustment steering gear.

With the normal vehicle weight resting on the front suspension, observe the flexible coupling or universal joint as an assistant turns the steering wheel 1/2 turn in each direction. If the vehicle has power steering, the engine should be running with the gear selector in park. The flexible coupling or universal joint must be replaced if there is free-play in this component.

## Flexible Coupling Replacement

If the flexible coupling must be replaced, loosen the coupling-to-steering gear stub shaft bolt. Disconnect the steering column from the instrument panel and move the column rearward until the flexible coupling can be removed from the steering column shaft. Remove the coupling-to-steering shaft bolts and disconnect the coupling from the shaft. When the new coupling and the steering column are installed on some vehicles, the clearance between the coupling clamp and the steering gear adjusting plug should be 1/16 inch (1.5 millimeters) **(Figure 30)**. This specification may vary depending on the vehicle. Always use the vehicle manufacturer's specifications in the service manual.

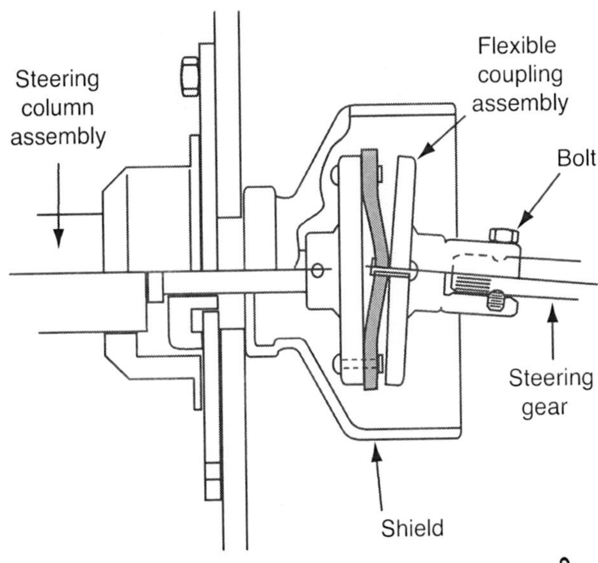


Figure 30. Flexible coupling installation.

## STEERING LINKAGE MAINTENANCE AND DIAGNOSIS

When servicing or replacing steering wheels, columns, or linkages, technicians actually have the customer's life in their hands. If steering components are not serviced properly or are not tightened to the specified torque, the steering may become disconnected, resulting in a complete loss of steering control. This condition may cause a collision resulting in vehicle damage, personal injury, and an expensive lawsuit for the technician and the shop where he or she is employed. Therefore, when performing any automotive service, always be sure the vehicle manufacturer's recommended service procedures and torque specifications are followed.

### Diagnosis of Center Link, Pitman Arm, and Tie-Rod Ends

The vehicle should be raised and safety stands must be positioned under the lower control arms to support the vehicle weight. Use vertical hand force to check for looseness in all the pivots on the tie-rod ends and the center link. Check the seals on each tie-rod end and pivot on the center link or pitman arm for damage and cracks. Cracked seals allow dirt to enter the pivoted joints which results in rapid wear. If looseness or damaged seals are found on any pivoted joint on the tie-rods and center link, these components must be replaced. Inspect rubber-encapsulated tie-rod ends for looseness of the ball stud in the rubber capsule and looseness of the stud and rubber capsule in the outer housing. If either of these conditions are present, replace the tie-rod end.

You Should Know *Never attempt to straighten steering linkage components. This action may weaken the metal and cause sudden component failure, vehicle damage, or personal injury.*

The second part of this diagnosis is done with the front wheels resting on the shop floor. If the vehicle is equipped with power steering, start the engine and allow the engine to idle with the transmission in park and the parking brake applied. While someone turns the steering wheel one-quarter turn in each direction from the straight ahead position, observe all the pivoted joints on the tie-rod ends and center link. This test allows the technician to check the steering linkage pivots under load. If any of the pivoted joints show a slight amount of play, they must be replaced.

Worn tie-rod ends result in excessive steering wheel free-play, incorrect front wheel toe setting, tire squeal on turns, tread wear on front tires, front wheel shimmy, rattling noise on road irregularities. **Front wheel toe** is the distance between the front edges of the front tires compared to the distance between the rear edges of these tires. **Front wheel shimmy** is a rapid oscillation of the front wheels to the right and the left.

### Idler Arm Diagnosis

Grasp the center link firmly near the idler arm and apply a 25-pound vertical load to the idler arm. If the total idler arm total vertical movement measured at the outer end of the arm exceeds 1/4 inch, idler arm replacement is necessary. If idler arm vertical movement is excessive, the tie-rod is not parallel to the lower control arm. Excessive idler arm vertical movement causes these steering problems: excessive toe change and front tire tread wear, excessive steering wheel free-play and reduced steering control, and front end shimmy.

You Should Know *A binding idler arm may suddenly break off and cause complete loss of steering control, vehicle damage, or personal injury.*

Binding idler arm bushings result in these complaints: hard steering, a squawking noise when the front wheels are turned, and poor steering wheel returnability.

## STEERING LINKAGE SERVICE

Proper steering linkage service is extremely important to maintain vehicle safety and provide normal tire life. For example, if the retaining nut on a tie-rod end is not tightened to the specified torque and retained with a cotter pin, the tie-rod may become disconnected, resulting in a loss of steering control. This action may result in a collision causing vehicle damage and personal injury.

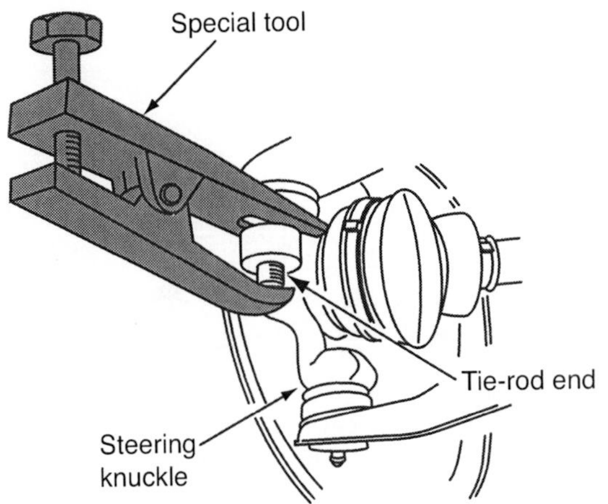


Figure 31. Tie-rod end removal.

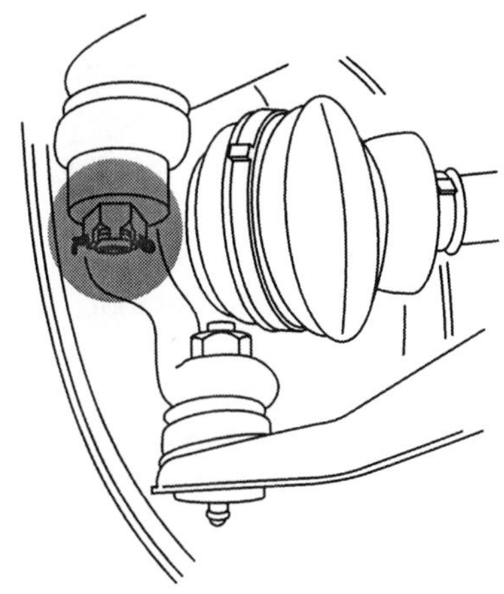

Figure 32. Tie-rod nut and cotter pin installation.

## Tie-Rod End Replacement

The cotter pin and nut must be removed prior to tie-rod end replacement. A puller is used to remove the tie-rod end from the steering arm **(Figure 31)**.

The tie-rod clamp must be loosened before the tie-rod is removed from the sleeve. Count the number of turns required to remove the tie-rod end from the sleeve and install the new tie-rod with the same number of turns. Even when this procedure is followed, the toe must be checked after the steering linkage components are replaced. Before the new tie-rod end is installed, center the stud in the tie-rod end. When the tie-rod stud is installed in the steering arm opening, only the threads should be visible above the steering arm surface. If the machined surface of the tie-rod stud is visible above the steering arm surface or the stud fits loosely in the steering arm opening, this opening is worn or the tie-rod end is not correct for that application. The tie-rod nut must be torqued to manufacturer's specifications and the cotter pin must be installed through the tie-rod end and nut openings **(Figure 32)**. When rubber-encapsulated tie-rod ends are installed and tightened, the front wheels must be in the straight-ahead position.

The tie-rod nut must never be loosened from the specified torque to install the cotter pin. Another method of positioning replacement tie-rod ends is to measure the distance from the center of the tie-rod stud to the end of the sleeve prior to removal. When the new tie-rod is installed, be sure this measurement is the same. The slots in the tie-rod sleeve must be positioned away from the opening in the sleeve clamps **(Figure 33)**. Leave the sleeve clamps loose until the front wheel toe is checked and then tighten

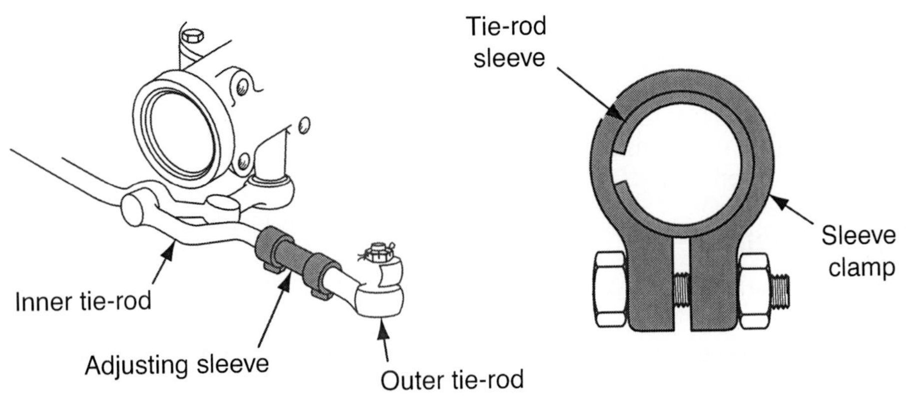


Figure 33. Proper slot and clamp position on tie-rod sleeves.

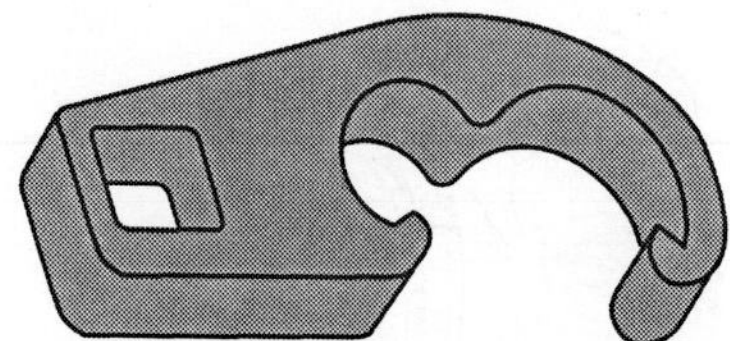

**Figure 34.** Tie-rod sleeve adjusting tool.

the sleeve clamp bolts to the specified torque. A special tool is available to rotate the tie-rod sleeves and set the front wheel toe **(Figure 34)**.

## Pitman Arm Diagnosis and Replacement

Some pitman arms contain a ball socket joint on the outer end. The threaded extension on this ball socket fits into the center link. On other steering linkages, the ball socket joint is in the center link and the threaded extension fits into the pitman arm opening. If the pitman arm is bent, it must be replaced. If the pitman arm is bent, the tie-rod is not parallel to the lower control arm. Under this condition, excessive front wheel toe change occurs on road irregularities and front tire wear may be excessive. The following is a typical pitman arm replacement procedure:

1. Position the front wheels straight ahead and remove the cotter pin and nut from the ball socket joint on the outer end of the pitman arm.
2. Remove the ball socket extension from the pitman arm or center link with a tie-rod end puller.
3. Loosen the pitman arm to pitman shaft nut.
4. Use a puller to pull the pitman arm loose on the shaft.
5. Remove the nut, lock washer, and pitman arm.
6. Check the pitman shaft splines. If the splines are damaged or twisted, the shaft must be replaced.
7. Reverse steps 1 through 5 to install the pitman arm. The pitman arm to shaft nut and the ball socket extension nut must be tightened to the manufacturer's specified torque. Be sure the pitman arm is installed in the correct position on the shaft splines. Install the cotter pin in the ball socket extension.

## Center Link Replacement

Follow these steps for a typical center link replacement:

1. Remove the cotter pins from the tie-rod to center link nuts and the idler arm and pitman arm to center link nuts.
2. Remove the nuts on the tie-rod inner ends, idler arm to center link ball socket extension, and the pitman arm to center link ball socket extension.
3. Use a tie-rod end puller to pull the inner tie-rods from the center link. Follow the same procedure to remove the center link to pitman arm ball socket extension.

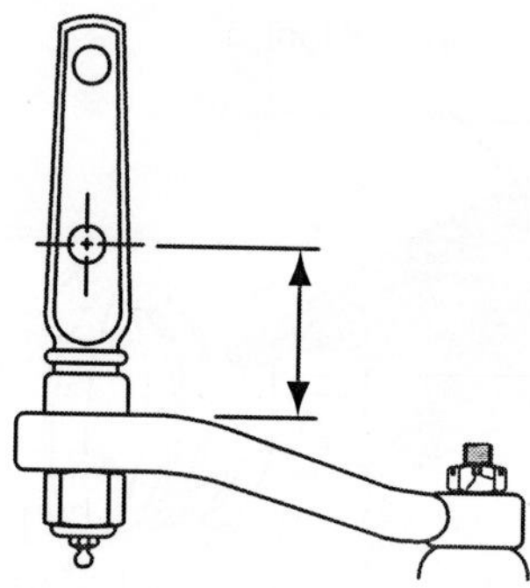

**Figure 35.** Specified clearance between center of lower bracket bolt hole and upper idler arm surface.

4. Remove the idler arm from the center link and remove the center link.
5. Reverse steps 1 through 4 to install the center link. Tighten all the ball socket nuts to the manufacturer's specified torque and install cotter pins in all the nuts. If the ball sockets have grease fittings, lubricate the ball sockets with a grease gun and chassis lubricant.

The following is a typical idler arm removal and replacement procedure:

1. Remove the idler arm to center link cotter pin and nut.
2. Remove the center link from the idler arm.
3. Remove the idler arm bracket mounting bolts and remove the idler arm.
4. If the idler arm has a steel bushing, thread the bracket into the idler arm bushing until the specified clearance is obtained between the center of the lower bracket bolt hole and the upper idler arm surface **(Figure 35)**.
5. Install the idler arm bracket to frame bolts and tighten the bolts to the specified torque. Be sure that lock washers are installed on the bolts.
6. Install the center link into the idler arm and tighten the mounting nut to the specified torque. Install the cotter pin in the nut.
7. If the idler arm steel bushing or bushings contain a grease fitting, lubricate as required.

The idler arm adjustment is very important. If this adjustment is incorrect, front wheel toe is affected. After idler arm replacement, the front wheel toe should be checked.

## Steering Damper Diagnosis and Replacement

Some steering systems have a damper connected between the center link and the chassis. A damper is similar to a small shock absorber. The purpose of the damper is to prevent the transmission of steering shock and vibrations to the steering wheel. A worn-out steering damper may cause excessive steering shock and vibration on the steering wheel, especially on irregular road surfaces. A rattling noise occurs if the damper mounting bolts or brackets are loose.

## Steering Arm Diagnosis

If the front rims have been damaged, the steering arms should be checked for a bent condition. Measure the distance from the center of the tie-rod stud to the edge of the rim on each side. Unequal readings may indicate a bent steering arm. Bent steering arms must be replaced.

# POWER STEERING PUMP MAINTENANCE AND DIAGNOSIS

Proper power steering pump maintenance and diagnosis is very important to maintain vehicle safety. A sudden loss of power steering pump pressure results in a large increase in steering effort. When turning a corner, a sudden loss of power steering pump pressure may cause the driver to lose steering control, resulting in a collision.

When performing power steering pump maintenance and diagnosis, one of the first steps is to inspect the pump drive belt and measure the belt tension. These procedures were explained previously in Chapter 37. A loose power steering belt causes low pump pressure, hard steering, and belt chirping or squealing, especially on acceleration.

## Fluid Level Checking

Most vehicle manufacturers recommend power steering fluid or automatic transmission fluid in power steering systems. Always use the type of fluid recommended in the vehicle manufacturer's service manual. If the power steering fluid level is low, steering effort is increased and may be erratic. A low fluid level may cause a growling noise in the power steering pump. Some vehicle manufacturers now recommend checking the power steering pump fluid level with the fluid at an ambient temperature of 176°F (80°C). Follow these steps to check the power steering fluid level:

1. With the engine idling at 1,000 rpm or less, turn the steering wheel slowly and completely in each direction several times to boost the fluid temperature.
2. If the vehicle has a remote power steering fluid reservoir, check for foaming in the reservoir which indicates a low fluid level or air in the system.
3. Observe the fluid level in the remote reservoir. This level should be at the hot full mark. Shut the engine off and remove dirt from the neck of the reservoir with a shop towel. If the power steering pump has an integral reservoir, the level should at the hot level on the dipstick. When an external reservoir is used, the dipstick is located in the external reservoir **(Figure 36)**.
4. Pour the required amount of the vehicle manufacturer's recommended power steering fluid into the reservoir to bring the fluid level to the hot full mark on the reservoir or dipstick with the engine idling.

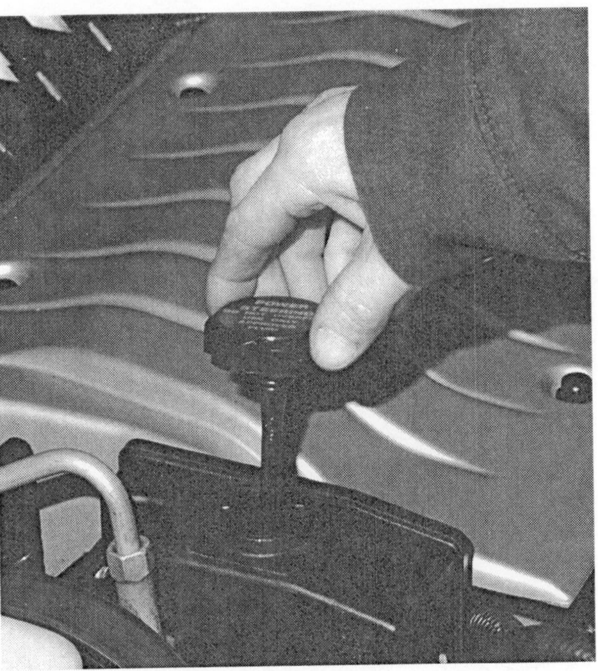

**Figure 36.** Power steering reservoir and dipstick.

## Power Steering Pump Oil Leak Diagnosis

The possible sources of power steering pump oil leaks are the drive shaft seal, reservoir O-ring seal, high-pressure outlet fitting, and the dipstick cap. If leaks occur at any of the seal locations, seal replacement is necessary. When a leak is present at the high-pressure outlet fitting, tighten this fitting to the specified torque **(Figure 37)**. If the leak

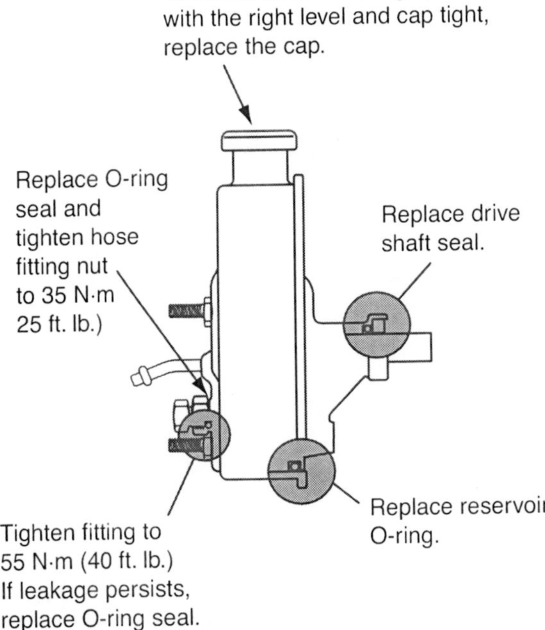
Check oil level. If leakage persists with the right level and cap tight, replace the cap.
Replace O-ring seal and tighten hose fitting nut to 35 N·m 25 ft. lb.)
Replace drive shaft seal.
Tighten fitting to 55 N·m (40 ft. lb.) If leakage persists, replace O-ring seal.
Replace reservoir O-ring.

**Figure 37.** Power steering pump oil leak diagnosis.

still occurs, replace the O-ring seal on the fitting and retighten the fitting.

## Power Steering System Draining and Flushing

If the power steering fluid is contaminated with moisture, dirt, or metal particles, the system must be drained and new fluid installed. Follow these steps to drain and flush the power steering system:

1. Lift the front of the vehicle with a floor jack and install jack stands under the suspension. Lower the vehicle onto the jack stands and remove the floor jack.
2. Remove the return hose from the remote reservoir that is connected to the steering gear. Place a plug on the reservoir outlet and position the return hose in an empty drain pan **(Figure 38)**.
3. With the engine idling, turn the steering wheel fully in each direction and stop the engine.
4. Fill the reservoir to the hot full mark with the manufacturer's recommended fluid.
5. Start the engine and run the engine at 1,000 rpm while observing the return hose in the drain pan. When fluid begins to discharge from the return hose, shut the engine off.
6. Repeat steps 4 and 5 until there is no air in the fluid discharging from the return hose.
7. Remove the plug from the reservoir and reconnect the return hose. Bleed the power steering system.

## Bleeding Air from the Power Steering System

When air is present in the power steering fluid, a growling noise may be heard in the pump and steering effort may be increased or erratic. When a power steering system has been drained and refilled, follow this procedure to remove air from the system:

1. Fill the power steering pump reservoir as outlined previously.
2. With the engine running at 1,000 rpm, turn the steering wheel fully in each direction three or four times. Each time the steering wheel is turned fully to the right or left, hold it there for 2 to 3 seconds before turning it the in the other direction.
3. Check for foaming of the fluid in the reservoir. When foaming is present, repeat steps 1 and 2.
4. Check the fluid level with the engine running and be sure it is at the hot full mark. Shut the engine off and make sure the fluid level does not increase more than 0.020 inch (0.5 millimeter).

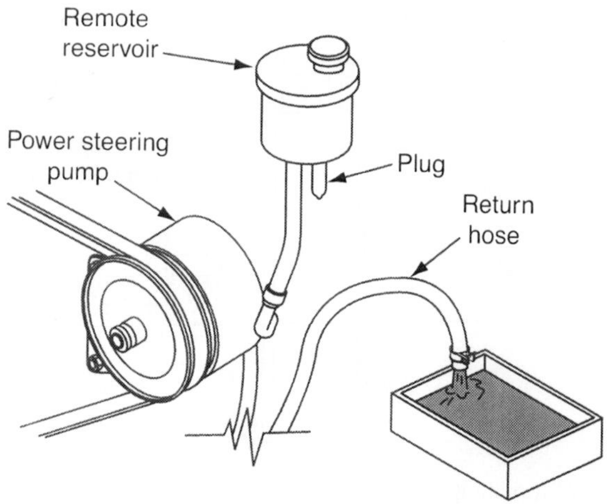


**Figure 38.** Return hose installed in drain pan for power steering draining and flushing.

## Power Steering Pump Pressure Test

Since there are some variations in power steering pump pressure test procedures and pressure specifications, the vehicle manufacturer's test procedures and specifications must be used. If the power steering pump pressure is low, steering effort is increased. Erratic power steering pump pressure causes variations in steering effort and the steering wheel may jerk as it is turned. Since a power steering pump will never develop the specified pressure if the belt is slipping, the belt tension must be checked and adjusted if necessary prior to a pump pressure test. The following is a typical power steering pressure test procedure:

1. With the engine stopped, disconnect the pressure line from the power steering pump and connect the gauge side of the pressure gauge to the pump outlet fitting. Connect the valve side of the gauge to the pressure line.
2. Start the engine and turn the steering wheel fully in each direction two or three times to bleed air from the system. Be sure the fluid level is correct and the fluid temperature is at least 176°F (80°C). A thermometer may be inserted in the pump reservoir fluid to measure the fluid temperature.

**You Should Know** *During the power steering pump pressure test, if the pressure gauge valve is closed for more than 10 seconds, excessive pump pressure may cause power steering hoses to rupture resulting in personal injury. Do not allow the fluid to become too hot during the power steering pump pressure test. Excessively high fluid temperature reduces pump pressure. Wear protective gloves and always shut the engine off before disconnecting gauge fittings because the hot fluid may cause burns.*

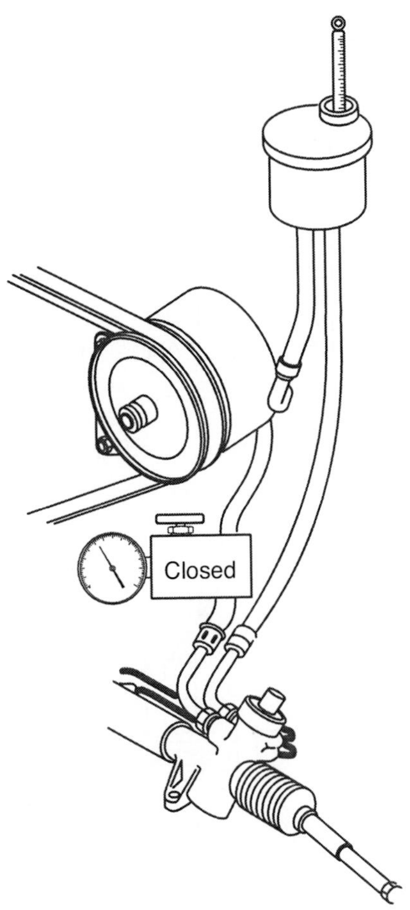


**Figure 39.** Power steering pump pressure test with gauge valve closed.

3. With the engine idling, close the pressure gauge valve for no more than 10 seconds and observe the pressure gauge reading **(Figure 39)**. Turn the pressure gauge valve to the fully open position. If the pressure gauge reading did not equal the vehicle manufacturer's specifications, repair or replace the power steering pump.
4. Check the power steering pump pressure with the engine running at 1,000 rpm and 3,000 rpm and record the pressure difference between the two readings. If the pressure difference between the pressure readings at 1,000 rpm and 3,000 rpm does not equal the vehicle manufacturer's specifications, repair or replace the flow control valve in the power steering pump.
5. With the engine running, turn the steering wheel fully in one direction and observe the steering pump pressure while holding the steering wheel in this position. If the pump pressure is less than the vehicle manufacturer's specifications, the steering gear housing has an internal leak and should be repaired or replaced.

**Figure 40.** Steering effort measurement.

6. Be sure the front tire pressures are correct and center the steering wheel with the engine idling. Connect a spring scale to the steering wheel and measure the steering effort in both directions **(Figure 40)**. If the power steering pump pressure is satisfactory and the steering effort is more than the vehicle manufacturer's specifications, the power steering gear should be repaired.

## POWER STEERING PUMP SERVICE

If a growling noise is present in the power steering pump after the fluid level is checked and air has been bled from the system, the pump bearings or other components are defective and pump replacement is required. When the power steering pump pressure is lower than specified and the flow control valve is operating normally, pump replacement is necessary.

When the power steering pump is replaced, proceed as follows:

1. Disconnect the power steering return hose from the remote reservoir or pump and allow the fluid to drain from this hose into a drain pan. Discard the used fluid.
2. Loosen the bracket or belt tension adjusting bolt and the pump mounting bolt.
3. Loosen the belt tension until the belt can be removed. On some cars it is necessary to lift the vehicle on a hoist to gain access to the power steering pump from underneath the vehicle.
4. Remove the hoses from the pump and cap the pump fittings and hoses.
5. Remove the belt tension adjusting bolt and the mounting bolt and remove the pump.

6. Check the pump mounting bolts and bolt holes for wear. Worn bolts must be replaced. If the bolt mounting holes in the pump are worn, pump replacement is necessary.
7. Reverse steps 1 through 5 to install the power steering pump. Tighten the belt as described previously and tighten the pump mounting and bracket bolts to the manufacturer's specifications. If O-rings are used on the pressure hose, replace the O-ring. Be sure the hoses are not contacting the exhaust manifold, catalytic converter, or exhaust pipe during or after pump replacement.
8. Fill the pump reservoir with the manufacturer's recommended power steering fluid and bleed air from the power steering system as previously described.

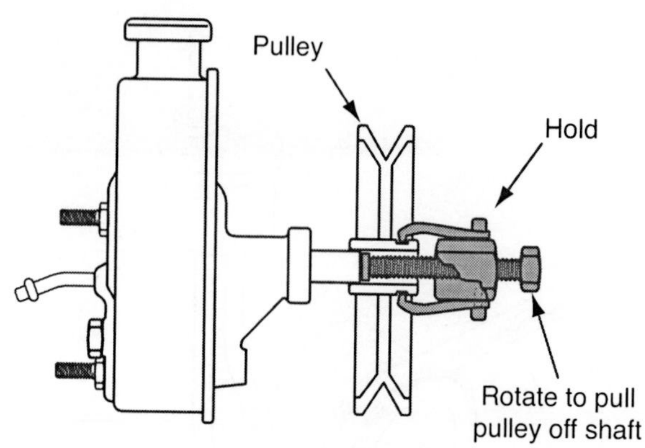


**Figure 41.** Press-on power steering pump pulley removal.

## Power Steering Pump Pulley Replacement

If the pulley wobbles while the pulley is rotating, the pulley is probably bent and pulley replacement is necessary. Worn pulley grooves also require pulley replacement. Always check the pulley for cracks. If this condition is present, pulley replacement is essential. A pulley that is loose on the pump shaft must be replaced. Never hammer on the pump drive shaft during pulley removal or replacement. This action will damage internal pump components. If the pulley is pressed onto the pump shaft, a special puller is required to remove the pulley **(Figure 41)** and a pulley installation tool is used to install the pulley **(Figure 42)**.

If the power steering pump pulley is retained with a nut, mount the pump in a vise. Always tighten the vise on one of the pump mounting bolt surfaces and do not tighten the vise with excessive force. Use a special holding tool to keep the pulley from turning and loosen the pulley nut with a box-end wrench. Remove the nut, pulley, and woodruff key. Inspect the pulley, shaft, and woodruff key for wear. Be sure the key slots in the shaft and pulley are not worn. Replace all worn components.

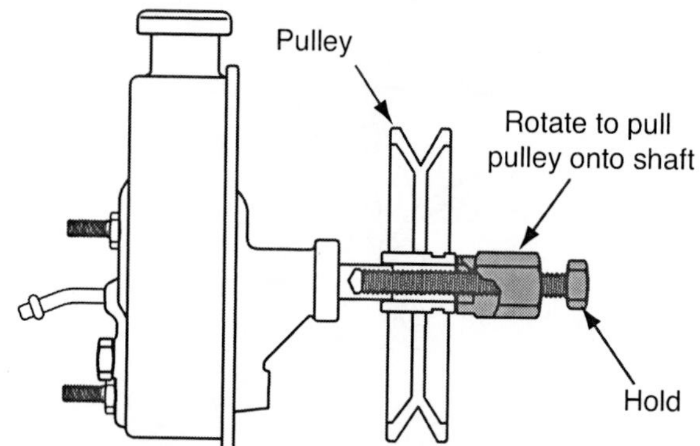


**Figure 42.** Press-on power steering pump pulley installation.

## Checking Power Steering Lines and Hoses

Power steering lines should be checked for leaks, dents, sharp bends, cracks, or contact with other components. Lines and hoses must not rub against other components. This action could wear a hole in the line or hose. Many high-pressure power steering lines are made from high-pressure steel-braided hose with molded steel fittings on each end **(Figure 43)**.

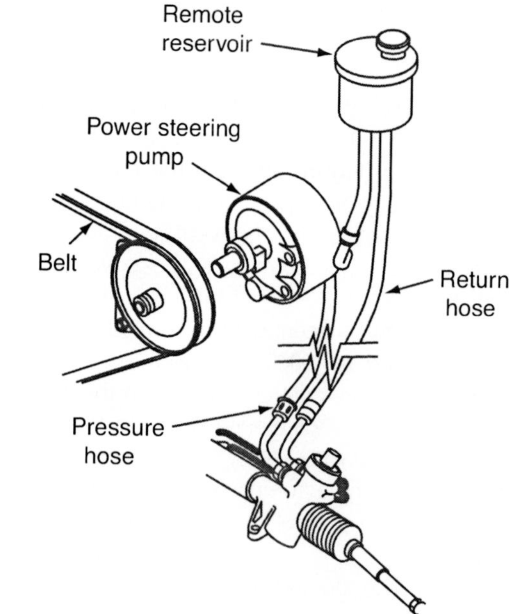


**Figure 43.** Power steering hoses and lines.

## Summary


- Steering columns help to provide steering control, driver convenience, and driver safety.
- Many steering columns provide some method of energy absorption to protect the driver protection during a frontal collision.
- Tilt steering columns provide increased driver convenience while driving and getting in or out of the driver's seat.
- The ignitions switch, dimmer switch, signal light switch, hazard switch, and the wipe/wash switch may be mounted in the steering column.
- When the ignition switch is in the lock position, a locking plate and lever in the upper steering column locks the steering wheel and the gear shift.
- In a parallelogram steering linkage, the tie-rods are parallel to the lower control arms.
- The parallelogram steering linkage minimizes toe change as the control arms move up and down on road irregularities.
- A rack-and-pinion steering linkage has reduced friction points and it is light weight and compact compared to a parallelogram steering linkage.
- A power steering pump may have a vane, roller, or slipper-type rotor assembly, but all three types of pumps operate on the same basic principle.
- The flow control valve in a power steering pump controls pump pressure.
- A loose power steering belt causes low pump pressure, hard steering, and belt chirping or squealing, especially on acceleration.
- Most vehicle manufacturers recommend power steering fluid or automatic transmission fluid in the power steering system.
- Air may be bled from the power steering system by turning the steering wheel fully in each direction several times with the engine running. Each time the wheel is turned fully to the right or left, it should be held in that position for 2 to 3 seconds.
- A pressure gauge and manual valve are connected in series in the power steering pump pressure hose to check pump pressure. Maximum pump pressure is checked by closing the manual valve for less than 10 seconds.
- After the power steering pump pressure is tested and proven to be satisfactory, steering gear leakage may be tested by turning the steering wheel fully in one direction and observing the pressure reading. If the power steering pump pressure is less than specified, the steering gear has an internal leak.
- With the engine idling, a spring scale may be attached at the outer end of the steering wheel crossbar to measure steering effort.


## Review Questions

1. While discussing energy absorbing or collapsible steering columns, Technician A says on some energy absorbing steering columns, the column-to-instrument panel mount is designed to allow column movement if the driver is thrown against the steering wheel in a frontal collision. Technician B says in some collapsible steering columns, steel pins shear off in the two-piece jacket and steering shaft to allow column collapse if the driver is thrown against the steering wheel in a frontal collision. Who is correct?
   A. Technician A
   B. Technician B
   C. Both Technician A and Technician B
   D. Neither Technician A nor Technician B
2. While discussing parallelogram steering linkages, Technician A says the tie-rods are parallel to the upper control arms. Technician B says the idler arm helps to maintain the proper center link and tie-rod height. Who is correct?
   A. Technician A
   B. Technician B
   C. Both Technician A and Technician B
   D. Neither Technician A nor Technician B
3. While discussing parallelogram steering linkages, Technician A says loose steering linkage components may cause excessive tire wear. Technician B says the tie-rod sleeves provide a method of toe adjustment. Who is correct?
   A. Technician A
   B. Technician B
   C. Both Technician A and Technician B
   D. Neither Technician A nor Technician B
4. While discussing types of steering linkages, Technician A says compared to a parallelogram steering linkage, a rack-and-pinion linkage has more friction points. Technician B says in a rack-and-pinion steering linkage, the tie-rods are parallel to the lower control arms. Who is correct?
   A. Technician A
   B. Technician B
   C. Both Technician A and Technician B
   D. Neither Technician A nor Technician B

5. All these statements about parallelogram steering linkages are true *except*:
   A. Parallelogram steering linkages may be mounted in front of or behind the front suspension.
   B. Parallelogram steering linkages have the tie-rods mounted parallel to the lower control arms.
   C. Parallelogram steering linkages have a pitman arm that connects the center link to the tie-rods.
   D. Parallelogram steering linkages have an idler arm that supports one end of the center link.
6. On a vehicle with a parallelogram steering linkage, when one front wheel strikes a bump on the road surface, the front wheel and suspension move upward. During this upward wheel action:
   A. the lower control arm moves through an arc.
   B. the tie-rod moves through an arc that is much different compared to the control arm arc.
   C. there is a considerable amount of front wheel toe change.
   D. during upward and downward wheel movement, toe change causes tire tread wear.
7. In a rack-and-pinion steering gear and linkage:
   A. The tie-rods are connected to the rack through inner tie-rod ends.
   B. The tie-rods are not parallel to the lower control arms.
   C. The pinion gear is connected to the sector shaft.
   D. The mounting position of the steering gear does not affect vehicle steering.
8. All these statements about power steering pump pressure testing are true *except*:
   A. The pressure gauge and valve should be connected in the power steering pump return hose to check pump pressure.
   B. The pressure gauge valve should be closed for 10 seconds during the power steering pump pressure test.
   C. The pressure test should be performed when the power steering fluid is cold.
   D. The engine should be running at 2,500 rpm during the pressure test.
9. While discussing power steering pump pressure, Technician A says if the maximum pump pressure is satisfactory with the pressure gauge valve closed but the pressure is lower than specified with the steering wheel turned fully in one direction, the steering gear may have an internal leak. Technician B says if the pump pressure is satisfactory and the front tires are properly inflated, if the steering effort is higher than specified, the steering gear may require replacing or repairing. Who is correct?
   A. Technician A
   B. Technician B
   C. Both Technician A and Technician B
   D. Neither Technician A nor Technician B
10. A loose power steering belt may cause all of these problems *except*:
    A. squealing during engine acceleration.
    B. low power steering pump pressure.
    C. damage to the power steering pump bearings.
    D. excessive steering effort.
11. In some collapsible steering columns ____________ ____________ in the outer column jacket and steering shaft shear off if the driver is thrown against the steering wheel in a frontal collision.
12. In a parallelogram steering linkage, the center link connects the pitman arm to the ____________ ____________.
13. A worn flexible coupling may cause a ____________ noise in the steering column.
14. Power steering fluid level should be checked with the fluid ____________.
15. Describe the proper clamp position on a tie-rod sleeve.
16. Describe the proper front wheel position when a rubber-encapsulated tie-rod end is installed and tightened.
17. Explain the results of a worn idler arm.
18. Explain the advantages of a rack-and-pinion steering linkage compared to a parallelogram steering linkage.

# Manual and Power Steering Gears Maintenance, Diagnosis, and Service

## Introduction

During the late 1970s and 1980s, the domestic automotive industry converted much of their production from larger RWD cars to smaller, lightweight, and more fuel efficient FWD cars. These FWD cars required smaller, lightweight components wherever possible. Manual and power rack-and-pinion steering gears are lighter and more compact compared to the recirculating ball steering gears and parallelogram steering linkages used on most RWD cars. Therefore, rack-and-pinion steering gears are ideally suited to these compact FWD cars.

Steering systems have not escaped the electronics revolution. Many cars are presently equipped with a variable effort steering (VES) that provides greater power assist during low-speed cornering and parking for increased driver convenience. Some vehicles are equipped with electronic power steering.

Steering gear maintenance, diagnosis, and service is extremely important to maintain vehicle safety. If a vehicle has excessive steering wheel free-play, this problem should be diagnosed and corrected as soon as possible. For example, if this problem is caused by worn tie-rod ends or loose steering gear mounting bolts, these conditions create a safety hazard. If a worn tie-rod end becomes disconnected, the result is a complete loss of steering control. This condition may cause a collision resulting in vehicle damage and/or personal injury.

### MANUAL RECIRCULATING BALL STEERING GEARS

In a recirculating ball steering gear, the steering wheel and steering shaft are connected to the worm shaft. Ball bearings support both ends of the worm shaft in the steering gear housing. A seal above the upper worm shaft bearing prevents oil leaks and an adjusting plug is provided on the lower worm shaft bearing to adjust worm shaft bearing preload. Proper preloading of the worm shaft bearing is necessary to eliminate worm shaft end-play and prevent steering gear free-play and vehicle wander. A ball nut is mounted over the worm shaft and internal threads or grooves on the ball nut match the grooves on the worm shaft. Ball bearings run in the ball nut and worm shaft grooves (**Figure 1**).

When the worm shaft is rotated by the steering wheel, the ball nut is moved up or down on the worm shaft. The gear teeth on the ball nut are meshed with matching gear teeth on the pitman shaft sector. Therefore, ball nut movement causes pitman shaft sector rotation. Since the pitman shaft sector is connected through the pitman arm and steering linkage to the front wheels, the front wheels are turned by the pitman shaft sector. The lower end of the pitman shaft sector is usually supported by a bushing or a needle bearing in the steering gear housing. A bushing in the sector cover supports the upper end of this shaft.

When the front wheels are straight ahead, an interference fit exists between the sector shaft and ball nut teeth.

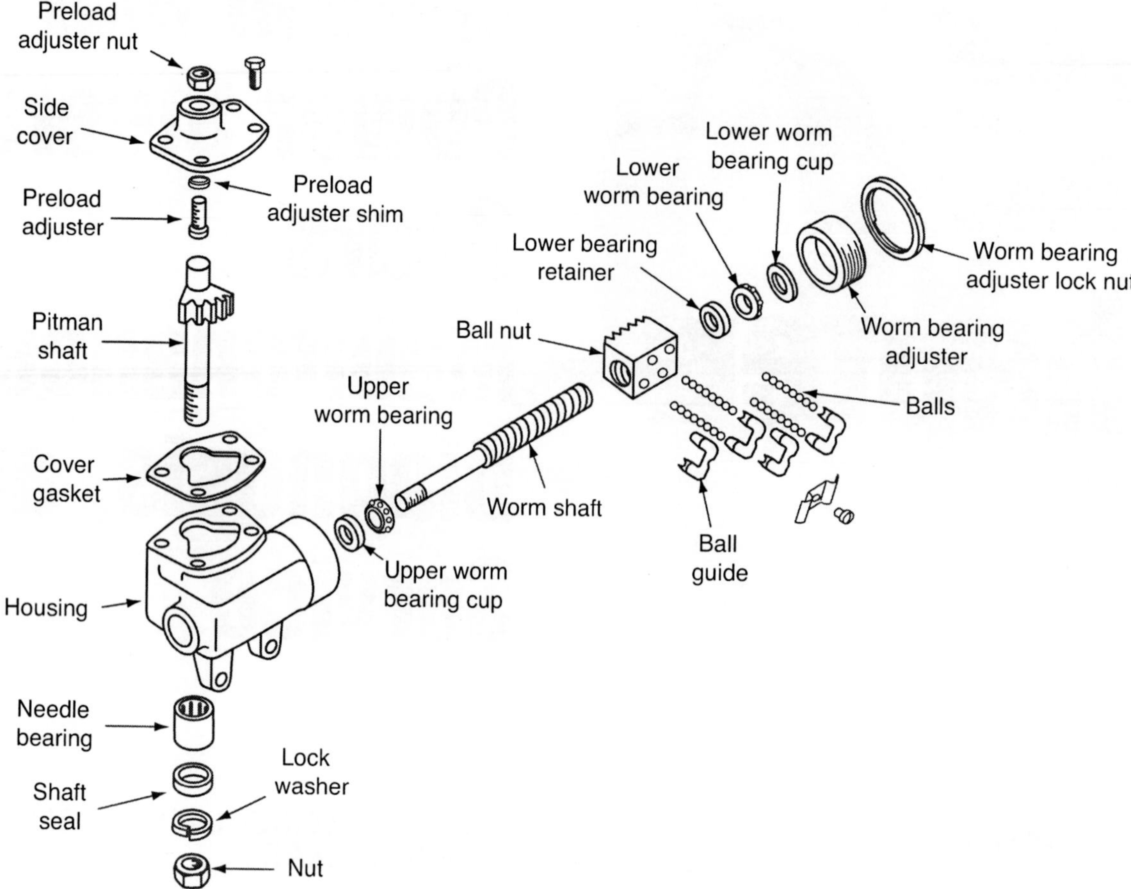


**Figure 1.** Manual recirculating ball steering gear.

This interference fit eliminates gear tooth lash when the front wheels are straight ahead and provides the driver with a positive feel of the road. Proper axial adjustment of the sector shaft is necessary to obtain the necessary interference fit between the sector shaft and worm shaft teeth. A sector shaft adjuster screw is threaded into the sector shaft cover to provide axial sector shaft adjustment **(Figure 2)**.

Manual recirculating ball steering gears have sector gear teeth designed to provide a constant ratio, whereas power recirculating ball steering gears usually have sector gear teeth with a variable ratio **(Figure 3)**. The sector gear teeth have equal lengths in a constant ratio steering gear, but the center sector gear tooth is longer compared to the other teeth in a variable ratio gear. The variable ratio steering gear varies the amount of mechanical advantage provided by the steering gear in relation to steering wheel position. This variable ratio provides "faster" steering. The steering gear ratio in a constant ratio manual steering gear is usually 15 or 16:1, whereas the average variable ratio steering gear ratio may be 13:1. When the same types of steering gears are compared, a higher numerical ratio provides reduced steering effort and increased steering wheel movement in relation to the amount of front wheel movement.

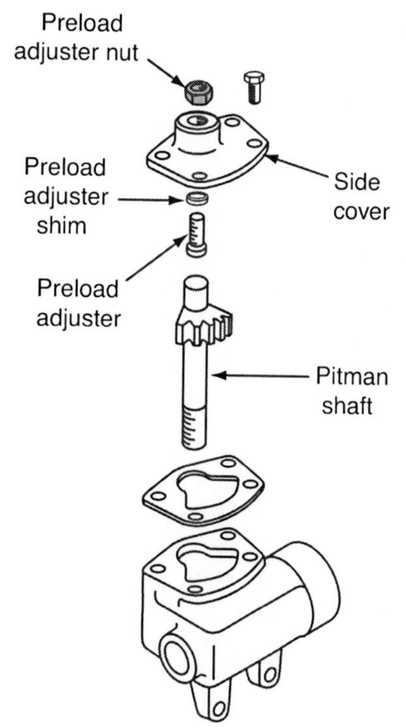


**Figure 2.** A sector shaft adjusting nut.

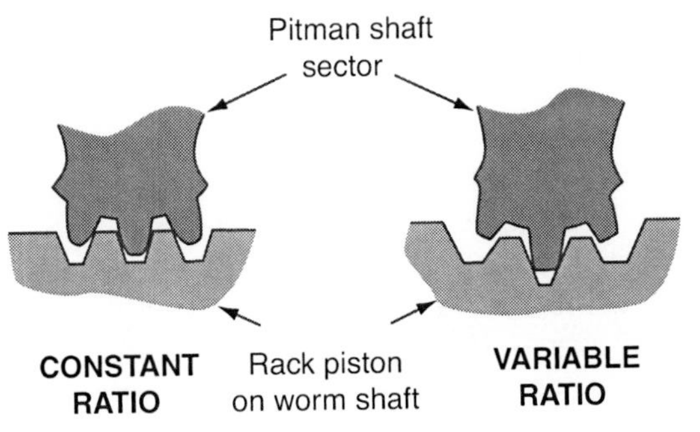


**Figure 3.** Constant and variable ratio steering gears.

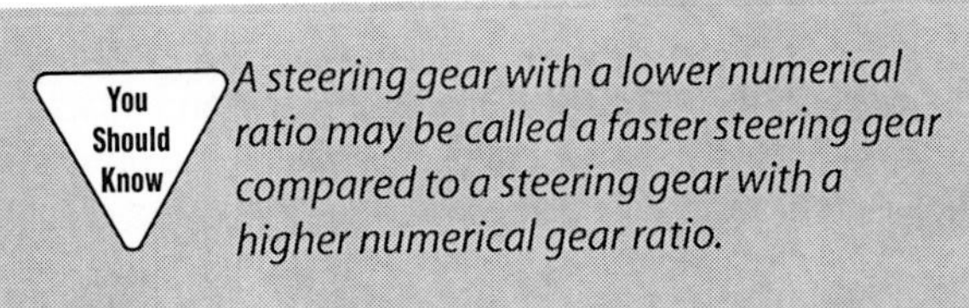


*A steering gear with a lower numerical ratio may be called a faster steering gear compared to a steering gear with a higher numerical gear ratio.*

Many recirculating ball steering gears are bolted to the frame with hard steel bolts **(Figure 4)**. These bolts must be tightened to the vehicle manufacturer's specified torque.

You Should Know

*Hard steel bolts may be used for steering gear mounting. Bolt hardness is indicated by ribs on the bolt head. Harder bolts have five, six, or seven ribs on the bolt heads. Never substitute softer steel bolts in place of the original harder bolts because these softer bolts may break allowing the steering box to detach from the frame which results in loss of steering control.*

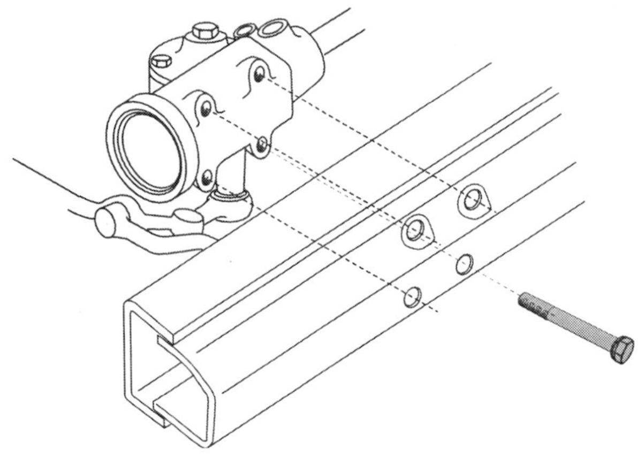

**Figure 4.** Steering gear mounting on the vehicle frame.

## MANUAL RACK-AND-PINION STEERING GEARS

The rack is a toothed bar which slides back and forth in a metal housing. The steering gear housing is mounted in a fixed position on the front crossmember or on the firewall. The rack takes the place of the idler and pitman arms in a parallelogram steering system and maintains the proper height of the tie-rods so they are parallel to the lower control arms. The rack may be compared to the center link in a parallelogram steering linkage. Bushings support the rack in the steering gear housing. Sideways movement of the rack pulls or pushes the tie-rods and steers the front wheels **(Figure 5)**.

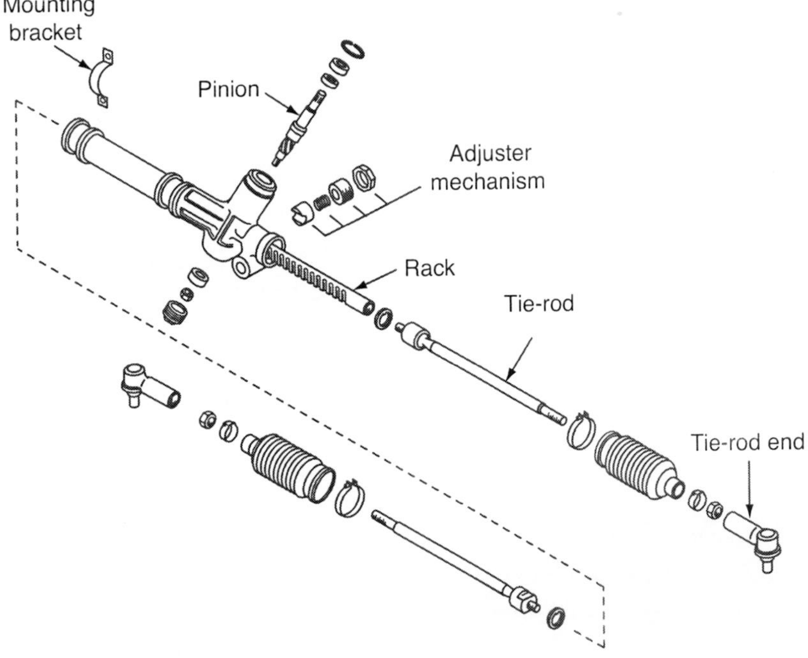


**Figure 5.** Manual rack-and-pinion steering gear components.

## Pinion

The pinion is a toothed shaft mounted in the steering gear housing so the pinion teeth are meshed with the rack teeth. Teeth on the rack and pinion may be spur or helical. The upper end of the pinion shaft is connected to the steering shaft from the steering column. Therefore, steering wheel rotation moves the rack sideways to steer the front wheels. The pinion is supported on ball bearings in the steering gear housing. **Spur** gear teeth are positioned parallel to the gear's rotational axis. **Helical** gear teeth are positioned at an angle in relation to the gear's rotational axis.

## Tie-Rods and Tie-Rod Ends

The tie-rods are similar to those used on parallelogram steering linkages. A spring-loaded ball socket on the inner end of the tie-rod is threaded onto the rack. When these ball sockets are torqued to the vehicle manufacturer's specification, a preload is placed on the ball socket. A bellows boot is clamped to the housing and tie-rod on each side of the steering gear. These boots keep contaminants out of the ball socket and rack.

A tie-rod end is threaded onto the outer end of each tie-rod. These tie-rod ends are similar to those used on parallelogram steering linkages. A jamb nut locks the outer tie-rod end to the tie-rod.

## Rack Adjustment

A rack bearing is positioned against the smooth side of the rack. A spring is located between the rack bearing and the rack adjuster plug that is threaded into the housing. This adjuster plug is retained with a locknut. The rack bearing adjustment sets the preload between the rack-and-pinion teeth which affects steering harshness, noise, and feedback.

## Steering Gear Ratio

When the steering wheel is rotated from lock to lock, the front wheels turn about 30 degrees each in each direction from the straight-ahead position. Therefore, the total front wheel movement from left to right is approximately 60 degrees. With a steering ratio of 1:1, 1 degree of steering wheel rotation would turn the front wheels 1 degree, and 30 degrees of steering wheel rotation in either direction would result in lock-to-lock front wheel movement. This steering ratio is much too extreme because the slightest steering wheel movement would cause the vehicle to swerve. The steering gear must have a ratio which allows more steering wheel rotation in relation to front wheel movement.

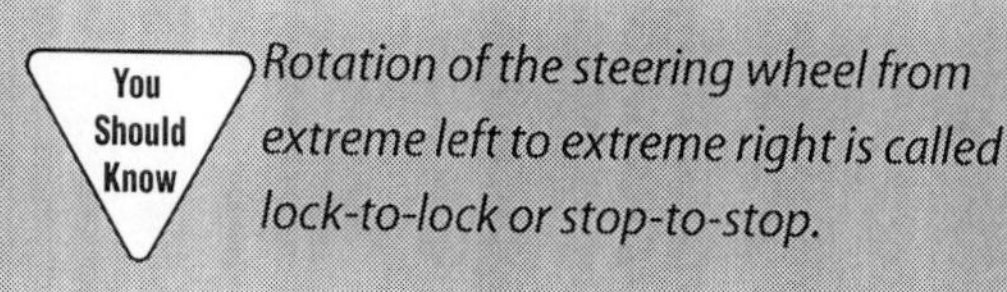


*Rotation of the steering wheel from extreme left to extreme right is called lock-to-lock or stop-to-stop.*

A steering ratio of 15:1 is acceptable. This ratio provides 1 degree of front wheel movement for every 15 degrees of steering wheel rotation. To calculate the steering ratio, divide the lock-to-lock steering wheel rotation in degrees by the total front wheel movement in degrees. For example, if the lock-to-lock steering wheel rotation is 3.5 turns, or 1260 degrees, and the total front wheel movement is 60 degrees, the steering ratio is $1260 \div 60 = 21:1$.

As a general rule, large, heavy cars have higher numerical steering ratios compared to small, lightweight cars.

## Manual Rack-and-Pinion Steering Gear Mounting

Large rubber insulating grommets are positioned between the steering gear and the mounting brackets. These bushings help prevent the transfer of road noise and vibration from the steering gear to the chassis and passenger compartment. The rack-and-pinion steering gear may be attached to the front crossmember or the cowl. Proper steering gear mounting is important to maintain the parallel relationship between the tie-rods and the lower control arms. The firewall is reinforced at the steering gear mounting locations to maintain the proper steering gear position.

## Advantages and Disadvantages of Rack-and-Pinion Steering

As mentioned previously, the rack-and-pinion steering gear is lighter and more compact compared to a recirculating ball steering gear and parallelogram steering linkage. Therefore, the rack-and-pinion steering gear is most suitable for FWD unibody vehicles.

Since there are fewer friction points in the rack-and-pinion steering compared to the recirculating ball steering gear with a parallelogram steering linkage, the driver has a greater feeling of the road with rack-and-pinion steering gear. However, fewer friction points reduce the steering system's ability to isolate road noise and vibration. Therefore, the rack-and-pinion steering system may have more complaints of road noise and vibration transfer to the steering wheel and passenger compartment.

# POWER RECIRCULATING BALL STEERING GEARS

The ball nut and pitman shaft sector are similar in manual and power recirculating ball steering gears. In the power steering gear, a torsion bar is connected between the steering shaft and the worm shaft. Since the front wheels are resting on the road surface, they resist turning and the parts attached to the worm shaft also resist turning. This turning resistance causes torsion bar deflection when the wheels are turned. This deflection is limited to a predetermined amount. The worm shaft is connected to the rotary valve body and the torsion bar pin also connects the torsion bar to the worm shaft. The upper end of the tor-

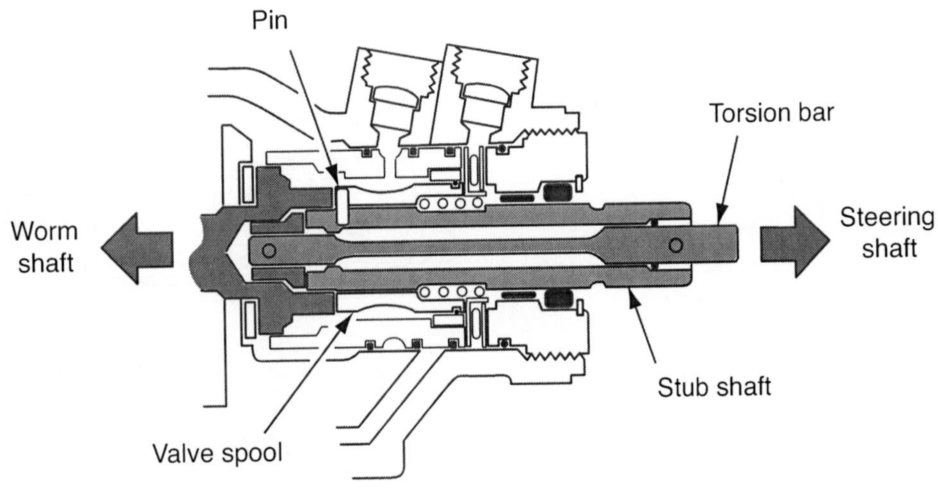


**Figure 6.** A torsion bar and stub shaft.

sion bar is attached to the steering shaft and wheel. A stub shaft is mounted inside the rotary valve and a pin connects the outer end of this shaft to the torsion bar. The pin on the inner end of the stub shaft is connected to the spool valve in the center of the rotary valve **(Figure 6)**.

When the car is driven with the front wheels straight ahead, oil flows from the power steering pump through the spool valve, rotary valve, and low-pressure return line to the pump inlet **(Figure 7)**. In the straight-ahead steering gear position, oil pressure is equal on both sides of the recirculating ball piston and the oil acts as a cushion that prevents road shocks from reaching the steering wheel.

If driver makes a left turn, torsion bar deflection moves the spool valve inside the rotary valve body so that oil flow is

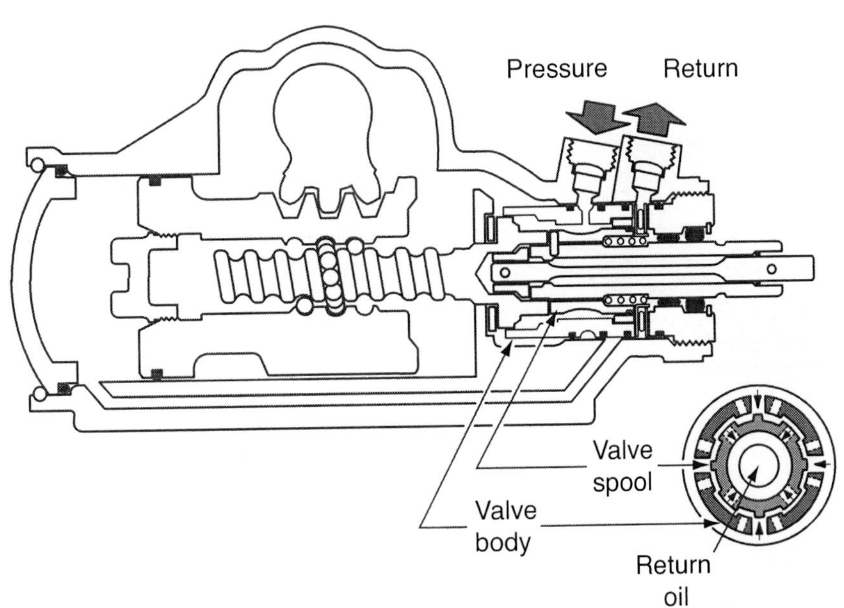


**Figure 7.** Power steering fluid flow with the wheels straight ahead.

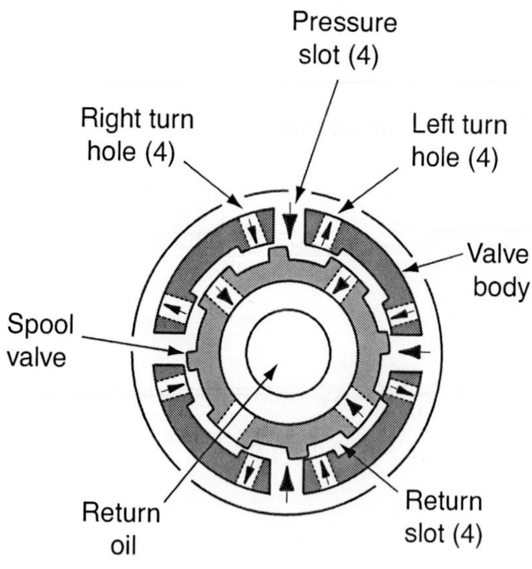


Figure 8. Spool valve position during a left turn.

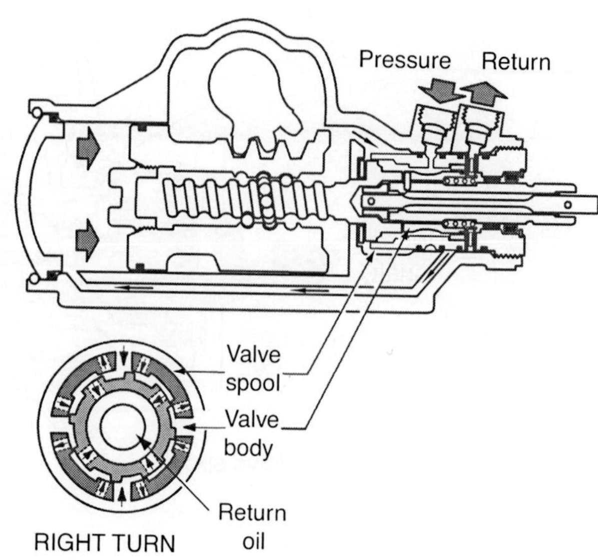


Figure 10. Power steering gear fluid flow during a right turn.

directed through the rotary valve to the left turn holes in the spool valve **(Figure 8)**. Since power steering fluid is directed from these left turn holes to the upper side of the recirculating ball piston **(Figure 9)**, this hydraulic pressure on the piston assists the driver in turning the wheels to the left.

When the driver makes a right turn, torsion bar deflection moves the spool valve so that oil flows through the spool valve, rotary valve, and a passage in the housing to the pressure chamber at the lower end of the ball nut piston **(Figure 10)**. During a right turn, hydraulic pressure applied to the lower end of the recirculating ball piston helps the driver to turn the wheels.

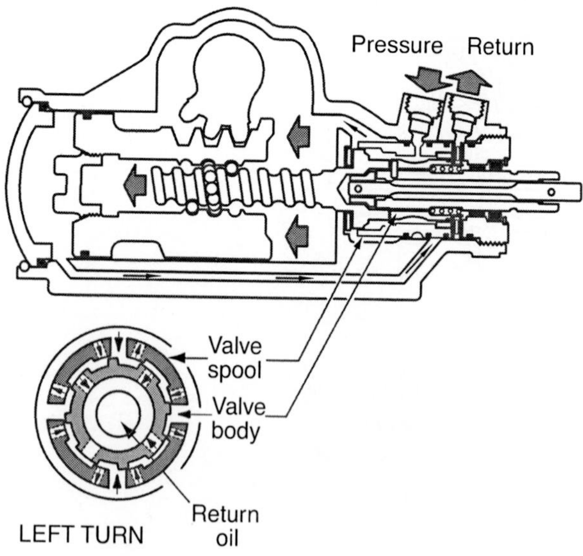


Figure 9. Power steering gear fluid flow during a left turn.

If a front wheel strikes a bump during a turn and the front wheels are driven in the direction opposite to the turning direction, the recirculating ball piston tends to move against the hydraulic pressure and force oil back out the pressure inlet port. This action would create a kickback on the steering wheel. Under this condition, a poppet valve in the pressure inlet fitting closes and prevents kickback action.

## POWER RACK-AND-PINION STEERING GEARS

A power assisted rack-and-pinion steering gear uses the same basic operating principles as a manual rack-and-pinion steering gear, but in the power assisted steering gear, hydraulic fluid pressure from the power steering pump is used to reduce steering effort. A rack piston is integral with the rack. This piston is located in a sealed chamber in the steering gear housing. Hydraulic fluid lines are connected to each end of this chamber and rack seals are positioned in the housing at ends of the chamber. A seal is also located on the rack piston **(Figure 11)**.

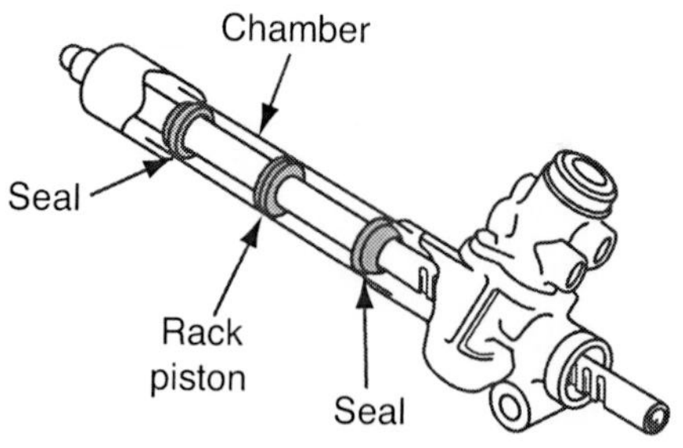


Figure 11. Hydraulic chamber in a power rack-and-pinion steering gear.

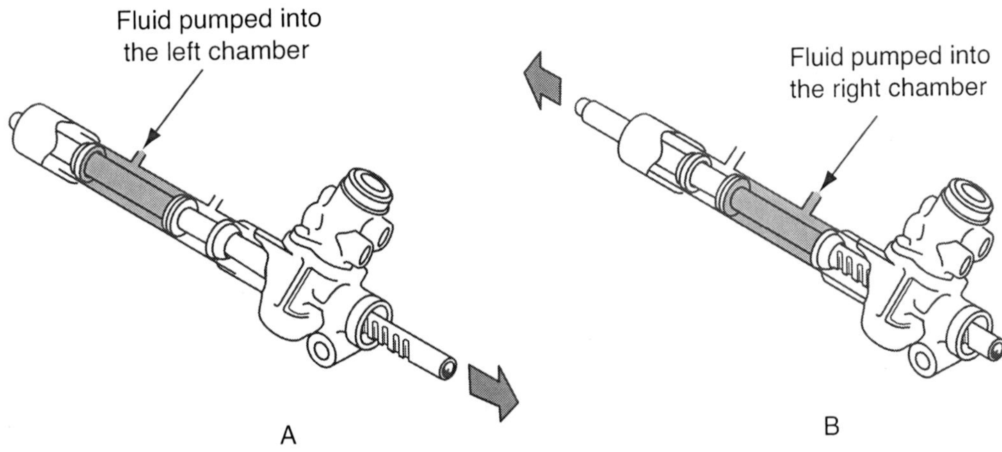


**Figure 12.** Rack movement during right and left turns.

The right and left side of a vehicle is determined from the driver's seat. If a left turn is completed, fluid is pumped into the left side of the fluid chamber and exhausted from the right chamber area. This hydraulic pressure on the left side of the rack piston helps the pinion to move the rack to the right **(Figure 12A)**.

When a right turn is made, fluid is pumped into the right side of the fluid chamber and fluid flows out of the left end of the chamber. Thus hydraulic pressure is exerted on the right side of the rack piston which assists the pinion gear in moving the rack to the left **(Figure 12B)**.

Since the steering gear is mounted behind the front wheels, rack movement to the left is necessary for a right turn, while rack movement to the right causes a left turn. Power rack-and-pinion steering gears may be classified as end take-off or center take-off.

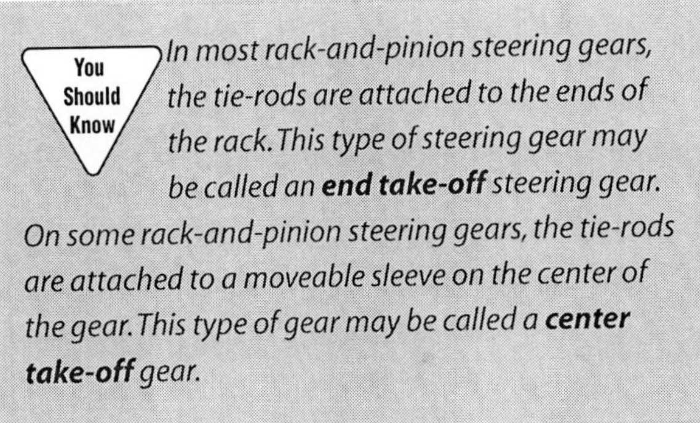


*In most rack-and-pinion steering gears, the tie-rods are attached to the ends of the rack. This type of steering gear may be called an* ***end take-off*** *steering gear. On some rack-and-pinion steering gears, the tie-rods are attached to a moveable sleeve on the center of the gear. This type of gear may be called a* ***center take-off*** *gear.*

## Rotary Valve and Spool Valve Operation

Fluid direction in the steering gear is controlled by a spool valve attached to the pinion assembly **(Figure 13)**. A

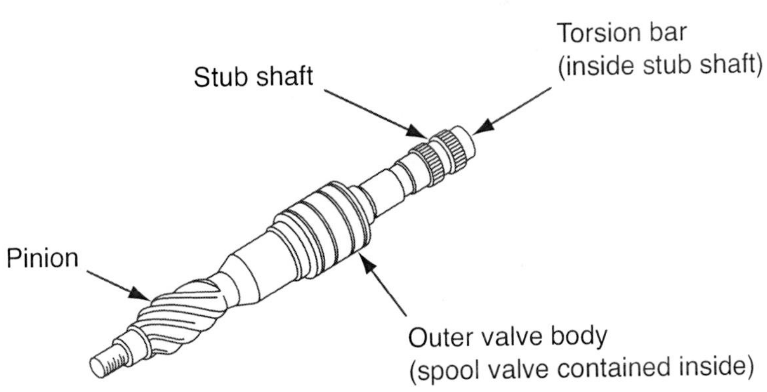


**Figure 13.** Pinion assembly for a power rack-and-pinion steering gear.

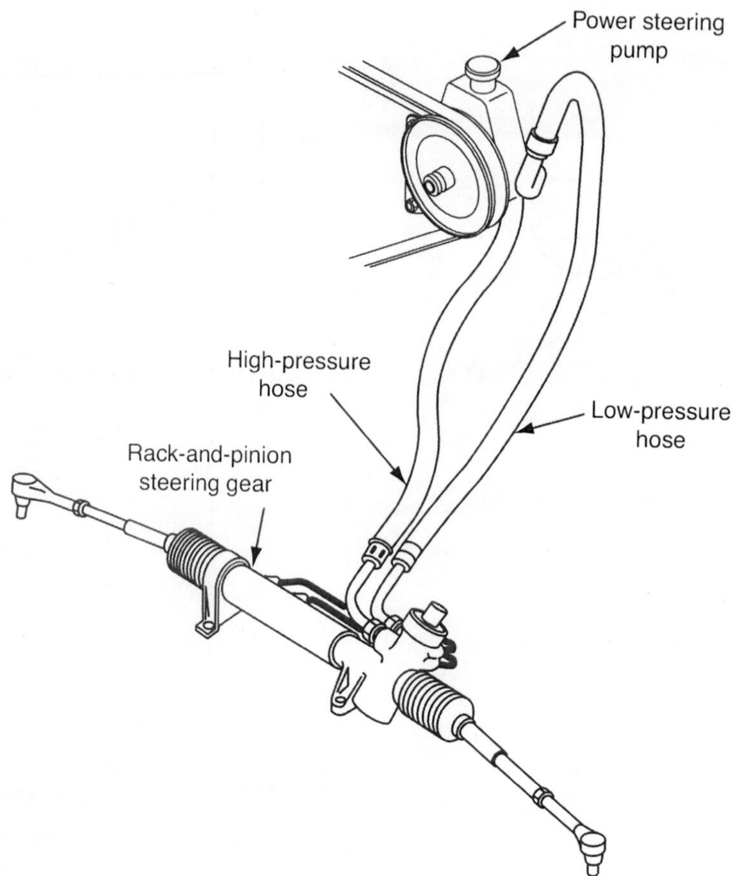


**Figure 14.** Power rack-and-pinion steering gear with connecting hoses.

stub shaft on the pinion assembly is connected to the steering shaft and wheel. The pinion is connected to the stub shaft through a torsion bar that twists when the steering wheel is rotated and springs back to the center position when the wheel is released. A rotary valve body contains an inner spool valve that is mounted over the torsion bar on the pinion assembly.

When the front wheels are in the straight-ahead position, fluid flows from the pump through the high-pressure hose to the center rotary valve body passage. Fluid is then routed through the valve body to the low-pressure return hose and the pump reservoir **(Figure 14)**.

Teflon rings, or O rings seal the rotary valve ring lands to the steering gear housing. A great deal of force is required to turn the pinion and move the rack because of the vehicle weight on the front wheels. When the driver turns the wheel, he or she forces the stub shaft to turn. However, the pinion resists turning because it is in mesh with the rack which is connected to the front wheels. This resistance of the pinion to rotation results in torsion bar twisting. During this twisting action, a pin on the torsion bar moves the spool valve with a circular motion inside the rotary valve. If the driver makes a left turn, the spool valve movement aligns the inlet center rotary valve passage with the outlet passage to the left side of the rack piston. Therefore, hydraulic fluid pressure applied to the left side of the rack piston assists the driver in moving the rack to the right.

When a right turn is made, twisting of the torsion bar moves the spool valve and aligns the center rotary valve passage with the outlet passage to the right side of the rack piston **(Figure 15)**. Under this condition, hydraulic fluid pressure applied to the rack piston helps the driver to move the rack to the left. The torsion bar provides a feel of the road to the driver.

When the steering wheel is released after a turn, the torsion bar centers the spool valve and power assist stops. If hydraulic fluid pressure is not available from the pump, the power steering system operates like a manual system, but steering effort is higher. When the torsion bar is twisted to a designed limit, tangs on the stub shaft engage with drive tabs on the pinion. This action mechanically transfers motion from the steering wheel to the rack and front

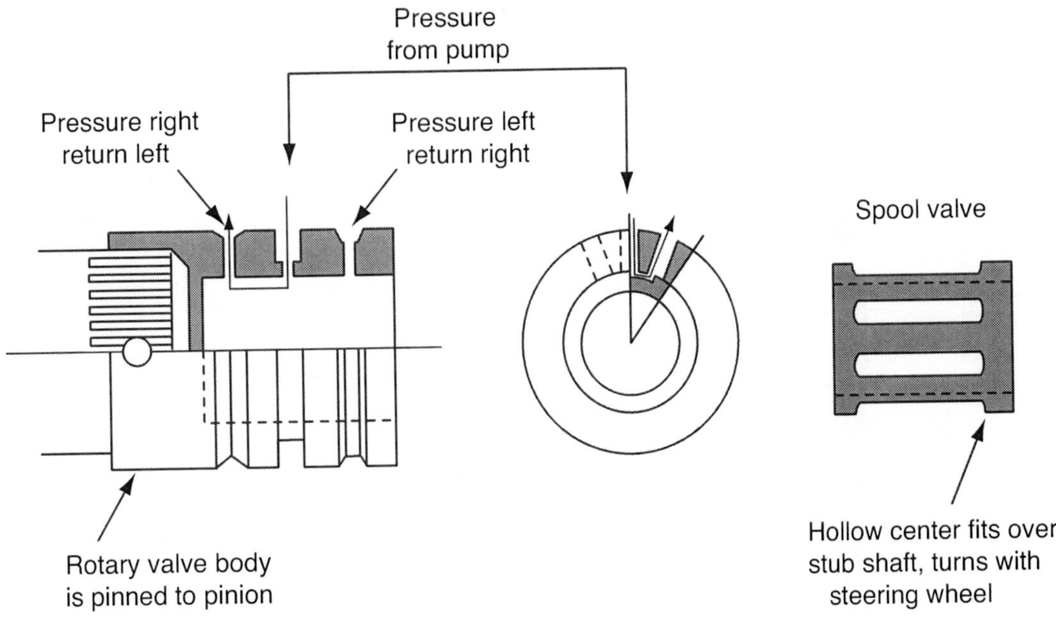


**Figure 15.** Spool valve movement inside the rotary valve.

wheels. Since hydraulic pressure is not available on the rack piston, greater steering effort is required. If a front wheel raises going over a bump or drops into a hole, the tie-rod pivots along with the wheel. However, the rack and tie-rod still maintain the left-to-right wheel direction.

The rack boots are clamped to the housing and the rack. Since the boots are sealed and air cannot be moved through the housing, a breather tube is necessary to move air from one boot to the other when the steering wheel is turned **(Figure 16)**. This air movement through the vent tube prevents pressure changes in the bellows boots during a turn.

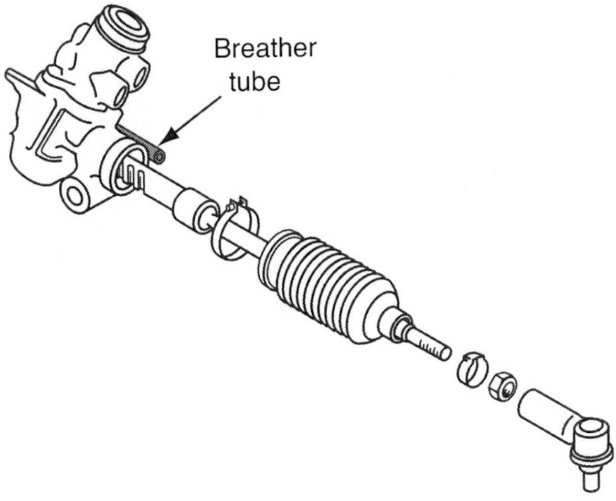


**Figure 16.** Breather tube and boot.

## ELECTRONIC POWER STEERING SYSTEMS

Some vehicles have a VES system. In the VES system, the vehicle speed sensor input is sent to the VES controller. This controller supplies a PWM voltage to the actuator solenoid in the power steering pump. The controller also provides a ground connection for the actuator solenoid **(Figure 17)**.

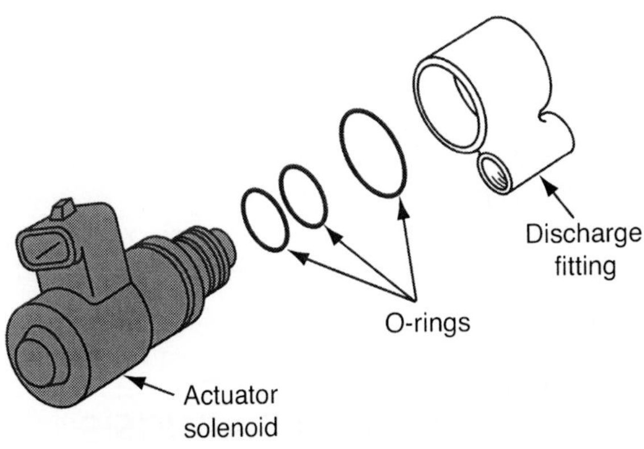


**Figure 17.** Actuator solenoid in a variable effort steering (VES) system.

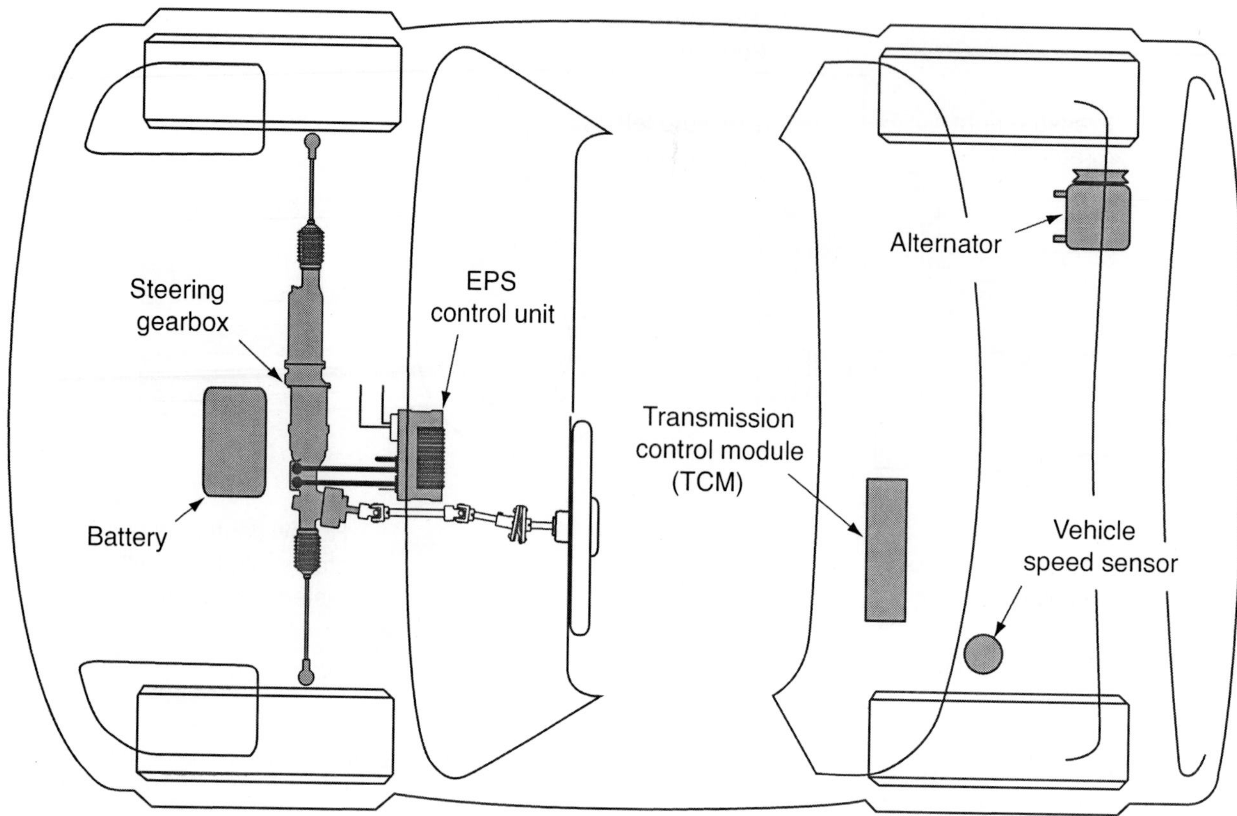


**Figure 18.** Electronic power steering (EPS) system components.

When the vehicle is operating at low speeds, the controller supplies a PWM signal to position the actuator solenoid plunger so the power steering pump pressure is higher. Under this condition, greater power assist is provided for cornering or parking. If the vehicle is operating at higher speed, the controller changes the PWM signal to the actuator solenoid and the solenoid plunger is positioned to reduce power steering pump pressure. This action reduces power steering assist to provide improved road feel for the driver.

A significant number of vehicles are equipped with electronic power steering. In these steering gears, an electric motor provides steering assist when the front wheels are turned **(Figure 18)**. Because the electronic power steering system does not have a power steering pump, there is no engine power required to drive this pump and fuel economy is increased slightly.

## MANUAL STEERING GEAR MAINTENANCE AND DIAGNOSIS

Manual steering gear maintenance is very important to provide normal steering gear life and proper steering gear operation. For example, if a manual steering gear is operated for a period of time with a low lubricant level, steering gear life is shortened.

### Manual Recirculating Ball Steering Gear Maintenance and Diagnosis

When the oil level plug is removed from the steering gear, the lubricant should be level with the bottom of the plug opening. If the oil level is low, fill the steering gear with the manufacturer's specified steering gear lubricant. If a steering gear provides binding and hard, uneven steering effort, the oil level is very low. If the oil level is low, visually check the sector shaft seal and the worm shaft seal area for leaks. Leaking seals must be replaced. Hard steering effort may also be caused by wheel alignment problems. Defective worm shaft bearings cause uneven turning effort and steering gear noise. Excessive steering effort may be caused by worn steering gears. A rattling noise from the steering gear may be caused by loose mounting bolts, worn steering shaft U-joints, or flexible coupling. Excessive steering wheel free-play may be caused by a loose worm shaft bearing preload adjustment, a loose sector shaft lash adjustment, worn steering gears, loose steering gear mounting bolts, or worn steering shaft U-joints or flexible coupling.

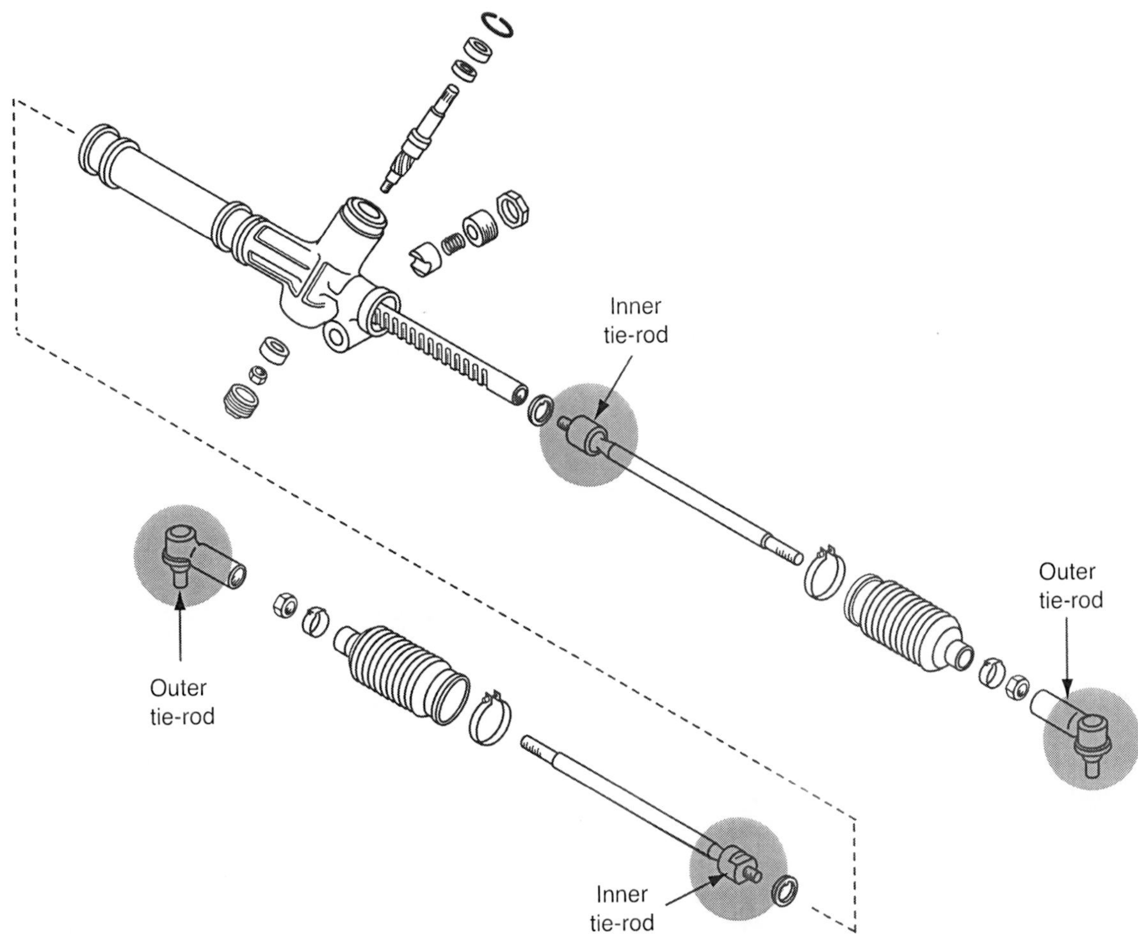


**Figure 19.** Wear points at the inner and outer tie-rod ends in a rack-and-pinion steering gear.

## Manual or Power Rack-and-Pinion Steering Gear On Car Inspection

The wear points are reduced to four in a rack-and-pinion steering gear. These wear points are the inner and outer tie-rod ends on both sides of the rack-and-pinion assembly **(Figure 19)**.

The first step in manual or power rack-and-pinion steering gear diagnosis is a very thorough inspection of the complete steering system. During this inspection, all steering system components, such as the inner and outer tie-rod ends, bellows boots, mounting bushings, couplings, or universal joints, ball joints, tires, and steering wheel free-play, must be checked. Follow these steps for manual or power rack-and-pinion steering gear inspection:

1. With the front wheels straight ahead and the engine stopped, rock the steering wheel gently back and forth with light finger pressure **(Figure 20)**. Measure the

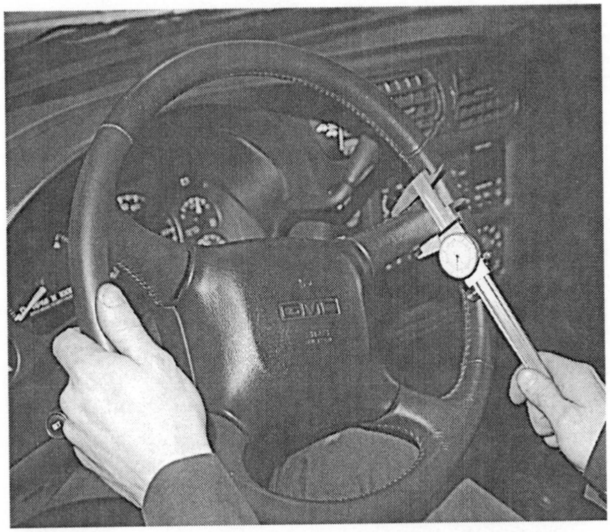

**Figure 20.** Measuring steering wheel free-play.

maximum steering wheel free-play. The maximum specified steering wheel free-play on some vehicles is 1.18 inches (30 millimeters). Always refer to the vehicle manufacturer's specifications in the service manual. Excessive steering wheel free-play indicates worn steering components.

2. With the vehicle sitting on the shop floor and the front wheels straight ahead, have an assistant turn the steering wheel about one quarter turn in both directions. Watch for looseness in the flexible coupling in the steering shaft. If looseness is observed, replace the coupling or universal joint.
3. While an assistant turns the steering wheel about one half turn in both directions, watch for movement of the steering gear housing in the mounting bushings. If there is any movement of the housing in these bushings, replace the bushings. The steering gear mounting bushings may be deteriorated by oil soaking, heat, or age.
4. Grasp the pinion shaft extending from the steering gear and attempt to move it vertically. If there is steering shaft vertical movement, a pinion bearing preload adjustment may be required. When the steering gear does not have a pinion bearing preload adjustment, replace the necessary steering gear components.
5. Road test the vehicle and check for excessive steering effort. A bent steering rack, tight rack bearing adjustment, or damaged front drive axle joints in a front-wheel drive car may cause excessive steering effort.
6. Visually inspect the bellows boots for cracks, splits, leaks, and proper clamp installation. Replace any boot that indicates any of these conditions. If the boot clamps are loose or improperly installed, tighten or replace the clamps as necessary. Since the bellows boots protect the inner tie-rod ends and the rack from contamination, boot condition is extremely important. Boots should be inspected each time undercar service such as oil and filter change or chassis lubrication is performed.
7. Loosen the inner bellows boot clamps and move each boot toward the outer tie-rod end until the inner tie-rod end is visible. Push outward and inward on each front tire and watch for movement in the inner tie-rod end. If any movement or looseness is present, replace the inner tie-rod end. An alternate method of checking the inner tie-rod ends is to squeeze the bellows boots and grasp the inner tie-rod end socket. Movement in the inner tie-rod end is then felt as the front wheel is moved inward and outward. Hard plastic bellows boots may be found on some applications. With this type of bellows boot, remove the ignition key from the switch to lock the steering column and push inward and outward on the front tire while observing any lateral movement in the tie-rod. When lateral movement is observed, replace the inner tie-rod end.

**You Should Know** *Bent steering components must be replaced. Never straighten steering components because this action may weaken the metal and result in sudden component failure, serious personal injury, and vehicle damage.*

8. Grasp each outer tie-rod end and check for vertical movement. While an assistant turns the steering wheel one quarter turn in each direction, watch for looseness in the outer tie-rod ends. If any looseness or vertical movement is present, replace the tie-rod end. Check the outer tie-rod end seals for cracks and proper installation of the nuts and cotter pins. Cracked seals must be replaced. Inspect the tie-rods for a bent condition. Bent tie-rods or other steering components must be replaced. Do not attempt to straighten these components.

## MANUAL STEERING GEAR SERVICE

Proper manual steering gear service is essential to supply normal steering control and vehicle safety. For example, loose manual steering gear worm shaft bearing preload and sector lash adjustments cause vehicle wander and reduced steering control.

### Manual Recirculating Ball Steering Gear Removal and Replacement

Follow these preliminary steps prior to steering gear removal or adjustment:

1. Disconnect the battery ground cable.
2. Raise the vehicle with the front wheels in the straight-ahead position. If the vehicle is lifted with a floor jack, place jack stands under the chassis or suspension and lower the vehicle onto the jack stands.
3. Remove the pitman arm nut and washer. Mark the pitman arm position in relation to the pitman shaft with a center punch and use a puller to remove the pitman arm.
4. Loosen the sector shaft backlash adjuster lock nut and back the adjuster screw off one-quarter turn.
5. Turn the steering wheel gently in one direction until it is stopped by the gear and then count the number of turns as the steering wheel is rotated in the opposite direction until it is stopped by the gear. Then turn the steering wheel back one-half the total number of turns toward the center position.
6. Remove the center steering wheel cover and place a socket and inch-pound torque wrench on the steering wheel nut. Do not use a torque wrench with a maximum scale reading above 50 in.-lb.
7. Rotate the steering wheel through a 90-degree arc and record the turning torque which indicates the worm shaft bearing preload.

Follow these steps for manual steering gear removal and replacement:

1. Disconnect the flexible coupling from the worm shaft.
2. Remove the steering gear-to-frame mounting bolts.
3. Remove the steering gear from the chassis. The steering gear may be cleaned externally with an approved cleaning solution.
4. Reverse steps 1 through 3 to re-install the steering gear. All bolts must be tightened to the specified torque. Be sure the steering gear is filled with the manufacturer's specified steering gear lubricant.

> **You Should Know** *Recirculating ball steering gears are often mounted near the exhaust manifold which may be extremely hot. Wear protective gloves and use caution when inspecting, adjusting, and servicing the steering gear.*

> **You Should Know** *Hard steel bolts may be used for steering gear mounting. Bolt hardness is indicated by ribs on the bolt head. Harder bolts have five, six, or seven ribs on the bolt heads. Never substitute softer steel bolts in place of the original harder bolts because these softer bolts may break allowing the steering box to detach from the frame, resulting in loss of steering control. When the steering linkage is disconnected from the gear, do not turn the steering wheel hard against the stops. This action may damage the ball guides in the steering gear.*

## Manual or Power Rack-and-Pinion Steering Gear Removal and Replacement

The replacement procedure is similar for manual or power rack-and-pinion steering gears. This removal and replacement procedure varies depending on the vehicle. On some vehicles, the front crossmember or engine support cradle must be lowered to remove the rack-and-pinion steering gear. Always follow the vehicle manufacturer's recommended procedure in the service manual. The following is a typical rack-and-pinion steering gear removal and replacement procedure:

1. Place the front wheels in the straight-ahead position and remove the ignition key from the ignition switch to lock the steering wheel. Place the driver's seat belt through the steering wheel to prevent wheel rotation if the ignition switch is turned on. This action maintains the clockspring electrical connector or spiral cable in the centered position on air bag-equipped vehicles.
2. Lift the front end with a floor jack and place jack stands under the vehicle chassis. Lower the vehicle onto the jack stands. Remove the left and right fender apron seals **(Figure 21)**.
3. Place punch marks on the lower universal joint and the steering gear pinion shaft so they may be reassembled in the same position

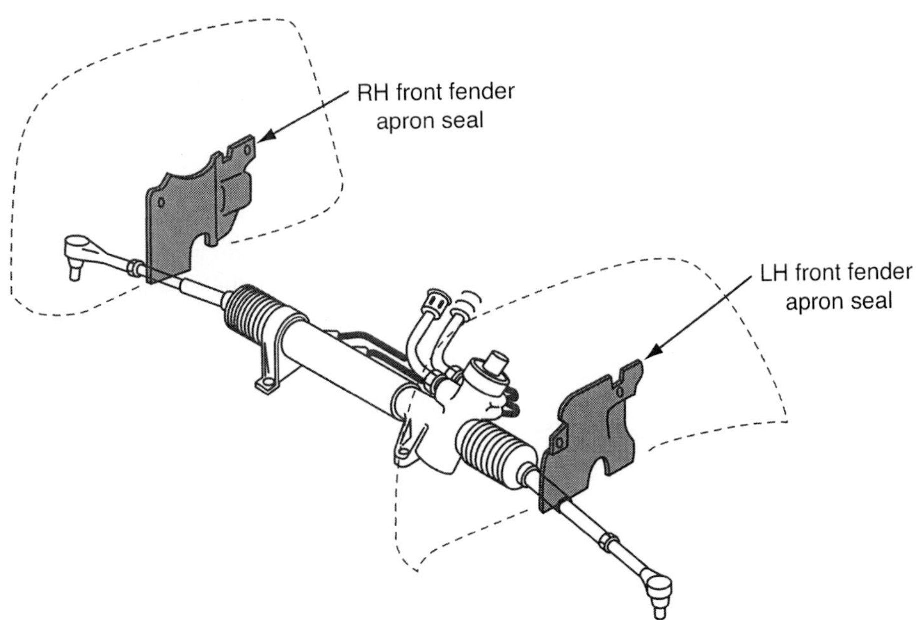


**Figure 21.** Left and right fender apron seals.

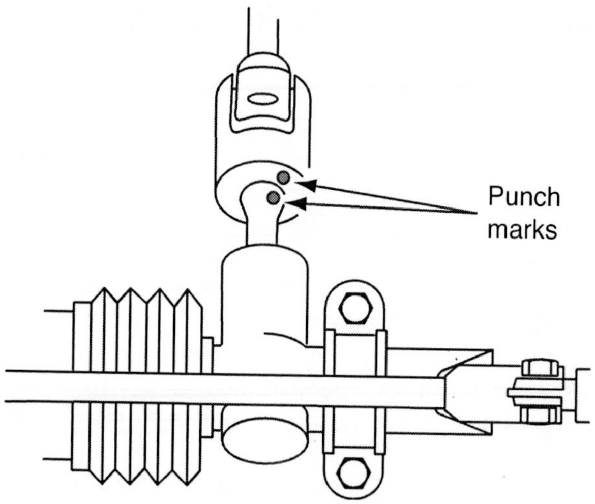


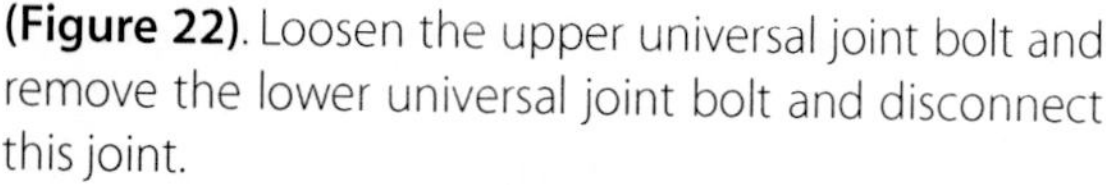

**Figure 22.** Punch marks on universal joint and pinion shaft.

**(Figure 22)**. Loosen the upper universal joint bolt and remove the lower universal joint bolt and disconnect this joint.

4. Remove the cotter pins from the outer tie-rod ends. Loosen but do not remove the tie-rod end nuts. Use a tie-rod end puller to loosen the outer tie-rod ends in the steering arms. Remove the tie-rod end nuts and remove the tie-rod ends from the arms.
5. Use the proper wrenches to disconnect the high-pressure hose and return hose from the steering gear **(Figure 23)**. This step is not required on a manual steering gear.
6. Remove the four stabilizer bar mounting bolts **(Figure 24)**.
7. Remove steering gear mounting bolts.
8. Remove the steering gear assembly from the right side of the car **(Figure 25)**.

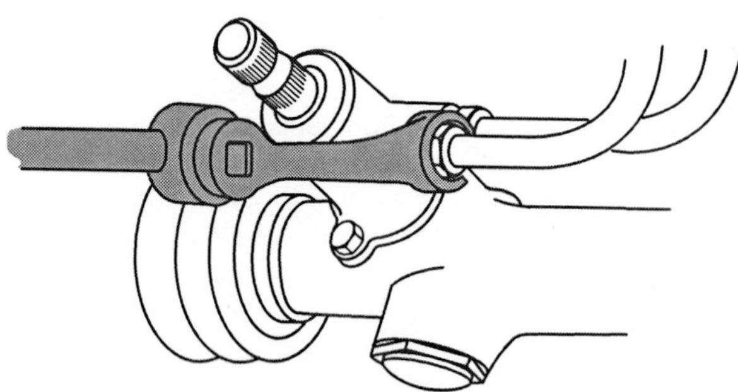

**Figure 23.** Removing high-pressure and return hoses from the steering gear.

**Figure 24.** Removing stabilizer bar mounting bolts.

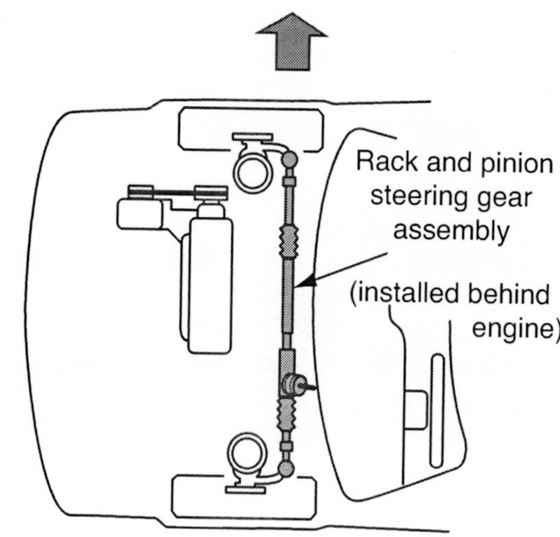


**Figure 25.** Removing steering gear from the car.

9. Position the right and left tie-rods the specified distance from the steering gear housing **(Figure 26)**. Install the steering gear through the right fender apron.
10. With the punch marks aligned, install the pinion shaft in the universal joint. Tighten the upper and lower universal joint bolts to the specified torque.
11. Install the steering gear mounting bolts and tighten these bolts to the specified torque.
12. Install the stabilizer bar mounting bolts and torque these bolts to specifications.

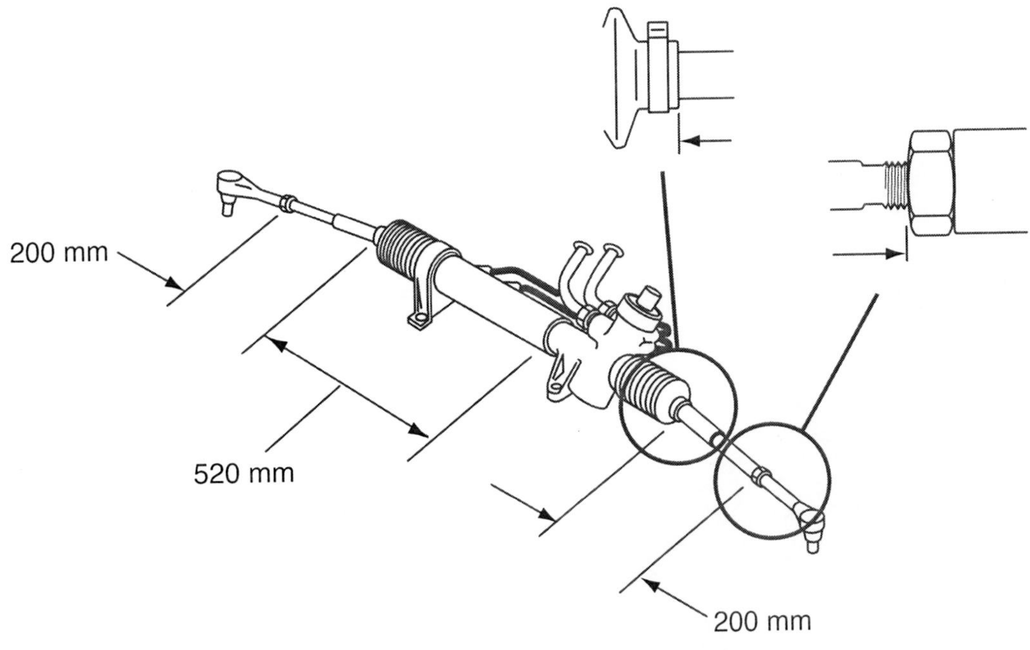


**Figure 26.** Right and left tie-rod position in relation to the steering gear housing prior to installation.

13. Install and tighten the high-pressure and return hoses to the specified torque. This step is not required on a manual rack-and-pinion steering gear.

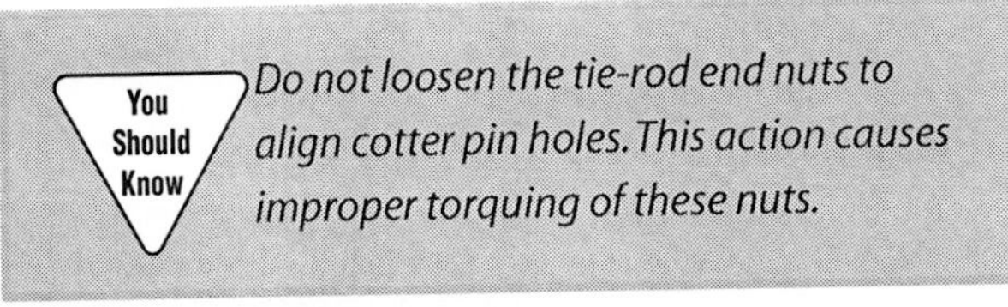


*Do not loosen the tie-rod end nuts to align cotter pin holes. This action causes improper torquing of these nuts.*

14. Install the outer tie-rod ends in the steering knuckles and tighten the nuts to the specified torque. Install the cotter pins in the nuts.
15. Check the front wheel toe and adjust as necessary. Tighten the outer tie-rod end jam nuts to the specified torque and tighten the outer bellows boot clamps.
16. Install the left and right fender apron seals and lower the vehicle with a floor jack.
17. Fill the power steering pump reservoir with the vehicle manufacturer's recommended power steering fluid and bleed air from the power steering system. This step is not required on a manual rack-and-pinion steering gear.
18. Road test the vehicle and check for proper steering gear operation and steering control.

## Manual Recirculating Ball Steering Gear Worm Shaft Bearing Preload Adjustment

If the worm shaft bearing preload adjustment is loose, steering wheel free-play is excessive. This results in steering wander when the vehicle is driven straight ahead. Steering effort is increased if the worm shaft bearing preload adjustment is too tight. When the worm shaft bearing preload is adjusted, use the following procedure:

1. Follow the preliminary steps listed previously.
2. Loosen the worm shaft adjuster plug locknut with a brass punch and a hammer. Tighten the adjuster plug until all the worm shaft end play is removed and then loosen the plug one-quarter turn.

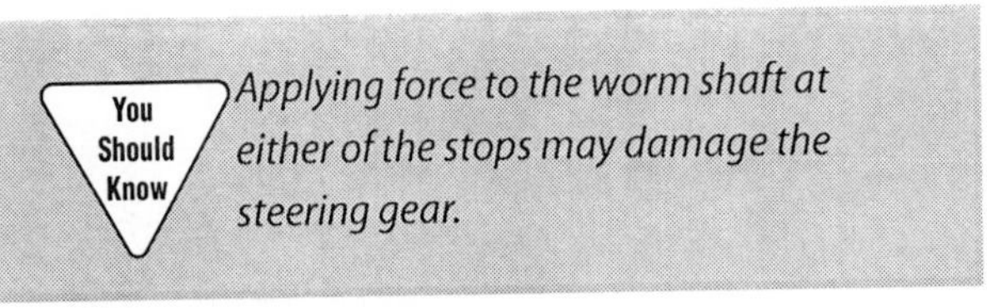


*Applying force to the worm shaft at either of the stops may damage the steering gear.*

3. Turn the worm shaft fully to the right with a socket and an inch-pound torque wrench. Then turn the worm shaft one-half turn toward the center position.

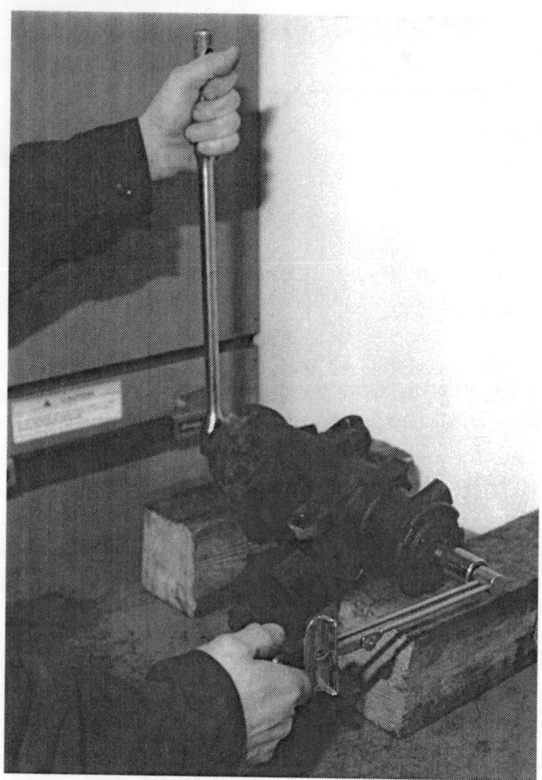

Figure 27. Adjusting worm shaft bearing preload, manual recirculating ball steering gear.

4. Tighten the adjuster plug until the specified bearing preload is indicated on the torque wrench as the worm shaft is rotated **(Figure 27)**. The specification on some steering gears is 5 to 8 in. lb. (0.56 to 0.896 N·m). Always use the vehicle manufacturer's specified preload.
5. Tighten the adjuster plug locknut to 85 ft. lb. (114 N·m).

## Manual Recirculating Ball Steering Gear Sector Lash Adjustment

When the sector shaft lash adjustment is too loose, steering wheel free-play is excessive and vehicle wander occurs when the vehicle is driven straight ahead. A loose sector shaft lash adjustment decreases driver road feel. If the sector lash adjustment is too tight, steering effort is increased, especially with the front wheels in the straight-ahead position.

The following procedure may be used when the pitman shaft sector lash is adjusted:

1. Turn the pitman backlash adjuster screw outward until it stops and then turn it in one turn.
2. Rotate the worm shaft fully from one stop to the other stop and carefully count the number of shaft rotations.
3. Turn the worm shaft back exactly one-half the total number of turns from one of the stops.
4. With the steering gear positioned as it was in step 3, connect an inch-pound torque wrench and socket to the worm shaft and note the steering gear turning torque while rotating the wormshaft 45 degrees in each direction.
5. Turn the pitman backlash adjuster screw until the torque wrench reading is 6 to 10 in. lb. (0.44 to 1.12 N·m) more than the worm shaft bearing preload torque in step 4 **(Figure 28)**.
6. Tighten the pitman backlash adjuster screw locknut to the specified torque.

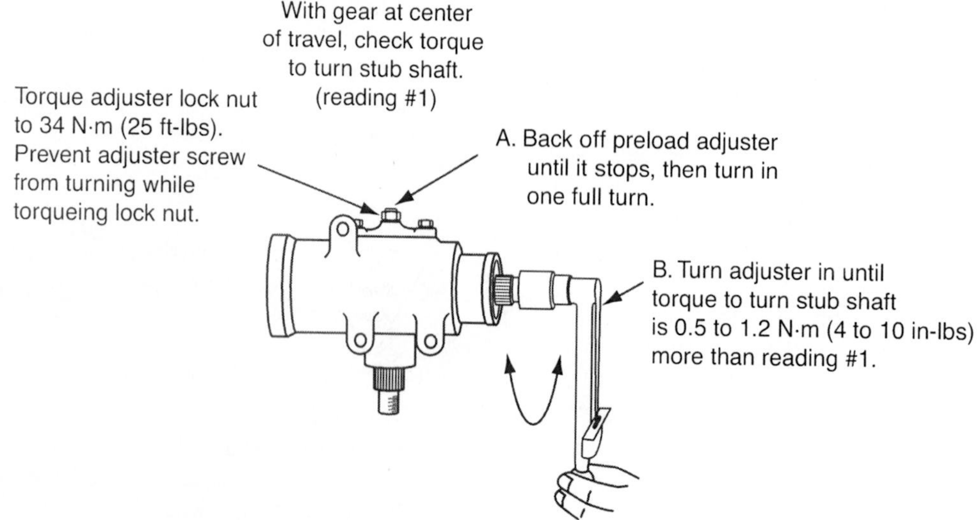
With gear at center of travel, check torque to turn stub shaft. (reading #1)
Torque adjuster lock nut to 34 N·m (25 ft-lbs). Prevent adjuster screw from turning while torqueing lock nut.
A. Back off preload adjuster until it stops, then turn in one full turn.
B. Turn adjuster in until torque to turn stub shaft is 0.5 to 1.2 N·m (4 to 10 in-lbs) more than reading #1.

Figure 28. Adjusting pitman backlash adjuster screw, manual recirculating ball steering gear.

## POWER STEERING GEAR MAINTENANCE AND DIAGNOSIS

If the steering gear is noisy, check these items:

1. A loose pitman shaft lash adjustment may cause a rattling noise when the steering wheel is turned.
2. Cut or worn dampener O-ring on the valve spool. When this defect is present, a squawking noise is heard during a turn.
3. Loose steering gear mounting bolts.
4. Loose or worn flexible coupling or steering shaft U-joints.

A hissing noise from the power steering gear is normal if the steering wheel is at the end of its travel or when the steering wheel is rotated with the vehicle standing still. If the steering wheel jerks or surges when the steering wheel is turned with the engine running, check the power steering pump belt condition and tension. When excessive kickback is felt on the steering wheel, check the poppet valve in the steering gear.

When the steering is loose, check these defects:

1. Air in the power steering system. To remove the air, fill the power steering pump reservoir and rotate the steering wheel fully in each direction several times.
2. Loose pitman lash adjustment.
3. Loose worm shaft thrust bearing preload adjustment.
4. Worn flexible coupling or universal joint.
5. Loose steering gear mounting bolts.
6. Worn steering gears.

A complaint of hard steering while parking could be caused by one of these defects:

1. Loose or worn power steering pump belt.
2. Low oil level in the power steering pump.
3. Excessively tight steering gear adjustments.
4. Defective power steering pump with low pressure output.
5. Restricted power steering hoses.

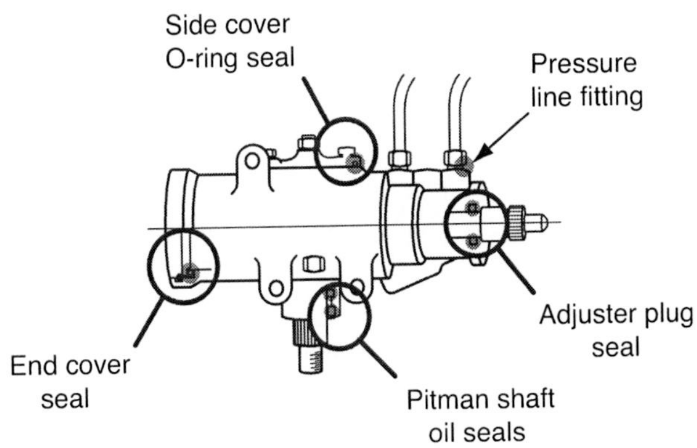


**Figure 29.** Power recirculating ball steering gear oil leak locations.

6. Defects in the steering gear such as:
   (a) Pressure loss in the cylinder because of scored cylinder, worn piston ring, or damaged back-up O-ring.
   (b) Excessively loose spool in valve body.
   (c) Defective or improperly installed gear check poppet valve.

### Power Recirculating Ball Steering Gear Oil Leak Diagnosis

Four locations where oil leaks may occur in a power steering gear are the following:

1. Side cover O-ring seal **(Figure 29)**.
2. Adjuster plug seal.
3. Pressure line fitting.
4. Pitman shaft oil seals.
5. Top cover seal.

If an oil leak is present at any of these areas, complete or partial steering gear disassembly and seal or O-ring replacement is necessary.

### Power Rack-and-Pinion Steering Gear Oil Leak Diagnosis

If power steering fluid leaks from the cylinder end of the power steering gear, the outer rack seal is leaking **(Figure 30)**.

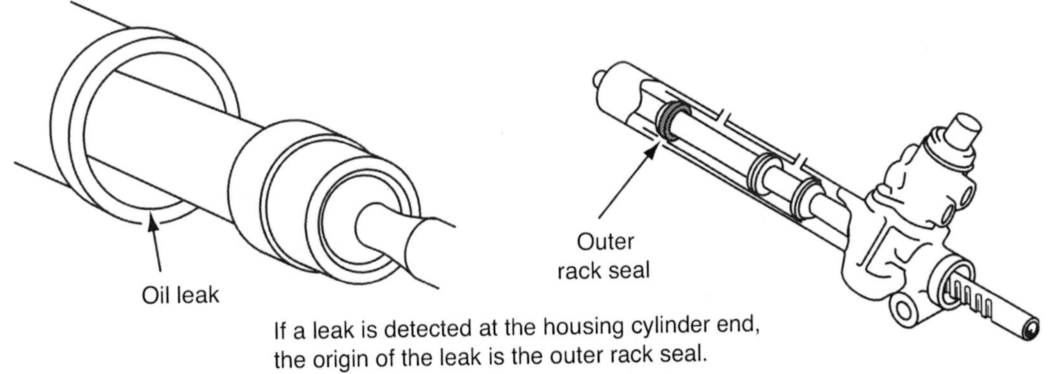


**Figure 30.** Oil leak diagnosis at outer rack seal.

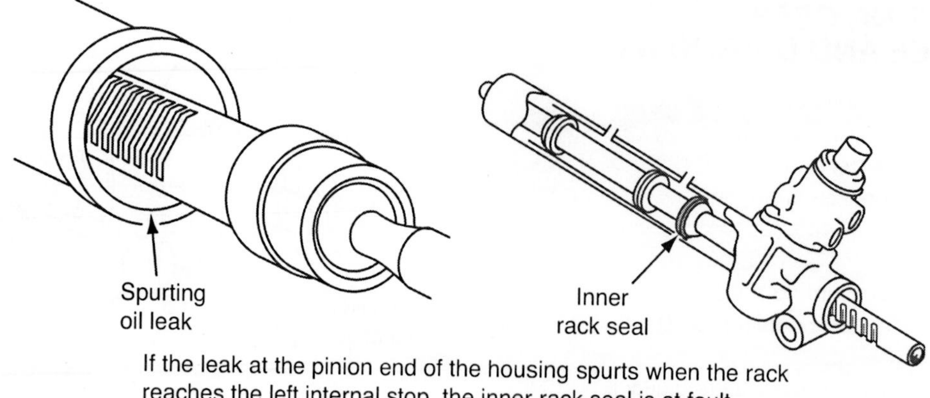
Spurting
oil leak
Inner
rack seal
If the leak at the pinion end of the housing spurts when the rack reaches the left internal stop, the inner rack seal is at fault.

**Figure 31.** Inner rack seal leak diagnosis.

The inner rack seal is defective if oil leaks from the pinion end of the housing when the rack reaches the left internal stop **(Figure 31)**. An oil leak at one rack seal may result in oil leaks from both boots because the oil may travel through the breather tube between the boots.

If an oil leak occurs at the pinion end of the housing and this leak is not influenced when the steering wheel is turned, the pinion seal is defective **(Figure 32)**.

If an oil leak occurs in the pinion coupling area, the input shaft seal is leaking **(Figure 33)**. This seal and the pinion seal will require replacement because the pinion seal must be replaced if the pinion is removed.

When the rack is removed, the inner and outer rack seals and the pinion seal must be replaced **(Figure 34)**. If oil leaks occur at fittings, these fittings must be torqued to the manufacturer's specifications. If the leak is still present, the

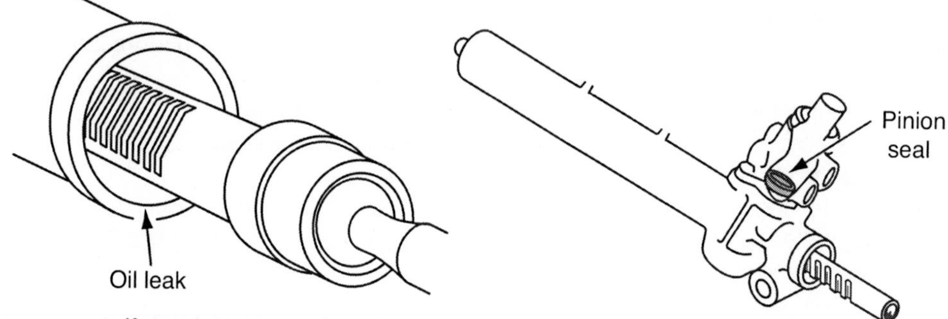
Pinion
seal
Oil leak
If you detect a leak at the pinion end of the housing and it is not influenced by the direction of the turn, the origin of the leak is the pinion seal.

**Figure 32.** Pinion seal leak diagnosis.

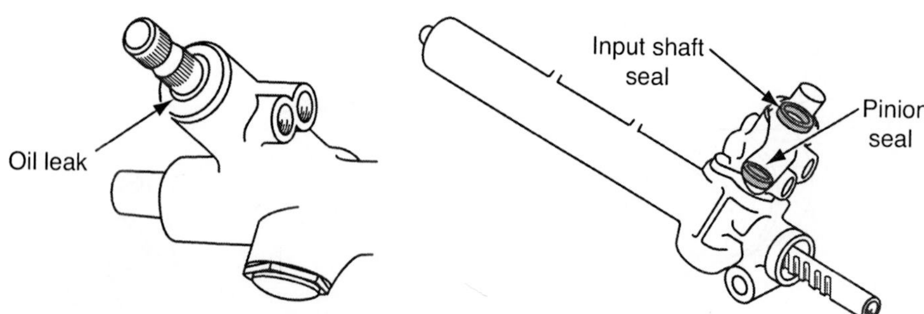
Input shaft
seal
Pinion
seal
Oil leak
If you discover a leak at the pinion coupling area, you will have to replace both the input shaft seal and the lower pinion seal.

**Figure 33.** Oil leak diagnosis in pinion coupling area.

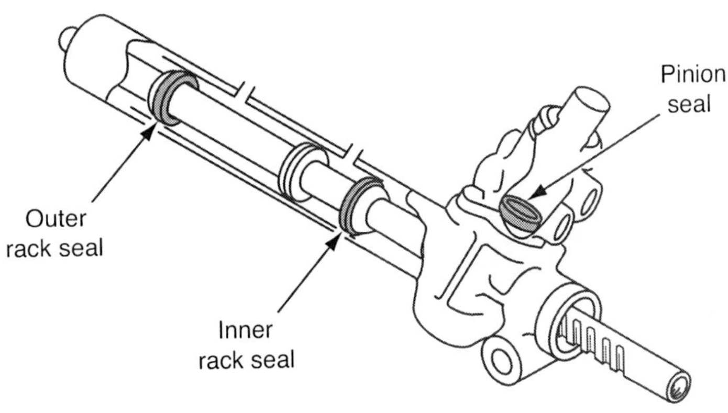


**Figure 34.** Rack seals and pinion seal.

line and fitting should be replaced. Leaks in the lines or hoses require line or hose replacement.

> **You Should Know** *If the engine has been running for a length of time, power steering gears, pumps, lines, and fluid may be very hot. Wear eye protection and protective gloves when servicing these components.*

## Power Rack-and-Pinion Steering Gear Turning Imbalance Diagnosis

The same amount of effort should be required to turn the steering wheel in either direction. A pressure gauge connected to the high-pressure hose should indicate the same pressure when the steering wheel is turned in each direction. Steering effort imbalance or lower power assist in each direction may be caused by defective rack seals **(Figure 35)**.

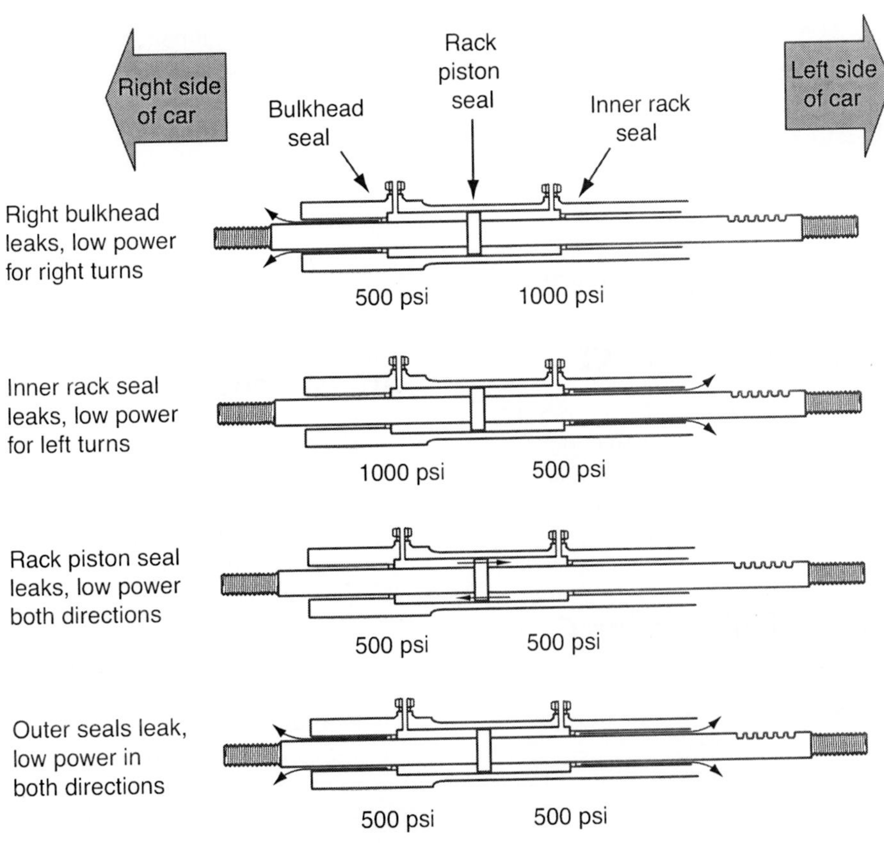


**Figure 35.** Effect of defective rack seals on steering effort imbalance and low power assist.

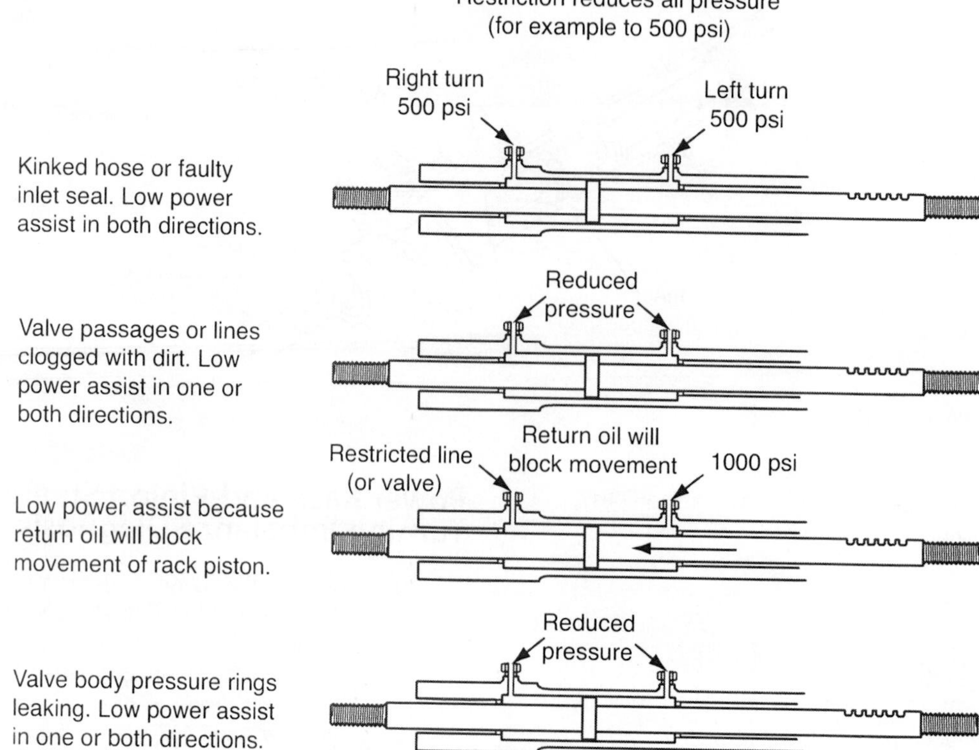
Restriction reduces all pressure
(for example to 500 psi)
Right turn
500 psi
Left turn
500 psi
Kinked hose or faulty inlet seal. Low power assist in both directions.
Reduced pressure
Valve passages or lines clogged with dirt. Low power assist in one or both directions.
Restricted line (or valve)
Return oil will block movement
1000 psi
Low power assist because return oil will block movement of rack piston.
Reduced pressure
Valve body pressure rings leaking. Low power assist in one or both directions.

**Figure 36.** Effect of worn rotary valve rings and seals or restricted lines or hoses on steering effort.

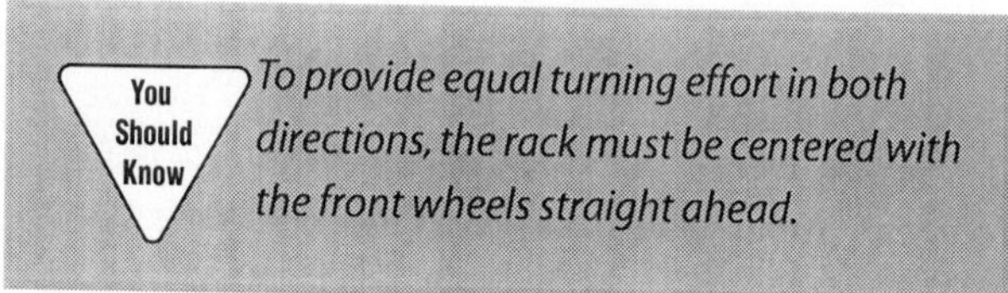
You Should Know

*To provide equal turning effort in both directions, the rack must be centered with the front wheels straight ahead.*

Steering effort imbalance or low power assist in both directions may also be caused by defective rotary valve rings and seals or restricted hoses and lines **(Figure 36)**.

## POWER STEERING GEAR SERVICE

Proper power steering gear service procedures are critical to maintain vehicle safety and normal steering control. For example, if steering gear mounting bolts are not tightened to the specified torque, these bolts may loosen and fall out resulting in a complete loss of steering control. This action may cause a collision and personal injury.

### Power Recirculating Ball Steering Gear Replacement

When the power steering gear is replaced, proceed as follows:

1. Disconnect the hoses from the steering gear and cap the lines and fittings to prevent dirt from entering the system.
2. Remove the pitman arm nut and washer and mark the pitman arm in relation to the shaft with a center punch. Use a puller to remove the pitman arm.
3. Disconnect the steering shaft from the worn shaft.
4. Remove the steering gear mounting bolts and remove the steering gear from the chassis.
5. Reverse steps 1 through 4 to install the steering gear. All bolts must be tightened to the specified torque. Be sure the pitman arm is installed in the original position.

### Power Recirculating Ball Steering Gear Adjustment

A loose worm shaft thrust bearing preload adjustment or sector lash adjustment causes excessive steering freeplay and steering wander. The power recirculating ball steering gear adjustment procedures may vary depending on the vehicle make and model year. Always follow the vehicle manufacturer's recommended procedure in the service manual. When the worm shaft thrust bearing preload adjustment is performed, use this procedure:

1. Remove the worm shaft thrust bearing adjuster plug locknut with a hammer and brass punch **(Figure 37)**.
2. Turn this adjuster plug inward or clockwise until it bottoms and tighten the plug to 20 ft. lbs. (27 N·m).
3. Place an index mark on the steering gear housing next to one of the holes in the adjuster plug **(Figure 38)**.

**Figure 37.** Removing worm shaft thrust bearing adjuster plug locknut.

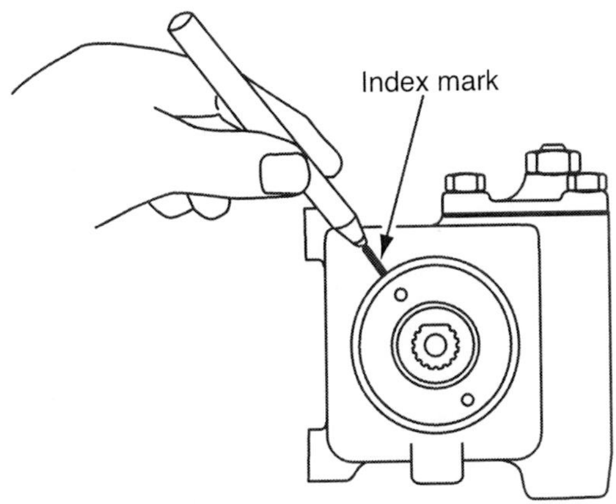


**Figure 38.** Placing index mark on steering gear housing opposite one of the adjuster plug holes.

4. Measure 0.50 inch (13 millimeters) counterclockwise from the index mark and place a second index mark at this position **(Figure 39)**.
5. Rotate the adjuster plug counterclockwise until the hole in the adjuster plug is aligned with the second index mark placed on the housing.

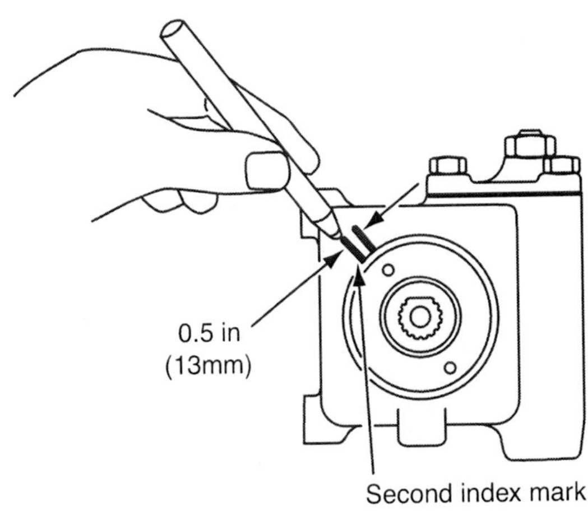


**Figure 39.** Measuring 0.50 inch (13 millimeters) counterclockwise from the index mark on the steering gear housing.

6. Install and tighten the adjuster plug locknut to the specified torque.

When the pitman sector shaft lash adjustment is performed, proceed as follows:

1. Rotate the stub shaft from stop to stop and count the number of turns.
2. Starting at either stop, turn the stub shaft back two-thirds of the total number of turns. In this position, the flat on the stub shaft should be facing upward and the master spline on the pitman shaft should be aligned with the pitman shaft backlash adjuster screw **(Figure 40)**.
2. Turn the pitman shaft backlash adjuster screw fully counterclockwise and then turn it clockwise one turn.
3. Use an inch-pound torque wrench to turn the stub shaft through a 45-degree arc on each side of the position in step 2. Read the over-center torque as the stub shaft turns through the center position.
4. Continue to adjust the pitman shaft adjuster screw until the torque is 6 to 10 in. lbs. ( 0.6 to 1.2 N·m) more than the torque in step 3.

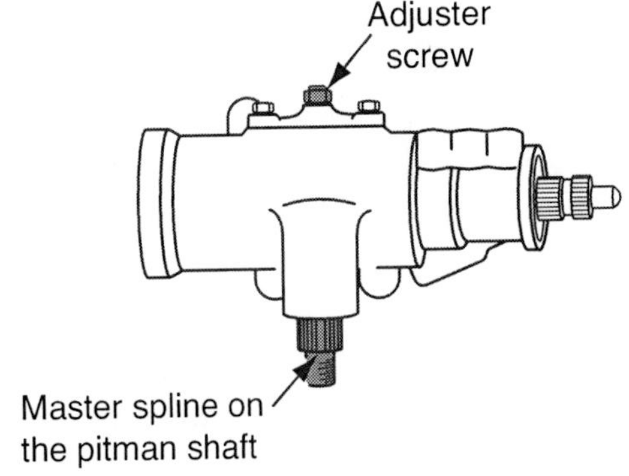


**Figure 40.** Pitman shaft master spline aligned with pitman backlash adjuster screw.

5. Hold the pitman shaft adjuster screw in this position and tighten the locknut to the specified torque.

## Power Recirculating Ball Steering Gear Side Cover O-Ring Replacement

Prior to any disassembly procedure, clean the steering gear with an approved cleaning solution. The steering gear service procedures vary depending on the make of gear. Always follow the vehicle manufacturer's recommended procedure in the service manual. The following is a typical side cover O-ring replacement procedure:

1. Loosen the pitman backlash adjuster screw locknut and remove the side cover bolts. Rotate the pitman backlash adjuster screw clockwise to remove the cover from the screw.
2. Discard the O-ring and inspect the side cover matching surfaces for metal burrs and scratches.
3. Lubricate a new O-ring with the vehicle manufacturer's recommended power steering fluid and install the O-ring.
4. Rotate the pitman backlash adjuster screw counterclockwise into the side cover until the side cover is properly positioned on the gear housing. Turn this adjuster screw fully counterclockwise and then one turn clockwise. Install and tighten the side cover bolts to the specified torque. Adjust the pitman sector shaft lash as explained previously.

## Power Recirculating Ball Steering Gear End Plug Seal Replacement

Follow these steps for end plug seal replacement:

1. Insert a punch into the access hole in the steering gear housing to unseat the retaining ring and remove the ring **(Figure 41)**.

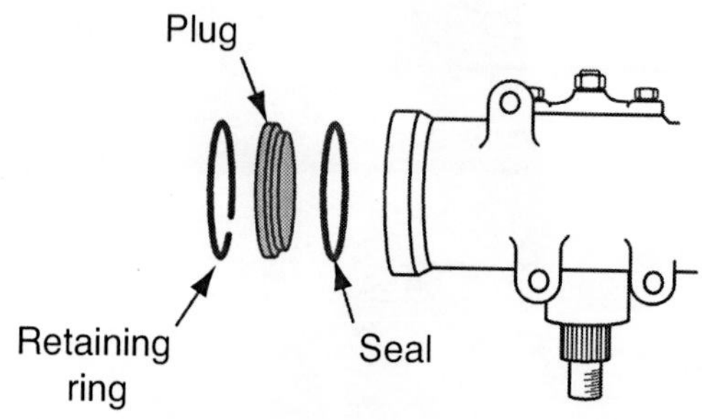


**Figure 41.** Removing steering gear end plug, retaining ring, and seal.

2. Remove the end plug and seal.
3. Clean the end plug and seal contact area in the housing with a shop towel.
4. Lubricate a new seal with the vehicle manufacturer's recommended power steering fluid and install the seal.
5. Install the end plug and retaining ring.

## Power Recirculating Ball Steering Gear Worm Shaft Bearing Adjuster Plug Seal and Bearing Replacement

Follow these steps for worm shaft bearing adjuster plug seal and bearing service:

1. Remove the adjuster plug locknut and use a special tool to remove the adjuster plug.
2. Use a screwdriver to pry at the raised area of the bearing retainer to remove this retainer from the adjuster plug **(Figure 42)**.
3. Place the adjuster plug face down on a suitable support, and use the proper driver to remove the needle bearing, dust seal, and lip seal.

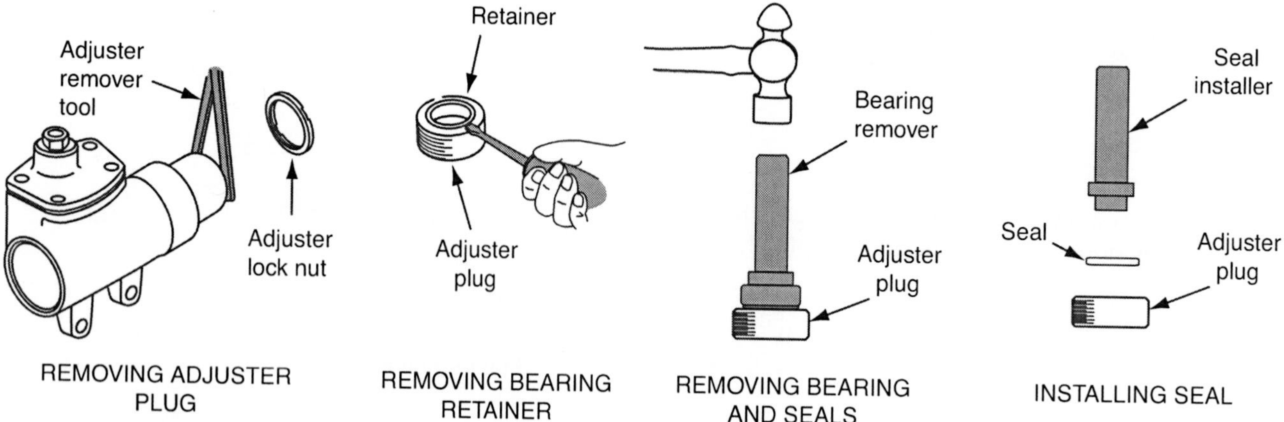


**Figure 42.** Removing and replacing worm shaft adjuster plug, bearing, and seal.

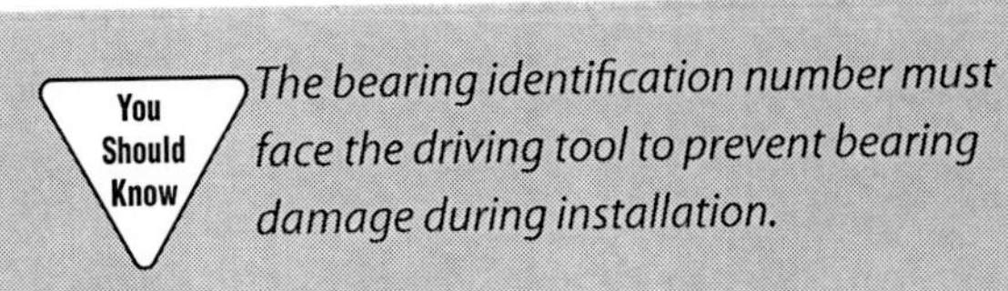

You Should Know
The bearing identification number must face the driving tool to prevent bearing damage during installation.


4. Place the adjuster plug outside face up on a suitable support and use the proper driver to install the needle bearing dust seal and lip seal.
5. Install the bearing retainer in the adjuster plug and lubricate the bearing and seal with the vehicle manufacturer's recommended power steering fluid.
5. Install the adjuster plug and locknut and adjust the worm shaft bearing preload as discussed previously.

## Power Rack-and-Pinion Steering Gear Adjustment

Follow this procedure to perform the rack bearing adjustment on a power rack-and-pinion steering gear.

1. Use the proper tool to rotate the rack spring cap 12 degrees counterclockwise.
2. Turn the pinion shaft fully in each direction and repeat this action.
3. Loosen the rack spring cap until there is no tension on the rack guide spring.
4. Place the proper turning tool and an inch-pound torque wrench on top of the pinion shaft.
5. Tighten the rack spring cap while rotating the pinion shaft back and forth **(Figure 43)**. Continue tightening the rack spring cap until the specified turning torque is indicated on the torque wrench.
6. Install and tighten the rack spring cap locknut.

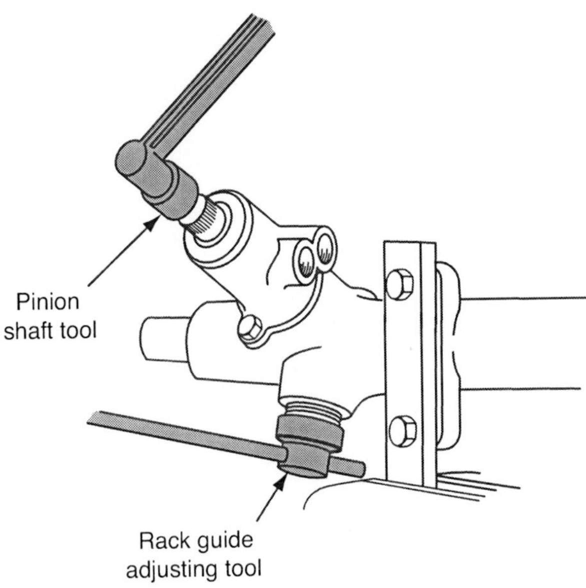


**Figure 43.** Adjusting pinion turning torque.

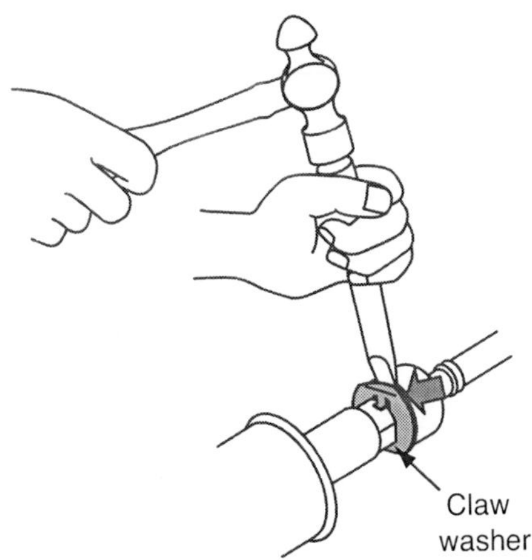


**Figure 44.** Removing claw washers from the inner tie-rod sockets.

## Removing and Replacing Tie-Rod Ends

To remove and replace the inner tie-rod ends on a power rack-and-pinion steering gear, follow this procedure:

1. Use a hammer and chisel to remove the claw washers from the inner tie-rod socket **(Figure 44)**.
2. Use a wrench to hold the rack from turning and loosen the inner tie-rod socket with the proper tool **(Figure 45)**.
3. Remove the inner tie-rod socket from the rack and install the new inner tie-rod socket and claw washer on the rack.
4. Hold the rack from turning and tighten the inner tie-rod socket to the specified torque.
5. Use a hammer and punch to stake the claw washer on the the inner tie-rod socket.

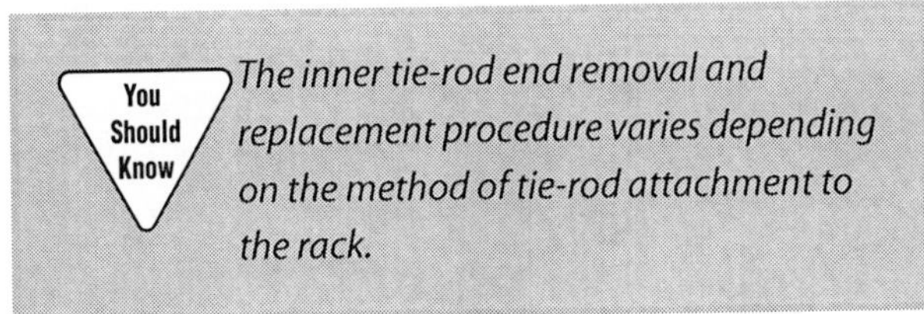

You Should Know
The inner tie-rod end removal and replacement procedure varies depending on the method of tie-rod attachment to the rack.


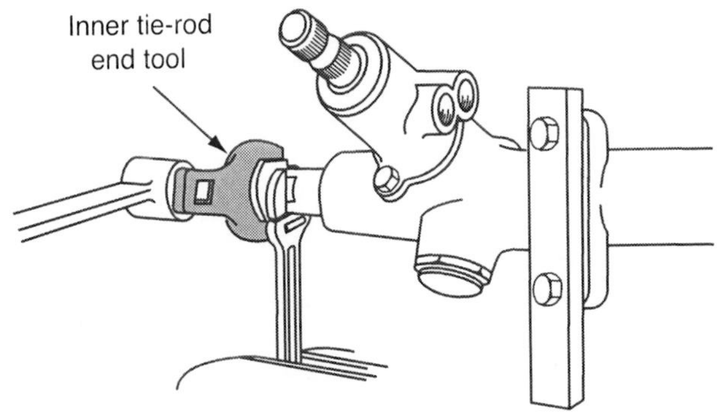


**Figure 45.** Loosening inner tie-rod sockets.

## Summary


- With the front wheels straight ahead in a recirculating ball steering gear, an interference fit exists between the ball nut teeth and the sector shaft teeth.
- When the front wheels are turned in, a power recirculating ball steering gear torsion bar deflection moves the spool valve inside the rotary valve. This valve movement directs the power steering fluid to the appropriate side of the recirculating ball piston to provide steering assist.
- Manual or power rack-and-pinion steering systems are lighter and more compact compared to recirculating ball steering gears and parallelogram steering linkages.
- The rack-and-pinion steering gear must be mounted so the rack maintains the tie-rods in a parallel position in relation to the lower control arms.
- The inner tie-rods are connected to the rack through spring-loaded inner ball sockets.
- The rack bearing and adjuster plug maintains proper preload between the rack and pinion teeth.
- Steering ratio is the relationship between steering wheel movement and front wheel movement to the right or left.
- A rattling noise and excessive steering free-play may be caused by loose gear mounting bolts, worn steering shaft U-joints, or worn flexible coupling in a recirculating ball steering gear system.
- A loose worm shaft bearing preload adjustment or sector lash adjustment on a manual or power recirculating ball steering gear causes excessive steering wheel free-play, steering wander, and reduced feel of the road.
- On a manual or power recirculating ball steering gear, the sector lash adjusting screw is tightened until the correct worm shaft turning torque is obtained as the worm shaft is rotated through the center position.
- Since bellows boots protect the inner tie-rod ends from contamination, boot condition should be checked each time undercar service such as oil and filter change or chassis lubrication is performed.
- Bellows boots that are cracked, split, leaking, or deteriorated must be replaced.
- Do not straighten bent steering components such as tie-rods.
- Always hold the rack while loosening the inner tie-rod ends.
- The rack adjuster plug must be adjusted until the correct turning torque is obtained on the pinion shaft.


## Review Questions

1. While discussing steering wheel free-play, Technician A says steering free-play refers to the amount of steering wheel rotation before the front wheels begin to move. Technician B says excessive steering free-play causes vehicle wander when the vehicle is driven straight ahead. Who is correct?
   A. Technician A
   B. Technician B
   C. Both Technician A and Technician B
   D. Neither Technician A nor Technician B
2. While discussing interference fit between the sector shaft teeth and ball nut teeth in a recirculating ball steering gear, Technician A says the interference fit between the sector shaft teeth and ball nut teeth is present when the front wheels are straight ahead. Technician B says the interference fit between the sector shaft teeth and the ball nut teeth provides the driver with a positive feel of the road. Who is correct?
   A. Technician A
   B. Technician B
   C. Both Technician A and Technician B
   D. Neither Technician A nor Technician B
3. While discussing steering ratio, Technician A says steering ratio is the relationship between the degrees of steering wheel rotation and the degrees of front wheel movement to the right or left. Technician B says if the lock-to-lock steering wheel rotation is 2.5 turns and the total front wheel movement is 60 degrees, the steering ratio is 17:1. Who is correct?
   A. Technician A
   B. Technician B
   C. Both Technician A and Technician B
   D. Neither Technician A nor Technician B
4. While discussing power rack-and-pinion steering gear operation, Technician A says fluid is directed to the appropriate side of the rack piston by the position of the rotary valve inside the spool valve. Technician B says fluid is directed to the appropriate side of the rack piston by the position of the rack in relation to the pinion. Who is correct?
   A. Technician A
   B. Technician B
   C. Both Technician A and Technician B
   D. Neither Technician A nor Technician B

5. Excessive steering wheel free-play may be caused by all of these problems *except*:
   A. a worn tie-rod end.
   B. a tight rack bearing adjustment.
   C. a loose steering gear mounting.
   D. a worn flexible coupling.
6. A vehicle with a power recirculating ball steering gear has a complaint of steering wander and reduced road feel. The most likely cause of this problem is:
   A. a loose sector lash adjustment.
   B. low power steering fluid level.
   C. a loose idler arm.
   D. a bent pitman arm.
7. A vehicle with a power recirculating ball steering gear requires excessive steering effort. All of these defects may be the cause of this complaint *except*:
   A. low power steering pump pressure.
   B. a restricted high-pressure power steering hose.
   C. underinflated front tires.
   D. loose outer tie-rod ends.
8. A power rack-and-pinion steering gear has a fluid leak at the pinion end of the housing and this leak is worse when the steering wheel is turned with the engine running. The most likely cause of this leak is:
   A. a worn input shaft seal.
   B. a leaking inner rack seal.
   C. a leaking outer rack seal.
   D. a leaking pinion seal.
9. A power steering gear has an excessive steering effort complaint and the power steering pump pressure is satisfactory. The most likely cause of this problem is:
   A. a leaking rack piston seal.
   B. a leaking inner rack seal.
   C. a leaking outer rack seal.
   D. a worn inner tie-rod end.
10. A manual rack-and-pinion steering gear has a loose mounting bushing. The most likely complaint caused by this problem is:
    A. reduced steering effort.
    B. reduced road feel.
    C. excessive steering wheel free-play.
    D. a rattling noise.
11. When the steering has free-play, there is some ____________ ______________ movement before the front wheels start to turn.
12. The worm shaft end plug provides a worm shaft bearing ____________ adjustment.
13. During a left turn, the power steering pump forces fluid into the __________ side of the rack chamber and fluid is removed from the __________ side of this chamber.
14. When adjusting a power rack-and-pinion steering gear, the rack ____________ ____________ is rotated while measuring the pinion ____________ ____________.
15. Explain the procedure for performing a rack bearing adjustment on a power rack-and-pinion steering gear.
16. Define the term "faster steering."
17. Explain the purpose of the bellows boots in a rack-and-pinion steering gear.
18. Explain how the spool valve is moved inside the rotary valve in a power rack-and-pinion steering gear.

# Bilingual Glossary

**Aiming pads** Small projections on the front of some headlights to which headlight aligning equipment may be attached.
***Patines de alineación*** *Pequeñas proyecciones en la parte delantera de algunos faros a los cuales se puede conectar el equipo de alineación de faros.*

**Air bag diagnostic monitor (ASDM)** An automotive computer responsible for air bag system operation.
***Monitor de bolsa de aire (ASD)*** *Una computadora automotiva que es responsable por la operación del sistema de la bolsa de aire.*

**Airflow restriction indicator** An indicator located in the air cleaner or intake duct to display air cleaner restriction by the color of the indicator window.
***Indicadora de restricción del aire*** *Un indicador ubicado en el limpiador de aire o en el ducto de entrada que indica la restricción del aire por medio del color de la ventanilla del indicador.*

**Alternating current** flows in one direction and then in the opposite direction.
***Corriente alterna*** *fluye en una dirección y luego en la dirección opuesta.*

**American Petroleum Institute (API) rating** A universal engine oil rating that classifies oils according to the type of service for which the oil is intended.
***Evaluación del Instituto Americano de Petroleo (API)*** *Una evaluación universal del aceite automotivo que clasifica a los aceites según el tipo del servicio que se le requiere.*

**Amperes** A measurement for the amount of current flowing in an electric circuit.
***Amperes*** *Una medida de la cantidad del corriente que fluye en un circuito eléctrico.*

**Analog meter** A meter with a pointer and a scale to indicate a specific reading.
***Medidora análoga*** *Un medidor con una indicadora y una gama para indicar una lectura específica.*

**Analog voltage signal** A signal that is continuously variable within a specific range.
***Señal análoga del voltaje*** *Una señal que es variable continuamente en una gama específica.*

**Angular bearing load** A load applied at an angle somewhere between the horizontal and vertical positions.
***Carga de soporte angular*** *Una carga aplicada en un ángulo que se encuentra entre las posiciones horizontales y verticales.*

**Atom** The smallest particle of an element.
***Átomo*** *La partícula más pequeña de un elemento.*

**Automatic Transmission Rebuilders Association (ATRA)** provides technical information to transmission shops and technicians.
***Asociación de Reconstrutores de Transmisiones Automáticas (ATRA)*** *Provea la información técnica a los talleres de transmisiones y a los técnicos.*

**Automotive dealership** sells and services vehicles produced by one or more vehicle manufacturers.
***Sucursal automotivo*** *vende y repara los vehículos producidos por un o varios fabricantes de automóviles.*

**Automotive Service Association (ASA)** promotes professionalism and excellence in the automotive repair industry through education, representation, and member services.
***Asociación de Servicio Automativo (ASA)*** *Promueve la profesionalidad y la excelencia en la industria de reparación automotiva por medio de la educación, la representación y los servicios de sus miembros.*

**Belleville spring** A diaphragm spring made from thin sheet metal that is formed into a cone shape.
***Resorte Belleville*** *Un resorte de tipo diafragma hecho de una hoja delgada de metal que es en la forma de un cono.*

**Bimetallic strip** contains two different metals fused together that expand at different rates and cause the strip to bend.
***Tira bimetálica*** *tiene dos metales distintos fundidos que expanden en dos velocidades causando que la tira se dobla.*

**Binary coding** The assignment of numeric values to digital signals.
***Codigo binario*** *La asignación de valores numéricos a las señales digitales.*

**Blowby** The amount of leakage between the piston rings and the cylinder walls.
***Fuga*** *La cantidad de fuga entre los anillos de pistones y las paredes del cilíndro.*

**Bolt diameter** The measurement across the threaded area of the bolt.
***Diámetro del perno*** *La medida a través del área enroscada del perno.*

**Bolt length** The distance from the bottom of the bolt head to the end of the bolt.
***Longitud del perno*** *La distancia de la parte inferior de la cabeza del perno a la extremidad del perno.*

**Bottom dead center (BDC)** The piston position in an engine when the piston is at the very bottom of its stroke.
***Punto muerto inferior (BDC)*** *La posición del pistón en un motor cuando el pistón esta en la punta más baja de su carrera.*

**Brake fade** Occurs when the brake pedal height gradually decreases during a prolonged brake application.
***Pérdida de adherencia de los frenos*** *Ocurre cuando la altura del pedal de frenos baja gradualmente durante la aplicación prolongada del freno.*

**Brake pedal free-play** The amount of brake pedal movement before the booster pushrod contacts the master cylinder piston.
***Juego del pedal del freno*** *La cantida del movimiento del pedal de freno antes de que la aumentadora de la varilla empujadora toca el piston del cilindro maestro.*

**Brake pressure modulator valve (BPMV)** An assembly containing the solenoid valves connected to each wheel in an antilock brake system (ABS).
***Válvula modulador del presión del freno (BPMV)*** *Una asamblea que contiene las válvulas del solenoide conectadas a cada rueda en un sistema de frenos antibloqueos (ABS).*

**British thermal unit** The amount of heat required to raise the temperature of 1 pound of water 1°F.
***Unedad termal (caloría) inglesa*** *La cantidad del calor requirido para subir la temperatura de un libra de agua por 1 grado F.*

**Canceling angles** are present when the vibration from one universal joint is canceled by an equal and opposite vibration from another universal joint.
***Ángulos de anulación*** *se presentan cuando la vibración de una junta universal se anula por una vibración igual e opuersta de otra junta universal.*

**Catalyst** A material that accelerates a chemical reaction without being changed itself.
***Catalizador*** *Una materia que acelera una reación química sín ser cambiada sí misma.*

**Catalytic converter** A component mounted in the exhaust system ahead of the muffler to reduce hydrocarbon (HC), carbon monoxide (CO), and oxides of nitrogen ($NO_x$) emissions.
***Convertidor catalítica*** *Un componente montado en el sistema de escape afrente del silenciador para reducir las emisiones del hidrocarburo (HC), el óxico de carbono y los óxidos de nitrógeno.*

**Cell group** A battery cell group contains alternately spaced positive and negative plates kept apart by porous separators.
***Grupo de células*** *Un grupo de células de batería que contiene las placas positivas y negativas puestas alternativamente y separadas por separadores porosos.*

**Central port injection (CPI)** A fuel injection system with a central port injector that supplies fuel to a mechanical poppet injector in each intake port.
***Inyección de puerta central (CPI)*** *Un sistema de inyección de combustible con una puerta de inyector central que suministra el combustible a un inyector mecánico de contrapunto en cada puerta de entrada.*

**Circuit breaker** A mechanical device that opens and protects an electric circuit if excessive current flow is present.
***Disyuntor*** *Un dispositivo mecánico que abre para proteјer un circuito eléctrico si se presenta un flujo de corriente excesivo.*

**Closed loop** A computer operating mode that occurs when the engine is partially warmed up. In this mode, the computer uses the oxygen sensor signal to control the air-fuel ratio.
***Bucle cerrado*** *Un modo de operación de computador que ocurre cuando el motor se ha calentado parcialmente. En este modo, la computador usa la señal del sensor de oxígeno para controlar la relación del aire al combustible.*

**Closed loop flushing procedure** flushes the refrigeration system with the system intact, without allowing refrigerant to escape to the atmosphere.
***Procedimiento de enjugado del bucle cerrado*** *Enjuega el sistema de refrigeración con el sistma intacto, sin permitir fugar el refrigerante a la atmósfera.*

**Clutch pedal free-play** The amount of clutch pedal movement before the release bearing contacts the pressure plate release levers.
***Juego del pedal del embrague*** *La cantidad del movimiento del pedal del embrague antes de que el cojinete de desembrague toca las palancas de desembrague de la placa de presión.*

**Cohesion** The tendency of engine oil to remain on the friction surfaces of engine components.
***Cohesión*** *La tendencia que tiene el aceite del motor de quedarse en las superficies de fricción de los componentes del motor.*

**Combination valve** A brake system valve that contains a metering valve, proportioning valve, and a switch to operate the brake system warning light.
***Válvula de combinación*** *Una válvula del sistema del frenado que tiene una válvula medidora, una válvula dosificante, y un interruptor que opera la lámpara testigo del sistema del frenado.*

**Compound** A compound is a material with two or more types of atoms.
***Compuesta*** *Una compuesta es una materia con dos tipos o más de átomos.*

**Compound gauge** has the ability to read two values.
***Indicador compuesta*** *tiene la habilidad de leer dos valores.*

**Compression ratio** The relationship between combustion chamber volume with the piston at TDC and the volume with the piston at BDC.
***Tasa de compresión*** *La relación entre el volumen de la cámara de combustión con el pistón en punto muerto superior y el volumen con el pistón en punto muerto inferior.*

**Conicity** A manufacturing defect in a tire caused by improperly wound plies that causes the tire to be slightly cone-shaped.
***Conicidad*** *Un defecto de fabricación del neumático debido a que los pliegues no sean envueltas correctamente causando que el neumático sea de una forma ligeramente cónica.*

**Connector position assurance (CPA)** pin holds two wiring connectors together so they cannot be separated until the pin is removed.
***Clavija aposicionadora de conectores (CPA)*** *sostiene juntos dos conectores de alambre para que no se separan hasta que se haya quitado la clavija.*

**Constant-running release bearing** makes light contact with the pressure plate release levers even when the clutch pedal is fully upward.
***Cojinete de desembrague contínuo*** *mantiene un contacto ligero con las palancas de desembrague de la placa de presión aún cuando el pedal del desembrague esta completamente en alto.*

**Constant velocity (CV) joints** transfer a uniform torque and a constant speed while operating at a wide variety of angles.
***Juntas de velocidad contínua*** *transfieren una torsión uniforme y una velocidad constante mientras que opera con una gran variedad de ángulos.*

**Conventional theory** states that current flows from positive to negative through an electric circuit.
***Teoría convencional*** *declara que el corriente fluye del positivo al negativo por medio de un circuito eléctrico.*

**Coolant hydrometer** A tester that measures coolant specific gravity to determine the antifreeze content of the coolant.
***Hidrómetro de refrigerante*** *Una probadora que mide la gravedad específica del refrigerante para determinar el contenido de anticogelante del refrigerante.*

**Corrosive** A material that is corrosive causes another material to be gradually worn away by chemical action.
***Corrosivo*** *Una materia que es corrosive causa que otra materia se deteriore por medio de acción química.*

**Coupling point** occurs in a torque converter when the impeller pump, turbine, and stator begin to rotate together.
***Punto de acoplamiento*** *ocurre en un convertidor de torsión cuando la bomba impelador, la turbina, y el estátor comienzan a girar juntos.*

**Critical speed** The rotational speed at which a component begins to vibrate.
***Velocidad crítica*** *La velocidad de rotación en la cual un componente comienza a vibrar.*

**Cross counts** The number of times the oxygen sensor signal switches from lean to rich in a given time period.
***Cuenta de cruzadas*** *El número de veces en que cambia el señal del sensor de oxígeno de pobre a rico en un periodo específico.*

**Crossflow radiator** A radiator in which the coolant flows horizontally from one radiator tank to the opposite tank.
***Radiador de flujo transversal*** *Un radiador en el cual el refrigerante fluye horizontalmente de un tanque de radiador al tanque opuesto.*

**Cross threaded** A defective thread condition caused by starting a fastener onto its threads when the fastener is tipped slightly to one side and the threads on the fastener are not properly aligned.
***Rosca corrida*** *Una condición defectuosa de rosca causada al poner un fijador en su rosca mientras que el fijador esté un poco inclinado hacia un lado y las roscas del fijador no estan alineadas correctamente.*

**Current diagnostic trouble codes (DTCs)** represent faults that are present during the diagnosis.
***Códigos diagnóstico de fallos corrientes (DTCs)*** *representan los fallos que se presentan durante el diagnósis.*

**Data link connector (DLC)** An electrical connector to which a scan tool may be connected to diagnose various electronic systems on a vehicle.
***Conector de enlace de datos (DLC)*** *Un conector eléctrico al cual se puede conectar una herramienta exploradora para diagnosticar varios sistemas electrónicas de un vehículo.*

**Detonation** The sudden explosion of the air-fuel mixture in the combustion chamber rather than a smooth burning action.
***Detonación*** *Una explosión repentina de la mezcla aire-combustible en la cámara de combustión en vez de una combustión sin variación.*

**Diagnostic procedure charts** located in service manuals to provide the necessary diagnostic steps in the proper order to diagnose specific vehicle problems.
***Gráfico de procedimientos diagnósticos*** *ubicados en los manuales de servicio para proveer los pasos necesarios de diagnóstico en su orden correcto para diagnosticar prolemas específicos del vehículo.*

**Digital meter** A meter with a digital reading that indicates a specific value.
***Medidor digital*** *Un medidor con una lectura digital que indica un valor específico.*

**Digital voltage signal** A voltage signal that is either high or low.
***Señal digital de voltaje*** *Un señal de voltaje que es alto o bajo.*

**Diode trio** A small device in some alternators that contains three diodes.
***Triple diodo*** *Un pequeño dispositivo en algunos alternadores que contiene tres diodos.*

**Direct current** flows only in one direction.
***Corriente directa*** *fluye en una sóla dirección.*

**Disc-type return spring** A beveled steel washer that flattens out when pressure is applied.
***Resorte de llamada en disco*** *Una arandela de acero biselado que se aplasta cuando se le aplica la presión.*

**Distributor ignition (DI)** An ignition system that uses a distributor to distribute the spark from the coil to the spark plugs.
***Encendido por distribuidor (DI)*** *Un sistema de encendido que usa un distribuidor para distribuir la chispa de la bobina a las bujías.*

**Dog teeth** A set of small gear teeth around the outer gear diameter with gaps between the teeth.
***Diente de sierra*** *Un grupo de dientes pequeños de engranaje alrededor del diámetro exterior del engranaje que tiene aberturas entre los dientes.*

**Downflow radiator** A radiator in which the coolant flows vertically downward from one radiator tank to the opposite tank.
***Radiador de flujo hacia abajo*** *Un radiador en el cual el refrigerante fluye verticalmente hacia abajo de un tanque del radiador al tanque opuesto.*

**Dual overhead cam (DOHC)** An engine with the intake and exhaust valves mounted in the cylinder heads and two camshafts are mounted on each cylinder head.
***Leva doble en cabeza (DOHC)*** *Un motor que tiene las válvulas de entrada e escape montadas en las cabezas de los cilíndros y dos árboles de leva montados en cada cabeza del cilíndro.*

**Electrolyte** A mixture of sulfuric acid and water in a lead acid battery.
***Electrólito*** *Una mezcla del ácido sulfúrico y el agua en una batería de plomo con ácido.*

**Electromagnetic induction** The process of inducing a voltage in a conductor by moving the conductor through the magnetic field or vice versa.
***Inducción electromagnético*** *El proceso de inducir un voltaje en un conductor moviendo el conductor al través de un campo magnético o vice versa.*

**Electromagnetic pickup coil** A pickup coil containing a permanent magnet surrounded by a coil of wire.
***Devanado detector electromagnético*** *Un devanado detector que contiene un imán permanente envuelto por un rollo de alambre.*

**Electrons** Negatively charged particles located on the various rings of an atom.
***Electrones*** *Las partículas de carga negativa que se encuentran en los varios anillos de un átomo.*

**Electron theory** states that electrons move from negative to positive through an electric circuit.
***Teoría atómica*** *declara que los electrones mueven del negativo al positivo al través de un circuito eléctrico.*

**Electronic ignition (EI)** An ignition system with a coil for each spark plug or pair of spark plugs.
***Encendido electrónico (EI)*** *Un sistema de encendido que tiene una bobina por cada bujía or par de bujías.*

**Element** An element is a liquid, solid, or gas with only one type of atom.
***Elemento*** *Un elemento es un líquido, un sólido o un gas con un sólo tipo de átomo.*

**Engine coolant temperature (ECT) sensor** A thermistor-type sensor that sends an analog voltage signal to the powertrain control module (PCM) in relation to coolant temperature.
***Sensor de temperatura del refrigerante de motor*** *Un sensor tipo termistor que manda un señal de voltaje analogo al módulo de control del trenmotor en relación a la temperatura del refrigerante.*

**Enhanced evaporative system** has the capability to check for leaks in the system and detect a 0.040-inch or 0.020-inch diameter leak depending on the vehicle model year.
***Sistema evaporativo intensificado*** *tiene la capacidad de buscar las fugas en el sistema y detectar una fuga con un diametro de 0.040- de una pulgada o 0.020 de una pulgada según el año del model del vehículo.*

**Environmental Protection Agency (EPA)** A United States government agency in charge of all aspects of environmental protection.
**Agencia de Protección del Medio Ambiente (EPA)** *Una agencia del gobierno de los Estados Unidos que se encarga de cada aspecto de la protección del medio ambiente.*

**EVAP emission leak detector (EELD)** produces a nontoxic smoke and blows this smoke into various components for leak-detection purposes.
***Detector de fuga de emisión EVAP (EELP)*** *produce un humo nontóxico y introduce el humo en varios componentes con el propósito de detectar las fugas.*

**Female-type quick disconnect coupling** A coupling attached to the end of a shop air hose. An opening in the center of the coupling is inserted over the male part of the quick disconnect coupling. The female part of the coupling contains the locking and release mechanism that allows easy locking and release action with the male part of the coupling.
***Acoplador de desconecta rápida tipo hembra*** *Un acoplador conectado a la extremidad de una manguera de aire del taller. Una abertura en el centro del acoplador se enchufa con el parte macho del acoplador de desconecta rápida. La parte hembra del acoplador contiene el mecanismo de conexión y desconexión que permite una acción fácil de conexión y desconexión con el parte macho del acoplador.*

**Fire and explosion data section** Part of an MSDS that informs employees regarding the flash point of hazardous materials.
***Sección de datos de fuegos e explosiones*** *Parte del MSDS que informa a los empleados sobre los puntos de explosión de las materiales peligrosas.*

**Fixed constant velocity (CV) joint** does not allow any inward or outward drive axle movement to compensate for changes in axle length.
***Junta de velocidad fija continua*** *no permite que el movimiento hacia adentro o afuera compensa los cambios de longitud del eje.*

**Flash diagnostic trouble codes (DTCs)** are displayed by the flashes of the malfunction indicator light (MIL) in the instrument panel.
***Códigos diagnósticos instantáneos de fallos (DTCs)*** *se exhiben por medio de los parpadeos de la lámpara indicadora de fallos (MIL)en el tablero de instrumentos.*

**Flash programming** The process of downloading computer software from a scan tool or PC into an onboard computer.
***Programación instantánea*** *El proceso de instalar las programaciones de computadora de una herramienta exploradora o de una computadora portátil (PC) a una computadora a bordo.*

**Flash-to-pass feature** allows a driver to move the signal light lever forward enough to operate the headlights on high beam to indicate he or she is going to pass the vehicle in front.
***Opción de señales rápidos*** *permite que un conductor mueva la palanca indicadora de señales hacia adelante lo bastante para operar los faros largos para indicar que el conductor va pasar el vehículo que va en frente.*

**Flex fan** A cooling fan with flexible blades that straighten out as engine speed increases to reduce engine power required to turn the fan.
***Ventilador flexible*** *Un ventilador de enfriamiento que tiene las aletas flexibles que se enderezan con más velocidad al motor para reducir la cantidad de potencia del motor requirido para operar el ventilador.*

**Flexible clutch** allows some movement between the clutch facings and the hub.
***Embrague flexible*** *permite algo de movimiento entre las caras del embrague y el cubo.*

**Floating caliper** A brake caliper that is designed so it slides sideways during a brake application.
***Calibre flotante*** *Un calibre de frenos diseñado para que se deslice horizontalmente durante la aplicación de los frenos.*

**Forward bias** An electrical connection between a voltage source and a diode that results in current flow through the diode.
***Polarización negativa frontal*** *Una conexión entre la fuente del voltaje y un diodo que resulta en un flujo de corriente por medio del diodo.*

**Four-channel antilock brake system (ABS)** has a pair of solenoids for each wheel to modulate the brake system pressure individually at each wheel.
***Sistema de frenos antibloqueantes de cuatro canales (ABS)*** *tiene un par de solenoides en cada rueda para modular la presión del sistema de frenos individualmente en cada rueda.*

**Fuel cells** electrochemically combine oxygen from the air with hydrogen from a hydrocarbon fuel to produce electricity.
***Acumuladores de combustible*** *combinan en un proceso electroquímico al oxígeno de la atmósfera con un combustible de hidrocarburo para producir la electricidad.*

**Full fielding** The process of supplying full field current to an alternator to obtain maximum output from the alternator.
***Campo completo*** *El proceso de proveer una corriente de campo completo a un alternador para obtener la salida máxima del alternador.*

**Fuse** A component that protects an electric circuit from excessive current flow.
***Fusible*** *Un componente que proteja un circuito eléctrico de un flujo excesivo de corriente.*

**Gear ratio** The ratio between the drive and driven gears.
***Relación de engranajes*** *La relación entre los engranajes de impulso y los de mando.*

**Good conductor** A good conductor has one, two, or three valence electrons that move easily from atom to atom.
***Buen conductor*** *Un buen conductor tiene uno, dos o tres electrones de valencia que viajan fácilmente de un átomo a otro átomo.*

**Grade marks** Radial lines on the head of a bolt in the USC system indicating the hardness of the steel in the bolt.
***Marcas de calidad*** *Las líneas radiales en la cabeza de un perno en el sistema USC indicando la dureza del acero contenido en el perno.*

**Graphing voltmeter** A voltmeter that indicates voltage readings in graph form.
***Volímetro gráfico*** *Un voltímetro que indica las lecturas del voltaje en una forma gráfica.*

**Grounded circuit** An unwanted copper-to-metal connection in an electric circuit.
***Circuito a tierra*** *Una conexión no deseada de cobre a metal en un circuito eléctrico.*

**Halogen** A term for a group of chemically treated nonmetallic elements including chlorine, fluorine, and iodine.
***Halógeno*** *Un término para un grupo de elementos nometálicos tratados quimicamente que incluyen al cloruro, la fluorina y el iodo.*

**Helical gears** have teeth that are cut at an angle to the center line of the gear.
***Engranaje helicoidal*** *tiene las dientes cortadas en un ángulo a la línea central del engranaje.*

**High impedance test light** A test light that contains a high resistance bulb.
***Lámpara de alta impedancia*** *Una lámpara de prueba que contiene un foco de alta resistencia.*

**History diagnostic trouble codes (DTCs)** are caused by intermittent faults and these faults are not present during the diagnosis.
***Códigos preexistentes de fallos diagnósticos (DTCs)*** *se causan por fallos intermitentes y éstos no están presentes durante un diagnósis.*

**Hotchkiss drive** An open drive shaft with two universal joints.
***Árbol Hotchkiss*** *Un árbol motor abierto que tiene dos juntas universales.*

**Hybrid organic additive technology (HOAT)** A special coolant additive package to help prevent cooling system corrosion.
***Tecnología de aditivos orgánicos híbrido (HOAT)*** *Un paquete de aditivos refrigerantes especiales para ayudar en la prevención de la corrosión del sistema de enfriamiento.*

**Hybrid vehicle** A vehicle with two power sources such as a gasoline engine and an electric motor.
***Vehículo híbridos*** *Un vehículo con dos fuentes de potencia tal como un motor de gasolina y un motor eléctrica.*

**Hydrometer** A tester that measures the specific gravity of a liquid.
***Hidrómetro*** *Un probador que mide la gravedad específica de un líquido.*

**Ignitable** A substance that is ignitable can be ignited spontaneously or by another source of heat or flame.
***Inflamable*** *Una substancia que es inflamable se puede encender espontáneamente o por medio de otro origen del calor o llama.*

**Incandescence** The process of changing electrical energy to heat energy in a light bulb to produce light.
***Incandencia*** *El proceso de cambiar la energía eléctrica a la energía de calor en una bombilla eléctrica para producir la luz.*

**Independent repair shop** An independent repair shop is privately owned and operated without being affiliated with a vehicle manufacturer, automotive parts manufacturer, or chain organization.
***Taller de reparaciones independiente*** *Un taller de reparaciones independient es poseído e operado sín afilicación con el fabricante de un vehículo, el fabricante de partes automotivos o una organización de sucursal.*

**Inductive pickup** A type of pickup that senses the amount of current flow from the magnetic strength surrounding a conductor.
***Captador inductivo*** *Un tipo de captador que percibe la cantidad del flujo del corriente por la fuerza del campo magnético que rodea un conductor.*

**Infinite ohmmeter reading** A resistance reading on an ohmmeter that is beyond measurement.
***Lectura infiníta de ohmímetro*** *Una lectura de la resistencia en un ohímetro que no se puede medir.*

**Injector pulse width** The length of time in milliseconds that an injector is open.
***Apertura del pulso del inyector*** *La duración del tiempo en milisegundos que se queda abierto un inyector.*

**In-phase universal joints** are all on the same plane.
***Juntas universales en fase*** *todos están en el mismo plano.*

**Insulator** An insulator has five or more valence electrons which do not move easily from atom to atom.
***Aislador*** *Un aislador tiene cinco o más electrones de valencia que no se mueven fácilmente de un átomo a otro átomo.*

**International Automotive Technicians Network (iATN)** A large group of automotive technicians in many countries that share technical knowledge with other members through the Internet.
***Red International de Técnico Automotivos (IATN)*** *Un grupo grande de técnicos automotivos en muchos paises que comparten la experiencia técnica con otros miembros por medio del Internet.*

**International System** A system of weights and measures in which every unit may be multiplied or divided by 10 to obtain larger or smaller units.
***Sistema Internacional*** *Un sistema de pesos y medidas en el cual cada unedad se puede multiplicar o dividirse por 10 para obtener una unedad más grande o chica.*

**Jounce travel** Upward tire and wheel movement.
***Sacudo*** *Un movimiento hace arriba del pneumático y la rueda.*

**Kinetic energy** Energy in motion.
***Energía kinética*** *La energía en movimiento.*

**Laboratory (lab) scope** A scope that displays various waveforms across the screen with a very fast trace.
***Aparato óptico del laboratorio*** *Un aparato que demuestra varias ondas en la pantalla con un registro muy rápido.*

**Latent heat of condensation** The heat released during condensation from a gas to a liquid.
***Calor latente del condensación*** *El calor soltado durante la condensación de un gas a un líquido.*

**Latent heat of vaporization** The amount of heat required to change a liquid to a gas after the liquid reaches the boiling point.
***Calor latente de la vaporización*** *La cantidad del calor requirido para cambiar un líquido a un gas después de que el líquido ha llegado a su punto de ebullición.*

**Lateral tire runout** The amount of sideways wobble in a rotating tire.
***Corrimiento lateral del pneumático*** *La cantidad de bamboleo en un pneumático en rotación.*

**Left-hand thread** A fastener with a left-hand thread must be rotated counterclockwise to tighten the fastener.
***Rosca con paso a izquierdas*** *Un fijador con un paso a izquierdas se tiene que girar en sentido contrario a la agujas del reloj para apretarlo.*

**Line pressure** The pressure delivered from the pressure regulator valve in an automatic transmission.
***Presión de la línea*** *La presión que se entrega de la válvula reguladora de presión en una transmisión automática.*

**Male-type quick disconnect coupling** The part of a coupling that is threaded into an air-operated tool. This male coupling contains grooves and ridges to provide a locking action with the female part of the coupling.
***Acoplador de desconecta rápida tipo macho*** *La parte de un acoplador enroscada a una herramienta pneumática. El acoplador macho contiene las ranuras y los bordes para proveer una acción de conexión con la parte hembra del acoplador.*

**Malfunction indicator light (MIL)** A warning light in the instrument panel that is illuminated by the PCM to indicate engine computer system defects.
***Lámpara indicadora de fallos (MIL)*** *Una lámpara de avisos en el tablero de instrumentos iluminada por el PCM para indicar los defectos del sistema de computadora del motor.*

**Material Safety Data Sheets** provide all the necessary data about hazardous materials.
***Hojas de Datos de Seguridad de los Materiales*** *proveen todos los datos necesarios pertinente a las materiales peligrosas.*

**Message center** A digital display in the instrument panel where various warning messages are displayed.
***Centro de avisos*** *Una presentación digital en el tablero de instrumentos en donde aparecen varios indicadores de aviso.*

**Microprocessor** The decision-making and calculating chip in a computer.
***Microprocesor*** *El componente que toma las desiciones y calcula en una computadora.*

**Mobile Air Conditioning Society Worldwide (MACS Worldwide)** provides technical and business information to automotive air conditioning shops and technicians.
***Asociación de Acondicionadores de Aire Mundial (MACS Worldwide)*** *provee la información técnica y de negocios para los talleres y los técnicos de aire acondicionado.*

**Molecule** A molecule is the smallest particle that a compound can be divided into and still retain its characteristics.
***Moécula*** *Una molécula es la partícula más pequeña al que se puede dividir una compuesta y aún mantener sus características.*

**Monolith-type catalytic converter** contains a catalyst-coated monolith that is similar to a honeycomb.
***Convertidor catalítico tipo monolítico*** *contiene un monolito parecido a una panal cubierto de una catalista.*

**Multiport fuel injection (MFI)** A fuel injection system that opens two or more injectors simultaneously.
***Inyección de combustible de puertas múltiples*** *Un sistema de inyección de combustible que abre dos o más inyectores a la vez.*

**National Automotive Technicians' Education Foundation (NATEF)** An independent affiliate of ASE responsible for evaluating and certifying automotive, autobody, and medium/heavy truck training programs.
***Fondación Nacional de Educación de Técnicos Automotivos (NATEF)*** *Un afiliado independiente del ASE que se encarga de la evaluación y certificación de las programas de entrenación automotivas, de carrocería y de camiones grandes/medianos.*

**National Institute for Automotive Service Excellence (ASE)** An independent organization responsible for testing and certification of automotive, autobody, and medium/heavy truck technicians in the United States.
***Instituto Nacional de Excelencia en Servicio Automotivo (ASE)*** *Una organización independiente que se encarga de la examinación y certificación de las programas de entrenación automotivas, de carrocería y de camiones grandes/medianos.*

**Neutrons** Particles with no electrical charge that are located with the protons in the nucleus of an atom.
***Neutrones*** *Las partículas que no tienen una carga eléctrica que se encuentran con los protones en el núcleo de un átomo.*

**Normally closed contacts** are closed when no pressure is supplied to the unit.
***Contactos normalmente cerrados*** *están cerrados mientras que no se les aplica la presión.*

**Normally open contacts** are open when no pressure is supplied to the unit.
***Contactos normalmente abiertos*** *están abiertos mientras que no se les aplica la presión.*

**Occupational Safety and Health Administration (OSHA)** regulates working conditions in the United States.
***Acta de Seguridad y Salud en el Lugar de Trabajo (OSHA)*** *Regula las condiciones del trabajo en los Estados Unidos.*

**Ohms** A measurement for resistance in an electric circuit.
***Ohmios*** *Una medida de la resistencia en un circuito eléctrico.*

**On-board diagnostic II (OBD II)** A type of computer system installed in all cars and light-duty trucks manufactured since 1996.
***Diagnóstico A bordoll (OBDII)*** *Un tipo de sistema de computadora que ha sido instalado en todos los coches y las camionetas de carga ligera fabricados des desde el 1996.*

**Open circuit** An unwanted break in a electric circuit.
***Circuito abierto*** *Un corte no deseado en un circuito eléctrico.*

**Open circuit voltage test** A voltage test that is performed with no electrical load on the battery.
***Prueba de circuito abierto*** *Una prueba de voltaje que se efectúa mientras que no tenga una carga eléctrica en la batería.*

**Open loop** A computer operating mode that occurs during engine warmup. In this mode, the computer ignores the oxygen sensor signal and the computer program and other parameters control the air-fuel ratio.
***Bucle abierto*** *Un modo de operación de una computadora que ocurre durante el calentamiento del motor. En este modo, la computadora no hace caso a la señal del sensor de oxígeno y la programa de computadora y otros parámetros controlan la relación de combustible al aire.*

**Opposed cylinder engine** An engine with the cylinder banks mounted at 180 degrees in relation to each other.
***Sistema de cilíndros opuestos*** *Un motor que tiene los bancos de cilíndros montados a los 180 grados en relación al uno al otro.*

**Out-of-round** A variation in drum diameter at different locations.
***Fuera de redondo (ovulado)*** *Una variación del diámetro del tambor en varios lugares.*

**Outside micrometer** A precision measuring instrument designed to measure the outside diameter of various components.
***Micrómetro del exterior*** *Un instrumento de medidas precisas diseñado a medir el diámetro exterior de varios componentes.*

**Overhead cam (OHC)** An engine with the intake valves, exhaust valves, and the camshaft mounted in the cylinder heads.
***Árbol de leva en cabeza (OHC)*** *Un motor con las válvulas de entrada, las válvulas de salida y el árbol de levas montado en la cabeza del cilíndro.*

**Overhead valve (OHV)** An engine with the valves mounted in the cylinder heads and the camshaft and valve lifters located in the engine block.
***Válvula en cabeza (OHV)*** *Un motor con las válvulas montadas en las cabezas del cilíndro y las levantaválvulas ubicadas en el bloque motor.*

**Oxidation** A process that occurs when some engine oil combines with oxygen in the air to form an undesirable compound.
***Oxidación*** *Un proceso que ocurre cuando un poco del aceite del motor combina con el oxígeno en el ambiente para formar una compuesta no deseada.*

**Oxides of nitrogen ($NO_x$)** An exhaust emission caused by nitrogen and oxygen combining at high temperature above 2,500°F in the combustion chambers.
***Óxidos de nitrógeno ($No_x$)*** *Una emisión causada por la combinación del nitrógeno y el oxígeno en altas temperaturas mas del 2,500 F en las cámaras de combustión.*

**Pellet-type catalytic converter** contains 100,000 to 200,000 small pellets coated with a catalyst material.
***Convertidor catalítico tipo pastilla*** *contiene 100,000 a 200,000 pastillitas cubiertas con una materia catalítica.*

**Periodic table** A listing of the known elements according to their number of protons and electrons.
***Tabla periódica*** *Una lista de los elementos conocidos según su número de protones e electrones.*

**Permissible exposure limit (PEL)** A section on MSDS sheets that indicates the maximum amount of hazardous material in the air to which a person may be exposed on a daily basis without harmful effects on the human body.
***Límite permisíble de exposición (PEL)*** *Una sección de las hojas de MSDS que indican la cantidad máxima delas materiales peligrosos en la atmósfera a la cual una person puede exponerse diariamente sín ocasionar efectos dañosos al cuerpo humano.*

**Photochemical smog** is formed by sunlight reacting with hydrocarbons (HC) and oxides of nitrogen ($NO_x$).
***Contaminación fotoquímico*** *formado por uaa reacción del luz del sol con los hidrocarburos (HC) y los óxidos de nitrógeno*

**Physical data section** Part of an MSDS that provides information about the hazardous material, such as appearance and odor of the material.
***Sección de datos físicos*** *Parte de un MSDS que provee la información sobre las materiales peligrosas, tal tomo su aparencia y el olor de la materia.*

**Plunging constant velocity (CV) joint** allows inward and outward drive axle movement to compensate for changes in axle length.
***Junta de velocidad descendida continua (CV)*** *Permite el movimiento hacia adentro y hacia afuera del eje para compensar la longitud del eje.*

**Ported manifold vacuum** is supplied from above the edge of the throttle valve(s).
***Vacío de la manívela de la puerta*** *se suministra por encima del borde de la válvula de toma de vapor.*

**Positive crankcase ventilation (PCV) valve** A valve that controls the amount of crankcase vapors flowing through the valve into the intake manifold.
***Válvula de ventilación positiva del cárter (PCV)*** *Una válvula que controla la cantidad de los vapores del cárter que fluyen por la válvula a la manívela de entrada.*

**Preignition** The ignition of the air-fuel mixture in the combustion chamber by means other than the spark plug, such as a hot piece of carbon.
***Pre encendido*** *El encendido de la mezcla del aire-combustible en la cámara de combustion por otro metodo que por la bujía, tal como por un pedazo caliente del carbono.*

**Printed circuit board** A thin, insulating board used to mount and connect various electronic components such as resistors, diodes, switches, capacitors, and microchips in a pattern of conductive lines.
***Placa de circuítos impresa*** *Una placa aisladora delgada que se usa para montar y conectar a varios componentes electrónicos tal como los resistores, los diodos, los interruptores, los capacitores, y los microchips en un dibujo de líneas conductivas.*

**Protons** Positively charged particles located in the nucleus of an atom.
***Protones*** *Las partículas de carga positiva ubicados en el núcleo de un átomo.*

**Pulse width modulation (PWM)** An on/off voltage signal with a variable on time.
***Modulación de separación de impulsos (PWR)*** *Una señal de apagado/prendido con un tiempo prendido variado.*

**Quad driver** A group of four transistors in a computer chip that operates some of the output controls.
***Exitadora de cuatro*** *Un grupo de cuatro transistores en un chip de computadora que operan algunos de los controles de salida.*

**Quench area** The area in the combustion chamber near the metal surfaces where the flame front is extinguished.
***Área extinctor*** *El área en la cámara de combustión cerca de las superficies de metal en dónde se extingue la llama delantera.*

**Quick-lube shop** specializes in automotive lubrication work.
***Taller de lubricación rápida*** *especializa en el trabajo de lubricacián automotivo.*

**Radial bearing load** A load applied in a vertical direction.
***Carga radial*** *Una carga aplicada en una dirección vertical.*

**Radial runout** The amount of diameter variation in a tire.
***Corrimiento radial*** *La cantidad de variación del diámetro de un pneumático.*

**Reactive** A material that is reactive reacts with some other chemicals and gives off a gas(es) during the reaction.
***Reactivo*** *Una material que es reactivo reacciona con otras sustancias químicas y crea un gas o los gases durante la reacción.*

**Reactivity and stability section** Part of an MSDS that informs employees regarding the mixing of other materials with a hazardous material.
***Sección de reacción e estabilidad*** *Parte de un MSDS que informa los empleados sobre la mezla de otras materiales con una material peligrosa.*

**Rebound travel** Downward tire and wheel movement.
***Viaje de rebote*** *Un movimiento hacia abajo de la rueda y el pnuemático.*

**Reciprocating** An up-and-down or back-and-forth motion.
***Recíproco*** *Un movimiento de arriba-y-abajo o de delante-y detrás.*

**Resource Conservation and Recovery Act (RCRA)** States that hazardous material users are responsible for hazardous materials from the time they become a waste until the proper waste disposal is completed.
***Acta de Conservación y Recobro de Recursos (RCRA)*** *Declara que los consumidores de materiales peligrosas son responsables por las materiales peligrosas del tiempo de que se convierten en deshechos hasta que se haya deshechado correctamente y completamente de ellas.*

**Reverse bias** An electrical connection between a voltage source and a diode that causes the diode to block voltage.
***Polarización reversa*** *Una conexión entre una fuente de voltaje y un diodo que causa que el diodo bloquea el voltaje.*

**Right-hand thread** A fastener with a right-hand thread must be rotated clockwise when tightening the fastener.
***Rosca a la derecha*** *Un fijador con una rosca a la derecha debe girarse en el sentido de las agujas del reloj para apretarse.*

**Rigid clutch** Does not allow any movement between the clutch facings and the hub.
***Embrague rígido*** *No permite ningún movimiente entre las caras del embrague y el cubo.*

**Rotary flow** occurs when the fluid in a torque converter moves in a circular direction with the impeller pump and turbine.
***Flujo rotatorio*** *ocurre cuando el fluido en un convertidor del par mueva en una dirección circular con la bomba impulsor y la turbina.*

**Semiconductor** A semiconductor has four valence electrons and these materials are used in the manufacture of diodes and transistors.
***Semiconductor*** *Un semiconductor tiene cuatro electrones de valencia y estas materiales se usan en la fabricación de los diodos y los transistores.*

**Separators** are positioned between each pair of battery plates to keep these plates from touching each other.
***Separadores*** *se posicionan entre cada par de placas de la batería para prevenir que estas placas se tocan.*

**Sequential fuel injection (SFI)** An injection system in which each injector is opened individually.
***Inyección de combustible en secuencia (SFI)*** *Un sistema de inyección en el cual cada inyector se abre individualmente.*

**Servo action** occurs when the operation of the primary brake shoe applies mechanical force to help apply the secondary shoe.
***Acción Servo*** *ocurre cuando la operación de la zapata principal del freno aplica la fuerza mecánica para asistir en la aplicación de la zapata segundaria.*

**Shorted circuit** An unwanted copper-to-copper connection in an electric circuit.
***Corto circuito*** *Una conexión no deseada de cobre-a-cobre en un circuito eléctrico.*

**Society of Automotive Engineers (SAE) rating** A universal oil rating that classifies oil viscosity in relation to that atmospheric temperature in which the oil will be operating.
***Grado de la Asociación de Ingenieros Automotivos (SAE)*** *Un grado universal que clasifica la viscosidad del aceite con relación a la temperatura del ambiente en la cual va operar el aceite.*

**Solenoid** An electro-mechanical device used to effect a push-pull mechanical operation using electric current.
***Solenoide*** *Un dispositivo electro-mecánico que sirve para efectuar una operación mecánica de tire-jale usando un corriente eléctrico.*

**Spark plug heat range** indicates the ability of a spark plug to dissipate heat.
***Gama de calor de la bujía*** *indica la habilidad de una bujía en dispersar el calor.*

**Spark plug reach** The distance from the lower end of the plug shell to the shoulder on the shell.
***Alcanze de la bujía*** *La distancia desde la extremidad inferior del cuerpo de la bujía hasta el collarín del cuerpo.*

**Specialty Shop** specializes in one type of repair work.
***Taller especializada*** *especializa en un tipo de reparación.*

**Specific gravity** is the weight of a liquid, such as battery electrolyte, in relation to an equal volume of water.
***Gravedad específica*** *es el peso de un líquido, tal como la electrólito de la batería, con relación a un volumen identico del agua.*

**Speed density system** A fuel injection system in which the two main signals used for air-fuel ratio control are engine speed and manifold absolute pressure (MAP).
***Sistema de densidad de velocidad*** *Un sistema de inyección de combustible en el cual las dos señales principales usados para el control de la relación del aire-combustible son la velocidad del motor y la presión absoluta del manívela (MAP).*

**Sprung weight** The weight of the chassis supported by the springs.
***Peso provisto de muelles*** *El peso de la carrocería apoyado por los muelles.*

**Stabilizer bar** A long, spring steel bar connected from the crossmember to each lower control arm to reduce body sway.
***Barra estabilizadora*** *Una barra larga de acero de resorte conectada desde el miembre transversal a cada brazo de control inferior para reducir la oscilación de la carrocería.*

**Staked hub nut** A nut that is secured to a drive axle by staking the nut lip into a recess in the axle.
***Tuerca de cubo con yunque*** *Una tuerca que se enclavija al eje de mando enclavijando el borde de la tuerca en una ranura en el eje.*

**Starter drive** A mechanical device that connects and disconnects the starter armature shaft and the flywheel ring gear.
***Acoplamiento del motor de arranque*** *Un dispositivo mecánico que conecta y disconecta el árbol de la armadura del arranque y el anillo dentado del volante.*

**Static pressure** The pressure in a system when the system is inoperative.
***Presión estática*** *La presión en un sistema cuando el sistema no está en operación.*

**Stoichiometric air-fuel ratio** The ideal air-fuel ratio of 14.7:1 on a gasoline engine that provides the best engine performance, economy, and emissions.
***Relación estequiométrico de aire-combustible*** *La relación ideal de aire-combustible de 14.7:1 en un motor de gasolina que provee lo mejor en ejecución, economía y emisiones de un motor.*

**Straight-cut gears** have teeth that are parallel to the gear centerline.
***Engranaje de corte recta*** *tiene los dientes que son paralelos a la línea central del engranaje.*

**Strut chatter** A clicking noise that is heard when the front wheels are turned to the right or left.
***Traqueteo de las manguetas*** *Un ruido de chasquido que se oye cuando las ruedas delanteras se giran a la derecha o a la izquierda.*

**Synchronizer** brings two components to the same speed to allow shifting without gear clashing.
***Sincronizador*** *iguala la velocidad de dos componentes para permitir el cambio de velocidad sin un choque entre los engranajes.*

**Synthetic lubricant** is developed in a laboratory.
***Lubricante sintético*** *se desarrolla en un laboratorio.*

**Thermistor** A special resistor that increases its resistance as the temperature decreases.
***Termistor*** *Un resistor especial que incrementa su resistencia al caer la temperatura.*

**Thermodynamics** The relationship between heat energy and mechanical energy.
***Termodinámica*** *La relación entre la energía de calor y la enería mecánica.*

**Thread depth** The height of the thread from its base to the top of its peak.
***Profundidad del hilado*** *La altura del filete desde su fondo a lo más alto de su pico.*

**Thread pitch** In the USC system, thread pitch is the number of threads per inch.
***Paso del filete*** *En el sistema USC, el paso del filete es el número de hilado por pulgada.*

**Three-channel antilock brake system (ABS)** has a pair of solenoids for each front wheel, but only one pair of solenoids for both rear wheels.
***Sistema de frenos antibloqueantes de tres canales (ABS)*** *tiene un par de solenoides en cada rueda delantera, pero sólo un par de solenoides para ambas ruedas traseras.*

**Threshold limit value** A section on MSDS sheets that indicates the maximum amount of hazardous material in the air to which a person may be exposed on a daily basis without harmful effects on the human body.
***Tasa umbral del límite*** *Una sección de las hojas MSDS que indican la cantidad máxima de la material peligrosa en el atmósfera a la cual una persona puede exponerse diariamente sín ocasionar efectos dañosos en el cuerpo humano.*

**Thrust bearing load** A load applied in a horizontal direction.
***Carga límite de empuje*** *Una carga aplicada en una dirección horizontal.*

**Top dead center (TDC)** The piston position in a engine when the piston is at the very top of its stroke.
***Punto muerto superior (TDC)*** *La posición del piston en un motor cuando el piston esta en la punta más alta de su carrera.*

**Torque** A twisting force.
***Torsión*** *Una fuerza que tuerce.*

**Torque steer** A condition in which the vehicle steering pulls to one side during hard acceleration.
***Dirección con torsión*** *Una condición en la cual la dirección del vehículo jala a un lado durante una aceleración fuerte.*

**Toxic** A toxic substance is poisonous to animal or human life.
***Tóxico*** *Una sustancia tóxica es venenosa a la vida de animales o humanos.*

**T-pin** A common pin that is T-shaped.
***Clavija de T*** *Una clavija común en forma de T.*

**Tracking bar** is attached from the chassis to the rear suspension to prevent rear lateral body movement.
***Barra de rectificación*** *se conecta desde el chasis a la suspensión trasera para prevenir el movimiento lateral trasera de la carrocería.*

**Transaxle vent** is located in the transaxle case to allow air to escape from the transaxle.
***Toma de aire*** *del transeje se encuentra en el cárter del transeje para permitir escapar el aire del transeje.*

**Transition time** The time required for the oxygen sensor to switch from lean to rich and rich to lean.
***Tiempo de transición*** *El tiempo requirido para que el sensor de oxígeno cambia de pobre a rico y de rico a pobre.*

**Transversely mounted engine** An engine that is mounted sideways in the vehicle.
***Motor montado transversalmente*** *Un motor que esta montado de lado en el vehículo.*

**Trim height** The distance between two specific chassis locations or between a chassis location and the road surface on a computer-controlled suspension system.
***Altura de centrado*** *La distancia entre dos puntos específicos en el chasis o entre un punto del chasis y la superficie del camino en un sistema de suspensión controlado por computadora.*

**Ultrasonic leak detector** An electronic device that emits a beeping noise when it senses hydrocarbons (HC).
***Detector de fuga ultrasónico*** *Un dispositivo electrónico que emite un ruido pipiante cuando percibe los hidrocarburos (HC).*

**United States Customary (USC)** A systems of weights and measures used in the United States.
***Consuetudinario de los Estados Unidos (USC)*** *Un sistema de pesos y medidas que se usan en los Estados Unidos.*

**Unsprung weight** The weight of the suspension that is not supported by the springs.
***Muelles sin peso*** *El peso de la suspensión que no se soporten por los muelles.*

**Valence ring** The outer ring on an atom.
***Anillo de valencia*** *El anillo exterior de un átomo.*

**Valve guides** Cylindrical components that support the valve stems in the cylinder heads.
***Guías de la válvula*** *Los componentes cilíndricos que soportan los vástagos de la válvula en las cabezas del cilíndro.*

**Valve overlap** The number of degrees of crankshaft rotation when both the intake and exhaust valves are open at the same time with the piston near TDC on the exhaust stroke.
***Superposición de la válvula*** *El número de grados de rotación del árbol de manívela cuando estan abiertas la válvula de entrada y la válvula de escape al mismo tiempo con el pistón cerca del TDC en la carrera del escape.*

**Vehicle identification number (VIN)** A number located in the top left-hand side of the dash indicating various vehicle data such as model, year, body style, engine type, and serial number.
***Número de Identificación del Vehículo (VIN)*** *Un número ubicado en la parte superior a mano izquierda del tablero indicando varios datos pertinentes al vehículo tal como el modelo, el año, el estilo de chasis, el tipo del motor y el número de serie.*

**Viscous coupling** A sealed chamber containing a thick, honey-like liquid.
***Acoplamiento viscoso*** *Una cámara sellada que contiene un líquido espeso parecido al miel.*

**Voltage** A measurement for electrical pressure difference.
***Voltaje*** *Una medida de la diferencia en presión eléctrica.*

**Voltage drop** The difference in voltage at two different locations in a circuit.
***Caída del voltaje*** *La diferencia en el voltaje en dos lugares distinctos en un circuito.*

**Volumetric efficiency** The actual amount of air-fuel mixture entering the engine cylinders compared to the amount of air-fuel mixture that could enter the cylinders under ideal conditions.
***Eficiencia volumétrica*** *La cantidad actual de la mezcla aire-combustible entrando a los cilíndros del motor comparada a la cantidad de la mezcla aire-combustible que podría entrar en los cilíndro bajo condiciones ideales.*

**Vortex flow** occurs in a torque converter when the stator is redirecting fluid from the turbine back into the impeller pump.
***Flujo de torbellino*** *ocurre en un convertidor del par cuando el estator esta volviendo a dirigir el fluido de la turbina hacia la bomba impulsor.*

**Waste spark system** An electronic ignition (EI) system that simultaneously fires a spark plug on the compression stroke and another spark plug on the exhaust stroke.
***Sistema de chispa resíduo*** *Un sistema de encendido electrónico (EI) que dispara una chispa de la bujía en la carrera de compresión simultáneamente con otra chispa de la bujía en la carrera de escape.*

**Watts** A measurement for electrical energy.
***Vatios*** *Una medida de la energía eléctrica.*

**Wheel shimmy** A rapid, repeated lateral wheel movement.
***Abaniqueo de las ruedas*** *Una vibración lateral rápida de la rueda.*

**Wheel tramp** The rapid, repeated lifting of the tire and wheel off the road surface.
***Salto de rueda*** *La elevación rápida y repetida del pneumático y la rueda de la superficie del camino.*

# Index